CALCIUM IN BIOLOGICAL SYSTEMS

CALCIUM IN BIOLOGICAL SYSTEMS

Edited by
Ronald P. Rubin
Medical College of Virginia
Virginia Commonwealth University
Richmond, Virginia

George B. Weiss
CIBA-GEIGY Corporation
Summit, New Jersey

and
James W. Putney, Jr.
Medical College of Virginia
Virginia Commonwealth University
Richmond, Virginia

PLENUM PRESS • NEW YORK AND LONDON

Library of Congress Cataloging in Publication Data

Main entry under title:

Calcium in biological systems.

Based in part upon the proceedings of the 67th Annual Meeting of the Federation of American Societies for Environmental Biology, held in Chicago, Ill., Apr. 10-15, 1983, and sponsored by the American Society for Pharmacology and Experimental Therapeutics and others.

Includes bibliographies and index.

1. Calcium—Physiological effect—Congresses. 2. Calcium—Metabolism—Congresses. 3. Calcium—Metabolism—Disorders—Congresses. 4. Calcification—Congresses. I. Rubin, Ronald P. II. Weiss, George B. III. Putney, James W. IV. Federation of American Societies for Experimental Biology. Meeting (67th: 1983: Chicago, Ill.) V. American Society for Pharmacology and Experimental Therapeutics. [DNLM: 1. Calcification, Physiological—congresses. 2. Calcinosis—congresses. 3. Calcium—congresses. 4. Calcium Channel Blockers—therapeutic use—congresses. 5. Muscle Contraction—drug effects—congresses. 6. Myocardial Contraction—drug effects—congresses. QV 276 C14297 1983]

QP535.C2C26325   1985   599′.019214   84-18246

ISBN-13: 978-1-4612-9453-5   e-ISBN-13: 978-1-4613-2377-8
DOI: 10.1007/978-1-4613-2377-8

©1985 Plenum Press, New York
Softcover reprint of the hardcover 1st edition 1985
A Division of Plenum Publishing Corporation
233 Spring Street, New York, N.Y. 10013

All rights reserved

No part of this book may be reproduced, stored in a retrieval system, or transmitted in any form or by any means, electronic, mechanical, photocopying, microfilming, recording, or otherwise, without written permission from the Publisher

Contributors

Paul R. Adams • Department of Neurobiology and Behavior, State University of New York, Stony Brook, New York 11794

Bernard W. Agranoff • Departments of Biological Chemistry and Psychiatry, University of Michigan, Ann Arbor, Michigan 48109

Myles H. Akabas • Departments of Physiology and Biophysics and Neuroscience, Albert Einstein College of Medicine, Bronx, New York 10461

W. Almers • Department of Physiology and Biophysics, University of Washington, Seattle, Washington 98195

H. Clarke Anderson • Department of Pathology and Oncology, University of Kansas Medical Center, Kansas City, Kansas 66103

Robert J. Bache • Cardiovascular Division, Department of Medicine, University of Minnesota Medical School, Minneapolis, Minnesota 55455

Edward M. Barr • Division of Pharmacology, College of Pharmacy, The University of Texas at Austin, Austin, Texas 78712

S. M. Baylor • Department of Physiology, University of Pennsylvania, Philadelphia, Pennsylvania 19104

Robert E. Beamish • Experimental Cardiology Section, Department of Physiology, University of Manitoba, Winnipeg, Manitoba R3E 0W3, Canada

Gustavo Benaim • Laboratory of Biochemistry, Swiss Federal Institute of Technology (ETH), 8092 Zurich, Switzerland; *permanent address:* Universidad Central de Venezuela, Facultad de Ciencias, Escuela de Biologia, Apartado 21201, Departamento de Biologia Celular, Caracas, Venezuela

Michael J. Berridge • A.F.R.C. Unit of Insect Neurophysiology and Pharmacology, Department of Zoology, University of Cambridge, Cambridge CB2 3EJ, England

John A. Bevan • Department of Pharmacology, School of Medicine, University of Vermont, Burlington, Vermont 05405

Mordecai P. Blaustein • Department of Physiology, University of Maryland, School of Medicine, Baltimore, Maryland 21201

Adele L. Boskey • Department of Biochemistry, Cornell University Medical College, and Department of Ultrastructural Biochemistry, The Hospital for Special Surgery, New York, New York 10021

Charles Eric Brown • Department of Biochemistry, The Medical College of Wisconsin, Milwaukee, Wisconsin 53226

David A. Brown • Department of Pharmacology, School of Pharmacology, London WCIN IAX, England

Hellmut Brunner • Biological Research Laboratories, Pharmaceuticals Division, Ciba-Geigy Corporation, CH 4000 Basel, Switzerland

Daniel B. Burnham • Department of Physiology, University of California, San Francisco, California 94143

Lewis Cantley • Department of Biochemistry and Molecular Biology, Harvard University, Cambridge, Massachusetts 02138

Ernesto Carafoli • Laboratory of Biochemistry, Swiss Federal Institute of Technology (ETH), 8092 Zurich, Switzerland

P. L. Carlen • Addiction Research Foundation Clinical Institute, Playfair Neuroscience Unit, Toronto Western Hospital, Departments of Medicine (Neurology), Physiology, and Institute of Medical Science, University of Toronto, Toronto, Ontario M5S 2S1, Canada

R. Casteels • Laboratory of Physiology, Catholic University of Louvain, Gasthuisberg Campus, B3000 Louvain, Belgium

L. Judson Chandler • Division of Pharmacology, College of Pharmacy, The University of Texas at Austin, Austin, Texas 78712

Herman S. Cheung • Division of Rheumatology, Department of Medicine, Medical College of Wisconsin, Milwaukee, Wisconsin 53226

Haing U. Choi • Orthopedic Research Laboratories, Montefiore Medical Center, Bronx, New York 10467

Robert B. Clark • Department of Physiology and Biophysics, University of Texas Medical Branch, Galveston, Texas 77550

Fredric S. Cohen • Department of Physiology, Rush Medical College, Chicago, Illinois 60612

Jay N. Cohn • Cardiovascular Division, Department of Medicine, University of Minnesota Medical School, Minneapolis, Minnesota 55455

Dennis L. Coleman • Department of Pharmaceutics, University of Utah, Salt Lake City, Utah 84112

Andrew Constanti • Department of Pharmacology, School of Pharmacology, London WCIN IAX, England

J. J. Corcoran • Department of Pharmacology, Duke University Medical Center, Durham, North Carolina 27710

Leoluca Criscione • Biological Research Laboratories, Pharmaceuticals Division, Ciba-Geigy Corporation, CH 4000 Basel, Switzerland

James L. Daniel • Temple University, Thrombosis Research Center, Philadelphia, Pennsylvania 19104

Laura C. Daniell • Division of Pharmacology, College of Pharmacy, The University of Texas at Austin, Austin, Texas 78712

Hector F. DeLuca • Department of Biochemistry, University of Wisconsin, Madison, Wisconsin 53706

Mrinal K. Dewanjee • Section of Nuclear Medicine, Mayo Clinic and Foundation, Rochester, Minnesota 55905

Naranjan S. Dhalla • Experimental Cardiology Section, Department of Physiology, University of Manitoba, Winnipeg, Manitoba R3E 0W3, Canada

Sue K. Bolitho Donaldson • Department of Physiology and Department of Medical Nursing, Rush University, Chicago, Illinois 60612

David E. Dong • Department of Pharmaceutics, University of Utah, Salt Lake City, Utah 84112

H. H. Draper • Department of Nutrition, College of Biological Science, University of Guelph, Guelph, Ontario N1G 2W1, Canada

G. Droogmans • Laboratory of Physiology, Catholic University of Louvain, Gathuisberg Campus, B-3000 Louvain, Belgium

Robert Dunn, Jr. • Department of Physiology and Department of Medical Nursing, Rush University, Chicago, Illinois 60612

Rosemary Dziak • Department of Oral Biology, State University of New York at Buffalo, Buffalo, New York 14226

Contributors

Joseph Eichberg • Department of Biochemical and Biophysical Sciences, University of Houston, Houston, Texas 77004

Leigh English • Department of Biochemistry and Molecular Biology, Harvard University, Cambridge, Massachusetts 02138

Alexandre Fabiato • Department of Physiology and Biophysics, Medical College of Virginia, Richmond, Virginia 23298

John N. Fain • Section of Biochemistry, Division of Biology and Medicine, Brown University, Providence, Rhode Island 02912

Marc Fallert • Biological Research Laboratories, Pharmaceuticals Division, Ciba-Geigy Corporation, CH 4000 Basel, Switzerland

Joseph J. Feher • Department of Physiology and Biophysics, Medical College of Virginia, Richmond, Virginia 23298

Clare Fewtrell • Department of Pharmacology, New York State College of Veterinary Medicine, Cornell University, Ithaca, New York 14853, and Section on Chemical Immunology, Arthritis and Rheumatism Branch, National Institute of Arthritis, Diabetes, Digestive and Kidney Diseases, National Institutes of Health, Bethesda, Maryland 20205

Alan Finkelstein • Departments of Physiology and Biophysics and Neuroscience, Albert Einstein College of Medicine, Bronx, New York 10461

A. Fleckenstein • Physiological Institute, University of Freiburg, 7800 Freiburg, West Germany

Gisa Fleckenstein-Grün • Physiological Institute, University of Freiburg, 7800 Freiburg, West Germany

Roger B. Galburt • Department of Restorative Dentistry, Tufts University, School of Dental Medicine, Boston, Massachusetts 02111

Stanley M. Garn • Center for Human Growth and Development and the Epidemiology Department of the School of Public Health, University of Michigan, Ann Arbor, Michigan 48109

Melvin J. Glimcher • Laboratory for the Study of Skeletal Disorders and Rehabilitation, Department of Orthopedic Surgery, Harvard Medical School, Children's Hospital Medical Center, Boston, Massachusetts 02115

M. E. Gnegy • Department of Pharmacology, University of Michigan, Ann Arbor, Michigan 48109

T. Godfraind • Laboratory of Pharmacodynamics and Pharmacology, Catholic University of Louvain, B-1200 Brussels, Belgium

P. P. Godfrey • Division of Cellular Pharmacology, Medical College of Virginia, Richmond, Virginia 23298

H. Gonzalez-Serratos • Department of Biophysics, University of Maryland School of Medicine, Baltimore, Maryland 21201

Paul Greengard • The Rockefeller University, New York, New York 10021

N. Gurevich • Addiction Research Foundation Clinical Institute, Playfair Neuroscience Unit, Toronto Western Hospital, Departments of Medicine (Neurology), Physiology, and Institute of Medical Science, University of Toronto, Toronto, Ontario M5S 2S1, Canada

Erik Gylfe • Department of Medical Cell Biology, Biomedicum, University of Uppsala, S-751 23 Uppsala, Sweden

T. J. Hallam • Physiological Laboratory, University of Cambridge, Cambridge CB2 3EG, England

Hiroaki Harasaki • Department of Artificial Organs, Cleveland Clinic Foundation, Cleveland, Ohio 44106

Charles A. Harrington • Analytical Neurochemistry Laboratory, Texas Research Institute of Mental Sciences, Houston, Texas 77030

R. Adron Harris • Denver Veterans Administration Hospital and Department of Pharmacology, University of Colorado, School of Medicine, Denver, Colorado 80262

Marguerite Hawley • Department of Cardiology and Laboratory of Human Biochemistry, Children's Hospital Medical Center, Boston, Massachusetts 02115

Victor M. Hawthorne • Center for Human Growth and Development and the Epidemiology Department of the School of Public Health, University of Michigan, Ann Arbor, Michigan 48109

Bo Hellman • Department of Medical Cell Biology, Biomedicum, University of Uppsala, S-751 23 Uppsala, Sweden

Kent Hermsmeyer • Department of Pharmacology, The Cardiovascular Center, University of Iowa, Iowa City, Iowa 52242

Holm Holmsen • Department of Biochemistry, University of Bergen, N-5000 Bergen, Norway

Susan L. Howard • Department of Cardiology and Laboratory of Human Biochemistry, Children's Hospital Medical Center, Boston, Massachusetts 02115

David S. Howell • Department of Medicine, University of Miami, School of Medicine, and U.S. Veterans Administration, Miami, Florida 33101

Daniel A. Huetteman • Department of Physiology and Department of Medical Nursing, Rush University, Chicago, Illinois 60612

Chiu Shuen Hui • Department of Biological Sciences, Purdue University, West Lafayette, Indiana 47907

Hwei-Chuen Hsu • Department of Pharmaceutics, University of Utah, Salt Lake City, Utah 84112

Joyce J. Hwa • Department of Pharmacology, School of Medicine, University of Vermont, Burlington, Vermont 05405

Shiro Kakiuchi • Department of Neurochemistry and Neuropharmacology, Institute of Higher Nervous Activity, Osaka University Medical School, Nakanoshima, Kita-ku, Osaka 530, Japan

N. Kirshner • Department of Pharmacology, Duke University Medical Center, Durham, North Carolina 27710

T. Kitazawa • The Pennsylvania Muscle Institute and Departments of Physiology and Pathology, University of Pennsylvania School of Medicine, Philadelphia, Pennsylvania 19104; *present address:* Department of Pharmacology, Juntendo University School of Medicine, Tokyo 113, Japan

Claude B. Klee • Laboratory of Biochemistry, National Cancer Institute, National Institutes of Health, Bethesda, Maryland 20205

Szloma Kowarski • Department of Physiology, Columbia University College of Physicians and Surgeons, New York, New York 10032

Lois B. Kramer • Metabolic Section, Veterans Administration Hospital, Hines, Illinois 60141

Dieter M. Kramsch • Cardiovascular Institute, Boston University School of Medicine, Boston, Massachusetts 02118

Hans Kühnis • Biological Research Laboratories, Pharmaceuticals Division, Ciba-Geigy Corporation, CH 4000 Basel, Switzerland

Yvonne Lai • The Rockefeller University, New York, New York 10021

Eduardo G. Lapetina • Department of Molecular Biology, The Wellcome Research Laboratories, Burroughs Wellcome Company, Research Triangle Park, North Carolina 27709

Steven W. Leslie • Division of Pharmacology, College of Pharmacy, The University of Texas at Austin, Austin, Texas 78712

Robert Levenson • Department of Biology and Center for Cancer Research, Massachusetts Institute of Technology, Cambridge, Massachusetts 02139

Judith T. Levy • Department of Chemistry, Wellesley College, Wellesley, Massachusetts 02025

Robert J. Levy • Department of Cardiology and Laboratory of Human Biochemistry, Children's Hospital Medical Center and Department of Pediatrics, Harvard Medical School, Boston, Massachusetts 02115

Jane B. Lian • Department of Biological Chemistry (Orthopedic Surgery), Laboratory for the Study of Skeletal Disorders and Rehabilitation, Harvard Medical School, Children's Hospital Medical Center, Boston, Massachusetts 02115

Sue-Hwa Lin • Section of Biochemistry, Division of Biology and Medicine, Brown University, Providence, Rhode Island 02912

Leona Ling • Department of Biochemistry and Molecular Biology, Harvard University, Cambridge, Massachusetts 02138

Contributors

Irene Litosch • Section of Biochemistry, Division of Biology and Medicine, Brown University, Providence, Rhode Island 02912

Ian G. Macara • Department of Biochemistry and Molecular Biology, Harvard University, Cambridge, Massachusetts 02138; *present address:* Department of Radiation Biology and Biophysics, University of Rochester School of Medicine, Rochester, New York 14642

David A. McCarron • Division of Nephrology and Hypertension, Oregon Health Sciences University, Portland, Oregon 97201

Daniel J. McCarty • Division of Rheumatology, Department of Medicine, Medical College of Wisconsin, Milwaukee, Wisconsin 53226

G. McClellan • The Pennsylvania Muscle Institue and Departments of Physiology and Pathology, University of Pennsylvania School of Medicine, Philadelphia, Pennsylvania 19104

E. W. McCleskey • Department of Physiology and Biophysics, University of Washington, Seattle, Washington 98195

Robert L. Macdonald • Departments of Neurology and Physiology, University of Michigan, Ann Arbor, Michigan 48109

Teresa L. McGuinness • The Rockefeller University, New York, New York 10021

James T. McMahon • Department of Pathology, Cleveland Clinic Foundation, Cleveland, Ohio 44106

Allan S. Manalan • Laboratory of Biochemistry, National Cancer Institute, National Institutes of Health, Bethesda, Maryland 20205

Gretchen S. Mandel • Department of Medicine and Research Service, Veterans Administration Medical Center, Medical College of Wisconsin, Milwaukee, Wisconsin 53193

Neil S. Mandel • Department of Medicine and Research Service, Veterans Administration Medical Center, Medical College of Wisconsin, Milwaukee, Wisconsin 53193

Johanne Martel-Pelletier • Rheumatic Disease Unit, Notre Dame Hospital, University of Montreal, Montreal, Quebec H2L 4K8, Canada

Max Meier • Biological Research Laboratories, Pharmaceuticals Division, Ciba-Geigy Corporation, CH 4000 Basel, Switzerland

Elias K. Michaelis • Neurobiology, Department of Human Development, and Center for Biomedical Research, University of Kansas, Lawrence, Kansas 66045

Mary L. Michaelis • Neurobiology, Department of Human Development, and Center for Biomedical Research, University of Kansas, Lawrence, Kansas 66045

Sara Morales • Department of Medicine, University of Miami, School of Medicine, Miami, Florida 33101

Ofelia Muniz • Department of Medicine, University of Miami, School of Medicine, Miami, Florida 33101

Paul H. Naccache • Department of Pathology, University of Connecticut Health Center, Farmington, Connecticut 06032

Abraham E. Nizel • Department of Oral Health Services, Tufts University, School of Dental Medicine, Boston, Massachusetts 02111

Yukihiko Nose • Department of Artificial Organs, Cleveland Clinic Foundation, Cleveland, Ohio 44106

M. O'Beirne • Addiction Research Foundation Clinical Institute, Playfair Neuroscience Unit, Toronto Western Hospital, Departments of Medicine (Neurology), Physiology, and Institute of Medical Science, University of Toronto, Toronto, Ontario M5S 2S1, Canada

Donald B. Olsen • Department of Surgery, Division of Artificial Organs, University of Utah, Salt Lake City, Utah 84112

Lauren Oshry • Department of Chemistry, Wellesley College, Wellesley, Massachusetts 02025

Charles C. Ouimet • The Rockefeller University, New York, New York 10021

Mary P. Owen • Department of Pharmacology, School of Medicine, University of Vermont, Burlington, Vermont 05405

P. T. Palade • Department of Physiology and Biophysics, University of Washington, Seattle, Washington 98195; *present address:* Department of Physiology and Biophysics, The University of Texas Medical Branch, Galveston, Texas 77550

Genaro M. A. Palmieri • Department of Internal Medicine, University of Tennessee Center for the Health Sciences, Memphis, Tennessee 38163

Vincenzo Panagia • Department of Oral Biology, University of Manitoba, Winnipeg, Manitoba R3E 0W3, Canada

C. Owen Parkes • Department of Physiology, University of British Columbia, Vancouver, British Columbia V6T 1W5, Canada

Jean Pierre Pelletier • Rheumatic Disease Unit, Notre Dame Hospital, University of Montreal, Montreal, Quebec H2L 4K8, Canada

A. Robin Poole • Joint Diseases Research Laboratory, Shriners Hospital for Crippled Children, Montreal, Quebec H3G 1A6, Canada

Paul A. Price • Department of Biology, University of California at San Diego, La Jolla, California 92093

Walter C. Prozialeck • Department of Physiology and Pharmacology, Philadelphia College of Osteopathic Medicine, Philadelphia, Pennsylvania 19131

J. W. Putney, Jr. • Division of Cellular Pharmacology, Medical College of Virginia, Richmond, Virginia 23298

M. E. Quinta-Ferreira • Department of Physiology, University of Pennsylvania, Philadelphia, Pennsylvania 19104; *present address:* Departmento de Fisica, Universidade de Coimbra, 3000 Coimbra, Portugal

Howard Rasmussen • Departments of Cell Biology and Internal Medicine, Yale University, New Haven, Connecticut 06510

T. J. Rink • Physiological Laboratory, University of Cambridge, Cambridge CB2 3EG, England

Harald Rogg • Biological Research Laboratories, Pharmaceuticals Division, Ciba-Geigy Corporation, CH 4000 Basel, Switzerland

Gary S. Rogoff • Department of Complete Denture Prosthetics, Tufts University, School of Dental Medicine, Boston, Massachusetts 02111

Lawrence C. Rosenberg • Orthopedic Research Laboratories, Montefiore Medical Center, Bronx, New York 10467

Philip M. Rosoff • Department of Biochemistry and Molecular Biology, Harvard University, Cambridge, Massachusetts 02138

Ronald P. Rubin • Division of Cellular Pharmacology, Medical College of Virginia, Richmond, Virginia 23298

A. Sanchez • Physiological Laboratory, University of Cambridge, Cambridge CB2 3EG, England

Kiriti Sarkar • Department of Pathology, School of Medicine, Faculty of Health Sciences, University of Ottawa, Ottawa, Ontario K1H 8M5, Canada

Leslie Satin • Department of Neurobiology and Behavior, State University of New York, Stony Brook, New York 11794

David Schachter • Department of Physiology, Columbia University College of Physicians and Surgeons, New York, New York 10032

Frederick J. Schoen • Department of Pathology, Brigham and Women's Hospital, Boston, Massachusetts 02115

Jeffrey S. Schwartz • Cardiovascular Division, Department of Medicine, University of Minnesota Medical School, Minneapolis, Minnesota 55455

Edward B. Seguin • Departments of Biological Chemistry and Psychiatry, University of Michigan, Ann Arbor, Michigan 48109

Ramadan I. Sha'afi • Department of Physiology, University of Connecticut Health Center, Farmington, Connecticut 06032

Wolfgang Siess • Department of Molecular Biology, The Wellcome Research Laboratories, Burroughs Wellcome Company, Research Triangle Park, North Carolina 27709; *present address:* Medizinische Klinik Innenstadt der Universität München, 8 Munich 2, West Germany

E. M. Silinsky • Department of Pharmacology, Northwestern University Medical School, Chicago, Illinois 60611

Pawan K. Singal • Experimental Cardiology Section, Department of Physiology, University of Manitoba, Winnipeg, Manitoba R3E 0W3, Canada

A. P. Somlyo • The Pennsylvania Muscle Institute and Departments of Physiology and Pathology, University of Pennsylvania School of Medicine, Philadelphia, Pennsylvania 19104

A. V. Somlyo • The Pennsylvania Muscle Institute and Departments of Physiology and Pathology, University of Pennsylvania School of Medicine, Philadelphia, Pennsylvania 19104

Herta Spencer • Metabolic Section, Veterans Administration Hospital, Hines, Illinois 60141

Paula H. Stern • Department of Pharmacology, Northwestern University Medical and Dental Schools, Chicago, Illinois 60611

Marguerite A. Stout • Department of Physiology, UMDNJ–New Jersey Medical School, Newark, New Jersey 07103

Edward Toggart • Cardiology Division, The Milton S. Hershey Medical Center, The Pennsylvania State University, Hershey, Pennsylvania 17033

G. Treisman • Department of Pharmacology, University of Michigan, Ann Arbor, Michigan 48109

Arnold G. Truog • Biological Research Laboratories, Pharmaceuticals Division, Ciba-Geigy Corporation, CH 4000 Basel, Switzerland

R. Y. Tsien • Department of Physiology–Anatomy, University of California, Berkeley, California 94720

Hans K. Uhthoff • Department of Surgery, School of Medicine, Faculty of Health Sciences, University of Ottawa, and Division of Orthopaedics, Ottawa General Hospital, Ottawa, Ontario K1H 8M5, Canada

Lucio A. A. Van Rooijen • Departments of Biological Chemistry and Psychiatry, University of Michigan, Ann Arbor, Michigan 48109

Michael Wallace • Section of Biochemistry, Division of Biology and Medicine, Brown University, Providence, Rhode Island 02912

Benjamin Weiss • Department of Pharmacology, Medical College of Pennsylvania, Philadelphia, Pennsylvania 19129

Mary Ann Werz • Department of Neurology, University of Michigan, Ann Arbor, Michigan 48109

John A. Williams • Department of Physiology, University of California, San Francisco, California 94143

S. P. Wilson • Department of Pharmacology, Duke University Medical Center, Durham, North Carolina 27710

E. E. Windhager • Department of Physiology, Cornell University Medical College, New York, New York 10021

Raymond J. Winquist • Cardiovascular Pharmacology, Merck Sharp & Dohme Research Laboratories, West Point, Pennsylvania 19486

Abraham M. Yaari • Department of Developmental Dentistry, University of Southern California School of Dentistry, Los Angeles, California 90089-0641

Hiro-aki Yamamoto • Faculty of Pharmaceutical Sciences, Fukuyama University, Fukuyama, Hiroshima 729-02, Japan

Li-An Yeh • Department of Biochemistry and Molecular Biology, Harvard University, Cambridge, Massachusetts 02138

S. Yoshikami • Laboratory of Chemical Physics, National Institute of Arthritis, Diabetes, and Digestive and Kidney Diseases, National Institutes of Health, Bethesda, Maryland 20205

Robert Zelis • Cardiology Division, The Milton S. Hershey Medical Center, The Pennsylvania State University, Hershey, Pennsylvania 17033

Mauro Zurini • Laboratory of Biochemistry, Swiss Federal Institute of Technology (ETH), 8092 Zurich, Switzerland

Preface

This volume is based in part upon the proceedings of the Calcium Theme held during the 67th Annual Meeting of the Federation of American Societies for Experimental Biology, which took place in Chicago, April 10–15, 1983. The American Society for Pharmacology and Experimental Therapeutics had the primary responsibility for organizing the scientific program with the assistance of other member societies, including the American Physiology Society, American Association of Pathologists, and American Institute of Nutrition.

The purpose of the Calcium Theme was to review progress in the diverse areas of investigation bearing on the ubiquitous role of calcium in biological systems. In addition to contributions from those participating in the Theme, this volume also includes a number of invited papers that were added to fill certain voids in topics covered. The authors were selected because they are investigators active in the mainstream of their particular research area, possessing the acumen to analyze cogently not only their own recent findings but also to relate these findings to their respective area. New information as well as reviews of current concepts generally highlight the individual contributions. Undoubtedly, some readers may argue with the emphasis made and/or the conclusions reached on individual topics. In such cases, other volumes will hopefully provide a forum for alternative points of view.

Due to the broad scope of subjects covered and the large number of contributions, the papers have been arranged in three sections. Section A deals with the interactions of calcium with cellular phospholipids and calcium-binding proteins (including calmodulin); the role of calcium in the regulation of ionic permeability and stimulus–secretion coupling; and the actions of this cation on blood-forming cells. This section is introduced by three general introductory chapters on the actions of calcium. Section B is devoted to the role of calcium in regulating contractility of skeletal, smooth, and cardiac muscle. Special emphasis is placed upon calcium antagonists and their role in elucidating physiological mechanisms and in applying them to diseases of the cardiovascular system. Section C deals with the nutritional and pathophysiological aspects of calcium's actions. including interactions of calcium with vitamin D and other calcemic agents; altered calcium homeostasis in animals and man; and normal and pathological calcification processes.

This volume clearly delineates the ubiquitous and pivotal role of calcium in diverse physiological and pathological states, and it is our fervent hope that these papers will stimulate even greater research activity in this vast and critical area. While emphasis has been placed on readability, rather than exhaustive analysis, the various accounts are sufficiently well documented to make the treatise valuable not only to those involved in all branches of calcium research, but also to teachers and students of general biology, physiology, biochemistry, and pathology.

The Editors would like to express their gratitude to the following Symposia Chairmen who organized individual sessions for the Theme and later invaluably assisted in collecting and editing the manuscripts of their respective sections: H. Clarke Anderson, Mordecai Blaustein, F. Norman Briggs, R. Casteels, Maurice Feinstein, Margaret Gnegy, R. Adron Harris, D. M. Hegsted, Jane Lian, Daniel McCarty, Eugene Silinsky, Elwood Speckmann, Paula Stern, Stuart Taylor, and Paul Vanhoutte. We would also like to thank Rebecca Harris and Deborah Slovenec for their valuable assistance in the editing and typing of the manuscripts. Finally, we are grateful to FASEB for providing the physical and financial resources to stage the Calcium Theme which spawned most of the chapters constituting this volume.

Ronald P. Rubin
George B. Weiss
James W. Putney, Jr.

Contents

A. METABOLIC AND FUNCTIONAL ASPECTS OF CALCIUM ACTION

I. General Aspects of Calcium and Cell Function

1. Historical and Biological Aspects of Calcium Action 5
   Ronald P. Rubin

2. Calcium Ion: A Synarchic and Mercurial But Minatory Messenger 13
   Howard Rasmussen

3. Intracellular Calcium as a Second Messenger: What's so Special about Calcium? 23
   Mordecai P. Blaustein

II. Roles of Phosphoinositides in Calcium-Regulated Systems

4. Calcium-Mobilizing Receptors: Membrane Phosphoinositides and Signal Transduction 37
   Michael J. Berridge

5. Platelet Response in Relation to Metabolism of Inositides and Protein Phosphorylation 45
   Eduardo G. Lapetina and Wolfgang Siess

6. Receptor-Mediated Changes in Hepatocyte Phosphoinositide Metabolism: Mechanism and Significance 53

Joseph Eichberg and Charles A. Harrington

7. Hormonal Regulation of Phosphoinositide Metabolism in Rat Hepatocytes ... 61

John N. Fain, Michael Wallace, Sue-Hwa Lin, and Irene Litosch

8. Implication of Phosphoinositides in Receptor-Mediated Stimulation of Phospholipid Labeling 67

Bernard W. Agranoff, Edward B. Seguin, and Lucio A. A. Van Rooijen

9. Calcium and Receptor-Mediated Phosphoinositide Breakdown in Exocrine Gland Cells 73

J. W. Putney, Jr. and P. P. Godfrey

III. Stimulus–Secretion Coupling

10. Calcium and Stimulus–Secretion Coupling in Pancreatic Acinar Cells ... 83

John A. Williams and Daniel B. Burnham

11. Glucose Regulation of Insulin Release Involves Intracellular Sequestration of Calcium 93

Bo Hellman and Erik Gylfe

12. Role of Calcium in Stimulus–Secretion Coupling in Bovine Adrenal Medullary Cells 101

N. Kirshner, J. J. Corcoran, and S. P. Wilson

13. Calcium and Transmitter Release: Modulation by Adenosine Derivatives ... 109

E. M. Silinsky

14. The Role of Calcium and Osmosis in Membrane Fusion 121

Fredric S. Cohen, Myles H. Akabas, and Alan Finkelstein

IV. Calcium Regulation of Hemopoietic Cells

15. Activation and Desensitization of Receptors for IgE on Tumor Basophils ... 129

Clare Fewtrell

16. Relationship between Calcium, Arachidonic Acid Metabolites, and
Neutrophil Activation 137

Ramadan I. Sha'afi and Paul H. Naccache

17. Receptor-Controlled Phosphatidate Synthesis during Acid Hydrolase
Secretion from Platelets 147

Holm Holmsen

18. Calcium and Diacylglycerol: Separable and Interacting Intracellular
Activators in Human Platelets 155

T. J. Rink, R. Y. Tsien, A. Sanchez, and T. J. Hallam

19. Protein Phosphorylation and Calcium as Mediators in Human Platelets .. 165

James L. Daniel

20. Na^+ and Ca^{2+} Fluxes and Differentiation of Transformed Cells 173

Lewis Cantley, Philip M. Rosoff, Robert Levenson, Ian G. Macara, Li-An Yeh,
Leona Ling, and Leigh English

V. Calcium as a Regulator of Membrane Permeability

21. Calcium-Activated Potassium Channels in Bullfrog Sympathetic Ganglion
Cells.. 181

Paul R. Adams, David A. Brown, Andrew Constanti, Robert B. Clark, and
Leslie Satin

22. Electrophysiological Evidence for Increased Calcium-Mediated Potassium
Conductance by Low-Dose Sedative-Hypnotic Drugs................. 193

P. L. Carlen, N. Gurevich, and M. O'Beirne

23. Neurochemical Studies on the Effects of Ethanol on Calcium-Stimulated
Potassium Transport 201

Hiro-aki Yamamoto

24. Role of Calcium in the Regulation of Sodium Permeability in the Kidney 207

E. E. Windhager

VI. Calcium as a Regulator of Neuronal Function

25. Neuronal Calcium as a Site of Action for Depressant Drugs 215

R. Adron Harris

26. Sedative-Hypnotic Drugs and Synaptosomal Calcium Transport 221
   Steven W. Leslie, Edward M. Barr, L. Judson Chandler, and Laura C. Daniell

27. Barbiturate and Opiate Actions on Calcium-Dependent Action Potentials and Currents of Mouse Neurons in Cell Culture 227
   Robert L. Macdonald and Mary Ann Werz

28. Effects of Depressant Drugs on Sodium–Calcium Exchange in Resealed Synaptic Membranes 237
   Elias K. Michaelis and Mary L. Michaelis

29. Calcium Extrusion by Retinal Rods Modifies Light Sensitivity of Dark Current 245
   S. Yoshikami

VII. Calmodulin

30. Mechanisms of Pharmacologically Altering Calmodulin Activity 255
   Walter C. Prozialeck and Benjamin Weiss

31. The Purified Calcium-Pumping ATPase of Plasma Membrane: Structure–Function Relationships 265
   Ernesto Carafoli, Mauro Zurini, and Gustavo Benaim

32. Calmodulin-Binding Proteins That Control the Cytoskeleton by a Flip-Flop Mechanism 275
   Shiro Kakiuchi

33. Calmodulin-Sensitive Adenylate Cyclase Activity: Interaction with Guanyl Nucleotides and Dopamine 283
   M. E. Gnegy and G. Treisman

34. Calcium/Calmodulin-Dependent Protein Phosphorylation in the Nervous System 291
   Teresa L. McGuinness, Yvonne Lai, Charles C. Ouimet, and Paul Greengard

35. Calcineurin, a Calmodulin-Stimulated Protein Phosphatase 307
   Allan S. Manalan and Claude B. Klee

B. CALCIUM AND THE REGULATION OF MUSCLE CONTRACTILITY

VIII. Calcium and Skeletal Muscle Contractility

36. Calcium Channels in Vertebrate Skeletal Muscle 321
    W. Almers, E. W. McCleskey, and P. T. Palade

37. Mechanisms of Calcium Release in Skinned Mammalian Skeletal Muscle Fibers 331
    Sue K. Bolitho Donaldson, Robert Dunn, Jr., and Daniel A. Huetteman

38. Isotropic Components of Antipyrylazo III Signals from Frog Skeletal Muscle Fibers 339
    S. M. Baylor, M. E. Quinta-Ferreira, and Chiu Shuen Hui

39. Ion Movements Associated with Calcium Release and Uptake in the Sarcoplasmic Reticulum 351
    A. V. Somlyo, T. Kitazawa, H. Gonzalez-Serratos, G. McClellan, and A. P. Somlyo

IX. Calcium and Cardiac Contractility

40. Role of Calcium in Heart Function in Health and Disease: 100 Years of Progress 361
    Pawan K. Singal, Vincenzo Panagia, Robert E. Beamish, and Naranjan S. Dhalla

41. Calcium Both Activates and Inactivates Calcium Release from Cardiac Sarcoplasmic Reticulum 369
    Alexandre Fabiato

42. Routes of Calcium Flux in Cardiac Sarcoplasmic Reticulum 377
    Joseph J. Feher

X. Calcium and Calcium Antagonists in Smooth Muscle

43. Calcium and Myogenic or Stretch-Dependent Vascular Tone 391
    John A. Bevan, Joyce J. Hwa, Mary P. Owen, and Raymond J. Winquist

44. Chemical Skinning: A Method for Study of Calcium Transport by
Sarcoplasmic Reticulum of Vascular Smooth Muscle 399

Marguerite A. Stout

45. Cellular and Subcellular Approaches to the Mechanism of Action of
Calcium Antagonists 411

T. Godfraind

46. Calcium Antagonist Effects on Vascular Muscle Membrane Potentials
and Intracellular Ca^{2+} 423

Kent Hermsmeyer

47. Suppression of Experimental Coronary Spasms by Major Calcium
Antagonists 431

Gisa Fleckenstein-Grün

48. CGP 28392, a Dihydropyridine Ca^{2+} Entry Stimulator 441

Arnold G. Truog, Hellmut Brunner, Leoluca Criscione, Marc Fallert,
Hans Kühnis, Max Meier, and Harald Rogg

XI. Calcium Entry Blockers and Disease

49. Role of Ca^{2+} in and Effect of Ca^{2+} Entry Blockers on Excitation–
Contraction and Excitation–Secretion Coupling 453

R. Casteels and G. Droogmans

50. Calcium Antagonists in the Treatment of Arrhythmias 459

A. Fleckenstein

51. Calcium Entry Blockers in Coronary Artery Disease 471

Jay N. Cohn, Robert J. Bache, and Jeffrey S. Schwartz

52. The Use of Calcium Entry Blockers in Congestive Heart Failure 479

Robert Zelis and Edward Toggart

C. NUTRITIONAL AND PATHOPHYSIOLOGIC ASPECTS OF CALCIUM ACTION

XII. Vitamin D and Other Calcemic Agents

53. The Vitamin D–Calcium Axis—1983 491

Hector F. DeLuca

54. Vitamin D and the Intestinal Membrane Calcium-Binding Protein 513
David Schachter and Szloma Kowarski

55. Calcium-Binding Protein in the Central Nervous System and Other Tissues 519
C. Owen Parkes

56. The Vitamin K-Dependent Bone Protein and the Action of 1,25-Dihydroxyvitamin D_3 on Bone 525
Paul A. Price

57. Prostaglandins as Mediators of Bone Cell Metabolism 533
Rosemary Dziak

58. Interactions of Calcemic Hormones and Divalent Cation Ionophores on Fetal Rat Bone *in Vitro* 541
Paula H. Stern

XIII. Alterations in Calcium Metabolism and Homeostasis

59. Altered Cell Calcium Metabolism and Human Diseases 551
Howard Rasmussen and Genaro M. A. Palmieri

60. Dietary Calcium in the Pathogenesis and Therapy of Human and Experimental Hypertension 561
David A. McCarron

61. Calcium Intake and Bone Loss in Population Context 569
Stanley M. Garn and Victor M. Hawthorne

62. Similarities and Differences in the Response of Animals and Man to Factors Affecting Calcium Needs 575
H. H. Draper

63. Factors Influencing Calcium Balance in Man 583
Herta Spencer and Lois B. Kramer

64. Role of Dietary Calcium and Vitamin D in Alveolar Bone Health: Literature Review Update 591
Gary S. Rogoff, Roger B. Galburt, and Abraham E. Nizel

XIV. Normal Biological Calcification

65. Normal Biological Mineralization: Role of Cells, Membranes, Matrix Vesicles, and Phosphatase 599

    H. Clarke Anderson

66. The Role of Collagen and Phosphoproteins in the Calcification of Bone and Other Collagenous Tissues................................ 607

    Melvin J. Glimcher

67. Biological Processes Involved in Endochondral Ossification 617

    Lawrence C. Rosenberg, Haing U. Choi, and A. Robin Poole

68. Role of Lipids in Mineralization: An Experimental Model for Membrane Transport of Calcium and Inorganic Phosphate...................... 625

    Abraham M. Yaari and Charles Eric Brown

XV. Pathological Calcification

69. Factors Contributing to Intracavitary Calcification 633

    Jane B. Lian

70. Mineral, Lipids, and Proteins Associated with Soft Tissue Deposits..... 645

    Adele L. Boskey

71. Polymer Properties Associated with Calcification of Cardiovascular Devices 653

    Dennis L. Coleman, Hwei-Chuen Hsu, David E. Dong, and Donald B. Olsen

72. Calcification of Cardiac Valve Bioprostheses: Host and Implant Factors . 661

    Robert J. Levy, Frederick J. Schoen, Susan L. Howard, Judith T. Levy, Lauren Oshry, and Marguerite Hawley

73. Pathogenesis of Valve Calcification: Comparison of Three Tissue Valves 669

    Hiroaki Harasaki, James T. McMahon, and Yukihiko Nose

74. Noninvasive Imaging of Dystrophic Calcification..................... 677

    Mrinal K. Dewanjee

75. Prevention of Arterial Calcium Deposition with Diphosphonates and Calcium Entry Blockers 685

    Dieter M. Kramsch

XVI. Crystal Deposition

76. Arthritis and Calcium-Containing Crystals: An Overview 699
    Daniel J. McCarty

77. Nucleoside Triphosphate (NTP) Pyrophosphohydrolase in
    Chondrocalcinotic and Osteoarthritic Cartilages 705
    David S. Howell, Jean Pierre Pelletier, Johanne Martel-Pelletier, Sara Morales,
    and Ofelia Muniz

78. Nucleation and Growth of CPPD Crystals and Related Species *in Vitro* .. 711
    Neil S. Mandel and Gretchen S. Mandel

79. Biological Effects of Calcium-Containing Crystals on Synoviocytes 719
    Herman S. Cheung and Daniel J. McCarty

80. Rotator Cuff Tendinopathies with Calcifications 725
    Kiriti Sarkar and Hans K. Uhthoff

Index 731

CALCIUM IN BIOLOGICAL SYSTEMS

A

METABOLIC AND FUNCTIONAL ASPECTS OF CALCIUM ACTION

I

General Aspects of Calcium and Cell Function

1

Historical and Biological Aspects of Calcium Action

Ronald P. Rubin

Historical Aspects

Our knowledge of the biological importance of calcium began about a century ago with the serendipitous discovery made by Sydney Ringer that a solution of sodium chloride plus tap water was more efficacious in maintaining cardiac contractility than a corresponding one constituted from distilled water [23]. Ringer concluded that the minute amount of calcium present in tap water antagonized the "injurious" effects of sodium. Locke and other investigators subsequently demonstrated that motor nerves exposed to pure sodium chloride lose their stimulant effects on muscle contractility which could be restored by adding calcium (and potassium) in the proper proportions [25]. So, by the early 20th century researchers had become well aware that normal biological activity depended on specific concentrations of calcium, sodium, and potassium, with calcium occupying a key position in the natural order of things.

In 1911 Mines, following up Locke's experiments, found that during perifusion of the frog gastrocnemius with 0.7% sodium chloride there was a gradual diminution in the response of the muscle to electrical stimulation of the sciatic nerve. The addition of calcium, strontium, or barium, but not magnesium, restored the response (Table I). Mines concluded that it was not mere electric charge, but some special chemical property that enabled calcium, barium, and/or strontium to interact selectively with unknown constituents within tissues [20]. This provided an important first step in our elucidation of calcium's role in biological systems, although we still do not understand the biochemical basis of this role. Nevertheless, evidence accumulated over the years generally substantiates the basic conclusion of Mines that only strontium, and at times barium, are effective substitutes for calcium in cell activation. Magnesium, on the other hand, acts as an inhibitor.

Another fundamental aspect of calcium's pivotal role in biological systems came to light when the early biologists noted the importance of calcium in maintaining the normal per-

---

Ronald P. Rubin • Division of Cellular Pharmacology, Medical College of Virginia, Richmond, Virginia 23298.

Table I. Influence of Divalent Cations on the Passage of Excitation from Nerve to Skeletal Muscle[a,b]

Calcium		Strontium		Barium		Magnesium	
+	−	+	−	+	−	+	−
6	1	11	4	7	4	0	5

[a]Modified from Mines [20].
[b]+, contraction of frog gastrocnemius in response to sciatic nerve stimulation; − lack of response.

meability of cells by its ability to protect against lysis by mechanical, pH, or osmotic effects. For example, McCutcheon and Lucke [19], using the sea urchin egg as an osmometer, found that after the addition of sodium or potassium, the permeability of the cell to water was even greater than in the presence of a nonelectrolyte such as dextrose (Table II). Calcium (and magnesium) had the opposite effect; in fact, calcium reduced permeability to approximately the same value observed with seawater.

This action of calcium on the barrier properties of cell membranes—whether or not the cells are electrically excitable—is viewed as a stabilizing action. The term *membrane stabilization* was coined by Guttman in 1940 to describe the ability of calcium to prevent depolarization by high potassium without altering the resting membrane potential [6]. Shanes expanded the concept to include any agent that interfered with alterations in the resting potential, such as local anesthetics, lanthanum, or magnesium [26]. While nerve and other tissues become less responsive to stimulation as the external calcium concentration is increased, the threshold for inducing electrical activity is markedly reduced and spontaneous firing is observed at low calcium concentrations, although the release of neurotransmitter is depressed (Fig. 1). Thus, we can invoke two major effects of calcium that may be viewed as reciprocal actions: the first, to stabilize the cell, making it more resistant to activators; and the second, as an activator itself, to promote cellular activity once inside the cell.

Testimony to the sometimes complex and multifaceted roles of calcium in biological systems was first provided by Heilbrunn in the three editions of *An Outline of General Physiology* published between 1937 and 1952. Heilbrunn [9] perceived that a basic property of all living cells was to utilize calcium. His "theory of cell stimulation" proclaimed that when a cell was stimulated by various means, the levels of ionized calcium within the cell—

Table II. The Effect of Certain Cations on Permeability of Sea Urchin Eggs to Water[a]

Medium	Permeability
Dextrose	0.093
NaCl + dextrose	0.129
KCl + dextrose	0.096
$CaCl_2$ + dextrose	0.054
$MgCl_2$ + dextrose	0.050
Seawater	0.048

[a]Modified from McCutcheon and Lucke [19].

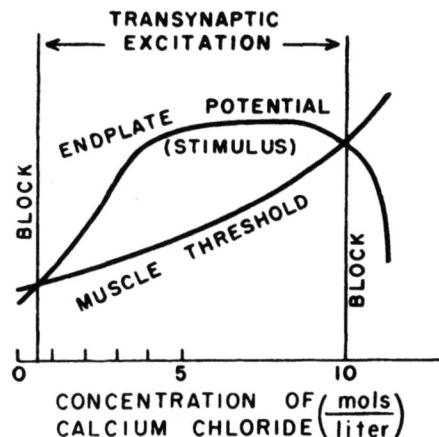

Figure 1. Diagrammatic construction of the relationship of endplate potential and excitability of muscle to concentration of calcium chloride in solution bathing a myoneural junction (frog). The excitation of the muscle by motor nerve impulses is blocked if the concentration is too high or too low. The magnitude of the endplate potential diminishes with the decrease in the calcium concentration due to a reduction in the amount of neurotransmitter delivered to the postsynaptic membrane. However, the threshold for excitation also decreases under calcium-deprived conditions. [Taken from Brink [1] with permission; as modified from G. Coppeé, *Arch. Int. Physiol.* **54**:334, 1946.]

which were very low—dramatically increased. Some of the putative functions that Heilbrunn proposed for ionic calcium included: (1) to promote cellular adhesion and/or intercellular communication, (2) effects on various enzyme systems, including ATPases and lipases, (3) regulation of permeability properties of cell membranes, (4) regulation of cell division, (5) control of metabolic activity of the cell—even before the relationship between mitochondrial calcium uptake and oxidative metabolism had been established, (6) effect of calcium on gel–sol transition states in the cytosol, and (7) a putative role in cell death. Heilbrunn even prophesied the pivotal role that calcium would play in the actions of drugs, e.g., cardiac glycosides, and proposed that many pharmacological agents express their effects by mobilizing this cation. His "theory of anesthesia," suggesting that the action of anesthetics may be associated with the loss of calcium from cell membranes, is presently an accepted theory for explaining the membrane actions of anesthetics. These agents are thought to displace membrane calcium, themselves bind to the membranes, and generally depress passive fluxes of cations [18].

Biological Importance of Calcium

Because of its unique properties, calcium is the most crucial cation in terms of diversity of function. While 95% of the calcium in the body is sequestered in bone, this reservoir is in dynamic equilibrium with the remaining pools of calcium serving to maintain a highly coordinated mechanism for calcium homeostasis. The cell is surrounded by fluid whose composition is kept constant by neural and endocrine mechanisms. There are large concentration differences between inside and outside the cell which are maintained by uphill movements, and the large electrochemical gradient enables the free calcium concentration of the cell to be increased rapidly and dramatically.

The functions of calcium in biological systems are not only diverse and multifaceted, but also fundamental in nature. In addition to providing rigidity to the organism by being a major constituent of bone, calcium exerts significant effects on the structural and functional properties of biomembranes. Membranes are simply less leaky in the presence of adequate amounts of calcium. Calcium also carries charge across biological membranes of excitable cells and influences excitability by affecting the kinetics of sodium and potassium permeabilities [22]. Pathways for excitation in biomembranes are channel discrete structures capable of selective ionic flux. Calcium is essential for the functioning of ionic channels by binding to fixed negative charges on the surface of the cell [10]. Calcium is also implicated in optimizing response of receptors to membrane-active agents [29].

The microtubules and microfilaments, constituting the cytoskeletal system, provide the template for maintaining the normal architecture of the cell. Microtubules, composed of tubulin, appear to participate in intracellular transport by providing directionality to intracellular movement. Calcium, by inducing microtubule depolymerization, may regulate microtubule assembly and disassembly [13]. Microfilaments, composed of actin and myosin, may provide the motive force for intracellular movement [4,11]. In this context, calcium may express its effects on cellular activity either by the activation of myosin light-chain kinase or by a direct action on F(fibrous)-actin [2,14,28].

Cellular expressions of motility—representing properties and interactions of filaments—are cytoplasmic streaming, cell shape changes, ciliary motion, and muscle contraction [4,11,28]. The cellular mobilization of secretory granules and their ultimate export from the cell by exocytosis is another prime example of this critical role that calcium plays in many fundamental cellular processes [25]. In a similar fashion, calcium plays a pivotal role in delivering nutrients to the cell, being a positive modulator of endocytosis and axoplasmic flow [8,21]. By analogy with the known effects of calcium in stimulating contractile processes and motility in somatic cells, calcium triggers the increase in motility associated with sperm–egg interactions (acrosome and cortical reactions) [7]. Calcium also positively modulates the proliferation of epithelial and mesenchymal cells by regulating DNA synthesis and mitotic activity in concert with other cellular mediators [30].

At another level, calcium is not only crucial for enabling cells to adhere to one another, but calcium deprivation causes cells to uncouple, compromising the ability of cells to communicate with one another [17]. This intercellular communication appears vital for optimizing cellular function by facilitating the exchange of ions and macromolecules through gap junctions. An excess of calcium is also associated with the loss of cell coupling so that uncoupling of electrical activity in adjacent cells is one parameter by which electrophysiologists monitor augmented levels of ionic cellular calcium [12].

The relatively high affinity of calcium for negatively charged sites on the surface of the cell [10] helps to explain the rather unique role that this cation plays in influencing excitable membranes, cell permeability, and cell communication. These sites may be constituted of acidic phospholipids. Displacement of calcium from such putative sites initiates cellular activation. A second stage may then involve the inward movement of extracellular calcium or the mobilization of cellular calcium to initiate cellular activation.

But just as important as calcium is in initiating and maintaining cellular activity by its protean actions, it may also be a putative mediator of cell injury and death [5]. The mechanism for calcium accumulation by mitochondria is quite sensitive to ischemic injury; this may constitute a primary insult during cell injury by raising the free ionic calcium concentration to toxic levels. Alternatively, a consequence of cell injury may include an increase in membrane permeability to calcium. It is therefore mandatory that there exist intracellular calcium buffering mechanisms to prevent calcium from accumulating to levels intolerable to the cell. So, cellular calcium is held at extremely low levels by several different mechanisms, including energy-dependent pumps, e.g., Ca^{2+}-ATPase, and intracellular binding. Calcium is bound to anions and macromolecules, as well as intracellular organelles such as mitochondria and endoplasmic reticulum. These intracellular stores not only serve as a calcium buffering mechanism but also as a reservoir for calcium mobilization during activation.

Cellular Mechanisms of Calcium Action

The molecular mechanisms by which calcium-dependent signals are translated into biological responses are now under intense scrutiny. These calcium-initiated signals may be expressed through several intracellular receptors, including proteins and/or phospholipids. A

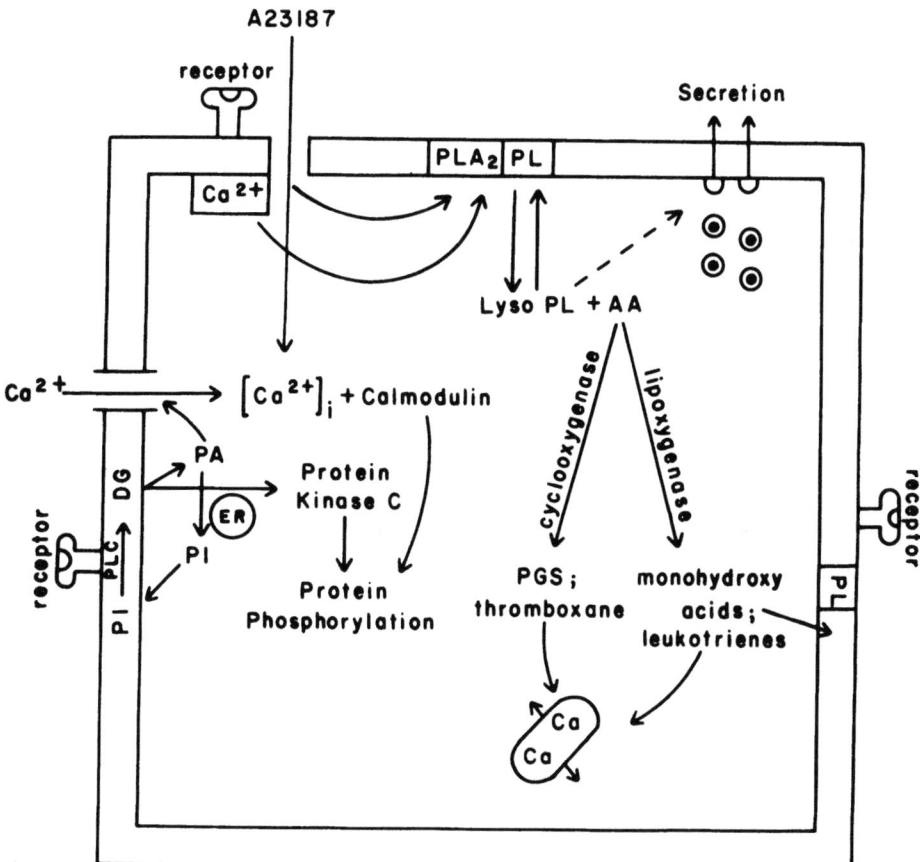

Figure 2. A hypothetical model depicting the role of calcium in cellular activation. Agonist- or ionophore-induced stimulation causes calcium to accumulate within the cytosol by entering from the extracellular fluid or by being released from cellular stores. The mobilization of calcium activates membranous phospholipase A_2 (PLA$_2$) to release from phospholipids (PL) arachidonic acid (AA), which serves as a substrate for synthesizing cyclooxygenase and lipoxygenase products. These metabolites may release calcium from intracellular organelles or incorporate into phospholipids of the cell membrane to modify subsequent responses to stimuli. The lysophospholipid (Lyso PL) may also participate in cellular activation by promoting membrane fusion reactions, e.g., exocytosis. However, a rapid reacylation of the lysophospholipid is deemed essential to prevent accumulation of these potentially cytotoxic substances. In addition to calcium mobilization, receptor-linked activation of protein kinase C, via phospholipase C-mediated phosphatidylinositol (PI) breakdown to diacylglycerol (DG), may be synergistic step in cell activation. Phosphatidic acid (PA), a component of the phosphatidate–PI cycle presumably taking place in the endoplasmic reticulum (ER), may serve as a physiological calcium ionophore.

number of calcium-binding proteins have been identified that have properties consistent with putative cellular calcium receptors, as for example troponin C of muscle; calcium-binding protein of the intestinal tract; and calmodulin [15]. Calmodulin evokes the most general interest since this ubiquitous protein is able to confer activation upon a number of enzymes [2,3]. In other words, calmodulin is able to act as a multifunctional calcium-dependent regulator. These calcium–protein interactions may be expressed by phosphorylation involving calmodulin which confers a calcium sensitivity upon specific protein kinases [2]. Such interactions may, for example, regulate cellular activity by activating adenylate cyclase or myosin light-chain kinase.

On the other hand, since many changes in phospholipid turnover are calcium-mediated, the actions of calcium may be expressed through enzymes that regulate phospholipid metabolism (Fig. 2). The relationship of phospholipase A_2 activity to calcium action may involve the release of arachidonic acid from position 2 of phospholipids, with the concomitant metabolism of arachidonic acid through the cyclooxygenase and lipoxygenase pathways [16,24]. The prostaglandins, hydroxy acids, and leukotrienes synthesized may serve as cellular mediators, or may be exported from cells to serve as autacoids. Alternatively, the agonist-induced activation of phospholipase C and the resulting cleavage of the polar head group of inositol phospholipids may provide the biochemical basis for calcium gating.

Yet, recent evidence indicates that a more holistic approach may be crucial for elucidating the molecular mechanism of calcium action (Fig. 2). A second species of calcium-dependent protein kinase (protein kinase C) has been described that is activated by diglyceride, a product of the phospholipase C-mediated reaction, rather than by calmodulin or cAMP for its activity [27]. Thus, phosphorylation of cellular proteins associated with calcium mobilization may be mediated through both calmodulin- and phospholipid-sensitive systems. The discovery of protein kinase C may have far-reaching implications because it enables us to generate new paradigms relative to the potential interactions of calcium, proteins, and phospholipids. Accepting the premise that protein phosphorylation is a crucial step in cell activation, then one may speculate that, along with calcium (and calmodulin), diacylglycerol—derived from the phospholipase C-mediated breakdown of inositol phospholipids—may play a key role in activation of protein kinases. Arachidonic acid metabolites may also participate in this series of molecular events by somehow interacting with protein kinase C. Thus, while calcium–protein interactions may help to explain the cellular actions of calcium, comprehension of the total picture requires that phospholipids also be taken into account. It is, therefore, apparent that insight into the molecular mechanisms mediating the diverse actions of calcium will stem from a multifaceted approach to this fundamental problem.

References

1. Brink, F. The role of calcium ions in neural processes. *Pharmacol. Rev.* **6**:243–298, 1954.
2. Brostrom, C. O.; Wolff, D. J. Properties and functions of calmodulin. *Biochem. Pharmacol.* **30**:1395–1405, 1981.
3. Cheung, W. Y. Calmodulin plays a pivotal role in cellular regulation. *Science* **207**:19–27, 1980.
4. Clarke, M.; Spudich, J. A. Nonmuscle contractile proteins: The role of actin and myosin in cell motility and shape determination. *Annu. Rev. Biochem.* **46**:797–822, 1977.
5. Farber, J. L. The role of calcium in cell death. *Life Sci.* **29**:1289–1295, 1981.
6. Guttman, R. Stabilization of spider crab nerve membranes by alkaline earths, as manifested in resting potential measurements. *J. Gen. Physiol.* **23**:343–364, 1940.
7. Gwatkin, R. B. L. *Fertilization Mechanisms in Man and Mammals*, New York, Plenum Press, 1977.

8. Hammerschlag, R. The role of calcium in the initiation of fast axonal transport. *Fed. Proc.* **39**:2809–2814, 1980.
9. Heilbrunn, L. V. *An Outline of General Physiology*, 3rd ed., Philadelphia, Saunders, 1952, pp. 105–740.
10. Hille, B. Gating in sodium channels of nerve. *Annu. Rev. Physiol.* **38**:139–152, 1976.
11. Hitchcock, S. E. Regulation of motility in non-muscle cells. *J. Cell Biol.* **74**:1–15, 1977.
12. Iwatsuki, N.; Petersen, O. H. Pancreatic acinar cells: Acetylcholine evoked electrical uncoupling and its ionic dependency. *J. Physiol. (London)* **274**:81–96, 1978.
13. Karr, T. L.; Kristofferson, D.; Purich, D. L. Calcium ions induce endwise depolymerization of bovine brain microtubules. *J. Biol. Chem.* **255**:11853–11856, 1980.
14. Korn, E. D. Actin polymerization and its regulation by proteins from nonmuscle cells. *Physiol. Rev.* **62**:672–737, 1982.
15. Kretsinger, R. H. Structure and evolution of calcium-modulated proteins. *CRC Crit. Rev. Biochem.* **8**:119–174, 1978.
16. Laychock, S. G.; Putney, J. W., Jr. Roles of phospholipid metabolism in secretory cells. In: *Cellular Regulation of Secretion and Release*, P. M. Conn, ed., New York, Academic Press, 1982, pp. 53–105.
17. Loewenstein, W. R.; Rose, B. Calcium in (junctional) intercellular communication and a thought on its behavior in intracellular communication. *Ann. N.Y. Acad. Sci.* **307**:285–307, 1978.
18. Low, P. S.; Lloyd, D. H.; Stein, T. M.; Rogers, J. A. Calcium displacement by local anesthetics. *J. Biol. Chem.* **254**:4119–4125, 1979.
19. McCutcheon, M.; Lucke, B. The effect of certain electrolytes and nonelectrolytes on permeability of living cells to water. *J. Gen. Physiol.* **12**:129–138, 1929.
20. Mines, G. R. On the replacement of calcium in certain neuro-muscular mechanisms by allied substances. *J. Physiol. (London)* **42**:251–266, 1911.
21. Prusch, R. D.; Hannafin, J. A. Sucrose uptake by pinocytosis in *Amoeba proteus* and the influence of external calcium. *J. Gen. Physiol.* **74**:523–535, 1979.
22. Putney, J. W., Jr. Stimulus–permeability coupling: Role of calcium in the regulation of membrane permeability. *Pharmacol. Rev.* **30**:209–245, 1979.
23. Ringer, S. A further contribution regarding the influence of the different constituents of the blood on the contraction of the heart. *J. Physiol. (London)* **4**:29–42, 1883.
24. Rubin, R. P. Calcium–phospholipid interactions in secretory cells: A new perspective on stimulus–secretion coupling. *Fed. Proc.* **41**:2181–2187, 1982.
25. Rubin, R. P. *Calcium and Cellular Secretion*, New York, Plenum Press, 1982.
26. Shanes, A. M. Electrochemical aspects in excitable cells. *Pharmacol. Rev.* **10**:59–273, 1958.
27. Takai, Y.; Minakuchi, R.; Kikkawa, U.; Sano, K.; Kaibuchi, K.; Yu, B.; Matsubara, T.; Nishizuka, Y. Membrane phospholipid turnover, receptor function, and protein phosphorylation. *Prog. Brain Res.* **56**:289–301, 1982.
28. Taylor, D. L.; Heiple, J.; Wang, Y.-L.; Luna, E. J.; Tanasugarn, L.; Brier, J.; Swanson, J.; Fechheimer, M.; Amato, P.; Rockwell, M.; Daley, G. Cellular and molecular aspects of amoeboid movement. *Cold Spring Harbor Symp. Quant. Biol.* **46**:101–111, 1982.
29. Triggle, D. J. Receptor–hormone interrelationships. In: *Membrane Structure and Function*, Volume 3, E. E. Bittar, ed., New York, Wiley, 1980, pp. 1–58.
30. Whitfield, J. F.; Boynton, A. L.; MacManus, J. P.; Silkorska, M.; Tsang, B. K. The regulation of cell proliferation by calcium and cyclic AMP. *Mol. Cell. Biochem.* **27**:155–179, 1979.

2

Calcium Ion
A Synarchic and Mercurial But Minatory Messenger

Howard Rasmussen

It was almost a century ago when Sidney Ringer demonstrated that extracellular Ca^{2+} is essential for the normal functioning of the isolated frog heart [19]. Although it was assumed for some time that Ca^{2+} stabilized the plasma membrane of the heart cell, the key Ca^{2+} pool in terms of cardiac contractile function is the small pool of free Ca^{2+} in the cell cytosol, $[Ca^{2+}]_c$. The concentration of Ca^{2+} in this pool rises during systole and falls during diastole. The rise precedes the contractile response, and the fall precedes the relaxation process. Troponin C has been identified as the calcium receptor component of the contractile protein system [5]. Thus, we have come a long way toward defining the cellular basis of Ca^{2+} action in the cardiac cell. However, in the past 10–15 years, we have also come to realize that this messenger or coupling function of Ca^{2+} is not confined to the heart, or other excitable tissues, but is a universal one [17]. Ca^{2+} serves as a coupling factor or second messenger in the evocation of the specific response of almost every type of differentiated cell by its appropriate extracellular messenger, which may be a hormone, circulating metabolite, or neurotransmitter.

Thousands of publications appear annually concerning the manner in which this calcium messenger system participates in one type of cellular response or another. The information available is abundant, redundant, bewildering, and contradictory. Yet, from this storehouse of data certain patterns and relationships emerge. As a consequence, it is possible to present an impressionistic portrait of how this system performs its role in cell activation. In this portrait, three themes are interwoven: Ca^{2+} as a synarchic, mercurial, and minatory messenger.

Ca^{2+} as Synarchic Messenger

The discovery of cAMP and the elucidation of its role as a second messenger in the action of peptide and amine hormones [23] coincided with the identification of Ca^{2+} as a

---

Howard Rasmussen • Departments of Cell Biology and Internal Medicine, Yale University, New Haven, Connecticut 06510.

coupling factor between excitation and response in excitable tissues [8]. For a brief time, it was thought that two fundamentally different modes of cell activation operated: in excitable tissues Ca^{2+} coupled stimulus to response and in nonexcitable tissues cAMP served this function. However, by the late 1960s, it became evident that a considerable overlap existed in the functions of these two intracellular messengers [16]. For example, it was clear that cAMP and the enzymes involved in its synthesis and degradation were abundant in neural tissues, and that Ca^{2+} played a role in hormone action in nonexcitable tissues [20].

Experimental evidence accumulated in the ensuing decade revealed that not only is there overlap between the two systems, but an intimate and universal interaction occurs between them [17]. Thus, Ca^{2+} and cAMP often serve together in regulating a particular process or response. Changes in [cAMP] influence cellular Ca^{2+} metabolism, and changes in [Ca^{2+}] influence cellular cAMP metabolism. From the experimental evidence, which demonstrates this widespread interaction between Ca^{2+} and cAMP, the concept of *synarchic* regulation was put forth in 1981 [17]. The concept avers that Ca^{2+} and cAMP nearly always serve together as intracellular messengers in modulating cellular responses to those extracellular messengers that provoke a particular cell type to perform its specific function, e.g., secrete glucose or a steroid hormone, transport H_2O, or contract.

The schematic representation of this universal cellular control system presented in Fig. 1 emphasizes that there are two interacting pathways by which information flows from cell surface to cell interior, and that there are multiple interactions between events in one pathway with those in the other. However, the figure depicts all known possible routes of interaction between the two, and in that sense is misleading because in actual operation not all possible routes are expressed in any particular cell type.

This means that in reality, this universal system exhibits various patterns of interactions between its two limbs when it expresses its function in any particular cell type. At least five patterns are clearly discernible simply by considering the fashion in which the initial event in each pathway is brought about [17]. In some cases, the pathways act in a *coordinate* fashion; both intracellular messengers increase in concentration in response to a rise in the concentration of a single extracellular messenger. In a slight variation on this theme, a small rise in concentration of an extracellular messenger causes information flow in one limb, and a larger

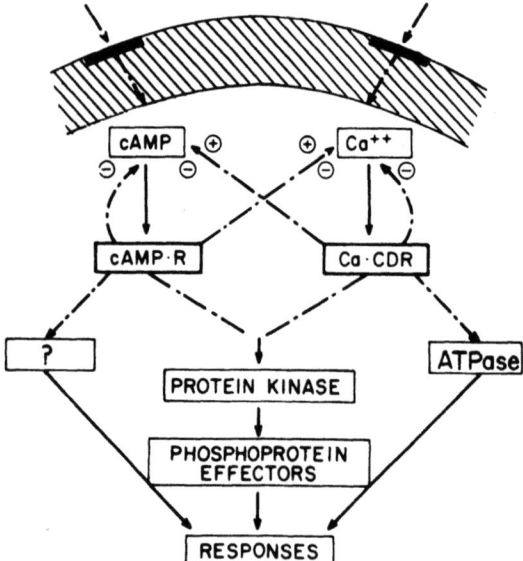

Figure 1. A schematic representation of the two interacting pathways of information flow from cell surface to cell interior which employ respectively cAMP and Ca^{2+} as synarchic messengers in the regulation of cell function.

rise in the concentration of the same extracellular messenger causes flow in the second limb. In such cases, the flow in the second limb reinforces that in the first and a greater cellular response results. This type of *hierarchical* pattern is also achieved by having different extracellular messengers control flow in the two different limbs.

Employing two different extracellular messengers that each produces flow in only one of the limbs, also produces an *antagonistic* pattern, if the effect of one intracellular messenger is opposed or blunted by the action of the other. A *redundant* pattern can also be observed in which each limb is controlled by a separate extracellular messenger, and a rise in concentration of either intracellular messenger alone leads to the same cellular response. Finally, a *sequential* pattern is seen when one of the intracellular messengers initiates the flow of information in the other pathway.

These various patterns of synarchic regulation are discussed more fully in a previous publication [17]. The importance of recognizing the nature and expression of this universal cellular control device lies in both practical and philosophic realms. From a practical point of view, it is not meaningful to study the role of cAMP in a particular cellular response without considering the role of Ca^{2+} and vice versa. From a philosophic point of view, an appreciation of the universal as well as the particular features of synarchic regulation allows one to understand both the evolutionary adaptability of this cellular control device, and the great plasticity of its expression in a given cell type under different metabolic circumstances.

Ca^{2+} as Mercurial Messenger

In order to appreciate the organizational elegance of the calcium messenger system, it is necessary to understand two related aspects of its function: the cellular metabolism of Ca^{2+} and the changes it undergoes during cell activation; and the nature of the molecular interactions between calcium ion, its intracellular receptor proteins, and the response elements whose functions are controlled by these interactions.

Cellular responses to extracellular messengers can be characterized as following one of three patterns. In the mast cell, for example, the response to a given stimulus is brief even in the sustained presence of the extracellular messenger [11]. The reason for this transient response is that the opening of the calcium gate in the plasma membrane is brief, so the rise of $[Ca^{2+}]_c$ is both brief and restricted to a small intracellular domain immediately beneath the site of Ca^{2+} influx. In a cardiac cell, there is an inherent contraction–relaxation cycle that employs Ca^{2+} even when the cell is in its basal state [5]. Extracellular stimuli alter either the amplitude and/or the frequency of this cyclic response. Many other cells, including hepatocytes, renal tubular cells, and exocrine or endocrine secretory cells, respond to the continued presence of the extracellular messenger with a sustained cellular response [4,17]. The remaining discussion will focus on cellular response of the latter type in the calcium messenger system.

The various cell types that display a maintained response to a sustained increase in the concentration of an extracellular messenger share a number of behavioral characteristics. First, upon activation there is both calcium release from an intracellular pool (plasma membrane and/or endoplasmic reticulum) and increased entry of calcium into the cell [21]. Second, there is a sharp, but transient, rise in the $[Ca^{2+}]_c$ followed by a fall to near basal values. Third, despite the transient rise in $[Ca^{2+}]_c$, there is a sustained high rate of calcium influx and enhanced cellular response [4]. Fourth, this augmented influx is nearly balanced by a high rate of calcium efflux, so that there is only a small rate of net accumulation of cellular calcium [4]. Fifth, the mitochondria serve as an intracellular sink for calcium [1], and the nonionic intramitochondrial pool of calcium provides a means by which the $[Ca^{2+}]_c$ can be stabilized at any desirable level.

Before considering the significance of this type of organizational arrangement, it is necessary to consider how small changes in $[Ca^{2+}]_c$ bring about dramatic changes in cell function. The most important answer to this question is that Ca^{2+} mediates its effects by binding to a specific class of intracellular proteins, termed *calcium receptor proteins*. The most ubiquitous and thoroughly studied is calmodulin [6], and its function will be considered in the ensuing discussion.

Cell activation may be viewed as a process of information flow from cell surface to interior [17]. One can identify at least seven steps in such an information flow sequence: (1) *recognition* of extracellular messenger by the cell surface receptor; (2) *transduction* of extracellular message into intracellular messenger at the plasma membrane; (3) *transmission* of messenger into the cell interior; (4) *recognition* of intracellular messenger by a specific receptor protein; (5) *modulation* of the function of other proteins, *response elements*, by the messenger–receptor protein complex; (6) *response* of the cell as a consequence of the altered functions of response elements; and (7) *termination* of response by effects at one of a variety of steps within this sequence. Viewed in this fashion, the classic model of cell activation in the calcium messenger system can be characterized as a process of *amplitude* modulation, i.e., a rise in the concentration (amplitude) of Ca^{2+} in the cell cytosol leads to an association of Ca^{2+} with receptor protein(s); and these Ca^{2+}–receptor complexes interact with response elements to modulate their function.

If the association reactions between Ca^{2+}, receptor protein, e.g., calmodulin, and response element, e.g., phosphodiesterase, were simple, noncooperative interactions, then the Ca^{2+}-binding properties of calmodulin would determine the $[Ca^{2+}]$ range in the cell cytosol over which amplitude modulation takes place. When Ca^{2+} binds to calmodulin, four Ca^{2+} are bound to one calmodulin molecule [15]. Binding begins at a $[Ca^{2+}]$ of 1 μM and is complete at 100 μM. These data imply that the $[Ca^{2+}]_c$ is below 1 μM in the nonactivated cell, and 100 μM in the fully activated cell. However, if one studies the Ca^{2+}-dependent activation of a response element such as phosphodiesterase *in vitro*, the results are quite

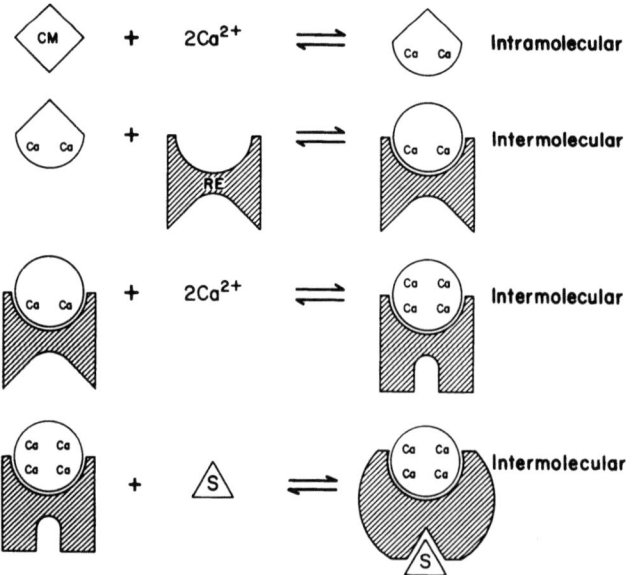

Figure 2. A stylized model of the four steps in the ordered, cooperative sequence of interactions between Ca^{2+}, calmodulin (CM), a response element (RE), and a substrate (S) of the response element when a rise in $[Ca^{2+}]$ in the cell cytosol leads to the activation of the response element. Each change in shape is meant to represent a change in conformation and is induced by either intramolecular or intermolecular events.

different [14]. Activation begins at 0.2 μM and is maximal at 1.0 μM. This dramatic difference between the Ca^{2+} binding profile of calmodulin and the Ca^{2+} activation profile of phosphodiesterase is due to two factors. First, calmodulin is in considerable excess to phosphodiesterase, and second, the interactions between Ca^{2+}, calmodulin, phosphodiesterase, and the substrate (cAMP) are ordered and highly cooperative.

A model depicting these ordered, cooperative association reactions is given in Fig. 2. Four separate steps can be identified. First is the cooperative binding of two Ca^{2+} to calmodulin as a consequence of a rise in $[Ca^{2+}]_c$; second, the association of $Ca_2 \cdot$calmodulin with the response element. The binding of the response element alters the affinity of $Ca_2 \cdot$calmodulin for Ca^{2+} at the two other Ca^{2+}-binding sites so that $Ca_2 \cdot$calmodulin·RE binds two more Ca^{2+}. Third, the binding of these last two Ca^{2+} leads to a conformational change in the response element converting it from a low-activity to a high-activity form; and fourth, the binding of substrate to the high-activity form leads to a further allosteric modification of response element structure so that the association of $Ca_4 \cdot$calmodulin to RE is even tighter, i.e., it induces a positively cooperative association [7]. These kinetic properties suggest that once the system is activated by calcium, it will remain activated even if the $[Ca^{2+}]_c$ falls from its peak value, i.e., it displays hysteresis [18].

The kinetic behavior of calmodulin-regulated response elements, and the changes in cellular calcium metabolism which occur upon activation of the cell, are two lines of

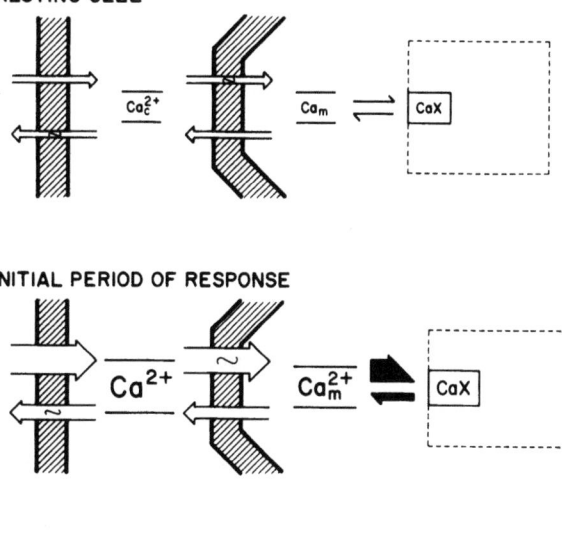
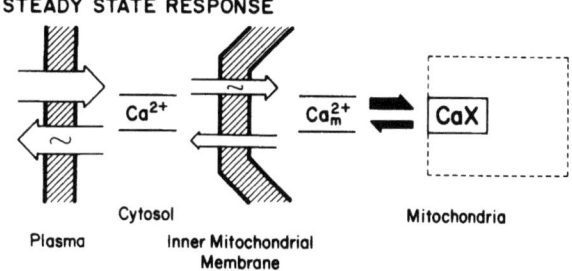

Figure 3. A schematic representation of cellular calcium metabolism during the process of amplitude modulation in the calcium messenger system. In the resting or nonactivated cell (top) the calcium concentration in the cell cytosol $\overline{Ca_c^{2+}}$ is low, as is that in the mitochondrial matrix space $\overline{Ca_m^{2+}}$. The nonionic pool of calcium (CaX) in the mitochondria is also relatively small, and the rates of calcium fluxes across both the plasma (left) and the mitochondrial (middle) membranes are small and balanced. During the initial period of cellular response to a hormone (middle), there is a release of Ca^{2+} from an intracellular pool (plasma membrane and/or endoplasmic reticulum—not shown) and an increased Ca^{2+} entry into the cell across the plasma membrane. These changes lead to an increase in the $[Ca^{2+}]$ in the cytosol. This rise in cytosolic $[Ca^{2+}]$ leads to the activation of the mitochondrial calcium pump resulting in the uptake of Ca^{2+} into both ionic and nonionic mitochondrial calcium pools. The rise in $[Ca^{2+}]_c$ leads to the allosteric (calmodulin) activation of the Ca^{2+} pump in the plasma membrane. The increase in activity of this pump, the cessation of further Ca^{2+} release from the intracellular pool, and the uptake of Ca^{2+} into the mitochondria, all lead to a fall in $[Ca^{2+}]_c$ so that during the sustained steady-state response, the $[Ca^{2+}]_c$ is only slightly higher than that in the resting cell, but rates of both Ca^{2+} influx and efflux across the plasma and mitochondrial membranes are increased, and nearly balanced, so that there is very little net accumulation of calcium by the cell.

converging evidence that lead to a similar conclusion, i.e., small and transient changes in $[Ca^{2+}]_c$ shift intracellular, calcium-regulated systems from a state of low to a state of high activity. A critical feature of this control system, that links calcium metabolism to the activation of cellular response elements, is the coupling of calmodulin to the Ca^{2+} pump in the plasma membrane [24]. In other words, the Ca^{2+} pump is one of the calmodulin-regulated response elements. This process contributes significantly to the secondary fall in $[Ca^{2+}]_c$ after its initial sharp increase during the initial phase of cell activation.

It is now possible to recapitulate and, with some speculation, present a profile of the events which transpire during amplitude modulation in the calcium messenger system. When extracellular messenger binds to receptor, there ensues a release of Ca^{2+} from the endoplasmic reticulum–plasma membrane pool, and an increase in the rate of Ca^{2+} influx. As a consequence, $[Ca^{2+}]_c$ rapidly rises from 0.1 μM to 1–4 μM (Fig. 3), which is sufficient to shift calmodulin-controlled response elements to a state of high activity. One of these elements is the plasmalemmal Ca^{2+} pump. Activation of this pump, along with calcium sequestration within the mitochondria (Fig. 3 and see below), diminishes the $[Ca^{2+}]_c$ from 1.0–4.0 μM to 0.4–0.6 μM, where it is maintained. In this new steady state, there is a continued high rate of Ca^{2+} influx and efflux (Fig. 3) and the response elements maintain their high-activity states. Efflux almost balances influx, but a small net accumulation of Ca^{2+} occurs within the nonionic pool in the mitochondrial matrix space.

This is the profile of a *mercurial* messenger: a small and transient rise in messenger concentration switches the cell from one functional state to another, which is then maintained at a much lower steady-state messenger concentration.

Calcium as Minatory Messenger

A question that arises is why has such a finely tuned system evolved, employing such small changes in $[Ca^{2+}]_c$, when such an abundance of Ca^{2+} exists in the extracellular fluid? To answer this question, it is necessary to consider an experiment done half a century ago by Heilbrunn [13], who found that if he inflicted a small wound in the surface membrane of a marine egg, the cellular contents would begin to leak out. However, if Ca^{2+} was present in the extracellular medium, then a "cortical precipitation reaction" took place, in which the wound healed, and the egg could then function normally. In the absence of Ca^{2+}, this reaction did not occur. The cellular contents continued to be extruded and the cell died.

Interesting as these findings were, another observation is more the focus of our attention. If excessive Ca^{2+} was present in the extracellular medium at the time the injury was inflicted, the cell died even though the wound healed. Heibrunn observed a wave of cytolysis which began at the site of injury and swept across the cell. If this wave swept completely across the cell, death ensued, but if it reached only part way, the cell recovered. These experiments were, I believe, the first to demonstrate that excess Ca^{2+} is toxic to the cell. Evidence obtained, largely in the past 15 years, has led to the recognition of this aspect of cellular calcium metabolism: calcium is a *minatory* messenger.

To put this feature of Ca^{2+} and cell function in perspective, it is necessary to consider the role of mitochondria in cellular metabolism. It has been known for 20 years that isolated mitochondria accumulate large amounts of calcium by an energy-dependent mechanism [1,4,9,17,25]. Yet, serious consideration has been given only recently to the physiological function of this mitochondrial activity.* Despite the considerable controversy that surrounds

---

*In recent years, a number of investigators have attempted to develop models and hypotheses concerning the mechanism and significance of Ca^{2+} metabolism by mitochondria. These range from the suggestion that in

this topic [2], several facts have become clear. The first is that *in situ* mitochondria contain considerably less calcium ($\sim$ 0.5–1 nmole/ng protein) than originally perceived [2]. The second is that the free [Ca^{2+}] in the mitochondrial matrix space ([Ca^{2+}]$_m$) approximates that in the cell cytosol, despite existence of a large electric potential (inside negative) across the mitochondrial membrane [12]. The third is that the free Ca^{2+} is less than 0.1% of the total calcium within the mitochondrial matrix space; much of the remainder exists in a nonionic exchangeable pool largely as a calcium–phosphate–ATP complex [12]. The fourth is that there are separate calcium influx and efflux pathways in the inner mitochondrial membrane [1]. The fifth is that during cell activation or injury the calcium content of the mitochondria often increases [10].

These various facts are clues which tempt speculation on the role of mitochondria in cellular calcium homeostasis. They can be put into sharper focus by considering the kinetic characteristics of the plasma membrane and mitochondrial membrane pumps. When the activities of these two pumps (determined in cell fractions) are plotted as a function of the [Ca^{2+}]$_c$, the activity of the plasma membrane pump increases dramatically at [Ca^{2+}]$_c$ between 0.1 and 0.5–0.6 μM, while activity of the mitochondrial pump increases gradually. However, the activity of the mitochondrial pump selectively increases dramatically at [Ca^{2+}]$_c$ above 0.6 μM, so above 1.0 μM this pump becomes the predominant pathway by which Ca^{2+} leaves the cell. These data indicate that in the [Ca^{2+}] range of 0.1 to 0.6 μM, the major route of calcium egress out of the cytosol is most likely through the plasmalemmal calcium pump, whereas above 0.8 μM the major route would be across the mitochondrial membrane. Thus, during prolonged activation, [Ca^{2+}]$_c$ is kept below 0.6 μM presumably to minimize net accumulation of Ca^{2+} by mitochondria and, thus, by the cell.

The two portions of the activity curve of the mitochondrial calcium pump imply that this pump functions in two different ways depending on whether the [Ca^{2+}]$_c$ is low (0.1–0.6 μM) or high (> 0.6 μM). The increase in net mitochondrial calcium uptake is gradual, between 0.1 and 0.6 μM Ca^{2+}. Both the inwardly directed pump and the outwardly directed efflux pathway across the inner mitochondrial membrane are functional. A rise in the [Ca^{2+}]$_c$ (e.g., from 0.1 μM to 0.5 μM) leads immediately to an increase in the rate of mitochondrial Ca^{2+} uptake. However, [Ca^{2+}]$_m$ changes slowly because it is in equilibrium with the large pool of nonionic exchangeable calcium ([Ca^{2+}]$_m$) [12]. If [Ca^{2+}]$_c$ remains at 0.5 μM, then this exchangeable mitochondrial pool will eventually fill, and the [Ca^{2+}]$_m$ will increase until the rate of calcium efflux from mitochondria equals influx. At this point a new steady state will be achieved. The value of this arrangement is that the nonionic exchangeable mitochondrial pool serves as a means of stabilizing [Ca^{2+}]$_c$. This stabilization is necessary because very small alterations in [Ca^{2+}]$_c$ produce very large changes in the cellular response, and because the pool of free Ca^{2+} in the cell cytosol is vanishingly small in comparison with the pools in the endoplasmic reticulum (100 μmoles), mitochondria (100 μmoles), and extracellular fluids (1000 μmoles).

This stabilizing function of the mitochondrial calcium sink at low [Ca^{2+}]$_c$, although vitally important to the proper functioning of the calcium messenger system, also may contribute considerable inertia in the system, making it difficult to rapidly alter the [Ca^{2+}]$_c$. In order to overcome this problem, it would be necessary to supply considerably more Ca^{2+} during cell activation than that required simply for raising [Ca^{2+}]$_c$ from 0.1 μM to 1.0 μM, and saturating the calmodulin pool in the cytosol. To accomplish these latter requirements

---

unperturbed cells, mitochondria contain insignificant quantities of Ca^{2+}, to the view that the mitochondria may be the primary source of internal Ca^{2+} involved in cell regulation. Interested readers are referred to a number of recent reviews which deal with these ideas [1,3,4,9,22,25].

(assuming no buffering or sequestering of the Ca^{2+}) would demand less than 10 μmoles of Ca^{2+}. At the time of activation, the amount of calcium released into or taken up into the cell cytosol may be of the order of 90–120 μmoles per liter of cell H_2O, or nearly 10 times the amount needed to activate the calmodulin-controlled response elements. This apparent large excess of Ca^{2+} may be required to overcome the inertia built into the system as a consequence of linking the nonionic exchangeable calcium pool in the mitochondrial matrix space to the free pool in the cytosol.

When the $[Ca^{2+}]_c$ exceeds 0.6 μM, the slope of the uptake profile of the mitochondrial pump is steep, and the predominant pathway operating is the influx (or pump) pathway. The efflux pathway operates maximally at values of approximately 0.5 μM. Under conditions of high $[Ca^{2+}]_c$ the mitochondria serve as a reservoir for storing excess Ca^{2+}. Their capacity to store calcium is considerable but finite; and when this capacity is exceeded, cell dysfunction and cell death follow. Calcium is truly a threatening or minatory messenger.

Plasticity of Metabolic Control by the Calcium Messenger System

A survey of our knowledge of the calcium messenger system would be incomplete without considering another feature of its organizational expression. In the preceding discus-

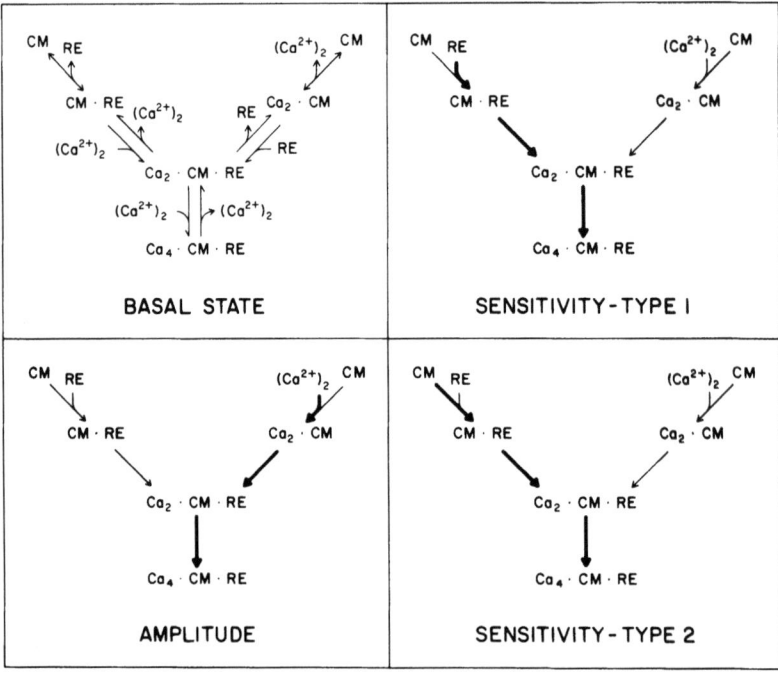

Figure 4. A schematic representation of amplitude and two types of sensitivity modulation of the function of a response element (RE) whose activity is regulated by Ca^{2+} and calmodulin (CM). The basal state of this system is depicted in the left upper panel. The affinity of CM for RE is low and the concentration of Ca^{2+} is low so there is very little Ca_4·CM·RE (the active form of the response element) present. In the left lower panel, the change seen during amplitude modulation is depicted. A rise in $[Ca^{2+}]$ leads to the formation of Ca_2·CM and this in turn leads to Ca_2·CM·RE and then to Ca_4·CM·RE. In the right upper panel, the changes in type 1 sensitivity modulation are represented. In this case, a change in RE structure leads to an increase in the affinity of RE for CM. Hence, Ca_2·CM·RE is formed without a rise in $[Ca^{2+}]$, and then Ca_4·CM·RE forms. In the right lower panel, the changes in type 2 sensitivity modulation are depicted. An increase in CM concentration leads to an increase in CM·RE formation, and then to Ca_2·CM·RE and Ca_4·CM·RE.

sion, the various features of amplitude modulation were considered. However, response element function can also be modulated by the process of *sensitivity* modulation [17]. By definition, sensitivity modulation is a change in the concentration of activated response element ($Ca_4 \cdot$calmodulin$\cdot$RE) without a change in $[Ca^{2+}]_c$. Such a change can occur as a consequence of: an increase in calmodulin or response element concentration; or a change in calmodulin or response element structure, leading to a change in their affinity for one another (Fig. 4).

The most thoroughly studied models of sensitivity modulation are ones in which cAMP-dependent phosphorylation of a response element leads to a change in the association of response element to calmodulin [17]. Two examples are phosphorylase b kinase which has an increase in its affinity for calmodulin, i.e., it displays positive sensitivity modulation; and myosin light-chain kinase, which has a decrease in its affinity for calmodulin, i.e., it displays negative sensitivity modulation.

Aside from supplying a molecular basis for some aspects of synarchic regulation, recognition of the process of sensitivity modulation prompts the realization that even though almost every cell employs this stereotyped control device to couple stimulus to response, the operation of sensitivity modulation permits variable cellular responses to a given change in $[Ca^{2+}]_c$: calcium-regulated cellular events manifest considerable *plasticity* of response. Sensitivity modulation can also allow the selective activation of only one of many calcium-regulated events, or prevent one or more calcium-regulated events from occurring even in the face of a rise in $[Ca^{2+}]_c$. Likewise, it can provide a means by which the different Ca^{2+}-regulated events will exhibit quantitatively varied responses to a change in $[Ca^{2+}]_c$.

In summary, the calcium messenger system functions as a link by which events at the cell surface are communicated to the cell interior. The system operates as an elegantly simple but highly plastic control device. In carrying out its role, calcium ion functions as a synarchic and mercurial, but minatory messenger.

ACKNOWLEDGMENTS. Work in the author's laboratory was supported by NIH Grant AM-19183 and by a grant from the Muscular Dystrophy Association.

References

1. Akerman, K. D. O.; Nicholls, D. G. Physiological and bioenergetic aspects of mitochondrial calcium transport. *Rev. Physiol. Biol. Pharmacol.* **95**:149–201, 1983.
2. Barnard, T. Mitochondrial matrix granules, dense particles and the sequestration of calcium by mitochondria. *Scanning Electron Microscopy* **11**:419–433, 1981.
3. Blaustein, M. P.; McGraw, C. F.; Somlyo, A. V.; Schweitzer, E. S. How is the cytoplasmic calcium concentration controlled in nerve terminals? *J. Physiol. (Paris)* **76**:459–470, 1980.
4. Borle, A. B. Control, modulation and regulation of cell calcium. *Rev. Physiol. Biochem. Pharmacol.* **90**:13–169, 1981.
5. Chapman, R. A. Excitation–contraction coupling in cardiac muscle. *Prog. Biophys. Mol. Biol.* **35**:1–52, 1979.
6. Cheung, W. Y. Calmodulin plays a pivotal role in cellular regulation. *Science* **207**:19–27, 1980.
7. Cheung, W. Y.; Lynch, T. J.; Wallace, R. W.; Tallant, E. C. cAMP renders Ca^{2+}-dependent phosphodiesterase refractory to inhibition by calmodulin-binding protein. *J. Biol. Chem.* **256**:4439–4443, 1981.
8. Ebashi, S. Excitation–contraction coupling. *Annu. Rev. Physiol.* **38**:293–313, 1976.
9. Fishkum, G.; Lehninger, A. L. Mitochondrial regulation of intracellular calcium. In: *Calcium and Cell Regulation*, W. Y. Cheung, ed., New York, Academic Press, 1982, pp. 39–79.
10. Fleckenstein, A. Drug-induced changes in cardiac energy. *Adv. Cardiol.* **12**:193–197, 1974.
11. Foreman, J. C.; Mongar, J. L. Calcium and the control of histamine secretion from mast cells. In: *Calcium Transport in Contraction and Secretion*, E. Carafoli, F. Clementi, W. Drabikowsky, and A. Margreth, eds., Amsterdam, North-Holland, 1975, pp. 175–184.
12. Hansford, R. G.; Castro, F. Intramitochondrial and extra-mitochondrial free calcium ion concentrations of

suspension of heart mitochondria with very low, plausibly physiological, contents of total calcium. *J. Bioenerg. Biomembr.* **14:**171–186, 1982.
13. Heilbrunn, L. V. *Dynamic Aspects of Living,* New York, Protoplasm Academic Press, 1956.
14. Huang, C. Y.; Chau, V.; Chock, P. G.; Wang, J. H.; Sharma, R. F. Mechanism of activation of cyclic nucleotide phosphodiesterase: Requirement of the binding of four Ca^{2+} to calmodulin for activation. *Proc. Natl. Acad. Sci. USA* **78:**871–874, 1981.
15. Klee, C. B. Conformational transition accompanying the binding of Ca^{2+} to the protein activator of 3'5'-cyclic adenosine monophosphate phosphodiesterase. *Biochemistry* **16:**1017–1024, 1977.
16. Rasmussen, H. Cell communication, calcium ion and cyclic adenosine monophosphate. *Science* **170:**404–412, 1970.
17. Rasmussen, H. *Calcium and cAMP as Synarchic Messengers,* New York, Wiley, 1981.
18. Rasmussen, H.; Waisman, D. M. Modulation of cell function in the calcium messenger system. *Rev. Physiol. Biochem. Pharmacol.* **95:**111–148, 1982.
19. Ringer, S. A further contribution regarding the influence of the different constituents of the blood on the contraction of the heart. *J. Physiol. (London)* **4:**29–49, 1883.
20. Rubin, R. P. *Calcium and Cellular Secretion,* New York, Plenum Press, 1982.
21. Schulz, I. Messenger role of calcium in function of pancreatic acinar cell. *Am. J. Physiol.* **239:**6335–6347, 1980.
22. Somlyo, A. P.; Somlyo, A. V.; Shuman, H.; Scarpa, A.; Endo, M.; Inesi, G. Mitochondria do not accumulate significant Ca concentrations in normal cells. In: *Calcium and Phosphate Transport across Biomembranes,* F. Bronner and M. Paterlik, eds., New York, Academic Press, 1981, pp. 87–93.
23. Sutherland, E. W.; Rall, T. W. Formation of cyclic adenine ribonucleotide by tissue particles. *J. Biol. Chem.* **232:**1065–1076, 1958.
24. Vincenzi, F. F.; Larsen, F. L. The plasma membrane calcium pump: Regulation by a soluble Ca^{2+} binding protein. *Fed. Proc.* **39:**2427–2431, 1980.
25. Williamson, J. R.; Cooper, R. H.; Hoek, J. B. Role of calcium in the hormonal regulation of liver metabolism. *Biochim, Biophys. Acta* **639:**243–295, 1981.

3

Intracellular Calcium as a Second Messenger
What's so Special about Calcium?

Mordecai P. Blaustein

Intracellular calcium serves as a second messenger for the control of a variety of cell functions, including secretion, contraction, phototransduction, cell division and differentiation, and potassium and sodium permeability. Since many aspects of calcium metabolism are strikingly similar in very diverse cell types, we will focus on certain general features of calcium metabolism that are applicable to a large variety of cells. Our objective is to gain some understanding of why calcium is such a good second messenger.

Intracellular Free Ca^{2+} and the Signal-to-Background Ratio

The resting intracellular free calcium concentration, $[Ca^{2+}]_{in(rest)}$, is very low in virtually all animal cells (10^{-7}–3×10^{-7} M) [1,13]. Kretsinger [36] has suggested that this situation may have initially evolved to avoid formation of insoluble $Ca_2(PO_4)_3$ in the cytosol, so that cells could utilize inorganic phosphate as a form of energy currency. However, even if the second messenger role of Ca^{2+} was only an evolutionary afterthought, the low $[Ca^{2+}]_{in(rest)}$ has been utilized to advantage in most cells. In addition to this low background, small changes in the absolute amount of Ca^{2+} in the cytosol can provide a very large signal-to-background (S-B) ratio:

$$\text{S-B ratio} = \frac{\Delta[Ca^{2+}]_{in}}{[Ca^{2+}]_{in(rest)}}$$

where $\Delta[Ca^{2+}]_{in}$ is the increment of $[Ca^{2+}]_{in}$ due to cellular activity. Since $\Delta[Ca^{2+}]_{in}$ may be as high as 10^{-6}–10^{-5} M (or perhaps even greater, in localized regions of cytoplasm; see below), the S-B ratio may range from 1 to 10 or even 100 or more, in some cells. The source of this Ca^{2+} may be the extracellular fluid or intracellular stores.

Mordecai P. Blaustein • Department of Physiology, University of Maryland, School of Medicine, Baltimore, Maryland 21201.

Normally, there is a very steep, inwardly directed electrochemical gradient for Ca^{2+} because the extracellular free Ca^{2+} is on the order of 10^{-3} M in vertebrates. The gradient is even higher in marine invertebrates [5], and resting membrane potentials are on the order of -40 to -90 mV (cytosol negative). Therefore, in cells with Ca^{2+}-selective ion channels [29,54], activation of channels by appropriate electrical or chemical signals allows Ca^{2+} to enter the cells rapidly. A good example is the presynaptic nerve terminal, in which much of the Ca^{2+} required to trigger neurotransmitter release comes from the extracellular fluid [24,33]. The Ca^{2+} enters the terminal through voltage-regulated Ca^{2+}-selective channels when the nerve terminals are depolarized [40,49,50].

Alternatively, in some cells most of the second messenger Ca^{2+} may be derived from intracellular stores during cell activation. A good example is the vertebrate skeletal muscle fiber, in which virtually all of the Ca^{2+} needed to activate contraction comes from the sarcoplasmic reticulum [18], a specialized form of smooth endoplasmic reticulum [e.g., 22].

In at least some types of cells, Ca^{2+} may have several second messenger functions. For example, in neurons, it is involved in neurotransmitter secretion [8,24,25,33,40], control of potassium conductance [45, and Chapter 21, this volume], and axoplasmic transport [15,30]. Since each of these activities must be regulated independently, and since they may occur simultaneously in different parts of the same cell, Ca^{2+} signals must exhibit spatial, as well as temporal, resolution (41,52). This occurs because the Ca^{2+} signals are well damped as a consequence of the fact that Ca^{2+} is strongly buffered and sequestered, and cannot readily diffuse through the cytoplasm [3,7,31,55].

A critical aspect of this regulatory function is the ability of cells to detect Ca^{2+} signals. Cells must have appropriate calcium-sensing molecules such as proteins that can rapidly detect Ca^{2+} signals in the physiological range (10^{-7}–10^{-5} M). Two good examples of such calcium-sensing proteins are the regulatory proteins, calmodulin [16,44] and troponin C of skeletal muscle [51]. These Ca^{2+}-dependent regulatory molecules have binding sites that wrap around a Ca^{2+} ion, thereby modifying molecular conformation (see below). This structural change is then translated into an alteration in the activities of regulated molecules such as myosin light-chain kinase or troponin I, to which the Ca^{2+}-dependent regulatory molecules are bound.

Terminating the Signal: Ca^{2+} Buffering and Sequestration

Not only must the signal be turned on, but it must also be terminated appropriately. In some situations the signal must be very crisp, and terminate abruptly. For instance, if Ca^{2+} is not removed from transmitter release sites at nerve terminals sufficiently rapidly, excess transmitter may accumulate in the synaptic cleft, and synaptic transmission may be significantly altered. Ca^{2+} may be removed from release sites at nerve terminals with a half-time of about 1 msec, as reflected in the decay of the transmitter release rate [34].

Spatial as well as temporal aspects are also important in termination of the signal. It may be necessary to prevent the Ca^{2+} signal from spreading to several different sensors in the same cell where Ca^{2+} may have multiple modulatory functions. A good example of a Ca^{2+}-binding suppressor (or Ca^{2+}-buffer) protein is the parvalbumin found in skeletal muscle [60,73]. Other cytoplasmic proteins may also be important for buffering the Ca^{2+} level in cells for other, perhaps special, purposes. For instance, the vitamin D-induced calcium-binding protein found in intestinal columnar epithelial cells, may help to maintain $[Ca^{2+}]_{in}$ low in the cytosol, while at the same time promoting Ca^{2+} transport from the mucosal to serosal border of intestinal cells [47,62,69]. This would be particularly important when demand for Ca^{2+} absorption is high and permeability of the mucosal (brush border) cell membrane to Ca^{2+} rises under the influence of vitamin D [20,53].

The binding capacity of cytoplasmic Ca^{2+}-binding proteins is limited, and other mechanisms have evolved for longer term and higher capacity cytoplasmic Ca^{2+} buffering, i.e., the intracellular Ca^{2+}-sequestering organelles. As mentioned above, these organelles may also serve as stores from which large amounts of Ca^{2+} can be rapidly released when cells are activated (as in the case of skeletal muscle).

Energized mitochondria accumulate Ca^{2+} with high affinity [39]. Therefore, it had long been assumed that these organelles are involved in the regulation of $[Ca^{2+}]_{in}$. However, several types of observations now indicate that the mitochondria from many types of cells do not accumulate much Ca^{2+} when the ambient free Ca^{2+} concentration is in the physiological range ($< 10^{-6}-10^{-5}$ M):

1. The affinity of mitochondria for Ca^{2+} is greatly reduced in the presence of physiological free Mg^{2+} concentrations which, for cytosol, are on the order of millimolar [32,68].
2. In the presence of physiological Ca^{2+} and Mg^{2+} concentrations, smooth endoplasmic reticulum appears to have a much higher affinity for Ca^{2+} than do mitochondria. Indeed, there appears to be a threshold for the mitochondrial Ca^{2+} uptake system, and these organelles do not appear to detect Ca^{2+} at $[Ca^{2+}]$ values below about 5–10 μM [10,14], which is already at the peak of the dynamic physiological range. These effects are illustrated in Fig. 1.
3. X-ray microprobe studies on appropriately prepared, quick-frozen tissue specimens

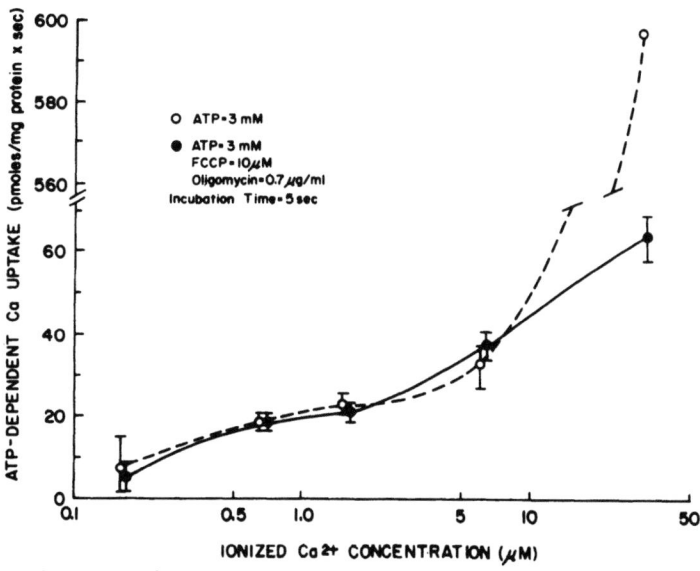

Figure 1. Effect of ionized Ca^{2+} concentration on ATP-dependent ^{45}Ca uptake by rat brain nerve terminals (synaptosomes) treated with saponin to expose the intraterminal organelles. ^{45}Ca uptake was measured in the absence of mitochondrial poisons to determine uptake into smooth endoplasmic reticulum (SER) and mitochondria (open circles), and in the presence of mitochondrial poisons (FCCP and oligomycin) to determine uptake into SER, alone (solid circles). The $[Ca^{2+}]$ was controlled with Ca^{2+}-EGTA buffers; synaptosomes were incubated with ^{45}Ca for 5 sec at 30 °C. [From Ref. 10, with permission.]

indicate that mitochondria from several types of cells contain very low levels of Ca^{2+} unless the tissue is damaged (as indicated by a high Na^+/K^+ ratio in the cytoplasm) [59,61].

Other studies have provided positive evidence that the smooth endoplasmic reticulum in various cell types (muscle, nerve, liver, adipocytes, etc.) can and, likely, does normally buffer free Ca^{2+} in the physiological range [4,14,41,42,56]. ATP-dependent Ca^{2+} transport mechanisms, virtually identical to those that operate in the sarcoplasmic reticulum, appear to

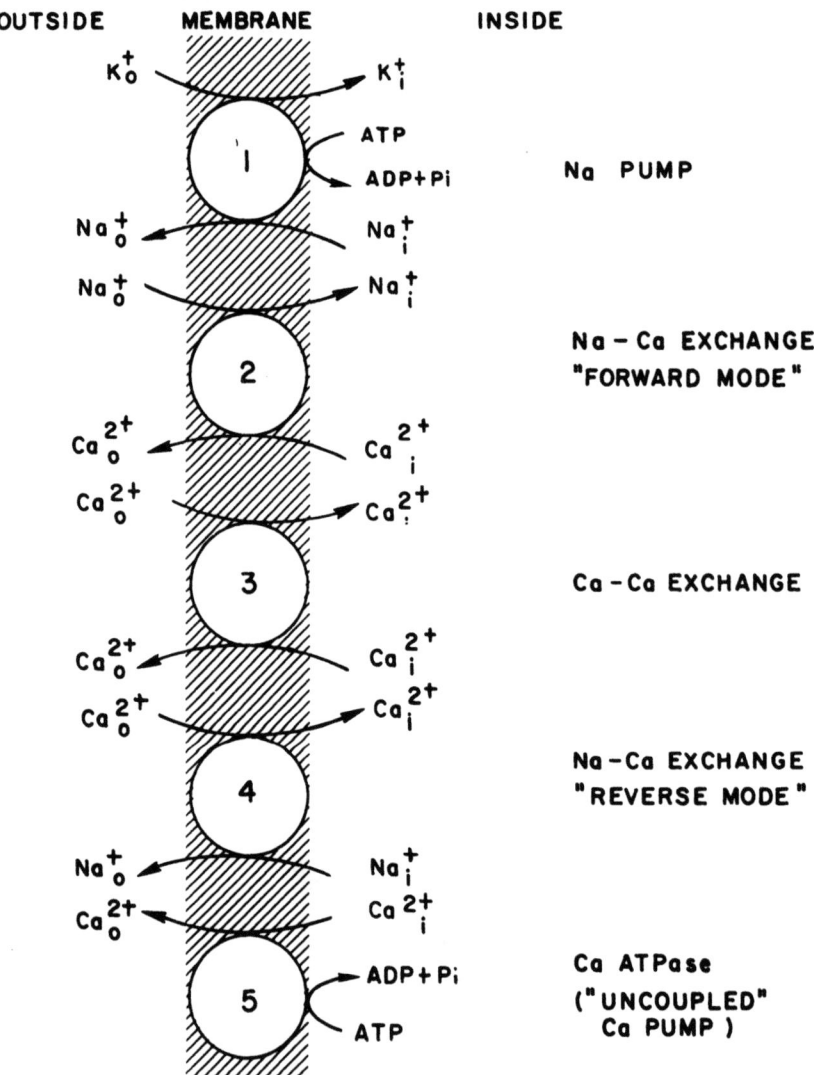

Figure 2. Diagrammatic representation of the Na^+–Ca^{2+} exchange (*2–4*) and Ca^{2+}-ATPase (*5*) modes of Ca^{2+} transport across the plasma membrane. Both transport systems may mediate Ca^{2+} extrusion (*2* and *5*, respectively). In addition, the Na^+–Ca^{2+} exchange system may operate in a "reverse mode"—bringing Ca^{2+} into the cell in exchange for exiting Na^+ (*4*), and it may also mediate Ca^{2+}–Ca^{2+} exchange (*3*). The Na^+ pump is shown at the top of the diagram (*1*). [From Ref. 6, with permission.]

be involved in sequestration of Ca^{2+} in the endoplasmic reticulum of nonmuscle cells [11,12,17,65].

Calcium Extrusion and Control of the Ca^{2+} Electrochemical Gradient

Ultimately, the cells must extrude the Ca^{2+} that enters passively, especially during periods of cell activity. Clearly, active transport mechanisms are required to maintain the very large electrochemical gradients for Ca^{2+} (mentioned above) that are observed in virtually all animal cells. Two general types of active transport mechanisms for Ca^{2+} have been described (see Fig. 2): (1) those that are fueled directly by ATP, and involve Ca^{2+}-ATPases, and (2) secondary active transport systems that are driven by energy from the Na^+ electrochemical gradient and involve Na^+-Ca^{2+} exchange. In some cells only a Ca^{2+}-ATPase is present and is responsible for maintaining the low $[Ca^{2+}]_{in}$. A classic example is the human red blood cell, which has no intracellular organelles to sequester Ca^{2+}. However, in nerve, muscle, secretory cells, intestinal and renal epithelial cells both Na^+-Ca^{2+} exchange and Ca^{2+}-ATPase mechanisms are present [5,6,9].

There is considerable uncertainty and controversy about which of the two mechanisms predominates in various cell types under normal physiological conditions. Nevertheless, in many of the aforementioned cell types, alterations in the Na^+ electrochemical gradient produce the types of changes in net Ca^{2+} fluxes and in $[Ca^{2+}]_{in}$ that are to be expected from the involvement of a Na^+-Ca^{2+} exchange mechanism [6,9]. This is seen, for example, in the heart, where twitch tension is directly correlated with the intracellular Na^+ level [37,38,58,70]; in neurons, where posttetanic potentiation of transmitter release is dependent on Na^+ accumulation in nerve terminals [2,46]; and in the transport of Ca^{2+} and Na^+ across intestinal and urinary tract epithelia [63,64, and see Chapter 24, this volume]. The implication is that $[Ca^{2+}]_{in}$ in these cells is regulated in part by Na^+-Ca^{2+} exchange, even though other Ca^{2+} extrusion systems may be present. It appears that Ca^{2+}-ATPase cannot compensate completely for the effects of changes in the Na^+ electrochemical gradient on the distribution of Ca^{2+} controlled by Na^+-Ca^{2+} exchange in these cells.

What's so Special about the Ca^{2+} Ion?

The aspects of Ca^{2+} metabolism that enable this cation to serve as an effective second messenger have been reviewed in the preceding part of this chapter and are summarized in Fig. 3. However, the main questions to be addressed are: Why is the Ca^{2+} ion so effective in this regard? Why not other ions? A logical answer is that, of all the abundant cations in biological systems, Ca^{2+} is particularly easy to select for. The combination of atomic size and charge density permits a high degree of selectivity for Ca^{2+} over Mg^{2+}, Na^+, and K^+, the other abundant cations. As a consequence of a few physicochemical constraints, molecules that bind Ca^{2+} much more effectively than other abundant cations, could be readily evolved. This is so despite the fact that Ca^{2+} and Mg^{2+} are both divalent, and that Ca^{2+} and Na^+ have virtually identical crystal radii (1.06 and 1.12 Å, respectively) [19,57].

In 1962, Eisenman [27,28] proposed a simple electrostatic model to explain monovalent cation selectivity in glass electrodes as a step toward understanding ionic selectivity in biological systems. The model was based on coulombic interactions between cations and anionic sites on the exchanger—either fully dissociated anions or the negative centers of

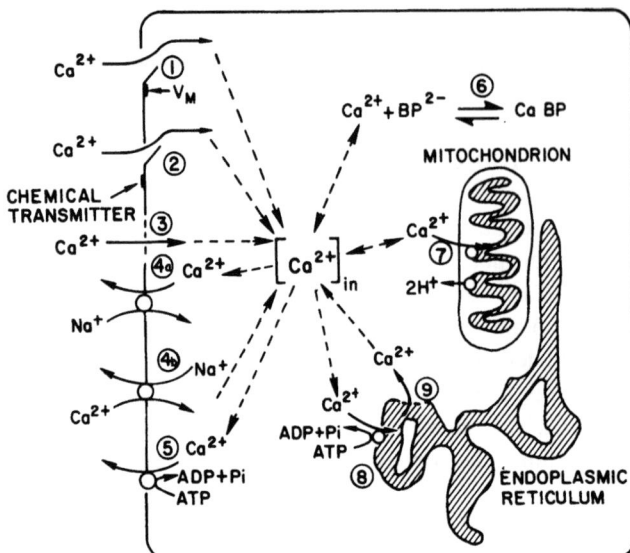

Figure 3. Diagrammatic summary of cellular Ca^{2+} metabolism. Ca^{2+} may enter cells via voltage-regulated (*1*) or chemical transmitter-activated (*2*) channels, through other (leakage?) pathways (*3*) or "reverse mode" Na^+-Ca^{2+} exchange (*4b*). Intracellular Ca^{2+} is buffered by cytoplasmic binding proteins (BP^{2-}; *6*) and can be sequestered in mitochondria (*7*) or endoplasmic reticulum (*8*); Ca^{2+} can also be triggered to enter the cytosol from the endoplasmic reticulum (*9*). Ca^{2+} extrusion takes place either via Na^+-Ca^{2+} exchange (*4a*) or a Ca^{2+}-ATPase (*5*).

dipoles such as carbonyl oxygens. The key parameter in this model was the "equivalent radius" of anionic sites (with these sites being treated as hard spheres). Selectivity was determined by the relative strengths of the cation–anion coulombic interactions. This simple model was very successful in predicting the allowable alkali metal ion selectivity sequences observed in ion-selective glasses and in biological systems.

Truesdell and Christ [66] extended this model to divalent cations, with the realistic assumption that the anionic binding sites in ion exchangers (both glasses and biological systems) are likely to be monovalent and multipolar rather than multivalent and monopolar. In this model, both the distance between the anionic sites, and the equivalent anionic site radii (or anionic field strength), are important for selectivity. The divalent-versus-monovalent cation selectivity critically depends on the distance between paired anionic sites; the divalent cation selectivity sequence depends in part on equivalent radii of the anionic sites. For example, a divalent cation will bind in preference to a monovalent cation of equal diameter if monovalent anionic sites are separated by a distance not much greater than the diameter of the cations. However, as the distance between the two negative charges increases, selectivity decreases and then reverses. Thus, with a large separation between two anionic sites (so that these anions act independently, and no interactions occur), binding of monovalent cations is preferred. This model predicts allowable alkaline earth ion selectivity sequences in biological systems [23]. Moreover, Nachshen [48] has recently applied the model successfully to Ca^{2+} channels in nerve membranes. The model helps to explain channel selectivity, as well as the relative potency of multivalent cation blockers.

The importance of steric factors and the distance between anionic sites is apparent from evidence that selectivity depends on how snugly a bound ion fits into ligand-binding sites. If the ion fits into the binding site too loosely, the coulombic interaction between cation and water oxygen will be much stronger than between cation and anionic sites on the ligand, so that the (latter) site will not substitute effectively for water of hydration. This is illustrated by the high degree of selectivity of arsenazo III for Ca^{2+} in preference to Mg^{2+} [35, and see Fig. 4]. The cavity formed by four of the arsenate oxygens and two hydroxyl oxygens (minimum cavity diameter = 2.15 Å) is sufficient to admit a Ca^{2+} ion (diameter = 2.12 Å), but too large to hold a Mg^{2+} ion (diameter = 1.56 Å) tightly. The cavity cannot be reduced

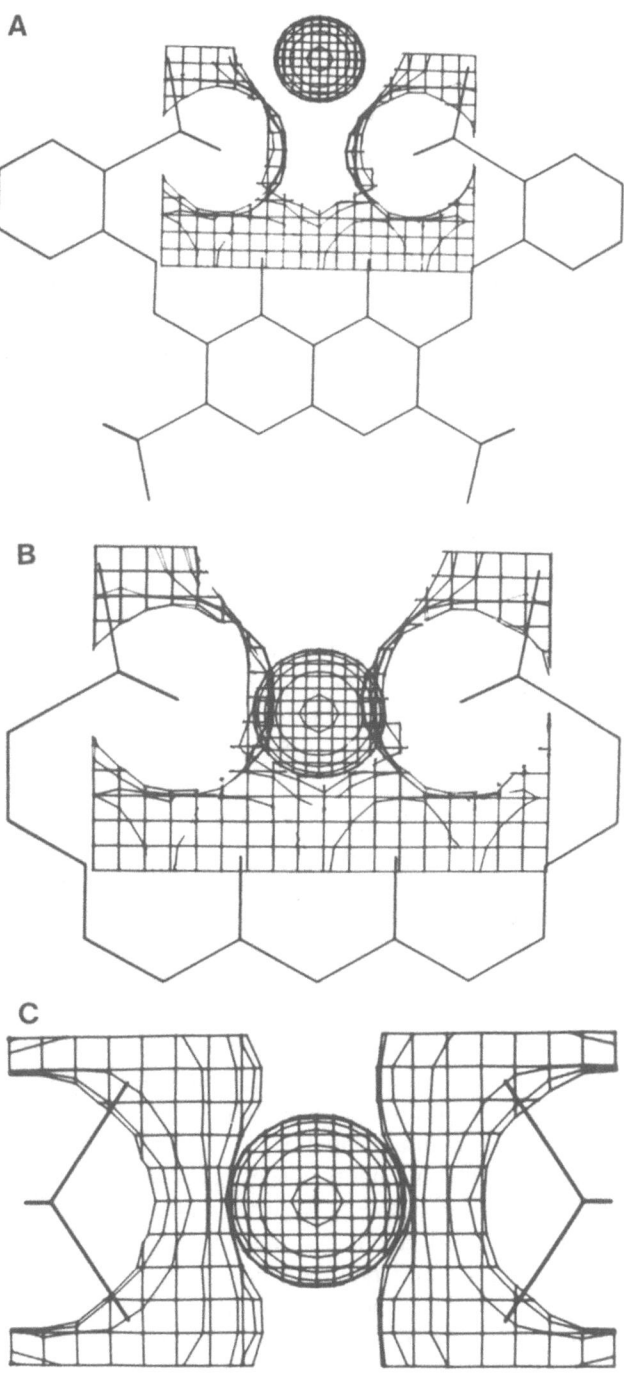

Figure 4. Molecular structure of arsenazo III planar conformation, calculated on a molecular modeling computer (MMS-X) developed at Washington University [71]. A and B show the molecule face-on. The interatomic bonds are seen distinctly. The cagelike region near the top of the molecule represents the (computed) surface formed around the van der Waals radii of the two hydroxyl oxygens that project upward from the naphthalene ring, and the oxygens from the two arsenate groups. The two oxygens from each arsenate that project downward, into the pocket, are superimposed on one another in this side view (A and B). However, one oxygen from each pair actually projects above the plane of the rest of the molecule, while the other oxygen projects below this plane; this relationship is readily observed when the molecule is viewed from on top (C). The circle above the arsenazo III, in A, corresponds to a Ca^{2+} ion with a radius of 1.0 Å [19,57]. In B, the Ca^{2+} ion has been moved into the pocket, to the position of closest approach to the hydroxyl oxygens. With the ion in the pocket, there is room for a water molecule to fit in over the Ca^{2+}. In C, the molecule is seen from on top, looking down into the pocket, with the Ca^{2+} ion touching the surface of the hydroxyl oxygens (as in B). The projecting van der Waals surfaces of the arsenate oxygens seem well designed to help hold the Ca^{2+} in the pocket with a near-perfect fit. The coordination number for Ca^{2+} in the pocket is 6 (octahedral)—or 7, if a water molecule covers the mouth of the pocket. [Unpublished data of C. D. Barry, N. C. Kendrick, and M. P. Blaustein.]

in size because of steric interference between adjacent carbons and oxygens that line the cavity. Thus, for Ca^{2+}, the hydration energy due to interaction between water oxygens and the Ca^{2+} can be compensated by the coulombic interaction between Ca^{2+}, and the oxygens of arsenazo III.

However, when a Mg^{2+} ion enters the cavity, it will sit too far away from some of the arsenazo III oxygens, so that electrostatic interaction will be weak and will not be able to compensate adequately for hydration energy (i.e., the energy required to strip away the inner hydration shell). A virtually identical explanation ("snugness" of fit) holds for the $Ca^{2+} \gg Mg^{2+}$ selectivity of EGTA [67]. Similar reasoning can be employed to explain why the neutral cyclic depsipeptide, valinomycin, exhibits a high degree of selectivity for K^+ over Na^+. The minimum cavity formed by the valinomycin molecule has a diameter of 2.7–3.1 Å, which is far too large for Na^+ (diameter = 2.24 Å). However, all six ester carbonyl oxygens can participate in ion–dipole interactions with K^+ [43].

Kretsinger [36] has described the high-affinity Ca^{2+}-binding sites of many proteins as "EF hands," whose essential features conform to the principles described above. The outstretched thumb and index finger each correspond to two turns of a helix, one ahead of, and one beyond the Ca^{2+}-binding site in the peptide chain. The coiled middle finger represents a 12-residue loop containing six Ca^{2+} coordinating ligands—oxygen-containing amino acid side chains and oxygens from peptide bonds. The oxygens provide octahedral coordination for a Ca^{2+}. This binding site structure, which has been highly conserved in evolution, is present in such diverse molecules as parvalbumin, calmodulin, troponin C, and the intestinal vitamin D-dependent Ca^{2+}-binding protein. As the Ca^{2+}-binding protein wraps around a Ca^{2+} ion, the protein conformation will be altered to confer regulatory properties on Ca^{2+}-binding proteins such as calmodulin and troponin.

The molecules that bind Ca^{2+} selectively with dissociation constants in the range of 10^{-7}–10^{-5} M must have similar ion-binding sites with the features mentioned above [36]—be they Ca^{2+}-dependent modulators, Ca^{2+} buffers, Ca^{2+} carriers, or Ca^{2+}-selective channels. In this context, the buffering and transport of Ca^{2+}, the generation of a large Ca^{2+} electrochemical gradient, and the possibility for large S-B ratios, are all consequences of the physicochemical principles outlined above. These features enable Ca^{2+} to serve as an effective second messenger.

In contrast, other abundant cations and Cl^- would be much less effective than Ca^{2+} as a relatively universal second messenger. The monovalent ions Na^+, K^+, and Cl^- all play an important role in cellular osmotic balance; however, their cytosolic concentrations are too high to generate large S-B ratios. On the other hand, Mg^{2+} does not normally play an important role in osmotic regulation. It is possible to develop ligands that select strongly for Mg^{2+} over Ca^{2+} by excluding ions significantly larger than Mg^{2+} from ligand-binding sites. However, even if the background concentration of free Mg^{2+} were very low ($\sim 10^{-6}$ M) Mg^{2+} would not be as effective as Ca^{2+}. Weakly acidic and neutral groups such as amines and imidazole bind Mg^{2+} in preference to Ca^{2+}, whereas Ca^{2+} is preferred by strongly acidic and multidentate ligands such as carboxylates, sulfonates, and phosphates [72]. Thus, a low $[Ca^{2+}]_{in(rest)}$ would be necessary to avoid having Ca^{2+} serve as a bridge between various anions and thereby inducing precipitation [72]. Furthermore, the greater charge density and larger hydration energy of Mg^{2+} would cause this ion to move on and off binding sites much more slowly than Ca^{2+} [26], so that Mg^{2+} "signals" would be much less "crisp" than Ca^{2+} "signals." Thus, Ca^{2+} stands alone among the abundant cations as a candidate for a relatively versatile second messenger role.

ACKNOWLEDGMENTS. I thank Drs. C. D. Barry and N. C. Kendrick for permission to use unpublished data; Drs. D. A. Nachshen and H. Rasgado-Flores for extremely fruitful discussions and for very helpful suggestions about the manuscript; and Mrs. A. Renninger for preparing the typescript. Supported by research grants from the Muscular Dystrophy Association, the NIH (NS-16106 and AM-32276), and the NSF (PCM-14027).

References

1. Ashley, C. C.; Campbell, A. K. (eds.) *Detection and Measurement of Free Ca^{2+} in Cells*, Amsterdam, Elsevier/North-Holland, 1979.
2. Atwood, H. L.; Charlton, M. P.; Thompson, C. S. Neuromuscular transmission in crustaceans is enhanced by a sodium ionophore, monensin, and by prolonged stimulation. *J. Physiol. (London)* **335**:179–195, 1983.
3. Baker, P. F.; Crawford, A. C. Mobility and transport of magnesium in squid giant axons. *J. Physiol. (London)* **227**:855–874, 1972.
4. Becker, G. L.; Fiskum, G.; Lehninger, A. L. Regulation of free Ca^{2+} by liver mitochondria and endoplasmic reticulum. *J. Biol. Chem.* **255**:9009–9012, 1980.
5. Blaustein, M. P. The interrelationship between sodium and calcium fluxes across cell membranes. *Rev. Physiol. Biochem. Pharmacol.* **70**:32–82, 1974.
6. Blaustein, M. P. The energetics and kinetics of sodium–calcium exchange in barnacle muscles, squid axons and mammalian heart: The role of ATP. In: *Electrogenic Transport: Fundamental Principles and Physiological Implications*, M. P. Blaustein and M. Lieberman, eds., New York, Raven Press, 1983, pp. 129–147.
7. Blaustein, M. P.; Hodgkin, A. L. The effect of cyanide on the efflux of calcium from squid axons. *J. Physiol. (London)* **200**:497–527, 1969.
8. Blaustein, M. P.; Nachshen, D. A.; Drapeau, P. Excitation–secretion coupling: The role of calcium. In: *Chemical Neurotransmission—75 Years*, L. Stjärne, P. Hedquist, A. Wennmalm, and H. Lagerkrantz, eds., New York, Academic Press, 1981, pp. 125–138.
9. Blaustein, M. P.; Nelson, M. T. Sodium–calcium exchange: Its role in the regulation of cell calcium. In: *Membrane Transport of Calcium*, E. Carafoli, ed., New York, Academic Press, 1982, pp. 217–236.
10. Blaustein, M. P.; Rasgado-Flores, H. The control of cytoplasmic free calcium in presynaptic nerve terminals. In: *Calcium and Phosphate Transport across Biomembranes*, F. Bronner and M. Peterlik, eds., New York, Academic Press, 1981, pp. 53–58.
11. Blaustein, M. P.; Ratzlaff, R. W.; Kendrick, N. C.; Schweitzer, E. S. Calcium buffering in presynaptic nerve terminals. I. Evidence for involvement of a nonmitochondrial ATP-dependent sequestration mechanism. *J. Gen. Physiol.* **72**:15–41, 1978.
12. Blaustein, M. P.; Ratzlaff, R. W.; Schweitzer, E. S. Calcium buffering in presynaptic nerve terminals. II. Kinetic properties of the nonmitochondrial Ca sequestration mechanism. *J. Gen. Physiol.* **72**:43–66, 1978.
13. Blinks, J. R.; Wier, W. G.; Hess, P.; Prendergast, F. G. Measurement of Ca^{2+} concentrations in living cells. *Prog. Biophys. Mol. Biol.* **40**:1–114, 1982.
14. Burgess, G. M.; McKinney, J. S.; Fabiato, A.; Leslie, B. A.; Putney, J. W., Jr. Calcium fluxes in saponin-permeabilized hepatocytes. *J. Biol. Chem.* **258**:15336–15345, 1983.
15. Chan, S. Y.; Ochs, S.; Worth, R. M. The requirement for calcium ions and the effect of other ions on axoplasmic transport in mammalian nerve. *J. Physiol. (London)* **301**:477–504, 1980.
16. Cheung, W. Y. Calmodulin plays a pivotal role in cellular regulation. *Science* **207**:19–27, 1980.
17. Colca, J. R.; McDonald, J. M.; Kotagal, N.; Patke, C.; Fink, C. J.; Greider, M. H.; Lacy, P. E.; McDaniel, M. L. Active calcium uptake by islet-cell endoplasmic reticulum. *J. Biol. Chem.* **257**:7223–7228, 1982.
18. Costantin, R. Activation in striated muscle. In: *Handbook of Physiology*, Section 1, *The Nervous System*, Volume 1, *Cellular Biology of Neurons*, Part I, E. Kandel, ed., Bethesda, American Physiological Society, 1975, pp. 215–259.
19. Cotton, F. A.; Wilkinson, G. *Advanced Inorganic Chemistry*, 3rd ed., New York, Interscience, 1972, pp. 189 and 206.
20. DeLuca, H. F. Vitamin D metabolism and function. *Arch. Intern. Med.* **138**:836–847, 1978.
21. Demaille, J. G. Calmodulin and calcium binding proteins: Evolutionary diversification of structure and function. In: *Calcium and Cell Function*, Volume II, W. Y. Cheung, ed., New York, Academic Press, 1982, pp. 111–144.
22. DeMeis, L.; Inesi, G. The transport of calcium by sarcoplasmic reticulum and various microsomal preparations. In: *Membrane Transport of Calcium*, E. Carafoli, ed., New York, Academic Press, 1982, pp. 141–186.
23. Diamond, J. M.; Wright, E. M. Biological membranes: The physical basis of ion and nonelectrolyte selectivity. *Annu. Rev. Physiol.* **31**:581–646, 1969.
24. Drapeau, P.; Blaustein, M. P. Calcium and neurotransmitter release: What we know and don't know. In: *Trends in Autonomic Pharmacology*, S. Kalsner, ed., Batimore, Urban & Schwarzenberg, 1982, pp. 117–130.
25. Drapeau, P.; Blaustein, M. P. Initial release of ^{3}H-dopamine from rat striatal synaptosomes: Correlation with calcium entry. *J. Neurosci.* **3**:703–713, 1983.
26. Eigen, M.; Winkler, R. Alkali ion carriers: Specificity, architecture, and mechanisms. *Neurosci. Res. Prog. Bull.* **9**:330–338, 1971.

27. Eisenman, G. On the elementary atomic origin of equilibrium ion specificity. In: *Membrane Transport and Metabolism*, A. Kleinzeller and A. Kotyk, eds., New York, Academic Press, 1962, pp. 169–179.
28. Eisenman, G. Cation selective electrodes and their mode of operation. *Biophys. J.* (Suppl.) 259–323, 1962.
29. Hagiwara, S.; Byerly, L. Calcium channel. *Annu. Rev. Neurosci.* **4**:69–125, 1981.
30. Hammerschlag, R. The role of calcium in the initiation of fast axoplasmic transport. *Fed. Proc.* **39**:2809–2814, 1980.
31. Hodgkin, A. L.; Keynes, R. D. Movements of labelled calcium in squid axons. *J. Physiol. (London)* **128**:28–60, 1957.
32. Hutson, S. M.; Pfeiffer, D. R.; Lardy, H. A. Effect of cations and anions on the steady state kinetics of energy-dependent Ca^{2+} transport in rat liver mitochondria. *J. Biol. Chem.* **251**:5251–5258, 1976.
33. Katz, B. *The Release of Neural Transmitter Substances*, Springfield, Ill., Thomas, 1968.
34. Katz, B.; Miledi, R. The role of calcium in neuromuscular facilitation. *J. Physiol. (London)* **195**:481–492, 1968.
35. Kendrick, N. C.; Ratzlaff, R. W.; Blaustein, M. P. Arenazo III as an indicator for ionized calcium in physiological salt solutions: Its use for determination of the CaATP dissociation constant. *Anal. Biochem.* **83**:433–450, 1977.
36. Kretsinger, R. H. Evolution of the informational role of calcium in eukaryotes. In: *Calcium Binding Proteins and Calcium Function*, R. Wasserman, R. Corradino, E. Carafoli, R. H. Kretsinger, D. MacLennan, and F. Siegel, eds., Amsterdam, Elsevier/North-Holland, 1977, pp. 63–72.
37. Lee, C. O.; Dagostino, M. Effect of strophanthidin on intracellular Na ion activity and twitch tension of constantly driven canine cardiac Purkinje fibers. *Biophys. J.* **40**:185–198, 1982.
38. Lee, C. O.; Vassalle, M. Modulation of intracellular Na^+ activity and cardiac force by norepinephrine and Ca^{2+}. *Am. J. Physiol.* **244**:C110–C114, 1983.
39. Lehninger, A. L. Mitochondria and calcium ion transport. *Biochem. J.* **119**:129–138, 1970.
40. Llinas, R.; Steinberg, I. Z.; Walton, K. Presynaptic calcium currents and their relation to synaptic transmission: Voltage clamp study in squid giant synapse and theoretical model for the calcium gate. *Proc. Natl. Acad. Sci. USA* **73**:2918–2922, 1976.
41. McGraw, C. F.; Nachshen, D. A.; Blaustein, M. P. Calcium movement and regulation in presynaptic nerve terminals. In: *Calcium and Cell Function*, W. Y. Cheung, ed., New York, Academic Press, 1982, pp. 81–110.
42. McGraw, C. F.; Somlyo, A. V.; Blaustein, M. P. Localization of calcium in presynaptic nerve terminals: An ultrastructural and electron microprobe analysis. *J. Cell Biol.* **85**:228–241, 1980.
43. McLaughlin, S.; Eisenberg, M. Antibiotics and membrane biology. *Annu. Rev. Biophys. Bioeng.* **4**:335–366, 1975.
44. Means, A. R.; Dedman, J. R. Calmodulin—An intracellular calcium receptor. *Nature (London)* **285**:73–77, 1980.
45. Meech, R. W. Calcium-dependent potassium activation in nervous tissues. *Annu. Rev. Biophys. Bioeng.* **7**:1–18, 1978.
46. Misler, S.; Hurlbut, W. P. Post-tetanic potentiation of acetylcholine release at the frog neuromuscular junction develops after stimulation in Ca^{2+}-free solutions. *Proc. Natl. Acad. Sci. USA* **80**:315–319, 1983.
47. Morrissey, R. L.; Empson, R. N., Jr.; Zolock, D. T.; Bickle, D. D.; Bucci, T. J. Intestinal response to $1\alpha,25$-dihydroxycholecalciferol. II. A timed study of the intracellular localization of calcium binding protein. *Biochim. Biophys. Acta* **538**:34–41, 1978.
48. Nachshen, D. A. Selectivity of the Ca binding site in synaptosome Ca channels: Inhibition of Ca influx by multivalent metal cations. *J. Gen Physiol.* **83**:941–967, 1984.
49. Nachshen, D. A.; Blaustein, M. P. Some properties of potassium-stimulated calcium influx in presynaptic nerve endings. *J. Gen. Physiol.* **76**:709–728, 1980.
50. Nachshen, D. A.; Blaustein, M. P. The influx of calcium, strontium and barium in presynaptic nerve endings. *J. Gen. Physiol.* **79**:1065–1087, 1982.
51. Potter, J. D.; Johnson, J. D. Troponin. In: *Calcium and Cell Function*, Volume II, W. Y. Cheung, ed., New York, Academic Press, 1982, pp. 145–173.
52. Putney, J. W., Jr.; Weiss, S. J.; Leslie, B. A.; Marier, S. A. Is calcium the final mediator of exocytosis in the rat parotid gland? *J. Pharmacol. Exp. Ther.* **203**:144–155, 1977.
53. Rasmussen, H.; Fontaine, O.; Max, E. E.; Goodman, D. B. P. The effect of 1α-hydroxyvitamin D_3 administration on calcium transport in chick intestine brush border membrane vesicles. *J. Biol. Chem.* **254**:2993–2999, 1979.
54. Reuter, H. Calcium channel modulation by neurotransmitters, enzymes and drugs. *Nature (London)* **301**:569–574, 1983.
55. Rose, B.; Loewenstein, W. R. Permeability of a cell junction and the local cytoplasmic free ionized calcium concentration: A study with aequorin. *J. Membr. Biol.* **28**:87–119, 1976.

56. Schweitzer, E. S.; Blaustein, M. P. Calcium buffering in presynaptic nerve terminals: Free calcium levels measured with arsenazo III. *Biochim. Biophys. Acta.* **600**:912–921, 1980.
57. Shannon, R. D. Revised effective ionic radii and systematic studies of interatomic distance in halides and chalcogenides. *Acta Crystallogr. Sect. A* **32**:751–767, 1976.
58. Sheu, S.-S.; Fozzard, H. A. Transmembrane Na^+ and Ca^{2+} electrochemical gradients in cardiac muscle and their relationship to force development. *J. Gen. Physiol.* **80**:325–351, 1982.
59. Somlyo, A. P.; Somlyo, A. V.; Shuman, H. Electron probe analysis of vascular smooth muscle: Composition of mitochondria, nuclei and cytoplasm. *J. Cell Biol.* **81**:316–335, 1979.
60. Somlyo, A. V.; Gonzales-Serratos, H.; Shuman, H.; McClellan, G.; Somlyo, A. P. Calcium release and ionic changes in the sarcoplasmic reticulum of tetanized muscle: An electron-probe study. *J. Cell Biol.* **90**:577–594, 1981.
61. Somlyo, A. V.; Shuman, H.; Somlyo, A. P. Elemental distribution in striated muscle and effects of hypertonicity: Electron probe analysis of cryosections. *J. Cell Biol.* **74**:828–857, 1977.
62. Spencer, R.; Charman, M.; Wilson, P. W.; Lawson, D. E. M. The relationship between vitamin D-stimulated calcium transport and intestinal calcium-binding protein in the chicken. *Biochem. J.* **170**:93–101, 1978.
63. Taylor, A. Role of cytosolic calcium and Na–Ca exchange in regulation of transepithelial sodium and water absorption. In: *Ion Transport by Epithelia*, S. G. Schultz, ed., New York, Raven Press, 1981, pp. 233–259.
64. Taylor, A.; Windhager, E. E. Possible role of cytosolic calcium and Na–Ca exchange in regulation of transepithelial sodium transport. *Am. J. Physiol.* **236**:F505–F512, 1979.
65. Terman, B. I.; Gunter, T. E. Characterization of the submandibular gland microsomal calcium transport system. *Biochim. Biophys. Acta* **730**:151–160, 1983.
66. Truesdell, A. H.; Christ, C. L. Glass electrodes for calcium and other divalent cations. In: *Glass Electrodes for Hydrogen and Other Cations*, G. Eisenman, ed., New York, Dekker, 1967, pp. 291–321.
67. Tsien, R. Y. New calcium indicators and buffers with high selectivity against magnesium and protons: Design, synthesis and properties of prototype structures. *Biochemistry* **19**:2396–2404, 1980.
68. Vinogradov, A.; Scarpa, A. The initial velocities of calcium uptake by rat liver mitochondria. *J. Biol. Chem.* **248**:5527–5531, 1973.
69. Wasserman, R. H.; Fullmer, C. S. Vitamin D-induced calcium binding protein. In: *Calcium and Cell Function*, Volume II, W. Y. Cheung, ed., New York, Academic Press, 1982, pp. 175–216.
70. Wasserstrom, J. A.; Schwartz, D. J.; Fozzard, H. A. Relation between intracellular sodium and twitch tension in sheep cardiac Purkinje strands exposed to cardiac glycosides. *Circ. Res.* **52**:697–705, 1983.
71. Weaver, D. C.; Barry, C. D.; McDaniel, M. L.; Marshall, G. R.; Lacy, P. E. Molecular requirements for recognition at a glucoreceptor for insulin release. *Mol. Pharmacol.* **16**:361–368, 1979.
72. Williams, R. J. P. The biochemistry of sodium, potassium, magnesium and calcium. *Q. Rev. Chem. Soc.* **24**:331–365, 1970.
73. Wnuk, W.; Cox, J. A.; Stein, E. A. Parvalbumins and other soluble high-affinity calcium binding proteins from muscle. In: *Calcium and Cell Function*, Volume II, W. Y. Cheung, ed., New York, Academic Press, 1982, pp. 243–278.

II

Roles of Phosphoinositides in Calcium-Regulated Systems

4

Calcium-Mobilizing Receptors
Membrane Phosphoinositides and Signal Transduction

Michael J. Berridge

When a hormone or a transmitter arrives at the surface of a cell, it must interact with a specific receptor in order to exert its effect. Each receptor is linked to a tranducing mechanism responsible for translating the chemical information extracted from the external signal into a meaningful intracellular signal often in the form of a second messenger such as cAMP or calcium. Although we now have detailed information on how receptors are coupled to adenylate cyclase to form cAMP there is less information concerning the way in which calcium-mobilizing receptors (e.g., muscarinic, α_1-adrenergic, H_1-histamine, V_1-vasopressin, etc.) bring about an increase in the intracellular level of calcium. Such receptors not only stimulate the influx of calcium across the plasma membrane, but can also mobilize calcium from intracellular reservoirs. It is this all-pervasive effect of agonists on membrane calcium permeability that is one of the most puzzling aspects of calcium-mobilizing receptors.

Another characteristic feature of these receptors is their multifunctional role in generating putative second messengers other than calcium [2]. They are responsible for activating guanylate cyclase to form cGMP, they release arachidonic acid which is converted into the eicosanoids (prostaglandins, prostacyclin, thromboxanes, leukotrienes), and they stimulate the hydrolysis of the phosphoinositides to form diacylglycerol that acts as a second messenger within the plane of the membrane to activate protein kinase C. It will be argued that one of these events, the hydrolysis of membrane phosphoinositides, represents a basic transducing mechanism responsible for forming all these signals. The aim of this review, therefore, is to describe how agonists stimulate the hydrolysis of these phosphoinositides and how the increased metabolism of these lipids may initiate a cascade of intracellular processes that sequentially release diacylglycerol to activate protein kinase C [18,27], mobilize calcium [21], release arachidonic acid [9], and activate guanylate cyclase [12].

Michael J. Berridge • A.F.R.C. Unit of Insect Neurophysiology and Pharmacology, Department of Zoology, University of Cambridge, Cambridge CB2 3EJ, England.

Table I. The Distribution, Specificity, and Function of Receptors That Use Membrane Phosphoinositides as a Transducing Mechanism[a]

Agonist	Receptor classification	Tissue
Acetylcholine	Muscarinic	Brain, parotid, pancreas, smooth muscle, adrenal, salt gland
Norepinephrine	α_1	Brain, parotid, fat cell, liver, smooth muscle, thyroid
Histamine	H_1	Brain, smooth muscle, endothelium
Vasopressin	V_1	Liver, smooth muscle
5-Hydroxytryptamine	$5\text{-}HT_1$	Fly salivary gland
ACTH	—	Adrenal, brain
Thrombin	—	Blood platelet
PTH	—	Kidney
Concanavalin A	—	Mast cell, lymphocyte
Substance P	—	Brain, parotid
Angiotensin II	—	Liver, adrenal, kidney
EGF	—	Epidermoid carcinoma
Caerulein	—	Pancreas, smooth muscle

[a] Data from Ref. 11, 17, 22–25.

The Agonist-Dependent Hydrolysis of Membrane Phosphoinositides

The ability of agonists to stimulate the hydrolysis of phosphoinositides is ubiquitous and receptor specific (Table I). There is an enormous variety of agonists that induce this effect during the activation of many different cellular processes including contraction, secretion, metabolism, and cell growth (Table I). Where receptor classifications are available, the hydrolysis of the phosphoinositides is always associated with one type of receptor (Table I). In the adrenergic system for example, the occupation of α_2- or β-receptors has no effect, whereas the activation of α_1 receptors stimulates the hydrolysis of this relatively minor group of membrane phospholipids with little or no change in the remaining phospholipids.

So far, the general term *phosphoinositide* has been used to refer to the three inositol-containing phospholipids (Fig. 1). The major component is phosphatidylinositol (PI), which exists in metabolic equilibrium with the two polyphosphoinositides [PI 4phosphate (PIP) and PI 4,5bisphosphate (PIP_2)] [8]. A small percentage of the PI is phosphorylated at the 4-position to form PIP which, in turn, is phosphorylated at the 5-position to form PIP_2. Both intermediates are of crucial importance because they represent the primary substrate for the receptor mechanism. Agonists interact with these receptors to increase the breakdown of all three phosphoinositides [for reviews see 2,21,22,25].

Since the major phosphoinositide is PI, it was reasonable to assume that it must be the primary substrate being degraded. More recent studies, however, have now identified polyphosphoinositides as those lipids hydrolyzed by the receptor mechanism. Tissues which have been incubated with [^{32}P]- or [^{3}H]inositol have a label incorporated into all three phosphoinositides and upon stimulation with agonists the first lipid to show a decrease in radioactivity is not PI but PIP_2. This rapid agonist-dependent breakdown of PIP_2 has been described in liver [19,31,37], parotid [39], blood platelets [6], and in the fly salivary gland [3]. Confirmation that PIP_2 is the primary substrate for the receptor mechanism was obtained by studying the appearance of water-soluble inositol phosphates [3,5]. Within 5 sec of stimulating the insect salivary gland with 5-HT, the level of inositol 1,4,5-trisphosphate (IP_3) derived

4. Calcium-Mobilizing Receptors

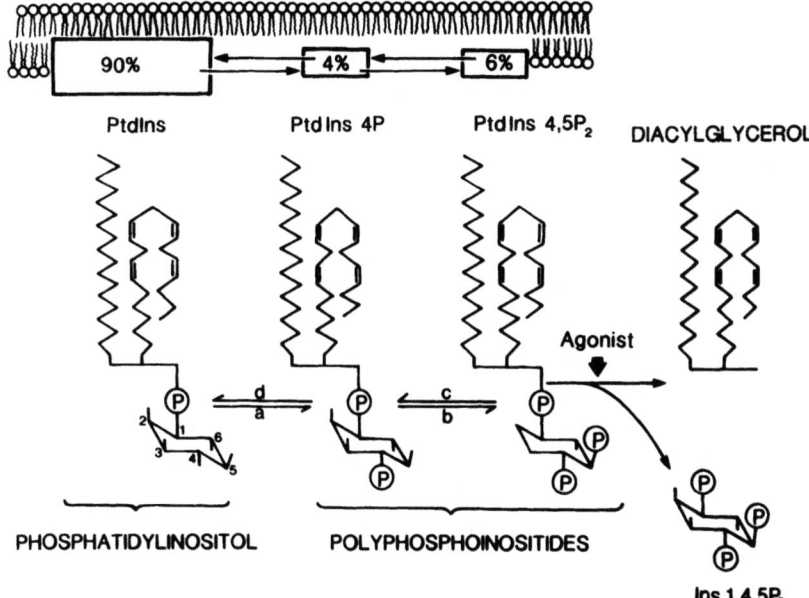

Figure 1. A summary of the major biochemical pathways responsible for the metabolism of phosphoinositides. The boxes at the top represent the proportion of PI (PtdIns) and the two polyphosphoinositides as measured in the insect salivary gland [3]. Agonists act by stimulating the hydrolysis of PIP_2 (PtdIns4,5P_2) by a phosphodiesterase to give diacylglycerol and IP_3 (Ins1,4,5P_3). a, PtdIns kinase; b, PtdIns4P kinase; c, PtdIns4,5P_2 phosphomonoesterase; d, PtdIns4P phosphomonoesterase.

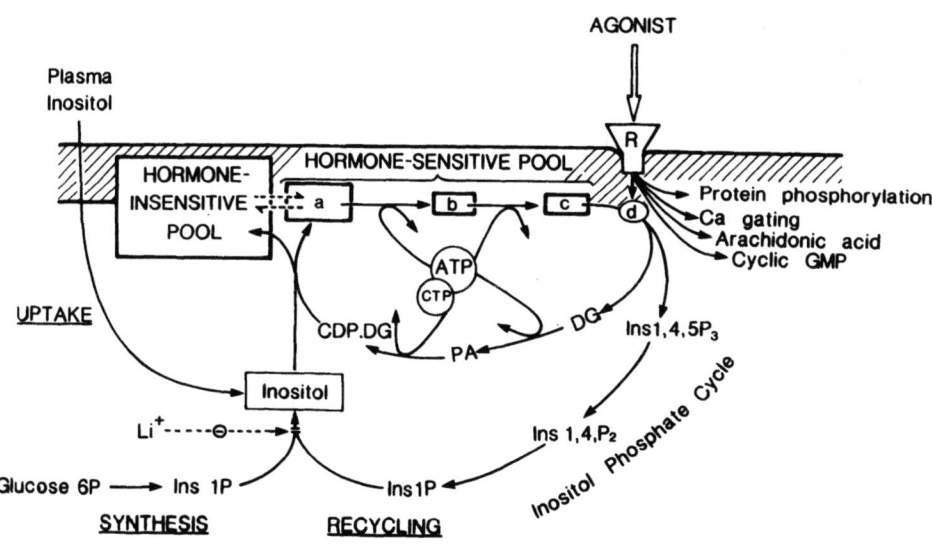

Figure 2. Inositol metabolism in relation to the operation of calcium-mobilizing receptors. Most of the PI in the membrane exists in a hormone-insensitive pool that equilibrates slowly (dashed lines) with a smaller hormone-sensitive pool (a) that is phosphorylated to PIP (b) and PIP_2 (c). The latter is hydrolyzed by a phosphodiesterase (d) when agonists interact with receptors (R). The IP_3 (Ins1,4,5P_3) is dephosphorylated progressively to inositol which recombines with CDP diacylglycerol (CDP.DG) to replenish the pool of PI (a). Inositol can also be taken up from plasma or it can be synthesized *de novo* from glucose-6-phosphate. The relationship of this hydrolysis of PIP_2 to other processes stimulated by this receptor is shown in Fig. 3.

from PIP$_2$ increased fourfold, whereas there was no change in inositol 1-phosphate or free inositol levels. A marked increase in the level of IP$_3$ has also been described in blood platelets 5 sec after stimulation with thrombin (1a). All these observations support the model suggesting that the primary action of agonists is to stimulate PIP$_2$ hydrolysis (Fig. 2). As this polyphosphoinositide is degraded, it will be constantly replaced by phosphorylation of PI (Fig. 2).

An interesting feature of this model is that the agonist is using the polyphosphoinositides, which are a minor component of the membrane pool of phosphoinositides (Fig. 2). As they are degraded as part of the receptor mechanism, their replacement by phosphorylation of PI requires 2 molecules of ATP (Fig. 2). This model thus explains why the agonist-dependent breakdown of PI in pancreas [15], liver [29], parotid [28], and blood platelets [16] is energy dependent. A constant supply of ATP is essential to convert PI into the PIP and PIP$_2$ that are hydrolyzed as part of the transducing mechanism of calcium-mobilizing receptors. The agonist-dependent generation of a calcium signal is thus an energy-dependent process which means that both the entry as well as the extrusion of calcium are coupled to ATP. Should ATP become limiting, this elegant autoregulatory device will restrict entry and so prevent the cell from being swamped with calcium.

Hydrolysis of PI 4,5-Bisphosphate Initiates a Cascade of Intracellular Signals

The hydrolysis of polyphosphoinositides has far-reaching implications because it initiates a cascade of events responsible for producing some if not all of the second messengers associated with calcium-mobilizing receptors. Because of the complexity of this system it is convenient to divide the cascade into three components (Fig. 3): (1) calcium-independent events; (2) calcium-dependent events; (3) an amplification loop mainly responsible for releasing arachidonic acid.

Calcium-independent events are defined as those processes that are triggered by the agonist without any change in the intracellular level of calcium. One such event is the agonist-dependent hydrolysis of the phosphoinositides, which is independent of calcium in many different cells [2,21,25]. One of the products of phosphoinositide breakdown is diacylglycerol (DG), which is a potent activator of protein kinase C [18,27]. It has been proposed that protein kinase C can phosphorylate specific proteins to induce certain cellular responses independently of any increase in the intracellular level of calcium [18,27]. Exogenous DG acts synergistically with low concentration of ionophore A23187 to stimulate aggregation and release in blood platelets [18]. Such a synergism may explain the observation that thrombin (acting together with a low concentration of calcium ionophore) can stimulate shape change and aggregation in the absence of any change in the intracellular level of calcium [32]. Similarly, exogenous DG or the phorbol ester PMA (4-phorbol-12-myristate-13-acetate) can stimulate platelet secretion without any change in the intracellular level of calcium (see Chapter 18).

Calcium-dependent events represent the most important consequence of activating calcium-mobilizing receptors. The very close relationship that exists between the agonist-dependent hydrolysis of phosphoinositides and the gating of calcium was first recognized by Michell [21]. But the biochemical link between the two processes has proved difficult to find. One interesting possibility is that phosphatidic acid (PA), which is formed by phosphorylation of DG, may act as an ionophore to facilitate the entry of extracellular calcium (Fig. 3) [30,34]. However, this model does not explain how agonists stimulate the release of

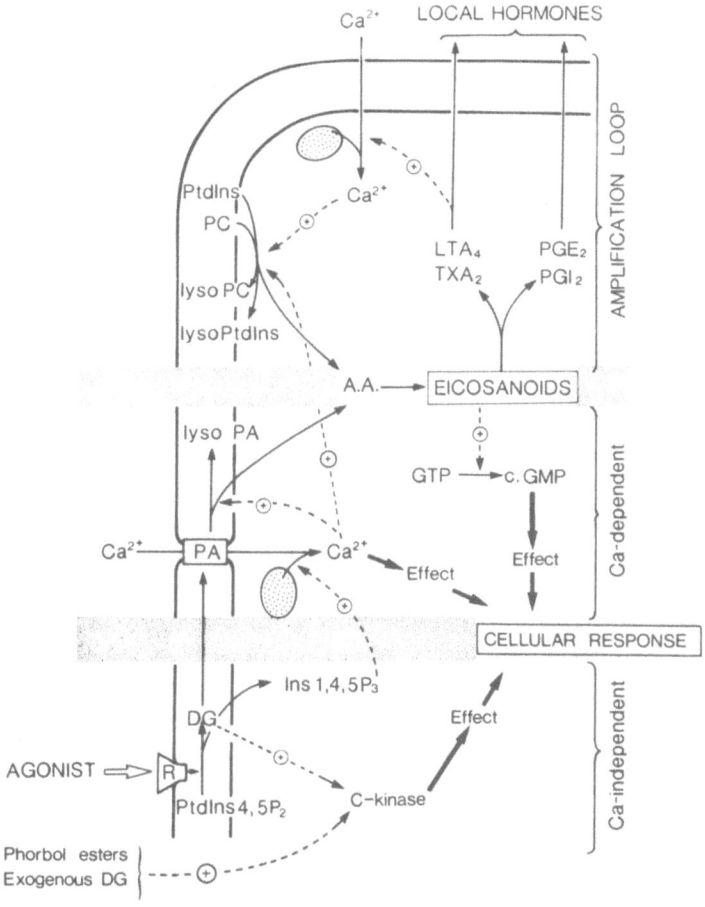

Figure 3. The intracellular signal cascade initiated by the agonist-dependent hydrolysis of PIP_2 ($PtdIns4,5P_2$) to diacylglycerol (DG) and IP_3 ($Ins1,4,5P_3$). PA, phosphatidic acid; PC, phosphatidylcholine; A.A., arachidonic acid. See text for further details.

intracellular calcium. An interesting possibility is that one of the water-soluble products of phosphoinositide hydrolysis, such as IP_3 or inositol 1,4-bisphosphate (IP_2) may function as a second messenger to release internal calcium (Fig. 3) [3]. Evidence to support this idea has been obtained using permeabilized pancreatic cells [36a], liver cells [7a, 17a], and smooth muscle cells [36b] that release calcium from a nonmitochondrial internal store upon addition of IP_3 [36a].

In addition to stimulating a variety of cellular responses by acting on calcium receptors (troponin C, calmodulin), the increase in intracellular calcium is responsible for activating lipases to release arachidonic acid [33]. The latter can be derived from two main sources. The hydrolysis of PIP_2 and PIP generates DG and PA that function as specific substrates for a release of arachidonic acid that is closely linked to the receptor mechanism (Fig. 3). Arachidonic acid is then metabolized via the cyclooxygenase and lipoxygenase pathways to form a plethora of eicosanoids, some of which are released from the cell to function as local hormones (Fig. 3). Arachidonic acid itself or one of its metabolites may be responsible for activating guanylate cyclase to raise the intracellular level of cGMP [12].

An amplification loop represents the other main source of arachidonic acid, which is derived from different lipids (PI, PC, and PE) through the activation of a nonspecific phospholipase A_2 (Fig. 3). The calcium necessary to activate this lipase may be derived either from the original receptor mechanism (as described above) or it may result from the action of eicosanoids such as the leukotrienes (LTA_4) or thromboxanes functioning through their own calcium-mobilizing receptors. This amplification loop is very much an optional extra that is particularly well developed in certain cells such as neutrophils and blood platelets that release large quantities of eicosanoids as part of their functional coordination with other cells. The agonist-dependent hydrolysis of PIP_2 initiates a cascade of events that reverberates throughout the cell to induce a whole host of second messengers.

Turnover of Phosphoinositides and Receptor Sensitivity—The Importance of Lithium

Not all cellular PI is immediately available to the transducing process [7,10], leading to the concept that the receptor mechanism depends on a small hormone-sensitive pool of rapidly turning over phosphoinositide (Fig. 2). The main features of this turnover are a lipid and an inositol phosphate cycle that process the products of the receptor mechanisms (DG, IP_2 and IP_3) so that they can be resynthesized to PI. Any reduction in the operation of these cycles will reduce the level of phosphoinositides in the hormone-sensitive pool which leads to a reduction in receptor sensitivity.

Resynthesis of PI is quite dependent on the maintenance of high intracellular inositol levels. The latter is derived from three main sources: from recycling of the inositol headgroup, *de novo* synthesis from glucose-6-P, or uptake of plasma inositol derived from the diet (Fig. 2). The latter source is not available to the brain because inositol does not readily cross the blood–brain barrier [20]. Since the nervous system must synthesize its own inositol, it is much more susceptible to metabolic defects or therapeutic alterations in inositol metabolism. A large number of brain neurotransmitters function to hydrolyze phosphoinositides (Table I) and their normal operation may be very dependent on a constant supply of inositol. Small changes in free inositol levels in the endoneurium can lead to profound changes in neural activity as measured through changes in oxygen consumption [36]. Galactosemia or streptozotocin-induced diabetes causes defects in inositol metabolism that could be responsible for the associated abnormalities in neural function [1b,13,26,38]. An even more exciting prospect is that the therapeutic action of Li^+ in controlling manic-depressive illness might be explained through changes in inositol metabolism [4,14].

Lithium acts by inhibiting the enzyme inositol-l-phosphatase (IP) [14], thus blocking the production of free inositol originating not only from turnover but also from *de novo* synthesis; these are the only two inositol-producing processes in the brain. As such, Li^+ levels raise IP levels while lowering free inositol levels [4,35] necessary to synthesize the PI required to replenish the hormone-sensitive pool (Fig. 2). By inhibiting the resynthesis of PI, Li^+ will gradually reduce phosphoinositide levels, which could result in a decline in the effectiveness of these multifunctional receptors. As a corollary to this hypothesis, manic-depressive illness may arise through an abnormal increase in activity or sensitivity of those receptors in brain (Table I) that use phosphoinositides as a part of their transducing mechanism.

References

1a. Agranoff, B. W.; Murthy, P.; Seguin, E. B. Thombin-induced phosphodiesteratic cleavage of phosphatidylinositol bisphosphate in human platelets. *J. Biol. Chem.* **258**:2076–2078, 1983.

1b. Bell, M. E.; Peterson, R. G.; Eichberg, J. Metabolism of phospholipids in peripheral nerve from rats with chronic streptozotocin-induced diabetes: Increased turnover of phosphatidylinositol-4,5-bisphosphate. *J. Neurochem.* **39:**192–200, 1982.
2. Berridge, M. J. Phosphatidylinositol hydrolysis: A multifunctional transducing mechanism. *Mol. Cell. Endocrinol.* **24:**115–140, 1981.
3. Berridge, M. J. Rapid accumulation of inositol trisphosphate reveals that agonists hydrolyse polyphosphoinositides instead of phosphatidylinositol. *Biochem. J.* **212:**849–858, 1983.
4. Berridge, M. J.; Downes, C. P.; Hanley, M. R. Lithium amplifies agonist-dependent phosphatidylinositol responses in brain and salivary glands. *Biochem. J.* **206:**587–595, 1982.
5. Berridge, M. J.; Dawson, R. M. C.; Downes, C. P.; Heslop, J. P.; Irvine, R. F. Changes in the levels of inositol phosphates after agonist-dependent hydrolysis of membrane phosphoinositides. *Biochem. J.* **212:**473–482, 1983.
6. Billah, M. M.; Lapetina, E. G. Rapid decrease of phosphatidylinositol 4,5-bisphosphate in thrombin-stimulated platelets. *J. Biol. Chem.* **257:**12705–12708, 1982.
7. Billah, M. M.; Lapetina, E. G. Evidence for multiple metabolic pools of phosphatidylinositol in stimulated platelets. *J. Biol. Chem* **257:**11856–11859, 1982.
7a. Burgess, G. M.; Godfrey, P. P.; McKinney, J. S.; Berridge, M. J.; Irvine, R. F.; Putney, J. W. The second messenger linking receptor activation to internal Ca release in liver. *Nature (London)* **309:**63–66, 1984.
8. Downes, C. P.; Michell, R. H. Phosphatidylinositol 4-phosphate and phosphatidylinositol 4,5-bisphosphate: Lipids in search of a function. *Cell Calcium* **3:**467–502, 1982.
9. Fain, J. N. Involvement of phosphatidylinositol breakdown in elevation of cytosol Ca^{2+} by hormones and relationship to prostaglandin formation. In: *Hormone Receptors*, L. D. Kohn, ed., New York, Wiley, 1982, pp. 237–276.
10. Fain, J. N.; Berridge, M. J. Relationship between phosphatidylinositol synthesis and recovery of 5-hydroxytryptamine-responsive Ca^{2+} flux in blowfly salivary glands. *Biochem. J.* **180:**655–661, 1979.
11. Farese, R. V. Phosphoinositide metabolism and hormone action. *Endocrinol. Rev.* **4:**78–95, 1983.
12. Gerzer, R.; Hamet, P.; Ross, A. H.; Lawson, J. A.; Hardman, J. G. Calcium-induced release from platelet membranes of fatty acids that modulate soluble guanylate cyclase. *J. Pharmacol. Exp. Ther.* **226:**180–186, 1983.
13. Greene, D. A.; De Jesus, P. V.; Winegrad, A. I. Effect of insulin and dietary myoinositol on impaired peripheral motor nerve conduction velocity in acute streptozotocin diabetes. *J. Clin. Invest.* **55:**1326–1336, 1975.
14. Hallcher, L. M.; Sherman, W. R. The effects of lithium ion and other agents on the activity of *myo*-inositol-l-phosphatase from bovine brain. *J. Biol. Chem.* **255:**10896–10901, 1980.
15. Hokin, M. R. Breakdown of phosphatidylinositol in the pancreas in response to pancreozymin and acetylcholine. In: *Secretory Mechanisms of Exocrine Glands*, N. A. Thorn and O. H. Petersen, eds., Copenhagen, Munksgaard, 1974, pp. 101–115.
16. Holmsen, H.; Kaplan, K. L.; Dangelmaier, C. A. Differential energy requirements for platelet responses. *Biochem. J.* **208:**9–18, 1982.
17. Jones, L. M.; Cockcroft, S.; Michell, R. H. Stimulation of phosphatidylinositol turnover in various tissues by cholinergic and adrenergic agonists, by histamine and by caerulein. *Biochem. J.* **182:**669–676, 1979.
17a. Joseph, S. K.; Thomas, A. P.; Williams, R. J.; Irvine, R. F.; Williamson, J. R. *myo*-Inositol 1,4,5-trisphosphate: A second messenger for the hormonal mobilization of intracellular Ca^{2+} in liver. *J. Biol. Chem.* **259:**3077–3081, 1984.
18. Kaibuchi, K.; Sano, K.; Hoshijima, M.; Takai, Y.; Nishizuka, Y. Phosphatidylinositol turnover in platelet activation; calcium mobilization and protein phosphorylation. *Cell Calcium* **3:**323–335, 1982.
19. Kirk, C. J.; Creba, J. A.; Downes, C. P.; Michell, R. H. Hormone-stimulated metabolism of inositol lipids and its relationship to hepatic receptor function. *Biochem. Soc. Trans.* **9:**377–379, 1981.
20. Margolis, R. V.; Press, R.; Altszuler, N.; Stewart, M. A. Inositol production by the brain in normal and alloxan-diabetic dogs. *Brain Res.* **28:**535–539, 1971.
21. Michell, R. H. Inositol phospholipids and cell surface receptor function. *Biochim. Biophys. Acta* **415:**81–147, 1975.
22. Michell, R. H. Inositol phospholipids in membrane function. *Trends Biochem. Res.* **4:**128–131, 1979.
23. Michell, R. H. Inositol lipid metabolism in dividing and differentiating cells. *Cell Calcium* **3:**429–440, 1982.
24. Michell, R. H.; Jones, L. M.; Jafferji, S. S. A possible role for phosphatidylinositol breakdown in muscarinic cholinergic stimulus–response coupling. *Biochem. Soc. Trans.* **5:**77–81, 1977.
25. Michell, R. H.; Kirk, C. J. Why is phosphatidylinositol degraded in response to stimulation of certain receptors? *Trends Pharmacol. Sci.* **2:**86–89, 1981.
26. Natarajan, V.; Dyck, P. J.; Schmid, H. H. O. Alterations of inositol lipid metabolism of rat sciatic nerve in streptozotocin-induced diabetes. *J. Neurochem.* **36:**413–419, 1981.

27. Nishizuka, Y. Phospholipid degradation and signal translation for protein phosphorylation. *Trends Biochem. Sci.* **8**:13–16, 1983.
28. Poggiolo, J.; Weiss, S. J.; McKinney, J. S.; Putney, J. W. Effects of antimycin A on receptor-activated calcium mobilization and phosphoinositide metabolism in rat parotid gland. *Mol. Pharmacol.* **23**:71–77, 1983.
29. Prpic, V.; Blackmore, P. F.; Exton, J. H. Phosphatidylinositol breakdown induced by vasopressin and epinephrine in hepatocytes is calcium-dependent. *J. Biol. Chem.* **257**:11323–11331, 1982.
30. Putney, J. W. Inositol lipids and cell stimulation in mammalian salivary gland. *Cell Calcium* **3**:369–383, 1982.
31. Rhodes, D.; Prpic, V.; Exton, J. H.; Blackmore, P. F. Stimulation of phosphatidylinositol 4,5-bisphosphate hydrolysis in hepatocytes by vasopressin. *J. Biol. Chem.* **258**:2770–2773, 1983.
32. Rink, T. J.; Smith, S. W.; Tsien, R. Y. Cytoplasmic free Ca^{2+} in human platelets: Ca^{2+} thresholds and Ca^{2+}-independent activation for shape-change and secretion. *FEBS Lett.* **148**:21–26, 1982.
33. Rubin, R. P. Calcium–phospholipid interactions in secretory cells: A new perspective on stimulus–secretion coupling. *Fed. Proc.* **41**:2181–2187, 1982.
34. Salmon, D. M.; Honeyman, T. W. Proposed mechanism of cholinergic action in smooth muscle. *Nature (London)* **284**:344–345, 1980.
35. Sherman, W. R.; Leavitt, A. L.; Honchar, M. P.; Hallcher, L. M.; Phillips, B. E. Evidence that lithium alters phosphoinositide metabolism: Chronic administration elevates primarily D-*myo*-inositol-l-phosphate in cerebral cortex of the rat. *J. Neurochem.* **36**:1947–1951, 1981.
36. Simmons, D. A.; Winegrad, A. I.; Martin, D. B. Significance of tissue *myo*-inositol concentrations in metabolic regulation in nerve. *Science* **217**:848–851, 1982.
36a. Streb, H.; Irvine, R. F.; Berridge, M. J.; Schulz, I. Release of Ca^{2+} from a nonmitochondrial intracellular store in pancreatic acinar cells by inositol-1,4,5-trisphosphate. *Nature (London)* **306**:67–69, 1983.
36b. Suematsu, E.; Hirata, M.; Hashimoto, T.; Kuriyama, H. Inositol 1,4,5-trisphosphate releases Ca^{2+} from intracellular store sites in skinned single cells of porcine coronary artery. *Biochem. Biophys. Res. Commun.* **120**:481–485, 1984.
37. Thomas, A. P.; Marks, J. S.; Coll, K. E.; Williamson, J. R. Quantitation and early kinetics of inositol lipid changes induced by vasopressin in isolated and cultured hepatocytes. *J. Biol. Chem.* **258**:5716–5725, 1983.
38. Warefield, A. S.; Segal, S. *Myo*-inositol and phosphatidylinositol metabolism in synaptosomes from galactose-fed rats. *Proc. Natl. Acad. Sci. USA* **75**:4568–4572, 1978.
39. Weiss, S. J.; McKinney, J. S.; Putney, J. W. Receptor mediated net breakdown of phosphatidylinositol-4,5-bisphosphate in parotid acinar cells. *Biochem. J.* **206**:555–560, 1982.

5

Platelet Response in Relation to Metabolism of Inositides and Protein Phosphorylation

Eduardo G. Lapetina and Wolfgang Siess

In the last few years our work has concentrated on lipid changes associated with platelet responses. We found initially that phosphatidic acid is rapidly formed in thrombin-stimulated platelets due to the breakdown of inositol phospholipids [1]. A specific phospholipase C degrades inositol phospholipids, and the 1,2-diacylglycerol so formed is phosphorylated to phosphatidic acid by 1,2-diacylglycerol kinase. The time course for the formation of phosphatidic acid differs from that for liberation of arachidonic acid from various phospholipids such as phosphatidylcholine, phosphatidylethanolamine, phosphatidylinositol, and also phosphatidic acid by phospholipase A_2. This suggested a sequential stimulation of phospholipase C and phospholipases A_2 [1,2].

Subsequently, the effects of calcium and cAMP on the "phosphatidylinositol cycle" were studied [3,4]. Calcium increases the amount of phosphatidic acid by inhibiting the resynthesis of phosphatidic acid into phosphatidylinositol [2–5]. The accumulated phosphatidic acid and lysophosphatidic acid (the product of degradation of phosphatidate by phospholipase A_2 activity) are claimed to have the properties of Ca^{2+} ionophore and if so, they could then facilitate the liberation of arachidonic acid from various phospholipids by a Ca^{2+}-sensitive phospholipase A_2. Alternatively, arachidonic acid could be liberated by a transacylation reaction between lysophosphatidic acid and other arachidonyl-containing phospholipids [6], or by a direct action of phosphatidic acid and lysophosphatidic acid on phospholipases A_2 [2]. cAMP decreases phosphatidic acid levels and, consequently, the liberation of arachidonic acid [2,3]. We have also studied various platelet enzymes related to these reactions such as phospholipase C, 1,2-diacylglycerol kinase, phosphatidate-specific phospholipase A_2, and phospholipases A_2 that degrade phosphatidylinositol, phosphatidylcholine, and phosphatidylethanolamine [7–10]. More recently, a rapid loss of phosphatidylinositol 4,5-bisphosphate by phosphodiesteratic cleavage has been demonstrated in

---

Eduardo G. Lapetina and Wolfgang Siess • Department of Molecular Biology, The Wellcome Research Laboratories, Burroughs Wellcome Company, Research Triangle Park, North Carolina 27709.   Present address of W.S.: Medizinische Klinik Innenstadt der Universität München, 8 Munich 2, West Germany.

stimulated platelets which is Ca^{2+}-independent and might precede the loss of phosphatidylinositol [11–13].

The sequential activations of phospholipases C and A_2 are related to different platelet responses [14–17]. The phosphodiesteratic cleavage of the inositides provides active products such as 1,2-diacylglycerol (activates protein kinase C), its phosphorylated product, phosphatidic acid, and inositol phosphates. These reactions are associated with an early platelet response, i.e., shape change [15,16]. Subsequently, phospholipases A_2 liberate phospholipid-associated arachidonic acid which is metabolized by cyclooxygenase and lipoxygenase. Arachidonic acid metabolites could amplify platelet activation. The formation of cyclooxygenase products parallels the release reaction and aggregation [18]. This mechanism of amplification could be due to activation of phospholipase C by metabolites derived from cyclooxygenase activity.

Cyclooxygenase-Dependent Mechanism for Phosphatidylinositol Degradation and Formation of 1,2-diacylglycerol and Its Phosphorylated Product, Phosphatidic Acid

Stimulation of platelets with physiological agents such as thrombin, collagen, ADP, or platelet-activating factor induces rapid changes in inositide metabolism [19–21]. These changes are associated with stimulation of phospholipase C which seems essential for platelet activation. Another platelet activator, i.e., arachidonic acid, also activates phospholipase C in intact platelets [16]. Addition of exogenous, unlabeled arachidonic acid (μM concentration) to washed human platelets prelabeled with [³H]arachidonic acid stimulates the rapid and transient formation of [³H]1,2-diacylglycerol and [³H]phosphatidic acid (Fig. 1). Maximal formation of 1,2-diacylglycerol precedes that of phosphatidic acid, indicating the sequential action of phospholipase C and 1,2-diacylglycerol kinase. The accumulation of [³H]phosphatidic acid (260% over control) is higher than that of [³H]1,2-diacylglycerol (80% over control) probably because 1,2-diacylglycerol kinase is very effective in transforming 1,2-diacylglycerol to phosphatidic acid.

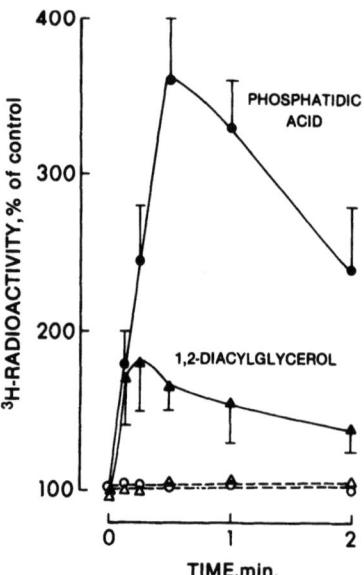

Figure 1. Exogenous arachidonic acid (5 μM) stimulates the formation of [³H]1,2-diacylglycerol (▲) and [³H]phosphatidic acid (●) in human platelets prelabeled with [³H]arachidonic acid. Results (mean ± S.D.) are expressed as percent of the control values from four different experiments. The open symbols refer to [³H]1,2-diacylglycerol (△) and [³H]phosphatidic acid (○) in aspirin-treated platelets as described in Ref. 16. Total radioactivity for [³H]1,2-diacylglycerol and [³H]phosphatidic acid in unstimulated controls was 205 ± 15 and 610 ± 10 cpm, respectively.

Table I. Effect of Exogenous, Unlabeled Arachidonic Acid on [³H]-Arachidonyl-Phospholipids of Human Platelets[a]

Phospholipid	Incubation time (sec)		
	0	15	30
[³H]-Phosphatidylinositol	84.7 ± 1.6 (100)	83.2 ± 0.8 (98.2)	79.2 ± 1.2* (93.6)
[³H]-Phosphatidylserine	57.5 ± 1.8 (100)	57.8 ± 1.1 (100)	56.7 ± 1.4 (98.5)
[³H]-Phosphatidylcholine	176.9 ± 4.2 (100)	176.6 ± 2.2 (99.8)	172.9 ± 6.4 (97.7)
[³H]-Phosphatidylethanolamine	54.9 ± 1.3 (100)	55.1 ± 0.8 (100)	54.5 ± 2.1 (99.3)

[a]Samples (0.5 ml) of human platelets (3×10^8 cells) prelabeled with [³H]-AA were stimulated with exogenous, unlabeled AA (5 μM). Measurements were done in quadruplicates. The lipids were extracted and analyzed as described in Ref. 16. Results are expressed as cpm (mean ± S.D.) $\times 10^{-3}$. Numbers in parentheses indicate values as percent of controls.
*$p < 0.01$ as compared to control.

Platelets stimulated with arachidonic acid also show a rapid breakdown of [³H]phosphatidylinositol which is statistically significant 30 sec after stimulation, and coincides with the maximal formation of phosphatidic acid (Fig. 1, Table I). Direct comparison of the loss in [³H]phosphatidylinositol after 30 sec (Table I) and the gain in [³H]1,2-diacylglycerol plus [³H]phosphatidic acid (Fig. 1) in the same time incubation, shows a recovery of about 50%. This low recovery could be explained by some phospholipase A_2 degradation of phosphatidylinositol and/or phosphatidic acid and the action of 1,2-diacylglycerol lipase on 1,2-diacylglycerol. There is no measurable breakdown of phosphatidylcholine, phosphatidylethanolamine, and phosphatidylserine. Pretreatment of platelets with aspirin, which irreversibly inhibits platelet cyclooxygenase, blocks the formation of 1,2-diacylglycerol and phosphatidic acid in arachidonic acid-stimulated platelets (Fig. 1). Thus, cyclooxygenase metabolites such as endoperoxides and thromboxane A_2 mediate arachidonic acid-induced stimulation of platelet phospholipase C [16].

Phosphatidic Acid Levels in Relation to Protein Phosphorylation, Shape Change, Serotonin Release, and Aggregation

Low concentrations of arachidonic acid (0.05–0.2 μM) induce shape change in human platelets without serotonin release or aggregation (Fig. 2, upper panel). Shape change induced by arachidonic acid occurs simultaneously with [³²P]phosphatidic acid formation (Fig. 2, lower panel) and with concurrent stimulation of phosphorylation of a 40K protein (Fig. 3). Concentrations of arachidonic acid greater than 0.5 μM induce serotonin release and aggregation, which is associated with accumulation of higher levels of phosphatidic acid and phosphorylation of the 40K protein (Fig. 4). Arachidonic acid at very high concentrations (0.1 mM) neither activates platelets nor induces phosphatidic acid formation (Fig. 4).

Endoperoxides and thromboxane A_2 may be responsible for arachidonic acid-induced stimulation of phospholipase C, since formation of 1,2-diacylglycerol and/or phosphatidic acid is blocked by aspirin, indomethacin, and N-methylimidazole [16]. Conversion to endoperoxides and thromboxane A_2 is a necessary, but not a sufficient, step for inducing platelet activation by exogenous arachidonic acid. Prostacyclin does not affect arachidonic acid metabolism by platelet cyclooxygenase and thromboxane synthetase, but does inhibit shape change, formation of phosphatidic acid, and protein phosphorylation [16]. The dependence of these various responses on prostacyclin concentration is almost identical, suggesting that a

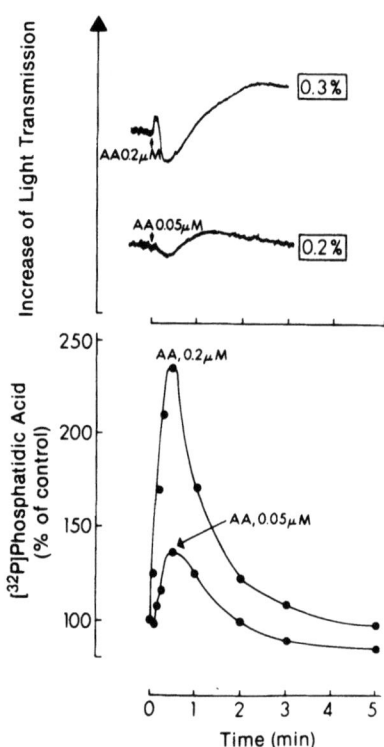

Figure 2. Arachidonic acid (AA) stimulates formation of [^{32}P]phosphatidic acid and shape change in human platelets. Platelet shape change is indicated by a decrease of light transmission and narrowing of the oscillation tracing which is produced by discoid platelets (upper panel). At various times as indicated, 0.1-ml aliquots were directly transferred into tubes containing 3.75 volume chloroform/methanol (1 : 2) for lipid extraction, and phosphatidic acid (lower panel) was estimated as described in Ref. 16. Release of [^{3}H]serotonin was measured after 3 min of stimulation and the values (% of total) are indicated inside the rectangles.

common site exists for prostacyclin in blocking each of these responses. Prostacyclin increases cAMP levels in platelets, and may prevent Ca^{2+} mobilization and, hence, phospholipase C stimulation induced by endoperoxides and thromboxane A_2 [16].

The action of collagen on phosphatidic acid formation in platelets also seems to be critically dependent on metabolites derived from cyclooxygenase and thromboxane synthetase [14]. Inhibitors of these two enzyme systems, such as indomethacin, aspirin, and N-methylimidazole, completely depress the action of all concentrations of collagen [14]. Endoperoxides may thus be involved in mediating collagen-induced formation of phosphatidic acid and platelet activation, since a direct action of the endoperoxide analog U-44069 has been detected [14].

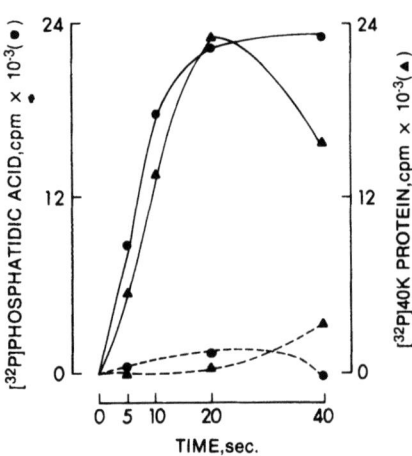

Figure 3. Time courses for the arachidonic acid-induced formation of [^{32}P]phosphatidic acid and phosphorylation of a 40K protein in the presence and absence of indomethacin. Washed human platelets prelabeled with ^{32}P (1.0 ml) were preincubated in the aggregometer tubes at 37°C for 3 min without (solid lines) or with 10 μM indomethacin (broken lines) and then stimulated with 0.5 μM arachidonic acid for different times as indicated. At the indicated times, a 0.1-ml sample was quenched for separation of proteins on polyacrylamide gels and the remaining sample (0.9 ml) was extracted with chloroform/methanol as described in Ref. 16. The radioactivity in phosphatidic acid (●) and the 40K protein (▲) is referred to 1 ml of platelet suspension which contains 7.5×10^8 platelets.

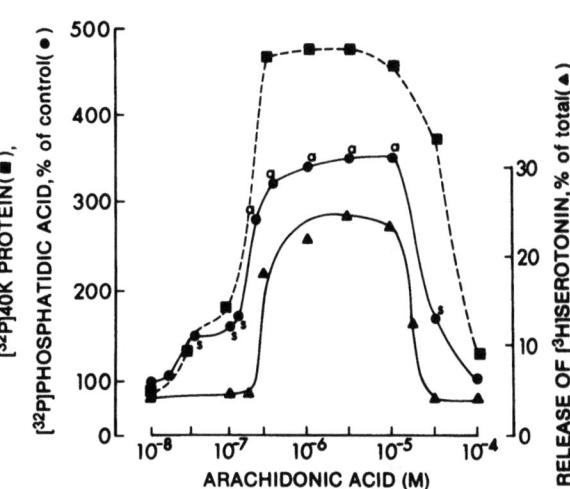

Figure 4. Differential effects of various concentrations of arachidonic acid on the formation of phosphatidic acid, protein phosphorylation, and platelet activation. Samples (1.0 ml) of platelets prelabeled with $^{32}P_i$ were placed into aggregometer tubes and exposed to various concentrations of arachidonic acid. [^{32}P]-Phosphatidic acid (●) and [^{32}P]-40K protein (■) were estimated before and 30 sec after addition of arachidonic acid as in Fig. 3 and Ref. 16. Platelet shape change (s) and aggregation (a) were recorded in the aggregometer. The release of serotonin was stimulated in platelets prelabeled with [^{3}H]serotonin. Samples (0.2 ml) of washed platelet suspension prelabeled with [^{3}H]serotonin were exposed to various concentrations of arachidonic acid and the [^{3}H]serotonin released (▲) into the medium was estimated as described in Ref. 16.

Thrombin and Platelet-Activating Factor Stimulate Phosphatidic Acid Formation by an Endoperoxide-Independent Mechanism

Formation of phosphatidic acid by thrombin may be the result of direct interaction of thrombin with its receptors and immediate activation of phospholipase C [14]. The action of platelet-activating factor is analogous to that of thrombin, since it does not require endoperoxides or thromboxanes to trigger stimulation of phospholipase C, formation of phosphatidic acid, and platelet shape change [15,21].

Platelet shape change and the accompanying formation of phosphatidic acid induced by platelet-activating factor show a similar sensitivity to inhibitors (Fig. 5). Neither response is inhibited by 50 μM trifluoperazine (a calmodulin antagonist which inhibits phospholipases A_2) nor 10 μM indomethacin (an inhibitor of cyclooxygenase). Examination by scanning electron microscopy also indicates that indomethacin does not prevent the shape change induced by platelet-activating factor [15]. On the other hand, prostacyclin prevents both platelet shape change [15] and phosphatidic acid formation induced by platelet-activating factor (Fig. 3).

The formation of phosphatidic acid induced by platelet-activating factor is also dependent on the concentration of platelet-activating factor in the range 1 nM to 1 μM, with a maximal effect at 0.1 μM [15]. Trifluoperazine or indomethacin does not inhibit phosphatidic acid formation induced by varied concentrations of platelet-activating factor. However, the formation of phosphatidic acid is sensitive to prostacyclin at all platelet-activating factor concentrations tested [15].

Phosphatidic acid is derived from 1,2-diacylglycerol formed as a consequence of phospholipase C activation. The generation of 1,2-diacylglycerol has been associated with the phosphorylation of a 40K protein by protein kinase C [22,23]. This series of metabolic events involving membrane phospholipids may be relevant to the mechanism of transmembrane signaling [22,23]. Stimulation of platelets by platelet-activating factor induces shape change with simultaneous formation of phosphatidic acid and phosphorylation of a 40K protein (Fig. 6). Both phosphorylated products reach maximal levels within 20 sec (Fig. 6).

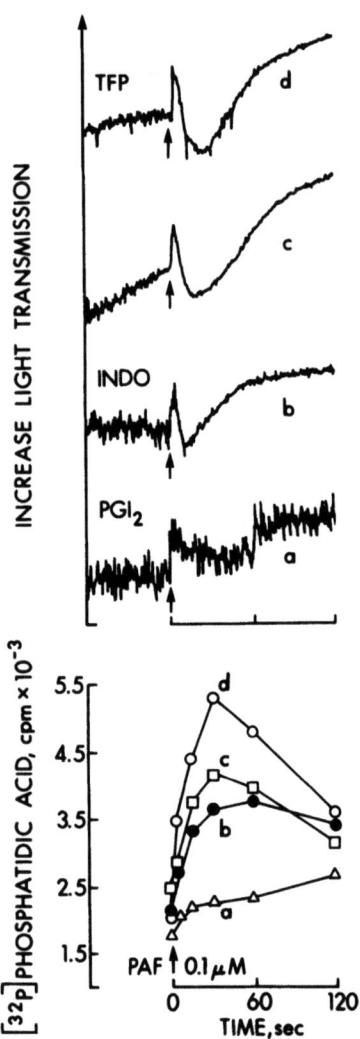

Figure 5. Effect of trifluoperazine (TFP), indomethacin (INDO), and prostacyclin (PGI$_2$) on platelet shape change and formation of phosphatidic acid. Washed human platelets prelabeled with ^{32}P (1.0-ml samples) were preincubated for 3 min in the aggregometer tubes at 37°C while stirring with 4 ng/ml prostacyclin (a) or with 10 μM INDO (b) without additions (c) or with 50 μM TFP (d) and then stimulated with 0.1 μM platelet-activating factor (PAF). Samples (0.1 ml) were directly taken from the aggregometer tube at different times for extraction of lipids and separation of phosphatidic acid. Radioactivity shown relates to phosphatidic acid in a 0.1-ml sample of platelet suspension. Upper panel shows recordings of shape change and lower panel the formation of [^{32}P]phosphatidic acid. [Other details as in Ref. 15.]

Pathways for Phospholipases C and A$_2$ Activation in Stimulated Platelets

Stimulation of phospholipase C leading to platelet phosphatidic acid formation can follow two distinct pathways. (1) Arachidonic acid and collagen stimulate phospholipase C and phosphatidic acid formation via the initial production of endoperoxides or thromboxanes [14,15]. Exogenous arachidonic acid is directly converted by cyclooxygenase to those metabolites [16], while collagen [14] first stimulates phospholipase A$_2$ to liberate arachidonic acid. (2) Thrombin or platelet-activating factor directly stimulates phosphatidic acid formation by an endoperoxide-independent mechanism [14,15].

These pathways for phospholipase C and A$_2$ activation may be considered putative positive "crossover" feedback mechanisms. Thrombin or platelet-activating factor stimulates phospholipase C that leads to the formation of 1,2-diacylglycerol and phosphatidic acid. These changes correlate with shape change of platelets. Subsequently, activation of phospholipases A$_2$ liberates arachidonic acid from various phospholipids, including phosphatidylcholine, phosphatidylinositol, phosphatidylethanolamine, and phosphatidic acid.

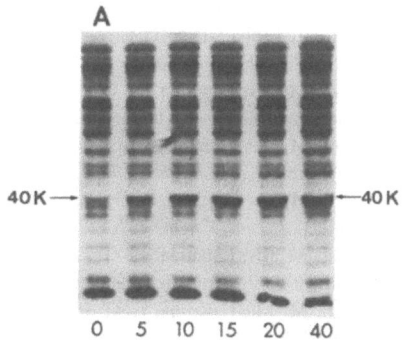

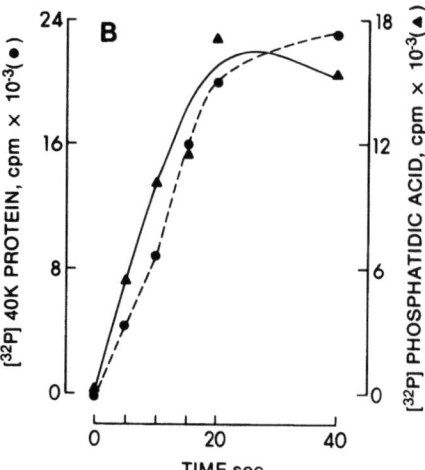

Figure 6. Time course of platelet-activating factor-induced phosphorylation of a 40K protein and formation of phosphatidic acid during platelet shape change. Washed human platelets prelabeled with ^{32}P (1.0 ml) were preincubated in the aggregometer tubes at 37°C for 3 min and then stimulated with 0.1 μM platelet-activating factor for different times as indicated. At the specific times, a 0.1-ml sample was quenched for separation of proteins on polyacrylamide gels and the remaining sample (0.9 ml) was extracted with chloroform/methanol as described in Ref. 15. The upper panel (A) shows an autoradiograph of the separation of ^{32}P-labeled proteins equivalent to 0.05 ml of platelet suspension. The radioactivity in phosphatidic acid (▲) and 40K protein (●) shown in the lower panel (B) is referred to 1 ml of platelet suspension which contains 7.5×10^8 platelets after subtraction of the basal values. In this particular experiment the basal values, corresponding to 1 ml of platelet suspension, for phosphatidic acid and for the 40K protein were 13,243 and 16,920 cpm, respectively.

These biochemical changes correlate with release of ADP and serotonin from dense granules. Endoperoxides and/or thromboxane A_2 could cause "feedback" stimulation of the phospholipase C, thereby amplifying platelet responses.

References

1. Lapetina, E. G.; Cuatrecasas, P. Stimulation of phosphatidic acid production in platelets precedes the formation of arachidonate and parallels the release of serotonin. *Biochim. Biophys. Acta* **573**:394–402, 1979.
2. Lapetina, E. G. Regulation of arachidonic acid production: Role of phospholipases C and A_2. *Trends Pharmacol. Sci.* **3**:115–118, 1982.
3. Lapetina, E. G.; Billah, M. M.; Cuatrecasas, P. The phosphatidylinositol cycle and the regulation of arachidonic acid production. *Nature (London)* **292**:367–369, 1981.
4. Lapetina, E. G.; Billah, M. M.; Cuatrecasas, P. The initial action of thrombin on platelets. Conversion of phosphatidylinositol to phosphatidic acid preceding production of arachidonic acid. *J. Biol. Chem.* **256**:5037–5040, 1981.
5. Billah, M. M.; Lapetina, E. G. Evidence for multiple metabolic pools of phosphatidylinositol in stimulated platelets. *J. Biol. Chem.* **257**:11856–11859, 1982.
6. Lapetina, E. G.; Billah, M. M.; Cuatrecasas, P. Lysophosphatidic acid potentiates the thrombin-induced production of arachidonate metabolites in platelets. *J. Biol. Chem.* **256**:11984–11987, 1981.
7. Billah, M. M.; Lapetina, E. G.; Cuatrecasas, P. Phosphatidylinositol-specific phospholipase-C of platelets: Association with 1,2-diacylglycerol-kinase and inhibition by cyclic-AMP. *Biochem. Biophys. Res. Commun.* **90**:92–98, 1979.

8. Billah, M. M.; Lapetina, E. G.; Cuatrecasas, P. Phospholipase A_2 and phospholipase C activities of platelets: Differential substrate specificity, Ca^{2+} requirement, pH dependence, and cellular localization. *J. Biol. Chem.* **255**:10227–10231, 1980.
9. Billah, M. M.; Lapetina, E. G.; Cuatrecasas, P. Phospholipase A_2 activity specific for phosphatidic acid: A possible mechanism for the production of arachidonic acid in platelets. *J. Biol. Chem.* **256**:5399–5403, 1981.
10. Billah, M. M.; Lapetina, E. G. Formation of lysophosphatidylinositol in platelets stimulated with thrombin or ionophore A_{23187}. *J. Biol. Chem.* **257**:5196–5200, 1982.
11. Billah, M. M.; Lapetina, E. G. Rapid decrease of phosphatidylinositol 4,5-bisphosphate in thrombin-stimulated platelets. *J. Biol. Chem.* **257**:12705–12708, 1982.
12. Billah, M. M.; Lapetina, E. G. Degradation of phosphatidylinositol 4,5-bisphosphate is insensitive to Ca^{2+}-mobilization in stimulated platelets. *Biochem. Biophys. Res. Commun.* **109**:217–222, 1982.
13. Billah, M. M.; Lapetina, E. G. Platelet-activating factor stimulates metabolism of phosphoinositides in horse platelets: Possible relationship to Ca^{2+}-mobilization during stimulation. *Proc. Natl. Acad. Sci. USA* **80**:965–968, 1983.
14. Siess, W.; Cuatrecasas, P.; Lapetina, E. G. A role for cyclooxygenase products in the formation of phosphatidic acid in platelets: Differential mechanisms of action of thrombin and collagen. *J. Biol. Chem.* **258**:4683–4686, 1983.
15. Lapetina, E. G.; Siegel, F. L. Shape change induced in human platelets by platelet activating factor: Correlation with the formation of phosphatidic acid and phosphorylation of a 40,000 dalton protein. *J. Biol. Chem.* **258**:7241–7244, 1983.
16. Siess, W.; Siegel, F. L.; Lapetina, E. G. Arachidonic acid stimulates the formation of 1,2-diacylglycerol and phosphatidic acid in human platelets: Degree of phospholipase C activation correlates with protein phosphorylation, platelet shape change, serotonin release and aggregation. *J. Biol. Chem.* **258**:11236–11242, 1983.
17. Lapetina, E. G. Metabolism of inosites and the activation of platelets. *Life Sci.* **32**:2069–2082, 1983.
18. Lapetina, E. G.; Siess, W. The role of phosphatidylinositol-specific phospholipase C and phospholipases A_2 in platelet responses. *Life Sci.* **33**:1011–1018, 1983.
19. Broekman, M. J.; Ward, J. W.; Marcus, A. J. Phospholipid metabolism in stimulated platelets: Changes in phosphatidylinositol, phosphatidic acid and lysophospholipids. *J. Clin. Invest.* **66**:275–283, 1980.
20. Lloyd, J. V.; Nishizawa, E. E.; Mustard, J. F. Effect of ADP-induced shape change on incorporation of ^{32}P into platelet phosphatidic acid and mono-, di- and triphosphatidyl inositol. *Br. J. Haematol.* **25**:77–99, 1973.
21. Lapetina, E. G. Platelet-activating factor stimulates the phosphatidylinositol cycle: Appearance of phosphatidic acid is associated with the release of serotonin in horse platelets. *J. Biol. Chem.* **257**:7314–7317, 1982.
22. Kishimoto, A.; Takai, Y.; Mori, T.; Kikkawa, U.; Nishizuka, Y. Activation of a calcium and phospholipid-dependent protein kinase by diacylglycerol, its possible relation to phosphatidylinositol turnover. *J. Biol. Chem.* **255**:2273–2276, 1980.
23. Sano, K.; Takai, Y.; Yamanishi, J.; Nishizuka, Y. A role of calcium-activated phospholipid-dependent protein kinase in human platelet activation. *J. Biol. Chem.* **258**:2010–2013, 1983.

6

Receptor-Mediated Changes in Hepatocyte Phosphoinositide Metabolism
Mechanism and Significance

Joseph Eichberg and Charles A. Harrington

Hepatocytes respond to a variety of extracellular stimuli which evoke intracellular metabolic changes. Among these, glucagon, isoproterenol, and other β-adrenergic receptor agonists act via elevation of cAMP levels, which leads to activation of glycogen phosphorylase and a resultant onset of glycogenolysis. However, another group of stimuli, which includes α_1-adrenergic agonists, vasopressin, and angiotensin II, each acting at a distinct cell surface receptor, enhances glycogen phosphorylase activity in rat hepatocytes. The mechanism is cAMP independent and involves a rapid rise in free cytosolic Ca^{2+} [6,10,31]. The mechanism of Ca^{2+} mobilization is unsettled, but cumulative evidence suggests that the ion is released from at least one and possible several intracellular sites. An increase in cytosolic Ca^{2+} may also constitute a step in the initiation of other cellular events such as inactivation of glycogen synthetase and stimulation of potassium fluxes across the plasma membrane [10]. The release of Ca^{2+} into the cytosol appears to be followed by extensive net uptake of the ion from the external medium, perhaps to replenish depleted intracellular stores.

The purpose of this brief review is to examine the evidence which implicates metabolism of inositol-containing phospholipids in these receptor-mediated changes in intracellular Ca^{2+} concentrations and to describe recent studies in our laboratory which indicate that rat liver plasma membrane preparations may be used to advantage in studies of the mechanism which links receptor–ligand interactions with stimulation of phosphoinositide metabolism.

Stimulated Phosphatidylinositol Metabolism in the Intact Hepatocyte

Enhanced ^{32}P incorporation into PI in liver slices in response to epinephrine was first reported by De Torrontegui and Berthet [9] and in isolated rat hepatocytes exposed to either

---

Joseph Eichberg • Department of Biochemical and Biophysical Sciences, University of Houston, Houston, Texas 77004.   *Charles A. Harrington* • Analytical Neurochemistry Laboratory, Texas Research Institute of Mental Sciences, Houston, Texas 77030.

epinephrine or vasopressin by Kirk *et al.* [19]. Within the past few years, convincing evidence has accumulated that PI metabolism is stimulated by α_1-adrenergic agonists, vasopressin, and angiotensin II, with each class of agents acting at a separate receptor [5,28]. As in many other tissues which demonstrate this metabolic effect, the initial step is considered to be a decrease in PI levels [by 5–10% in the hepatocyte (18,21)], and stimulated incorporation of radioactivity into the molecule then reflects its subsequent compensatory resynthesis. The disappearance of PI has until recently been universally interpreted as a consequence of phosphodiesteratic cleavage by a PI-specific phospholipase C to yield 1,2-diacylglycerol and a mixture of D-inositol-1-phosphate and D-inositol-cyclic-1,2-phosphate. Diacylglycerol is then utilized to reform PI via phosphatidic acid and CMP-phosphatidic acid (CDP-diacylglycerol) as intermediates, as described more fully in Chapter 4 by Berridge. The entire metabolic process is known as the PI cycle.

A strong correlation has been established between those receptor-mediated responses which elicit stimulated PI turnover and those which elevate cytosolic Ca^{2+} concentrations [21,22]. However, a number of incompletely answered questions remain concerning the relationship between these phenomena. One long-standing issue is whether PI disappearance precedes or follows the rise in cytosolic Ca^{2+}. This problem has usually been examined by manipulations which either lower or raise cytosolic Ca^{2+}. The stimulation of phosphorylase activity upon receptor activation in intact hepatocytes is readily inhibited upon Ca^{2+} depletion. Removal of Ca^{2+} with EGTA in the incubation medium either reduces [13,18] or abolishes [25] stimulated PI metabolism brought about by α_1-adrenergic agonists or vasopressin. In our hands, diminished but still measurable stimulated ^{32}P labeling of PI in the presence of phenylephrine was evident following treatment of hepatocytes with EGTA and was not increased by Ca^{2+} addition to depleted cells (Fig. 1). The Ca^{2+} ionophore A23187 also elevates cytosolic Ca^{2+} to an extent sufficient to activate phosphorylase, without affecting PI metabolism [5,25,28].

Michell and co-workers have argued that since the dose–response curve for activation of glycogen phosphorylase occurs at nearly two orders of magnitude lower agonist con-

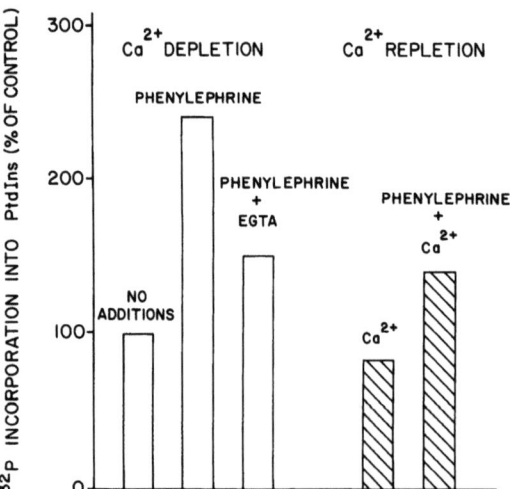

Figure 1. Effect of Ca^{2+} depletion and repletion on phenylephrine-stimulated incorporation of ^{32}P into phosphatidylinositol in rat hepatocytes isolated by a modification of the method of Pesch and co-workers [16]. Bars 1 and 2: Hepatocytes were preincubated with ^{32}P for 20 min in a medium composed of 136 mM NaCl, 3 mM KCl, 1.2 mM $MgCl_2$, 1 mM $CaCl_2$, 3 mM Hepes buffer pH 7.4, and 5 mM glucose and hereafter called Hepes-glucose buffer. Aliquots of the suspension (2×10^6 cells) were added to incubation vessels with Hepes-glucose buffer, ^{32}P, and 100 μM phenylephrine, if desired, and incubations continued for 60 min. Bar 3: To deplete cells of Ca^{2+}, hepatocytes were incubated for 30 min in Hepes-glucose buffer with Ca^{2+} omitted and in the presence of 2 mM EGTA. The cells were washed at 0°C with Ca^{2+}-free buffer. Aliquots (2×10^6 cells) were incubated with ^{32}P in Hepes-glucose buffer for 20 min. Phenylephrine was then added and incubation continued for 30 min. Bars 4 and 5 (hatched): After Ca^{2+} depletion, as described, the cation was added back to washed cells to a final concentration of 1 mM, and the suspension incubated for 30 min at 37°C. Aliquots (2×10^6 cells) were incubated for 20 min with ^{32}P and then for a further 60 min with or without 100 μM phenylephrine as described for Bar 3. Results are the average of two incubations.

centrations than that for stimulated inositol phospholipid breakdown, then the latter event, unlike the physiological response with which it is associated, requires virtually full occupancy of available receptors by agonist [23]. These authors suggest that since a critical level of cytosolic Ca^{2+} must be reached for phosphorylase to be stimulated, an analogous Ca^{2+} requirement, if it existed, for PI degradation, would require a much higher cytosolic level of cation, proportional to the far greater number of receptors occupied. Hence, enhanced PI metabolism should be much more sensitive to Ca^{2+} deprivation than phosphorylase activity, whereas in fact the opposite is seen. Thus, under suitable conditions of EGTA treatment, stimulation of phosphorylase is abolished without effect on inositol phospholipid disappearance [27]. It should also be kept in mind that depletion of Ca^{2+} using chelators may partially or wholly remove a pool of membrane-bound Ca^{2+} on which the PI-specific phospholipase C is dependent, or which is otherwise critical for retention of cellular physiological integrity.

Another series of unresolved questions involves the subcellular site of the responsive pool of inositol phospholipid, as well as of those enzyme molecules which catalyze the steps of the PI cycle. Clearly, if changes in PI metabolism are intimately connected with receptor occupancy, the responsive pool would most logically be associated with the plasma membrane. Efforts to identify a subcellular location by prelabeling hepatocytes with [^{3}H]inositol and then exposing the cells to vasopressin prior to fractionation have produced conflicting results. Thus, Kirk *et al.* [18] and Rhodes *et al.* [26] found essentially the same percent decrease in [^{3}H]-PI in all subcellular fractions, whereas Lin and Fain [20] reported a nearly 20% loss of PI only from a fraction they identified as enriched in plasma membranes.

The subcellular localizations of other steps in the PI cycle are also uncertain. Most speculative schemes depict released diacylglycerol being phosphorylated in the plasma membrane, phosphatidic acid then being transferred to the endoplasmic reticulum where PI biosynthesis is completed; this latter molecule is then transferred back to the plasma membrane via the PI-specific exchange protein [15]. A difficulty with this mechanism is that no protein has been described which promotes phosphatidic acid exchange between cell organelles.

Receptor-Mediated PI Disappearance in Rat Liver Plasma Membranes

Agonist-stimulated PI metabolism, which was discovered 30 years ago, has only been observed in intact cells or preparations that retain properties of whole cells (e.g., synaptosomes). Previous efforts to reproduce receptor-mediated PI depletion in subcellular fractions have not been successful and the usual explanation has been that the preservation of a closed membranous system with native vectorial organization of membrane constituents is essential for expression of the effect.

Recently, our laboratory has investigated conditions which might allow receptor-mediated PI disappearance to occur in plasma membrane-enriched preparations from rat liver [12]. In doing this, the amount of PI in the membranes was determined by chemical measurement using a method involving rapid separation of phospholipids by high-performance thin-layer chromatography, and quantification of lipid P by scanning densitometry after the chromatogram was sprayed with a molybdate-containing reagent [14]. When partially purified rat liver plasma membranes [24] were incubated for 30 min with or without added norepinephrine (50 μM), the amount of PI remained constant. The inclusion of rat liver cytosol had no effect in control incubations, but the addition of norepinephrine caused nearly a 30% decrease in membranous PI (Fig. 2). The decrease was not seen when boiled cytosol was used, whereas EGTA (3 mM) failed to prevent the disappearance of PI. The extent of PI

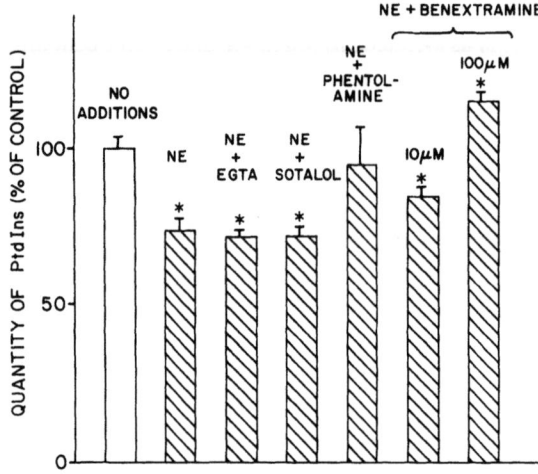

Figure 2. Rat liver plasma membranes and cytosol were prepared and incubated for 30 min in a medium consisting of Hepes buffer containing Mg^{2+} (10 mM) and Ca^{2+} (1 mM), as described in Ref. 12. The concentrations of pharmacological agents were: norepinephrine 50 µM; EGTA 3.0 mM; sotalol and phentolamine 500 µM; benextramine as shown. $*p < 0.001$ different from no additions.

depletion depended on the concentration of norepinephrine (Table I). Evidence that the decrease was mediated by α_1-adrenergic receptors was obtained by the finding that phentolamine, benextramine, and prazosin blocked PI loss, whereas sotalol had no effect (Fig. 2 and Table I). Although decrease of PI mediated by α_1-adrenergic receptors occurs in rat liver plasma membrane preparations supplemented with cytosol, we have thus far been unable to demonstrate an analogous decrease with vasopressin, possibly due to the lability of the receptors involved.

In these experiments, we did not identify the products derived from PI, although the expectation was that the responsible enzyme might be the soluble Ca^{2+}-dependent PI-specific phospholipase C. If this is so, it is surprising that the decrease in PI persisted in the presence of EGTA. However, the Ca^{2+} requirement of the enzyme has been documented using dispersed substrate where much of it may be needed to associate with the substrate and render it susceptible to enzymatic attack. Thus, cytosolic phospholipase C may degrade membrane-bound PI in the presence of much lower concentrations of Ca^{2+}, and this minimal

Table I. Effect of Varying Concentrations of Norepinephrine and Prazosin on Phosphatidylinositol Levels in Rat Liver Plasma Membranes[a]

Addition	nmoles PI present
None	2.9 ± 0.3
Norepinephrine	
1 µM	3.0 ± 0.1
10 µM	2.2 ± 0.1
50 µM	1.4 ± 0.2
50 µM norepinephrine + prazosin	
0.1 µM	1.4 ± 0.2
1 µM	1.5 ± 0.1
10 µM	2.8 ± 0.2
100 µM	3.0 ± 0.1

[a]Rat liver plasma membranes and cytosol were prepared and incubated for 30 min as described in Ref. 13.

quantity may be resistant to chelation. The finding that after treatment with EGTA intact hepatocytes will still exhibit some stimulation of PI metabolism upon receptor activation is consistent with this explanation. Our results are also in harmony with views of others that receptor-mediated PI degradation can take place in the virtual absence of cytosolic Ca^{2+} [2,23].

Recently, Fain and co-workers found that vasopressin caused an 11% increase in PI levels in rat liver plasma membranes (treated with deoxycholate) in the absence of Ca^{2+} [30] (see Chapter 7). This group observed a drop in PI concentration of similar magnitude when detergent- and Ca^{2+}-free plasma membranes obtained from inositol-deficient rats and prelabeled with [^{3}H]inositol were incubated with either vasopressin or epinephrine [11]. A norepinephrine-induced loss of approximately 10% of PI from isolated rat liver plasma membranes occurred in 1 min and was blocked by phentolamine [29]. Depletion of divalent cations mimicked this action of norepinephrine. In none of these studies was the effect of cytosol on these phenomena investigated. Hence, the results cannot be compared directly with our own; conceivably, variable amounts of cytosol adsorbed to the membranes could have supplied enzyme. Alternatively, endogenous enzyme activity in the plasma membranes might have been responsible for the modest loss of PI following receptor activation.

Receptor-Activated Polyphosphoinositide Degradation

The formulation of proposals that consider stimulated PI breakdown an obligatory event preceding intracellular Ca^{2+} mobilization has encountered a particularly serious difficulty; namely, an appreciable decrease of this phospholipid, whether measured chemically or by radioassay, is usually detectable only after several minutes, whereas the increase in cytosolic Ca^{2+} concentration occurs within a few seconds [7]. In this regard, recent observations in hepatocytes, which implicate polyphosphoinositides in Ca^{2+} mobilization [8,17,23,26,27], have cast an entirely new light on the significance of enhanced PI metabolism. In these experiments, all three hepatic phosphoinositides were labeled either with [^{3}H]inositol to equilibrium or with ^{32}P until the monoesterified phosphate groups of polyphosphoinositides were in equilibrium with cellular ATP. On exposure of hepatocytes to vasopressin, radioactivity in PI 4-phosphate (PIP) and PI 4,5-bisphosphate (PIP_2) was detectably decreased within a few seconds and declined to 50–80% of the initial quantity within as little as 20 sec, well before any decrease in PI was apparent. Within a short time therafter, labeled polyphosphoinositide returned toward its initial level. In general, the response was elicited by the same range of concentrations of those agonists which caused depletion of PI. That the decline in polyphosphoinositide content is indeed due to phosphodiesteratic cleavage has not yet been shown in liver, but the appearance of expected water-soluble products, inositol 1,4,5-trisphosphate and inositol 1,4-bisphosphate, has been documented in other stimulated tissues [1,3,4]. Moreover, the rate at which these compounds were generated was comparable to the rate of disappearance of PIP and PIP_2.

The inositol polyphosphates are acted on by phosphomonoesterases sequentially to yield inositol 1-phosphate and eventually myoinositol, products which could also arise from the degradation of PI. Further, the absolute rate of disappearance of both PI and the polyphosphoinositides in hepatocytes following stimulation appears to be similar [23]. These findings have led to the proposal that PI disappearance is caused, not by breakdown of the molecule, but by its utilization for resynthesis of the depleted store of polyphosphoinositides which undergoes rapid degradation.

In the light of this hypothesis, the decrease of PI observed under appropriate conditions in plasma membrane preparations incubated with cytosol may have resulted from its conversion at least in part to one or both of the polyphosphoinositides, which in turn were rapidly hydrolyzed. For this to occur, adequate ATP must be present in the cytosol, and this has not yet been demonstrated. Nonetheless, the failure of many investigators to demonstrate stimulated phosphoinositide metabolism in cell-free preparations could be due to absence of a sufficient supply of ATP needed for PIP and PIP_2 synthesis.

Conclusion

It now seems established that receptor activation in hepatocytes is followed by rapid breakdown of a pool of polyphosphoinositides (Fig. 3). The slower disappearance of PI may be due either to conversion of this molecule to PIP and PIP_2, its degradation by phosphodiesteratic cleavage, or a combination of the two processes. At present, the preponderance of evidence seems to support the conclusion that elevation of cytosolic Ca^{2+} does not precede and therefore does not regulate inositol phospholipid breakdown, although this point is not completely settled. However, a pool of Ca^{2+}, quite possibly membrane-bound, may be required for activation of one or more enzymes, especially the PI-specific phospholipase C or perhaps a polyphosphoinositide-specific phospholipase C. Despite the strong circumstantial association between stimulated phosphoinositide turnover and a rise in intracellular Ca^{2+}, there is as yet no direct evidence for a mechanistic link between these two events. Possibly a labile hydrolysis product of polyphosphoinositides, such as inositol 1,4,5-trisphosphate, will emerge as a vital link in these phenomena. The liver plasma membrane preparation may prove to be a valuable model system in future studies concerned with the relationship between phosphoinositide metabolism and Ca^{2+} mobilization.

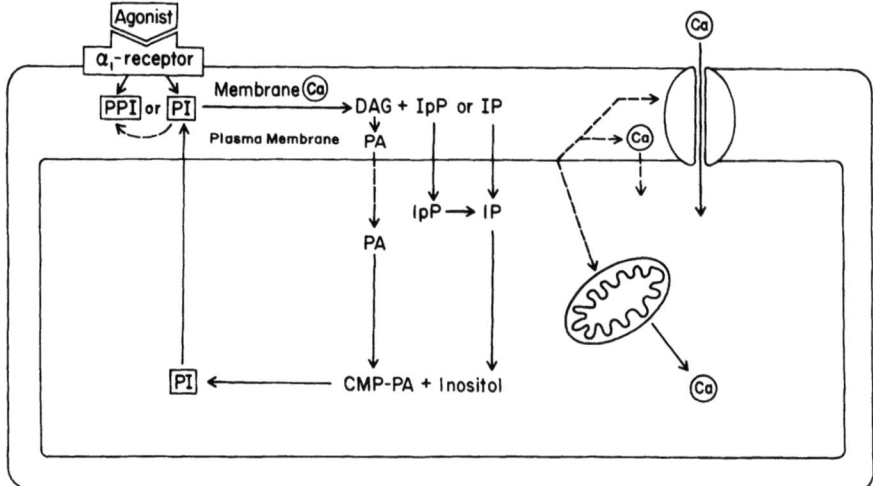

Figure 3. Proposed relationship between receptor activation, stimulated phosphoinositide metabolism, and Ca^{2+} mobilization in the hepatocyte. The hypothetical scheme shown for an α_1-adrenergic-mediated mechanism is applicable to other effective agonists and their receptors as well. PI, phosphatidylinositol; PPI, phosphatidylinositol 4-phosphate and/or phosphatidylinositol 4,5-bisphosphate; IP, inositol 1-phosphate; IpP, inositol polyphosphates, i.e., inositol 1,4-bisphosphate and/or inositol 1,4,5-trisphosphate; DAG, diacylglycerol; PA, phosphatidic acid; CMP-PA, cytidine monophosphate-phosphatidic acid (CDP-diacylglycerol). [Modified from Ref. 11.]

ACKNOWLEDGMENTS. The research described in this chapter was supported by grants from the National Institutes of Health (NS-12493) and from the Robert A. Welch Foundation (E-675).

References

1. Agranoff, B. W.; Murthy, P.; Sequin, E. P. Thrombin-induced phosphodiesteratic cleavage of phosphatidylinositol bisphosphate in human platelets. *J. Biol. Chem.* **258:**2076–2078, 1983.
2. Berridge, M. J. Phosphatidylinositol hydrolysis: A multifunctional hypothesis. *Mol. Cell. Endocrinol.* **24:**115–140, 1981.
3. Berridge, M. J. Rapid accumulation of inositol trisphosphate reveals that agonists hydrolyse polyphosphoinositides instead of phosphatidylinositol. *Biochem. J.* **212:**849–858, 1983.
4. Berridge, M. J.; Dawson, R. M. C.; Downes, C. P.; Heslop, J. P.; Irvine, R. F. Changes in the levels of inositol phosphates after agonist-dependent hydrolysis of membrane phsophoinositides. *Biochem. J.* **212:**473–482, 1983.
5. Billah, M. M.; Michell, R. H. Phosphatidylinositol metabolism in rat hepatocytes stimulated by glycogenolytic hormones. *Biochem. J.* **182:**661–668, 1979.
6. Blackmore, P. F.; Hughes, B. P.; Shuman, E. A.; Exton, J. H. α-Adrenergic activation of phosphorylase in liver cells involves mobilization of intracellular calcium without influx of extracellular calcium. *J. Biol. Chem.* **257:**190–197, 1982.
7. Charest, R.; Blackmore, P. F.; Berthon, B.; Exton, J. H. Changes in free cytosolic Ca^{2+} in hepatocytes following α_1-adrenergic stimulation. *J. Biol. Chem.* **258:**8769–8773, 1983.
8. Creba, J. A.; Downes, C. P.; Hawkins, P. T.; Brewster, G.; Michell, R. H.; Kirk, C. J. Rapid breakdown of phosphatidylinositol-4-phosphate and phosphatidylinositol-4,5-bisphosphate in rat hepatocytes stimulated by vasopressin and other Ca^{2+}-mobilizing hormones. *Biochem. J.* **212:**733–747, 1983.
9. De Torrontegui, G.; Berthet, J. The action of adrenaline and glucagon on the metabolism of phospholipids in the rat liver. *Biochim. Biophys. Acta* **116:**467–476, 1966.
10. Exton, J. H. Molecular mechanisms involved in α_1-adrenergic responses. *Mol. Cell. Endocrinol.* **23:**233–264, 1981.
11. Fain, J. N.; Lin, S.-H.; Randazzo, P.; Robinson, S.; Wallace, M. Hormonal regulation of glycogen phosphorylase in rat hepatocytes: Activation of phosphatidylinositol breakdown by vasopressin and alpha$_1$ catecholamines. In: *Isolation, Characterization and Use of Hepatocytes*, R. A. Harris and N. W. Cornell, eds., New York, Elsevier Biomedical, 1983, pp. 411–418.
12. Harrington, C. A.; Eichberg, J. Norepinephrine causes α_1-adrenergic receptor-mediated decrease of phosphatidylinositol in isolated rat liver plasma membranes supplemented with cytosol. *J. Biol. Chem.* **258:**2087–2090, 1983.
13. Harrington, C. A.; Davis, C. M.; Eichberg, J. α_1-Adrenergic receptor-mediated phosphatidylinositol metabolism in rat hepatocytes and purified plasma membranes. Abstr. 63rd Annual Endocrine Meeting. p. 113, 1981.
14. Harrington, C. A.; Fenimore, D. C.; Eichberg, J. Fluorometric analysis of polyunsaturated phosphatidylinositol and other phospholipids in the picomole range using high-performance thin layer chromatography. *Anal. Biochem.* **106:**307–313, 1980.
15. Helmkamp, G. M.; Harvey, M. S.; Wirtz, K. W. A.; Van Deenen, L. L. M. Phospholipid exchange between membranes: Purification of bovine brain proteins that preferentially catalyze the transfer of phosphatidylinositol. *J. Biol. Chem.* **249:**6382–6389, 1974.
16. Howard, R. B.; Lee, J. C.; Pesch, L. R. A. The fine structure, potassium content, and respiratory activity of isolated rat liver parenchymal cells prepared by improved enzymatic techniques. *J. Cell Biol.* **57:**642–658, 1973.
17. Kirk, C. J. Ligand-stimulated inositol lipid metabolism in the liver: Relationship to receptor function. *Cell Calcium* **3:**399–411, 1982.
18. Kirk, C. J.; Michell, R. H.; Hems, D. A. Phosphatidylinositol metabolism in rat hepatocytes stimulated by vasopressin. *Biochem. J.* **194:**155–165, 1981.
19. Kirk, C. J.; Verrinder, T. R.; Hems, D. A. Rapid stimulation, by vasopressin and adrenaline, of inorganic phosphate incorporation into phosphatidylinositol in isolated hepatocytes. *FEBS Lett.* **83:**267–271, 1977.
20. Lin, S.-H.; Fain, J. N. Vasopressin and epinephrine stimulation of phosphatidylinositol breakdown in the plasma membrane of rat hepatocytes. *Life Sci.* **29:**1905–1912, 1981.

21. Michell, R. H. Inositol phospholipids and cell surface receptor function. *Biochim. Biophys. Acta* **415**:81–147, 1975.
22. Michell, R. H. Inositol phospholipids in membrane function. *Trends Biochem. Sci.* **4**:128–131, 1979.
23. Michell, R. H.; Kirk, C. J.; Jones, L. M.; Downes, C. P.; Creba, J. A. The stimulation of inositol lipid metabolism that accompanies calcium mobilization in stimulated cells: Defined characteristics and unanswered questions. *Philos. Trans. R. Soc. London Ser. B* **296**:123–137, 1981.
24. Pohl, S. L.; Birnbaumer, L.; Rodbell, M. J. The glycogen-sensitive adrenyl cyclase system in plasma membranes of rat liver. *J. Biol. Chem.* **246**:1849–1856, 1971.
25. Prpic, V.; Blackmore, P. F.; Exton, J. H. Phosphatidylinositol breakdown induced by vasopressin and epinephrine in hepatocytes is calcium-dependent. *J. Biol. Chem.* **257**:11323–11331, 1982.
26. Rhodes, D.; Prpic, V.; Exton, J. H.; Blackmore, P. F. Stimulation of phosphatidylinositol-4,5-bisphosphate hydrolysis in hepatocytes by vasopressin. *J. Biol. Chem.* **258**:2770–2773, 1983.
27. Thomas, A. P.; Marks, J. S.; Coll, K. E.; Williamson, J. R. Quantitation and early kinetics of inositol lipid changes induced by vasopressin in isolated and cultured hepatocytes. *J. Biol. Chem.* **258**:5716–5725, 1983.
28. Tolbert, M. E. M.; White, A. C.; Aspry, K.; Curtis, J.; Fain, J. N. Stimulation by vasopressin and α-catecholamines of phosphatidylinositol formation in isolated rat liver parenchymal cells. *J. Biol. Chem.* **257**:1938–1944, 1980.
29. Wallace, M. A.; Poggioli, J.; Gireaud, F.; Claret, M. Norepinephrine-induced loss of phosphatidylinositol from isolated rat liver plasma membranes: Effect of divalent cations. *FEBS Lett.* **156**:239–243, 1983.
30. Wallace, M. A.; Randazzo, P.; Li, S.-Y.; Fain, J. Direct stimulation of phosphatidylinositol degradation by addition of vasopressin to purified rat liver plasma membranes. *Endocrinology* **111**:341–343, 1982.
31. Williamson, J. R.; Cooper, R. H.; Hoek, J. B. Role of calcium in the hormonal regulation of liver metabolism. *Biochim. Biophys. Acta* **639**:243–295, 1981.

7

Hormonal Regulation of Phosphoinositide Metabolism in Rat Hepatocytes

John N. Fain, Michael Wallace, Sue-Hwa Lin, and Irene Litosch

This chapter reviews recent advances on the hormonal regulation of phosphoinositide metabolism in rat hepatocytes. The pioneering work of Hokin and Hokin on pancreatic slices demonstrated that acetylcholine stimulated the uptake of ^{32}P into phosphatidic acid and phosphatidylinositol (PI) but not into other phospholipids [1]. Durell *et al.* [2] suggested that the increased turnover of PI and phosphatidic acid was secondary to an initial hormone-stimulated breakdown of PI.

Our current view of the phosphoinositide cycle is depicted in Fig. 1. Agonist-stimulated breakdown of phosphoinositides occurs through a phospholipase C to produce inositol phosphates and diacylglycerol. The diacylglycerol may be subsequently phosphorylated to form phosphatidic acid. In the presence of CTP the phosphatidic acid is converted to CDP-diacylglycerol which combines with inositol to form PI. PI may be phosphorylated by PI kinase and PI 4-phosphate kinase to generate PI 4-phosphate and PI 4,5-bisphosphate.

Michell [3] postulated that PI breakdown is linked to the mechanism by which hormones stimulate influx of extracellular Ca^{2+}. An excellent correlation exists between agonist-stimulated phosphoinositide turnover and elevation of intracellular Ca^{2+} [3–5]. In hepatocytes, α_1-adrenergic agonists and vasopressin stimulate selective degradation of the phosphoinositides and increase intracellular Ca^{2+} levels [3–5]. Elevations in cytosolic Ca^{2+} probably result from Ca^{2+} release from intracellular sites [6–9] and probably from Ca^{2+} entry via receptor-regulated channels. The signal generated by hormone to trigger intracellular Ca^{2+} mobilization may be diacylglycerol or inositol phosphate derived from phosphoinositide breakdown. The mobilization of intracellular Ca^{2+} appears sufficient for α-adrenergic activation of glycogen phosphorylase [6–9].

The resynthesis of phosphatidic acid and PI may also be a consequence of intracellular Ca^{2+} mobilization. Release of Ca^{2+} from the endoplasmic reticulum could relieve an inhibitory constraint on phospholipid synthesis resulting in a stimulated $^{32}P_i$ uptake into PI [4]. The stimulated resynthesis of PI is seen only with $^{32}P_i$ label, but not with [^3H]inositol.

John N. Fain, Michael Wallace, Sue-Hwa Lin, and Irene Litosch • Section of Biochemistry, Division of Biology and Medicine, Brown University, Providence, Rhode Island 02912.

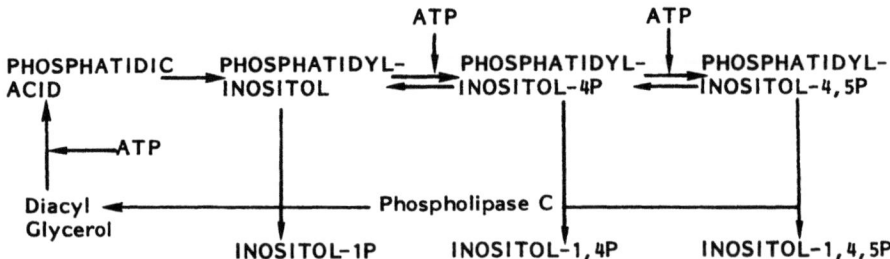

Figure 1. Pathways involved in the phosphoinositide cycle. Diacylglycerol is converted to phosphatidic acid, the precursor of phosphatidylinositol. The figure indicates that there are kinases and phosphatases involved in the interconversion of the three phosphoinositides. The phosphodiesterase cleavage of the three phosphoinositides gives the appropriate inositol phosphate plus diacylglycerol and this enzyme(s) has been referred to as phospholipase C.

While there is appreciable incorporation of [^{3}H]inositol into PI in hepatocytes, there is no further stimulation of [^{3}H]inositol uptake with α_1-adrenergic activation [10]. This suggests that labeled inositol does not equilibrate with the inositol pool used for the compensatory resynthesis of [^{32}P]-PI seen in the presence of vasopressin. Furthermore, vasopressin does not stimulate $^{32}P_i$ or [^{3}H]inositol incorporation into the polyphosphoinositides [10] even at short time periods [11,12], suggesting that the metabolic regulation of PI synthesis differs from that of PI 4,5-bisphosphate.

Lin and Fain [13] found selective degradation of plasma membrane PI in hepatocytes incubated for only 5 min with vasopressin or epinephrine. PI was labeled by injection of [^{3}H]inositol into rats 18 hr prior to isolation of hepatocytes. In these studies it was necessary to homogenize the hepatocytes at 5°C and use conditions which minimized activation of lysosomal phospholipases as well as redistribution of PI through phospholipid transfer proteins.

Wallace *et al.* [14] reported that vasopressin increases total PI degradation in rat liver membranes. This effect of vasopressin was not dependent on Ca^{2+} and was seen in the presence of deoxycholate [14]. Subsequently, a similar effect on breakdown of labeled PI was found in hepatic plasma membranes obtained from rats injected 18 hr previously with myo[2-^{3}H]inositol [11]. There was a 12% decrease in labeled PI over a 20-min incubation of membranes with 50 mU/ml of vasopressin in buffer containing 1 mM $CaCl_2$ and 17% breakdown in Ca^{2+}-free buffer containing 0.5 mM EGTA [11]. It was not necessary to add deoxycholate to these membranes, which were prepared from inositol-deficient rats. An 8% breakdown of total rat hepatic plasma membrane PI was found under conditions in which norepinephrine released bound Ca^{2+} from hepatocyte membranes [15,16]. Much work is still required to determine whether Ca^{2+} release and PI breakdown are concurrent events or causally linked.

The direct effect of norepinephrine on PI degradation in rat liver plasma membranes was independently reported by Harrington and Eichberg [17]. Their results were similar to those reported here in that Ca^{2+} was not required, but they found effects only in the presence of cytosol. The localization and nature of the enzyme activated by hormone remains to be elucidated. There is a hepatic cytosolic phospholipase C which acts on exogenous PI in the presence of appreciable amounts of Ca^{2+} [18]. The role of Ca^{2+}, *in vitro,* might be to neutralize the negative charge on PI allowing the enzyme to interact with its substrate. In contrast, when PI is present in the native membrane there may be no requirement for added Ca^{2+}. The activity of the cytosolic enzyme on PI may be modulated by other phospholipids and diacylglycerol [19,20].

7. Phosphoinositides, Hepatocytes, Hormones

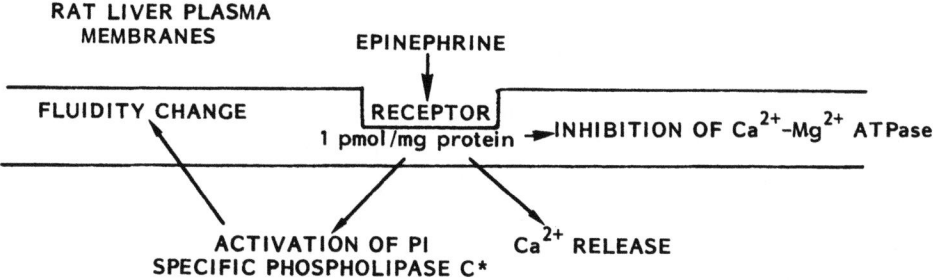

Figure 2. A speculative model for the initial events which occur after the interaction of epinephrine with rat liver plasma membranes. *One thousand picomoles of phosphatidylinositol (PI) are degraded per milligram of membrane protein per 1 min.

A schematic diagram for events occuring at the plasma membrane depicts epinephrine interacting with α_1-adrenoceptors [21] (Fig. 2). This interaction causes phospholipase C activation which cleaves the phosphoinositides. The rate of PI breakdown in hepatic plasma membranes is estimated to be about 1000 pmoles/mg protein per min [15]. More molecules of PI are degraded than can be accounted for by the hypothesis that receptor activation simply exposes to degradation PI molecules in the vicinity of the receptor. This suggests that an amplification step, possibly involving phospholipase C activation, must occur.

Three other effects of epinephrine on rat hepatocytes are depicted in Fig. 2. Epinephrine triggers $^{45}Ca^{2+}$ release from isolated rat liver plasma membranes and induces a change in membrane fluidity [16]. Additionally, vasopressin inhibits Ca^{2+}-Mg^{2+}-ATPase activity of hepatocyte plasma membrane [22]. Whether the changes in Ca^{2+} binding, membrane fluidity, and Ca^{2+}-Mg^{2+}-ATPase activity are related to phosphoinositide breakdown remains to be established.

The direct activation of PI breakdown by hormone addition to membranes is difficult to reconcile with the proposal by Michell and his associates [5,23] that hormones only increase PI 4,5-bisphosphate degradation. Loss of PI is presumed to occur solely through its phosphorylation to replenish PI 4,5-bisphosphate. So little polyphosphoinositide is present in isolated rat plasma membranes that its degradation cannot be measured. However, membranes incubated with [^{32}P]-ATP incorporate label into polyphosphoinositides. The uptake is rapid, with half- and near-maximal uptake at 10 and 30 sec, respectively [11]. Vasopressin does not affect incorporation of [^{32}P]-ATP into polyphosphoinositides [17]. Under the conditions used for measuring polyphosphoinositide formation, a 6% breakdown of total PI is observed over 30 min in the presence of vasopressin [11].

Our laboratory [11,12] has confirmed the rapid vasopressin-stimulated breakdown of polyphosphoinositides in hepatocytes reported by Michell *et al.* [5,23]. Four groups have reported on phosphoinositide metabolism in hepatocytes and have come to at least three different conclusions. Exton's laboratory found a rapid degradation of polyphosphoinositides but claimed that breakdown was secondary to an elevation in Ca^{2+} [24,25]. Michell proposed that polyphosphoinositide breakdown is the primary event in hormone action and loss of PI occurs through its phosphorylation to PI 4,5-bisphosphate [5,23,26]. Our group [11,12] and that of Williamson and co-workers [27] found that hormones directly accelerate the breakdown of all hepatic phosphoinositides. We postulate that phospholipase(s) are activated by hormones which degrade all three phosphoinositides (Fig. 1).

In hepatocytes prelabeled for 45 min with either [^3H]inositol or $^{32}P_i$, or both, an appreciable degradation of all phosphoinositides occurred in response to hormones [11]. Vasopressin (without removal of the labeled precursor) produced a transient loss of label from PI and polyphosphoinositides in the presence of Ca^{2+}. Loss of label was augmented if

EGTA was added to cells 10 min prior to hormone addition [11]. This effect disappeared by 120 sec, as uptake of label increased sufficiently to balance the breakdown of labeled phosphoinositides [11].

Rhodes et al. [25], using hepatocytes isolated from rats injected with [^{3}H]inositol, found that the rapid breakdown of labeled PI 4,5-bisphosphate induced by hormone was abolished by prolonged washing and incubation of hepatocytes in buffer containing 1 mM EGTA. They concluded that phosphoinositide breakdown was secondary to the elevation in cytosolic Ca^{2+}. However, these same investigators found that the calcium ionophore which increases cytosolic Ca^{2+} and stimulates glycogen phosphorylase, did not initiate phosphoinositide breakdown [24,25]. Since prolonged exposure to EGTA may have detrimental effects on cellular physiology, the finding that EGTA reduced the hormone effect may be due to factors other than Ca^{2+} depletion. Other laboratories have reported that neither PI synthesis [10,28] nor breakdown [11] was abolished by incubating hepatocytes in Ca^{2+}-free medium, under conditions in which activation of glycogen phosphorylase was abolished [10].

In hepatocytes previously incubated with $^{32}P_i$ or [^{3}H]inositol, near-maximal breakdown of PI and PI 4,5-bisphosphate occurs very rapidly in response to vasopressin [12]. The percent disappearance of ^{32}P label at 30 sec was 10% for PI, 15% for PI 4-phosphate, and 40% for PI 4,5-bisphosphate. With [^{3}H]inositol label, PI lost approximately five times as much label as did phosphatidylinositol 4,5-bisphosphate over 30 sec [12]. These results demonstrate that the hormone-sensitive pool of hepatocyte phosphoinositides can be labeled *in vitro* with both [^{3}H]inositol and $^{32}P_i$.

Measurements of total phospholipid content 30 sec after addition of vasopressin produced a major surprise. Despite the loss of 30–50% of labeled PI 4,5-bisphosphate, total content increased by 40% [12], indicating that vasopressin increases turnover of this phosphoinositide. The increase in content appears to result from a marked stimulation of PI 4,5-bisphosphate synthesis from pools of ATP and inositol which equilibrate poorly with [^{32}P]-ATP and [^{3}H]inositol. The increase in polyphosphoinositide content supports the concept that hormones also enhance PI conversion to PI 4,5-bisphosphate.

References

1. Hokin, M. R.; Hokin, L. E. Enzyme secretion and the incorporation of ^{32}P into phospholipids of pancreas slices. *J. Biol. Chem.* **203**:967–977, 1953.
2. Durell, J.; Garland, J. T.; Friedel, R. O. Acetylcholine action: Biochemical aspects. *Science* **165**:862–866, 1969.
3. Michell, R. H. Inositol phospholipids and cell surface receptor function. *Biochim. Biophys. Acta* **415**:81–147, 1975.
4. Fain, J. N. Involvement of phosphatidylinositol breakdown in elevation of cytosol Ca^{2+} by hormones and relationship to prostaglandin formation. *Horizons Biochem. Biophys.* **6**:237–276, 1982.
5. Downes, C. P.; Michell, R. H. Phosphatidylinositol 4-phosphate and phosphatidylinositol 4,5-bisphosphate lipids in search of a function. *Cell Calcium* **3**:467–502, 1982.
6. Blackmore, P. F.; Brumley, F. T.; Marks, J. L.; Exton, J. H. Studies on alpha-adrenergic activation of hepatic glucose output. *J. Biol. Chem.* **253**:4851–4858, 1978
7. Chen, J. L.; Babcock, D. F.; Lardy, H. A. Norepinephrine, vasopressin, glucagon, and A-23187 induce efflux of calcium from an exchangeable pool in isolated rat hepatocytes. *Proc. Natl. Acad. Sci. USA* **75**:2234–2238, 1978.
8. Althaus-Salzmann, M.; Carafoli, E.; Jakob, A. Ca^{2+}, K^+ redistributions and alpha-adrenergic activation of glycogenolysis in perfused rat livers. *Eur. J. Biochem.* **106**:241–248, 1980.
9. Malbon, C. C.; Gilman, H. R.; Fain, J. N. Hormonal stimulation of cyclic AMP accumulation and glycogen phosphorylase activity in calcium-depleted hepatocytes from euthyroid and hypothyroid rats. *Biochem. J.* **188**:593–599, 1980.

10. Tolbert, M. E. M.; White, A. C.; Aspry, K.; Cutts, J.; Fain, J. N. Stimulation by vasopressin and α-catecholamines of phosphatidylinositol formation in isolated rat liver parenchymal cells. *J. Biol. Chem.* **255**:1938–1944, 1980.
11. Fain, J. N.; Lin, S.-H.; Randazzo, P.; Robinson, S.; Wallace, M. Hormonal regulation of glycogen phosphorylase in rat hepatocytes: Activation of phosphatidylinositol breakdown by vasopressin and alpha$_1$ catecholamines. In: *Isolation, Characterization and Use of Hepatocytes*, R. A. Harris and N. W. Cornell, eds., New York, Elsevier, 1983, pp. 411–418.
12. Litosch, I.; Lin, S.-H.; Fain, J. N. Rapid changes in hepatocyte phosphoinositides induced by vasopressin. *J. Biol. Chem.* **258**:13727–13732, 1983.
13. Lin, S.-H.; Fain, J. N. Vasopressin and epinephrine stimulation of phosphatidylinositol breakdown in the plasma membrane of rat hepatocytes. *Life Sci.* **18**:1905–1912, 1981.
14. Wallace, M. A.; Randazzo, P.; Li, S.-Y.; Fain, J. N. Direct stimulation of phosphatidylinositol degradation by addition of vasopressin to purified rat liver plasma membranes. *Endocrinology* **111**:341–343, 1982.
15. Wallace, M. A.; Giraud, F.; Poggioli, J.; Claret, M. Norepinephrine-induced loss of phosphatidylinositol from isolated rat liver plasma membrane. *FEBS Lett.* **156**:239–243, 1983.
16. Burgess, G. M.; Giraud, F.; Poggioli, J.; Claret, M. α-Adrenergically mediated changes in membrane lipid fluidity and Ca^{2+} binding in isolated rat liver plasma membranes. *Biochim. Biophys. Acta* **731**:387–396, 1983.
17. Harrington, C. A.; Eichberg, J. Norepinephrine causes α$_1$-adrenergic receptor-mediated decrease of phosphatidylinositol in isolated rat liver plasma membranes supplemented with cytosol. *J. Biol. Chem.* **258**:2087–2090, 1983.
18. Kemp, P.; Hübscher, G.; Hawthorne, J. N. Phosphoinositides. 3. Enzymic hydrolysis of inositol-containing phospholipids. *Biochem. J.* **79**:193–200, 1960.
19. Dawson, R. M. C.; Hemington, N.; Irvine, R. F. The inhibition and activation of Ca^{2+}-dependent phosphatidylinositol phosphodiesterase by phospholipids and blood plasma. *Eur. J. Biochem.* **112**:33–38, 1980.
20. Hofmann, S. J.; Majerus, P. W. Modulation of phosphatidylinositol-specific phospholipase C activity by phospholipid interactions, diglycerides and calcium ions. *J. Biol. Chem.* **257**:14359–14364, 1982.
21. Goodhardt, M.; Ferry, N.; Geynet, P.; Hanoune, J. Hepatic α$_1$-adrenergic receptors show agonist-specific regulation by guanine nucleotides. *J. Biol. Chem.* **257**:11577–11583, 1982.
22. Lin, S.-H.; Wallace, M. A.; Fain, J. N. Regulation of Ca^{2+}-Mg^{2+}-ATPase activity in hepatocyte plasma membranes by vasopressin and phenylephrine. *Endocrinology* **113**:2268–2275, 1983.
23. Michell, R. H.; Kirk, C. J.; Jones, L. M.; Downes, C. P.; Creba, J. A. The stimulation of inositol lipid metabolism that accompanies calcium mobilization in stimulated cells: Defined characteristics and unanswered questions. *Philos. Trans. R. Soc. London Ser. B* **296**:123–137, 1981.
24. Prpic, V.; Blackmore, P. F.; Exton, J. H. Phosphatidylinositol breakdown induced by vasopressin and epinephrine in hepatocytes is calcium-dependent. *J. Biol. Chem.* **257**:11323–11331, 1982.
25. Rhodes, D.; Prpic, V.; Exton, J. H.; Blackmore, P. F. Stimulation of phosphatidylinositol 4,5-bisphosphate hydrolysis in hepatocytes by vasopressin. *J. Biol. Chem.* **258**:2770–2773, 1983.
26. Kirk, C. J.; Creba, J. A.; Downes, C. P.; Michell, R. H. Hormone-stimulated metabolism of inositol lipids and its relationship to hepatic receptor function. *Biochem. Soc. Trans.* **9**:377–379, 1981.
27. Thomas, A. P.; Marks, J. S.; Coll, K. E.; Williamson, J. R. Quantitation and early kinetics of inositol lipid changes induced by vasopressin in isolated and cultured hepatocytes. *J. Biol. Chem.* **258**:5716–5725, 1983.
28. Kirk, C. J.; Verrinder, T. R.; Hems, D. A. The influence of extracellular calcium concentration on the vasopressin-stimulated incorporation of inorganic phosphate into phosphatidylinositol in hepatocyte suspensions. *Biochem. Soc. Trans.* **6**:1031–1033, 1978.

8

Implication of Phosphoinositides in Receptor-Mediated Stimulation of Phospholipid Labeling

Bernard W. Agranoff, Edward B. Seguin, and Lucio A. A. Van Rooijen

"Stimulated phospholipid labeling," "the phosphatidylinositol effect," and similar terms have been in use for some years to describe the enhanced turnover of phospholipid that occurs when a variety of intact cells are exposed to ligands for which they have appropriate cell surface receptors. Hokin and Hokin initially observed [19] that acetylcholine not only causes release of amylase from pigeon pancreas slices, but stimulates incorporation of $^{32}P_i$ into two quantitatively minor phospholipids, eventually identified as phosphatidate (PA) and phosphatidylinositol (PI) [10,17,18]. Furthermore, the stimulated incorporation was blocked in the presence of atropine [19], establishing the pharmacological specificity of the phenomenon. In the intervening 30 years, similar or related observations have been made in a multitude of cell systems, suggesting the possible existence of a second messenger effector system in stimulus–secretion coupling, and in which PA and PI play a role [22]. In the interim, the cAMP mechanism has unfolded and its biochemical basis has been largely elucidated, while the basis of the "phosphatidylinositol effect" remains largely unresolved. While there have been implications that calcium gating, protein kinase, and prostanoid metabolism are somehow involved, they are at this writing still in the speculative realm.

One of the problems encountered in reviewing the various effects on ligand turnover is that model systems currently in use differ significantly, not only in specific tissue and ligands used, but in the biochemical paradigm employed. Thus, in some labeling studies, incorporation of labeled inositol rather than of $^{32}P_i$ is examined, while in others alterations in the absolute amounts of PA and PI are studied, without the use of radioisotopes. In many experiments, the ligand is added to prelabeled tissue and accelerated degradation rather than *de novo* incorporation is studied. This latter procedure, i.e., stimulated phospholipid degradation, has been employed extensively in studies on polyphosphoinositides, emphasized in this review. The reader should bear these differences in mind when comparing what might be considered contradictory or confirmatory studies.

Bernard W. Agranoff, Edward B. Seguin, and Lucio A. A. Van Rooijen • Departments of Biological Chemistry and Psychiatry, University of Michigan, Ann Arbor, Michigan 48109.

Biochemical Mechanisms: The PI Cycle

The finding that PA and PI were linked biosynthetically by a pathway that conserves radiophosphate [3] demonstrated that PA labeling preceded that of PI in ligand-stimulation experiments. The discovery of a PI-specific phosphodiesterase leading to inositol phosphate and diacylglycerol production [11] and the further finding that diacylglycerol could be rephosphorylated to PA [20], made possible the proposal of a closed cycle in which the ligand-receptor interaction leads to phosphodiesteratic cleavage of PI [13,21]. While the breakdown of prelabeled PI has been demonstrated in some cell systems in which enhanced phospholipid turnover follows ligand addition, this has proven not to be the case in others. To identify alternative sources of the putative diacylglycerol released upon ligand-receptor interaction, we initiated experiments that have led us to suggest that polyphosphoinositides may serve as the donor.

The Polyphosphoinositides

Phosphorylated derivatives of PI, PI 4-phosphate and PI 4,5-bisphosphate, are referred to collectively as the polyphosphoinositides. Why polyphosphoinositides have only recently been implicated in the stimulated labeling effect is largely attributable to technical factors. Standard lipid solvents, if not acidified, may not extract PI 4-phosphate (PIP) or PI 4,5-bisphosphate (PIP_2). Furthermore, unless special TLC conditions are employed, the two polyphosphoinositides will not migrate from the origin.

PIP and PIP_2 are synthesized sequentially from PI by specific kinases and are dephosphorylated by specific phosphomonoesterases, such that PI 5-phosphate is neither synthesized from PI nor formed from PIP_2 degradation [8,9]. The simultaneous action of these enzymes quickly equilibrates the phosphomonoester radioactivity of the 4' and 5' positions of D-myoinositol with that of $[\gamma\text{-}^{32}P]$-ATP in tissue incubations. Stimulated labeling of the polyphosphoinositides as the result of receptor activation has not been observed. In fact, when preparations are prelabeled prior to addition of ligand, loss of radioactivity from these lipids can be demonstrated. Studies in our laboratory employing nerve ending preparations indicated a potentiation of stimulated labeling of PA and PI by Ca^{2+} ionophore A23187 with that produced by muscarinic ligands. This was accompanied by a potentiated loss in radioactivity from prelabeled polyphosphoinositides [15]. These results focused our attention on the possible participation of polyphosphoinositides in stimulated lipid labeling.

Phosphodiesteratic Breakdown of Polyphosphoinositides

The stimulated breakdown of polyphosphoinositides as a result of receptor-ligand interaction has now been demonstrated in a number of systems, including muscarinic stimulation of iris muscle [1] and parotid gland [25], thrombin stimulation of platelets [7], and vasopressin stimulation of hepatocytes [23]. Stimulated turnover is generally seen within seconds, and there is a greater breakdown of PIP_2 than of PIP. The question then arises as to the nature of the degradation. PIP_2 can be dephosphorylated to the 4-phosphate via phosphomonoesterase, or cleaved to inositol trisphosphate (IP_3) and diacylglycerol via phosphodiesterase. There is at present little experimental evidence to support other possibilities such as phosphodiesteratic cleavage to PA and inositol bisphosphate ("phospholipase D"), or deacylation to lysolipids.

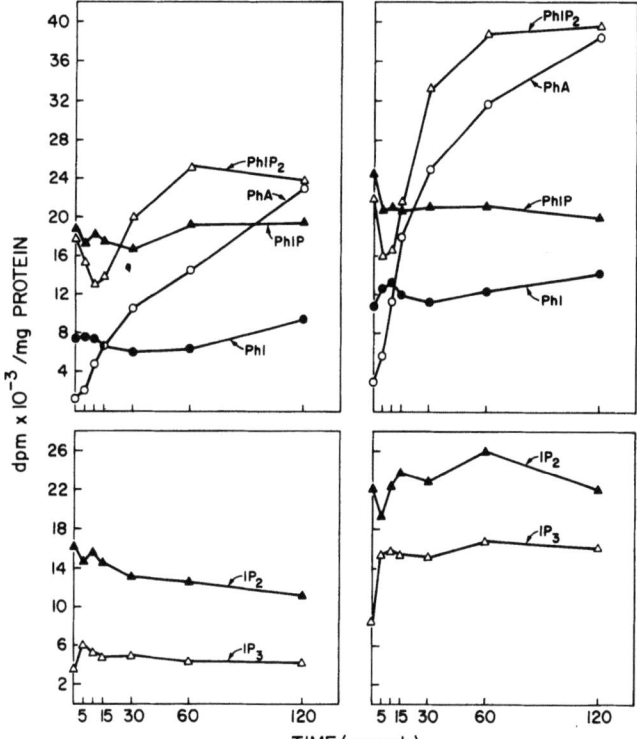

Figure 1. The effect of the addition of thrombin on the levels of phospholipids and inositol phosphates in two platelet preparations prelabeled with $^{32}P_i$. These data demonstrate that the initial event following platelet activation by thrombin is a diesteratic breakdown of PI 4,5-bisphosphate (PhIP$_2$), generating IP$_3$. DAG is also formed, and is converted to labeled PA. [From Agranoff et al. [4].]

The turnover of PIP$_2$ in intact nerve endings is so rapid, even in the absence of ligands, that this system appeared impractical for the study of stimulated polyphosphoinositide breakdown. We turned then to human platelets, a preparation known to support a brisk stimulation of PA and PI labeling in response to thrombin. By using a technique whereby both phospholipid and labeled water-soluble derivatives could be measured, the rapid loss of prelabeled PIP$_2$ was demonstrated, with the appearance of approximately equal amounts of radioactivity in a substance comigrating in high-voltage electrophoresis with IP$_3$ (Fig. 1). A smaller but consistent decrease of PIP was seen, which was not accompanied by an increase in IP$_2$. No marked changes in labeled PI were observed. While receptor-mediated breakdown of PIP$_2$ had been previously described in a number of tissues, the platelet studies were the first to demonstrate that the mediated breakdown was indeed phosphodiesteratic and of the phospholipase C type. Release of IP$_2$ and IP$_3$ from serotonin-stimulated salivary gland has also recently been described by Berridge (see Chapter 4).

The Nerve Ending Preparation

We have documented the muscarinic nature of stimulated labeling in nerve endings and summarized evidence that receptors are localized to postsynaptic sites [2]. As indicated above, ionophore A23187 mimics the stimulatory effect of muscarinic ligands, although unlike the latter, the effect of ionophore is not blocked by atropine. On the other hand, EGTA blocks stimulated labeling of PA and PI induced by either muscarinic agents or ionophore, thereby suggesting a role for Ca^{2+} in stimulated labeling [14]. The synergistic action of ionophore and carbachol indicates separate mechanisms for their respective effects.

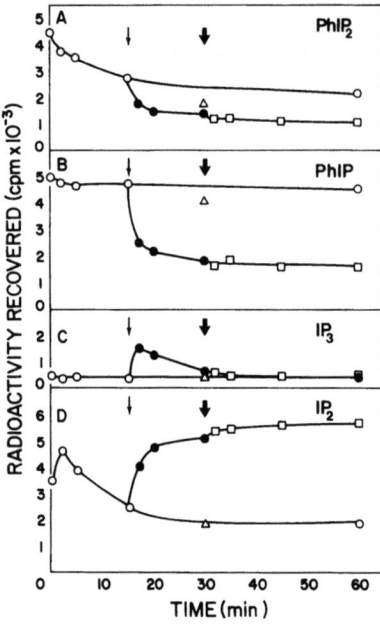

Figure 2. The effect of Ca^{2+} addition on $^{32}P_i$-labeled polyphosphoinositides and inositol phosphates. Nerve ending membranes prelabeled with $^{32}P_i$ were prepared by freeze-thawing and subsequent washing. An initial loss of label from the polyphosphoinositides was observed, which was not accompanied by an increase in inositol phosphates. The addition of 1.5 mM Ca^{2+} (○, ↓) elicited a rapid loss of label from both PI 4-phosphate (PhIP) and PI 4,5-bisphosphate with a concomitant increase in IP_2 and IP_3. The effect of excess EGTA (●, ◆) indicates that the increase in the water-soluble products was not due to inhibition of their phosphatases. These data demonstrate the presence of a Ca^{2+}-stimulated phosphodiesterase acting on endogenous polyphosphoinositides in nerve ending membranes. [From Van Rooijen et al. [24].]

Thus, Ca^{2+} is required for, but does not mediate, cholinergic stimulation of PA and PI labeling. While the role of Ca^{2+} in initiation of enhanced phospholipid turnover is unclear, Ca^{2+} eventually does play a regulatory role in systems where ligand addition results in a measurable biological effect such as secretion.

The loss of radioactivity from $^{32}P_i$-prelabeled nerve ending membranes has recently been investigated in our laboratory [24]. Addition of Ca^{2+} to washed membranes promotes a

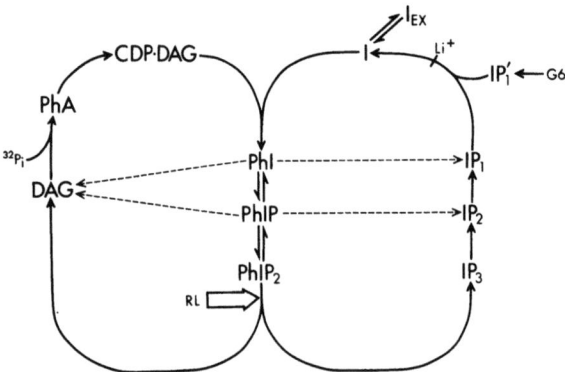

Figure 3. Integrated cycles in phosphoinositide turnover. The cycle proposed on the left is inferred from $^{32}P_i$ labeling experiments. The receptor–ligand interaction (RL) leads to phosphodiesteratic cleavage of PI 4,5-bisphosphate ($PhIP_2$) to DAG and IP_3. The DAG is phosphorylated by $[^{32}P]$-ATP to form PhA. In most tissues, the steady-state levels of labeled CDP·DAG are too low to detect. The reaction of CDP·DAG with inositol leads to the formation of PhI which is sequentially phosphorylated to the polyphosphoinositides, PhIP and $PhIP_2$. The observed stimulated labelings of PhA and PhI with various ligands are then reflections of a restorative phase in phospholipid turnover, whereby $PhIP_2$ is eventually resynthesized.

The cycle on the right is based on results with both $^{32}P_i$ prelabeling experiments and, in tissues where it is feasible to do so, with labeled inositol. The IP_3 produced is presumed to be D-myoinositol 1,4,5-trisphosphate and is dephosphorylated to IP_2 (D-myoinositol 1,4-bisphosphate) which is in turn dephosphorylated to IP_1 (D-myoinositol 1-phosphate). L-Myoinositol 1-phosphate (IP'_1) is formed from cyclization of glucose 6-phosphate. Both IP_1 and IP'_1 are degraded by a specific phosphatase that is inhibited by Li^+ [16]. That breakdown of IP_1 contributes significantly to cellular inositol levels is suggested by the finding that Li^+-treated rats having high brain IP levels also have reduced amounts of free inositol [5]. Extracellular inositol (I_{ex}) may play a limited role in cellular inositol levels. The integration of the two cycles in cells is supported by the finding that cholinergic ligands added to Li^+-treated salivary glands greatly enhance accumulation of IP [6].

phosphodiesteratic cleavage of both PIP and PIP$_2$ (Fig. 2). Enzyme activity appears specific for the polyphosphoinositides, since prelabeled PI is not degraded, even in the presence of excess Ca^{2+}. While PIP$_2$ is preferentially degraded in intact tissues, the endogenous enzyme in nerve ending membranes appears to degrade both PIP and PIP$_2$ equally. Thus, the phosphodiesteratic activity observed in membranes may be mediated by an enzyme that is distinct from the one that mediates the effect of stimulated labeling. It is more likely, however, that the endogenous enzyme mediates stimulated breakdown and that this discrepancy provides a clue regarding regulation of phosphodiesterase activity under *in vivo*, compared with *in vitro*, conditions. Since we are dealing with endogenous enzymes and substrates, apparent substrate specificities may be a reflection of the physical state of the enzyme or substrate. Thus, judgements on specificity should be reserved until definitive studies can be performed in well-defined systems.

In any event, these results, taken together with recent evidence in a number of other systems that phosphodiesteratic breakdown of PIP$_2$ and release of IP$_3$ are early consequences of receptor–ligand activation, indicate that the "PA–PI cycle" should include PIP and PIP$_2$ as obligatory intermediates (Fig. 3, left side). It is of some historical interest that this possibility was raised previously [12].

References

1. Abdel-Latif, A. A.; Akhtar, R. A.; Hawthorne, J. N. Acetylcholine increases the breakdown of triphosphoinositide of rabbit iris muscle prelabelled with [^{32}P]phosphate. *Biochem. J.* **162**:61–73, 1977.
2. Agranoff, B. W. Biochemical mechanisms in the phosphatidylinositol effect. *Life Sci.* **32**:2047–2054, 1983.
3. Agranoff, B. W.; Bradley, R. M.; Brady, R. O. The enzymatic synthesis of inositol phosphatide. *J. Biol. Chem.* **233**:1077–1083, 1958.
4. Agranoff, B. W.; Murthy, P.; Seguin, E. B. Thrombin-induced phosphodiesteratic cleavage of phosphatidylinositol bisphosphate in human platelets. *J. Biol. Chem.* **258**:2076–2078, 1983.
5. Allison, J. H.; Blisner, M. E.; Holland, W. H.; Hipps, P. P.; Sherman, W. R. Increased brain myo-inositol 1-phosphate in lithium-treated rats. *Biochem. Biophys. Res. Commun.* **71**:664–670.
6. Berridge, M. J.; Downes, C. P.; Hanley, M. R. Lithium amplifies agonist-dependent phosphatidylinositol responses in brain and salivary glands. *Biochem. J.* **206**:587–595, 1982.
7. Billah, M. M.; Lapetina, E. G. Rapid decrease of phosphatidylinositol 4,5-bisphosphate in thrombin-stimulated platelets. *J. Biol. Chem.* **257**:12705–12708, 1982.
8. Brockerhoff, H.; Ballou, C. E. The structure of the phosphoinositide complex of beef brain. *J. Biol. Chem.* **236**:1907–1911, 1961.
9. Chang, M.; Ballou, C. E. Specificity of ox brain triphosphoinositide phosphomonoesterase. *Biochem. Biophys. Res. Commun.* **26**:199–205, 1967.
10. Dawson, R. M. C. The measurement of ^{32}P labelling of individual kephalins and lecithin in a small sample of tissue. *Biochim. Biophys. Acta* **14**:374–379, 1954.
11. Dawson, R. M. C. Studies on the enzymic hydrolysis of monophosphoinositide by phospholipase preparations from *P. notatum* and ox pancreas. *Biochim. Biophys. Acta* **33**:68–77, 1959.
12. Durell, J.; Sodd, M. A.; Friedel, R. O. Acetylcholine stimulation of the phosphodiesteratic cleavage of guinea pig brain phosphoinositides. *Life Sci.* **7**:363–368, 1968.
13. Durell, J.; Garland, J. T.; Friedel, R. O. Acetylcholine action: Biochemical aspects. *Science* **165**:862–866, 1969.
14. Fisher, S. K.; Agranoff, B. W. Calcium and the muscarinic synaptosomal phospholipid labeling effect. *J. Neurochem.* **34**:1231–1240, 1980.
15. Fisher, S. K.; Agranoff, B. W. Enhancement of the muscarinic synaptosomal phospholipid labeling effect by the ionophore A23187. *J. Neurochem.* **37**:968–977, 1981.
16. Hallcher, L. M.; Sherman, W. R. The effects of lithium ion and other agents on the activity of *myo*-inositol-1-phosphatase from bovine brain. *J. Biol. Chem.* **255**:10896–10901, 1980.
17. Hokin, L. E.; Hokin, M. R. Effects of acetylcholine on the turnover of phosphoryl units in individual phospholipids of pancreas slices and brain cortex slices. *Biochim. Biophys. Acta* **18**:102–110, 1955.

18. Hokin, L. E.; Hokin, M. R. Phosphoinositides and protein secretion in pancreas slices. *J. Biol. Chem.* **233**:805–810, 1958.
19. Hokin, M. R.; Hokin, L. E. Enzyme secretion and the incorporation of P^{32} into phospholipids of pancreas slices. *J. Biol. Chem.* **203**:967–977, 1953.
20. Hokin, M. R.; Hokin, L. E. The synthesis of phosphatidic acid from diglyceride and adenosine triphosphate in extracts of brain microsomes. *J. Biol. Chem.* **234**:1381–1386, 1959.
21. Hokin, M. R.; Hokin, L. E. Interconversions of phosphatidylinositol and phosphatidic acid involved in the response to acetylcholine in the salt gland. In: *Metabolism and Physiological Significance of Lipids*, R. M. C. Dawson and D. N. Rhodes, eds., New York, Wiley, 1964, pp. 423–434.
22. Michell, R. H. Inositol phospholipids and cell surface receptor function. *Biochim. Biophys. Acta* **415**:81–147, 1975.
23. Rhodes, D.; Prpic, V.; Exton, J. H.; Blackmore, P. F. Stimulation of phosphatidylinositol 4,5-bisphosphate hydrolysis in hepatocytes by vasopressin. *J. Biol. Chem.* **258**:2770–2773, 1983.
24. Van Rooijen, L. A. A.; Seguin, E. B.; Agranoff, B. W. Phosphodiesteratic breakdown of endogenous polyphosphoinositides in nerve ending membranes. *Biochem. Biophys. Res. Commun.* **112**:919–926, 1983.
25. Weiss, S. J.; McKinney, J. S.; Putney, J. W., Jr. Receptor-mediated net breakdown of phosphatidylinositol 4,5-bisphosphate in parotid acinar cells. *Biochem. J.* **206**:555–560, 1982.

9

Calcium and Receptor-Mediated Phosphoinositide Breakdown in Exocrine Gland Cells

J. W. Putney, Jr., and P. P. Godfrey

Exocrine glands provide excellent models for examining the relationship of phosphoinositide metabolism to receptors and ion movements. Homogeneous preparations of tissue fragments, isolated acini, and isolated cells are well characterized and extensively studied as to the cellular events involved in the secretory response; specifically, the roles of Ca^{2+} and cAMP as second messengers have been examined in some detail.

Calcium-Mobilizing Receptors

The exocrine glands most extensively studied as to receptor control of ion movements and phospholipid metabolism are the exocrine pancreas, parotid salivary gland, and lacrimal glands of the rat. The calcium-mobilizing pathways of all three glands are activated through muscarinic–cholinergic receptors [12], i.e., by acetylcholine and congeners (carbachol, methacholine). The parotid and lacrimal glands have α-adrenoceptors similarly coupled to calcium mobilization, and the salivary gland and pancreas have peptide receptors for substance P and cholecystokinin, respectively. Available evidence suggests that activation of these receptors causes an elevation in the intracellular concentration of Ca^{2+}, which in turn triggers a number of cellular responses, including secretion of enzymes and alterations in monovalent ion fluxes.

Figure 1 shows the pattern of response obtained with carbachol as stimulus in rat lacrimal gland slices [14]. A technique was used whereby unidirectional efflux of ^{86}Rb was monitored as an indicator of membrane permeability to K^+. Addition of carbachol causes an abrupt increase in the rate of ^{86}Rb efflux which subsequently declines to a lesser but still significantly elevated level. The appropriate receptor blocking drug, in this case atropine,

---

J. W. Putney, Jr., and P. P. Godfrey • Division of Cellular Pharmacology, Medical College of Virginia, Richmond, Virginia 23298.

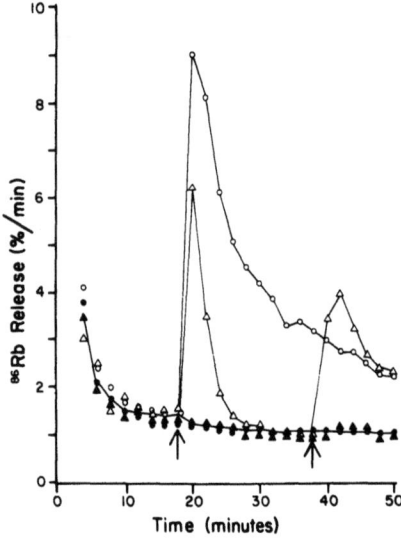

Figure 1. Release of ^{86}Rb from slices of rat lacrimal gland by carbachol. Slices were incubated in media containing ^{86}Rb and then carried through a series of brief nonradioactive incubations so that rate coefficients for efflux could be determined as previously described [14]. ●, control; ○, 10^{-5} M carbachol 18–50 min; ▲, 10^{-4} M atropine 12–50 min, 10^{-5} M carbachol 18–15 min; △, no added Ca^{2+} + 10^{-4} M EGTA 0–50 min, 10^{-5} M carbachol 18–50 min, 3.1 mM Ca^{2+} 38–50 min.

completely blocks the response. When experiments were carried out in a calcium-deficient medium, however, the response was only partially inhibited. The initial rise in ^{86}Rb efflux still occurred, although with some attenuation. The response subsequently returned to baseline, indicating that the more tonic component of the response is absolutely dependent on the presence of extracellular calcium. This inhibitory effect of calcium omission was readily reversible (Fig. 1). Reintroduction of extracellular calcium rapidly restored the elevated rate of ^{86}Rb efflux.

This biphasic response pattern has also been observed in parotid gland and exocrine pancreas (measuring amylase secretion), and the basic conclusion drawn by investigators has been the same in each case. Agonists are believed to act initially by mobilizing Ca^{2+} from some cellular depot into the cytosol. The resulting increase in intracellular $[Ca^{2+}]$ then triggers a variety of cellular responses such as ^{86}Rb efflux (opening of calcium-sensitive K^+ channels) or amylase secretion. Following this Ca^{2+} release (or concurrent with it), plasma membrane permeability to Ca^{2+} is increased, presumably due to activation of receptor-regulated Ca^{2+} channels or gates. These two modes of calcium mobilization, internal Ca^{2+} release and plasmalemmal Ca^{2+} gating, are believed to mediate respectively the (apparently) Ca^{2+}-independent and Ca^{2+}-dependent phases of responses (Fig. 1).

Alterations in Phosphoinositide Metabolism

These same calcium-mobilizing receptors are also associated with alterations in phosphoinositide metabolism. In fact, the first published account of agonist-induced effects on phosphatidylinositol (PI) was the 1953 report of Hokin and Hokin [7] of increased turnover ($^{32}PO_4$ incorporation) of PI in exocrine pancreas fragments treated with acetylcholine.

In the rat parotid gland, a number of specific observations have been made relevant to understanding the pathways involved in the phosphoinositide effect. Cholinergic agonists stimulate labeling of PI [6] and phosphatidic acid (PA) [18], cause net breakdown of PI [8] and net synthesis of PA [16], and increase the cellular levels of diacylglycerol (DG) (unpublished observation) and soluble inositol phosphates [2,3]. Taken together, these findings suggest that activation of cholinergic receptors causes breakdown of PI (or one of its phosphorylated derivatives; see below) by a phosphodiesteratic or phospholipase C type of mechanism. The resulting DG is rapidly phosphorylated by DG kinase to PA from which PI is resynthesized.

Most eukaryotic cells are capable of phosphorylating the inositol ring of PI by a specific kinase to form PI 4-phosphate (PIP), which can be further phosphorylated to form PI 4,5-bisphosphate [10]. These polyphosphoinositides are generally present in relatively low concentration relative to PI (< 1 nmole/mg protein) [5,19]. In ^{32}P-labeling experiments, the rapid turnover of monoester phosphates results in apparent equilibration of labeled PIP and PIP_2 with ATP (γ-phosphate) in about 1 hr [5,19]. Thus, net changes in ^{32}P-labeled polyphosphoinositides can be used to indicate net changes in cellular mass of these minor lipids.

In experiments in which ^{32}P-labeled polyphosphoinositides were examined in parotid acinar cells, methacholine induced a rapid ($t_{1/2} < 1$ min) decline in PIP_2 but had no effect on PIP [19]. Qualitatively similar results were obtained in pancreas and lacrimal gland [5,13]. This decline in PIP_2 level is generally more rapid than PI breakdown, suggesting to some investigators that polyphosphoinositide breakdown may be the initial reaction activated by calcium-mobilizing agonists [9,11].

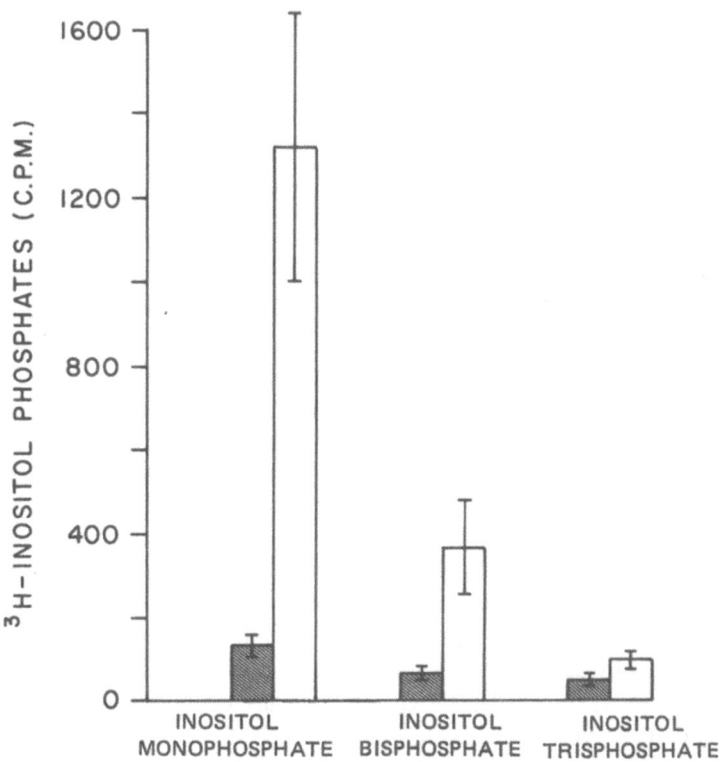

Figure 2. Effects of methacholine on the formation of [^{3}H]inositol phosphates in lacrimal cells. Cells were incubated with [^{3}H]inositol for 90 min, washed, and resuspended in cold medium. 10 mM LiCl was then added 5 min prior to the addition of 0.1 mM methacholine (white bar) or control (hatched bar). After 30 min the incubations were terminated by the addition of perchloric acid. The individual inositol phosphates were separated by anion-exchange chromatography. Results are means ± S.E.M. for five experiments. All three are increased significantly in the methacholine-treated samples compared to controls ($p < 0.05$ by paired t test). [For details see Ref. 5.]

Recently, a technique has been described whereby formation of soluble inositol phosphates, the putative products of inositide breakdown, can be analyzed [2,3]. Rat lacrimal cells were incubated with [^{3}H]inositol to label head groups of the inositides and, after experimental manipulations, soluble radioactive products were separated by anion-exchange chromatography. These experiments were carried out in the presence of 10 mM LiCl which blocks enzymatic degradation of inositol phosphate to inositol [2,3]. Methacholine caused a statistically significant increase in the levels of all three species of inositol phosphates, the effect being most pronounced for the case of inositol phosphate (Fig. 2). The slight but significant stimulation of inositol trisphosphate formation confirms the supposition that PIP_2 breakdown occurs by a phospholipase C type of mechanism.

As the other inositol phosphates could be formed either from breakdown of their respective inositol lipids or from sequential enzymatic dephosphorylation of inositol trisphosphate, it is not possible to conclude from these data whether direct breakdown of inositol lipids other than PIP_2 also occurs. In the parotid gland, however, inositol trisphosphate and bisphosphate formation precedes the appearance of inositol phosphate, and the quantity of inositol trisphosphate formed exceeds the cellular content of PIP_2 [1a]. These findings are consistent with the previous suggestion that the lipids primarily acted upon by phospholipase C are the polyphosphoinositides rather than PI.

Calcium and Phosphoinositide Breakdown

The physiological significance of altered phosphoinositide metabolism to the stimulus–response pathway in exocrine glands is not known with certainty. One attractive hypothesis was put forth by Michell in 1975 [10], whereby inositide breakdown is envisioned as an early reaction in the stimulus–response pathway which may serve to couple surface membrane receptor activation to cellular calcium mobilization. Evidence supporting this hypothesis is largely circumstantial. One point put forth in the original argument was that the PI effect appeared in most instances not to be a calcium-mediated reaction, suggesting that it preceded calcium mobilization in the sequence of reactions following receptor occupation. However, exceptions to this generalization have been noted [4]. Recent studies, therefore, have sought to examine more carefully the relationship between calcium and phosphoinositide metabolism specifically with reference to the suspected initial reaction, the breakdown of PIP_2 [19].

Figure 3 depicts the time course of agonist-induced alterations in [^{32}P]-PIP_2 as well as associated changes in ^{86}Rb efflux in the parotid gland. In both cases, the experiments were carried out in the presence (filled circles) or absence (open circles) of extracellular calcium [19]. The preparations were both treated sequentially with methacholine, atropine (a muscarinic–cholinergic receptor antagonist which blocks the effects of methacholine), and substance P (a calcium-mobilizing peptide). Methacholine induces a rapid breakdown of PIP_2 which is unaffected by calcium deprivation. This finding alone does not rule out calcium as a mediator of this effect, since, as the ^{86}Rb data indicate, the initial phase of calcium mobilization in these cells comes from internal pools which are not depleted by removing extracellular calcium. The ^{86}Rb data also show, however, that the second stimulus, substance P, fails to elicit any ^{86}Rb response in the calcium-deficient medium. This is presumably due to the fact that the internal pool of calcium, once mobilized by methacholine, cannot be replenished in the absence of external calcium [1,15]. This protocol did not result in significant inhibition of the ability of substance P to cause PIP_2 breakdown in the absence of calcium. These findings strongly argue that stimulation of PIP_2 breakdown in the parotid gland occurs independently of both calcium influx and internal calcium release.

9. Ca^{2+} and Receptor-Mediated Phosphoinositide Breakdown

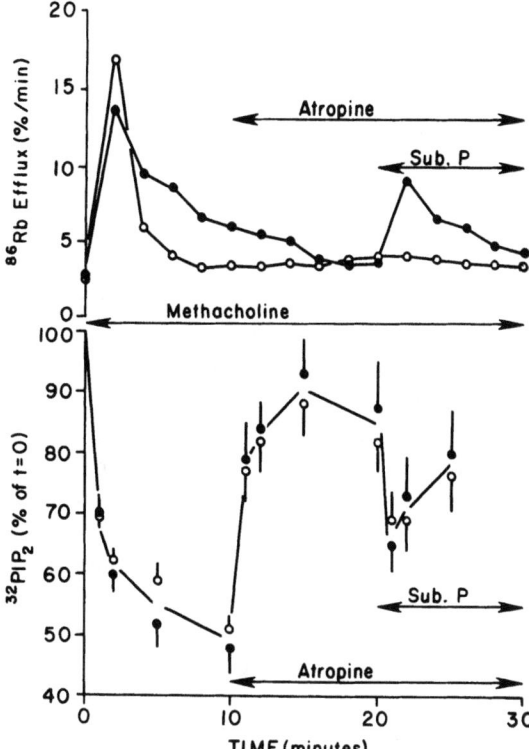

Figure 3. Effect of Ca^{2+} omission on changes in parotid gland ^{86}Rb flux (top) and $[^{32}P]$-PIP_2 levels (bottom) induced by agonists and antagonists. Methacholine (0.1 mM), atropine (10 μM), and substance P (0.1 μM) were added as indicated. ●, control; ○, excess EGTA added 5 min prior to methacholine. [For details see Ref. 19.]

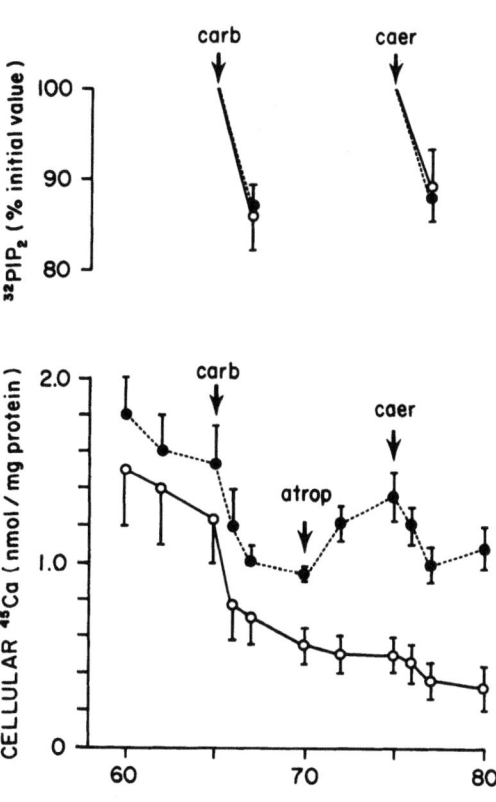

Figure 4. Effect of Ca^{2+} omission on ^{45}Ca content (bottom) and $[^{32}P]$-PIP_2 levels (top) in pancreatic acini in response to agonists and antagonists. Carbachol (carb, 10 μM), atropine (atrop, 10 μM), and caerulein (caer, 0.1 μM) were added as indicated (●), control; (○), excess EGTA added 10 min prior to carbachol. [For details see Ref. 13.]

Similar experiments have been carried out for the exocrine pancreas [13], and are summarized in Fig. 4. Cellular calcium release in the pancreas is of sufficient size that it can be measured as a net change in cell-associated ^{45}Ca in isolated cell or acini preparations. If acini are stimulated by carbachol, net efflux of ^{45}Ca occurs (Fig. 4), which is believed to reflect internal calcium release [17]. When atropine is added, calcium is reaccumulated and subsequently can be released by the peptide agonist, caerulein. When the calcium-chelating agent, EGTA, is added in excess of external calcium (open circles), no calcium can be sequestered when atropine is added and no calcium is released with caerulein. However, [^{32}P]-PIP$_2$ breakdown due to both carbachol and caerulein is unaffected. Again, the conclusion is that neither external calcium nor internal calcium release affects agonist-induced PIP$_2$ breakdown in exocrine pancreas.

Qualitatively similar results have been obtained for the lacrimal gland by measuring formation of inositol phosphates rather than inositide breakdown [15]. Collectively, these findings strongly suggest that in the parotid gland, lacrimal gland, and exocrine pancreas, phosphoinositide breakdown is an early event in the stimulus–response coupling pathway. Changes in cell lipid metabolism could affect calcium movements at least in part by the actions of inositol trisphosphate to release internal calcium [17a]. Clearly much more experimental work is necessary before definitive conclusions can be drawn regarding the precise role of the phosphoinositides in exocrine cell function.

ACKNOWLEDGMENT. Supported in part by National Institutes of Health Grants DE-05764 and EY-03533.

References

1. Aub, D. L.; McKinney, J. S.; Putney, J. W., Jr. Nature of the receptor-regulated calcium pool in the rat parotid gland. *J. Physiol. (London)* **331**:557–565, 1982.
1a. Aub, D. L.; Putney, J. W., Jr. Metabolism of inositol phosphates in parotid cells: Implications for the pathways of the phosphoinositide effect and the possible messenger role of inositol trisphosphate. *Life Sci.* **34**:1347–1355, 1984.
2. Berridge, M. J.; Dawson, R. M. C.; Downes, C. P.; Heslop, J. P.; Irvine, R. F. Changes in the levels of inositol phosphates following agonist-dependent hydrolysis of membrane phosphoinositides. *Biochem. J.* **212**:473–482, 1983.
3. Berridge, M. J.; Downes, C. P.; Hanley, M. R. Lithium amplifies agonist-dependent phosphatidylinositol responses in brain and salivary glands. *Biochem. J.* **206**:587–595, 1982.
4. Cockcroft, S. Does phosphatidylinositol breakdown control the Ca^{2+}-gating mechanism? *Trends Pharmacol. Sci.* **2**:340–342, 1981.
5. Godfrey, P. P.; Putney, J. W., Jr. Receptor-mediated metabolism of the phosphoinositides and phosphatidic acid in rat lacrimal acinar cells. *Biochem. J.* **218**:187–195, 1984.
6. Hokin, L. E.; Sherwin, A. L. Protein secretion and phosphate turnover in the phospholipids in salivary glands in vitro. *J. Physiol. (London)* **135**:18–29, 1957.
7. Hokin, M. R.; Hokin, L. E. Enzyme secretion and the incorporation of P^{32} into phospholipides of pancreas slices. *J. Biol. Chem.* **203**:967–977, 1953.
8. Jones, L. M.; Michell, R. H. Breakdown of phosphatidylinositol provoked by muscarinic cholinergic stimulation of rat parotid gland fragments. *Biochem. J.* **142**:583–590, 1974.
9. Kirk, C. J.; Creba, J. A.; Downes, C. P.; Michell, R. H. Hormone-stimulated metabolism of inositol lipids and its relationship to hepatic receptor function. *Biochem. Soc. Trans.* **9**:377–379, 1981.
10. Michell, R. H. Inositol phospholipids and cell surface receptor function. *Biochim. Biophys. Acta* **415**:81–147, 1975.
11. Poggioli, J.; Weiss, S. J.; McKinney, J. S.; Putney, J. W., Jr. Actions of antimycin A on receptor-activated calcium mobilization and phosphoinositide metabolism in rat parotid gland. *Mol. Pharmacol.* **23**:71–77, 1983.
12. Putney, J. W., Jr. Stimulus–permeability coupling: Role of calcium in the receptor regulation of membrane permeability. *Pharmacol. Rev.* **30**:209–245, 1978.

13. Putney, J. W., Jr.; Burgess, G. M.; Halenda, S. P.; McKinney, J. S.; Rubin, R. P. Effects of secretagogues on [^{32}P]phosphatidylinositol 4,5-bisphosphate metabolism in the exocrine pancreas. *Biochem. J.* **212**:483–488, 1983.
14. Putney, J. W., Jr.; Parod, R. J.; Marier, S. H. Control by calcium of exocytosis and membrane permeability to potassium in the rat lacrimal gland. *Life Sci.* **20**:1905–1912, 1977.
15. Putney, J. W., Jr.; Poggioli, J.; Weiss, S. J. Receptor regulation of calcium release and calcium permeability in parotid gland cells. *Philos. Trans. R. Soc. London Ser. B* **296**:37–45, 1981.
16. Putney, J. W., Jr.; Weiss, S. J.; Van De Walle, C. M.; Haddas, R. A. Is phosphatidic acid a calcium ionophore under neurohumoral control? *Nature (London)* **284**:345–347, 1980.
17. Stolze, H.; Schulz, I. Effect of atropine, ouabain, antimycin A, and A23187 on "trigger Ca^{2+} pool" in exocrine pancreas. *Am. J. Physiol.* **238**:G338–G348, 1980.
17a. Streb, H.; Irvine, R. F.; Berridge, M. J.; Schulz, I. Release of Ca^{2+} from a nonmitochondrial intracellular store in pancreatic acinar cells by inositol-1,4,5-trisphosphate. *Nature (London)* **306**:67–68, 1983.
18. Weiss, S. J.; McKinney, J. S.; Putney, J. W., Jr. Regulation of phosphatidate synthesis by secretagogues in parotid acinar cells. *Biochem. J.* **204**:587–592, 1982.
19. Weiss, S. J.; McKinney, J. S.; Putney, J. W., Jr. Receptor mediated net breakdown of phosphatidylinositol-4,5-bisphosphate in parotid acinar cells. *Biochem. J.* **206**:555–560, 1982.

III

Stimulus–Secretion Coupling

10

Calcium and Stimulus–Secretion Coupling in Pancreatic Acinar Cells

John A. Williams and Daniel B. Burnham

Pancreatic acinar cells synthesize and secrete a variety of digestive enzymes with the major physiological secretagogues being the neurotransmitter acetylcholine and the gut hormone cholecystokinin. These secretagogues also bring about other changes in acinar cell function, including depolarization, secretion of Cl^--rich pancreatic juice, and increased glucose uptake, O_2 consumption, and protein synthesis [14,34]. Most if not all of these secretagogue effects are mediated at least in part by Ca^{2+} as a second messenger. Douglas and colleagues originally coined the term *stimulus–secretion coupling* to refer to the role of Ca^{2+} as an intracellular messenger in the control of secretion by a number of tissues [10]. Although this term has also been applied to pancreatic acinar cells, it might be more appropriate to refer to this sequence as stimulus–response coupling since as mentioned above, secretagogues affect a number of acinar cell functions.

Evidence Supporting a Role for Ca^{2+} in Pancreatic Stimulus–Secretion Coupling

The evidence supporting an intracellular messenger role for Ca^{2+} in pancreatic acinar cells, which is summarized in Table I, has been the subject of several recent reviews [30,34]. This evidence emphasizes the ultimate importance of a rise in cytoplasmic free Ca^{2+}, which is believed to then activate secretion. However, compared to other secretory cells such as neurons, chromaffin cells, and pancreatic islet beta cells, stimulus–secretion coupling in acinar cells is prominently dependent on Ca^{2+} derived from intracellular stores. Thus, removal of extracellular Ca^{2+}, which immediately inhibits secretion by the aforementioned secretory cells, does not inhibit acinar cell secretion until after a 10- to 20-min period, during which time intracellular Ca^{2+} stores become depleted [28,34]. Depletion of acinar intracellular Ca^{2+} stores by combined use of an extracellular chelator and a calcium ionophore

---

John A. Williams and Daniel B. Burnham • Department of Physiology, University of California, San Francisco, California 94143.

Table I. Evidence Supporting a Role for Ca^{2+} in
Pancreatic Acinar Cell Stimulus–Secretion Coupling
Stimulated by CCK and ACh

Depletion of extracellular and intracellular Ca^{2+} inhibits the action of secretagogues

Secretagogues release bound or sequestered intracellular Ca^{2+} and increase the influx of Ca^{2+}

Artificial increases in cytoplasmic Ca^{2+} by use of ionophores mimic the action of secretagogues

Secretagogues increase the cytoplasmic concentration of Ca^{2+}

---

will, however, rapidly depress secretion [24]. The observations that secretion eventually falls in the absence of extracellular Ca^{2+} and that secretagogues increase $^{45}Ca^{2+}$ uptake into acinar cells indicate the existence of a membrane Ca^{2+} channel regulated by secretagogues. However, acinar cells do not appear to have voltage-dependent Ca^{2+} channels and neither secretion nor Ca^{2+} fluxes are blocked by calcium channel blockers such as D600 [8,22]. A recent suggestion is that Ca^{2+} may enter via the secretagogue-controlled Na^+–K^+ channel [22]. In summary, pancreatic secretion takes place using intracellular Ca^{2+} stores but extracellular Ca^{2+} is necessary to maintain cellular Ca^{2+} and ultimately secretion. The existence of receptor-mediated membrane Ca^{2+} channels in acinar cells requires further investigation.

Until quite recently there has been no direct evidence that free ionized calcium increases in acinar cells in response to secretagogues. Utilizing Ca^{2+}-sensitive microelectrodes, O'Doherty and Stark [21] measured a resting cytoplasmic Ca^{2+} concentration of 0.4 µM which was increased by acetylcholine in a dose dependent manner to about 1 µM. A higher basal level of Ca^{2+}, but a similar twofold increase in intracellular Ca^{2+} in response to secretagogues, was also estimated in a report using the Ca^{2+}-sensitive photoprotein aequorin, which was introduced by reversible hypotonic hemolysis [7]. Using the fluorescence of quin-2 as an intracellular Ca^{2+} indicator [32] also indicates a rise in cytoplasmic Ca^{2+} in acinar cells responding to secretagogues [20]. Thus, these recent data document a rise in intracellular Ca^{2+} in response to secretagogues. Whether or not this rise in Ca^{2+} is sufficient to explain all of the effects of secretagogues on acinar cells is currently under investigation.

Secretagogue-Induced Calcium Mobilization

The major alteration in acinar cell calcium induced by secretagogues is the mobilization of sequestered intracellular calcium. This was originally demonstrated by labeling intracellular Ca^{2+} stores with $^{45}Ca^{2+}$ and showing that the efflux of the radioactive Ca^{2+} was increased by acetylcholine and CCK [3,18]. Although initial studies were negative, it was subsequently demonstrated that secretagogues also increase $^{45}Ca^{2+}$ uptake [8,17]. However, the dominant initial effect was Ca^{2+} mobilization shown by determining the cellular content of $^{45}Ca^{2+}$ in the presence of a constant level of extracellular $^{45}Ca^{2+}$ [13]. The loss of total acinar cell Ca^{2+} in response to a secretagogue was shown definitively by measuring cellular calcium content by atomic absorption spectroscopy (Fig. 1).

Studies with the fluorescent chelate probe chlorotetracycline revealed that calcium was

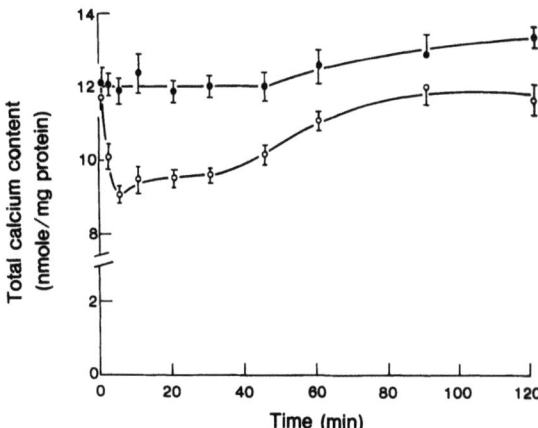

Figure 1. Total calcium content of mouse pancreatic acini in presence or absence of the cholinergic agonist bethanechol. ●, control; ○, 30 μM bethanechol. [From Ref. 8.]

being lost from a membrane-associated or sequestered pool dependent on cellular ATP [4]. That Ca^{2+} is sequestered in an ATP-dependent pool has also been shown by measuring effects of metabolic inhibitors on $^{45}Ca^{2+}$ fluxes [8,31]. This pool of bound Ca^{2+} can be discharged with a mitochondrial inhibitor. However, this does not necessarily imply that mitochondria are the source of the Ca^{2+}, since mitochondria are the predominate source of ATP in acini. Both metabolic inhibitors and secretagogues release calcium from the same pool, as determined with chlorotetracycline and $^{45}Ca^{2+}$ [4,8,31]. The source of mobilized calcium is readily accessible from the plasma membrane. This supposition is supported by the finding that this calcium pool rapidly refills when secretagogue action is blocked; this Ca^{2+} can then be rereleased by a second secretagogue [31].

The intracellular source of mobilized calcium and its mode of control are currently under study but unresolved. The major sites of Ca^{2+} sequestration in acinar cells, as in other cells, are mitochondria and endoplasmic reticulum. Both isolated pancreatic microsomes and mitochondria show distinct ATP-dependent Ca^{2+} uptake systems consistent with their role as an intracellular Ca^{2+} store [23]. *In vitro* microsomal Ca^{2+} uptake was recently localized to rough endoplasmic reticulum [25]. These data on isolated organelles are also supported by a study in acinar cells permeabilized with saponin in which Ca^{2+} uptake into endoplasmic reticulum and mitochondria was observed following precipitation with oxalate [33]. Subcellular fractionation of isolated pancreatic acini labeled with $^{45}Ca^{2+}$ demonstrated that carbachol stimulation reduced $^{45}Ca^{2+}$ content in all fractions, but most prominently in the endoplasmic reticulum [9]. Moreover, the endoplasmic reticulum was the only cell fraction to exhibit a secretagogue-induced loss of total calcium content (Table II). By contrast, however, an earlier study in pancreatic lobules with a different protocol suggested mitochondria as the source of mobilized $^{45}Ca^{2+}$ [5].

The mechanism by which occupancy of plasma membrane secretagogue receptors leads to Ca^{2+} mobilization is unknown. If the source of Ca^{2+} is an intracellular organelle, then an unidentified mediator must be postulated. For this reason, some investigators have suggested the plasma membrane as a Ca^{2+} source [31]. However, it seems unlikely that enough Ca^{2+} is normally bound to the plasma membrane to account entirely for the amount of Ca^{2+} lost (2–3 nmoles/mg acinar protein; Fig. 1). Alternatively, based on a mechanism proposed for cardiac muscle [11], release of a small amount of Ca^{2+} from the plasma membrane could produce a localized Ca^{2+} concentration sufficient to trigger further Ca^{2+} release from intracellular stores.

Table II. Effect of Carbachol on the Calcium Content of Pancreatic Acini and Subcellular Fractions[a]

	Ca^{2+} content (nmoles/mg protein)	
	Control	Carbachol (1 μM)
Whole acini	11.1 ± 0.8	8.5 ± 0.6*
Mitochondrial fraction	34.3 ± 1.5	30.8 ± 1.4
Zymogen granule fraction	38.4 ± 2.3	37.4 ± 1.8
Microsomal fraction	47.4 ± 3.7	37.7 ± 2.2*

[a]Acini were incubated with or without carbachol for 10 min before homogenization in 0.3 M sucrose containing 1 mM benzamidine and 10 μM ruthenium red.
*$p < 0.05$ difference from control.
Data from Ref. 9.

Considerable interest has recently focused on the possible role of phosphatidylinositide breakdown and Ca^{2+} mobilization in transmembrane signal transduction [19,26]. As originally described in pancreas, secretagogues lead to phosphatidylinositol breakdown by a phospholipase C-like mechanism, with increased production of phosphatidic acid [15]. In addition, secretagogues also induce breakdown of the polyphosphoinositides, phosphatidylinositol 4-phosphate and phosphatidylinositol 4,5-bisphosphate [27]. As polyphosphoinositides bind Ca^{2+} with relatively high affinity, Ca^{2+} release might arise from their breakdown. However, in most cells polyphosphoinositides account for only a few percent of total phosphoinositides. As another possible mechanism, the increased amount of phosphatidic acid could act as a Ca^{2+} ionophore [26]. Lastly, some other product of the breakdown of phosphatidylinositides may function as the second messenger leading to Ca^{2+} mobilization.

Ca^{2+}-Activated Changes in Protein Phosphorylation

In contrast to recent advances in our knowledge of the regulation of cellular calcium, much less is known about the mechanism by which Ca^{2+} brings about secretion in acinar and other secretory cells. Three general observations must be explained: (1) a small rise in Ca^{2+} must be able to exert its effect in the presence of a large excess of Mg^{2+}, (2) the process must require ATP, and (3) it should be a general process activated by cAMP, since Ca^{2+} may be replaced or act in parallel with cAMP as a second messenger. Currently, considerable interest centers on the possibility that this Ca^{2+}-dependent process involves changes in phosphorylation of specific regulatory proteins. Changes in protein phosphorylation would account for the aformentioned three general observations. Indeed, several classes of Ca^{2+}-activated protein kinases and phosphatases have been described and related to various cell functions, including metabolic and contractile events [6,29]. However, in contrast to intermediary metabolism, there is little evidence as to which enzymes or structural proteins may be involved in the control of secretion. Therefore, a more phenomenological approach to describing the role of protein phosphorylation in calcium action has initially been necessary in the acinar cell.

Acinar Cell Protein Phosphorylation

When isolated mouse pancreatic acini were incubated with $^{32}P_i$ to label cellular ATP, a calcium-mediated secretagogue such as carbachol then added for 1–10 min, and the radioac-

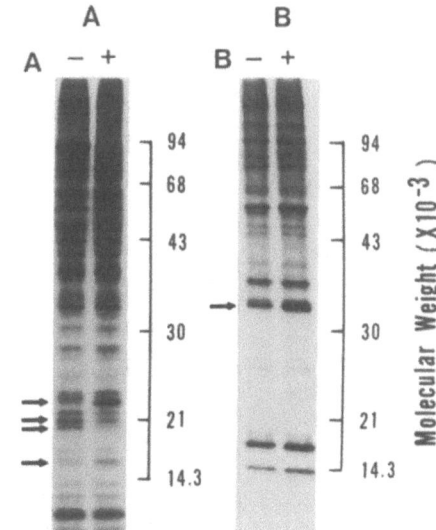

Figure 2. Autoradiograph of soluble (A) and particulate (B) fractions of ^{32}P-labeled pancreatic acini incubated with (+) or without (−) 3 μM carbachol for 5 min. Arrows indicate bands altered by carbachol. [From Ref. 1.]

tivity of soluble and particulate proteins evaluated by polyacrylamide gel electrophoresis and autoradiography, changes in five phosphoprotein bands were consistently observed (Fig. 2). Carbachol increased labeling of 23,000-M_r (23K-S) and 16,000-M_r (16K-S) proteins in the soluble fraction and reduced protein labeling in the 21,000- and 20,500-M_r regions (21K-S and 20.5K-S, respectively). Carbachol also stimulated labeling of a 32,500-M_r protein (32.5K-P) in the particulate fraction. Changes in protein phosphorylation occurred over the carbachol concentration range of 10^{-7} to 10^{-5} M, which is the range over which this stimulus induces various biological effects in acini. All these changes in phosphorylation were blocked by atropine [1]. Similar changes were also induced by cholecystokinin-octapeptide (CCK$_8$), which were blocked by the CCK receptor antagonist dibutyryl cGMP and the Ca^{2+} ionophore A23187. These results suggest that the changes in protein phosphorylation are induced by interaction of secretagogues with specific receptors, and are secondary to the rise in intracellular Ca^{2+}, rather than being involved in its generation.

The time course of changes in protein phosphorylation induced by carbachol revealed that increased phosphorylation of the 32.5K-P, 23K-S, and 16K-S proteins reached a maximum after 5 to 10 min, whereas maximal dephosphorylation of the 20.5 and 21K-S proteins occurred within 1 min (Fig. 3). Under similar conditions, amylase release was increased at 1 min and maximally stimulated by 3 min (Fig. 3). Addition of atropine to carbachol-stimulated acini inhibited amylase release, with the effect initially observed at 3 min and maximal after 5–10 min. Atropine reversed the dephosphorylation of the 20.5 and 21K-S proteins over a similar time course; and 10 min after treatment with atropine, phosphorylation had returned to control levels [1]. By contrast, atropine reversal of increased phosphorylation of the 32.5K-P protein was only partially achieved and the 23K-S protein was unaltered by 10-min atropine treatment [1]. Thus, closely decreased phosphorylation of the 20.5 and 21K-S proteins correlates more closely with onset of secretion than does increased phosphorylation of the 32.5K-P, 23K-S, and 16K-S proteins.

This correlation was also substantiated by studies with insulin, which shares certain actions exhibited by secretagogues, such as regulation of protein synthesis and glucose transport, although it does not act as a secretagogue [14]. Analysis of the effects of insulin on protein phosphorylation revealed that this hormone increased phosphorylation of the 32.5K-P, 23K-S, and 16K-S proteins but had no effect on the 20.5 and 21K-S proteins [1]. The lack of relation of altered phosphorylation of the 32.5K-P protein to secretion is also indicated by

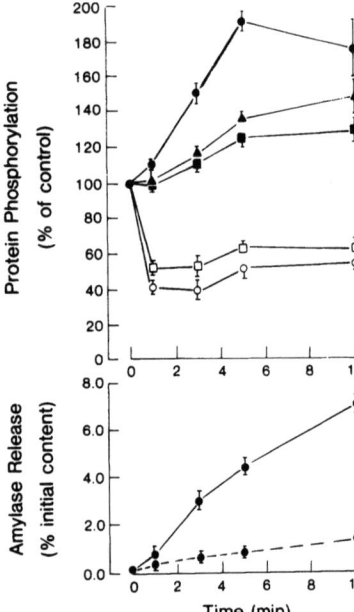

Figure 3. Time course of the effects of carbachol on protein phosphorylation (upper panel) and amylase release (lower panel). Upper panel: ●, 23K-S; ■, 16K-S; ▲, 32.5K-P; □, 21K-S; ○, 20.5K-S bands. Lower panel: ———, carbachol (3 μM); -----, control. [From Ref. 1.]

the work of Freedman and Jamieson, who observed secretagogue-induced phosphorylation of a similar ribosomal protein in rat pancreas (M_r 29,000) [12]. This protein was identified as ribosomal protein S-6 involved in the control of protein synthesis.

Acinar Cell Protein Kinase and Phosphatase Activity

Calcium-mediated changes in acinar cell protein phosphorylation *in situ* imply the existence of calcium-regulated kinase and phosphatase activities in acinar cells. Previously described Ca^{2+}-activated kinases are dependent on either calmodulin or phospholipids. To study acinar cell kinase activity, acini were homogenized in hypotonic buffer containing EGTA and the resultant cytosol incubated with [γ-^{32}P]-ATP followed by gel electrophoresis. The addition of calcium during the incubation led to intense phosphorylation of a 92,000-M_r protein and lesser labeling of a 50,000- to 52,000-M_r band [2]. This phosphorylation was dependent on the presence of endogenous calmodulin, since upon removal of endogenous calmodulin by phenothiazine-Sepharose, the readdition of calcium had no effect, but the addition of calcium plus calmodulin led to phosphorylation of the aforementioned proteins (Fig. 4). When phosphatidylserine was present along with calcium, a completely different set of proteins was phosphorylated [2]. Neither calmodulin nor phosphatidylserine affected protein phosphorylation in the absence of calcium.

Thus, separate calmodulin- and phosphatidylserine-dependent kinases exist in acini. In addition, a cAMP-activated kinase can also be demonstrated which phosphorylates a number of proteins, including the 92,000-M_r protein [2]. These different kinases can also be distinguished using inhibitors. Trifluoperazine at low micromolar concentrations inhibits the calmodulin-dependent kinase. Low concentrations of this inhibitor enhanced activity of the phosphatidylserine-dependent kinase; but 100–300 μM trifluoperazine inhibited this enzyme with only a small effect on the cAMP-activated kinase [2]. By contrast, the cAMP kinase inhibitor from beef heart was selective for cAMP-activated kinase.

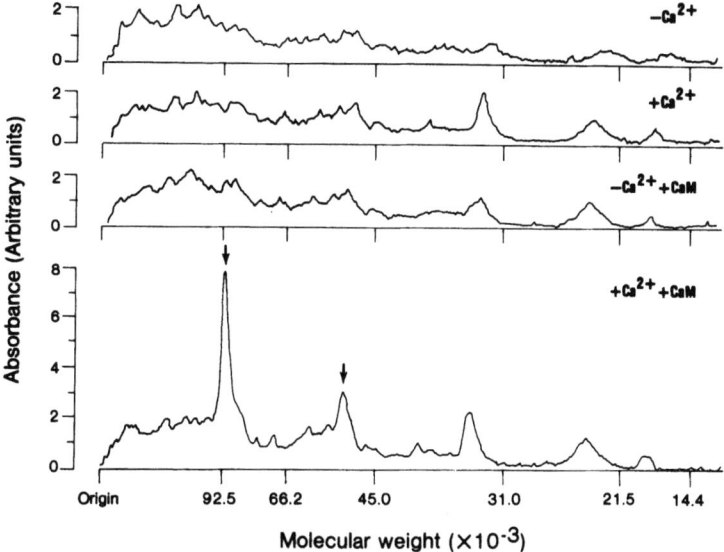

Figure 4. Densitometric tracing of autoradiograph of pancreatic acinar cytosol incubated with [γ-^{32}P]-ATP. Where indicated, Ca^{2+} was added at 300 μM and calmodulin (CaM) at 10 μg/ml. Arrows indicate bands altered by Ca^{2+} and calmodulin. Endogenous calmodulin was previously removed using phenothiazine-coupled Sepharose.

These Ca^{2+}-activated kinases are responsive to micromolar concentrations of Ca^{2+}. The calmodulin-dependent and the phosphatidylserine-dependent kinases were half-maximally activated by 6 and 12 μM Ca^{2+}, respectively, when measured on endogenous substrates [2]. While these concentrations are probably higher than that reached *in situ* in response to secretagogues, the kinases are clearly not in their native environment. This is also revealed by the fact that endogenous proteins phosphorylated in response to Ca^{2+} in the broken cell preparation differ from those phosphorylated *in situ*.

In other tissues, Ca^{2+}-activated kinases are generally substrate-specific, and several including phosphorylase kinase and myosin light-chain kinase have been identified. It is not yet clear whether the acinar calmodulin-dependent kinase(s) represents a kinase that operates on a different substrate or a distinct secretion-related kinase. In this context, mouse pancreatic cytosol will not readily phosphorylate rabbit skeletal muscle phosphorylase (D. B. Burnham, unpublished data).

Table III. Ca^{2+}-Activated Protein Phosphatase Activity in Pancreatic Acinar Cytosol Using Phosphorylase Kinase as Substrate[a]

Addition	Phosphatase activity (pmole ^{32}P released/min/mg protein)	
	Control	300 μM Ca^{2+}
None	0.14	0.32
Calmodulin (10 μg/ml)	0.18	0.61
Calmodulin plus trifluoperazine (100 μM)	0.12	0.17

[a]Rabbit skeletal muscle phosphorylase kinase was labeled using the catalytic subunit of cAMP protein kinase. Acinar cytosol was centrifuged through a Sephadex G-50 column and then incubated (0.5–1 mg protein/ml) for 1 min at 30°C with [^{32}P]phosphorylase kinase (40,000 cpm) and the amount of radioactivity in TCA-soluble material determined.

The data on Ca^{2+}-activated dephosphorylation of acinar protein *in situ* also suggest that a specific Ca^{2+}-activated phosphatase should be present. Such a phosphatase has recently been described in skeletal muscle and brain, using inhibitor 1 and phosphorylase kinase as substrates and designated type 2B phosphatase [6,16]. In preliminary studies using phosphorylase kinase, casein, and lysine-rich and mixed histones as substrates, we have identified both Ca^{2+}-independent and -dependent phosphatase activity in pancreas. Calcium-activated phosphatase activity is most clearly demonstrated using phosphorylase kinase as substrate, and is enhanced by exogenous calmodulin and inhibited by trifluoperazine (Table III). In addition, this activity is half-maximally activated by Ca^{2+} in the submicromolar range (0.3–0.8 μM). Whether this phosphatase is responsible for the selective dephosphorylation observed in intact acinar cells remains to be established.

Conclusion

Although it is now well established that mobilization of intracellular Ca^{2+} occupies a central role in stimulus–secretion coupling in exocrine pancreas, certain issues concerning the regulation of cellular Ca^{2+} remain unresolved. In particular, does occupancy of secretagogue receptors lead to Ca^{2+} mobilization through generation of an intracellular mediator? Clues to solving this problem may come from determining whether Ca^{2+} is released from the plasma membrane, in addition to one or more intracellular organelles. In conjunction with the study of intracellular calcium mobilization, the regulation of calcium influx through secretagogue-regulated membrane Ca^{2+} channels should be further evaluated. Although altered protein phosphorylation is currently a favored mode of action of Ca^{2+} in pancreatic stimulus–secretion coupling, much more work is needed to link changes in phosphorylation of specific proteins with regulation of acinar function. This will be accomplished by identifying these proteins, with regard to their subcellular localization, function, and regulation by specific protein kinases and phosphatases.

ADKNOWLEDGMENT. The research carried out in the authors' laboratory was supported by NIH Grant GM-19998.

References

1. Burnham, D. B.; Williams, J. A. Effects of carbachol, cholecystokinin, and insulin on protein phosphorylation in isolated pancreatic acini. *J. Biol. Chem.* **257**:10523–10528, 1982.
2. Burnham, D. B.; Williams, J. A. Activation of protein kinase activity in isolated pancreatic acini by calcium and cyclic AMP. *Am. J. Physiol.* **246**:G500–G508, 1984.
3. Case, R. M.; Clausen, T. The relationship between calcium exchange and enzyme secretion in the isolated rat pancreas. *J. Physiol. (London)* **235**:75–102, 1973.
4. Chandler, D. E.; Williams, J. A. Intracellular divalent cation release in pancreatic acinar cells during stimulus–secretion coupling. II. Subcellular localization of the fluorescent probe chlorotetracycline. *J. Cell Biol.* **76**:386–399, 1978.
5. Clemente, F.; Meldolesi, J. Calcium and pancreatic secretion-dynamics of subcellular calcium pools in resting and stimulated acinar cells. *Br. J. Pharmacol.* **55**:369–379, 1975.
6. Cohen, P. The role of protein phosphorylation in the neural and hormonal control of cellular activity. *Nature (London)* **296**:613–620, 1982.
7. Dormer, R. L. Direct demonstration of increases in cytosolic free Ca^{2+} during stimulation of pancreatic enzyme secretion. *Biosci. Rep.* **3**:233–240, 1983.
8. Dormer, R. L.; Poulsen, J. H.; Licko, V.; Williams, J. A. Calcium fluxes in isolated pancreatic acini: Effects of secretagogues. *Am. J. Physiol.* **240**:G38–G49, 1981.

9. Dormer, R. L.; Williams, J. A. Secretagogue-induced changes in subcellular Ca^{2+} distribution in isolated pancreatic acini. *Am. J. Physiol.* **240**:G130–G140, 1981.
10. Douglas, W. W. Stimulus–secretion coupling: The concept and clues from chromaffin and other cells. *Br. J. Pharmacol.* **34**:451–474, 1968.
11. Fabiato, A.; Fabiato, F. Calcium-induced release of calcium from sarcoplasmic reticulum of skinned cells from human, dog, cat, rabbit, rat, and frog hearts and from fetal and newborn rat ventricles. *Ann. N.Y. Acad. Sci.* **307**:491–521, 1978.
12. Freedman, S. D.; Jamieson, J. D. Hormone-induced protein phosphorylation. II. Localization to the ribosomal fraction from rat exocrine pancreas and parotid of a 29,000-dalton protein phosphorylated in situ in response to secretagogues. *J. Cell Biol.* **95**:909–917, 1982.
13. Gardner, J. D.; Conlon, T. P.; Klaeveman, H. L.; Adams, T. D.; Ondetti, M. A. Action of cholecystokinin and cholinergic agents on calcium transport in isolated pancreatic acinar cells. *J. Clin. Invest.* **56**:366–375, 1975.
14. Goldfine, I. D.; Williams, J. A. Receptors for insulin and CCK in the acinar pancreas: Relationship to hormone action. *Int. Rev. Cytol.* **85**:1–38, 1983.
15. Hokin-Neaverson, M. Acetylcholine causes a net decrease in phosphatidylinositol and a net increase in phosphatidic acid in mouse pancreas. *Biochem. Biophys. Res. Commun.* **58**:763–768, 1974.
16. Ingebritsen, T. S.; Cohen, P. Protein phosphatases: Properties and role in cellular regulation. *Science* **221**:331–338, 1983.
17. Kondo, S.; Schulz, I. Calcium ion uptake in isolated pancreas cells induced by secretagogues. *Biochim. Biophys. Acta* **419**:76–92, 1976.
18. Matthews, E. K.; Petersen, O. H.; Williams, J. A. Pancreatic acinar cells: Acetylcholine-induced membrane depolarization, calcium efflux and amylase release. *J. Physiol. (London)* **234**:689–701, 1973.
19. Michell, R. H.; Jafferji, S. S.; Jones, L. M. The possible involvement of phosphatidylinositol breakdown in the mechanism of stimulus–response coupling at receptors which control cell surface calcium gates. *Adv. Exp. Biol. Med.* **83**:447–464, 1977.
20. Ochs, D. L.; Korenbrot, J. I.; Williams, J. A. Intracellular free calcium concentrations in isolated pancreatic acini: Effects of secretagogues. *Biochem. Biophys. Res. Commun.* **117**:122–128, 1983.
21. O'Doherty, J.; Stark, R. J. Stimulation of pancreatic acinar secretion: Increases in cytosolic calcium and sodium. *Am. J. Physiol.* **242**:G513–G521, 1982.
22. Petersen, O. H.; Maruyama, Y. What is the mechanism of the calcium influx to pancreatic acinar cells evoked by secretagogues? *Pfluegers Arch.* **396**:82–84, 1983.
23. Ponnappa, B. C.; Dormer, R. L.; Williams, J. A. Characterization of an ATP dependent Ca^{2+} uptake system in mouse pancreatic microsomes. *Am. J. Physiol.* **240**:G122–G129, 1981.
24. Ponnappa, B. C.; Williams, J. A. Effects of ionophore A23187 on calcium flux and amylase release in isolated mouse pancreatic acini. *Cell Calcium* **1**:267–278, 1980.
25. Preissler, M.; Williams, J. A. Localization of ATP-dependent calcium transport activity in mouse pancreatic microsomes. *J. Membr. Biol.* **73**:137–144, 1983.
26. Putney, J. W. Recent hypotheses regarding the phosphatidylinositol effect. *Life Sci.* **29**:1183–1194, 1981.
27. Putney, J. W.; Burgess, G. M.; Halenda, S. P.; McKinney, J. S.; Rubin, R. P. Effects of secretagogues on [^{32}P] phosphatidylinositol 4,5-bisphosphate metabolism in the exocrine pancreas. *Biochem. J.* **212**:483–488, 1983.
28. Scheele, G.; Haymovits, A. Cholinergic and peptide-stimulated discharge of secretory protein in guinea pig pancreatic lobules: Role of intracellular and extracellular calcium. *J. Biol. Chem.* **254**:10346–10353, 1979.
29. Schulman, H. Calcium-dependent protein phosphorylation. In: *Handbook of Pharmacology*, J. A. Nathanson and J. W. Kebabian, eds., Berlin, Springer-Verlag, 1982, pp. 425–470.
30. Schulz, I. Messenger role of calcium in function of pancreatic acinar cells. *Am. J. Physiol.* **239**:G335–G337, 1980.
31. Stolze, H.; Schulz, I. Effect of atropine, ouabain, antimycin A and A23187 on the "trigger Ca^{2+} pool" in the exocrine pancreas. *Am. J. Physiol.* **238**:G338–G348, 1980.
32. Tsien, R. Y.; Pozzan, T.; Rink, T. J. Calcium homeostasis in intact lymphocytes: Cytoplasmic free calcium monitored with a new, intracellularly trapped fluorescent indicator. *J. Cell Biol.* **94**:325–334, 1982.
33. Wakasugi, H.; Kimura, T.; Haase, W.; Kribben, A.; Kaufmann, R.; Schulz, I. Calcium uptake into acini from rat pancreas: Evidence for intracellular ATP-dependent calcium sequestration. *J. Membr. Biol.* **65**:205–220, 1982.
34. Williams, J. A. Regulation of pancreatic acinar cell function by intracellular calcium. *Am. J. Physiol.* **238**:G269–G279, 1980.

11

Glucose Regulation of Insulin Release Involves Intracellular Sequestration of Calcium

Bo Hellman and Erik Gylfe

Like other secretory processes, insulin release from the pancreatic β cells is thought to be triggered by an increase in cytosolic Ca^{2+}. Glucose stimulation of insulin release involves increased Ca^{2+} entry after opening of potential-dependent channels in the plasma membrane [7,10,12,16,21]. However, glucose also has effects on intracellular calcium distribution which are of significance for the secretory response [7,10]. This chapter reviews evidence for the concept of a dual action of glucose on cytosolic Ca^{2+}, postulating the secretory signal to be the result of a balance between stimulated entry of Ca^{2+} and enhanced sequestration of the ion in organelle sinks.

Extrusion of Ca^{2+}

Most of our understanding of how glucose affects Ca^{2+} extrusion from the pancreatic β cells has been obtained by following the washout of radioactivity from islets loaded with ^{45}Ca [2,7,9,10,12,21]. Using this approach, glucose has a dual action in terms of stimulation and inhibition of ^{45}Ca efflux (Fig. 1). Both effects represent rapidly reversible processes with different sensitivities to glucose. The dose–response relationship for the stimulatory effect mimics that of insulin release in being sigmoidal with a half-maximal response at about 9 mM glucose. The enhanced ^{45}Ca efflux is also similar to the release process in being associated with depolarization of the β cells and requiring the presence of extracellular Ca^{2+}. However, the relationship between stimulation of ^{45}Ca efflux and insulin release represents only to a minor extent the release of calcium from secretory granules together with hormone. The stimulatory component in the action of glucose on ^{45}Ca efflux can instead be attributed to displacement of ^{45}Ca from intracellular binding sites following increased entry of nonradioactive Ca^{2+}. The extent of intracellular $^{40}Ca/^{45}Ca$ exchange is consequently sufficient to overcome the competitive inhibition of the outward transport of ^{45}Ca exerted by

---

Bo Hellman and Erik Gylfe • Department of Medical Cell Biology, Biomedicum, University of Uppsala, S-751 23 Uppsala, Sweden.

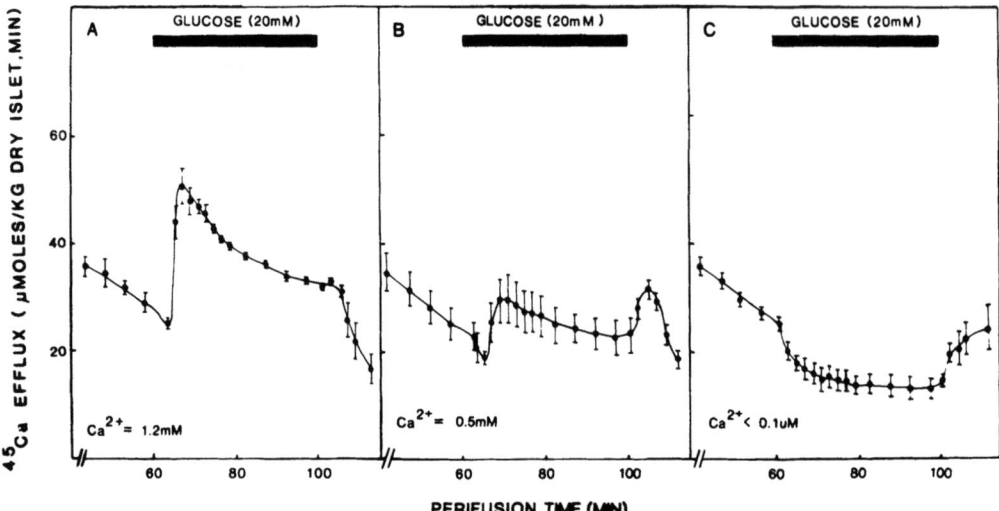

Figure 1. Effects of glucose on ^{45}Ca efflux from *ob/ob* mouse islets perifused with different concentrations of Ca^{2+}. The islets were loaded with ^{45}Ca in the presence of 20 mM glucose and perifused with media containing 1.2 mM (A), 0.5 mM (B), or < 0.1 μM (C) Ca^{2+}. The lowest concentration of Ca^{2+} was obtained by including 0.5 mM EGTA in the medium. The amounts of ^{45}Ca in the medium were expressed in terms of calcium with the same specific radioactivity as in the loading medium. Mean values ± S.E.M ($n = 4$).

the entering ^{40}Ca. Isolated pancreatic islets are not unique in responding with a stimulated ^{45}Ca efflux when exposed to agents which increase the entry of nonradioactive Ca^{2+}. Similar effects have been observed during perifusion of the posterior pituitary and the adrenal medulla [6].

In accordance with the idea that stimulation of ^{45}Ca efflux follows from an increased Ca^{2+} entry, a lowering of the extracellular concentration of Ca^{2+} unmasked the inhibitory component of the glucose action. Thus, when the extracellular Ca^{2+} concentration was < 0.1 μM, only inhibition was observed (Fig. 1C). The inhibitory component not only precedes the stimulatory phase but also disappears more rapidly with omission of glucose. Thus, at moderately reduced concentrations of Ca^{2+} the result is a substantial "off-induced" increase of ^{45}Ca efflux (Fig. 1B). In the absence of extracellular Ca^{2+}, the inhibitory effect of glucose reflects a true decrease of Ca^{2+} efflux. Inhibition is seen also when islets are loaded with ^{45}Ca to isotopic equilibrium, implying that it cannot simply be due to preferential mobilization of calcium stores with low specific radioactivity. Since part of the escaping ^{45}Ca is presumably bound to negatively charged sites on the β-cell surface, stimulated reentry of Ca^{2+} might contribute to the inhibitory action of glucose. However, this factor has proved to be of minor importance, as indicated from the absence of a clear inhibition in the presence of excessive K^+, or other agents with depolarizing effects on pancreatic β cells.

Although it is generally agreed that glucose inhibits Ca^{2+} efflux from β cells, there are divergent views as to the mechanism involved. Apart from the original concept that glucose interferes with the active extrusion of Ca^{2+} from β cells [12,20], impaired efflux may result from trapping of the cation in cellular organelles [7,10]. It is fundamental for the understanding of the glucose regulation of insulin release to discriminate between these alternatives, since they would have opposite effects on the cytosolic concentration of Ca^{2+}. The outward transport of Ca^{2+} from the pancreatic β cells seems to be mediated both by a specific Ca^{2+}-ATPase [17] and a Na^+/Ca^{2+} countertransport mechanism [8]. Although islets have been reported to contain Ca^{2+}-ATPase activity subject to glucose inhibition [14], the significance of this observation for the outward transport of Ca^{2+} remains to be elucidated. There is no evidence that glucose inhibits Ca^{2+} efflux by increasing Na^+ activity in β cells. It has instead been postulated that glucose inhibition of Ca^{2+} efflux reflects an increased production of H^+, competing with Ca^{2+} for exit by Na^+/Ca^{2+} countertransport [12]. Various observations make it difficult to accept this view. Exposure to glucose makes β cells more alkaline, rather than reducing the intracellular pH [13,15]. Moreover, glucose is equally effective in inhibiting ^{45}Ca efflux in the absence of Na^+ when the depletion of cytosolic K^+ is prevented by substituting Na^+ for K^+ [9]. Also the observation that the magnitude and even direction of the Ca^{2+} gradient across the plasma membrane has little effect on inhibition of ^{45}Ca efflux obtained in the presence of 4 mM glucose [2] is difficult to reconcile with the idea that glucose interferes with the active extrusion of Ca^{2+}.

Deposition of Calcium into Intracellular Stores

If the glucose-induced suppression of Ca^{2+} efflux from β cells is due to reduction of the cytosolic Ca^{2+} activity by intracellular sequestration, it should be possible to demonstrate a stimulated calcium uptake into organelles different from that following depolarization-induced Ca^{2+} entry. Addition of 25 mM K^+ was as effective as the maximally stimulating concentration of glucose in promoting the intracellular uptake of ^{45}Ca into the β-cell-rich pancreatic islets of *ob/ob* mice [3]. In both cases, enhanced uptake was associated with the appearance of increased amounts of ^{45}Ca in subcellular fractions containing mitochondria and secretory granules. Nevertheless, there were definite differences manifested in terms of a lower mobility of the ^{45}Ca incorporated in response to glucose.

By comparing ^{45}Ca efflux from islets loaded with isotope in the presence and absence of glucose, it has been possible to obtain additional information about the intracellular calcium sensitive to glucose. Calcium incorporated in response to glucose was readily mobilized, when the β cells were exposed to metabolic inhibitors [7]. Part of this effect could be attributed to increased intracellular Na^+. Raised cytoplasmic Na^+ proved to be effective in mobilizing ^{45}Ca incorporated in response to glucose also in the presence of an intact metabolism, as indicated from experiments where Na^+ permeability was increased or Na^+/K^+-ATPase inhibited [11].

Depolarization-Independent Net Uptake of Ca^{2+}

A major obstacle in studies of pancreatic β cells is the difficulty in obtaining them in sufficient amounts without significant contamination from other cell types. In the exploration of Ca^{2+} movements related to insulin release, we have therefore not only used the large and β-cell-rich pancreatic islets isolated from *ob/ob* mice but also a clonal cell line (RINm5F) established from a transplantable rat islet tumor. The insulin release pattern of the latter cells is characterized by a significant secretory response to excess K^+ but not to glucose [5,18]. After attachment to fibronectin-coated beads, the cells were loaded with ^{45}Ca and efflux of the isotope recorded in the same perifusion system as employed when studying isolated islets. Using this experimental approach, it was possible to demonstrate that differences in the effects of excess K^+ and glucose on insulin release paralleled the promotion of ^{45}Ca efflux [1]. The inability of glucose to stimulate insulin release is therefore probably due to impaired depolarization rather than due to a lack of voltage-dependent Ca^{2+} channels. Despite the absence of a distinct glucose stimulation of ^{45}Ca efflux, the sugar also inhibited isotope washout from the RINm5F cells during perifusion with Ca^{2+}-deficient medium.

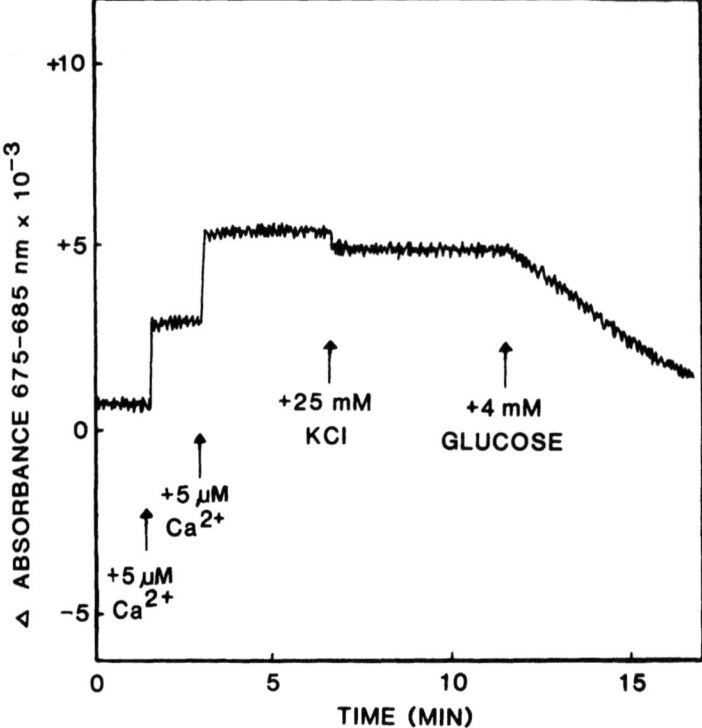

Figure 2. Effects of K$^+$ and glucose on net fluxes of Ca^{2+} in RINm5F cells. Cells equivalent to 4.1 mg protein were suspended in 0.3 ml medium buffered to pH 7.4 with 25 mM Hepes. The medium contained 20 µM arsenazo III and at the beginning of the experiment 6 µM Ca^{2+} as determined by EGTA titration. The Ca^{2+} activity of the medium was monitored continuously be measuring the absorbance difference 675–685 nm of arsenazo III.

It is difficult to evaluate from the radioisotope studies to what extent alterations of ^{45}Ca fluxes reflect net changes in calcium content. We have therefore made a direct comparison of how glucose and high K$^+$ affect net fluxes of Ca^{2+} in RINm5F cells by monitoring alterations of the Ca^{2+} concentration in a suspending medium containing micromolar concentrations of the cation [5]. The results of such an analysis are shown in Fig. 2. Whereas the addition of 25 mM K$^+$ failed to affect net uptake of Ca^{2+} in the suspended cells, the addition of 4 mM glucose resulted in a clear stimulation. The data indicate that in the presence of micromolar concentrations of the cation, glucose promotes net uptake of Ca^{2+} by a mechanism different from depolarization.

Concentration of Cytosolic Ca^{2+}

Since RINm5F cells exhibit both the characteristic glucose-induced reduction of ^{45}Ca efflux as well as a glucose-stimulated net uptake of Ca^{2+}, they should be suitable for directly testing whether the sugar can lower cytosolic Ca^{2+}. We have performed such studies with quin-2 [19], an indicator especially designed for monitoring intracellular Ca^{2+} [20]. Unstimulated RINm5F cells have a cytosolic Ca^{2+} concentration of about 100 nM, which was more than tripled when the cells were depolarized with excess K$^+$. Technical difficulties have prevented valid conclusions as to how glucose affects the cytosolic Ca^{2+} when cation entry is inhibited by using a Ca^{2+}-deficient medium. However, the quin-2 technique made it possible to demonstrate a lowering of cytosolic Ca^{2+} in the presence of glucose. The addition of 4 mM glucose to an incubation medium containing 1 mM Ca^{2+} results in a significant suppression of cytosolic Ca^{2+} in RINm5F cells (Fig. 3).

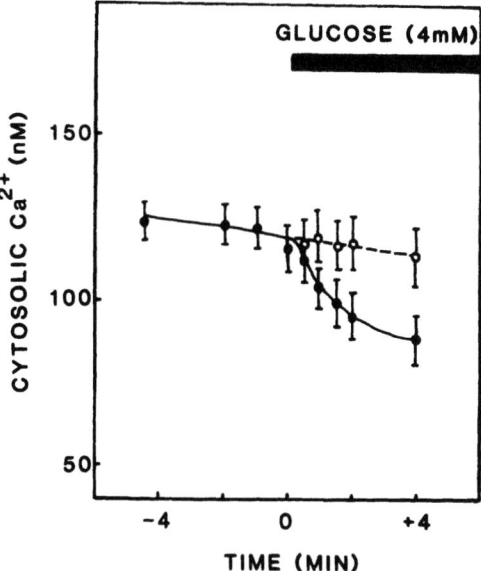

Figure 3. Effects of glucose on the concentration of cytosolic Ca^{2+} in RINm5F cells as estimated with the fluorescent indicator quin-2. Cells equivalent to 5–10 mg protein and containing about 3 mM quin-2 were incubated with 1 mM Ca^{2+} in 1 ml medium. Glucose (4 mM) was introduced as shown by the horizontal bar. The open symbols represent the Ca^{2+} activities expected without modification of the medium. Mean values ± S.E.M ($n = 6$).

Rate of Insulin Release

Glucose is the major physiological stimulus of insulin release. However, it follows from the notion of glucose lowering cytosolic Ca^{2+} that it might be possible to find conditions when the sugar inhibits insulin release. Indeed, we have recently reported such a situation [11]. During perifusion with a Ca^{2+}-deficient medium, insulin release induced by raising

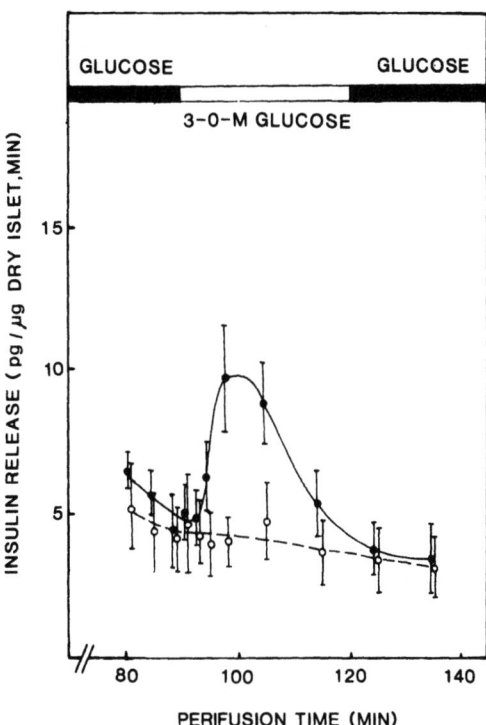

Figure 4. Effect of glucose removal on insulin release from isolated islets exposed to Ca^{2+}-deficient medium. Islets were microdissected from *ob/ob* mice and perifused with medium containing 1 mg/ml albumin. After 30 min of exposure to 1.28 mM Ca^{2+} and 20 mM glucose, Ca^{2+} was omitted and 0.5 mM EGTA added. The filled symbols refer to experiments where glucose was temporarily replaced with equimolar 3-*O*-methylglucose between 90 and 120 min of perifusion. The open symbols represent control experiments with similar medium changes maintaining the presence of glucose. Mean values ± S.E.M. ($n = 5$).

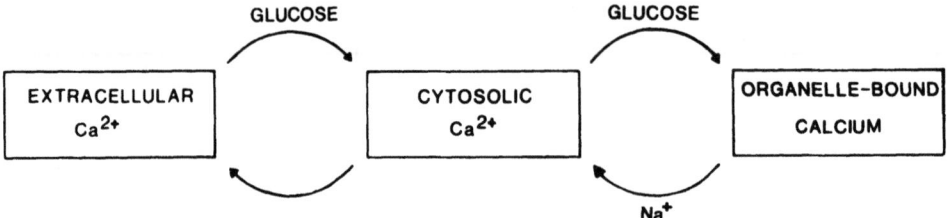

Figure 5. Diagram illustrating the postulated dual action of glucose on cytosolic Ca^{2+}. The effect of glucose on cytosolic Ca^{2+} reflects the balance between a stimulated entry of Ca^{2+} and the trapping of the ion in intracellular stores sensitive to Na^+

intracellular Na^+ was suppressed by glucose. Another example of the inhibitory action of glucose on insulin release under conditions of impaired Ca^{2+} entry is shown in Fig. 4. The mere replacement of glucose by an equimolar concentration of the nonmetabolizable analog 3-O-methylglucose results in transient stimulation of insulin release in a Ca^{2+}-deficient medium.

Functional Significance of Intracellular Ca^{2+} Trapping

The experiments described have shown that glucose not only promotes the entry of Ca^{2+} into the pancreatic β cells but also stimulates trapping of the cation in cellular organelles. The action of glucose on cytosolic Ca^{2+} activity, and consequently on the rate of insulin release, reflects the balance between these processes (Fig. 5). Under physiological conditions, the net effect of glucose over time is to enhance the cytosolic concentration of Ca^{2+}. Nevertheless, it seems likely from the dynamics of the ^{45}Ca efflux that exposure to glucose causes an initial reduction of Ca^{2+} activity. Since K^+ conductance of the β-cell membrane appears to be under the control of cytosolic Ca^{2+} [4], a reduction of Ca^{2+} activity by glucose might be the process that initiates depolarization. The intracellular buffering of Ca^{2+} may become less pronounced with time due to a limited capacity for Ca^{2+} sequestration. Such a limitation might account for certain unexplained phenomena, such as the slowly increasing, second phase of glucose-stimulated insulin release and the fact that the secretory response is enhanced after priming with the sugar.

ACKNOWLEDGMENTS. Supported by grants from the Swedish Medical Research Council (12x-562 and 12x-6240), the Swedish Diabetes Association, and the Nordic Insulin Foundation.

References

1. Abrahamsson, H.; Berggren, P.-O.; Hellman, B. Mobilisation of ^{45}Ca from clonal insulin-releasing cells (RINm5F). Submitted for publication.
2. Abrahamsson, H.; Gylfe, E.; Hellman, B. Influence of external calcium ions on labelled calcium efflux from pancreatic β-cells and insulin granules in mice. *J. Physiol. (London)* **311**:541–550, 1981.
3. Andersson, T.; Betsholtz, C.; Hellman, B. Calcium and pancreatic β-cell function. 12. Modification of ^{45}Ca fluxes by excess of K^+. *Acta Endocrinol. (Copenhagen)* **96**:87–92, 1981.
4. Atwater, I.; Dawson, C. M.; Ribalet, B.; Rojas, E. Potassium permeability activated by intracellular calcium ion concentration in the pancreatic β-cell. *J. Physiol. (London)* **288**:575–588, 1979.
5. Gylfe, E.; Andersson, T.; Rorsman, P.; Abrahamsson, H.; Arkhammar, P.; Hellman, P.; Hellman, B.; Oie, H.

K.; Gazdar, A. F. Depolarization-independent net uptake of Ca^{2+} into clonal insulin-releasing cells exposed to glucose. *Biosci. Rep.* **3**:927–937, 1983.
6. Gylfe, E.; Hellman, B. Lack of Ca^{2+} ionophoretic activity of hypoglycemic sulfonylureas in excitable cells and isolated secretory granules. *Mol. Pharmacol.* **22**:715–720, 1982.
7. Hellman, B.; Andersson, T.; Berggren, P.-O.; Flatt, P.; Gylfe, E.; Kohnert, K.-D. The role of calcium in insulin secretion. *Horm. Cell Regul.* **3**:69–96, 1979.
8. Hellman, B.; Andersson, T.; Berggren, P.-O.; Rorsman, P. Calcium and pancreatic β-cell function. 11. Modification of ^{45}Ca fluxes by Na^+ removal. *Biochem. Med.* **24**:143–152, 1980.
9. Hellman, B.; Gylfe, E. Glucose inhibits ^{45}Ca efflux from pancreatic β-cells also in the absence of Na^+/Ca^{2+} countertransport. *Biochim. Biophys. Acta* **770**:136–141, 1984.
10. Hellman, B.; Gylfe, E.; Berggren, P.-O.; Andersson, T.; Abrahamsson, H.; Rorsman, P.; Betsholtz, C. Ca^{2+} transport in pancreatic β-cells during glucose stimulation of insulin secretion. *Upsala J. Med. Sci.* **85**:321–329, 1980.
11. Hellman, B.; Honkanen, T.; Gylfe, E. Glucose inhibits insulin release induced by Na^+ mobilisation of intracellular calcium. *FEBS Lett.* **148**:289–292, 1982.
12. Herchuelz, A.; Malaisse, W. J. Calcium movements and insulin release in pancreatic islet cells. *Diabete Metab. (Paris)* **7**:283–288, 1981.
13. Lebrun, P.; Malaisse, W. J.; Herchuelz, A. Effect of the absence of bicarbonate upon intracellular pH and calcium fluxes in pancreatic islet cells. *Biochim. Biophys. Acta* **721**:357–365, 1982.
14. Levin, S. R.; Kasson, B. G.; Driessen, J. F. Adenosine triphosphatases of rat pancreatic islets. Comparison with those of rat kidney. *J. Clin. Invest.* **62**:692–701, 1978.
15. Lindström, P.; Sehlin, J. Glucose increases the intracellular pH in isolated pancreatic islets. *Acta Endocrinol. (Copenhagen)* **257**:43, 1983.
16. Meissner, H. P.; Schmelz, H. Membrane potential of beta cells in pancreatic islets. *Pfluegers Arch.* **351**:195–206, 1974.
17. Pershadsingh, H. A.; McDaniel, M. L.; Landt, M.; Bry, C. G.; Lacy, P. E.; McDonald, J. M. Ca^{2+}-activated ATPase and ATP-dependent calmodulin-stimulated Ca^{2+} transport in islet cell plasma membrane. *Nature (London)* **288**:492–495, 1980.
18. Praz, G. A.; Halban, P. A.; Wollheim, C. B.; Blondel, B.; Strauss, A. J.; Renold, A. E. Regulation of immunoreactive insulin release from a rat cell line (RINm5F). *Biochem. J.* **210**:345–352, 1983.
19. Rorsman, P.; Berggren, P.-O.; Gylfe, E.; Hellman, B. Reduction of the cytosolic Ca^{2+} activity in clonal insulin-releasing cells exposed to glucose. *Biosci. Rep.* **3**:939–946, 1983.
20. Tsien, R. Y.; Pozzan, T.; Rink, T. J. Calcium homeostasis in intact lymphocytes; cytoplasmic free calcium monitored with a new intracellularly trapped fluorescent indicator. *J. Cell Biol.* **94**:325–334, 1982.
21. Wollheim, C. B.; Sharp, G. W. G. Regulation of insulin release by calcium. *Physiol. Rev.* **61**:914–973, 1981.

12

Role of Calcium in Stimulus–Secretion Coupling in Bovine Adrenal Medullary Cells

N. Kirshner, J. J. Corcoran, and S. P. Wilson

The requirement of extracellular calcium for secretion of catecholamines from the adrenal medulla was first demonstrated by Douglas and Rubin in 1961 [10]. They proposed that acetylcholine stimulated secretion by promoting the influx of calcium into chromaffin cells, thus initiating intracellular events resulting in the release of adrenaline and noradrenaline. Evidence consistent with this hypothesis was obtained by Douglas and Poisner [9] who showed that perfused cat adrenal glands treated with acetylcholine retained more of a loading dose of $^{45}Ca^{2+}$ than did untreated glands. Because of the limited experimental flexibility of the perfused gland system it was not possible to carry out a detailed analysis of calcium uptake, but continued studies of stimulus–secretion coupling provided evidence that entry of calcium into the cell was an early step in secretion [8]. With the advent of methods for maintaining primary cultures of adrenal medulla chromaffin cells, it became possible to determine directly the effects of stimulation on the entry of calcium as well as other ions into the cells. Several laboratories have undertaken such studies [20,23,28], and the essential features are described here.

Time Course of Calcium Uptake and Catecholamine Secretion

Stimulation of adrenal medullary cell cultures with various types of secretagogues leads to the rapid uptake of $^{45}Ca^{2+}$ (Fig. 1). The time course for $^{45}Ca^{2+}$ uptake depends on the mode of stimulation and precedes catecholamine secretion [23]. This is seen in Fig. 2 and is more apparent upon stimulation with veratridine than with nicotine. The maximal rates of calcium uptake and catecholamine secretion occur within the first minute after stimulation with nicotine. Thereafter the rates of both processes decline, the rate of calcium uptake more rapidly than the rate of catecholamine secretion. Holz et al. [20] obtained similar results using carbachol and reported that significant calcium uptake could be detected 15 sec after

---

N. Kirshner, J. J. Corcoran, and S. P. Wilson • Department of Pharmacology, Duke University Medical Center, Durham, North Carolina 27710.

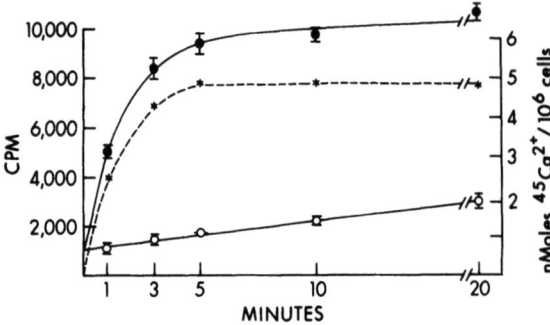

Figure 1. Time course for the accumulation of $^{45}Ca^{2+}$ in the presence (●) and absence (○) of 50 μM acetylcholine (+ 10 μM eserine). Asterisks mark ACh-stimulated $^{45}Ca^{2+}$ accumulation after background subtraction.

onset of stimulation, at which time there was no significant catecholamine secretion. Upon stimulation with veratridine, the maximal rate of $^{45}Ca^{2+}$ uptake occurs during the first to second minute of stimulation while the maximal rate of catecholamine secretion occurs during the third to fourth minute of stimulation. Rat pheochromocytoma cells also take up $^{45}Ca^{2+}$ upon stimulation with carbamylcholine, veratridine, or high K^+ concentrations [4,33].

An increased influx of calcium is sufficient to cause catecholamine secretion. This has been demonstrated by the use of calcium ionophores [5] and by rendering isolated cells permeable to small ions by exposing them to short bursts of intense electrical fields [25] or by treatment with digitonin [14,35]. Knight and Baker [25] have shown that addition of Mg-ATP and calcium to medium containing cells which had been exposed to electrical fields results in secretion of catecholamines. The required concentrations of calcium and Mg-ATP for half-maximal release were 1 μM and 1 mM, respectively. Cells treated with digitonin also secrete catecholamines upon addition of calcium [14,35]. The effects of Mg-ATP with low concentrations of calcium on digitonin-treated cells are variable [14,35]. This may be related to the fact that little or no time elapsed between digitonin treatment and addition of calcium; thus, variable amounts of endogenous ATP may have diffused out of cells. In the

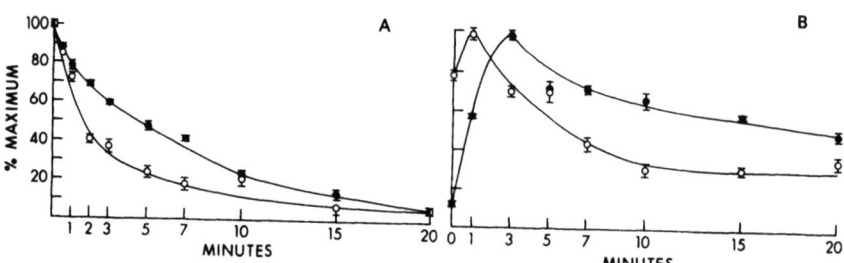

Figure 2. (A) Rate of nicotine-stimulated $^{45}Ca^{2+}$ uptake (○) and endogenous catecholamine release (●) as a function of time. Cells were incubated in 0.3 ml of unlabeled medium containing 10 μM nicotine. At the indicated times after addition of nicotine, 30 μl of medium containing $^{45}Ca^{2+}$ was added and incubation was continued for 1 min. The amounts of $^{45}Ca^{2+}$ taken up during the 1-min period, as well as the amounts of endogenous catecholamines released, were then determined. Maximal rates of $^{45}Ca^{2+}$ uptake and catecholamine release were 2.1 nmole/min and 4.6% release of total content/min per 10^6 cells, respectively. Nonstimulated controls have been subtracted. Controls were also run in which cells were first incubated with nicotine for 20 min, after which d-tubocurarine (100 μM final concentration) was added, prior to incubation with $^{45}Ca^{2+}$. Prior nicotine stimulation did not increase control $^{45}Ca^{2+}$ uptake. (B) Rates of $^{45}Ca^{2+}$ uptake (○) and catecholamine release (●) stimulated by veratridine (100 μM). The experiment was carried out as described in (A). Maximal rates of $^{45}Ca^{2+}$ uptake and catecholamine release were 0.77 nmole/min per 10^6 cells and 1.2% release of total content/min per 10^6 cells, respectively.

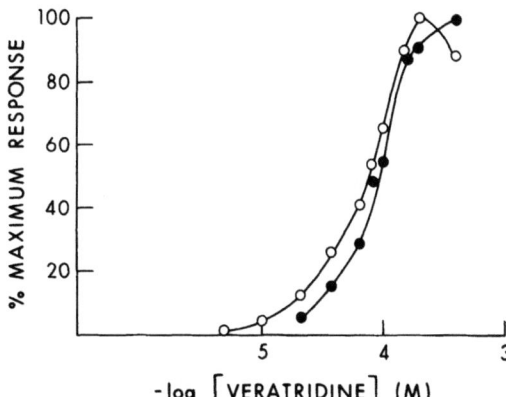

Figure 3. Veratridine dose–response curves for the release of catecholamines (○) and $^{45}Ca^{2+}$ uptake (●). Cells were stimulated for 10 min. Maximal catecholamine release and $^{45}Ca^{2+}$ uptake were, respectively, 25% of total content released and 2.1 nmole/10^6 cells.

experiments reported by Knight and Baker [25], 15 to 30 min or longer generally elapsed between electrical treatment and addition of calcium, allowing ample time for Mg-ATP to diffuse out of cells. Addition of Mg-ATP was not required for secretion in digitonin-treated cells with higher concentrations of calcium (> 100 μM), which may be attributed to negligible loss of cellular ATP [35].

Dose–Response Relationships

The dose–response curves for $^{45}Ca^{2+}$ uptake and catecholamine secretion are similar using veratridine as secretagogue (Fig. 3) [23]. However, with nicotine as the agonist, the dose–response curve for $^{45}Ca^{2+}$ uptake is shifted to the right of that for catecholamine secretion (Fig. 4A); the EC_{50} for $^{45}Ca^{2+}$ uptake is about threefold greater than that for catecholamine secretion. Thus, 10 μM nicotine elicits 75% of the maximal secretory response, but only 15% of the maximal $^{45}Ca^{2+}$ uptake (Fig. 4A). Comparable results have

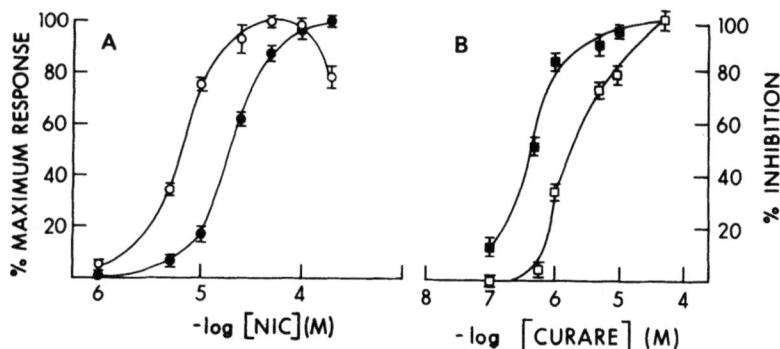

Figure 4. (A) Nicotine dose–response curves for release of endogenous catecholamines (○) and stimulation of $^{45}Ca^{2+}$ (●). Maximal percent release of total catecholamines and $^{45}Ca^{2+}$ uptake were 15.5% and 5.4 nmole/10^6 cells, respectively. Incubation was carried out for 5 min. (B) Dose–response curves for the inhibition of nicotine-stimulated catecholamine release (□) and $^{45}Ca^{2+}$ uptake (■) by d-tubocurarine. Cells were stimulated for 5 min with 10 μM nicotine in the presence of the indicated concentrations of d-tubocurarine. Maximal rate of nicotine-stimulated catecholamine release and $^{45}Ca^{2+}$ uptake in the absence of curare are 18% of total cellular catecholamine content and 3.4 nmole/10^6 cells, respectively.

been reported using carbachol [20], but the dose–response curves for $^{45}Ca^{2+}$ uptake and catecholamine secretion are similar with acetylcholine as the agonist [28].

Nicotine-stimulated $^{45}Ca^{2+}$ uptake and catecholamine secretion are both inhibited by curare, and again there is a marked difference in sensitivity (Fig. 4B). $^{45}Ca^{2+}$ uptake is inhibited to a much greater extent at low concentrations of curare than is catecholamine secretion; 1 µM curare inhibited $^{45}Ca^{2+}$ uptake and catecholamine secretion by 85 and 35%, respectively. High concentrations of nicotine promote $^{45}Ca^{2+}$ uptake in excess of that required for secretion, but a high correlation exists between $^{45}Ca^{2+}$ uptake and catecholamine secretion with low concentrations of nicotine or carbachol [20,23].

Dependency of Calcium Uptake and Catecholamine Secretion upon External Calcium Concentration

The dependencies of $^{45}Ca^{2+}$ uptake and catecholamine secretion upon the external calcium concentration are very similar with veratridine and 50 mM K^+ as secretagogues; both uptake and secretion are maximal between 1 and 2 mM external calcium [23]. However, the situation is quite different with nicotine as the stimulus. Catecholamine secretion is maximal at 1 to 3 mM external calcium, but $^{45}Ca^{2+}$ uptake does not reach a plateau at external calcium concentrations as high as 10 mM. Holz et al. [20] also found that K^+-stimulated $^{45}Ca^{2+}$ uptake was maximal at 2 mM calcium, while carbachol-stimulated uptake of $^{45}Ca^{2+}$ continued to increase up to 15 mM calcium, the highest concentration tested. However, in contrast to our results, the calcium dependency for carbachol-induced catecholamine secretion was the same as that for $^{45}Ca^{2+}$ uptake and a plateau for secretion was not seen at 2 mM calcium [20]. An explanation for this difference is not apparent.

The fact that veratridine- or K^+-stimulated $^{45}Ca^{2+}$ uptake was maximal at 2 mM calcium could be due either to saturation of intracellular sequestration of $^{45}Ca^{2+}$ (which appears unlikely), or to saturation of one or more steps in the uptake process, implying a specific binding site involved in calcium transport. To test these alternatives, the calcium concentration dependency for $^{45}Ca^{2+}$ uptake at two different K^+ concentrations was determined [23]. If saturation of intracellular binding was responsible, then similar levels of $^{45}Ca^{2+}$ uptake should occur with the two different concentrations of K^+, but at different external calcium concentrations. However, the results clearly show that the same levels of $^{45}Ca^{2+}$ uptake do not occur and that with both concentrations of K^+, maximal uptake of $^{45}Ca^{2+}$ occurs at the same extracellular calcium concentration. These studies indicate that calcium uptake stimulated by veratridine and high K^+ involves interaction of calcium with specific saturable sites.

Differences in the dependency upon the external concentration of calcium for veratridine- and nicotine-stimulated $^{45}Ca^{2+}$ uptake may be explained by utilization of different channels for calcium entry. The gated entry of calcium into excitable cells occurs through two different types of channels, voltage-sensitive calcium channels and receptor-operated channels. Opening of the former channel is regulated by the membrane potential, while the latter channel is dependent on receptor occupancy and perhaps secondarily on membrane potential. Depolarizing agents such as 50 mM K^+ and veratridine promote calcium entry through voltage-sensitive calcium channels, while nicotine promotes entry through the receptor-operated channel. Depolarization of the chromaffin cell elicited by nicotine or acetylcholine [3,13,22] is not mandatory for $^{45}Ca^{2+}$ uptake or catecholamine secretion, since both events can be elicited in isotonic sucrose [23]. Ritchie [30] and Stallcup [33] in earlier studies on $^{45}Ca^{2+}$ uptake in pheochromocytoma cells similarly concluded that utilization of different channels for calcium entry depends on the mode of stimulation.

Differential Inhibition of Calcium Uptake and Catecholamine Secretion by D600

Effects of D600 on nicotine- and veratridine-stimulated calcium uptake and catecholamine secretion were investigated in an attempt to differentiate further the channels used for calcium entry. D600 blocks the slow inward current carried by calcium and sodium in myocardial cells [26,27], as well as the fast sodium channels activated by veratridine in cultured heart and neuroblastoma cells [18]. D600 also inhibits catecholamine secretion by perfused bovine adrenal glands and medullary cell cultures [6,29]. In adrenal medullary cell cultures D600 is about 10-fold more potent in inhibiting nicotine-induced $^{45}Ca^{2+}$ uptake than in inhibiting veratridine-induced $^{45}Ca^{2+}$ uptake [6]. Moreover, inhibition of the veratridine-induced effects is competitive with veratridine, while inhibition of the nicotine-induced effects is noncompetitive with nicotine. Inhibition of nicotine-stimulated $^{45}Ca^{2+}$ uptake and catecholamine secretion by D600 is qualitatively similar to inhibition by curare; $^{45}Ca^{2+}$ uptake is more sensitive to inhibition than is catecholamine secretion. On the other hand, veratridine-induced $^{45}Ca^{2+}$ uptake and catecholamine secretion are equally sensitive to inhibition by D600.

D600 also inhibits $^{22}Na^{+}$ uptake stimulated by nicotine or veratridine in adrenal medullary cell cultures, and again nicotine-stimulated uptake is threefold more sensitive to inhibition by D600 than is veratridine-stimulated uptake [6]. Veratridine-stimulated sodium uptake occurs through tetrodotoxin-sensitive channels and nicotine-stimulated sodium uptake occurs through tetrodotoxin-insensitive channels, presumably the ion channel associated with the nicotinic receptor. These studies thus suggest that D600 inhibits nicotine-induced $^{45}Ca^{2+}$ uptake and catecholamine secretion by interfering with the receptor-operated ion channel; while inhibition of veratridine-induced $^{45}Ca^{2+}$ uptake is due to an action on the voltage-dependent calcium channel.

Mediation of Secretion by Other Divalent Cations

Although calcium is the physiological ion which mediates stimulus–secretion coupling, it can be replaced by other divalent cations. Strontium appears to freely substitute for calcium in the perfused gland [11], in isolated adrenal medullary cells [7], and in cells permeabilized with digitonin [35]. Barium by itself causes intense secretion of catecholamines by perfused adrenals [12] and isolated medullary cells [7]. Strontium can also elicit secretion from cell cultures but is a much weaker agonist than barium [7]. Manganese ion can also replace calcium in stimulus–secretion coupling in perfused glands, cell cultures, and digitonin-permeabilized cells [2,7,35]. In the presence of manganese the onset of secretion is delayed and the duration of secretion is prolonged relative to that observed in the presence of calcium [2,7]. Cations which replace calcium in stimulus–secretion coupling may do so by directly activating the calcium-dependent reaction or by increasing calcium availability by displacing it from intracellular storage sites.

Termination of Secretion

Once secretion is in progress, what are the conditions necessary to maintain secretion? Is it necessary to have continuous stimulation and continuous influx of Ca^{2+}? A partial answer to these questions was obtained in a study in which nicotine-stimulated catecholamine secretion and calcium uptake were interrupted after 5 min by the addition of

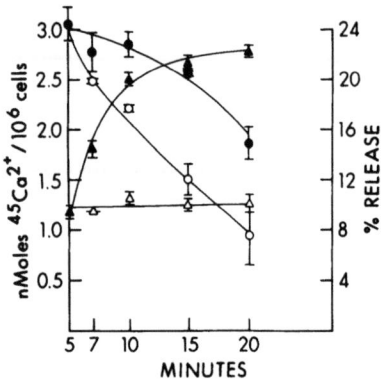

Figure 5. Effect of *d*-tubocurarine on nicotine-stimulated $^{45}Ca^{2+}$ accumulation (circles) and on endogenous catecholamine release (triangles). Cells were incubated with 0.3 ml of 10 μM nicotine for 5 min, at which time 30 μl of 1 mM *d*-tubocurarine in Locke's medium (○, △) or Locke's medium alone (●, ▲) was added. Net accumulation of $^{45}Ca^{2+}$ and secretion of endogenous catecholamines was examined at various times after the addition of curare.

curare [23], which caused an immediate cessation of secretion and a rapid efflux of $^{45}Ca^{2+}$ (Fig. 5). Although one cannot distinguish the effects of receptor occupancy from calcium influx in this experiment, it appears that substantial cytoplasmic levels of calcium do not accumulate during stimulation and that continued Ca^{2+} influx is necessary for continued secretion. Similar results were obtained in a second experiment in which the addition of D600 during 50 mM K^+-stimulated $^{45}Ca^{2+}$ uptake and catecholamine secretion produced an immediate cessation of secretion and loss of previously accumulated $^{45}Ca^{2+}$ from the cell [23].

Intracellular Events Involved in Secretion

The entry of calcium into the cell as an early event in stimulus–secretion coupling is now well established, but additional work is required to define better the routes of entry. The events which occur subsequent to the entry of calcium are poorly understood, but the development of permeable cell models provides systems to investigate proposed intracellular events, and significant new information has already been reported [14,25,35].

Early studies on catecholamine secretion demonstrated a requirement for metabolic energy [24,31], but it was not determined whether the energy was utilized for the entry of calcium, for the intracellular events subsequent to the entry of calcium, or for both processes. The demonstration [25] that ATP is required for secretion by leaky cells shows that the intracellular event or events subsequent to the entry of calcium utilize energy. This, of course, does not rule out the possibility that the events leading to the stimulation-induced calcium uptake may also require ATP.

The ATP requirement for secretion suggests that a phosphorylation reaction occurs, and attempts have been made to demonstrate specific protein phosphorylation associated with adrenal medullary secretion. Increased phosphorylation of two protein bands following stimulation has been reported [1]. The protein band of 60,000 M_r is probably tyrosine hydroxylase [19], while little is known of the other band of 95,000 M_r. Phosphorylation of both bands occurred very rapidly after onset of stimulation, was calcium-dependent, and was inhibited by pharmacological agents which block catecholamine secretion [1]. The relationship between protein phosphorylation and secretion remains to be elucidated.

A possible role of calmodulin-mediated reactions in stimulus–secretion coupling has been investigated, and conflicting results have been reported [21,22,24]. Pimozide and a number of phenothiazine-like drugs cause a similar inhibition of both $^{45}Ca^{2+}$ uptake and catecholamine secretion induced by nicotine, veratridine, and 50 mM K$^+$ [32]. The IC$_{50}$ values for inhibition of nicotine- and veratridine-induced effects were in the range of 1 to 10 µM while the IC$_{50}$ values for the K$^+$-induced effects were in the 20 to 65 µM range. The phenothiazines were generally less effective inhibitors when the calcium ionophore ionomycin was used to elicit secretion rather than high K$^+$. Thus, these drugs act at diverse sites, and a primary effect is inhibition of calcium transport into the cell.

One cannot determine from these experiments whether the intracellular events utilize a calmodulin-dependent reaction because calcium uptake is affected by the phenothiazines to the same extent as secretion. Wada *et al.* [34] similarly concluded that trifluoperazine depressed catecholamine secretion mainly by inhibiting calcium uptake. However, Kenigsberg *et al.* [21] found that trifluoperazine, at concentrations that markedly inhibited catecholamine secretion, had little or no effect on calcium uptake and concluded that the drug blocked secretion distal to calcium entry. They suggested that a calmodulin-dependent reaction is involved in secretion. Knight and Baker [25] found that 10 µM trifluoperazine elicited a 40 to 50% inhibition of catecholamine secretion from permeabilized cells, while 10 µM pimozide had no effect. However, these same concentrations of trifluoperazine and pimozide have no effect on secretion from digitonin-treated cells [35].

It was shown a number of years ago that *N*-ethylmaleimide inhibits catecholamine secretion evoked by acetylcholine or KCl in perfused adrenal glands by 90 to 95% [17]. In digitonin-treated adrenal medullary cell cultures *N*-ethylmaleimide also inhibits catecholamine release evoked by the addition of calcium, indicating the role of a sulfhydryl group in the intracellular secretory event [35].

In conclusion, the studies described here support the concept proposed more than 20 years ago [8] that stimulation of the adrenal medulla results in the influx of calcium which triggers the subsequent events resulting in secretion of catecholamines. Our understanding of how calcium gains entry into the cell is still superficial, but the combination of tissue culture and advanced electrophysiological techniques [15,16] should provide new insights into this process. The intracellular events leading to secretion initiated by the entry of calcium are unknown, but permeabilized cells provide a promising model system for investigating the molecular events in this process.

ACKNOWLEDGMENT. The work reported from this laboratory was supported by NIH Grant AM-05427.

References

1. Amy, C. M.; Kirshner, N. Phosphorylation of adrenal medulla cell proteins in conjunction with stimulation of catecholamine secretion. *J. Neurochem.* **36**:847–854, 1981.
2. Arqueros, L.; Daniels, A. J. Manganese as agonist and antagonist of calcium ions: Dual effect upon catecholamine release from adrenal medulla. *Life Sci.* **28**:1535–1540, 1981.
3. Biales, B.; Dichter, M.; Tischler, A. Electrical excitability of cultured adrenal chromaffin cells. *J. Physiol. (London)* **262**:743–753, 1976.
4. Chalfie, M.; Hoodley, D.; Pastan, S.; Perlman, R. L. Calcium uptake into rat pheochromocytoma cells. *J. Neurochem.* **27**:1405–1409, 1976.
5. Conn, P. M.; Kilpatrick, D.; Kirshner, N. Ionophoretic Ca^{2+} mobilization in rat gonadotropes and bovine adrenomedullary cells. *Cell Calcium* **1**:29–133, 1980.
6. Corcoran, J. J.; Kirshner, N. Inhibition of calcium uptake, sodium uptake and catecholamine secretion by methoxyverapamil (D600) in primary cultures of adrenal medulla cells. *J. Neurochem.* **40**:1106–1109, 1983.
7. Corcoran, J. J.; Kirshner, N. Effects of manganese and other divalent cations on calcium uptake and catecholamine secretion by primary cultures of bovine adrenal medulla. *Cell Calcium* **4**:127–137, 1983.
8. Douglas, W. W. Involvement of calcium in exocytosis and the exocytosis vesiculation sequence. *Biochem. Soc. Symp.* **39**:1–28, 1974.

9. Douglas, W. W.; Poisner, A. M. On the mode of action of acetylcholine in evoking adrenal medullary secretion: Increased uptake of calcium during the secretory response. *J. Physiol. (London)* **162**:385–392, 1962.
10. Douglas, W. W.; Rubin, R. P. The role of calcium in the secretory response of the adrenal medulla to acetylcholine. *J. Physiol. (London)* **159**:40–57, 1961.
11. Douglas, W. W.; Rubin, R. P. The effects of alkaline earths and other divalent cations on adrenal medullary secretion. *J. Physiol. (London)* **175**:231–241, 1964.
12. Douglas, W. W.; Rubin, R. P. Stimulant action of barium on the adrenal medulla. *Nature (London)* **203**:305–307, 1964.
13. Douglas, W. W.; Kanno, T.; Sampson, S. R. Influence of the ionic environment on the membrane potential of adrenal chromaffin cells and on the depolarizing effect of acetylcholine. *J. Physiol. (London)* **191**:107–121, 1967.
14. Dunn, L. A.; Holz, R. W. Catecholamine secretion from digitonin-treated adrenal medullary chromaffin cells. *J. Biol. Chem.* **258**:4989–4993, 1983.
15. Fenwick, E. M.; Marty, A.; Neher, E. A patch-clamp study of bovine chromaffin cells and of their sensitivity to acetylcholine. *J. Physiol. (London)* **331**:577–597, 1982.
16. Fenwick, E. M.; Marty, A.; Neher, E. Sodium and calcium channels in bovine chromaffin cells. *J. Physiol. (London)* **331**:599–635, 1982.
17. Ferris, R. M.; Viveros, O. H.; Kirshner, N. Effects of various agents on the Mg^{2+}-ATP stimulated incorporation and release of catecholamines by isolated bovine adrenomedullary storage vesicles and on secretion from the adrenal medulla. *Biochem. Pharmacol.* **19**:505–514, 1970.
18. Galper, J. B.; Catterall, W. A. Inhibition of sodium channels by D600. *Mol. Pharmacol.* **15**:174–178, 1979.
19. Haycock, J. W.; Meligeni, J. A.; Bennett, W. F.; Waymire, J. C. Phosphorylation and activation of tyrosine hydroxylase mediate the acetylcholine-induced increase in catecholamine biosynthesis in adrenal chromaffin cells. *J. Biol. Chem.* **257**:12641–12648, 1982.
20. Holz, R. W.; Sentor, R. A.; Frye, R. A. Relationship between Ca^{2+} uptake and catecholamine secretion in primary dissociated cultures of adrenal medulla. *J. Neurochem.* **39**:635–646, 1982.
21. Kenigsberg, R. L.; Cote, A.; Trifaro, J. M. Trifluoperazine, a calmodulin inhibitor, blocks secretion in cultured chromaffin cells at a step distal from calcium entry. *Neuroscience* **7**:2277–2286, 1982.
22. Kidikoro, Y.; Miyazaki, S.; Ozawa, S. Acetylcholine-induced membrane depolarization and potential fluctuations in the rat adrenal chromaffin cell. *J. Physiol. (London)* **324**:221–237, 1982.
23. Kilpatrick, D. L.; Slepetis, R. J.; Corcoran, J. J.; Kirshner, N. Calcium uptake and catecholamine secretion by cultured bovine adrenal medulla cells. *J. Neurochem.* **38**:427–435, 1982.
24. Kirshner, N.; Smith, W. J. Metabolic requirements for secretion from the adrenal medulla. *Life Sci.* **8**:799–803, 1969.
25. Knight, D. E.; Baker, P. F. Calcium-dependence of catecholamine release from bovine adrenal medullary cells after exposure to intense electric fields. *J. Membr. Biol.* **68**:107–140, 1982.
26. Kohlhardt, M.; Bauer, B.; Krause, H.; Fleckenstein, A. Differentiation of the transmembrane Na and Ca channels in mammalian cardiac fibers by the use of specific inhibitors. *Pfluegers Arch.* **335**:309–322, 1972.
27. Nawrath, H.; Ten Eick, R. E.; McDonald, T. F.; Trautwein, W. On the mechanism underlying the action of D600 on slow inward current and tension in mammalian myocardium. *Circ. Res.* **40**:408–414, 1977.
28. Oka, M.; Isosaki, M.; Watanabe, J. Calcium flux and catecholamine release in isolated bovine adrenal medullary cells: Effects of nicotinic and muscarinic stimulation. In: *Advances in the Biosciences*, Volume 36, *Synthesis, Storage and Secretion of Adrenal Catecholamines*, F. Izumi, K. Kumakura, and M. Oka, eds., Elmsford, N.Y.; Pergamon Press, 1982, pp. 29–36.
29. Pinto, J. E. B.; Trifaro, J. M. The different effects of D600 (methoxyverapamil) on the release of adrenal catecholamines induced by acetylcholine, high potassium or sodium deprivation. *Br. J. Pharmacol.* **57**:127–132, 1976.
30. Ritchie, A. K. Catecholamine secretion in a rat pheochromocytoma cell line: Two pathways for calcium entry. *J. Physiol. (London)* **286**:541–561, 1979.
31. Rubin, R. P. The role of energy metabolism in calcium-evoked secretion from the adrenal medulla. *J. Physiol. (London)* **206**:181–192, 1970.
32. Slepetis, R.; Kirshner, N. Inhibition of $^{45}Ca^{2+}$ uptake and catecholamine secretion by phenothiazines and pimozide in adrenal medulla cell cultures. *Cell Calcium* **3**:183–190, 1982.
33. Stallcup, W. B. Sodium and calcium fluxes in a clonal nerve cell line. *J. Physiol. (London)* **286**:525–540, 1979.
34. Wada, A.; Yanagihara, N.; Izumi, F.; Sakuroi, S.; Kobayashi, H. Trifluoperazine inhibits $^{45}Ca^{2+}$ uptake and catecholamine secretion and synthesis in adrenal medullary cells. *J. Neurochem.* **40**:481–486, 1983.
35. Wilson, S. P.; Kirshner, N. Calcium-evoked secretion from digitonin-permeabilized adrenal medullary chromaffin cells. *J. Biol. Chem.* **258**:4994–5000, 1983.

13

Calcium and Transmitter Release
Modulation by Adenosine Derivatives

E. M. Silinsky

Overview of Calcium-Dependent Transmitter Release

The currently accepted scheme of evoked transmitter secretion may be summarized as follows (for recent reviews, see [44,46]):

(1) A wave of depolarization (the action potential) invades the nerve terminal and opens voltage-sensitive calcium channels. (2) Extracellular calcium, preequilibrated with the extracellular surface of the nerve terminal membrane, enters the nerve terminal down its concentration gradient through the open calcium channels. (3) Calcium, once near the internal face of the nerve membrane, causes synaptic vesicles with encapsulated neurotransmitter to fuse with the nerve terminal at specific releasing sites. (4) The transmitter contents of the fused vesicles are discharged into the synaptic cleft by exocytosis.

The suggested quantitative relationships between occupancy by calcium of extracellular sites near calcium channels and transmitter release are summarized schematically in Fig. 1. In this figure, occupied binding sites at calcium channels are blackened and transmitter exocytosis is illustrated as asterisks. Fig. 1a illustrates the simplest condition, a *linear* relationship between the fraction of the total number of extracellular calcium channel binding sites occupied (Y_{Ca}) and the fraction of total number of releasable packets of transmitter that are actually released (Y_R). In Fig. 1a, where $Y_{Ca} = 0.7$, then $Y_R = 0.7$; i.e., 7 of the 10 vesicles discharge their transmitter contents. Such a linear relationship has been reported on occasion (e.g., in crayfish or in some ganglia), but most nerve endings show more complex behavior. For example, it has been proposed that several calcium ions must cooperate to release transmitter; specifically Y_{Ca} must be raised to a *power* (the fourth or possibly even the fifth power) to describe secretion at neuromuscular junctions [9]. As shown in Fig. 1b, where $Y = 0.7$, by a fourth power relationship only 20% of the available transmitter parcels would be released with 70% of the calcium binding sites occupied.

E. M. Silinsky • Department of Pharmacology, Northwestern University Medical School, Chicago, Illinois 60611.

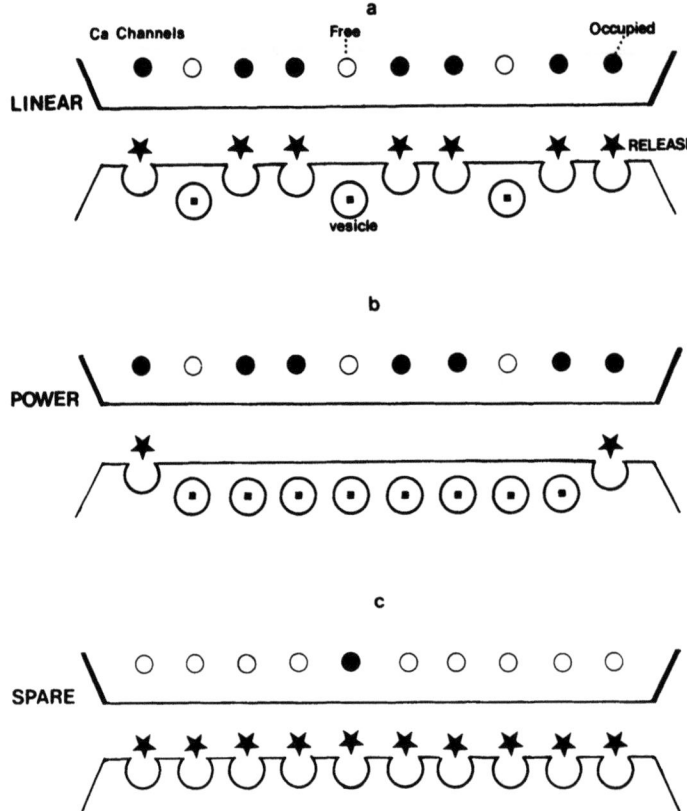

Figure 1. Suggested quantitative relationships between occupied extracellular calcium binding sites (filled circles) and transmitter release (asterisks). See text for further details.

More recently, however, the mathematical framework of receptor theory has been applied to the motor nerve ending; the results suggest a different conclusion, namely, that the majority of calcium channels are "*spare*" ([43,44,50] Fig. 1c). As illustrated in Fig. 1c, only about 10% of the extracellular calcium channel binding sites need to be occupied to produce maximal acetylcholine (ACh) secretion, in turn leaving the majority of these sites spare or reserve [50]. Such a result is commonly observed in complex receptor–effector systems. This is because the initiation signal, i.e., the full agonist binding to receptors (in this instance calcium binding to the external orifice of the calcium channel), need *not* be large to produce maximal effects (secretion) because of the enormous amplification inherent in the sequence of reaction intermediates that intervenes between binding and response [2]. The published experimental results of the motor nerve ending are in accord with the notion of spare channels for calcium, the full agonist for synchronous evoked ACh release, but not for the partial agonist strontium [43,44]. Spare calcium channels have also been observed in mammalian brain [10–12]. Specifically, it has been shown that maximal release of neurosecretory products in response to depolarization in the mammalian neurohypophysis occurs at an extracellular calcium concentration where calcium entry is only a modest fraction of the maximum [10,11]. Similarly, at nerve terminals in rat brain, antagonism of Ca^{2+} fluxes by La^{3+} does not inhibit the output of transmitter at higher extracellular Ca^{2+} concentrations [12]. The properties of alkaline earth cations in supporting transmitter release will be discussed more fully below.

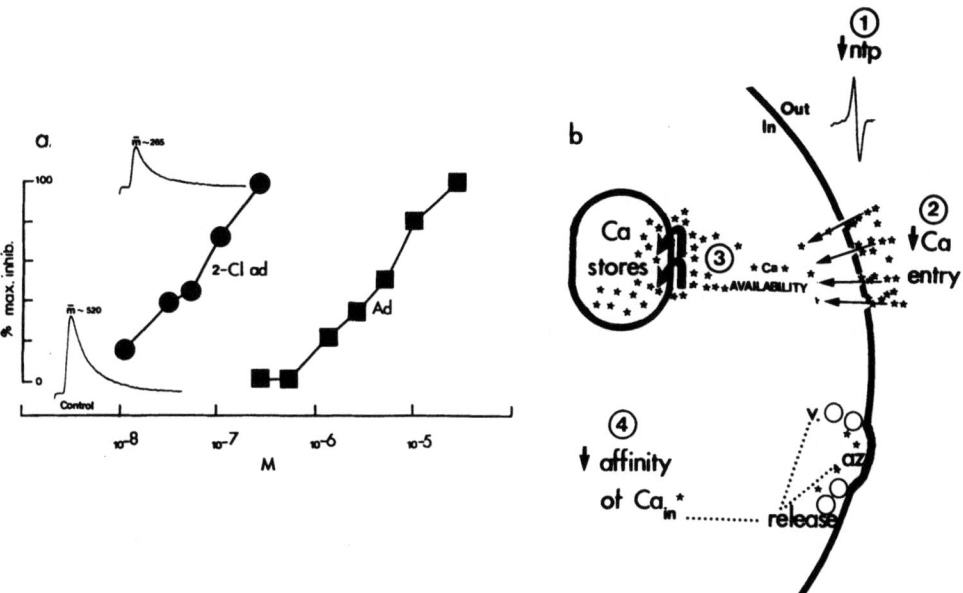

Figure 2. The observations (a) and the possible mechanisms (b) for the inhibitory effects of adenosine derivatives on acetylcholine (ACh) release in the frog cutaneous nerve–muscle preparation.

(a) The relationship between concentration of adenosine receptor agonist and the mean number of ACh packets (m) released synchronously in normal calcium solutions. The control end-plate potential (epp) (m = 520 quanta) and the maximally inhibited epp (m = 265) are shown as insets. Each epp is the mean response to four stimuli delivered at a frequency of 0.05 Hz. The concentrations are plotted on semilogarithmic coordinates against the fraction of a maximal level of inhibition. ●, 2-chloroadenosine (2-Cl ad); ■, adenosine (ad). Ringer solution contained 1.8 mM Ca^{2+} and 6.4 mg/liter tubocurarine. In a total of seven experiments made in low-Ca^{2+} (0.35–0.7 mM) high-Mg^{2+} (5–6 mM) Ringer where m < 10 or in 1.8 mM Ca^{2+} Ringer, adenosine was one to two orders of magnitude more potent than 2-chloroadenosine as an inhibitor of m. (For further details of these methods and results see Ref. 8, 23, and 45.)

(b) Possible sites at which adenosine derivatives may interfere with transmitter release. Asterisks = calcium. See text for details.

The Effects of Adenosine Derivatives on Calcium-Dependent Transmitter Release

Adenosine derivatives are released by synaptic activation at a number of sites in the peripheral and central nervous system [5,13,14,19,28,39,47,52,54] and exert depressant effects on the release of transmitter substances at the majority of synapses investigated [4,13–15,19,28,32–36,42,45,52]. This has led to the suggestion that adenosine derivatives may be negative feedback modulators of transmitter secretion [32,39]. Despite the current interest in adenosine, however, little is known as to how this substance inhibits evoked transmitter release. The general effects of adenosine derivatives on acetylcholine (ACh) release from motor nerve terminals are illustrated in Fig. 2a. Both adenosine (ad) and 2-chloroadenosine (2-Cl ad) depress the mean number of ACh packets released synchronously (m) in response to a nerve impulse [reflected electrophysiologically as the end-plate potential (epp); see insets] by a maximum of approximately 50%. However, 2-chloroadenosine is one to two orders of magnitude more potent than adenosine as an inhibitor of calcium-dependent ACh release [2-chloroadenosine is not a substrate for uptake or for deamination [7,28]]. These inhibitory effects are blocked by the adenosine receptor antagonist theophylline [15,45].

Suggested Mechanisms for the Inhibitory Effects of Adenosine Derivatives at Motor Nerve Endings

Various suggested mechanisms to explain the inhibitory effects of adenosine are illustrated in Fig. 2b as follows. Site [1]-adenosine depresses the nerve terminal action potential (ntp) [1]. Site [2]-adenosine depresses calcium entry into the nerve terminal [19,32,35]. Site [3]-adenosine increases the rate of clearance of calcium into storage sites such as mitochondria [4,33]. (Note that mechanisms [2] and [3] together may be viewed as a decreased availability of intracellular calcium for transmitter release.) Site [4]-adenosine decreases the apparent intracellular affinity of calcium for a structural component of the secretory apparatus [43,45], e.g., synaptic vesicles (v) or active zones of release (az).

Evidence That Adenosine Derivatives Decrease the Intracellular Affinity of Alkaline Earth Cation Activators of ACh Release

Figures 3a–d suggest that mechanisms [1]–[3] (Fig. 2b) probably cannot explain the effects of adenosine derivatives at the motor nerve ending. First, the focal ntp is not depressed by 2-chloroadenosine, whereas the end-plate current (epc), which reflects ACh release, is decreased (Figs. 3a,b). Next, calcium channels (site [2], Fig. 2b) may be eliminated as the primary target of adenosine by the observation that when ACh release is activated by calcium-containing liposomes that bypass calcium channels [18,25,27,29], 2-chloroadenosine has its characteristic inhibitory effect (Fig. 3c). Finally, it is unlikely that adenosine impairs ACh release by increasing the rate of clearance of divalent cations (site [3], Fig. 2b). A form of evoked release that is a sensitive indicator of cation clearance rate, namely asynchronous neurally evoked release in barium (see inset, Fig. 3d), is inhibited by 2-chloroadenosine and adenosine without a change in the rate constant of decay (= slope of the regression lines in Fig. 3d).* The rate constant of decay of asynchronous ACh release in barium solutions (Fig. 3d) reflects the clearance of the activating cation from the site of release [41,53].

Direct evidence consistent with a decreased affinity of intracellular calcium (site [4], Fig. 2b) is provided in Figs. 3e and f. Specifically, Fig. 3e shows that a competitive relationship [51] exists between calcium and 2-chloroadenosine; the site, however, must be beyond the calcium channel (e.g., Fig. 3c) and is presumably intracellular. This *intracellular* competition with calcium is reflected in the relationship between *extracellular* calcium and 2-chloroadenosine because of the presence of spare calcium channels [43,44,50,51]. Thus, increasing the extracellular calcium concentration overcomes inhibition produced by 2-chloroadenosine, because spare calcium channels can now deliver additional calcium to intracellular sites. This behavior, when taken in conjunction with the high apparent intracellular affinity of calcium [25,48], allows the fraction of intracellular sites occupied by calcium to be restored to the high level needed to produce maximal ACh release.† There are

---

*Asynchronous ACh release in barium solutions is initiated by the entry of barium through calcium channels. It is reflected electrophysiologically as a slow depolarizing of the postsynaptic membrane (V in Fig. 3d) produced by the large discharge of miniature end-plate potentials (see Refs. 40 and 41 and inset to Fig. 3d).

†Evoked ACh release as a fraction of maximal levels is proportional to the fraction of total number of intracellular divalent cation binding sites occupied at the secretory apparatus Y (e.g., Ref. 45). By the adsorption isotherm [51], $Y = [Me_{intracellular}]K_{Me}/(1 + [Me_{intracellular}]K_{Me})$, where Me = Ca or Sr and K_{Me} is the intracellular affinity of the metal ion for the secretory apparatus. This equation shows that Y can be restored after a depression of K_{Me} (e.g., caused by adenosine or 2-chloroadenosine) by increasing $[Me]_{intracellular}$, e.g., by using spare calcium channels. Even though strontium moves through the same channel as calcium [48], there are no spare channels for strontium.

13. Calcium and Transmitter Release

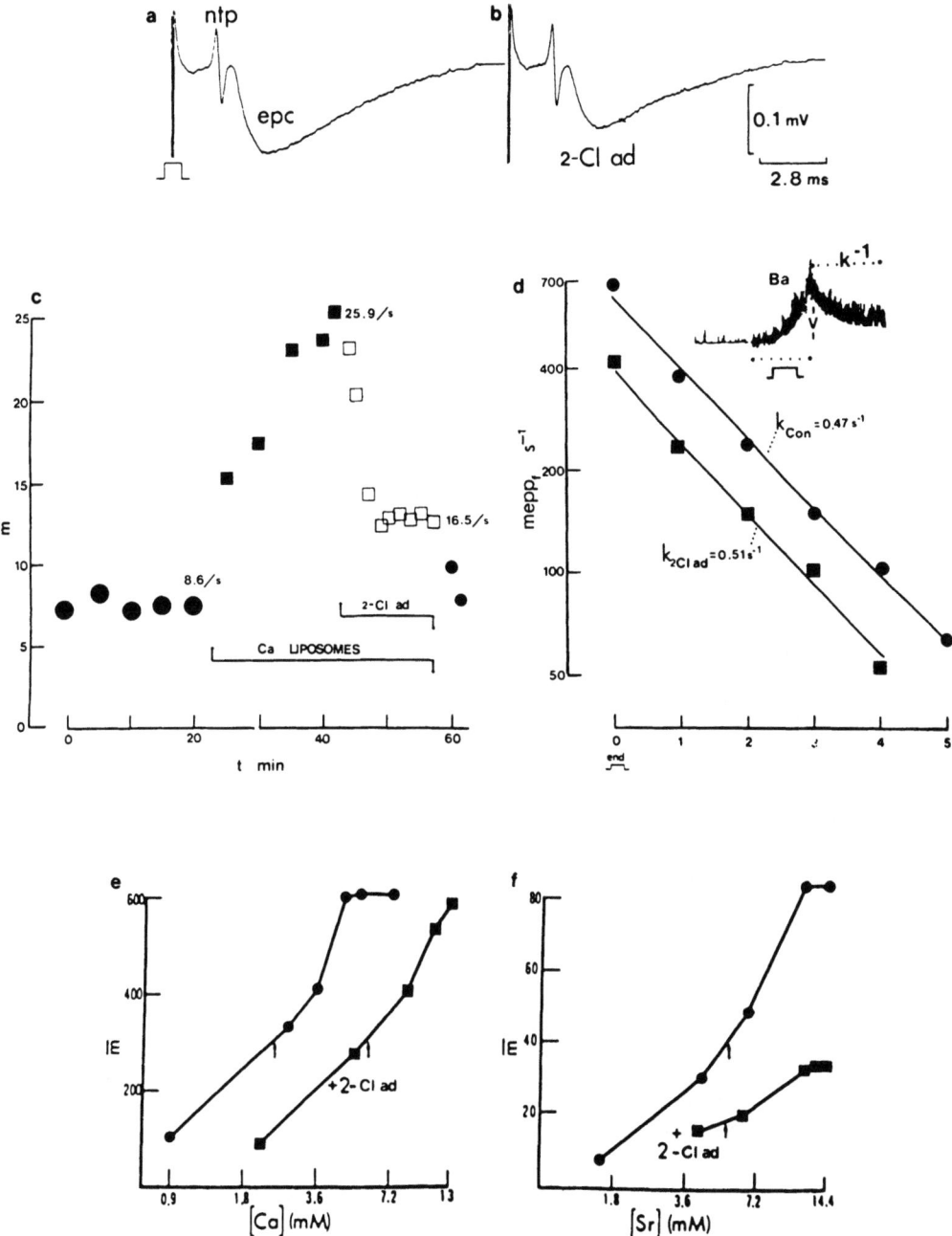

Figure 3. Experimental evaluation of sites (1)–(4) in Fig. 1b. Parts a–d provide evidence against sites (1)–(3); c, e and f provide evidence in favor of site [4].

(a, b) Absence of effect of 2-chloroadenosine on nerve terminal action potentials (ntp). (a) Averaged control response. epc, end-plate current. (b) Averaged response after 8 min in 10 μM 2-chloroadenosine. Each record is the mean response to 128 stimuli delivered at 1 Hz. Ringer solution contained 0.5 mM Ca^{2+}, 6 mM Mg^{2+}. In all five experiments (three with 10 μM 2-chloroadenosine, two with 100 μM adenosine) release was depressed by a mean of 65% without a decrease in the size of the ntp.

(c) Antagonism by 2-chloroadenosine (10 μM) of ACh release evoked by intracellular calcium. ●, m in control solution = 0.35 mM Ca^{2+}, 1 mM Mg^{2+} Ringer. ■, m in control + Ca^{2+} liposomes. □, m in control + Ca^{2+} liposomes + 2-chloroadenosine. Insets show mean mepp frequency (per sec). Ca^{2+} liposomes were prepared as described in Ref. 25. Large symbols: averaged m in response to 128 or 64 stimuli; small symbols: response to 32 stimuli. Stimulation frequency = 0.48 Hz. [For further details see Ref. 45.]

continued

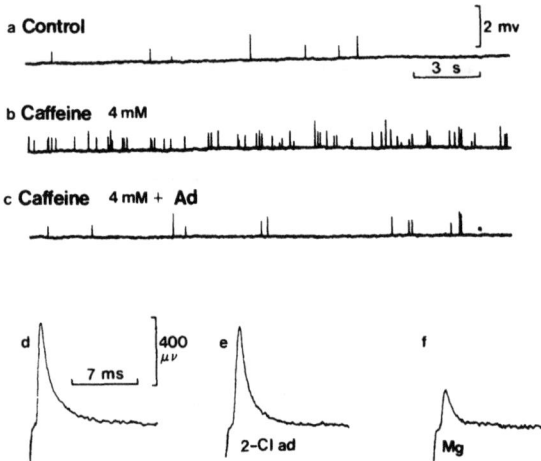

Figure 4. Possible involvement of adenylate cyclase in mediating the inhibitory effects of adenosine derivatives: in (a–c), inhibition by adenosine is increased in magnitude in the presence of caffeine. Ca^{2+}-free Ringer contained 3 mM Mg^{2+}. In (a), control mepp frequency = 0.46 ± 0.06 sec^{-1} (mean ± S.E.M., n = 124). In (b), mepp frequency in 4 mM caffeine Ringer = 2.46 ± 0.17 (n = 76). In (c), mepp frequency in 4 mM caffeine + 100 μM adenosine = 0.66 ± 0.11 (n = 62). In (d–f), antagonism of the inhibitory effects of 2-chloroadenosine (e) but not of Mg^{2+} (f) by an adenylate cyclase inhibitor (RMI 12,330A) is shown. Preparation was pretreated for approximately 45 min with 70 μM RMI 12,330A (a cyclo-alkyl lactamide) and then washed with normal Ringer without drug for another 45 min. (d) Control responses after RMI 12,330A treatment in normal (1.8 mM Ca^{2+}) Ringer without tubocurarine. (e) 0.8 μM 2-chloroadenosine. (f) 5 mM Mg^{2+}. Each epp is the mean response to 16 stimuli delivered at a frequency of 0.1 Hz. The very small epp's are due to the irreversible postsynaptic depression of end-plate sensitivity to ACh produced by RMI 12,330A. [For further details see Ref. 45.]

no spare channels for the partial agonist strontium [24,43]. If 2-chloroadenosine reduced the intracellular strontium affinity, then increasing the extracellular strontium concentration should not restore ACh release, as little additional strontium can be delivered to the cytoplasm in the absence of spare channels. The strontium–2-chloroadenosine relationship should thus appear noncompetitive; Fig. 3f shows this to be the case.

Coupling of Adenosine Receptor Binding to Inhibition: The Role of Adenylate Cyclase

The results presented here suggest that adenosine receptor agonists reduce the intracellular affinity of calcium (and other divalent cations) for sites involved in the process of ACh release. Further experimental evidence in support of this conclusion has previously been described [45]. The adenosine receptors responsible for these effects appear to be of the R variety [21] (R indicates that an intact ribose moiety is generally required for agonist behavior [6,20,21]). Such R sites are situated at the external surface of the cell membrane, blocked by theophylline, and linked to adenylate cyclase. As a working model, let us suppose that adenosine derivatives stimulate adenylate cyclase [37] and the subsequent

---

(d) Depression of Ba-mediated asynchronous ACh release (mepp frequency) by 2-chloroadenosine (10 μM) without a change of the rate constant of decay of mepp frequency (k). Stimulation was delivered at a frequency of 30 Hz for 2 sec [dotted line in inset which shows response in control Ringer (0.7 mM Ba^{2+}, 6 mM Mg^{2+})]. mepp frequency was determined first by measuring V at 1-sec intervals after the termination of nerve stimulation and then using the following equation: mepp frequency = V/τ (mepp), where V is the amplitude of the slow depolarization and τ is the time constant of the mepp (the amplitude of the miniature potentials) [41]. Each symbol is the mean of four determinations of mepp frequency at each particular time. Circles show decay of mepp frequency under control conditions, which was described by the equation: mepp frequency $(t) = 627e^{-0.47t}$. Squares show decay in 2-chloroadenosine where mepp frequency $(t) = 402e^{-0.51t}$.

(e, f) Log concentrations–m curves for Ca^{2+} (e) and Sr^{2+} (f) before (circles) and after (squares) antagonism with 2-chloroadenosine (25 μM). [Reprinted from Ref. 43 with permission of the British Pharmacological Society.]

increase in local intracellular cAMP level inhibits release [cf. 15]. If this were true, and *if the local cAMP concentration were the limiting factor as to the maximal level of inhibition produced by adenosine*, then agents which allow for larger increases in the local concentration of cAMP, e.g., phosphodiesterase inhibitors, should augment inhibition by adenosine (Figs. 4a–c). Indeed, caffeine (4 mM) increases mepp frequency fivefold in the virtual absence of extracellular calcium (Fig. 4b). This effect is presumably due to the calcium-translocating action of methylxanthines [30]. Increased release is restored to control levels by 100 μM adenosine (Fig. 4c). This level of inhibition is far greater than normally observed in the absence of phosphodiesterase inhibition (e.g., Figs. 2 and 3), and this enhanced effect of adenosine may be attributed to phosphodiesterase inhibition by caffeine as similar effects were observed with other phosphodiesterase inhibitors (e.g., Ref. 45).* Another prediction of this working model is that an irreversible inhibitor of adenylate cyclase would blunt the inhibitory effects of adenosine derivatives but not of a calcium channel blocker such as magnesium [18]. After treatment with the adenylate cyclase inhibitor RMI 12,330A [16,17] 2-chloroadenosine has no effect but magnesium (5 mM) still inhibits ACh release in its characteristic manner (e.g., Ref. 45) (Figs. 4d–f).

The results described above are in accord with the view that local increases in cAMP in the vicinity of secretory sites are responsible for the inhibitory effects of adenosine derivatives. Other evidence consistent with this interpretation is as follows. (1) Adenosine enhances cAMP levels in peripheral cholinergic neurons [37]. (2) Imidazole, which stimulates phosphodiesterases, prevents the effects of adenosine derivatives [34]. (3) Imidazole [34], RMI 12,330A (unpublished), and P-site agonists (P sites require an unsubstituted purine ring and are exclusively associated with an inhibition of adenylate cyclase [21,42]) all increase ACh release. This would be expected if resting levels of cAMP are producing tonic inhibition of ACh release, since all of these agents would be predicted to reduce resting cAMP levels.

If one accepts the premise that R-site *subtypes* exist [6,20], then the sites on motor nerve endings which inhibit ACh release are more likely to be R_a (= A_2) in which adenosine receptor agonists stimulate adenylate cyclase than R_i (= A_1) in which adenylate cyclase is inhibited [6]. The [adenosine]–ACh release relationship shown in Fig. 2a is also consistent with this suggestion [28], although apparent potencies may be skewed by uptake and deamination [7]. (Although these results are consistent with the working model, the evidence thus far is circumstantial.)†

Summary

Figure 5 summarizes these results with adenosine in "cartoon" form. In (a), adenosine (ad) is shown prior to binding to its receptor ([1]R). Two other components constitute the adenylate cyclase triad [3], namely the nucleotide-binding regulatory subunit ([2]N), which acts as a shuttle between the receptor and the catalytic site, and the catalytic site ([3]C) of adenylate cyclase (cyclops) (Fig. 5). Adenosine binding to R causes the N subunit to

---

*These results with caffeine are of interest in that this agent has been suggested to act as an adenosine receptor blocker [30]; it is evident that this is not the case at the neuromuscular junction (see also Ref. 45).

†The effects of exogenous cAMP derivatives are curious in this regard; increases [49] and/or small decreases (unpublished) and no effect [26] have all been observed in response to dibutyryl cAMP. Exogenously applied cAMP derivatives might phosphorylate multiple cellular sites and produce complex effects in many cell compartments, while only a small compartment of cAMP near the membrane is involved in modulation of secretion by adenosine. Although exogenous cAMP could increase mepp frequency under certain conditions [26,49], cAMP is an unlikely mediator of the physiologically functional form of ACh release detected as m.

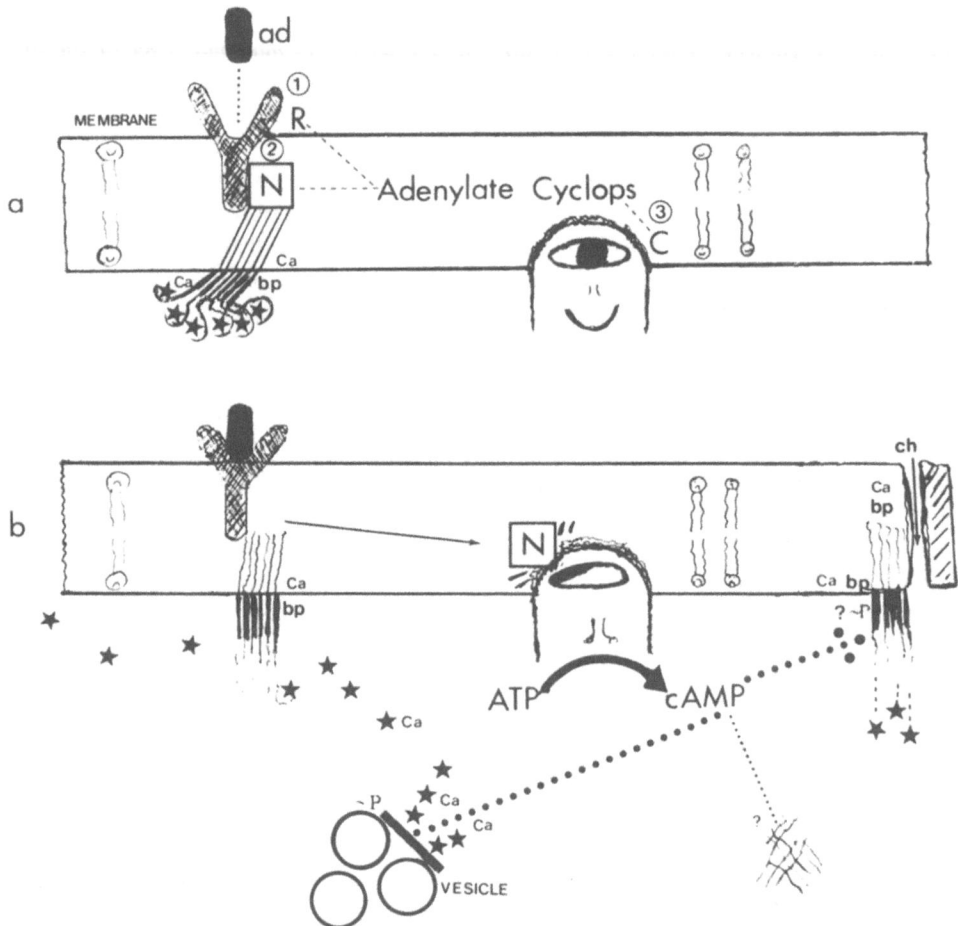

Figure 5. A "cartoon" summary of the mechanism by which adenosine receptor agonists inhibit ACh release. In (a), adenosine (ad) is shown prior to binding to the R-site receptor subunit ([1]R) of adenylate cyclase (cyclops). The other subunits are the guanosine nucleotide-binding regulatory subunit ([2]N) and the catalytic subunit ([3]C). In (b), adenosine binds to R, N dissociates from R and bonks C in the head; this provokes C to generate cAMP from ATP (cAMP can be viewed as cyclic AMP or cyclops AMP). The local increase in cAMP may interfere with calcium-dependent membrane phosphorylations such as that shown at the level of the synaptic vesicle (calcium = asterisks) or cAMP may phosphorylate (~P) calcium-binding proteins (Ca bp) in the presynaptic membrane (or vesicle) and reduce the affinity of these proteins for calcium [compare the right-hand part of (b) with the left of (a)]. These interactions are likely to be determined at the level of the protein kinases. The possibility that such events might control channel function (ch) in some systems or interact at the cytoskeletal level (?, e.g., at a vesicle docking protein) is shown as well. The N subunit may be structurally connected to cytoskeletal elements or to Ca bp (e.g., Ref. 31) and raises the possibility that departure of N from its original resting position alters the conformation of Ca bp apart from any subsequent activation of the catalytic subunit. It may be speculated that the G subunit could modulate m while subsequent alterations in cAMP concentrations might modulate mepp frequency.

"bonk" the C unit of the cyclops, provoking it to make cAMP from ATP (Fig. 5b). (cAMP may be viewed as cyclic AMP or cyclops AMP, depending on the reader's current state of light-heartedness.) The local increase in cAMP may then produce a decreased affinity of a structural component of the release process for calcium. For example, a protein component of synaptic vesicles (protein 1 or synapsin I) has specific peptides phosphorylated by calcium (asterisks) and this calcium-dependent phosphorylation is inhibited by cAMP [38]. Such an interaction at the vesicle or at the presynaptic membrane [22] at the level of phosphorylation could explain the reduction of the intracellular affinity for calcium and other cation activators of ACh release by adenosine receptor agonists. The cAMP-induced phosphorylation of vesicles or active zones of release could also reduce the affinity of calcium (see legend to Fig. 5 and Ref. 46a for further speculation). Regardless of the precise involvement of protein phosphorylation, extracellular adenosine receptors on adenylate cyclase inhibit the physiologically functional, synchronous release of ACh from nerve endings by reducing the apparent affinity for calcium of an essential component of the secretory apparatus.

ACKNOWLEDGMENTS. This work was supported in part by a Fogarty Senior International Fellowship from the USPHS and a USPHS research grant.

References

1. Akasu, T.; Hirai, K.; Koketsu, K. Increase of acetylcholine receptor sensitivity by adenosine triphosphate: A novel action of ATP on ACh sensitivity. *Br. J. Pharmacol.* **74**:505–507, 1981.
2. Ariens, E. J. Receptors: From fiction to fact. *Trends Pharmacol. Sci.* **1**:11–15, 1979.
3. Birnbaumer, L.; Iyengar, R. Coupling of receptors to adenylate cyclases. In: *Cyclic Nucleotides I: Handbook of Experimental Pharmacology*, J. A. Nathanson and J. E. Kebabian, eds., Berlin, Springer-Verlag, 1982, pp. 153–183.
4. Branisteanu, D. D.; Haulica. N. P.; Proca, B.; Nhue, B. G. Adenosine effects upon transmitter release parameters in the Mg^{2+}-paralyzed neuromuscular junction of the frog. *Naunyn-Schmiedebergs Arch. Pharmakol.* **308**:273–279, 1979.
5. Burnstock, G. Past and current evidence for the purinergic nerve hypothesis. In: *Physiological and Regulatory Functions of Adenosine and Adenine Nucleotides*, H. P. Baer and G. I. Drummond, eds., New York, Raven Press, 1979, pp. 3–32.
6. Van Calker, D.; Muller, M.; Hamprecht, B. Adenosine regulates via two different types of receptors, the accumulation of cyclic AMP in cultured brain cells. *J. Neurochem.* **33**:999–1005, 1979.
7. Clarke, D. A.; Davoll, J., Phillips, F. S.; Brown, B. G. Enzymatic deamination and vasopressor effects of adenosine analogues. *J. Pharmacol. Exp. Ther.* **106**:291–302, 1952.
8. Del Castillo, J.; Katz, B. Quantal components of the end-plate potential. *J. Physiol. (London)* **124**:560–573, 1954.
9. Dodge, F. A., Jr.; Rahamimoff, R. Cooperative action of Ca ions in transmitter release at the neuromuscular junction. *J. Physiol. (London)* **193**:419–432, 1967.
10. Douglas, W. W.; Poisner, A. M. Stimulus-secretion coupling in a neurosecretory organ: The role of calcium in the release of vasopressin from the neurohypophysis. *J. Physiol. (London)* **172**:1–18, 1964.
11. Douglas, W. W.; Poisner, A. M. Calcium movements in the neurohypophysis of the rat and its relation to the release of vasopressin. *J. Physiol. (London)* **172**:19–30, 1964.
12. Drapeau, P.; Blaustein, M. P. Initial release of ^{3}H-dopamine from rat striatal synaptosomes: Correlation with calcium entry. *J. Neurosci.* **3**:703–713, 1983.
13. Dunwiddie, T. V.; Hoffer, B. J. The role of cyclic nucleotides in the nervous system. In: *Cyclic Nucleotides II: Handbook of Experimental Pharmacology*, J. E. Kebabian and J. A. Nathanson, eds., Berlin, Springer-Verlag, 1982, pp. 389–463.
14. Fredholm, B. B.; Hedqvist, P. Modulation of neurotransmission by purine nucleotides and nucleosides. *Biochem. Pharmacol.* **29**:1635–1643, 1980.
15. Ginsborg, B. L.; Hirst, G. D. S. The effect of adenosine on the release of the transmitter from the phrenic nerve of the rat. *J. Physiol. (London)* **224**:629–645, 1972.
16. Grupp, G.; Grupp, I. L.; Johnson, C. L.; Matlib, M. A.; Rouslin, W.; Schwartz, A.; Wallick, E. T.; Wang, T.; Wisler, P. Effect of RMI 12,330A, a new inhibitor of adenylate cyclase on myocardial function and subcellular activity. *Br. J. Pharmacol.* **70**:429–442, 1980.
17. Guellaen, G.; Mahu, J. L.; Mavier, P.; Berthelot, P.; Hanoune, J. RMI 12,330A, an inhibitor of adenylate cyclase in rat liver. *Biochim. Biophys. Acta* **484**:465–475, 1977.
18. Kharasch, E. D.; Mellow, A. M.; Silinsky, E. M. Intracellular magnesium does not antagonize calcium-dependent acetylcholine secretion. *J. Physiol. (London)* **314**:255–263, 1981.

19. Kuroda, Y. Physiological roles of adenosine derivatives which are released during neurotransmission in mammalian brain. *J. Physiol. (Paris)* **74**:463–470, 1978.
20. Londos, C.; Cooper, D. M. F.; Wolff, J. Subclasses of external adenosine receptors. *Proc. Natl. Acad. Sci. USA* **77**:2551–2554, 1980.
21. Londos, C.; Wolff, J. Two distinct adenosine sensitive sites on adenylate cyclase. *Proc. Natl. Acad. Sci. USA* **74**:5482–5486, 1977.
22. de Lorenzo, R. J. Calmodulin in neurotransmitter release and synaptic function. *Fed. Proc.* **41**:2265–2272, 1982.
23. McLachlan, E. M.; Martin, A. R. Non-linear summation of end-plate potentials in the frog and mouse. *J. Physiol. (London)* **311**:307–324, 1981.
24. Meiri, U.; Rahamimoff, R. Activation of transmitter release by strontium and calcium ions at the neuromuscular junction. *J. Physiol. (London)* **215**:709–726, 1971.
25. Mellow, A. M.; Perry, B. D.; Silinsky, E. M. Effects of calcium and strontium in the process of acetylcholine release from motor nerve endings. *J. Physiol. (London)* **328**:547–562, 1982.
26. Miyamoto, M. D.; Breckenridge, B. M. A cyclic adenosine monophosphate link in the catecholamine enhancement of transmitter release at the neuromuscular junction. *J. Gen. Physiol.* **63**:609–624, 1974.
27. Pagano, R. E.; Weinstein, J. N. Interactions of liposomes with mammalian cells. *Annu. Rev. Biophys. Bioeng.* **7**:435–468, 1978.
28. Phillis, J. W.; Wu, P. H. Adenosine and adenosine triphosphate as neurotransmitter/neuromodulator in the brain: The evidence is mounting. In: *Trends in Autonomic Pharmacology*, Volume 2, S. Kalsner, ed., Baltimore, Urban & Schwarzenberg, 1982, pp. 237–261.
29. Rahamimoff, R.; Meiri, H.; Erulkar, S. D.; Barenholz, Y. Changes in transmitter release induced by ion-containing liposomes. *Proc. Natl. Acad. Sci. USA* **75**:5214–5216, 1978.
30. Rall, T. W. Central nervous system stimulants: The xanthines. In: *The Pharmacological Basis of Therapeutics*, A. G. Gilman, L. S. Goodman, and A. Gilman, eds., New York, Macmillan Co., 1980, pp. 592–607.
31. Rasenick, M. M.; Stein, P. J.; Bitensky, M. W. The regulatory subunit of adenylate cyclase interacts with cytoskeletal components. *Nature (London)* **294**:560–562, 1981.
32. Ribeiro, J. A. Purinergic modulation of transmitter release. *J. Theor. Biol.* **80**:259–270, 1979.
33. Ribeiro, J. A.; Dominguez, M. L. Mechanisms of depression of neuromuscular transmission by ATP and adenosine. *J. Physiol. (Paris)* **74**:491–496, 1978.
34. Ribeiro, J. A.; Dominguez, M. J.; Goncalves, M. L. Purine effects at the neuromuscular junction and their modification by theophylline, imidazole and verapamil. *Arch. Int. Pharmacodyn. Ther.* **238**:206–219, 1979.
35. Ribeiro, J. A.; Sa-Almeida, A. M.; Namorado, J. M. Adenosine and adenosine triphosphate decrease ^{45}Ca uptake by synaptosomes stimulated by potassium. *Biochem. Pharmacol.* **28**:1297–1300, 1979.
36. Ribeiro, J. A.; Walker, J. The effects of adenosine triphosphate and adenosine diphosphate on transmission at the rat and frog neuromuscular junctions. *Br. J. Pharmacol.* **54**:213–218, 1975.
37. Roch, P.; Salamin, A. Adenosine-promoted accumulation of cyclic AMP in rabbit vagus nerve. *Experientia* **32**:1419–1421, 1976.
38. Schulman, H. Calcium-dependent protein phosphorylation. In: *Cyclic Nucleotides I: Handbook of Experimental Pharmacology*, J. A. Nathanson and J. W. Kebabian, eds., Berlin, Springer-Verlag, 1982, pp. 425–478.
39. Silinsky, E. M. On the association between transmitter secretion and the release of adenine nucleotides from mammalian motor nerve terminals. *J. Physiol. (London)* **247**:145–162, 1975.
40. Silinsky, E. M. Can barium support the release of acetylcholine by nerve impulses? *Br. J. Pharmacol.* **59**:215–217, 1977.
41. Silinsky, E. M. On the role of barium in supporting the asynchronous release of acetylcholine quanta by motor nerve impulses. *J. Physiol. (London)* **274**:157–171, 1978.
42. Silinsky, E. M. Evidence for specific adenosine receptors at cholinergic nerve endings. *Br. J. Pharmacol.* **71**:191–194, 1980.
43. Silinsky, E. M. On the calcium receptor that mediates depolarization–secretion coupling at cholinergic motor nerve terminals. *Br. J. Pharmacol.* **73**:413–429, 1981.
44. Silinsky, E. M. Properties of calcium receptors that initiate depolarization–secretion coupling. *Fed. Proc.* **41**:2172–2180, 1982.
45. Silinsky, E. M. On the mechanism by which adenosine receptor activation inhibits the release of acetylcholine from motor nerve endings. *J. Physiol. (London)* **346**:243–256, 1984.
46. Silinsky, E. M. The biophysical pharmacology of calcium-dependent acetylcholine secretion. *Pharmacol. Rev.* in press.
47. Silinsky, E. M.; Hubbard, J. I. Release of ATP from rat motor nerve terminals. *Nature (London)* **243**:404–405, 1973.
48. Silinsky, E. M.; Mellow, A. M. The relationship between strontium and other divalent cations in the process of

transmitter release from cholinergic nerve endings. In: *Handbook of Stable Strontium,* S. Skoryna, ed., New York, Plenum Press, 1981, pp. 263–285.
49. Standaert, F. G.; Dretchen, K. L. Cyclic nucleotides and neuromuscular transmission. *Fed. Proc.* **38:**2182–2192, 1979.
50. Stephenson, R. P. A modification of receptor theory. *Br. J. Pharmacol.* **11:**379–393, 1956.
51. Stephenson, R. P.; Barlow, R. B. Concepts of drug action, quantitative pharmacology and biological assay. In: *A Companion to Medical Studies,* R. Passmore and J. S. Robson, eds., Oxford, Blackwell, 1970, pp. 1–19.
52. Stone, T. W. Physiological roles for adenosine and adenosine 5'-triphosphate in the nervous system. *Neuroscience* **6:**523–555, 1981.
53. Zengel, J. E.; Magleby, K. L. Changes in miniature end-plate potential frequency during repetitive nerve stimulation in the presence of Ca^{2+}, Ba^{2+} and Sr^{2+} at the frog neuromuscular junction. *J. Gen. Physiol.* **77:**503–529, 1981.
54. Zimmermann, H. Vesicle recycling and transmitter release. *Neuroscience* **4:**1773–1804, 1979.

14

The Role of Calcium and Osmosis in Membrane Fusion

Fredric S. Cohen, Myles H. Akabas, and Alan Finkelstein

Although exocytosis is triggered by a rise in intracellular levels of Ca^{2+}, it is not known how Ca^{2+} produces this effect. Because Ca^{2+} regulates so many cellular functions, it is difficult to establish what role any one of these plays in exocytosis. An approach for determining the mechanisms underlying fusion is to study those aspects of the process that can be attributed solely to phospholipid bilayer interactions, independent of membrane proteins and cytoplasmic factors. Thus, considerable effort has gone into studying fusion between artificial phospholipid bilayer membranes, with most of this effort directed to fusion between vesicles [8,9,11]. On the other hand, we have studied the fusion of phospholipid vesicles to planar phospholipid bilayer membranes. Here, we describe the physical principles that govern fusion in this system and discuss their possible applicability to fusion and exocytosis as they occur biologically.

Osmotic Swelling

Bona fide fusion between phospholipid vesicles and planar membrane can occur, and it was suggested that swelling of vesicles in contact with the planar membrane results in fusion [3,12]. The following experiment subsequently demonstrated directly that this hypothesis is correct [4].

Vesicles were formed with the channel-forming protein porin incorporated in their membranes; they were made in a solution containing 200 mM stachyose, a tetrasaccharide not permeant through porin channels. These vesicles were then added to one side (*cis*) of a voltage-clamped planar membrane. The aqueous phase of the *cis* side contained the same stachyose solution as was used to prepare the vesicles; the *trans* side of the planar membrane contained an isosmotic solution. Following the addition of vesicles to the *cis* side, 15 mM

Fredric S. Cohen • Department of Physiology, Rush Medical College, Chicago, Illinois 60612. *Myles H. Akabas and Alan Finkelstein* • Departments of Physiology and Biophysics and Neuroscience, Albert Einstein College of Medicine, Bronx, New York 10461.

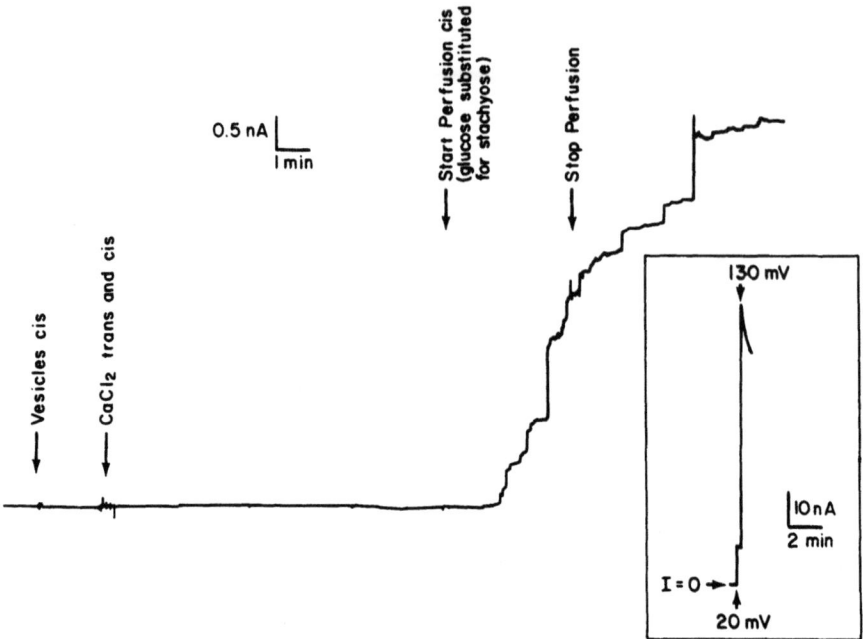

Figure 1. Osmotic swelling of vesicles leads to fusion. A planar membrane was voltage-clamped at 20 mV, *cis* side positive, and separated 100 mM KCl, 200 mM stachyose, 10 mM MES (2[*N*-morpholino]ethane sulfonic acid), 2 mM $MgCl_2$, 0.1 mM EDTA, pH 6.0 on the *cis* side (same solution as in the interior of the vesicles) and 100 mM KCl, 200 mM glucose, 10 mM MES, 2 mM $MgCl_2$, 0.1 mM EDTA, pH 6.0 on the *trans* side. Multimellar vesicles [egg PC (phosphatidylcholine): bovine PS (phosphatidylserine), 4 : 1] reconstituted with porin were added to the *cis* side of a hydrocarbon-free asolectin membrane. Two minutes later 15 mM $CaCl_2$ was added to the *cis* and *trans* compartments. Fusion did not occur over the next 10 min. The *cis* side was then perfused with a vesicle-free solution of the same composition as that of the original solution, except that the 200 mM stachyose was replaced with 200 mM glucose. A round of fusion lasting several minutes occurred. (Inset) After the burst of fusion, the membrane potential was switched from 20 mV to 130 mV. The decay of current is characteristic of the closing of porin channels at high voltages. [Reprinted with permission from Ref. 4.]

$CaCl_2$ was added to both sides of the membrane (*cis* and *trans*). No fusion occurred under these conditions, as assayed by conductance increases in the planar membrane. (When fusion occurs, it must, by definition, result in the total incorporation of the vesicular membrane into the planar membrane, channels included. The incorporation of porin channels into the planar membrane is manifested by characteristic jumps in conductance.) If the *cis* side is now perfused with a vesicle-free solution, in which isosmolal glucose is substituted for stachyose, vesicles free in solution will be removed, and glucose will enter any vesicle in contact with the planar membrane through their porin channels. Water will follow glucose into the vesicles, and they will osmotically swell. Under these conditions, fusion occurs.

Perfusion of the *cis* side, with glucose replacing stachyose, results in jumps in membrane conductance (Fig. 1). [These conductance jumps result from porin incorporation rather than nonspecific conductance leaks in the planar membrane, as the current (conductance) decreases with time upon switching the membrane voltage from 20 mV to 130 mV. This decrease is characteristic of the closing of porin channels at high voltage [10].] The conductance jumps are due to fusion and not simply to transfer of porin channels from vesicular to planar membrane. This is evident from the sizes of the conductance jumps, which correspond to several porin channels being incorporated simultaneously into the planar membrane (i.e., within 200 μ sec), rather than being inserted individually as would be expected for a transfer process. Thus, osmotic swelling of vesicles in contact with the planar membrane results in fusion.

Ca^{2+} and the Pre-fusion State

In the experiment just described, swelling of the vesicles was accomplished without establishing osmotic gradients across the planar membrane. Osmotic swelling of the vesicles, and therefore fusion, can also be induced by establishing such gradients.

An osmotic gradient across the membrane, *cis* side hyperosmotic, results in water flow from the *trans* to the *cis* side, and if the vesicles are sufficiently close to the planar membrane, water will enter the vesicles, causing them to swell and fuse. After vesicles and 15 mM $CaCl_2$ are added to the *cis* side, the addition of 400 mM urea to the *cis* side causes fusion (Fig. 2). Because urea is permeable through the porin channel (and through the vesicular bilayer), there is no net shrinkage or swelling of vesicles due to osmotic gradients between the *cis* compartment and the vesicular interior. Water flow from the *trans* compartment into the vesicular interior, however, causes vesicular swelling. Abolishing the osmotic gradient by adding 400 mM urea *trans* results in the cessation of fusion. Reestablishment of an osmotic gradient by adding 400 mM urea *cis* causes resumption of fusion.

The existence of a pre-fusion state can be readily demonstrated by establishing osmotic gradients in this manner. In Fig. 3, vesicles are added to the *cis* compartment and 15 mM $CaCl_2$ is added to both *cis* and *trans* compartments. The *cis* compartment is then perfused free of vesicles by using the identical $CaCl_2$-containing solution. Ten minutes later, 400 mM urea is added to the *cis* compartment and a burst of fusion occurs. An additional 400 mM increment of urea concentration causes another round of fusion, etc. Thus, $CaCl_2$ has clearly caused vesicles to associate tightly with the membrane, as evidenced by a population of vesicles (not removed by perfusion of the *cis* compartment) that are capable of fusing with

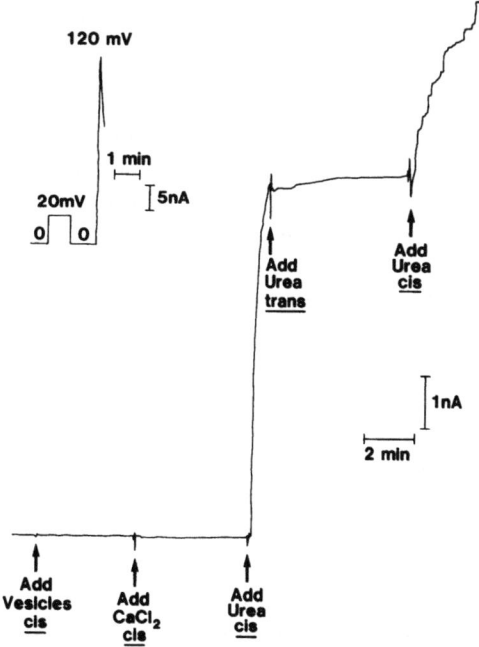

Figure 2. Osmotic gradients across the planar membrane cause fusion. Phospholipid vesicles (egg PC: bovine PS, 4 : 1) reconstituted with porin by a freeze-thaw method were added to the *cis* side ($\approx 10^{11}$ vesicles/ml) of a solvent-free asolectin membrane voltage-clamped at 20 mV. The addition of 15 mM $CaCl_2$ to the *cis* side did not result in conductance increases. Adding 400 mM urea *cis* resulted in fusion, whereas abolishing the gradient by adding 400 mM *trans* abolished fusion. Recreating the osmotic gradient by adding 400 mM urea *cis* caused fusion to resume. From the inset, it is seen that the conductance increases were due to incorporation of porin—the current decreased with time when the membrane potential was switched to 120 mV. The membrane was initially bathed in symmetrical 100 mM NaCl, 10 mM MES, 3 mM $MgCl_2$, 0.1 mM EDTA, pH 6.0.

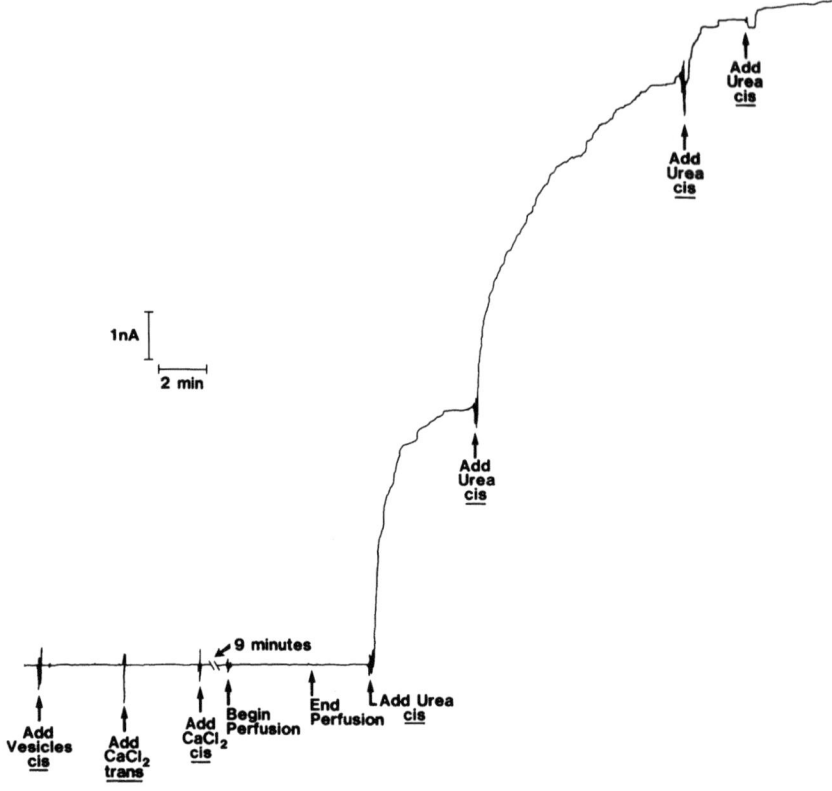

Figure 3. Ca^{2+} promotes the formation of the prefusion state. Phospholipid vesicles (same as Fig. 2) were added to the *cis* side of an asolectin membrane voltage-clamped at 20 mV. 15 mM $CaCl_2$ was added *trans* and *cis*. Ten minutes later the *cis* side was perfused with five times its volume of a Ca^{2+}-free solution (100 mM NaCl, 10 mM MES, 1 mM EDTA, pH 6.0). Adding 400 mM urea *cis* resulted in a burst of fusion. An additional 400 mM urea *cis* caused another burst of fusion. A third addition of 400 mM urea to the *cis* side promoted a small round of fusion. Thereafter, addition of urea *cis* did not stimulate fusion. The membrane was initially bathed in 100 mM NaCl, 10 mM MES, 3 mM $CaCl_2$, 0.1 mM EDTA, pH 6.0.

the planar membrane. Different populations of vesicles associated with the planar membrane apparently exist, since increasing osmotic gradients produce successive bursts of fusion.

If the above experiment is performed without addition of $CaCl_2$ to the *cis* compartment, no fusion occurs when osmotic gradients are established. Thus, divalent cations are required for the formation of the pre-fusion state. On the other hand, if the pre-fusion state is first formed by adding vesicles and $CaCl_2$, and the vesicles in solution are then removed by perfusing the *cis* side with a divalent cation-free 1 mM EDTA buffer, the addition of urea to the *cis* side still results in some fusion. Thus, whereas divalent cations are necessary to form the pre-fusion state, they are not necessary for maintaining some of the vesicles in this long-lived state.

Because establishing an osmotic gradient across the planar membrane results in vesicular swelling, the vesicles are presumably within a few angstroms of the planar membrane in the pre-fusion state. If they were not, then almost all of the water flowing into the aqueous corridor between the vesicle and the planar membrane would be shunted around the vesicle (rather than flow into the vesicle), and no fusion would occur [1].

For experiments of the type illustrated in Fig. 2, where vesicles, $CaCl_2$, and an osmoticant are added to the *cis* side of a planar membrane, the fusion rate decreases with time and eventually ceases (usually within 1 hr). After fusion has ceased, if the vesicles in solution are removed via perfusion and new vesicles are now added, no new fusion is observed. (Control experiments show that planar membranes bathed in $CaCl_2$ and osmoticants for 90 min without vesicles will still support fusion when vesicles are added.) Relatedly, planar membranes bathed with vesicles and $CaCl_2$ for 2 hr in the absence of an osmotic gradient will manifest a single large burst of fusion, when an osmotic gradient is created. Both of the above phenomena are explained by the presence of vesicles which have adsorbed to but not fused with the planar membrane, thereby occluding the planar membrane so that "fusible" vesicles cannot approach. The reduction in free area of planar membrane accounts for the decrease in the rate of fusion with time and in its eventual cessation.

Possible Biological Relevance

As already stated, one of the reasons for studying membrane fusion in artificial systems is to determine which forces might be involved in the biological process. In using the model system, our first goal is to achieve fusion and then to describe the observed process in physicochemical terms. What is conceptually most important is that a physicochemical force can lead to fusion in the model system. The concentration of Ca^{2+} used in the system is of secondary concern. We are studying a simplified system, not a reconstituted synapse. Proteins which lead to exocytotic release, and are activated by submicromolar concentrations of Ca^{2+} in an intact cell, are absent in the model system.

There are two steps in the fusion process of the model system that have been described. The first is the close association of vesicles with the planar membrane, and the second is the swelling of vesicles leading to fusion. Ca^{2+} affects only the first step in this system. In exocytosis there are two analogous steps: the close apposition of vesicles with plasma membrane, and the actual fusion process. Calcium could affect either or both of these steps.

The postulate that Ca^{2+} leads to a closer apposition of vesicle and plasma membrane in analogy with the pre-fusion state has previously been advanced [5,6]. That it is possible for micromolar concentrations of Ca^{2+} to promote the pre-fusion state in the model system is suggested by the observation that 10 μM Ca^{2+} plus appropriate osmotic gradients produced fusion when a calcium-binding protein was incorporated into the planar membrane [13]. Mg^{2+} did not substitute for Ca^{2+}, just as it does not replace Ca^{2+} in the secretory process. Other mechanisms such as activation of contractile elements by a rise in intracellular Ca^{2+} might be responsible for bringing vesicles to fusion sites on the plasma membrane in the intact cell [2,7].

Our findings with the model system clearly show that vesicular swelling can lead to fusion. An obvious implication of the results is that an elevation in intracellular ionic Ca^{2+} triggers fusion by causing vesicular swelling. This could be accomplished in a number of ways. Calcium could alter ion permeability by opening Ca^{2+}-activated channels, resulting in an influx of ions into the vesicle with water following; alternatively, calcium could modulate pump activity or activate osmotically inactive vesicular constituents. On the other hand, synaptic vesicles could be sustaining a hydrostatic pressure. When the vesicles and plasma membrane come into intimate contact as a result of Ca^{2+} stimulation, vesicles burst and fuse because the radius of curvature in the region of contact is greater than elsewhere, and thus these vesicles cannot sustain the preexisting hydrostatic pressure.

These projected mechanisms for osmotically induced fusion mechanisms are only illustrative and admittedly highly speculative. But the importance of vesicular swelling in promoting fusion in this model system should not be ignored. While the relevance of osmotic forces in biological fusion and exocytosis remains to be tested directly, osmotic forces have been implicated in several exocytotic systems [1,3]. Osmotic forces are well regulated from the simplest cells to the most complex organisms. In view of the importance of osmotic swelling of vesicles for fusion in this simple model system, it is quite possible that osmoregulation was employed for fusion between membranes early in evolution.

ACKNOWLEDGMENTS. The work reported in this chapter was supported by NIH Grants GM-27367-03, GM-29210-06, and 5T32-GM-7288.

References

1. Akabas, M. H.; Cohen, F. S.; Finkelstein, A. Separation of the osmotically driven fusion event from vesicle-planar membrane attachment in a model system for exocytosis. *J. Cell Biol.* **98**:1063–1071, 1984.
2. Berl, S.; Psuszkin, S.; Nicklas, W. J. Actomyosin-like protein in brain. *Science* **179**:441–446, 1973.
3. Cohen, F. S.; Zimmerberg, J.; Finkelstein, A. Fusion of phospholipid vesicles with planar phospholipid bilayer membranes. II. Incorporation of a vesicular membrane marker into the planar membrane. *J. Gen. Physiol.* **75**:251–270, 1980.
4. Cohen, F. S.; Akabas, M. H.; Finkelstein, A. Osmotic swelling of phospholipid vesicles causes them to fuse with a planar phospholipid bilayer membrane. *Science* **217**:458–460, 1982.
5. Creutz, C. E.; Pazoles, C. J.; Pollard, H. B. Identification and purification of an adrenal medullary protein (synexin) that causes calcium-dependent aggregation of isolated chromaffin granules. *J. Biol. Chem.* **253**:2858–2866, 1978.
6. Dean, P. M. Exocytosis modelling: An electrostatic function for calcium in stimulus–secretion coupling. *J. Theor. Biol.* **54**:289–308, 1975.
7. Howell, S. L.; Tyhurst, M. Microtubules, microfilaments, and insulin secretion. *Diabetologia* **22**:301–308, 1982.
8. Kendall, D. A.; MacDonald, R. C. A fluorescence assay to monitor vesicle fusion and lysis. *J. Biol. Chem.* **257**:13892–13895, 1982.
9. Miller, C.; Arvan, P.; Telford, J. N.; Racker, E. Ca^{2+}-induced fusion of proteoliposomes: Dependence on transmembrane osmotic gradients. *J. Membr. Biol.* **30**:271–282, 1976.
10. Schindler, H.; Rosenbusch, J. P. Matrix protein from *Escherichia coli* outer membranes forms voltage-controlled channels in lipid bilayers. *Proc. Natl. Acad. Sci. USA* **75**:3751–3755, 1978.
11. Wilschut, J.; Papahadjopolous, D. Ca^{2+}-induced fusion of phospholipid vesicles monitored by mixing of aqueous contents. *Nature (London)* **281**:690–692, 1979.
12. Zimmerberg, J.; Cohen, F. S.; Finkelstein, A. Fusion of phospholipid vesicles with planar phospholipid bilayer membranes. I. Discharge of vesicular contents across the planar membrane. *J. Gen. Physiol.* **75**:241–250, 1980.
13. Zimmerberg, J.; Cohen, F. S.; Finkelstein, A. Micromolar Ca^{2+} stimulates fusion of lipid vesicles with planar bilayers containing a calcium-binding protein. *Science* **210**:906–908, 1980.

IV

Calcium Regulation of Hemopoietic Cells

15

Activation and Desensitization of Receptors for IgE on Tumor Basophils

Clare Fewtrell

The aggregation of receptors for IgE on the surface of mast cells and basophils leads to secretion of a variety of mediators of immediate hypersensitivity. Many of these mediators, including histamine and serotonin, are stored in granules within the cytoplasm, while others such as arachidonic acid and its metabolites (e.g., prostaglandins and leukotrienes) are newly formed when cells are stimulated to secrete. In common with other secretory systems, mast cell granule exocytosis is presumed to occur as a result of a rise in cytoplasmic free Ca^{2+} and is also dependent on a source of metabolic energy. The mechanism of IgE receptor activation and the events leading to exocytosis are beginning to be unraveled [1] and we now know a considerable amount about the molecular structure of the first component in the sequence, namely the receptor for IgE [2].

Tumor Basophils

An important breakthrough in the study of the molecular components involved in stimulus–secretion coupling in mast cells and basophils was the discovery of a basophilic leukemia [3]. These tumor basophils can be grown in culture or as a solid tumor in rats [4,5], and this model provided sufficient material for isolation and detailed molecular characterization of the receptor for IgE [2]. Furthermore, a particularly responsive secreting subline (2H3) of these basophilic leukemia (RBL) cells has been cloned [6]. The development of these models has opened the way for the study of the entire sequence of events leading from receptor activation to the final secretory response of the cells and for the isolation and characterization of the molecular components involved.

While the fundamental mechanisms involved in the initiation of secretion from mast

---

Clare Fewtrell • Department of Pharmacology, New York State College of Veterinary Medicine, Cornell University, Ithaca, New York 14853, and Section on Chemical Immunology, Arthritis and Rheumatism Branch, National Institute of Arthritis, Diabetes, Digestive and Kidney Diseases, National Institutes of Health, Bethesda, Maryland 20205.

cells and basophils are probably the same, a number of differences exist between the different cell types. It was therefore imperative to characterize the secretory response of the tumor basophils in some detail and to compare it with that seen with normal mast cells and basophils. These studies, which are summarized in Ref. 7, reveal that while certain differences do indeed exist, secretion from the tumor basophils fundamentally resembles that from normal mast cells and basophils. We have continued to study the properties of secretion from RBL cells and have examined the basis for some of the apparent differences observed between different basophil and mast cell types. In addition, a number of the unique properties of the tumor cells have been exploited in order to learn more about the mechanism of IgE receptor-mediated stimulus–secretion coupling.

Activation by Multivalent Antigens

When mast cells or basophils are stimulated with increasing concentrations of antigen or anti-IgE, secretion increases and ultimately reaches a peak or a plateau [8,9]. At higher concentrations of stimulating ligand, secretion declines, often returning to baseline levels [8,9]. RBL cells respond similarly when stimulated with polyclonal anti-IgE; however, with an antigen such as bovine serum albumin (BSA) to which dinitrophenyl groups have been covalently attached (DNP antigen), secretion reaches a plateau which is often maintained over four orders of magnitude of antigen concentration (Fig. 1, curve a). The apparent failure to obtain a high-dose inhibition is due to the number of receptors that have been sensitized with IgE. In curve a (Fig. 1), 100% of the receptors are occupied with IgE. However, as the number of sensitized receptors is reduced, the shape of the dose–response curves changes dramatically. When 10% of the receptors are occupied with IgE, the maximal response is virtually unaltered, but at high antigen concentrations there is a very marked reduction in the secretory response (Fig. 1, curve b). When the surface IgE is reduced still further (Fig. 1, curve c), secretion at all concentrations of antigen is reduced.

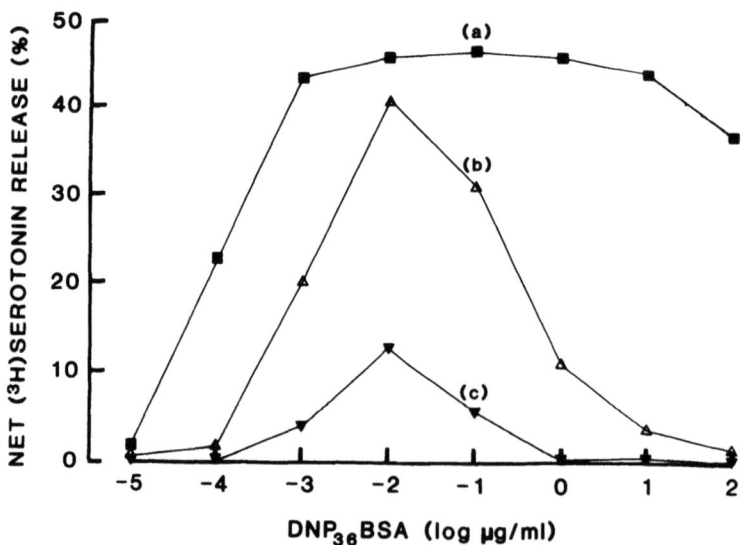

Figure 1. Effect of IgE receptor number on the dose–response curve to the multivalent antigen $DNP_{36}BSA$. Tumor basophils were sensitized such that 100% (curve a), 10% (curve b), or 3% (curve c) of their receptors were occupied with mouse IgE anti-DNP. The cell concentration in this and other experiments was 2×10^6 cells/ml.

These results are qualitatively similar to those predicted for secretion induced by a bivalent hapten [10]. However, the curves should be symmetrical and have their maxima at the same concentration of antigen [10]. This was clearly not the case when a multivalent antigen was used to trigger secretion from tumor basophils (Fig. 1). The ascending part of the curve is always steeper than the descending phase. This is most apparent in curve b, and is a regular feature of similar experiments. Furthermore, the inflection point of the curves is gradually shifted to the left as the IgE receptor occupancy is decreased (Fig. 1). There may, however, be a trivial explanation for the skewed nature of these curves; at low antigen concentrations there may be a significant depletion of free antigen due to binding to cells. So the free antigen concentration will therefore be considerably lower than the added concentration, which is plotted on the abscissa. Thus, the ascending part of the curve will be too steep, since it will be shifted to the right at low concentrations of antigen. Binding studies with labeled antigen are necessary to completely exclude this possibility altogether; however, experiments using 10- and 100-fold lower cell concentrations gave qualitatively similar results, suggesting some alternative explanation

One intriguing possibility is that the asymmetry of the curves is related to the magnitude of the unit signals generated by receptor aggregates of different sizes. It is clear from studies utilizing chemically cross-linked oligomers of IgE that the formation of receptor dimers is a sufficient unit signal for the initiation of secretion [11–13]. However, existing evidence suggests that larger receptor aggregates generate a qualitatively different and more effective unit signal [12,14]. At low concentrations of multivalent antigen, large receptor aggregates may readily form, and this process may not be markedly affected by the number of receptors occupied by IgE. However, at high concentrations of antigen, when binding is limited by the availability of antigen-combining sites on IgE, monomeric and dimeric binding of antigen will predominate, and this may be more sensitive to variations in the density of cell surface IgE. Since larger receptor aggregates are more efficient at inducing secretion from RBL cells [12], this postulate would explain the asymmetry of the dose–response curves in Fig. 1. Further studies are necessary to determine the validity of this hypothesis.

High-Dose Inhibition

There are two potential causes of the inhibition of secretion seen with high concentrations of antigen (Fig. 1). The first is due to excess cross-linking and is typically observed with stimulating ligands such as concanavalin A and anti-IgE [8]. These ligands bind to multiple sites on the IgE molecule to generate highly cross-linked IgE–receptor aggregates on the cell surface that are apparently unable to trigger secretion. The second type of high-dose inhibition is due to a reduction in cross-links and is typically seen with bivalent hapten antigens [10]. As the concentration of bivalent hapten increases, unoccupied antigen-combining sites on IgE will become extremely scarce, causing an increasing number of bivalent hapten molecules to bind monomerically.

It is not easy to predict which type of high-dose inhibition will occur with a multivalent hapten antigen. However, the two types of inhibition can be distinguished by using a monovalent ligand to compete with the multivalent ligand and reduce the extent of cross-linking. When excess cross-linking is the cause, addition of a monovalent ligand will reverse the inhibitory effect and restore the secretory response [8]. However, when high-dose inhibition is due to a reduction in cross-links, the pattern depicted in Fig. 2 will be seen. The monovalent hapten competes with the bi- or multivalent hapten antigen on the ascending part of the curve to reduce the response. At higher antigen concentrations, where a significant

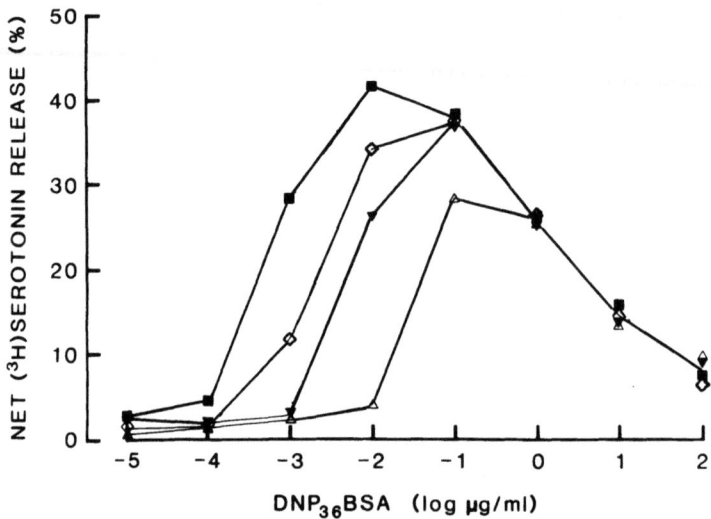

Figure 2. Effect of the monovalent hapten DNP-lysine on the dose–response curve to the multivalent antigen DNP$_{36}$BSA. Tumor basophils were sensitized such that 15% of their receptors were occupied with mouse IgE anti-DNP. ■, control (no DNP-lysine); ◊, 3×10^{-8} M; ▼, 10^{-7} M; △, 3×10^{-7} M DNP-lysine.

amount of monomeric binding of antigen is already occurring, the monovalent hapten will have little or no additional inhibitory effect. The net result will therefore be to reduce the maximal response and to increase the antigen concentration required to achieve it [10]. Thus, Fig. 2 clearly demonstrates that the high-dose inhibition seen with a multivalent DNP antigen having an average of 36 DNP groups, is due to a reduction in cross-links. These results are consistent with data obtained with rabbit basophils using a similar multivalent antigen [15].

The difference between high-dose inhibition induced by anti-IgE and by DNP antigen is confirmed by the experiment shown in Table I. Tumor basophils were sensitized such that 93% of the receptors bound rat IgE and 7% bound mouse IgE anti-DNP. The cells were then exposed to a supramaximal concentration of anti-mouse IgE or DNP antigen, neither of which induced a significant amount of secretion (Table I, first column). Cells exposed to

Table I. Ability of Antigen-Specific IgE–Receptor Complexes on Tumor Basophils[a] to Respond to a Second Stimulus after Exposure to a Supramaximal Concentration of Stimulating Ligand for 90 min

Supramaximal primary stimulus	Optimal secondary stimulus			
	None	DNP antigen (0.1 μg/ml)	Anti-mouse IgE (1 μg/ml)	Anti-rat IgE (10 μg/ml)
None	13.3[b]	38.8	31.2	34.2
DNP antigen[c] (100 μg/ml)	12.5	—	28.6	39.1
Anti-mouse IgE (30 μg/ml)	19.6	16.3	—	33.9

[a] Cells were sensitized such that 7% of the receptors were occupied with mouse IgE anti-DNP and the remaining 93% with rat myeloma IgE.
[b] % release of incorporated [^{3}H]serotonin.
[c] Bovine γ-globulin to which an average of 15 dinitrophenyl groups had been coupled.

excess DNP antigen were subsequently able to secrete normally in response to an optimal concentration of anti-mouse IgE. However, cells previously exposed to excess anti-mouse IgE failed to respond to a normally optimal concentration of DNP antigen. The last column in Table I shows that cells respond well when receptors occupied with rat IgE are stimulated with anti-rat IgE. These data demonstrate that cells were not nonspecifically desensitized after exposure to a supramaximal concentration of anti-mouse IgE.

Maintenance of Receptor Cross-Links

For secretion to occur in human basophils, IgE receptors must remain cross-linked. If antigen-induced receptor aggregates are disrupted with a monovalent hapten, secretion ceases immediately [16]. However, in view of the relatively rapid rate of secretion and desensitization seen with these cells, generation of a putative intermediate in the activation process might be overlooked. Since both secretion and desensitization (see below) occur much more slowly in RBL cells, one might be able to detect a more stable intermediate step in the activation process that did not require the maintenance of cross-linked receptors. However, receptor aggregates are clearly necessary for continued secretion from RBL cells (Fig. 3). Addition of the monovalent hapten DNP-lysine to cells at various times after stimulation completely halts secretion, as does addition of the calcium chelator EGTA (Fig. 3a). Since washing away unbound antigen had no effect on the subsequent response of the cells (Fig. 3b), the monovalent hapten is exerting its effect by disrupting preformed receptor aggregates and not simply by preventing binding of additional multivalent antigen.

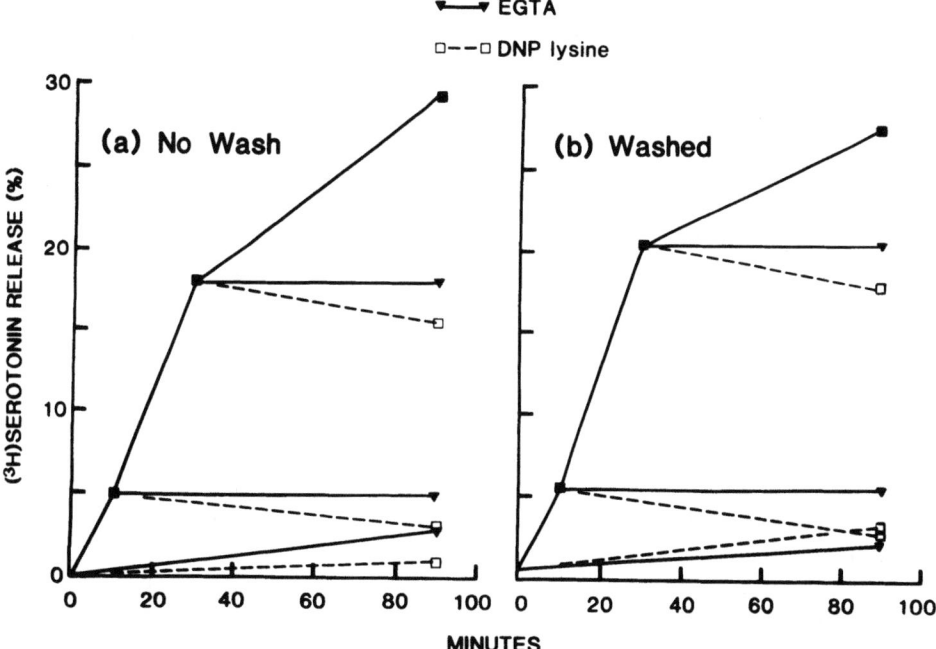

Figure 3. Secretion from tumor basophils at 37°C is halted immediately when receptor cross-links are broken with 10^{-5} M DNP-lysine (□---□) or when extracellular calcium is chelated with 5 mM EGTA (▼—▼). In the control curve (■—■), 5 mM EGTA was added and the cells were cooled to 0°C at the indicated times. Cells were preincubated with 1 μg/ml DNP antigen (DNP$_{15}$ bovine γ-globulin) for 30 min at 0°C before warming to 37°C (zero time). (b) As in (a) except that cells were washed to remove unbound DNP antigen before warming to 37°C.

Similar experiments in which cells were preincubated with multivalent antigen at 37°C in the absence of Ca^{2+} and then washed and incubated with calcium, gave essentially the same results. It was not possible to achieve a state of cell activation that did not require the maintenance of receptor cross-links for continued secretion.

Desensitization

When receptors for IgE on normal rat mast cells are aggregated in the absence of extracellular calcium, they rapidly lose their ability to secrete when calcium is subsequently restored [17]. The rate at which this desensitization occurs varies, depending on the stimulus used [18]. When cells are exposed to an antigen such as ovalbumin, which will only aggregate receptors bearing IgE molecules that recognize ovalbumin, the rate of desensitization is relatively rapid ($t_{1/2} = \sim 1$ min). However, using stimuli such as Concanavalin A or anti-IgE, which will aggregate all IgE molecules regardless of their antigenic specificity, the rate of desensitization may be considerably slower ($t_{1/2} = \sim 10$ min). Desensitization in human basophils has been analyzed in considerable detail by Lichtenstein and his colleagues [16,19]. The rate of desensitization is generally slower ($t_{1/2} = \sim 20$ min) than that observed in mast cells; however, it can also vary significantly depending on the cell donor, and the nature and concentration of the antigenic stimulus.

Preliminary experiments suggested that the phenomenon of desensitization may not occur in tumor basophils [20]. Even when calcium was added back 90 min after cells had been stimulated, the secretory response to the readdition of calcium was almost as large as that obtained when calcium was added together with the primary stimulus [21]. However, these experiments were carried out with cells in which all receptors were occupied with IgE. It is clear from studies with human basophils [22] and RBL cells [21, and Fig. 1] that a maximal secretory response can be elicited even when only a small fraction of the receptors (>10% in Fig. 1) is occupied with IgE. It is therefore possible that desensitization is taking place, but that a sufficient number of activated receptors remains for maximal secretion to occur. The effect of reducing the number of receptors occupied with IgE on the rate of desensitization was therefore examined. When the number of sensitized receptors drops below that required for a maximal secretory response, desensitization does indeed occur, and its rate increases as the receptor number is lowered further (Fig. 4).

Two types of desensitization have been described in human basophils: "specific," in which cells are desensitized to one antigen, but can still respond to anti-IgE or another antigen, and "nonspecific," in which exposure to one antigen renders the cells unresponsive to both anti-IgE and other antigens [16]. The type of desensitization observed is dependent on the number of antigen-specific IgE molecules involved. When this number is low, desensitization is specific; but as the number of antigen-specific IgE molecules on the cell surface is increased, a gradual transition to nonspecific desensitization occurs [19]. Both specific and nonspecific desensitization are also seen with rat mast cells (Fewtrell and Gomperts, unpublished).

Since we were only able to observe significant desensitization in tumor basophils when the number of DNP-specific IgE molecules was small (3% or less; see Fig. 4), desensitization should be specific; this was indeed found to be the case. Cells were sensitized with a mixture of rat and DNP-specific mouse IgE and then desensitized to DNP antigen (see Fig. 4). The secretion obtained when cells were subsequently stimulated with anti-rat IgE in the presence of calcium was as large as that obtained with cells not exposed to DNP antigen, thus clearly demonstrating that desensitization was specific for DNP antigen.

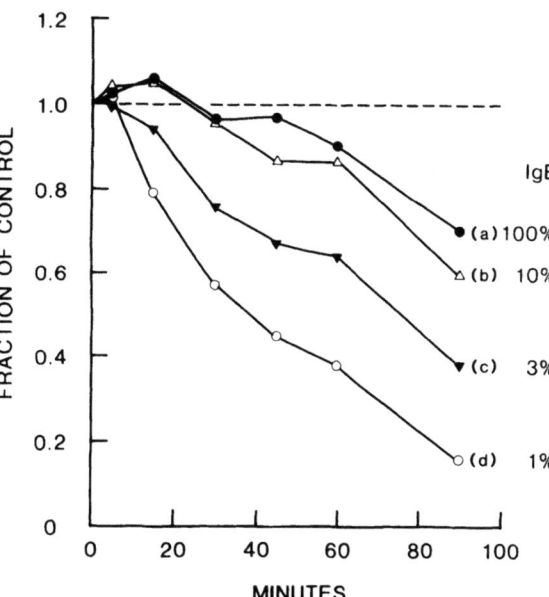

Figure 4. Effects of IgE receptor number on the rate of desensitization of tumor basophils to DNP antigen (0.01 μg/ml DNP$_{15}$ bovine γ-globulin). The % values on the right of the figure indicate the proportion of receptors occupied with mouse IgE anti-DNP in each case. Results are expressed as a fraction of the [^{3}H]serotonin secretion obtained when calcium and DNP antigen were added to the cells simultaneously (zero time). These values were (a) 36%, (b) 35%, (c) 27%, and (d) 18%. In all cases cells were deprived of calcium for approximately 2 hr and then incubated with calcium for another hour to allow secretion to occur. DNP antigen was added at the indicated times prior to the addition of calcium to the cells.

The molecular basis for desensitization is still poorly understood, and more than one mechanism may be involved. Since secretion is instantly halted when receptor aggregates are disrupted [16, and Fig. 3], it is clear that internalized receptor aggregates (which would be insensitive to disruption) are unable to induce secretion. The slow desensitization seen with tumor basophils may be due to removal of activated receptors from the cell surface, since internalization occurs rapidly after cell stimulation [23]. More detailed comparative studies of the kinetics of the two processes will be required to confirm or exclude this possibility.

In conclusion, a considerable amount is now known about the receptor for IgE and the manner in which it is activated. However, the molecular details of receptor activation and the events leading to exocytosis are still poorly understood. In particular, the way(s) in which cytoplasmic Ca^{2+} is elevated in mast cells and basophils, and the potential involvement of other pathways such as phosphatidylinositol turnover [24] remain a mystery. Studies with tumor basophils promise to be invaluable in elucidating these mechanisms.

ACKNOWLEDGMENT. I would like to thank Dr. Henry Metzger for his advice, encouragement, and many stimulating discussions during the course of this work.

References

1. Metzger, H.; Ishizaka, T. Symposium: Transmembrane signaling by receptor aggregation: The mast cell receptor for IgE as a case study. *Fed. Proc.* **41**:7–34, 1982.
2. Metzger, H. The receptor on mast cells and related cells with high affinity for IgE. *Contemp. Top. Mol. Immunol.* **9**:115–145, 1983.
3. Eccleston, E.; Leonard, B. J.; Lowe, J.; Welford, H. Basophilic leukaemia in the albino rat and a demonstration of the basopoietin. *Nature New Biol.* **244**:73–76, 1973.
4. Kulczycki, A., Jr.; Isersky, C.; Metzger, H. The interaction of IgE with rat basophilic leukemia cells. I. Evidence for specific binding of IgE. *J. Exp. Med.* **139**:600–616, 1974.
5. Isersky, C.; Metzger, H.; Buell, D. N. Cell cycle-associated changes in receptors for IgE during growth and differentiation of a rat basophilic leukemia cell line. *J. Exp. Med.* **141**:1147–1162, 1975.
6. Barsumian, E. L.; Isersky, C.; Petrino, M. G.; Siraganian, R. P. IgE-induced histamine release from rat

basophilic leukemia cell lines—Isolation of releasing and nonreleasing clones. *Eur. J. Immunol.* **11**:317–323, 1981.
7. Fewtrell, C.; Metzger, H. Stimulus–secretion coupling in rat basophilic leukemia cells. In: *Biochemistry of the Acute Allergic Reactions*, E. L. Becker, A. S. Simon, and K. F. Austen, eds., New York, Liss, 1981, pp. 295–314.
8. Magro, A. M.; Alexander, A. Histamine release: In vitro studies of the inhibitory region of the dose–response curve. *J. Immunol.* **112**:1762–1765, 1974.
9. Marone, G.; Kagey-Sobotka, A.; Lichtenstein, L. M. IgE-mediated histamine release from human basophils: Differences between antigen E- and anti-IgE-induced secretion. *Int. Arch. Allergy Appl. Immunol.* **65**:339–348, 1981.
10. Dembo, M.; Goldstein, B.; Sobotka, A. K.; Lichtenstein, L. M. Degranulation of human basophils: Quantitative analysis of histamine release and desensitization due to a bivalent penicilloyl hapten. *J. Immunol.* **123**:1864–1872, 1979.
11. Segal, D. M.; Taurog, J. D.; Metzger, H. Dimeric immunoglobulin-E serves as a unit signal for mast cell degranulation. *Proc. Natl. Acad. Sci. USA* **74**:2993–2997, 1977.
12. Fewtrell, C.; Metzger, H. Larger oligomers of IgE are more effective than dimers in stimulating rat basophilic leukemia cells. *J. Immunol.* **125**:701–710, 1980.
13. Kagey-Sobotka, A.; Dembo, M.; Goldstein, B.; Metzger, H.; Lichtenstein, L. M. Qualitative characteristics of histamine release from human basophils by covalently cross-linked IgE. *J. Immunol.* **127**:2285–2291, 1981.
14. MacGlashan, D. W., Jr.; Schleimer, R. P.; Lichtenstein, L. M. Qualitative differences between dimeric and trimeric stimulation of human basophils. *J. Immunol.* **130**:4–6, 1983.
15. DeLisi, C.; Siraganian, R. P. Receptor cross-linking and histamine release. *J. Immunol.* **122**:2293–2299, 1979.
16. Sobotka, A. K.; Dembo, M.; Goldstein, B.; Lichtenstein, L. M. Antigen-specific desensitization of human basophils. *J. Immunol.* **122**:511–517, 1979.
17. Foreman, J. CL; Garland, L. G. Desensitization in the process of histamine secretion induced by antigen and dextran. *J. Physiol. (London)* **239**:381–391, 1974.
18. Gomperts, B. D.; Cockroft, S.; Bennett, J. P.; Fewtrell, C. M. S. Early events in the activation of Ca^{2+} dependent secretion: Studies with rat peritoneal mast cells. *J. Physiol. (Paris)* **76**:383–393, 1980.
19. MacGlashan, D. W., Jr.; Lichtenstein, L. M. The transition from specific to nonspecific desensitization in human basophils. *J. Immunol.* **127**:2410–2414, 1981.
20. Taurog, J. D.; Mendoza, G. R.; Hook, W. A.; Siraganian, R. P.; Metzger, H. Noncytotoxic IgE-mediated release of histamine and serotonin from murine mastocytoma cells. *J. Immunol.* **119**:1757–1761, 1977.
21. Fewtrell, C.; Kessler, A.; Metzger, H. Comparative aspects of secretion from tumor and normal mast cells. In: *Advances in Inflammation Research*, Volume 1, G. Weissmann, B. Samuelsson, and R. Paoletti, eds., New York, Raven Press, 1979, pp. 205–221.
22. Conroy, M. C.; Adkinson, N. F., Jr.; Lichtenstein, L. M. Measurement of IgE on human basophils: Relation to serum IgE and anti-IgE-induced histamine release. *J. Immunol.* **118**:1317–1321, 1977.
23. Isersky, C.; Rivera, J.; Segal, D. M.; Triche, T. The fate of IgE bound to rat basophilic leukemia cells. II. Endocytosis of IgE oligomers and effect on receptor turnover. *J. Immunol.* **131**:388–396, 1983.
24. Cockcroft, S. Phosphatidylinositol metabolism in mast cells and neutrophils. *Cell Calcium* **3**:337–349, 1982.

16

Relationship between Calcium, Arachidonic Acid Metabolites, and Neutrophil Activation

Ramadan I. Sha'afi and Paul H. Naccache

The neutrophils represent the first line of defense against foreign and pathogenic elements, and in so doing must perform several functions. In response to chemotactic factors, they detect, move toward, and accumulate at the site of infection. This phenomenon is termed *chemotaxis*. Once at the site of infection, they ingest or phagocytize the appropriate particles. The phagocytic process is followed by the discharge of the neutrophils' various lysosomes and granules into the phagocytic vacuole thereby promoting degradation and digestion of the engulfed particle or organism. Thus, most neutrophil functions such as motility, cell shape change, phagocytosis, and degranulation that are activated by chemotactic factors depend on the mechanical displacement of part of, or all of, the cell. Mechanochemical coupling is thus an essential component of neutrophil responsiveness. In addition to eliciting various neutrophil functions, chemotactic factors activate several biochemical events, including stimulation of certain membrane-bound enzymes, oxidative metabolism, and actin associated with the cytoskeleton [4,22,30,31,33].

In this chapter available experimental evidence will be summarized and current status of some of the major elements will be discussed regarding neutrophil activation by the synthetic chemotactic peptide formyl-methionyl-leucyl-phenylalanine (f-Met-Leu-Phe). These elements include: (1) membrane receptors, (2) the role of extracellular sodium and potassium, (3) the nature of the second messenger, (4) the role of arachidonic acid metabolites in cell activation, (5) the mechanism for calcium mobilization by lipid mediators, and (6) some of the main questions remaining to be investigated.

Membrane Receptors

Neutrophil activation by f-Met-Leu-Phe is mediated through plasma membrane receptors [4,11,19]. This conclusion is supported by several experimental observations. (1) The

---

Ramadan I. Sha'afi • Department of Physiology, University of Connecticut Health Center, Farmington, Connecticut 06032. *Paul H. Naccache* • Department of Pathology, University of Connecticut Health Center, Farmington, Connecticut 06032.

biological activity of a series of closely related synthetic oligopeptides displays an extreme degree of structural specificity. (2) High- and low-affinity sites for radiolabeled formylated peptides have been detected on whole cells and isolated plasma membranes. (3) These binding sites, and the functions elicited upon their occupation, display the characteristics of down-regulation and receptor internalization. These receptors behave in much the same manner as classical peptide hormone receptors.

Nature of the Involvement of Na^+ and K^+ Ions in Neutrophil Activation

While f-Met-Leu-Phe stimulates Na^+ and K^+ movements in neutrophils [22], removal of Na^+ or K^+ only shifts the chemotaxis and degranulation dose–response curves to the right; and the addition of Na^+ or K^+ ionophores does not activate cells [17,22,24]. Furthermore, amiloride at concentrations which totally inhibit stimulated Na^+ movement only marginally inhibits most of the f-Met-Leu-Phe-stimulated neutrophil responses. One must therefore conclude that the roles of Na^+ and K^+ fluxes in neutrophil activation are not directly and intimately related to the mechanism of signal transduction.

Nature of the Second Messenger in Neutrophil Activation by Chemotactic Factors

Identifying the nature of the second messenger and defining the mechanisms responsible for the transduction process are essential for elucidating the overall mechanism of "excitation–response coupling" in neutrophil activation. The most frequently mentioned second messengers in cell physiology are (1) membrane potential, (2) levels of intracellular cyclic nucleotides, (3) intracellular pH, and (4) levels of intracellular ionized calcium.

Although membrane potential changes elicited by various chemotactic factors have been reported in neutrophils [21], it seems unlikely that these potential changes per se are causally involved in neutrophil activation. Many functional activities can be dissociated from changes in membrane potential under several experimental conditions. First, incubating neutrophils in high K^+, which should depolarize the cell membrane, does not activate or inhibit responsiveness to chemotactic stimuli. Second, neutrophils from chronic granulomatous disease patients lack a membrane potential change even though their chemotactic and secretory responses are normal [21].

Second messenger functions are fulfilled by cAMP in many cells. However, this is probably not the case in neutrophils, since elevation of cAMP levels diminishes cell responsiveness. It is also possible to dissociate the chemotactic factor-induced change in cAMP levels from cell activation by the same stimulus [26]. The f-Met-Leu-Phe-induced rapid and transient increases in intracellular cAMP are most probably related to modulation, and not initiation, of neutrophil responsiveness [15,22,25].

Similarly, f-Met-Leu-Phe-stimulated increase in cell pH is not likely to be the neutrophils' second messenger since amiloride, which totally abolishes the pH increase, only marginally inhibits neutrophil activation by the same chemotactic factor [22,23, unpublished data). Furthermore, stimulus-dependent increases in cell pH can be dissociated in time from cell activation by the same stimulus [22,23].

Regulation of the intracellular concentration of calcium in neutrophils is achieved by pump–leak systems at the plasma membrane and by Ca^{2+} binding to cytoplasmic constituents and plasma membrane. Calcium ions enter the cells by membrane potential-dependent and -independent channels. Only the presence of the latter channels can be demonstrated experimentally in neutrophils. In addition, there are two different energy-dependent mechanisms that control the rate and extent of Ca^{2+} efflux: a specific calcium pump, the presence of which has been demonstrated in the plasma membrane of neutrophils and inside-out membrane vesicles; and a Na^+ influx, Ca^{2+} efflux exchange pump, the presence of which cannot be experimentally demonstrated in neutrophil membranes [28]. In addition to these membrane events, control of cytosol Ca^{2+} may also be dependent on membrane binding, buffering by cytoplasmic constituents, and accumulation into intracellular organelles and/or granules.

The conclusion that calcium ion plays the role of second messenger in neutrophil activation by chemotactic factors [4,22,30,33] is based on the following observations: (1) cell responsiveness is modulated by the concentration of calcium in the suspending medium; (2) introduction of calcium into the cytosol by means other than the first messenger activates cells; (3) inhibitors which antagonize the activity of an intracellular calcium receptor (calmodulin) inhibit cell activation; and (4) receptor occupancy by the first messenger causes a translocation of Ca^{2+} from certain sites (increased influx from the outside medium, release from membranous stores and intracellular organelles) to other sites (calmodulin, Ca^{2+}-sensitive enzymes). This translocation is usually reflected in an increased level of intracellular exchangeable Ca^{2+}. This rise could be accomplished in either of two basic ways: (1) stimulation of the mechanisms responsible for calcium movements into the cytoplasm (in the neutrophils these include uptake from the extracellular medium and translocation from intracellular stores and/or cytoplasmic constituents), or (2) inhibition of the mechanisms of calcium extrusion from the cytoplasm. The available evidence strongly favoring the first possibility includes: (1) demonstration of intracellular calcium redistribution, (2) extracellular calcium dependency of stimulated neutrophil responsiveness, (3) stimulation of ^{45}Ca efflux from preloaded cells, (4) lack of inhibition of the Ca^{2+}-ATPase by f-Met-Leu-Phe, and (5) rapid decrease in the steady-state level of cell-associated ^{45}Ca when the cells are stimulated in the absence of extracellular calcium.

The increases in the level of intracellular free calcium due to the chemotactic factor have been deduced indirectly by using radioactive tracer and other indirect techniques. Recently, however, an increase in intracellular ionic calcium in rabbit and human neutrophils stimulated by chemotactic factors has been directly demonstrated using the calcium-sensitive fluorescent probe quin-2 [27,32]. The increase comes mainly from intracellular stores and not from the extracellular medium. It was also possible to investigate the relationship between intracellular Ca^{2+} and cAMP in human neutrophils and platelets by using this probe. Addition of agents that stimulate the cellular level of cAMP reduces the intracellular level of free Ca^{2+} (unpublished data).

The issue of stimulated calcium translocation from cytoplasmic stores to calmodulin and/or Ca^{2+}-sensitive enzymes raises another point of experimental significance. In view of the general deleterious metabolic effects of calcium, it is highly likely that the increase in calcium would be localized to the immediate vicinity of desired calcium-binding sites. Therefore, at physiologically relevant concentrations of stimuli, translocated calcium is likely to be absorbed locally and would not diffuse to the cytoplasm proper. Because of this contingency and the competition for calcium between calcium-sensitive probes and intracellular binding sites, the probes located in the cytoplasm, dependent on their affinity for calcium relative to other binding sites, may not detect physiologically relevant calcium, but only excess calcium mobilized by higher levels of stimulation. A corollary of this is that a failure to detect experimentally an increase in cell ionized Ca^{2+} following stimulation does not imply that Ca^{2+} is not involved in cell activation. Similarly, the commonly observed stimulation of superoxide production *in vitro* by high concentrations of chemotactic factors may be due to calcium mobilization in excess of the binding capacity of the physiologically relevant binding sites. The findings that superoxide production occurs only at high concentrations of the stimulus and is inhibited by the removal of Ca^{2+} from the suspension medium, are consistent with this view.

In discussing the nature of the second messenger in neutrophil activation, we have recently dissociated cell activation from a rise in intracellular free calcium as measured by quin-2 fluorescence. Phorbol myristate acetate (PMA) (0.1 μM) causes neutrophil aggregation and degranulation, increased oxygen consumption, and actin association with the cytoskeleton without detectable increases in intracellular free calcium (unpublished data). In view of the high affinity of quin-2 [27] for calcium and of the high intracellular concentration of quin-2 [27] achieved during loading, the inability to detect an increase in fluorescence following stimulation may indeed reflect a lack of calcium mobilization by low concentrations of PMA. This may suggest the presence of an additional second messenger. Alternatively, PMA may activate at a point distal to calcium mobilization or increase intracellular cGMP levels. PMA activates protein kinase C, and it is also known that diacylglycerol formed by the hydrolysis of phosphoinositides [phosphatidylinositol (PI), PI 4-phosphate (PIP), and PI 4,5-bisphosphate (PIP_2)] by phospholipase C also activates protein kinase C (for review and original references, see Ref. 9). In view of these findings, diacylglycerol, in addition to calcium, may serve as a second messenger in neutrophil activation. This hypothesis is also supported by our recent findings that occupancy of receptors by f-Met-Leu-Phe leads to a rapid increase in the levels of 1,2-diacylglycerol and monoacylglycerol in rabbit neutrophils (unpublished data).

Role of Arachidonic Acid Metabolites in Neutrophil Activation

Arachidonic acid is the precursor of a number of extremely important molecules such as leukotrienes, prostacyclin, thromboxanes, and prostaglandins. Although the steps involved in the release of arachidonic acid from membrane phospholipids following cell stimulation are complex and not fully understood, one or more of the following pathways are thought to be responsible for releasing arachidonic acid (for review see Refs. 5 and 9). First, arachidonic acid can be cleaved from phosphatidylcholine (PC), PI, and phosphatidylethanolamine (PE) by phospholipase A_2. Second, diacylglycerol lipase can cleave arachidonic acid from diacylglycerol formed by the hydrolysis of PI through phospholipase C. Third, phosphatidic acid-specific phospholipase A_2 can cleave arachidonic acid from phosphatidic acid which is formed by phosphorylation of diacylglycerol.

Arachidonic acid metabolism generates three families of compounds with biological activities. The synthesis of the prostaglandins and the thromboxanes is initiated by the enzyme prostaglandin endoperoxide-synthetase, and that of the leukotrienes (leukotrienes B_4, C_4, D_4, etc.) by the various lipoxygenase pathways. Arachidonic acid is metabolized by the three pathways in platelets, while in neutrophils, metabolism predominantly takes place by the lipoxygenase pathway. It is reasonable to conclude from available data that mobilization and metabolism of arachidonic acid are involved in neutrophil activation, and that this involvement is mediated through the action of certain arachidonic acid metabolites on calcium homeostasis. This conclusion is based on several experimental findings [22,9]. First, the biological activities of exogenously added arachidonic acid and some of its metabolites have been demonstrated. Second, the responsiveness of neutrophil activation by chemotactic factors can be modulated by certain lipase and lipoxygenase inhibitors. Third, chemotactic factors cause the release of arachidonic acid previously incorporated in phospholipids and stimulate its subsequent metabolism [7,13,16,20]. Fourth, leukotriene B_4 mimics the effects of chemotactic factors on calcium homeostasis, i.e., intracellular Ca^{2+} redistribution and net influx.

An important question which arises at this point concerns the physiological roles of lipid mediators such as leukotrienes. Are they intercellular first messengers designed to amplify the activation or recruitment of nearby cells? Or are they intracellular mediators that express the effects of receptor activators? Existing experimental results favor the first alternative. That is, the role of arachidonic acid pathway is modulatory in nature and is not an intracellular mediator of the action of the first messenger, i.e., it is not necessary for cell activation. This view is based on three main experimental findings. First and most importantly, leukotrienes and thromboxane can escape from cells of origin and stimulate other tissues. A compound produced by cells for an essential intracellular role would be conserved by these cells and not be rapidly lost to the external environment. Second, specific binding sites for radiolabeled leukotriene B_4 have been detected on neutrophils. Third, preincubation of cells with high concentrations of arachidonic acid or leukotriene B_4, followed by washing, inhibits subsequent responses of neutrophils to arachidonic acid and leukotriene B_4, but not to f-Met-Leu-Phe. On the other hand, preincubation with high concentrations of f-Met-Leu-Phe inhibits the subsequent responses to all three stimuli. This view, which is discussed in terms of leukotriene B_4 and neutrophil activation, can be extended to other lipid mediators such as thromboxane A_2, prostaglandin endoperoxides, and platelet-activating factors and to other types of cells, including platelets (see Ref. 9).

Mechanisms for Calcium Mobilization by Lipid Mediators

The proposed mechanisms by which lipid mediators such as leukotrienes increase plasma membrane permeability to Ca^{2+} fall into two main categories. The view that they act as Ca^{2+} ionophores is supported by the finding that they transport Ca^{2+} in model systems (albeit at high concentrations), release Ca^{2+} from membrane fractions, and promote calcium uptake by intact cells. Despite this evidence, it is unlikely that they act as ionophores at physiological concentrations. This conclusion is based on many experimental observations (for review see Ref. 9). First, the effects of these mediators are cell-specific at physiological concentrations. Second, they are extremely potent and their effects are sensitive to relatively minor stereochemical modifications. Third, specific binding sites for several of these mediators (leukotriene B_4, platelet-activating factor, and prostaglandins) have been demonstrated. Fourth, neutrophils become deactivated after exposure to leukotriene B_4. Fifth, depending on the cell type, reactions that ensue after stimulation generate several lipid mediators. If the latter act as ionophores, they would offer significant danger to the cell. The second view is that lipid mediators, as protein hormones, act through membrane receptors and that there is a linkage of receptor-protein with an existing ion conductance channel protein. The generally accepted view regarding lipid mediators and calcium transport, at least in neutrophils, is that at physiological concentrations they express their effects through receptors. On the other hand, at relatively high concentrations ($> 10^{-6}$ M), these mediators, as well as many other unsaturated fatty acids, may behave operationally as calcium ionophores.

Some of the Major Questions Remaining to Be Answered

Although the second messenger role in neutrophils and in many other cells is fulfilled by calcium ion, there are many questions related to this role that remain to be answered. They fall into two main categories. One relates to the mechanism by which calcium becomes available after stimulation, and the other to the biochemical and biophysical changes initiated by the rise in Ca^{2+}. The first category includes the nature of the signal responsible for the increased membrane permeability to calcium following stimulation, the identification of the cellular pools from which calcium is released, and the nature of the signal for calcium release.

The following mechanisms have been advanced to account for stimulus–dependent increase in membrane permeability and/or release of calcium from intracellular stores.

1. It has been proposed that stimulus-induced mobilization of calcium is mediated by phospholipid and/or protein methylation [3]. According to this view, the stimulus induces a conversion of PE, which is usually located in the half of the membrane bilayer facing the cytoplasmic side, into PC, which is located in the half of the bilayer facing the outside [13]. This conversion and redistribution of certain phospholipids may alter membrane properties and thus lead to calcium mobilization. This view is based primarily on the observation that methyltransferase inhibitors which inhibit the methylation reaction partially block f-Met-Leu-Phe-stimulated calcium movement, arachidonic acid release, and chemotaxis in rabbit neutrophils [3]. Although the methylation response may turn out to be an important step in the overall sequence of events in the excitation–response coupling, a role in calcium mobilization is uncertain since the experimental support is correlative in nature and based on studies using inhibitors. In addition, inhibition of methylation reactions is not sufficient to account for impairment of chemotaxis by certain inhibitors of methylation [10]. Furthermore, the methylation response is not associated with all receptors that function through calcium [14].

2. It has been hypothesized that a specific stimulus-induced breakdown of PIP_2 is closely involved in the regulation of calcium gating. This hypothesis holds that the initial lipolytic response to stimulation does not require a rise in the basal level of intracellular free calcium; and PIP_2 breakdown is viewed as directly receptor-linked and ultimately responsible in turn for the mobilization of calcium necessary for cell activation. This proposal is based on the observation that there are a variety of cell surface receptors that, when activated, increase exchangeable cell calcium and PIP_2 metabolism in target cells [5,6,9,18,29].

Certain investigators have even proposed that occupancy of the receptor by an agonist elicits a breakdown of PIP_2 and/or PIP causing calcium release from the phospholipid [6]. Although an attractive hypothesis, based on the low binding constant for calcium to phospholipids, it is unlikely that sufficient calcium can be released from PIP_2 to activate the cell. Furthermore, the amounts of PIP_2 and PIP and the changes brought about by stimulation are extremely small. This does not rule out the possibility that calcium released from phospholipids following stimulation recruits more calcium, which in turn activates cellular processes. The turnover rates of the phosphate but not glycerol moieties in both PIP_2 and PIP under unstimulated conditions are very high in neutrophils, so it is unlikely that small changes in this turnover are used as a signal [29, unpublished data]. If calcium mobilization is indeed mediated by PIP_2 turnover, then it must be through the action on the intracellular calcium stores of products of the P/P_2 breakdown, such as inositol 1,4,5-trisphosphate, and not by calcium release from this phospholipid.

Two experimental findings remain unresolved with respect to the relationship between PIP_2 and calcium mobilization by chemotactic factors. First, PIP_2 turnover requires the presence of high concentrations of the stimulus ($> 10^{-9}$ M f-Met-Leu-Phe in rabbit peritoneal neutrophils), whereas significant calcium mobilization can be detected at much lower concentrations (10^{-11} M f-Met-Leu-Phe). Second, stimulus-dependent PI and PIP_2 turnovers are affected by removal of external calcium [8,20,29], whereas calcium mobilization is only slightly affected by removal of external calcium [32]. Calcium sensitivity of phosphoinositide metabolism has been observed in other systems [1,2,9,12], and the apparent dependence on external calcium may be due to the inhibitory effect of calcium on PI resynthesis [5]. A second likely explanation is that incubation of cells in the absence of calcium (+ EGTA) for 5 min reduces basal levels of intracellular free calcium so that the enzyme polyphosphatidylinositol 4,5-bisphosphate phosphodiesterase cannot be fully activated upon receptor occupancy. This may explain the apparent contradictory experimental findings with respect to calcium sensitivity of phosphoinositide's metabolism, since the final level of intracellular free calcium, following the incubation of cells in calcium-free medium, will depend on membrane permeability to Ca^{2+} and duration of incubation.

3. A linkage between receptor-protein and an existing ion conductance channel protein has also been hypothesized. A receptor-linked ion channel provides very high specificity both with respect to the nature of agonists capable of opening channel "gates" and to sites of action, since only membranes containing receptor-linked channels can be affected. While it is commonly assumed that the "channels" are proteinaceous, they may also be phospholipid in nature. Direct experimental support for this hypothesis awaits the isolation and reconstitution of the receptor in artificial membranes.

In summary, very little information is available concerning not only the identification of pools from which functional calcium is released but also the nature of the release mechanisms. Occupancy of the receptor by the agonist is thought to effect calcium release from membranous sites such as the plasma membrane and intracellular organelles. The major piece of experimental evidence supporting this view is the finding that the fluorescence signal of the cell-associated chelate probe chlorotetracycline decreases upon cell stimulation. The biophysical and biochemical changes triggered by the rise in the cell level of exchangeable calcium are also still a mystery. Since mechanochemical coupling is an essential component of neutrophil activities, it is reasonable to assume that phosphorylation of certain proteins (myosin light chain, actin-binding protein) takes place and actin–myosin interactions are initiated.

ACKNOWLEDGMENT. Supported by NIH Grants AI-13734-05 and AM-31000-01.

References

1. Abdel-Latif, A. A.; Luke, B.; Smith, J. P. Studies on the properties of the soluble phosphatidylinositol-phosphodiesterase of rabbit iris smooth muscle. *Biochim. Biophys. Acta* **614:**425–434, 1980.
2. Akhtar, R. A.; Abdel-Latif, A. A. Studies on the properties of triphosphoinositide phosphodiesterase and phosphodiesterase of rabbit iris smooth muscle. *Biochim. Biophys. Acta* **527:**150–170, 1978.
3. Bareis, D. L.; Hirata, F.; Schiftmann, E.; Axelrod, J. Phospholipid metabolism, calcium flux and the receptor-mediated induction of chemotaxis in rabbit neutrophils. *J. Cell Biol.* **93:**690–697, 1982.
4. Becker, E. L.; Naccache, P. H.; Showell, H. J.; Walenga, R. W. Early events in neutrophil activation: Receptor stimulation, ionic fluxes and arachidonic acid metabolism. In: *Lymphokines*, Volume 4, E. Pick and M. Land, eds., New York, Academic Press, 1981, pp. 194–334.
5. Berridge, N. J. A novel cellular signaling system based on the integration of phospholipid and calcium metabolism. In: *Calcium and Cell Function*, Volume 3, W. Y. Cheung, ed., New York, Academic Press, 1983, pp. 1–36.

6. Billah, M. M.; Lapetina, E. G. Rapid decrease of phosphatidylinositol 4,5-bisphosphate in thrombin-stimulated platelets. *J. Biol. Chem.* **257**:12705-12708, 1982.
7. Bokoch, G.; Reed, P. W. Stimulation of arachidonic metabolism in the polymorphonuclear leukocyte by an N-formylated peptide. *J. Biol. Chem.* **255**:10223-10226, 1980.
8. Cockcroft, S.; Bennett, J. P.; Gomperts, B. D. Stimulus-secretion coupling in rabbit neutrophils is not mediated by phosphatidylinositol breakdown. *Nature (London)* **288**:275-277, 1980.
9. Feinstein, M. B.; Sha'afi, R. I. The role of calcium in arachidonic acid metabolism and in the actions of arachidonic acid-derived metabolites. In: *Calcium and Cell Function*, Volume 4, W. Y. Cheung, ed., New York, Academic Press, 1983, pp. 337-376.
10. Garcia-Castro, I.; Mato, J. M.; Vasanthakumas, G.; Wiesmann, W. P.; Schiffmann, E.; Chiang, P. K. Paradoxical effects of adenosine on neutrophil chemotaxis. *J. Biol. Chem.* **258**:4345-4349, 1983.
11. Goetzl, E. J.; Foster, D. W.; Goldman, D. W. Specific effects on human neutrophils of antibodies to a membrane protein constituent of neutrophil receptors for chemotactic formyl-methionyl peptides. *Immunology* **45**:1-8, 1982.
12. Griffin, H. D.; Hawthorne, J. N. Calcium-activated hydrolysis of phosphatidyl-myo-inositol 4-phosphate and phosphatidyl-myo-inositol 4,5-bis phosphate in guinea pig synaptosomes. *Biochem. J.* **176**:541-552, 1978.
13. Hirata, F.; Corcoran, B. A.; Venkatasubramanian, K.; Schiffmann, E.; Axelrod, J. Chemoattractants stimulate degradation of methylated phospholipids and release of arachidonic acid in rabbit leukocytes. *Proc. Natl. Acad. Sci. USA* **76**:2640-2643, 1979.
14. Hotchkiss, A.; Jordan, J. V.; Hirata, F.; Schulman, N. R.; Axelrod, J. Phospholipid methylation and human platelet function. *Biochem. Pharmacol.* **30**:2089-2095, 1981.
15. Jackowski, S.; Sha'afi, R. I. Response of adenosine cyclic 3',5'-monophosphate levels in rabbit neutrophils to the chemotactic peptide formyl-methionyl-phenylalanine. *Mol. Pharmacol.* **16**:473-481, 1979.
16. Jubiz, W.; Radmark, O.; Malmsten, C.; Hansson, G.; Lindgrent, J. A.; Palmblad, J.; Uden, A.-M.; Samuelsson, B. A novel leukotriene produced by stimulation of leukocytes with formylmethionylleucylphenylalanine. *J. Biol. Chem.* **257**:6106-6110, 1982.
17. Korchak, H. M.; Weissmann, G. Stimulus-response coupling in the human neutrophil: Transmembrane potential and the role of extracellular Na^+. *Biochim. Biophys. Acta* **601**:180-194, 1980.
18. Michell, R. H.; Kirk, C. J. The unknown meaning of receptor-stimulated inositol lipid metabolism. *Trends. Pharmacol. Sci.* **3**:1490-1491, 1982.
19. Niedel, J. E.; Cuatrecasas, P. Formyl peptide chemotactic receptors of leukocytes and macrophages. *Curr. Top. Cell. Regul.* **27**:137-170, 1982.
20. Rubin, R. P.; Sink, L. E.; Freer, R. J. On the relationship between formyl methionyl-leucyl-phenylalanine stimulation of arachidonyl phosphatidylinositol turnover and lysosomal enzyme secretion by rabbit neutrophils. *Mol. Pharmacol.* **19**:31-73, 1981.
21. Seligmann, B. D.; Gallin, J. L. Use of lipophilic probes of membrane potential to assess human neutrophil activation: Abnormality in chronic granulomatous disease. *J. Clin. Invest.* **66**:493-503, 1980.
22. Sha'afi, R. I.; Naccache, P. H. Ionic movements in neutrophil chemotaxis and secretion. In: *Advances in Inflammation Research*, Volume 2, G. Weissmann, ed., New York, Raven Press, 1981, pp. 115-148.
23. Sha'afi, R. I.; Naccache, P. H.; Molski, T. F. P.; Volpi, M. Chemotactic stimuli-induced changes in the pH of rabbit neutrophils. In: *Intracellular pH: Its Measurements, Regulation and Utilization in Cellular Functions*, Kroc Foundation Series, Volume 15, R. Nuccitelli and D. W. Deamer, New York, Liss, 1982, pp. 513-525.
24. Showell, H. J.; Naccache, P. H.; Sha'afi, R. I.; Becker, E. L. The effects of extracellular K^+, Na^+, and Ca^{2+} on lysosomal enzyme secretion from polymorphonuclear leukocytes. *J. Immunol.* **119**:804-811, 1977.
25. Simchowitz, L.; Fischbein, L. L.; Spilberg, I.; Atkinson, J. P. Induction of a transient elevation in intracellular levels of adenosine-3',5'-cyclic monophosphate by chemotactic factors: An early event in human neutrophil activation. *J. Immunol.* **124**:1482-1491, 1980.
26. Simchowitz, L.; Spilberg, I.; Atkinson, J. P. Evidence that functional responses of human neutrophils occur independently of transient elevations in cAMP levels. *67th Annual Meeting of FASEB* **42**:1080, 1983.
27. Tsien, R. Y.; Pozzan, T.; Rink, T. J. Calcium homeostasis in intact lymphocytes: Cytoplasmic free calcium monitored with a new, intracellular trapped fluorescent indicator. *J. Cell Biol.* **94**:325-334, 1982.
28. Volpi, M.; Naccache, P. H.; Sha'afi, R. I. Calcium transport in inside-out membrane vesicles prepared from rabbit neutrophils. *J. Biol. Chem.* **258**:4153-4158, 1983.
29. Volpi, M.; Yassin, R.; Naccache, P. H.; Sha'afi, R. I. Chemotactic factor causes rapid decreases in phosphatidylinositol 4,5 bis-phosphate and phosphatidylinositol 4-monophosphate in rabbit neutrophils. *Biochem. Biophys. Res. Commun.* **112**:957-964, 1983.
30. Weissmann, G.; Smolen, J.; Korchak, H.; Hoffstein, S. The secretory code of the neutrophils. In: *Cellular Interactions*, J. T. Dingle and J. L. Gordon, eds., Amsterdam, Elsevier/North-Holland, 1980, pp. 15-31.

31. White, J. R.; Naccache, P. H.; Sha'afi, R. I. The synthetic chemotactic peptide formyl-methionyl-leucyl-phenylalanine causes an increase in actin associated with the cytoskeleton in rabbit neutrophils. *Biochem. Biophys. Res. Commun.* **108**:1144–1149, 1982.
32. White, J. R.; Naccache, P. H.; Molski, T. F. P.; Borgeat, P.; Sha'afi, R. I. Direct demonstration of increased intracellular concentration of free calcium in rabbit and human neutrophils following stimulation by chemotactic factors. *Biochem. Biophys. Res. Commun.* **113**:44–50, 1983.
33. Wilkinson, D. C. *Chemotaxis and Inflammation*, 2nd ed., Edinburgh, Churchill Livingstone, 1982.

17

Receptor-Controlled Phosphatidate Synthesis during Acid Hydrolase Secretion from Platelets

Holm Holmsen

Thrombin induces a number of biochemical and cellular responses* in platelets. Binding of this ligand to its putative receptors (for review see Ref. 12) produces an immediate decrease in phosphatidylinositol (PI) and PI 4,5-bisphosphate (PIP_2) formation of diacylglycerol (DG), phosphatidate (PA), and free arachidonate (AA) (and its oxygenation products); and profound stimulation of ATP consumption, glycogenolysis, and protein phosphorylation. Concomitantly, platelets change shape, aggregate, form pseudopods, and release constituents from three different types of storage granules by exocytosis. These biochemical and cellular responses are induced by elevation of cytoplasmic Ca^{2+}. Also, thrombin-induced responses are usually accompanied by an increase in the concentration of free cellular Ca^{2+} [17]. Hence, the interaction of thrombin with its receptors causes an enhanced availability of cytoplasmic Ca^{2+}, which subsequently triggers various responses, except formation of DG and PA, and decrease in PIP_2 levels [1,11,18]. DG may act directly as a second messenger so that responses may be triggered without mobilization of Ca^{2+} (see Chapter 18).

These observations suggest that those processes that are not induced by elevation of cytoplasmic Ca^{2+}, i.e., DG and PA formation, PIP_2 reduction, are involved in Ca^{2+} mobilization. Rapid breakdown of PI to DG and phosphorylation of the latter to PA occurs in a vast number of cells in response to agonists. A subsequent, slow conversion of PA to PI also occurs in these cells, thus completing the so-called PI cycle (PI response) (Fig. 1). The two first steps (PI hydrolysis and DG phosphorylation) occur early enough in the stimulus–response coupling sequence to be involved in Ca^{2+} mobilization. The divalent cationophoretic properties of PA [19] make its synthesis a plausible candidate for a calcium-gating process. There is very little free DG, PA, and cytidine diphosphodiacylglycerol

---

*The behavior of whole cells or subcellular fractions that cannot be described at the molecular level are called "cellular responses" (e.g., exocytosis, aggregation, shape change) while those that can be described at the molecular level are called "biochemical responses" (e.g., formation of PA: DG + ATP → PA + ADP; AA liberation: phospholipid + H_2O → lysophospholipid + AA).

Holm Holmsen • Department of Biochemistry, University of Bergen, N-5000 Bergen, Norway.

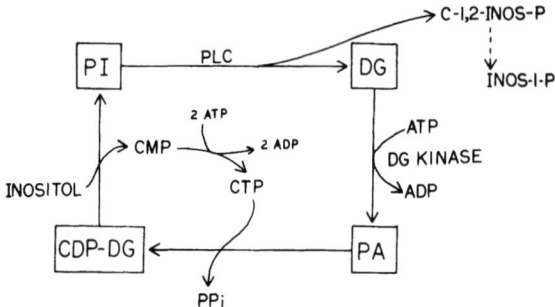

Figure 1. The phosphatidylinositol cycle.

(CDP-DG) in resting cells, while PI exists in considerable amounts on the inner leaflet of the plasma membrane and in intracellular membranes [3]. Perhaps this is the reason that it is generally believed that breakdown of PI to DG and cyclic inositol 1,2-phosphate is the first step in the PI cycle, and hence controlled by the agonist–receptor complex. No direct experimental evidence exists for such receptor-controlled PI hydrolysis, although a large body of circumstantial evidence is available [13,14,16]. Thus, time courses of PI disappearance and cellular responses, as well as dose–response curves for PI disappearance, agonist binding and response have all been shown in several cell systems to support the idea of PI hydrolysis as the receptor-controlled step in the PI cycle. Also, in several systems, agonist-induced PI hydrolysis is independent of cellular calcium, which is a prerequisite for a receptor-controlled reaction that could lead to mobilization of cytoplasmic Ca^{2+}.

On the other hand, there are observations that appear to mitigate against PI hydrolysis as the initial receptor-controlled step. In most cells phospholipase C, which catalyzes this step, is not associated with membranes (for review see Ref. 6), whereas a receptor-controlled enzyme would be expected to be membrane-bound. Furthermore, the Ca^{2+} requirement for phospholipase C activity [6] is difficult to reconcile with a process primarily involved in Ca^{2+} mobilization. Several reports have demonstrated disappearance of PI from intact cells which is dependent on cytoplasmic Ca^{2+} [4,15]; however, the PI level could have been decreased by a mechanism other than diesteratic hydrolysis by phospholipase C, such as phosphorylation and deacylation. Finally, the thrombin-induced loss of PI from platelets is inhibited by ATP deprivation [10], indicating either that PI is removed by energy-requiring processes (phosphorylation) or that activation of phospholipase C is energy-requiring.

We have previously shown that thrombin-induced responses in platelets can be subdivided into two groups: some require only a short period of thrombin–platelet interaction and proceed after thrombin has been removed, while others require the sustained presence of thrombin. This subdivision was based on the effects of hirudin* on thrombin-induced platelet responses. Surprisingly, thrombin-stimulated PA formation,† but not PI disappearance, ceased instantaneously upon rapid removal of thrombin. Acid hydrolase secretion also stopped abruptly (Fig. 2), as did AA production [8]. These responses, therefore, require the sustained presence of thrombin. On the other hand, both dense granule secretion (Fig. 2) and aggregation, as well as PI disappearance and synthesis, continue after removal of thrombin

---

*Hirudin is an antithrombin which rapidly combines with thrombin to form a high-affinity complex [5]. This complex formation abolishes thrombin's proteolytic and platelet-stimulating activity sites on the platelet surface [20].

†PA is measured as total radioactivity of [^{32}P]-PA in platelets prelabeled with $^{32}P_i$ and transferred to a P_i-free medium. [^{32}P]-PA formed has the same specific radioactivity as labeled phosphate groups of ATP, ADP, and P_i [7,9]; thus, total radioactivity of PA is directly proportional to its mass.

17. PA Synthesis during Acid Hydrolase Secretion

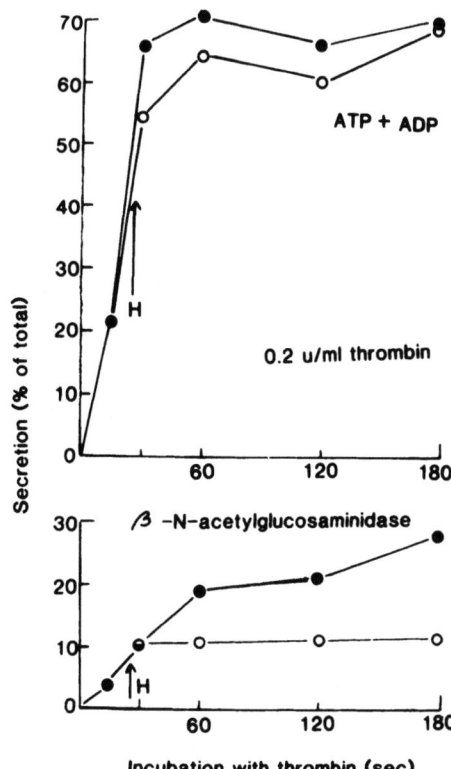

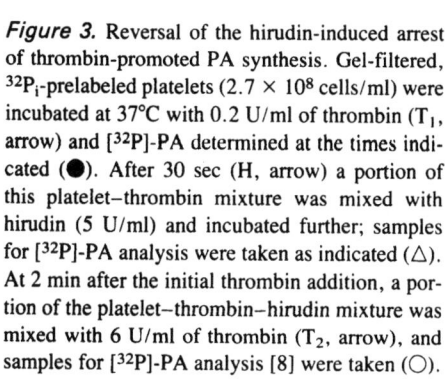

Figure 2. Effect of hirudin on thrombin-induced dense granule and acid hydrolase secretion. Two portions of gel-filtered platelets (3–8 × 10^8 cells/ml) were incubated with 0.2 U/ml of thrombin at 37°C and measurement of secretion of ATP + ADP and β-N-acetylglucosaminidase was done at times indicated [8]. To one portion 10 U/ml of hirudin was added (○) as indicated by the arrows, while no hirudin was added to the other portion (●).

with hirudin [8]. These responses thus require a short interaction between platelets and thrombin and continue unaffected by its subsequent removal.

The abrupt cessation of PA formation (Fig. 3) and acid hydrolase secretion [9] was reversible, since they could be restored by addition of thrombin in amounts exceeding those of hirudin. These effects of sequential addition of thrombin–hirudin–thrombin on PI disap-

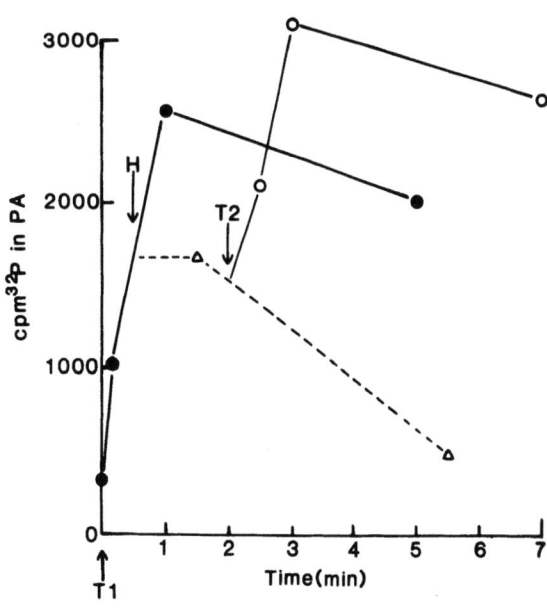

Figure 3. Reversal of the hirudin-induced arrest of thrombin-promoted PA synthesis. Gel-filtered, $^{32}P_i$-prelabeled platelets (2.7 × 10^8 cells/ml) were incubated at 37°C with 0.2 U/ml of thrombin (T$_1$, arrow) and [^{32}P]-PA determined at the times indicated (●). After 30 sec (H, arrow) a portion of this platelet–thrombin mixture was mixed with hirudin (5 U/ml) and incubated further; samples for [^{32}P]-PA analysis were taken as indicated (△). At 2 min after the initial thrombin addition, a portion of the platelet–thrombin–hirudin mixture was mixed with 6 U/ml of thrombin (T$_2$, arrow), and samples for [^{32}P]-PA analysis [8] were taken (○).

pearance were complex [9]. Thrombin always produced an immediate removal of radioactive PI, which had been prelabeled by incubating platelets with [^{3}H]-AA and then transferring them to a fatty acid-free medium. Subsequent addition of excess hirudin either had no effect on the decline in [^{3}H]-PI or caused a slow inhibition. The posthirudin addition of thrombin, which caused a profound reinitiation of PA formation and acid hydrolase secretion, had virtually no effect on [^{3}H]-PI disappearance [9]. Under the conditions used, maximal binding of [^{125}I]thrombin initially and after hirudin occurred in less than 13 sec, and more than 95% of the bound thrombin was removed from platelets within 13 sec [9].

This novel approach of studying the effects of association–dissociation–reassociation of the agonist–receptor complex thus demonstrated a close correlation of agonist binding with PA formation and with acid hydrolase secretion. It should, however, be emphasized that this correlation only holds for that group of cellular responses in platelets that requires sustained receptor occupancy. These data do not give information as to possible correlations between individual steps in the PI cycle and the more intensively studied platelet responses, dense granule secretion and aggregation.

We have therefore pursued the apparent coupling between receptor occupancy, PA production, and acid hydrolase secretion. If PA production were controlled directly through receptor occupancy and PA acted as a cationophore to provide Ca^{2+} for the exocytotic event, acid hydrolase secretion might be expected to occur after PA production, have the same dose–response relationship as PA formation, and be induced by ionophores with PA production. These three criteria have been met experimentally. Figure 4 shows that PA formation precedes acid hydrolase secretion, and demonstrates that dense granule secretion is complete when less than 50% of the final PA level has been reached. The variation in PA production was almost superimposable on acid hydrolase secretion, while dense granule secretion was maximal at 50% of maximal PA production and acid hydrolase secretion [9]. Calcium ionophore A23187 induces little PA formation in untreated platelets [11], and triggers acid hydrolase secretion in platelets treated with indomethacin and aspirin without PA formation [Holmsen, unpublished].

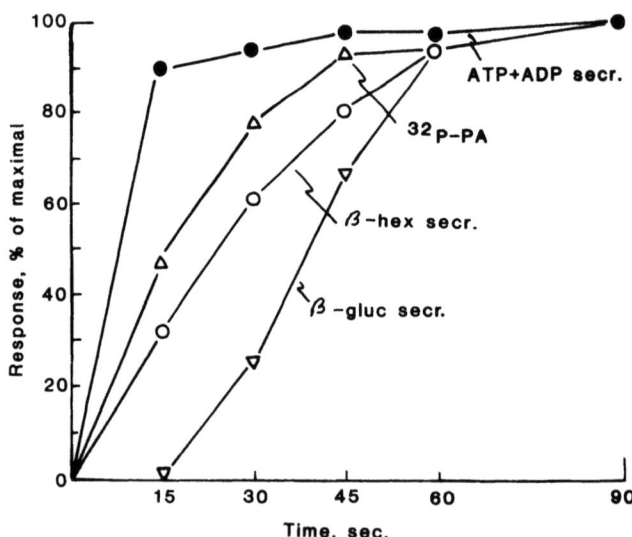

Figure 4. Time course of secretion and PA formation in platelets. Gel-filtered, ^{32}P$_i$-labeled platelets (4.3 × 10^8 cells/ml) were incubated with 0.2 U/ml of thrombin at 37°C. Samples for measurement of dense granule secretion (ATP + ADP, ●), acid hydrolase secretion (β-hexosaminidase, ○; β-glucuronidase, ▽), and [^{32}P]-PA (△) were prepared at times shown and analyzed as described elsewhere [8].

17. PA Synthesis during Acid Hydrolase Secretion

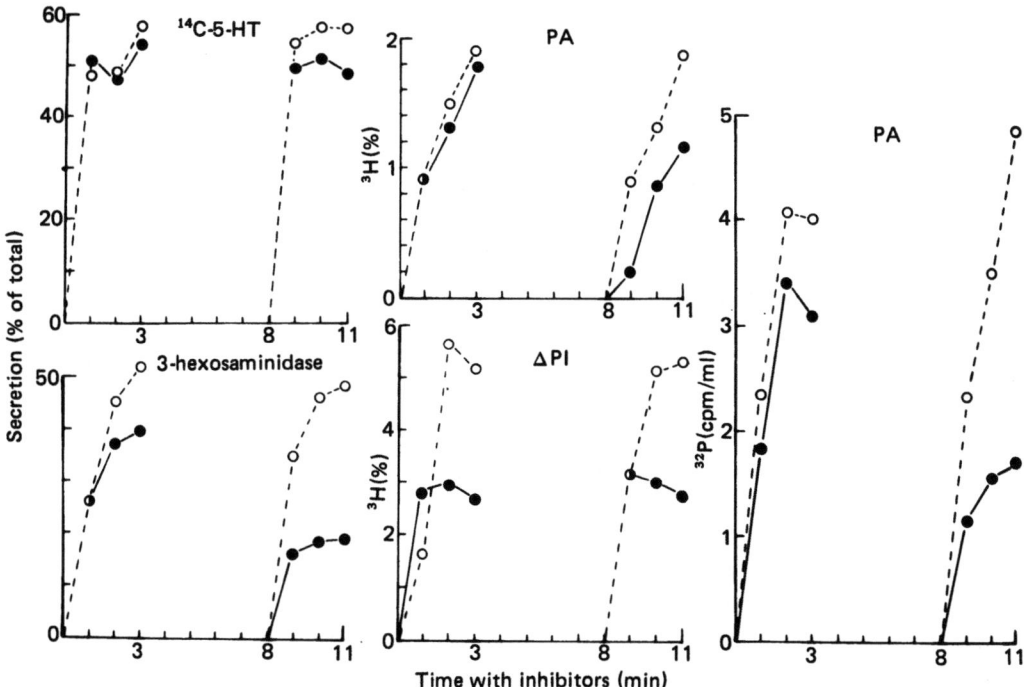

Figure 5. Effect of incubation with metabolic inhibitors on secretion, PA formation, and PI disappearance. Platelets labeled with [^{3}H-arachidonate, [^{14}C]serotonin, and ^{32}P$_i$ were gel-filtered and split in two portions. One portion was incubated with 4 μg/ml of antimycin A and 30 mM 2-deoxyglucose and the other portion was incubated with the solvents for the inhibitors (1.7 μl ethanol and 33 μl 0.15 M NaCl per ml cell suspension). Immediately (0 min on abscissa) and 8 min (8 min on abscissa) after addition of inhibitors or solvents a sample was taken from each incubation mixture and incubated further with 0.5 U/ml of thrombin; samples for determination of the various parameters were taken at times shown and analyzed [8]. The fall in the level of [^{3}H]-PI is expressed as ΔPI, the difference in ^{3}H radioactivity of PI between the thrombin-treated aliquot and a corresponding sample of unstimulated platelets. The ^{3}H radioactivity is expressed for each metabolite as its percentage of ^{3}H relative to the total amount of ^{3}H in the sample.

Additional evidence favoring a coupling of PA formation, but not PI disappearance, to thrombin binding and acid hydrolase secretion came from studies on the dependence of ATP availability for individual biochemical and cellular responses. Figure 5 shows that there was a progressive inhibition of acid hydrolase secretion and PA formation (both [^{32}P]-PA and [^{3}H]-PA) as the incubation time proceeded in the presence of metabolic inhibitors. In sharp contrast, inhibition of PI disappearance was not progressive, as the same degree of inhibition was found immediately after addition of inhibitors as after 8 min. Dense granule secretion (monitored with preabsorbed [^{14}C]serotonin) was hardly affected by the metabolic inhibitors during this short incubation.

The results presented in this chapter suggest a tight coupling of thrombin receptor occupancy, PA formation, and acid hydrolase secretion. In light of the association of DG kinase with platelet membranes [2], the absolute requirement of phospholipase C for Ca^{2+} [6], and the ionophoretic properties of PA [19], the present results are compatible with the tentative scheme given in Fig. 6. The thrombin–receptor complex (TR*) directly activates Mg^{2+}-dependent DG kinase in the platelet membrane, and trace amounts of DG present are converted to PA. These minute amounts of newly formed PA mobilize sufficient Ca^{2+} to activate phospholipase C close to the PI-rich inner layer of the plasma membrane, so that DG

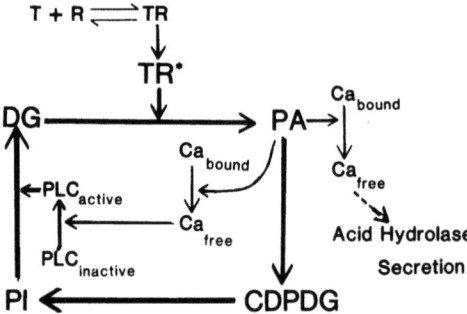

Figure 6. Tentative scheme of coupling of thrombin receptor occupancy, PA formation, and acid hydrolase secretion. For explanation, see text.

production is initiated. This mechanism constitutes a positive feedback on PA formation (still controlled by the postulated TR*-activated DG kinase), so that increasing amounts of Ca^{2+} are mobilized. This serves both to maintain an active phospholipase C and to stimulate acid hydrolase secretion. This scheme (Fig. 6) explains the dependence of PI disappearance in response to thrombin on energy metabolism by ATP-requiring PA formation. Most importantly, our hypothesis sharply differs from the generally accepted view that receptor occupancy triggers phospholipase C activity, or modulates the membrane so that PI becomes available to phospholipase C. Our data suggest that the correlation of acid hydrolase secretion to PA formation is *vectorial*. Thus, there is no correlation of secretion with the (scalar) concentration of PA, since a high level of PA is present after secretion is terminated by hirudin. The correlation is with the *process of PA formation*. It may reflect a situation in which only PA is ionophoretically active at its site of synthesis. PA is then rapidly removed, thereby losing its ability to mobilize Ca^{2+}.

ACKNOWLEDGMENTS. This work was performed at the Thrombosis Research Center, Temple University, Philadelphia, and supported by NIH Grant HL-14217.

References

1. Billah, M. M.; Lapetina, E. G. Rapid decrease of phosphatidylinositol 4,5-bisphosphate in thrombin-stimulated platelets. *J. Biol. Chem.* **257**:12705–12708, 1982.
2. Call, F. L.; Rubert, M. Diglyceride kinase in human platelets. *J. Lipid Res.* **14**:466–474, 1973.
3. Chap, H.; Perret, B.; Mauco, G.; Plantavid, M.; Laffont, F.; Simon, M. F.; Douste-Blazy, L. Organization and role of platelet membrane phospholipids as studied with phospholipases A_2 from various venoms and phospholipases C from bacterial origin. *Toxicon* **20**:291–298, 1982.
4. Cockcroft, S.; Bennet, J. P.; Gomperts, B. D. The dependence on Ca^{2+} of phosphatidylinositol breakdown and enzyme secretion in rabbit neutrophils stimulated by formyl-methionyl-leucyl-phenylalanine or ionomycin. *Biochem. J.* **200**:501–508, 1981.
5. Fenton, J. W. Trombin specificity. *Ann. N.Y. Acad. Sci.* **370**:468–495, 1981.
6. Hofmann, S. L.; Majerus, P. W. Identification and properties of two distinct phosphatidylinositol-specific phospholipase C enzymes from sheep seminal vesicular glands. *J. Biol. Chem.* **257**:6461–6469, 1982.
7. Holmsen, H.; Dangelmaier, C. A.; Akkerman, J. W. N. Determination of levels of glycolytic intermediates and nucleotides in platelets by pulse-labeling with ^{32}P-orthophosphate. *Anal. Biochem.* **131**:266–272, 1983.
8. Holmsen, H.; Dangelmaier, C. A.; Holmsen, H. K. Thrombin-induced platelet responses differ in requirement for receptor occupancy: Evidence for tight coupling of occupancy and compartmentalized phosphatidic acid formation. *J. Biol. Chem.* **256**:9393–9396, 1981.

9. Holmsen, H.; Dangelmaier, C. A.; Rongved, S. Tight coupling of thrombin-induced acid hydrolase secretion and phosphatidate synthesis to receptor occupancy in human platelets. *Biochem. J.* **221**:in press, 1985.
10. Holmsen, H.; Kaplan, K. L.; Dangelmaier, C. A. Differential energy requirement for platelet responses: A simultaneous study of dense granule, alpha-granule and acid hydrolase secretion, arachidonate liberation, phosphatidylinositol turnover and phosphatidate formation. *Biochem. J.* **208**:9–18, 1982.
11. Lapetina, E. G.; Billah, M. M.; Cuatrecasas, P. The phosphatidylinositol cycle and the regulation of arachidonic acid production. *Nature (London)* **292**:367–369, 1981.
12. Larsen, N. E.; Simons, E. R. Preparation and application of a photoreactive thrombin analogue: Binding to human platelets. *Biochemistry* **20**:4141–4147, 1981.
13. Michell, R. H. Inositol phospholipids and cell surface receptor function. *Biochim. Biophys. Acta* **415**:81–147, 1975.
14. Michell, R. H.; Kirk, C. J.; Jones, L. M.; Downes, C. P.; Creba, J. The stimulation of inositol lipid metabolism that accompany calcium mobilization in stimulated cells: Defined characteristics and unanswered questions. *Philos. Trans. R. Soc. London Ser. B* **296**:123–144, 1981.
15. Prpić, V.; Blackmore, P. F.; Exton, J. H. Phosphatidylinositol breakdown induced by vasopressin and epinephrine in hepatocytes is calcium-dependent. *J. Biol. Chem.* **257**:11323–11331, 1982.
16. Putney, J. W. Minireview: Recent hypothesis regarding the phosphatidylinositol effect. *Life Sci.* **29**:1183–1194, 1981.
17. Rink, T. J.; Smith, S. W.; Tsien, R. Y. Cytoplasmic free Ca^{2+} in human platelets: Ca^{2+} thresholds and Ca-independent activation for shape change and secretion. *FEBS Lett.* **148**:21–26, 1982.
18. Rittenhouse-Simmons, S. Differential activation of platelet phospholipases by thrombin and ionophore A23187. *J. Biol. Chem.* **256**:4153–4155, 1981.
19. Serhan, C.; Anderson, P.; Goodman, E.; Dunham, P.; Weissmann, G. Phosphatidate and oxidized fatty acids are calcium ionophores. *J. Biol. Chem.* **256**:2736–2741, 1981.
20. Tam, S. W.; Fenton, J. W.; Detwiler, T. C. Dissociation of thrombin from platelets by hirudin: Evidence for receptor processing. *J. Biol. Chem.* **254**:8723–8725, 1979.

18

Calcium and Diacylglycerol
Separable and Interacting Intracellular Activators in Human Platelets

T. J. Rink, R. Y. Tsien, A. Sanchez, and T. J. Hallam

Platelet responses to many agonists involve rapid stimulation of calcium fluxes and turnover of phosphoinositides. A rise in cytoplasmic free calcium, $[Ca^{2+}]_i$, has been regarded as the final common pathway for shape change, secretion, and aggregation; phosphoinositide turnover has been viewed mainly as a means of mobilizing calcium [and as one route for releasing arachidonate for formation of thromboxane A_2 (TxA_2)]. However, much of the evidence is indirect and interpretation necessarily speculative. Two recent developments have allowed a more informed analysis of second messenger pathways.

1. $[Ca^{2+}]_i$ can actually be measured in intact, functioning platelets, by the intracellularly trapped fluorescent indicator quin-2 [24,29–31]. This technique allows a much more direct assessment of the role of calcium than previously possible. It also greatly aids the study of other putative messengers by permitting measurement and monitored manipulation of one known activator, i.e., calcium.

2. Nishizuka and his colleagues have shown that an immediate product of phosphoinositide breakdown, diacylglycerol, has activator properties in its own right, as a direct stimulator for C kinase, and appears to play a key role in the response to several platelet agonists [14].

We outline here some of our current ideas on the separate and potentiating roles of $[Ca^{2+}]_i$ and diacylglycerol, together with indications that other intracellular activator pathways, as yet unidentified, are also at work.

Primary Effects of Agonists

It may be helpful to summarize what we presently suppose are the primary actions of platelet agonists, even at the expense of anticipating the experimental evidence to be presented below.

T. J. Rink, A. Sanchez, and T. J. Hallam • Physiological Laboratory, University of Cambridge, Cambridge CB2 3EG, England.   *R. Y. Tsien* • Department of Physiology–Anatomy, University of California, Berkeley, California 94720.

1. Elevation of $[Ca^{2+}]_i$. One initial effect of agonist receptor interaction is stimulated calcium influx which may well be independent of stimulated phosphoinositide breakdown.

2. Stimulated phosphoinositide breakdown by phospholipase C, with formation of diacylglycerol, is a rapid response to many agonists [14,25]. This process is probably independent of calcium in that it does not necessarily require or cause a change in $[Ca^{2+}]_i$.

Different agonists may activate one or both of these processes to different degrees and so produce varying changes of $[Ca^{2+}]_i$ and diacylglycerol.

3. TxA_2 formation. One cannot properly discuss platelet activation without considering TxA_2, although we see it more as an internally generated agonist that reinforces the effect of other agonists, rather than an intracellular second messenger. Most agonists can cause release of arachidonate from phospholipids, which is then converted to TxA_2 by cyclooxygenase and thromboxane synthetase. The release may follow either elevation of $[Ca^{2+}]_i$, which stimulates phospholipase A_2, or formation of diacylglycerol from which diglyceride lipase can liberate arachidonate [25]. [We suspect there may be other pathways. For instance, epinephrine stimulates TxA_2 formation but does not appear itself to elevate $[Ca^{2+}]_i$ (see below) nor is there evidence that it directly causes phosphoinositide breakdown.] Whatever the mechanisms for TxA_2 production, it is exceedingly important to realize that TxA_2 itself, acting as an agonist, elevates $[Ca^{2+}]_i$ [9] and promotes breakdown of phosphoinositides [17,27]. One therefore has to distinguish between direct responses to other agonists and those due to consequent formation of TxA_2, a distinction that has often been overlooked.

4. Release of intracellular calcium. One can imagine receptor-occupancy opening a calcium channel in the surface membrane to increase calcium influx. It has been more difficult to discern how a surface receptor (or membrane depolarization in excitable cells) causes calcium release from internal stores, as occurs in platelets and many other cell types. It might be directly mediated via the close apposition between membranes of the open canalicular system and the dense tubular system (an endoplasmic reticular membrane system similar to sarcoplasmic reticulum), or there may be an intracellular trigger substance (see Chapter 4). Whatever the mechanism, this internal release adds to the elevation in $[Ca^{2+}]_i$ that is produced by stimulated influx.

What Can Calcium Do?

Calcium ionophores can stimulate essentially normal responses including the cytoskeletal rearrangements of shape change, exocytotic secretion, and aggregation [6–8]. These studies have been extended in our laboratory by measurement of the levels of $[Ca^{2+}]_i$ required for these responses.

Figure 1 shows the response of aspirin-treated, quin-2-loaded cells to graded doses of the calcium ionophore ionomycin. Shape change occurs at several hundred nanomolar $[Ca^{2+}]_i$, while secretion and aggregation have an apparent threshold of about 1 μM, and substantial secretion of ATP requires several micromolar $[Ca^{2+}]_i$. These results fit well with the calcium secretion relation found in platelets made permeable to calcium buffers by high-voltage discharge [15]. It is of course impossible to be certain that increasing $[Ca^{2+}]_i$ is the only intracellular activator pathway operating, but one can eliminate production of TxA_2 by blocking cyclooxygenase. Under these conditions, diacylglycerol formation in response to ionophore will likely be small or negligible [25], though this point needs further experimental confirmation. The key point is that a secretory response triggered by calcium requires an approximately 10-fold increase in $[Ca^{2+}]_i$ from the basal level.

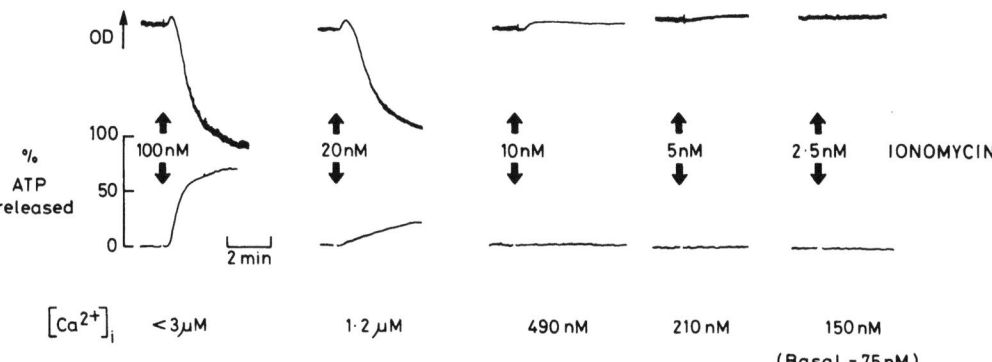

Figure 1. Responses of quin-2-loaded human platelets to different concentrations of ionomycin. The cells were resuspended in Hepes-buffered saline, 1 mM calcium with 100 μM aspirin to block cyclooxygenase. The upper trace shows the optical density of the stirred suspension to indicate shape change and aggregation. The lower trace shows ATP release (as percent of the maximum releasable by thrombin), monitored by luciferin luminescence, as a measure of dense granule secretion. The figures under the trace show the $[Ca^{2+}]_i$ levels achieved with each concentration of ionomycin measured from the quin-2 fluorescence as previously described (24). All experiments were done at 37°C.

Responses Independent of Raised $[Ca^{2+}]_i$

As we examined responses to natural agonists it became clear that shape change and, perhaps more surprisingly, secretion could be stimulated while $[Ca^{2+}]_i$ remained at or near basal levels (24, and see e.g. Fig. 4), and well below the $[Ca^{2+}]_i$ threshold observed with calcium ionophore. The next step was to identify the trigger, if not elevated $[Ca^{2+}]_i$.

Diacylglycerol Activates While $[Ca^{2+}]_i$ Remains at Basal Levels

Exogenous oleoylacetyl-glycerol promotes phosphorylation of the 40K substrate of C kinase and some secretion in human platelets [14]. Typical responses of quin-2-loaded cells to this diacylglycerol reveal a substantial secretion of ATP which shows a characteristic delayed, slow time course (Figs. 2A,B). Secretion is independent of external calcium and TxA_2 formation, and it is not accompanied by any rise in $[Ca^{2+}]_i$. The optical density trace indicates a partial shape change and aggregation when there is external Ca^{2+}. These results suggest that diacylglycerol could be the calcium-dependent trigger for secretion. (They also suggest that it is not the main stimulus for the shape-change which is independent of Ca^{2+} flux or the discharge of internal calcium.)

Support for the idea that activation of C kinase can initiate secretion comes from responses to 12-*O*-tetradecanoyl-13-phorbol acetate (TPA). This agent, long known to activate platelets [23,32] and many other cells, activates C kinase *in vitro* [2], and promotes phosphorylation of the 40K substrate in intact platelets [2,3]. TPA produces responses in quin-2-loaded platelets strikingly similar to those produced by diacylglycerol, and also without any elevation of $[Ca^{2+}]_i$ (Figs. 2C,D). Our data thus show that diacylglycerol and TPA stimulate secretion without increasing $[Ca^{2+}]_i$. However, we have been unable to reduce $[Ca^{2+}]_i$ below 50 nM and therefore cannot state whether a "constitutive" requirement for some calcium exists in intact cells.

Our reading of Kaibuchi *et al.* [13,14] is that in the presence of phosphatidylserine and phosphatidylethanolamine (ubiquitous in platelet membranes), diacylglycerol can give substantial activation of isolated C kinase at very low $[Ca^{2+}]_i$, i.e., in a medium containing

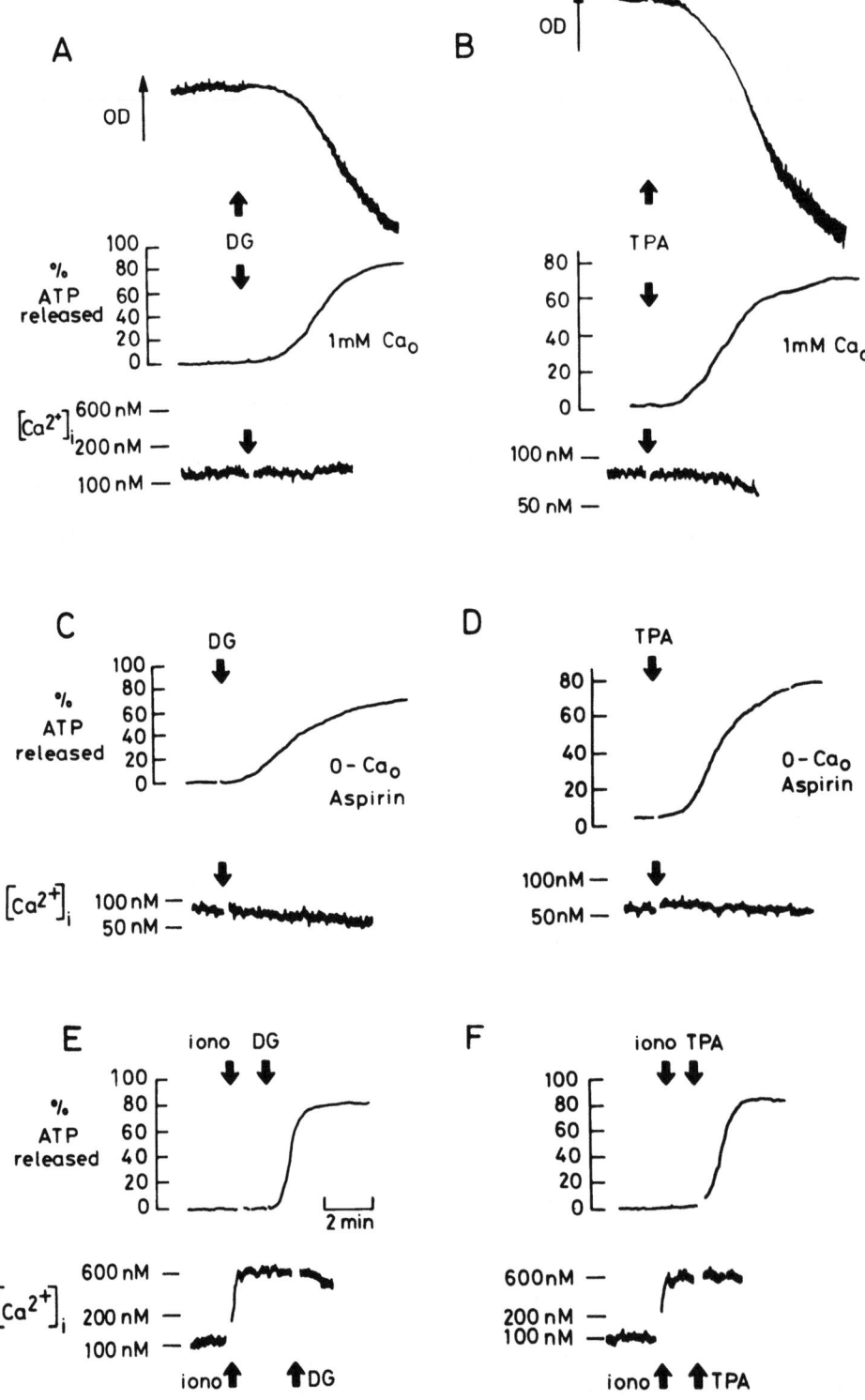

Figure 2. Responses to exogenous diacylglycerol and TPA. In A, C, and E, 60 μg/ml oleoylacetylglycerol was added as indicated (DG). In B, D, and F, 20 nM TPA was added. In panels A and B, the upper traces show optical density of a stirred suspension of quin-2-loaded platelets, the middle traces secretion of ATP, and the lower traces $[Ca^{2+}]_i$. In C, D, and E, and F, only ATP secretion and $[Ca^{2+}]_i$ are shown. The external Ca was 1 mM in A and B, and also in E and F. In C and D there was 1 mM EGTA and no added calcium; the cells were also preincubated with 100 μM aspirin. In E and F, ionomycin, 5 and 10 nM, respectively, was added prior to the diacylglycerol and TPA, as indicated.

EGTA with no added calcium. TPA can also apparently activate C kinase *in vitro* in calcium-free conditions [5]. These observations suggest that these agents can activate C kinase without any absolute requirement for calcium.

Interaction between Calcium and Diacylglycerol

Evidence for interaction between calcium and diacylglycerol in intact platelets comes from results such as those shown in Figs. 2E and F. Here $[Ca^{2+}]_i$ was raised to 600 nM by ionomycin, a level too low either to cause secretion or to presumably activate C kinase (cf. Ref. 14). However, subsequently added diacylglycerol or TPA now gave a much faster response than that seen when they were added alone, with $[Ca^{2+}]_i$ remaining near 100 nM. The site of this accelerating action of "subthreshold" rises in $[Ca^{2+}]_i$ is still an open question, but one that should be resolvable with presently available techniques. The interaction could be on C kinase itself. Alternatively, modest elevation of $[Ca^{2+}]_i$ may accelerate an otherwise rate-limiting step in the secretory process, subsequent to the 40K phosphorylation. Either way, the results highlight an important effect of apparently subthreshold elevations of $[Ca^{2+}]_i$ by revealing one way that one agonist, which produces a modest rise in $[Ca^{2+}]_i$, sensitizes platelets to another agonist which induces diacylglycerol formation.

The question also arises as to whether secretion triggered by large increases in $[Ca^{2+}]_i$ in the 1–10 µM range involves C kinase activation. This seems likely since: (1) $[Ca^{2+}]_i$ in the 1–50 µM range activates C kinase *in vitro*, without needing diacylglycerol, and (2) concentrations of ionophore that stimulate secretion promote phosphorylation of the 40K substrate [11,13,14].

What Do Agonists Do to $[Ca^{2+}]_i$?

With the introduction of quin-2, it has been possible to obtain direct quantitative answers to this question. Different agonists produce varied patterns of calcium in quin-2-loaded platelets (Fig. 3). The responses in calcium-free medium are taken to reflect stimulus-evoked discharge of intracellular calcium stores, and the responses in the presence of 1 mM calcium are presumed to reflect both calcium influx and internal release. The difference between responses should therefore represent calcium influx.

Thrombin, PAF, or ADP rapidly raise $[Ca^{2+}]_i$ manyfold with 1 mM external calcium. Thrombin is the most effective: an optimal concentration produces an elevation to several micromolar. With PAF the rise is more transient, probably due to receptor desensitization, and rarely exceeds 1 µM, while the ADP-evoked rise seldom exceeds 700–800 nM.

Each of these agonists gives a rapid but much reduced response in calcium-free medium, indicating that they discharge internal stores, but that normally the response is mainly due to calcium influx. This was expected, at least for thrombin and PAF, from previously documented stimulation of tracer calcium influx [e.g., 16,18]. The presence of quin-2 is expected to reduce the response in calcium-free medium by increasing the calcium buffering capacity of the cytoplasm [21,24]. In cells not loaded with quin-2, maximum internal release may elevate calcium to about 600 nM [24]. The major source of internal calcium may be the dense tubular system, rather than the mitochondria, since uncoupling agents neither mimic nor prevent the thrombin-evoked rise in $[Ca^{2+}]_i$ seen in calcium-free medium [28].

Three substances have been widely considered as triggers for internal release of calcium, but seem to us unlikely to have a major role. TxA_2 can cause some internal release (see Fig. 3D). However, the increased quin-2 fluorescence produced by thrombin, ADP, and

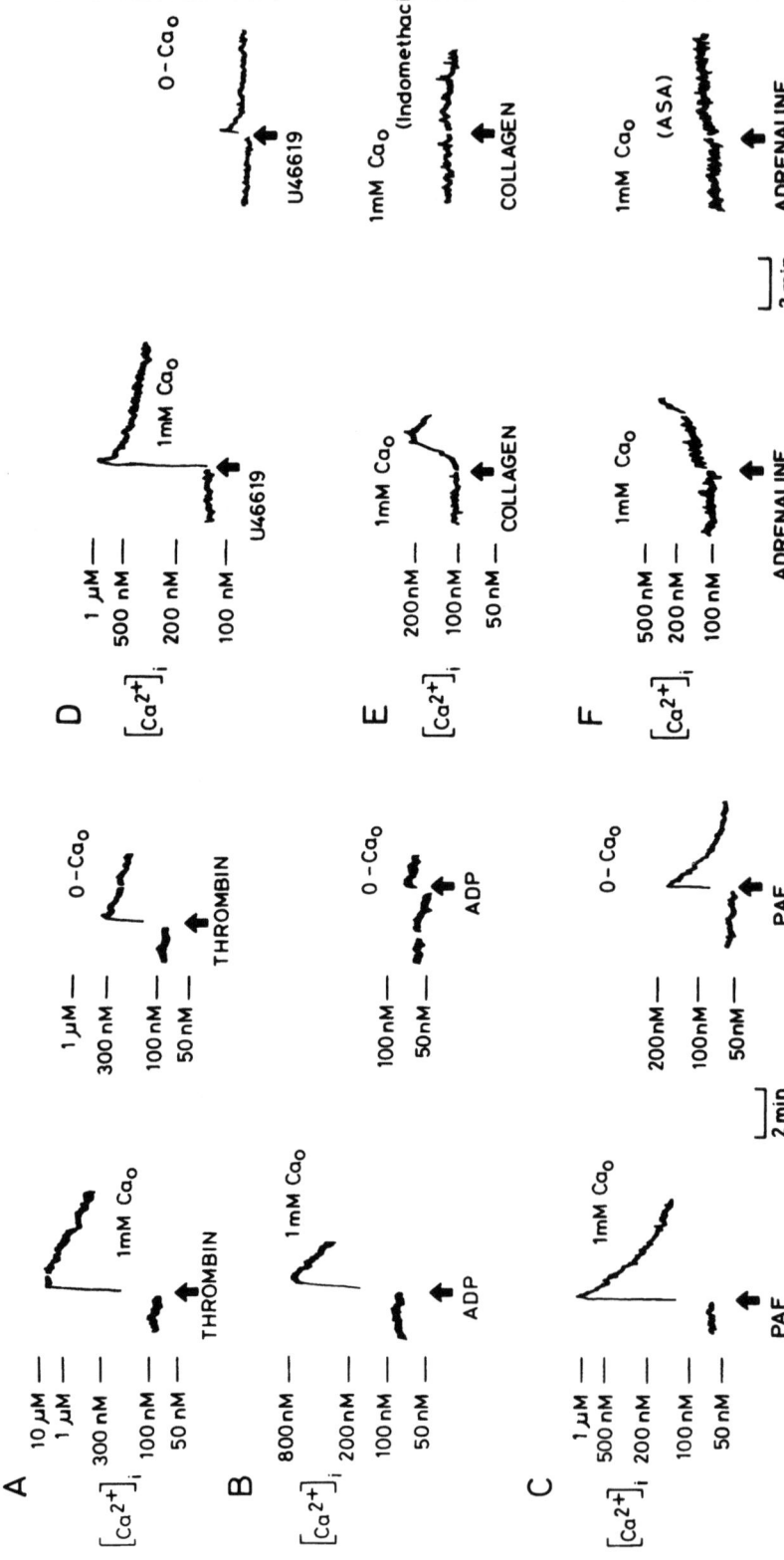

Figure 3. $[Ca^{2+}]_i$ changes produced by different agonists. Panels A, B, C, and D show responses to 0.5 U/ml thrombin, 40 nM PAF, 10 μM ADP, and 1 μM U46619, in first 1 mM and then 50 nM external Ca^{2+}. Panels E and F show response to 10 μg/ml collagen and 40 μM epinephrine in 1 mM Ca^{2+} solution, first without, and then with, blockade of cyclooxygenase.

PAF, in calcium-free medium, is only slightly reduced by complete blockade of TxA_2 production. TxA_2 seems to *reinforce* rather than to *mediate* the effects of these other agonists. Phosphatidic acid and lysophosphatidic acid appear to be produced too slowly to account for the very rapid, internal release, which is complete in 2–3 sec with PAF and thrombin. Moreover, there is considerable doubt over their alleged ability to act as calcium ionophores [12].

By contrast, collagen and epinephrine do not seem to elevate directly $[Ca^{2+}]_i$ but do so only via TxA_2 formation. When TxA_2 production is blocked, collagen and epinephrine produce little or no change in $[Ca^{2+}]_i$ (Figs. 3E,F). Without cyclooxygenase blockade these agonists produce a delayed rise in $[Ca^{2+}]_i$, presumably due to the slow formation of TxA_2. These seemingly surprising results fit in with the inability of collagen and epinephrine to stimulate shape change in aspirin-treated cells. An apparent discrepancy exists between our failure to observe a rise in $[Ca^{2+}]_i$ with epinephrine in aspirin-treated cells and the (small)

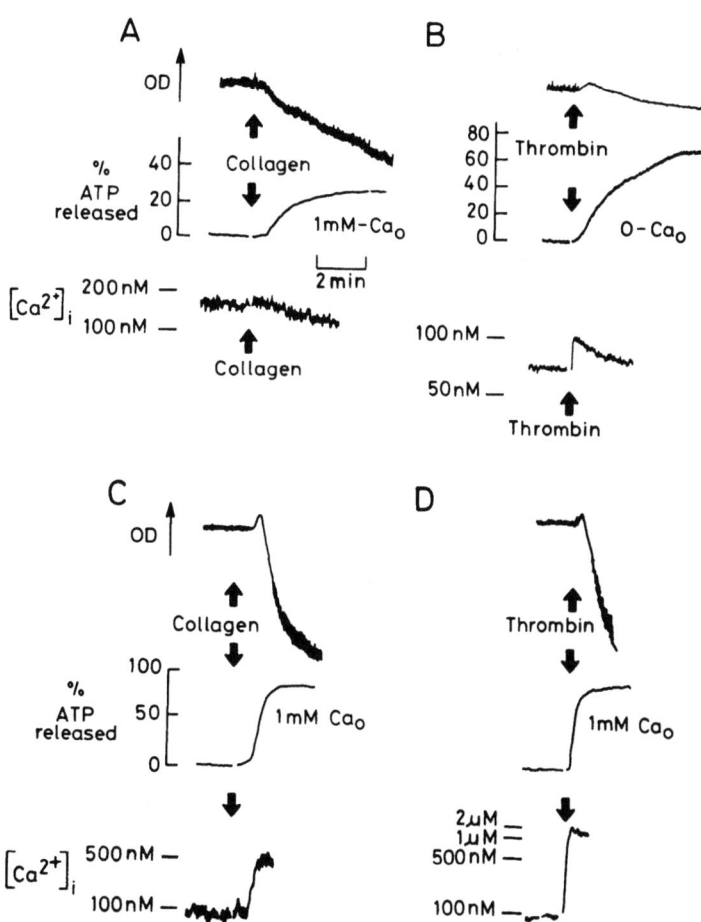

Figure 4. Response to collagen and thrombin when $[Ca^{2+}]_i$ stays near basal levels, A and B, and when $[Ca^{2+}]_i$ can increase, C and D. (A) 10 μg/ml collagen added, 1 mM external Ca^{2+}, aspirin-treated cells. (B) 0.05 U/ml thrombin added, no added external Ca^{2+}, 1 mM EGTA, aspirin-treated cells. (C) 10 μg/ml collagen added, 1 mM external Ca^{2+}, cells not treated with aspirin. (D) 0.5 U/ml thrombin added, 1 mM external calcium, aspirin-treated cells.

increase in $^{45}Ca^{2+}$ uptake previously reported [20]. However, others have been unable to find such an uptake under these conditions [4], which fits with our findings. Of all these agonists only thrombin can readily raise $[Ca^{2+}]_i$ to levels sufficient for calcium alone to stimulate secretion and aggregation. Yet all of these agonists can aggregate quin-2-loaded platelets and cause substantial secretion, suggesting the crucial involvement of other activators, such as diacylglycerol.

Phosphoinositide Breakdown, Independent of Elevated $[Ca^{2+}]_i$

Thrombin, PAF, and collagen stimulate the hydrolysis of phosphoinositides via phospholipase C, with the formation of diacylglycerol [14,25]. In many systems phosphoinositide turnover can be stimulated independently of a rise in $[Ca^{2+}]_i$ [e.g., 1,10,22]. In platelets the evidence is strongest for collagen stimulation. In the absence of TxA_2 formation, collagen causes no increase in $[Ca^{2+}]_i$ (Figs. 3E and 4A). Yet phosphoinositide breakdown occurs under these conditions [25], which we have confirmed in quin-2-loaded platelets (Irvine, Hallam, Sanchez, and Rink, unpublished observations). Responses to natural agonists could thus be triggered by diacylglycerol formed *in situ* without any change in $[Ca^{2+}]_i$.

Responses to Natural Agonists

Figures 4A and B show responses to collagen which fit reasonably well with what we know of calcium and diacylglycerol. Figure 4A shows no rise in $[Ca^{2+}]_i$ in aspirin-treated cells, but a delayed secretory response with a time course resembling that seen with exogenous diacylglycerol or TPA. Recently we have found [26] that chlorpromazine blocks the response to diacylglycerol, TPA, and collagen in aspirin-treated cells, at concentrations (20–40 μM) that only slightly reduce secretion and $[Ca^{2+}]_i$ changes produced by thrombin. Since chlorpromazine inhibits isolated C kinase and this inhibition can be overcome by calcium ions [19], our findings add support to the proposition that collagen acts via C kinase, presumably by promoting the formation of diacylglycerol.

When TxA_2 formation is permitted and $[Ca^{2+}]_i$ rises, the secretory response to collagen is larger and very briskly follows the delayed elevation of $[Ca^{2+}]_i$ (Fig. 4C). This enhanced secretory response presumably reflects the effect of a greater phosphoinositide breakdown together with the elevated $[Ca^{2+}]_i$, and is rather similar to that seen in Figs. 2E and F, where a subthreshold elevation of $[Ca^{2+}]_i$ enhanced the responses to diacylglycerol and TPA.

Figure 4B shows responses to thrombin in conditions where $[Ca^{2+}]_i$ rise due to internal release was minimized, by loading with more quin-2 than usual and incubating for a prolonged period in calcium-free solution. TxA_2 production was also suppressed. Secretion had a time course similar to that seen with exogenous diacylglycerol or TPA. The secretion response shown in Fig. 4B could be largely attributable to thrombin-evoked formation of diacylglycerol. A large rise in $[Ca^{2+}]_i$ and very rapid secretion ensues when the experiment is repeated in the presence of calcium (Fig. 4D).

Results like those in Fig. 4B demonstrate that thrombin produces extra effects at basal $[Ca^{2+}]_i$, not produced by diacylglycerol or TPA, including a discharge of internal calcium and shape change. Although the triggers for these responses are not yet known, possible candidates include one or more of the inositol phosphates which are water-soluble products released during breakdown of phosphoinositides (see Chapters 4 and 9).

Conclusions

This chapter is intended as a brief progress report. As a result of the discovery of diacylglycerol as an activator of C kinase, and the ability to measure $[Ca^{2+}]_i$, we have much more data than were available 2 years ago; we clearly need much more. R. A. Steinhardt has succinctly stated what is needed to pin down the role of a second messenger as "show it, move it, block it." One can more or less do these things for calcium. With diacylglycerol there are some data showing it, i.e., measuring its formation, and some results of moving it, i.e., observing its effects or those of its mimic, TPA. Many more experiments are needed, particularly measurements of diacylglycerol formation correlated with monitored $[Ca^{2+}]_i$, protein phosphorylation, and cellular responses. We still lack selective blockers of diacylglycerol formation and its action; these agents are urgently needed as experimental tools and they could have significant therapeutic implications.

The role of diacylglycerol and C kinase in other cell types is doubtlessly now being investigated in numerous laboratories, and may bring together much of the hitherto uninterpretable data showing the very widespread correlation of phosphoinositide turnover with cell stimulation, and numerous effects of activation by TPA.

Our data in particular suggest that a critical reappraisal is needed regarding the role of $[Ca^{2+}]_i$ in many aspects of cell activation. For instance, secretory exocytosis can no longer be assumed to imply a rise in $[Ca^{2+}]_i$. The fact that elevated $[Ca^{2+}]_i$ can trigger a process does not necessarily infer that a natural agonist must act this way. We are at an exciting stage in the unraveling of an "intracellular endocrinology," where new messengers are still being discovered and their interactions and molecular targets are beginning to be defined.

ACKNOWLEDGMENTS. This work was supported by grants from the Science and Engineering Research Council, U.K., and the National Institutes of Health, U.S.A. A.S. is a recipient of a long-term EMBO Fellowship. The U46619 was a gift from the Upjohn Company, Kalamazoo, Michigan.

References

1. Berridge, M. J. 5-Hydroxytryptamine stimulation of phosphatidylinositol hydrolysis and calcium signalling in the blowfly salivary gland. *Cell Calcium* 3:385–397, 1982.
2. Castagna, M.; Takai, Y.; Kaibuchi, K.; Sano, K.; Kikkawa, U.; Nishizuka, Y. Direct activation of calcium-activated, phospholipid-dependent protein kinase by tumour-promoting phorbol esters. *J. Biol. Chem.* 257:7847–7851, 1982.
3. Chiang, T. M.; Cagen, L. M.; Kang, A. H. Effects of 12-O-tetradecanoyl phorbol 13-acetate on platelet aggregation. *Thromb. Res.* 21:611–622, 1981.
4. Clare, K. A.; Scrutton, M. C. The role of Ca^{2+} uptake in the response of human platelets to adrenaline and to 1-O-alkyl-2-acetyl-Sn-glycero-3-phosphorylcholine (platelet-activating factor). *Eur. J. Biochem.* in press, 1984.
5. Donolly, T. E.; Jensen, R. Activation of calcium-stimulated, phospholipid protein kinase by tumour-promoting phorbol ester does not require calcium. *Fed. Proc.* 42:783, 1983.
6. Feinman, R. D.; Detwiler, T. C. Platelet secretion induced by divalent cation ionophores. *Nature (London)* 249:172–173, 1974.
7. Feinstein, M. B.; Fraser, C. Human platelet secretion and aggregation induced by calcium ionophores. *J. Gen. Physiol.* 66:561–581, 1975.
8. Gerrard, J. M.; White, J. G.; Rao, G. H. R. Effects of the ionophore A23187 on blood platelets. II. Influence on ultrastructure. *Am. J. Pathol.* 77:151–166, 1974.
9. Hallam, T. J.; Rink, T. J.; Sanchez, A. Effects of arachidonate and a thromboxane A_2 analogue on human

platelets studied with a fluorescence indicator for cytoplasmic free calcium. *J. Physiol. (London)* **343**:97–98, 1983.
10. Harrington, C. A.; Eichberg, J. Norepinephrine causes α1-adrenergic receptor-mediated decrease of phosphatidylinositol in isolated rat liver plasma membranes supplemented with cytosol. *J. Biol. Chem.* **258**:2087–2090, 1983.
11. Haslam, R. J.; Lynham, J. A. Relationship between phosphorylation of blood platelet proteins and secretion of platelet granule constituents. I. Effects of different aggregatory agents. *Biochem. Biophys. Res. Commun.* **77**:714–722, 1977.
12. Holmes, R. P.; Yoss, N. L. Failure of phosphatidic acid to translocate calcium across phosphatidylcholine membranes. *Nature* **305**:637–638, 1983.
13. Kaibuchi, K.; Takai, Y.; Nishizuka, Y. Cooperative roles of various membrane phospholipids in the activation of calcium-activated, phospholipid-dependent protein kinase. *J. Biol. Chem.* **256**:7146–7149, 1981.
14. Kaibuchi, K.; Sano, K.; Hoshijima, M.; Takai, Y.; Nishizuka, Y. Phosphatidylinositol turnover in platelet activation; calcium mobilization and protein phosphorylation. *Cell Calcium* **3**:323–335, 1982.
15. Knight, D. E.; Hallam, T. J.; Scrutton, M. C. Agonist selectivity and second messenger concentration in Ca^{2+}-mediated secretion. *Nature (London)* **296**:256–257, 1982.
16. Lee, T. C.; Malone; B.; Blank, M. L.; Snyder, F. 1-Alkyl-2-acetyl-sn-glycero-3-phosphocholine (PAF) stimulates calcium influx in rabbit platelets. *Biochem. Biophys. Res. Commun.* **102**:1262–1268, 1981.
17. MacIntyre, D. E.; and Pollock, W. K. Platelet-activating factor, U44069 and vasopressin stimulate phosphatidylinositol turnover in human blood platelets. *Br. J. Pharmacol.* **77**:466P, 1982.
18. Massini, P.; Luscher, E. F. On the significance of influx of Ca^{2+} ions into stimulated human blood platelets. *Biochim. Biophys. Acta* **435**:652–663, 1976.
19. Mori, T.; Takai, Y.; Minakuchi, B. Y.; Nishizuka, Y. Inhibitory action of chlorpromazine, dibucaine, and other phospholipid-interacting drugs on calcium-activated, phospholipid-dependent protein kinase. *J. Biol. Chem.* **255**:8378–8380, 1980.
20. Owen, N. E.; Feinberg, H.; and Le Breton, G. C. Epinephrine induces Ca^{2+} uptake in human blood platelets. *Am. J. Physiol.* **239**:H483–H488, 1980.
21. Pozzan, T.; Arslan, P.; Tsien, R. Y.; Rink, T. J. Anti-immunoglobulin, cytoplasmic free calcium, and capping in B-lymphocytes. *J. Cell Biol.* **94**:335–340, 1982.
22. Putney, J. W., Jr. Inositol lipids and cell stimulation in mammalian salivary gland. *Cell Calcium* **3**:369–383, 1982.
23. Rao, G. H. R.; White, J. G. Influence of esterase inhibitors on platelet aggregation and release induced by phorbol myristate acetate. *Biochem. Pharmacol.* **24**:293–295, 1974.
24. Rink, T. J.; Smith, S. W.; Tsien, R. Y. Cytoplasmic free calcium in human blood platelets: Calcium thresholds and calcium-independent activation for shape-change and secretion. *FEBS Lett.* **148**:21–26, 1982.
25. Rittenhouse, S. E. Inositol lipid metabolism in the responses of stimulated platelets. *Cell Calcium* **3**:311–322, 1982.
26. Sanchez, A.; Hallam, T. J.; Rink, T. J. Trifluoperazine and chlorpromazine block secretion from human platelets evoked at basal cytoplasmic free calcium by activators of C-kinase. *FEBS Lett.* **164**:43–46, 1983.
27. Siess, W.; Cuatrecasas, P.; Lapetina, E. G. A role for cyclooxygenase products in the formation of phosphatidic acid in stimulated human platelets. *J. Biol. Chem.* **258**:4683–4686, 1983.
28. Smith, S. W. Intracellular free calcium and human platelet activation. M.Sc. thesis, University of Cambridge, 1982.
29. Tsien, R. Y. New calcium indicators and buffers with high selectivity against magnesium and protons: Design, synthesis and properties of prototype structures. *Biochemistry* **19**:2396–2404, 1980.
30. Tsien, R. Y. A non-disruptive technique for loading calcium buffers and indicators into cells. *Nature (London)* **290**:527–528, 1981.
31. Tsien, R. Y.; Pozzan, T.; Rink, T. J. Calcium haemostasis in intact lymphocytes: Cytoplasmic free calcium monitored with a new, intracellularly trapped fluorescent indicator. *J. Cell Biol.* **94**:325–334, 1982.
32. Zucker, M. B.; Troll, W.; Belman, S. The tumour-promoter phorbol ester (12-O-tetradecanoyl-phorbol-13-acetate), a potent aggregating agent for blood platelets. *J. Cell Biol.* **60**:325–336, 1974.

19

Protein Phosphorylation and Calcium as Mediators in Human Platelets

James L. Daniel

Protein phosphorylation is a major mechanism by which hormonal stimulation of cells is coupled to cell function [7,30]. Protein kinase and phosphatases regulate phosphorylation of cellular proteins, many of which are enzymes. In turn, phosphorylations regulate or modulate function or activity of these proteins. Activity of protein kinases and phosphatases is regulated by intracellular messengers such as cAMP, cGMP and calcium. Two types of cAMP-dependent protein kinases have been described [31]. In both cases cAMP activates kinases by binding to and dissociating a regulatory subunit. This dissociation removes the inhibition that the regulatory subunit imposes on the catalytic subunit. Binding of cGMP to cGMP-dependent protein kinase activates the kinase without dissociation of a regulatory subunit [13]. Two types of calcium-activated protein kinases have been described. One type requires the complex action of calmodulin and calcium for activation [34]. The other type, called protein kinase C, requires three factors for full activation: micromolar levels of calcium, phospholipid, and diacylglycerol [29]. The regulation of enzymes that dephosphorylate proteins, phosphoprotein phosphatases, is less clear but some phosphatases that may be regulated through calcium–calmodulin have been described [7].

Elevated levels of cyclic nucleotides are associated with inhibition of cell function in human blood platelets [16]. Inhibitory agonists such as adenosine and prostaglandins E_1 and I_2 act through elevation of platelet cAMP level [16]. Kaulen and Gross [25] partially purified cAMP-dependent protein kinase from platelets and later Haslam et al. [18] showed that both types of cAMP-dependent protein kinases are present in platelets. Photoaffinity labels reveal that regulatory cAMP-binding subunits of the kinases are present [18]. In experiments in which platelets were incubated with $^{32}PO_4$ and then treated with PGE_1 to elevate cAMP, Haslam et al. [17] demonstrated several substrates of cAMP-dependent protein kinase including polypeptides of apparent molecular weight of 50,000, 36,000, 24,000, and 20,000. Increases in labeling of the 24K and 50K proteins were detectable after 30 sec. The authors concluded that these latter two phosphorylations might be of more physiological significance based on the fact that inhibition of platelet function is complete by this time.

James L. Daniel • Temple University, Thrombosis Research Center, Philadelphia, Pennsylvania 19104.

One possible regulatory role for certain of these cAMP-dependent phosphorylations comes from studies of Käser-Glanzmann et al. [23,24] on calcium-sequestering membrane vesicles. Calcium seems to be a mediator of platelet responses such as shape change, aggregation, and secretion (see Ref. 11 for a review). When an exogenous cAMP-dependent protein kinase was added to vesicles, the rate of calcium uptake was increased [23]. It was later shown that an endogenous protein kinase was present which could cause a similar increase in calcium uptake by the vesicles [24]. When [γ-^{32}P]-ATP was added to this preparation, radioactivity was incorporated into proteins of molecular weights in the region of 22,000–24,000 [24]. In related experiments, 50K and 36K proteins were found in the cytosolic fraction of platelets that had been labeled ^{32}PO$_4$ and then treated with PGE$_1$ [12]. A membrane fraction contained the 24K and 22K proteins. The membrane fraction from the PGE$_1$-treated cells manifested a faster rate of ATP-dependent, oxalate-stimulated ^{45}Ca^{2+} uptake than control membranes. When the ^{45}Ca^{2+}-labeled vesicles were centrifuged on sucrose gradients, the ^{45}Ca^{2+} comigrated with the ^{32}P-labeling of the 22K and 24K proteins. These results suggest that one mechanism for cAMP-dependent inhibition of platelet function might be through increasing activity of a Ca^{2+} pump localized either to the platelet membrane or in sarcoplasmic reticulum-like vesicles. The pump in its activated form can more rapidly translocate cytoplasmic calcium, and as a result effects of platelet agonists are diminished or prevented.

Adelstein and Hathaway [20] demonstrated another possible substrate for cAMP-dependent protein kinase by showing that myosin light-chain kinase, which is the enzyme that phosphorylates the 20K light chain of platelet myosin, is phosphorylated by a cAMP-dependent protein kinase. Myosin light-chain kinase requires that calmodulin–Ca^{2+} complex for activity [19]. When the kinase is phosphorylated, its ability to bind calmodulin–Ca^{2+} is diminished [20]. There may also be a slight decrease in the maximum velocity of the enzyme. The phosphorylation of this kinase has yet to be demonstrated in an intact platelet, probably because it is present in low concentrations. However, these data suggest that elevated cAMP may inhibit by interfering with myosin light chain phosphorylation which regulates activation of the contractile system (see below).

The role of cGMP in platelet function is less certain. Based on studies with sodium nitroprusside, a known activator of guanylate cyclase in other cells [5] and inhibitor of platelet function [5,16], Haslam et al. [18] concluded that sodium nitroprusside inhibits platelet function through elevation of cGMP and not through a secondary elevation of cAMP levels. However, some conflicting and inconsistent data have made this conclusion somewhat tentative [18,35]. Addition of sodium nitroprusside to ^{32}PO$_4$-labeled platelets showed preferential labeling of a 49K polypeptide, and the 50K polypeptide labeled when cAMP was elevated [18].

Two unidentified proteins of molecular weight of 20K and 47K on SDS polyacrylamide gels showed enhanced incorporation of ^{32}PO$_4$ when ^{32}PO$_4$-labeled platelets were treated with thrombin [28]. Nishizuka and Takai [29] have shown that the 47K protein is phosphorylated by a protein kinase from brain which they have called C-kinase. The kinase requires three factors for full activity: micromolar concentrations of calcium, phospholipid, and small amounts of diacylglycerol (DG). Lyons and Atherton [27] isolated the 47K protein in the denatured state, and more recently Imaoka et al. [22] purified this protein under nondenaturing conditions. In spite of the intense interest in this protein, its function remains unknown.

Adelstein et al. [3] were the first to show that the 20K myosin light chain could be phosphorylated when crude platelet actomyosin was incubated with [γ-^{32}P]-ATP. Later, platelet 20K myosin light chain was shown to be phosphorylated by the specific calmodulin–Ca^{2+}-dependent protein kinase discussed above. A phosphatase that dephosphorylates the light chain has been isolated from platelets; however, it is not known whether this phosphatase is a regulated enzyme [2]. The actin-activated ATPase of platelet myosin is very low when it is dephosphorylated and enhanced as much as 20-fold when myosin is fully phosphorylated [1,32]. A model system for the study of isometric contraction is the actomyosin thread. The tension generated by a particular thread depends directly on the level of light chain phosphorylation of the myosin from which the thread was made [26]. Taken together, these results suggest that phosphorylation regulates contractile activity of myosin.

Platelet myosin phosphorylation is the focus of our attention. The biochemical regulation of this process is well defined and phosphorylation appears to be a major regulatory mechanism for contractile activity of platelet actomyosin. A study was undertaken in our laboratory to determine whether myosin phosphorylation occurs in intact platelets and in concert with physiological responses. Our studies led to the demonstration that myosin light chain is phosphorylated in intact platelets and showed that the 20K band of Lyons and coworkers' experiments [28] was myosin light chain [8]. Later, using alkaline urea polyacrylamide gel electrophoresis to separate phosphorylated and unphosphorylated light chain by charge difference, the change in radioactivity that occurred upon platelet activation was shown to represent a true change in myosin phosphorylation rather than a change in specific radioactivity of a constant amount of light-chain phosphate [9]. These studies also demonstrated that the change in the phosphorylation state of myosin in intact cells was significant (i.e., thrombin stimulation of the cells caused an increase in myosin phosphorylation from 10% to 90%). The phosphorylation of these proteins occurs in concert with platelet dense granule secretion [10,14,17]. Haslam and Lynham [15] showed that so-called "Ca^{2+} antagonists" blocked phosphorylation of both proteins. Although they did not detect phosphorylation during platelet aggregation, they only studied phosphorylation after relatively long incubations with agonist (about 1 min) [14].

To examine the association between myosin phosphorylation and platelet secretion, the temporal relationship between platelet dense granule secretion and myosin phosphorylation at temperatures below 37°C was investigated. In many of these experiments, secretion seemed to lag behind myosin phosphorylation. In one experiment, the time for myosin phosphorylation and secretion to reach 50% of its final level was less than 5 sec and 50 sec, respectively (Fig. 1). These data suggest that either the two phenomena are unrelated or there is another step after myosin phosphorylation that is rate limiting for secretion.

Based on these experiments, we decided to determine whether myosin phosphorylation could occur in the absence of secretion. Addition of ADP to gel-filtered platelets in the absence of fibrinogen and divalent metal ions produces a rapid shape change without aggregation or secretion [6]. Figure 2 shows an experiment in which the time course of myosin phosphorylation was measured in platelets undergoing shape change in response to ADP. Phosphorylation occurred very rapidly (< 10 sec), which is in agreement with the time course of shape change under these conditions [6]. No aggregation was detected by light microscopy. Secretion of either ATP or [^{14}C]serotonin was not enhanced, although shape change was maintained during the 5-min period of study. However, by this time, the level of myosin phosphorylation had returned to basal values. When cells fixed 5 min after adding ADP were studied by scanning electron microscopy, the shape of the cells correlated with the optical record, i.e., the cells still appeared the same as those fixed 10 sec after the addition of ADP. This experiment suggests that contractile proteins are necessary to produce, but not to maintain, the shape change response.

Pretreatment of cells with indomethacin did not affect the extent or time course of myosin phosphorylation. In other experiments (not shown), cells treated with ADP under conditions where aggregation and shape change occurred, manifested a similar myosin phosphorylation response to that shown in Fig. 2. Also, the dose–response curve of the

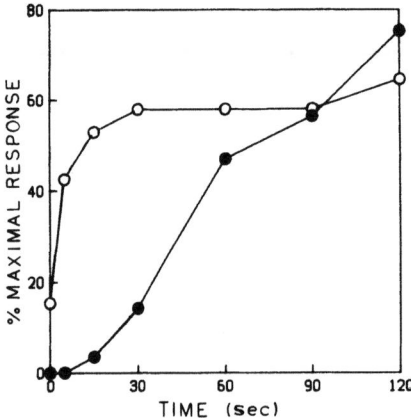

Figure 1. Relationship between the phosphorylation of myosin light chain and secretion of [^{14}C]serotonin. Gel-filtered platelets which had been labeled with [^{14}C]serotonin were incubated with 0.5 U/ml of thrombin at 30°C for the times indicated. Myosin phosphorylation (○) and secreted serotonin (●) were measured as previously described [9]. Myosin phosphorylation is expressed as a percent of phosphorylated light chain as determined on alkaline urea PAGE. Secretion is expressed as a percent of that induced by 5 U/ml of thrombin for 30 sec.

ADP-induced shape change was the same as that for myosin phosphorylation. These data suggest that contractile proteins play an important role in the shape change response, but not directly in aggregation.

The phosphorylation of the 47K peptide has also been measured under the same conditions. While myosin phosphorylation is substantial (about 40–60%) 5 sec after the addition of ADP, phosphorylation of the 47K protein was only slightly enhanced (Table I). However, when platelets are stimulated with thrombin and secretion occurs, phosphorylation of the 47K protein is more apparent. Preliminary experiments have revealed that by using conditions similar to those shown in Fig. 1, the time course of 47K phosphorylation could be well correlated with that of secretion.

According to the hypothesis advanced by Nishizuka and Takai [29], phosphorylation of the 47K protein requires DG. In order to test this hypothesis, gel-filtered platelets were labeled with [^{3}H]arachidonic acid and stimulated with ADP so that only shape change occurred. Since the 47K protein was not phosphorylated under these conditions, the amount of DG present was expected to be similar to that found in unstimulated cells. Indeed, ADP stimulation does not result in a significant increase in [^{3}H]-DG above the control level (Table II). By contrast, thrombin causes as much as a 10-fold increase in radioactivity associated with DG.

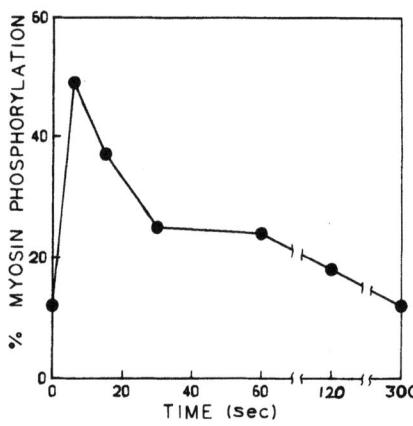

Figure 2. Myosin phosphorylation during platelet shape change. Gel-filtered platelets were incubated at 37°C with stirring and 10 μM ADP in the presence of 5 mM EDTA. NaOH was added with the EDTA to maintain the pH at 7.4. Samples were taken at the indicated times for measurement of myosin phosphorylation. Since alkaline PAGE was used to measure light chain phosphorylation, the values are absolute [9].

Table I. Phosphorylation of Myosin Light Chain and the 47,000-Dalton Polypeptide during Platelet Shape Change[a]

	Control	ADP	Thrombin
Myosin light chain	12%	54%	80%
47,000-dalton polypeptide	3%	5%	100%

[a] Myosin phosphorylation was measured as in Fig. 1 after 5-sec treatment with each agent or with saline for the control. Phosphorylation of the 47,000-dalton polypeptide was measured for densitometry of autoradiograms of 13.5% SDS PAGE as described previously [9].

Figure 3 shows a working model which can be used as a basis for future experiments. Weak agonists such as ADP cause an increase in cytoplasmic calcium. The increase in cytoplasmic calcium results in microtubule depolymerization [33], which may result in rounding of the cell. Actin filaments either polymerize or rearrange. Calcium is bound by calmodulin (CM), myosin light-chain kinase (MLCK) is activated, and myosin becomes phosphorylated. Myosin, after binding as filaments at the membrane, interacts with actin by a sliding filament mechanism similar to that of skeletal muscle. This actin–myosin interaction may cause extension of filopodia. Later, myosin light chain becomes dephosphorylated due to a fall in cytoplasmic calcium or an activation of the protein phosphatase which dephosphorylates the light chain. However, the filopodia are maintained, and after some time microbutules are found in the filopodia.

A stronger stimulus, such as thrombin, may mobilize more calcium, either directly or by activating the formation of such agents as thromboxane A_2 (TXA_2), phosphatidic acid (PA), or lysophosphatidic acid (LPA), which may open calcium gates. Also, thrombin causes DG production either by activating phospholipase C (PLC) or by exposing phosphatidylinositol (PI) to more efficient cleavage by PLC. DG and Ca^{2+} activate protein kinase C (PK-C) which phosphorylates the 47K polypeptide (P47). The phosphorylated form of this protein may promote secretion by an as yet unknown mechanism. The greater level of cytoplasmic Ca^{2+} may also lead to phosphorylation of another pool of myosin which, in concert with 47K, might be involved in triggering secretion. While most of the aspects of this model have not been proven experimentally, it does provide a basis for further investigation of stimulus–response coupling in platelets.

Table II. Comparison of ADP and Thrombin-Induced [³H]-Diacylglycerol Production in [³H]-Arachidonate-Labeled Platelets[a]

Expt	Control		ADP	Thrombin
1	78,	268	139	1033
2	376,	305	276	2612

[a] Platelets were labeled in plasma with 1 μCi/ml of [³H]arachidonic acid. After gel filtration, the cells were treated for 5 sec with each agent or saline and then the lipids extracted (4). Diacylglycerol was measured as ³H cpm after diacylglycerol was separated from other lipids on a two-dimensional TLC plate. The solvent in the first dimension was petroleum ether/diethyl ether/acetic acid (70/20/4). The second dimension was the system of Hong and Levine [21].

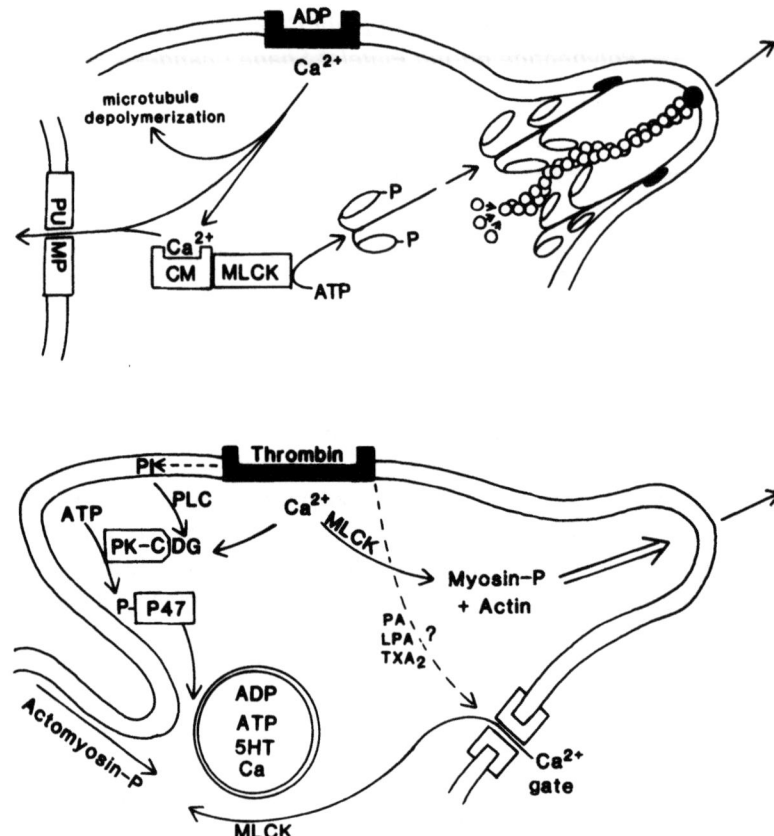

Figure 3. Hypothetical model of stimulus–response coupling in platelets. See text for further details and explanation of abbreviations.

References

1. Adelstein, R. S.; Conti, M. A. Phosphorylation of platelet myosin increases actin-activated myosin ATPase activity. *Nature (London)* **256**:597–598, 1975.
2. Adelstein, R. S.; Chacko, S.; Barylko, B.; Scordilis, S.; Conti, M. A. The role of myosin phosphorylation in the regulation of platelet and smooth muscle contractile proteins. In: *Current Topics in Intracellular Regulation II. Contractile Systems in Non-muscle Tissue* S. V. Perry, A. Margeth, and R. S. Adelstein, eds., Amsterdam, Elsevier/North-Holland, 1976, pp. 153–163.
3. Adelstein, R. S.; Conti, M. A.; Anderson, W., Jr. Phosphorylation of human platelet myosin. *Proc. Natl. Acad. Sci. USA* **70**:3115–3119, 1973.
4. Bligh, E. G.; Dyer, W. J. A rapid method of total lipid extraction and purification of phospholipid. *Can. J. Biochem.* **37**:911–917, 1959.
5. Bohme, E.; Graf, H.; Schultz, G. Effects of sodium nitroprusside and other smooth muscle relaxants on cyclic GMP formation in smooth muscle and platelets. *Adv. Cyclic Nucleotide Res.* **9**:131–143, 1978.
6. Born, G. V. R.; Dearnly, R.; Foulks, J. G.; Sharp, D. E. Quantification of the morphological reaction of platelets to aggregating agents and of its reversal by aggregation inhibitors. *J. Physiol. (London)* **280**:193–212, 1978.
7. Cohen, P. The role of protein phosphorylation in neural and hormonal control of cellular activity. *Nature (London)* **296**:613–620, 1982.
8. Daniel, J. L.; Holmsen, H.; Adelstein, R. S. Thrombin-stimulated myosin phosphorylation in intact platelets and its possible involvement in secretion. *Thromb. Haemost.* **38**:984–989, 1977.
9. Daniel, J. L.; Molish, I. R.; Holmsen, H. Myosin phosphorylation in intact platelets. *J. Biol. Chem.* **256**:7510–7514, 1981.

10. Daniel, J. L.; Molish, I. R.; Holmsen, H.; Salganicoff, L. Phosphorylation of myosin light chain in intact platelets: Possible role in platelet secretion and clot retraction. *Cold Spring Harbor Conf. Cell Prolif.* **8**:913–928, 1981.
11. Feinstein, M. B. The role of calcium in hemostasis. In: *Progress in Hemostasis and Thrombosis*, Volume 6, T. H. Spaet, ed., New York, Grune & Stratton, 1982, pp. 25–61.
12. Fox, J. E. B.; Say, A. K.; Haslam, R. J. Subcellular distribution of the different proteins phosphorylated on exposure of intact platelets to ionophore A23187 or to prostaglandin E_1. *Biochem. J.* **184**:651–661, 1979.
13. Gill, G. N.; Holdy, K. E.; Walton, G. M.; Kanstein, C. B. Purification and characterization of 3':5'-cyclic GMP-dependent protein kinase. *Proc. Natl. Acad. Sci. USA* **73**:3918–3923, 1976.
14. Haslam, R. J.; Lynham, J. A. Relationship between phosphorylation of blood platelet proteins and secretion of platelet granule constituents. I. Effects of different aggregating agents. *Biochem. Biophys. Res. Commun.* **77**:714–722, 1977.
15. Haslam, R. J.; Lynham, J. A. Relationship between phosphorylation of blood platelet proteins and secretion of platelet granule constituents. II. Effects of different inhibitors. *Thromb. Res.* **12**:619–628, 1978.
16. Haslam, R. J.; Davidson, M. M. L.; Desjardins, J. V. Inhibition of adenylate cyclase by adenosine analogues in preparations of broken and intact human platelets. *Adv. Cyclic Nucleotide Res.* **9**:533–552, 1978.
17. Haslam, R. J.; Lynham, J. A.; Fox, J. E. B. Effects of collagen, ionophore A23187 and prostaglandin E_1 on the phosphorylation of specific proteins in blood platelets. *Biochem. J.* **178**:397–406, 1979.
18. Haslam, R. J.; Salma, S. E.; Fox, J. B.; Lynham, J. A.; Davidson, M. M. L. Roles of cyclic nucleotides and of protein phosphorylation in the regulation of platelet function. In: *Platelets: Cellular Response Mechanisms and Their Biological Significance*, M. M. L. Davidson, A. Rotman, F. A. Meyer, C. Gitler, and A. Silverberg, eds., New York, Wiley, 1980, pp. 213–231.
19. Hathaway, D. R.; Adelstein, R. A. Human platelet myosin light chain kinase requires the calcium-binding protein calmodulin for activity. *Proc. Natl. Acad. Sci. USA* **76**:1653–1657, 1979.
20. Hathaway, D. R.; Eaton, C. R.; Adelstein, R. S. Phosphorylation of human platelet myosin light chain kinase by the catalytic subunit of cyclic AMP-dependent protein kinase. *Nature (London)* **291**:252–254, 1981.
21. Hong, S. L.; Levine, L. Inhibition of arachidonic acid release from cells as the biochemical action of anti-inflammatory corticosteroids. *Proc. Natl. Sci. USA* **73**:1730–1734, 1976.
22. Imaoka, T.; Lynham, J. A.; Haslam, R. J. Purification and characterization of the 47,000 dalton protein that is phosphorylated during the platelet release reaction. *Fed. Proc.* **42**:660, 1983.
23. Käser-Glanzmann, R.; Gerber, E.; Luscher, E. F. Regulation of the cellular calcium level in human platelets. *Biochim. Biophys. Acta* **558**:344–347, 1979.
24. Käser-Glanzmann, R.; Jakabova, M.; George, J. N.; Luscher, E. F. Stimulation of Ca^{2+} in platelet membrane vesicles by adenosine 3',5'-cyclic monophosphate and protein kinase. *Biochim. Biophys. Acta* **466**:429–440, 1977.
25. Kaulen, H. D.; Gross, R. Purification and properties of a soluble cyclic AMP dependent protein kinase of human platelets. *Hoppe-Seylers Z. Physiol. Chem.* **355**:471–480, 1975.
26. Lebowitz, E. A.; Cooke, R. Contractile properties of actomyosin from human blood platelets. *J. Biol. Chem.* **253**:5443–5447, 1978.
27. Lyons, R. M.; Atherton, R. M. Characterization of a platelet protein phosphorylated during the thrombin-induced release reaction. *Biochemistry* **18**:544–552, 1979.
28. Lyons, R. M.; Stanford, N.; Majerus, P. W. Thrombin-induced protein phosphorylation in human platelets. *J. Clin. Invest.* **5**:924–936, 1975.
29. Nishizuka, Y.; Takai, Y. Calcium and phospholipid turnover in a new receptor function for protein phosphorylation. *Cold Spring Harbor Conf. Cell Prolif.* **8**:237–250, 1981.
30. Rosen, O. M.; Krebs, E. G. Protein phosphorylation. *Cold Spring Harbor Conf. Cell Prolif.* Volume 8, 1981.
31. Rubin, C. S.; Rosen, O. M. Protein phosphorylation. *Annu. Rev. Biochem.* **44**:831–896, 1975.
32. Sellers, J. R.; Pato, M. D.; Adelstein, R. S. Reversible phosphorylation of smooth muscle myosin, heavy meromyosin, and platelet myosin. *J. Biol. Chem.* **256**:13137–13142, 1981.
33. Steiner, M.; Ikeda, Y. Quantitative assessment of polymerized and depolymerized microtubules. *J. Clin. Invest.* **63**:443–448, 1979.
34. Wang, J. H.; Tam, S. W.; Lewis, W. G.; Sharma, R. K. The role of calmodulin in the regulation of protein phosphorylations in skeletal muscle. *Cold Spring Harbor Conf. Cell Prolif.* **8**:357–371, 1981.
35. Weiss, A.; Baenziger, N. L.; Atkinson, J. P. Platelet release reaction and intracellular cGMP. *Blood* **52**:524–531, 1978.

20

Na^+ and Ca^{2+} Fluxes and Differentiation of Transformed Cells

Lewis Cantley, Philip M. Rosoff, Robert Levenson, Ian G. Macara, Li-An Yeh, Leona Ling, and Leigh English

Alterations in cation fluxes across the plasma membrane appear to play a central role in the regulation of cellular growth and differentiation [1,2,4,6–8,17,19]. Oocyte maturation is stimulated by Ca^{2+} release into the cytoplasm [7], and serum factors which stimulate growth of quiescent cells in culture cause rapid changes in Na^+, H^+, Ca^{2+}, and K^+ fluxes across the plasma membrane [4,8,17]. However, it has not been conclusively demonstrated that these changes in cation fluxes are essential for the subsequent acceleration of growth. In this context, a limited number of established cell lines, which are capable of undergoing differentiation in culture, provide an opportunity for investigating the role of cation fluxes in growth regulation and signaling terminal differentiation.

We have been investigating cation fluxes in two established cell lines that are capable of differentiation in culture. Friend murine erythroleukemia cells, cultured from mouse spleen infected with Friend virus, undergo terminal erythroid differentiation when treated with dimethylsulfoxide (DMSO) or a variety of other agents. Agents that induce differentiation in Friend cells have the common ability to depress K^+ uptake through the Na^+-K^+ pump [13]. In fact, the specific Na^+-K^+-ATPase inhibitor ouabain induces erythropoiesis in a Friend cell line selected for a partial ouabain resistance [1]. These observations suggest an important role for the Na^+-K^+ pump in differentiation.

Our laboratory is also investigating a pre-B-lymphocyte cell line (70Z/3) isolated from spleens of mice which had been subjected to chemical mutagens [14]. These cells, like Friend cells, divide indefinitely in suspension. Their immunoglobulin genes have rearranged and the heavy chain of IgM is synthesized but no light chain is produced and no IgM appears on the surface of the cells. When treated with lipopolysaccharide (LPS) or supernatants from

---

Lewis Cantley, Philip M. Rosoff, Ian G. Macara, Li-An Yeh, Leona Ling, and Leigh English • Department of Biochemistry and Molecular Biology, Harvard University, Cambridge, Massachusetts 02138.   *Robert Levenson* • Department of Biology and Center for Cancer Research, Massachusetts Institute of Technology, Cambridge, Massachusetts 02139. Present address of I.G.M.: Department of Radiation Biology and Biophysics, University of Rochester School of Medicine, Rochester, New York 14642.

cloned helper T cells, light chain is produced and the complete IgM molecule is expressed on the surface [15]. This process apparently represents differentiation to a more mature B cell [14,15].

In this chapter evidence will be summarized in support of the idea that in both Friend cells and 70Z/3 cells, commitment to differentiation results from an elevated cytoplasmic Na^+ concentration initiated by changes in Na^+ flux across the plasma membrane.

Inducing Agents Alter Cation Fluxes across the Plasma Membrane of Friend Cells

The first evidence that alterations in cation fluxes may be an essential signal in Friend cell differentiation was reported by Mager and Bernstein [13]. They showed that ouabain-sensitive $^{86}Rb^+$ uptake (an assay of the Na^+-K^+ pump) into Friend cells decreased by approximately 40% 6–12 hr after the addition of DMSO, hypoxanthine, or N,N-dimethyl-acetamide (all agents that induce erythropoiesis). Xanthine, which does not induce erythropoiesis, had no effect on $^{86}Rb^+$ uptake.

In addition to the decrease in $^{86}Rb^+$ uptake, we observed an increase in $^{45}Ca^{2+}$ flux [10], a rise in cellular Na^+ (I. Macara, unpublished observation), and a decrease in cyanine dye uptake into Friend cell mitochondria (an indication of mitochondrial membrane potential [11]). All of these flux changes are presumed to be a consequence of decreased activity of the Na^+-K^+ pump because of the following observations: (1) treatment with ouabain raises the cellular Na^+ concentration and enhances $^{45}Ca^{2+}$ flux into the cells (probably via a Na^+/Ca^{2+} exchange system) [18]; (2) addition of amiloride or EGTA prevents the DMSO enhancement of Ca^{2+} influx and impairment of cyanine dye uptake [10,11]; (3) addition of Ca^{2+} ionophore A23187 decreases mitochondrial cyanine dye uptake [11]. Thus, in Friend cells inhibition of the Na^+-K^+ pump causes enhancement of Ca^{2+} flux, which in turn brings about the decrease in cyanine dye uptake. However, these experiments do not rule out the possibility that DMSO may also affect other cation transport systems by a mechanism independent of the effect on the Na^+-K^+ pump.

Because of the implied primary role of the Na^+-K^+ pump in altering cation fluxes in differentiating Friend cells, the mechanism by which DMSO mediates inhibition was investigated. The 100,000-M_r catalytic subunit of this Na^+-K^+-ATPase was phosphorylated in living cells and in purified plasma membranes by a highly specific plasma membrane-bound protein kinase [20]. In vitro, this kinase adds approximately three phosphates per 100,000-M_r peptide within 10 min at room temperature [3,20]. It is not regulated by cAMP or Ca^{2+} and is not inhibited by EGTA or heparin. When Friend cells are treated with DMSO, phosphorylation of the Na^+-K^+-ATPase decreases in parallel with the decrease in Na^+-K^+ pump activity, suggesting that DMSO induces the change in cation fluxes by altering the phosphorylation level of the Na^+-K^+ pump (L.-A. Yeh, unpublished observations).

LPS Alters Na^+ and K^+ Fluxes in 70Z/3 Cells

Although the mechanism by which LPS alters cation fluxes in 70Z/3 cells is different from effects of DMSO on Friend cells, an increase in cellular Na^+ is observed in both cases. LPS enhances an amiloride-sensitive Na^+ influx system in 70Z/3 cells to raise cellular Na^+ [16]. The Na^+-K^+ pump is activated, rather than inhibited, as an apparent consequence of elevated cellular Na^+ [16]. The effects of LPS on Ca^+ fluxes are not yet clear; however, preliminary results suggest that cytoplasmic Ca^{2+} levels increase.

An Essential Role for Ca^{2+} in Differentiation of Friend Cells and 70Z/3 Cells

The first evidence that a change in Ca^{2+} flux in Friend cells is important for differentiation came from studies with amiloride, which at concentrations (40 μM) having little effect on Na^+ influx or cell viability, prevent induction of erythropoiesis by DMSO, butyric acid, or hypoxanthine [10,12]. This block of commitment can be overcome if the culture is supplemented with ionophore A23187 [10,12]. The inhibition by amiloride of DMSO-induced $^{45}Ca^{2+}$ influx [10] occurs inside the cell and may be either a direct or an indirect effect on Na^+/Ca^{2+} exchange [18]. Also consistent with the importance of enhanced Ca^{2+} flux for differentiation is the observation that EGTA blocks differentiation at concentrations that permit cell growth [2,10].

Although Ca^{2+} flux studies have not yet been conducted on 70Z/3 cells, effects of amiloride and EGTA on surface IgM expression in these cells are similar to their effects on Friend cell erythropoiesis. Amiloride blocks LPS induction of surface IgM expression at concentrations (66 μM) that do not affect Na^+ uptake or cell viability [16]. EGTA also prevents surface IgM expression, but only if added several hours prior to LPS, suggesting that internal Ca^{2+} pools are important for differentiation of 70Z/3 cells.

Specific Ionophores and Ion Transport Inhibitors Induce Differentiation

Convincing evidence that cation flux changes are important signals for differentiation comes from studies with specific ionophores. Although we have been unable to induce erythropoiesis of Friend cells with ionophores alone, a significant acceleration of commitment is observed when ionophores are added either together with or prior to other inducers. Regardless of the inducer used, a latency of about 12 hr occurs between addition of drug and appearance of a significant fraction of Friend cells capable of inducer-independent erythropoiesis [9]. Following this initial latency, commitment occurs stochastically. We found that treatment of Friend cells with A23187 for 1 hr prior to DMSO addition allowed stochastic commitment to occur without the 12-hr lag [2]. A23187 alone did not cause commitment and prolonged treatment with this drug produced cell death. This result suggests that a change in Ca^{2+} flux is essential for commitment but may not alone be sufficient to induce differentiation. With all of the inducers employed, approximately a 12-hr latency was observed prior to commitment, and in all cases treatment of Friend cells with A23187 eliminated the latency [2,12]. Table I summarizes these results by showing the fraction of cells committed 18 hr after various drug additions. In the absence of ionophores, less than 5% of the cells commit to erythropoiesis, while in their presence, about 20% of the population is committed by 18 hr.

Commitment to erythropoiesis can also be accelerated by agents which raise cellular Na^+. Treatment of Friend cells with ouabain plus monensin for 4 hr was as effective as A23187 in eliminating the latency for commitment (Table I and Ref. 18). This result is not surprising since this treatment enhances $^{45}Ca^{2+}$ flux into Friend cells [18]. The time required for DMSO or hypoxanthine to reduce Na^+-K^+ pump activity by 40% (about 12 hr) is approximately equal to the latency for commitment [2,5,13]. These data are therefore consistent with the hypothesis that the rate-limiting event in commitment is a decrease in the activity of the Na^+-K^+ pump activity and that enhanced Ca^{2+} flux is a consequence of this inhibition.

Table I. Acceleration of Commitment of Friend Cells to Erythropoiesis by Specific Ion Transport Effectors

Drug[a]	% of cells committed[b]		Reference
	18 hr	50 hr	
Control	<5	<5	2,10–12,18
DMSO	<5	60	10
A23187 (1 hr)	<5	<5	2
FCCP (1 hr)	<5	<5	11
Monensin (4 hr)	<5	<5	18
Ouabain (4 hr)	<5	<5	18
Ouabain + monensin (4 hr)	<5	<5	18
DMSO + A23187 (1 hr)	20	60–65	2,11
DMSO + FCCP (1 hr)	18	60–65	11
DMSO + monensin (4 hr)	9	60	18
DMSO + ouabain (4 hr)	8	59	18
DMSO + ouabain (4 hr) + monensin (4 hr)	24	70	18
Butyric acid	7	41	12
Butyric acid + A23187 (1 hr)	20	60	12
Butyric acid + FCCP (1 hr)	18	~55	12
Hypoxanthine	5	45	12
Hypoxanthine + A23187 (1 hr)	20	65	12
Hypoxanthine + FCCP (1 hr)	21	55	12

[a]Unless otherwise indicated the drug was present for the full 18 or 50 hr. The drug concentrations used were: DMSO, 1.5%; A23187, 1 µg/ml; FCCP, 10 µg/ml; monensin, 5 µg/ml; ouabain, 150 µM; butyric acid, 1 mM; hypoxanthine, 0.5 mg/ml.
[b]Commitment was assayed by the fraction of cells that form benzidine-positive colonies when subcloned in a plasma clot [9]. The cells were subcloned in the absence of drugs at either 18 or 50 hr after initial drug addition.

A third way to accelerate commitment is by adding FCCP to Friend cells for 1 hr prior to DMSO treatment. FCCP probably accelerates commitment by altering Ca^{2+} fluxes, since this agent uncouples mitochondria and releases mitochondrial Ca^{2+} into the cytoplasm [11].

Expression of surface IgM on 70Z/3 cells can be induced with either ouabain or monensin. Induction of surface IgM requires continuous exposure of these cells to LPS for 18 hr (Ref. 16 and Table II). However, a 4-hr exposure to ouabain plus monensin is as effective as continuous exposure to LPS in inducing surface IgM. Ouabain or monensin alone also induces surface IgM, although less effectively than the drug combination. As discussed above, LPS raises the Na^+ level in 70Z/3 cells. These results taken together

Table II. Ouabain and Monensin Induce Differentiation of 70Z/3 Cells

Drug	% sIg$^+$ cells at 24 hr[a]
Control	1
LPS (10 µg/ml; 24 hr)	71
LPS (10 µg/ml; 4 hr)	4
Ouabain (400 µM; 4 hrs)	22
Monensin (2.5 µg/ml; 4 hr)	27
Ouabain (400 µM) + monensin (2.5 µg/ml; 4 hr)	75

[a]Surface IgM expression was assayed by counting the number of fluorescent cells after labeling with fluorescein-conjugated goat anti-mouse IgM [16].

indicate that the essential function of LPS in inducing surface IgM expression is the activation of Na^+ entry.

In summary, during differentiation of virally and chemically transformed cell lines, agents raise cellular Na^+ in both cell lines. This is accomplished in Friend cells by decreasing the rate of Na^+ efflux through the Na^+-K^+ pump, while in 70Z/3 cells an amiloride-sensitive Na^+ influx is activated. The increase in cellular Na^+ appears to be the rate-limiting event for commitment in both cell lines. Inhibition of the Na^+-K^+ pump in Friend cells produces an enhanced Ca^{2+} influx which appears essential for commitment.

ACKNOWLEDGMENTS. This research was supported by NIH Grant GM-28538 and by a Grant-in-Aid from the American Heart Association. P.M.R. was supported by USPHS Training Grant CA-09172-08 awarded to the Division of Pediatric Oncology, Dana–Farber Cancer Institute, Department of Pediatrics, Harvard Medical School. L.-A.Y. was supported by a postdoctoral fellowship from the Massachusetts Affiliate of the American Heart Association. L.C. is the recipient of an Established Investigatorship from the American Heart Association. L.E. is an NIH Postdoctoral Fellow.

References

1. Bernstein, A.; Hunt, V.; Crichley, V.; Mak, T. W. Induction by ouabain of hemoglobin synthesis in cultured Friend erythroleukemia cells. *Cell* **9**:375–391, 1976.
2. Bridges, K.; Levenson, R.; Housman, D.; Cantley, L. Calcium regulates the commitment of murine erythroleukemia cells to terminal differentiation. *J. Cell Biol.* **90**:542–544, 1981.
3. Cantley, L.; Yeh, L.-A.; Ling, L.; Schulz, J.; English, L. Characterization of a plasma membrane kinase which specifically phosphorylates the (Na,K) pump. In: *Structure and Function of Membrane Proteins*, F. Palmieri, ed., Amsterdam, Elsevier/North-Holland, 1983, pp. 73–79.
4. Deutsch, C.; Price, M. A.; Johansson, C. A sodium requirement for mitogen-induced proliferation in human peripheral blood lymphocytes. *Exp. Cell Res.* **136**:359–369, 1981.
5. English, L. H.; Macara, I. G.; Cantley, L. Vanadium stimulates the (Na,K) pump in Friend erythroleukemia cells and blocks erythropoiesis. *J. Cell Biol.* **97**:1299–1302, 1983.
6. Hesketh, T. R.; Smith, G. A.; Houslay, G. B.; Warren, G. B.; Metcalfe, J. A. Is an early calcium flux necessary to stimulate lymphocytes? *Nature (London)* **267**:490–494, 1977.
7. Jaffe, L. F. Calcium explosions as triggers of development. *Ann. N.Y. Acad. Sci.* **339**:86–101, 1980.
8. Kaplan, J. G. Activation of cation transport during lymphocyte stimulation: The molecular theology of spinning metabolic wheels. *Trends Biochem. Sci.* **4**:N147–N149, 1979.
9. Levenson, R.; Housman, D. Memory of MEL cells to a previous exposure to inducer. *Cell* **17**:485–490, 1979.
10. Levenson, R.; Housman, D.; Cantley, L. Amiloride inhibits murine erythroleukemia cell differentiation: Evidence for a Ca^{2+} requirement for commitment. *Proc. Natl. Acad. Sci. USA* **77**:5948–5952, 1980.
11. Levenson, R.; Macara, I. G.; Smith, R. L.; Cantley, L.; Housman, D. Role of mitochondrial membrane potential in the regulation of murine erythroleukemia cell differenitation. *Cell* **28**:855–863, 1982.
12. Levenson, R.; Macara, I.; Cantley, L.; Housman, D. Ionic regulation of MEL cell commitment. *J. Cell. Biochem.* **21**:1–8, 1983.
13. Mager, D.; Bernstein, A. Early transport changes during erythroid differentiation of Friend erythroleukemia cells. *J. Cell. Physiol.* **94**:275–285, 1978.
14. Paige, C. J.; Kincaide, P. W.; Ralph, P. Murine B cell leukemia with inducible surface immunoglobulin expression. *J. Immunol.* **121**:641–647, 1978.
15. Paige, C. J.; Schreier, M. H.; Sidman, C. L. Mediators from cloned T helper cell lines effect immunoglobulin expression in B cells. *Proc. Natl. Acad. Sci. USA* **79**:4756–4760, 1982.
16. Rosoff, P. M.; Cantley, L. Increasing cellular Na^+ induces differentiation in a pre-B lymphocyte cell line. *Proc. Natl. Acad. Sci. USA* **80**:7547–7550, 1983.
17. Rozengurt, E.; Gelehrter, T. D.; Legg, A.; Pettican, P. Mellitin stimulates Na^+ entry, Na-K pump activity and DNA synthesis in quiescent cultures of mouse cells. *Cell* **23**:781–788, 1981.
18. Smith, R. L.; Macara, I. G.; Levenson, R.; Housman, D.; Cantley, L. Evidence that a Na^+/Ca^{2+} antiport system regulates murine erythroleukemia cell differentiation. *J. Biol. Chem.* **257**:773–780, 1982.

19. Szamel, M.; Schneider, S.; Resch, K. Functional interrelationship between (Na,K)ATPase and lysolecithin acyltransferase in plasma membranes of mitogen stimulated rabbit thymocytes. *J. Biol. Chem.* **256**:9198–9204, 1981.
20. Yeh, L.-A.; Ling, L.; English, L.; Cantley, L. Phosphorylation of the (Na,K)ATPase by a plasma membrane bound protein kinase in Friend erythroleukemia cells. *J. Biol. Chem.* **258**:6567–6574, 1983.

V

Calcium as a Regulator of Membrane Permeability

21

Calcium-Activated Potassium Channels in Bullfrog Sympathetic Ganglion Cells

Paul R. Adams, David A. Brown, Andrew Constanti, Robert B. Clark, and Leslie Satin

The existence of a calcium activated potassium current (I_c) in the somata of vertebrate sympathetic ganglion cells was postulated to account for the calcium-sensitive spike afterhyperpolarizations present in these cells [19,22,26]. We have studied I_c in bullfrog ganglion cells more directly by using various voltage-clamp techniques, partly in order to understand better the role this current plays in spike repolarization, spike afterhyperpolarization, and spontaneous hyperpolarizations, and partly to define the difference between I_c and the M-current I_m [4]. Both I_c and I_m are voltage-sensitive potassium currents sensitive to transmitters, the former being activated by internal calcium and the latter inactivated by external acetylcholine. Despite these superficial similarities, it turns out that the two currents have virtually nothing in common.

Our understanding of I_c in bullfrog cells comes from various types of experiments. We can separately measure calcium influx and the effects of artificially applied internal calcium on potassium currents. In principle, and to some extent in practice, one can predict what will happen when calcium is induced to flow in across the cell membrane by appropriate depolarizing voltage pulses to activate I_c "naturally." These voltage-clamp results are then used to predict some qualitative features of action potentials in these cells. Lastly, we will describe an unusual spontaneous event which we believe reflects the packeted action of calcium acting as a true intracellular transmitter.

Calcium Influx

Voltage-dependent calcium influx into bullfrog neurons has been directly demonstrated using arsenazo [28] but we have used an indirect approach [2,3,5], hoping that the clamp

---

Paul R. Adams and Leslie Satin • Department of Neurobiology and Behavior, State University of New York, Stony Brook, New York 11794.   *David A. Brown and Andrew Constanti* • Department of Pharmacology, School of Pharmacology, London WCIN IAX, England.   *Robert B. Clark* • Department of Physiology and Biophysics, University of Texas Medical Branch, Galveston, Texas 77550.

current observed after eliminating most of the sodium and potassium currents is calcium current. External sodium was replaced by TEA, 4AP was added, and internal and external potassium replaced by cesium, using the nystatin loading technique [29]. The residual voltage-dependent currents are probably indeed carried by calcium, because they are greatly reduced in nominally zero external calcium solutions, and are completely blocked by cadmium [100–500 μM]. Of course, these tests do not exclude the possibility of contamination by residual calcium-mediated I_c or outward cesium-current through calcium channels [14]. Such currents were indeed shown to be present, by injecting calcium iontophoretically into cells subjected to the procedures outlined above; but they were about 1000 to 2000-fold smaller than in normal cella. The main effect of this contamination is to make inactivation of I_{Ca} look more rapid than it really is.

Using -40 mV holding potential the calcium current first turned on at -21 ($\pm$ 1.4) mV, and reached a peak at $+10$ ($\pm$ 0.5) mV ($n = 16$). Its peak amplitude (with 10 mM external calcium) was 5.8 ($\pm$ 0.5) nA. At physiological calcium levels its peak amplitude could be estimated at 2.6 nA. Near threshold, the peak calcium current increases e-fold for about 8-mV depolarization. The turn-on and turn-off (tail current) of I_{Ca} were both approximately exponential, with a time constant that was a symmetrical bell-shaped function of voltage centered at -2 mV. The midpoint of the steady-state conductance curve determined from tail-current amplitude was about 10 mV positive to the midpoint of the time-constant bell. The time constant at the peak of the bell was 4 msec.

The calcium current appeared to inactivate if the cell was kept depolarized for at least 100 msec. The apparent inactivation was most pronounced at potentials giving the largest inward current. True inactivation was assessed from tail-current measurements and appeared to eliminate about half the calcium current by 1 sec at $+10$ to $+20$ mV. Unfortunately, the demanding protocol required for testing true inactivation prevented an investigation of the voltage dependence of its magnitude and time course. The time constant of recovery from inactivation was several seconds, and became faster at more negative potentials.

These data are all relevant to calculating the size and shape of I_c transients generated by voltage steps in normal Ringer solution. Perhaps the most useful number is the size of the calcium load entering during a 50-msec pulse to $+10$ mV—about 130 pC.

Calcium Injections

Iontophoretic injection of calcium into voltage-clamped ganglion cells evoked an outward current (6; Fig. 1a). The size of the outward current increases nonlinearly as the holding potential is made more positive. Part of the increase is due to the increased driving force on potassium movement. In high potassium the current is inward at negative potentials, is zero near the expected K-equilibrium potential, and becomes outward at more positive membrane potentials (Fig. 1c). The nonlinearity arises because the calcium-evoked conductance is itself voltage dependent (Fig. 1d; see also Refs. 6,12,32), increasing initially e-fold for about 12-mV membrane depolarization (Fig. 1d), until saturation is reached at positive potentials.

The effect of membrane voltage on the calcium-evoked potassium conductance (G_c) does not appear instantly, so that following a voltage step, I_c shows an instantaneous change (due to the changed driving force), followed by a relaxation (reflecting opening or closing of potassium channels [6] (Fig. 1b). Such voltage-jump relaxation responses are well-known features of transmitter-operated channels [1,25,27]. In principle, the voltage sensitivity of G_c could reflect an effect of potential on the channel open time, on the opening frequency, or

21. Ca^{2+}-Activated K$^+$ Channels in Sympathetic Ganglion

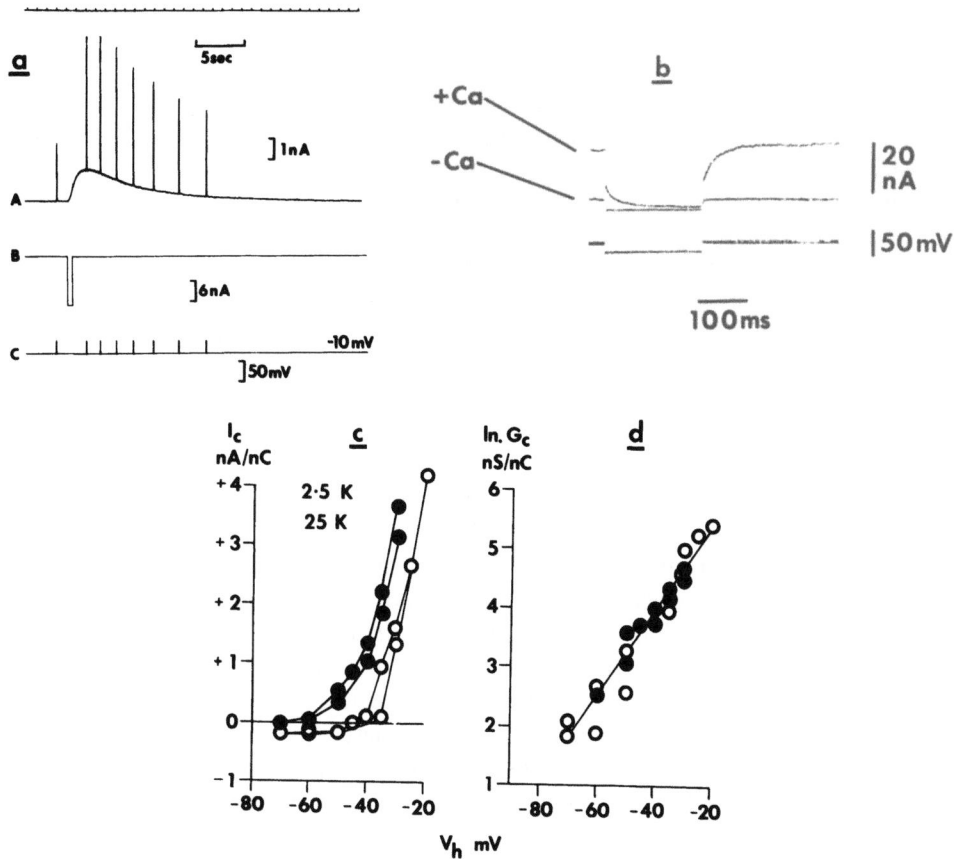

Figure 1. Membrane currents elicited by calcium injection into bullfrog neurons. (a) The outward current responses (row A) to an iontophoretic calcium injection (monitored in row B) in a cell voltage-clamped with two microelectrodes (potential monitored in row C). Holding potential −10 mV, clamp steps to 0 mV. (b) Expanded timebase records (from another similar experiment) of membrane currents elicited by hyperpolarizing clamp steps before (−Ca trace) or during (+Ca trace) iontophoretic calcium injection via a third intracellular microelectrode. (c) The amplitude of the peak current elicited by constant calcium injection into a bullfrog neuron by using a single-electrode voltage clamp. Two runs each were made in normal Ringer and in raised potassium. (d) The calcium-evoked conductance increase (on a log scale) calculated from the data in (c), as a function of membrane potential. The slope of the line corresponds to an *e*-fold change in conductance for 11-mV depolarization.

combination of both. In theory, this could be determined by analyzing the effect of voltage on the relaxation time course. However, though the relaxations were clearly faster at more negative potentials, especially when the calcium injection was small, they did not always show a single exponential time course. This may reflect multiple open states or nonuniform calcium concentration, and prevents any simple analysis. The relaxation time constant at −10 mV was about 15 msec.

The calcium-evoked current is noisy, as if it were due to the random activation of highly conductive channels. We estimated the unitary current through the channels by taking the ratio of the current variance and mean. The value we obtained, 6 pA at holding potentials of −20 to 0 mV, should be viewed with caution because the power spectrum of the current noise was non-Lorentzian, but it is compatible with our other estimates (see below). The response to calcium injections was not blocked by external cadmium (100 μM) but was blocked by millimolar concentrations of external TEA.

I_c Evoked by "Natural" Calcium Entry

In normal Ringer solution, voltage steps to -20 mV or more positive evoke large outward currents that are presumably carried by potassium. These large outward currents can be separated into calcium-independent and calcium-dependent components by treating the cells with low-calcium Ringer or calcium blockers such as cadmium, cobalt, manganese, or nickel. The calcium-independent component is probably mostly Hodgkin–Huxley-type delayed rectifier current. Prolonged depolarizations (30 sec or more) cause complete inactivation of this component, so that the current surviving such long depolarizations is leakage current plus I_c. Thus, in the absence of external calcium, short hyperpolarizing voltage pulses applied after keeping the cell depolarized for 30 sec or more do not evoke any time-dependent currents. However, in the presence of calcium such short voltage pulses reveal relaxing currents which strongly resemble those seen during internal calcium injection experiments [6] (Fig. 2).

Our interpretation of these relaxing currents is that during prolonged voltage steps the internal calcium levels are raised and fairly uniform in the presence of external calcium. During short probing voltage pulses there is little change in the internal calcium level, and the relaxations reveal the voltage-dependent behavior of an ensemble of I_c channels subjected to a low, constant, and uniform (but unknown) internal calcium concentration. The time constant of these relaxations is a function of membrane potential (cf. Fig. 2). Its value at -10 mV is 16 msec, in agreement with that measured by calcium injection. The time constant changes exponentially with 27-mV membrane hyperpolarization. If this time constant represents the lifetime of I_c channels, this result suggests that about half of the voltage dependence of the mean calcium-evoked conductance reflects a voltage-dependent channel-opening frequency.

Prolonged depolarizations were also used to study the properties of the calcium-dependent noise. In this case, probably because the attained intracellular calcium levels were low but uniform, Lorentzian spectra were usually obtained. The conductance of a single I_c channel was estimated to be 100 pS [6].

The outward currents evoked by *brief* positive steps from a negative holding potential are more complex and variable, possibly reflecting very rapidly changing internal calcium levels, both in space and in time. First, the size of the outward current produced by repeating

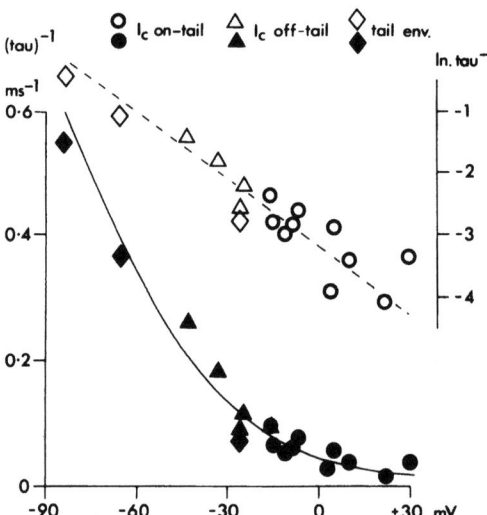

Figure 2. Reciprocal relaxation time constants of I_c elicited by prolonged depolarization in calcium-containing Ringer solution. The time constants were determined by imposing short hyperpolarizations during the maintained depolarizations, and measuring relaxations due to closing (triangles) or opening (circles) of I_c channels. Diamonds show measurements from the envelope of turn-on tails elicited by variable-duration hyperpolarizations to the test potential. The straight line in the semilog plot has a slope of 27 mV.

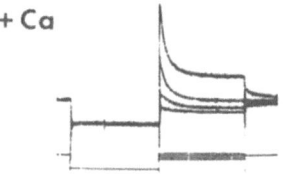

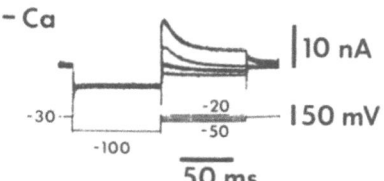

Figure 3. Transient outward currents triggered by hyperpolarizing–depolarizing sequences in a bullfrog neuron. The top traces were recorded in our normal frog Ringer, the bottom traces in the same cell after switching to a nominally calcium-free Ringer. Note that part of the transient outward current triggered by steps to −20 or −30 mV is calcium sensitive. The calcium-insensitive transient outward current is thought to be A current.

such a step, say to +20 mV for 50 msec, usually declines gradually with time, even if the cell is not stimulated between tests. Second, the size of the current increases and then decreases as the step is made more positive, giving an N-shaped current–voltage relation [4,5]. However, the rate of decline and the depth of N are very variable. Both are correlated with the susceptibility of the large outward current to calcium blockers or calcium deprivation, and therefore both phenomena reflect the participation of I_c. The decline probably occurs because I_{Ca} is itself labile (I_c elicited by direct calcium injection is robust). The N shape is generally thought to reflect decreasing calcium entry at positive potentials [20] though a contribution from rectification of the single-channel current–voltage relation also seems possible (see below).

The initial size of the peak calcium-dependent outward current for a step from −30 mV to +10 mV is about 30 nA [4]. The calcium influx caused by such a pulse would be expected to evoke an outward current of about 18 nA (correcting the sensitivity to calcium reported in Ref. 6 at −30 mV to +10 mV by using the observed voltage dependence of I_c). However, this rough agreement with theory is probably misleading, because the theory predicts that the evoked outward current should grow linearly with increasing pulse duration at least until inactivation of I_{Ca} becomes prominent. On the other hand, the calcium-sensitive component of outward current usually decreases as the pulse is prolonged beyond about 10 msec [6,15]. This behavior is particularly pronounced when the cell is briefly stepped to −100 mV before depolarizing to the I_c threshold (Fig. 3). The transient outward current generated by this protocol resembles the A current which can be recorded in these cells at more negative potentials [4]; this led us to postulate that the outward current is generated by a transient overshoot in internal calcium acting on the I_c system [8].

Single I_c Channels

Armed with an estimate of the conductance of a single I_c channel, we began patch-clamping cultured adult bullfrog ganglion cells. Depolarized, cell-attached patches exhibit a cornucopia of outward current events (Fig. 4A), which seem to fall into three general classes: (1) very large events (10 pA at 0 mV) exhibiting brief interruptions; (2) medium-sized events of rather scattered amplitude; and (3) very prolonged but small events present in a very small fraction of patches. The medium-sized events are initially quite numerous but then typically

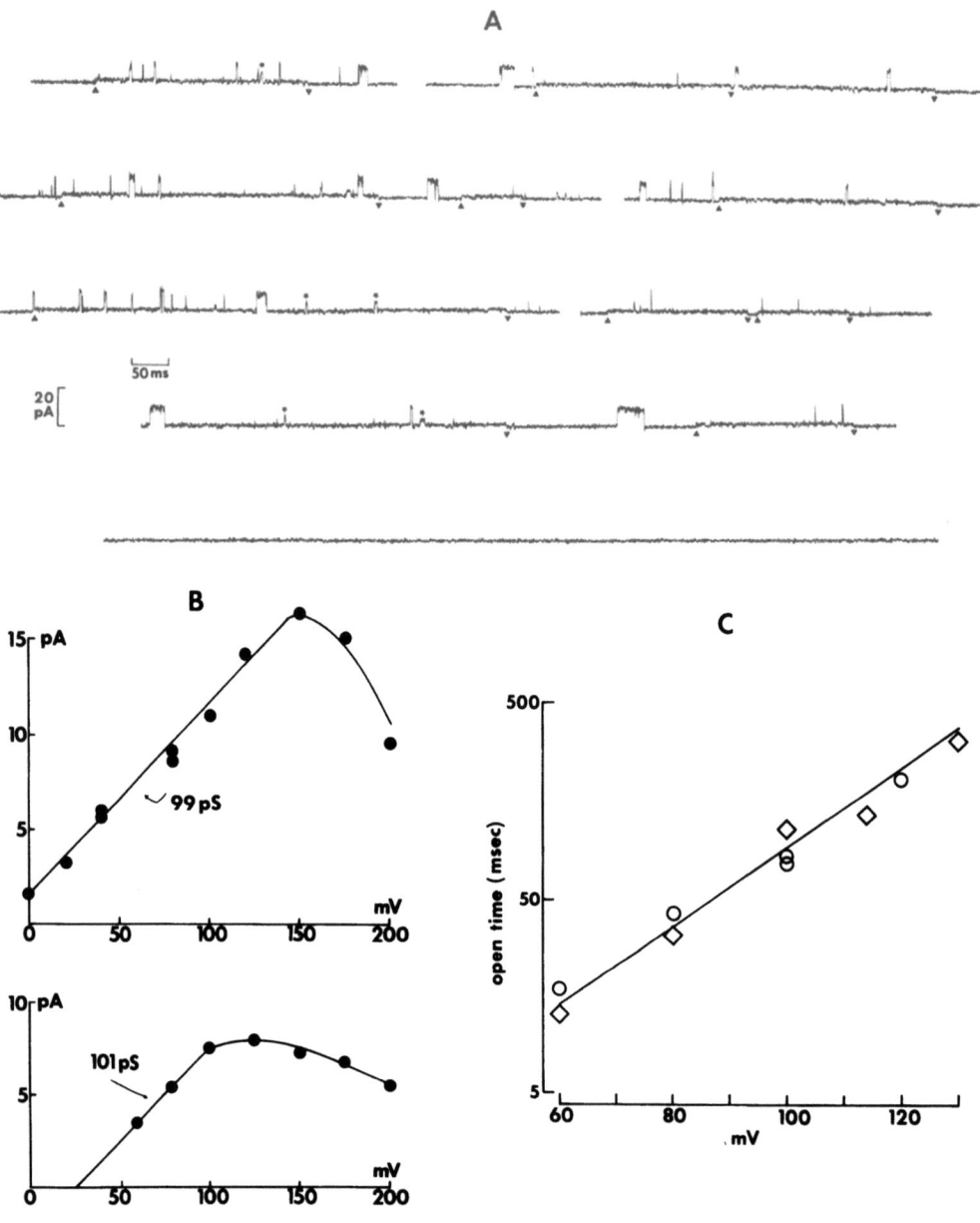

Figure 4. Properties of single I_c channels. (A) Single-channel currents recorded with a patch pipette on the surface of a cultured bullfrog neuron. The bottom trace shows background noise recorded at the resting potential, while the remaining traces show records when the patch was depolarized by 60 mV. The large outward I_c channel currents are not marked. Stars mark putative delayed rectifier channel currents; each arrowhead marks the turn-on or turn-off of a long-lived outward current channel that may be an M channel. (B) Two examples of the single-channel current–voltage relation for large outward events of on-cell patches in cultured bullfrog neurons. The potentials indicated are displacements from rest. (C) A semilog plot showing mean open time of large outward events in two on-cell patches. The slope of the line drawn is 22mV.

inactivate over several seconds, though they appear singly and sporadically thereafter. They are, therefore, probably delayed rectifier channels.

The large events are usually brief and infrequent unless the patch is clamped to very positive potentials. However, in some cells they are present even at potentials between rest and zero; and if the other channel types are absent, they can be studied at leisure. Their current voltage relation is shown in Fig. 4B. The initial linear region has a slope of 100 pS; this, coupled with their ability to be blocked by external TEA, led us to identify then as I_c channels. The single-channel current becomes flat and eventually falls during extreme depolarization. The negative resistance characteristic may explain the N shape of the whole-cell macroscopic outward current (see above). The decline in the single-channel current at positive potentials is paralleled by increasing noisiness of the open-channel current. Both probably reflect vibration of a blocking moiety or molecule into the channel. The mean open time of these channels increases exponentially for about 22-mV depolarization; the opening frequency also increases as the membrane is depolarized. The total time channels are open thus increases steeply (e-fold for 17 mV) before leveling off.

Several laboratories are examining the interrelationship of calcium concentration, membrane potential, and channel kinetics using excised membrane patches [7,13,17,21,30] to define a kinetic scheme for the I_c channel. This information, coupled with quantitative descriptions of calcium diffusion, buffering, sequestration, and transport, should eventually allow the prediction of I_c transients generated by voltage or current steps.

Roles of I_c in Bullfrog Cells

The classical role for I_c in ganglion cells and other neurons is the generation of spike afterhyperpolarizations (AHP). Our data, in fact, predict a two-component decay of the AHP. The initial component would reflect voltage-dependent closure of I_c channels triggered by spike repolarizaton. Since the time constant for this process is only a few milliseconds, it might be severely filtered by the membrane capacity. The second component (probably lasting several seconds) reflects dissipation of the internal calcium load that entered during the action potential. The transient nature of the current evoked by brief calcium injections shows that mechanisms exist for rapidly disposing of such calcium loads. These mechanisms can be overloaded, for example by a massive sustained calcium injection, thereby greatly prolonging evoked currents. The reported effects of various transmitters on spike AHP [16,24], which have been ascribed to a reduction in I_c, may reflect an altered ability of the cell to handle calcium loads.

The effect of steady polarizing currents on the AHP in current-clamp experiments is often interpreted in terms of resultant changes in driving force for potassium movements. However, two other effects will also be important. The steady currents modify both the membrane potential level from which the spike takes off, and the membrane potential to which it returns. The former will mainly affect the size of the calcium load, while the latter affects the size of the G_c response to this load.

The exquisite sensitivity of G_c to membrane potential in vertebrate neurons [6,13] is rather surprising, since it implies that the G_c observed during the spike AHP is a mere shadow of that developed during the spike. A highly voltage-sensitive G_c seems much better adapted to produce spike *repolarization*. We have obtained evidence for this in bullfrog neurons [6].

A third quite unexpected role for G_c in these cells emerged when intracellular recordings from the cultured bullfrog ganglion cells were made. About half of these cells exhibited

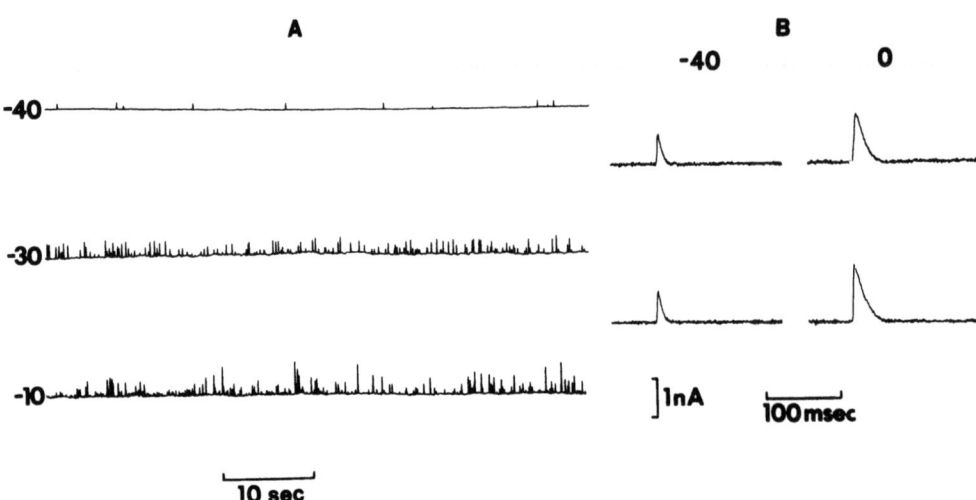

Figure 5. Spontaneous miniature outward currents (s.m.o.c.) recorded in a bullfrog cultured ganglion cell. The traces in (A) show slow timebase records at three different potentials. The traces in (B) show fast sweep records at two different holding potentials (left and right). Note the increased s.m.o.c. frequency and decreased decay rate at depolarized potentials. Two-microelectrode voltage clamp. The Ringer solution contained 200 µM Cd^{2+}.

spontaneous hyperpolarizations that resembled miniature synaptic potentials. However, recordings from pairs of cells failed to reveal any sign of synaptic connection between cells. The spontaneous hyperpolarizations resembled those observed in fresh mudpuppy cardiac principal cells [9] and cultured dorsal root ganglion cells [18]. Under voltage clamp they were seen as spontaneous miniature outward currents (s.m.o.c.; Fig. 5) with an abrupt rising phase, lasting about 3 msec, and an exponential decaying phase. S.m.o.c. amplitudes were very variable, both from cell to cell and within a given cell. The amplitudes were exponentially distributed. The interval between s.m.o.c. also seemed random, though distributions have not yet been examined. However, frequency was very steeply dependent on potential, rising abruptly between -50 mV and -10 mV, sinking between -10 mV and $+10$ mV, before rising again at more positive potentials. Voltage-dependent increases in s.m.o.c. frequency were still seen in cells exposed to 200 µM cadmium to block calcium current. The striking effect of cell potential on s.m.o.c. frequency suggests that they are generated somewhere in the cell itself, rather than by an input to the cell, especially since the effect persists in the presence of cadmium.

Depolarization also reduced the rate of decay of s.m.o.c., with no effect on rise time. The s.m.o.c. decay time constant increased e-fold for 33-mV depolarization, with an average decay time constant of 8 msec at -40 mV. Similar values were inferred from spectral data for spontaneous events in dorsal root ganglion cells [18]. The voltage dependence and decay time constant seemed uncannily similar to the lifetime of I_c channels determined by noise, voltage jump, and single-channel recording. This could be explained if a s.m.o.c. were due to the internal release of calcium "packet" near the inner membrane surface, whose contents then rapidly bound to the calcium receptors presumably associated with I_c channels. This explanation fits well with Mathers and Barker's [18] observation that s.m.o.c. are quite sensitive to external TEA. In the risetime of a s.m.o.c., calcium could only diffuse about 1 µM, suggesting that subsurface cisterns of the endoplasmic reticulum, which store calcium, might be the origin of such packets [10].

In *Amphioxus* muscle cells and cerebellar neurons, fuzz connects the subsurface endoplasmic reticulum to the surface membrane [11]. These structures could communicate the

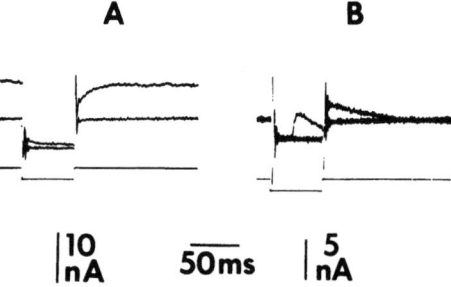

Figure 6. Comparison of relaxations during calcium injection (A) or a s.m.o.c. (B). In each case two traces were superimposed, during one of which neither calcium injection nor s.m.o.c. occurred (square current responses). In all four traces a 30-mV hyperpolarizing voltage command (monitored in bottom beam) was applied. During calcium injection (A) the baseline moves up due to the development of an outward current of about 4 nA. During hyperpolarization I_c turns off almost completely, and turns back on with an exponential time course on returning to the holding potential (−10 mV). The relaxation time constant was 17 msec. In (B) (recorded prior to insertion of calcium electrode, hence lower leakage) a voltage jump was applied during the decay of a s.m.o.c. (holding potential +10 mV).

surface voltage to the calcium stores, in a manner similar to the coupling between the T system and sarcoplasmic reticulum of muscle. The exponential size distribution would be obtained if calcium packets were generated by subsurface calcium channels opening for random periods. During the decay of a s.m.o.c. free calcium would again be low, so random I_c channel closure would generate the observed exponential time course. This idea was examined by injecting calcium iontophoretically into cells from which s.m.o.c. were also recorded. The voltage jump method was employed to determine the I_c channel lifetime (Fig. 6). Measurements of s.m.o.c. decay time constants and I_c relaxation time constants agreed over a wide voltage range (Fig. 7), supporting our interpretation.

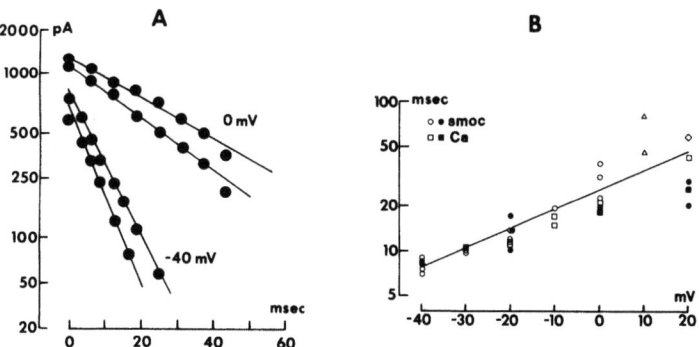

Figure 7.(A) Semilog plots of the decay phase (starting at the peak) of four s.m.o.c., two recorded at −40 mV holding potential and two at 0 mV. The time constants obtained from the slopes of the straight lines were 7.0, 9.0, 22.9, and 31.3 msec. The plots were made from oscilloscope records of events like those shown as chart records in Fig. 1. In (B), the time constants from these and other plots are shown on a logarithmic scale as a function of holding potential. Open circles are data from the cell illustrated in Fig. 1; filled circles are data from another cell in another culture. Squares show relaxation time constants determined as shown in Fig. 4A in the same cell, after insertion of a calcium pipette. Triangles refer to relaxations of s.m.o.c. following voltage jumps (see Fig. 4B). Diamond shows time constant of I_c-like relaxation determined without calcium injection, probably reflecting an appreciable resting intracellular calcium level after prolonged impalement with three electrodes. 200 μM Cd^{2+} present for all experiments. The solid line has a slope of 33 mV, for an e-fold change of the time constant.

Table I. Features of Chemical Transmission

Transmitter storage
Quantal release
Depolarization-triggered release
Action on membrane receptors
Synchronous opening of ion channels
Removal of transmitter
Random closing of ion channels

Table I summarizes some of the properties usually associated with conventional synaptic transmission. Most of these properties are also exhibited by s.m.o.c.s., reinforcing the parallel between calcium as an intracellular transmitter and conventional extracellular transmitters. Spontaneous hyperpolarizations of much longer duration observed under certain conditions in bullfrog ganglion cells [23] may represent internal calcium release activating I_c [13,14]. Indeed, calcium release during slow spontaneous hyperpolarizations has recently been measured directly [24]. Possibly s.m.o.c. represent the constituent subunits of slow hyperpolarizations.

Light-evoked miniature current events of similar time course are commonly recorded in photoreceptors [30,33]. These events may reflect release of discrete packets of internal transmitter near the inner membrane surface. The responses in bullfrog ganglion cells as described here may thus provide a useful model for widespread transduction phenomena.

ACKNOWLEDGMENTS. Supported by NIH Grants NS-14986 and NS-18579 and by grants to D.A.B. and A.C. from the Medical Research Council and Wellcome Trust.

References

1. Adams, P. R. Kinetics of agonist conductance changes during hyperpolarization at frog end-plates. *Br. J. Pharmacol.* **53**:308–310, 1975.
2. Adams, P. R. The calcium current of a vertebrate neurone. In: *Advances in Physiological Sciences*, Volume 4, J. Salanki, ed., Proc. 28th Int. Congre. Physiol. Sci. Budapest, Akademiai Kiado, 1981.
3. Adams, P. R. Activation of calcium current in bullfrog sympathetic neurones. *J. Physiol. (London)* Submitted for publication.
4. Adams, P. R.; Brown, D. A.; Constanti, A. M-currents and other potassium currents in bullfrog sympathetic neurones. *J. Physiol. (London)* **330**:537–572, 1982.
5. Adams, P. R.; Brown, D. A.; Constanti, A. Voltage clamp analysis of membrane currents underlying repetitive firing of bullfrog sympathetic neurons. In: *Physiology and Pharmacology of Epileptogenic Phenomena*, M. R. Klee, ed., New York, Raven Press, 1982.
6. Adams, P. R.; Constanti, A.; Brown, D. A.; Clark, R. B. Intracellular Ca^{2+} activates a fast voltage-sensitive K^+-current in vertebrate sympathetic neurones. *Nature (London)* **296**:746–749, 1982.
7. Barrett, J. N.; Magleby, K. L.; Pallotta, B. S. Properties of single calcium-activated potassium channels in cultured rat muscle. *J. Physiol. (London)* **331**:211–230, 1982.
8. Brown, D. A.; Constanti, A.; Adams, P. R. Calcium-dependence of a component of transient outward current in bullfrog ganglion cells. *Soc. Neurosci. Abstr.* **8**:252, 1982.
9. Hartzell, H. C.; Kuffler, S. W.; Stickgold, R.; Yoshikami, D. Synaptic excitation and inhibition resulting from direct actions of acetylcholine on two types of chemoreceptors on individual amphibian parasympathetic neurones. *J. Physiol. (London)* **271**:817–846, 1977.
10. Henkart, M. Identification and function of intracellular calcium stores in axons and cell bodies of neurons. *Fed. Proc.* **39**:2783–2789, 1980.

11. Henkart, M.; Landis, D. M. D.; Reese, T. S. Similarity of junctions between plasma membranes and endoplasmic reticulum in muscle and neurons. *J. Cell Biol.* **70**:388–347, 1976.
12. Hermann, A.; Hartung, K. Noise and relaxation measurements of the Ca^{2+} activated K^+ current in *Helix* neurones. *Pfluegers Arch.* **393**:254–261, 1982.
13. Latorre, R.; Vergara, C.; Hidalgo, C. Reconstitution in planar lipid bilayers of a Ca^{2+}-dependent K^+ channel from transverse tubule membranes isolated from rabbit skeletal muscle, *Proc. Natl. Acad. Sci. USA* **79**:805–809, 1982.
14. Lee, K. S.; Tsien, R. W. Reversal of current through calcium channels in dialysed single heart cells. *Nature (London)* **297**:498–501, 1982.
15. MacDermott, A. B.; Weight, F. F. Action potential repolarization may involve a transient Ca^{2+}-sensitive outward current in a vertebrate neurone. *Nature (London)* **300**:185–188, 1982.
16. Madison, D. V.; Nicoll, R. A. Noradrenaline blocks accommodation of pyramidal cell discharge in hippocampus. *Nature (London)* **299**:636–638, 1982.
17. Marty, A. Ca-dependent K channels with large unitary conductance in chromaffin cell membranes. *Nature (London)* **291**:497–500, 1981.
18. Mathers, D. A.; Barker, J. L. Spontaneous hyperpolarizations at the membrane of cultured mouse dorsal root ganglion cells. *Brain Res.* **211**:451–455, 1981.
19. McAfee, D. A.; Yarowsky, P. J. Calcium-dependent potentials in the mammalian sympathetic neurone. *J. Physiol. (London)* **290**:507–523, 1974.
20. Meech, R. W.; Standen, N. B. Potassium activation in *Helix aspersa* neurones under voltage clamp: A component mediated by calcium influx. *J. Physiol. (London)* **249**:211–239, 1975.
21. Methfessel, C.; Boheim, G. The gating of single calcium-dependent potassium channels is described by an activation/blockade mechanism. *Biophys. Struct. Mech.* **9**:35–60, 1982.
22. Minota, S. Calcium ions and the post-tetanic hyperpolarization of bullfrog sympathetic ganglion cells. *Jpn. J. Physiol.* **24**:501–512, 1974.
23. Morita, K.; Koketsu, K.; Kuba, K. Oscillation of $[Ca^{2+}]$ linked K^+ conductances in bullfrog sympathetic ganglion cell is sensitive to intracellular axons. *Nature (London)* **283**:204–205, 1980.
24. Morita, K.; North, R. A.; Tokimasa, T. Muscarinic agonists inactivate potassium conductance of guinea pig myenteric neurones. *J. Physiol. (London)* **333**:125–139, 1982.
25. Neher, E.; Sakmann, B. Voltage-dependence of drug-induced conductance in frog neuromuscular junction. *Proc. Natl. Acad. Sci. USA* **72**:2140–2144, 1975.
26. North, R. A. The calcium-dependent slow afterhyperpolarization in myenteric plexus neurones with tetrodotoxin-resistant action potentials. *Br. J. Pharmacol.* **49**:709–711, 1973.
27. Sheridan, R.; Lester, H. A. Rates and equilibrium at acetylcholine receptor of electrophorus electroplaques: A study of neurally evoked postsynaptic currents and of voltage-jump relaxations. *J. Gen. Physiol.* **70**:187–219, 1977.
28. Smith, S. J.; MacDermott, A. B.; Weight, F. F. Intracellular calcium transients elicited by synaptic and electrical membrane activation and by theophylline measured in bullfrog neurons using arsenazo III. *Soc. Neurosci. Abstr.* **7**:15, 1981.
29. Tillotson, D. Inactivation of Ca conductance is dependent on entry of Ca ions in molluscan neurons. *Proc. Natl. Acad. Sci. USA* **76**:1497–1500, 1979.
30. Wong, B. S.; Lecar, H.; Adler, M. Single calcium-dependent potassium channels in clonal anterior pituitary cells. *Biophys. J.* **39**:313–317, 1982.
31. Wong, F. Nature of light-induced conductance changes in ventral photoreceptors of *Limulus*. *Nature (London)* **275**:76–79, 1978.
32. Woolum, J. C.; Gorman, A. L. F. Time dependence of the calcium-activated potassium current. *Biophys. J.* **36**:297–302, 1981.
33. Yau, K. W.; Lamb, D. A.; Baylor, D. A. Light-induced fluctuations in membrane current of single toad rod outer segments. *Nature, (London)* **269**:78–80, 1977.

22

Electrophysiological Evidence for Increased Calcium-Mediated Potassium Conductance by Low-Dose Sedative-Hypnotic Drugs

P. L. Carlen, N. Gurevich, and M. O'Beirne

In this chapter evidence is presented favoring the hypothesis that low-dose actions (i.e., mild sedation and anxiolysis) of ethanol, a water-soluble benzodiazepine (midazolam), and pentobarbital are due to enhanced calcium-mediated potassium conductance (Ca-gK). This hypothesis, which developed from results of electrophysiological experiments recording intracellularly from CA1 and CA3 cells in mammalian hippocampal slices, centers around the fact that injection of Ca^{2+} into excitable cells induces a membrane hyperpolarization by a selective increase in gK [20,24,30,32,33]. Physiologically, Ca-gK is usually triggered by a depolarization-induced influx of Ca^{2+} from the surrounding medium. On the other hand, persistently raised intracellular free Ca^{2+} concentration, $[Ca^{2+}]_i$, will actually reduce the depolarization-induced inward Ca^{2+} current [14,16,20,24,37,43]. However, it was shown in voltage-clamped dorid neurons that the intraneuronal free $[Ca^{2+}]_i$ and not the amount of the Ca^{2+} current is related to the degree of activation of the Ca-gK [15]. Therefore, Ca-gK could be increased even though the inward Ca^{2+} current is reduced if the source of increased $[Ca^{2+}]_i$ is intracellular. The exact way that Ca-gK is activated by a depolarizing current pulse or by injected Ca^{2+} is unclear. We have used the size of the Ca^{2+} spikes evoked in neurons perfused with tetrodotoxin (TTX), which blocks Na^+-dependent action potentials, as an indirect monitor of the free $[Ca^{2+}]_i$.

Ethanol is thought to act mainly by interacting with the lipid of cell membranes in a physical rather than a chemical manner [18]. There is also evidence for ethanol enhancement of GABA inhibitory action in cortical [34] and locus coeruleus neurons [42]. Barbiturates and benzodiazepines are classically thought to enhance GABA actions by specifically binding to neuronal receptors.

The *in vitro* mammalian (rat and guinea pig) hippocampal slice (Fig. 1D) is well suited

---

P. L. Carlen, N. Gurevich, and M. O'Beirne • Addiction Research Foundation Clinical Institute, Playfair Neuroscience Unit, Toronto Western Hospital, Departments of Medicine (Neurology), Physiology, and Institute of Medical Science, University of Toronto, Toronto, Ontario M5S 2S1, Canada.

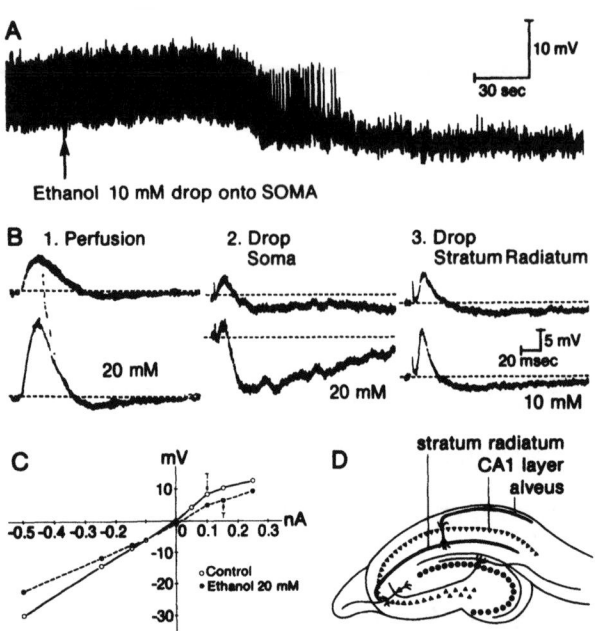

Figure 1. Electrophysiological effects of low doses of ethanol measured intracellularly in CA1 cells of a hippocampal slice preparation.

(A) Spontaneous activity and resting membrane potential obtained with a potassium acetate electrode were monitored on a d.c. chart recorder (Gould). Approximately 1.5 min after focal application of a drop containing 10 mM ethanol onto the somatic region of the CA1 cell, the cell was hyperpolarized; this was followed by cessation of the spontaneous spiking. Spike heights were attenuated by the chart recorder.

(B) An EPSP–IPSP sequence elicited by stratum radiatum stimulation. (1) Bath perfusion of 20 mM ethanol caused an increased EPSP and IPSP. The stimulation current for both records was 18 μA. The EPSP sometimes reached threshold causing the neuron to fire (lower trace). (2) Drop application of ethanol (20 mM) onto the somatic area caused a greatly enhanced IPSP after 5 min. (3) Ethanol (10 mM) focally applied to the stratum radiatum increased the EPSP measured 2 min after ethanol application. Stimulation currents for (2) and (3) were 22 and 100 μA, respectively.

(C) Injected current plotted against the peak voltage response. Cell input resistance (R_{in}) was decreased after ethanol exposure; R_{in} was measured in all cells with 100-msec constant-current pulses injected by means of an active bridge amplifier. With no injected current, the ethanol plot crosses the y axis at −1 mV, reflecting the degee of resting membrane hyperpolarization seen 5 min after ethanol exposure in this cell. Control R_{in} was 60 megaohms, reduced with ethanol perfusion to 44 megaohms.

(D) Diagram of hippocampal slice preparation. The stratum radiatum contains excitatory afferents to the apical dendrites of CA1 cells. The stratum oriens, which lies between the alveus and CA1 soma layers, contains excitatory afferents to the basilar dendrites of CA1 cells.

for examining the cellular neurophysiological mechanisms of drugs. It is a central mammalian neuronal preparation with intact local circuitry and well-defined neuronal somatic and dendritic layers. The extracellular fluid composition and perfusate drug concentrations can be precisely controlled. Also, stable intracellular recordings can be achieved and known concentrations of drugs or putative neurotransmitters can be focally applied to soma or dendritic regions using pressure ejection. Electrophysiological recording techniques from brain slices are described elsewhere [9,10,12]. In our hands, low doses of these drugs cause neuronal inhibition, which is not due to enhanced GABA action. The results of these experiments are briefly summarized below.

Ethanol

Many *in vitro* experiments have used ethanol in anesthetic, as opposed to sedative or intoxicating, concentrations. In most jurisdictions, the "legal limit" for blood alchol concentrations is 100 mg% (21.7 mM) or less. Lethal blood concentrations range above 100 mM. We initially noted that extracellularly measured population spike field potentials evoked orthodromically from stratum radiatum stimulation and measured in the CA1 soma layer were reversibly depressed at bath perfusate threshold concentrations of 100 mM ethanol

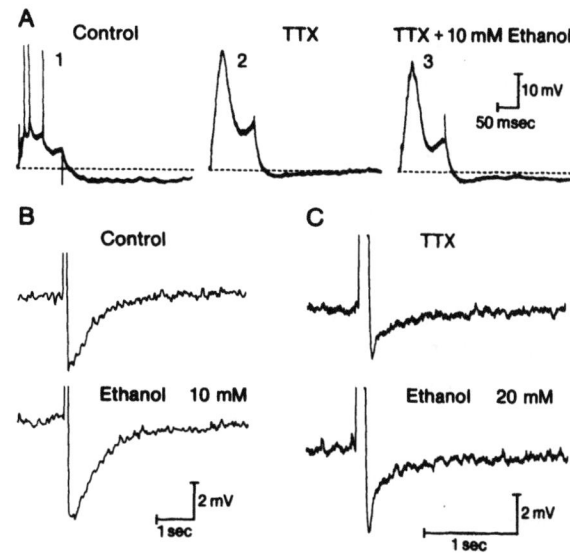

Figure 2. Augmentation of AHPs by ethanol.
(A) (1) Post-repetitive-firing AHP in control medium with a 0.35-nA 100-msec constant-current pulse. Spikes were retouched and full heights (not shown) were 80 mV. (2) TTX (20 μM) was applied in large drops to the stratum radiatum, pyramidale, and oriens, blocking fast Na^+ spikes. A higher injected depolarizing current (3.5 nA) elicited a Ca^{2+} spike followed by an AHP which was probably mediated by increased K^+ conductance (19,24). (3) Three minutes after drop application of 10 mM ethanol to the soma, the AHP (at 3.5 nA) was increased, but the Ca^{2+} spike was unaffected. Bridge balance was not corrected after ethanol application.
(B) Increases in depth and length of AHP after drop application of 10 mM ethanol to the soma in control solution. The current pulse for control and ethanol measurements was 0.75 nA, and each pulse caused four spikes. A chart recorder was used for (B) and (C), and depolarizing responses to 100-msec current pulses were cut off because of high gain.
(C) In the presence of TTX, the AHP following the Ca^{2+} spike increased in depth and was markedly prolonged after drop application of ethanol (20 mM) to the soma. A 1.0-nA constant-current pulse was used.

[8]. On interacting a prior antidromically elicited population spike (alveus stimulation), the orthodromically elicited field potential was depressed at much lower bath ethanol concentrations (20 mM) [13]. This conditioning antidromic pulse causes both synaptically evoked recurrent inhibition and intrinsic neuronal postspike inhibition (enhanced gK).

A study of ethanol actions using intracellular recording techniques was then undertaken on more than 80 CA1 cells [9]. Ethanol (5–20 mM), applied by bath perfusion or by focal drop application, generally elicited a hyperpolarization of the recorded neuron, a moderate increase in membrane conductance (Fig. 1C), and a decrease in spontaneous spiking when present (Fig. 1A). These effects were resistant to intracellular chloride ion injection (i.e., the reversal potential of this process did not seem to be altered), leaving enhanced K^+ conductance (gK) as the most reasonable alternative. When synaptic transmission was blocked by TTX or by low or zero extracellular Ca^{2+} concentrations (with 2.4 mM $MnCl_2$ added), the effects persisted.

To further explore the hypothesis of enhanced gK, the postspike afterhyperpolarization (AHP) was examined. The very early portion of the AHP is thought to be a voltage-dependent gK [25,39], and the later portion, a calcium-activated gK [19,25]. Also, hyperpolarization following the stratum radiatum-evoked EPSP in CA1 cells is thought to be due in part to Ca-gK [35] and to GABA-mediated chloride conductance. The AHP following a train of spikes elicited by a depolarizing constant-current pulse was consistently prolonged and usually deepened following low-dose ethanol application (Fig. 2), even though the cell often hyperpolarized. TTX was used to block Na^+-dependent spikes, bringing out higher threshold Ca^{2+} spikes [40]. Surprisingly, low-dose ethanol either had no effect, or increased the threshold and latency and decreased the size of these Ca^{2+} spikes, even though the ensuing AHPs were prolonged and deepened (Fig. 2A and 2C).

These data, coupled with the persistence of ethanol-induced hyperpolarization and conductance increase in zero Ca^{2+} perfusate, suggested that Ca-gK was enhanced by an intracellular mechanism such as increased $[Ca^{2+}]_i$. As previously discussed, increased $[Ca^{2+}]_i$ is an effective inhibitor of inward Ca^{2+} currents [14,16,20,24,37,43] and also enhances gK [20,24,30,32,33]. There is also biochemical evidence that ethanol directly augments Ca-gK without requiring changes in the concentration of $[Ca^{2+}]_i$ (44; Yamamoto, this volume). However, this effect was seen only with ethanol concentrations greater than 25 mM (Yamamoto, this volume). Interestingly, focal application of ethanol either onto stratum radiatum, wherein lie excitatory synaptic terminals innervating CA1 cells, or onto the soma where the inhibitory terminals lie, respectively, enhanced orthodromically elicited excitatory postsynaptic potentials (EPSPs) (Fig. 1B3) and inhibitory postsynaptic potentials (IPSPs) (Fig. 1B2). EPSPs from stratum oriens were enhanced by somatic drop and not by stratum radiatum drop application of ethanol. These data suggest a presynaptic action of ethanol, enhancing nerve-evoked transmitter release, possibly due to raised intraterminal free calcium. Both distal dendritic and somatic drop applications of ethanol caused the usual postsynaptic hyperpolarization and moderate conductance increase. At these low doses (i.e., 5–20 mM) no enhancement of focally applied pressure-ejected GABA was noted. However, in some preliminary experiments, 100 mM ethanol sometimes enhanced GABA responses in CA1 neurons and the early, presumably GABA-mediated, part of the orthodromically elicited IPSP. Shefner et al. [42] also noted enhanced GABA responses in locus coeruleus neurons using 30 mM ethanol.

Midazolam

A series of experiments was instituted using pressure-ejected drops of midazolam, a water-soluble benzodiazepine [10]. The proper concentration to simulate *in vivo* effects when applied to neurons *in vitro* is not clear. Clinically effective serum [17] and concomitant brain [28] concentrations are typically in the micromolar range, whereas cerebrospinal fluid concentrations are in the nanomolar range [27]. Binding affinities of benzodiazepines to central neurons are also typically in the low nanomolar range [6]. The most consistent inhibitory results were, in our hands, noted at 10^{-9} to 10^{-8} M, recording intracellularly from more than 40 CA1 cells.

Within 1 min following drop application to the CA1 soma region, a hyperpolarization and moderate conductance increase occurred in most cells, with a decrease in spontaneous activity, when present. These effects were, as with ethanol, resistant to intracellular chloride injection and synaptic blockade with TTX. However, unlike ethanol, midazolam actions were blocked in zero Ca^{2+} solution (with 2.4 mM $MnCl_2$ added). Orthodromic IPSPs tended to be enhanced by somatic drop application. These IPSPs are in part mediated by Ca-gK [35].

To delineate further the interaction of midazolam with Ca-gK, AHPs and Ca^{2+} spikes were examined. Like ethanol, midazolam enhanced the AHPs, in spite of frequently accompanying tonic hyperpolarization and conductance increase. Unlike ethanol, midazolam enhanced Ca^{2+} spikes, decreasing their threshold and latency. In five recent experiments, midazolam (5×10^{-9} M) perfused onto the slice, instead of being focally pressure-ejected onto the soma, caused a significant tonic hyperpolarization and conductance increase in every case. AHPs were also prolonged and EPSPs were increased in height, sometimes much more than expected from the degree of hyperpolarization.

No enhancement of GABA action was noted with low nanomolar concentrations of midazolam, but using a 10^{-7} M drop, GABA action was enhanced as noted by Jahnsen and Laursen [26]. Also, Alger and Nicoll [2] showed enhanced hyperpolarizing somatic responses to iontophoresed GABA by perfusing 10^{-6} M diazepam.

With a water-soluble specific benzodiazepine antagonist, Ro14-7437, in similar nanomolar concentrations, opposite effects were found following somatic drop application [11]. Ro14-7437 caused a tonic depolarization, conductance decrease, and increased spontaneous spiking. These effects were also resistant to intracellular injection of Cl^- ions or synaptic blockade by TTX, and were prevented in Ca^{2+}-free medium. The responses occurred with or without prior application of midazolam, raising the possibility of an endogenous benzodiazepine-like inhibitory ligand in the slice itself. Importantly, AHPs and Ca^{2+} spikes in the presence of TTX were diminished by this benzodiazepine antagonist.

The above results suggest that the benzodiazepine agonist midazolam acts by enhancing Ca-gK, whereas the antagonist tends to block Ca-gK. Midazolam acts differently from ethanol in that midazolam enhances Ca^{2+} spikes and is ineffective in a calcium-free perfusate. Our interpretation of these data is that midazolam augments Ca-gK by enhancing transmembrane Ca^{2+} currents, both spike-evoked and a Ca^{2+} current present at the resting membrane potential. The end result is increased $[Ca^{2+}]_i$, which in turn augments gK. However, ethanol may raise $[Ca^{2+}]_i$ by an intracellular mechanism rather than by enhancing transmembrane Ca^{2+} currents.

Pentobarbital

Pentobarbital enhances IPSPs and GABA action in *in vitro* hippocampal CA1 cells using concentrations of 100 μM or higher [1], which are coma-producing in man and animals. Mildly sedative concentrations are in the low micromolar range (unpublished observations). A series of experiments in over 20 CA1 and 20 CA3 cells have been conducted to date using both bath-applied and focally pressure-ejected sodium pentobarbital (10^{-6}–10^{-5} M). The results obtained resemble those obtained with ethanol. Within 1 to 2 min of somatic drop-application, there developed a tonic hyperpolarization (> 70% of cells), moderate conductance increase (54% of cells), and decreased spontaneous spiking (18/23 cells) when present, which was resistant to intracellular Cl^- leakage or synaptic blockade by TTX. AHPs and often IPSPs were enhanced, but with these low doses, GABA responses were not increased. Ca^{2+} spikes evoked in the presence of TTX were slightly attenuated or unchanged with these low doses. Heyer and MacDonald [23] have also shown reduction of Ca^{2+}-dependent action potentials in cultured mouse spinal cord neurons with threshold effects at 25 μM pentobarbital. The data accumulated to date suggest that low-dose pentobarbital enhances gK, possibly by elevating $[Ca^{2+}]_i$ by releasing it from an intracellular store.

Calcium-Mediated Potassium Conductance: A Hypothesis for Low-Dose Sedative-Anxiolytic Drug Action

Krnjevic [29] proposed over 10 years ago that enhanced Ca-gK might be the mechanism of general anesthesia in vertebrate central neurons. Nicoll and Madison [36] recently showed that several general anesthetics, including ether and halothane, augment gK in rat hippocampal neurons. Our results with low-dose sedative-hypnotic drugs showed enhanced Ca-gK, although gK was augmented by different mechanisms for ethanol as compared to midazolam. These drugs augment GABA-mediated inhibition at higher doses. The low-dose sedative-anxiolytic actions of these drugs could be due to enhanced Ca-gK, and the higher-dose anesthetic actions could also include increased GABA-mediated inhibition. In the case of benzodiazepines and to a lesser extent pentobarbital, the appropriate doses that should be applied onto neurons *in vitro* to mimic sedation or general anesthesia are not clearly known, but they are crucial for proper understanding of these drug-induced phenomena.

This section of the volume emphasizes the novel but important role that Ca^{2+} plays in sedative drug action. Neuronal Ca^{2+} metabolism is still poorly understood but obviously closely linked to many intracellular physiological and biochemical processes. Intracellular free Ca^{2+} is tightly regulated and is below micromolar concentrations, whereas the extracellular Ca^{2+} concentration is in the low millimolar range. There are several mechanisms for regulation of $[Ca^{2+}]_i$. $[Ca^{2+}]_i$ may be rapidly buffered by cytoplasmic high-affinity Ca^{2+}-binding proteins [3], or more slowly by intracellular Ca^{2+}-sequestering organelles, such as the endoplasmic reticulum [4] or mitochondria [31]. However, the physiological significance of the role of mitochondria in buffering $[Ca^{2+}]_i$ has been questioned [5,7]. Ca^{2+} "pumps" in the neuronal membrane may also play a pivotal role. In human erythrocytes, $[Ca^{2+}]_i$ is regulated by a Ca^{2+}-Mg^{2+}-dependent ATPase [38], and this pump does not exchange Ca^{2+} for any other ions. Another pump exchanges intracellular Ca^{2+} for extracellular Na^+, and this system seems to be powered by energy from the Na^+ electrochemical gradient (5; Michaelis and Michaelis, this volume). Finally, the actual $[Ca^{2+}]_i$ helps to regulate inward Ca^{2+} currents [14,16,20,24,37,43]

The decreased Ca^{2+} spikes observed with ethanol [9], and possibly pentobarbital, in hippocampal slice cells, and with pentobarbital in cultured mouse spinal cord neurons (23; Macdonald and Werz, this volume), suggest that both of these drugs raise $[Ca^{2+}]_i$. Ethanol and barbiturates alter neuronal Ca^{2+} metabolism [21], and inhibit intrasynaptosomal sequestration of Ca^{2+} [22]; and pentobarbital inhibits synaptosomal Ca^{2+} transport (Leslie et al., this volume). Erythrocyte ghost membrane $[Ca^{2+}]_i$ rises by about 60% in the presence of 20 mM ethanol, although pentobarbital has no effect [41]; and furthermore, low doses of ethanol inhibit the Na^+–Ca^{2+} membrane exchange pump, suggesting a possible mechanism for raising $[Ca^{2+}]_i$ (Michaelis and Michaelis, this volume).

Unlike ethanol and pentobarbital, midazolam appears to increase Ca-gK by enhancing transmembrane Ca^{2+} since this latter drug augments Ca^{2+} spikes and is ineffective in a Ca^{2+}-free perfusate. All three drugs show various degrees of behavioral cross-tolerance and all have similar sedative actions. Of course, the available data do not rule out the possibility that all three drugs directly activate gK rather than acting through Ca^{2+}-mediated mechanisms. To generalize our theory of increased Ca-gK being the common mechanism of action for sedation, similar intracellular electrophysiological investigations using other sedative drugs and examining other brain regions are required. Further experimental manipulations are indicated, including intracellular injection of Ca^{2+}-chelating agents and gK blockers, as well as voltage-clamp and patch-clamp recordings. The anesthetic action of these drugs at higher doses may be due to activation of other inhibitory mechanisms, particularly enhancement of GABA-mediated inhibition.

ACKNOWLEDGMENTS. This work was supported in part by the U.S. National Institutes of Health, the Medical Research Council of Canada, the Alcoholic Beverage Medical Research Foundation, and the Ontario Mental Health Foundation. Typing was done by Mary Cairoli and Cathy Van Der Geissen. Figures 1 and 2 reproduced with permission from *Science*, Volume 215, pp. 306–309, 1982. Copyright 1982 by the American Association for the Advancement of Science.

References

1. Alger, B. E.; Nicoll, R. A. Feed-forward dendritic inhibition in rat hippocampal pyramidal cells studied in-vitro. *J. Physiol. (London)* **328**:105–123, 1982.
2. Alger, B. E.; Nicoll, R. A. Pharmacological evidence for two kinds of GABA receptor on rat hippocampal pyramidal cells studied in-vitro. *J. Physiol. (London)* **328**:125–141, 1982.

3. Baker, P. F.; Schlaepfer, W. W. Uptake and binding of calcium by axoplasm isolated from giant axons of LOLIGO and MYXICOLA. *J. Physiol. (London)* **276**:103–125, 1978.
4. Blaustein, M. P.; McGraw, C. F.; Somlyo, A. V.; Schweitzer, E. S. How is the cytoplasmic calcium concentration controlled in nerve terminals? *J. Physiol. (Paris)* **76**:459–470, 1980.
5. Blaustein, M. P.; Nelson, M. T. Sodium–calcium exchange: Its role in the regulation of cell calcium. In: *Calcium Transport across Biological Membranes*, E. Carafoli, ed., New York, Academic Press, 1982, pp.217–236.
6. Braestrup, C.; Squires, R. F. Specific benzodiazepine receptors in rat brain characterized by high-affinity [^{3}H]diazepam-binding. *Proc. Natl. Acad. Sci. USA* **74**:3805–3809, 1977.
7. Brinley, F. J.,Jr. Regulation of intracellular calcium in squid ions. *Fed. Proc.* **39**:2778–2782, 1980.
8. Carlen, P. L.; Corrigall, W. A. Ethanol tolerance measured electrophysiologically in hippocampal slices and not in neuromuscular junctions from chronically ethanol-fed rats. *Neurosci. Lett.* **17**:95–100, 1980.
9. Carlen, P. L.; Gurevich, N., Durand, D. Ethanol in low doses augments calcium mediated mechanisms measured intracellularly in hippocampal neurons. *Science* **215**:306–309, 1982.
10. Carlen, P. L.; Gurevich, N.; Polc, P. Low dose benzodiazepine neuronal inhibition: Enhanced Ca^{++}-mediated K^+ conductance. *Brain Res.* **271**: 358–364, 1983.
11. Carlen, P. L.; Gurevich, N.; Polc, P. The excitatory effects of the specific benzodiazepine antagonist Ro-14-7437 measured intracellularly in CA1 cells. *Brain Res.* **271**:115–119, 1983.
12. Dingledine, R.; Dodd, J,; Kelly, J. S. The *in-vitro* brain slice as a useful neurophysiological preparation for intracellular recording. *J. Neurosci. Methods* **2**:323–362, 1980.
13. Durand, D.; Corrigall, W. A.; Kujtan, P.; Carlen, P. L. Effects of low concentrations of ethanol on CA1 hippocampal neurons in vitro. *Can. J. Physiol. Pharmacol.* **59**:979–984, 1981.
14. Eckert, R.; Ewald, D. Residual calcium ions depress activation of calcium dependent current. *Science* **216**:730–733, 1982.
15. Eckert, R.; Tillotson, D. Potassium activation associated with intraneuronal free calcium. *Science* **200**:437–439, 1978.
16. Eckert, R.; Tillotson, D. L. Calcium-mediated inactivation of the calcium conductance in caesium-loaded giant neurones of *Aplysia californica*. *J. Physiol. (London)* **314**:265–280, 1981.
17. Garattini, S.; Mussini, E.; Marucci, F.; Guaitani, A. Metabolic studies on benzodiazepines in various animal species. In: *The Benzodiazepines*, S. Garattini, E. Mussini, and L. O. Randall, eds., New York, Raven Press, 1973, pp.75–97.
18. Goldstein, D. B.; Chin, J. H. Interaction of ethanol with biological membranes. *Fed. Proc.* **40**:2073–2076, 1981.
19. Gustafasson, B.; Wigstrom, H. Evidence for two types of afterhyperpolarizations in CA1 pyramidal cells in the hippocampus. *Brain Res.* **206**:462–468, 1981.
20. Hagiwara, S.; Byerly, L. Calcium Channels. *Annu. Rev. Neurosci.* **4**:69–125, 1981.
21. Harris, R. A. Psychoactive drugs as antagonists of actions of calcium. In: *Calcium Antagonists*, G. Weiss, ed., Bethesda, American Physiological Society, pp.223–231.
22. Harris, R. A. Ethanol and pentobarbital inhibition of intrasynaptosomal sequestration of calcium. *Biochem. Pharmacol.* **30**:3209–3215, 1981.
23. Heyer, E. J.; MacDonald, R. L. Barbiturate reduction of calcium-dependent action potentials: Correlation with anesthetic action. *Brain Res.* **236**:157–171, 1982.
24. Hofmeier, G.; Lux, H. D. The time courses of intracellular free calcium and related electrical effects after injection of $CaCl_2$ into neurons of the snail, *Helix pomatia*. *Pfluegers Arch.* **391**:242–217, 1981.
25. Hotson, J. R.; Prince, D. A. A calcium-activated hyperpolarization follows repetitive firing in hippocampal neurons. *J. Neurophysiol.* **43**:409–419, 1980.
26. Jahnsen, H.; Laursen, A. M. The effects of benzodiazepine on the hyperpolarizing and the depolarizing responses of hippocampal cells to GABA. *Brain Res.* **207**:214–217, 1981.
27. Kanto, J.; Kangas, L.; Siirotola, T. Cerebrospinal-fluid concentrations of diazepam and its metabolites in man. *Acta. Pharmacol. Toxicol.* **36**:328–334, 1975.
28. Klotz, U. Effect of age on levels of diazepam in plasma and brain of rats. *Naunyn-Schmiedeberg's Arch. Pharmacol.* **307**:167–169, 1979.
29. Krnjevic, K. Excitable membranes and anesthetics. In: *Cellular Biology and Toxicity of Anesthetics*, B. R. Fink, ed., Baltimore, Williams & Wilkins, 1972, pp.3–9.
30. Krnjevic, K.; Lisiewicz, A. Injections of calcium-ions into spinal motoneurones. *J. Physiol. (London)* **225**:363–390, 1972.
31. Lehninger, A. L. Mitochondria and calcium ion in transport. *Biochem. J.* **119**:129–138, 1970.
32. Meech, R. W. Intracellular calcium injection causes increased potassium conductance in *Aplysia* nerve cells. *Comp. Biochem. Physiol. A* **42**:493–499, 1972.

33. Meech, R. W. The sensitivity of *Helix aspersa* neurones to injected calcium ions. *J. Physiol. (London)* **237**:259–277, 1974.
34. Nestoros, J. N. Ethanol specifically potentiates GABA-mediated neurotransmission in feline cerebral cortex. *Science* **209**:708–710, 1980.
35. Nicoll, R. A.; Alger, B. E. Synaptic excitation may activate a calcium dependent potassium conductance in hippocampal pyramidal cells. *Science* **212**:957–959, 1981.
36. Nicoll, R. A.; Madison, D. V. General anesthetics hyperpolarize neurons in the vertebrate central nervous system. *Science* **217**:1055–1057, 1982.
37. Plant, T. D.; Standen, N. B. Calcium current inactivation in identified neurones of *Helix aspersa*. *J. Physiol. (London)* **321**:273–285, 1981.
38. Schatzmann, H. J.; Burgin; H. Calcium in human red blood cells. *Ann. N.Y. Acad. Sci.* **307**:125–147, 1978.
39. Schwartzkroin, P. A.; Prince, D. A. Effects of TEA on hippocampal neurons. *Brain Res.* **185**:169–181, 1980.
40. Schwartzkroin, P. A.; Slawsky, M. Probable calcium spikes in hippocampal neurons. *Brain Res.* **135**:157–161, 1977.
41. Seeman, P.: Chau, M.; Goldberg, M.; Sauks, T.; Sax, L. The binding of Ca^{2+} to the cell membrane by volatile anesthetics (alcohols, acetone, ether) which induce sensitization of nerve or muscle. *Biochim. Biophys. Acta* **225**:185–193, 1971.
42. Shefner, S. A.; Chiu, T. H.; Anderson, E. G. Intracellular measurements of ethanol effects on rat locus coeruleus neurons in a brain slice preparation. *Soc. Neurosci. Abstr.* **8**:651, 1982.
43. Standen, N. B. Ca channel inactivation by intracellular Ca injection into *Helix* neurones. *Nature (London)* **293**:158–159, 1981.
44. Yamamoto, H. A.; Harris, R. A. Calcium-dependent ^{86}Rb efflux and ethanol intoxication: Studies of human red blood cells and rodent brain synaptosomes. *Eur. J. Pharmacol.* **88**:357–363, 1983.

23

Neurochemical Studies on the Effects of Ethanol on Calcium-Stimulated Potassium Transport

Hiro-aki Yamamoto

To understand the actions of ethanol at the synaptic level the dynamics of neuronal calcium must be considered. One action of calcium is to regulate the efflux of potassium from nerve cells. Electrophysiological studies have demonstrated a long-lasting neuronal afterhyperpolarization due to calcium-dependent potassium conductance in neurons of the central nervous system [1,12,13,15]. These studies indicate that this process plays an important role in control of neuronal excitability.

Recent evidence indicates that ethanol enhances this calcium-dependent potassium current in hippocampal slices [4,8]. There are few other studies of the effects of drugs or toxins on the calcium-stimulated potassium potential, although apamin, a bee venom polypeptide of 18 amino acids [6], selectively blocks this potassium conductance in nerve cells [2,8,11,12].

Alcohols Increase Calcium-Dependent ^{86}Rb Efflux from Erythrocytes

Biochemical studies indicate that ^{86}Rb efflux is activated by intracellular calcium in human erythrocytes (RBC) [9,14,18]. Accordingly, the effect of alcohols on calcium-dependent potassium efflux was investigated, using ^{86}Rb as a tracer for potassium in RBC extensively washed and metabolically depleted to eliminate active transport. In agreement with other studies [9,14], ^{86}Rb efflux was dependent on the calcium concentration, with little activity at 0.15 μM free calcium and maximal activity at 0.7 μM calcium (Fig. 1). In vitro addition of 100 mM ethanol significantly increased calcium-dependent ^{86}Rb efflux at lower concentrations of calcium (< 0.4 μM), but had no effect at higher concentrations of calcium (> 0.6 μM). These results suggest that ethanol enhances the action of low calcium concentrations. Two other alcohols, n-propanol and n-butanol, also stimulate calcium-dependent ^{86}Rb efflux, but the longer chain-length alcohols are effective at lower concentrations than

---

Hiro-aki Yamamoto • Faculty of Pharmaceutical Sciences, Fukuyama University, Fukuyama, Hiroshima 729-02, Japan.

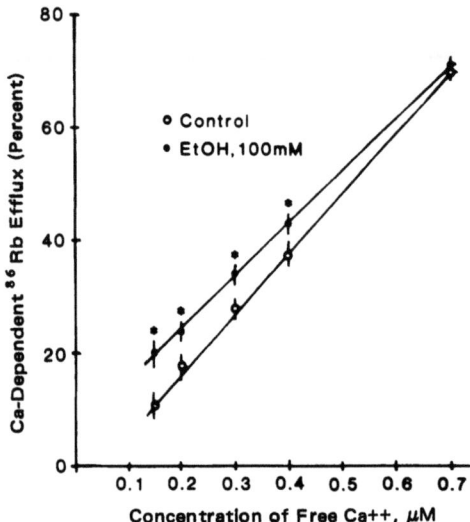

Figure 1. Effect of free calcium concentration on calcium-dependent efflux of ^{86}Rb from human RBC in the presence and absence of ethanol. Fresh RBC were depleted of their endogenous energy stores and lysed and resealed with ^{86}Rb and calcium–EGTA buffers to give varying concentrations of free calcium [9,13,14]. The cells were then incubated at 37°C for 2 hr and the ^{86}Rb released (supernatant) was determined by scintillation spectrometry. Calcium-independent ^{86}Rb efflux was calculated as ^{86}Rb efflux in the absence of internal calcium (EGTA alone), and calcium-dependent ^{86}Rb efflux was calculated as ^{86}Rb efflux in the presence of calcium (total efflux) minus calcium-independent efflux [19].

The abscissa represents the calcium-dependent efflux of ^{86}Rb expressed as a percentage of the total amount of ^{86}Rb in the tube at the beginning of the final incubation. Data indicated by filled circles were obtained in the presence of 100 mM ethanol; open circles represent efflux in the absence of ethanol. Each point is the average of four determinations. Vertical bars signify ± S.E.M. Asterisks indicate significant difference from control; $p < 0.01$ by t test for paired observations.

those required for ethanol action [19]. The relative potency of these alcohols in enhancing ^{86}Rb efflux is consistent with their lipid solubility [16].

When RBC were preincubated at 60°C, the calcium-dependent ^{86}Rb efflux was completely abolished, whereas the calcium-independent ^{86}Rb efflux was unaffected (Table I). These results indicate that this mild heat treatment did not affect resealing of the RBC or alter membrane "leakiness." The heat denaturation experiment also demonstrates that the effect of ethanol is specific for the calcium-dependent component of efflux and suggests that this efflux was mediated by a protein that was easily denatured.

Effect of Apamin and Ethanol on Calcium-Dependent ^{86}Rb Efflux in Synaptosomes

Efflux of ^{86}Rb from brain synaptosomes of rats and mice [5,20] was enhanced by low concentrations of free calcium [19]. The calcium sensitivity of ^{86}Rb efflux is similar for both

Table I. Effect of Ethanol and Heat Pretreatment on Ca-Dependent and Ca-Independent Efflux of ^{86}Rb from Human RBC

		^{86}Rb efflux[b]	
Drug	Preincuation[a]	Ca-dependent	Ca-independent
None	37°C	21 ± 1	18 ± 2
Ethanol, 100 mM	37°C	26 ± 1*	17 ± 3
Ethanol, 200 mM	37°C	31 ± 2*	16 ± 2
None	60°C	1 ± 1	18 ± 2
Ethanol, 100 mM	60°C	2 ± 2	17 ± 2
Ethanol, 200 mM	60°C	1 ± 2	17 ± 3

[a]Preincubation time was 5 min.
[b]Values represent percent efflux, mean ± SEM, $n = 4$. Efflux was determined as described previously [19].
*Significant effect of ethanol, $p < 0.01$, t test for paired observations.

Table II. Effect of Ethanol and Apamin on the Ca-Dependent Efflux of ^{86}Rb from Mouse Brain Synaptosomes

Addition	Ca-dependent ^{86}Rb efflux
None	5.0 ± 1.5^a
Apamin (10^{-7} M)	3.8 ± 0.9*
Ethanol (100 mM)	9.8 ± 2.0†
Ethanol (100 mM) + apamin (10^{-7} M)	5.5 ± 1.6*

^{a}Values represent percent Ca-dependent efflux, mean $\pm$ SEM, $n = 6$. Efflux was determined as described previously (19).
*Significant effect of apamin, $p < 0.01$, t test for paired observations.
†Significant effect of ethanol, $p < 0.01$, t test for paired observations.

synaptosomes and RBC, with 0.1 μM free calcium producing a detectable efflux and 1 μM producing near-maximal efflux. The calcium-dependent ^{86}Rb efflux was significantly increased by *in vitro* addition of 100 mM ethanol. The stimulatory effect of ethanol is more pronounced at low concentrations of free calcium. These results are similar to those obtained with RBC (Fig. 1.).

Apamin (0.1 μM) significantly inhibited calcium-dependent ^{86}Rb efflux from synaptosomes and blocked the stimulatory effect of ethanol on calcium-dependent ^{86}Rb efflux (Table II). Despite the similarities of calcium-dependent ^{86}Rb efflux and the effect of ethanol in synaptosomes and RBC, the response to apamin may differ in these two systems. Electrophysiological evidence indicates that apamin inhibits calcium-stimulated potassium efflux of nerve cells [2,12], but it does not alter this K^+ efflux in RBC [3]. Similarly, biochemical studies show apamin to be a potent inhibitor of calcium-stimulated ^{86}Rb efflux from synaptosomes. Specific binding sites for apamin may exist on nerve membranes [8], and these sites may be present on a component of the transport system found in brain cells, but it is unlikely that the apamin-binding component is involved in the action of ethanol. The observation that apamin antagonizes the stimulatory effects of ethanol on the calcium-dependent efflux of ^{86}Rb from synaptosomes may be useful in examining the importance of potassium efflux in alcohol action. If increased apamin-dependent ^{86}Rb efflux is responsible for ethanol intoxication, then administration of apamin should reduce the effects of ethanol.

Effect of Apamin on Ethanol-Induced Narcosis

Intracerebroventricular injection of apamin reduced ethanol sleep time (Table III). A dose of 0.01 μg reduced sleep time by about 50% and doses of 0.05 and 0.1 μg shortened sleep time even further. It was not possible, however, to prevent the loss of righting reflex with apamin. These results suggest that stimulation of calcium-dependent potassium efflux is one of several neurochemical changes responsible for alcohol narcosis.

It is of interest to consider possible mechanisms for stimulation of calcium-dependent ^{86}Rb efflux by ethanol. The effect of ethanol was greatest at a low concentration of free calcium, indicating that ethanol may enhance cellular responsiveness to small changes in intracellular calcium. A possible explanation for ethanol's effect is that it increases calcium binding to membrane sites that regulate efflux. Indeed, ethanol increases calcium binding to the cytoplasmic surface of human RBC [10,17] and mouse brain synaptosomes [7]. These findings support the postulate that these drugs affect ^{86}Rb efflux by altering calcium binding to the membrane surface.

Table III. Effects of Apamin on the Duration of Loss of Righting Reflex (Sleep Time) Produced by Ethanol

Pretreatment[a]	Ethanol sleep time (min)
Saline	53 ± 6[b]
Apamin	
0.01 μg/brain	29 ± 3*
0.05 μg/brain	10 ± 1*
0.1 μg/brain	9 ± 1*

[a] Mice were pretreated with saline or apamin (intracerebroventricular injection) and 1 min later all mice were injected i.p. with 4 g/kg ethanol. Sleep time is defined as the time from the loss of righting reflex to the return of the reflex. Assessment of the reflex required that the mice right themselves onto all four paws twice within 30 sec.
[b] Values represent mean ± SEM, $n = 6$.
*Significant effect of apamin, $p < 0.01$.

In summary, biochemical and behavioral observations, in conjunction with the electrophysiological findings of Carlen et al. [4] (see also Chapter 22), suggest that stimulation of calcium-dependent potassium efflux is one of the neurochemical mechanisms responsible for the pharmacological and toxicological actions of ethanol.

ACKNOWLEDGMENT. I thank Dr. R. Adron Harris for his assistance.

References

1. Alger, B. E.; Nicoll, R. A. Epileptiform burst afterpolarization: Calcium-dependent potassium potential in hippocampal CA 1 pyramidal cells. *Science* 210:1125–1126, 1980.
2. Banks, B. E.; Brown, C.; Burgess, G. M.; Cocks, T. M.; Jenkinson, D. H. Apamin blocks certain neurotransmitter-induced increases in potassium permeability. *Nature (London)* 282:415–417, 1979.
3. Burgess, G. M.; Claret, M.; Jenkinson, D. H. Effects of quinine and apamin on the calcium-dependent potassium permeability of mammalian hepatocytes and red cells. *J. Physiol. (London)* 317:67–90, 1981.
4. Carlen, P. L.; Gurevich, N.; Durand, D. Ethanol in low doses augments calcium-mediated mechanisms intracellularly in hippocampal neurons. *Science* 215:306–309, 1982. (See also Carlen et al., this volume.)
5. Cotman, C. W.; Matthews, D. A. Synaptic plasma membranes from rat brain synaptosomes: Isolation and partial characterization. *Biochim. Biophys. Acta* 249:380–394, 1971.
6. Habermann, E. Bee and wasp venoms, *Science* 177:314–322, 1972.
7. Harris, R. A.; Fenner, D. Ethanol and synaptosomal calcium binding. *Biochem. Pharmacol.* 31:1790–1792, 1982.
8. Hugues, M.; Romey, G.; Duval, D.; Vincent, J. P.; Lazdunski, M. Apamin as a selective blocker of the calcium-dependent potassium channel in neuroblastoma cells: Voltage-clamp and biochemical characterization of the toxin receptor. *Proc. Natl. Acad. Sci. USA* 79:1308–1312, 1982.
9. Lewi, V. L.; Ferreira, H. G. Calcium transport and the properties of a calcium-activated potassium channel in red cell membranes. *Curr. Top. Memb. Transp.* 10:217–277, 1978.
10. Low, P. S.; Lloyd, D. H.; Stein, T. M.; Rogers, J. A. Calcium displacement by local anesthetics: Dependence on pH and anesthetic charge. *J. Biol. Chem.* 254:4119–4125, 1979.
11. Meech, R. W. Calcium-dependent potassium activation in nervous tissue. *Annu. Rev. Biophys. Bioeng.* 7:1–18, 1978.
12. Moolinaar, W. H.; Spector, I. Ionic currents in cultured mouse neuroblastoma cells under voltage-clamp conditions. *J. Physiol. (London)* 278:265–286, 1978.
13. Nicoll, R. A.; Alger, B. E. Synaptic excitation may activate a calcium-dependent potassium conductance in hippocampal pyramidal cells. *Science* 212:957–959, 1981.

14. Porzig, H. Studies on the cation permeability of human red cell ghosts: Characterization and biological significance of two membrane sites with high affinities for Ca. *J. Memb. Biol.* **31**:317–349, 1977.
15. Schwartzkroin, P. A.; Stafstrom, C. E. Effects of EGTA on the calcium-activated after-hyperpolarization in hippocampal CA3 pyramidal cells. *Science* **210**:1125–1126, 1980.
16. Seeman, P. The membrane actions of anesthetics and tranquilizers. *Pharmacol. Rev.* **24**:583–655, 1972.
17. Seeman, P.; Chau, M.; Goldberg, M.; Sauks, T.; Sax, L. The binding of Ca^{2+} to the cell membrane increased by volatile anesthetics (alcohols, acetone, ether) which induce sensitization of nerve or muscle. *Biochim. Biophys. Acta* **225**:185–193, 1971.
18. Simons, T. J. B. Calcium-dependent potassium exchange in human red cell ghosts. *J. Physiol. (London)* **256**:227–244, 1976.
19. Yamamoto, H.; Harris, R. A. Calcium-dependent ^{86}Rb efflux and ethanol intoxication: Studies of human red blood cells and rodent brain synaptosomes. *Eur. J. Pharmacol.* **88**:357–363, 1983.
20. Yamamoto, H.; McCain, H. W.; Izumi, K.; Misawa, S.; Way, E. L. Effect of amino acid, especially taurine and 2-aminobutyric acid (GABA), on analgesia and calcium depletion induced by morphine in mice. *Eur. J. Pharmacol.* **71**:177–184, 1981.

24

Role of Calcium in the Regulation of Sodium Permeability in the Kidney

E. E. Windhager

Calcium ions have long been known to influence membrane permeability to ions in excitable tissues and erythrocytes. Recent evidence also indicates that calcium plays an important role in regulating epithelial net transport of ions and H_2O. Specifically, there is mounting evidence that increases in cytosolic Ca^{2+} activity reduce sodium permeability of the apical membrane of sodium-absorbing epithelial cells [2,6,11].

An early indication that calcium plays such a role was obtained by Taylor [9], who observed that quinidine, a drug known to increase cytosolic Ca^{2+} activity in other tissues, inhibits net sodium transport in toad bladders. This inhibition was reversed by vasopressin, which increases luminal sodium permeability. Similar results were obtained when calcium ionophores were used to increase cytosolic Ca^{2+} concentrations. Subsequently, Erlij and Grinstein [3,6] reported that a reduction in the sodium concentration of the medium bathing the inner surface of frog skins inhibits the rate of sodium transport in a calcium-dependent manner. Their interpretation of this finding was based on the concept that a Na^+–Ca^{2+} exchange mechanism operates across the basolateral cell membrane of epithelial cells as originally proposed by Blaustein [1] for renal tubules. These and other studies led to the hypothesis that cytosolic Ca^{2+} ions are involved in a feedback regulation of sodium transport (Figs. 1 and 2; also see Ref. 10 for review).

According to this model, a primary change in active sodium transport alters the intracellular Na^+ concentration and hence the electrochemical driving force for passive sodium backflux into the cell across the basolateral cell membrane. If this passive sodium entry is coupled to calcium efflux via the operation of a Na^+–Ca^{2+} countertransport mechanism, changes in intracellular sodium concentration lead to parallel changes in cytosolic Ca^{2+} activity. When Na^+ activity rises, Ca^{2+} activity will also increase. The change in intracellular calcium is then thought to modify luminal membrane permeability to sodium in such a way that the rate of sodium entry across the apical membrane is kept in step with its rate of

---

E. E. Windhager • Department of Physiology, Cornell University Medical College, New York, New York 10021.

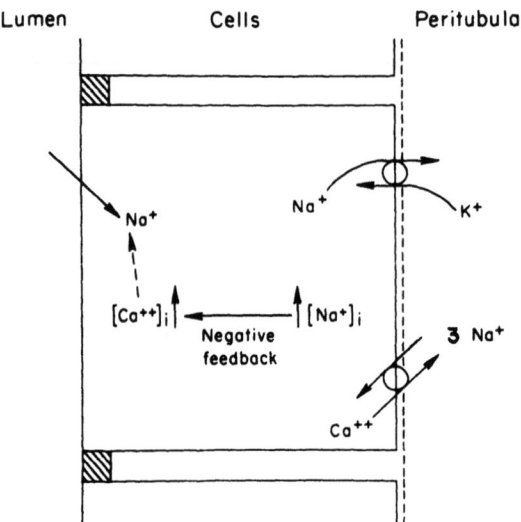

Figure 1. Intracellular feedback hypothesis. [Modified from Ref. 10.]

extrusion across the basolateral cell membrane. Thus, a primary reduction in Na^+-pump activity, as during ouabain administration, increases cytosolic Ca^{2+} levels, and this in turn induces a decrease in the rate of luminal entry of sodium. The major prediction of this model is that any increase in cytosolic Ca^{2+} activity reduces net sodium transport.

To test the applicability of this model to renal epithelia, initial studies were carried out on isolated proximal tubules of rabbit kidneys perfused with a solution containing 145 mM sodium [4]. This same sodium concentration was present in the contraluminal bath during control and recovery periods. During the experimental period, peritubular sodium levels were reduced to about 40 mEq/liter by substitution of sodium with lithium, resulting in a reduction in fluid reabsorption to 46% of the control values. Fluid reabsorption recovered to near control values when sodium concentrations were restored to their physiological level in the peritubular bath. This same study [4] also examined whether the observed reduction in fluid reabsorption is caused by decreased efflux of sodium from the lumen or by increased

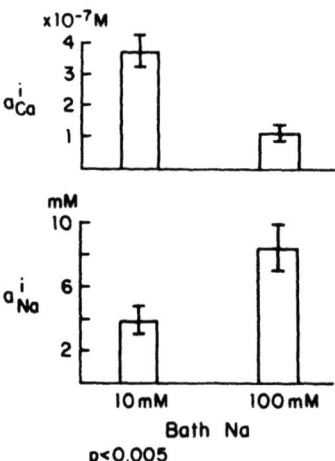

Figure 2. Measurements of cytosolic Na^+ and Ca^{2+} activities [11]. Microelectrode impalements of proximal tubular cells of *Necturus* were done on different nephrons.

backflux. ^{22}Na was added to the luminal perfusate, and when NaCl in the bath was partially replaced by LiCl, sodium efflux fell by 28%. This reduction in efflux corresponds to 51% of the observed drop in fluid reabsorption.

These results indicate that inhibition of fluid reabsorption induced by low peritubular sodium concentrations is to a large extent caused by a decrease in unidirectional sodium transport from lumen to bath. The decrease in sodium efflux occurred despite a sodium concentration gradient of 105 mEq/liter favoring sodium movement in the reabsorptive direction. Significantly, in separate experiments, inhibition of sodium transport was found to depend on the extracellular calcium concentration. The least inhibition was observed at the lowest calcium levels in the peritubular fluid.

Recently, Frindt and Windhager [5] measured isotopic sodium efflux from isolated perfused rabbit collecting tubules during periods in which the peritubular sodium concentration was reduced from 145 to 5 mM by replacement with choline. Sodium reduction resulted in a significant decline in ^{22}Na efflux, which could be reversed by addition of vasopressin to the bath. Sodium backflux permeability was not significantly affected under these conditions. The results obtained were consistent with the existence of Na^+-Ca^{2+} exchange in proximal and collecting tubules and inhibition of sodium transport by high cytosolic calcium levels. Experiments were also performed on proximal [4] and cortical collecting [5] tubules in which the effect of quinidine in sodium transport was tested. In both nephron segments, 5×10^{-5} M quinidine inhibited ^{22}Na efflux by about 30%. Passive paracellular permeability was unaffected. Also, the calcium ionophore A23187 inhibited proximal tubular sodium reabsorption by approximately 30% [4]. Thus, in proximal and distal tubules, a variety of experimental conditions thought to increase cytosolic Ca^{2+} activity resulted in a decrease in sodium transport.

Interpretation of these experiments is hampered by the fact that cytosolic Ca^{2+} activities have not been directly measured in epithelial cells of mammalian kidneys. However, such information was obtained on proximal tubular cells of *Necturus* kidney, employing calcium-selective glass microelectrodes [7]. In 11 successful measurements on proximal tubular cells of perfused *Necturus* kidneys, the cytosolic Ca^{2+} activity averaged 116 ($\pm$ 32) nM (mean $\pm$ S.D.), which is the same order of magnitude as in other tissues.

To test whether cytosolic calcium levels fluctuate in a manner predicted by operation of Na^+-Ca^{2+} exchange process in the basolateral cell membrane, measurements of cytosolic activity were carried out during a control period, during reduction of and then during restoration of a normal sodium concentration in peritubular fluid [7]. The luminal fluid sodium concentration was kept constant at 100 mM. When the sodium concentration in peritubular fluid was reduced from 100 to 10 mM by substitution with choline, there was a small depolarization of the membrane potential and a simultaneous increase in calcium activity from 115 to 371 nM (n = 6–8). After restoration of the normal sodium concentration, cytosolic calcium and membrane potential returned toward their original values (118 nM). In four successful measurements of this type, cytosolic Ca^{2+} activity recovered to 106 from 376 nM during the experimental periods.

Net fluid reabsorption was also measured by the split droplet technique. The half-time, about 65 min for net fluid transport at low peritubular sodium concentrations, was about 2 times longer than that found at normal sodium concentrations. These results indicate that lowering peritubular sodium concentration in *Necturus* results in a similar inhibition of sodium reabsorption as previously found in the proximal tubule of rabbit kidneys [4]. These findings also support the view that a Na^+-Ca^{2+} exchange process is located in the peritubular cell membrane and that an increase in cytosolic calcium levels inhibits net sodium transport in proximal tubules.

More recently, we have attempted to evaluate changes in cytosolic Na^+ activity in isolated perfused proximal tubules of the *Necturus* kidney [8,11] under different experimental conditions. Cytosolic Na^+ activity measured with recessed-tip sodium-selective glass microelectrodes filled with 1.5 M NaCl averaged 11.6 mM and the peritubular membrane potential -52.0 mV ($n = 23$). A lowering of peritubular sodium from 100 to 10 mM, by replacement of sodium with choline, depolarized the basolateral membrane potential by 7.6 mV and decreased cytosolic Na^+ activity from 8.3 to 3.9 mM ($n = 5$). These results are shown in the lower part of Fig. 2, while the upper diagram summarizes previously discussed calcium data. The low cytosolic Na^+ activity at a low peritubular sodium concentration (with 10 mM sodium in the contraluminal fluid) is consistent with the hypothesis that increased cytosolic Ca^{2+} activity reduces Na^+ influx into cells. This finding makes it also unlikely that elevated cytosolic calcium levels observed during periods of low peritubular sodium inhibit active sodium transport out of the cell.

Support for this view has also been obtained by Dr. Frindt in studies with amphotericin B on cortical collecting tubules. These experiments were done on rabbit tubules pretreated with mineralocorticoids, thereby achieving a near-maximal rate of sodium transport. Reducing the peritubular bath sodium concentration from 145 to 5 mM, depressed ^{22}Na efflux from 21.7 to 14.4 pM/cm per sec ($n = 5$); this reduction was presumably due to increased cytosolic calcium levels. When amphotericin B, a compound known to increase sodium permeability, was added to the luminal perfusate (5.4×10^{-6} M), sodium efflux rose to levels similar to control values at normal bath sodium concentrations. The fact that amphotericin B was able to drive sodium efflux to its apparent pump maximum suggests that a primary inhibition of the sodium pump was not responsible for the reduction in sodium efflux. Rather, we conclude that a decrease in the sodium permeability of the luminal cell membrane is responsible for reducing sodium efflux in these experiments.

These findings are supported by studies on proximal tubules of *Necturus* kidneys in which intracellular Ca^{2+} and Na^+ activities were measured during administration of quinidine (10^{-4} M). Calcium activity rose from 62 to 648 nM after addition of the drug [11]. In all cases, this effect was at least partially reversible. In contrast to the rise in intracellular Ca^{2+}, the cytosolic Na^+ activity decreased significantly from 15 to 12 mM after administration of quinidine [11]. This is not what one would expect if the sodium pumps were poisoned. The decrease in sodium efflux produced by this drug is best explained by a reduction in luminal sodium permeability, presumably due to a drug-induced increase in cytosolic Ca^{2+} activity.

In sum, our data support the view that a Na^+–Ca^{2+} exchange mechanism operates at the peritubular cell membranes of renal tubules and that the cytosolic Ca^{2+} activity plays a regulatory role for renal tubular sodium transport.

References

1. Blaustein, M. P. The interrelationship between sodium and calcium fluxes across cell membranes. *Rev. Physiol. Biochem. Pharmacol.* **70**:33–82, 1974.
2. Chase, H. S., Jr.; Al-Awqati, Q. Calcium reduces the sodium permeability of luminal membrane vesicles from toad urinary bladder: Studies using a fast-reaction apparatus. *J. Gen. Physiol.* **81**:643–666, 1983.
3. Erlij, D.; Grinstein, S. Intracellular calcium regulates transepithelial sodium transport in the frog skin. *Biophys. J.* **17**:23a, 1977.
4. Friedman, P. A.; Figueiredo, J. F.; Maack, T.; Windhager, E. E. Sodium–calcium interactions in the renal proximal convoluted tubule of the rabbit. *Am. J. Physiol.* **240**:F588–F568, 1981.
5. Frindt, G.; Windhager, E. E. Effect of quinidine, low peritubular [Na] or [Ca] on Na transport in isolated perfused rabbit cortical collecting tubules. *Fed. Proc.* **42**:305, 1983.

6. Grinstein, S.; Erlij, F. Intracellular calcium and the regulation of sodium transport in the frog skin. *Proc. R. Soc. London Ser. B* **202:**353–360, 1978.
7. Lee, C. O.; Taylor, A.; Windhager, E. E. Cytosolic calcium ion activity in epithelial cells of *Necturus* kidney. *Nature (London)* **287:**859–861, 1980.
8. Lorenzen, M.; Sackin, H.; Lee, C. O., Windhager, E. E. Intracellular Na^+ activity, basolateral cell membrane potential, and transepithelial voltage in isolated perfused proximal tubules of *Necturus* kidney. *Fed. Proc.* **40:**394a, 1981.
9. Taylor, A. Effect of quinidine in the action of vasopressin. *Fed. Proc.* **34:**285, 1975.
10. Taylor, A.; Windhager, E. E. Possible role of cytosolic calcium and Na–Ca exchange in regulation of transepithelial sodium transport. *Am. J. Physiol.* **236:**F505–F512, 1979.
11. Windhager, E. E.; Taylor, A.; Maack, T.; Lee, C. O., Lorenzen, M. Studies on renal tubular function. In: *Functional Regulation at the Cellular and Molecular Levels,* R. A. Corradino, ed., Amsterdam, Elsevier/North-Holland, 1982, pp. 299–316.

VI

Calcium as a Regulator of Neuronal Function

25

Neuronal Calcium as a Site of Action for Depressant Drugs

R. Adron Harris

The crucial role of calcium in neurotransmission and neuronal function has attracted attention to this ion as a possible mediator of the effects of depressant drugs. Calcium was first implicated in the action of opiates in a report published in 1936 showing that injection of calcium gluconate decreased development of morphine tolerance and injection of a calcium chelator (oxalate) enhanced tolerance [36]. Thirty years later, evidence was provided that intracerebral injections of calcium antagonized the analgesic effects of opiates and administration of calcium chelators produced the opposite effect [27]. These results were confirmed and extended by Way's group [5,6,21], and provided the conceptual basis for the hypothesis that opiates act by reducing availability of neuronal calcium (see also MacDonald and Werz, this volume).

Investigations on the role of calcium in the actions of ethanol were initiated by the observation of Ehrenpreis in 1965 that isopropyl alcohol increases calcium binding to an extract of sciatic nerve [9]. This observation was extended to demonstrate that relatively low concentrations of ethanol (10–50 mM) increase calcium binding to the cytoplasmic surface of erythrocyte and brain synaptosomal membranes [18,28,32]. This effect of alcohols led to comparative studies with barbiturates. Although barbiturates increase the binding of calcium to phospholipids [2], their predominant effect on biological membranes appears to be to decrease calcium binding [18]. Despite their opposing effects on calcium binding, nerve blockage produced by both alcohols and barbiturates may be reversed by elevated concentrations of calcium [33]. These and other observations were made in the 1960s and early 1970s and provide indirect evidence for a role of calcium in the actions of opiates, barbiturates, and alcohols. These results, together with a growing body of literature (reviewed in Refs. 5,6,15,24, and Macdonald and Werz, this volume) demonstrating that these three classes of drugs inhibit the calcium-dependent release of certain neurotransmitters, provided the impetus for detailed studies of effects of these drugs on neuronal calcium fluxes.

Progress in neuropharmacology is often dependent on, and limited by, technical ad-

---

R. Adron Harris • Denver Veterans Administration Hospital and Department of Pharmacology, University of Colorado, School of Medicine, Denver, Colorado 80262.

Table I. Summary of the Acute Effects of Depressant Drugs on Neuronal Calcium Function Measured in Vitro[a]

Cellular function	Drug effect and references		
	Opiates	Alcohols	Barbiturates
Calcium binding to plasma membranes	0 (26; unpublished observations)	↑ (18, 28, 32)	↓ (18)
Voltage-dependent calcium influx	↓ or 0 (15; Chapter 27)	↓ (20, 29, 30, 34; Chapter 26)	↓ (3, 10; Chapters 26 and 27)
Calcium efflux: Na^+-Ca^{2+} exchange	?	↓ (Chapter 28)	?
Calcium efflux: Ca^{2+}-ATPase	?	↑ (40)	↑ (23, 25)
Calcium sequestration: mitochondrial	0 (13; unpublished observations)	sl ↓ (unpublished observations)	↓ and ↑ (31, 35, 38)
Calcium sequestration: nonmitochondrial	↓ (13)	sl ↓ (14, 25)	sl ↓ (14, 25)
Calcium-dependent potassium efflux	↑ (37)	↑ (39; Chapters 22 and 23)	↑ (Chapter 22)

[a] ↑ indicates increased activity; ↓ indicates decreased activity; 0 indicates no effect; ? indicates drug effect not known; sl indicates effect is small and requires high concentration. Selected references are given in parentheses. Only results obtained from studies of acute, *in vitro* drug application are presented.

vances in neurophysiology and neurochemistry. Although the importance of calcium in neuronal transmission has long been recognized, only in the last few years have techniques evolved for rigorous electrophysiological and biochemical analyses of calcium transport in the nerve cell. These studies, which are discussed in detail in this volume, suggest seven processes as potential sites for interactions between calcium and depressant drugs (Table I). The ability of these drugs to alter multiple components of the calcium homeostatic system raises three important questions: What are the biochemical mechanisms responsible for these actions? What are the functional consequences of these alterations? Which, if any, of these changes are responsible for the observed pharmacological effects?

None of these questions can be answered definitively at the present time, but some information is available for each of them. Regarding mechanism of action, the opiates act through specific receptors, but it is not clear whether these receptors directly affect ion channels or act through intermediate messengers (5,6; Macdonald and Werz, this volume). At least some opiate receptors are coupled to adenylate cyclase; and it is possible that changes in cAMP mediate effects of opiates on ion channels, as cAMP-dependent protein phosphorylation has been implicated in the regulation of neuronal ion fluxes [1,7]. Further studies are needed to define the steps between opiate–receptor interaction and changes in ion fluxes. The effects of barbiturates and alcohols may be due to their ability to disorder membrane lipids [19,22] or to interact directly with membrane proteins [12]. To distinguish between these two possibilities, investigators have compared effects of alcohols, barbiturates, and other membrane perturbants on membrane order (as judged by the fluorescence polarization of diphenylhexatriene, a probe of the membrane core) with their actions on ion transport. These studies demonstrate that the chemically diverse membrane perturbants increase synaptic Ca^{2+}-ATPase activity, which is correlated with the degree of membrane fluidization [40]. The drug effects on voltage-dependent calcium influx and calcium-stimulated potassium efflux are not, however, correlated with their membrane-disordering effect

[16,17]. Thus, the actions of alcohols and barbiturates on neuronal ion transport appear to be due to both lipid perturbation and a direct action on membrane proteins.

Evaluation of the functional consequences of the alterations given in Table I requires that results obtained from isolated systems under specialized conditions be integrated to reflect brain function. At present, our knowledge of the relative importance of each of the processes is not sufficient to allow definitive conclusions. For example, ethanol may increase intraneuronal calcium by inhibiting $Na^+ - Ca^{2+}$ exchange and thereby enhance potassium efflux. On the other hand, the effect of ethanol on voltage-dependent calcium influx and on Ca^{2+}-ATPase might decrease intracellular calcium, possibly inhibiting potassium efflux. *In vivo* there is a dynamic interplay of these processes on a millisecond time scale that is difficult to evaluate from biochemical data obtained *in vitro* with a time scale of seconds to minutes.

Despite these reservations, the consistency of electrophysiological and biochemical data tempts speculation that the effect of depressant drugs is to impair synaptic transmission by reducing calcium influx and by enhancing calcium-dependent potassium efflux. This does not imply that equal effects occur at all synapses. At a given nerve terminal, the drug might alter one, both, or neither of these transport mechanisms. For the opiates, selective effects are easily envisioned because not all nerve endings contain opiate receptors and not all opiate receptors are coupled to the same effector (Macdonald and Werz, this volume). Such selectivity is not necessarily expected in the case of barbiturates and alcohols. But differences in brain regional sensitivity to effects of these drugs on calcium influx [10,34], and their selectivity in inhibiting the release of certain neurotransmitters [4], suggest that synaptic specificity does occur with regard to effects of these drugs on neuronal ion transport. The determinants of such specificity are a challenging and important area for future research.

The last, and most difficult, question concerns the pharmacological importance of observed changes in ion fluxes. Three approaches have been used to address this question: (1) determine if the drug concentration required to alter ion transport in an isolated system is consistent with concentrations achieved *in vivo;* (2) determine if the effects and potencies obsered *in vivo* for pharmacologically related agents are correlated with these observed *in vitro;* and (3) determine if agents which antagonize the drug effect *in vitro* also block the effect *in vivo* (or vice versa). The subsequent articles and others [11,15,23,34,35] discuss these approaches in detail, and indicate, with some qualifications, that the effects presented in Table I may be related to the pharmacological action of the drugs. One inconsistency is that biochemical changes produced by pharmacologically relevant concentrations of the drugs are often quite small. The lack of sensitivity may merely indicate that conditions *in vitro* are different from those *in vivo*.

Another problem is that good antagonists of the actions of alcohols and barbiturates are not available. The opiate and benzodiazepine receptor antagonists have proven invaluable for sorting out the specific and nonspecific actions of these drugs (Carlen *et al.*, this volume; Macdonald and Werz, this volume); and it is unfortunate that analogous pharmacological tools do not exist for other depressant drugs. One approach to this problem is to use agents which act directly on ion transport processes. A promising example is apamin, a polypeptide that blocks calcium-stimulated potassium efflux and also reduces the anesthetic action of ethanol (39; Yamamoto, this volume). Development of specific blocking agents for neuronal calcium transport is proceeding [8] and these compounds will undoubtedly help us to understand the role of ion fluxes in the action of depressant drugs.

In conclusion, investigations of the role of calcium in the action of depressant drugs span five decades, but only during the last decade have we learned enough about the physiology and biochemistry of neuronal calcium homeostasis to begin to define drug actions.

ACKNOWLEDGMENTS. Supported in part by funds from the Veterans Administration and USPHS Grant DA-02855.

References

1. Alkon, D. L.; Acosta-Urquidi, J.; Olds, J.; Kuzma, G.; Neary, J. T. Protein kinase injection reduces voltage-dependent potassium currents. *Science* **219**:303–306, 1983.
2. Blaustein, M. P. Phospholipids as ion exchangers: Implications for a possible role in biological membrane excitability and anesthesia. *Biochim. Biophys. Acta* **135**:653–668, 1967.
3. Blaustein, M. P.; Ector, A. C. Barbiturate inhibition of calcium uptake by depolarized nerve terminals *in vitro*. *Mol. Pharmacol.* **11**:369–378, 1975.
4. Carmichael, F. J.; Israel, Y. Effects of ethanol on neurotransmitter release by rat brain cortical slices. *J. Pharmacol. Exp. Ther.* **193**:824–834, 1975.
5. Chapman, D. B.; Way, E. L. Metal ion interactions with opiates. *Annu. Rev. Pharmacol.* **20**:553–579, 1980.
6. Chapman, D.; Way, E. L. Pharmacologic consequences of calcium interactions with opioid alkaloids and peptides. In: *Calcium Regulation by Calcium Antagonists*, R. G. Rahwan and D. T. Witiak, eds., Washington, D.C., American Chemical Society, 1982, pp. 119–142.
7. dePeyer, J. E.; Cachelin, A. B.; Levitan, I. B.; Reuter, H. Ca^{2+}-activated K^+ conductance in internally perfused snail neurons is enhanced by protein phosphorylation. *Proc. Natl. Acad. Sci. USA* **79**:4207–4211, 1982.
8. Dethmers, J. K.; Cragoe, E. J.; Kaczorowski, G. J. Inhibition of Na^+/Ca^{++} exhange in bovine pituitary plasma membrane vesicles by analogs of amiloride. *Fed. Proc.* **42**:2245, 1983.
9. Ehrenpreis, S. An approach to the molecular basis of nerve activity. *J. Cell. Comp. Physiol.* **66**:159–164, 1965.
10. Elrod, S. V.; Leslie, S. W. Acute and chronic effects of barbiturates on depolarization-induced calcium influx into synaptosomes from rat brain regions. *J. Pharmacol. Exp. Ther.* **212**:131–136, 1980.
11. Faber, D. S.; Klee, M. R. Actions of ethanol on neuronal membrane properties and synaptic transmission. In: *Alcohol and Opiates: Neurochemical and Behavioral Mechanisms*, K. Blum, ed., New York, Academic Press, 1977, pp. 41–63.
12. Franks, N. P.; Lieb, W. R. Molecular mechanisms of general anaesthesia. *Nature (London)* **300**:487–492, 1982.
13. Guerrero-Munoz, F.; Guerrero, M.; Way, E. L. Effect of morphine on calcium uptake by lysed synaptosomes. *J. Pharmacol. Exp. Ther.* **211**:370–374, 1979.
14. Harris, R. A. Ethanol and pentobarbital inhibit intrasynaptosomal sequestration of calcium. *Biochem. Pharmacol.* **30**:3209–3215, 1981.
15. Harris, R. A. Psychoactive drugs as antagonists of actions of calcium. In: *New Perspectives on Calcium Antagonists*, G. B. Weiss, ed., Washington D.C., American Physiological Society, 1981, pp.223–231.
16. Harris, R. A. Ethanol membrane perturbation, and synaptosomal ion transport. *Proc. West. Pharmacol. Soc.* **26**:255–257, 1983.
17. Harris, R. A. Differential effects of membrane perturbants on voltage-activated sodium and calcium channels and calcium-dependent potassium channels. *Biophys. J.* **45**:132–134, 1984.
18. Harris, R. A.; Fenner, D. Ethanol and synaptosomal calcium binding. *Biochem. Pharmacol.* **31**:1790–1792, 1982.
19. Harris, R. A.; Hitzemann, R. J. Membrane fluidity and alcohol actions. In: *Currents in Alcoholism*, Volume VIII, M. Galanter, ed., New York, Grunne and Stratton, 1981, pp.379–404.
20. Harris, R. A.; Hood, W. F. Inhibition of synaptosomal calcium uptake by ethanol. *J. Pharmacol. Exp. Ther.* **213**:562–568, 1980.
21. Harris, R. A.; Loh, H. H.; Way, E. L. Effects of divalent cations, cation chelators and an ionophore on morphine analgesia and tolerance. *J. Pharmacol, Exp. Ther.* **195**:488–498, 1975.
22. Harris, R. A.; Schroeder, F. Effects of barbiturates and ethanol on the physical properties of brain membranes. *J. Pharmacol. Exp. Ther.* **223**:424–431, 1982.
23. Harris, R. A.; Stokes, J. A. Effects of a sedative and a convulsant barbiturate on synaptosomal calcium transport. *Brain Res.* **242**:157–163, 1982.
24. Ho, I. K.; Harris, R. A. Mechanism of action of barbiturates. *Annu. Rev. Pharmacol.* **21**:83–111, 1981.
25. Hood, W. F.; Harris, R. A. Effects of depressant drugs and sulfhydryl reagents on the transport of calcium by isolated nerve endings. *Biochem. Pharmacol.* **29**:957–959, 1980.

26. Hoss, W.; Okumura, K.; Formaniak, M.; Tanaka, R. Relation of cation binding sites on synaptic vesicles to opiate action *Life Sci.* **24:**1003–1010, 1979.
27. Kakunaga, T.; Kaneto, H.; Hano, K. Pharmacologic studies on analgesics. VII. Significance of the calcium ion in morphine analgesia. *J. Pharmacol. Exp. Ther.* **153:**134–141, 1966.
28. Michaelis, E. K.; Myers, S. L. Calcium binding to brain synaptosomes: Effects of chronic ethanol intake. *Biochem. Pharmacol.* **28:**2081–2087, 1979.
29. Oakes, S. G.; Pozos, R. B. Electrophysiologic effects of acute ethanol exposure. I. Alterations in the action potentials of dorsal root ganglia neurons in dissociated culture. *Dev. Brain Res.* **5:**243–249, 1982.
30. Oakes, S. G.; Pozos, R. S. Electrophysiologic effects of acute ethanol exposure. II. Alterations in the calcium component of action potentials from sensory neurons in dissociated culture. *Dev. Brain Res.* **5:**251–255, 1982.
31. Pincus, J. H.; Hsiao, K. Calcium uptake mechanisms affected by some convulsant and anticonvulsant drugs. *Brain Res.* **217:**119–127, 1981.
32. Seeman, P.; Chau, M.; Goldberg, M.; Sauks, T.; Sax, L. The binding of Ca^{2+} to the cell membrane increased by volatile anesthetics (alcohols, acetone, ether) which induce sensitization of nerve or muscle. *Biochim. Biophys. Acta* **225:**185–193, 1971.
33. Seeman, P.; Chen, S. S.; Chau-Wong, M.; Staiman, A. Calcium reversal of nerve blockage by alcohols, anesthetics, tranquilizers, and barbiturates. *Can. J. Physiol. Pharmacol.* **52:**526–534, 1974.
34. Stokes, J. A.; Harris, R. A. Alcohols and synaptosomal calcium transport. *Mol. Pharmacol.* **22:**99–104, 1982.
35. Sweetman, A. J.; Esmail, A. F. Effect of the general anaesthetics, alphaxalone, hexobarbitone and halothane on calcium uptake into rat brain mitochondria *in vitro* and *in vivo*. *Methods Find. Exp. Clin. Pharmacol.* **4:**299–305, 1982.
36. Weger, P.; Amsler, C. Weiteres zum Problem der Gewöhnung Morphin. *Arch. Exp. Pathol. Pharmakol.* **181:**489–493, 1936.
37. Williams, J. T.; Egan, T. M.; North, R. A. Enkephalin opens potassium channels on mammalian central neurones. *Nature (London)* **299:**74–77, 1982.
38. Willow, M.; Bygrave, F. L. Effects of pentobarbitone on $^{45}Ca^{2+}$ transport by rat brain mitochondria. *J. Neurochem.* **39:**557–562, 1982.
39. Yamamoto, H.-A.; Harris, R. A. Calcium-dependent ^{86}Rb efflux and ethanol intoxication: Studies of human red blood cells and rodent brain synaptosomes. *Eur. J. Pharmacol.* **88:**357–363, 1983.
40. Yamamoto, H.-A.; Harris, R. A. Effects of ethanol and barbiturates on Ca^{++}-ATPase activity of erythrocyte and brain membranes. *Biochem. Pharmacol.* **32:**2787–2791, 1983.

26

Sedative-Hypnotic Drugs and Synaptosomal Calcium Transport

Steven W. Leslie, Edward M. Barr, L. Judson Chandler, and Laura C. Daniell

Sedative-hypnotic drugs exert potent actions within the synapse and on neuronal cell bodies in the central nervous system. Barbiturates block synaptic transmission at concentrations within their anesthetic range [12,14], whereas much larger concentrations are required to suppress axonal conduction. Both pre- and postsynaptic functions are altered by low barbiturate concentrations. Postsynaptically, barbiturates depress excitatory postsynaptic potentials [2,21,22] and enhance postsynaptic inhibition [10,24]. Presynaptically, anesthetic concentrations of pentobarbital decrease neurotransmitter release from cerebral cortical slices [11] and brain prisms from different brain regions [26]. With regard to the effects of sedative-hypnotic drugs on neuronal cell bodies, barbiturates inhibit voltage-dependent calcium uptake [8] and facilitate chloride conductance [10] in neuronal cell cultures.

Over the past several years we have been interested in the possibility that sedative-hypnotic drugs may inhibit neurotransmitter release by blocking voltage-dependent calcium uptake into presynaptic nerve terminals. Calcium enters the neuron through voltage-dependent or receptor-operated calcium channels [1,4,13,23,25,27] (Fig. 1). Since calcium entry into the presynaptic nerve terminal is essential for exocytotic neurotransmitter release, agents that block presynaptic calcium uptake will suppress neurotransmitter release and, therefore, synaptic function.

Sedative-Hypnotic Drugs Inhibit Synaptosomal Calcium Uptake

Anesthetic concentrations of barbiturates inhibit voltage-dependent calcium uptake into isolated synaptosomes [3,6,9,16]. In these studies synaptosomes were depolarized with potassium chloride and $^{45}Ca^{2+}$ uptake was measured for 1 to 2 min. Studies in our laboratory revealed that brain regions vary in their sensitivity to the inhibitory effects of pentobar-

---

Steven W. Leslie, Edward M. Barr, L. Judson Chandler, and Laura C. Daniell • Division of Pharmacology, College of Pharmacy, The University of Texas at Austin, Austin, Texas 78712.

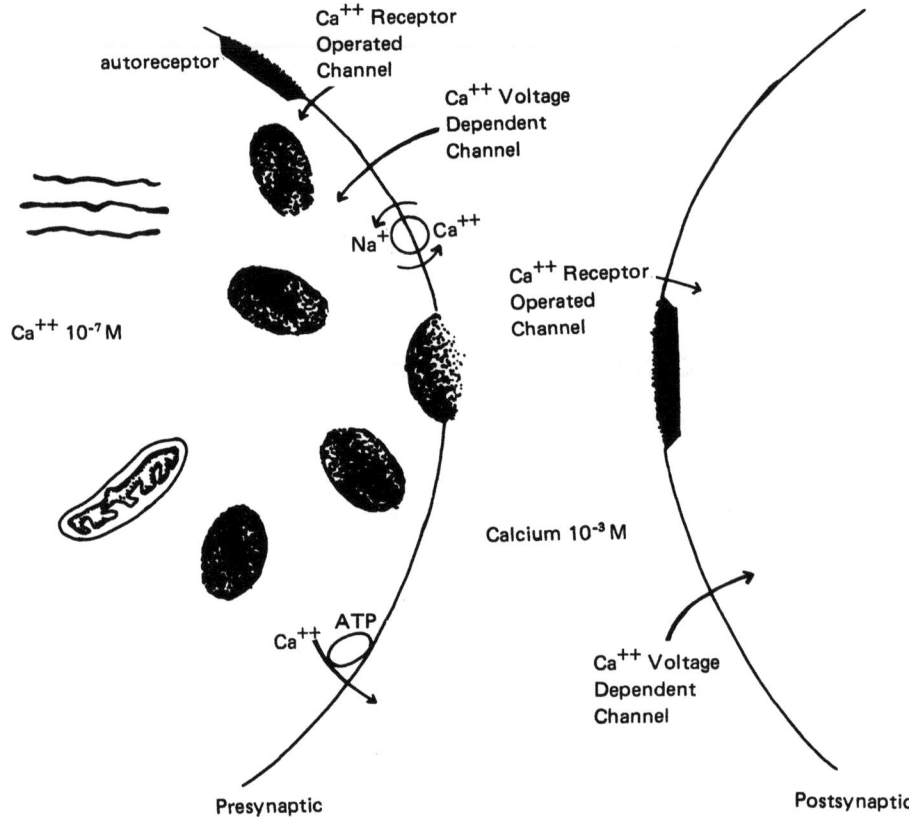

Figure 1. Representation of voltage-dependent and receptor-operated calcium channels in pre- and postsynaptic terminals. Calcium enters passively because of the large concentration gradient across the cell membrane and is eventually removed from the cell interior by Ca^{2+}-ATPase and Na^+–Ca^{2+} exchange activation.

bital. A greater inhibition of synaptosomal calcium uptake was observed in brain stem and cerebellum than in cerebral cortex (77, 73, and 39%, respectively, using 0.3 mM pentobarbital). $^{45}Ca^{2+}$ uptake into synaptosomes from midbrain, striatum, or hypothalamus was only slightly inhibited by pentobarbital. These differences in regional sensitivity to barbiturates are in agreement with electrophysiological studies [5,12].

Tolerance Development to Inhibition of Calcium Uptake

Chronic barbiturate administration results in the development of functional tolerance. Therefore, a biochemical mechanism responsible for barbiturate sedation should undergo adaptation after chronic exposure resulting in less responsivity to depressant actions of the drug. Thus, phenobarbital was administered in a lab chow diet to male Sprague–Dawley rats for 14 days. $^{45}Ca^{2+}$ uptake by synaptosomes isolated from these drug-treated rats was compared to uptake observed in synaptosomes isolated from control rats and rats treated acutely with phenobarbital. Chronic phenobarbital administration resulted in the development of biochemical tolerance to the inhibitory effect of pentobarbital on calcium uptake [6]. Although the addition of pentobarbital to control synaptosomes from cerebral cortex, brain stem, and cerebellum resulted in a marked inhibition of KCl-induced $^{45}Ca^{2+}$ uptake (39, 77, and 73%, respectively), chronic barbiturate administration resulted in an adaptation such that $^{45}Ca^{2+}$ uptake by these brain regions was inhibited by only 21, 6, and 16%, respectively.

Thus, barbiturates inhibit voltage-dependent calcium uptake into presynaptic nerve terminals in a manner consistent with a cellular mechanism of anesthesia. In addition, chronic barbiturate administration results in tolerance development to inhibition of calcium uptake. This adaptive process may represent a cellular mechanism responsible for expression of functional tolerance. Similar studies have been performed with chlordiazepoxide [17] and chlorpromazine [15]. Both of these drugs inhibited calcium uptake in concentrations which produce sedation, and chronic administration resulted in tolerance development to inhibition of calcium uptake similar to that observed with barbiturates.

Inhibition of Fast- and Slow-Phase Calcium Uptake

Recent evidence suggests that calcium entry into nerve terminals occurs through two separate processes, referred to as "fast-" and "slow-phase" uptake [7,20]. The fast component may modulate phasic release of neurotransmitters subsequent to an action potential whereas the slow component may be relevant to conditions associated with more prolonged stimulation such as tetanic stimulation [20]. Our work shows that the fast-phase component for calcium uptake continues for approximately 3 sec (Table I). This observation agrees with the findings of Gripenberg *et al.* [7], but differs slightly from results of Nachshen and Blaustein [20] which suggest that the fast-phase process terminates within 1 sec. During the 5- to 15-sec time interval, $^{45}Ca^{2+}$ uptake rates were diminished to 10% or less of the values observed at 0–1 sec, suggesting that only calcium channels involved in the slow-phase process remain open after 5 sec of depolarization (Table I). Thus, the experiments described above investigating the effects of barbiturates and other sedative drugs on calcium uptake at 1–2 min examined slow-phase uptake parameters.

The inhibitory effects of 0.2 mM pentobarbital on fast- and slow-phase calcium uptake into cerebrocortical synaptosomes are illustrated in Table II. Pentobarbital inhibited voltage-dependent $^{45}Ca^{2+}$ uptake at all time points examined but produced its most potent inhibitory

Table I. Rates of Depolarization-Dependent $^{45}Ca^{2+}$ Uptake by Synaptosomes from Rat Brain Regions[a]

Uptake time intervals (sec)	nmoles $^{45}Ca^{2+}$/sec/mg protein			
	Cerebral cortex ($n = 6$)	Cerebellum ($n = 6$)	Midbrain ($n = 5$)	Brain stem ($n = 6$)
0–1	1.02 ± 0.14	1.08 ± 0.10	1.21 ± 0.12	0.41 ± 0.05*
1–3	0.72 ± 0.10	0.40 ± 0.04	0.33 ± 0.04	0.15 ± 0.04**
3–5	0.22 ± 0.09	0.17 ± 0.09	0.27 ± 0.07	0.18 ± 0.06
5–15	0.07 ± 0.02	0.09 ± 0.02	0.12 ± 0.02	0.05 ± 0.01
15–30	0.05 ± 0.02			

[a]Uptake rates (mean ± S.E.M.) were determined by least-squares linear regression analysis for each time interval. The data represent net $^{45}Ca^{2+}$ uptake rates after 65 mM KCl depolarization. Resting $^{45}Ca^{2+}$ uptake (5 mM KCl) was subtracted from total $^{45}Ca^{2+}$ uptake (after 65 mM KCl addition) to obtain net $^{45}Ca^{2+}$ uptake values for linear aggression analysis.
*Statistically significant ($p < 0.05$; one-way analysis of variance; Scheffe's post hoc test) from cerebral cortex, cerebellum, and midbrain values.
**Statistically significant ($p < 0.05$) from cerebral cortex values.

Table II. Time-Dependent Inhibition of Synaptosomal $^{45}Ca^{2+}$ Uptake by Pentobarbital[a]

Time (sec)	μmoles $^{45}Ca^{2+}$/g protein		% inhibition
	Control	Pentobarbital (0.2 mM)	
1	1.95 ± 0.30	0.61 ± 0.33	69
3	3.50 ± 0.28	1.77 ± 0.03	49
5	3.66 ± 0.43	2.23 ± 0.16	39
15	5.35 ± 0.69	3.53 ± 0.09	34
30	5.86 ± 1.24	4.27 ± 0.01	28
60	6.91 ± 1.71	5.26 ± 0.25	24

[a]Rates of $^{45}Ca^{2+}$ uptake values calculated as described in footnote a of Table I. Data represent the mean of two experiments performed in duplicate using cerebrocortical syntaptosomes from male Sprague–Dawley rats.

effect on the fast component of $^{45}Ca^{2+}$ uptake. A concentration–response study with pentobarbital showed that inhibition of fast-phase calcium uptake (3 sec) was concentration-dependent. Pentobarbital, 0.05, 0.10, 0.20, and 0.30 mM, inhibited $^{45}Ca^{2+}$ uptake by 6, 21, 36, and 51% respectively.

Time–response and concentration–response studies were also performed with ethyl alcohol [18]. Addition of 80 mM ethanol to synaptosomes isolated from rat cerebral cortex resulted in significant reduction of $^{45}Ca^{2+}$ uptake at 1–3 sec. Uptake of $^{45}Ca^{2+}$ by cerebrocortical synaptosomes was not altered at 5 sec or later. Ethanol inhibited fast-phase calcium uptake (3-sec measurement time) over its sedative-anesthetic range (25–100 mM) in a concentration-dependent manner [18]. These results suggest that in cerebrocortical synaptosomes both ethanol and pentobarbital are potent blockers of the fast-phase, voltage-dependent calcium channel. Pentobarbital also blocked slow-phase calcium uptake into cerebrocortical synaptosomes while ethanol did not. Studies using synaptosomes isolated from rat midbrain, cerebellum, and brain stem showed that ethanol produced a more generalized inhibition of fast- and slow-phase calcium uptake [18]. Ethanol, 80 mM, significantly inhibited $^{45}Ca^{2+}$ uptake into cerebellum and brain stem synaptosomes at 3–5 sec, while $^{45}Ca^{2+}$ uptake into midbrain synaptosomes was inhibited by ethanol at 1, 3, 5, and 15 sec. Ethanol also differentially inhibits slow-phase synaptosomal $^{45}Ca^{2+}$ uptake, depending upon the brain region studied [28].

Since ethanol inhibits $^{45}Ca^{2+}$ uptake into isolated presynaptic nerve terminals, studies were designed to examine whether chronic administration of ethanol resulted in the development of tolerance to inhibition of $^{45}Ca^{2+}$ uptake as has been demonstrated for other sedative-hypnotic drugs (6,15–17). A summary of these results is shown in Table III and a more complete description may be obtained elsewhere [18]. Table III shows that *in vitro* addition of ethanol, 80 mM, significantly inhibited $^{45}Ca^{2+}$ uptake into brain synaptosomes by 3 sec in *ad libitum* control and *pair-fed* control groups. Ethanol, 80 mM, did not significantly inhibit $^{45}Ca^{2+}$ uptake into synaptosomes from the chronic ethanol-treated rats. These results are quite similar to those obtained with chronic barbiturate [6,16], chlordiazepoxide [17], and chlorpromazine [15] administration and provide further support for the suggestion that functional tolerance development after chronic sedative-hypnotic drug administration may occur as a result of cellular adaptation to inhibition of calcium transport at the presynaptic nerve terminal.

Table III. Effects of Chronic Ethanol Consumption on Depolarization-Dependent $^{45}Ca^{2+}$ Uptake by Synaptosomes Isolated from Cerebral Cortex[a,b]

	Assay conditions	nmoles $^{45}Ca^{2+}$/3 sec/mg protein	% inhibition
Ad libitum control	Net uptake	3.94 ± 0.36	
	Net uptake with 80 mM ethanol	2.73 ± 0.27*	30.7
Pair-fed control	Net uptake	3.75 ± 0.23	
	Net uptake with 80 mM ethanol	2.88 ± 0.23*	23.2
Chronic ethanol diet	Net uptake	3.19 ± 0.22	
	Net uptake with 80 mM ethanol	2.75 ± 0.22	13.8

[a] Initial weight of rats in each group was 225–250 g. Final weights after the 8-week diet period were 357 ± 3, 468 ± 20, and 346 ± 9 g for pair-fed control, ad libitum control, and chronic ethanol-treated rats, respectively. Blood ethanol levels for the chronic ethanol-treated group measured at 7 PM on the day before sacrifice averaged 146 ± 20 mg/dl. Each value in this table represents the mean ± S.E. of six to seven experiments. Net uptake was determined as for Tables I and II. Asterisks indicate statistically significant ($p < 0.05$; one-way analysis of variance) inhibition produced by 80 mM ethanol addition in vitro as compared with the respective control value.

[b] Ethanol was administered to Sprague–Dawley rats in a nutritionally adequate liquid diet for 8 weeks (19). A pair-fed group received a liquid diet in which ethanol was replaced isocalorically with dextrin. A third, ad libitum control group which was maintained on liquid diet containing dextrin was also included. Synaptosomes were isolated from cerebral cortex of animals in each of these groups and $^{45}Ca^{2+}$ uptake was examined.

Conclusions

Sedative-hypnotic drugs are potent inhibitors of voltage-dependent calcium uptake by presynaptic nerve endings and neuronal cell bodies [3,6,8,9,15-18,28], whereas voltage-dependent calcium is unaltered by these drugs [3,6,8]. The concentrations required to block calcium uptake are in the anesthetic range. These observations suggest that sedative-hypnotic drugs produce some of their pharmacological actions through inhibition of calcium uptake. Further support for this suggestion is derived from the observation that, at least for barbiturates and ethanol, brain regional differences exist for their inhibitory potency on voltage-dependent calcium uptake. These brain regional differences agree closely with previous electrophysiological findings. In addition, the demonstration that chronic administration of sedative-hypnotic drugs results in tolerance development to inhibition of calcium uptake lends support to the possibility that functional tolerance develops as a result of an adaptive process involving the neuronal calcium channel.

ACKNOWLEDGMENTS. This work was supported by National Institute on Alcoholism and Alcohol Abuse Research Grant AA-05809, a grant from the Alcoholic Beverage Medical Research Foundation, and Research Scientist Development Award AA-00044 to S.W.L.

References

1. Baker, P. F. Transport and metabolism of calcium ions in nerve. *Prog. Biophys. Mol. Biol.* **24**:177–223, 1972.
2. Barker, J. L.; Gainer, H. Pentobarbital: Selective depression of excitatory postsynaptic potentials. *Science* **182**:720–722, 1973.

3. Blaustein, M. P.; Ector, A. C. Barbiturate inhibition of calcium uptake by depolarized nerve terminals *in vitro*. *Mol. Pharmacol.* **11**:369–378, 1975.
4. Bolton, T. B. Mechanism of action of transmitters and other substances on smooth muscle. *Physiol. Rev.* **59**:606–718, 1979.
5. Brazier, M. The electrophysiological effects of barbiturates on the brain. In: *Physiological Pharmacology*, Volume 1, W. Root and F. Hoffman, eds., New York Academic Press 1963, pp. 219–238.
6. Elrod, S. V.; Leslie, S. W. Acute and chronic effects of barbiturates on depolarization-induced calcium influx into synaptosomes from rat brain regions. *J. Pharmacol. Exp. Ther.* **212**:131–136, 1980.
7. Gripenberg, J.; Heinonen, E.; Jansson, S.-E. Uptake of radiocalcium by nerve endings isolated from rat brain: Kinetic studies. *Br. J. Pharmacol.* **71**:265–271, 1980.
8. Heyer, E. J.; MacDonald, R. L. Barbiturate reduction of calcium-dependent action potentials: Correlation with anesthetic action. *Brain Res.* **236**:157–171, 1982.
9. Hood, W. F.; Harris, R. A. Effects of depressant drugs and sulfhydryl reagents on the transport of calcium by isolated nerve endings. *Biochem. Pharmacol.* **29**:957–959, 1980.
10. Huang, L.-Y. M.; Barker, J. L. Pentobarbital: Stereospecific actions of (+) and (−) isomers revealed on cultured mammalian neurons. *Science* **207**:195–197, 1980.
11. Kalant, H.; Grose, W. Effects of ethanol and pentobarbital on release of acetylcholine from cerebral cortex slices. *J. Pharmacol. Exp. Ther.* **158**:386–393, 1967.
12. King, E. E. Differential action of anesthetics and interneuron depressants upon EEG arousal and recruitment responses. *J. Pharmacol. Exp. Ther.* **116**:404–417, 1956.
13. Kostyuk, P. G. Calcium ionic channels in electrically excitable membranes. *Neuroscience* **5**:945–959, 1980.
14. Larrabee, M.; Posternak, J. Selective action of anesthetics on synapses and axons in mammalian sympathetic ganglia. *J. Neurophysiol.* **15**:91–114, 1952.
15. Leslie, S. W.; Elrod, S. V.; Coleman, R.; Belknap, J. K. Tolerance to barbiturate and chlorpromazine-induced central nervous system sedation—Involvement of calcium mediated stimulus–secretion coupling. *Biochem. Pharmacol.* **28**:1437–1440, 1979.
16. Leslie, S. W.; Friedman, M. B.; Wilcox, R. E.; Elrod, S. V. Acute and chronic effects of barbiturates on depolarization-induced calcium influx into rat synaptosomes. *Brain Res.* **185**:409–417, 1980.
17. Leslie, S. W.; Friedman, M. B.; Coleman, R. R. Effects of chlordiazepoxide on depolarization-induced calcium influx into synaptosomes. *Biochem. Pharmacol.* **29**:2439–2443, 1980.
18. Leslie, S. W.; Barr, E.; Chandler, J.; Farrar, R. P. Inhibition of fast- and slow-phase depolarization-dependent synaptosomal calcium uptake by ethanol. *J. Pharmacol. Exp. Ther.* **225**:571–575, 1983.
19. Miller, S. S.; Goldman, M. E.; Erickson, Ç. K; Shorey, R. L. Induction of physical dependence on and tolerance to ethanol in rats fed a nutritionally complete and balanced liquid diet. *Psychopharmacology* **68**:55–59, 1980.
20. Nachshen, D. A.; Blaustein, M. P. Some properties of potassium-stimulated calcium influx in presynaptic nerve endings. *J. Gen Physiol.* **76**:709–728, 1980.
21. Nicoll, R. A. Pentobarbital: Differential postsynaptic actions on sympathetic ganglion cells. *Science* **199**:451–452, 1978.
22. Nicoll, R. A.; Iwamoto, E. T. Action of pentobarbital on sympathetic ganglion cells. *J. Neurophysiol.* **41**:977–985, 1978.
23. Putney, J. W., Jr. Stimulus–permeability coupling—Role of calcium in the receptor regulation of membrane permeability. *Pharmacol. Rev.* **30**:209–245, 1978.
24. Ransom, B. R.; Barker, J. L. Pentobarbital selectively enhances GABA-mediated post-synaptic inhibition in tissue cultured mouse spinal neurons. *Brain Res.* **114**:530–535, 1976.
25. Reuter, H. Divalent cations as charge carriers in excitable membranes. *Prog. Biophys. Mol. Biol.* **26**:1–43, 1973.
26. Richter, J. A.; Waller, M. B. Effects of pentobarbital on the regulation of acetylcholine content and release in different regions of rat brain. *Biochem. Pharmacol.* **26**:609–615, 1977.
27. Rosenberger, L. B.; Triggle, D. J. Calcium, calcium translocation and specific calcium antagonists. In: *Calcium in Drug Action*, G. B. Weiss, ed., New York, Plenum Press, 1978, p. 3.
28. Stokes, J. A.; Harris, R. A. Alcohols and synaptosomal calcium transport. *Mol. Pharmacol.* **22**:99–104, 1982.

27

Barbiturate and Opiate Actions on Calcium-Dependent Action Potentials and Currents of Mouse Neurons in Cell Culture

Robert L. Macdonald and Mary Ann Werz

The purpose of this chapter is to consider the actions of depressant drugs on calcium entry into neurons. Since presynaptic calcium entry is an essential link in excitation–secretion coupling of synaptic transmission, a reduction of calcium entry by depressant drugs should produce a reduction in neurotransmitter release and therefore a reduction in synaptic transmission. In this chapter we will review evidence obtained by using intracellular recordings from mouse neurons in primary dissociated cell culture which indicates that barbiturate and opiate drugs reduce presynaptic calcium entry. Evidence will also be presented which suggest that there are substantial differences in the manner in which these two drug classes modify calcium movement.

Mouse Neurons in Cell Culture

Experiments were performed on mouse spinal cord and dorsal root ganglion neurons grown in primary dissociated cell culture which were prepared using methods previously described [28]. Briefly, spinal cords and attached dorsal root ganglia were removed from 12.5- to 14-day-old mouse fetuses, mechanically disrupted to a single-cell suspension, and plated on 35-mm collagen-coated dishes in a horse serum/fetal calf/Eagle's minimum essential medium (MEM) growth medium. After 3 to 5 days in culture, the rapid division of nonneuronal cells was suppressed by an antimitotic agent, and the cultures were changed to a horse serum/MEM growth medium for 4 to 12 weeks. All experiments were performed in a protein-free salt solution.

Spinal cord neurons in cell culture are multipolar and electrically excitable, develop an extensive synaptic connectivity, and have chemosensitivity to a variety of neurotransmitters

---

Robert L. Macdonald • Departments of Neurology and Physiology, University of Michigan, Ann Arbor, Michigan 48109. *Mary Ann Werz* • Department of Neurology, University of Michigan, Ann Arbor, Michigan 48109.

[17,24,28]. Intracellular impalement of spinal cord neurons bathed in a physiological salt solution reveals spontaneous activity consisting of a random mixture of excitatory and inhibitory postsynaptic potentials and irregular firing of action potentials. Dorsal root ganglion neurons in cell culture are round and smooth with one or more axons emerging from the soma. The cells are electrically excitable, and action potentials have both sodium and calcium components. The neurons are not spontaneously active and receive no synaptic input.

Barbiturate Actions

Numerous barbiturates are used clinically as anticonvulsants, sedative hypnotics, and anesthetics. Each of the clinical effects of the barbiturates is likely due to different cellular mechanisms of action. Barbiturates have both synaptic and nonsynaptic (membrane) actions in the central nervous system. These agents: (1) alter postsynaptic neurotransmitter action including enhancement of GABA-mediated inhibition and antagonism of excitatory synaptic transmission, (2) reduce presynaptic calcium entry and neurotransmitter release, (3) enhance membrane chloride ion conductance, (4) alter membrane sodium and potassium conductances, and (5) modify repetitive firing [see 10,18,27,31 for reviews]. The correspondence of particular barbiturate actions to the clinical effects of barbiturates remains controversial. Evidence will be presented below which suggests that barbiturate-induced reduction of calcium entry into presynaptic terminals may relate to sedative-hypnotic and anesthetic, but not anticonvulsant, actions of these agents.

Barbiturates reduce neurotransmitter release from group Ia afferent fibers, vagal terminals, and cortical, striatal, hippocampal, midbrain and pons–medulla slices [16,29,30,34,35]. Furthermore, barbiturates reduce calcium uptake into cortical synaptosomes at concentrations similar to those effective in reducing neurotransmitter release [1] and in blocking calcium action potentials in R2 neurons of *Aplysia* abdominal ganglia [8]. Thus, it is likely that barbiturates block release of neurotransmitter by reducing presynaptic entry of calcium.

To obtain detailed information on presynaptic actions of barbiturates, one would like to directly record inward calcium movements from presynaptic terminals as a function of differing barbiturate concentrations. Unfortunately, this is not feasible in the mammalian nervous system. However, it is possible to measure neuronal calcium entry by recording calcium-dependent action potentials from both spinal cord and dorsal root ganglion neuron somata in cell culture following significant blockade of repolarizing potassium conductance with the potassium channel blockers tetraethylammonium and/or 3-aminopyridine [5,9]. Such calcium-dependent action potentials have also been recorded from presynaptic terminals in the giant synapse of squid stellate ganglion [13].

The effect of phenobarbital, pentobarbital, and barbituric acid on the duration of calcium-dependent action potentials of mouse spinal cord neurons in cell culture [10] has been investigated (Fig. 1). Pentobarbital produced a dose-dependent reduction of calcium-dependent action potential duration from 30 to 500 μM, with a 50% reduction occurring at about 200 μM. Within the effective range of concentrations, all neurons studied were affected by both barbiturates. Barbituric acid, which is clinically inactive, did not modify calcium-dependent action potential duration at 2 mM. Hence, clinically active barbiturates reduce calcium-dependent action potential duration and thus calcium entry. Pentobarbital reduction of calcium entry occurs at concentrations which are similar to those achieved during deep barbiturate anesthesia. Phenobarbital shortened calcium-dependent action potentials only at

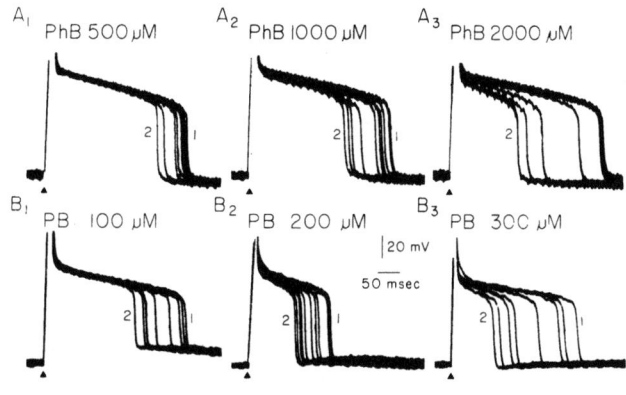

Figure 1. Phenobarbital (PhB) and pentobarbital (PB) shortening of calcium-dependent action potentials was dose-dependent. PhB and PB were applied to spinal cord neurons by superfusion. Superimposed calcium-dependent action potentials stimulated with 10-msec depolarizing pulses every 30 sec prior [1] and after [2] superfusion of 500 (A_1), 1000 (A_2), 2000 μM (A_3) PhB, and 100 (B_1), 200 (B_2), 300 μM (B_3) PB; [1] and [2] are to the right and left of their action potentials, respectively. A_1, A_2, A_3 were recorded from the same neuron (RMP −58 mV); B_1, B_2, B_3 were recorded from the same neuron (RMP −55 mV). The bathing solutions were control bathing solution with 25 mM TEA, 3 μM TTX in both A_1, A_2, A_3 and B_1, B_2, B_3 and 5 mM 3-AP only in A_1, A_2, A_3. [From Ref. 10.]

concentrations which exceed those relevant for clinical anticonvulsant action, but which are relevant for sedation and anesthesia.

It is possible for barbiturates to reduce calcium-dependent action potential duration either by decreasing calcium conductance, and therefore directly blocking calcium entry, or by enhancing a membrane potassium conductance which hastens repolarization. To study this, we applied pentobarbital and phenobarbital to spinal cord neurons at resting membrane potential [20]. No effects on resting membrane potential or resting membrane conductance were seen at concentrations which produce action potential shortening. Thus, barbiturates do not affect a resting calcium or potassium conductance. To investigate barbiturate actions on voltage- and calcium-dependent potassium conductances, the effect of barbiturate on potassium-dependent afterhyperpolarization which follows calcium-dependent action potentials was studied. Rather than enhancing afterhyperpolarization, barbiturates reduced it, which is consistent with a reduction of calcium entry and therefore a secondary reduction of calcium-activated potassium conductance. It is therefore likely that barbiturates reduce calcium conductance rather than enhance potassium conductance.

This phenomenon was studied more directly by injecting dorsal root ganglion neurons with the potassium channel blocker cesium and demonstrating that phenobarbital and pentobarbital shortened calcium-dependent action potential duration with unaltered potency [20] (Fig. 2A). Single-electrode voltage-clamp studies of the inward currents produced by depolarizing step clamps were also performed (Fig. 2B). Both phenobarbital and pentobarbital reduced the inward calcium current in this system. In the presence of the calcium channel blocker cadmium, the barbiturates did not alter the remaining outward current. Thus, clinically active barbiturates modify calcium entry by reducing calcium conductance and not by enhancing potassium conductance.

While this effect may be quite similar to divalent cation calcium channel blockers such as cadmium, these experiments do not reveal whether the reduction in calcium entry is due to direct blockade of calcium channels or to modification of activation or inactivation kinetics of the channels. Barbiturate action would appear to be non-receptor-mediated and involve an interaction of barbiturates directly with calcium channels. Thus, barbiturates may produce sedative-hypnotic and anesthetic actions, at least in part, by a nonspecific (or nonselective) reduction of presynaptic calcium entry and subsequent reduction of transmitter release from synapses throughout the nervous system.

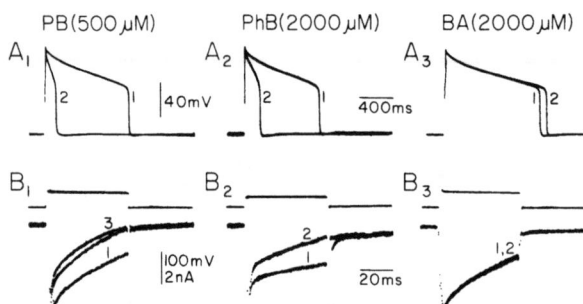

Figure 2. Barbiturates decrease calcium-dependent action potential duration and calcium dependent inward currents following blockade of potassium conductance by intracellular cesium injection. Action potentials (A) and calcium inward currents (B) evoked from dorsal root ganglion neurons bathed in medium with cesium substituted for potassium. Cesium was also injected intracellularly with a recording micropipette filled with 3 M CsCl. (A) Substantial blockade of potassium conductance was achieved, suggested by the prolongation of calcium-dependent action potentials to durations of 500 msec to 2 sec and by loss of the afterhyperpolarization. Pentobarbital (A_1) and phenobarbital (A_2) decreased the calcium component of the action potential under these conditions while the inactive barbiturate, barbituric acid, was without effect (A_3). (B) Pentobarbital (B_1), phenobarbital (B_2), but not barbituric acid (B_3) decreased the amplitude of inward currents evoked by depolarizing voltage commands from a holding potential of -60 mV.

Opiate Actions

Opiate alkaloids and opioid peptides produce their major clinical actions by interacting with specific opiate receptors [26,32,33]. However, the mechanisms of opiate action remain unclear. Opiates have been suggested to possess: (1) presynaptic actions to block neurotransmitter release, reduce presynaptic calcium entry, reduce action potential invasion of synaptic terminals, or enhance GABA-mediated presynaptic inhibition, (2) postsynaptic actions to reduce glutamic acid, acetylcholine, GABA, and glycine responses, and (3) direct actions on neuronal membranes to produce membrane hyperpolarization secondary to enhancement of membrane potassium conductance (see 25 for review).

Opiate receptors have been demonstrated on primary afferent terminals in the spinal cord [6,7,11,15], and opiate alkaloids and opioid peptides may produce their analgesic actions by interacting with presynaptic receptors to reduce transmitter release [12,19,23]. We have studied opiate actions on dorsal root ganglion neuron to spinal cord neuron synaptic connections in culture [19]. When opiate alkaloids were applied to the site of termination of monosynaptic connections between dorsal root ganglion and spinal cord neurons, a reduction in excitatory postsynaptic potentials was seen (Fig. 3). Etorphine effects on synaptic transmission were blocked by the opiate antagonist naloxone, consistent with mediation by opiate receptors. Quantal analysis was then performed using the variance technique to identify whether or not this reduction in excitatory postsynaptic potential amplitude was due to a direct action of the opiate alkaloid on presynaptic terminals to block transmitter release or on postsynaptic receptors for the neurotransmitter to produce an antagonism of transmitter action. Using this technique, etorphine was shown to act presynaptically to reduce synaptic transmission. Subsequent to our demonstration of this presynaptic action of the opiates, Mudge et al. [23] demonstrated that the opioid peptide D-Ala²-Met enkephalinamide reduced the calcium component of the action potential of chick dorsal root ganglion neurons. Thus, it was suggested that reduction in transmitter release was due to an opiate receptor-mediated reduction of presynaptic calcium entry.

This action of opioid peptides and opiate alkaloids was recently reinvestigated on calcium-dependent action potentials recorded from mouse dorsal root ganglion neurons in cell culture. The assumption underlying these studies was the existence of opiate receptors on the somata of dorsal root ganglion neurons similar to those on presynaptic terminals. Both

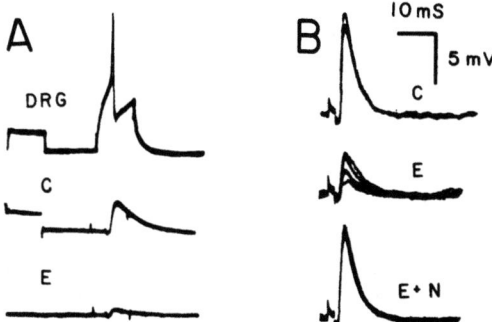

Figure 3. Etorphine reduces the amplitude of dorsal root ganglion (DRG) to spinal cord (SC) neuron monosynaptic excitation postsynaptic potentials (EPSP). Stimulation of a DRG neuron produced a typical action potential (DRG) which was followed by a 12-mV EPSP in the SC neuron with a 2.5-msec latency (A-C, B-C). Iontophoresis of etorphine (1 mM, 5 nA) near the SC neuron produced a reduction in EPSP amplitude (A-E, B-E). Iontophoresis of naloxone during the etorphine application reversed the antagonism (B-E + N). The calibration pulse in (A) is 10 mV and 5 msec. [From Ref. 19.]

opioid peptides and opiate alkaloids reduced calcium-dependent action potential duration, and the reductions were reversed by naloxone (Fig. 4) [36]. Thus, opiate alkaloids and opioid peptides interact with specific opiate receptors on presynaptic terminals to modify the entry of calcium and thus to reduce transmitter release.

There are subtypes of opiate receptors in both the central and autonomic nervous systems [2,14,22]. The three most extensively studied are μ receptors, which are morphine-like; δ receptors, which are enkephalin-like; and κ receptors, which are ethoketocyclazocine-like. The question was raised whether the actions of enkephalins and opiate alkaloids on dorsal root ganglion action potential duration were due to interactions with μ, δ, and/or κ receptors. To investigate this, the actions on calcium-dependent action potentials of morphiceptin, a μ-selective opioid peptide [3], and leucine-enkephalin, an opioid peptide which prefers δ receptors but significantly cross-reacts with μ receptors [14], were compared [37,39]. Several different classes of neurons could be identified. First, a number of neurons responded to leucine-enkephalin at 1 μM but did not respond to morphiceptin at 1 μM (Fig. 4B). Second, other neurons responded equipotently to morphiceptin and to leucine-en-

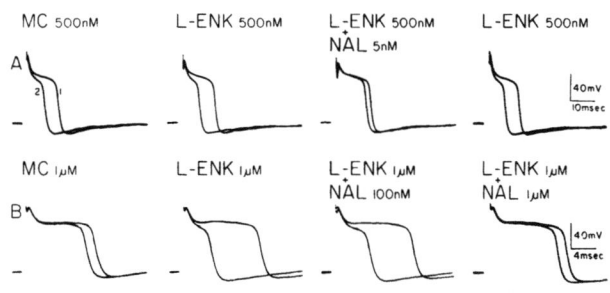

Figure 4. Morphiceptin and leucine-enkephalin decrease DRG neuron calcium-dependent action potentials but with a heterogeneous response pattern and differential naloxone sensitivity. (A) and (B) illustrate DRG neurons with a different response pattern to leucine-enkephalin and morphiceptin and different naloxone sensitivity. (A) In a DRG neuron that responded to both morphiceptin and leucine-enkephalin with a large decrease of calcium-dependent action potential duration, the response to leucine-enkephalin and morphiceptin was highly sensitive to naloxone antagonism with naloxone at a concentration 1/100th the opioid peptide concentration producing substantial antagonism. (B) A DRG neuron that responded well to leucine-enkephalin but not morphiceptin. A much higher concentration of naloxone was required for antagonism of the opioid peptide response with approximately equimolar concentrations of naloxone and leucine-enkephalin needed. [From Ref. 37]

kephalin (Fig. 4A). Third, certain neurons responded somewhat more briskly to leucine-enkephalin than to morphiceptin. A large fourth population of neurons responded to neither peptide. This suggested that those neurons responding to leucine-enkephalin, but not to morphiceptin, likely possessed predominately δ receptors on their somata. Neurons responding equipotently to leucine-enkephalin and to morphiceptin were likely to contain predominately μ receptors. The remaining two neuronal populations had mixed μ and δ receptors, or were devoid of μ or δ receptors.

To further clarify this point, naloxone antagonism of opioid peptide-induced shortening of calcium-dependent action potential duration was studied. Naloxone has a much more potent action on μ receptors than on δ receptors [2,14]. Therefore, if opiate action on neurons which responded to morphiceptin and leucine-enkephalin was due to an interaction of the peptides with μ receptors, naloxone should potently antagonize the opiate effect. Conversely, the opiate effect should have been relatively insensitive to naloxone on neurons which only responded to leucine-enkephalin. Responses on neurons to morphiceptin and leucine-enkephalin could be antagonized by naloxone, with an agonist to antagonist concentration ratio of up to 1000 to 1 (Fig. 4A). However, responses on those neurons which only responded to leucine-enkephalin, were antagonized only by a 1 to 1 agonist to antagonist concentration ratio. Thus, there is a heterogeneous distribution of μ and δ receptors among dorsal root ganglion neurons.

Recently, these observations have been extended to actions of the opioid peptide dynorphin A, which may preferentially bind to κ receptors [4,40]. Cells were found that responded to 1 μM dynorphin A but did not respond to either leucine-enkephalin or morphiceptin [21] (Fig. 5A). These responses to dynorphin A were naloxone reversible. Thus, there are specific μ, δ, and κ (dynorphin A) receptors present on dorsal root ganglion neuron somata. Furthermore, individual neurons may express either μ, δ, or κ receptors alone or in varying combinations.

Having pharmacologically characterized the response of dorsal root ganglion neurons to μ, δ, and κ ligands, we investigated the ionic basis for opioid peptide actions. Neither leucine-enkephalin, morphiceptin, nor dynorphin A modified resting membrane potential or resting membrane conductance, suggesting that none of the peptides acted on a resting membrane conductance [21,38]. The effects of these peptides on the potassium-dependent

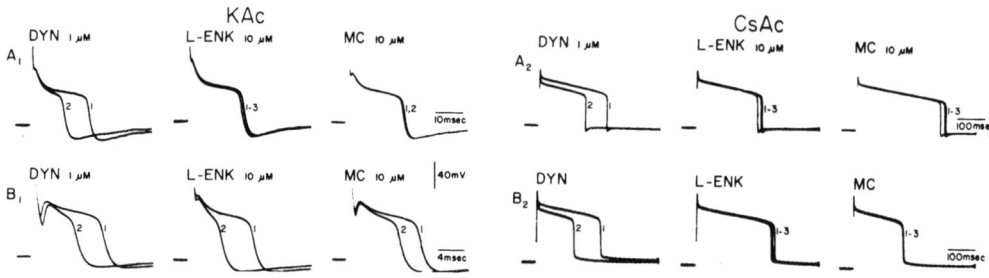

Figure 5. Intracellular injection of cesium blocks neuronal responses to morphiceptin and leucine-enkephalin but not to dynorphin A. Opioid peptides were tested on somatic calcium-dependent action potentials during recording with KAc-filled micropipette. (A) A DRG neuron that responded only to dynorphin A during recording with a KAc-filled micropipette. The dynorphin A response persisted following intracellular injection of cesium. (B) A DRG neuron that responded to dynorphin A, leucine-enkephalin, and morphiceptin during recording with KAc-filled micropipettes. Intracellular injection of cesium blocked neuronal responses to leucine-enkephalin and morphiceptin but not to dynorphin A. [From Ref. 21.]

action potential after hyperpolarization produced in these neurons were also investigated. Both leucine-enkephalin and morphiceptin increased and dynorphin A reduced the afterhyperpolarization, suggesting that leucine-enkephalin and morphiceptin, acting through μ and δ receptors, had a somewhat different action than dynorphin A.

To further investigate this possibility, the effect of intracellular cesium injection to block membrane potassium conductance was examined (Fig. 5). The paradigm followed was to impale a dorsal root ganglion neuron, determine its sensitivity to the opioid peptides in varying combinations, and then remove the recording micropipette and reimpale the cell with a cesium-containing micropipette. Following leakage of cesium into the cells, the majority of potassium channels were blocked. The opioid peptide action was then reassessed. In all cases, the actions of morphiceptin and leucine-enkephalin were blocked by cesium injection (Fig. 5B), supporting the idea that these opioid peptides are coupled to potassium channels. However, the action of dynorphin A was not blocked (Fig. 5), and in many cases was enhanced, by the cesium injection. This result with dynorphin A was similar to that seen with pentobarbital, cadmium, and norepinephrine, which clearly block calcium conductance. Therefore, not only is there a heterogeneous distribution of μ, δ, and κ receptors among dorsal root ganglion neurons, but there may be a heterogeneous coupling of opiate receptors to ion channels. μ and δ receptors may be coupled to membrane potassium channels, whereas dynorphin A receptors may be coupled to a membrane calcium channel.

Thus, on dorsal root ganglion neurons in cell culture, opioid peptides and opiate alkaloids interact with μ, δ, and κ opiate receptors to modify calcium entry either by increasing potassium conductance (μ and δ receptors) or by decreasing calcium conductance (κ receptors). Such an action occurring at presynaptic terminals would decrease calcium entry, and thus reduce transmitter release only on terminals which contain specific opiate receptors involved.

In conclusion, we have demonstrated that two classes of compounds, barbiturates and opiates, modify calcium entry into mouse neurons in cell culture. While exhibiting this similarity, many differences in the actions of these two groups of agents are apparent. Barbiturates act nonselectively with calcium channels on all cells to directly reduce calcium entry. Such an effect would be ideally suited to produce a nonselective suppression of neuronal functioning and therefore sedation or anesthesia. In contrast, opiates act selectively on a specific receptor subtype on those nerve terminals that contain the receptor. This action to reduce calcium entry and subsequent transmitter release is accomplished by two different mechanisms, depending upon opiate receptor subtype: either to enhance potassium conductance or to block calcium conductance. These differences between binding sites and ionic mechanisms of action between these two classes of compounds may have direct relevance for differences in their clinical actions.

References

1. Blaustein, M. P.; Ector, A. C. Barbiturate inhibition of calcium uptake by depolarized nerve terminals *in vitro*. *Mol. Pharmacol.* **11**:369–378, 1975.
2. Chang, K.-J.; Hazum, E.; Cuatrecasas, P. Novel opiate binding sites selective for benzomorphan drugs. *Proc. Natl. Acad. Sci. USA* **78**:4141–4145, 1981.
3. Chang, K.-J.; Kittan, A.; Hazum, E.; Cuatrecasas, P.; Change, J.-K. Morphiceptin (NH$_4$-Tyr-Pro-Phe-Pro-Pro-CONH$_2$): A potent and specific agonist for morphine (μ) receptors. *Science* **212**:75–77, 1981.
4. Chavkin, C.; James, I. F.; Goldstein, A. Dynorphin is a specific endogenous ligand of the κ opioid receptor. *Science* **215**:413–415, 1982.
5. Dichter, M. A.; Fischbach, G. D. The action potential of chick dorsal root ganglion neurones maintained in cell culture *J. Physiol. (London)* **267**:281–298, 1977.

6. Fields, H. L.; Emson, P. C.; Leigh, B. K.; Gilbert, R. F. T.; Iversen, L. L. Multiple opiate receptor sites on 1° afferent fibres. *Nature (London)* **284**:351–353, 1980.
7. Gämse, R.; Holzer, P.; Lembeck, F. Indirect evidence for presynaptic location of opiate receptors on chemosensitive primary sensory neurons. *Naunyn–Schmiedeberg's Arch. Pharmacol.* **308**:281–285, 1979.
8. Goldring, J. M.; Blaustein, M. P. Effect of pentobarbital on Na and Ca action potentials in an invertebrate neuron. *Brain Res.* **240**:273–283, 1982.
9. Heyer, E. J.; Macdonald, R. L. Calcium- and sodium-dependent action potentials of mouse spinal cord and dorsal root ganglion neurons in cell culture. *J. Neurophysiol.* **97**:641–655, 1982.
10. Heyer, E. J.; Macdonald, R. L. Barbiturate reduction of calcium dependent action potentials: Correlation with anesthetic action. *Brain Res.* **236**:157–171, 1982.
11. Hiller, J. M.; Simon, E. J.; Crain, S. M.; Peterson, E. R. Opiate receptors in cultures of fetal mouse DRG and spinal cord: Predominance in DRG neurites. *Brain Res.* **145**:396–400, 1978.
12. Jessell, T. M.; Iversen, L. L. Opiate analgesics inhibit substance P release from rat trigeminal nucleus. *Nature (London)* **268**:549–551, 1977.
13. Katz, M.; Miledi, R. Tetrodotoxin-resistant electric activity in presynaptic terminals. *J. Physiol. (London)* **203**:459–487, 1969.
14. Kosterlitz, H. W.; Lord, J. A. H.; Paterson, S. J.; Waterfield, A. A. Effects of changes in the structures of enkephalins of narcotic analgesic drugs on their interactions with μ- and α-receptors. *Br. J. Pharmacol.* **68**:333–342, 1980.
15. Lamotte, C.; Pert, C. B.; Snyder, S. H. Opiate receptor binding in primate spinal cord: Distribution and changes after dorsal root section. *Brain Res.* **112**:407–412, 1976.
16. Lindmar, R.; Löffelholz, K.; Weide, W. Inhibition by pentobarbital of the acetylcholine release from the postganglionic parasympathetic neuron of the heart. *J. Pharmacol. Exp. Ther.* **210**:166–173, 1979.
17. Macdonald, R. L.; Barker, J. L. Neuropharmacology of spinal cord neurons in primary dissociated cell culture. In: *Excitable Cells in Tissue Culture*, P. G. Nelson and M. Lieberman, eds., New York, Plenum Press, 1981, pp. 81–109.
18. Macdonald, R. L.; McLean, M. J. Cellular bases of barbiturate and phenytoin anticonvulsant drug action. *Epilepsia* **23**(Suppl. 1):S7–S18, 1982.
19. Macdonald, R. L.; Nelson, P. G. Specific opiate-induced depression of transmitter release from dorsal root ganglion cells in culture. *Science* **199**:1449–1451, 1978.
20. Macdonald, R. L.; Werz, M. A. Barbiturates decrease voltage-dependent calcium conductance of mouse neurons in dissociated cell culture. *Soc. Neurosci. Abstr.* **8**:568, 1982.
21. Macdonald, R. L.; Werz, M. A. Dynorphin decreases calcium conductance of mouse cultured dorsal root ganglion neurons. *Soc. Neurosci. Abstr.* **9**:1129, 1983.
22. Miller, R. J. Multiple opiate receptors for multiple opioid peptides. *Med. Biol.* **60**:1–6, 1982.
23. Mudge, A. W.; Leeman, S. E.; Fischbach, G. D. Enkephalin inhibits release of substance P from sensory neurons in culture and decreases action potential duration. *Proc. Natl. Acad. Sci. USA* **76**:526–530, 1979.
24. Nelson, P. G.; Neale, E. A.; Macdonald, R. L. Electrophysiological and structural studies of neurons in dissociated cell cultures of the central nervous system. In: *Excitable Cells in Tissue Culture*, P. G. Nelson and M. Lieberman, eds., New York, Plenum Press, 1980, pp.50–80.
25. North, R. A. Opiates, opioid peptides, and single neurones. *Life Sci.* **24**:1527–1546, 1979.
26. Pert, C. B.; Snyder, S. H. Properties of opiate receptor binding in rat brain. *Proc. Natl. Acad. Sci. USA* **70**:2243–2247, 1973.
27. Prichard, J. W. Barbiturates: Physiological effects I. In: *Antiepileptic Drugs: Mechanisms of Action*, G. H. Glasser, J. K. Penry, and D. M. Woodbury, eds., New York, Raven Press, 1980, pp. 505–522.
28. Ransom, B. R.; Neale, E.; Henkart, M.; Bullock, P. N.; Nelson, P. G. Mouse spinal cord in cell culture. I. Morphology and intrinsic neuronal electrophysiologic properties. *J. Neurophysiol.* **40**:1132–1150, 1977.
29. Richter, J. A.; Waller, M. B. Effects of pentobarbital on the regulation of acetylcholine content and release on different regions of rat brain. *Biochem. Pharmacol.* **26**:609–615, 1977.
30. Richter, J. A.; Werling, L. L. K-stimulated acetylcholine release: Inhibition by several barbiturates and chloral hydrate but not by ethanol, chlordiazepoxide or 11-OH-9-tetrahydrocannabinol. *J. Neurochem.* **32**:935–941, 1979.
31. Schulz, D. W.; Macdonald, R. L. Barbiturate enhancement of GABA-mediated inhibition and activation of chloride ion conductance: Correlation with anticonvulsant and anesthetic actions. *Brain Res.* **209**:177–188, 1981.
32. Simon, E. J.; Hiller, J. M.; Edelman, I. Stereospecific binding of the potent narcotic analgesic [^{3}H] etorphine to rat brain homogenate. *Proc. Natl. Acad. Sci. USA* **70**:1947–1949, 1973.
33. Terenius, L. Characteristics of the "receptor" for narcotic analgesics in synaptic plasma fraction of rat brain. *Acta Pharmacol. Toxicol.* **33**:377–384, 1973.

34. Waller, M. B.; Richter, J. A. Effects of pentobarbital and Ca^{2+} on the resting and K^+-stimulated release of several endogenous neurotransmitters from rat midbrain slices. *Biochem. Pharmacol.* **29**:2189–2198, 1980.
35. Weakly, J. N. Effect of barbiturates on "quantal" synaptic transmission in spinal motoneurones. *J. Physiol. (London)* **204**:63–77, 1969.
36. Werz, M. A.; Macdonald, R. L. Opioid peptides decrease calcium-dependent action potential duration of mouse dorsal root ganglion neurons in cell culture. *Brain Res.* **239**:315–321, 1982.
37. Werz, M. A.; Macdonald, R. L. Heterogeneous sensitivity of cultured dorsal root ganglion neurones to opioid peptides selective for μ- and δ-opiate receptors. *Nature (London)* **299**:730–733, 1982.
38. Werz, M. A.; Macdonald, R. L. Opioid peptides selective for mu- and delta-opiate receptors reduce calcium-dependent action potential duration by increasing potassium conductance. *Neurosci. Lett.* **42**:173–178, 1983
39. Werz, M. A.; Macdonald, R. L. Opioid peptides with differential affinity for mu- and delta-receptors decrease sensory neuron calcium-dependent action potentials. *J. Pharmacol. Exp. Ther.* **227**:394–402, 1983.
40. Wüster, M.; Schulz, R.; Herz, A. Opiate activity and receptor selectivity of dynorphin$_{1-13}$ and related peptides. *Neurosci. Lett.* **20**:79–83, 1980.

28

Effects of Depressant Drugs on Sodium–Calcium Exchange in Resealed Synaptic Membranes

Elias K. Michaelis and Mary L. Michaelis

Many drugs that act on the central nervous system (CNS) to produce neuronal depression may do so through alterations in the regulation of Ca^{2+} fluxes across neuronal membranes [17]. Such actions could involve decreases in either the amount of Ca^{2+} that is bound to the neuronal plasma membrane or enters the neuronal compartment through activation of voltage-sensitive channels [4,18,21,25,26,28,42]. Decreases in depolarization-induced Ca^{2+} conductance produced by anionic or neutral drugs such as barbiturates or alcohols [25,26] would be expected to cause decreased neurotransmitter release and, secondarily, a failure in synaptic transmission. This assumption is based on the involvement of Ca^{2+} in the initiation of exocytotic release of certain transmitters. Such decreases in neurotransmitter release have indeed been observed following acute administration of these depressant drugs *in vivo* or following the *in vitro* application of these agents to neuronal preparations [14,20,34,41]. Yet, in certain preparations such as isolated striatal nerve endings, the neuromuscular junction, or the hippocampal slice preparation, an agent such as ethanol increases the release of endogenous transmitter substance [11,13,22,39]. This apparent discrepancy in the actions of a depressant drug such as ethanol may be due to the existence of multiple targets for its actions, and these may include cellular entities which maintain intracellular free calcium ion concentrations ($[Ca^{2+}]_i$) at a very low level (10^{-7} M).

The low $[Ca^{2+}]_i$ is maintained against a 10^4-fold gradient ($[Ca^{2+}]_o/[Ca^{2+}]_i$) and requires the expenditure of metabolic energy to sequester free Ca^{2+} into intracellular organelles or to pump Ca^{2+} out of the cell. Some of the Ca^{2+}-handling processes appear to exist in most cell types as, for example, the transport of Ca^{2+} into mitochondria [9,10]. Other processes, such as the Na^+–Ca^{2+} exchange carrier, have been primarily studied in excitable cells such as neurons and myocytes [e.g., 1,3,35,36]. The various systems for regulating intracellular Ca^{2+} concentrations in neurons are shown in diagrammatic form in

---

Elias K. Michaelis and Mary L. Michaelis • Neurobiology, Department of Human Development, and Center for Biomedical Research, University of Kansas, Lawrence, Kansas 66045.

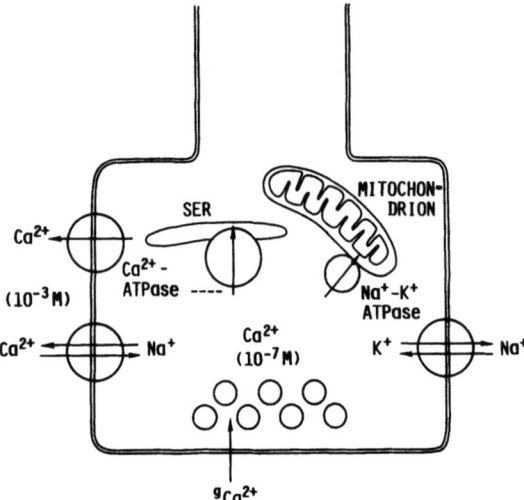

Figure 1. Schematic representation of proposed Ca^{2+}-buffering activities in the nerve ending region. Depolarization-induced Ca^{2+} influx is indicated as g_{Ca}.

Fig. 1. Interference with activity of any of the Ca^{2+}-regulating systems in neurons could lead to progressive intracellular accumulation of this cation. An increase in free $[Ca^{2+}]_i$ may modify neurotransmitter release, the conductance of Ca^{2+}-sensitive K^+ channels, and several Ca^{2+}-regulated metabolic activities such as protein phosphorylation, and calmodulin-dependent activation of enzymes and transport carriers [8,11,12,40].

Characteristics of Neuronal Plasma Membrane Na^+–Ca^{2+} Exchange Activity and Effects of Ethanol

The plasmalemmal Na^+–Ca^{2+} exchange carrier is thought to be the most active, relatively high-affinity Ca^{2+}-transporting system in neuronal membranes [6]. Although Ca^{2+}-stimulated ATPase activity of axonal and synaptic membranes and of the intraterminal endoplasmic reticulum has a higher affinity for Ca^{2+} (0.2–0.5 μM), it has a much lower capacity than the Na^+-dependent Ca^{2+} transport mechanism [2,6,23,29,31]. Activity of the Na^+–Ca^{2+} antiport has been studied in perfused squid giant axon and in intact nerve ending particles or synaptosomes [1,3,5]. Isolated synaptic plasma membrane vesicles obtained following osmotic rupturing of brain synaptosomes exhibit high activity of this Na^+–Ca^{2+} exchange carrier [15,30].

The basic characteristics of this carrier activity in synaptic membranes are summarized in Table I and are compared with those previously determined for the intact synaptosomal preparation and the internally dialyzed squid giant axon. On the basis of their activity, these membrane preparations can be very useful for measuring the effects of various drug agents on this transport system. In addition, the study of this transport activity in isolated synaptic plasma membranes affords an opportunity to study changes in this carrier without substantial contamination by other neuronal Ca^{2+}-handling activities. This transport system is not directly dependent on available metabolic energy but instead utilizes an existing Na^+ gradient across the neuronal membrane; therefore, it is not affected by metabolic inhibitors. It is very sensitive, however, to ionophores that collapse the Na^+ gradient across the plasma membrane (Table I). Finally, it appears to be a distinct entity from voltage-sensitive Ca^{2+} channels, since neither verapamil (0.02–0.2 mM) nor D600 (0.2 mM) inhibits Na^+–Ca^{2+} exchange activity (Table I).

Table I. Characteristics of Synaptic Plasma Membrane, Synaptosome, and Axon Na^+–Ca^{2+} Exchange Transport[a]

	Synaptic membranes[b]	Synaptosomes[c]	Axons[d]
Kinetic constants[e]			
K_{Na^+}	6.6 mM	18 mM	50–120 mM
$K_{Ca^{2+}}$	38.6 μM	200 μM	2–3 mM
V_{max}	1.52 nmoles/mg/15 sec	1.59 nmoles/mg/min	2 pmoles/cm²/sec
Ionophores and other blockers			
Nigericin	Inhibition	—	—
Verapamil	No effect	—	—
D600[f]	No effect	No effect	No effect
Oligomycin[f]	No effect	No effect	No effect
Ouabain	No effect	No effect	No effect

[a]All characteristics summarized are for the Na^+-dependent Ca^{2+} uptake in these preparations except for the K_{Na^+} of synaptosomes and synaptic membranes.
[b]Data from Refs. 30, 31, and unpublished observations.
[c]Data from Ref. 5.
[d]Data from Ref. 1.
[e]K_{Na^+}, $K_{Ca^{2+}}$, and V_{max} are the apparent activation constants for Na^+ and Ca^{2+} and the estimated maximum transport activity for Ca^{2+}.
[f]Data from Ref. 15.

Ethanol was the first central depressant drug explored for its activity on the Ca^{2+} transport system. Ethanol had biphasic effects on the activity of the Na^+-dependent Ca^{2+} transport of synaptic membranes measured at 23°C (Fig. 2). At subanesthetic concentrations (< 50 mM), ethanol caused a small enhancement in Na^+-dependent Ca^{2+} transport, while at higher concentrations it inhibited carrier activity. The inhibitory activity of aliphatic alcohols was directly related to their aliphatic chain length (Fig. 3), suggesting that the actions of alcohols were due to adsorption or partitioning of the aliphatic portion of each alcohol into the lipid environment of the membrane bilayer. The standard free energy ($\Delta G°$) for the partitioning of methylene groups of alcohols into synaptic membranes was estimated

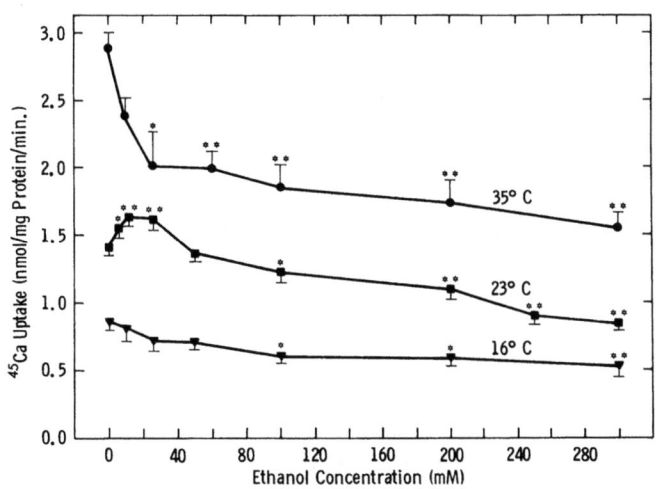

Figure 2. The effects of ethanol on Na^+-dependent Ca^{2+} uptake into synaptic membrane vesicles measured at three different incubation temperatures. NaCl-loaded vesicles were incubated in the presence or absence of an outward-directed Na^+ gradient ($[Na^+]_i > [Na^+]_o$) with 10 μM $^{45}CaCl_2$ and indicated ethanol concentration in the external medium. Membranes were preincubated for 120 sec with ethanol prior to being transferred to appropriate salt solutions where incubations were carried out for 15 sec at 16 and 23°C, and for 8 sec at 35°C in order to measure transport during the initial linear phase. All values were extrapolated to nmoles Ca^{2+} taken up/min. Calcium uptake values which differ significantly from those obtained for a given temperature in the absence of ethanol are indicated with asterisks: *$p < 0.05$, **$p < 0.01$. [From Ref. 31.]

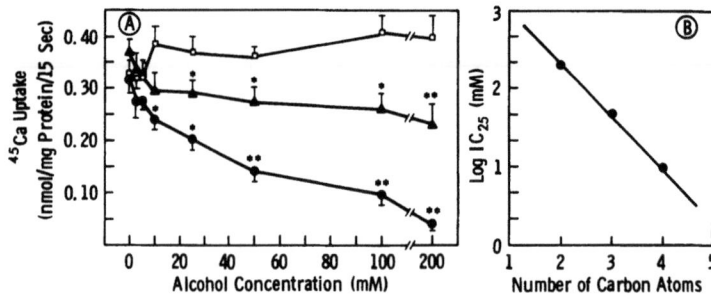

Figure 3. The effects of short-chain aliphatic alcohols on Na$^+$-dependent Ca^{2+} influx. (A) The NaCl-loaded vesicles were preincubated for 120 sec with the indicated concentrations of methanol (□), 1-propanol (▲), and 1-butanol (○) prior to transfer of the vesicles to incubation media containing 10 μM ^{45}CaCl$_2$. Samples which differed significantly from the zero alcohol values are indicated with asterisks: *$p < 0.05$, **$p < 0.01$. (B) Alcohol concentrations required to inhibit Na$^+$-dependent Ca^{2+} uptake by 25% (IC$_{25}$) plotted against the number of carbon atoms of n-alcohols. These values were obtained from Fig. 2 and Fig. 3A. [From Ref. 31.]

from the plot shown in Fig. 3B and was found to be −889 cal/mole [31]. This value is very similar to the $\Delta G°$ for the transfer of each methylene group of aliphatic alcohols from an aqueous medium to an organic phase ($\Delta G° = -907$ cal/mole) [37], or from buffer medium to dipalmitoylphosphatidylcholine vesicle membranes ($\Delta G° = -819.6$ cal/mole) [24].

The inhibition of Na$^+$–Ca^{2+} exchange activity in synaptic membranes was not only produced by ethanol and related aliphatic alcohols, but was also observed following exposure of synaptic membranes to cationic local anesthetic agents and to benzyl alcohol. The IC$_{50}$ values for inhibition of the exchange carrier in synaptic membranes by dibucaine (0.5 mM), tetracaine (1.8 mM), and benzyl alcohol (2.8 mM) were within the range of their local anesthetic activity [38]. Unlike local anesthetic agents, the central depressant and anxiolytic agent desmethyl diazepam (0.5 nM to 5 μM) did not inhibit Na$^+$–Ca^{2+} exchange transport activity.

Inhibition of Na$^+$-dependent Ca^{2+} transport of synaptic membranes by *in vitro* exposure of membranes to ethanol was significant only at concentrations that were within the anesthetic range when transport activity was measured at 23°C. The K_I for ethanol-induced inhibition was estimated to be 1.0 M [31]. If this were the only observed effect of ethanol on this transport system, it would be difficult to ascribe any of the actions of ethanol at subanesthetic concentrations to inhibition of the Na$^+$–Ca^{2+} exchange carrier. However, when the effects of ethanol on the exchange carrier activity were determined at temperatures near the physiological range, this alcohol, even at concentrations as low as 2.5 and 5 mM, inhibited transport carrier activity without any evidence of enhancement of transport function (Fig. 2). A continuous trend toward inhibition of transport activity with increasing ethanol concentrations was also observed at 16°C (Fig. 2). Thus, the biphasic effects of ethanol on Na$^+$–Ca^{2+} exchange carriers can only be detected at temperatures in the range of 23°C.

Such an unusual pattern of activity may be due to two basic factors: the sensitivity of the carrier system to the lipid environment of the membrane and the effect of ethanol on the thermotropic behavior of these lipids [32]. The influence of the lipid environment on the activity of these carriers would presumably be manifested as a marked sensitivity of transport to changes in the physical state of membrane lipids brought about by such treatments as changes in incubation temperature or exposure to agents that disrupt membrane lipid organization. Ethanol-induced alteration in the thermotropic behavior of synaptic membrane lipids might be detectable as a shift in the temperature-dependent phase transition of the membrane lipids following exposure of synaptic membranes to this alcohol.

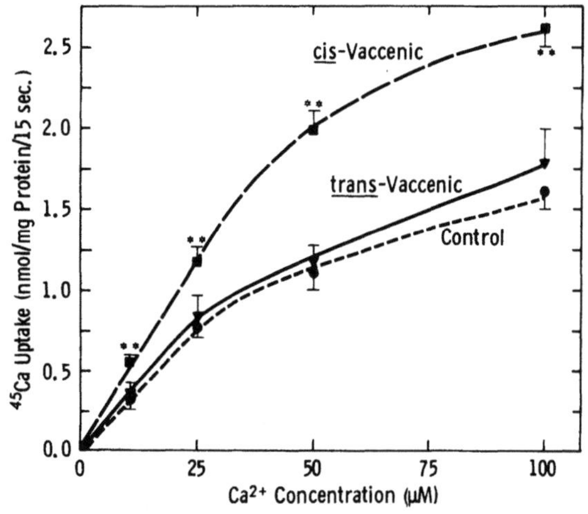

Figure 4. Effects of pretreatment of membranes with *cis*- or *trans*-vaccenic acid on Na$^+$-dependent Ca^{2+} uptake. Membrane vesicles were preincubated with 1 mM *cis*- or *trans*-vaccenic acid or appropriate amount of methanol for the controls. Na$^+$-dependent Ca^{2+} uptake was measured for 15 sec at 23°C. Calcium uptake values which differ significantly from controls are indicated by asterisks: **$p < 0.01$. [From Ref. 32.]

Bulk Lipid Phase Changes in Synaptic Membranes and Function of the Na$^+$–Ca^{2+} Exchange Carrier

A clear demonstration of the sensitivity of synaptic membrane exchange transport activity to the state of membrane lipids was obtained when the two isomers of vaccenic acid were introduced into the synaptic membrane (Fig. 4). The *cis*-vaccenic isomer, which causes increased fatty acid motion within biological membranes [33], produced a dramatic and consistent increase in activity of the exchange carrier (Fig. 4). On the other hand, the *trans*-vaccenic isomer had no effect on transport carriers. In addition, the saturated fatty acid stearic acid had no effect on the Na$^+$–Ca^{2+} exchange transport activity, whereas oleic acid, which has a Δ^9 *cis* unsaturated conformation, produced stimulation very similar to that observed with *cis*-vaccenic acid. Examination of the physical state of synaptic membrane lipids by means of electron paramagnetic resonance (EPR) techniques revealed a large increase in membrane lipid motion following treatment with *cis*-vaccenic acid, whereas *trans*-vaccenic had no effect on membrane lipid motion [32].

These observations indicate a high degree of sensitivity of transport carriers to the physical state of membrane lipids and, consequently, to drug agents that may alter membrane lipid organization. It is well known that both the aliphatic alcohols, as well as benzyl alcohol and the cationic local anesthetics, affect membrane lipid organization. [7,16]. Although there is still considerable uncertainty about the precise effects of these agents on lipid fatty acid motion and the manner in which changes in lipid or protein organization contribute to their biological effects, the fact that they affect both membrane lipid and protein motion in biological membranes is beyond question.

The activity of the Na$^+$–Ca^{2+} exchange carrier in synaptic membranes exhibits a high degree of sensitivity to the incubation temperature (Fig. 2). When temperature-dependent changes in this activity were examined in the form of an Arrhenius plot, there were clear transition phases in the temperature range 22–28°C that coincided with lipid phase transitions in synaptic membranes detected by EPR [32] and fluorescence polarization [19] techniques. The ATP-dependent synaptic membrane carrier Na$^+$ + K$^+$-ATPase also exhibits a phase transition at 24°C [27]. The addition of 25 mM ethanol caused inhibition of carrier activity at all temperatures in the range 8–36°C, except for the transition phase region of 22–28°C, where it enhanced ($\simeq 20\%$) transport. These findings indicate that ethanol even at relatively low concentrations affects thermotropic transitions within synaptic membranes and that its actions on Na^+–Ca^{2+} exchange transport are probably the result of its effects on membrane lipid organization.

Involvement of Ca^{2+} Transport Processes in the Actions of Central Depressants

The inhibitory actions of alcohols on Na^+-dependent Ca^{2+} transport of synaptic membranes provide a plausible mechanism for the observed enhancement of neurotransmitter release and prolongation of the open time of Ca^{2+}-activated K^+ channels in central neurons following exposure to ethanol [11]. The net accumulation of intracellular Ca^{2+} in hippocampal neurons by a mechanism other than depolarization-induced Ca^{2+} influx led to increased release of both excitatory and inhibitory neurotransmitters, as well as prolongation of afterhyperpolarization, a Ca^{2+}-mediated increase in K^+ efflux [11]. Activation of the latter process may be more directly responsible for suppression of central neuronal excitability than are the observed increases or decreases in both excitatory and inhibitory transmitter release. An inhibitory effect on Na^+-dependent Ca^{2+} transport has only been demonstrated so far for alcohols and local anesthetics. Thus, it is still premature to hypothesize that other central depressant drugs produce neuronal suppression through a similar process. In addition, the link between inhibition of the Na^+–Ca^{2+} exchange carrier activity and increased transmitter release or prolongation of the afterhyperpolarization is only tentative at this stage. Since there are no specific inhibitors for this transport system, it is not yet possible to test the idea that inhibition of this transport system alone produces these changes in neuronal activity.

An inhibitory effect on the Na^+–Ca^{2+} exchange carrier was exhibited by all agents tested that are known to interact with the lipid milieu of biological membranes. Inhibition was not observed with agents that are believed to interact with specific receptor sites, including diazepam and verapamil. Such observations tentatively suggest that it may be possible to predict which classes of central depressants will influence activity of this plasma membrane transport carrier and possibly enhance Ca^{2+}-activated K^+ efflux following neuronal depolarization. Included within this class would be drugs that primarily interact with membrane lipids or lipid–protein domains.

ACKNOWLEDGMENTS. The research reported herein was supported by Grant NS–16364 from the National Institute of Neurological and Communicative Disorders and Stroke, and Grant AA–04732 from the National Institute of Alcoholism and Alcohol Abuse. We thank Tim Tehan for expert technical assistance and Kathy Wright for typing the manuscript. We acknowledge the support of the Center for Biomedical Research, University of Kansas.

References

1. Baker, P. F. The regulation of intracellular calcium in giant axons of *Loligo* and *Myxicola*. *Ann. N.Y. Acad. Sci.* **307**:250–267, 1978.
2. Baugé, L.; DiPolo, R.; Osses, L,; Barnola, F.; Campos, M. A (Ca^{2+}, Mg^{2+})-ATPase activity in plasma membrane fragments isolated from squid nerves. *Biochim. Biophys. Acta* **644**:147–152, 1981.
3. Blaustein, M. P. The interrelationship between sodium and calcium fluxes across cell membranes. *Rev. Physiol. Biochem. Pharmacol.* **7**:33–82, 1974.
4. Blaustein, M. P.; Ector, A. C. Barbiturate inhibition of calcium uptake by depolarized nerve terminals *in vitro*. *Mol. Pharmacol.* **11**:369–378, 1975.
5. Blaustein, M. P.; Oborn, C. J. The influence of sodium on calcium fluxes in pinched-off nerve terminals *in vitro*. *J. Physiol. (London)* **247**:657–686, 1975.

6. Blaustein, M. P.; Ratzlaff, R. W.; Kendrick, N. K. The regulation of intracellular calcium in presynaptic nerve terminals. *Ann. N.Y. Acad. Sci.* **28**:195–212, 1978.
7. Boulanger, Y.; Schreier, S.; Leitch, L. C.; Smith, I. C. P. Multiple binding sites for local anesthetics in membranes: Characterization of the sites and their equilibria by deuterium NMR of specifically deuterated procaine and tetracaine. *Can. J. Biochem.* **58**:986–996, 1980.
8. Brostrom, C. O.; Huang, Y. C.; Breckenridge, B. M.; Wolff, D. J. Identification of a Ca^{2+}-binding protein as a calcium-dependent regulator of brain adenylate cyclase. *Proc. Natl. Acad. Sci. USA* **72**:64–68, 1975.
9. Carafoli, E. In vivo effect of uncoupling agents on the incorporation of calcium and strontium into mitochondria and other subcellular fractions of rat liver. *J. Gen. Physiol.* **58**:1849–1864, 1967.
10. Carafoli, E.; Malmstrom, K.; Capano, M.; Sigel, E.; Crompton, M. Mitochondria and the regulation of cell calcium. In: *Calcium Transport in Contraction and Secretion*, E. Carafoli, F. Clementi, W. Drabikowski, and A. Margreth, eds., Amsterdam, North-Holland, 1975, pp. 53–64.
11. Carlen, P. L.; Gurevich, N.; Durand, D. Ethanol in low doses augments calcium-mediated mechanisms measured intracellularly in hippocampal neurons. *Science* **215**:306–309, 1982. (See also Carlen *et al.*, this volume.)
12. Cheung, W. Y.; Bradham, L. S.; Lynch, T. J.; Lin, Y. M.; Tallant, E. A. Protein activator of cyclic-3',5'-nucleotide phosphodiesterase of bovine rat brain also activates its adenylate cyclase. *Biochem. Biophys. Res. Commun.* **65**:1055–1062, 1975.
13. Curran, M.; Seeman, P. Alcohol tolerance in a cholinergic nerve terminal: Relation to the membrane expansion–fluidization theory of ethanol action. *Science* **197**:910–911, 1977.
14. Erickson, C. K.; Graham, D. T. Alteration of cortical and reticular acetylcholine release by ethanol *in vivo. J. Pharmacol. Exp. Ther.* **185**:583–593, 1973.
15. Gill, D. L.; Grollman, E. F.; Kohn, L. D. Calcium transport mechanisms in membrane vesicles from guinea pig brain synaptosomes. *J. Biol. Chem.* **256**:184–192, 1981.
16. Gordon, L. M.; Sauerheber, R. D.; Esgate, J. A.; Dipple, I.; Marchmont, R. J.; Houslay, M. The increase in bilayer fluidity of rat liver plasma membranes achieved by the local anesthetic benzyl alcohol affects the activity of intrinsic membrane enzymes. *J. Biol. Chem.* **255**:4519–4527, 1980.
17. Harris, R. A. Psychoactive drugs as antagonists of actions of calcium. In: *Calcium Antagonists*, Bethesda, American Physiological Society, 1981, pp. 223–231.
18. Harris, R. A.; Fenner, D. Ethanol and synaptosomal calcium binding. *Biochem. Pharmacol.* **31**:1790–1792, 1982.
19. Harris, R. A.; Schroeder, F. Ethanol and the physical properties of brain membranes: Fluorescence studies. *Mol. Pharmacol.* **20**:128–137, 1981.
20. Haycock, J. W.; Levy, W. B.; Cotman, C. W. Pentobarbital depression of stimulus–secretion coupling in brain-selective inhibition of depolarization-induced calcium-dependent release. *Biochem. Pharmacol.* **26**:159–161, 1977.
21. Hood, W. F.; Harris, R. A. Effects of depressant drugs and sulfhydryl reagents on the transport of calcium by isolated nerve endings. *Biochem. Pharmacol.* **29**:957–959, 1980.
22. Inoue, F.; Frank, G. B. Effects of ethyl alcohol on excitability and on neuromuscular transmission in frog skeletal muscle. *Br. J. Pharmacol. Chemother.* **30**:186–193, 1967.
23. Javors, M. A.; Bowden, C. L.; Ross, D. H. Kinetic characterization of Ca^{2+} transport in synaptic membranes. *J. Neurochem.* **37**:381–387, 1981.
24. Kamaya, H.; Kaneshina, S.; Ueda, I. Partition equilibrium of inhalation anesthetics and alcohols between water and membranes of phospholipids with varying acyl chain-lengths. *Biochim. Biophys. Acta* **646**:135–142, 1981.
25. Leslie, S. W.; Elrod, S. V.; Coleman, R.; Belknap, J. K. Tolerance to barbiturate and chlorpromazine-induced central nervous system sedation: Involvement of calcium-mediated stimulus–secretion coupling. *Biochem. Pharmacol.* **28**:1437–1440, 1979.
26. Leslie, S. W.; Friedman, M. B.; Wilcox, R. E.; Elrod, S. V. Acute and chronic effects of barbiturates on depolarization-induced calcium influx into rat synaptosomes. *Brain Res.* **185**:409–417, 1980.
27. Levental, M.; Tabakoff, B. Sodium-potassium-activated adenosine triphosphatase as a measure of neuronal membrane characteristics in ethanol-tolerant mice. *J. Pharmacol. Exp. Ther.* **212**:315–319, 1980.
28. Michaelis, E. K.; Myers, S. L. Calcium binding to brain synaptosomes: Effects of chronic ethanol intake. *Biochem. Pharmacol.* **28**:2081–2087, 1979.
29. Michaelis, E. K.; Michaelis, M. L.; Chang, H. H.; Kitos, T. E. High affinity Ca^{2+}-stimulated Mg^{2+}-dependent ATPase in rat brain synaptosomes, synaptic membranes, and microsomes. *J. Biol. Chem.* **258**:6101–6108, 1983.
30. Michaelis, M. L.; Michaelis, E. K. Ca^{2+} fluxes in resealed synaptic plasma membrane vesicles. *Life Sci.* **28**:37–45, 1981.
31. Michaelis, M. L.; Michaelis, E. K. Alcohol and local anesthetic effects on Na^+-dependent Ca^{2+} fluxes in brain synaptic membrane vesicles. *Biochem. Pharmacol.* **32**:963–969, 1983.

32. Michaelis, M. L.; Michaelis, E. K.; Tehan, T. Alcohol effects on synaptic membrane Ca^{2+} fluxes. *Pharmacol. Biochem. Behav.* **18[Suppl. 1]**: 19–23, 1983.
33. Orly, J.; Schramm, M. Fatty acids as modulators of membrane function: Catecholamine activated adenylate cyclase of the turkey erythrocyte. *Proc. Natl. Acad. Sci. USA* **72**:3433–3437, 1975.
34. Phillis, J. W.; Jhamandas, K. The effects of chlorpromazine and ethanol on *in vivo* release of acetylcholine from the cerebral cortex. *Comp. Gen. Pharmacol.* **2**:306–310, 1971.
35. Pitts, B. J. R. Stoichiometry of sodium–calcium exchange in cardiac sarcolemmal vesicles. *J. Biol. Chem.* **254**:6232–6235, 1979.
36. Reeves, J. P.; Sutko, J. L. Sodium–calcium exchange activity generates a current in cardiac membrane vesicles. *Science* **208**:1461–1464, 1980.
37. Rytting, J. H.; Houston, L. P.; Higuchi, T. Thermodynamics group contribution for the hydroxyl, amino and methylene groups. *J. Pharm. Sci.* **67**:615–618, 1978.
38. Seeman, P. The membrane actions of anesthetics and tranquilizers. *Pharmacol. Rev.* **24**:583–655, 1972.
39. Seeman, P.; Lee, T. The dopamine-releasing actions of neuroleptics and ethanol. *J. Pharmacol. Exp. Ther.* **190**:131–140, 1974.
40. Shulman, H.; Greengard, P. Ca^{2+}-dependent protein phosphorylation system in membranes from various tissues, and its activation by "calcium-dependent regulator." *Proc. Natl. Acad. Sci. USA* **75**:5432–5436, 1978.
41. Sunahara, G. I.; Kalant, H. Effect of ethanol on potassium-stimulated and electrically-stimulated acetylcholine release *in vitro* from rat cortical slices. *Can. J. Physiol. Pharmacol.* **58**:706–711, 1980.
42. Yamamoto, H.; Harris, R. A.; Loh, H. H.; Way, E. L. Effects of acute and chronic morphine treatments on calcium localization and binding in brain. *J. Pharmacol. Exp. Ther.* **205**:255–264, 1978.

29

Calcium Extrusion by Retinal Rods Modifies Light Sensitivity of Dark Current

S. Yoshikami

A chemical mode for signal transmission in biological systems occurs whenever the sender and the receiver are physically and electrically distinct. For transmission within a cell it involves (1) the creation or release of a specific chemical upon stimulation, (2) the diffusion of the chemical to the receptor and interaction with it, and (3) the removal of the chemical signal by either transportation away from the intracellular receptor by extrusion through the plasma membrane (3a), uptake by intracellular organelles (3b), by reversible reactions with binding sites (3c), or degradation (3d). The amount of transmitter required for its production and rapidity of appearance and removal by a sensory system depends primarily on the substance's migration rate, the distance spanned, and the system noise [4,9].

In photoexcitation of vertebrate retinal rods, a chemical transmission step is necessary because light is absorbed in disk membranes that are discontinuous with the plasma membrane of the outer segment where the dark current is controlled [3,8]. The minimum signal strength is of the order of hundreds of transmitters released per photon absorbed [9,23] and the bandwidth of the dark current modulation is d.c. to 0.2 kHz [17]. The signal gain in the receiver is greater than 10^6 per photoisomerization (5% current suppression), which is sufficient to readily exceed membrane noise [17].

Calcium ion has been proposed as the transmitter in rods [8,23]. The transmitter release, reaction, and recovery steps in the calcium cycle hypothesis are schematically shown in Fig. 1. Step (1), revealed through (3a) [7,27], step (2) [18,23,24], and step (3a) [8,25,27] of the calcium cycle hypothesis have been observed in live rods. Isolated rod outer segment disks have been shown to perform steps (1) and (3b) [6]. When the Ca^{2+} chelator EGTA is introduced into rod cytoplasm, the light sensitivity of the dark current to dim light flashes is diminished as if cytoplasmic calcium changes were necessary for rod photoexcitation [10,18]. The EGTA in the rod cytoplasm acts like an additional component (c) of step (3), serving to stabilize the cytoplasmic Ca^{2+}. In a similar way, by changing the rates of other

---

S. Yoshikami • Laboratory of Chemical Physics, National Institute of Arthritis, Diabetes, and Digestive and Kidney Diseases, National Institutes of Health, Bethesda, Maryland 20205.

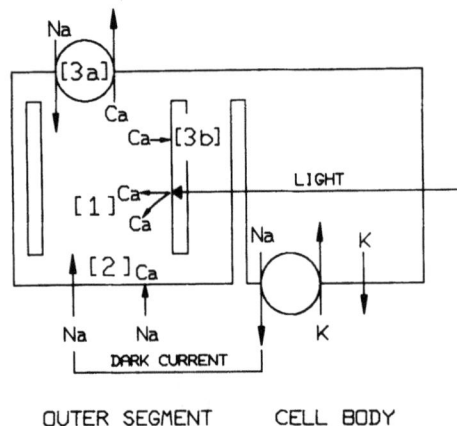

Figure 1. The calcium cycle hypothesis for rod photoexcitation. [1] Light causes a disk to release Ca^{2+} into the cytoplasm of the rod outer segment. [2] Ca^{2+} diffuses to the plasma membrane and stops the dark current. [3] Ca^{2+} is removed from the cytoplasm by the Na^+-Ca^{2+} exchange pump in the plasma membrane [3A] and Ca^{2+} pumps in the rod disk [3b]. ATP-driven Na^+-K^+ pumps power the dark current and indirectly power the Na^+-Ca^{2+} exchange pump through the transmembrane Na^+ gradient they create.

mechanisms comprising step (3) that remove the light-elevated cytoplasmic Ca^{2+}, the light sensitivity of the dark current should be altered.

A Na^+-Ca^{2+} exchange pump that can be perturbed readily at the cell exterior [1,16] regulates cytoplasmic Ca^{2+} in a variety of cells (1), including rods [9,25,27]. The contribution of the cellular extrusion component (a) of step (3) in the calcium hypothesis to the shape of rod dark current responses to light is examined here by perturbing the membrane potential suddenly to change the driving force on the Na^+-Ca^{2+} pump. The removal rate of cytoplasmic Ca^{2+} through the plasma membrane is found to influence the kinetics of the rod's responses to light flashes. This result agrees qualitatively, at least, with the predictions of the Ca^{2+} cycle model.

Membrane Depolarization Enhances the Effect of Light on the Dark Current

The driving force of Na^+-Ca^{2+} exchange pump in the plasma membrane can be changed by altering the membrane voltage. Thus, a rapid alteration of cellular membrane potential, caused for example by large and sudden changes in external $[K^+]$, enhances or retards Ca^{2+} extrusion from a cell using such a mechanism [1,16]. If the dark current were directly controlled by light through an increase in cytoplasmic Ca^{2+}, then the efficacy of Ca^{2+} removal by Na^+-Ca^{2+} exchange in the outer segment should modify the kinetics and amplitude of the light-induced changes of the dark current. Depolarization with high-K^+ Ringer's solution should reduce the Ca^{2+} ejection rate by the plasma membrane. The light-elevated level of cytoplasmic Ca^{2+} would be prolonged thus increasing the duration of dark current suppression produced by a given stimulus flash.

The results from such an experiment are shown in Fig. 2. The recovery time of the dark current (shown as dark voltage) from 0 to 50% of its dark value following a flash (100 photons absorbed per rod) is increased from 1 to 5 sec by depolarization. The duration of light-altered change of transmitter level in the cytoplasm appears to be lengthened, because Ca^{2+} efflux was concurrently decreased. The changes in $[Ca^{2+}]$ shown in Fig. 2B were transformed to Ca^{2+} fluxes through the plasma membrane of outer segments with the method described in Ref. 27. The maximum Ca^{2+} efflux at the peak of the light response of the dark current was decreased by the depolarization from 24 ($\pm$ 1.2) $\times$ 10^3 to 12.5 ($\pm$ 1.1) $\times$ 10^3 Ca^{2+}/rod per sec. The integrated Ca^{2+} output decreased with depolarization from

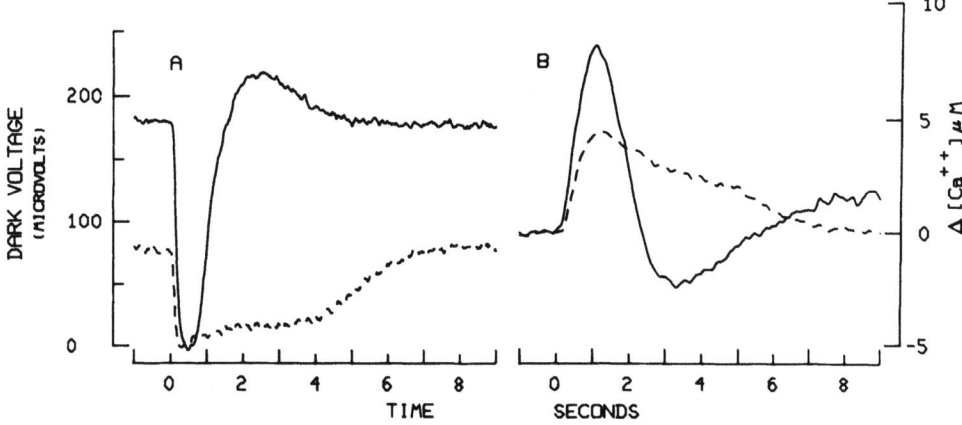

Figure 2. Effect of a rapid depolarization of the rod membrane on the light response of the dark voltage gradient produced by the dark current and the light-stimulated release of Ca^{2+} by rod outer segments. 2 × 3-mm pieces of isolated retinas from dark-adapted Sprague–Dawley rats were sustained at 32°C in either of two kinds of Ringer's solutions differing only in [K^+] or [choline$^+$]. The low-K^+ Ringer's solution contained 2.5 mM potassium isethionate and 25 mM choline isethionate; the high-K^+ Ringer's solution contained 25 mM K isethionate and 2.5 mM choline isethionate. In addition both solutions contained 110 mM sodium isethionate, 0.5 mM $MgCl_2$, 0.1 mM Na_2HPO_4, 10 mM Hepes buffer at pH 7.13, 11 mM glucose and were saturated with calcium oxalate to equalize the [Ca^{2+}] at about 90 μM. Two Ringer's-filled glass pipette electrodes, one placed at the tips of the outer segments and the other 60 μm into the rod layer, were used to measure the voltage gradient produced by the dark current in the interstitial space [10]. A Ca^{2+}-sensitive micropipette using the ionphore ETH1001 [19] was placed 25 μm into the interstitial space of the rod layer to determine the efflux of Ca^{2+} from the rod outer segments [27]. A flowing solution bathed the retina, and the solution composition in the region where the measurements were made was changed rapidly by displacing it with a jet of another solution [27]. In this manner the solution about the rod outer segment could be changed in 0.25 sec. The liquid junction potential transient created by the change in solution composition was removed by subtracting the response of the system in darkness to [K^+] change from that obtained with the change superimposed on a light flash. (A) Dark voltage response of rods absorbing 100 photons each per flash at time zero in Ringer's containing 2.5 mM K^+ (———). One second before a similar flash, the rods were depolarized with a jet of Ringer's containing 25 mM K^+ (-----). The [Na^+] was kept constant at 110 mM. (B) Corresponding changes of Ca^{2+} in the interstitial space of the rod outer segment layer.

2.47×10^4 to 1.67×10^4 Ca^{2+}/rod per sec 1.5 sec after the flash. The reduction in membrane potential induced by K^+ diminishing light-stimulated Ca^{2+} efflux to 52% of its original rate, was estimated to be between 7.4 and 13 mV.* A 26 mV depolarization of squid axon membrane with a 10-fold rise in external [K^+] depressed Ca^{2+} efflux to 64% [1].

The K^+-induced depolarization also halved the maximum light-suppressible dark current by decreasing the driving force of the dark current. The transient overshoot of the dark

---

*The lower and upper bound estimates of the change in rod membrane potential induced by the 10-fold increase in external [K^+] were determined by using the lowest and highest membrane potentials of rods from several published values (-15 to -30 mV) of axolotl, frog, and gecko [20–22] and the rod cytoplasmic [Na^+] of 40 mM that was obtained with X-ray microbeam analysis on rat rods [26]. Because isethionate was the principal anion in the Ringer's solution, only the contributions of Na^+ and K^+ to the membrane potential were considered. The change of external [K^+] in the interstitial space of the rod outer segments in 1 sec was sufficiently rapid to allow one to consider that the rod [K^+] and [Na^+] were not significantly altered by the dark current [23]. Thus, the internal and external cation concentrations were assumed equal for the Na^+ and K^+ gradient across the rod membrane and the independence relation equation: $V_m = RT/F \ln(([K^+]_{ext.} + b[Na^+]_{ext.})/([K^+]_{int.} + b[Na^+]_{int.}))$ [11] was used to calculate the b of 0.605 and 0.296 that respectively yielded -15 and -30 mV for the range of resting membrane potentials of the rod. The range of the membrane potential change resulting from the 10-fold increase in external [K^+] was then determined from the b values.

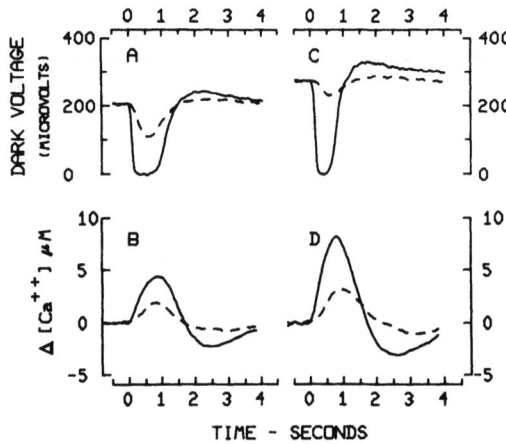

Figure 3. Dark voltage responses to light resulting in 100 (——) or 15 (-----) photons being absorbed per rod per flash and the associated changes of Ca^{2+} in the interstitial space surrounding the outer segment. Experimental method was as for Fig. 2. (A,B) Retina in 25 mM K^+. (C, D) Retina repolarized by a 10-fold drop in [K^+] at -2 sec.

current following the flash in low-K^+ Ringer's and the parallel influx of Ca^{2+} into rods will be considered below.

After 2 min in the depolarizing high-K^+ Ringer's solution, the dark current recovered to 96% of that in low-K^+ medium. Although its recovery time to a light flash was shortened, it remained longer than the recovery time in low-K^+ medium. The Ca^{2+} efflux also remained correspondingly slower.

Repolarizing Rods Reduce the Effect of Light on the Dark Current

Increasing the driving force on the Na^+–Ca^{2+} pump by repolarization should have an effect opposite to that described above. Increased pump efflux rate of Ca^{2+} should decrease the lifetime of the light-released Ca^{2+} in the cytoplasm thus rendering it less potent as a suppressor of the dark current.

Repolarizing the rod membrane by decreasing external [K^+] to 1/10th of the initial level of 25 mM caused the dark current to increase and to speed its recovery to a flash of light (Figs. 3A,C). A flash of 15 photons absorbed per rod transiently reduced the dark current by 50%, but when the cell was repolarized, a flash of the same intensity now reduced it by only

Table I. Comparison of Light-Stimulated Ca^{2+} Efflux Rates from Rod Outer Segments before and after Membrane Repolarization

Solution composition	Percent dark current suppressed		Ca^{2+} efflux rate at peak of dark current suppression (Ca^{2+}/rod/sec)	Integrated Ca^{2+} ejected/rod after 1.5 sec
	15 photons absorbed/ rod/flash	100 photons absorbed/ rod/flash		
25 mM K^+ unchanged	50	100	$(7.3 \pm 1.2) \times 10^3$ $(14.5 \pm 1.1) \times 10^3$	4.9×10^3 12.3×10^3
25 to 2.5 mM [K^+] jump 2 sec before the flash	17	100	$(11.5 \pm 1.7) \times 10^3$ $(26 \pm 1.3) \times 10^3$	8.9×10^3 21×10^3

17%, while the Ca^{2+} efflux rate was almost doubled (Figs. 3C,D, and Table I). However, at a constant $[K^+]$ of 2.5 mM, 25% of the dark current was reduced by a flash of the same intensity. The light energy in a flash needed to suppress the dark current by 50% in 2.5 mM K^+ is twice that needed in 25 mM K^+. The response times of the rod between the 5-μsec flash (15 photons absorbed per rod per flash) and the resultant decreases in the dark current just exceeding the noise level were 10 and 100 msec for depolarized and repolarized rods, respectively (Figs. 3A,C).

For a given amount of Ca^{2+} released by disks on light stimulation, an enhanced cellular efflux of Ca^{2+} will cause the cytoplasmic $[Ca^{2+}]$ in rods to attain a lower level and to fall more rapidly. When Ca^{2+} efflux was enhanced, the dark current responded to a given light intensity with a smaller amplitude change and recovered more rapidly.

Interpretations

Though Ca^{2+} efflux from rods is affected by manipulating the membrane potential with external $[K^+]$, could the change in external $[K^+]$ shape the rod dark current response by other means? Potassium ion by itself might act on the Na^+ conductance [2] on step (1), or through the membrane potential effects on step (1) or (2). The increase in K^+ alone does not seem to be involved because after 2 min in the high-K^+ Ringer's solution the dark current recovers almost fully and the kinetics of the dark current and the light-stimulated Ca^{2+} efflux rate recover to lesser extents. The growth of the dark current in the high-K^+ Ringer's solution indicates that the rod internal $[K^+]$ and membrane potential have reequilibrated. Thus, the direct influence of K^+ on the Na^+ conductance appears negligible when compared to its effect through membrane potential on step (3a) and the resulting change in the Ca^{2+} efflux rate.

The rise in external $[Ca^{2+}]$ at the outer segment layer following a light flash is not due simply to an effect of membrane hyperpolarization (dark current suppression) stimulating Ca^{2+} efflux and reducing internal $[Ca^{2+}]$. It is also not due to a buildup of external $[Ca^{2+}]$ created by a suppression of an inward Ca^{2+} current using the entryway of Na^+ for the dark current, because more Ca^{2+} can appear extracellularly during a smaller light suppression of the dark current than during a larger one (Figs. 3B,D, and Table I). It is a light-stimulated rise in cytoplasmic Ca^{2+} that leads to Ca^{2+} extrusion from rods.

Ca^{2+} applied to the outer segments shuts the dark current [18,23,24]. With the aid of Ca^{2+} ionophores put into the plasma membrane of rods, it was estimated that at an internal $[Ca^{2+}]$ of no more than 1 μM, 50% of the dark current would be suppressed [9]. The internal $[Ca^{2+}]$ can be raised 1.5 μM by considering the retention of the Ca^{2+} ejected by step (3a) from the rod in 1.5 sec after a flash (Fig. 2B). Similarily the Ca^{2+} ejected by the rod to a flash causing the absorption of 15 photons per rod (Table 1) would suppress at least 14% of the dark current when depolarized (4900 Ca^{2+}) and 26% of the dark current when repolarized (8900 Ca^{2+}). But removing Ca^{2+} more quickly from rods after a flash causes a given amount of light to suppress less dark current than otherwise (Table I).

When the rod membrane is repolarized by reducing external $[K^+]$ the resulting smaller response of the dark current to the same light energy is not due to less Ca^{2+} being released by the disk per photon absorbed in step (1); in fact, more Ca^{2+} is extruded by the rod during this response. The total Ca^{2+} released by a disk per photoisomerization in rat rods is not known (it is on the order of thousands per photoisomerization in frog disk [6]), but by increasing the fraction of the Ca^{2+} removed from the cytoplasm the action of light on the dark current is reduced. It is not likely that there is some other transmitter controlling the dark current because Ca^{2+} is released by light in sufficient speed and quantity to suppress the dark current by itself.

The calcium cycle model requires that the Ca^{2+} ejected by the cell be recovered, and the disk be recharged with Ca^{2+} following illumination. Ca^{2+} fluxes determined from the records in Figs. 3B and D show that following a flash of light, half of the ejected Ca^{2+} reenters the rod within 5 sec (cf. Fig. 3 in [27]). Because the dark current is not suppressed during the reentry of Ca^{2+}, cytoplasmic Ca^{2+} activity must be kept low during the uptake process. A high uptake of Ca^{2+} by the illuminated disk during recovery permits a large dark current as Ca^{2+} passes through the cytoplasm. This is similar to the effect noted when enhanced efflux of light-released Ca^{2+} diminished its effect on the dark current. The high uptake rate would contribute to the dark current overshoot already quantitatively attributed to a change in the Na^+ content of rods during illumination (W. A. Hagins and S. Yoshikami, unpublished). Moreover, light-stimulated Ca^{2+} uptake by disks would serve to reduce the dark current reduction produced by a subsequent illumination of the same intensity [17] during the period of enhanced Ca^{2+} uptake, because the second stimulus would appear to release fewer transmitters.

A second light stimulus produces a diminished response. Exposing rods to a steady light (450 photons absorbed/rod per sec) causes full suppression of the dark current and a surge of Ca^{2+} from the rods. Within a few seconds, the dark current level and the Ca^{2+} efflux rate diminish to a steady level of one-half their initially high levels and remain reduced for the duration of the illumination. When a light flash (11,000 photons absorbed/rod per flash) that can fully suppress the dark current of an unilluminated retina for several seconds is superimposed on the steady light, only half of the remaining dark current is suppressed. The additional Ca^{2+} efflux created by the superimposing flash is also proportionately smaller than that from a retina not exposed to the steady light (unpublished). From this observation one cannot distinguish between the effects of an increase in Ca^{2+} uptake by light or a decrease in Ca^{2+} release by a second stimulus. The contributions of each to the photoexcitatory process could probably be determined by examining the influence of background light on uptake and light-stimulated efflux of ^{45}Ca by isolated rod disks.

Isolated rod disks accumulate Ca^{2+} actively and release it upon illumination [6]. cGMP but not cAMP enhances Ca^{2+} uptake by disks if cellular nucleotide triphosphates exceed diphosphates or if ATP (or GTP) and phosphocreatine are present [6]. Is the energy for recharging the disks with transmitter derived from cGMP hydrolysis and is light-activated phosphodiesterase in rods [15] involved in this process [14]? The GTP-binding protein that activates the phosphodiesterase is membrane bound [5,12] and poorly activates phosphodiesterase in adjacent disk membranes [13]. Such a localization would help reduce the stimulation of Ca^{2+} uptake by adjacent unilluminated disks and so decrease their scavenging of the Ca^{2+} released by activated disks. A preferential recharging of the activated disks with the transmitter would thus be assured.

ACKNOWLEDGMENTS. I am grateful to W. A. Hagins for his valuable suggestions and discussions. I also thank Ronald Fico and John I. Powell of the Computer Systems Laboratory, and W. H. Jennings of the Laboratory of Chemical Physics, NIH, for development of the computerized laboratory data acquisition system used in this study.

References

1. Blaustein, M. P.; Russell, J. M.; De Weer, P. Calcium efflux from internally dialyzed squid axons: The influence of external and internal cations. *J. Supramol. Struct.* 2:558–581, 1974.

2. Brown, J. E.; Pinto, L. H. Ionic mechanism for the photoreceptor potential of the retina of *Bufo marinus*. *J. Physiol. (London)* **236**:575–591, 1974.
3. Cohen, A. I. Further studies on the question of patency of saccules in outer segments of vertebrate photoreceptors. *Vision Res.* **10**:445–453, 1970.
4. Cone, R. A. The internal transmitter model for visual excitation: Some quantitative implications. In: *Biochemistry and Physiology of Visual Pigments*, H. Langer, ed., Berlin, Springer-Verlag, 1973, pp. 275–282.
5. Fung, B. K. K.; Hurley, J. B.; Stryer, L. Flow of information in the light-triggered cyclic nucleotide cascade of vision. *Proc. Natl. Acad. Sci. USA* **78**:152–156, 1981.
6. George, J. S.; Hagins, W. A. Control of Ca^{2+} in rod outer segment disks by light and cyclic GMP. *Nature (London)* **303**:344–348, 1983.
7. Gold, G. H.; Korenbrot, J. I. Light-induced calcium release by intact retinal rods. *Proc. Natl. Acad. Sci. USA* **77**:5557–5561, 1980.
8. Hagins, W. A. The visual process: Excitatory mechanisms in the primary receptor cells. *Annu. Rev. Biophys. Bioeng.* **1**:131–158, 1972.
9. Hagins, W. A.; Yoshikami, S. Ionic mechanisms in excitation of photoreceptors. *Ann. N.Y. Acad. Sci.* **264**:314–325, 1975.
10. Hagins, W. A.; Yoshikami, S. Intracellular transmission of visual excitation in photoreceptors: Electrical effects of chelating agents introduced into rods by vesicle fusion. In: *Vertebrate Photoreception*, P. Fatt and H. Barlow, eds., New York, Academic Press, 1977, pp. 97–139.
11. Hodgkin, A. L.; Katz, B. The effect of sodium ions on the electrical activity of the giant axon of the squid. *J. Physiol. (London)* **108**:37–77, 1949.
12. Kuhn, H. Light- and GTP-regulated interaction of GTPase and other proteins with bovine photoreceptor membrane. *Nature (London)* **283**:587–589, 1980.
13. Liebman, P. A.; Sitaramayya, A.; Puch, E. N., Jr. Origin of delays in PDE activation of rod disk membranes (RDM). *Biophys. J.* **37**:88a, 1982.
14. Lipton, S. A.; Rasmussen, H.; Dowling, J. E. Electrical and adaptive properties of rod photoreceptors in *Bufo marinus*. II. Effects of cyclic nucleotides and prostaglandins. *J. Gen. Physiol.* **70**:771–791, 1977.
15. Miki, N.; Keirns, J. J.; Marcus, F. R.; Freeman, J.; Bitensky, M. W. Regulation of cyclic nucleotide concentration in photoreceptors: An ATP dependent stimulation of cyclic nucleotide phosphodiesterases by light. *Proc. Natl. Acad. Sci. USA* **70**:3820–3824, 1973.
16. Mullins, L. J.; Brindley, F. J., Jr. The sensitivity of calcium efflux from squid axons to change in membrane potential. *J. Gen. Physiol.* **65**:135–152, 1975.
17. Penn, R. A.; Hagins, W. A. Kinetics of the photocurrent of retinal rods. *Biophys. J.* **12**:1073–1094, 1972.
18. Pinto, L. N.; Brown, J. E.; Coles, J. A. Mechanism for the generation of the receptor potential of rods of *Bufo marinus*. In: *Vertebrate Photoreception*, P. Fatt, and H. Barlow, eds., New York, Academic Press, 1977, pp. 159–167.
19. Simon, W.; Ammann, D.; Oehme, M.; Morf, W. E. Calcium-selective electrodes. *Ann. N.Y. Acad. Sci.* **307**:52–70, 1978.
20. Toyoda, J.; Hashimoto, H.; Tomita, T. The rod response in the frog as studied by intracellular recording. *Vision Res.* **10**:1093–1100, 1970.
21. Toyoda, J.; Nosaki, H.; Tomita, T. Light-induced resistance changes in single photoreceptors of *Necturus* and gecko. *Vision Res.* **9**:453–463, 1969.
22. Werblin, F. S. Regenerative hyperpolarization in rods. *J. Physiol. (London)* **244**:53–81, 1975.
23. Yoshikami, S.; Hagins, W. A. Control of the dark current in vertebrate rods and cones. In: *Biochemistry and Physiology of Visual Pigments*, H. Langer, ed., Berlin, Springer-Verlag, 1973, pp. 245–255.
24. Yoshikami, S.; Hagins, W. A. Kinetics of control of the dark current of retinal rods by Ca^{++} and by light. *Fed. Proc.* **39**:1814, 1980.
25. Yoshikami, S.; Hagins, W. A. Sodium–calcium exchange and the kinetics of the rod response. *Biophys. J.* **33**:288a, 1981.
26. Yoshikami, S.; Foster, M. C.; Hagins, W. A. Ca^{++} regulation, dark current control, and Na^+ gradient across the plasma membrane of retinal rods. *Biophys. J.* **41**:342a, 1983.
27. Yoshikami, S.; George, J. S.; Hagins, W. A. Light-induced calcium fluxes from outer segment layer of vertebrate retinas. *Nature (London)* **286**:395–398, 1980.

VII

Calmodulin

30

Mechanisms of Pharmacologically Altering Calmodulin Activity

Walter C. Prozialeck and Benjamin Weiss

Considerable evidence has now accumulated indicating that calmodulin is the principal mediator of the effects of Ca^{2+} in most eukaryotic cells [for reviews see 3,8,24,25]. Since calmodulin plays such a fundamental role in cell biology, agents that inhibit its activity should produce important pharmacological effects. An understanding of the mechanisms by which drugs alter calmodulin activity may suggest new approaches for modifying various physiological or pathological processes. Furthermore, the development of selective calmodulin antagonists may provide a useful means for further studying the biological roles of calmodulin.

In the mid 1970s, phenothiazine antipsychotics were reported to inhibit calmodulin-induced activation of a Ca^{2+}-dependent form of phosphodiesterase [28,31,54]. Subsequent studies have shown that a vast array of compounds belonging to diverse chemical and pharmacological classes inhibit calmodulin's actions in a variety of systems [for reviews, see 36,56,57]. Since calmodulin regulates a multitude of processes, and many different classes of drugs inhibit its activity, the question arises as to whether it is possible to alter selectively calmodulin-dependent events. The purpose of this review is to summarize the various mechanisms by which the presently available calmodulin antagonists act and to suggest several approaches that might be used to develop more selective calmodulin inhibitors.

General Mechanisms for Altering Calmodulin-Dependent Processes

To consider the mechanisms by which drugs might modify the activity of calmodulin, it is first necessary to review the mechanisms by which calmodulin produces its physiological effects. Most of the effects of calmodulin result from activation of specific enzymes [3,24]. The general mechanism by which calmodulin regulates these enzymes has been reviewed

---

Walter C. Prozialeck • Department of Physiology and Pharmacology, Philadelphia College of Osteopathic Medicine, Philadelphia, Pennsylvania 19131. *Benjamin Weiss* • Department of Pharmacology, Medical College of Pennsylvania, Philadelphia, Pennsylvania 19129.

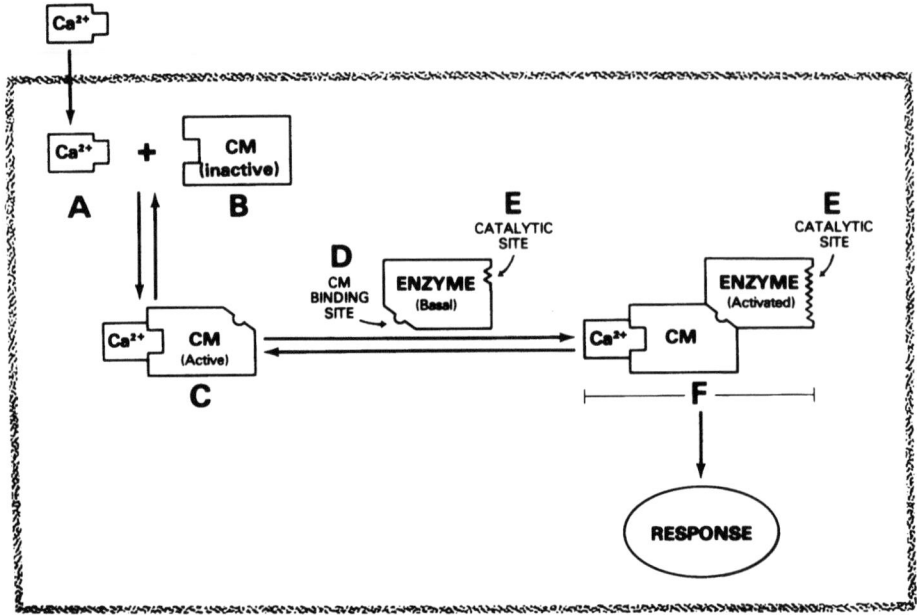

Figure 1. Mechanisms for altering calmodulin activity. See text for description.

elsewhere [3,8,17,24,25]. In brief, each calmodulin molecule can bind up to four Ca^{2+} ions; the binding of Ca^{2+} alters the conformation of calmodulin, increasing its helical content and exposing hydrophobic regions. In this conformation, the Ca^{2+}–calmodulin complex can bind to target enzymes and, through an unknown mechanism, alter their activities.

A variety of mechanisms exist by which drugs might inhibit the effects of calmodulin (see Fig. 1). Agents might act by: (A) reducing the availability of Ca^{2+} (this could be accomplished by inhibiting the entry of Ca^{2+} into cells, by inhibiting its release from intracellular pools, or by chelating it); (B) binding to calmodulin and altering its ability to bind Ca^{2+}; (C) binding to the Ca^{2+}–calmodulin complex and modifying its activity; (D) binding to the calmodulin–recognition site on the calmodulin-sensitive enzyme and thereby preventing interaction of the Ca^{2+}–calmodulin complex with the enzyme; (E) interacting with the catalytic portion of the calmodulin-sensitive enzyme; and (F) interacting with the ternary Ca^{2+}–calmodulin-enzyme complex. Each of these mechanisms is discussed in more detail below.

Agents That Alter the Availability of Ca^{2+}

Agents that inhibit the actions of calmodulin by decreasing the concentration of available Ca^{2+} include Ca^{2+}-chelating agents, such as EGTA [28,59], and the calcium entry blockers, such as verapamil [14]. Since these agents do not interact directly with calmodulin or its binding sites, they display little specificity for calmodulin-regulated systems.

Agents That Interact with Calmodulin and Alter Its Ability to Bind Ca^{2+}

Relatively little is known about agents in this category. They might act by binding directly to the Ca^{2+}-binding sites on calmodulin or by binding to some other site on the molecule thereby inducing changes in the Ca^{2+}-binding regions. Although a variety of di-

and trivalent cations interact with the Ca^{2+}-binding sites on calmodulin [7,32,47], it is not clear whether the agents can antagonize or potentiate the effects of Ca^{2+}. Mg^{2+} at millimolar concentrations inhibits activation of calmodulin-dependent enzymes by competing with Ca^{2+} for ion-binding sites on calmodulin [2,3,23]. Recently, several metals, including Hg, Cd, Zn, Co, and Sr, have been shown to inhibit the calmodulin-induced activation of phosphodiesterase [11]. However, the mechanism underlying this effect has yet to be determined.

Although examinations of the nuclear magnetic resonance spectrum of calmodulin suggest that binding of phenothiazine antipsychotics alters the Ca^{2+}-binding domains of calmodulin [15,45], the pharmacological significance of this is unclear. Drug binding requires Ca^{2+} [29,59] and the inhibitory effects of drugs cannot be overcome by increasing the concentration of Ca^{2+} [28].

Agents That Interact with the Ca^{2+}–Calmodulin Complex

The vast majority of calmodulin antagonists described thus far act by binding directly to the Ca^{2+}–calmodulin complex and thereby alter its activity [36,59]. Of these agents, the phenothiazine antipsychotics have been studied most extensively. The phenothiazines bind to two distinct classes of sites on calmodulin: a class of saturable, high-affinity, Ca^{2+}-dependent sites and a class of low-affinity, Ca^{2+}-independent sites [29–31]. The Ca^{2+}-dependent sites appear to be the biochemically important sites, since there is an excellent correlation between the Ca^{2+}-dependent binding of various drugs and their anticalmodulin potencies [31,59]. There are two or three Ca^{2+}-dependent drug-binding sites per calmodulin molecule, with the most potent agents displaying dissociation constants of 10 μM or less [31,56].

A variety of other agents belonging to diverse pharmacological and chemical classes also interact with the Ca^{2+}–calmodulin complex (see Table I). Many of these seemingly unrelated agents compete with the phenothiazines for binding sites on calmodulin, although they exhibit different affinities for calmodulin [19,31,42,56].

The fact that such a wide variety of drugs can interact with the Ca^{2+}–calmodulin

Table I. Pharmacological Classes of Drugs That Inhibit Calmodulin by Binding to the Ca^{2+}–Calmodulin Complex

Class	References
α-Adrenergic antagonists	13, 51
Antianginal agents	4, 14
Antianxiety agents	31
Antidepressants	28, 31, 39
Antidiarrheals	60
Antihistaminics	28, 30
Antipsychotics	28, 31, 39
Cancer chemotherapeutics	22, 51, 52
Local anesthetics	46, 49, 50
Neuropeptides	34, 42, 43, 58
Insect venom peptides	9, 43
Smooth muscle relaxants	18, 19
Miscellaneous agents	
Calmidazolium (R 24 571)	48
Compound 48/80	16
Triton X-100	44

complex might suggest that this interaction is not associated with any particular structure or pharmacological activity and thus represents a nonspecific effect. However, the drugs that interact with calmodulin do, in fact, exhibit certain physicochemical and pharmacological similarities [36,39,57]. Chemically, the most potent calmodulin-binding drugs contain a large hydrophobic region and a positively charged amino group. The hydrophobic group should consist of two or more aromatic rings—presumably to interact with hydrophobic residues on calmodulin—and must be separated from the amino group by four or more atoms. The charged amino group presumably interacts with negatively charged aspartic or glutamic acid residues on calmodulin. This general structure is found in several classes of antipsychotic drugs as well as in the local anesthetics, antihistaminics, antimalarials, α-adrenergic antagonists, and certain smooth muscle relaxants. The structural similarities among these agents may not only explain why they interact with calmodulin, but also why there is overlap in their pharmacological activities. This overlap is illustrated by the fact that the phenothiazine antipsychotics also display antihistaminic, local anesthetic, smooth muscle relaxant, and α-blocking properties.

There are two possible explanations for the commonality of structure and pharmacological activity among these agents. One possibility is that the various physiological systems on which the drugs act utilize different biochemical components that happen to have similar structural requirements for pharmacological manipulation. The other possibility is that some common biochemical component, such as calmodulin, is involved in regulating a variety of systems; thus, drugs acting on that component would produce a variety of effects. Perhaps, the actions of the calmodulin-binding drugs will be ultimately explained by some combination of these two possibilities.

In addition to drugs, there are several naturally occurring compounds that interact with the Ca^{2+}-calmodulin complex. These include a group of calmodulin-binding proteins and a variety of peptides [34,42,43,58]. Several neuropeptides and certain insect venom peptides are among the most potent calmodulin inhibitors yet discovered; some of them are orders of magnitude more potent than the most potent antipsychotic drugs [9,43]. Mellitin [43] inhibits calmodulin-stimulated phosphodiesterase activity without affecting basal activity of the enzyme. The IC_{50} value of mellitin is approximately 200 nM and its K_i 30 nM. Therefore, this compound is about 50 times more potent than trifluoperazine in inhibiting calmodulin activity. When mellitin is acetylated, its anticalmodulin activity decreases by approximately fivefold. Since acetylation of the lysine residues in mellitin does not substantially alter its hydrophobicity but does decrease the net positive charge on the molecule, these results again show the importance of ionic interactions between calmodulin inhibitors and calmodulin [43]. This finding that certain neuropeptides inhibit calmodulin suggests that these or perhaps some as yet undiscovered peptides may function as endogenous regulators of calmodulin activity and may themselves provide potential sites for pharmacological intervention.

Early studies showed that binding of phenothiazines and related agents to calmodulin is a reversible process, since the binding could be reversed by removing Ca^{2+} or by dialyzing against an excess of competing drug [29,31,59]. However, recent findings have revealed that some agents can bind irreversibly to calmodulin under certain conditions. Upon irradiation with ultraviolet light [37,56] or treatment with peroxidase–hydrogen peroxide [56], chlorpromazine and trifluoperazine bind irreversibly to calmodulin, presumably through a free radical mechanism. Like the reversible binding of these agents to calmodulin, the irreversible binding is enhanced by Ca^{2+}. The Ca^{2+}-dependent binding is saturable, with one Ca^{2+}-dependent drug-binding site per molecule [37]. The irreversible binding of phenothiazines to calmodulin results in the irreversible inactivation of calmodulin; that is, calmodulin that is irreversibly linked to the drug by ultraviolet irradiation cannot activate phosphodiesterase

[37], although the drug–calmodulin complex inhibits phosphodiesterase activation by native calmodulin [38].

The α-adrenergic antagonists phenoxybenzamine and dibenamine also inhibit calmodulin irreversibly, probably by binding directly to calmodulin in a Ca^{2+}-dependent manner [13]. These agents generate ethyleneimmonium and carbonium ion intermediates that might bind irreversibly to electron-rich functional groups on calmodulin. Although this may be a nonselective effect, several alkylating agents that are not α-adrenergic antagonists display meager anticalmodulin activity [13,51].

Besides providing a potentially useful tool for studying drug–calmodulin interactions, irreversible binding of phenothiazines and α-adrenergic antagonists to calmodulin *in vitro* raises the possibility that they might also interact irreversibly with calmodulin *in vivo*. Perhaps these agents can be photochemically or metabolically activated *in vivo* to yield free radicals that might bind irreversibly to calmodulin. This could have important implications regarding mechanisms by which they produce certain of their long-term effects.

Agents That Act at Calmodulin-Binding Sites on Calmodulin-Sensitive Enzymes

Thus far, relatively little attention has been given to developing agents that bind to recognition sites on calmodulin-sensitive enzymes. However, recent findings [38] suggest that these sites may prove useful for pharmacological intervention. Results of one such study are shown in Fig. 2. Calmodulin that had been linked to chlorpromazine by ultraviolet irradiation (CPZ-CM) had no effect on basal phosphodiesterase activity but did inhibit enzyme activation by native calmodulin in a concentration-dependent manner (IC_{50} 450 ng/sample). Assuming that the CPZ-CM complex has a molecular weight of about 17,000, its K_i would be approximately 100 nM. Another study [38] showed that CPZ-CM increased

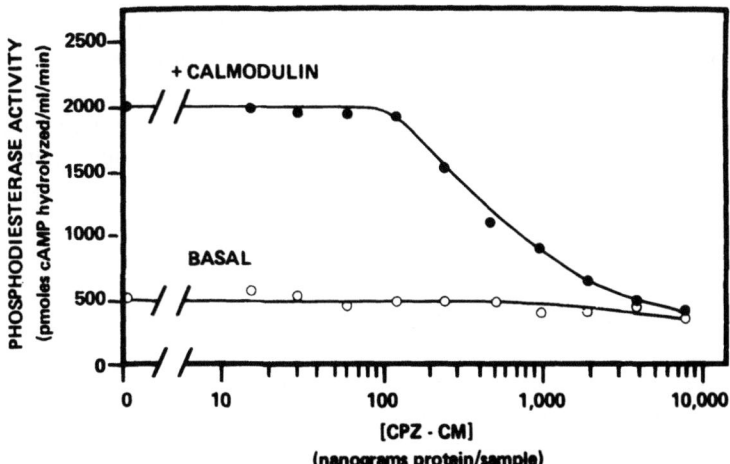

Figure 2. Inhibition of the calmodulin-induced activation of phosphodiesterase by chlorpromazine-linked calmodulin. Chlorpromazine-linked calmodulin (CPZ-CM) was prepared by irradiating purified bovine brain calmodulin with ultraviolet light (λ_{max} = 366 nm), in the presence of 100 μM chlorpromazine and 100 μM $CaCl_2$ [37,38]. The sample was then dialyzed extensively to remove reversibly bound chlorpromazine. Phosphodiesterase activity was measured in the absence and presence of calmodulin (1 unit; approximately 10 nM) along with varied amounts of CPZ-CM and 400 μM cAMP, in a final volume of 125 μl [39]. Each point represents the mean value of three samples.

the activation constant (K_a) for the interaction of calmodulin with phosphodiesterase but did not affect maximal enzyme activation (V_{max}) by calmodulin. Neither calmodulin nor CPZ-CM altered the K_m for the interaction between phosphodiesterase and cAMP.

These results indicate that CPZ-CM inhibits the calmodulin-induced activation of phosphodiesterase by competing with calmodulin for regulatory sites on the enzyme and not by blocking the interaction of cAMP with the enzyme. Thus, it may be possible to develop a new class of calmodulin antagonists directed at calmodulin-binding sites on calmodulin-sensitive enzymes. Since various calmodulin-sensitive enzymes may have somewhat different calmodulin-binding sites (for review see 24), agents directed at these sites might display greater selectivity than agents that interact with calmodulin itself.

Agents That Act on the Catalytic Portion of Calmodulin-Sensitive Enzymes

Like Ca^{2+} chelators and Ca^{2+}-entry blockers, agents that interact with catalytic portions of calmodulin-sensitive enzymes, such as methylxanthine phosphodiesterase inhibitors, could be considered indirect calmodulin antagonists, since they do not interact with calmodulin or its binding sites. These agents are relatively nonspecific because they inhibit the nonstimulated form of the enzyme as well as the calmodulin-stimulated form [32,53,55]. In addition, they inhibit enzyme activation by agents other than calmodulin or calmodulin-insensitive forms of the enzyme [53].

Agents That Act on the Ternary Ca^{2+}–Calmodulin–Enzyme Complex

No agents that interact specifically with the ternary Ca^{2+}–calmodulin–enzyme complex have yet been described, although agents acting on the enzyme or calmodulin itself might remain bound after the ternary complex is formed.

Agents That Act at Multiple Sites

Certain agents may act by more than one of the mechanisms described above. For example, calmodulin-binding drugs such as the phenothiazines may inhibit the calmodulin-induced activation of Ca^{2+}-transport ATPase by interacting with both calmodulin and the enzyme itself [1,17,33]. Likewise, the antiarrhythmic and antianginal agent nimodipine inhibits calmodulin-induced phosphodiesterase activation by interacting with both calmodulin and the catalytic portion of the enzyme [14]. The benzodiazepine antianxiety agents not only bind directly to calmodulin [31] but also interact with calmodulin-sensitive protein kinase of rat brain membranes [12].

Approaches to Developing Selective Calmodulin Antagonists

If calmodulin inhibitors are to be used to modify a particular physiological process or to probe calmodulin effects, a high degree of specificity would be desirable. Unfortunately, most presently available calmodulin inhibitors are relatively nonspecific. Calmodulin-binding drugs such as the phenothiazines and related agents interact with a variety of biochemical sites other than calmodulin [41,48] and produce nonspecific alterations in biological membranes [40,41]. Furthermore, since calmodulin regulates so many physiological processes, drugs which inhibit its activity would be expected to produce a wide range of pharmacological effects.

How is it possible then to develop selective calmodulin antagonists? First, the structures

30. Pharmacological Alterations

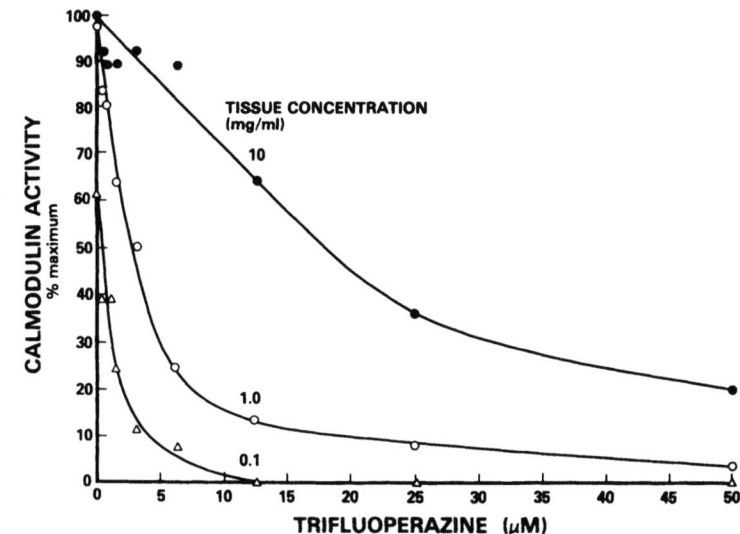

Figure 3. Effect of trifluoperazine on calmodulin activity as a function of the concentration of tissue. Rat caudate nuclei were homogenized and boiled to destroy any endogenous phosphodiesterase activity. Calmodulin activity in the homogenates was assessed by its ability to activate a calmodulin-sensitive form of phosphodiesterase. The calmodulin-inhibitory effects of varying concentrations of trifluoperazine were then determined in the presence of varying concentrations of caudate tissue homogenate (0.1, 1.0, and 10 mg tissue/ml).

of presently available calmodulin-binding drugs might be altered to increase their anticalmodulin activities, while decreasing other drug actions. For example, structure–activity studies suggest that anticalmodulin activities of phenothiazine derivatives may be enhanced while decreasing anticholinergic, antihistaminic, or α-blocking properties [39]. A presently available calmodulin inhibitor that may offer this type of specificity is the compound calmidazolium (R 24 571), which is a potent calmodulin antagonist that displays little affinity for several neurotransmitter receptors with which other classes of calmodulin-binding drugs interact [48].

There are a number of possible approaches that might be used to develop agents directed at calmodulin or at a specific calmodulin-sensitive enzyme. First, the specificity with which calmodulin inhibitors act depends on the same factors that determine specificity of any drugs, including drug disposition. It may be possible to develop drugs with pharmacokinetic properties that would allow them to interact selectively with calmodulin in a particular tissue or subcellular fraction.

Another approach might be to take advantage of the fact that the degree to which drugs inhibit calmodulin depends on the tissue concentration of calmodulin. Figure 3 shows that at high concentrations of caudate nucleus (i.e., at high concentrations of calmodulin), higher levels of trifluoperazine are required to inhibit calmodulin activity. When the concentration of caudate homogenate was 0.1 mg/ml, the IC_{50} was about 1 μM; at a tissue concentration of 10 mg/ml, the IC_{50} was about 20 μM. Since the concentration of calmodulin varies markedly from tissue to tissue [6,21], calmodulin inhibitors may produce greater effects in those tissues having lower concentrations of calmodulin.

Similarly, different calmodulin-sensitive enzymes may require different concentrations of calmodulin or calmodulin in different degrees of Ca^{2+} occupancy for activation [2,10,20,24,35]. If this is the case, a calmodulin-binding drug might exert a greater effect on one enzyme than another, even though in each instance it was acting on calmodulin. For example, consider the theoretical situation in which one calmodulin-sensitive enzyme (A) is maximally activated at a calmodulin concentration of 10 nM; whereas another enzyme (B) requires 100 nM calmodulin for full activation. If a calmodulin inhibitor were to reduce the effective calmodulin concentration to 10 nM, it would greatly inhibit the activation of B but would have little effect on A. In this manner, a nonselective calmodulin inhibitor might, in fact, produce a selective effect.

Drug specificity might also depend on the existence of more than one form of calmodulin [5,26,27]. Perhaps the structural differences that exist in calmodulin from various sources could be utilized to develop more selective drugs. The recent finding that calmodulin from mutant strains of *Volvox carteri* displays a lower sensitivity to inhibition by fluphenazine than calmodulin from the wild-type strain [26,27] suggests that this may be a fruitful approach.

Finally, the recent finding [38] that chlorpromazine-linked calmodulin (discussed above) inhibits calmodulin-induced phosphodiesterase activation by competing with calmodulin for a regulatory site on the enzyme suggests that a new class of calmodulin inhibitors directed at regulatory sites on calmodulin-sensitive enzymes might be developed. Since various calmodulin-sensitive enzymes may have somewhat different calmodulin-regulatory sites [24], it may be possible to produce a more selective pharmacological effect by developing agents that interact with one calmodulin-sensitive enzyme but not another.

ACKNOWLEDGMENTS. We thank Ms. Becky Simon and Patrice Hart for their excellent technical assistance and Ms. Chris Cirillo for her help with the manuscript. Portions of this research were supported by a Research Starter Grant from the Pharmaceutical Manufacturers Association Foundation to W.C.P. and by Grant MH–30096 from the National Institute of Mental Health and funds from the Department of Public Welfare, Commonwealth of Pennsylvania, to B.W.

References

1. Adunyah, E. S.; Niggli, V.; Carafoli, E. The anticalmodulin drugs trifluoperazine and R 24 571 remove the activation of the purified erythrocyte Ca^{2+}-ATPase by acidic phospholipids and by controlled proteolysis. *FEBS Lett.* **143**:65–68, 1982.
2. Blumenthal, D. K.; Stull, J. T. Activation of skeletal muscle myosin light chain kinase by calcium $(2+)$ and calmodulin. *Biochemistry* **19**:5608–5614, 1980.
3. Brostrom, C. O.; Wolff, D. J. Properties and functions of calmodulin. *Biochem. Pharmacol.* **30**:1395–1405, 1981.
4. Broström, S.-L.; Bengt, L.; Mårdh, S.; Forsen, S.; Thulin, E. Interaction of the antihypertensive drug felodipine with calmodulin. *Nature (London)* **292**:777–778, 1981.
5. Burgess, W. H. Characterization of calmodulin and calmodulin isotypes from sea urchin gametes. *J. Biol. Chem.* **257**:1800–1804, 1982.
6. Chafouleas, J. G.; Dedman, J. R.; Munjaal, R. P.; Means, A. R. Calmodulin: Development and application of a sensitive radioimmunoassay. *J. Biol. Chem.* **254**:10262–10267, 1979.
7. Chao, S. H.; Suzuki, Y.; Zysk, J. R.; Cheung, W. Y. Metal cation-induced activation of calmodulin is a function of ionic radii. *Fed. Proc.* **42**:1087, 1983.
8. Cheung, W. Y. Calmodulin plays a pivotal role in cellular regulation. *Science* **207**:19–27, 1980.
9. Comte, M.; Maulet, Y.; Cox, J. A. Ca^{2+}-dependent high-affinity complex formation between calmodulin and melittin. *Biochem. J.* **209**:269–272, 1983.
10. Cox, J. A.; Comte, M.; Stein, E. A. Activation of human erythrocyte Ca^{2+}-dependent Mg^{2+}-activated ATPase by calmodulin and calcium: Quantitative analysis. *Proc. Natl. Acad. Sci. USA* **79**:4265–4269, 1982.
11. Cox, J. L.; Harrison, S. D. Correlation of metal toxicity with in vitro calmodulin inhibition. *Biochem. Biophys. Res. Commun.* **115**:106–111, 1983.
12. DeLorenzo, R. J.; Burdette, S.; Holderness, J. Benzodiazepine inhibition of the calcium–calmodulin protein kinase system in brain membrane. *Science* **213**:546–549, 1981.

13. Earl, C. Q.; Prozialeck, W. C.; Weiss, B. Interaction of alpha adrenergic antagonists with calmodulin. *Life Sci.* **35**:525–534, 1984.
14. Epstein, P. M.; Fiss, K.; Hachisu, R.; Andrenyak, D. M. Interaction of calcium antagonists with cyclic AMP phosphodiesterases and calmodulin. *Biochem. Biophys. Res. Commun.* **105**:1142–1149, 1982.
15. Forsén, S.; Thulin, E.; Drakenberg, T.; Krebs, J.; Seamon, K. A ^{113}Cd NMR study of calmodulin and its interaction with calcium, magnesium and trifluoperazine. *FEBS Lett.* **117**:189–194, 1980.
16. Gietzen, K.; Delgado, E. S.; Bader, H. Compound 48/80: A powerful and specific inhibition of calmodulin-dependent Ca^{2+}-transport ATPase. *IRCS Med. Sci. Biochem.* **11**:12–13, 1983.
17. Gietzen, K.; Sadorf, I.; Bader, H. A model for the regulation of the calmodulin-dependent enzymes erythrocyte Ca^{2+}-transport ATPase and brain phosphodiesterase by activators and inhibitors. *Biochem. J.* **207**:541–548, 1982.
18. Hidaka, H.; Asano, M.; Tanaka, T. Activity–structure relationship of calmodulin antagonists: Naphthalenesulfonamide derivatives. *Mol. Pharmacol.* **20**:571–578, 1981.
19. Hidaka, H.; Yamaki, T.; Naka, M.; Tanaka, T.; Hayashi, H.; Kobayashi, R. Calcium-regulated modulator protein interacting agents inhibit smooth muscle calcium-stimulated protein kinase and ATPase. *Mol. Pharmacol.* **17**:66–72, 1980.
20. Huang, C. Y.; Chau, V.; Chock, P. B.; Wang, J. H.; Sharma, R. K. Mechanism of activation of cyclic nucleotide phosphodiesterase: Requirement of the binding of four Ca^{2+} to calmodulin for activation. *Proc. Natl. Acad. Sci. USA* **78**:871–874, 1981.
21. Kakiuchi, S.; Yasuda, S.; Yamazaki, R.; Teshima, Y.; Kanda, K.; Kakiuchi, R.; Sobue, K. Quantitative determinations of calmodulin in the supernatant and particulate fractions of mammalian tissues. *J. Biochem. (Tokyo)* **92**:1041–1048, 1982.
22. Katoh, N.; Wise, B. C.; Wrenn, R. W.; Kuo, J. F. Inhibition by adriamycin of calmodulin-sensitive and phospholipid-sensitive calcium-dependent phosphorylation of endogenous proteins from heart. *Biochem. J.* **198**:199–205, 1981.
23. Kilimann, M.; Heilmeyer, L. M. G. The effect of Mg^{2+} on the Ca^{2+}-binding properties of non-activated phosphorylase kinase. *Eur. J. Biochem.* **73**:191–197, 1977.
24. Klee, C. B. Calmodulin: Structure–function relationships. In: *Calcium and Cell Function*, volume 1, W. Y. Cheung, ed., New York, Academic Press, 1980, pp. 59–77.
25. Klee, C. B.; Crouch, T. H.; Richman, P. G. Calmodulin. *Annu. Rev. Biochem.* **49**:489–515, 1980.
26. Kurn, N. Inhibition of phosphate uptake by fluphenazine, a calmodulin inhibitor: Analysis of *Volvox* wild-type and fluphenazine resistant mutant strains. *FEBS Lett.* **144**:68–72, 1982.
27. Kurn, N.; Sela, B.-A. Altered calmodulin activity in fluphenazine-resistant mutant strains: Pleiotropic effect on development and cellular organization in *Volvox carteri*. *Eur. J. Biochem.* **121**:53–57, 1981.
28. Levin, R. M.; Weiss, B. Mechanism by which psychotropic drugs inhibit adenosine cyclic 3',5'-monophosphate phosphodiesterase of brain. *Mol. Pharmacol.* **12**:581–589, 1976.
29. Levin, R. M.; Weiss, B. Binding of trifluoperazine to the calcium-dependent activator of cyclic nucleotide phosphodiesterase. *Mol. Pharmacol.* **13**:690–697, 1977.
30. Levin, R. M.; Weiss, B. Specificity of the binding of trifluoperazine to the calcium-dependent activator of phosphodiesterase and to a series of other calcium-binding proteins. *Biochim. Biophys. Acta* **540**:197–204, 1978.
31. Levin, R. M.; Weiss, B. Selective binding of antipsychotics and other psychoactive agents to the calcium-dependent activator of cyclic nucleotide-phosphodiesterase. *J. Pharmacol. Exp. Ther.* **208**:454–459, 1979.
32. Lin, Y. M.; Liu, Y. P.; Cheung, W. Y. Cyclic 3':5'-nucleotide phosphodiesterase: Purification, characterization and active form of the protein activator from bovine brain. *J. Biol. Chem.* **249**:4943–4954, 1974.
33. Luthra, M. G. Trifluoperazine inhibition of calmodulin-sensitive Ca^{2+}-ATPase and calmodulin-insensitive $(Na^+ + K^+)$- and Mg^{2+}-ATPase activities of human and rat red blood cells. *Biochim. Biophys. Acta* **692**:271–277, 1982.
34. Malencik, D. A.; Anderson, S. R. Binding of hormones and neuropeptides by calmodulin. *Biochemistry* **22**:1995–2001, 1983.
35. Malnoë, A.; Cox, J. A.; Stein, E. A. Ca^{2+}-dependent regulation of calmodulin binding and adenylate cyclase activation in bovine cerebellar membranes. *Biochim. Biophys. Acta* **714**:84–92, 1982.
36. Prozialeck, W. C. Structure–activity relationships of calmodulin antagonists. *Annu. Rep. Med. Chem.* **18**:203–212, 1983.
37. Prozialeck, W. C.; Cimino, M.; Weiss, B. Photoaffinity labeling of calmodulin by phenothiazine antipsychotics. *Mol. Pharmacol.* **19**:264–269, 1981.
38. Prozialeck, W. C.; Wallace, T. L.; Weiss, B. Chlorpromazine-linked calmodulin: A novel calmodulin antagonist. *Fed. Proc.* **42**:1087, 1983.

39. Prozialeck, W. C.; Weiss, B. Inhibition of calmodulin by phenothiazines and related drugs: Structure–activity relationships. *J. Pharmacol. Exp. Ther.* **222**:509–516, 1982.
40. Roufogalis, B. D. Phenothiazine antagonism of calmodulin: A structurally-nonspecific interaction. *Biochem. Biophys. Res. Commun.* **98**:607–613, 1981.
41. Seeman, P. Anti-schizophrenic drugs: Membrane receptor sites of action. *Biochem. Pharmacol.* **26**:1741–1748, 1977.
42. Sellinger-Barnette, M.; Weiss, B. Interaction of β-endorphin and other opioid peptides with calmodulin. *Mol. Pharmacol.* **21**:86–91, 1982.
43. Sellinger-Barnette, M.; Weiss, B. Interaction of various peptides with calmodulin. *Adv. Cyclic Nucleotide Protein Phosphorylation Res.* **16**:261–276, 1984.
44. Sharma, R. K.; Wang, J. H. Inhibition of calmodulin-activated cyclic nucleotide phosphodiesterase by Triton X-100. *Biochem. Biophys. Res. Commun.* **100**:710–715, 1981.
45. Shimizu, T.; Hatano, M.; Nagao, S.; Nozawa, Y. ^{43}Ca NMR studies of Ca^{2+}–*Tetrahymena* calmodulin complexes. *Biochem. Biophys. Res. Commun.* **106**:1112–1118, 1982.
46. Tanaka, T.; Hidaka, H. Interaction of local anesthetics with calmodulin. *Biochem. Biophys. Res. Commun.* **101**:447–453, 1981.
47. Teo, T. S.; Wang, J. H. Mechanism of activation of a cyclic adenosine 3′:5′ monophosphate phosphodiesterase from bovine heart by Ca^{2+} ions. *J. Biol. Chem.* **248**:5950–5955, 1973.
48. Van Belle, H. R 24 571: A potent inhibitor of calmodulin-activated enzymes. *Cell Calcium* **2**:483–494, 1981.
49. Volpi, M.; Sha'afi, R. I.; Epstein, P. M.; Andrenyak, D. M.; Feinstein, M. B. Local anesthetics, mepacrine and propranolol are antagonists of calmodulin. *Proc. Natl. Acad. Sci. USA* **78**:795–799, 1981.
50. Volpi, M.; Sha'afi, R. I.; Feinstein, M. B. Antagonism of calmodulin by local anesthetics: Inhibition of calmodulin-stimulated calcium transport of inside-out membrane vesicles. *Mol. Pharmacol.* **20**:363–370, 1981.
51. Watanabe, K.; West, W. L. Calmodulin, activated cyclic nucleotide phosphodiesterase, microtubules, and vinca alkaloids. *Fed. Proc.* **41**:2292–2299, 1982.
52. Watanabe, K.; Williams, E. F.; Law, J. S.; West, W. L. Effects of vinca alkaloids on calcium–calmodulin regulated cyclic adenosine 3′,5′-monophosphate phosphodiesterase activity from brain. *Biochem. Pharmacol.* **30**:335–340, 1981.
53. Weiss, B. Differential activation and inhibition of the multiple forms of cyclic nucleotide phosphodiesterase. *Adv. Cyclic Nucleotide Res.* **5**:195–211, 1975.
54. Weiss, B.; Fertel, R.; Figlin, R.; Uzunov, P. Selective alteration of the activity of the multiple forms of adenosine 3′,5′-monophosphate phosphodiesterase of rat cerebrum. *Mol. Pharmacol.* **10**:615–625, 1974.
55. Weiss, B.; Hait, W. N. Selective cyclic nucleotide phosphodiesterase inhibitors as potential therapeutic agents. *Annu. Rev. Pharmacol. Toxicol.* **17**:441–477, 1977.
56. Weiss, B.; Prozialeck, W. C.; Cimino, M.; Barnette, M. S.; Wallace, T. L. Pharmacological regulation of calmodulin. *Ann. N.Y. Acad. Sci.* **356**:319–345, 1980.
57. Weiss, B.; Prozialeck, W. C.; Wallace, T. L. Interaction of drugs with calmodulin: Biochemical, pharmacological and clinical implications. *Biochem. Pharmacol.* **31**:2217–2226, 1982.
58. Weiss, B.; Sellinger-Barnette, M. Effects of antipsychotic dopamine antagonists and polypeptide hormones on calmodulin. In: *Apomorphine and Other Dopaminomimetics*, Volume 1, G. L. Gessa and G. U. Corsini, eds., New York, Raven Press, 1981, pp. 179–192.
59. Weiss, B.; Wallace, T. L. Mechanisms and pharmacological implications of altering calmodulin activity. In: *Calcium and Cell Function*, Volume 1, W. Y. Cheung, ed., New York, Academic Press, 1980, pp. 329–379.
60. Zavecz, J. H.; Jackson, T. E.; Limp, G. L.; Yellin, T. O. Relationship between anti-diarrheal activity and binding to calmodulin. *Eur. J. Pharmacol.* **78**:375–377, 1982.

31

The Purified Calcium-Pumping ATPase of Plasma Membrane
Structure–Function Relationships

Ernesto Carafoli, Mauro Zurini, and Gustavo Benaim

The Ca^{2+}-pumping ATPase of plasma membrane, known to exist since 1966 [13], has now been characterized as a high Ca^{2+} affinity enzyme, present in all likelihood in all eukaryotic plasma membranes (for a recent review, see Ref. 11). The essential properties of this enzyme, as they can be extracted from a very large number of studies, are summarized in Table I. The ATPase is perhaps less well characterized than other members of the E_1, E_2 class of transport ATPases like the Na^+/K^+-ATPase and the Ca^{2+}-ATPase of sarcoplasmic reticulum. For example, it is still not known whether the acyl phosphate formed during the enzyme cycle is an aspartyl phosphate. It is also not established to the satisfaction of all specialists whether the Ca^{2+}/ATP stoichiometry is, as indicated in Table I, always and invariably 1. Recent reviews on the ATPase have appeared, the most comprehensive being perhaps the one by Schatzmann [14].

One interesting property of the plasma membrane Ca^{2+}-ATPase is its sensitivity to calmodulin which was discovered in 1977 [3,4]. The stimulation is due to the direct interaction of calmodulin with the enzyme [7,10] and corresponds to a shift of the affinity for Ca^{2+} toward lower K_m values. However, the liver plasma membrane Ca^{2+}-pumping ATPase is insensitive to calmodulin [5,6]. It now appears probable that this ATPase belongs to a distinct class of plasma membrane Ca^{2+}-ATPases.

The interaction of the ATPase with calmodulin was utilized by Niggli *et al.* [9] to devise a successful method to isolate and purify the enzyme from membranes. Previous attempts [see e.g. Ref. 12] based on conventional protein purification procedures had met with formidable difficulties, the most serious being the very minute amounts of enzyme present in plasma membranes (about 0.02% of the total membrane protein). Niggli *et al.* [9] used a calmodulin affinity chromatography column, on which a Triton X-100 solubilizate of EDTA-

---

Ernesto Carafoli, Mauro Zurini, and Gustavo Benaim • Laboratory of Biochemistry, Swiss Federal Institute of Technology (ETH), 8092 Zurich, Switzerland.   Permanent address for G. B.: Universidad Central de Venezuela, Facultad de Ciencias, Escuela de Biologia, Apartado 21201, Departamento de Biologia Celular, Caracas, Venezuela.

Table I. Properties of the Plasma Membrane
Ca^{2+}-Pumping ATPase

Type of enzyme	E_1, E_2, forms acyl phosphate
Affinity for Ca^{2+} (K_m)	<1 µM
V_{max} of Ca^{2+} transport	0.5 nmole/mg membrane protein/sec
Ca^{2+}/ATP stoichiometry	1:1?
Inhibition	Vanadate, $K_{1/2}$ < 1 µM

washed erythrocyte membranes was applied. After extensive washing of the column with Ca^{2+} removed most of the unbound protein, elution with EDTA produced a sharp protein peak that contained a very high Ca^{2+}-ATPase activity. The EDTA-eluted peak possessed essentially a single protein of M_r 138,000, which formed acyl phosphate upon incubation with [γ-^{32}P]-ATP and Ca^{2+}, and was completely inhibited by low concentrations of vanadate.

The purified ATPase has now been characterized in the laboratories of Carafoli and Penniston (for a comprehensive review see Ref. 2). Such studies reproduce the properties of the enzyme *in situ* (Table II) and have helped settle some of the issues which *in situ* studies had left undecided. These issues include the ATP/Ca^{2+} stoichiometry, which now appears to be almost certainly 1, and the charge balance during the transport cycle: the enzyme functions as an electroneutral Ca^{2+}–$2H^+$ exchanger.

One of the particularly important observations made on the purified enzyme is that calmodulin as an activator can be replaced by a limited proteolytic treatment of the enzyme [8]. As is the case with activation by calmodulin, that produced by proteolysis is also essentially a K_m effect, i.e., the activated enzyme has a much higher affinity for Ca^{2+} (K_m about 0.5 µM) than the enzyme in the absence of activating treatments (K_m 10–20 µM). Controlled proteolysis, coupled to activity measurements and to a variety of affinity-label procedures, has recently been used to map functional domains in the isolated enzyme molecule. In the following pages, a brief account of this work will be presented.

Controlled Proteolysis of the Purified Plasma Membrane Ca^{2+}-ATPase by Trypsin

Trypsin splits the purified plasma membrane ATPase according to a pattern which is considerably more complex than that of the Ca^{2+}-ATPase of sarcoplasmic reticulum. However, under controlled experimental conditions the pattern is reproducible and presents a

Table II. The Purified Ca^{2+}-ATPase
of the Erythrocyte Membrane

M_r	138,000
Affinity for Ca^{2+} (K_m)	0.4–0.5 µM
V_{max} of Ca^{2+} transport	180–1200 nmoles/mg protein/sec
Ca^{2+}/ATP stoichiometry	1
Inhibition	Vanadate, $K_{1/2}$ < 1 µM
Activation	Calmodulin, acidic phospholipids, fatty acids, limited proteolysis

31. Ca^{2+}-Pumping ATPase of Plasma Membrane

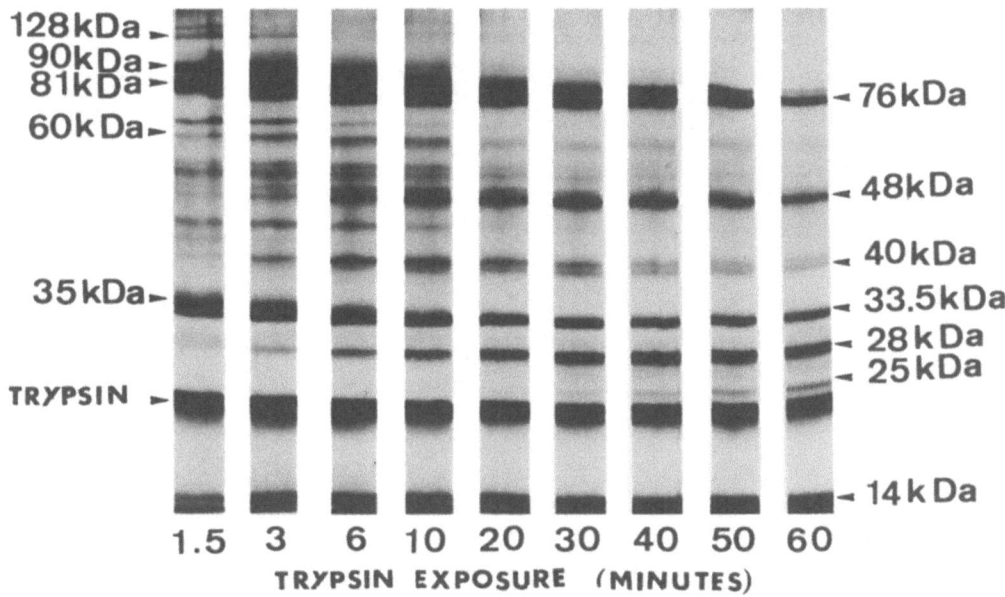

Figure 1. Controlled proteolysis and activation of the purified Ca^{2+}-ATPase. Aliquots (50 μl) of Triton X-100-solubilized, purified Ca^{2+}-ATPase (corresponding to about 7 μg of protein) were exposed at 0°C to 1 μg of trypsin for the times indicated. The reaction was stopped by the addition of threefold concentrated electrophoresis buffer followed by 5 min boiling, and 40 μl of each sample was submitted to 7% SDS-polyacrylamide slab gel electrophoresis. The gels were stained with a silver impregnation method. Standards: myosin (200K); β-galactosidase (116.25K); phosphorylase B (92.5K); bovine serum albumin (66.2K); ovalbumin (45K); carbonic anhydrase (31K).

number of well-recognizable basic features (Fig. 1). Some of the fragments produced are limit polypeptides, like those of M_r 14,000, 28,000, 33,500, 48,000. Others, such as those of M_r 124,000 and 90,000, are transient. A group of fragments in the 85,000–75,000 region gradually condenses into a doublet of M_r 81,000–76,000. But it seems evident that the latter products, although certainly less transient than, for example, the 90,000-M_r polypeptide, tend to be degraded further after prolonged proteolysis. The pattern shown in Fig. 1 can be

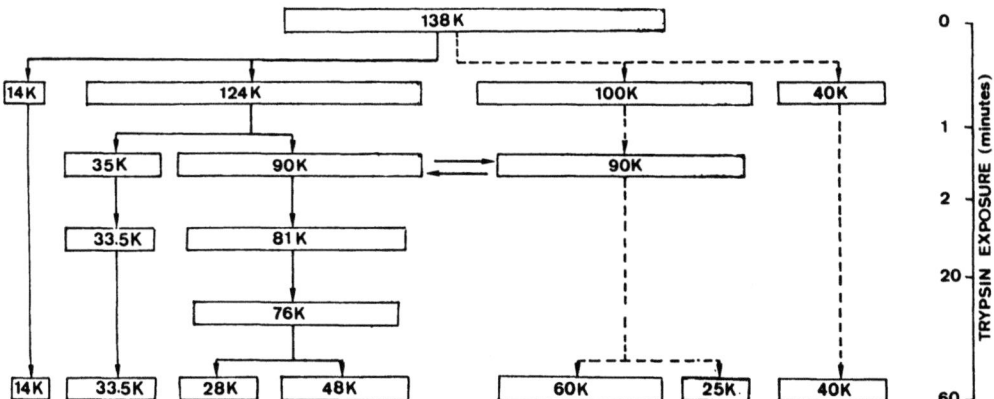

Figure 2. Model for trypsin proteolysis. It is proposed that attack by trypsin at a few points, with some molecules being resistant at some points, will give rise to the observed patterns. The cleavage points shown and the pattern indicated by the solid lines will produce the main features seen by protein staining. The dashed lines indicate the minor proteolysis pathway (see text).

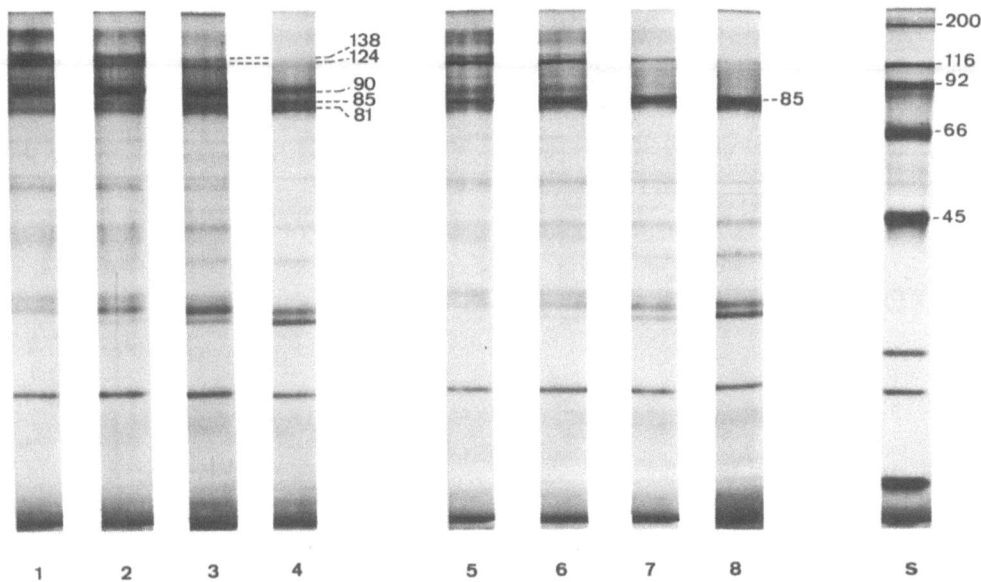

Figure 3. Effect of calmodulin (plus Ca^{2+}) on the pattern of trypsin proteolysis of the purified erythrocyte Ca^{2+}-ATPase. The basic conditions of the experiments were those of the experiment shown in Fig. 1, except that the temperature of the proteolysis medium was 37°C, and the trypsin/ATPase ratio was much lower (0.25 μg trypsin, about 7 μg ATPase). Ca^{2+} was 100 μM, calmodulin 1 μg. Concentration of polyacrylamide in the gels, 10%.

provisionally interpreted according to a scheme in which a major proteolytic pathway produces in sequence most of the (major) fragments seen (Fig. 2). A second pathway, probably reflecting a minor conformer of the ATPase, is postulated to produce a number of other minor products.

The fractionation scheme in Fig. 1 reflects the splitting of the enzyme at 0°C, in a

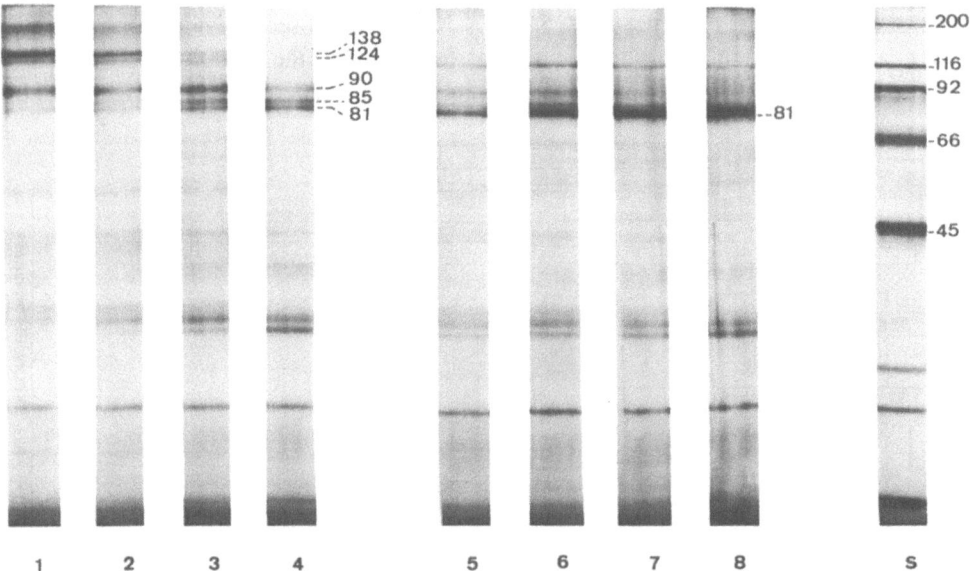

Figure 4. Effect of vanadate on the pattern of trypsin proteolysis of the purified erythrocyte ATPase. Conditions as in the experiment of Fig. 3, except that calmodulin and Ca^{2+} were replaced by 10 μM vanadate and 10 mM Mg^{2+}. Concentration of polyacrylamide in the gels, 10%.

medium containing 130 mM KCl, 20 mM Hepes, pH 7.2, 1 mM $MgCl_2$, 100 mM $CaCl_2$, 2 mM dithiothreitol, 0.05% Triton X-100, 0.05% phosphatidylcholine, 1 µg trypsin, and 7 µg purified ATPase. Presumably, under the conditions chosen, the ATPase was not "activated" and neither of its two configurations (E_1 and E_2) was privileged. On the other hand, very evident differences in the fragmentation scheme were visible when the enzyme was exposed to trypsin in the activated (presumably E_1), as compared to the E_2 configuration [1]. The former conformation was obtained by adding Ca^{2+} and calmodulin under conditions of limiting trypsin at 37°C, and produced a pattern that proceeded until a major 85,000-M_r polypeptide was formed and accumulated (Fig. 3). In the E_2 configuration, which was obtained by the addition of vanadate and Mg^{2+}, the proteolysis proceeded instead to a fragment of M_r 81,000, which under the experimental conditions was relatively protected from further digestion (Fig. 4). This suggests conformational changes were induced by Ca^{2+}-calmodulin, or by vanadate, which make different portions of the ATPase accessible to the attack by trypsin.

Functional Domains in the Plasma Membrane Ca^{2+}-ATPase

Among the domains of functional interest in the ATPase molecule are the calmodulin-binding region and the region of ATP-binding and/or acyl phosphate formation. Experiments on the trypsinized enzyme have now identified the calmodulin receptor in the (transient) 90,000-M_r polypeptide [15]. The conclusion rests on affinity labeling studies with azido-modified, radioactive calmodulin (not shown), and especially on the isolation of the pure 90,000-M_r fragment from calmodulin affinity columns. The 90,000-M_r fragment had calmodulin-stimulated, Ca^{2+}-dependent ATPase activity, and yielded the 81,000-M_r polypeptide upon further proteolysis (Fig. 5). It is of great interest that the 90,000-M_r calmodulin receptor could be reconstituted into liposomes, and shown to pump Ca^{2+} inwardly by a calmodulin-sensitive process (15; Fig. 6). This raises the question as to the role of the ~ 50,000-M_r portion of the ATPase molecule, which is lost as proteolysis progresses to the 90,000-M_r polypeptide. Possibly, this region, which evidently is not involved in the reaction mechanism proper, could be responsible for some regulatory function.

The second domain of interest in the ATPase molecule is the site where ATP is bound and where the acyl phosphate bond is formed. Work is now in progress on the identification of the fragment which bears the acyl phosphate function [15]. So far, it can be concluded from experiments with [γ-^{32}P]-ATP that acyl phosphate is formed on all fragments of M_r higher than 76,000–81,000. Association of radioactivity from [γ-^{32}P]-ATP with lower M_r products has not yet been demonstrated, but this could be due to difficulties relative to the experimental system. Indeed, separation of the lower M_r fragments under polyacrylamide gel conditions that would not be harmful to the acyl phosphate bond has proven very difficult.

As for the ATP-binding site, it need not be the same as the acyl phosphate site, although the two sites, even if separated by a (long) sequence of amino acids, may be very close in space due to the folding of the polypeptide chains. The Ca^{2+}-ATPase of plasma membranes, unlike other membrane ATPases, apparently bears the acyl phosphate and the ATP-binding function in the same fragments produced by the proteolysis (Fig. 7). For this experiment, radioactive dialdehyde ATP, produced by oxidation of [^{14}C]-ATP by periodic acid, was used. The autoradiograph shows radioactivity in the intact enzyme and in the same fragments of M_r down to 76,000–81,000 which also became labeled by ^{32}P radioactivity from [γ-^{32}P]-ATP. In addition, the experiment shows that one of the two fragments which the scheme in Fig. 1 depicts as deriving from the splitting of the 81,000–76,000-M_r product, that having M_r of 48,000, also becomes labeled. By contrast, no radioactivity is visible in the region

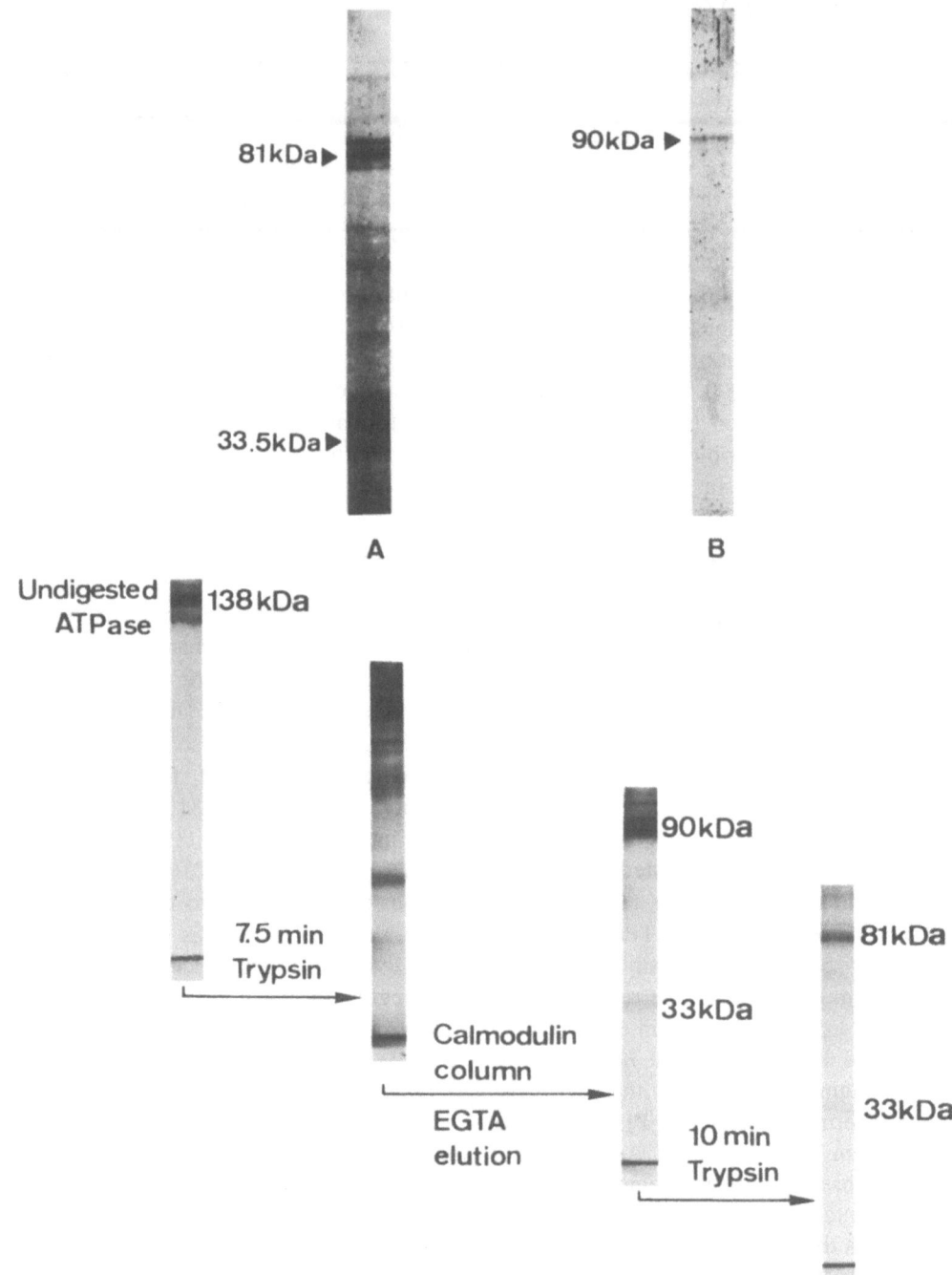

Figure 5. Calmodulin affinity chromatography of the trypsinized purified Ca^{2+}-ATPase, and proteolytic degradation of the purified 90,000-M_r fragment.
(Top panel) Trypsinized purified Ca^{2+}-ATPase was passed through a calmodulin affinity column in the presence of 50 μM $CaCl_2$. After several washes with a Ca^{2+}-containing buffer, the column was eluted with 2 mM EDTA. Details are given in Ref. 15. Fractions of the Ca^{2+}-wash and of the EDTA-eluate were submitted to 7% SDS-polyacrylamide slab gel electrophoresis. The gels were stained with a silver impregnation method. (A) Ca^{2+}-wash; (B) EDTA-eluate.
(Bottom panel) After trypsinization and calmodulin affinity chromatography, 50 μl of column eluate was digested for 10 min at 0°C with 0.5 μg trypsin. A small amount of the 33,500-M_r component was found after the second trypsinization but this was not due to proteolysis at this step, since the same amount appeared as a contaminant of the 90,000-M_r fragment. [From Ref. 15.]

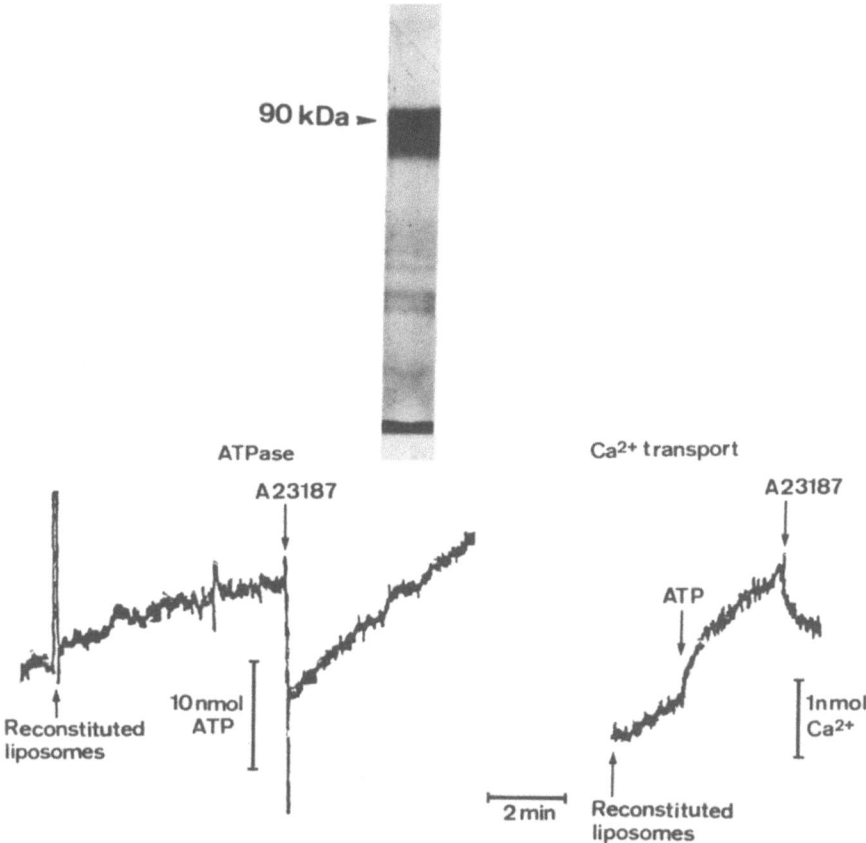

Figure 6. Ca^{2+}-ATPase and Ca^{2+} uptake by the 90,000-M_r fragment reconstituted into liposomes. The 90,000-M_r fragment shown in the gel was reconstituted and its ATPase and Ca^{2+} uptake measured. The ATPase reaction was measured by a coupled enzyme assay [15]. The reaction was started by adding 50 μl of reconstituted vesicles containing 0.2 μg of ATPase protein. The Ca^{2+} uptake was monitored using a Ca^{2+}-selective electrode; the total reaction volume was 1 ml and contained 130 mM KCl, 20 mM Tris-Cl, pH 6.6, 0.5 mM $MgCl_2$, 10 μM $CaCl_2$ 1 mM dithiothreitol, and 50 μl of reconstituted vesicles. The reaction was started by adding ATP to a concentration of 50 μM. Ionophore A23187 concentration was 2 μM. [From Ref. 15.]

corresponding to the polypeptide of M_r 28,000. The label in the 48,000-M_r product increases with trypsinization time, whereas the label in the 81,000–76,000-M_r doublet decreases (Fig. 7). This supports the proposal (see the scheme of Fig. 1) that the limited trypsin products of M_r 28,000–48,000 arise from the degradation of the 81,000–76,000-M_r polypeptides (the derivation of the 81,000-M_r product from the 90,000-M_r calmodulin-binding region has been directly demonstrated by the experiment of Fig. 5).

A summary of the results discussed above on the definition of some of the functional domains of the ATPase is presented in Fig. 8; only the relevant trypsin proteolysis fragments are shown. Work now in progress includes the splitting of the ATPase by proteolytic agents different from trypsin, the identification of the acyl phosphate-forming site among the low-M_r products of proteolysis, and the identification of the domain(s) of the ATPase which interacts with effectors of activity different from calmodulin.

ACKNOWLEDGMENT. The work described here was made possible by the financial contribution of the Swiss Nationalfonds (Grant 3.189–0.82).

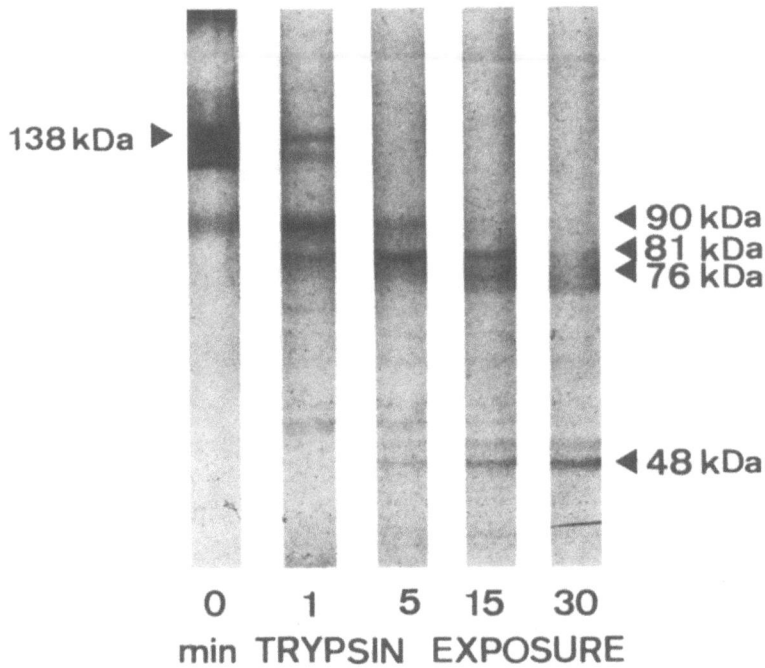

Figure 7. Localization of ATP-binding sites by labeling with ATP-dialdehyde. Autoradiogram showing binding of ATP dialdehyde to ATPase and to selected fragments. 200 μl of 100 μM [U-^{14}C]-ATP (specific activity 0.55 Ci/nmole) was oxidized for 1 hr on ice and in the dark with 10 μl of a 2.2 mM NaIO$_4$ solution. At the end of the incubation, excess NaIO$_4$ was removed by addition of 1 μl 100% ethylene glycol. Samples of purified ATPase were incubated for 1 hr at 37°C with 6–7 M o-ATP, transferred to ice, and submitted to trypsin treatment for times varying from 0 to 30 min. Electrophoresis was carried out on 7% gels. The gels were dehydrated in 100% Me$_2$SO, and immersed for 3 hr in 20% 2,5-diphenyloxazole (PPO) in Me$_2$SO (w/w), washed 1 hr with several changes of water, dried, and autoradiographed prior to staining by a silver impregnation procedure [From Ref. 15.]

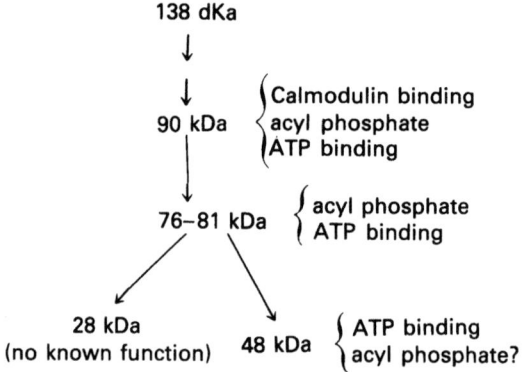

Figure 8. Functional domains in the Ca^{2+}-ATPase as revealed by controlled trypsin proteolysis.

References

1. Benaim, G.; Zurini, M.; Carafoli, E. *J. Biol. Chem.* in press, 1984.
2. Carafoli, E.; Zurini, M. The Ca^{2+}-pumping ATPase of plasma membranes: Purification, reconstitution and properties. *Biochim. Biophys. Acta* **683**:279–301, 1982.
3. Gopinath, R. M.; Vincenzi, F. M. Phosphodiesterase protein activator mimics red blood cell cytoplasmic activator of (Ca^{2+}-Mg^{2+}) ATPase. *Biochem. Biophys. Res. Commun.* **77**:1203–1209, 1977.
4. Jarrett, H. W.; Penniston, J. T. Partial purification of the Ca^{2+}-Mg^{2+} ATPase activator from human erythrocytes: Its similarity to the activator of 3':5'-cyclic nucleotide phosphodiesterase. *Biochem. Biophys. Res. Commun.* **77**:1210–1216, 1977.
5. Kraus-Friedmann, M.; Biber, J.; Murer, H.; Carafoli, E. Calcium uptake in isolated hepatic plasma membrane vesicles. *Eur. J. Biochem.* **129**:7–12, 1983.
6. Loterzstein, S.; Hanoune, J.; Pecker, F. A high affinity calcium stimulated magnesium-dependent ATPase in rat liver plasma membranes. *J. Biol. Chem.* **256**:11209–11215, 1981.
7. Lynch, T. J.; Cheung, W. Y. Human erythrocyte Ca^{2+}-Mg^{2+}-ATPase: Mechanism of stimulation by Ca^{2+}. *Arch. Biochem. Biophys.* **194**:165–170, 1979.
8. Niggli, V.; Adunyah, E. S.; Carafoli, E. Acidic phospholipids, unsaturated fatty acids, and limited proteolysis mimic the effect of calmodulin on the purified erythrocyte Ca^{2+}-ATPase. *J. Biol. Chem.* **256**:8588–8592, 1981.
9. Niggli, V.; Penniston, J. T.; Carafoli, E. Purification of the (Ca^{2+}-Mg^{2+})-ATPase from human erythrocyte membranes using a calmodulin affinity column. *J. Biol. Chem.* **254**:9955–9958, 1979.
10. Niggli, V.; Ronner, P.; Carafoli, E.; Penniston, J. T. Effects of calmodulin on the (Ca^{2+}-Mg^{2+})-ATPase partially purified from erythrocyte membranes. *Arch. Biochem. Biophys.* **198**:124–130, 1977.
11. Penniston, J. T. Plasma membrane Ca^{2+}-ATPase as active Ca^{2+} pumps. In: *Calcium and Cell Function*, Volume 4, W. J. Cheung, ed., New York, Academic Press, in press.
12. Peterson, S.; Ronner, P.; Carafoli, E. Partial purification and reconstitution of the (Ca^{2+}-Mg^{2+})-ATPase of erythrocyte membranes. *Arch. Biochem. Biophys.* **186**:202–210, 1978.
13. Schatzmann, H. ATP-dependent Ca^{2+}-extrusion from human red cells. *Experientia* **22**:364–368, 1966.
14. Schatzmann, H. The plasma membrane calcium pump of erythrocytes and other animal cells. In: *Membrane Transport of Calcium*, E. Carafoli, ed., New York, Academic Press, 1982, pp. 41–108.
15. Zurini, M.; Krebs, J.; Penniston, J. T.; Carafoli, E. Controlled proteolysis of the purified Ca^{2+}-ATPase of the erythrocyte membrane. *J. Biol. Chem.* **259**:618–627, 1984.

32

Calmodulin-Binding Proteins That Control the Cytoskeleton by a Flip-Flop Mechanism

Shiro Kakiuchi

Calmodulin is thought to be a Ca^{2+}-dependent regulator of the contractile apparatus and cytoskeleton of smooth muscle and nonmuscle tissues. Recent results from our laboratory have led to the view that regulatory actions of Ca^{2+} and calmodulin in this connection are mediated by a number of specific calmodulin-binding proteins that are able to bind to F-actin filaments or tubulin [1]. For instance, caldesmon [2] from smooth muscle and tau factor [3] from brain microtubules are calmodulin-binding proteins able to bind to F-actin or tubulin, respectively. The Ca^{2+}-dependent binding of calmodulin to these proteins eliminates the interaction between these proteins and F-actin or tubulin. Therefore, the binding of these proteins to calmodulin or cytoskeletal protein alternates depending on the concentration of Ca^{2+}: at an elevated Ca^{2+} concentration, they bind to calmodulin and, at a reduced concentration ($< 10^{-6}$ M), they bind to F-actin or tubulin (flip-flop binding; depicted schematically in Fig. 1). The binding of these proteins to target cytoskeletal proteins alters the function of the latter proteins, and this flip-flop mechanism appears to be a general principle through which the Ca^{2+}-dependent regulatory effect of calmodulin is transmitted to the cytoskeleton. In this chapter the calmodulin-binding proteins related to this flip-flop mechanism are described.

Caldesmon

Caldesmon is a major Ca^{2+}-dependent calmodulin-binding protein in smooth muscle and platelets. It was originally purified from chicken gizzard smooth muscle [2], but its immunologically cross-reactive forms were also demonstrated in bovine aorta and uterus and human platelets [4]. They all show flip-flop regulated binding to F-actin (Fig. 1a). Upon SDS-polyacrylamide gel electrophoresis, gizzard and platelet caldesmon resolved into two closely related bands of M_r 150,000 and 147,000, but caldesmon in aorta and uterus gave a

---

Shiro Kakiuchi • Department of Neurochemistry and Neuropharmacology, Institute of Higher Nervous Activity, Osaka University Medical School, Nakanoshima, Kita-ku, Osaka 530, Japan.

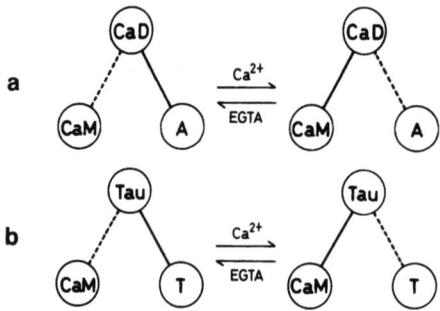

Figure 1. Calmodulin-binding protein-linked flip-flop regulation of the cytoskeleton. CaM, calmodulin; CaD, caldesmon; A, F-actin filaments; Tau, tau factor; T, tubulin.

single band of M_r 150,000. The concentration of caldesmon in chicken gizzard was estimated to be 240 mg/100 g tissue, which is about 8 μM taking 300,000 for its molecular weight. This concentration of caldesmon binds 16 μM calmodulin, which is about 70% of the total amount of calmodulin present in this tissue.

The binding of caldesmon to F-actin did not alter the physical state of the actin filaments, i.e., it does not produce gel formation or shortening of the filaments, nor does it influence reannealing of fragmented actin filaments [5]. Instead, it alters the interaction of F-actin filaments with other proteins such as filamin or myosin. Thus, the filamin-induced gelation of actin filaments was inhibited by the binding of caldesmon to actin filaments, and this inhibition was abolished by flip-flop regulated binding of calmodulin to caldesmon [5] (Fig. 2). The inhibition of caldesmon on the F-actin–filamin interaction is not due to a simple competition between caldesmon and filamin for the same binding site on F-action filaments since the amount of filamin associated with F-actin remained unchanged by addition of increasing concentrations of caldesmon. The molar ratio of filamin(dimer) to actin(monomer) of a gel complex was 1:240, and the molar ratio of caldesmon(dimer) to actin(monomer) that completely inhibited the filamin-induced gelation was 1:60 ~ 70.

Actin–myosin interaction of smooth muscle, as determined by superprecipitation or actomyosin ATPase activity, is also controlled by the caldesmon-linked flip-flop mechanism [6] (Fig. 3). Phosphorylation of myosin light chains and the presence of tropomyosin are prerequisites for actin–myosin interactions. The mechanism of smooth muscle contraction is still controversial. A generally accepted view is that phosphorylation of two of the 20,000-M_r light chains of myosin catalyzed by a specific Ca^{2+}- and calmodulin-dependent myosin light-chain kinase is responsible for activating myosin–actin interaction (for reviews see Refs. 7–9). However, the presence of a thin filament-linked control has also been claimed [10]. Our results suggest that caldesmon-linked flip-flop regulation of F-actin filaments may be implicated in the regulation of smooth muscle contraction.

135,000-M_r Protein

Another calmodulin-binding protein having an estimated M_r of 135,000 was purified from chicken gizzard smooth muscle [11]. Like caldesmon, this protein binds to F-actin

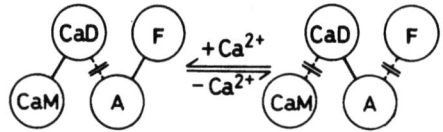

Figure 2. Schematic representation of regulation of filamin-induced gelation of actin filaments by caldesmon-linked flip-flop mechanism. CaM, calmodulin; CaD, caldesmon; A, F-actin filaments; F, filamin.

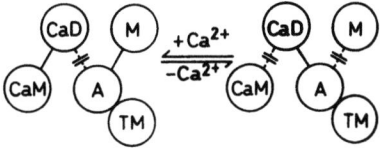

Figure 3. Schematic representation of regulation of myosin–F-actin interaction by caldesmon-linked flip-flop mechanism. CaM, calmodulin; CaD, caldesmon; A, F-actin filaments; M, myosin; TM, tropomyosin.

filaments in a flip-flop manner. This protein has myosin light-chain kinase activity, and its substrate specificity, specific activity value, and molecular weight estimated from SDS-PAGE agree with that reported for turkey gizzard myosin light-chain kinase [12]. Future work is needed to explore why this enzyme interacts with actin filaments in a flip-flop fashion.

Spectrin

The Ca^{2+}-pump ATPase of the erythrocyte membrane is known to require calmodulin for its activity [13,14]. While the concentration of this enzyme in erythrocytes is only 4500 molecules/cell [15], the concentration of calmodulin is approximately 160,000 molecules/cell (2.5 μM) [16]. This discrepancy was resolved when we found that spectrin in erythrocytes is a Ca^{2+}-dependent calmodulin-binding protein [16], whose concentration is about 200,000 molecules/cell [17]. Spectrin, composed of α (M_r 240,000) and β (M_r 220,000) subunits, is the major constituent of the erythrocyte cytoskeleton. Together with actin oligomers, spectrin tetramers ($\alpha_2\beta_2$) line the cytoplasmic surface of the erythrocyte membrane; this structure is thought to be responsible for controlling cell shape, elasticity of the cell, and lateral mobility of membrane proteins [17]. Therefore, an interesting possibility is the involvement of the Ca^{2+}-dependent spectrin–calmodulin interaction in the above roles of spectrin. Indeed, exposure of human erythrocytes to calmodulin-binding agents such as trifluoperazine or W-7 results in the deformation of cells to spherocytes [18].

Calspectin (Spectrin-Related Protein; Fodrin)

Following the discovery that erythrocyte spectrin is a Ca^{2+}-dependent calmodulin-binding protein (see above), a calmodulin-binding protein having an estimated M_r of 240,000 was purified from a membrane fraction of brain homogenates [19]. This protein was identical to the 240,000-M_r subunit (α subunit) of erythrocyte spectrin (Fig. 4a). We then isolated and purified a heterodimeric protein composed of 240,000 M_r (α subunit) and 235,000 M_r (β subunit) (Fig. 4b) from the membrane fraction of brain and the 240,000-M_r polypeptide

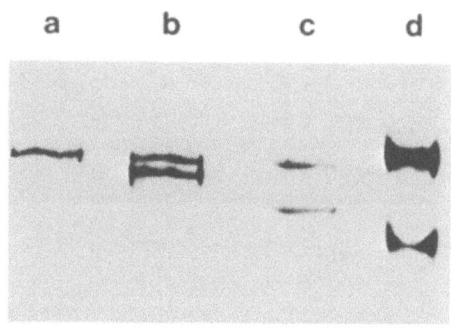

Figure 4. SDS-polyacrylamide gel electrophoresis of high-molecular-weight calmodulin-binding proteins. (a) 240,000-M_r calmodulin-binding protein purified as reported earlier [19]; (b) calspectin purified from bovine brain [20]; (c) spectrin of human erythrocytes; (d) chicken gizzard smooth muscle filamin (upper band) and myosin heavy chain (lower band) as controls. The concentration of polyacrylamide was 5%.

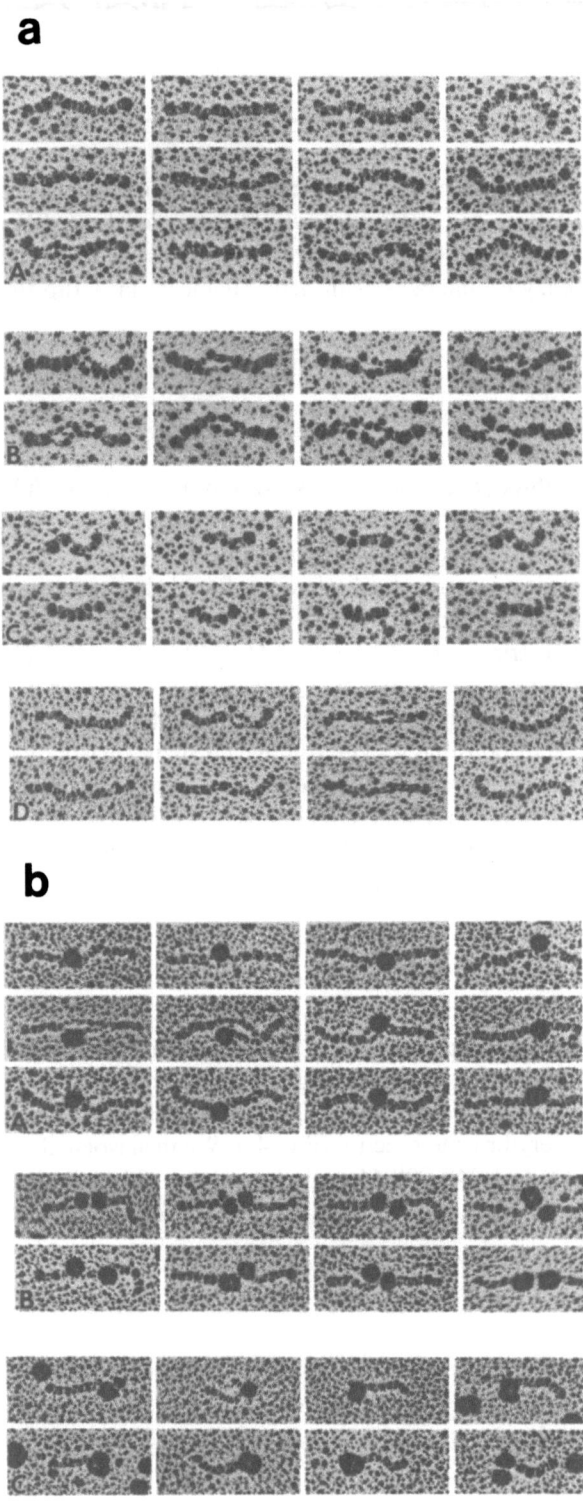

Figure 5. Rotary-shadowing electron micrography of calspectin molecules [31]. (a) Morphology of calspectin molecules. A and B, calspectin tetramers; C, calspectin dimers; D, calspectin tetramers reconstituted from dimers. (b) Calspectin molecules in which calmodulin-binding sites are labeled with ferritin particles. Biotin-conjugated calmodulin molecules that had been bound to calspectin were decorated by avidin–ferritin complexes. A, individual calspectin tetramer labeled with a single ferritin particle; B, individual calspectin tetramer labeled with two ferritin particles; C, individual calspectin dimer labeled with a ferritin particle. Bar, 0.1 μM (× 168,000).

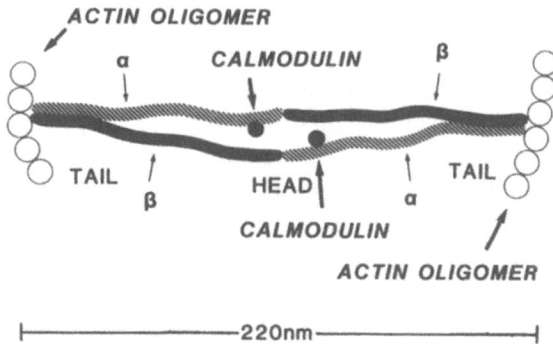

Figure 6. Schematic representation of calspectin molecule (tetramer) and its binding sites for calmodulin and actin oligomers.

obtained earlier was identified as the α subunit of a dimer [20]. We designated this spectrin analog (see below) *calspectin* (from calmodulin-binding spectrinlike protein). Quite independently, the same protein has been identified by several other laboratories [21–23]. Levine and Willard [22] termed the protein *fodrin*, as it was identified as a specific component of cortical cytoplasm of many cells, including nerve cells. Later, Weber and associates purified the same protein from brain [24] and a similar but distinct protein from brush borders of chicken intestinal epithelial cells [25].

Earlier attempts to detect spectrin in nonerythroid cells by immunological means were unsuccessful, and spectrin was thought to be confined to erythrocytes (reviewed in Ref. 17). However, this view had to be modified by the discovery of calspectin, which was concluded to be a spectrin analog because of the following properties: (1) dimer–tetramer interconversion depending on conditions (pH, ionic strength, temperature) [20]; (2) similar molecular forms [flexible rods of about 200 nm (tetramer) in length] on rotary-shadowing electron microscopy of those of spectrin [20,24,25], presumably formed by head-to-head association of pairs of heterodimers (Figs. 5, 6); (3) potential to bind to F-actin filaments at their tail ends (Fig. 6) [20,23–25]; (4) capability of inducing the polymerization of G-actin to F-actin by increasing nucleation [26]; (5) Ca^{2+}-dependent binding to calmodulin [19,20]; (6) capability of binding to erythrocyte ankyrin [27].

The concentration of calspectin in brain was about 1 mg/g tissue, of which about 70% was associated with the particulate fraction [28]. This value is comparable to that of spectrin in erythrocyte (1.4 mg/g). Upon subcellular fractionation, its highest concentration was found in a fraction rich in synaptic membranes [28]. The same fraction contained a protein kinase activity that phosphorylated the β subunit of calspectin [29]. Solubilization from the membranes and partial purification of this activity yielded two activity peaks upon gel filtration column chromatography (800,000 and 88,000 M_r) [30]. The activity of the 800,000-M_r peak was dependent on the presence of both Ca^{2+} and calmodulin [30]. These results suggest that calspectin and its phosphorylation may by implicated in synaptic functions, including neurotransmitter release.

Recently, calmodulin molecules bound to calspectin were directly visualized by labeling calmodulin with ferritin [31]. These results revealed that calmodulin was attached to the head parts of calspectin dimers at a position 10–20 nm from the top of the head (Fig. 5b). The number of calmodulin-binding sites is one for each dimer and two for each tetramer.

This information, together with the fact that the isolated α subunit of calspectin is a calmodulin-binding polypeptide [19], establishes unequivocally that only the α subunit, but not the β subunit, is responsible for the calmodulin binding of calspectin. Calspectin is also an F-actin-binding protein (Fig. 6). In previous studies using purified calspectin, binding of calspectin to F-actin was not influenced by either calmodulin or Ca^{2+} or both [20]. However, in the presence of a factor contained in the supernatant of brain homogenates, calspectin alternately bound to calmodulin or F-actin, depending on the presence or absence, respectively, of Ca^{2+} (flip-flop binding) [32]. This factor (DE-200 factor), which was partially purified from a brain homogenate by DEAE-cellulose column chromatography, is nondialyzable, heat-labile, and devoid of calspectin kinase activity. The nature of this factor is yet to be elucidated.

340,000-M_r Protein

Recently, a 340,000-M_r calmodulin-binding protein was found in supernatant fractions from brain, adrenal gland, and pituitary gland [32]. This protein is also an F-actin-binding protein, and its binding to F-actin is regulated by Ca^{2+}-dependent binding of calmodulin in a flip-flop manner. The 340,000-M_r protein is distinct from microtubule-associated proteins (MAPs) 1 or 2 on SDS-PAGE [32]. However, the possibility was raised that the MAP2 reported in the literature as a 300,000-M_r polypeptide doublet may be derived from the 340,000-M_r protein because, upon aging for 3 days on ice, the 340,000-M_r protein is converted to a doublet that migrates with MAP2 on SDS gel.

Tau Factor

Cytoplasmic microtubules, purified by cycles of assembly and disassembly *in vitro*, are composed of tubulin and several MAPs. The presence of these MAPs may be necessary for the regulation of assembly–disassembly of tubulin because when tubulin is freed of MAPs, it no longer assembles into microtubules under standard polymerization conditions. Addition of MAPs back to the system fully restores the capacity of tubulin to polymerize. We recently found that tau factor, one species of MAPs, is a Ca^{2+}-dependent calmodulin-binding protein [3]. Subsequently, we were able to reconstitute a Ca^{2+}-sensitive tubulin polymerization system using purified tubulin, tau factor, and calmodulin [33]. In this system, in the absence of Ca^{2+}, tau factor–tubulin interaction took place, thereby leading to assembly of tubulin. In the presence of Ca^{2+}, tau factor bound to calmodulin, thereby leading to impaired tau factor–tubulin interactions with the consequent inhibition of microtubule assembly (a flip-flop regulation; Fig. 1b).

Conclusions

The calmodulin-binding proteins described here are key proteins through which regulatory actions of calmodulin are transmitted to the contractile apparatus and cytoskeleton of smooth muscle and nonmuscle tissues. These proteins include: caldesmon and 135,000-M_r protein from smooth muscles; spectrin from erythrocytes; calspectin from brain; 340,000-M_r protein from adrenal gland, pituitary gland, and brain; and tau factor from brain microtubules. These proteins alternately bind to calmodulin or cytoskeletal proteins (actin filaments or tubulin) in the presence or absence, respectively, of Ca^{2+} (flip-flop binding). This binding to target cytoskeletal proteins modulates their function; and Ca^{2+}-dependent binding to calmodulin eliminates the interaction between the calmodulin-binding proteins and target cytoskeletal proteins. We have proposed the name *cytocalbin* for these calmodulin-binding proteins (from cytoskeleton-related calmodulin-binding proteins) [32].

References

1. Kikiuchi, S.; Sobue, K. Control of the cytoskeleton by calmodulin and calmodulin-binding proteins. *Trend Biochem. Sci.* **8**:59–62, 1983.
2. Sobue, K.; Muramoto, Y.; Fujita, M.; Kakiuchi, S. Purification of a calmodulin-binding protein from chicken gizzard that interacts with F-actin. *Proc. Natl. Acad. Sci. USA* **78**:5652–5655, 1981.
3. Sobue, K.; Fujita, M.; Muramoto, Y.; Kakiuchi, S. The calmodulin-binding protein in microtubules is tau factor. *FEBS Lett.* **132**:137–140, 1981.
4. Kakiuchi, R.; Inui, M.; Morimoto, K.; Kanda, K.; Sobue, K.; Kakiuchi, S. Caldesmon, a calmodulin-binding, F actin-interacting protein, is present in aorta, uterus and platelets. *FEBS Lett.* **154**:351–356, 1983.
5. Sobue, K.; Morimoto, K.; Kanda, K.; Maruyama, K.; Kakiuchi, S. Reconstitution of Ca^{2+}-sensitive gelation of actin filaments with filamin, caldesmon and calmodulin. *FEBS Lett.* **138**:289–292, 1982.
6. Sobue, K.; Morimoto, K.; Inui, M.; Kanda, K.; Kakiuchi, S. Control of actin–myosin interaction of gizzard smooth muscle by calmodulin- and caldesmon-linked flip-flop mechanism. *Biomed. Res.* **3**:188–196, 1982.
7. Hartshorne, D. J.; Siemankowski, R. F. Regulation of smooth muscle actomyosin. *Annu. Rev. Physiol.* **43**:519–530, 1981.
8. Adelstein, R. Calmodulin and the regulation of the actin–myosin interaction in smooth muscle and nonmuscle cells. *Cell* **30**:349–350, 1982.
9. Stull, J. T. Phosphorylation of contractile proteins in relation to muscle function. *Adv. Cyclic Nucleotide Res.* **13**:39–93, 1980.
10. Ebashi, S.; Nonomura, Y.; Nakamura, S.; Nakasone, H.; Kohama, K. Regulatory mechanism in smooth muscle: Actin-linked regulation. *Fed. Proc.* 41:2863–2867, 1982.
11. Sobue, K.; Morimoto, K.; Kanda, K.; Fukunaga, K.; Miyamoto, E.; Kakiuchi, S. Interaction of 135000-M_r calmodulin-binding protein (myosin kinase) and F-actin: Another Ca^{2+}- and calmodulin-dependent flip-flop switch. *Biochem. Int.* **4**:503–510, 1982.
12. Adelstein, R. S.; Klee, C. B. Purification and characterization of smooth muscle myosin light chain kinase. *J. Biol. Chem.* **256**:7501–7509, 1981.
13. Gopinath, R. M.; Vincenzi, F. F. Phosphodiesterase protein activator mimics red blood cell cytoplasmic activator of $(Ca^{2+}-Mg^{2+})$ATPase. *Biochem. Biophys. Res. Commun.* **77**:1203–1209, 1977.
14. Jarrett, H. W.; Penniston, J. T. Partial purification of the $Ca^{2+}-Mg^{2+}$ ATPase activator from human erythrocytes: Its similarity to the activator of 3′,5′-cyclic nucleotide phosphodiesterase. *Biochem. Biophys. Res. Commun.* **77**:1210–1216, 1977.
15. Jarrett, H. W.; Kyte, J. Human erythrocyte calmodulin: Further chemical characterization and the site of its interaction with the membrane. *J. Biol. Chem.* **254**:8237–8244, 1979.
16. Sobue, K.; Muramoto, Y.; Fujita, M.; Kakiuchi, S. Calmodulin-binding protein of erythrocyte cytoskeleton. *Biochem Biophys. Res. Commun.* **100**:1063–1070, 1981.
17. Branton, D.; Cohen, C. M.; Tyler, J. Interaction of cytoskeletal proteins on the human erythrocyte membrane. *Cell* **24**:24–32, 1981.
18. Kidoguchi, K.; Hayashi, A.; Sobue, K.; Kakiuchi, S.; Hidaka, H. A possible involvement of calmodulin-spectrin interaction in the regulation of erythrocyte shape: A study with calmodulin antagonists. In: *Calmodulin and Intracellular Ca^{2+} Receptors,* S. Kakiuchi, H. Hidaka, and A. R. Means, eds., New York, Plenum Press, 1982, p. 438.
19. Kakiuchi, S.; Sobue, K.; Fujita, M. Purification of a 240000 M_r calmodulin-binding protein from a microsomal fraction of brain. *FEBS Lett.* **132**:144–148, 1981.
20. Kakiuchi, S.; Sobue, K.; Kanda, K.; Morimoto, K.; Tsukita, S.; Tsukita, S.; Ishikawa, H.; Kurokawa, M. Correlative biochemical and morphological studies of brain calspectin: A spectrin-like calmodulin-binding protein. *Biomed. Res.* **3**:400–410; 1982.
21. Davies, P. J. A.; Klee, C. B. Calmodulin-binding proteins: A high molecular weight calmodulin-binding protein from bovine brain. *Biochem. Int.* **3**:203–212, 1981.
22. Levine, J.; Willard, M. Fodrin: Axonally transported polypeptides associated with the internal periphery of many cells. *J. Cell Biol.* **90**:631–643, 1981.

23. Shimo-oka, T.; Watanabe, Y. Stimulation of actomyosin Mg^{2+}-ATPase activity by a brain microtubule-associated protein fraction: High-molecular-weight actin-binding protein is the stimulating factor. *J. Biochem.* **90**:1297–1307, 1981.
24. Glenney, J. R., Jr.; Glenney, P.; Weber, K. F-actin-binding and cross-linking properties of porcine brain fodrin, a spectrin-related molecule. *J. Biol. Chem.* **257**:9781–9787, 1982.
25. Glenney, J. R., Jr.; Glenney, P.; Osborn, M.; Weber, K. An F-actin- and calmodulin-binding protein from isolated intestinal brush borders has a morphology related to spectrin. *Cell* **28**:843–854, 1982.
26. Sobue, K.; Kanda, K.; Inui, M.; Morimoto, K.; Kakiuchi, S. Actin polymerization induced by calspectin, a calmodulin-binding spectrin-like protein. *FEBS Lett.* **148**:221–225, 1982.
27. Bennett, V.; Davis, J.; Fowler, W. Brain spectrin, a membrane-associated protein related in structure and function to erythrocyte spectrin. *Nature (London)* **299**:126–131, 1982.
28. Kakiuchi, S.; Sobue, K.; Morimoto, K.; Kanda, K. A spectrin-like calmodulin-binding protein (calspectin) of brain. *Biochem. Int.* **5**:755–762, 1982.
29. Sobue, K.; Kanda, K.; Yamagami, K.; Kakiuchi, S. Ca^{2+}- and calmodulin-dependent phosphorylation of calspectin (spectrin-like calmodulin-binding protein; fodrin) by protein kinase systems in synaptosomal cytosol and membranes. *Biomed. Res.* **3**:561–570, 1982.
30. Sobue, K.; Kanda, K.; Kakiuchi, S. Solubilization and partial purification of protein kinase systems from brain membranes that phosphorylate calspectin, a spectrin-like calmodulin-binding protein (fodrin). *FEBS Lett.* **150**:185–190, 1982.
31. Tsukita, S.; Tsukita, S.; Ishikawa, H.; Kurokawa, M.; Morimoto, K.; Sobue, K.; Kakiuchi, S. Binding sites of calmodulin and actin on the brain spectrin, calspectin. *J. Cell Biol.* **97**:574–578, 1983.
32. Sobue, K.; Kanda, K.; Adachi, J.; Kakiuchi, S. Calmodulin-binding proteins that interact with actin filaments in a Ca^{2+}-dependent flip-flop manner: Survey in brain and secretory tissues. *Proc. Natl. Acad. Sci. USA* **80**:6868–6871, 1983.
33. Kakiuchi, S.; Sobue, K. Ca^{2+}- and calmodulin-dependent flip-flop mechanism in microtubule assembly-disassembly. *FEBS Lett.* **132**:141–143, 1981.

33

Calmodulin-Sensitive Adenylate Cyclase Activity
Interaction with Guanyl Nucleotides and Dopamine

M. E. Gnegy and G. Treisman

Adenylate cyclase activity in mammalian brain differs from that in many peripheral tissues in that it can be stimulated by Ca^{2+}. The stimulation by Ca^{2+} is mediated by the endogenous Ca^{2+}-binding protein, calmodulin (CaM) [3,6]. In the striatum, several studies suggest a relationship between dopaminergic activity and the content and activity of CaM [7-9,12,15]. Rats treated chronically with antipsychotic drugs developed supersensitivity of dopamine (DA)-stimulated adenylate cyclase activity as well as increased CaM content in striatal membranes [20]. Similarly, Hanbauer *et al.* [11] found that striatal slices treated with amphetamine exhibited a decrease in both sensitivity of adenylate cyclase to DA and membrane content of CaM. We found that CaM increased adenylate cyclase sensitivity to DA in striatal membranes [10]. Thus, both *in vivo* and *in vitro* evidence suggests that CaM affects DA-stimulated adenylate cyclase activity in rat striatum.

The molecular mechanism by which CaM modulates adenylate cyclase activity is not understood. CaM may stimulate adenylate cyclase activity by acting on the catalytic subunit, with an action independent of the GTP-binding protein [13,17]. Stimulation of adenylate cyclase activity in bovine cerebral cortex by CaM and guanyl nucleotides may be either simple additivity between the two activators [16,17] or synergism [4,13]. Actually, there appear to be two components of adenylate cyclase activity, one sensitive and another insensitive to CaM [2,20].

In order to explain the effect of CaM on DA stimulation, we have explored the interrelationship between CaM and guanyl nucleotides. Our studies demonstrate that the nature of the interaction between CaM and guanyl nucleotides in rat striatum depends on the type and concentration of the guanyl nucleotide. In examining the interaction of CaM with guanyl nucleotides, our findings support the existence of two components of adenylate cyclase activity, which could be differentiated by their sensitivities to guanyl nucleotides, as well as to CaM. The difference in response of adenylate cyclase to CaM in the presence of GTP or the nonhydrolyzable analog guanosine-5-(β-γ-imido) triphosphate (GppNHp) could also be explained by an effect of Ca^{2+} and CaM on GTPase activity.

M. E. Gnegy and G. Treisman • Department of Pharmacology, University of Michigan, Ann Arbor, Michigan 48109.

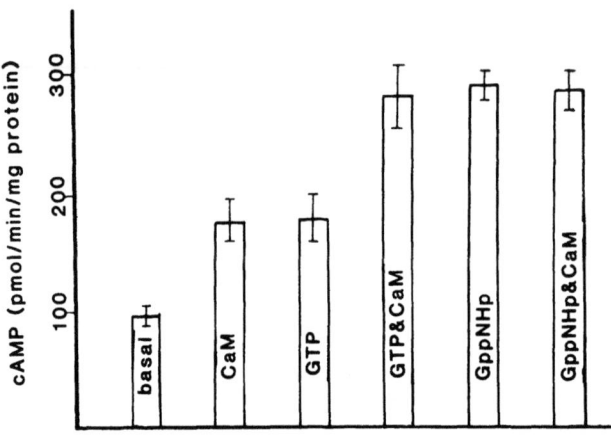

Figure 1. Stimulation of adenylate cyclase activity in a rat striatal particulate fraction by guanyl nucleotides. Concentrations of reactants used in the assay were: CaM, 140 nM with 100 μM CaCl$_2$ (0.12 μM free Ca^{2+}); GTP, 1 μM; GppNHp, 10 μM. Adenylate cyclase activity was assayed and free Ca^{2+} calculated as described by Gnegy and Treisman [10]. The results are the average of five separate experiments ± S.E.M.

Interaction of Calmodulin with Guanyl Nucleotides in Rat Striatum

In investigating stimulation of striatal adenylate cyclase activity by CaM and guanyl nucleotides, a difference was found between the ability of CaM to stimulate adenylate cyclase in the presence of GTP as opposed to GppNHp. Striatal adenylate cyclase activity was stimulated by CaM, GTP, and GppNHp in a striatal particulate fraction (27,000 g) washed twice with 1.2 mM EGTA to deplete endogenous Ca^{2+} and CaM (Fig. 1). CaM, in the presence of 0.1 μM Ca^{2+}, enhanced adenylate cyclase activity approximately twofold, as did a maximal concentration of GTP. Adenylate cyclase activation in the presence of 1 μM GTP and CaM was additive. CaM, however, was unable to stimulate enzyme activity in the presence of 10 μM GppNHp. Further, enzyme activity in the presence of GppNHp alone was the same as activity attained with GTP and CaM.

The interaction between calmodulin and guanyl nucleotides was examined further by

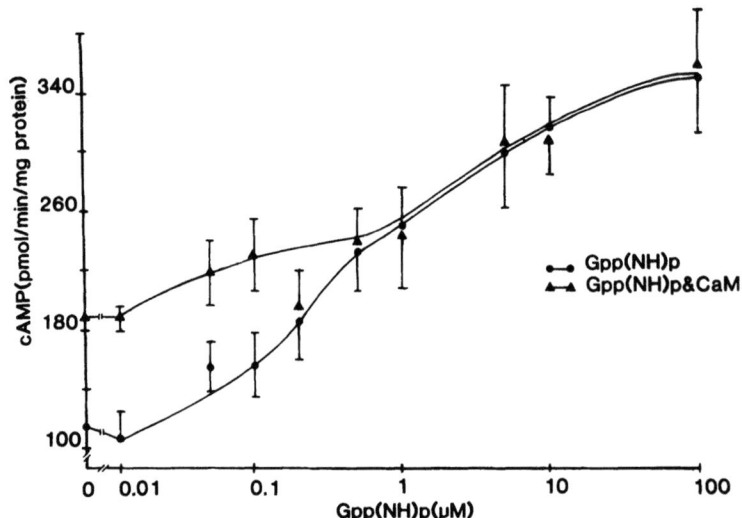

Figure 2. Effect of Ca^{2+} and CaM on GppNHp activation of adenylate cyclase in a rat striatal particulate fraction. Adenylate cyclase activity was determined in a rat striatal particulate fraction (27,000 g) depleted of CaM and Ca^{2+} that was prepared as described by Gnegy and Treisman [10]. The concentration dependence to GppNHp was measured in the absence (●) and presence of 120 nM Ca^{2+} (100 μM CaCl$_2$ added to the assay) and 500 ng CaM (▲). The results are the average of five separate experiments ± S.E.M.

demonstrating that GppNHp stimulated striatal adenylate cyclase activity over several orders of magnitude (Fig. 2). Enzyme stimulation of adenylate cyclase over such a wide concentration range suggests the existence of more than one component in the system. Eadie Hofstee analysis of adenylate cyclase activity in the presence of GppNHp gave a biphasic line which was compatible with the presence of two components of catalytic activity. Approximate kinetic constants for GppNHp were calculated assuming two catalytic components [18]. One component had an apparent K_a for GppNHp for 30 nM and an apparent V_{max} of 223 pmoles/min per mg protein. The second, less-sensitive component had an apparent K_a for GppNHp of 3.4 μM and an apparent V_{max} of 131 pmoles/min per mg protein.

CaM-induced activation of basal adenylate cyclase activity (from 80 pmoles/min per mg protein to 190 pmoles/min per mg protein) was slightly potentiated (225 pmoles/min per mg protein) by 10 nM GppNHp (Fig. 2). Enzyme activity was not further enhanced with GppNHp concentrations as high as 0.5 μM. At concentrations of GppNHp above 0.5 μM, adenylate cyclase activity in the presence of GppNHp plus CaM was the same as that with GppNHp alone. CaM activation was not additive with that of GppNHp at any concentration of GppNHp. These findings indicate that CaM and low GppNHp concentrations seem to competitively activate a component of adenylate cyclase activity that plateaus at approximately 220 pmoles/min per mg protein (Fig. 2).

This idea is further substantiated by an experiment that was performed in a particulate preparation that had not been subjected to EGTA extraction so that it still contained Ca^{2+} and CaM. In these membranes, basal activity was 190 pmoles/min per mg protein, which is nearly twice that found in EGTA-extracted membranes (100 pmoles/min per mg protein). Adenylate cyclase activity in this preparation was not stimulated by the addition of exogenous CaM or Ca^{2+}. Under these conditions, the dose–response curve to GppNHp was superimposable with results obtained in EGTA-washed membranes in the presence of Ca^{2+}, CaM, and GppNHp (upper line, Fig. 2). The GppNHp dose–response curve appeared to have only one component; and GppNHp stimulation did not occur at concentrations of 0.5 μM or less. Maximal activity (310 pmoles/min per mg protein) occurred at 100 μM GppNHp and was equivalent to maximal activity seen in EGTA-washed membranes at 100 μM GppNHp.

CaM may not provide further stimulation of adenylate cyclase over that observed with higher concentrations of GppNHp because GppNHp maximally activates the enzyme. Experiments performed at 25°C suggested that this was not the case, because adenylate cyclase activity stimulated by 280 nM CaM at 25°C was similar to that at 37°C (233 pmoles/min per mg protein). Activation induced by 10 μM GppNHp (185 pmoles/min per mg protein) was much reduced from that observed at 37°C. However, activity measured in the presence of 0.12 μM free Ca^{2+}, CaM, and GppNHp was equal to that of GppNHp (185 pmoles/min per mg protein) and followed the lag period for GppNHp. This suggests that GppNHp interferes with activation by Ca^{2+} and CaM.

Stimulation of striatal adenylate cyclase activity by various concentrations of GTP was also determined in the presence and absence of CaM (280 nM) and Ca^{2+} (0.12 μM). Adenylate cyclase activity reached a maximum value at 1 μM GTP, and Eadie Hofstee analysis was consistent with the presence of a single kinetic component. CaM and Ca^{2+} stimulated adenylate cyclase activity in an additive manner at every concentration of GTP (data not shown). Maximum adenylate cyclase activity detected in the presence of CaM and GTP (310 pmoles/min per mg protein) was similar to that found in the presence of 100 μM GppNHp. Adenylate cyclase stimulation by CaM and GTP was also additive when CaM concentrations were varied in the presence and absence of a maximal GTP concentration (1 μM).

These studies demonstrate that the ability of CaM to stimulate adenylate cyclase in the presence of guanyl nucleotide depends on the type and concentration of guanyl nucleotide. Adenylate cyclase activity measured in the presence of CaM and GTP reflected additivity at every concentration of these reactants. On the contrary, when the activating guanyl nucleotide was GppNHp, CaM further activated adenylate cyclase only at concentrations less than 0.2–0.5 μM GppNHp. Kinetic analysis of our data suggests that there are two different components of adenylate cyclase activity in striatal membranes which can be distinguished by their response to CaM and guanyl nucleotides. One component is activated by Ca^{2+} and CaM, and has a higher affinity for GppNHp. A second, CaM-insensitive component is activated by higher concentrations of GppNHp, as well as GTP.

Effect of CaM on Striatal GTPase Activity

Cassel et al. [5] have suggested that GTPase activity is the "turn-off" reaction of hormone-stimulated adenylate cyclase activity. In accordance with this model, hydrolysis of GTP to GDP restores the GTP-binding protein and catalytic subunit to a resting state which can then be restimulated by hormone binding to its receptor. The differential action of CaM on adenylate cyclase activation in the presence of GTP or GppNHp might be mediated by an effect on GTPase activity. If GTP hydrolysis is part of the function of CaM, then CaM would be expected to be inactive in the presence of GppNHp, which also inhibits GTPase activity.

We next examined whether Ca^{2+} and CaM affect GTPase activity in the EGTA-washed striatal particulate fraction. Ca^{2+} and CaM significantly inhibited GTPase activity at concentrations from 10 nM to 1 μM but not at 10 μM (Table I). The fact that CaM affected only low concentrations of GTP suggested that the GTPase activity detected corresponded to adenylate cyclase activation. The K_a of GTPase related to adenylate cyclase activity is below 1 μM [14]. When the activity was corrected for high-K_m GTPase, inhibition of low-K_m

Table I. Inhibition of Striatal GTPase Activity by CaM at Various Concentrations of GTP

Concentration of GTP (μM)	GTPase activity (cpm ± S.E.M.)[a]	
	− CaM	+ Ca^{2+} + CaM
10^{-8}	8436 ± 346	6793 ± 372*
10^{-7}	7058 ± 379	6060 ± 387*
10^{-6}	5236 ± 393	4321 ± 440*
10^{-5}	4088 ± 388	4188 ± 663
10^{-4}	3070 ± 374	3821 ± 405

[a]GTPase activity was measured in an EGTA-washed striatal particulate fraction from male Sprague–Dawley rats prepared as described by Gnegy and Treisman [10]. GTPase activity was assayed using [γ-^{32}P]-GTP according to the method of Cassel et al. [5]. The assay contents were modified to be more similar to those used in the adenylate cyclase assay. Assay contents in a volume of 100 μl were: 12.5 mM Tris HCl buffer, pH 7.5, 5 mM $MgCl_2$, 1 mM ATP, 1 mM AppNHp, 1 mM creatine phosphate, 5 U creatinine phosphokinase, 1 mM ouabain, 2 mM dithiothreitol, 5 μg membrane protein, 150 μM EGTA (carried over from the membranes), and 50,000 cpm [γ-^{32}P]-GTP plus various concentrations of cold GTP. When present, 500 ng CaM was added to the assay and 100 μM $CaCl_2$ (0.12 μM free Ca^{2+} in the assay). Incubations were carried out for 10 min at 37°C. Assays were terminated and free $^{32}P_i$ determined as described by Cassel et al. [5]. N = 4.
*$p < 0.05$ as compared to activity in the absence of Ca^{2+} and CaM.

Table II. Ca^{2+} Dependence of the CaM Effect on GTPase Inhibition and Adenylate Cyclase Activation in a Rat Striatal Particulate Fraction

pCa[c]	GTPase activity[a]		Adenylate cyclase activity[b]	
	− CaM	+ CaM	− CaM	+ CaM
	pmoles/min per mg protein ± S.E.M.			
—	520 ± 26	520 ± 25	97 ± 12	91 ± 15
8.75	ND[d]	519 ± 41	ND	ND
8.18	538 ± 35	489 ± 24	ND	107 ± 19
7.32	470 ± 45	443 ± 45	94 ± 9	157 ± 16
6.92	ND	348 ± 34	86 ± 10	153 ± 14
5.79	468 ± 27	362 ± 35	ND	136 ± 13
4.29	485 ± 29	314 ± 35	80 ± 9	105 ± 16

[a]GTPase activity was determined as described in footnote a of Table I. 500 ng of CaM was used in the assays. $N = 4$.
[b]Adenylate cyclase activity was determined in the EGTA-washed particulate fraction as described in Ref. 10. When present, 500 ng CaM was used in the assays. $N = 5$.
[c]Free Ca^{2+} was calculated from the equilibrium constant for Ca · EGTA as described in Ref. 10.
[d]ND, not determined.

GTPase by CaM was 35–40%, which is similar to inhibition found with other agents, including opiates [5].

The inhibition by CaM was dependent on the presence of Ca^{2+} (Table II). Ca^{2+} itself (10^{-8}–10^{-5} M) did not affect GTPase activity. In the presence of CaM, inhibition was greater with increasing Ca^{2+}; a maximum effect was achieved at 0.12 μM Ca^{2+}. The Ca^{2+} concentration dependence for inhibition of GTPase by CaM corresponded to the Ca^{2+} requirement for stimulation of adenylate cyclase (Table II). However, there was no biphasic effect of free Ca^{2+} on GTPase activity as there was for adenylate cyclase activation. GTPase inhibition was dependent on the concentration of CaM. With 10^{-7} M free Ca^{2+}, inhibition of GTPase was detected at 50 ng CaM and was maximal at 2 μg CaM, which is the concentration range effective for CaM stimulation of adenylate cyclase at the same Ca^{2+} concentration [10]. To establish further a relationship between the effect of CaM on GTPase activity and its ability to stimulate adenylate cyclase, the effect of CaM on GTPase activity in a rat liver particulate fraction was examined. CaM and Ca^{2+} had no effect on GTPase activity in EGTA-washed hepatic membranes; and CaM and Ca^{2+} were unable to enhance adenylate cyclase activity in liver membranes.

These data suggest that at least part of the action of CaM in stimulating adenylate cyclase activity is effected by an inhibition of GTPase activity. CaM may inhibit GTPase activity by modifying the interaction of a GTP-binding protein with the catalytic subunit. However, it is unlikely that the sole effect of CaM on cyclase stimulation is mediated by an inhibition of GTPase, because GTP is probably not a requirement for adenylate cyclase activation by CaM [13,17]. Since there are many types of GTPase in this preparation, however, the effect of CaM on a single species of GTPase might be greater than the apparent effect on a mixture of GTPase enzymes. In any event, in order to determine whether CaM directly affects the enzyme, it will be necessary to solubilize GTPase.

In conclusion, we have investigated the interaction of CaM and guanyl nucleotides in activating adenylate cyclase in rat striatum. Our studies indicate that CaM does interact with guanyl nucleotides since stimulation was not additive with the nonhydrolyzable GTP analog, GppNHp. This action can be partially explained by an effect of CaM in inhibiting a low-K_m GTPase activity in striatal membranes. Kinetic analysis of adenylate cyclase activation demonstrated two components of adenylate cyclase activity, one sensitive to CaM and low concentrations of GppNHp, the other stimulated by GTP and high concentrations of GppNHp, but not by CaM. Since GTP, rather than GppNHp, is the physiologically relevant guanyl nucleotide, it is very likely that GTP activates the CaM-sensitive component *in vivo*. There may be greater GTPase activity at the CaM-sensitive component or an alteration in coupling of the GTP-binding protein to this component. Our two-component model cannot delineate whether there are two separate catalytic subunits or one catalytic subunit with two GTP-binding proteins. Other studies similarly suggest, however, that there are two catalytic components of adenylate cyclase, one sensitive and the other insensitive to CaM [2,20]. After partial purification of the CaM-stimulated adenylate cyclase, the molecular weight of the catalytic subunit was determined to be 150,000 [1]. Tissues containing adenylate cyclase activity that do not respond to CaM may lack the CaM-sensitive component found in brain.

CaM affects DA sensitivity in the presence of either GTP or GppNHp [10], suggesting that DA stimulates adenylate cyclase activity through both components, using GTP as the guanyl nucleotide. Adenylate cyclase activation by DA, CaM, and low concentrations of GppNHp is increased in particulate fractions of striatum from rats chronically treated with haloperidol [19]. Thus, the CaM-sensitive component of adenylate cyclase manifests increased responsiveness under conditions of dopaminergic supersensitivity.

ACKNOWLEDGMENT. This work was supported by Grant R01 MH-36044-01 from the National Institute of Mental Health.

References

1. Andreason, T. J.; Heideman, W.; Rosenberg, G. B.; Storm, D. R. Photoaffinity labeling of brain adenylate cyclase preparations with azido[^{125}I]calmodulin. *Fed. Proc.* **42**:1852, 1983.
2. Brostrom, C. O.; Brostrom, M. A.; Wolff, D. J. Calcium-dependent adenylate cyclase from rat cerebral cortex: Reversible activation by sodium fluoride. *J. Biol. Chem.* **252**:5677–5685, 1977.
3. Brostrom, C. O.; Huang, Y.-C.; Breckenridge, B. M.; Wolff, D. J. Identification of a calcium-binding protein as a calcium-dependent regulator of brain adenylate cyclase. *Proc. Natl. Acad. Sci. USA* **72**:64–68, 1975.
4. Brostrom, M. A.;Brostrom, C. O.; Wolff, D. J. Calcium-dependent adenylate cyclase from rat cerebral cortex: Activation by guanine nucleotides. *Arch. Biochem. Biophys.* **191**:341–350, 1978.
5. Cassel, D.; Levkovitz, H.; Selinger, Z. The regulatory GTPase cycle of turkey erythrocyte adenylate cyclase. *J. Cyclic Nucleotide Res.* **3**:393–406, 1977.
6. Cheung, W. Y.; Bradham, L. S.; Lynch, T. J.; Lin, Y. M.; Tallant, E. A. Protein activator of cyclic 3′:5′-nucleotide phosphodiesterase of bovine or rat brain also activates its adenylate cyclase. *Biochem. Biophys. Res. Commun.* **66**:1055–1062, 1975.
7. Gnegy, M. E. Relationship of calmodulin and dopaminergic activity in the striatum. *Fed. Proc.* **41**:2273–2277, 1982.
8. Gnegy, M. E.; Lau, Y. S.; Treisman, G. Role of calmodulin in states of altered catecholamine sensitivity. *Ann. N.Y. Acad. Sci.* **356**:304–318, 1980.
9. Gnegy, M. E.; Lucchelli, A.; Costa, E. Correlation between drug-induced supersensitivity of dopamine-dependent striatal mechanisms and the increase in striatal content of the Ca^{2+}-regulated protein activator of cAMP phosphodiesterase. *Naunyn-Schmiedebergs Arch. Pharmacol.* **301**:121–127, 1977.
10. Gnegy, M.; Treisman, G. Effect of calmodulin on dopamine sensitive adenylate cyclase activity in rat striatal membranes. *Mol. Pharmacol.* **19**:256–263, 1981.
11. Hanbauer, I.; Gimble, J.; Lovenberg, W. Changes in soluble calmodulin following activation of dopamine receptors in rat striatal slices. *Neuropharmacology* **18**:851–857, 1979.
12. Hanbauer, I.; Pradham, S.; Yang, H.-Y.T. Role of calmodulin in dopaminergic transmission. *Ann. N.Y. Acad. Sci.* **356**:292–303, 1980.
13. Heideman, W.; Wierman, B. M.; Storm, D. R. GTP is not required for calmodulin stimulation of bovine brain adenylate cyclase. *Proc. Natl. Acad. Sci. USA* **79**:1462–1465, 1982.

14. Koski, G.; Klee, W. A. Opiates inhibit adenylate cyclase by stimulating GTP hydrolysis. *Proc. Natl. Acad. Sci. USA* **78**:4185–4189, 1981.
15. Lau, Y. S.; Gnegy, M. E. Chronic haloperidol treatment increased calcium-dependent phosphorylation in rat striatum. *Life Sci.* **30**:21–28, 1982.
16. Salter, R. S.; Krinks, M. H.; Klee, C. B.; Neer, E. J. Calmodulin activates the isolated catalytic unit of brain adenylate cyclase. *J. Biol. Chem.* **256**:9830–9833, 1981.
17. Seamon, K. B.; Daly, J. W. Calmodulin stimulation of adenylate cyclase in rat brain membranes does not require GTP. *Life Sci.* **30**:1457–1464, 1982.
18. Spears, G.; Sneyd, J. G. T.; Loten, E. G. A method for deriving kinetic constants for two enzymes acting on the same substrate. *Biochem. J.* **125**:1149–1151, 1971.
19. Treisman, G.; Muirhead, N.; Gnegy, M. E. Calmodulin-stimulated adenylate cyclase activity is increased in striatum from chronic haloperidol treated rats. *Soc. Neurosci. Abstr.* **9**:86, 1983.
20. Wescott, K. R.; LaPorte, D. C.; Storm, D. R. Resolution of adenylate cyclase sensitive and insensitive to Ca^{++} and calcium-dependent regulatory protein (CDR) by CDR-Sepharose affinity chromatography. *Proc. Natl. Acad. Sci. USA* **76**:204–208, 1979.

34

Calcium/Calmodulin-Dependent Protein Phosphorylation in the Nervous System

Teresa L. McGuinness, Yvonne Lai, Charles C. Ouimet, and Paul Greengard

The central role of Ca^{2+} in the physiology of the nervous system is well documented [2]. Ca^{2+} regulates electrical excitability of nerve cells [15,26], as well as synthesis [43] and release [3,46] of neurotransmitters. Unlike the cyclic nucleotide second messengers, which are thought to act exclusively on protein kinases [19,20,32,34,39,40], intracellular Ca^{2+} interacts with a variety of protein systems, including, but not limited to, protein kinases [9]. The cyclic nucleotides and Ca^{2+} also differ in the number of protein kinases they stimulate. cAMP and cGMP each activate one known catalytic component, while Ca^{2+} activates several different protein kinases. An exciting area of research is to identify those physiological actions of Ca^{2+} that are mediated through protein phosphorylation and to determine, by identification of the specific protein kinases and substrate proteins involved, the precise molecular pathways by which those actions of Ca^{2+} are achieved.

The possibility that one or more of the effects of Ca^{2+} at the presynaptic nerve terminal are mediated or modulated by protein phosphorylation was investigated by Krueger et al. [33], who showed that depolarization of intact synaptosomes increases the phosphorylation state of a number of proteins. Subsequently, Schulman and Greengard [47] demonstrated that phosphorylation of synaptosomal membrane proteins is dependent on the presence of the Ca^{2+}-binding protein calmodulin, which was previously shown to modulate the activities of cyclic nucleotide phosphodiesterase [8,25] and adenylate cyclase [6,10]. Ca^{2+} plus calmodulin increased ^{32}P incorporation into a number of membrane proteins, with those having molecular weights of about 50,000 and 60,000 showing the highest levels of incorporation.

Simultaneously with the findings that Ca^{2+}/calmodulin-dependent protein kinase activity existed in brain membrane fractions, several investigators demonstrated that calmodulin conferred Ca^{2+} sensitivity to a known Ca^{2+}-dependent protein kinase, myosin light-chain kinase [12,53,58]. Shortly thereafter, Cohen et al. [11] showed that calmodulin was a subunit of the Ca^{2+}-dependent enzyme, phosphorylase kinase, and Schulman and Greengard

---

Teresa L. McGuinness, Yvonne Lai, Charles C. Ouimet, and Paul Greengard • The Rockefeller University, New York, New York 10021.

[48] demonstrated Ca^{2+}/calmodulin-dependent protein kinase activity in every animal tissue examined. The latter study showed that the enzyme(s) phosphorylated a tissue-specific set of endogenous substrates.

Following these initial reports, an enormous amount of work has been done in the field of Ca^{2+}-dependent protein phosphorylation. It is clear from this work that Ca^{2+} activates a number of protein kinases, several by complexing with calmodulin (see below) and one in concert with phospholipid [51]. Yamauchi and Fujisawa [59] demonstrated three distinct Ca^{2+}/calmodulin-dependent protein kinases in rat brain. Similarly, Kennedy and Greengard [30] showed that at least four Ca^{2+}/calmodulin-dependent protein kinases exist in rat brain. In addition to myosin light-chain kinase and phosphorylase kinase, brain homogenates contain two enzymes that phosphorylate the substrate synapsin I (previously referred to as protein I). Several of the calmodulin-dependent protein kinases observed in these and other studies have recently been purified to apparent homogeneity. These include: a brain kinase that phosphorylates smooth muscle myosin light chain and microtubule protein [16]; a kinase that phosphorylates tubulin [17]; and a brain enzyme that phosphorylates tyrosine- and tryptophan-5-monooxygenases [60]. Two brain kinases that phosphorylate synapsin I (see below) have also been purified to near homogeneity from brain [31,35,38], and a Ca^{2+}/calmodulin-dependent protein kinase that phosphorylates glycogen synthase has been purified to apparent homogeneity from liver [1,44,45] and skeletal muscle [56,57]. These various enzymes are distinct from both myosin light-chain kinase and phosphorylase kinase, but some of them appear to be related to each other (see Concluding Remarks). A direct comparison of the physical and enzymatic properties of these various enzymes will be required before the total number of distinct Ca^{2+}/calmodulin-dependent protein kinases can be determined.

Purification and Characterization of a Ca^{2+}/Calmodulin-Dependent Protein Kinase That Phosphorylates Synapsin I

Synapsin I Phosphorylation

The studies of Krueger *et al.* [33] demonstrated that Ca^{2+} influx into rat brain synaptosomes greatly stimulated the phosphorylation of two protein bands of 86,000 and 80,000 M_r which are also phosphorylated by cAMP-dependent protein kinase [49,52]. These peptides have similar properties and are collectively designated synapsin I (synapsin Ia and synapsin Ib). Synapsin I is a neuron-specific protein, associated with synaptic vesicles in virtually all neurons [13,14,24]. Because phosphorylation of synapsin I may be involved in certain Ca^{2+}-dependent processes that modulate nerve terminal function, studies have been carried out in our laboratory to establish the mechanism by which Ca^{2+} regulates synapsin I phosphorylation. For this purpose, purified synapsin I has been used as a substrate to identify, purify, and characterize those brain Ca^{2+}-dependent protein kinases responsible for its phosphorylation.

Calmodulin Kinases I and II

Synapsin I can be phosphorylated on serine residues in two different regions of the molecule [23]. cAMP-dependent protein kinase and a Ca^{2+}/calmodulin-dependent protein kinase (calmodulin kinase I) both phosphorylate one serine residue (site 1) in one region of the molecule, and a second Ca^{2+}/calmodulin-dependent protein kinase (calmodulin kinase II) phosphorylates two serine residues (sites 2 and 3) in a different region of the molecule

[22,30]. Calmodulin kinase I is present in the cytosolic compartment of rat brain homogenates, while calmodulin kinase II is present in both cytosolic and particulate fractions [30].

Purification of Calmodulin Kinase II

Evidence that the calmodulin kinase II activities detected in the soluble and particulate fractions of brain homogenates were due to a single enzyme rather than two distinct enzymes was obtained by partially purifying and comparing the enzymes from the two fractions [31]. Extraction of the particulate-associated calmodulin kinase II activity was achieved by dilution of the particulate material into low-ionic-strength buffer. The extracted particulate kinase and the soluble kinase were partially purified by DEAE-cellulose chromatography followed by calmodulin-Sepharose affinity chromatography. The two partially purified enzymes, when combined and applied to an analytical DEAE-cellulose column, were eluted as a single peak of activity. Moreover, the two kinases were indistinguishable with respect to all physicochemical and kinetic properties examined and showed virtually identical substrate specificities (Table I). Therefore, the soluble and extracted particulate fractions were combined in the further purification of calmodulin kinase II.

Calmodulin kinase II has been purified to near homogeneity in our laboratory [35,37 McGuinness, Lai, and Greengard, in press; also see Fig. 1]. The purified kinase exhibited a

Table I. Comparison of Soluble and Particulate Calmodulin Kinase II Activities[a]

Property	Physicochemical and kinetic properties[b] Kinase	
	Soluble	Particulate
Elution from DEAE-cellulose	0.13–0.17 M NaCl	0.13–0.18 M NaCl
K_m for synapsin I	0.4 μM	0.4 μM
K_m for ATP	3.0 μM	3.9 μM
K_a for Ca^{2+}	~4 μM	~4 μM
K_a for calmodulin	0.4 μM	0.4 μM
IC_{50} for trifluoperazine	18 μM	18 μM
pH optimum	Broad (> pH 7.5)	Broad (> pH 7.5)

Substrate	Substrate specificities[c] Relative rates of phosphorylation (%)	
	Soluble	Particulate
Synapsin I (0.1 mg/ml)	100	100
Histone H3 (0.1 mg/ml)	14	13
Myosin light chain (0.3 mg/ml) (skeletal muscle)	2	2
Phosphorylase b (0.3 mg/ml)	1.4	1.1
Casein (0.4 mg/ml)	1.1	1.0
Phosvitin (0.4 mg/ml)	ND[d]	ND
Gelatin (0.2 mg/ml)	ND	ND

[a]Partially purified preparations of calmodulin kinase II were used for this study.
[b]Adapted from data in Ref. 31.
[c]Data from Ref. 31.
[d]ND, not detectable.

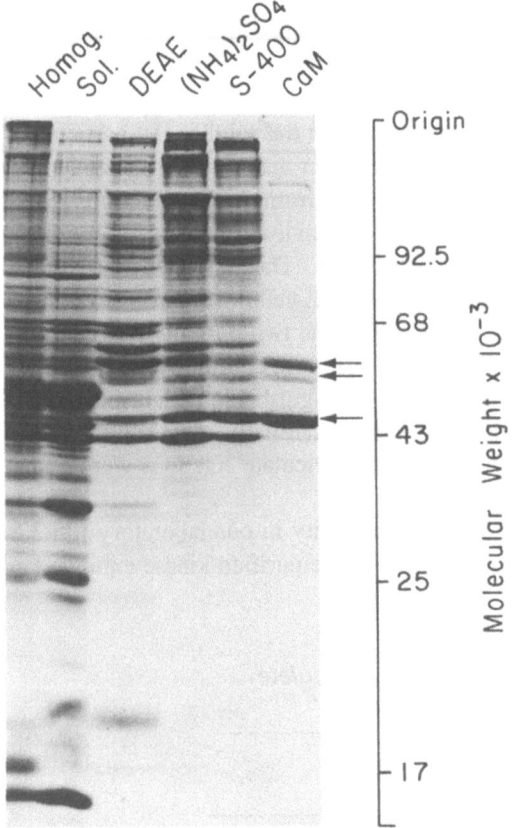

Figure 1. Protein staining pattern of calmodulin kinase II preparation at each stage of the purification: Homogenate (Homog.), 75 μg; pooled soluble and solubilized fractions (Sol.), 75 μg; DEAE-cellulose chromatography pool (DEAE), 60 μg; 35% ammonium sulfate pellet [$(NH_4)_2SO_4$], 45 μg; Sephacryl S-400 gel filtration pool (S-400), 37 μg; calmodulin-Sepharose affinity chromatography pool (CaM), 15 μg. Proteins were separated by SDS-polyacrylamide gel electrophoresis and stained with Coomassie brilliant blue. Arrows point to the 60,000-, 58,000-, and 50,000-M_r protein staining bands.

Calmodulin kinase II was purified according to the following procedure. Fifty rat forebrains were homogenized [31] and centrifuged (100,000g for 60 min) to separate soluble and particulate components. The particulate activity was extracted [31] and the solubilized protein collected by a second centrifugation (100,000g for 60 min). The two supernatant fractions, containing 75–95% of the original activity in the homogenate, were pooled and subjected to ion-exchange chromatography on DEAE-cellulose, followed by precipitation with 35% ammonium sulfate. The resuspended pellet was subjected to gel filtration chromatography using Sephacryl S-400, and the enzyme was eluted as a single peak of activity, corresponding to 600,000–650,000 M_r. The fractions containing enzyme activity were pooled and precipitated with 35% ammonium sulfate. The precipitated material was collected by centrifugation, resuspended, and applied to a calmodulin-Sepharose 4B affinity column in the presence of Ca^{2+}; the kinase activity was eluted with buffer containing EGTA. The fractions containing enzyme activity were pooled, brought to 50% glycerol, and stored at −20°C.

specific activity of approximately 4 μmoles min^{-1} mg^{-1} using synapsin I as substrate. The protein staining patterns at the various stages of the purification are shown in Fig. 1. The most highly purified preparation exhibited a major 50,000-M_r protein staining band and a less intense protein staining doublet of 60,000/58,000 M_r on SDS-polyacrylamide gels. Faintly stained bands could occasionally be seen at 43,000 and 170,000/140,000 M_r. The 50,000- and 60,000/58,000-M_r polypeptides of the highly purified calmodulin kinase II were shown to undergo Ca^{2+}/calmodulin-dependent phosphorylation [31; Fig. 2]. Analysis by two-dimensional gel electrophoresis confirmed that the radioactive bands comigrated with the protein staining bands of corresponding molecular weight [31; data not shown].

Association of the 50,000- and 60,000/58,000-M_r Polypeptides with Calmodulin Kinase II Activity

The high specific activity of the most purified enzyme preparation suggested that the most prominent band(s) (50,000 and 60,000/58,000 M_r) were components of the kinase. Additional evidence that these phosphoproteins were associated with calmodulin kinase II

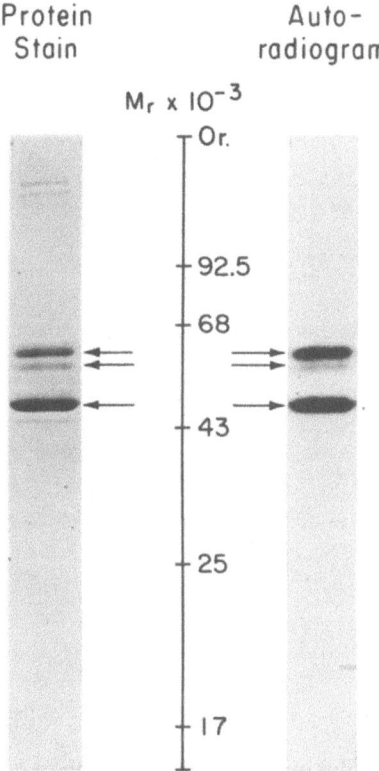

Figure 2. Protein staining pattern and autoradiogram of the purified calmodulin kinase II preparation. The kinase preparation (4 μg) was incubated with [γ-^{32}P]-ATP (10 μM), Ca^{2+} (300 μM), and calmodulin (0.03 mg/ml) as described in Ref. 31. 0.1 μg of the phosphorylated kinase was added to 14.9 μg of unlabeled kinase prior to electrophoresis. The kinase preparation was the same as that used for Fig. 1 (last lane). Proteins were subsequently separated by SDS-polyacrylamide gel electrophoresis and stained with Coomassie brilliant blue, and autoradiography was performed. Arrows point to the 60,000-, 58,000-, and 50,000-M_r polypeptides.

activity was obtained in studies showing that they coeluted with enzyme activity through a variety of purification procedures. The 50,000- and 60,000/58,000-M_r polypeptides, identified by protein staining and autoradiography, coincided with the peak of kinase activity during gel filtration, ammonium sulfate precipitation, sucrose density gradient centrifugation, nondenaturing gel electrophoresis, as well as during chromatography on DEAE-cellulose, hydroxylapatite, phosphocellulose, calmodulin-Sepharose affinity, and dye-ligand affinity columns [31,35].

The [^{125}I]calmodulin gel overlay technique [7], which was used to identify the calmodulin-binding component(s) of the enzyme, revealed that the 50,000- and 60,000/58,000-M_r polypeptides bind calmodulin. In addition, in photoaffinity labeling studies using [α-^{32}P]-8-azido ATP, preliminary evidence suggested that each of these polypeptides binds ATP and, thus, may contain active sites of the enzyme [35].

To further characterize the kinase, polyclonal and monoclonal antibodies were obtained that selectively precipitated the 50,000-M_r polypeptide and 60,000/58,000-M_r polypeptide doublet [27,36; McGuinness and Greengard, unpublished results]. In order to determine whether the antibodies independently recognized each of the polypeptides, immunoreactivity was analyzed on Western immunoblots. One mouse monoclonal antibody (C42.1) selectively labeled each of the three polypeptides in a partially purified calmodulin kinase II preparation, indicating immunological similarity between these polypeptides (Fig. 3). The relative intensities of the ^{125}I-labeled immunoreactive bands agreed with the relative intensities of the 50,000- and 60,000/58,000-M_r protein staining bands in the purified enzyme preparation. In

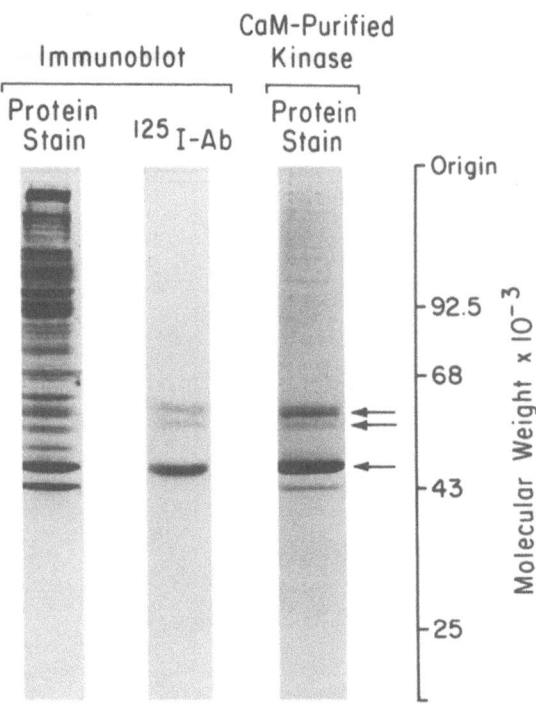

Figure 3. Immunoblot showing protein staining pattern (Protein Stain) and ^{125}I-labeled antibody reactivity (^{125}I-Ab) of a partially purified preparation of calmodulin kinase II. The preparation (30 μg), purified through the gel filtration step of the purification procedure, was subjected to SDS-polyacrylamide gel electrophoresis, transferred to nitrocellulose, and either stained for protein or analyzed for immunoreactivity. For the latter purpose, the nitrocellulose sheet was incubated with mouse monoclonal antibody C42.1 as described in Ref. 27. The sheet was then incubated in a 1:400 dilution of ^{125}I-labeled goat anti-mouse IgG F(ab')^2 fragment, washed, dried, and exposed to X-ray film. For comparison, the protein staining pattern of a highly purified preparation (15 μg) of calmodulin kinase II (CaM-purified kinase, Protein Stain) is shown. Arrows point to the 60,000-, 58,000-, and 50,000-M_r polypeptides.

similar immunoblots of brain homogenates, these bands were the only ones detectably labeled by this antibody [41]. Peptide mapping data [27,31] suggest that the 50,000-M_r polypeptide is probably not derived from the 60,000- or 58,000-M_r polypeptides, but the results are consistent with the 58,000-M_r component being derived from the 60,000-M_r polypeptide.

Substrate Specificity of Calmodulin Kinase II

Examination of the substrate specificity of calmodulin kinase II revealed that the enzyme phosphorylated a number of substrate proteins, with synapsin I being the most effective substrate examined (Table II). Calmodulin kinase II also phosphorylated cardiac phospholamban (Lai, McGuinness, and Greengard, unpublished results). The ability of calmodulin kinase II to recognize several different substrate proteins distinguishes it from myosin light chain kinase and phosphorylase kinase, and suggests that calmodulin II kinase is a multifunctional kinase involved in several different Ca^{2+}-regulated functions [37].

Comparison of Calmodulin Kinase II Polypeptides with Polypeptides of Corresponding Molecular Weight Present in Synaptic Junction Fractions

A calmodulin-dependent protein kinase has been demonstrated in postsynaptic density (PSD) preparations isolated from canine cerebral cortex [18]. An endogenous 50,000-M_r protein is a prominent substrate for this enzyme [18] and binds calmodulin [7]. This 50,000-M_r polypeptide is the most abundant PSD protein, and hence has been designated the major PSD protein (mPSDp) [4,5,29]. PSD fractions also contain a substrate protein [18] and a calmodulin-binding protein [7] in the 60,000-M_r range. Calmodulin-binding [21] and calmodulin-dependent phosphorylation [28] of 50,000-, 58,000-, and 60,000-M_r polypeptides have also been demonstrated in synaptic junction (SJ) fractions isolated from rat

Table II. Substrate Specificity of Highly Purified Calmodulin Kinase II[a]

Substrate	Concentration (mg/ml)	Relative rates of phosphorylation (%)
Synapsin I	0.1	100
Myosin P-light chain (smooth muscle)	0.8	16
Glycogen synthase	0.4	13
Histone H1	0.2	3.0
Microtubule-associated protein-2	0.8[b]	2.2
Acetyl-CoA carboxylase	0.3	1.0
ATP-citrate lyase	0.4	0.7
Myosin P-light chain (skeletal muscle)	0.2	0.1
Phosphorylase b	1.5	<0.02
Phosvitin	2.0	<0.02

[a]Adapted from data in Ref. 37.
[b]Concentration refers to concentration of microtubule protein carried through three cycles of *in vitro* assembly–disassembly.

forebrain. The SJ 50,000-M_r polypeptide has been shown to be identical to the mPSDp [21,29].

The similarities in apparent molecular weight, phosphorylation, and calmodulin-binding properties of the 50,000-, 58,000-, and 60,000-M_r polypeptides in SJ preparations and calmodulin kinase II suggested that they may be related. Therefore, the kinase polypeptides were further compared to their molecular weight counterparts present in SJ preparations [27,36]. One-dimensional proteolytic phosphopeptide maps, two-dimensional ^{125}I-labeled tryptic/chymotryptic peptide fingerprints, and cross-reactivity on Western immunoblots using polyclonal and monoclonal antibodies indicated that the three polypeptides in the purified calmodulin kinase II preparation were indistinguishable from their counterparts in the SJ preparation. One demonstration of the similarity of the 50,000-M_r polypeptides from the two sources is shown in Fig. 4. ^{125}I-labeled peptide tryptic/chymotryptic fingerprints of the 50,000-M_r polypeptide from calmodulin kinase II (Fig. 4A), from the SJ preparation (Fig. 4C), and from an equal mixture (cpm) of the two (Fig. 4B) indicated that the kinase and SJ 50,000-M_r polypeptides are virtually identical. Moreover, when purified calmodulin kinase II was added to SJ fractions, it increased the state of phosphorylation of the same endogenous proteins that were phosphorylated by the endogenous SJ enzyme [27,36]. No additional proteins were phosphorylated. When purified synapsin I was added to SJ fractions, the endogenous kinase phosphorylated the same region of the synapsin I molecule that is phosphorylated by calmodulin kinase II, further indicating similar substrate specificity for the two kinases.

Immunocytochemical Localization of Calmodulin Kinase II Polypeptides in Rat Brain

The above data suggest that calmodulin kinase II, of which the major PSD protein appears to be a component, may be involved in certain Ca^{2+}-dependent processes at SJ membranes. The enzyme, however, is not localized exclusively to SJ or PSD fractions.

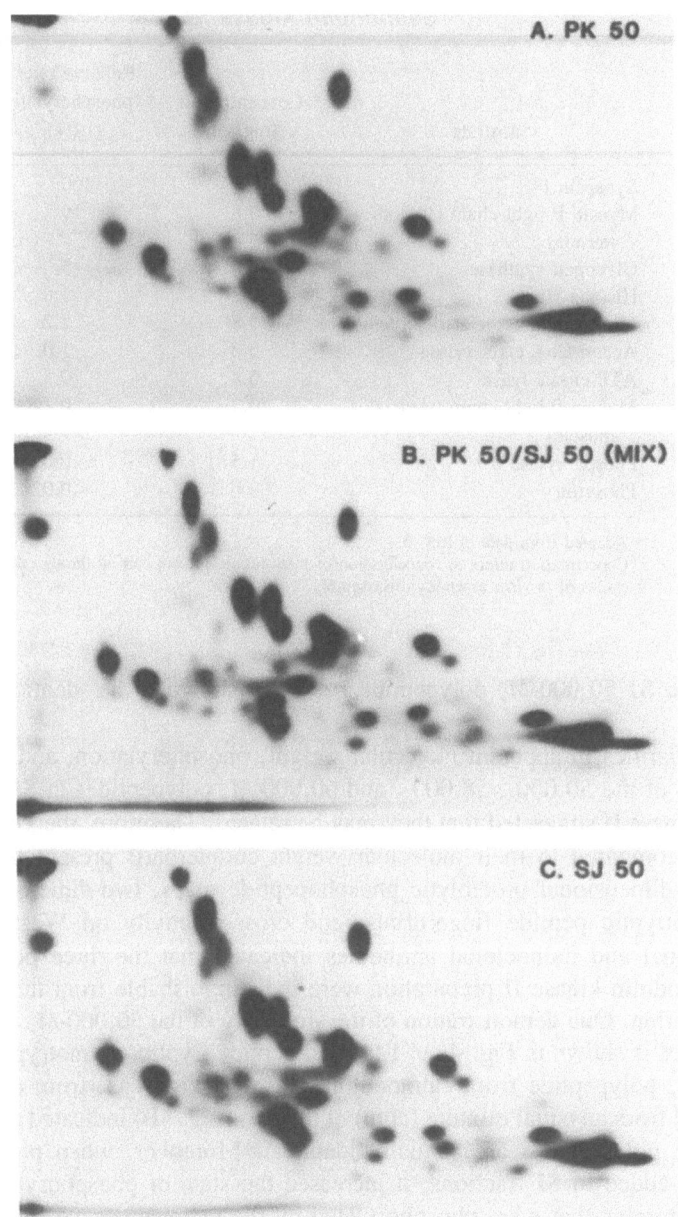

Figure 4. Two-dimensional ^{125}I-labeled peptide fingerprints obtained by tryptic/chymotryptic digestion of (A) 50,000-M_r polypeptide from calmodulin kinase II (PK 50), (C) 50,000-M_r polypeptide from an SJ preparation (SJ 50), and (B) a mixture of equal amounts (cpm) of the two 50,000-M_r polypeptides. Electrophoresis was from left to right in the first dimension and chromatography was from bottom to top in the second dimension. [Data from Ref. 27.]

Approximately 50% of the calmodulin kinase II activity detected in brain homogenates is present in the soluble fraction and much of the remaining particulate activity can be readily extracted [30,31]. It was, therefore, of interest to attempt to determine the localization of calmodulin kinase II by immunocytochemistry. Light and electron microscopic immunocytochemistry, using a monoclonal antibody (C42.1) that reacts with the 50,000- and 60,000/58,000-M_r polypeptides of the kinase, revealed immunoreactivity throughout neu-

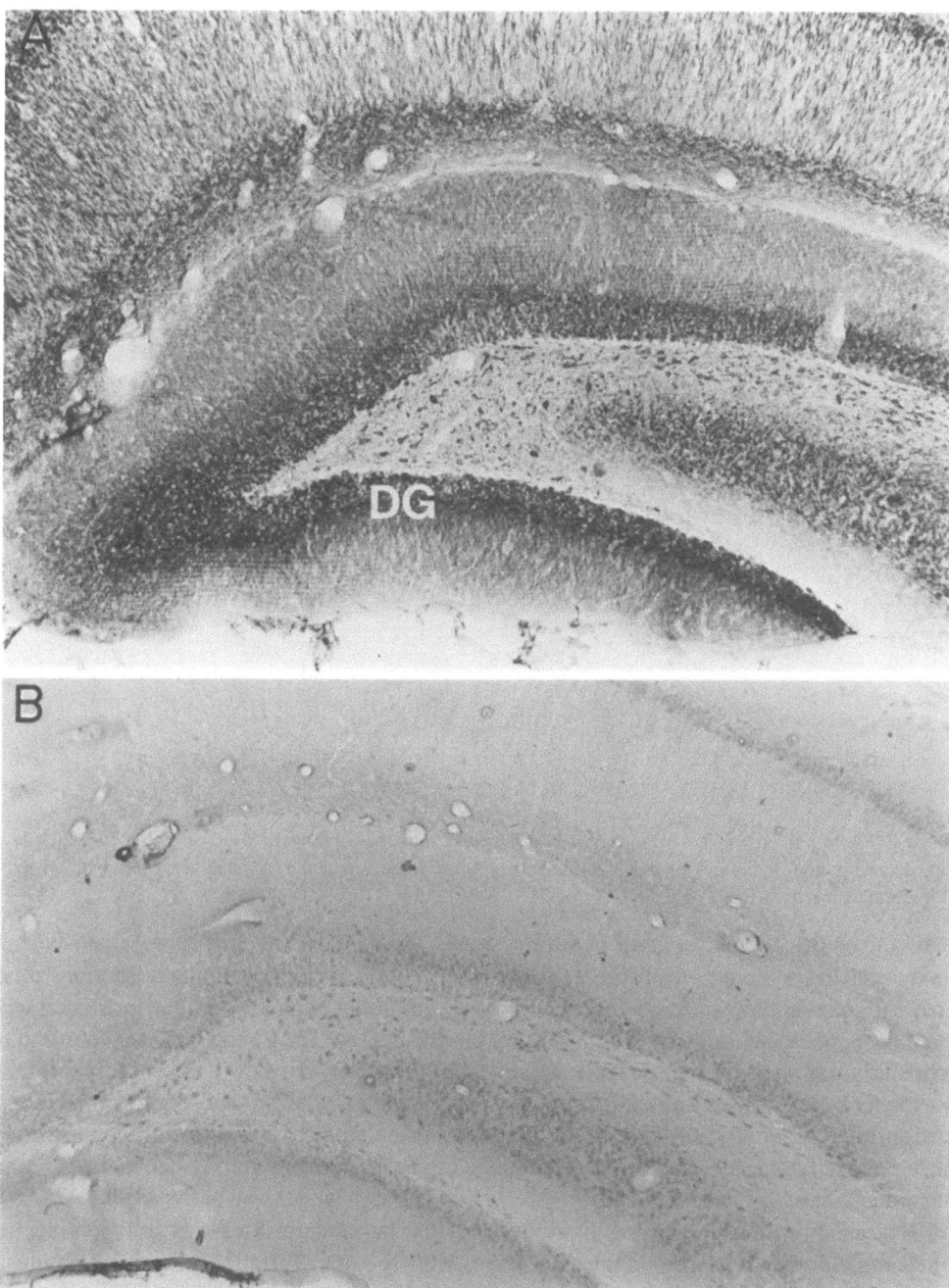

Figure 5. (A) Light micrograph showing calmodulin kinase II immunoreactivity in a coronal section through the hippocampal formation. Immunoreactivity is greatest in the dentate gyrus (DG). (B) Light micrograph of an adjacent section incubated with a control mouse monoclonal antibody (directed against dinitrophenol). The results shown in this figure and in Figs. 6 and 7 were obtained by the following procedures. Adult male rats were perfused with 500 ml of fixative containing 0.2% glutaraldehyde and 4% formaldehyde in 0.1 M phosphate buffer (pH 7.4). After 5-min postfixation *in situ*, brains were removed, cut into 1- to 2-mm slabs, and sectioned at 20 μm in a vibratome. The sections were then incubated successively with normal rabbit serum (1:100) for 1 hr, with monoclonal antibody C42.1 directed against calmodulin kinase II (1:2000) overnight, with rabbit anti-mouse IgG (1:20) for 1 hr, and with mouse peroxidase antiperoxidase (1:100) for 1 hr. The sections were then processed according to the peroxidase antiperoxidase method of Sternberger [50]. Tissue used for light microscopy was then mounted on microscope slides without counterstaining. Tissue for electron microscopy was osmicated (2% OsO_4 in 0.1 M phosphate buffer, pH 7.4) for 1 hr, dehydrated, embedded in epon-araldite, thin-sectioned (80 nm), and counterstained with aqueous uranyl acetate (1%) for 5 min.

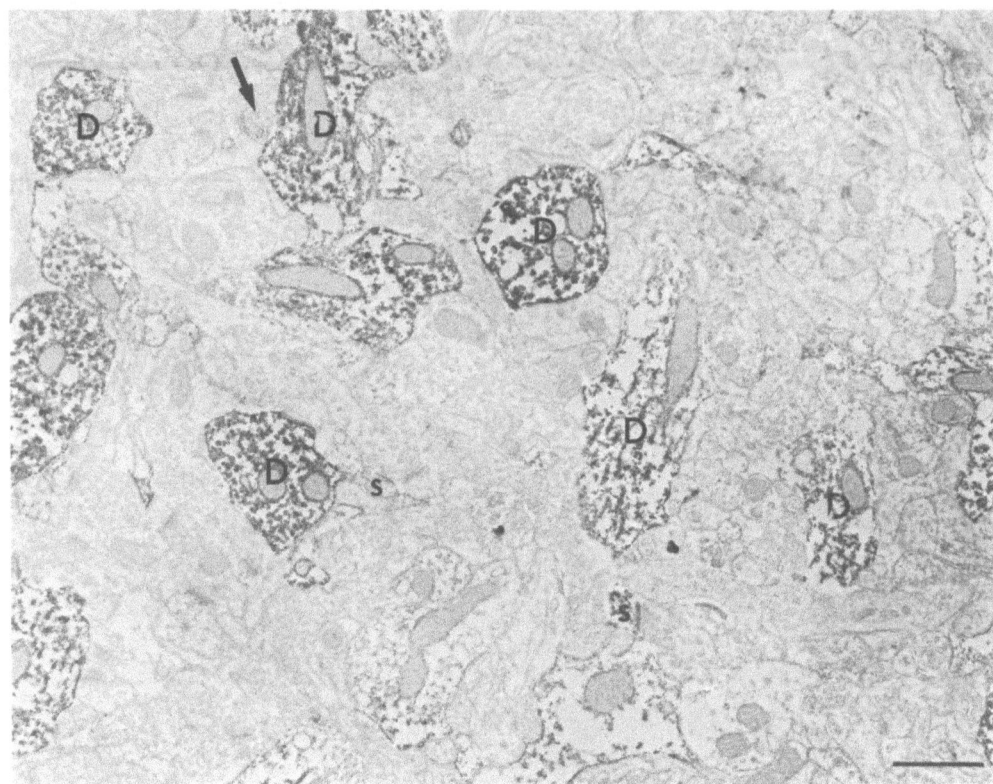

Figure 6. Electron micrograph showing calmodulin kinase II immunoreactivity in a section of the molecular layer of the dentate gyrus. Dendrites (D) and dendritic spines (s) are immunoreactive. A small nerve terminal (arrow) is lightly immunoreactive. Bar = 1 μm.

ronal cell bodies, dendrites, and dendritic spines (36,41). Horseradish peroxidase-labeled immunoreactive staining was present in the dentate gyrus of hippocampus in sections incubated with monoclonal antibody C42.1 (Fig. 5A), but not in adjacent sections incubated with a control monoclonal antibody (Fig. 5B). Within the dentate gyrus, staining was found to be particularly intense in granule cell bodies and dendrites. Electron microscopic immunocytochemistry showed limited reactivity in axons and axon terminals. Figures 6 and 7 are electron micrographs of ultrathin sections through the dentate gyrus showing dendrites (D) that are heavily labeled and nerve terminals (arrows) that are lightly labeled. No glial components have yet been labeled.

Examination of the regional distribution of the calmodulin kinase II polypeptides revealed immunoreactivity throughout the brain, with regional variation in staining intensity [36,41]. Areas with high immunoreactivity included: dentate gyrus, amygdala, septum, indusium griseum, outer layers of the neocortex, and the substantia gelatinosa of the spinal trigeminal nucleus. Large areas with lower-than-average staining intensity included: midbrain, cerebellum, pons, and medulla. This regional distribution qualitatively agrees with regional calmodulin kinase II activity as measured by phosphorylation of exogenous synapsin I and endogenous 50,000-, 58,000-, and 60,000-M_r polypeptides [54, 55].

Concluding Remarks

A Ca^{2+}/calmodulin-dependent protein kinase, calmodulin kinase II, purified on the basis of its ability to phosphorylate the neuron-specific protein synapsin I is present in

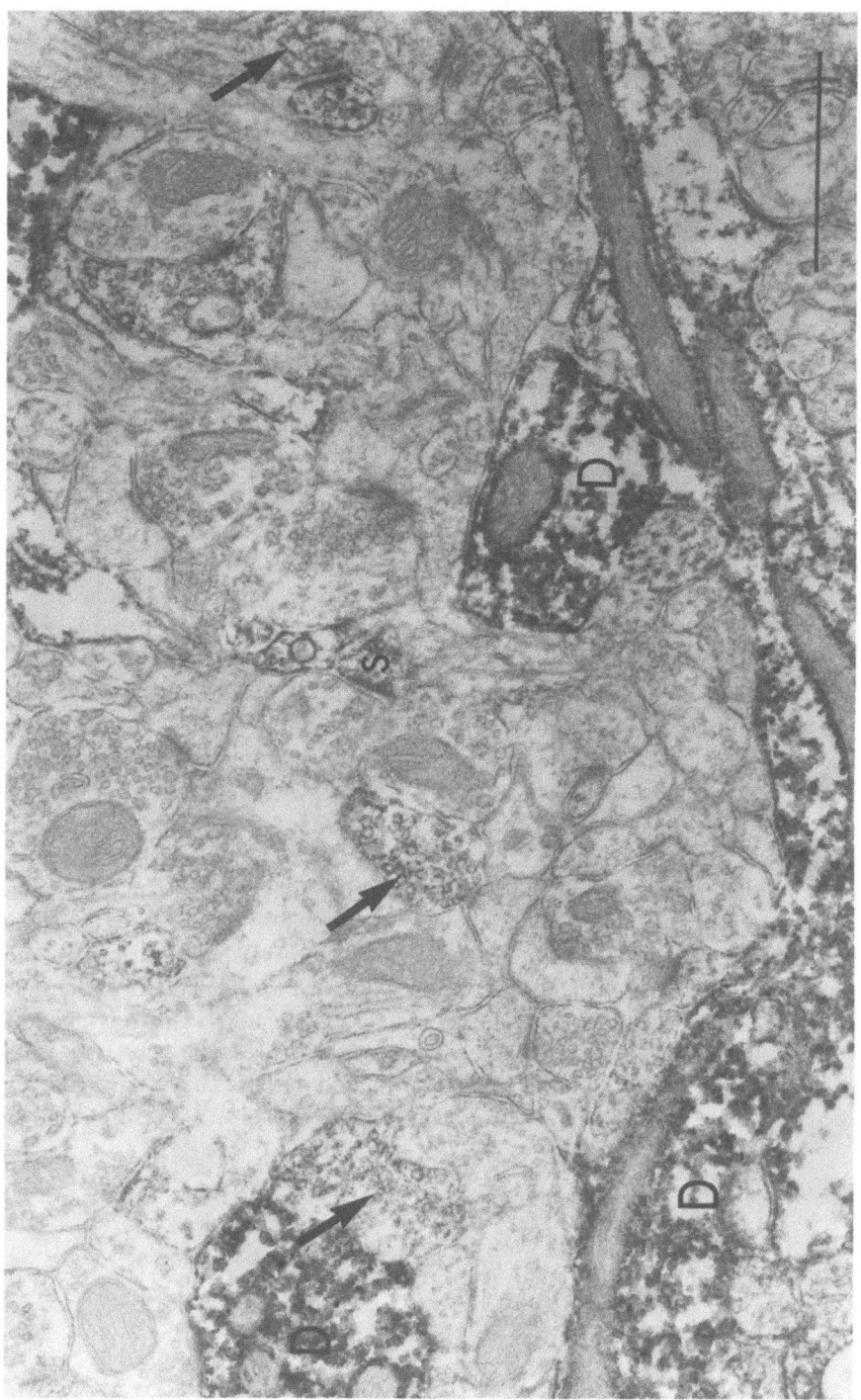

Figure 7. Higher magnification electron micrograph showing calmodulin kinase II immunoreactivity in a section of the molecular layer of the dentate gyrus. Dendrites (D) and dendritic spines (s) are immunoreactive. Nerve terminals (arrows) are lightly immunoreactive. Bar = 1 μm.

neurons and is localized in both soluble and particulate compartments of rat brain homogenates. It appears to be responsible for calmodulin-dependent protein kinase activity observed in SJ and PSD preparations and, by immunocytochemical evidence, is present within neuronal cell bodies, dendrites, axons, and terminals as well. Calmodulin kinase II activity, as measured by phosphorylation of exogenous synapsin I, exists in tissues other than brain, but in much lower apparent concentrations [30]. Preliminary experiments indicate that calmodulin kinase II represents approximately 0.4% of total rat brain protein (data not shown), suggesting that this enzyme is a relatively abundant brain protein. In agreement with this interpretation, the 50,000- and 60,000-M_r phosphoproteins were the most prominent substrates observed upon Ca^{2+}/calmodulin-dependent protein phosphorylation in a crude synaptosomal lysate preparation [31,47].

The ability of calmodulin kinase II to phosphorylate a number of substrate proteins suggests that it is a multifunctional enzyme, involved in several different Ca^{2+}-regulated processes. This would distinguish calmodulin kinase II from myosin light-chain kinase and phosphorylase kinase, each of which phosphorylates primarily one substrate protein. In support of the idea of a multifunctional role for calmodulin kinase II, the enzyme has properties similar to several calmodulin-dependent protein kinases purified from various tissues by several investigators. These include: a glycogen synthase kinase purified from liver [1,44,45] and from skeletal muscle [56,57]; a kinase purified from brain that phosphorylates smooth muscle myosin light chain and microtubule-associated protein-2 [16]; a kinase partially purified from brain that phosphorylates tubulin [17]; a kinase purified from brain that phosphorylates tyrosine-5-monooxygenase, tryptophan-5-monooxygenase, and microtubule-associated protein-2 [60]; and a kinase purified from *Torpedo californica* electric organ that phosphorylates synapsin I [42]. Most of these calmodulin-dependent kinases appear to have autophosphorylatable subunits in the 50,000- to 60,000-M_r range, are high-M_r oligomers of 500,000 to 800,000, and have purification properties similar to those of calmodulin kinase II. Moreover, calmodulin kinase II phosphorylates most of the substrates reported for these enzymes.

Recent studies [37] comparing calmodulin kinase II with the calmodulin-dependent glycogen synthase kinase isolated from skeletal muscle [56,57] indicate that these two enzymes share a number of properties including: substrate specificity, location in substrate of amino acid residue phosphorylated, immunological cross-reactivity, and phosphopeptide maps following limited proteolysis. Thus, these two enzymes may represent isozymes of a multifunctional calmodulin-dependent protein kinase. It will be of interest to determine whether the other enzymes listed above are closely related to calmodulin kinase II. Identification of the total number of calmodulin-dependent protein kinases, the substrates phosphorylated by each, and the tissue and subcellular distribution of these enzymes should help to elucidate the molecular pathways by which some of the physiological effects of Ca^{2+} are mediated.

ACKNOWLEDGMENTS. This work was supported by U.S. Public Health Service Grants MH–39327 and NS–21550, a grant from the McKnight Foundation, a contract from the USAF School of Aerospace Medicine, training grants GM–07205 (to T.L.M.) and GM–07324 (to Y.L.), and NIMH Fellowship MH–08601 (to C.C.O.).

References

1. Ahmad, Z.; DePaoli-Roach, A. A.; Roach, P. J. Purification and characterization of a rabbit liver calmodulin-dependent protein kinase able to phosphorylate glycogen synthase. *J. Biol. Chem.* **257**:8348–8355, 1982.

2. Baker, P. F. Transport and metabolism of calcium ions in nerve. *Prog. Biophys. Mol. Biol.* **24**:177–223, 1972.
3. Baker, P. F.; Hodgkin, A. L.; Ridgway, E. B. Depolarization and calcium entry in squid giant axons. *J. Physiol. (London)* **218**:709–755, 1971.
4. Banker, G.; Churchill, L.; Cotman, C. W. Proteins of the postsynaptic density. *J. Cell Biol.* **63**:456–465, 1974.
5. Blomberg, F.; Cohen, R. S.; Siekevitz, P. The structure of postsynaptic densities isolated from dog cerebral cortex. II. Characterization and arrangement of some of the major proteins within the structure. *J. Cell Biol.* **74**:204–225, 1977.
6. Brostrom, C. O.; Huang, Y.-C.; Breckenridge, B. M.; Wolff, D. J. Identification of a calcium-binding protein as a calcium-dependent regulator of brain adenylate cyclase. *Proc. Natl. Acad. Sci. USA* **72**:64–68, 1975.
7. Carlin, R. K.; Grab, D. J.; Siekevitz, P. Function of calmodulin in postsynaptic densities. III. Calmodulin-binding proteins of the postsynaptic density. *J. Cell Biol.* **89**:449–455, 1981.
8. Cheung, W. Y. Cyclic 3',5'-nucleotide phosphodiesterase: Demonstration of an activator. *Biochem. Biophys. Res. Commun.* **38**:533–538, 1970.
9. Cheung, W. Y. Calmodulin plays a pivotal role in cellular regulation. *Science* **207**:19–27, 1980.
10. Cheung, W. Y.; Bradham, L. S.; Lynch, T. J.; Lin, Y. M.; Tallant, E. A. Protein activator of cyclic 3':5'-nucleotide phosphodiesterase of bovine or rat brain also activates its adenylate cyclase. *Biochem. Biophys. Res. Commun.* **66**:1055–1062, 1975.
11. Cohen, P.; Burchell, A.; Foulkes, J. G.; Cohen, P. T. W.; Vanaman, T. C.; Nairn, A. C. Identification of the Ca^{2+}-dependent modulator protein as the fourth subunit of rabbit skeletal muscle phosphorylase kinase. *FEBS Lett.* **92**:287–293, 1978.
12. Dabrowska, R.; Sherry, J. M. F.; Aromatorio, D. K.; Harthshorne, D. J. Modulator protein as a component of the myosin light chain kinase from chicken gizzard. *Biochemistry* **17**:253–258, 1978.
13. De Camilli, P.; Cameron, R.; Greengard, P. Synapsin I (protein I), a nerve terminal-specific phosphoprotein. I. Its general distribution in synapses of the central and peripheral nervous system demonstrated by immunofluorescence in frozen and plastic sections. *J. Cell Biol.* **96**:1337–1354, 1983.
14. De Camilli, P.; Harris, S. M., Jr.; Huttner, W. B.; Greengard, P. Synapsin I (protein I), a nerve terminal-specific phosphoprotein. II. Its specific association with synaptic vesicles demonstrated by immunocytochemistry in agarose-embedded synaptosomes. *J. Cell Biol.* **96**:1355–1373, 1983.
15. Frankenhaeuser, B. The effect of calcium on the myelinated nerve fiber. *J. Physiol. (London)* **137**:245–260, 1957.
16. Fukunaga, K.; Yamamoto, H.; Matsui, K.; Higashi, K.; Miyamoto, E. Purification and characterization of a Ca^{2+}- and calmodulin-dependent protein kinase from rat brain. *J. Neurochem.* **39**:1607–1617, 1982.
17. Goldenring, J. R.; Gonzalez, B.; DeLorenzo, R. J. Isolation of brain Ca^{2+}–calmodulin tubulin kinase containing calmodulin binding proteins. *Biochem. Biophys. Res. Commun.* **108**:421–428, 1982.
18. Grab, D. J.; Carlin, R. K.; Siekevitz, P. Function of calmodulin in postsynaptic densities. II. Presence of a calmodulin-activatable protein kinase activity. *J. Cell. Biol.* **89**:440–448, 1981.
19. Greengard, P. Phosphorylated proteins as physiological effectors. *Science* **199**:146–152, 1978.
20. Greengard, P. Intracellular signals in the brain. *Harvey Lect.* **75**:277–331, 1981.
21. Groswald, D. E.; Montgomery, P. R.; Kelly, P. T. Synaptic junctions isolated from cerebellum and forebrain: Comparisons of morphological and molecular properties. *Brain Res.* **278**:63–80, 1983.
22. Huttner, W. B.; DeGennaro, L. J.; Greengard, P. Differential phosphorylation of multiple sites in purified protein I by cyclic AMP-dependent and calcium-dependent protein kinases. *J. Biol. Chem.* **256**:1482–1488, 1981.
23. Huttner, W. B.; Greengard, P. Multiple phosphorylation sites in protein I and their differential regulation by cyclic AMP and calcium. *Proc. Natl. Acad. Sci. USA* **76**:5402–5406, 1979.
24. Huttner, W. B.; Schiebler, W.; Greengard, P.; De Camilli, P. Synapsin I (protein I), a nerve terminal-specific phosphoprotein. III. Its association with synaptic vesicles studied in a highly purified vesicle preparation. *J. Cell Biol.* **96**:1374–1388, 1983.
25. Kakiuchi, S.; Yamazaki, R. Calcium dependent phosphodiesterase activity and its activating factor (PAF) from brain: Studies on cyclic 3',5'-nucleotide phosphodiesterase (III). *Biochem. Biophys. Res. Commun.* **41**:1104–1110, 1970.
26. Kelly, J. S.; Krnjevic, K.; Somjen, G. J. Divalent cations and electrical properties of cortical cells. *J. Neurobiol.* **1**:197–208, 1969.
27. Kelly, P. T.; McGuinness, T. L.; Greengard, P. Evidence that the major postsynaptic density protein is a component of a Ca^{2+}/calmodulin-dependent protein kinase. *Proc. Natl. Acad. Sci. USA* **81**:945–949, 1984.
28. Kelly, P.; McGuinness, T.; Greengard, P. Calcium/calmodulin-dependent phosphorylation in synaptic junctions. *Soc. Neurosci. Abstr.* **9**:1030, 1983.
29. Kelly, P. T.; Montgomery, P. R. Subcellular localization of the 52,000 molecular weight major postsynaptic density protein. *Brain Res.* **233**:265–286, 1982.

30. Kennedy, M. B.; Greengard, P. Two calcium/calmodulin-dependent protein kinases, which are highly concentrated in brain, phosphorylate protein I at distinct sites. *Proc. Natl. Acad. Sci. USA* **78**:1293–1297, 1981.
31. Kennedy, M. B.; McGuinness, T.; Greengard, P. A calcium/calmodulin-dependent protein kinase from mammalian brain that phosphorylates synapsin I: Partial purification and characterization. *J. Neurosci.* **3**:818–831, 1983.
32. Krebs, E. G. Protein kinases. *Curr. Top. Cell. Regul.* **5**:99–133, 1972.
33. Krueger, B. K.; Forn, J.; Greengard, P. Depolarization-induced phosphorylation of specific proteins, mediated by calcium ion influx, in rat brain synaptosomes. *J. Biol. Chem.* **252**:2764–2773, 1977.
34. Kuo, J. F.; Greengard, P. Cyclic nucleotide-dependent protein kinases, IV. Widespread occurrence of adenosine 3′,5′-monophosphate-dependent protein kinase in various tissues and phyla of the animal kingdom. *Proc. Natl. Acad. Sci. USA* **64**:1349–1355, 1969.
35. Lai, Y.; McGuinness, T. L.; Greengard, P. Purification and characterization of brain Ca^{2+}/calmodulin-dependent protein kinase II that phosphorylates synapsin I. *Soc. Neurosci. Abstr.* **9**:1029, 1983.
36. McGuinness, T. L.; Kelly, P. T.; Ouimet, C. C.; Greengard, P. Studies on the subcellular and regional distribution of calmodulin-dependent protein kinase II in rat brain. *Soc. Neurosci. Abstr.* **9**:1029, 1983.
37. McGuinness, T. L.; Lai, Y.; Greengard, P.; Woodgett, J. R.; Cohen, P. A multifunctional calmodulin-dependent protein kinase: Similarities between skeletal muscle glycogen synthase kinase and a brain synapsin I kinase. *FEBS Lett.* **163**:329–334, 1983.
38. Nairn, A. C.; Greengard, P. Purification and characterization of brain Ca^{2+}/calmodulin-dependent protein kinase I that phosphorylates synapsin I. *Soc. Neurosci. Abstr.* **9**:1029, 1983.
39. Nestler, E. J.; Greengard, P. Protein phosphorylation in the brain. *Nature (London)* **305**:583–588, 1983.
40. Nestler, E. J.; Greengard, P. *Protein Phosphorylation in the Nervous System*, New York, Wiley, 1984, pp. 17–42.
41. Ouimet, C. C.; McGuinness, T. L.; Greengard, P. Immunocytochemical localization of calcium/calmodulin-dependent protein kinase II in rat brain. *Proc. Natl. Acad. Sci. USA* **81**:5604–5608, 1984.
42. Palfrey, H. C.; Rothlein, J. E.; Greengard, P. Calmodulin-dependent protein kinase and associated substrates in *Torpedo* electric organ. *J. Biol. Chem.* **258**:9496–9503, 1983.
43. Patrick, R. L.; Barchas, J. D. Stimulation of synaptosomal dopamine synthesis by veratridine. *Nature (London)* **250**:737–739, 1974.
44. Payne, M. E.; Schworer, C. M.; Soderling, T. R. Purification and characterization of rabbit liver calmodulin-dependent glycogen synthase kinase. *J. Biol. Chem.* **258**:2376–2382, 1983.
45. Payne, M. E.; Soderling, T. R. Calmodulin-dependent glycogen synthase kinase. *J. Biol. Chem.* **255**:8054–8056, 1980.
46. Rubin, R. P. The role of calcium in the release of neurotransmitter substances and hormones. *Pharmacol. Rev.* **22**:389–428, 1972.
47. Schulman, H.; Greengard, P. Stimulation of brain membrane protein phosphorylation by calcium and an endogenous heat-stable protein. *Nature (London)* **271**:478–479, 1978.
48. Schulman, H.; Greengard, P. Ca^{2+}-dependent protein phosphorylation system in membranes from various tissues, and its activation by "calcium-dependent regulator." *Proc. Natl. Acad. Sci. USA* **75**:5432–5436, 1978.
49. Sieghart, W.; Forn, J.; Greengard, P. Ca^{2+} and cyclic AMP regulate phosphorylation of same two membrane-associated proteins specific to nerve tissue. *Proc. Natl. Acad. Sci. USA* **76**:2475–2479, 1979.
50. Sternberger, L. A. *Immunocytochemistry*, New York, Wiley, 1979, pp. 104–129.
51. Takai, Y.; Kishimoto, A.; Iwasa, Y.; Kawahara, Y.; Mori, T.; Nishizuka, Y. Calcium-dependent activation of a multifunctional protein kinase by membrane phospholipids. *J. Biol. Chem.* **254**:3692–3695, 1979.
52. Ueda, T.; Greengard, P. Adenosine 3′:5′-monophosphate-regulated phosphoprotein system of neuronal membranes. I. Solubilization, purification, and some properties of an endogenous phosphoprotein. *J. Biol. Chem.* **252**:5155–5163, 1977.
53. Waisman, D. M.; Singh, T. J.; Wang, J. H. The modulator-dependent protein kinase. *J. Biol. Chem.* **253**:3387–3390, 1978.
54. Walaas, S. I.; Nairn, A. C.; Greengard, P. Regional distribution of calcium- and cyclic adenosine 3′:5′-monophosphate-regulated protein phosphorylation systems in mammalian brain. I. Particulate systems. *J. Neurosci.* **3**:291–301, 1983.
55. Walaas, S. I.; Nairn, A. C.; Greengard, P. Regional distribution of calcium- and cyclic adenosine 3′:5′-monophosphate-regulated protein phosphorylation systems in mammalian brain. II. Soluble systems. *J. Neurosci.* **3**:302–311, 1983.
56. Woodgett, J. R.; Davison, M. T.; Cohen, P. The calmodulin-dependent glycogen synthase kinase from rabbit skeletal muscle: Purification, subunit structure and substrate specificity. *Eur. J. Biochem.* **136**:481–487, 1983.
57. Woodgett, J. R.; Tonks, N. K.; Cohen, P. Identification of a calmodulin-dependent glycogen synthase kinase in rabbit skeletal muscle, distinct from phosphorylase kinase. *FEBS Lett.* **148**:5–11, 1982.

58. Yagi, K.; Yazawa, M.; Kakiuchi, S.; Ohshima, M.; Uenishi, K. Identification of an activator protein for myosin light chain kinase as the Ca^{2+}-dependent modulator protein. *J. Biol. Chem.* **253**:1338–1340, 1978.
59. Yamauchi, T.; Fujisawa, H. Evidence for three distinct forms of calmodulin-dependent protein kinases from rat brain. *FEBS Lett.* **116**:141–144, 1980.
60. Yamauchi, T.; Fujisawa, H. Purification and characterization of the brain calmodulin-dependent protein kinase (kinase II), which is involved in the activation of tryptophan 5-monooxygenase. *Eur. J. Biochem.* **132**:15–21, 1983.

35

Calcineurin, a Calmodulin-Stimulated Protein Phosphatase

Allan S. Manalan and Claude B. Klee

Calcineurin, a major calmodulin-binding protein of brain [15], was initially identified as a heat-labile inhibitor of calmodulin-stimulated cyclic nucleotide phosphodiesterase [17,36]. Although the function of this protein was not known, because of its Ca^{2+}-binding properties [15] and its apparent neural localization [34], it was named *calcineurin* [15]. The recent discovery of a calmodulin-stimulated protein phosphatase with similar subunit structure [29] led to the identification of calcineurin as a phosphoprotein phosphatase [30]. Calcineurin subunit interactions can now be correlated with observed regulatory properties of the enzyme, providing the opportunity to explore the relationship between subunit structure and the mechanisms of phosphatase regulation.

Structure

Cross-linking experiments with dimethylsuberimidate have demonstrated that calcineurin is a heterodimer composed of a 61,000-M_r subunit, calcineurin A, and a 19,000-M_r subunit, calcineurin B [15–17].* The interaction between the two subunits of calcineurin is independent of Ca^{2+} or other divalent cations. Calcineurin interacts with calmodulin only in the presence of Ca^{2+}. Thus, in the presence of Ca^{2+} and calmodulin, treatment with dimethylsuberimidate produces additional cross-linked species of apparent M_r 92,000 and 76,000, corresponding to a complex of calmodulin, calcineurin A, and calcineurin B and either a complex of calmodulin and calcineurin A or a complex of calcineurin B and calcineurin A (Table I). No cross-linking of calmodulin with calcineurin B is observed. Thus, Calcineurin A is the calmodulin-binding subunit. These findings of cross-linking experiments

---

*Calcineurin B exhibits apparent M_r of 15,000–16,000 on SDS gel electrophoresis. The value of 19,000 is based on the amino acid sequence of calcineurin B [3].

Allan S. Manalan and Claude B. Klee • Laboratory of Biochemistry, National Cancer Institute, National Institutes of Health, Bethesda, Maryland 20205.

Table I. Physical and Enzymatic Properties of Calcineurin

	Calcineurin			Trypsin-treated calcineurin		
	EGTA	Ca^{2+}	Ca^{2+}·CaM	EGTA	Ca^{2+}	Ca^{2+}·CaM
$M_r{}^a$	76,000	76,000	92,000			
$s_{20,w}$ (S)[b]	4.5		5.0	4.3		4.3
Subunit composition[b]	A·B		CaM·A·B	A'·B[c]		A'·B[c]
Phosphatase[d] (nmole/min/mg)	0.4	0.5	2.1	3.2	2.5	1.8

[a]M_r was assessed by electrophoretic mobility of cross-linked products after treatment with dimethylsuberimidate [15].
[b]Sedimentation coefficients and subunit composition were determined by glycerol gradient centrifugation (Fig. 3).
[c]A' refers to the 44,000-M_r proteolytic fragment of calcineurin generated by trypsin treatment in the absence of calmodulin.
[d]Determined as previously described [23].

have been corroborated using the [^{125}I]calmodulin gel overlay method [22,23]. Only calcineurin A is labeled by [^{125}I]calmodulin, and this binding is dependent on Ca^{2+}. In the presence of Ca^{2+}, calcineurin exhibits high affinity for calmodulin ($K_i = 3 \times 10^8$ M^{-1}), allowing it to compete effectively with other calmodulin-binding proteins [8,17,20,21,33,36].

Calcineurin is itself a Ca^{2+}-binding protein, capable of binding 4 moles Ca^{2+}/mole, with dissociation constants in the micromolar range [15]. The mobility of the small subunit, calcineurin B, is altered by Ca^{2+} on SDS gel electrophoresis [15]. Such anomalous electrophoretic behavior is typical of "E-F hand" Ca^{2+}-binding proteins such as troponin C, calmodulin, and parvalbumins, consistent with the identification of calcineurin B as the Ca^{2+}-binding subunit.

The two subunits of calcineurin were separated from each other by DEAE-cellulose chromatography in the presence of 6 M urea, and the small subunit was subsequently purified by reverse-phase HPLC. The UV absorption spectrum of calcineurin B is similar to that of calmodulin. Its amino acid sequence revealed the presence of four "E-F" Ca^{2+}-binding domains analogous but not identical to those of calmodulin [3]. Calcineurin B contains as an amino-terminal blocking group myristic acid [2], which has also been identified on the catalytic subunit of cAMP-dependent protein kinase [7]. Although the function of this blocking group is not known, possibilities include [1] stabilization of the complex of the two subunits of calcineurin or [2] anchoring of calcineurin in membranes [22].

Enzymatic Activity

A Ca^{2+}- and calmodulin-regulated phosphatase activity has recently been found associated with calcineurin [30]. Because of the presence of Ca^{2+}-independent phosphatases in crude brain fractions, a Ca^{2+}- and calmodulin-stimulated phosphatase can be detected only after selective purification by calmodulin-Sepharose affinity chromatography. However, calcineurin can be quantified even in crude extracts by [^{125}I]calmodulin gel overlay [6,18,22]. Throughout the purification of calcineurin, accomplished by sequential column chromatography, a Ca^{2+}- and calmodulin-stimulated phosphatase activity copurifies with calcineurin [18,22]. Calcineurin and its associated phosphatase activity are retained by a substrate affinity column (thiophosphorylated myosin P-light chain-Sepharose) in the presence of Ca^{2+}, with specific elution by EGTA [31].

Calcineurin dephosphorylates the α subunit of phosphorylase b kinase, protein phosphatase inhibitor-1, myosin light chains [12,29], and the regulatory subunit of type II cAMP-dependent protein kinase [4,18]. The V_{max} value for these substrates is between 0.4 and 2 μmole/min per mg [29]. Histones IIa, VS, and the β subunit of phosphorylase kinase are dephosphorylated by calcineurin at a rate that is 1/10th to 1/1000th that of inhibitor-1. While [^{32}P]phosphorylated myosin light chains are a convenient substrate for phosphatase assay, practical concentrations of light chains (1–10 μM) lie well below the K_m value. Therefore, activities observed under these assay conditions are well below the maximal achievable velocity (0.4 μmole/min per mg). With [^{32}P]phosphorylated myosin light chains or histones IIa or VS as substrate, phosphatase activity of calcineurin is slightly stimulated by Ca^{2+}, and further stimulated (two- to fourfold) by calmodulin [18,23]. Maximal phosphatase activation is achieved at a calmodulin/calcineurin molar ratio of 1:1.

Calcineurin is therefore considered a member of the group of protein phosphatases designated 2B, defined on the basis of characteristic substrate specificity, stimulation by Ca^{2+} and calmodulin, and inhibition by phenothiazines [12]. Members of this group have been detected in several tissues [13] and purified to apparent homogeneity from skeletal muscle [29] and brain [17,28,33]. Low levels of a calcineurin-like protein are also present in heart [19,37]. These phosphatases exhibit differences in apparent M_r of their large subunits which, although suggestive of tissue specificity, may merely result from differing amounts of limited proteolysis.

Activation by Limited Proteolysis

The effects of limited proteolysis of calcineurin on the Ca^{2+}-, calmodulin-stimulated protein phosphatase have strengthened identification of calcineurin as a protein phosphatase. Like other calmodulin-regulated enzymes [9,11,19,21,24–26,32,35], calcineurin is irreversibly activated by limited proteolysis [23]. Treatment of calcineurin with trypsin activates the phosphatase assayed in the presence of EGTA. The protease-stimulated enzyme has a higher activity in the absence of Ca^{2+} than that of the undigested enzyme assayed in the presence of Ca^{2+} and calmodulin. Ca^{2+} or Ca^{2+}–calmodulin exerts an inhibitory effect on enzyme activity, which is lost upon storage. This results in an enzyme with specific activity similar to that of the intact calcineurin–calmodulin complex, but which is no longer affected by EGTA, Ca^{2+}, or calmodulin. Trypsin-induced activation of calcineurin is accompanied by conversion of the 61,000-M_r subunit, calcineurin A, to a 44,000-M_r fragment which cannot bind calmodulin (Fig. 1) [23]. The absence of other detectable polypeptides on SDS gels suggests that formation of this 44,000-M_r fragment, which is generated within 4 min of trypsin treatment, is the result of multiple cleavages involving a region of calcineurin A highly susceptible to proteolysis. Loss of this fragment is accompanied by enzyme activation and loss of the ability to bind calmodulin.

The addition of calmodulin protects calcineurin from proteolysis and enzyme activation (Fig. 1) [23]. Loss of enzyme stimulation by Ca^{2+} and calmodulin was observed only after 40 min of digestion at a higher concentration of trypsin. The pattern of digestion of calcineurin was also altered. Fragments of calcineurin A of 57,000, 55,000, 54,000, 46,000, and 40,000 M_r were generated over the course of 40 min of trypsin treatment (Figs. 1 and 2). Each of the four largest fragments retained the ability to bind calmodulin on [^{125}I]calmodulin gel overlay. Calcineurin B, identified on SDS gels by its differing mobility in the presence of Ca^{2+} or EGTA, was not significantly degraded (Fig. 2). [^{14}C]guanidinated calmodulin also appeared resistant to proteolysis when bound to calcineurin, since no calmodulin peptides

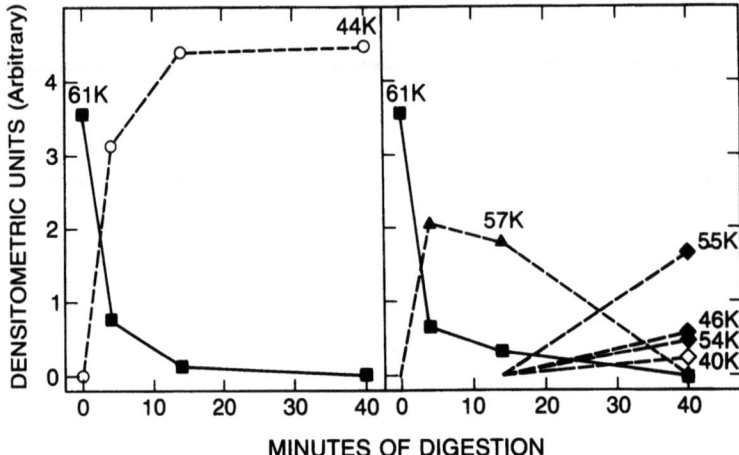

Figure 1. Effect of limited proteolysis of calcineurin on subunit size and ability to interact with calcineurin. Limited tryptic digestion of calcineurin was conducted in the absence (left panel) or presence (right panel) of 3 μM calmodulin. Trypsin concentration was 0.35 μg/ml (left panel) or 1.0 μg/ml (right panel). After digestion at 30°C for the indicated times, reactions were stopped by addition of soybean trypsin inhibitor at 100 μg/ml, and aliquots were subjected to SDS-polyacrylamide gel electrophoresis and [^{125}I]calmodulin gel overlay [23]. This figure presents the time course of digestion of calcineurin A, assessed by densitometric analysis of the Coomassie-stained gel. Those bands exhibiting [^{125}I]calmodulin labeling are indicated by the closed symbols. In the presence of calmodulin, the pattern of digestion of calcineurin is altered, with preservation of the ability of the large fragments to bind calmodulin.

were detected by autoradiography (Fig. 2). Even when [^{14}C]guanidinated calmodulin was present in excess over calcineurin, only very faint bands with M_r corresponding to that of fragments 78–148 and 1–77 were detected after 45 min of digestion (Fig. 2).

The effects of limited proteolysis on calcineurin subunit interactions and on the ability of calcineurin to interact with calmodulin were also evaluated. On centrifugation of calcineurin in a glycerol gradient containing Ca^{2+} and [^{14}C]guanidinated calmodulin [23], a 1:1:1 complex of calcineurin A, calcineurin B, and [^{14}C]guanidinated calmodulin was detected, with a sedimentation coefficient of 5 S (Fig. 3, Table I). Ca^{2+}-, calmodulin-stimulated protein phosphatase activity cosedimented with this complex [23]. When trypsin-treated calcineurin, prepared by trypsin treatment of calcineurin in the absence of calmodulin, was subjected to glycerol gradient centrifugation under the same conditions, no complex with labeled calmodulin was detected (Table I). However, calcineurin B cosedimented with the 44,000-M_r fragment of calcineurin A (Fig. 3). The interaction of calcineurin A and calcineurin B was unaffected by calmodulin binding or proteolysis, thus excluding subunit dissociation as a mechanism of calcineurin activation. Phosphatase activity cosedimented with the two subunits of trypsin-treated calcineurin ($s_{20,w}$ = 4.3 S) (Table I). In this case, phosphatase activity was higher in the presence of EGTA than in the presence of Ca^{2+} and calmodulin.

When glycerol gradients were used in the presence of EGTA, phosphatase activity again cosedimented with calcineurin or its proteolytic derivative [23]. Calcineurin had a sedimentation coefficient of 4.5 S, lower than that of the calmodulin–calcineurin complex (Table I). The trypsin-treated enzyme had a sedimentation coefficient of 4.3 S, whether in the presence of EGTA or in the presence of Ca^{2+} and calmodulin. Once again, calcineurin B cosedimented with the 44,000-M_r fragment of calcineurin A, demonstrating preservation of Ca^{2+}-independent interaction of the two subunits (Fig. 3, Table I).

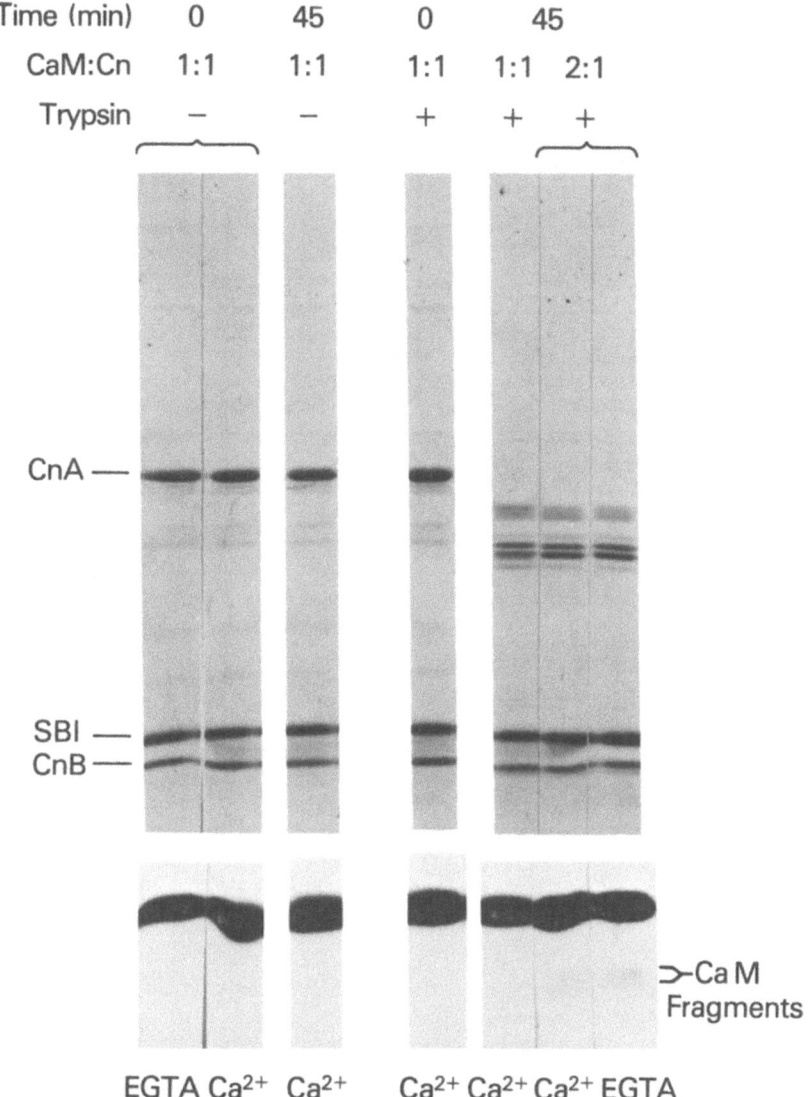

Figure 2. Effects of limited proteolysis on the calmodulin–calcineurin complex. Limited tryptic digestion of calcineurin was performed in the presence of [^{14}C]guanidinated calmodulin at a calmodulin/calcineurin ratio of 1:1 or 2:1. Digestion was conducted at 30°C for the indicated times at a trypsin concentration of 1 μg/ml. Reactions were stopped by addition of soybean trypsin inhibitor (100 μg/ml), and digests subjected to SDS-polyacrylamide gel electrophoresis, Coomassie blue staining, and autofluorography. Trypsin treatment of the calmodulin–calcineurin complex (1:1) resulted in generation of fragments of calcineurin A as shown in Fig. 1. Neither calcineurin B, identified by its characteristic increased mobility in the presence of Ca^{2+}, nor [^{14}C]guanidinated calmodulin, identified by autofluorography and Ca^{2+}-induced increased mobility, was detectably altered by trypsin treatment. When [^{14}C]guanidinated calmodulin was present in excess, very faint bands of M_r corresponding to calmodulin fragments 78–148 and 1–77 were detected by autofluorography.

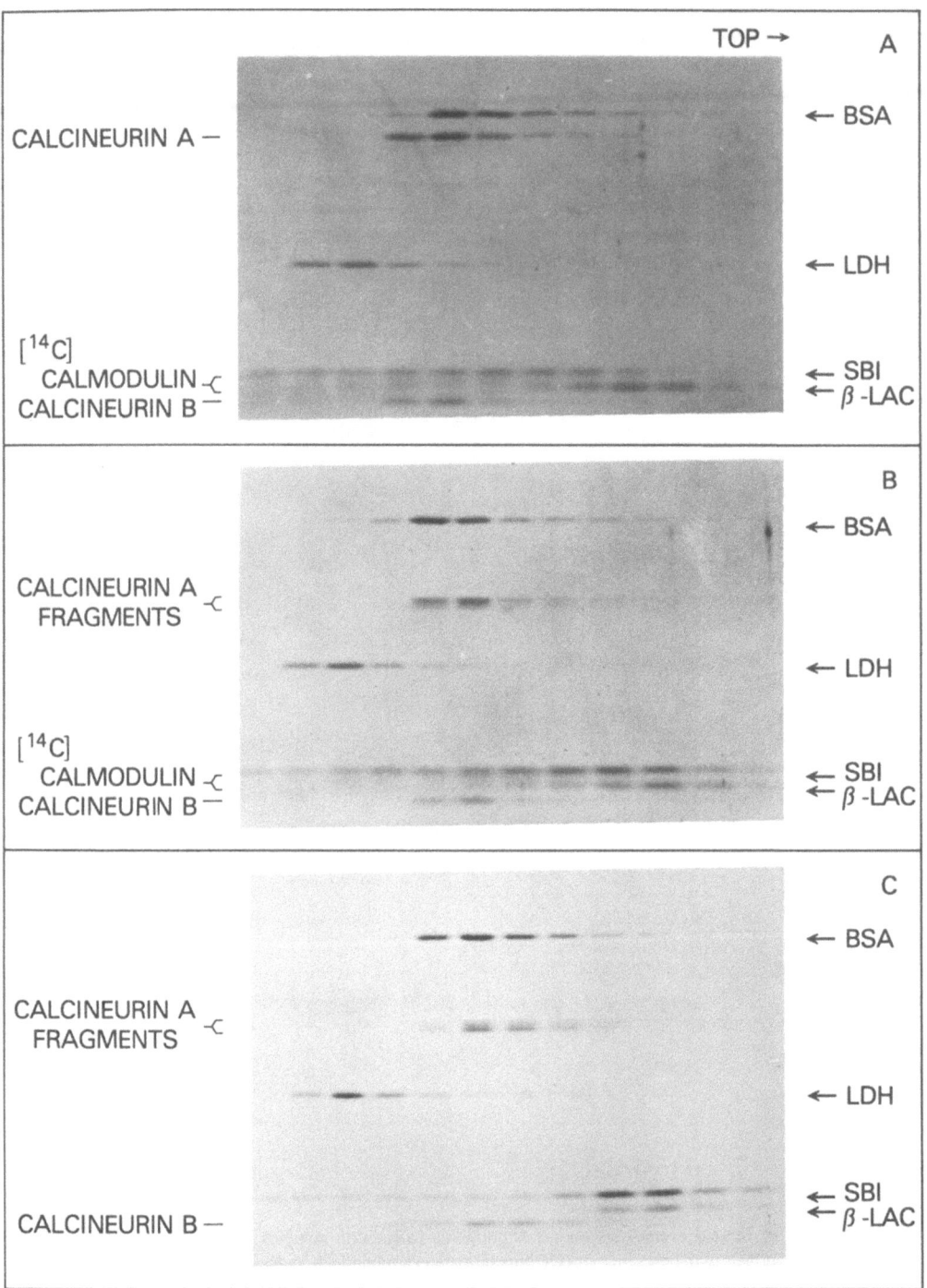

Figure 3. Glycerol gradient centrifugation of calcineurin and trypsin-treated calcineurin. Treatment with trypsin (0.35 μg/ml) was for 40 min at 30°C in the absence of calmodulin. Digestion was stopped by addition of soybean trypsin inhibitor (100 μg/ml). Soybean trypsin inhibitor (SBI) was also included in glycerol gradients to inhibit further proteolysis during sedimentation. Calcineurin, sedimented in the presence of Ca^{2+} and [^{14}C]guanidinated calmodulin (upper panel). Trypsin-treated calcineurin, sedimented in the presence of Ca^{2+} and [^{14}C]guanidinated calmodulin (middle panel) or in the presence of EGTA (lower panel). Aliquots of gradient fractions were analyzed by SDS-polyacrylamide gel electrophoresis, for comparison with results of phosphatase assay [23]. Calcineurin B cosedimented with calcineurin A or its proteolytic fragments both in the presence and absence of Ca^{2+}-calmodulin. [^{14}C]guanidinated calmodulin, identified by autofluorography as a diffuse band, was not well visualized on the Coomassie blue-stained gel. Sedimentation markers were lactate dehydrogenase (LDH), bovine serum albumin (BSA) and β-lactoglobulin (β-LAC).

Conclusions

Recent results further document the identification of calcineurin as a protein phosphatase. The findings of limited proteolysis suggest that at least two domains exist within calcineurin A. One domain contains the catalytic site as well as the site for interaction with calcineurin B. This region is relatively resistant to proteolysis, and further resolution of potentially distinct catalytic and calcineurin B-binding domains was not observed. The other domain, which is highly susceptible to proteolysis, appears essential for interaction with calmodulin. Removal of this calcineurin A domain produces relatively little change in the sedimentation coefficient of calcineurin (Table I), suggesting that it may contribute to asymmetry in the intact molecule. This observation suggests that the calmodulin-binding domain may be partially unfolded, consistent with its observed susceptibility to proteolysis. By analogy with other calmodulin-regulated enzymes (1,9,11,14,19,24–26,32,35), the present results are consistent with the presence of a calmodulin-binding domain, whose inhibitory effect on enzyme activity is nullified by either proteolysis or conformational changes induced by calmodulin binding.

Calcineurin appears to be a member of a class of calmodulin-binding proteins that contains an integral Ca^{2+}-binding subunit. Thus, like phosphorylase b kinase [27] and $Ca^{2+} + Mg^{2+}$-ATPase [5,10], calcineurin exhibits two mechanisms of Ca^{2+} regulation. The primary mode is mediated directly by Ca^{2+} binding to integral sites on calcineurin B, with an additional avenue for regulation by calmodulin when Ca^{2+} levels are increased.

Biochemical responses to regulatory stimuli are mediated by at least two major intracellular signals, cAMP and Ca^{2+}. Effects of Ca^{2+} may be synergistic or antagonistic, depending on the nature and distribution of cAMP- and Ca^{2+}-regulated effectors in a given tissue. Immunohistochemical studies have demonstrated localization of calcineurin to postsynaptic densities [6,38], suggesting a role for this phosphatase in neurotransmission. While the presence of calcineurin in the cytosol is not excluded, these results are compatible with the recent finding that a substantial portion of calcineurin in brain is membrane-bound (manuscript in preparation). Thus, it is possible that the myristyl NH_2-terminal acylation of calcineurin B [2] is involved in anchoring calcineurin in the membrane. Although the physiological substrate(s) of calcineurin is unknown, studies *in vitro* demonstrate that calcineurin is capable of dephosphorylating several substrates of cAMP-dependent protein kinase, including phosphorylase kinase, phosphatase inhibitor-1, as well as the regulatory subunit of type II cAMP-dependent protein kinase. These results suggest a role for calcineurin in antagonizing the biochemical effects of cAMP, thereby providing another modality for interaction by the two second messengers, cAMP and Ca^{2+}.

References

1. Adelstein, R. S.; Pato, M. D.; Sellers, J. R.; deLanerolle, P.; Conti, M. A. Regulation of actin–myosin interaction by reversible phosphorylation of myosin and myosin kinase. *Cold Spring Harbor Symp. Quant. Biol.* **46**:921–928, 1981.
2. Aitken, A.; Cohen, P.; Santikarn, S.; Williams, D. H.; Calder, A. G.; Smith, A.; Klee, C. B. Identification of the NH_2-terminal blocking group of calcineurin B as myristic acid. *FEBS Lett.* **150**:314–318, 1982.
3. Aitken, A.; Klee, C. B.; Cohen, P. The structure of the B-subunit of calcineurin. *Eur. J. Biochem.*, **139**:663–671, 1984.
4. Blumenthal, D. K.; Krebs, E. G. Dephosphorylation of cAMP-dependent protein kinase regulatory subunit (type II) by calcineurin (protein phosphatase 2B). *Biophys. J.* **41**:409a, 1983.
5. Carafoli, E.; Niggli, V.; Malmstrom, K.; Caroni, P. Calmodulin in natural and reconstituted calcium transport systems. *Ann. N.Y. Acad. Sci.* **356**:258–266, 1980.
6. Carlin, R. K.; Grab, D. J.; Siekevitz, P. Function of calmodulin in postsynaptic vesicles. III. Calmodulin-binding proteins of the postsynaptic density. *J. Cell Biol.* **89**:449–455, 1981.
7. Carr, S. A.; Biemann, K.; Shoji, S.; Parmelee, D. C.; Titani, K. n-Tetradecanoyl is the NH_2-terminal blocking

group of the catalytic subunit of cAMP-dependent protein kinase from bovine cardiac muscle. *Proc. Natl. Acad. Sci. USA* **79**:6128–6131, 1982.
8. Cohen, P.; Picton, C.; Klee, C. B. Activation of phosphorylase kinase from rabbit skeletal muscle by calmodulin and troponin. *FEBS Lett.* **104**:25–30, 1979.
9. Gietzen, K.; Sardorf, Il; Bader, H. A model for the regulation of the calmodulin-dependent enzymes erythrocyte Ca^{2+}-transport ATPase and brain phosphodiesterase by activators and inhibitors. *Biochem. J.* **207**:541–548, 1982.
10. Gietzen, K.; Seiler, S.; Fleischer, S.; Wolf, H. U. Reconstitution of the Ca^{2+}-transport system of human erythrocytes. *Biochem. J.* **188**:47–54, 1980.
11. Graf, E.; Verma, A. K.; Gorski, J. P.; Lopaschuk, G.; Niggli, V.; Zurini, M.; Carafoli, E.; Penniston, J. T. Molecular properties of calcium-pumping ATPase from human erythrocytes. *Biochemistry* **21**:4511–4516, 1982.
12. Ingebritsen, T. S.; Cohen, P. The protein phosphatases involved in cellular regulation. 1. Classification and substrate specificities. *Eur. J. Biochem.* **132**:255–261, 1983.
13. Ingebritsen, T. S.; Stewart, A. A.; Cohen, P. The protein phosphatases involved in cellular regulation. 6. Measurement of type-1 and type-2 protein phosphatases in extracts of mammalian tissues; an assessment of their physiological roles. *Eur. J. Biochem.* **132**:297–307, 1983.
14. Klee, C. B. Calmodulin: The coupling factor of the two second messengers Ca^{2+} and cAMP. In: *Protein Phosphorylation and Bio-regulation*, G. Thomas, E. J. Podesta, and J. Gordon, eds., Basel, Karger, 1980, pp. 61–69.
15. Klee, C. B.; Crouch, T. H.; Krinks, M. H. Calcineurin: A calcium- and calmodulin-binding protein of the nervous system. *Proc. Natl. Acad. Sci. USA* **76**:6270–6273, 1979.
16. Klee, C. B.; Haiech, J. Concerted role of calmodulin and calcineurin in calcium regulation. *Ann. N.Y. Acad. Sci.* **356**:43–54, 1980.
17. Klee, C. B.; Krinks, M. H. Purification of cyclic 3′,5′-nucleotide phosphodiesterase inhibitory protein by affinity chromatography of activator protein coupled to Sepharose. *Biochemistry* **17**:120–126, 1978.
18. Klee, C. B.; Krinks, M. H.; Manalan, A. S.; Cohen, P.; Stewart, A. A. Isolation and characterization of bovine brain calcineurin: A calmodulin-stimulated protein phosphatase. *Methods Enzymol.* **102**:227–244, 1983.
19. Krinks, M. H.; Haiech, J.; Rhoads, A.; Klee, C. B. Reversible and irreversible activation of cyclic nucleotide phosphodiesterase: Separation of the regulatory and catalytic domains by limited proteolysis. *Adv. Cyclic Nucleotide Res.* **16**:31–47, 1984.
20. Larsen, F. L.; Raess, B. U.; Hinds, T. R.; Vincenzi, F. F. Modulator binding protein antagonized activation of $(Ca^{2+} + Mg^{2+})$-ATPase and Ca^{2+} transport of red blood cell membranes. *J. Supramol. Struct.* **9**: 269–274, 1978.
21. Lynch, T. J.; Cheung, W. Y. Human erythrocyte Ca^{2+}-Mg^{2+}-ATPase: Mechanism of stimulation by Ca^{2+}. *Arch. Biochem. Biophys.* **194**:165–170, 1979.
22. Manalan, A. S.; Klee, C. B. Interaction of calmodulin with its target proteins. *Chem. Scr.* **21**:139–144, 1983.
23. Manalan, A. S.; Klee, C. B. Activation of calcineurin by limited proteolysis. *Proc. Natl. Acad. Sci. USA* **80**:4291–4295, 1983.
24. Meijer, L.; Guerrier, P. Activation of calmodulin-dependent NAD^+ kinase by trypsin. *Biochim. Biophys. Acta* **702**:143–146, 1982.
25. Meyer, W. L.; Fischer, E. H.; Krebs, E. G. Activation of skeletal muscle phosphorylase b kinase by Ca^{2+}. *Biochemistry* **3**:1033–1039, 1964.
26. Niggli, V.; Adunyah, E. S.; Carafoli, E. Acidic phospholipids, unsaturated fatty acids, and limited proteolysis mimic the effect of calmodulin on the purified erythrocyte Ca^{2+}-ATPase. *J. Biol. Chem.* **256**:8588–8592, 1981.
27. Picton, C.; Klee, C. B.; Cohen, P. Phosphorylase kinase from rabbit skeletal muscle: Identification of the calmodulin binding subunits. *Eur. J. Biochem.* **111**:553–561, 1980.
28. Sharma, R. K.; Desai, R.; Waisman, D. M.; Wang, J. H. Purification and subunit structure of bovine brain modulator binding protein. *J. Biol. Chem.* **254**:4276–4282, 1979.
29. Stewart, A. A.; Ingebritsen, T. S.; Cohen, P. The protein phosphatases involved in cellular regulation. 5. Purification and properties of a Ca^{2+}/calmodulin-dependent protein phosphatase (2B) from rabbit skeletal muscle. *Eur. J. Biochem.* **132**:289–295, 1983.
30. Stewart, A. A.; Ingebritsen, T. S.; Manalan, A.; Klee, C. B.; Cohen, P. Discovery of a Ca^{2+}- and calmodulin-dependent protein phosphatase: Probable identity with calcineurin (CaM-BP_{80}). *FEBS Lett.* **137**:80–84, 1982.
31. Tonks, N. K.; Cohen, P. Calcineurin is a calcium ion dependent, calmodulin stimulated protein phosphatase. *Biochim. Biophys. Acta* **747**:191–193, 1983.
32. Tucker, M. M.; Robinson, J. B., Jr.; Stellwagen, E. The effect of proteolysis on the calmodulin activation of cyclic nucleotide phosphodiesterase. *J. Biol. Chem.* **256**:9051–9058, 1981.

33. Wallace, R. W.; Lynch, T. J.; Tallant, E. A.; Cheung, W. Y. Purification and characterization of an inhibitor protein of brain adenylate cyclase and cyclic nucleotide phosphodiesterase. *Arch. Biochem. Biophys.* **187**:328–334, 1978.
34. Wallace, R. W.; Tallant, E. A.; Cheung, W. Y. High levels of a heat labile calmodulin-binding protein (CaM-BP$_{80}$) in bovine neostriatum. *Biochemistry* **19**:1831–1837, 1980.
35. Walsh, M. P.; Dabrowska, R.; Hinkins, S.; Hartshorne, D. J. Calcium-independent myosin light chain kinase of smooth muscle: Preparation by limited chymotryptic digestion of the calcium ion dependent enzyme, purification and characterization. *Biochemistry* **21**:1919–1925, 1982.
36. Wang, J. H.; Desai, R. Modulator binding protein: Bovine brain protein exhibiting the Ca^{2+}-dependent association with the protein modulator of cyclic nucleotide phosphodiesterase. *J. Biol. Chem.* **252**:4175–4184, 1977.
37. Wolf, H.; Hofmann, F. Purification of myosin light chain kinase from bovine cardiac muscle. *Proc. Natl. Acad. Sci. USA* **77**:5852–5855, 1980.
38. Wood, J. G.; Wallace, R.; Whitaker, J.; Cheung, W. Y. Immunocytochemical localization of calmodulin and a heat labile calmodulin-binding protein (CaM-BP$_{80}$) in basal ganglia of mouse brain. *J. Cell Biol.* **84**:66–76, 1980.

B

CALCIUM AND THE REGULATION OF MUSCLE CONTRACTILITY

VIII

Calcium and Skeletal Muscle Contractility

36

Calcium Channels in Vertebrate Skeletal Muscle

W. Almers, E. W. McCleskey, and P. T. Palade

In crustacean skeletal muscle [33] as well as in vertebrate cardiac muscle [e.g., 19], cell membrane depolarization opens voltage-dependent Ca^{2+} channels. Influx of Ca^{2+} through these channels contributes to both cell membrane depolarization and the Ca^{2+} supply to the contractile filaments. In vertebrate skeletal muscle, however, activity appears to be independent of the influx (or presence) of extracellular Ca^{2+} [14]. Instead, as in nerve axons, cell membrane excitation is mediated by voltage-dependent Na^+ and K^+ channels [2,13,61] and the Ca^{2+} needed to activate contractile proteins comes entirely from an intracellular store, the sarcoplasmic reticulum. Thus, there appeared to be no physiological reason why the cell membrane of vertebrate skeletal muscle should possess Ca^{2+} channels, and this may explain why their existence was not discovered until recently [18,63,66]. In this review, we will discuss the properties of Ca^{2+} channels in vertebrate skeletal muscle, and will speculate about their physiologic role.

Location, Voltage Dependence, and Kinetics

In frog skeletal muscle, most or all Ca^{2+} channels reside in the transverse tubular system (TTS). This conclusion derives from the finding that when the TTS is partially disrupted by glycerol treatment, Ca^{2+} current is diminished in amplitude [59]. It also follows from results indicating that in fibers loaded with the Ca^{2+} chelator EGTA, the decline of I_{Ca} under maintained depolarization is due to Ca^{2+} depletion from the TTS (see later). In EGTA-loaded fibers, more than 93% of all Ca^{2+} channels are calculated to reside in the TTS [8]. However, it is not known whether these Ca^{2+} channels reside in the membrane bordering on the myoplasm or in contact with the terminal cisternae.

Since it is difficult to control the potential of the TTS [1], especially in the presence of a regenerative inward current [3,40], present data on voltage dependence and kinetics of Ca^{2+} channels must be interpreted with caution. Nevertheless, there is agreement that gating of

---

W. Almers, E. W. McCleskey, and P. T. Palade • Department of Physiology and Biophysics, University of Washington, Seattle, Washington 98195.   Present address of P.T.P.: Department of Physiology and Biophysics, The University of Texas Medical Branch, Galveston, Texas 77550.

Table I. Activation of Ca^{2+} Channels

Preparation	$t_{1/2}$ (msec)[a]	T (°C)	Reference
Paramecium	0.7	21	24
Presynaptic terminal of squid giant synapse	1	18	50
Limnea neuron	2.5	23–27	25
Bovine chromaffin cell	2	20–22	34
Barnacle muscle	3–10	18	46
Crayfish muscle	3	17	41
Guinea pig dissociated heart cells	2	21	48
Rat skeletal muscle	110	22	29
Frog skeletal muscle	100–300	20–24	8
Frog skeletal muscle	150	22–26	64

[a] $t_{1/2}$: time to reach half the maximal value of I_{Ca} during steps to approximately 0 mV. Values are approximate and were read off selected figures from the papers cited. Note that Ca^{2+} channels in vertebrate skeletal muscle activate one or two orders of magnitude more slowly than those in most other cell types.

Ca^{2+} channels is a relatively slow process. During step depolarization to 25–30 mV, I_{Ca} rises to half its peak value within 35–70 msec, and upon repolarization to −70 mV I_{Ca} declines with a time constant of about 10 msec [12] [see also 64]. These values apply at temperatures of 20–24°C and at an external Ca^{2+} activity about six times that of Ringer's solution. They indicate that activation and deactivation of Ca^{2+} channels under step changes in potential occur about 1000 times more slowly than activation and deactivation of Na^+ channels at similar temperatures. Vertebrate skeletal muscle Ca^{2+} channels also gate much more slowly than Ca^{2+} channels in most other tissues (Table I).

Decline of I_{Ca} under Maintained Depolarization

The basis for the decline of inward Ca^{2+} current is controversial, and has been attributed either to Ca^{2+} depletion from the TTS [8] or to a voltage-dependent gating process [26]. Depletion of Ca^{2+} would be expected if Ca^{2+} channels are in the TTS. Under conditions where external Ca^{2+} is the only significantly permeable ion, inward current across the tubular wall would be carried predominantly or entirely by Ca^{2+}, whereas at the mouth of the tubules, other ions in solution would contribute in approximate proportion to their concentrations. Therefore, the amount of Ca^{2+} leaving the TTS would exceed the amount entering from the outside. The Ca^{2+} concentration in the TTS would fall and current would decline until a steady state is reached and Ca^{2+} diffusion into the TTS balances I_{Ca}. When current is turned off (e.g., by closure of Ca^{2+} channels under repolarization), [Ca^{2+}] in the TTS would be expected to recover slowly due to Ca^{2+} diffusion.

In EGTA-loaded fibers investigated with the "Vaseline gap" technique, the following findings argue in favor of depletion causing the decline of inward current [8]. (1) The decline appears to be a current-dependent process. The rate constant of decline is strongly correlated with I_{Ca}, whether I_{Ca} is varied by altering the membrane potential, applying drugs, or changing the permeant ion. (2) The rate of decline is markedly slowed when external [Ca^{2+}] (and presumably also [Ca^{2+}] in the TTS) is buffered with malate. (3) The charge carried by the transient portion of I_{Ca} is similar or equal to that expected to be carried by the Ca^{2+} ions dissolved in the TTS. This is expected if I_{Ca} empties an extracellular compartment of the same volume as the TTS (0.3% of the fiber volume; see 58,62). (4) Recovery of inward current occurs at approximately the rate expected for diffusion of Ca^{2+} into the tubular network, i.e., 2–3 times slower than recovery of K^+ inward current from K^+ depletion [8]. (5) The decline is incomplete, with current declining to 20–30% of peak I_{Ca}. The final value of I_{Ca} is quite similar to the product of $a \times Q$, where a is the initial rate constant of recovery and Q is the charge under the I_{Ca} transient [8]. This similarity or equality suggests that final current and the initial rate of recovery are both determined by the same process (e.g., Ca^{2+} diffusion into the TTS) and generally is not expected from an "inactivation" mechanism gated by the membrane potential.

When intact fibers are investigated and muscle contraction is abolished by hypertonic solutions rather than by EGTA loading, a different picture emerges [26]. (1) Decline of current is generally complete at all potentials. (2) The rate of decline is not strongly dependent on potential. Between −25 and +35 mV, it may increase less than twofold with potential (Fig. 2A of Ref. 26) or remain constant [64]. This contrasts with a fourfold decrease with potential in EGTA-loaded fibers (Fig. 1 of Ref. 8). (3) Decline is on the whole slower than in EGTA-loaded fibers and occurs with a time constant of about 1 sec at 0 mV and 23°C. (4) The correlation between current and the rate of decline is not observed. In particular, temperature affects the rate of decline much more strongly than the magnitude of I_{Ca}. Point (4) particularly indicates that mechanisms other than depletion operate when internal EGTA is absent. The differences between EGTA-loaded fibers and intact fibers in hypertonic solutions raise two puzzling questions which will next be considered.

1. *Why does I_{Ca} not cause depletion in the absence of EGTA?* Although currents in osmotically shrunken fibers are as large as those in EGTA-loaded fibers, they decline more slowly and apparently do not cause depletion. Yet severe depletion is expected if I_{Ca} flows across the membrane of a TTS of normal morphology. On the basis of the findings of Cota *et al.* [26], the charge under the transient at low temperatures may be expected to be 5–10 times larger than that contained on the Ca^{2+} ions in a TTS occupying the normal 0.3% of fiber volume; yet most or all Ca^{2+} presumably passes across the tubular wall. Cota *et al.* [26] offer two explanations. (1) The TTS in hypertonic solutions may be swollen, as was suggested by Franzini-Armstrong *et al.* [36]. (2) The tubular membrane may contain a Ca^{2+} pump that causes a rapid and electrically silent extrusion of the Ca^{2+} entering as I_{Ca}. Both hypotheses need additional experimental confirmation. It is unclear whether there is sufficient swelling to prevent significant depletion, and also whether the tubule swelling observed morphologically is consistent with the rather small increase in membrane capacity observed (20–30%; Ref. 37, Almers, unpublished). Measured Ca^{2+} efflux rates from the cytoplasm of resting fibers are of the order of 0.1–0.2 pmole/cm² per sec [28], equivalent to 20–40 nA/cm². Even if efflux is increased 2-fold after 5 min of intense stimulation [28], it is still much too small to maintain a normal tubular $[Ca^{2+}]$ in the face of nearly 1000-fold larger I_{Ca}. Unless Ca^{2+} efflux rises two or more orders of magnitude while I_{Ca} flows, the question remains: Why does depletion not occur?

2. *What is the mechanism of Ca^{2+} channel "inactivation," and why is it not observed in EGTA-loaded fibers?* In EGTA-loaded fibers, "inactivation" at −10 to +10 mV must occur with a time constant of 7–10 sec or more at 20–24°C (Table 1 of Ref. 8). This is much slower than the decline of current attributed to voltage-dependent inactivation in intact fibers under otherwise identical conditions (∼ 1 sec) [26]. One possibility is that normal Ca^{2+} channels possess a relatively rapid (∼ 1 sec) voltage-dependent inactivation mechanism, which is lost or slowed when internal EGTA is present at high concentrations.

However, the "inactivation" observed in intact fibers [26,65] has some features that are unusual for a voltage-dependent gating process. While the degree of "inactivation" investigated with conditioning prepulses is steeply voltage dependent over the range of -40 to -20 mV, the rate constant for decline of I_{Ca} is nearly independent of voltage [65]. It is important to examine the idea that in the absence of EGTA, "inactivation" is influenced by intracellular $[Ca_2^+]$, as described in *Aplysia* neurons [69], stick insect muscle [15], and *Paramecium* [24]. Initially, this may seem unlikely because inactivation of frog skeletal muscle in EGTA-free solution appears to be independent of I_{Ca}. However, inactivation may be induced not by Ca^{2+} influx, but by Ca^{2+} release from the sarcoplasmic reticulum (SR). If so, the "h_∞ IF curve" of Sánchez and Stefani [65] may reflect the cell membrane potential-dependent control of myoplasmic Ca^{2+} by the SR. This hypothesis would explain why inactivation is absent in EGTA-loaded fibers. However, the rate constants for "inactivation" still would be unusually slow for such a mechanism, especially at lower temperatures.

Selectivity

Under physiological conditions, Ca^{2+} channels are highly effective in excluding the majority ions Na^+ and K^+ in skeletal muscle [11] as well as in other tissues. Yet Ca^{2+} channels appear highly permeable also to several foreign divalent cations such as Ba^{2+} and Sr^{2+}, sometimes more so than to Ca^{2+}. In skeletal muscle, Ba^{2+}, Sr^{2+}, Ca^{2+}, Mn^{2+}, and even Mg^{2+} may all carry inward current and are therefore permeant; Co^{2+} and Ni^{2+} are not measurably permeant [12].

Recent results [10,11,9] show that when external $[Ca^{2+}]$ is reduced to submicromolar levels by Ca^{2+} chelators, Ca^{2+} channels in skeletal muscle become freely permeable also to Li^+, Na^+, K^+, Rb^+, and Cs^+. The ability to exclude these monovalent cations is restored by external Ca^{2+}, which apparently binds to Ca^{2+} channels with high affinity and a dissociation constant of about 0.7 μM. Similarly, the permeabilities to Ba^{2+} [9] and Mg^{2+} (manuscript in preparation) are largest in the absence of Ca^{2+}, and are strongly reduced when Ca^{2+} is added externally. Evidently Ca^{2+} channels are intrinsically unselective and reject small cations other than Ca^{2+} only when Ca^{2+} is present.

These findings can be explained quantitatively by a model [9] wherein the Ca^{2+} channel is pictured as an aqueous "single-file" pore [43] containing two sites to which permeating ions must bind in succession. Ions inside the pore cannot pass each other but may dislodge one another by electrostatic repulsion. When more than one type of permeant cation is present, such a pore will be occupied preferentially by the ion with the highest affinity for the sites. In a pore occupied by one high-affinity ion, only another high-affinity ion will be able to bind to the second site sufficiently long to dislodge the resident ion, thereby generating net flux. Hence in ion mixtures, current through such a pore is carried predominantly by the ion with the highest affinity. On the basis of the effectiveness of various divalent cations (Ca^{2+}, Sr^{2+}, Co^{2+}, Mn^{2+}, Cd^{2+}, Ni^{2+}, Mg^{2+}) in blocking the permeability to monovalent cations, the ion with the highest affinity is Ca^{2+}. We believe that the specific affinity between the binding sites and Ca^{2+} forms the basis for selective ion transport in Ca^{2+} channels. By attributing selectivity to selective binding, the model explains why Ca^{2+} channels appear nonselective in the absence of Ca^{2+} or other high affinity cations. A similar model has recently been applied also to cardiac muscle [42].

Selectivity by affinity was previously proposed by Eisenmann [31] in order to explain the function of ion-sensitive glass electrodes. The application of this concept to aqueous pores has encountered resistance, however (see appendix of Ref. 21), because tight binding in pores seemed inconsistent with the high single-channel fluxes observed experimentally. In the model described here, this difficulty is overcome by allowing interaction between ions inside the pore. If we assume diffusion-limited access of Ca^{2+} to the pore, the model easily accounts for the high single-channel fluxes observed in other tissues [34].

Pharmacology

Divalent cations. As in other tissues [39], the Ca^{2+}-current of frog skeletal muscle is blocked by several divalent cations. The most potent blocker is Cd^{2+} with half-blockage concentration of approximately 0.4 mM. Other effective blockers are Co^{2+}, Ni^{2+}, and Mn^{2+}. Even Ca^{2+} itself can block Ba^{2+} currents through the Ca^{2+} channel with a half-blockage concentration of less than 1 mM. Thus, certain ions, such as Mn^{2+} and Ca^{2+}, are effective blockers, but are also highly permeant.

Organic blockers. The dihydropyridine derivative, nifedipine, is the most potent of the organic blockers investigated so far. In cell-free preparations, nifedipine and other dihydropyridines, most notably nitrendipine, bind with high affinity and specificity to Ca^{2+} channels. In a rabbit transverse tubule membrane preparation, dissociation constants of 4 and 1.4 nM have been calculated for nifedipine and nitrendipine, respectively [35]. These values are about 100 times lower than the electrophysiological half-blockage concentration obtained

Table II. Block of Ca^{2+} Channels in Frog Muscle Cell Membrane[a]

Substance	Half-blockage concentration (mM) ± S.D.	n
Nifedipine	0.00033 ± 0.00005	9
D600	0.010 ± 2	6
Dibucaine	0.008 ± 0.004	2
Tetracaine	0.14 ± 0.02	7
Lidocaine	1.4 ± 0.3	6
Procaine	2.0 ± 0.4	6
QX-314	27 ± 8	3
Secobarbital	0.22 ± 0.04	6
Pentobarbital	0.44 ± 0.06	10
Phenobarbital	1.3 ± 0.3	4
Cd^{2+}	0.36 ± 0.07	7
Co^{2+}	1.28 ± 0.23	6
Ni^{2+}	1.28 ± 0.28	4
Mn^{2+}	12.5 ± 2.7	4
Mg^{2+}	36.6 ± 11.6	3

[a]Fibers were soaked overnight in a solution containing 80 mM K_2EGTA plus ATP, as in Ref. 12. Vaseline-gap method was used throughout, with fiber ends cut in 100 mM K_2EGTA or 100 mM Cs_2EGTA. External solution contained 10 mM Ca^{2+} and 120 mM tetraethyl- (or tetramethyl-) ammonium, with the anion being methanesulfonate. Divalent blocking cations Cd^{2+}, Co^{2+}, and Ni^{2+} were added as the chloride salt. In experiments with Mg^{2+} or Mn^{2+}, Mg- or Mn-methanesulfonate replaced TEA-methanesulfonate on an isosmotic basis. Nifedipine experiments were performed under red light. Temperature 20–24°C throughout. All data from Palade and Almers (unpublished).

by us for nifedipine. A similar discrepancy seems to exist in cardiac muscle [20,49]. The reason for this disparity between electrophysiological and biochemical results is presently unclear, but it may reflect different states of the Ca^{2+} channel.

Various other substances, among them certain local anesthetics, also block effectively and reversibly; and dose–response curves are readily obtained. Table II lists a number of agents along with the concentrations at which they reduce peak current to 50%. All values were obtained at an external $[Ca^{2+}]$ of 10 mM [60].

When the myoplasm contains K^+ or Cs^+, inward I_{Ca} under maintained depolarization is followed by an outward current, presumably carried by intracellular K^+ or Cs^+. This outward current is seen despite 120 mM external TEA^+ [12], a substance known to block most K^+ channels [67]. Curiously, all agents that block current through Ca^{2+} channels also block this outward current with equal or greater effectiveness. Even Ca^{2+} appears to block, since outward current is increased more than twofold when external Ca^{2+} is completely replaced by Mg^{2+} [12]. Among the 15 agents investigated by us, not a single exception was found. The basis for this striking pharmacological parallel is obscure. The most obvious explanation, namely that the outward current flows through channels that are activated by Ca^{2+} entering through Ca^{2+} channels, is ruled out because outward current *increases* when Mg^{2+} replaces external Ca^{2+}. Furthermore, replacement of Ca^{2+} with other cations thought to be ineffective in activating Ca^{2+}-activated K^+ channels (such as Sr^{2+} and Ba^{2+}) does not abolish this outward current (Palade and Almers, unpublished).

Ca^{2+} Channels and Excitation–Contraction Coupling

Despite their location in the TTS, Ca^{2+} channels in frog skeletal muscle have no known role in excitation–contraction coupling. The channels appear to respond so slowly to potential changes that only a vanishingly small number of them are expected to open during a single action potential. In any event, action potentials and muscle twitches are still possible at nanomolar external $[Ca^{2+}]$ [14], so that an influx of "trigger Ca^{2+}" from the extracellular fluid appears unnecessary for initiation of contraction. K^+ contractures are greatly shortened at low external Ca^{2+} [e.g., 68], but Mg^{2+} can partially or entirely restore the duration of K^+ contractures [51].

A direct way to explore the involvement of Ca^{2+} channels in excitation–contraction coupling is to test for an effect of Ca^{2+} channel blockers on the ability of the SR to release Ca^{2+} in response to cell membrane depolarization. Changes in myoplasmic $[Ca^{2+}]$ can now be monitored optically [16,22,47,55], but for a pharmacological survey, it is more convenient to use the myofilaments of a voltage-clamped muscle fiber as an assay for myoplasmic $[Ca^{2+}]$. The procedure is to apply strong (e.g., to 50 mV), voltage-clamp depolarizations, and to determine the smallest duration ["minimum stimulus duration" (msd)] needed to elicit a threshold contraction visible under a microscope. The threshold contraction is taken to signal the rise of myoplasmic Ca^{2+} to an unknown value that is constant for a given fiber.

Recent experiments with the metallochromic dye arsenazo III [56] have confirmed that threshold contraction is accompanied by a stereotyped threshold rise in myoplasmic $[Ca^{2+}]$. When Ca^{2+} release from the SR is impaired by maintained depolarization, for example, the msd is increased, because release from the SR must be continued longer to establish a threshold myplasmic $[Ca^{2+}]$. Similarly, pharmacological blockers of Ca^{2+} release would be expected to increase msd. Local anesthetics, for example, are a class of membrane-permeant small molecules known to block a variety of ionic channels, including Ca^{2+} channels. Indeed, tetracaine, procaine, and dibucaine all lengthen the msd as though they interfered with Ca^{2+} release [4,7]. However, neither Ni^{2+}, nifedipine, nor D600 has any such effect, even when applied at concentrations sufficient to block more than 90% of the Ca^{2+} channels in the cell membrane [30,53,54]. Furthermore, excitation–contraction coupling is not impaired by diltiazem [38], a drug known to block Ca^{2+} channels in some tissues. These results strongly argue against a role for these Ca^{2+} channels in excitation–contraction coupling.

Recently, it has been demonstrated that frog muscle fibers become paralyzed after a K^+ contracture in the presence of 30 μM D600 at 7°C or below. During paralysis, cell membrane depolarization causes neither contraction [52] nor the "charge movement" [44] believed to be an early electrical event in excitation–contraction coupling [5,65]. Kaumann and Uchitel [45] had previously shown a decrease in peak tension of successive K^+ contractures by tonic frog muscle at room temperature. These findings might support a role for Ca^{2+} channels in excitation–contraction coupling. However, D600 is notoriously nonspecific, and K^+ contractures in the presence of nifedipine (20 μM) or Ni^{2+} (20 mM) do not lead to paralysis (McCleskey, in preparation). Muscle paralysis by D600 is evidently not mediated by the Ca^{2+} channels that are the subject of this review.

Apparently, excitation–contraction coupling does not depend upon Ca^{2+} entry from the extracellular space, be it via voltage-dependent Ca^{2+} channels or otherwise. However, Ca^{2+} release may occur via ionic channels in the SR membrane that are similar to Ca^{2+} channels in cell membranes. Electrical studies of SR membrane vesicles have not as yet shown the presence of Ca^{2+} channels [e.g., 57], but this does not rule out their existence, as Ca^{2+} channels in the SR may be rare [6]. They need occur only at very low density in order to generate fluxes equivalent to the rates of Ca^{2+} release observed experimentally [17]. Furthermore, Ca^{2+} channels in the SR are expected to have at least one property that is unusual for the Ca^{2+} channels studied so far: their opening and closing is determined in a membrane other than the one in which they reside. Nevertheless, it seemed of interest to ask whether "Ca^{2+} channels" in the SR are pharmacologically similar to those in the T membrane. Tetracaine, for example, blocks Ca^{2+} channels both in the T membrane and in the SR [7]. However, Ni^{2+}, D600, and nifedipine all fail to lengthen the msd, even if they are pressure-injected into muscle fibers or if muscles are exposed for extended periods (24 hr) to high concentrations of the lipid-soluble drugs, D600 and nifedipine [53]. Clearly, Ca^{2+} release from the SR occurs by mechanisms that are pharmacologically dissimilar from those exhibited by ordinary Ca^{2+} channels.

Our negative result with nifedipine is consistent with the failure of others [e.g., 35] to find nitrendipine binding to rabbit sarcoplasmic reticulum (see, however, 32).

Physiological Role

With a direct involvement in excitation–contraction coupling unlikely, the physiological role of Ca^{2+} channels in vertebrate skeletal muscle cannot be defined precisely at this stage. They may be involved in regulating the Ca^{2+} content of muscle over times considerably longer than the duration of a single twitch. Like all other cells, skeletal muscle fibers are capable of extruding Ca^{2+} against a large electrochemical gradient, with at least some of the extrusion sites located in the transverse tubules [23]. When myoplasmic [Ca^{2+}] rises, as it does during activity, Ca^{2+} efflux also rises [28]. To maintain cellular Ca^{2+} reserves, there must, therefore, be a pathway for Ca^{2+} entry, which is linked in some way to muscle activity. Depolarization-activated Ca^{2+} channels would be suitable for providing Ca^{2+} entry during activity. Moreover, since Ca^{2+} channels reside predominantly in the tubules, Ca^{2+} entering through them would uniformly reach the myoplasm and SR; this would clearly be desirable but could not be guaranteed if Ca^{2+} channels solely resided in the sarcolemma.

The question is still open as to how Ca^{2+} channels are activated in living muscle. Present voltage-clamp data give the impression that a single impulse activates only a few percent of all available Ca^{2+} channels, and that any activated channels close quickly upon repolarization (time constant of $\sim$ 10 msec). Judging by presently published kinetics, Ca^{2+} channels seem designed to ignore action potentials. However, all voltage-clamp studies so far have been carried out under unphysiological conditions, and the observed gating kinetics may be similarly unphysiological. Ca^{2+} channels may gate more rapidly at lower external $[Ca^{2+}]$ (see Fig. 10 of Ref. 8). Also, the observed Ca^{2+} influx per action potential (1 pmole/cm^2 at 0.7-Hz stimulation frequency [27]) is an order of magnitude larger than influx through Ca^{2+} channels predicted on the basis of voltage-clamp studies [12,64]. Furthermore, it is not clear to what extent summation of open Ca^{2+} channels occurs under tetanic stimulation. While a deactivation time constant of 10 msec does not suggest significant summation at stimulation frequencies under 50–100 Hz, for a short interval after closing, Ca^{2+} channels may need less time to reopen under a new depolarization. Finally, Donaldson and Beam [29] have suggested that in rat skeletal muscle, Ca^{2+} channels may gate at sufficient speed at 37°C to enable a significant number of them to respond to a single action potential.

In summary, more experiments are needed in order to answer the question as to whether a significant number of Ca^{2+} channels open under tetanic stimulation at physiological frequencies. At the same time, one should keep in mind the possibility that Ca^{2+} channels are regulated by factors other than action potentials. For example, Ca^{2+} channels in cardiac muscle are hormone-responsive [70].

ACKNOWLEDGMENT. Supported by USPHS grant AM-17803.

References

1. Adrian, R. H.; Almers, W. Membrane capacity measurements on frog skeletal muscle in media of low ion content. *J. Physiol. (London)* **237**:573–605, 1974.
2. Adrian, R. H.; Chandler, W. K.; Hodgkin, A. L. Voltage clamp experiments in striated muscle fibres. *J. Physiol. (London)* **208**:607–644, 1970.
3. Adrian, R. H.; Peachey, L. D. Reconstruction of the action potential of frog sartorius muscle. *J. Physiol. (London)* **235**:103–131, 1973.
4. Almers, W. Local anesthetics and excitation–contraction coupling in skeletal muscle: Effects on a Ca^{++} channel. *Biophys. J.* **18**:355–357, 1977.
5. Almers, W. Gating currents and charge movements in excitable membranes. *Rev. Physiol. Biochem. Pharmacol.* **82**:96–190, 1978.
6. Almers, W.; Adrian, R. H.; Levinson, S. R. Some dielectric properties of muscle membrane and their possible importance for excitation–contraction coupling. *Ann. N.Y. Acad. Sci.* **264**:278–292, 1975.
7. Almers, W.; Best, P. M. Effects of tetracaine on displacement currents and contraction of frog skeletal muscle. *J. Physiol. (London)* **262**:583–611, 1976.
8. Almers, W.; Fink, R.; Palade, P. T. Calcium depletion in frog muscle tubules: The decline of calcium current under maintained depolarization. *J. Physiol. (London)* **312**:177–207, 1981.
9. Almers, W.; McCleskey, E. W. Nonselective conductance in calcium channels of frog muscle: Calcium selectivity in a single-file pore. *J. Physiol. (London)* **353**:in press, 1984.
10. Almers, W.; McCleskey, E. W.; Palade, P. T. Frog muscle membrane: A cation-permeable channel blocked by micromolar external $[Ca^{++}]$. *J. Physiol. (London)* **332**:52P–53P, 1982.
11. Almers, W.; McCleskey, E. W.; Palade, P. T. A nonselective cation conductance in frog muscle membrane blocked by micromolar external calcium ions. *J. Physiol. (London)* **353**:in press, 1984.

12. Almers, W.; Palade, P. T. Slow calcium and potassium currents across frog muscle membrane: Measurements with a Vaseline gap technique. *J. Physiol. (London)* **312**:159–176, 1981.
13. Almers, W.; Roberts, W. M.; Ruff, R. L. Voltage clamp of rat and human skeletal muscle: Measurements with an improved loose-patch technique. *J. Physiol. (London)* **347**: 751–768, 1984.
14. Armstrong, C. M.; Bezanilla, F. M.; Horowicz, P. Twitches in the presence of ethylene glycol bis(β-aminoethyl ether)-N,N′-tetraacetic acid. *Biochim. Biophys. Acta* **267**:605–608, 1972.
15. Ashcroft, F. M.; Stanfield, P. R. Calcium and potassium currents in muscle fibres of an insect (*Carausius morosus*). *J. Physiol. (London)* **323**:93–115, 1982.
16. Baylor, S. M.; Chandler, W. K.; Marshall, M. W. Use of metallochromic dyes to measure changes in myoplasmic Ca^{++} during activity in frog skeletal muscle fibres. *J. Physiol. (London)* **331**:139–177, 1982.
17. Baylor, S. M.; Chandler, W. K.; Marshall, M. W. Sarcoplasmic reticulum calcium release in frog skeletal muscle fibres estimated from arsenazo III calcium transients. *J. Physiol. (London)* **344**:525–666, 1983.
18. Beaty, G. N.; Stefani, E. Calcium-dependent electrical activity in twitch muscle fibres of the frog. *Proc. R. Soc. London Ser. B.* **194**:141–150, 1976.
19. Beeler, G. W., Jr.; Reuter, H. Membrane calcium current in ventricular myocardial fibres. *J. Physiol. (London)* **207**:191–209, 1970.
20. Bellemann, P.; Ferry, D.; Lübbecke, F.; Glossmann, H. [³H]-nitrendipine, a potent calcium antagonist, binds with high affinity to cardiac membranes. *Arzneim. Forsch.* **31**:2064–2067, 1981.
21. Bezanilla, F. M.; Armstrong, C. M. Negative conductance caused by entry of sodium and cesium ions into the potassium channels of squid giant axons. *J. Gen. Physiol.* **60**:588–608, 1972.
22. Blinks, J. R.; Rüdel, R.; Taylor, S. R. Calcium transients in isolated amphibian skeletal muscle fibres: Detection with aequorin. *J. Physiol. (London)* **277**:291–323, 1978.
23. Brandt, N. R.; Caswell, A.; Brunschwig, J. P. ATP-energized Ca-pump in isolated transverse tubules of skeletal muscle. *J. Biol. Chem.* **255**:6290–6298, 1980.
24. Brehm, P.; Eckert, R. Calcium entry leads to inactivation of calcium channel in *Paramecium*. *Science* **202**:1203–1206, 1978.
25. Byerly, L.; Hagiwara, S. Calcium currents in internally perfused nerve cell bodies of *Limnea stagnalis*. *J. Physiol. (London)* **322**:503–528, 1982.
26. Cota, G.; Nicola Siri, L.; Stefani, E. Calcium-channel gating in frog skeletal muscle membrane: Effect of temperature. *J. Physiol. (London)* **338**:395–412, 1983.
27. Curtis, B. A. Ca fluxes in single twitch muscle fibers. *J. Gen. Physiol.* **50**:255–267, 1966.
28. Curtis, B. A. Calcium efflux from frog twitch muscle fibers. *J. Gen. Physiol.* **55**:243–253, 1970.
29. Donaldson, P. L.; Beam, K. G. Calcium currents in a fast twitch skeletal muscle of the rat. *J. Gen. Physiol.* **82**:449–468, 1983.
30. Dörrscheidt-Käfer, M. The action of D-600 on frog skeletal muscle: Facilitation of excitation–contraction coupling. *Pfluegers Arch.* **369**:259–267, 1977.
31. Eisenmann, G. Cation-selective glass electrodes and their mode of operation. *Biophys. J.* **2**:259–323, 1962.
32. Fairhurst, A. S.; Thayer, S. A.; Colker, J. E.; Beatty, D. A. A calcium antagonist drug binding site in skeletal muscle sarcoplasmic reticulum: Evidence for a calcium channel. *Life Sci.* **32**:1331–1339, 1983.
33. Fatt, P.; Ginsborg, B. L. The ionic requirements for the production of action potentials in crustacean muscle fibres. *J. Physiol. (London)* **142**:516–543, 1958.
34. Fenwick, E. M.; Marty, A.; Neher, E. Sodium and calcium channels in bovine chromaffin cells. *J. Physiol. (London)* **331**:599–635, 1982.
35. Fosset, M.; Jaimovich, E.; Delpont, E.; Lazdunski, M. [³H] Nitrendipine receptors in skeletal muscle. *J. Biol. Chem.* **258**:6086–6092, 1983.
36. Franzini-Armstrong, C.; Heuser, J. E.; Reese, T. S.; Somlyo, A. P.; Somlyo, A. V. T-tubule swelling in hypertonic solutions: A freeze substitution study. *J. Physiol. (London)* **283**:133–140, 1978.
37. Freygang, W. H., Jr.; Rapoport, S. I.; Peachey, L. D. Some relations between changes in the linear electrical properties of striated muscle fibers and changes in ultrastructure. *J. Gen. Physiol.* **50**:2437–2458, 1967.
38. Gonzalez-Serratos, H.; Valle-Aguilera, R.; Lathrop, D. A.; del Carmen Garcia, M. Slow inward calcium currents have no obvious role in muscle excitation–contraction coupling. *Nature (London)* **298**:292–294, 1982.
39. Hagiwara, S. Ca-dependent action potential. In: *Membranes*, Volume 3, G. Eisenman, ed., New York, Dekker, 1975, pp. 359–381.
40. Heiny, J. A.; Vergara, J. Optical signals from surface and T system membranes in skeletal muscle fibers: Experiments with the potentiometric dye NK2367. *J. Gen. Physiol.* **80**:203–230, 1982.
41. Henček, M.; Zachar, J. Calcium currents and conductances in the muscle membrane of the crayfish. *J. Physiol. (London)* **268**:51–71, 1977.
42. Hess, P.; Tsien, R. W. Mechanism of ion permeation through calcium channels. *Nature* **309**:453–456, 1984.
43. Hille, B.; Schwarz, W. Potassium channels as multi-ion, single-file pores. *J. Gen. Physiol.* **72**:409–442, 1978.

44. Hui, C. S.; Milton, R. L.; Eisenberg, R. S. Elimination of charge movement in skeletal muscle by a calcium antagonist. *Biophys. J.* **41**:178a, 1983.
45. Kaumann, A. J.: Uchitel, O. D. Reversible inhibition of potassium contractures by optical isomers of verapamil and D600 on slow muscle fibres of the frog. *Naunyn-Schmiedeberg's Arch. Pharmacol.* **292**:21–27, 1976.
46. Keynes, R. D.; Rojas, E.; Taylor, R. E.; Vergara, J. Calcium and potassium systems of a giant barnacle muscle fibre under membrane potential control. *J. Physiol. (London)* **229**:409–455, 1973.
47. Kovács, L.; Ríos, E.; Schneider, M. F. Calcium transients and intramembrane charge movement in skeletal muscle fibres. *Nature (London)* **279**:391–396, 1979.
48. Lee, K. S.; Tsien, R. W. Reversal of current through calcium channels in dialysed single heart cells. *Nature (London)* **297**:498–501, 1982.
49. Lee, K. S.; Tsien, R. W. Mechanism of calcium channel blockade by verapamil, D600, diltiazem, and nitrendipine in single dialysed heart cells. *Nature (London)* **302**:790–794, 1983.
50. Llinás, R.; Steinberg, I. Z.; Walton, K. Presynaptic calcium currents in squid giant synapse. *Biophys. J.* **33**:289–322, 1981.
51. Lüttgau, H.-C.; Spiecker, W. The effects of calcium deprivation upon mechanical and electrophysiological parameters in skeletal muscle fibres of the frog. *J. Physiol. (London)* **296**:411–429, 1979.
52. McCarthy, R. T.; Milton, R. L.; Eisenberg, R. S. Paralysis of frog skeletal muscle fibres by the calcium antagonist D600. *J. Physiol. (London)* **341**:495–506, 1983.
53. McCleskey, E. W. Ca^{++} channels in frog cell membrane and intracellular Ca^{++} release: A pharmacological comparison. Ph.D. thesis, University of Washington, Seattle, 1983.
54. McCleskey, E. W.; Almers, W. Pharmacological comparison of e.c. coupling and the skeletal muscle Ca^{++} channel. *Biophys. J.* **33**:33a, 1981.
55. Miledi, R.; Parker, I.; Schalow, G. Measurement of calcium transients in frog muscle by the use of arsenazo III. *Proc. R. Soc. London Ser. B* **198**:201–210, 1977.
56. Miledi, R.; Parker, I.; Zhu, P. H. Calcium transients studied under voltage-clamp control in frog twitch muscle fibres. *J. Physiol. (London)* **340**:649–680, 1983.
57. Miller, C. Voltage-gated cation conductance channel from fragmented sarcoplasmic reticulum: Steady-state electrical properties. *J. Membr. Biol.* **40**:1–23, 1978.
58. Mobley, B. A.; Eisenberg, B. R. Sites of components in frog skeletal muscle by methods of stereology. *J. Gen. Physiol.* **66**:31–46, 1975.
59. Nicola Siri, L.; Sánchez, J. A.; Stefani, E. Effect of glycerol treatment on the calcium current of frog skeletal muscle. *J. Physiol. (London)* **305**:87–96, 1980.
60. Palade, P. T.; Almers, W. The Ca^{++}-channel in frog muscle cell membrane: A pharmacological profile. *Biophys. J.* **33**:151a, 1981.
61. Pappone, P. A. Voltage-clamp experiments in normal and denervated mammalian skeletal muscle fibres. *J. Physiol. (London)* **306**:377–410, 1980.
62. Peachey, L. D. The sarcoplasmic reticulum and transverse tubules of the frog's sartorius. *J. Cell Biol.* **25**:209–231, 1965.
63. Sánchez, J. A.; Stefani, E. Inward calcium current in twitch muscle fibres of the frog. *J. Physiol. (London)* **283**:197–209, 1978.
64. Sánchez, J. A.; Stefani, E. Kinetic properties of calcium channels of twitch muscle fibres of the frog. *J. Physiol. (London)* **337**:1–17, 1983.
65. Schneider, M. F.; Chandler, W. K. Voltage dependent charge movement in skeletal muscle: A possible step in excitation–contraction coupling. *Nature (London)* **242**:244–246, 1973.
66. Stanfield, P. R. A calcium-dependent inward current in frog skeletal muscle fibres. *Pfluegers Arch.* **368**:267–270, 1977.
67. Stanfield, P. R. Tetraethylammonium ions and the potassium permeability of excitable cells. *Rev. Physiol. Biochem. Pharmacol.* **97**:1–67, 1983.
68. Stefani, E.; Chiarandini, D. J. Skeletal muscle: Dependence of potassium contractures on extracellular calcium. *Pfluegers Arch.* **343**:143–150, 1973.
69. Tillotson, D. Inactivation of Ca conductance dependent on entry of Ca ions in molluscan neurones. *Proc. Natl. Acad. Sci. USA* **76**:1497–1500, 1979.
70. Tsien, R. W. Calcium channels in excitable membranes. *Annu. Rev. Physiol.* **45**:341–358, 1983.

37

Mechanisms of Calcium Release in Skinned Mammalian Skeletal Muscle Fibers

Sue K. Bolitho Donaldson, Robert Dunn, Jr., and Daniel A. Huetteman

Contraction of intact skeletal muscle fibers is initiated and governed by depolarization of the sarcolemma and transverse tubules (TTs) [3,5,16,17]. This excitation signal elicits from the sarcoplasmic reticulum (SR) Ca^{2+} release which activates the contractile apparatus [11,23]. However, the nature of the TT–SR junctional communication and the mechanism of SR Ca^{2+} release are not fully understood. The mechanically skinned fiber preparation, in which the sarcolemma has been removed completely (peeled fibers) [20] or partially (longitudinally split fibers) [12], is a rather unique preparation for the study of these critical excitation-contraction (EC) coupling steps. In mechanically skinned fibers, the environment of the TT–SR junction and the SR can be controlled and, in addition, Ca^{2+} fluxes readily monitored. While the earliest studies of Ca^{2+} release mechanisms in skinned fibers were performed on frog skeletal fibers [11,23], a peeled mammalian skeletal fiber preparation has recently been introduced [6–8]. The physiological Ca^{2+} release mechanism is expected to be the same in frog and mammalian skeletal fibers. However, peeled mammalian and frog fibers do not respond identically to all stimuli that elicit the release of Ca^{2+} [8,19]. The purpose of this chapter is to present evidence that these discrepancies may be due to varied responses to either SR volume changes or directional changes in bathing solution $[K^+] \times [Cl^-]$.

Cl^- Stimulation of Ca^{2+} Release

Since it is not possible to mimic *in vivo* depolarization of the TTs in skinned skeletal muscle fibers, other stimuli are used to elicit Ca^{2+} release from the SR. Constantin and Podolsky [4] elicited localized contraction of skinned frog skeletal fibers in oil by applying electric current pulses. But this method of stimulation has not been used successfully with the skinned fibers in aqueous solutions [23]. Since aqueous environment is necessary in order to

---

Sue K. Bolitho Donaldson, Robert Dunn, Jr., and Daniel A. Huetteman • Department of Physiology and Department of Medical Nursing, Rush University, Chicago, Illinois 60612.

take full advantage of the accessibility of the internal environment with skinned fibers, a means of ionic depolarization has been used to stimulate SR Ca^{2+} release from skinned fibers in aqueous solutions. Ionic depolarization of internal membranes is achieved by abruptly substituting a membrane-permeant anion (Cl^-) for a membrane-impermeant anion (e.g., propionate) [4,23] in the bathing solutions. Simultaneous substitution of a permeant cation (K^+) for an impermeant one (e.g., choline), such that bathing solution $[K^+] \times [Cl^-]$ does not vary, should limit volume changes of the affected membrane systems [2,19] during ionic depolarization. Amphibian skinned fibers release Ca^{2+} when stimulated with Cl^- at constant [19] and increased $[K^+] \times [Cl^-]$ [12], although a significant part of the response to the latter stimulus appears to be due to osmotic swelling of SR rather than ionic depolarization [19]. Cl^--induced Ca^{2+} release can be detected as net ^{45}Ca efflux from the fiber or as transient activation of force generation [23]. Caffeine also elicits Ca^{2+} release, but the mechanism involved, Ca^{2+}-induced release of Ca^{2+}, is probably not a physiological one in skeletal fibers [11].

Mammalian peeled fibers have a highly variable response to ionic depolarization with Cl^-. Peeled mammalian cardiac cells do not have Cl^--induced Ca^{2+} release [14]. A significant number of peeled and all split mammalian skeletal fibers do not respond to any type of Cl^- stimulation with a measurable tension transient, even though they have sizable caffeine contractures and thus functional SR [8]. Furthermore, the entire Cl^--induced release of Ca^{2+} in mammalian peeled skeletal fibers can be eliminated by treating the fiber in a manner that facilitates ouabain binding to Na^+-K^+-ATPase sites on the interior of the TTs [8,15]. These data suggest that sealed, polarized TTs mediate the Cl^--induced response in peeled mammalian skeletal fibers.

We have explored the Ca^{2+} release response of peeled rabbit adductor fibers as a result of ionic depolarization with Cl^- at constant and varied $[K^+] \times [Cl^-]$. The methods used for peeling the mammalian skeletal fibers and preparing the bathing solutions are as published previously [8–10]. The fiber was peeled, mounted in the tension transducer, and initially immersed in a relaxing solution with 0.05 mM EGTA, 2 mM $MgATP^{2-}$, 15 mM CP^{2-} (added as $tris_2CP$), 70 mM Na^+, 1 mM Mg^{2+}, propionate, imidazole (concentration varied to make ionic strength 0.15 M), 15 U/ml CPK, pH 7, and no added Ca^{2+} (pCa approximately 7.0). The relaxing solution in which the fibers were peeled should have been trapped initially in sealed TTs. The high Na^+ content of this solution facilitated ouabain binding to TT Na^+-K^+-ATPase sites or, in the absence of ouabain, helped to establish a polarizing TT diffusion potential when the peeled fiber was later immersed in bathing solutions with high $[K^+]$ and low $[Na^+]$. Furthermore, the high propionate and low Cl^- concentrations of the relaxing solution trapped in TTs maximized the depolarizing effect of Cl^- stimulation by creating the largest possible $[Cl^-]$ gradient across the TT membrane.

The first tension trace (A) in Fig. 1 displays the ionic depolarization protocol and a Cl^--induced tension transient in one half of a peeled rabbit adductor fiber. Transferring the fiber from the relaxing solution to a caffeine (10 mM) solution at low Mg^{2+} concentration caused Ca^{2+} release from the SR. The magnitude of the caffeine contracture gives a measure of the fiber's SR Ca^{2+} content [11]. However, it does not empty the fiber of Ca^{2+} since detergent (Brij-58) treatment, which should destroy all membranes, causes another tension transient (not shown in Fig. 1) following the one induced by caffeine [8]. The fiber was then rinsed in the relaxing solution and transferred to a Ca^{2+} load solution for 2 min (Fig. 1). The load solution contained 4 mM EGTA and sufficient total calcium for pCa = 5.6. The monovalent ionic composition of the load solution was 4 mM Na^+, 66 mM K^+, 4 mM Cl^-, and at least 66 mM propionate. Thus, the load solution provides for entry and sequestration of Ca^{2+} into the fiber's internal membrane systems (e.g., SR) and allows for equilibration of these

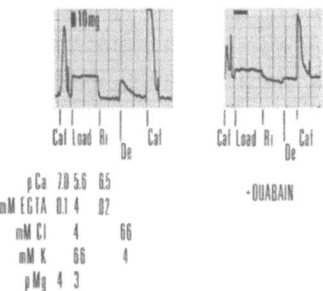

Figure 1. Elimination of Cl^--induced tension transient by ouabain in a peeled rabbit adductor magnus fiber. The fibers were immersed in relaxing solution unless otherwise indicated below the time marks on the tension records. The temperature of the solutions was $22 \pm 1°C$. The complex equilibrium for the bathing solution was determined using binding constants from the literature and a computer program [10]. Oligomycin (1 mg/ml) was added to prevent Ca^{2+} uptake by mitochondria [18]. Tension traces A and B are from peeled halves of the same fiber mounted separately in the same tension tranducer (A before B) and exposed to the same bathing solutions after peeling; the fiber half for record B was treated with ouabain prior to peeling as described in the text. Both fiber halves respond to caffeine (10 mM, pMg 4) stimulation demonstrating that the SR was functional, but only half A responded to Cl^- stimulation at constant $[K^+] \times [Cl^-]$ (De) with an observable tension transient.

systems with a low $[Cl^-]$, high $[K^+]$ bathing solution. Except for the relaxing solution, the $[Na^+]$ was 4 mM in all solutions in order to provide a constant stimulus for any Na^+-K^+-ATPase in the TTs. The fiber was then rinsed with a solution identical in monovalent ionic content to the load solution but with very low EGTA concentration (0.02 mM) and no added Ca^{2+}; this removed excess calcium and EGTA from the fiber's interior. Next, the fiber was stimulated with Cl^- by transfer to a solution that differed from the rinse solution only in monovalent ionic composition. The ionic depolarization solution created an abrupt substitution of Cl^- for propionate and choline for K^+ such that $[K^+] \times [Cl^-]$ remained constant (4 $\times$ 66). The ionic depolarization solution elicited a tension transient. The subsequent caffeine contracture is larger than the original one because the fiber had been loaded with Ca^{2+}. Thus, the peeled mammalian fiber responded to the Cl^- application at constant $[K^+] \times [Cl^-]$ with a Cl^--induced tension transient that is qualitatively similar to that of the frog peeled fibers [19].

However, Cl^- application without concurrent cation substitution (i.e., with increasing $[K^+] \times [Cl^-]$) frequently did not induce Ca^{2+} release in peeled mammalian fibers, and when it did the resulting tension transient was smaller than the Cl^--induced one at constant $[K^+] \times [Cl^-]$ [8]. These findings are opposite of those for frog peeled fibers [19].

Ouabain Elimination of Cl^--Induced Ca^{2+} Release

The second tension trace (B) of Fig. 1 was obtained from the other half of the adductor fiber. This half of the fiber was soaked, prior to peeling, in a relaxing solution containing 1 mM ouabain at 0°C for 2.5 hr. This presoaking in ouabain virtually eliminated the Cl^--induced transient at constant $[K^+] \times [CL^-]$ without greatly affecting SR function, assayed as the caffeine-induced release. As shown previously [8], adding ouabain to the bathing solutions after peeling does not affect either type of Ca^{2+} release, as would be expected. Ouabain is not very lipid soluble and its binding sites are on the interior of the TTs [13,22].

These results along with other data from previous studies [6-8] confirm the conclusions that: (1) the Cl^--induced (constant $[K^+] \times [Cl^-]$) Ca^{2+} release in peeled mammalian skeletal fibers is the result of depolarization of sealed, polarized TTs and (2) SR depolarization does not elicit Ca^{2+} release. The second conclusion is also consistent with the lack of evidence for SR depolarization preceding Ca^{2+} release in intact fibers [21]. Similar data have not been accumulated for frog peeled skeletal fibers. The Cl^--induced Ca^{2+} releases observed by Endo and Nakajima [12] in split skeletal fibers may have been triggered by the increase in $[K^+]\times[Cl^-]$, rather than ionic depolarization, or perhaps by stimulation of some undetected polarized TTs. Thus, Cl^--induced Ca^{2+} release from skinned frog fibers may also be due to TT depolarization [23]. The discrepancies in Cl^--induced responses of frog and mammalian peeled skeletal fibers are more likely due to variable responses to changes in $[K^+]\times[Cl^-]$ during ionic depolarization than to inherent differences in physiological mechanisms of Ca^{2+} release.

Ca^{2+} Release Induced by Changes in $[K^+]\times[Cl^-]$

Cl^- stimulation with a concurrent increase in $[K^+]\times[Cl^-]$ (and subsequent swelling of internal membrane systems) enhances Cl^--induced release in frog peeled skeletal fibers [19] but depresses it in mammalian peeled fibers [8]. In order to determine whether mammalian peeled skeletal fibers respond differently to several types of SR volume changes, fibers that did not appear to have a TT type of mechanism operant (i.e., they did not have an observable tension transient in response to ionic depolarization at constant $[K^+]\times[Cl^-]$) were tested using the protocol shown in Fig. 2.

The peeled adductor fiber was loaded, rinsed, and then ionically depolarized with Cl^- at constant $[K^+]\times[Cl^-]$ but did not respond with a Cl^--induced tension transient. However, following subsequent transfer to a bathing solution with decreased $[K^+]\times[Cl^-]$ (4 × 4), which should shrink the SR, the fiber responded with transient tension generation even though the potential of internal membrane systems should not have been altered during the decrease in $[K^+]\times[Cl^-]$. Since mammalian fibers do not release Ca^{2+} when $[Cl^-]$ is decreased and $[K^+]$ increased at constant $[K^+]\times[Cl^-]$ [8], the stimulation used in this study is probably not due to the diffusion potential created by the $[Cl^-]$ gradient.

Following a caffeine contracture and a second load and rinse, the fiber was transferred directly to the low $[K^+]$ solution at $[K^+]\times[Cl^-] = 16$, which should shrink the SR while simultaneously depolarizing it via a $[K^+]$ gradient instead of a $[Cl^-]$ gradient. This procedure also elicited Ca^{2+} release, although the resulting tension transient is slightly smaller, suggesting that prior ionic depolarization may enhance the Ca^{2+} release.

The final load and rinse of the fiber was followed with stimulation by Cl^- at constant $[K^+]$ and thus with increasing $[K^+]\times[Cl^-]$ (66 × 66). This stimulation did not yield a tension transient, suggesting that SR swelling does not elicit Ca^{2+} release in mammalian peeled fibers. However, when the $[K^+]\times[Cl^-]$ was decreased from 66 × 66 to 4 × 4,

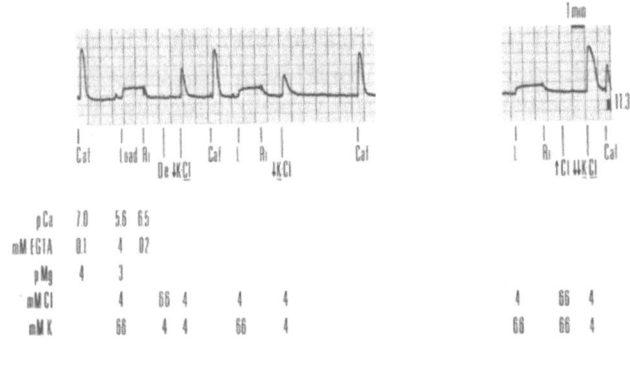

Figure 2. Tension transients in a peeled rabbit adductor magnus fiber induced by decreases and increases in $[K^+]\times[Cl^-]$. The fiber was selected because of its nonresponsiveness to ionic depolarization (De) (shown after the first load and rinse marked on the tension trace), thus lack of TT tension transient. As can be seen in the first two responses to decreased $[K^+]\times[Cl^-]$ (↓ K·Cl; ↓ K·Cl), the fiber responded with a slightly larger tension transient when it was ionically depolarized (De) prior to rather than during stimulation with decreased $[K^+]\times[Cl^-]$. The peeled fiber did not respond to ionic depolarization with increased $[K^+]\times[Cl^-]$ (↑ Cl), but it generated the largest tension transient to the final and largest decrease in $[K^+]\times[Cl^-]$ (↓ ↓ K·Cl).

Figure 3. Osmotic stimulation of a peeled adductor magnus fiber by sucrose, imidazole, and decreased $[K^+]\times[Cl^-]$. Tension traces A, B, and C were recorded in order of presentation. Part A: Tension trace shows failure of the fiber to respond to ionic depolarization at constant $[K^+]\times[Cl^-]$ (De) even though caffeine induced a tension transient. Adding and removing 50 mM sucrose (S) from the depolarizing (De) solution did not elicit an observable tension transient. Part B: The fiber did release Ca^{2+} in response to depolarization with a decrease in $[K^+]\times[Cl^-]$ ($\downarrow$ K·Cl) but this response was inhibited by the presence of sucrose (S) beginning in the rinse solution. Removing the sucrose without reloading the fiber restored the response to the decrease in $[K^+]\times[Cl^-]$. However, sucrose addition or removal did not elicit tension transients. Part C: The fiber was stimulated by a decrease in $[K^+]\times[Cl^-]$ and then caffeine (10 mM, pMg 4) both of which induced tension transients. The protocol was repeated but 50 mM imidazole was added at the end of the rinse. The fiber still responds to the decrease in $[K^+] \times [Cl^-]$ in the presence of imidazole with a large tension transient; imidazole itself did not trigger Ca^{2+} release. The remainder of the tension trace is a repeat of the first of Part C except that the rinse time period where imidazole was added is shortened to match those preceding the nonimidazole trials.

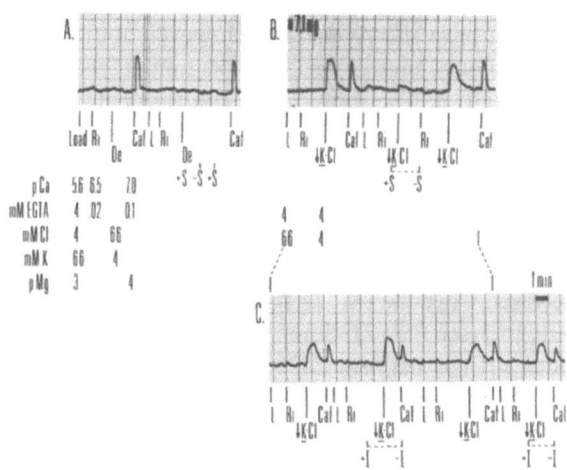

which should have induced the largest decrease in SR volume of the three ionic changes, the largest osmotic tension transient of the trace was generated.

The Ca^{2+} release in response to decreasing $[K^+]\times[Cl^-]$ results in alterations of SR calcium content, as evidenced by its effect on the magnitude of the caffeine contractures [11] (Fig. 2). Thus, mammalian peeled skeletal fibers differ from frog fibers [19] in that they respond to changes in $[K^+]\times[Cl^-]$ which should shrink, rather than swell, SR. However, it is not clear that shrinkage per se was the trigger for Ca^{2+} release. The addition of 50 mM sucrose or 50 mM imidazole (pH 7.0) to the rinse solution did not cause release of Ca^{2+}, even though these solutes should osmotically shrink the SR to a greater extent than the decrease in $[K^+]\times[Cl^-]$ produced by transfer from the rinse to decreased $[K^+]$ solution (Fig. 3).

Figure 3A shows that the peeled adductor fiber, after loading and rinsing, did not respond to ionic depolarization by Cl^- at constant $[K^+]\times[Cl^-]$ and thus did not have an observable TT-dependent component of Ca^{2+} release. The caffeine contracture demonstrates that the SR was functional. The same bathing solution protocol was again repeated except that 50 mM sucrose was included in the ionic depolarization solution. The fiber still did not release Ca^{2+}. Similarly, removing and replacing the sucrose, all in the ionic depolarization solution, failed to elicit Ca^{2+} release; however, caffeine still induced Ca^{2+} release. Thus, osmotic stimulation with sucrose, which should have caused shrinkage and swelling of the SR during ionic depolarization, did not elicit SR Ca^{2+} release.

Following a second load and rinse (Fig. 3B) stimulating the same fiber by decreasing the $[K^+]\times[Cl^-]$ elicited a large tension transient. This treatment did not empty the SR since caffeine subsequently stimulated a transient contraction (Fig. 3B). This series of bathing solution changes was repeated with 50 mM sucrose added to the rinse and to the decreased $[K^+]\times[Cl^-]$ solutions. Sucrose did not itself elicit Ca^{2+} release but inhibited the response triggered by a decrease in $[K^+]\times[Cl^-]$.

The protocol used in Fig. 3B was repeated using 50 mM imidazole (pH 7.0) to increase the osmolarity (Fig. 3C). Imidazole is not likely to permeate quickly, if at all, and so initially it should have shrunk the SR. However, it did not elicit Ca^{2+} release in the rinse solution. Furthermore, imidazole did not inhibit the Ca^{2+} release induced by a decrease in $[K^+] \times [Cl^-]$. The protocol for the first half of the Fig. 3C trace is then repeated.

Since imidazole and sucrose did not themselves elicit Ca^{2+} release, even when the SR should have been depolarized during osmotic stimulation, the decrease in $[K^+] \times [Cl^-]$ probably does not trigger Ca^{2+} release osmotically. Similarly, depolarization created by the decrease in $[K^+] \times [Cl^-]$ or the $[K^+]$ gradient cannot be the trigger for Ca^{2+} release, since ionic depolarization at constant $[K^+] \times [Cl^-]$ did not trigger Ca^{2+} release and the response occurred when the $[K^+] \times [Cl^-]$ was decreased by lowering the concentration of either $[K^+]$ or $[Cl^-]$ (see Fig. 2). Some as yet unknown parameter of stimulation inherent in a decreased $[K^+] \times [Cl^-]$ triggers Ca^{2+} release in mammalian peeled skeletal fibers.

The inhibitory effect of sucrose on the Ca^{2+} release induced by a decrease in $[K^+] \times [Cl^-]$ either may be a specific effect of sucrose or may be due to reduction of SR volume by sucrose, as has been observed for frog skinned fibers [1]. If imidazole is SR impermeant, the former conclusion is supported; whereas, if imidazole slowly permeates the SR in the rinse solution, the latter conclusion is more representative of the data.

Based on earlier data from peeled mammalian skeletal fibers, ionic depolarization of TTs, but not of SR membrane, triggers Ca^{2+} release [6–8]. Analogous studies of peeled frog fibers have not been reported, but existing data do not preclude this interpretation [11,14,23]. Indeed, electrical depolarization of frog peeled fibers in oil was blocked by ouabain [4], suggesting that TTs rather than SR were stimulated. Thus, there is evidence that frog and mammalian peeled skeletal fibers respond similarly to stimulation of Ca^{2+} release via a TT–SR junctional step initiated by TT depolarization. Differences in the Cl^--induced Ca^{2+}-release properties of frog and mammalian peeled skeletal fibers may be accounted for either by differences in the fibers' responses to SR volume changes or by an as yet unidentified parameter related to variation in $[K^+] \times [Cl^-]$. Since there is no evidence to indicate that SR volume changes trigger Ca^{2+} release during twitches in intact fibers, the differences between frog and mammalian peeled fiber responses to ionic depolarization may not be physiologically relevant. However, these differences must be accounted for in comparing amphibian and mammalian skinned fiber SR Ca^{2+}-release data.

ACKNOWLEDGMENTS. We thank Susan Bickham for typing the manuscript and Dr. Christine Kasper for helpful comments. Supported by grants from the NIH (AM-31511, HL-23128) and the Muscular Dystrophy Association of America.

References

1. Assayama, J.; Ford, L. E.; Surdyk-Droske, M. F. Relationship between sarcoplasmic reticulum volume and calcium capacity in skinned frog skeletal muscle fibres. *J. Muscle Res. Cell Motil.* **4**:307–319, 1983.
2. Boyle, P. J.; Conway, E. J. Potassium accumulation in muscle and associated change. *J. Physiol. (London)* **100**:1–63, 1941.
3. Costantin, L. L. Contractile activation in skeletal muscle. *Prog. Biophys. Mol. Biol.* **29**:197–224, 1975.
4. Costantin, L. L.; Podolsky, R. J. Depolarization of the internal membrane system in the activation of frog skeletal muscle. *J. Gen. Physiol.* **50**:1101–1124, 1967.
5. Costantin, L. L.; Taylor, S. R. Graded activation in frog muscle fibers. *J. Gen. Physiol.* **61**:424–443, 1973.
6. Donaldson, S. K. Mammalian skinned muscle fibers: Evidence of an ouabain-sensitive component of Cl^--stimulated Ca^{2+} release. *Biophy. J.* **37**:23a, 1982.
7. Donaldson, S. K. Mammalian peeled fibers: Ca^{2+} release is graded by Cl^- and blocked by ouabain. *Biophys. J.* **41**:231a, 1983.

8. Donaldson, S. K. B. Peeled mammalian skeletal muscle fibers: Possible stimulation of Ca^{2+} release via the TT-SR junction. *J. Gen. Physiol.* (in press), 1984.
9. Donaldson, S. K. B.; Hermansen, L. Differential direct effects of H^+ on Ca^{2+}-activated tension generation of skinned fibers from the soleus, cardiac, and adductor magnus muscles of rabbits. *Pfluegers Arch.* **376**:55–65, 1978.
10. Donaldson, S. K. B.; Kerrick, W. G. L. Characterization of the effects of Mg^{2+} on Ca^{2+}- and Sr^{2+}-activated tension generation of skinned skeletal muscle fibers. *J. Gen. Physiol.* **66**:427–444, 1975.
11. Endo, M. Calcium release from the sarcoplasmic reticulum. *Physiol. Rev.* **57**:71–108, 1977.
12. Endo, M.; Nakajima, J. Release of calcium induced by 'depolarization' of the sarcoplasmic reticulum membrane. *Nature New Biol.* **246**:216–218, 1973.
13. Erdmann, E.; Krawietz, W.; Presek, P. Receptor for cardiac glycosides. In: *Myocardial Failure*, G. Riecker, A. Weber, and J. Goodwin, eds., Berlin, Springer-Verlag, 1977, pp. 120–131.
14. Fabiato, A.; Fabiato, F. Calcium release from the sarcoplasmic reticulum. *Circ. Res.* **40**:119–129, 1977.
15. Hegyvary, C. Ouabain-binding and phosphorylation of $(Na^+ + K^+)$ ATPase treated with N-ethylmaleimide or oligomycin. *Biochim. Biophys. Acta* **422**:365–379, 1976.
16. Hodgkin, A. L.; Horowicz, P. Potassium contractures in single muscle fibers. *J. Physiol. (London)* **153**:386–403, 1960.
17. Huxley, A. F.; Taylor, R. E. Local activation of striated muscle fibers. *J. Physiol. (London)* **144**:426–441, 1958.
18. Lardy, H.; Reed, P.; Lin, H. C. Antibiotic inhibitors of mitochondrial ATP synthesis. *Fed. Proc.* **34**:1707–1710, 1975.
19. Mobley, B. A. Chloride and osmotic contractures in skinned frog muscle fibers. *J. Membr. Biol.* **46**:315–329, 1979.
20. Natori, R. The property and contraction process of isolated myofibrils. *Jikeikai Med. J.* **1**:119–126, 1954.
21. Oetliker, H. An appraisal of the evidence for a sarcoplasmic reticulum membrane potential and its relation to calcium release in skeletal muscle. *J. Muscle Res. Cell Motil.* **3**:247–272, 1982.
22. Schwartz, A.; Lindenmayer, G. E.; Allen, J. A. The sodium-potassium adenosine tri-phosphatase: Pharmacological, physiological and biochemical aspects. *Pharmacol. Rev.* **27**:3–134, 1975.
23. Stephenson, E. W. Activation of fast skeletal muscle: Contributions of studies on skinned fibers. *Am. J. Physiol.* **240**:C1–C19, 1981.

38

Isotropic Components of Antipyrylazo III Signals from Frog Skeletal Muscle Fibers

S. M. Baylor, M. E. Quinta-Ferreira, and Chiu Shuen Hui

The activation of the contractile response in a vertebrate twitch muscle fiber is known to depend on a transient rise in myoplasmic free Ca^{2+}. In order to learn more about the details of the cellular processes controlling and controlled by Ca^{2+}, it would be useful to be able to accurately measure the amplitude and time course of the myoplasmic Ca^{2+} transient. Recent work in a number of laboratories has shown that any of several metallochromic "calcium" indicator dyes, once introduced into the myoplasm, may be useful for this purpose in frog twitch fibers [2,4,5,8,18,19,21–25,31]. For example, arsenazo III and antipyrylazo III at a myoplasmic concentration in the range 0.5–1.5 mM give large and easily measured Ca^{2+} signals in response to a single action potential or voltage-clamp pulse.

Recently, a detailed comparison of the signals from intact single fibers injected with either of these two dyes suggested that the antipyrylazo III Ca^{2+} signal may be simpler and more reliably interpreted than the arsenazo III Ca^{2+} signal [8]. Nevertheless, the intracellular use of antipyrylazo III cannot be considered to be without complexity, since its signal during activity is known to include at least two other components [4,8] besides the Ca^{2+} component (ΔCa):

1. An anisotropic (dichroic) component, dependent upon activity in a subpopulation of dye molecules that are likely bound to oriented structures accessible to myoplasm, e.g., the myofilaments or the sarcoplasmic reticulum (SR).
2. A maintained isotropic component, best resolved after the Ca^{2+} and dichroic components have returned to baseline, which likely reflects a maintained rise in myoplasmic Mg^{2+} or pH ($\Delta Mg/\Delta pH$).

This chapter is concerned with the two major isotropic components (the ΔCa and the $\Delta Mg/\Delta pH$) of the antipyrylazo III muscle signal and includes a description of how they are

---

S. M. Baylor and M. E. Quinta-Ferreira • Department of Physiology, University of Pennsylvania, Philadelphia, Pennsylvania 19104.   *Chiu Shuen Hui* • Department of Biological Sciences, Purdue University, West Lafayette, Indiana 47907.   Present address of M.E.Q.-F.: Departmento de Fisica, Universidade de Coimbra, 3000 Coimbra, Portugal.

measured and the type of physiological information that they may give. A new conclusion suggested is that the maintained component may predominantly reflect a rise in Mg^{2+} rather than a fall in H^+. If so, a likely source of the elevation would be the Mg^{2+} bound to the Ca^{2+}, Mg^{2+} sites on parvalbumin [14,27]. Following stimulation, a portion of this Mg^{2+} is presumed to be released to the myoplasmic pool in exchange for a fraction of activator Ca^{2+} that temporarily is bound to parvalbumin [6,10,11,15,16,26,30], before the Ca^{2+} is resequestered in the SR by a means of the SR Ca^{2+} pump. The time course of decay of the maintained antipyrylazo III signal might therefore reflect the movement of Mg^{2+} back to the Ca^{2+}, Mg^{2+} sites on parvalbumin, as the fraction of released Ca^{2+} captured by parvalbumin is pumped back into the SR.

Use of Antipyrylazo III as an Intracellular Indicator

The methods employed in this study have been described in detail elsewhere [3,7,8]. Briefly, an intact single twitch fiber from the semitendinosus or iliofibularis muscle of the frog (*Rana temporaria*) was mounted in a normal Ringer solution at a long sarcomere spacing on an optical bench. Dye was pressure-injected into myoplasm following penetration by a dye-filled micropipette. A small region of fiber near the injection site was then transilluminated with a spot of quasi-monochromatic light, 35–65 μm in diameter, in order to measure absorbance. The optical data were collected sequentially in time using a series of wavelengths, at each wavelength using two forms of polarized light, denoted 0° (electric vector parallel to the fiber axis) and 90° (electric vector perpendicular to the fiber axis). Fiber activity was initiated by means of a single action potential elicited by a brief shock from a pair of extracellular electrodes. Any remaining twitch response was recorded by a sensitive tension transducer attached to one tendon end of the fiber. Records were obtained only from fibers showing stable, all-or-nothing optical responses.

The optical signals measured experimentally were proportional to light intensities or fractional changes in light intensity but have been calibrated in terms of absorbances A and changes in absorbance ΔA. For this purpose the following relationships were used [cf. 6]:

$$A(\lambda) = \log_{10}(J/I) \tag{1}$$

where J is the intensity of the transmitted light of wavelength λ measured in the absence of the fiber and I in its presence, and

$$\Delta A(\lambda) = -(\Delta I/I)/\log_e 10 \tag{2}$$

where Δ indicates small changes. In general, the values of A and ΔA include contributions from both the fiber itself (the "intrinsic" component) and the injected dye molecules. The intrinsic component is usually a small fraction of the total, for which correction can be made [cf. 3] to get the dye-related component. This corrected value of A can then be related to the concentration c of dye molecules in the fiber that contribute to the signal by means of Beer's law.

$$A(\lambda) = \epsilon(\lambda) \, c \, l \tag{3}$$

where l is path length through the fiber and $\epsilon(\lambda)$ is the molar extinction coefficient for the dye at wavelength λ. An analogous equation is applicable to the situation during activity, when dye changes from one state (e.g., Ca^{2+}-free) to a second state (e.g., Ca^{2+}-bound):

Resting Absorbance of Antipyrylazo III in Muscle Cells

$$\Delta A(\lambda) = \Delta\epsilon(\lambda)\, \Delta c\, l \qquad (4)$$

In this case $\Delta A(\lambda)$ is the change in dye-related absorbance, $\Delta\epsilon(\lambda)$ the change in molar extinction coefficient between the two states, and Δc the concentration of dye molecules that have changed states.

Resting Absorbance of Antipyrylazo III in Muscle Cells

The open circles in Fig. 1A show measurements of polarized absorbance in a resting fiber injected with approximately 1.2 mM antipyrylazo III. In order to obtain the dye-related

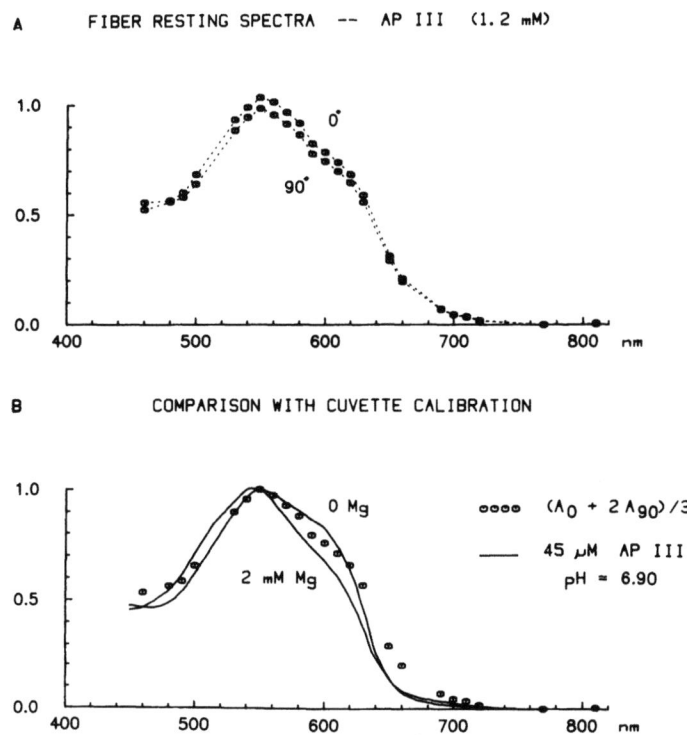

Figure 1. (A) Polarized antipyrylazo III absorbance (open circles) as a function of wavelength (indicated in nm on the abscissa) from a resting muscle fiber just prior to stimulation. The 90° data at wavelength λ have been corrected for the intrinsic absorbance by scaling up the value of $A_{90}(810) = 0.0118$ by the factor $(810/\lambda)^{1.3}$ and subtracting it from $A_{90}(\lambda)$; a similar correction was made to the $A_0(\lambda)$ absorbance data using an $(810/\lambda)^{1.1}$ scaling factor applied to $A_0(810) = 0.0183$. During the run, which took 9.5 min to complete, the value of $A_{90}(550)$ decreased from 0.335 to 0.286 while $A_0(550)$ decreased from 0.352 to 0.301. At each wavelength the data points have been normalized by the amplitude of the signal $(A_0 + 2A_{90})/3$ that would have been measured during that sweep using 550-nm light, as judged from bracketing measurements during the run. For all $\lambda < 700$ nm, $A_0(\lambda)$ is greater than $A_{90}(\lambda)$—see dashed lines connecting the two data sets. (Note: the observation that dye-related 0° absorbance is significantly greater than 90° absorbance is quite insensitive to the exact procedure used to correct the fiber's intrinsic absorbance. A justification for the particular method used here will be given in detail elsewhere.)
(B) Isotropic absorbance (open circles) calculated from (A) using the weighted average $(A_0 + 2A_{90})/3$. The mean isotropic value of $A(550)$ during the run was 0.317, which, using equation (3) with $\epsilon(550) = 2.55 \times 10^4$ M^{-1} cm^{-1} [Ref. 29], corresponds to an average dye concentration of 1.2 mM for this 104-μm-diameter fiber. Sarcomere spacing, 3.4 μm; spot diameter, 65 μm; fiber ID, 061082.2; 17.1°C. Continuous curves are cuvette calibrations measured at room temperature on the optical bench apparatus in a pH 6.90 solution containing 45 μM antipyrylazo III, 150 mM KCl, 10 mM Pipes, 0 or 2 mM MgCl$_2$, 0.25 mM EGTA, no added Ca^{2+}. Continuity in the curves was obtained from an interpolation by eye between measurements taken every 10 nm. The curves cross at $\lambda = 460, 550,$ and 660 nm.

component shown in Fig. 1A, the raw absorbance data were corrected for the small contribution to the total absorbance that is attributable to the fiber itself in the absence of dye. In addition, both data sets have been normalized by the value of $A(550)$ that would have been measured using either polarized form if all the dye molecules were isotropically (randomly) oriented. The 0° measurements in Fig. 1A are consistently higher, by about 5.4% on average, than the 90° measurements, indicating that some fraction of the dye molecules in the fiber is anisotropically (nonrandomly) oriented. The most likely explanation is that this fraction is bound to one or more oriented intracellular structures. A lower limit for the bound fraction is given by the ratio $0.054/3 = 0.018$ [see Ref. 5]. The fraction actually bound could be substantially larger. For comparison, Kovács et al. [19], who made measurements of the diffusion of antipyrylazo III into cut fibers, estimated a bound fraction of about 0.3.

The open circles in Fig. 1B show a weighted average, $(A_0 + 2A_{90})/3$, of the two data sets in Fig. 1A. As shown in [5], this averaging procedure, when applied to a cylindrical structure with radial symmetry such as a muscle fiber, permits calculation of the isotropic absorbance signal, i.e., the net absorbance that would be measured if all of the dye molecules were randomly oriented (which they are not). Although the circles in Fig. 1B reflect contributions from at least two different subpopulations of dye molecules (the oriented and the randomly distributed ones), with possibly different underlying $\epsilon(\lambda)$ dependencies, it is of interest to see if the isotropic spectrum can be fit by an *in vitro* calibration curve. Since myoplasmic free Ca^{2+} in a resting muscle fiber should be negligibly small [12,32] in comparison with the effective dissociation constant of the dye [29], the simplest situation would be that the resting antipyrylazo III spectrum is determined by myoplasmic pH and Mg^{2+} alone. Therefore, also shown in Fig. 1B are calibration curves obtained in a cuvette solution at zero Ca^{2+}, pH 6.90 [which is the estimated myoplasmic pH in these experiments (Ref. 3)], and at two different free Mg^{2+} levels, 0 and 2 mM. For $480 < \lambda < 620$ nm, the muscle data are bracketed by the calibration curves, suggesting that free Mg^{2+} in myoplasm may be approximately 1 mM [cf. 3]. However, some doubt is cast on this suggestion from Fig. 1B because of the disagreement observed between the muscle data and the calibration curves for wavelengths greater than 620 nm. The source of this discrepancy is not clear at present but is consistently observed in all fibers examined. It cannot be explained by assuming a nonzero resting Ca^{2+} level, but may relate to the contribution that the oriented dye molecules make to the average isotropic spectrum. This suggestion needs further experimental investigation.

The conclusion that follows from Fig. 1B and other muscle experiments and *in vitro* calibrations (not shown) is that the resting isotropic spectrum of antipyrylazo III in an intact fiber cannot be satisfactorily explained by any reasonable pair of values assumed for resting myoplasmic pH and Mg^{2+} alone. This conclusion therefore differs somewhat from that of Kovács et al. [19], who made less complete measurements in the range $\lambda > 620$ nm and used a different method to calculate resting dye absorbance.

Absorbance Changes in Antipyrylazo III during Activity

Figure 2A shows original records of changes in polarized absorbance (upper seven pairs of traces) and tension (lowermost trace) in response to a single propagated action potential. The wavelengths used to make the measurements shown in Fig. 2 are of interest because of the spectral features of the Ca^{2+}–antipyrylazo III reaction observed in cuvette (see continuous curve, Fig. 4A). For example, relative peaks are expected for a ΔCa signal at $\lambda = 460$, 550, 650, and 710 nm, whereas small to negligible changes are expected at $\lambda = 490$, 590,

38. Isotropic Components of Antipyrylazo III Signals

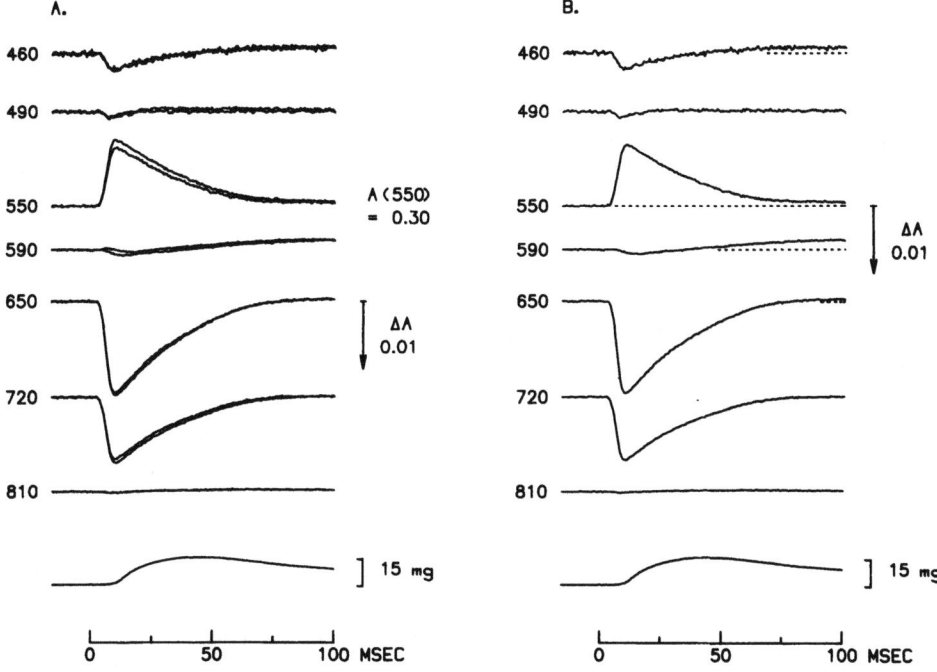

Figure 2. (A) Original records of polarized absorbance changes (upper seven pairs of traces) and tension (lowest trace) in response to an action potential initiated at zero time by an external shock. The records at different wavelengths (indicated in nm at the left) were taken sequentially over a 4.7-min period in response to single shocks separated by 0.2 to 0.8 min. At each wavelength, the 0° and 90° absorbance traces (recorded simultaneously by two seperate photodetectors) are shown superimposed. In the 550- and 590-nm records the 0° trace is above the 90° trace during the time of the early signal, whereas it is below in the 650- and 720-nm records. The 590-, 650-, and 720-nm records are single sweeps, while signal-averaging of two to five sweeps was used for the other records. The mean isotropic value of $A(550)$ during the run is shown to the right of the 550-nm pair of traces. An upward deflection in the optical traces corresponds to an increase in transmitted intensity (or a decrease in fiber absorbance). Data collected within same time period and from same fiber region as that shown in Fig. 1. (B) Isotropic absorbance changes calculated from (A) using the weighted average $(\Delta A_0 + 2\Delta A_{90})/3$. Extended baselines are shown as dashes in some cases in this and subsequent figures. No correction for the intrinsic absorbance change has been made to either (A) or (B).

and 810 nm. The signal at 810 nm is particularly important as a control, since 810 nm is beyond the absorbance band of the dye, and this signal therefore assesses the magnitude of any intrinsic components of the signals at the other wavelengths. Since the amplitude of the 810-nm signal is quite small relative to the shorter wavelength signals, the latter signals primarily reflect dye-related changes [cf. 3]. Furthermore, the close similarity of the members of each pair of polarized traces indicates that the dye-related activity may arise principally from molecules that are randomly oriented and therefore may be in the myoplasmic solution rather than bound to oriented structures. (The small differences detectable by polarization in the 550-, 590-, 650-, and 720-nm signals represent activity in a subpopulation of dye molecules that are oriented. This dichroic signal will be described in detail elsewhere.)

The traces in Fig. 2B show the weighted average, $(\Delta A_0 + 2\Delta A_{90})/3$, of the records in part A. This averaging procedure is directly analogous with that carried out in Fig. 1 for analysis of the resting absorbance. Thus, the records in Fig. 2B reflect the isotropic absorbance changes, i.e., the average changes in $\epsilon(\lambda)$ that occur for all subpopulations of dye molecules, independent of their orientation [see 5]. These signals consist primarily of two dye-related components that are distinguishable by their time course and spectral dependence.

The earlier temporal component in Fig. 2B is largest in the 550-, 650-, and 720-nm traces, which all have approximately the same waveform, reaching a peak about 10.4 msec after stimulation. Under the assumption that the later temporal component (see next paragraph) is relatively small at this time, the spectral dependence of the earlier component can be obtained by plotting the normalized amplitude of the absorbance change (at 10.4 msec) as a function of wavelength. This is shown as the open square data in Fig. 4A. The continuous curve in the same figure is a scaled Ca^{2+}-antipyrylazo III difference spectrum obtained from cuvette measurements. The close agreement between the muscle data and the cuvette curve indicates that the early signal component primarily reflects the transient formation of Ca^{2+}-dye complex. This reaction in turn undoubtedly reflects the rise and fall of myoplasmic free Ca^{2+} that serve to drive the twitch response. If one uses the calibration constants of Ríos and Schneider [29] for the Ca^{2+}-antipyrylazo III reaction involving 1 Ca^{2+} ion and 2 dye molecules, the peak value of the myoplasmic free Ca^{2+} transient in this experiment is estimated to be 2.2 μM [see also 4,8,19]. In view of the unexplained features of the muscle resting spectra in Fig. 1, this estimate must be considered somewhat tentative.

At late times in Fig. 2B, the calcium transient returns nearly to baseline, as judged by the 650- and 720-nm signals. However, at other wavelengths, for example at 460 and 590 nm, the final levels are offset from the baseline, indicating the presence of a second temporal component. This later component can be seen more clearly by examining the traces at a higher gain and slower sweep speed (Fig. 3A). All the signals, except for the 490-nm signal,

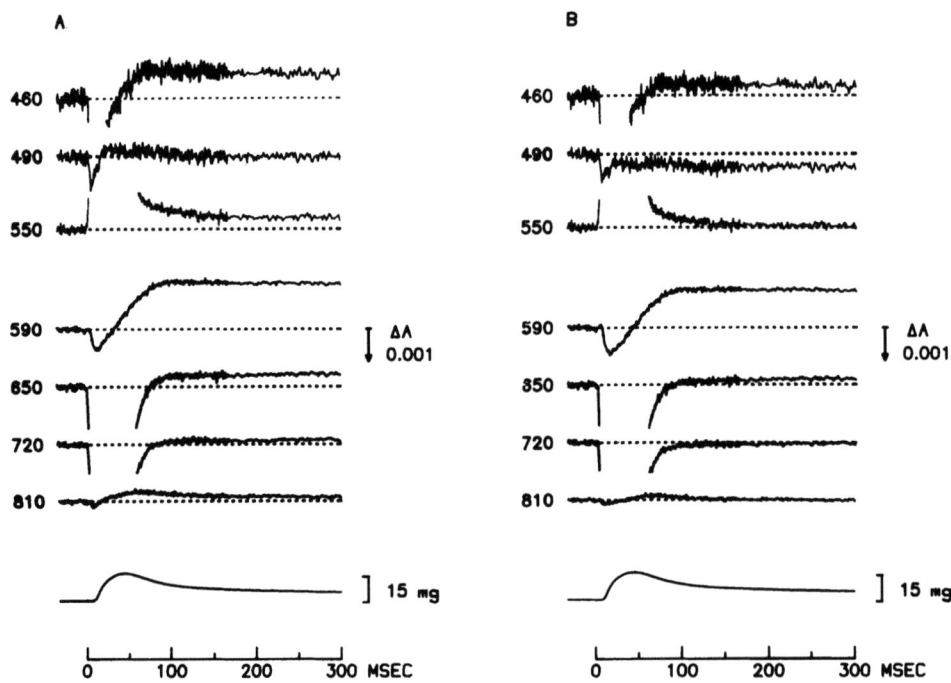

Figure 3. (A) Same traces as Fig. 2B, but on a longer time base and at 5 times the vertical gain. (B) Same traces as in (A), except corrected for the intrinsic component. The basis of the intrinsic correction was to use a standard waveform and a $(1/\lambda)^2$-dependence [cf. Ref. 3] as determined from other fibers stimulated in the absence of dye. The correction procedure forced the 810-nm trace to have an average value of zero between 168 and 300 msec, but not otherwise.

show a baseline offset at the end of the records (Fig. 3A). This maintained signal is not entirely attributable to a change in dye-related absorbance, since some offset is also seen in the 810-nm record. In order to examine in detail the dye-related component, it is necessary to correct the records in Fig. 3A for the non-dye-related contribution. The estimated dye-related signals are shown in Fig. 3B. The procedure for removing the non-dye-related component in this experiment (see Legend of Fig. 3) apparently worked satisfactorily at late times, as evidenced by the return of the 810-nm record in Fig. 3B to baseline. However, the procedure did not work perfectly at early times, as a small component, probably a movement artifact, is evident in this record during the time when the tension response is greatest.

The spectral dependence of the maintained dye-related component was determined from

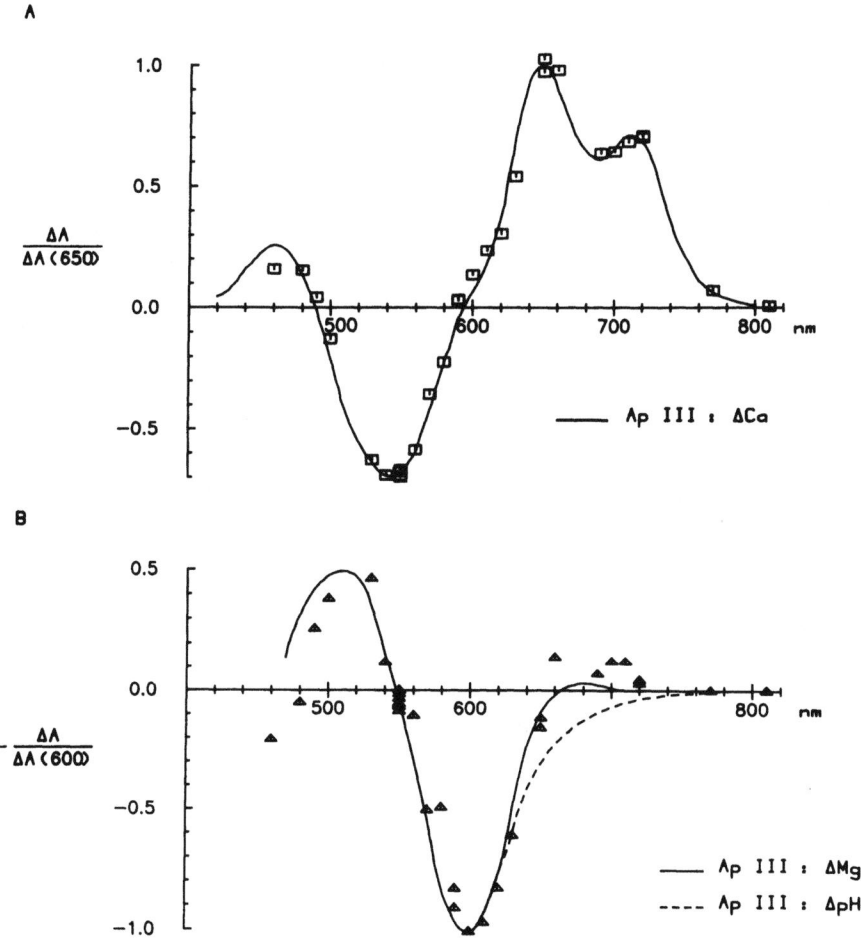

Figure 4. Comparison of the wavelength dependence of the early isotropic signal from a muscle fiber (open squares, part A) or the late isotropic signal (open triangles, part B) with cuvette calibration curves (continuous and dashed lines). At each wavelength the muscle data points were normalized by the amplitude of the signal that would have been measured during that sweep using light of 650 nm (part A) or 600 nm (part B), as judged from time-interpolated bracketing measurements during the run. Data were collected within same time period and from same fiber region as that shown in Fig. 1A. Calibration curves were taken from Figs. 10 and 11 of Ref. 4 and scaled to have a value of unity at 650 nm (part A) or 600 nm (part B). (A) Amplitude of muscle signal was measured from traces as in Fig. 2B, 10.4 msec after stimulation. $\Delta A(650)$ averaged 0.0141 over a time period when $A(550)$ averaged 0.3115. (B) Amplitude of muscle signal was measured from traces as in Fig. 3B, taken as the average value over the period 168–328 msec after stimulation. $\Delta A(600)$ was -0.0015 at a time when $A(550)$ was 0.321.

the final portions of the records illustrated in Fig. 3B and plotted in Fig. 4B (open triangle data). This dependence is clearly different from that of the ΔCa signal determined at early times (Fig. 4A), but is quite similar to the cuvette difference spectra shown in Fig. 4B reflecting an increase in either Mg^{2+} (continuous curve) or pH (dashed curve). Thus, the maintained isotropic signal likely reflects an increase in myoplasmic Mg^{2+} and/or pH resulting from events associated with the twitch response. A fraction of the maintained signal could result from a small maintained elevation in free Ca^{2+}; for example, such an elevation might account for the discrepancy between the muscle data points and the ΔMg calibration curve for $\lambda > 660$ nm. Alternatively, this discrepancy might be related to the same mechanism that is responsible for the discrepancy between the muscle data points and the calibration curves in Fig. 1B for $\lambda > 620$ nm.

Because of the similarity of the Mg^{2+} and pH difference spectra in Fig. 4B, it is not possible to assign by means of spectral information alone the fractional contributions of ΔMg and ΔpH to the total maintained signal in Fig. 3B. However, calibration of the possible amplitudes of the ΔpH or ΔMg signals suggests that the absorbance change may primarily reflect an increase in Mg^{2+} rather than a fall in H^+. For example, based on the calibration information for antipyrylazo III given in Refs. 4 and 29 the amplitude of the maintained change, if due to a change in free Mg^{2+} alone, requires an increase of 77 µM (relative to an assumed resting level of 2 mM) or, if due to a change in pH alone, requires an increase of 0.014 units (relative to an assumed resting level of 6.90). This value for the pH change is 7 times larger than the 0.002 units measured at a similar time after stimulation in fibers injected with the pH indicator dye phenol red [3]. Thus it appears quantitatively unlikely that a pH change can be the sole or major determinant of the second isotropic component of the antipyrylazo III signal, although it would appear reasonable to assign a small fraction (perhaps 14% = 0.002/0.014) of the maintained signal of Fig. 4B to the measured myoplasmic alkalization.

As discussed in Ref. 3, a likely source of the alkalization is the removal of protons required by the hydrolysis of phosphocreatine during a twitch. This mechanism predicts a pH increase of approximately 0.002 units if one assumes that the amount of phosphocreatine breakdown is 0.17 mM [3], a value estimated from measurements of activation heat [28]. In addition, because phosphocreatine is a Mg^{2+} buffer [13], hydrolysis of 0.17 mM phosphocreatine is expected to increase free Mg^{2+} by about 0.24% [3], or approximately 5 µM if resting free Mg^{2+} is 2 mM. Thus the combined effects of the pH and Mg^{2+} changes expected from phosphocreatine hydrolysis would appear to explain no more than 20–25% of the maintained antipyrylazo III absorbance change in Fig. 4B.

On the other hand, the kinetic modeling in Ref. 6 predicts that 300 msec after stimulation the Ca^{2+}–Mg^{2+} sites on parvalbumin should have released a total of approximately 100 to 150 µM Mg^{+2} to the myoplasmic pool in exchange for a similar amount of Ca^{2+} captured primarily from the activator sites on troponin before the Ca^{2+} is returned to the interior of the SR by the SR Ca^{2+} pump. As a result of the release of Mg^{2+} from parvalbumin there should be an increase in myoplasmic free $[Mg^{2+}]$ which, due to buffering, will be less than the total released. The nonbuffered fraction for Mg^{2+} may be roughly estimated as 0.4 in the following way. Under the assumption that a single myoplasmic site, S, of total concentration $[S_T] = [MgS] + [S]$ is the principal Mg^{2+} buffer, where $[MgS]$ and $[S]$ represent the concentrations of bound and free site, repectively, then

$$[MgS] = \frac{[S_T][Mg^{2+}]}{[Mg^{2+}] + K_D} \quad (5)$$

if K_D is the dissociation constant of the site for Mg^{2+}. Given some small change in total

Figure 5. Dye-related ΔA (upper), intrinsic ΔA (middle), and tension (lower) on a very slow sweep speed. The middle trace was taken from the 810-nm record shown in Fig. 2B (or Fig. 3A). The upper trace is the difference between the 590-nm and 490-nm traces shown in Fig. 3B. Both 490- and 590-nm traces individually returned toward (in the case of the 590-nm signal) or overshot (in the case of the 490-nm signal) the baseline at the end of the sweep. Since the two signals had nonidentical waveforms, the presence of another dye-related component is postulated. This component may be a very slowly time-dependent decrease in absorbance at all wavelengths, possibly related to the mechanism that accounts for the maintained change seen in the middle trace. The best estimate of the final $\Delta Mg/\Delta pH$ time course in this experiment may therefore come from the difference trace shown here, in which common mode signals on either side of an isosbestic point at 550 nm would tend to cancel, whereas the $\Delta Mg/\Delta pH$ signal would summate. The computer least-squares fit of an exponential function (without baseline offset) to the last 1.2 sec of the difference record gave a good fit with a time constant of 1.60 sec, corresponding to a rate of 0.63 sec^{-1}.

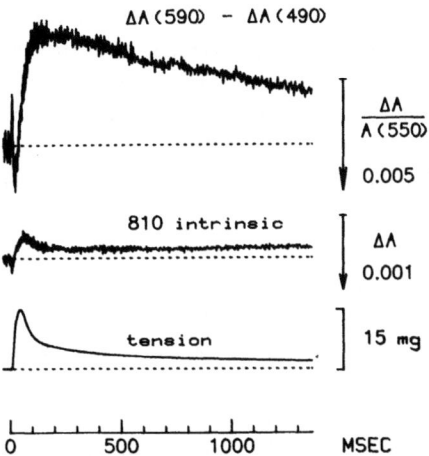

Mg^{2+} allotted between $\Delta[Mg^{2+}]$ and $\Delta[MgS]$, the nonbuffered fraction for Mg^{2+} then satisfies:

$$\frac{\Delta[Mg^{2+}]}{\Delta[Mg^{2+}] + \Delta[MgS]} = \left[1 + \frac{[S_T]K_D}{([Mg^{2+}] + K_D)^2} \right]^{-1} \quad (6)$$

If phosphocreatine is assumed to be the major myoplasmic buffer for Mg^{2+} other than parvalbumin, then substitution in equation (6) with $[Mg^{2+}] = 2$ mM, $[S_T] = 43$ mM and $K_D = 25$ mM (see Refs. 3, 13) yields a nonbuffered fraction of 0.4 for the Mg^{2+} released by parvalbumin. Hence, the estimated increase in free $[Mg^{2+}]$ 300 msec after stimulation expected from the exchange of Ca^{2+} for Mg^{2+} on parvalbumin is 0.4 times 100 to 150 μM, or 40 to 60 μM. Such an increase is presumed to account for between 52% (40/77) and 78% (60/77) of the maintained signal observed in Fig. 4B.

The overall conclusion from these considerations is that the maintained isotropic component of the antipyrylazo III signal is likely to arise from both the increase in myoplasmic pH and the increase in myoplasmic free Mg^{2+} that result from both phosphocreatine hydrolysis and Ca^{2+}–Mg^{2+} exchange on parvalbumin, although the latter process, involving Mg^{2+} alone, appears to be quantitatively the most significant. It will therefore be of interest in future experiments to measure the time course of the absorbance signal on a time scale of seconds so that its falling phase may be resolved and possibly used to set limits on the time course of return of the released Mg^{2+} to its binding sites on parvalbumin. This latter time course may give information concerning which of two processes are more important as the rate-limiting step in restoration of the resting state following a twitch: (1) the rate of Ca^{2+} pumping by the SR Ca^{2+} pump at low levels of free Ca^{2+}, or (2) the rate of unbinding of Ca^{2+} from parvalbumin [17].

Summary and Conclusion

The experiments of this and previous reports [4,8,24,25] indicate that the use of metallochromic dyes such as antipyrylazo III and arsenazo III as myoplasmic Ca^{2+} indicators involves a number of complexities. In the case of antipyrylazo III, a detectable fraction of the dye molecules is oriented and presumably bound in a resting muscle fiber. Further, the resting isotropic spectrum for $\lambda > 620$ nm is significantly different from a cuvette calibration determined by pH and Mg^{2+} alone. This discrepancy implies that a significant fraction of the dye molecules behaves differently inside a muscle fiber than in a cuvette. Moreover, a component of the absorbance change measured during a twitch reflects activity in an oriented fraction of the dye molecules. (These may be the same molecules that are oriented in the resting state.) This component of the absorbance change may arise from a change either in molar extinction coefficient or in orientation (or both) of these dye molecules [5]. Finally, the data indicate the existence of at least two dye-related components to the isotropic absorbance change measured during a twitch—an early one reflecting a rise in myoplasmic Ca^{2+}, and a later one reflecting a rise in myoplasmic Mg^{2+} (and probably a fall in H^+). If in the instances described above the fractions of bound molecules are large, the calibration of the Ca^{2+}, Mg^{2+}, and pH signals (at rest or during activity) on the basis of cuvette measurements lacking muscle components could involve significant quantitative error.

Nevertheless, the use of antipyrylazo III in intact muscle fibers to monitor changes during activity in myoplasmic free Ca^{2+} and Mg^{2+} (and H^+) holds considerable promise as a new experimental technique. The Ca^{2+}–antipyrylazo III signal represents the fastest time course so far determined by any method for estimating the myoplasmic Ca^{2+} transient in response to a single action potential [see, e.g., 1,4,8,9,25]. These findings suggest that the antipyrylazo III signal is the most accurate temporal monitor of the average myoplasmic free Ca^{2+} transient and should provide the most reliable information in kinetic models of Ca^{2+} cycling in muscle [cf. 6,15,19,20,30]. Also, the close agreement between the muscle Ca^{2+} signal and a cuvette calibration (Fig. 4A) indicates that, in spite of the complications enumerated above, antipyrylazo III may monitor free Ca^{2+} in a fairly straightforward way. Finally, the presence of the later isotropic component, although adding complexity to a complete description of the antipyrylazo III signal in muscle, may also contribute important new physiologically relevant information. For example, this signal may prove to be a useful experimental tool for investigations on the intact fiber concerning the role that the Ca^{2+}, Mg^{2+} sites on parvalbumin play in promoting muscle relaxation.

ACKNOWLEDGMENTS. Financial support was provided by grants from the National Institutes of Health (NS-17620 to S.M.B. and NS-15375 to C.S.H.) and the American Heart Association (Established Investigatorship to S.M.B. and Research Grant to C.S.H.).

References

1. Baylor, S. M. Optical studies of excitation–contraction coupling using voltage-sensitive and calcium-sensitive probes. In: *Handbook of Physiology: Section on Skeletal Muscle*, L. D. Peachey and R. H. Adrian, eds., Bethesda, American Physiological Society, 1983, pp. 355–379.
2. Baylor, S. M.; Chandler, W. K.; Marshall, M. W. Arsenazo III signals in singly dissected frog twitch fibres. *J. Physiol. (London)* **287**:23P-24P, 1979.
3. Baylor, S. M.; Chandler, W. K.; Marshall, M. W. Optical measurement of intracellular pH and magnesium in frog skeletal muscle fibres. *J. Physiol. (London)* **331**:105–137, 1982.
4. Baylor, S. M.; Chandler, W. K.; Marshall, M. W. Use of metallochromic dyes to measure changes in myoplasmic calcium during activity in frog skeletal muscle fibres. *J. Physiol. (London)* **331**:139–177, 1982.
5. Baylor, S. M.; Chandler, W. K.; Marshall, M. W. Dichroic components of arsenazo III and dichlorophosphonazo III signals in skeletal muscle fibres. *J. Physiol. (London)* **331**:179–210, 1982.
6. Baylor, S. M.; Chandler, W. K.; Marshall, M. W. Sarcoplasmic reticulum calcium release in frog skeletal muscle fibres estimated from arsenazo III calcium transients. *J. Physiol. (London)* **344**:625–666, 1983.

7. Baylor, S. M.; Oetliker, H. A large birefringence signal preceding contraction in single twitch fibres of the frog. *J. Physiol. (London)* **264**:141–162, 1977.
8. Baylor, S. M.; Quinta-Ferreira, E. M.; Hui, C. S. Comparison of isotropic calcium signals from intact frog muscle fibers injected with arsenazo III or antipyrylazo III. *Biophys. J.* **44**:107–112, 1983.
9. Blinks, J. R.; Rudel, R.; Taylor, S. R. Calcium transients in isolated amphibian skeletal muscle fibres: Detection with aequorin. *J. Physiol. (London)* **277**:291–323, 1978.
10. Blum, H. E.; Lehky, P.; Kohler, L.; Stein, E. A.; Fisher, E. H. Comparative properties of vertebrate parvalbumins. *J. Biol. Chem.* **252**:2834–2838, 1977.
11. Briggs, N. Identification of the soluble relaxing factor as a parvalbumin. *Fed. Proc.* **34**:540, 1975.
12. Coray, A.; Fry, C. H.; Hess, P.; McGuigan, J. A. S.; Weingart, R. Resting calcium in sheep cardiac tissue and in frog skeletal muscle measured with ion-selective microelectrodes. *J. Physiol. (London)* **305**:60P–61P, 1980.
13. Curtin, N. A.; Woledge, R. C. Energy changes and muscle contraction. *Physiol. Rev.* **58**:690–761, 1978.
14. Gillis, J. M.; Piront, A.; Gosselin-Rey, C. Parvalbumins: Distribution and physical state inside the muscle cell. *Biochim. Biophys. Acta* **585**:444–450, 1979.
15. Gillis, J. M.; Thomason, D.; Lefevre, J.; Kretsinger, R. H. Parvalbumins and muscle relaxation: A computer simulation study. *J. Muscle Res. Cell Motil.* **3**:377–398, 1982.
16. Gosselin-Rey, C.; Gerday, C. Parvalbumins from frog skeletal muscle (*Rana temporaria*): Isolation and characterization, structural modifications associated with calcium binding. *Biochim. Biophys. Acta* **492**:53–63, 1977.
17. Johnson, J. D.; Robinson, D. E.; Robertson, S. P.; Schwartz, A.; Potter, J. D. Ca^{2+} exchange with troponin and the regulation of muscle contraction. In: *Regulation of Muscle Contraction: Excitation–Contraction Coupling,* A. D. Grinnell and M. A. B. Brazier, eds., New York, Academic Press, 1981, pp. 241–259.
18. Kovács, L.; Ríos, E.; Schneider, M. F. Calcium transients and intramembrane charge movement in skeletal muscle fibres. *Nature (London)* **279**:391–396, 1979.
19. Kovács, L.; Ríos, E.; Schneider, M. F. Measurement and modification of free calcium transients in frog skeletal muscle fibres by a metallochromic indicator dye. *J. Physiol. (London)* **343**:161–196, 1983.
20. Meltzer, W.; Ríos, E.; Schneider, M. F. Rate of calcium release in frog skeletal muscle. *Biophys. J.* **41**:396a, 1983.
21. Miledi, R.; Nakajima, S.; Parker, I.; Takahashi, T. Effects of membrane polarization on sarcoplasmic calcium release in skeletal muscle. *Proc. R. Soc. London Ser. B* **213**:1–13, 1981.
22. Miledi, R.; Parker, J.; Schalow, G. Measurement of changes in intracellular calcium in frog skeletal muscle fibres using arsenazo III. *J. Physiol. (London)* **269**:11P–13P, 1977.
23. Miledi, R.; Parker, I.; Zhu, P. H. Calcium transients evoked by action potentials in frog twitch muscle fibres. *J. Physiol. (London)* **333**:655–679, 1982.
24. Palade, P.; Vergara, J. Detection of Ca^{++} with optical methods. In: *Regulation of Muscle Contraction: Excitation–Contraction Coupling,* A. D. Grinnell and M. A. B. Brazier, eds., New York, Academic Press, 1981, pp. 143–160.
25. Palade, P.; Vergara, J. Arsenazo III and antipyrylazo III calcium transients in single skeletal muscle fibers. *J. Gen. Physiol.* **79**:679–707, 1982.
26. Pechère, J. F.; Derancourt, J.; Haiech, J. The participation of parvalbumins in the activation–relaxation cycle of vertebrate fast skeletal muscle. *FEBS Lett.* **75**:111–114, 1977.
27. Potter, J. D.; Johnson, J. D.; Dedman, J. R.; Schreiber, W. E.: Mandel, F.; Jackson, R. L.; Means, A. R. Calcium-binding proteins: Relationship of binding, structure, conformation and biological function. In: *Calcium-Binding Proteins and Calcium Function,* R. H. Wasserman, R. A. Corradino, E. Carafoli, R. H. Kretsinger, D. H. MacLennan, and F. L. Siegel, eds., Amsterdam, North-Holland, 1977, pp. 239–250.
28. Rall, J. A. Effects of temperature on tension, tension-dependent heat, and activation heat in twitches of frog skeletal muscle. *J. Physiol.* **291**:265–275, 1979.
29. Ríos, E.; Schneider, M. F. Stoichiometry of the reaction of calcium with the metallochromic indicator dyes antipyrylazo III and arsenazo III. *Biophys. J.* **36**:607–621, 1981.
30. Robertson, S. P.; Johnson, J. D.; Potter, J. D. The time-course of Ca^{2+} exchange with calmodulin, troponin, parvalbumin, and myosin in response to transient increases in Ca^{2+}. *Biophys. J.* **34**:559–569, 1981.
31. Schneider, M. F.; Ríos, E.; Kovács, L. Calcium transients and intramembrane charge movement in skeletal muscle. In: *Regulation of Muscle Contraction: Excitation–Contraction Coupling,* A. D. Grinnell and M. A. B. Brazier, eds., New York, Academic Press, 1981, pp. 131–141.
32. Snowdowne, K. W. Aequorin luminescence in frog skeletal muscle fibers at rest. *Fed. Proc.* **38**:1443, 1979.

39

Ion Movements Associated with Calcium Release and Uptake in the Sarcoplasmic Reticulum

A. V. Somlyo, T. Kitazawa, H. Gonzalez-Serratos, G. McClellan, and A. P. Somlyo

The release of Ca^{2+} from the sarcoplasmic reticulum (SR) during the activation of contraction and its uptake during relaxation are essential steps in the excitation–contraction coupling process of vertebrate skeletal muscle. Depolarization of the T-tubule membrane is thought to induce Ca^{2+} release through a trigger mechanism transmitted to the SR at the junction between the T-tubules and the terminal cisternae (TC), traversed by "feet," "bridging structures," or "pillars" [3,4,20]. The specific trigger mechanism of release, however, is not known, although two general classes of coupling mechanism, electrical and chemical messenger, have been proposed.

In order to assess the feasibility of these mechanisms, and particularly those involving electrical coupling, it was necessary to obtain an estimate of the trans-SR ion distribution that would indicate whether there is an electrical potential across the SR membrane and to measure the movement of other ions accompanying Ca^{2+} release and uptake. Because the release, unlike the uptake, process has been, at best, difficult to study in isolated SR, and the concentration of permeant monovalent ions in the SR is altered through fractionation procedures, it was important to determine *in situ* the compositional changes in the SR that accompany the release and the uptake of Ca^{2+}. Electron probe analysis, in conjunction with rapid freezing and cryoultramicrotomy, permits the measurement of elemental concentrations in microvolumes having the diameter of a 50-nm focused electron beam. Therefore, with these techniques the concentration of various elements in the TC and in adjacent regions of the I-band can be studied in small bundles of 15–20 frog semitendinosus fibers frozen at various stages of contraction and relaxation.

A. V. Somlyo, T. Kitazawa, G. McClellan, and A. P. Somlyo • The Pennsylvania Muscle Institute and Departments of Physiology and Pathology, University of Pennsylvania School of Medicine, Philadelphia, Pennsylvania 19104.   *H. Gonzalez-Serratos* • Department of Biophysics, University of Maryland School of Medicine, Baltimore, Maryland 21201.   Present address of T. K.: Department of Pharmacology, Juntendo University School of Medicine, Tokyo 113, Japan.

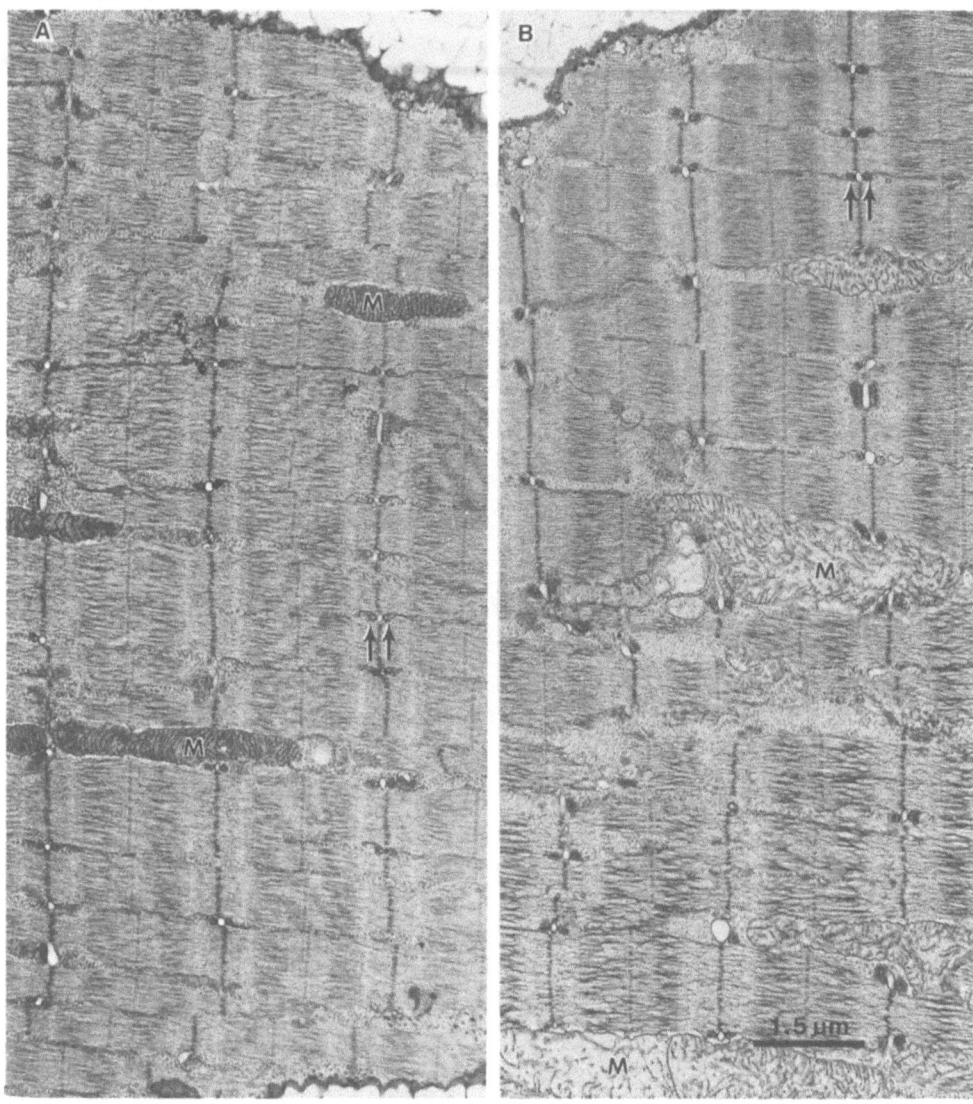

Figure 1. Images of freeze-substituted control (A) and valinomycin-treated (B) fibers showing swollen mitochondria (M) in the valinomycin-treated muscle compared with the control. In some regions only remnants of the mitochondrial cristae are observed in the valinomycin-treated (for 2 hr at 2–5°C) muscle. The triads of the sarcoplasmic reticulum (arrows) appear normal and, unlike the mitochondria, do not show evidence of swelling in the valinomycin-treated muscles. [From Ref. 13.]

The procedures for rapid freezing—which occurs in milliseconds—enable one to obtain tissue free of visible ice crystal damage to a depth of 5–10 μm from the surface (Fig. 1), while maintaining ionic gradients across cell membranes [21,22]. The specific elements occupying a given region of dry, unstained cryosections are recognized and quantitated through the characteristic X-ray peaks generated by these elements, when irradiated by a focused electron beam of the electron microscope. Details of the methods used for electron probe microanalytic quantitation [7,8,19] and for rapid freezing and cryoultramicrotomy [22,24] have been published. The optimal spatial resolution of the method with a field emission gun is better than 10 nm and with an LaB_6 gun operated in conventional transmission mode is about 100 nm; the minimal, detectable concentration of Ca is about 1 mmole/kg dry wt [13,19].

We shall summarize here our results with electron probe analysis showing the elemental composition of the SR at rest and during activation, including the effect of the K^+ ionophore valinomycin [17] on the SR, and the implication of these results concerning the trans-SR electrical potential and excitation–contraction coupling. Preliminary results describing the time course of the return of Ca to the TC during relaxation and the delayed trans-SR movements of Mg will also be provided.

The Composition of the Terminal Cisternae at Rest and during a Tetanus

The elemental composition of the terminal cisternae and of paired regions of adjacent I-band cytoplasm, in 10 paired frog muscles frozen at rest and during a 1.2-sec tetanus, was examined. Calcium was localized to the terminal cisternae using a focused 50- to 100-nm probe. As shown in Table I and reported elsewhere [21], the amount of Ca that is released during a 1.2-sec tetanus is 59% (69 mmoles/kg dry TC) of the resting Ca content. The release from this compartment—which represents about 4% of the volume of the whole fiber—would lead to a large (~ 1 mM) increase in total calcium in fiber water. These Ca measurements represent free as well as bound Ca. Frog muscle contains approximately 0.35 mM parvalbumin [5], which has two high-affinity Ca^{2+}-binding sites. Therefore, the total concentration of potential cytoplasmic Ca^{2+}-binding sites includes the Ca^{2+}-binding sites (0.7 mmole/liter) on parvalbumin, the Ca^{2+}-specific (0.2 mmole/liter) and the Ca^{2+}/Mg^{2+} sites (0.2 mmole/liter) of troponin. This total agrees well with the electron probe measurements of the amount of Ca^{2+} released from the TC [21].

Ca release was associated with a significant uptake of K and Mg into the TC (Table I). But the amount of cations taken up by the SR did not fully compensate for the total charge released in the form of Ca^{2+} [21]; this "apparent" charge deficit (62 mEq), if real, would

Table I. Elemental Composition of the Terminal Cisternae and of I-Band Cytoplasm[a,b]

	n	Na	Mg	P	S	Cl	K	Ca
		\multicolumn{7}{c}{Elemental Composition of the Terminal Cisternae Mean ± S.D., mmoles/kg dry wt}
Control	229	56 ± 39	59 ± 22	415 ± 82	214 ± 40	43 ± 19 (17)	554 ± 138	117 ± 48
Tetanus	222	58 ± 37	72 ± 23*	413 ± 82	225 ± 56	42 ± 20 (16)	604 ± 103*	48 ± 20*
		\multicolumn{7}{c}{Effect of tetanus (mEq): -138 Ca $+$ 50 K $+$ 26 Mg $= -62$ mEq*}
		\multicolumn{7}{c}{Elemental composition of I-band cytoplasm (analysis with small probes)}
Control	229	45 ± 44 (11)	54 ± 18	339 ± 80	263 ± 52	55 ± 27 (14)	510 ± 112 (128)	4.5 ± 7.1
Tetanus	222	40 ± 33 (10)	47 ± 18*	338 ± 76	256 ± 65	54 ± 26 (14)	505 ± 108 (126)	8.0 ± 7.5*
								Δ+3.5 ± 1.3

[a]Modified from Ref. 21.
[b]The concentrations expressed as mmoles/liter H_2O are shown in parentheses and are based on a value of 80% H_2O in the I-band and 72% H_2O in the TC.
*$p < 0.001$, statistical comparisons between control and tetanus.

give rise to an unrealistically high trans-SR membrane potential. Most of this charge deficit is probably apparent, and complete charge compensation is achieved by the uptake of protons and/or organic ions (see below) that cannot be measured by energy-dispersive electron probe analysis.

The concentrations of Na and Cl in the TC and adjacent I-band were not significantly different (Table I), ruling out the hypothesis, based on Na and Cl fluxes, that the ionic content of the SR is similar to that of the extracellular space (for review see 22). The Cl concentrations in the TC of resting and tetanized muscles, respectively, were not significantly different. The absence of an alteration in Cl^- concentration during a tetanus argues against a large and/or sustained electrical potential change across the SR membrane accompanying Ca^{2+} release. If the SR membranes are as permeable to Cl^- as estimated in isolated SR fractions and inferred from experiments on skinned skeletal muscle fibers, then Cl^- should distribute electrophoretically according to the potential change across this membrane. A trans-SR potential of -56 mV due to E_{Ca} should result in a readily detectable 10-fold Cl^- concentration gradient across the SR membrane. The absence of Cl^- movements toward such a gradient, suggests either that a large, sustained potential change is absent during Ca^{2+} release or that the SR membrane *in situ* is not as permeable to Cl^- as indicated by other experiments.

The Effects of Valinomycin on K^+ Movements across the SR Membrane during a Tetanus

In view of the high K^+ concentration in the cytoplasm and the reported high permeability of the SR membranes for K^+ [2; for review see 9,13] it was surprising that K^+ was not the sole or dominant cation moving into the SR to compensate for the charge loss due to Ca^{2+} release. The effect of valinomycin 5 μM (this concentration applied at 2–4°C for 2 hr has no effect on tetanus tension) on the composition of the TC at rest and during a 1.2-sec tetanus was recently determined [13]. Penetration of valinomycin into the fibers was apparent from the marked mitochondrial swelling seen in cryosections and in freeze-substituted muscles (Fig. 1); the K content of these swollen mitochondria was significantly increased in both the resting and the tetanized valinomycin-treated muscles. Morphometric analysis of rapidly frozen freeze-substituted muscles revealed that there was no significant change in the volume of the SR of untreated and valinomycin-treated tetanized muscles. Whole fiber composition measured over several sarcomeres with large-diameter electron probes showed no differences in the concentrations of Na, Mg, P, S, Cl, or Ca between untreated and valinomycin-treated muscles at rest or during a tetanus.

Analysis of the TC showed that valinomycin had no effect on the K content of *resting* muscle, providing further evidence against the existence of a significant electrical potential across the SR in resting muscle [13,21–23; for review see 14]. In the absence of a measurable K^+ gradient across this membrane, a large potential could only exist if the membrane were impermeable to K^+, in which case valinomycin would be expected to cause a large influx of K^+ and water, resulting in massive swelling. Neither measurements of K^+ permeability of isolated SR nor SR volume measurements in frozen muscles are consistent with this hypothesis.

Valinomycin had a significant effect on the composition of the SR during tetanus and Ca^{2+}-release (Table II). The K uptake that normally accompanies Ca release during a tetanus was approximately doubled ($p < 0.001$) in the presence of valinomycin, while Mg uptake also remained highly significant. Thus, under these conditions, valinomycin abol-

Table II. Effect of 5 μM Valinomycin on the Elemental
Composition of Terminal Cisternae
in Tetanized Muscles[a,b]

Element	Mean ± S.E.M., mmoles/kg dry TC wt	
	Control ($n = 65$)	Valinomycin ($n = 154$)
Na	98 ± 5.2	71 ± 2.9
Mg*	81 ± 2.4	92 ± 2.3
P	524 ± 7.4	510 ± 5.9
S	218 ± 4.5	217 ± 3.3
Cl	38 ± 1.6	45 ± 1.7
K**	654 ± 10.4	712 ± 10.3
Ca***	47 ± 2.3	53 ± 1.6

[a]Modified from Ref. 13.
[b]Statistical comparisons between control and valinomycin-treated TC: *$p < 0.005$, **$p < 0.001$, ***$p < 0.05$.

ished the apparent charge deficit: the electrical equivalents of Ca^{2+} released and Mg^{2+} plus K^+ taken up were approximately equal. The observation that K^+ transport is increased in the presence of valinomycin, a neutral carrier [17], implies that Ca^{2+} release is electrogenic. The persistent and even greater Mg uptake observed after valinomycin treatment, as compared with control tetanized muscles, is not understood. It may be explained by the presence, in the lumen of the TC, of Ca^{2+}-binding proteins (predominantly calsequestrin) with a higher affinity for Mg^{2+} than for K^+ [10,16]. The inward movement of K^+ into the SR during a tetanus and its enhancement with valinomycin treatment are likely to be due to both the electrical driving forces generated by outward Ca^{2+} movement and the presence of the acidic cation-binding proteins in the lumen of the SR. Furthermore, the increased K^+ uptake induced by valinomycin implies that the K^+ conductance of the SR in untreated muscles is insufficient for short-circuiting the potential generated by the Ca^{2+} current during release. These results are also consistent with the suggestion [21] that the apparent charge deficit observed during a tetanus reflects the movement of protons and, possibly, organic cations into the SR.

Ca^{2+} and Mg^{2+} Movements across the SR Membrane during Recovery from a Tetanus

The amount of Ca^{2+} released during a tetanus (approximately 1 mM in fiber H_2O) is localized to the cytoplasm and *not* trapped in the longitudinal SR [21]. The cytoplasmic localization of released Ca^{2+} is consistent with a large fraction being bound to parvalbumin. In view of the high ratio of (free) Mg^{2+} to Ca^{2+} in resting muscle and the known affinities of parvalbumin for these cations, it can be inferred that Mg^{2+} is associated with most of the parvalbumin in resting muscle. (Furthermore, during a twitch, Ca^{2+} binding to these sites would be rate limited by the relatively slow off-rates of Mg^{2+} from parvalbumin [6,18].) The half-time for the removal of Mg^{2+} from parvalbumin is approximately 550 msec [18], and, thus, there is sufficient time, during a 1.2-sec tetanus, for these sites to be saturated with Ca^{2+}.

The half-time of Ca^{2+} off-rate from parvalbumin is also rather slow, approximately 700

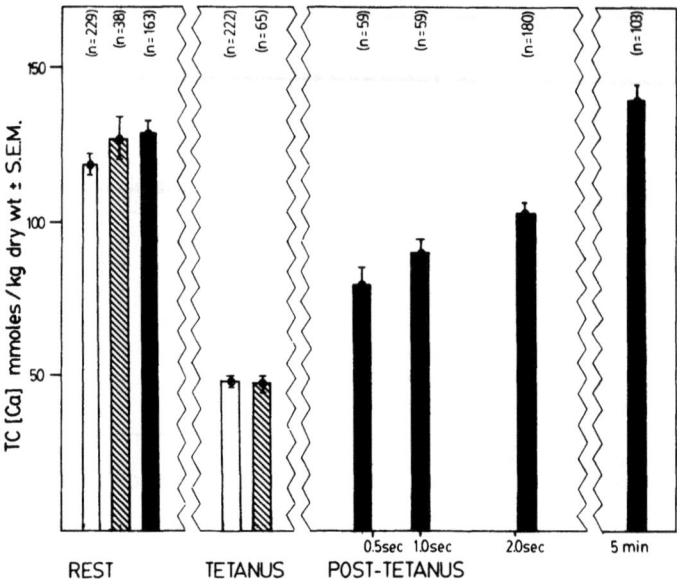

Figure 2. Histogram showing changes in the Ca concentrations of the terminal cisternae (TC) in resting muscles, during a 1.2-sec tetanus and at various intervals following a tetanus shown on the abscissa. The open bars represent the muscles published in Somlyo et al. [21] from 10 paired frog muscles and the hatched bars are taken from paired muscles published in Kitazawa et al. [13]. The solid bars represent a large group of frogs that were analyzed over a period of three years alternating pairs for the various experimental points. The number of TC analyzed are indicated at the top of each bar.

msec [18]. If these slow off-rates are also applicable to the protein *in vivo* and if Ca^{2+} binds to these sites *in vivo*, then the half-time of the return of Ca^{2+} to the TC following a tetanus should be as great or greater than the Ca^{2+} off-rate from parvalbumin. The following results show this, in fact, to be the case. Muscles were frozen at various intervals following a brief tetanus, and the composition of the TC and adjacent I-band cytoplasm were measured. By 0.5 sec after the end of a tetanus, the muscles had relaxed to baseline, and 31 mmoles Ca^{2+}/kg dry wt TC (Fig. 2) had returned to the TC. Based on a TC volume of 2.3 to 5% of the fiber volume and cell H_2O content of 77%, the Ca^{2+} uptake at the end of relaxation (about 0.2–0.45 mmole/kg wet fiber) was equal to or exceeded the amount of Ca^{2+} that can be bound to the Ca^{2+}-specific sites of troponin. This finding is consistent with observations on fragmented SR [11,15] showing that rates of Ca^{2+}-pumping are sufficient to account for Ca^{2+} removal during relaxation. At 2.1 sec after tetanus, 68% of the Ca had returned to the TC (Fig. 2), but the removal of Mg accumulated in the SR during Ca release showed a longer time course of recovery. There was no significant change in the Mg content of the TC at 2.1 sec posttetanus, compared with the tetanus value. However, by 5 min following a tetanus, the Mg content of the TC had decreased to the control resting levels.

The mechanism underlying the slow Mg^{2+} movements is not understood nor is its relationship (if any) to other slow recovery phenomena, such as posttetanic potentiation, repriming after a KCl contracture or the slow recovery of the full aequorin signal during a twitch [1]. A similar uptake of magnesium into the SR has been observed during caffeine contractures [26] and during quinine contractures [27]. These results imply that magnesium uptake is associated with the increased calcium permeability during release, rather than with calcium pumping. The influx of Mg^{2+} into the TC during tetanus, as well as caffeine and quinine contractures and with the slow removal of Mg from the TC after relaxation, suggest that there is a transient increase in the Mg^{2+}-permeability of the SR during activation.

The half-time for the return of Ca to the TC is approximately 1 sec (Fig. 2), considerably slower than a simple diffusion process, faster than measured by autoradiography [25], and similar to the Ca^{2+} off-rates determined for isolated (carp) parvalbumin [18]. This leads to the conclusion that the rate-limiting step for the return of Ca into the TC of the SR is the (rather slow) off-rate of Ca^{2+} from parvalbumin.

Summary and Conclusions

Our findings reveal that there is no significant electrical potential across the SR membrane in resting muscle: the concentrations of (unbound) K, Na, and Cl in the SR are similar to those in the cytoplasm; and valinomycin has no effect on the ionic composition of the SR in resting muscle. The membrane potential across the SR during tetanus is small and/or not sustained, as indicated by the lack of Cl redistribution across the SR during tetanus. This conclusion, however, is based on the assumption that the Cl^- permeability of the SR is high, as observed in fragmented SR preparations.

During a tetanus in frog muscle, 60% of the Ca in the TC is released into the fiber; this concentration is sufficient to saturate the Ca^{2+}-binding sites on troponin and on parvalbumin. Release of Ca^{2+} is electrogenic and compensated in part by the uptake of K^+ and Mg^{2+} into the SR and by the movement of a charged component (possibly protons) that is not detected by electron probe analysis.

The return of calcium to the TC during relaxation (400–500 msec after tetanus) indicates that the pumping rate in the SR is sufficiently high to account for the removal of an amount of Ca^{2+} equivalent to that bound to regulatory sites on troponin. The rate of return after relaxation is consistent with the time of Ca^{2+} release from parvalbumin. There is no evidence of delayed transit of Ca from the longitudinal tubules of the SR to the TC. The increase in Mg content of the TC during tetanus suggests a transient state of relatively high Mg^{2+} permeability, as the Mg content remains high at 2.1 sec after tetanus, by which time most (78%) of the Ca has returned to the TC.

ACKNOWLEDGMENT. Supported by Grant HL-15835 to the Pennsylvania Muscle Institute.

References

1. Blinks, J. R.; Rudel, R.; Taylor, S. R. Calcium transients in isolated amphibian skeletal muscle fibres: Detection with aequorin. *J. Physiol. (London)* **277**:291–323, 1978.
2. Duggan, P. B.; Martonosi, A. Sarcoplasmic reticulum. IX. The permeability of sarcoplasmic reticulum membranes. *J. Gen. Physiol.* **56**:147–167, 1970.
3. Eisenberg, B. R.; Gilai, A. Structural changes in single muscle fibers after stimulations at a low frequency. *J. Gen. Physiol.* **74**:1–16, 1979.
4. Franzini-Armstrong, C. Studies of the triad. I. Structure of the junction in frog twitch muscles. *J. Cell Biol.* **47**:488–499, 1970.
5. Gosselin-Rey, C.; Gerday, C. Parvalbumins from frog skeletal muscle (*Rana temporaria* L.): Isolation and characterization: Structural modifications associated with calcium binding. *Biochim. Biophys. Acta* **492**:53–63, 1977.
6. Haiech, J.; Derancourt, J.; Pechere, J.; Demaille, J. G. Magnesium and calcium binding to parvalbumins: Evidence for differences between parvalbumins and an explanation of their relaxing function. *Biochemistry* **13**:2752–2758, 1979.

7. Hall, T. A. The microprobe assay of chemical elements. In: *Physical Techniques in Biological Research*, Volume IA, G. Oster, ed., New York, Academic Press, 1971, pp. 151–267.
8. Hall, T. A.; Gupta, B. L. EDS quantitation and applications to biology. In: *Introduction to Analytical Electron Microscopy*, J. J. Hren, J. I. Goldstein, and D. C. Joy, eds., New York, Plenum Press, 1979, pp. 169–197.
9. Hasselbach, W.; Oetliker, H. Energetics and electrogenicity of the sarcoplasmic reticulum calcium pump. *Annu. Rev. Physiol.* **45**:325–339, 1983.
10. Ikemoto, N.; Nagy, G.; Bhatnagar, G. M.; Gergely, J. Studies on a metal-binding protein of the sarcoplasmic reticulum. *J. Biol. Chem.* **249**:2357–2365, 1974.
11. Inesi, G.; Scarpa, A. Fast kinetics of adenosine triphosphate dependent Ca^{2+} uptake by fragmented sarcoplasmic reticulum. *Biochemistry* **11**:356–359, 1972.
12. Kitazawa, T.; Shuman, H.; Somlyo, A. P. Quantitative electron probe analysis: Problems and solutions. *Ultramicroscopy* **11**:251–262, 1983.
13. Kitazawa, T.; Somlyo, A. P.: Somlyo, A. V. The effects of valinomycin on ion movements across the sarcoplasmic reticulum in frog muscle. *J. Physiol. (London)* **350**:253–268, 1984.
14. Oetliker, H. An appraisal of the evidence for a sarcoplasmic reticulum membrane potential and its relation to calcium release in skeletal muscle. *J. Muscle Res. Cell Motil.* **3**:247–272, 1982.
15. Ogawa, Y. Some properties of fragmented frog sarcoplasmic reticulum with particular reference to its response to caffeine. *J. Biochem. (Tokyo)* **67**:667–683, 1970.
16. Ostwald, T. J.; MacLennan, D. H. Isolation of a high affinity calcium-binding protein from sarcoplasmic reticulum. *J. Biol. Chem.* **249**:974–979, 1974.
17. Pressman, B. C. Biological applications of ionophores. *Annu. Rev. Biochem.* **45**:501–530, 1976.
18. Robertson, S. P.; Johnson, J. D.; Potter, J. D. The time-course of Ca^{2+} exchange with calmodulin, troponin, parvalbumin, and myosin in response to transient increases in Ca^{2+}. *Biophys. J.* **34**:559–569, 1981.
19. Shuman, H.; Somlyo, A. V.; Somlyo, A. P. Quantitative electron probe microanalysis of biological thin sections: Methods and validity. *Ultramicroscopy* **1**:317–339, 1976.
20. Somlyo, A. V. Bridging structures spanning the junctional gap at the triad of skeletal muscle. *J. Cell Biol.* **80**:743–750, 1979.
21. Somlyo, A. V.; Gonzalez-Serratos, H.; Shuman, H.; McClellan, G.; Somlyo, A. P. Calcium release and ionic changes in the sarcoplasmic reticulum of tetanized muscle: An electron probe study. *J. Cell Biol.* **90**:577–594, 1981.
22. Somlyo, A. V.; Shuman, H.; Somlyo, A. P. Elemental distribution in striated muscle and effects of hypertonicity: Electron probe analysis of cryo sections. *J. Cell Biol.* **74**:828–857, 1977.
23. Somlyo, A. V.; Shuman, H.; Somlyo, A. P. The composition of the sarcoplasmic reticulum *in situ:* Electron probe X-ray microanalysis of cryo sections. *Nature (London)* **268**:556–558, 1977.
24. Somlyo, A. V.; Silcox, J. Cryoultramicrotomy for electron probe analysis. In: *Microbeam Analysis in Biology*, C. Lechene and R. Warner, eds., New York, Academic Press, 1979, pp. 535–555.
25. Winegrad, S. The intracellular site of calcium activation of contraction in frog skeletal muscle. *J. Gen. Physiol.* **55**:77–88, 1970.
26. Yoshioka, T.; Somlyo, A. P. Composition of sarcoplasmic reticulum and mitochondria during caffeine contracture. *Biophys. J.* **45**:319a, 1984.
27. Yoshioka, T.; Somlyo, A. P. Calcium and magnesium contents and volume of the terminal cisternae in caffeine-treated skeletal muscle. *J. Cell Biol.* in press.

IX

Calcium and Cardiac Contractility

40

Role of Calcium in Heart Function in Health and Disease
100 Years of Progress

Pawan K. Singal, Vincenzo Panagia, Robert E. Beamish, and Naranjan S. Dhalla

Early History and Introduction

> A small quantity of calcium, added to saline solution with small amounts of potassium chloride, makes a good artificial circulating fluid and the ventricle will continue beating perfectly for more than four hours [Sydney Ringer, 1883].

In his experiments on isolated frog hearts, Ringer [34] first reported that "saline solution, to which is added one ten-thousandth part of potassium chloride, makes an excellent circulating fluid in experiments with the detached heart." However, in these misleading experiments, the saline solution (0.75% NaCl) was prepared with pipe water contaminated with traces of many inorganic salts including Ca^{2+}. When Ringer repeated these experiments with all solutions made using distilled water [35], he discovered that Ca^{2+} has an important role in heart function.

Over the past 100 years, considerable investigative effort has focused on the mechanism by which Ca^{2+} initiates cardiac contraction and maintains heart cell function. During the past two decades, a definite though poorly defined role for Ca^{2+} in the maintenance of cardiac cell structure has been identified. There has been an evolution of newer concepts concerning regulation of intracellular Ca^{2+} concentration within a very narrow range (10^{-7} to 10^{-5} M) by a closely regulated exchange between different subcellular Ca^{2+} stores and myofibrils. Furthermore, an understanding of the roles of Ca^{2+} in excitation–contraction coupling and in various metabolic processes under physiological as well as pathophysiological conditions has increased considerably. Due to space limitations only those observations which we believe are crucial for the demonstration of Ca^{2+} involvement in physiological and

---

Pawan K. Singal, Robert E. Beamish, and Naranjan S. Dhalla • Experimental Cardiology Section, Department of Physiology, University of Manitoba, Winnipeg, Manitoba R3E OW3, Canada. Vincenzo Panagia • Department of Oral Biology, University of Manitoba, Winnipeg, Manitoba R3E OW3, Canada.

Table I. Experiments Supporting the Role of Ca^{2+} in Heart Function in Health and Disease

Excitation–contraction coupling
  Loss of force development in a Ca^{2+}-free medium
  Initiation of a focal contraction by microinjection of Ca^{2+} in the myocardial cell
  Relationship between force development and extracellular Ca^{2+} concentration
  Contractile force development is preceded by an increase in intracellular free Ca^{2+}
  Contractile failure with lanthanum–an extracellular electron-dense tracer and a competitor of Ca^{2+}
  Relationship of contractile force with sarcolemmal Ca^{2+} binding
  Contractile force depression with Ca^{2+} antagonists
Pathophysiological aspects
  Loss of structural integrity during altered Ca^{2+} homeostasis in the cell
  Ca^{2+}-paradox phenomenon
  Intracellular Ca^{2+} deficiency and Ca^{2+} overload
  Abnormalities in subcellular Ca^{2+} transport systems
  Cardiac arrhythmias and Ca^{2+} antagonists
  Ca^{2+}-induced activation of phospholipases, proteases, and lysosomes

pathophysiological aspects of heart function will be emphasized. Events and experiments especially relevant to our present understanding of the role of Ca^{2+} in heart function in health and disease are listed in Table I.

Role of Ca^{2+} in Myocardial Physiology

Almost 25 years after Ringer's observations, Locke and Rosenheim [25] and Mines [27] demonstrated that surface electrical activity remained unaltered when contractile activity completely disappeared in hearts perfused with a Ca^{2+}-free medium. Microinjection of Ca^{2+} into the muscle fiber by Heilbrunn and Wiercinski [17] in 1947 initiated contraction. In 1955, Niedergerke [32] reported that the Ca^{2+} in the cell causing contraction was in ionic form and was proportional to the magnitude of the developed force. In more recent experiments carried out with cardiac fibers and aequorin (a Ca^{2+}-sensitive protein) the increase in intracellular free Ca^{2+} occurred before the contractile response [3]. These experiments revealed that Ca^{2+} is an essential link between cardiac muscle excitation and contraction.

Although many observations indicated the importance for contractility of membranous Ca^{2+}, it was Sandow [37] who proposed that Ca^{2+} released from membranes may be responsible for the activation of contraction. Support for this concept was provided by studies with lanthanum (La^{3+}), an electron-dense tracer that is restricted to the extracellular space [10] and depresses contractile force in a dose-dependent manner [36,39]. Since La^{3+} both displaces membrane-bound Ca^{2+} and reduces Ca^{2+} binding to the sarcolemmal membrane, it is considered to act by decreasing the supply of intracellular Ca^{2+}. More conclusive evidence for this is the demonstration of a direct relationship between Ca^{2+} binding and contractile force [9]. Antagonists of Ca^{2+}, which interfere with Ca^{2+} entry through sarcolemma, also depress contractile force development. Although these observations indicate that intracellular free Ca^{2+} levels required for contraction are modified by Ca^{2+} movements across the sarcolemma, the exact mechanism of Ca^{2+} influx as well as of its regulatory control remains to be defined. In this regard, Ca^{2+} influx through electrically and metabolically controlled slow channels as well as Na^+–Ca^{2+} exchange mechanisms in the sarcolemma have been identified [9]. The missing biochemical link between the increase in intracellular free Ca^{2+} and Ca^{2+} sensitivity of the contractile apparatus was supplied by the discovery of new muscle protein, troponin, by Ebashi and co-workers [11]. This protein provided Ca^{2+} sensitivity to actomyosin in the presence of tropomyosin in skeletal [11] and cardiac muscle [21].

Another intracellular membrane system important in the maintenance of cytoplasmic Ca^{2+} levels was first detected in a preparation isolated from skeletal muscle homogenates. This was called *soluble relaxing factor* because its *in vitro* effect on actomyosin proteins was reversible upon addition of Ca^{2+}. This soluble relaxing factor was later found to constitute membrane vesicles derived from the sarcoplasmic reticulum (SR) [4,19,44]. It is now known that Ca^{2+} transport into SR vesicles is associated with Ca^{2+}-pump ATPase, and this, in turn, is stimulated by cAMP-dependent protein kinase-mediated phosphorylation of phospholamban present in these membranes [45]. The SR Ca^{2+} release mechanisms, on the other hand, are not as well understood, although Ca^{2+}-pump ATPase protein may be involved [22]. Mitochondria have also been shown to accumulate as well as to release Ca^{2+}, but the question of their beat-to-beat role in altering intracellular Ca^{2+} levels remains controversial [6].

The Ca^{2+} level inside the cell during a contraction–relaxation cycle fluctuates approximately between 0.1 and 10 μM [26]. These changes in cytoplasmic free Ca^{2+} levels are generally attributed to Ca^{2+} influx and efflux across the sarcolemma, as well as to Ca^{2+} release and uptake mechanisms in the SR. However, the biochemical control mechanisms involved in these Ca^{2+} movements are far from clear. Two of the regulatory control agents identified (cAMP-dependent protein kinase and calmodulin) influence Ca^{2+} transport systems in these membranes [9]. The effects of these two mediators on Ca^{2+}-pump activities in SR and sarcolemma are summarized in Table II. Both Ca^{2+} uptake and Ca^{2+}-stimulated ATPase activities in the SR were markedly increased by protein kinase and calmodulin. Sarcolemmal Ca^{2+}-pump activities were also stimulated by these two substances. These observations illustrate the type of metabolic controls regulating membrane Ca^{2+} transport.

From the information accumulated to date, the biochemical basis of contraction can be

Table II. Ca^{2+}-Pump Activities in Rat Heart Sarcoplasmic Reticular and Sarcolemmal Fractions in the Absence and Presence of cAMP-Dependent Protein Kinase or Calmodulin[a]

	Control	Protein kinase (10 μg)	Calmodulin (2 μg)
Sarcoplasmic reticulum			
Ca^{2+}-uptake activity (nmoles Ca^{2+}/mg/5 min)	121 ± 7.1	165 ± 6.9*	184 ± 8.2*
Ca^{2+}-stimulated ATPase (μmoles P_i/mg/hr)	11.3 ± 1.0	15.0 ± 0.6*	16.8 ± 1.2*
Sarcolemma			
Ca^{2+}-binding activity (nmoles Ca^{2+}/mg/5 min)	4.6 ± 0.3	7.9 ± 0.5*	10.5 ± 0.7*
Ca^{2+}-stimulated ATPase (μmoles P_i/mg/hr)	5.3 ± 0.4	8.2 ± 0.7*	11.7 ± 0.5*

[a]Each value is a mean ± S.E. of four experiments. The effect of protein kinase was studied in the presence of 5 μM cAMP. The concentration of free Ca^{2+} employed in these experiments was 10 μM. The activities of sarcoplasmic reticulum (unlike sarcolemma) were studied in the presence of 2 mM potassium oxalate. *, significantly ($p < 0.05$) different from control. Methods for the isolation of sarcoplasmic reticulum and sarcolemma as well as for the determination of their activities were described elsewhere [43, 47].

described as follows. Membrane depolarization causes an increase in the intracellular free Ca^{2+} concentration through Ca^{2+} influx and release mechanisms. This free Ca^{2+} then binds to troponin and removes the inhibitory effect of troponin–tropomyosin on the actin–myosin interaction for initiating contraction. Intracellular free Ca^{2+} concentration is then lowered by efflux and uptake mechanisms to obtain cardiac relaxation. The effects of a

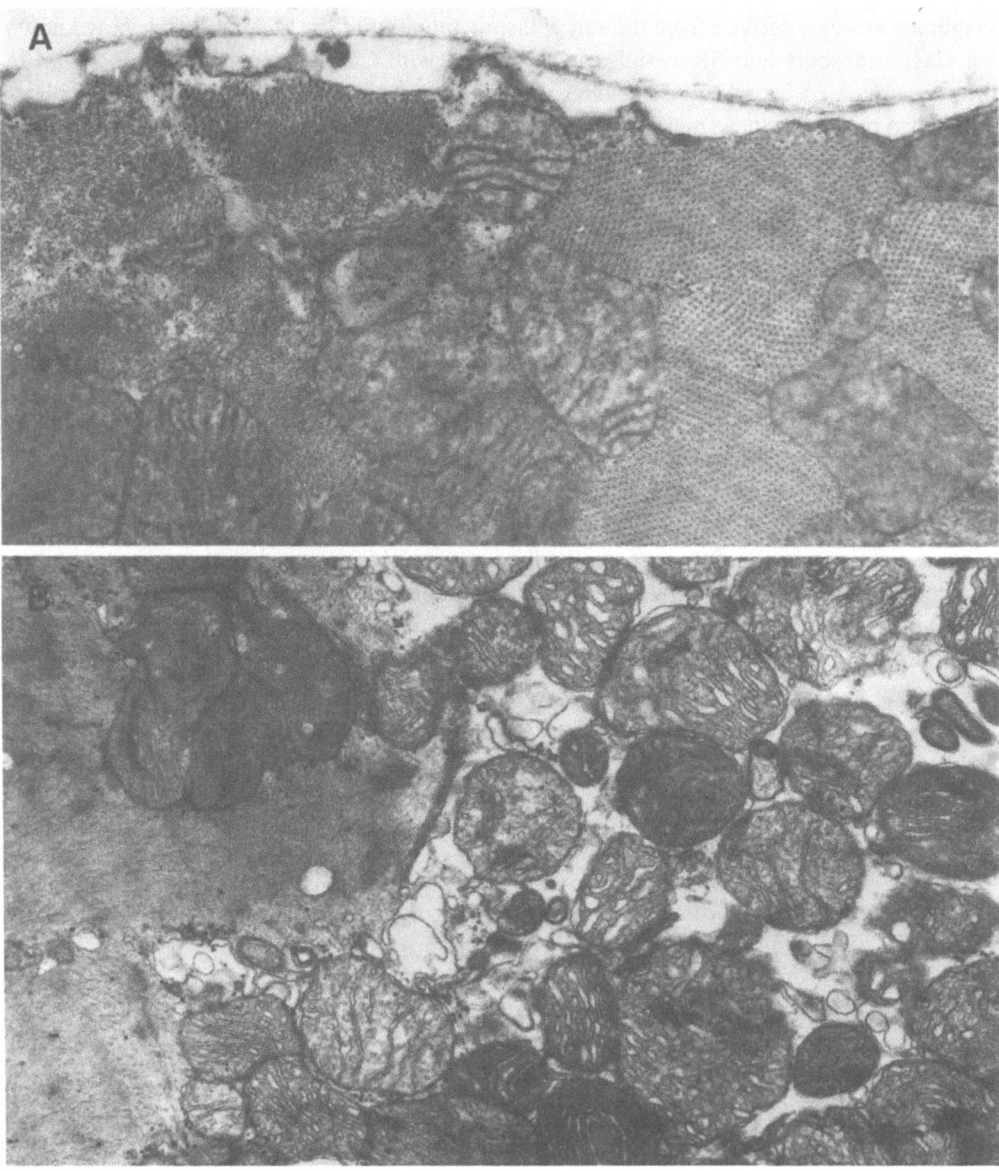

Figure 1. (A) Separation of basal lamina from the bilayer membrane of the myocardial cell in an isolated rat heart perfused with Ca^{2+}-free medium. × 15,000. (B) Isolated rat heart perfused with a Ca^{2+}-free medium for 5 min followed by a 5-min perfusion with a medium containing 1.25 mM Ca^{2+}. Cell damage is indicated by the formation of a contractile mass, extrusion of mitochondria, and loss of typical appearance of a cardiac cell. × 15,000. The conditions for heart perfusions as well as techniques for the processing of the myocardium for ultrastructural studies were described elsewhere [40].

variety of cardiotonic and cardiodepressant drugs can now be understood in terms of their actions on various Ca^{2+} transport systems; and these actions form the basis of many current therapeutic approaches to treatment of heart disease.

Role of Ca^{2+} in Myocardial Pathophysiology

The role of Ca^{2+} in maintenance of structural integrity of the cardiac cell was emphasized by experiments in which gross structural heart damage was reported after perfusion with Ca^{2+}-free medium [29,48]. Perfusion of hearts with Ca^{2+}-free media also separates the basal lamina from the plasma membrane bilayer (Fig. 1A). This cell surface component of the sarcolemma is critically important in controlling membrane Ca^{2+} transport [23,24], as well as other membrane enzyme functions [28]. Recently, changes in the membrane bilayer have also been reported in hearts perfused with a Ca^{2+}-free medium [14,40]. Alterations in membrane Ca^{2+} content have been associated with conformational changes in the bilayer of a variety of membranes [18,33]. While these studies clearly established the role of Ca^{2+} in the maintenance of cell structure, the exact mechanism of Ca^{2+} involvement remains to be determined. One possibility is the cross-linking of various membrane molecules by a nonionic binding through Ca^{2+} [5]. Thus, a lack of Ca^{2+} would result in membrane permeability changes.

Another extreme form of altered Ca^{2+} homeostasis is the intracellular Ca^{2+} overload which accompanies myocardial cell damage and contractile failure. This overload occurs in a variety of experimental models. Zimmerman and Hülsman [49] reported that successive heart perfusion with Ca^{2+}-free and Ca^{2+}-containing media results in massive structural damage (Fig. 1B), as well as in derangement of contractile and metabolic functions. Since perfusion with Ca^{2+}-free medium lowers myocardial Ca^{2+} content and reperfusion with Ca^{2+}-containing medium increases myocardial Ca^{2+} levels [1,2], this model of "calcium paradox" is of great interest to experimental cardiologists for studying the role of altered Ca^{2+} homeostasis in heart disease. In this regard, heart failure responsive to positive inotropic drugs can be associated with an intracellular Ca^{2+} deficiency [7], whereas intracellular Ca^{2+} overload accompanies prolonged hypoxic and chronic ischemic conditions [9,30,38]. Occurrence of Ca^{2+} overload has also been implicated in a wide variety of myocardial diseases, including catecholamine-induced myocardial necrosis [13]. Since myocardial changes in these Ca^{2+} overload situations can be reduced by lowering the Ca^{2+} concentrations in pre- or postperfusion media as well as by the use of drugs effecting Ca^{2+} influx [2,13,31,40], the concept that Ca^{2+} is involved in heart disease appears to be well established.

Two important questions underscore the pathophysiology of heart disease and the intracellular Ca^{2+} overload. First, what are the mechanisms of membrane change that lead to Ca^{2+} overload? Second, what processes are activated by the increase in intracellular Ca^{2+}? Answers to these questions are not available, but several suggestions have been advanced. Depending on the type of insult, the initial membrane injury could be due to a lack of Ca^{2+} support for the microarchitecture of the membrane, as in the case of the Ca^{2+} deficiency condition produced by the Ca^{2+}-free perfusion. Membrane injury can also be influenced by peroxidative changes in membrane lipids, as has been suggested for catecholamine-induced cardiomyopathy [41,42]. Once the semipermeable characteristics of the membrane are compromised, Ca^{2+} migrates down its concentration gradient from the extracellular space into the cytoplasm. High intracellular Ca^{2+} levels can have a multitude of effects, including depletion of high-energy phosphates and stimulation of proteases, phospholipases, and

Table III. ATP-Dependent Ca^{2+}-Binding with Heart Subcellular Membrane Fractions during the Development of Catecholamine-Induced Cardiomyopathy[a]

	ATP-dependent Ca^{2+} binding (nmoles Ca^{2+}/mg protein/min)		
Fraction	Control	3 hr after isoproterenol	24 hr after isoproterenol
Sarcoplasmic reticulum	18.4 ± 1.3	27.6 ± 2.1*	10.2 ± 0.9*
Mitochondria	9.2 ± 0.7	8.9 ± 0.8	13.5 ± 1.2*
Sarcolemma	3.6 ± 0.3	5.6 ± 0.4*	2.1 ± 0.1*

[a] Each value is a mean ± S.E. of four experiments. The concentration of free Ca^{2+} employed here was 10 μM. *, significantly ($p < 0.05$) different from control. Rats were injected with 40 mg/kg isoproterenol i.p. and hearts were removed 3 or 24 hr later. Control group of animals received a saline injection. Methods for the isolation of subcellular organelles as well as for the determintion of ATP-dependent Ca^{2+} binding and the assay conditions were described elsewhere [43, 45, 46].

lysosomal enzymes [9]. Lower energy states as well as stimulation of hydrolytic enzymes have been implicated in pathophysiology of the heart [8,13,16,20].

Gertz et al. [15] were first to recognize that Ca^{2+} transport properties of the SR vesicles were altered in a spontaneously failing heart. This observation led to examination of other membrane systems (e.g., mitochondria and sarcolemma) in different experimental models of heart failure. One or more of the subcellular membrane systems involved in the maintenance of intracellular Ca^{2+} levels in the myocardium may be modified in all experimental models of Ca^{2+} overload [7]. However, the nature and extent of the subcellular membrane involvement varies with the type and stage of the heart failure. This point is documented in Table III by data on ATP-dependent Ca^{2+}-binding activities of subcellular membranes isolated from hearts of rats injected with high doses of isoproterenol. ATP-dependent Ca^{2+}-binding activities in the SR and sarcolemma are considered to be involved in the removal of Ca^{2+} from the cytoplasm [6]. After 3 hr of drug treatment, these activities were significantly increased (indicating a "physiological adaptation" of these membranes to reduce intracellular Ca^{2+} overload known to occur in this model) [13]. However, at 24 hr both activities were depressed below control values, whereas mitochondrial Ca^{2+} uptake was stimulated. This depression probably represents a "pathological adaptation," since increased mitochondrial Ca^{2+} content impairs energy-producing capability and thus lowers the energy state. Increased Ca^{2+} in the cell in this model of cardiomyopathy is followed by severe pathological changes [41,42]. Catecholamine-induced alterations in the myocardium are reduced by Ca^{2+} antagonists [13], which are capable of blocking Ca^{2+} influx through slow Ca^{2+} channels [12]. The beneficial effects of Ca^{2+} antagonists in the treatment of arrhythmias and anginal pain give additional testimony to the fact that Ca^{2+} is intimately involved in pathophysiological aspects of heart function.

From the foregoing discussion, it is clear that during the past 100 years appreciable progress has been made in furthering our understanding of the role of Ca^{2+} in heart function in health and disease. The increase in our knowledge related to Ca^{2+} homeostasis in the myocardium has aided in the development of improved therapy for heart disease. It is hoped that remaining unsolved Ca^{2+}-related problems in the field of cardiac biology and pathology will be resolved in the near future.

ACKNOWLEDGMENTS. The research reported in this chapter was supported by a grant from the Medical Research Council. P.K.S. is a Research Scholar of the Canadian Heart Foundation.

References

1. Alto, L. E.; Dhalla, N. S. Myocardial cation contents during induction of calcium paradox. *Am. J. Physiol.* **237**:H713–H719, 1979.
2. Alto, L. E.; Singal, P. K.; Dhalla, N. S. Calcium paradox: Dependence of reperfusion-induced changes on the extracellular calcium concentration. *Adv. Myocardiol.* **2**:177–185, 1980.
3. Blinks, J. R.; Wier, W. G., Hess, P.; Prendergast, F. G. Measurement of Ca^{2+} concentrations in living cells. *Prog. Biophys. Mol. Biol.* **40**:1–114, 1982.
4. Carsten, M. E. The cardiac calcium pump. *Proc. Natl. Acad. Sci. USA* **52**:1456–1462, 1964.
5. Cook, W. J.; Bugg, C. E. Calcium–carbohydrate bridges composed of uncharged sugars: Structure of a hydrated calcium bromide complex of alpha-fucose. *Biochim. Biophys. Acta* **389**:428–435, 1975.
6. Dhalla, N. S.; Ziegelhoffer, A.; Harrow, J. A. C. Regulatory role of membrane systems in heart function. *Can. J. Physiol. Pharmacol.* **55**:1211–1234, 1977.
7. Dhalla, N. S.; Das, P. K.; Sharma, G. P. Subcellular basis of cardiac contractile failure. *J. Mol. Cell. Cardiol.* **10**:363–385, 1978.
8. Dhalla, N. S.; Singal, P. K.; Dhillon, K. S. Mitochondrial functions and drug-induced heart disease. In: *Drug-Induced Heart Disease*, M. R. Bristow, ed., Amsterdam, Elsevier/North-Holland, 1980, pp. 39–61.
9. Dhalla, N. S.; Pierce, G. N.; Panagia, V.; Singal, P. K.; Beamish, R. E. Calcium movements in relation to heart function. *Basic Res. Cardiol.* **77**:117–139, 1982.
10. Doggenweiler, C. F.; Frenk, S. Staining properties of lanthanum on cell membranes. *Proc. Natl. Acad. Sci. USA* **53**:425–430, 1965.
11. Ebashi, S.; Kodama, A. A new protein factor promoting aggregation of tropomyosin. *J. Biochem. (Tokyo)* **58**:107–108, 1965.
12. Fleckenstein, A. Specific inhibitors and promoters of Ca^{2+} action in the excitation–contraction coupling of heart muscle and their role in the prevention or production of myocardial lesions. In: *Calcium and the Heart*, P. Harris and L. Opie, eds., New York, Academic Press, 1976, pp. 135–188.
13. Fleckenstein, A.; Janke, J.; Doring, H. J.; Pachinger, O. Ca overload as the determinant factor in the production of catecholamine-induced myocardial lesions. *Rec. Adv. Stud. Card. Struct. Metab.* **2**:455–466, 1973.
14. Frank, J. S.; Rich, T. L.; Baydler, S.; Kreman, M. Calcium depletion in rabbit myocardium: Ultrastructure of the sarcolemma and correlation with the calcium paradox. *Circ. Res.* **51**:117–130, 1982.
15. Gertz, E. W.; Hess, M. L.; Lain, R.; Briggs, F. N. Activity of vesicular calcium pump in the spontaneously failing heart–lung preparation. *Circ. Res.* **20**:477–484, 1967.
16. Hearse, D. J. Oxygen deprivations and early myocardial contractile failure: A reassessment of the possible role of adenosine triphosphate. *Am. J. Cardiol.* **44**:1115–1121, 1979.
17. Heilbrunn, L. V.; Wiercinski, F. J. The action of various cations on muscle protoplasm. *J. Cell. Comp. Physiol.* **29**:15–32, 1947.
18. Hunter, D. R.; Haworth, R. A.; Southard, J. H. Relationship between configuration, function and permeability in calcium-treated mitochondria. *J. Biol. Chem.* **251**:5069–5077, 1976.
19. Inesi, G.; Ebashi, S.; Watanabe, S. Preparation of vesicular relaxing factor from bovine heart muscle. *Am. J. Physiol.* **207**:1339–1344, 1964.
20. Katz, A. M.; Reuter, H. Cellular calcium and cardiac cell death. *Am. J. Cardiol.* **44**:188–190, 1979.
21. Katz, A. M., Repke, D. I.; Cohen, B. R. Control of the activity of highly purified cardiac actomyosin by Ca^{2+}, Na^+ and K^+. *Circ. Res.* **19**:1062–1070, 1966.
22. Katz, A. M.; Repke, D. I.; Fudyma, G.; Shigekawa, M. Control of calcium efflux from sarcoplasmic reticulum vesicles by external calcium. *J. Biol. Chem.* **252**:4210–4214, 1977.
23. Langer, G. A. The structure and function of the myocardial cell surface. *Am. J. Physiol.* **235**:H461–H468, 1978.

24. Langer, G. A.; Frank, J. S. Lanthanum in heart cell culture: Effects on calcium exchange correlated with its localization. *J. Cell Biol.* **54**:441–445, 1972.
25. Locke, T. S.; Rosenheim, O. Contributions to the physiology of the isolated heart: The consumption of dextrose by mammalian cardiac muscle. *J. Physiol. (London)* **36**:205–220, 1907.
26. Marban, E.; Rink, T. J.; Tsien, R. W.; Tsien, R. Y. Free calcium in heart muscle at rest and during contraction measured with Ca^{2+}-sensitive microelectrodes. *Nature (London)* **286**:845–850, 1980.
27. Mines, G. R. On functional analysis of the action of electrolytes. *J. Physiol. (London)* **46**:188–235, 1913.
28. Moffat, M. P.; Singal, P. K.; Dhalla, N. S. Differences in sarcolemmal preparations: Cell surface material and membrane sidedness. *Basic Res. Cardiol.* **78**:451–461, 1983.
29. Muir, A. R. The effects of divalent cations on the ultrastructure of the perfused rat heart. *Am. J. Anat.* **101**:239–261, 1967.
30. Nayler, W. G.; Poole-Wilson, P. A.; Williams, A. Hypoxia and calcium. *J. Mol. Cell. Cardiol.* **11**:683–706, 1979.
31. Nayler, W. G.; Ferrari, R.; Williams, A. Protective effect of pretreatment with verapamil, nifedipine and propranolol on mitochondrial function in the ischemic and reperfused myocardium. *Am. J. Cardiol.* **46**:242–248, 1980.
32. Niedergerke, R. Local muscular shortening by intracellularly applied calcium. *J. Physiol. (London)* **128**:12–13, 1955.
33. Rand, R. P.; Sengupta, S. Cardiolipin forms hexagonal structures with divalent cations. *Biochim. Biophys. Acta* **255**:484–492, 1972.
34. Ringer, S. Concerning the influence exerted by each of the constituents of the blood on the contraction of the ventricle. *J. Physiol. (London)* **3**:380–393, 1882.
35. Ringer, S. A further contribution regarding the influence of the different constituents of the blood on the contraction of the heart. *J. Physiol. (London)* **4**:29–42, 1883.
36. Sanborn, W. D.; Langer, G. A. Specific uncoupling of excitation and contraction in mammalian cardiac tissue by lanthanum. *J. Gen. Physiol.* **56**:191–217, 1970.
37. Sandow, A. Excitation–contraction coupling in muscular response. *Yale J. Biol. Med.* **25**:176–201, 1952.
38. Shen, A. C.; Jennings, R. B. Myocardial calcium and magnesium in acute ischemic injury. *Am. J. Pathol.* **67**:417–440, 1972.
39. Singal, P. K.; Prasad, K. Extracellular calcium and positive inotropy of ionophore (X537-A) in cardiac muscle. *Jpn. J. Physiol.* **26**:529–535, 1976.
40. Singal, P. K.; Matsukubo, M. P.; Dhalla, N. S. Calcium-related changes in the ultrastructure of mammalian myocardium. *Br. J. Exp. Pathol.* **60**:96–106, 1979.
41. Singal, P. K.; Dhillon, K. S.; Beamish, R. E.; Dhalla, N. S. Protective effect of zinc against catecholamine-induced myocardial changes: Electrocardiographic and ultrastructure studies. *Lab. Invest.* **44**:426–433, 1981.
42. Singal, P. K.; Kapur, N.; Dhillon, K. S.; Beamish, R. E.; Dhalla, N. S. Role of free radicals in catecholamine-induced cardiomyopathy. *Can. J. Physiol. Pharmacol.* **60**:1390–1397, 1982.
43. Sulakhe, P. V.; Dhalla, N. S. Excitation–contraction coupling in heart. VII. Calcium accumulation in subcellular particles in congestive heart failure. *J. Clin. Invest.* **50**:1019–1027, 1971.
44. Tada, M.; Katz, A. M. Phosphorylation of the sarcoplasmic reticulum and sarcolemma. *Annu. Rev. Physiol.* **44**:401–423, 1982.
45. Tada, M.; Kirchberger, M. A.; Katz, A. M. Phosphorylation of a 22000 dalton component of the cardiac sarcoplasmic reticulum by adenosine 3′,5′-monophosphate-dependent protein kinase. *J. Biol. Chem.* **250**:2640–2657, 1975.
46. Takeo, S.; Duke, P.; Taam, G. M. L.; Singal, P. K.; Dhalla, N. S. Effects of lanthanum on the heart sarcolemmal ATPase and calcium binding activities. *Can. J. Physiol. Pharmacol.* **57**:496–503, 1979.
47. Tuana, B. S.; Dzurba, A.; Panagia, V.; Dhalla, N. S. Stimulation of heart sarcolemmal calcium pump by calmodulin. *Biochem. Biophys. Res. Commun.* **100**:1245–1250, 1981.
48. Weiss, D. L.; Surawicz, B.; Rubenstein, I. Myocardial lesions of calcium deficiency causing irreversible myocardial failure. *Am. J. Pathol.* **48**:653–666, 1966.
49. Zimmerman, A. N. E.; Hülsman, W. C. Paradoxical influence of calcium ions on the permeability of the cell membranes of the isolated rat heart. *Nature (London)* **211**:646–647, 1966.

41

Calcium Both Activates and Inactivates Calcium Release from Cardiac Sarcoplasmic Reticulum

Alexandre Fabiato

Introduction

Ringer reported that ventricular contraction ceases within 20 min in the presence of a Ca^{2+}-free saline solution and is restored by the addition of a Ca^{2+} salt [30]. Although this most frequently quoted article does not indicate the animal species from which the ventricle was taken, a previous paper indicated that the frog ventricle was used [29]. These articles by Ringer have been widely quoted to demonstrate the importance of transsarcolemmal Ca^{2+} influx in controlling cardiac contraction, not only in the frog ventricular tissue but in mammalian cardiac tissue as well. Repeating Ringer's experiments on small trabeculae from the frog ventricle, but with observation under the high-power microscope, confirmed Ringer's findings. Not only was contraction abolished within 2–3 min by Ca^{2+} removal, but microscopic examination found the preparation to be completely quiescent. In contrast, when the experiments were done in a mammalian (e.g., adult rat) ventricular tissue, Ca^{2+} removal resulted in an almost complete abolition of the contraction detected from the ends of the trabecula with a highly sensitive transducer but observation under the microscope showed localized cyclic contractions. This very prominent but desynchronized contractile behavior persisted for many hours in a Ca^{2+}-free solution (A. Fabiato, unpublished observations). This observation is one of many pieces of evidence suggesting that, in addition to the transsarcolemmal Ca^{2+} influx, an intracellular Ca^{2+} store participates in the excitation–contraction coupling of the adult mammalian cardiac tissue. Other experiments have demonstrated that the Ca^{2+} store is located in the sarcoplasmic reticulum (SR) [7,15].

The outstanding problem of cardiac excitation–contraction coupling is that of the link between these two sources of activator Ca^{2+}. One hypothesis, initially proposed for skeletal muscle [4,21], is that of a Ca^{2+}-induced release of Ca^{2+} from the SR. The source of the Ca^{2+} that triggers Ca^{2+} release is uncertain in skeletal muscle [15], but for mammalian cardiac muscle the hypothesis can be phrased in the following way: the transsarcolemmal

---

Alexandre Fabiato • Department of Physiology and Biophysics, Medical College of Virginia, Richmond, Virginia 23298.

influx of Ca^{2+} does not activate the myofilaments directly but through the induction of a Ca^{2+} release from the SR.

One of the major arguments against the hypothesis of Ca^{2+}-induced release of Ca^{2+} from the SR is that this phenomenon should *a priori* be all or none inasmuch as it would include a positive feedback: the Ca^{2+} released from the SR should cause more Ca^{2+} release. An all-or-none Ca^{2+}-induced release of Ca^{2+} would be incompatible with the observation of a gradation of cardiac contraction with the extracellular $[Ca^{2+}]$. This criticism was strongly emphasized [10] soon after the preliminary experiments suggesting the possibility of a Ca^{2+}-induced release of Ca^{2+} from the SR of skinned (sarcolemma mechanically removed) fibers from skeletal muscle [11,18–20]. This argument has also been deemed sufficient for rejecting this hypothesis for cardiac muscle [31] even though it had been previously demonstrated that Ca^{2+}-induced release of Ca^{2+} was graded with the level of [free Ca^{2+}] trigger in skinned cardiac cells [16].

In principle, as pointed out in the first demonstration of the gradation of the Ca^{2+}-induced release of Ca^{2+} with the level of [free Ca^{2+}] trigger [16], the argument that Ca^{2+}-induced release of Ca^{2+} from the SR must be all or none is far from compelling. In biology as well as in electronics, an amplifier with positive feedback does not necessarily produce an amplification up to saturation if the system is complex.

Subsequently, the gradation of the Ca^{2+}-induced release of Ca^{2+} from the SR has been further documented. It has been shown that Ca^{2+}-induced release of Ca^{2+} is not only graded with the peak level of [free Ca^{2+}] used as a trigger, but also with the level of preload of the SR with Ca^{2+} and the rate of change of [free Ca^{2+}] at the outer surface of the SR [17]. These observations have been made not only with tension recording but also directly by monitoring with chlorotetracycline the changes in amount of Ca^{2+} bound inside the SR [14,17], or by monitoring with aequorin the changes of myoplasmic [free Ca^{2+}] resulting from Ca^{2+} release from the SR [12,13].

None of the previous experiments directly demonstrated any mechanism for the gradation of the Ca^{2+}-induced release of Ca^{2+}. The experiment reported in the present communication directly demonstrates that a rapid increase of [free Ca^{2+}] at the outer surface of the SR to a relatively low peak level can induce Ca^{2+} release from the SR whereas an increase of [free Ca^{2+}] to a higher level inactivates Ca^{2+} release. Hence, the increase of myoplasmic [free Ca^{2+}] resulting from Ca^{2+}-induced release of Ca^{2+} from the SR should inactivate further Ca^{2+} release from the SR. Therefore, the process of Ca^{2+}-induced release of Ca^{2+} contains a negative feedback loop which helps to explain why this is a graded process. This unforeseen property [15] is just the opposite of the positive feedback that had been initially assumed by others [10,31].

Methods

The methods used for the experiment took advantage of the fact that the skinned cardiac cell has a much smaller diameter than the skinned skeletal muscle fiber, which permits the rapid diffusion of the externally applied solutions within the preparation. In addition, the changes of [free Ca^{2+}] in the intact cardiac cell are much less rapid than in the intact skeletal muscle fiber. Hence, it is possible to change the [free Ca^{2+}] at the outer surface of the SR during the course of an aequorin-detected Ca^{2+} transient caused by Ca^{2+} release [12,15]. The preparation was obtained from the adult rat ventricle by homogenization and microdissection as previously described [12,16]. The experiment has been repeated in 15 cells of width 8 ± 0.5 μm, thickness 6 ± 0.5 μm, and length 30–60 μm. These preparations were taken from the adult rat ventricle because Ca^{2+}-induced release of Ca^{2+} is obtained with a particularly low [free Ca^{2+}] in this tissue [13].

A low [free Ca^{2+}] was required because the detection of Ca^{2+} release was done with aequorin. The protocol required that aequorin be included in all the solutions to be used for the experiment. Each solution was contained in a micropipette at one of three different $-\log_{10}$ [free Ca^{2+}] (i.e., pCa) values: 7.40, 6.25, and 5.50. In the presence of a pMg 2.50, the time constant of the use of aequorin is about 2 hr at pCa 5.50. This time constant is sufficient for permitting the completion of the experiment without too much drift of the baseline. As will be shown, pCa 5.50 is also sufficient to inactivate Ca^{2+}-induced release of Ca^{2+} in a skinned cardiac cell from the adult rat ventricle. If skinned cardiac cells from other animal species had been used [13], too high a [free Ca^{2+}] may have been required to inactivate Ca^{2+}-induced release of Ca^{2+}, which would have rendered the monitoring of the aequorin bioluminescence more difficult, if not impossible, during this inactivation.

To permit very rapid changes of solutions, an artificial bathing chamber of 0.001- to 0.002-µl volume was created around the skinned cardiac cell. This was done by transferring the skinned cell in ion-free mineral oil and injecting the aqueous solution containing aequorin, ions, and substrates around it [12].

The change of solutions was done with a microprocessor-controlled system of microinjection–aspiration. This system, which is to be described in detail elsewhere, uses a very small volume (< 10 µl) of transmission fluid that is almost perfectly incompressible: degassed water. The degassed water was separated from the aequorin-containing solution by 0.1 µl of ion-free mineral oil.

Solution changes were done by aspiration of the previous solution into the micropipette in which it was originally contained, followed by injection of the new solution. There was no mixing between the two solutions. The tips of the micropipettes containing the solutions with aequorin in the presence of pCa 7.40, 6.25, and 5.50 were permanently in the field of view of the photomultiplier tube so that the large resting glow was constant [12]. Accordingly, the solution changes did not produce any shift of resting light other than that caused by the Ca^{2+} release within the skinned cardiac cell. The resting light produced a large voltage at the anode of the photomultiplier tube that was compensated by a counterbias voltage [12]. This permitted the definition of the zero of the aequorin light trace (Fig. 1). The cutoff frequency of the double-pole active filter was 30 Hz for the bioluminescence signal. Tension was recorded simultaneously with a filtration at a cutoff frequency of 5 Hz.

The recording also included a display of the signals of injections and aspirations of the three microprocessor-controlled channels. The amplitude of each signal was made different only to facilitate the identification of each of the microprocessor-controlled microsyringes. In contrast, the timing and duration of a signal were exact measures of the timing and duration of aspiration or injection. Injection and aspiration signals were first displayed simultaneously with tension and light transients. After the end of the experiment the signals were redisplayed at a higher speed to facilitate their analysis (bottom trace in Fig. 1).

At the end of the experiment the last aequorin-containing solution was reaspirated so that only the aequorin contained in the skinned cell remained present. This aequorin was entirely inactivated by the application of a solution at pCa 2.50 in the presence of 0.050 mM ethyleneglycol-bis(β-aminoethylether)-N,N'-tetraacetic acid (EGTA) and the same pMg of 2.50 as in the experimental solutions. The same [total EGTA] was used for this calibration as for the experiment because of the suggestion of an effect of EGTA, probably [free EGTA], on the calcium–aequorin reaction [28]. The inactivation of all the aequorin contained in the skinned cell produced a large bioluminescence signal. The area under the curve of this signal was treated with a microcomputer so that the amplitude of the maximum light was directly displayed according to a previously described computer program [12]. The maximum light corresponded to that which would be produced by the instantaneous inactivation of all the aequorin contained in the skinned cell by a saturating [free Ca^{2+}] in a rapid mixing chamber [12]. The ratio of peak light recorded during the Ca^{2+} transient to maximum light, and a cuvette calibration of aequorin bioluminescence under conditions very close to those used for the experiment, permitted the inference of the myoplasmic [free Ca^{2+}] reached during the Ca^{2+} transient [1].

Results

All of the 15 experiments were done under identical conditions. In all cases the solutions contained 0.050 mM EGTA, a pMg of 2.50, pMgATP 2.50, 12 mM phosphocreatine, 15 U/ml creatine phosphokinase, 0.160 M ionic strength, with K^+ and Cl^- as the major ionic species, and 30 mM N,N'-bis(2-hydroxyethyl)-2-aminoethanesulfonic acid (BES) at pH 7.10 and 22°C.

Initially the skinned cardiac cell was bathed in an aequorin-containing solution at pCa 7.40 (Fig. 1). At this pCa the preparation was quiescent. No cyclic contractions caused by Ca^{2+} overload of the SR were observed. The control recording consisted of the aspiration of the solution at pCa 7.40 and the injection of the solution at pCa 6.25. This pCa 6.25 corresponds to the optimum trigger for Ca^{2+}-induced release of Ca^{2+} [12]. A Ca^{2+} release was then induced, resulting in a transient of aequorin light followed by a tension transient. This step ended on the reaspiration of the solution at pCa 6.25 and the reinjection of the solution at pCa 7.40. Because of the return to the initial solution at low [free Ca^{2+}] after a delay of less than 1 sec, repeated, cyclic releases of Ca^{2+} were not observed (Fig. 1). The

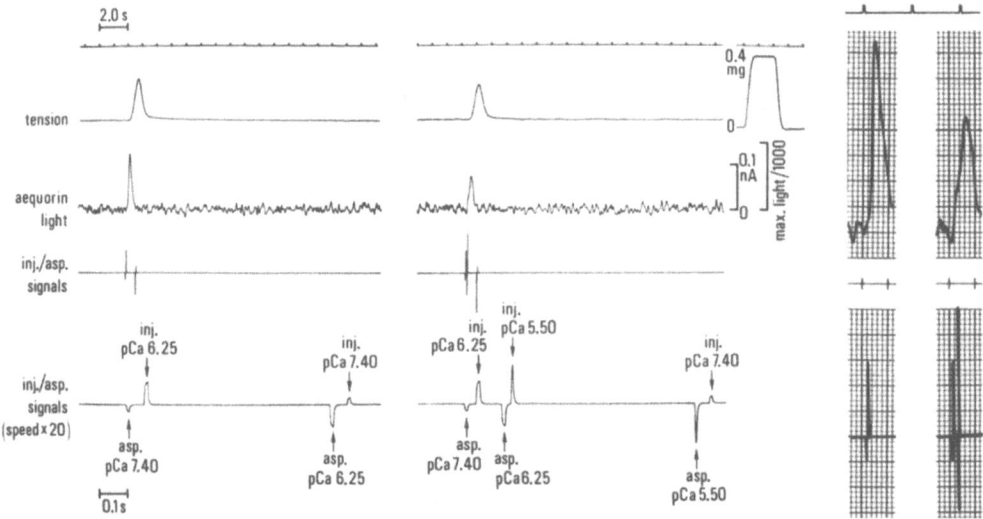

Figure 1. Induction of a Ca^{2+} release from the SR with an optimum increase of [free Ca^{2+}], premature inactivation of the Ca^{2+} release by a supraoptimum additional increase of [free Ca^{2+}], and prevention of the cyclic repetition of Ca^{2+}-induced release of Ca^{2+} by return to the initial low [free Ca^{2+}]. The experiment was done in an 8-μm-wide, 6-μm-thick, 54-μm-long skinned cardiac cell from the adult rat ventricle. Note that the abbreviation used for "second" is "s" in this figure legend, whereas "sec" is used in the text for consistency with other chapters in this book. [Reproduced with a slight modification from Ref. 15, with permission of the American Physiological Society.]

aequorin calibration at the end of the experiment showed that the peak [free Ca^{2+}] reached during the Ca^{2+} transient was pCa 5.48 ± 0.09 (S.D.). This control step was repeated three times at regular intervals of 25 sec.

A new Ca^{2+} release was induced after the same interval and with the same increase of the [free Ca^{2+}] to pCa 6.25. However, after 70–120 msec, the solution at pCa 6.25 was reaspirated and the solution at pCa 5.50 was injected. The experimental step ended, as for the control step, on a reaspiration of the solution at pCa 5.50 and a reinjection of the solution at pCa 7.40. This step was generally repeated three times, and the effects were qualitatively consistent in the 15 experiments. The increase of [free Ca^{2+}] to pCa 5.50 resulted in a decrease of the amplitude of the aequorin transient and in an accompanying decrease of the amplitude of the tension transient. The changes of amplitude in the light transient were always much more pronounced than those in the tension transient because of the stoichiometry of the calcium–aequorin binding [1]. A detail of the effect of the increase of myoplasmic [free Ca^{2+}] on the time course of the light transient is shown in the photographic enlargement at the right of Fig. 1. The injection of the solution at pCa 5.50 caused, after a delay of less than 25 msec, a marked decrease of the rate of the ascending phase of the Ca^{2+} transient. The peak of the aequorin transient was decreased and its duration curtailed.

A pCa of 5.50 was not significantly different from the average pCa reached at the peak of the Ca^{2+} transient. Thus, the injection of the pCa 5.50 solution had prematurely exposed the outer surface of the SR to a [free Ca^{2+}] level that would have been reached only at the peak of the Ca^{2+} transient if the release induced by pCa 6.25 had been allowed to follow its spontaneous course. Hence, an early inactivation of the Ca^{2+}-induced release of Ca^{2+} was produced by this experimental intervention.

From this experiment it is concluded that the inactivation of the Ca^{2+}-induced release of Ca^{2+} may be caused, at least in part, by the increase of myoplasmic [free Ca^{2+}] at the outer surface of the SR resulting from Ca^{2+}-induced release of Ca^{2+}.

Discussion

This experiment demonstrates for the first time a negative feedback loop in the process of Ca^{2+}-induced release of Ca^{2+} from the cardiac SR. This helps to explain why this process is not all or none and, thus, removes a major obstacle to the hypothesis of a physiological role for Ca^{2+}-induced release of Ca^{2+} in cardiac excitation–contraction coupling.

Other experiments demonstrate that the inactivation of the Ca^{2+}-induced release of Ca^{2+} is also time-dependent (A. Fabiato, series of three manuscripts in preparation). It has been previously demonstrated that the activation of the Ca^{2+}-induced release of Ca^{2+} is not only dependent on the level of [free Ca^{2+}] used as a trigger but also is inversely related to the time taken for the increase of [free Ca^{2+}] [13–15,17]. Thus, both activation and inactivation of the putative Ca^{2+} channel controlling Ca^{2+} release from the SR appear to be time- and Ca^{2+}-dependent.

The time- and Ca^{2+}-dependent activation and inactivation of the putative Ca^{2+} channel across the SR membrane may be amenable to an analytical description of the type done by Hodgkin and Huxley [23], by replacing the voltage parameters by Ca^{2+} concentration parameters in the equations. Such a model may have important implications for understanding the control of other types of channels. In particular, there is evidence that the inactivation of the Ca^{2+} channel across the cardiac sarcolemma is dependent on voltage, intracellular [free Ca^{2+}], and time [3,5,6,24–26,32].

The nature of the gating molecule(s) remains unknown. Although there is evidence against any direct involvement of the Ca^{2+} accumulation mechanism in the release process [15], it cannot be excluded that the high-affinity site of the Ca^{2+} pump in the SR may be the site gating the activation of the Ca^{2+}-induced release of Ca^{2+}. In fact, this is the only known site having an affinity for Ca^{2+} sufficiently high to be in the same range as the [free Ca^{2+}] required to induce Ca^{2+} release from the SR [8,9,27]. Low-affinity binding sites at the outer surface of the SR have been described [22]. These might participate in the inactivation process.

The inactivation of Ca^{2+}-induced release of Ca^{2+} by Ca^{2+} explains why too high a transsarcolemmal Ca^{2+} influx may cause a smaller Ca^{2+}-induced release of Ca^{2+} and consequently have a negative inotropic effect [2].

Summary

A microprocessor-controlled microinjection–aspiration system has been developed for experiments in skinned (sarcolemma removed by microdissection) single cardiac cells. This system permits rapid changes of the myoplasmic [free Ca^{2+}] at the outer surface of the SR during the course of a Ca^{2+} transient. The amount of Ca^{2+} released from the SR by Ca^{2+}-induced release of Ca^{2+}, which was detected with aequorin, increased when the peak [free Ca^{2+}] trigger was increased and the time for [free Ca^{2+}] change was decreased. An optimum Ca^{2+}-induced release of Ca^{2+} was elicited by an increase of [free Ca^{2+}] from $-\log_{10}$ [free Ca^{2+}] (i.e., pCa) 7.40 to pCa 6.25 within a few milliseconds in a skinned cardiac cell from the rat ventricle. The peak myoplasmic [free Ca^{2+}] reached during the Ca^{2+} transient was about pCa 5.50. A prematurely imposed increase of myoplasmic [free Ca^{2+}] in the solution to pCa 5.50 during the early part of the ascending phase of the Ca^{2+} transient curtailed the Ca^{2+} transient and decreased its maximum amplitude. Therefore, the putative Ca^{2+} channel used by Ca^{2+}-induced release of Ca^{2+} is inactivated by an increase of [free Ca^{2+}] to about the level reached at the peak of the Ca^{2+} transient. This negative feedback helps to explain why Ca^{2+}-induced release of Ca^{2+} is not an all-or-none process but instead is a graded process.

ACKNOWLEDGMENTS. This study was supported by Grant R01 HL-19138 from the National Heart, Lung and Blood Institute. I wish to thank Dr. F. N. Briggs for organizing this symposium and for his continued support and encouragement, Dr. E. B. Ridgway and Professor O. Shimomura for the gift of aequorin, Dr. E. B. Ridgway for many helpful discussions and review of the manuscript, and Dr. J. J. Feher who also kindly reviewed the manuscript.

References

1. Allen, D. G.; Blinks, J. R. The interpretation of light signals from aequorin-injected skeletal and cardiac muscle cells: A new method of calibration. In: *Detection and Measurement of Free Ca^{2+} in Cells*, C. C. Ashley and A. K. Campbell, eds., Amsterdam, Elsevier/North-Holland, 1979, pp. 159–174.
2. Aronson, R. S.; Capasso, J. M. Negative inotropic effect of elevated extracellular calcium in rat myocardium. *J. Mol. Cell. Cardiol.* **12**:1305–1309, 1980.
3. Ashcroft, F. M.; Stanfield, P. R. Calcium dependence of the inactivation of calcium currents in skeletal muscle fibers of an insect. *Science* **213**:224–226, 1981.
4. Bianchi, C. P.; Shanes, A. M. Calcium influx in skeletal muscle at rest, during activity, and during potassium contracture. *J. Gen. Physiol.* **42**:803–815, 1959.

5. Brehm, P.; Eckert, R. Calcium entry leads to inactivation of calcium channel in *Paramecium*. *Science* **202**:1203–1206, 1978.
6. Brown, A. M.; Morimoto, K.; Tsuda, Y.; Wilson, D. L. Calcium current-dependent and voltage-dependent inactivation of calcium channels in *Helix aspersa*. *J. Physiol. (London)* **320**:193–218, 1981.
7. Chapman, R. A. Excitation–contraction coupling in cardiac muscle. *Prog. Biophys. Mol. Biol.* **35**:1–52, 1979.
8. Chiesi, M.; Ho, M. M.; Inesi, G.; Somlyo, A. V.; Somlyo, A. P. Primary role of sarcoplasmic reticulum in phasic contractile activation of cardiac myocytes with shunted myolemma. *J. Cell Biol.* **91**:728–742, 1981.
9. Chiesi, M.; Inesi, G. The use of quench reagents for resolution of single transport cycles in sarcoplasmic reticulum. *J. Biol. Chem.* **254**:10370–10377, 1979.
10. Costantin, L. L.; Taylor, S. R. Graded activation in frog muscle fibers. *J. Gen. Physiol.* **61**:424–443, 1973.
11. Endo, M.; Tanaka, M.; Ogawa, Y. Calcium induced release of calcium from the sarcoplasmic reticulum of skinned skeletal muscle fibres. *Nature (London)* **228**:34–36, 1970.
12. Fabiato, A. Myoplasmic free calcium concentration reached during the twitch of an intact isolated cardiac cell and during calcium-induced release of calcium from the sarcoplasmic reticulum of a skinned cardiac cell from the adult rat or rabbit ventricle. *J. Gen. Physiol.* **78**: 457–497, 1981.
13. Fabiato, A. Calcium release in skinned cardiac cells: Variations with species, tissues, and development. *Fed. Proc.* **41**:2238–2244, 1982.
14. Fabiato, A. Fluorescence and differential light absorption recordings with calcium probes and potential-sensitive dyes in skinned cardiac cells. *Can. J. Physiol. Pharmacol.* **60**:556–567, 1982.
15. Fabiato, A. Calcium-induced release of calcium from the cardiac sarcoplasmic reticulum. *Am. J. Physiol.* **245**:C1–C14, 1983.
16. Fabiato, A.; Fabiato, F. Contractions induced by a calcium-triggered release of calcium from the sarcoplasmic reticulum of single skinned cardiac cells. *J. Physiol. (London)* **249**:469–495, 1975.
17. Fabiato, A.; Fabiato, F. Use of chlorotetracycline fluorescence to demonstrate Ca^{2+}-induced release of Ca^{2+} from the sarcoplasmic reticulum of skinned cardiac cells. *Nature (London)* **281**:146–148, 1979.
18. Ford, L. E.; Podolsky, R. J. Regenerative calcium release within muscle cells. *Science* **167**:58–59, 1970.
19. Ford, L. E.; Podolsky, R. J. Calcium uptake and force development by skinned muscle fibres in EGTA buffered solutions. *J. Physiol. (London)* **223**: 1–19, 1972.
20. Ford, L. E.; Podolsky, R. J. Intracellular calcium movements in skinned muscle fibres. *J. Physiol. (London)* **223**:21–33, 1972.
21. Frank, G. B. The current view of the source of trigger calcium in excitation–contraction coupling in vertebrate skeletal muscle. *Biochem. Pharmacol.* **29**:2399–2406, 1980.
22. Hasselbach, W.; Koenig, V. Low affinity calcium binding sites of the calcium transport ATPase of sarcoplasmic reticulum membranes. *Z. Naturforsch. Teil C* **35**:1012–1018, 1980.
23. Hodgkin, A. L.; Huxley, A. F. A quantitative description of membrane current and its application to conduction and excitation in nerve. *J. Physiol. (London)* **117**:500–544, 1952.
24. Isenberg, G.; Klöckner, U. Calcium currents of isolated bovine ventricular myocytes are fast and of large amplitude. *Pfluegers Arch.* **395**:30–41, 1982.
25. Kass, R. S.; Scheuer, T. Slow inactivation of calcium channels in the cardiac Purkinje fiber. *J. Mol. Cell. Cardiol.* **14**:615–618, 1982.
26. Marban, E.; Tsien, R. W. Is the slow inward calcium current of heart muscle inactivated by calcium? *Biophys. J.* **33**:143a, 1981.
27. Martonosi, A. N. Sarcoplasmic reticulum of skeletal muscle: The mechanism of calcium release. In: *Myology*, B. Q. Banker and A. W. Engel, eds., New York, McGraw–Hill, in press.
28. Ridgway, E. B.; Snow, A. E. Effects of EGTA on aequorin luminescence. *Biophys. J.* **41**:244a, 1983.
29. Ringer, S. Concerning the influence of season and of temperature on the action and on the antagonisms of drugs. *J. Physiol. (London)* **3**:115–124, 1882.
30. Ringer, S. A further contribution regarding the influence of the different constituents of the blood on the contraction of the heart. *J. Physiol. (London)* **4**:29–42, 1883.
31. Sommer, J. R.; Johnson, E. A. Ultrastructure of cardiac muscle. In: *Handbook of Physiology*, Section 2, *The Cardiovascular System*, Volume I, *The Heart*, R. M. Berne, N. Sperelakis, and S. R. Geiger, eds., Bethesda, American Physiological Society, 1979, pp. 113–186.
32. Tillotson, D. Inactivation of Ca conductance dependent on entry of Ca ions in molluscan neurons. *Proc. Natl. Acad. Sci. USA* **76**:1497–1500, 1979.

42

Routes of Calcium Flux in Cardiac Sarcoplasmic Reticulum

Joseph J. Feher

Net release of Ca^{2+} from the sarcoplasmic reticulum (SR) plays an important role in coupling excitation to contraction in cardiac muscle cells, and the uptake of Ca^{2+} by SR plays a major role in effecting relaxation of the contractile apparatus [9,20,26]. There are several possible routes of Ca^{2+} influx and efflux in SR that could account for the net uptake and release of Ca^{2+}. These putative pathways include passive diffusion, pump-mediated Ca^{2+} influx and efflux, carrier-mediated facilitated diffusion, and Ca^{2+} efflux through a gated channel. One goal of research in this field is to identify the routes of Ca^{2+} flux in SR and to determine their magnitude, time course, and roles in physiological regulation of Ca^{2+} flux. In this presentation, the routes of Ca^{2+} flux in SR vesicles isolated from dog hearts are examined. The results suggest that Ca^{2+} fluxes in cardiac SR occur through only three routes: (1) forward pump-mediated Ca^{2+} influx; (2) reverse pump-mediated Ca^{2+} efflux; and (3) passive efflux.

The Pump-Leak Model

Cardiac SR vesicles incubated in the presence of Ca^{2+}, ATP, and Mg^{2+} can attain a steady-state Ca^{2+} accumulation that persists for some time [12]. The actual steady-state Ca^{2+} uptake depends on the reaction conditions and varies with temperature, pH, and Mg^{2+}, nucleotide, and extravesicular Ca^{2+} concentrations [4,12,23]. After steady-state uptake is reached, ATPase hydrolysis by SR vesicles continues at a fairly rapid rate [12]. These initial observations can be explained by at least three different SR models. In the simple fixed-stoichiometry pump-leak model, the Ca^{2+}-ATPase couples Ca^{2+} influx to hydrolysis of ATP with a fixed stoichiometry (Fig. 1A). At steady state, Ca^{2+} influx mediated by the pump would exactly balance efflux through a passive pathway. A consequence of this model is that steady-state Ca^{2+} influx would equal passive Ca^{2+} efflux (Jf =

Joseph J. Feher • Department of Physiology and Biophysics, Medical College of Virginia, Richmond, Virginia 23298.

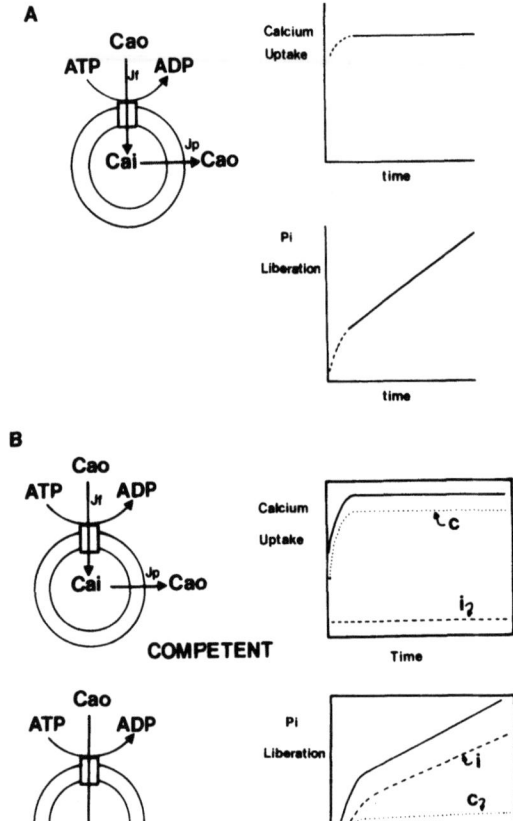

Figure 1. Pump-leak models of cardiac SR with their predictions for Ca^{2+} uptake and ATPase activity. (A) Fixed-stoichiometry pump-leak model predicts that the net ATPase activity (JNET) at steady state is stoichiometrically coupled to a Ca^{2+} influx (Jf) that balances a passive efflux (Jp). (B) Competent/incompetent pump-leak model allows for Ca^{2+} uptake contributed mainly by competent vesicles while ATPase activity may be contributed mainly by incompetent vesicles. In this case JNET would be much greater than Jf or Jp.

Jp) and that the net rate of ATP hydrolysis (JNET) would be stoichiometrically related to both Jf and Jp (JNET = Jp/α = Jf/α, where α is the coupling ratio generally regarded as an integer which may be 1.0 for cardiac SR and 2.0 for skeletal SR) [18].

In a second model (Fig. 1B), the SR preparation is considered a mixture of sealed competent vesicles and incompletely sealed, incompetent vesicles. Sealed vesicles would achieve a steady-state Ca^{2+} uptake much like that described for the fixed-stoichiometry, pump-leak model. Accumulated Ca^{2+} would inhibit the pump so that at steady state only a low rate of ATP hydrolysis remains in these competent vesicles. This slow ATP hydrolysis would be stoichiometrically related to the Ca^{2+} influx necessary to balance the passive efflux. On the other hand, incompletely sealed or incompetent vesicles are characterized by a very high permeability and a very low level of Ca^{2+} uptake due only to surface binding. Because incompetent vesicles would not accumulate Ca^{2+}, internal Ca^{2+} would be unable to inhibit the pump, and thus net ATP hydrolysis would be rapid. Presumably, this rapid ATP hydrolysis would remain coupled to a rapid Ca^{2+} influx which would be experimentally undetectable because the pump would merely transport Ca^{2+} from one region of the solution to another.

If both competent and incompetent vesicles were present in the same preparation, then total Ca^{2+} uptake and ATPase activity would be the sum of the activities present in the two vesicle types. Incompetent vesicles would make no contribution to the total Ca^{2+} influx or total passive efflux, so that a consequence of this model is Jf = Jp. Because incompetent vesicles would contribute to JNET but not to Jf, the consequence for this model is that JNET would be greater than Jf/α.

A third pump-leak model makes no restrictions on the coupling ratio of the pump. This variable-stoichiometry pump-leak model suggests that the pump may "slip" and thereby operate with a variable stoichiometry. While the predictions for the pump-leak model with incompetent vesicle contamination are indistinguishable from those for the pump-leak model with variable stoichiometry, they both can be distinguished from the prediction for the fixed-stoichiometry pump-leak model. Distinction between the models can be made by measuring Ca^{2+} influx (Jf), passive Ca^{2+} efflux (Jp), and net ATP hydrolysis (JNET).

The total Ca^{2+} influx at steady state (Jf) was evaluated from the rate of Ca^{2+} exchange. Cardiac SR vesicles were allowed to come to a previously determined steady-state uptake in the absence of tracer ^{45}Ca. After steady state was reached, ^{45}Ca tracer was added and the rate of ^{45}Ca equilibration across the membrane was measured by rapid sequential filtration of the reaction mixture. From the time course of ^{45}Ca equilibration, Ca^{2+} influx can be calculated [12]. The passive Ca^{2+} efflux was evaluated by quenching the Ca^{2+} pump after attainment of steady-state uptake. The pump was quenched by adding EGTA, which complexes activator Ca^{2+}; or by adding glucose plus hexokinase, which removes substrate ATP. In both cases the initial passive efflux was similar [11]. The net efflux measured in this way is the diffusional flux referable to the steady state which occurs prior to quenching the pump-mediated Ca^{2+} influx [12].

The results of several measurements of Jf, Jp, and JNET clearly show that Ca^{2+} influx at steady state and net ATP hydrolysis are much larger than the passive efflux (Table I). All pump-leak models, regardless of fixed or variable stoichiometry or contaminating incompletely sealed vesicles, are inconsistent with these data. A simple fixed-stoichiometry pump-leak model has also been ruled out for skeletal SR [1].

Table I. Calcium Influx (Jf), Passive Calcium Efflux (Jp), and Net ATPase Rate (JNET) in Cardiac Sarcoplasmic Reticulum Vesicles[a]

Experiment	Jf (nmoles/min/per mg)	Jp (nmoles/min/per mg)	JNET (nmoles/min per mg)
1	807	56	257
2	453	46	128
3	125	18	103
4	533	26	122
5	671	49	295
6	417	42	220
7	478	41	94

[a]Each experiment was performed using a separate preparation of cardiac SR vesicles. Reaction conditions were 100 mM KCl, 20 mM imidazole buffer, pH 7.0, 10 mM NaN_3, 100 μM ATP, 2.1 mM $MgCl_2$, from 47 to 125 μg cardiac SR protein/ml, and from 7.3 to 11.8 μM total Ca^{2+}, as determined by atomic absorption spectrometry. JNET in experiments 1-4 was determined using [γ-^{32}P]-ATP and in experiments 5-7 using [^{3}H]-ATP. Net ATP hydrolysis was measured by extracting $^{32}P_i$ after hydrolysis of [γ-^{32}P]-ATP [10] or by counting the ATP, ADP, and AMP fractions following thin-layer chromatography after hydrolysis of [2, 8-^{3}H]-ATP [28, 32].

Nondiffusional Routes of Calcium Efflux

Since passive Ca^{2+} efflux is much less than Ca^{2+} influx at steady state, there may be some nondiffusional route(s) of efflux which, together with passive efflux, balances Ca^{2+} influx. There are at least three possible routes for this nondiffusional efflux: (1) carrier-mediated facilitated diffusion; (2) Ca^{2+} efflux through the pump; and (3) Ca^{2+} efflux through a gated Ca^{2+} channel distinct from the pump. If the total Ca^{2+} efflux is Je, then at steady state we have:

$$Jf = Je = Jp + Jr$$

Where Jp is the passive, diffusional efflux and Jr is the flux through some other, nondiffusional pathway. Efflux through this nondiffusional pathway requires external free Ca^{2+} (Ca_o), since complexation of Ca_o by EGTA lowers the Ca^{2+} efflux from that present at steady state to the diffusional efflux at steady state [12]. Flux through this route also requires ATP since glucose plus hexokinase, which removes ATP, also reduces efflux from Je to Jp [12]. Addition of an ATP-regenerating system also reduces Je, even though the level of steady-state Ca^{2+} uptake and Jp are increased. These results show that Ca^{2+} efflux through the nondiffusional pathway requires ATP, ADP, and Ca_o. Of the three pathways mentioned above, only Ca^{2+} efflux through the pump appears to be consistent with these requirements. However, efflux through a gated channel may appear to have these requirements because flux through the gated channel would be linked by constraints of the steady-state condition to flux through the pump.

Asynchronous Uptake and Release Model (Gated Channel)

In the asynchronous uptake and release model, Ca^{2+} uptake occurs in distinct phases (see Fig. 2A). In the uptake phase, a gated Ca^{2+} channel is closed and Ca^{2+} influx by the pump results in net Ca^{2+} accumulation. Sometime after loading, the gated channel opens and net Ca^{2+} release ensues. The flux of Ca^{2+} through this channel is noted as Jg. Because the vesicles are not synchronized, some vesicles will be in the uptake phase, while others are in the release phase. The observed Ca^{2+} uptake would be the sum for the populations of vesicles and would be quite steady due to the large number of vesicles in different phases of uptake and release. These phases are drawn as sinusoids in Fig. 2A, although they could assume a different shape.

The asynchronous uptake and release model suggests that the net rate of ATP hydrolysis during steady state is rapid in order to balance efflux through the gated channel and diffusional pathway. This net ATP hydrolysis would be the sum of rapid ATP hydrolysis during the late release and early uptake phases, and slower ATP hydrolysis when intravesicular Ca^{2+} is high. According to this model, quenching of the Ca^{2+} pump prevents partially loaded vesicles from entering the release phase. Quenching of the pump would thus indirectly quench efflux through the gated channel. Vesicles already in the release phase might or might not continue to release, depending on whether or not the quenching agents have effects on the channel. In either case, if the release phase is normally fast compared to the uptake phase, then only a small fraction of the vesicles would be in this phase at any time. Even if the release phase continued after quenching the pump, only a small fraction of the total Ca^{2+} taken up would be released. In this way, any effect on Ca^{2+} influx by the pump would be linked to an effect on efflux through the gated channel.

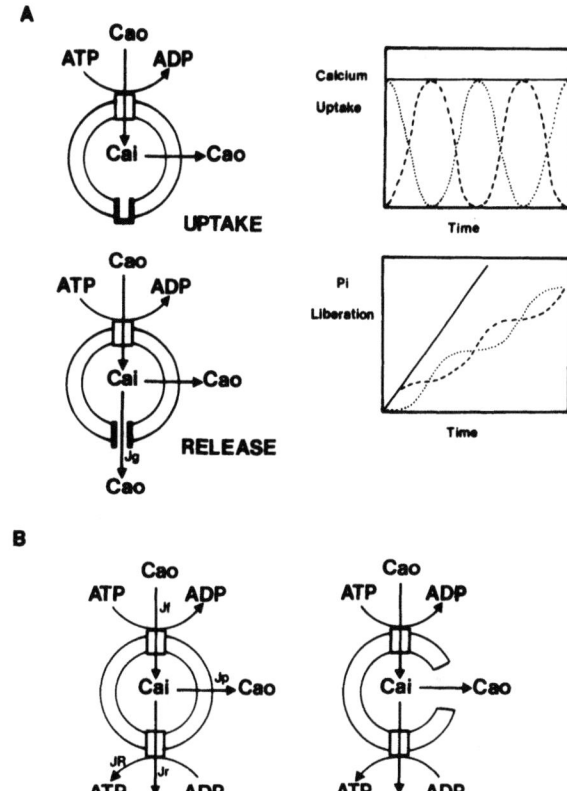

Figure 2. (A) Asynchronous uptake and release model of cardiac SR; (B) pump, reverse-pump, leak model of cardiac SR. (A) In the asynchronous uptake and release model, some vesicles (-----) take up Ca^{2+} while others (······) release Ca^{2+}. The uptake and release phases are governed by the closing and opening of a gated channel. Total Ca^{2+} uptake and P_i liberation (———) are steady. (B) In the pump, reverse-pump, leak model, Ca^{2+} influx (Jf) is balanced by the sum of passive efflux (Jp) and efflux by partial reversal of the ATPase (Jr). Efflux by partial reversal of the ATPase should be related to the reverse nucleotide flux (JR).

Variations of this model have been suggested previously. Spontaneous Ca^{2+} release by cardiac SR has been reported to occur even in the absence of a net observable release, and the released Ca^{2+} is subsequently reaccumulated [6]. Ryanodine increases oxalate-supported Ca^{2+} uptake but does not affect the rate of ATP hydrolysis, suggesting that ryanodine blocks Ca^{2+} efflux through a path distinct from the pump [19]. Ample evidence in skinned cardiac muscle fibers suggests that Ca^{2+} release from the SR occurs by a Ca^{2+}-gated mechanism [7,8].

The asynchronous uptake and release model should be particularly intriguing to physiologists because it suggests that Ca^{2+} efflux, rapid enough to explain *in vivo* Ca^{2+} release, may occur in isolated SR vesicles. Because of the asynchrony, Ca^{2+} efflux does not appear as a net efflux. However, this model does not explain why Ca^{2+} efflux through the nondiffusional pathway is reduced when an ATP-regenerating system is used. Removal of ADP is expected to activate Ca^{2+} influx and, because of the constraints of steady state, the increased influx should be associated with an increased efflux. However, Ca^{2+} influx is depressed when an ATP-generating system is used in cardiac [12] or skeletal SR [33].

Pump, Reverse-Pump, Leak Model

An alternative explanation for the data shown in Table I is that the nondiffusional Ca^{2+} efflux is mediated by a partial reversal of the Ca^{2+} pump (Fig. 2B). In skeletal SR at steady

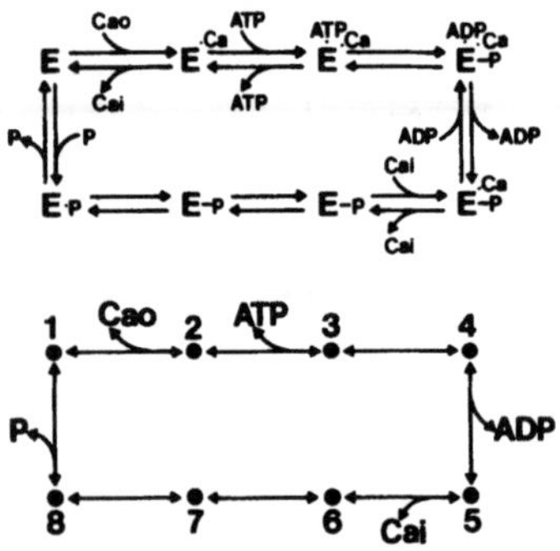

Figure 3. Reaction mechanism for Ca^{2+}-ATPase of sarcoplasmic reticulum [14,17,30].

state there is a rapid Ca^{2+}–Ca^{2+} exchange [22] and a rapid ATP–ADP exchange [5,15]. Ca^{2+} efflux can be stoichiometrically coupled to ATP synthesis from ADP and P_i in both skeletal [3,21] and cardiac [29,34] SR. There are a number of postulated mechanisms for the Ca^{2+}-ATPase of skeletal SR [14,17,30], and a simplified version of one such reaction mechanism is shown in Fig. 3. This scheme is shown primarily for illustration purposes and may be either incorrect or incomplete. At steady state, each of the enzyme states 1–8 has a nonzero population. Flux between adjacent states would be determined by the state populations, the rate constants governing the transitions, and the concentrations of any ligands involved in the transition [16]. Flux over nonadjacent states would be governed by the intervening rate constants; but the relation can be complicated [27]. When an enzyme makes a transition from state 1 to state 6 (Fig. 3), Ca^{2+} is transported from the outside to the inside of the SR vesicle. Similarly, transition from state 6 to state 1 corresponds to a pump-mediated Ca^{2+} efflux. In this scheme, Ca^{2+} influx mediated by a partial turnover of the pump is obligatorily linked to conversion of ATP to ADP and phosphoenzyme. Similarly, Ca^{2+} efflux by the pump is linked to conversion of ADP and phosphoenzyme to ATP. Net Ca^{2+} uptake corresponds to a completed cycle which, in this scheme, requires hydrolysis of ATP to ADP and liberated P_i.

Because the reaction scheme shown in Fig. 3 consists of a single cycle, net ATPase activity and net Ca^{2+} transport occurs with only fixed stoichiometry in this model. The observation is that net ATPase, which at steady state balances the leak, is much larger than the leak (Table I). The pump, reverse-pump, leak model is consistent with this observation only if some ATPase activity can occur uncoupled to the observed net Ca^{2+} uptake. This may happen in two ways. First, the reaction cycle may be branched so that Ca^{2+}-activated ATP hydrolysis occurs without Ca^{2+} transport. This could occur if either activator Ca^{2+} is not transported or transported Ca^{2+} does not desorb on the inside of vesicles. A second source of apparent uncoupling is ATPase activity from incompetent vesicles (Fig. 2B). The consequence of incompetent vesicles was discussed earlier. Basically, incompetent vesicles exhibit ATPase activity without contributing Ca^{2+} fluxes.

Distinction between Asynchronous Uptake/Release and Pump, Reverse-Pump, Leak Models

The asynchronous uptake and release model can be experimentally distinguished from the pump, reverse-pump, leak model. The asynchronous uptake and release model predicts that Ca^{2+} influx, Jf, will be greater than the gated channel flux, Jg, by an amount equal to

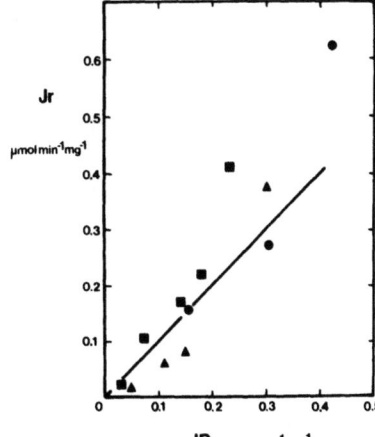

Figure 4. Relation between nondiffusional Ca^{2+} efflux (Jr) and the reverse nucleotide flux (JR). The points shown represent data obtained with three separate cardiac SR preparations corresponding to the three different symbols shown. The line shown in the figure corresponds to a slope of unity and is presented for comparison. Conditions were similar to those described in Table I.

the passive efflux, Jp. Ca^{2+} influx should be stoichiometrically related to the forward nucleotide flux, JF, the rate at which ATP is converted to ADP.

A consequence of the pump, reverse-pump, leak model is that Ca^{2+} efflux through the nondiffusional pathway is related to the reverse nucleotide flux, (JR), which is the rate at which ADP is converted to ATP. Influx of Ca^{2+} should be related to JF, the forward nucleotide flux. The pump, reverse-pump, leak model predicts that the nondiffusional flux, calculated as Jr = Jf − Jp, will be related to JR, while the asynchronous uptake and release model predicts that Jr will be much larger than JR. Distinction between the models can be made by comparing Jr to JR.

As mentioned above, Jr was calculated as Jr = Jf − Jp, where Jf was measured as Ca^{2+} exchange at steady state and Jp was measured as the net diffusional efflux following quench of the Ca^{2+} pump. The forward and reverse nucleotide fluxes were determined by adding [^{3}H]-ATP or [^{3}H]-ADP during steady state hydrolysis and then counting the ATP and ADP

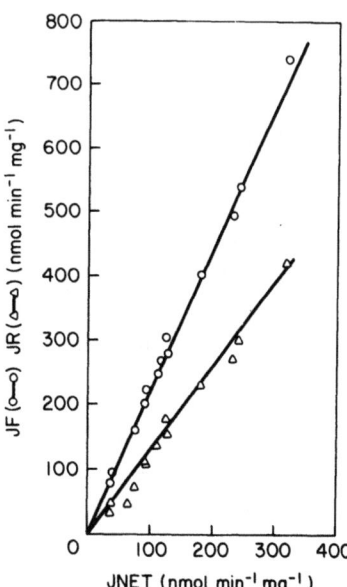

Figure 5. Relation between unidirectional nucleotide fluxes and net ATP hydrolysis in cardiac SR vesicles. Data shown are the composite for three separate SR preparations.

fractions at various times after adding the labeled nucleotide. The results indicate that the nondiffusional efflux is highly correlated with the reverse nucleotide flux and that these fluxes are nearly equal (Fig. 4). Both JF and JR are larger than JNET, suggesting that the pump works at a pseudo-equilibrium condition (Fig. 5). The similar magnitude of JF and JR has been observed in skeletal SR [25,31], where JR was much greater than the nondiffusional efflux [31,33]. It is clear from the results obtained in cardiac SR vesicles that the nondiffusional efflux is as large as JR (Fig. 4).

Further Tests of the Pump, Reverse-Pump, Leak Model

The observation that the nondiffusional efflux is highly correlated with the reverse nucleotide flux suggests that this efflux is mediated by the reverse pump, rather than by a gated channel distinct from the pump. Assignment of this efflux to a reverse pump is also consistent with the observations that the nondiffusional efflux requires ATP, ADP, and Ca_o.

There are several predictions that can be made for the pump, reverse-pump model. First, the model is consistent with the observed steady-state ATP hydrolysis only if the pump has a variable stoichiometry, or if there are incompetent vesicles contributing ATPase activity but not observable Ca^{2+} fluxes. The second prediction concerns Ca^{2+} fluxes when acetylphosphate is used as a substrate for the pump. Acetylphosphate supports Ca^{2+} uptake [24], but should terminate Ca^{2+} efflux mediated by the pump since it is a poor substrate for reversal of the ATPase [13,31]. If pump-mediated Ca^{2+} efflux is entirely responsible for the nondiffusional efflux, then acetylphosphate should turn the system into a pump-leak system (cf. Fig. 1B).

Both of the predictions for the pump, reverse-pump, leak model are observed. When cardiac SR vesicles are maximally loaded with calcium oxalate and then subjected to sucrose gradient centrifugation, about 14% of the recovered protein fails to show the density augmentation associated with calcium oxalate loading (Fig. 6). Ca^{2+}-ATPase activity of these unloaded vesicles was almost as high as that of native vesicles. A similar observation has also been reported using skeletal SR [2]. When acetylphosphate is used to support Ca^{2+} accumulation, Ca^{2+} influx is reduced to the level of passive efflux (Fig. 7). In this case, the Ca^{2+} fluxes can be described by the pump-leak model in which Ca^{2+} influx by the pump balances the parallel passive leak.

Significance of the Results

The reported results suggest that all of the Ca^{2+} fluxes in isolated cardiac SR vesicles can be explained by the pump, reverse-pump, leak model. The inability to detect fluxes through a gated channel distinct from the pump may be explained in several different ways. First, there may be no gated channel distinct from the pump, and Ca^{2+} efflux is mediated by the pump. An asynchronous uptake/release model may occur in this case if the pump undergoes certain slow transitions associated with the Ca^{2+} gradient. Second, the gated channel may not open under the steady-state conditions used. In these studies the external Ca^{2+} concentration ranged from 0.2 μM to 10 μM, which encompasses the range of Ca_o required for Ca^{2+} release in skinned cardiac cells [7]. However, free Ca^{2+} was constant during steady-state uptake, while a gated channel may respond to the rate of change of Ca_o [7]. Thus, a gated Ca^{2+} channel may be silent in these experiments because of the conditions employed. Third, the gated channel may be sparsely represented in the SR so that relatively

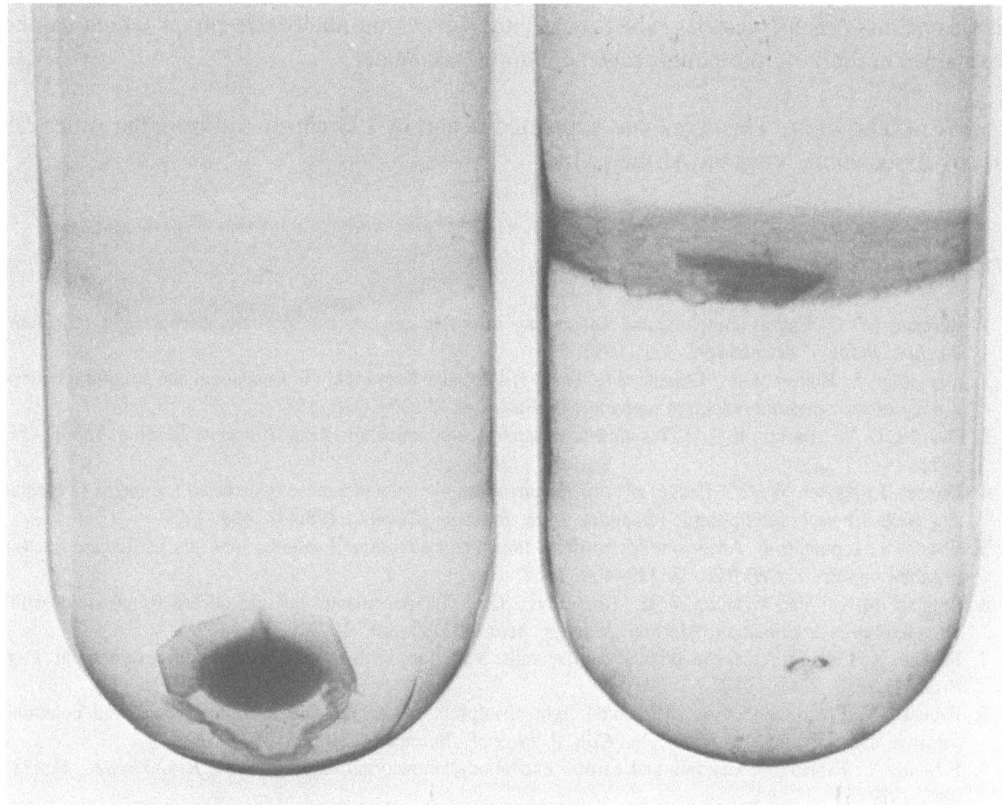

Figure 6. Competency of cardiac SR vesicles. Cardiac SR vesicles were loaded in the presence (left) or absence (right) of oxalate, layered over 40% sucrose, and then centrifuged. Calcium oxalate-loaded vesicles penetrated the 40% sucrose and were collected as a pellet. A small proportion (about 14%) of material failed to load with calcium oxalate.

few vesicles possess a channel. Fourth, the isolated SR preparation used in these studies may not contain the Ca^{2+} channel because the SR preparation is a subpopulation of the total tissue SR. In particular, it may be enriched in longitudinal SR that may be devoid of the Ca^{2+} channel.

It is not yet possible to distinguish among these various possible explanations from the

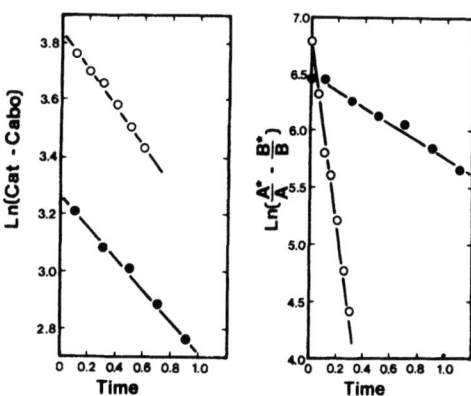

Figure 7. Effect of acetylphosphate on passive efflux (left) and total Ca^{2+} influx (right). Substrates used to support Ca^{2+} uptake were ATP (○) and acetylphosphate (●). The first-order plots of intravesicular Ca^{2+} (Cat − Cabo) give the passive efflux as the product of the slope and initial value of Cat − Cabo [11]. The Ca^{2+} influx can be calculated from the slope of the logarithm of the difference in specific activity of intravesicular (B*/B) and extravesicular (A*/A) Ca^{2+} as described previously [12]. With acetylphosphate, passive efflux was 15 nmoles/min per mg and Ca^{2+} influx was 14 nmoles/min per mg. When ATP was used, passive efflux was 29 nmoles/min per mg and Ca^{2+} influx was 191 nmoles/min per mg.

observations made on isolated cardiac SR. However, at this juncture, there appears to be at least one class of SR vesicles which contribute Ca^{2+} and nucleotide fluxes which can be explained entirely by the pump, reverse-pump, leak model.

ACKNOWLEDGMENT. This work was supported in part by a Grant-in-Aid from the American Heart Association, Virginia Affiliate, Inc.

References

1. Berman, M. C. Energy coupling and uncoupling of active calcium transport by sarcoplasmic reticulum. *Biochim. Biophys. Acta* **694**:95–121, 1982.
2. Chevallier, J.; Bonnet, J.-P.; Galante, M.; Tenu, J.-P.; Gulik-Krzywicki, T. Functional and structural heterogeneity of sarcoplasmic reticulum preparations. *Biol. Cell.* **30**:103–110, 1977.
3. Deamer, D. W.; Baskin, R. J. ATP synthesis in sarcoplasmic reticulum. *Arch. Biochem. Biophys.* **153**:47–54, 1972.
4. Dunnet, J.; Nayler, W. G. Effect of pH on calcium accumulation and release by isolated fragments of cardiac and skeletal muscle sarcoplasmic reticulum. *Arch. Biochem. Biophys.* **198**:434–438, 1979.
5. Ebashi, S.; Lipmann, F. Adenosine triphosphate-linked concentration of calcium ions in a particulate fraction of rabbit muscle. *J. Cell Biol.* **14**:389–400, 1962.
6. Entman, M. L.; Van Winkle, W. B.; Bornet, E.; Tate, C. Spontaneous calcium release from sarcoplasmic reticulum: A re-examination. *Biochim. Biophys. Acta* **551**:382–388, 1979.
7. Fabiato, A. Calcium release in skinned cardiac cells: Variations with species, tissues and development. *Fed. Proc.* **41**:2238–2244, 1982.
8. Fabiato, A. Fluorescence and differential light absorption recordings with calcium probes and potential-sensitive dyes in skinned cardiac cells. *Can. J. Physiol. Pharmacol.* **60**:556–567, 1982.
9. Fabiato, A.; Fabiato, F. Calcium and cardiac excitation–contraction coupling. *Annu. Rev. Physiol.* **41**:473–484, 1979.
10. Feher, J. J.; Briggs, F. N. The effect of calcium oxalate crystallization kinetics on the kinetics of calcium uptake and calcium ATPase activity of sarcoplasmic reticulum vesicles. *Cell Calcium* **1**:105–118, 1980.
11. Feher, J. J.; Briggs, F. N. The effect of calcium load on the calcium permeability of sarcoplasmic reticulum. *J. Biol. Chem.* **257**:10191–10199, 1982.
12. Feher, J. J.; Briggs, F. N. Determinants of calcium loading at steady state in sarcoplasmic reticulum. *Biochim. Biophys. Acta* **727**:389–402, 1983.
13. Friedman, Z.; Makinose, M. Phosphorylation of skeletal muscle microsomes by acetyl phosphate. *FEBS Lett.* **11**:69–72, 1970.
14. Guimaraes-Motta, H.; DeMeis, L. Pathway for ATP synthesis by sarcoplasmic reticulum ATPase. *Arch. Biochem. Biophys.* **203**:395–403, 1980.
15. Hasselbach, W.; Makinose, M. ATP and active transport. *Biochem. Biophys. Res. Commun.* **7**:132–136, 1962.
16. Hill, T. L. *Free Energy Transduction in Biology*, New York, Academic Press, 1977.
17. Inesi, G.; Kurzmack, M.; Kosk-Kosicka, D.; Lewis, D.; Scofano, H.; Guimaraes-Motta, H. Equilibrium and kinetic studies of calcium transport and ATPase activity in sarcoplasmic reticulum. *ZNaturforsch. Teil C* **37**:685–691, 1982.
18. Jones, L. R.; Besch, H. R. Calcium handling by cardiac sarcoplasmic reticulum. *Tex. Rep. Biol. Med.* **39**:19–35, 1979.
19. Jones, L. R.; Besch, H. R.; Sutko, J. L.; Willerson, J. T. Ryanodine-induced stimulation of net Ca uptake by cardiac sarcoplasmic reticulum vesicles. *J. Pharmacol. Exp. Ther.* **209**:48–55, 1979.
20. Levitsky, D. O.; Benevolensky, D. S.; Levchenko, T. S.; Smirnov, V. N.; Chazov, E. I. Calcium-binding rate and capacity of cardiac sarcoplasmic reticulum. *J. Mol. Cell. Cardiol.* **13**:785–796, 1981.
21. Makinose, M.; Hasselbach, W. ATP synthesis by the reverse of the sarcoplasmic calcium pump. *FEBS Lett.* **12**:271–272, 1971.
22. Martonosi, A.; Feretos, R. Sarcoplasmic reticulum. I. The uptake of Ca by sarcoplasmic reticulum fragments. *J. Biol. Chem.* **239**:648–658, 1964.
23. Penpargkul, S. Effects of adenine nucleotides on calcium binding by rat heart sarcoplasmic reticulum. *Cardiovasc. Res.* **13**:243–253, 1979.
24. Pucell, A.; Martonosi, A. Sarcoplasmic reticulum. XIV. Acetylphosphate and carbamylphosphate as energy sources for Ca transport. *J. Biol. Chem.* **246**:3389–3397, 1971.

25. Ronzani, N.; Migala, A.; Hasselbach, W. Comparison between ATP-supported and GTP-supported phosphate turnover of the calcium-transporting sarcoplasmic reticulum membranes. *Eur. J. Biochem.* **101**:593–606, 1979.
26. Solaro, J.; Briggs, F. N. Estimating the functional capabilities of sarcoplasmic reticulum in cardiac muscle. *Circ. Res.* **34**:531–540, 1974.
27. Stein, W. D. An algorithm for writing down flux equations for carrier kinetics, and its application to cotransport. *J. Theor. Biol.* **62**:467–478, 1976.
28. Suko, J.; Hasselbach, W. Characterization of cardiac sarcoplasmic reticulum ATP–ADP phosphate exchange and phosphorylation of the calcium transport adenosine triphosphatase. *Eur. J. Biochem.* **64**:123–130, 1976.
29. Suko, J.; Hellman, G.; Winkler, F. The reversal of the calcium pump of cardiac sarcoplasmic reticulum. *Basic Res. Cardiol.* **72**:147–152, 1977.
30. Takakuwa, Y.; Kanazawa, T. Reaction mechanism of (Ca,Mg)-ATPase of sarcoplasmic reticulum vesicles. *J. Biol. Chem.* **256**:2696–2700, 1981.
31. Takenaka, H.; Adler, P. N.; Katz, A. M. Calcium fluxes across the membrane of sarcoplasmic reticulum vesicles. *J. Biol. Chem.* **257**:12649–12656, 1982.
32. Verjovski-Almeida, S.; Kurzmack, M.; Inesi, G. Partial reactions in the catalytic and transport cycle of sarcoplasmic reticulum ATPase. *Biochemistry* **17**:5006–5013, 1978.
33. Waas, W.; Hasselbach, W. Interference of nucleoside diphosphates and inorganic phosphate with nucleoside triphosphate-dependent calcium fluxes and calcium-dependent nucleoside-triphosphate hydrolysis in membranes of sarcoplasmic reticulum vesicles. *Eur. J. Biochem.* **116**:601–608, 1981.
34. Winkler, F.; Suko, J. Phosphorylation of the calcium transport adenosine triphosphatase of cardiac sarcoplasmic reticulum by orthophosphate. *Eur. J. Biochem.* **77**:611–619, 1977.

X

Calcium and Calcium Antagonists in Smooth Muscle

43

Calcium and Myogenic or Stretch-Dependent Vascular Tone

John A. Bevan, Joyce J. Hwa, Mary P. Owen, and Raymond J. Winquist

Peripheral vascular bed resistance and capacitance is related not only to the architecture and passive physical properties (both relatively static components) of the bed, but also to the active vascular tone of its component blood vessels. This tone may be extrinsic or intrinsic in origin. Extrinsic tone results from influences that originate outside the smooth muscle cells, such as circulating vasoactive substances and locally released material from such cellular elements as neurons, mast cells, and platelets. Intrinsic tone, on the other hand, is considered to originate from within the smooth muscle cells themselves. This may not be a useful classification since all evidence suggests that myogenic (intrinsic) tone appears in response to stretch. Perhaps a classification of tone based on its primary cause would be more satisfactory.

Myogenic tone developing in response to stretch was first described by Bayliss [1] and is of considerable physiological interest, although there are no satisfactory estimates of its contribution to total vascular resistance nor to whole body regulation. It seems quite likely that it is a major factor in autoregulation [8] and that it occurs to some degree in most vascular beds [16]. The effect occurs more often in smaller than in larger vessels [8,20,26] and in arterial rather than venous vessels, although there are exceptions to this rule [2,14,43]. Maintained tone predominates in small-resistance vessels and rhythmic tone in small veins. It is clearly more than an optimization of the length of the smooth muscle cell leading to more optimal contraction [18]. In the microcirculation it maintains the constancy of capillary perfusion pressure, although this is not its only role in the circulation [4,43].

Analysis of the responses of a number of blood vessels to a variety of agonists indicates that the increased free intracellular Ca^{2+} level associated with increased extrinsic tone may result from Ca^{2+} originating from a number of sources. Most important in blood vessels is the Ca^{2+} entering from the extracellular space. However, there are contributions from a

---

John A. Bevan, Joyce J. Hwa, and Mary P. Owen • Department of Pharmacology, School of Medicine, University of Vermont, Burlington, Vermont 05405   *Raymond J. Winquist* • Cardiovascular Pharmacology, Merck Sharp & Dohme Research Laboratories, West Point, Pennsylvania 19486.

variety of cell-bound or sequestered Ca^{2+} pools in smooth muscle cells. These sites are at least two in number: (1) a relatively small depletable store responsible for the initial phasic response of a blood vessel and (2) a store whose size or functional importance differs remarkably in different vessels [5]. This latter store is of little or no importance in cerebral blood vessels [23], but is of major importance in the renal vasculature [32]. The coupling systems between the receptor and these several Ca^{2+} sources are independent [6], and their anatomical correlates have not been defined. Extrinsic vascular tone depends on both varying proportions and amounts of Ca^{2+} entering the cell from the extracellular space through receptor-operated or potential-sensitive channels and Ca^{2+} released from complex sequestered sites by an as yet undefined transmembrane coupling mechanism.

Intrinsic Tone, Wall Stress, and Ca^{2+}

Myogenic tone has been studied *in vitro* in only a few preparations. Uchida and Bohr [36] perfused small-resistance vessel segments, described the tone that occurred in different regional beds, and concluded that it was dependent on the presence of extracellular Ca^{2+}. The most extensive quantitative study has been undertaken in the circular fibers of the buccal segment of the rabbit facial vein [27,40]. A segment with similar features also occurs in man [24]. In the rabbit, the magnitude of developed tone is positively related to the degree of applied stretch (Fig. 1). The slope of the relationship is a function of external Ca^{2+} concentration and the Ca^{2+} gradient across the cell membrane so that the lower the external concentration and the gradient, the smaller the myogenic response to a given applied stress. This conclusion is consistent with exhaustive studies of this process *in vivo*, particularly in the microcirculation [18,19].

Johnson [17] has proposed a closed-loop gain model of myogenic tone which involves an in-series tensor-sensor. This model encompasses the results of a number of observations made after a variety of manipulations (e.g., hemorrhage). Speden [31] concluded from a careful mechanical analysis that an increase in the vascular smooth muscle cell membrane radius or length was the primary stimulus for myogenic response.

A brief word regarding the experimental methodology for measuring these parameters *in vitro* is appropriate. A response cannot be recorded *in vitro* until stress is applied to a vessel wall. However, as soon as stress is initiated, tone begins to develop and stress relaxation also takes place. Both take different times to reach equilibrium. The problem that this poses is circumvented experimentally by applying tension when external Ca^{2+} is reduced below that which supports contraction (25 μM). Once stress relaxation is complete, applied stress can be measured and the associated myogenic tone observed upon the introduction of a normal Ca^{2+} level into the tissue bath. Alternatively, myogenic tone can be measured once stress relaxation is complete by measuring the dilation that occurs on adding a maximum concentration of a vasodilator drug such as a nitrite or papaverine. These approaches yield similar results. Maintained myogenic tone only develops in the presence of external Ca^{2+}. It does not develop if the tissue is bathed in Mn^{2+} and/or La^{3+} [43]. The $t_{1/2}$ for the *loss* of myogenic tone is consistent with the calculated rate of loss of Ca^{2+} from the extracellular space by diffusion (Hwa, personal communication; 23).

By contrast, the rate of *onset* of myogenic tone found on introduction of Ca^{2+} into the prestressed tissue (see above) (Fig. 1) is much slower and is measured in minutes. The main limiting factor in this response is clearly not Ca^{2+} entry into the tissue extracellular space, nor is it due to some inherent property of the contractile machinery (the response due to agonists that release Ca^{2+} from a sequestered pool is much faster), but it presumably is due to the relatively slow entry of Ca^{2+} into the cell.

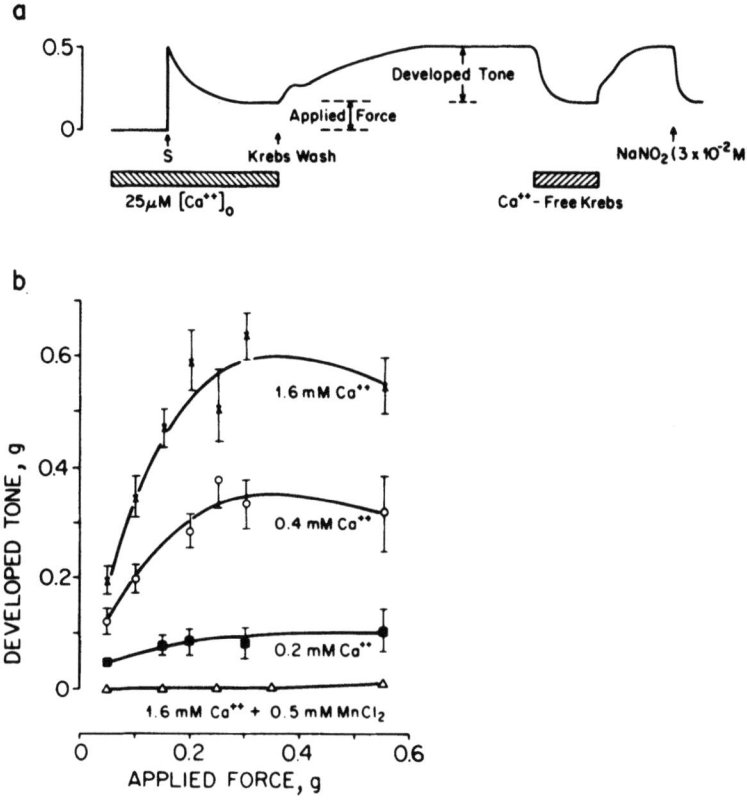

Figure 1. (a) Development of myogenic tone following stretch (S) in the rabbit facial vein with the experimental protocol designed to quantify the relationship of applied force to developed tone (see text). Rings are initially incubated in 25 μM Ca^{2+}, which is subthreshold for tone development after stretch. The equilibrated level of the stress relaxation was designated as the value of applied force. Washing with normal Ca^{2+} Krebs produced the corresponding value for developed tone. Tone is rapidly lost when rings are washed in Ca^{2+}-free Krebs. Exogenously added sodium nitrite ($NaNO_2$, 3×10^{-2} M) also relaxes tissues maximally. The units of the vertical axis are grams. (b) Relationship of myogenic tone developed in facial vein rings and applied force. As the applied force on the vessel wall increases, there is a corresponding increase in the myogenic developed tone. There is no further increase in developed tone with values of applied force greater than 0.35 g. Decreasing the external Ca^{2+} concentration (to 0.4 and 0.2 mM) causes progressively less tone to develop. Addition of manganous chloride (0.5 mM) along with the Krebs solution wash completely inhibits the development of myogenic tone. Each data point represents the mean (+ S.E.M.) of at least three experiments.

Rhythmic myogenic tone has been studied almost exclusively in the longitudinal muscle of the mammalian portal vein and those veins that converge upon it. It is dependent on the presence of Ca^{2+} in the bathing solution [12,39]. Calcium channel blocking agents reveal that the magnitude of periodic contraction but not the frequency of the spontaneous intrinsic activity depends on the presence of Ca^{2+} in the external environment [3].

Other Characteristics of Myogenic Tone

Magnitude. In the ear circulation of the rabbit, only smaller vessels develop tone. The terminal branch of the central ear artery, approximately 75-μm inside diameter, develops an equilibrium level of tone at l_0 of about 9% of the maximum possible contraction. This level approximated that obtained with sympathetic nerve stimulation at 2 Hz and is equivalent to that recorded in response to norepinephrine (NE) (4×10^{-8} M). Plasma NE levels in man

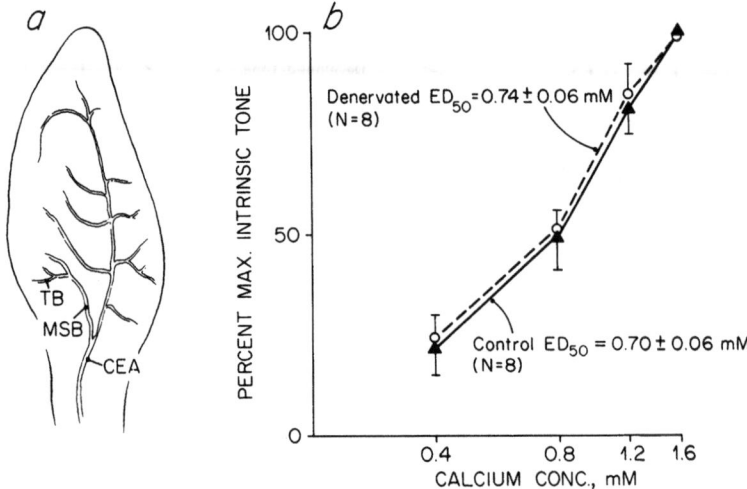

Figure 2. (a) Distribution of myogenic tone in vasculature of the rabbit ear. Tone is not observed in segments of the central ear artery studied *in vitro*, but it is seen in both the main side branch (MSB) and the terminal branch (TB) and is proportionately greater in the latter. [Figure courtesy of Dr. J. Walmsley.] (b) Relationship between myogenic tone as a percent of maximum obtained at l_0 and Ca^{2+} concentration in physiological saline solution before and after chronic sympathetic denervation.

during exercise rise as high as 7×10^{-8} M [10]. The stretch-induced tone developing in the buccal segment of the facial vein, which is considered to have a sphincter-like role in the circulation not related to peripheral resistance changes [42], was quantitatively much greater. The maximum level would be expected to occlude the venous channel.

Independence of innervation and known local tone-producing mechanisms. Myogenic tone still occurred in vessels taken from vascular beds after chronic sympathetic denervation [36], a finding similar to that reported for umbilical vessels [30,44]. Spontaneous rhythmic activity in the rabbit portal vein was unaffected by phenoxybenzamine [13].

In a careful quantitative study, Hwa has shown that the Ca^{2+} dose–response curve of a small branch of the ear artery is unaltered by chronic sympathectomy. Thus, the sensitivity of intrinsic tone to changes in Ca^{2+} concentration was unchanged (Fig. 2). There was, however, some evidence that the magnitude of the contraction was increased by denervation.

Rate dependency. The level of myogenic tone that develops in the umbilical artery was initially rate-dependent [30]. The electrical activity that accompanies tone in the rabbit portal vein is related to the velocity of stretch [14]. If stress was maintained, both myogenic tone and electrical phenomena diminished but never returned to the original level. The dynamic characteristics of tone development were concluded to be more important than the static ones. However, it must be pointed out that the type of tone that occurs in the portal vein may not be common in the circulation. Grande and Mellander [9] concluded after their study of the circulation to a muscle in the lower leg, that arterial vessels greater than 20-μm inside diameter responded to static rather than dynamic stretch. The rate sensitivity of the response was primarily evident in smaller vessels. Large arteries and veins seemed to lack myogenic activity. An interesting computer model of this system has been prepared [7].

Electrophysiology. In the rabbit portal vein rhythmic muscular activity is related to membrane potential changes similar to those recorded from intestinal smooth muscle. They consist of pacemaker potentials each with multiple spikes [15]. The electrophysiological correlates of maintained intrinsic tone are not as well known. In the rabbit facial vein no regenerative changes in the membrane potential were seen when the level of tone was altered [28]. Hyperpolarization occurred when stretch-induced tone that developed at optimum length was inhibited by high doses of isoproterenol or sympathetic nerves acting through postsynaptic β-receptors. Changes in membrane potential always preceded changes in tension. Relaxation appeared to be a function of hyperpolarization, but whether small increases in relaxation took place without some associated change in membrane potential could not be resolved. Phasic changes in the level of intrinsic tone could be induced by increasing the external K^+ concentration. This was accompanied by oscillations in the membrane potential in phase with contraction, but without regenerative-like activity.

Spontaneous electrical activity in the smooth muscle cells of guinea pig small mesenteric arteries has been recorded with intracellular electrodes when the arteries were distended by an increase in intramural pressure [45]. Resting membrane potentials were lower in vessels subjected to higher rather than lower pressures. Each action potential consisted of a prepotential (pacemaker potential) followed by a spike of variable amplitude. These spikes were very sensitive to a reduction in temperature (see below). A similar picture has been described for the cerebral arteries of the rat (Halpern, Harder, and Mongeon, 1983, personal communication). Until more experiments are performed, any generalizations and useful conclusions regarding electrophysiological correlates of myogenic tone cannot be made.

Temperature sensitivity. Many effects related in one way or another to contraction of vascular smooth muscle change only slightly (if at all) when temperature is increased from 20°C to about 40°C. By contrast, myogenic tone seems to emerge only at temperatures in the low to mid thirties and then to change remarkably with temperature increases up to 40°C and above. In the rabbit facial vein where this phenomenon has been studied in greatest detail [42], tone is first seen at 32–33°C. This is also true for tone in the small blood vessels of the ear artery, although quantitatively the changes are much smaller than in the facial vein. Rhythmic activity of the rabbit portal vein is absent below 30°C [38], and guinea pig mesenteric artery appears to be similar [45]. However, not all smooth muscle rhythmic activity possesses these characteristics and, often, a gradual decrease in rhythmic activity is seen [37]. It should be emphasized that there are many forms of rhythmic activity in vascular smooth muscle and that type observed in the rabbit portal vein may not be the most common.

A survey of the vascular smooth muscle literature does not reveal a new event in the lower (30°) temperature range. Many experiments have been conducted at two temperatures, one above and one below 33°C; the small difference in the magnitude of the responses observed indicates that no dramatic change occurs in this function with temperature. Johansson and Somlyo [15], discussing the effect of stretch on the membrane potential, assumed a gradual change in vascular smooth muscle with temperature increase. An extensive analysis of the effects of temperature change on the different components of contraction in the blood vessel wall does not identify a discontinuity or a new event between 30 and 35°C [37]. A distortion or stretch receptor may be an analogous tissue of interest in regard to stretch or distortion of vascular smooth muscle. Changes in electrical properties of the stretch receptor of Crustacea [33] and the magnitude of the generator potential in the Pacinian corpuscle [11] show no unusual changes at that temperature level.

Mechanism of Myogenic Tone

Perhaps the most intriguing feature of myogenic tone (and one that might provide insight into its mechanism) is its unique temperature dependence. The temperature of the endothermic transition between the gel and liquid-crystalline phase of phospholipids in a cell

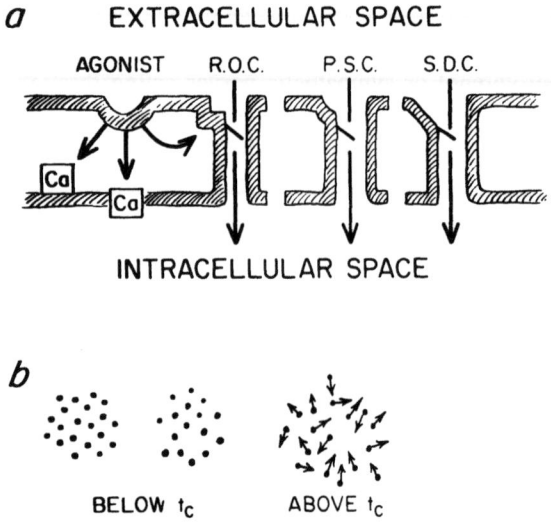

Figure 3. Diagrammatic representation of proposed Ca^{2+} channels in vascular smooth muscle. Agonist binding to a specific receptor can lead to Ca^{2+} mobilization from sequestered Ca^{2+} storage sites and entry of Ca^{2+} from the extracellular space via receptor-operated channels (R.O.C.). Also proposed are potential-sensitive channels (P.S.C.) and stretch-dependent channels (S.D.C.). The latter are dependent on the longitudinal stretch of the cell membrane. Such a mechanism may only be effective when at least parts of the cell membrane are held above their critical temperature for transition change. Above such temperatures, functionally important components of the cell membrane are no longer restricted but are capable of lateral movement and of association with each other to form multimolecular combinations.

membrane has been measured by differential scanning calorimetry. In some experimental systems in which the membrane has a composition similar to that found in some living cells, this transition occurs abruptly in the middle 30° range [21]. Evidence from analysis of electron spin resonance of steroid labels incorporated into lipid membranes indicates a change in the organization of the membrane system within this temperature range [35]. A complex mosaic molecular structure may exist in biological membranes below the transition temperatures: many steroids and other molecules may be clustered together into large, organized groups. However, when temperature is raised, the membrane organization changes and steroid molecules are free to undergo lateral diffusion within the phase of the membrane. A diffusional process may be necessary for the formation of specific groupings or functional complexes. This may lead to changes in the cellular electrical surface charge, permeability, and the affinity of the membrane surface to both ions and large molecules [35]. In the membrane of *Mycoplasma laidlawii* [29] and *E. coli* [25], correlations between phase transition and change in physiological properties have been found. The transduction of stretch to contraction may not occur below the transition temperature. However, above that level, the membrane is sufficiently labile so that changes in permeability can ensue.

It is well known that in some systems stretch leads to cellular distortion and to membrane depolarization [22]. It is unlikely that stretch-dependent tone is mediated through a change in the membrane potential. Firstly, muscle membrane constants change relatively little with stretch [11]. Secondly, depolarization would probably involve potential-sensitive channels, but these may differ from those insensitive to high-K^+ conditions. In several systems Ca^{2+} channel blockers (specifically, diltiazem and nimodipine) antagonize the K^+-induced contraction of smooth muscle cells in concentrations much lower than those that influence stretch [5,41], although this is not invariably the case [34].

In conclusion, we propose that there are specific stretch-operated channels in vascular smooth muscle of certain vessels that probably differ from those sensitive to potential change or operate through specific agonist receptors (Fig. 3). We suggest that stretch is coupled to Ca^{2+} entry through local changes in membrane properties and that these only occur when certain cellular membrane components are held above the membrane transition temperature.

ACKNOWLEDGMENT. Supported by Grants HL-20581 and HL-26414.

References

1. Bayliss, W. M. On the local reaction of the arterial wall to changes of internal pressure. *J. Physiol. (London)* **28**:220–231, 1902.
2. Bevan, J. A. Transient responses of rabbit cerebral blood vessels to norepinephrine: Correlation with intrinsic myogenic tone. *Circ. Res.* **45**:556–572, 1979.
3. Bevan, J. A. The selective action of diltiazem on cerebral vascular smooth muscle in the rabbit: Antagonism of extrinsic, but not intrinsic, maintained tone. *Am. J. Cardiol.* **49**:519–524, 1982.
4. Bevan, J. A.; Bevan, R. D.; Duckles, S. P. Adrenergic regulation of vascular smooth muscle. In: *Handbook of Physiology,* Section 2, *The Cardiovascular System,* Volume 2, *Vascular Smooth Muscle,* D. F. Bohr, A. P. Somlyo, and H. V. Sparks, Jr., eds., Baltimore, American Physiological Society, 1980, pp. 515–566.
5. Bevan, J. A.; Bevan, R. D.; Hwa, J. J.; Owen, M. P.; Tayo, F. M.; Winquist, R. J. Calcium, extrinsic and intrinsic (myogenic) vascular tone. In: *Calcium Modulators, Symposia of the Giovanni Lorenzini Foundation,* Volume 15, T. Godfraind, A. Albertini, and R. Paoletti, eds., Amsterdam, Elsevier/North-Holland, 1982, pp. 125–128.
6. Bevan, J. A.; McCalden, T. A.; Rapoport, R. M. Receptor-activated calcium mechanisms and their antoganism in cerebrovascular muscle. In: *New Perspectives on Calcium Antagonists,* G. B. Weiss, ed., Baltimore, Williams & Wilkins, 1981, pp. 123–129.
7. Borgstrom, P.; Grande, P.-O. Myogenic microvascular responses to change of transmural pressure: A mathematical approach. *Acta Physiol. Scand.* **106**:411–423, 1979.
8. Folkow, B. Description of the myogenic hypothesis. *Circ. Res.* **15**(Suppl. 1): 279–287, 1964.
9. Grande, P.-O.; Mellander, S. Characteristics of static and dynamic regulatory mechanisms in myogenic microvascular control. *Acta Physiol. Scand.* **102**:231–245, 1978.
10. Haggendal, J.; Harley, L. H.; Saltin, B. Arterial noradrenaline concentration during exercise in relation to the relative work levels. *Scand. J. Clin. Lab. Invest.* **26**:337–342, 1970.
11. Ishiko, N.; Loewenstein, W. R. Effects of temperature on the generator and action potentials of a sense organ. *J. Gen. Physiol.* **45**:105–124, 1962.
12. Johansson, B. Processes involved in vascular smooth muscle contraction and relaxation. Arthur C. Corcoran memorial lecture. *Circ. Res.* **43**(Suppl. 1):14–20, 1978.
13. Johansson, B.; Bohr, D. F. Rhythmic activity in smooth muscle from small subcutaneous arteries. *Am. J. Physiol.* **210**:801–806, 1966.
14. Johansson, B.; Mellander, S. Static and dynamic components in the vascular myogenic response to passive changes in length as revealed by electrical and mechanical recordings from the rat portal vein. *Circ. Res.* **36**:76–83, 1975.
15. Johansson, B.; Somlyo, A. P. Electrophysiology and excitation–contraction coupling. In: *Handbook of Physiology,* Section 2, *The Cardiovascular System,* Volume 2, *Vascular Smooth Muscle,* D. F. Bohr, A. P. Somlyo, and H. V. Sparks, Jr., eds., Baltimore, American Physiological Society, 1980, pp. 301–323.
16. Johnson, P. C. Review of previous studies and current theories of autoregulation. *Circ. Res.* **15**(Suppl. 1): 2–9, 1964.
17. Johnson, P. C. The myogenic response. In: *Handbook of Physiology,* Section 2, *The Cardiovascular System,* Volume 2, *Vascular Smooth Muscle,* D. F. Bohr, A. P. Somlyo, and H. V. Sparks, Jr., eds., Baltimore, American Physiological Society, 1980, pp. 409–442.
18. Johnson, P. C.; Intaglietta, M. Contributions of pressure and flow sensitivity to autoregulation in mesenteric arterioles. *Am. J. Physiol.* **231**:1686–1698, 1976.
19. Johnson, P. C.; Wayland, H. Regulation of blood flow in single capillaries. *Am. J. Physiol.* **212**:1405–1415, 1967.
20. Kontos, H. A.; Wei, E. P.; Navari, R. M.; Levasseur, J. E.; Rosenblum, W. I.; Patterson, J. L., Jr. Responses of cerebral arteries and arterioles to acute hypotension and hypertension. *Am. J. Physiol.* **23**:H371–H383, 1978.
21. Ladbrooke, B. D.: Williams, R. M.; Chapman, D. Studies on lecithin–cholesterol–water by differential scanning calorimetry and X-ray diffraction. *Biochim. Biophys. Acta* **150**:333–340, 1968.
22. Loewenstein, W. R. (ed.) Mechano-electric transduction in the Pacinian corpuscle: Initiation of sensory impulses in mechanoreceptors. In: *Principles of Receptor Physiology,* Berlin, Springer-Verlag, 1971, pp. 269–290. In: *Handbook of Sensory Physiology,* Volume I, H. Autrum, R. Jung, W. R. Loewenstein, D. M. MacKay, and H. L. Teuber, eds., Berlin, Springer-Verlag, 1971.
23. McCalden, T. A.; Bevan, J. A. Sources of activator calcium in rabbit basilar artery. *Am. J. Physiol.* **241**:H129–H133, 1981.
24. Mellander, S.; Andersson, P.-O.; Afzelius, L.-E.; Hellstrand, P. Neural beta-adrenergic dilatation of the facial vein in man: Possible mechanism in emotional blushing. *Acta Physiol. Scand.* **114**:393–399, 1981.

25. Overath, P.; Shairer, H. U.; Stoffel, W. Correlation of *in vivo* and *in vitro* phase transitions of membrane lipids in *Escherichia coli*. *Proc. Natl. Acad. Sci. USA* **67**:606–614, 1970.
26. Owen, M. P.; Walmsley, J. G.; Mason, M. F.; Bevan, R. D.; Bevan, J. A. Adrenergic control in three artery segments of diminishing diameter in the rabbit ear. *Am. J. Physiol.* **245**:H320–H326, 1983.
27. Pegram, B. L.; Bevan, R. D.; Bevan, J. A. Facial vein of the rabbit: Neurogenic vasodilation mediated by beta-adrenergic receptors. *Circ. Res.* **39**:854–860, 1976.
28. Prehn, J. L.; Bevan, J. A. Facial vein of the rabbit: Intracellularly recorded hyperpolarization of smooth muscle cells induced by beta-adrenergic receptor stimulation. *Circ. Res.* **52**:465–470, 1983.
29. Reinert, J. C.; Steim, J. M. Calorimetric detection of a membrane-lipid phase transition in living cells. *Science* **168**:1580–1582, 1970.
30. Sparks, H. V., Jr. Effect of quick stretch on isolated vascular smooth muscle. *Circ. Res.* **Suppl. I**:I254–I260, 1964.
31. Speden, R. M. The maintenance of arterial constriction at different transmural pressures. *J. Physiol. (London)* **229**:361–381, 1973.
32. Tayo, F. M.; Bevan, J. A. Resistance of the rabbit renal artery to calcium withdrawal and calcium entry blockers. Personal communication.
33. Terzuolo, C. A.; Washizu, Y. Relation between stimulus strength, generator potential and impulse frequency in stretch receptor of *Crustacea*. *J. Neurophysiol.* **25**:56–66, 1962.
34. Towart, R. The selective inhibition of serotonin-induced contractions of rabbit cerebral vascular smooth muscle by calcium-antagonistic dihydropyridines. *Circ. Res.* **48**:650–657, 1981.
35. Trauble, H.; Sackmann, E. Studies of the crystalline–liquid crystalline phase transition of lipid model membranes. III. Structure of a steroid–lecithin system below and above the lipid-phase transition. *J. Am. Chem. Soc.* **94**:4499–4510, 1972.
36. Uchida, E.; Bohr, D. F. Myogenic tone in isolated perfused resistance vessels from rats. *Eur. J. Physiol.* **216**:1343–1350, 1969.
37. Vanhoutte, P. M. Physical factors of regulation. In: *Handbook of Physiology*, Section 2, *The Cardiovascular System*, Volume 2, *Vascular Smooth Muscle*, D. F. Bohr, A. P. Somlyo, and H. V. Sparks, Jr., eds., Baltimore, American Physiological Society, 1980, pp. 443–474.
38. Vanhoutte, P. M.; Lorenz, R. R. Effect of temperature on reactivity of saphenous, mesenteric, and femoral veins of the dog. *Am. J. Physiol.* **218**:1746–1750, 1970.
39. Van Neuten, J. M.; Vanhoutte, P. M. Calcium entry blockers and vascular smooth muscle heterogeneity. *Fed. Proc.* **40**:2862–2865, 1981.
40. Winquist, R. J. Intrinsic myogenic tone and the adrenergic responses of the rabbit facial vein. Ph.D. doctoral dissertation, 1979.
41. Winquist, R. J.; Baskin, E. P. Calcium translocation through channels resistant to organic calcium entry blockers in a rabbit vein. *Am. J. Physiol.* **245**:H1024–H1030, 1983.
42. Winquist, R. J.; Bevan, J. A. Temperature sensitivity of tone in the rabbit facial vein: Myogenic mechanism for cranial thermoregulation. *Science* **207**:1001–1002, 1980.
43. Winquist, R. J.; Bevan, J. A. *In vitro* model of maintained myogenic vascular tone. *Blood Vessels* **18**:134–138, 1981.
44. Zaitev, N. D. Development of neural elements in the umbilical cord. *Arkh. Anat. Gistol. Embriol.* **37**:81–88, 1959.
45. Zelcer, E.; Sperelakis, N. Spontaneous electrical activity in pressurized small mesenteric arteries. *Blood Vessels* **19**:301–310, 1982.

44

Chemical Skinning
A Method for Study of Calcium Transport by Sarcoplasmic Reticulum of Vascular Smooth Muscle

Marguerite A. Stout

Free intracellular Ca^{2+} regulates force development and contraction in smooth muscle cells. Shifts in extracellular and intracellular Ca^{2+} pools increase cytosolic Ca^{2+} which then interacts with contractile and regulatory proteins to cause contraction. Relatively little is known about the sequence of molecular events which regulate these Ca^{2+} pools. In intact tissues the interpretation of data has been complicated by the intervening plasma membrane and the complex distribution of bound and free Ca^{2+} in the cytosol and in subcellular organelles. This has led to considerable controversy about the source of the activating Ca^{2+} and about the site of the rate-limiting step for Ca^{2+} transport.

As a result of these difficulties, other techniques were developed which made the cytosolic space and its components more accessible to experimental manipulation. Two of these techniques have made important contributions to understanding the cytosolic events essential for contraction and relaxation. The first is the isolation of vesicles of sarcoplasmic reticulum (SR) or mitochondria separated from other cellular components by fractionation procedures. With this method the distribution of Ca^{2+} in subcellular organelles has been measured and the kinetics of Ca^{2+} regulation studied. However, the data obtained on SR vesicles have been controversial due to questions raised about suitable markers to assess purity of the fraction, sidedness and integrity of the vesicles, and the applicability of the results to *in vivo* function [22].

The second technique uses detergents such as saponin to increase smooth muscle plasma membrane permeability to ions and some larger-molecular-weight solutes. Saponin is believed to increase membrane permeability by complexing with cholesterol [9,20,21]. The plasma membrane has a relatively high fraction of cholesterol and is readily made hyperpermeable by saponin [2]. The membranes of the SR and inner membrane of the mitochondria contain proportionally more protein and less than 2% cholesterol [3,16,17]. Since these membranes remain functionally intact after saponin treatment, changes in force development

---

Marguerite A. Stout • Department of Physiology, UMDNJ–New Jersey Medical School, Newark, New Jersey 07103.

have been used as an indicator of Ca^{2+} uptake and release by the SR [7,8,11,19]; however, this method may not be able to separate the effects of manipulations on the contractile and regulatory proteins from those on the SR.

I have developed a procedure which combines the advantages of each of these techniques. The technique uses saponin to increase plasma membrane permeability, thereby permitting the chemical composition of the cytosolic compartment to be manipulated and enabling SR Ca^{2+} transport to be isolated and studied *in situ* with radioactive tracers.

Establishing Optimal Skinning Conditions

Saponin-induced membrane permeability depends on the size of the muscle strip, the saponin concentration, the length of exposure, and the temperature [6]. An essential step was, therefore, to identify conditions which would skin the plasma membrane but leave the contractile proteins and membranes of the subcellular organelles functionally intact.

Optimal skinning conditions were established for rat caudal artery [23]. Exposure of strips to 0.1 mg/ml saponin for 60 min gave rapidly rising force responses to increasing Ca^{2+} and ATP concentrations, indicating that these substances readily entered the cytosolic space. The magnitude of the tension produced by a combination of 10^{-5} M Ca^{2+} and 5 mM MgATP was equivalent to or greater than the KCl-induced tension in the intact strip. This response represents approximately 60–90% of the maximal force produced in the intact strip with a combination of norepinephrine (NE) and KCl. The skinned strips contracted in 10 mM caffeine, suggesting that SR Ca^{2+} uptake and release were functional. The caffeine-induced contraction could be prevented by increasing the EGTA buffer from 0.5 mM to 3 mM. Thus, an EGTA buffer concentration of 3 mM or higher was sufficient to control rapid changes in cytoplasmic free Ca^{2+} levels, thereby enabling the Ca^{2+} concentration at the membrane boundaries to be controlled during isotopic equilibration and washout.

^{45}Ca Distribution

Caudal artery strips consist of a multicompartmental system of extracellular and cytosolic space and cytoplasmic organelles arranged in series. Ca^{2+} distribution and exchange can be studied in this system provided the Ca^{2+} compartments can be separated and identified [5]. To achieve this, free and bound Ca^{2+} must be rapidly removed from the extracellular and cytosolic space while simultaneously retaining the Ca^{2+} sequestered by organelles; therefore, two desaturation solutions having different free Ca^{2+} concentration were tested for their effects on Ca^{2+} distribution and rate of washout.

The procedure for incubation and desaturation of skinned caudal artery strips has previously been reported in detail [23]. Strips of caudal artery of approximately 5-cm length and 2-mm width were incubated for 30 min in solutions containing 1 mM EGTA, 5.24×10^{-7} M free Ca^{2+} (0.62 mM total Ca^{2+}), 140 mM K^+-acetate, 2 mM Mg^{2+}-acetate, 2 mM ATP, and 20 mM imidazole (pH 6.9 ± 0.05). The strips were mounted above a fraction collector and superfused with a desaturation solution containing 5 mM EGTA and a free Ca^{2+} concentration of 4.24×10^{-9} M or of 1.08×10^{-6} M. The other solute concentrations were identical to the incubation solution.

From the residual activity remaining in the tissue and the radioactivity in each sample, curves representing Ca^{2+} content as a function of time were constructed (Fig. 1). For each experiment the total Ca^{2+} content amounts to about 525 μmoles/kg wet tissue. Initially, the

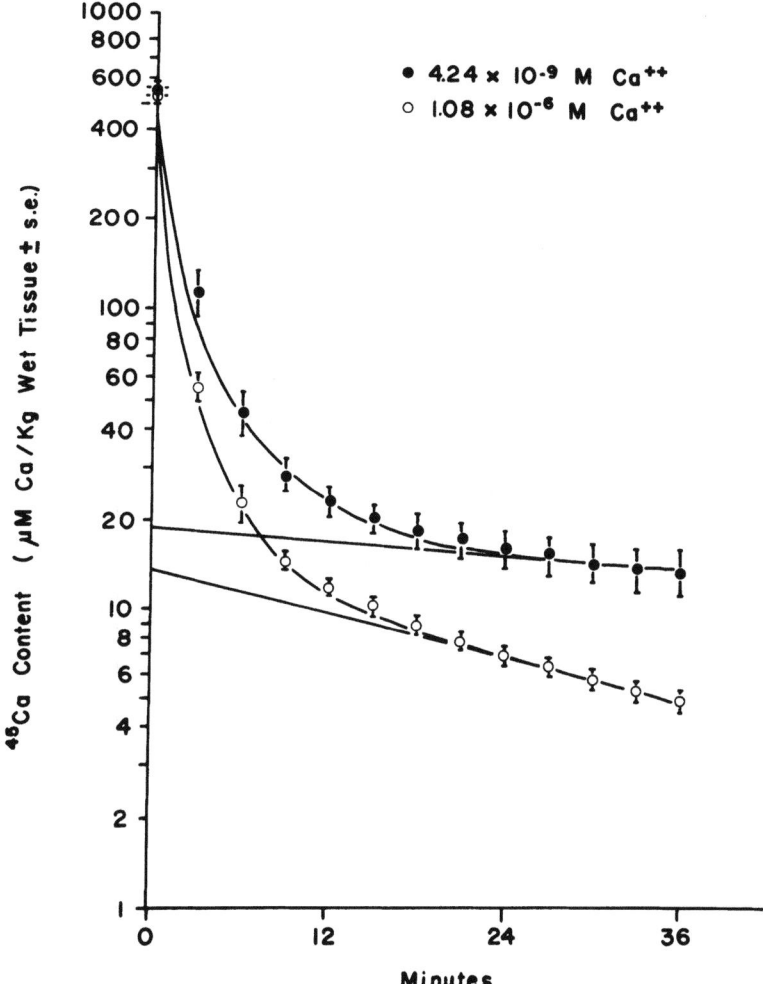

Figure 1. ^{45}Ca desaturation of skinned caudal artery strips. Strips were loaded with 5.24×10^{-7} M free Ca^{2+} and desaturated at 4.24×10^{-9} M (●) ($n = 3$) or 1.08×10^{-6} M (○) ($n = 3$) free Ca^{2+}. The solid curve connecting the experimental points was fit to the following equation: $y = Ae^{(k_1 t)} + Be^{-(k_2 t)} + Ce^{-(k_3 t)}$. The contents of each compartment (A, B, and C) and the rate constants (k) for each curve are given in Table I. The solid straight line represents the extrapolation of the Ca^{2+} content of the slow component to zero time. [Reproduced from *J. Pharmacol. Exp. Ther.* **255**:106, 1983.]

washout of Ca^{2+} continually changes at a rapid rate until about 24 min when it converges to a constant rate.

With curve peeling techniques [14], three Ca^{2+} compartments were identified. The Ca^{2+} content of each compartment and the rate coefficients are summarized in Table I. The Ca^{2+} content extrapolated for the fastest component under each condition amounts to 404 μmoles/kg. Calcium ion desaturates with a rate coefficient of the order of 1 min^{-1}. The Ca^{2+} content of the second (intermediate) component extrapolates to approximately 100 μmoles/kg. The rates of release range from 0.245 to 0.360 min^{-1} and correspond to time constants of 4.1 and 2.8. Thus, neither the Ca^{2+} content nor the rates of release of the fast or intermediate components are significantly changed by the free Ca^{2+} concentration of the desaturation solution.

The slow component has the smallest Ca^{2+} content of the three compartments. Since

Table I. Effect of Desaturation Solution on Ca^{2+} Content and Washout of Skinned Caudal Artery[a]

Desaturating free Ca^{2+}	Fast component		Intermediate component		Slow component	
	Ca^{2+} content (μmoles/kg wet wt)	RC^a (min^{-1})	Ca^{2+} content (μmoles/kg wet wt)	RC (min^{-1})	Ca^{2+} content (μmoles/kg wet wt)	RC (min^{-1})
4.24×10^{-9} M(3)[b]	404	1.05	110	0.245	19.0	0.0095
1.08×10^{-6} M(3)	404	1.40	100	0.360	13.6	0.0277

[a] RC, rate coefficient. [Adapted from Stout and Dieke [23] with permission from the American Society for Pharmacology and Experimental Therapeutics.]
[b] Number of tissues in parentheses.

the rate of release of the smallest component is significantly less than that of the fast and intermediate components, it can be easily distinguished from the other two. The rate coefficients, 0.0095 and 0.0277 $\min^{-1}$, correspond to time constants of 105 and 36 min, respectively. In contrast to the other two components, the rate of Ca^{2+} release of the slow component is sensitive to the free Ca^{2+} concentration in the desaturation solution. The markedly reduced rate of Ca^{2+} efflux in the presence of low free Ca^{2+} indicates that this solution produces the best separation of components.

Origin of the Fast and Intermediate Components

Next, the origin of each Ca^{2+} component was identified by comparing the Ca^{2+} content of skinned strips incubated at constant free Ca^{2+}, but varying EGTA buffer concentrations. The Ca^{2+} content of the fast and intermediate components are a linear function of the EGTA buffer concentrations (Fig. 2). Since nearly all of the Ca^{2+} in solution is complexed with EGTA, the compartmental contents are also directly proportional to the total Ca^{2+} concentration. The contents of these compartments are not a function of the free Ca^{2+} concentration. The fast and intermediate components, thus, behave as if the compartments are readily accessible to Ca^{2+} and CaEGTA; that is to say, extracellular compartments not confined by continuous and intact membranes. In contrast, the Ca^{2+} concentration of the slow component (not shown) remains constant with increasing EGTA concentration but varies as a function of the free Ca^{2+} concentration. This finding suggests that the slow component originates from an entirely different location than the fast or intermediate component.

Further evidence for the origin of the fast and intermediate components can be obtained from a comparison of the experimentally observed rates of diffusion with those predicated by a model approximating the geometry of the smooth muscle strips. The diffusion from muscle strips, therefore, was modeled as originating from an infinite sheet with diffusion through four sides. The diffusion out of this structure would be described by a series of exponential

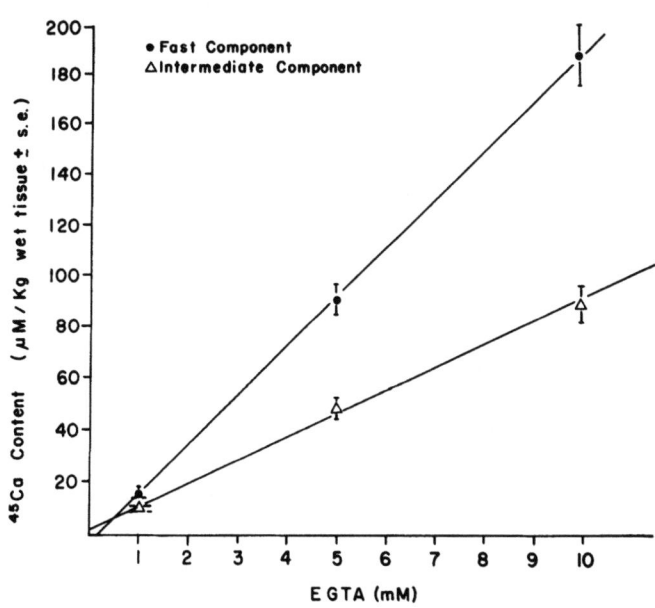

Figure 2. Ca^{2+} content of the fast and intermediate components as a function of EGTA. For each series of experiments tissues were loaded at constant free Ca^{2+} (1.12×10^{-8} M) but varying EGTA. The solid lines represent a linear regression fitted by the least-squares method. The correlation coefficients are 1.00 for each component. Each point represents the mean of four tissues. [Reproduced from *J. Pharmacol. Exp. Ther.* **255**:107, 1983.]

equations which rapidly converge to a single exponential. Using this model, the predicted rate of diffusion from the extracellular space would proceed with a rate constant of approximately 6.7 min^{-1}. By comparison, the rate coefficients obtained for the fast components of Ca^{2+} desaturation were significantly less than this value and were of the order of 1 min^{-1}. The difference between predicted and observed rates may be explained by the presence of impermeable or slightly permeable structures, such as connective tissue and smooth muscle cells, which increase the effective path length of diffusion. The rate coefficient of the fast component is, therefore, consistent with Ca^{2+} desaturated from the extracellular space.

The desaturation of the intermediate component is considerably slower than that of the fast component and does not appear to emanate from the extracellular space. Furthermore, the Ca^{2+} content of the intermediate component is approximately one-fourth that of the fast component. These observations are consistent with the Ca^{2+} originating from another compartment, the intracellular space, which is still confined by a hyperpermeable membrane. To test these assumptions, the distribution of $^{51}CrEDTA$ in intact and saponin skinned strips was compared (Stout, unpublished). The CrEDTA space measured in intact strips was approximately 55% and, after skinning, increased to about 75–80%. These data imply that the solute space of the cytoplasmic compartment comprises roughly 25% of the strip volume. The size of the water space for the intermediate component is consistent with the observation that the Ca^{2+} content of the intermediate component is approximately one-fourth that of the fast component (Table I).

Furthermore, the data indicate that, with saponin treatment, the cytosolic space is made readily accessible to CrEDTA (a molecule of $M_r \sim 300$). The permeability of the smooth muscle membrane to small molecules was further confirmed by exposing intact and skinned strips to 2% procion yellow for 10 min (Stout and Farnsworth, unpublished). In intact tissues the fluorescent dye (M_r 630) was confined to the extracellular space. By contrast, the dye distributed in both the extracellular and the cytoplasmic space of skinned strips. The distribution of procion yellow in skinned strips is a further confirmation that molecules as large as ATP, which has a similar molecular weight, enter the cytosolic space following saponin treatment.

Since small molecules enter the cytosolic space, estimates about the changes in membrane permeability following skinning were made using the rate coefficients of 0.24 min^{-1} (Table I) extrapolated for the intermediate component and the V/A ratio of 0.6×10^{-4} cm given by Jones for rat aorta [15]. The permeability of the skinned smooth muscle membrane to CaEGTA was determined to be 2.5×10^{-7} cm/sec. This value is an order of magnitude greater than the K^+ permeability of the intact smooth muscle membrane; however, it also indicates that the skinned plasma membrane still represents a partial barrier to solute diffusion. These results are consistent with the intermediate component having a slower rate of Ca^{2+} washout than the fast component.

Origin of the Slow Component

The Ca^{2+} content of the slow component was not directly proportional to the EGTA concentration described for the fast and intermediate components (Fig. 2). It was, however, a function of the free Ca^{2+} concentration, suggesting that it originated from a different compartment. The effects of free Ca^{2+} and ATP on the slow component are summarized in Fig. 3. Uptake of Ca^{2+} in solutions containing 10^{-8} M Ca^{2+} amounted to approximately 3 μmoles/kg wet tissue. The content increased to about 50 μmoles/kg at pCa 4.5. In the absence of Mg^{2+} and ATP, Ca^{2+} uptake is significantly reduced, but there is some increase

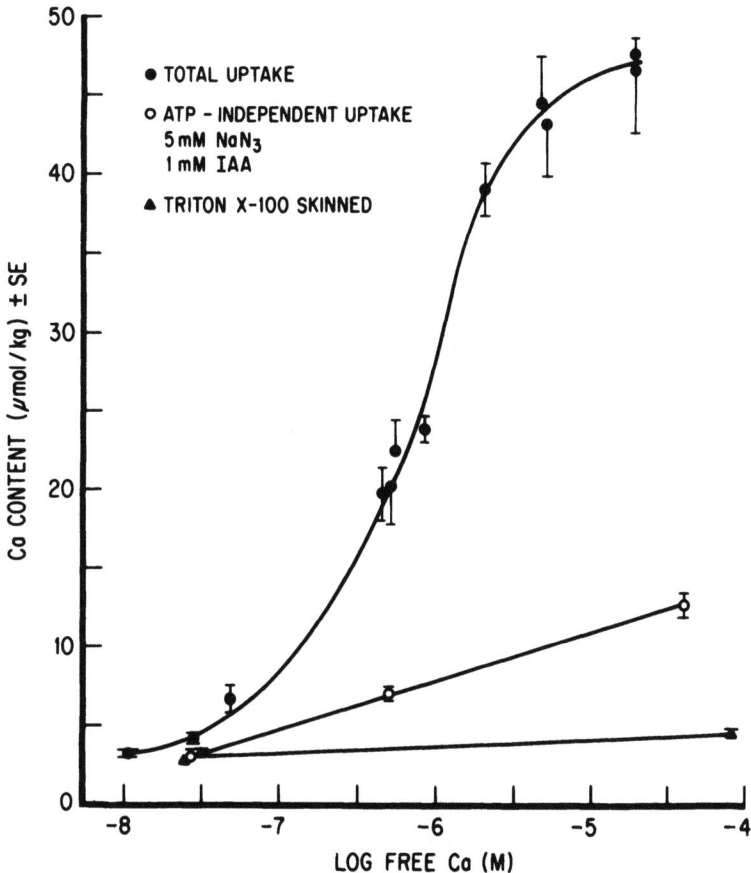

Figure 3. Factors affecting Ca^{2+} content in saponin skinned caudal artery. Control tissues (●) were loaded for 30 min in solutions containing 2 mM Mg^{2+} and 2 mM ATP. The ATP-independent component (○) was measured in the absence of Mg^{2+} and ATP. Na^+ azide and iodoacetate were added to incubation and desaturation solutions to prevent resynthesis of ATP. (▲), Ca^{2+} content of strips which were skinned for 60 min in 0.5% Triton X-100 but otherwise treated similarly to controls. Each point represents the mean of three or four strips.

in content as a function of the free Ca^{2+} concentration. It is not clear whether this increase represents Ca^{2+} binding, Ca^{2+} sequestration by an energy-independent mechanism, such as $Ca^{2+}-Ca^{2+}$ exchange, or incomplete depletion of ATP stores. When subcellular organelles are destroyed with Triton X-100, the Ca^{2+} content is reduced to approximately 3 μmoles/kg (Fig. 3). These data indicate that Ca^{2+} uptake by the slow component is Ca^{2+}-sensitive and energy-dependent. The absence of Ca^{2+} uptake in Triton suggests that the major fraction of the slow Ca^{2+} component originates in the mitochondria and SR. A minor fraction, amounting to about 3 μmoles/kg, may represent bound Ca^{2+}.

The components of energy-dependent Ca^{2+} sequestration represented by the slow component were examined in further detail. In isolated smooth muscle vesicles, azide inhibits nearly all Ca^{2+} sequestration by the mitochondria, without affecting the SR [10,12]. In skinned caudal artery, Ca^{2+} uptake measured in the presence of azide at pCa 6.3 and 5.7 does not differ statistically from control values (Fig. 4). Only at high Ca^{2+} concentrations (pCa 4.7 or lower) is Ca^{2+} uptake significantly reduced by azide. Further verification that Ca^{2+} uptake at pCa 8 to 5 represents SR sequestration was obtained with oxalate. Hess and Ford [12] have demonstrated that oxalate rapidly enters the SR but not the mitochondria of

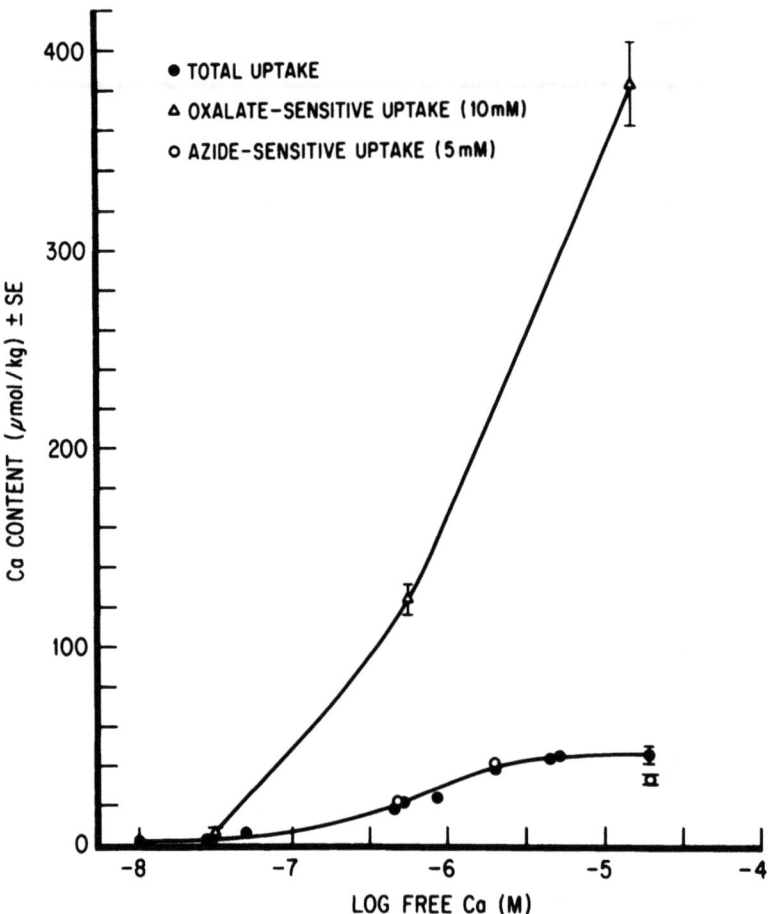

Figure 4. The effects of oxalate and azide on Ca^{2+} content of the slow component. The controls (●) are replotted from Fig. 3. The oxalate- (△) and azide- (○) sensitive components were measured in solution with Mg^{2+} and ATP concentrations equivalent to controls. Each point represents the mean of three or four strips.

smooth muscle vesicles. Skinned arterial strips were incubated in 10 mM oxalate-containing solutions in which free Ca^{2+}, Mg^{2+}, and MgATP were equivalent to controls. The upper tracing (Fig. 4) shows that Ca^{2+} uptake was increased over 500% at pCa 6.3 ($p \leq 0.001$) and over 700% ($p \leq 0.001$) at pCa 4.6. These studies provide convincing evidence that the SR sequestering mechanisms are functional and that in the absence of azide, the SR is mainly responsible for Ca^{2+} uptake above pCa 5.7.

The Ca^{2+} content of skinned strips was converted to units similar to those used for the Ca^{2+} concentration sequestered by the SR, in order to compare Ca^{2+} sequestration of vesicles with that of skinned caudal artery. Smooth muscle cells represent approximately 34% of the total weight or volume of the caudal artery strips. The volume of SR per volume of smooth muscle ranges from 2 to 5%, depending on the smooth muscle type [4]; thus, the SR of caudal artery would occupy 0.68–2% of the total volume or weight of the total strip. The Ca^{2+} content of skinned strips at pCa 6.3 amounts to 20.13 μmoles/kg wet tissue and represents Ca^{2+} sequestered by the SR with no mitochondrial contamination. Assuming that 1 mg of protein corresponds to 5 μl of SR volume, the Ca^{2+} content of smooth muscle SR ranges from 4.9 to 14.25 μM Ca^{2+}/g protein. This range has been compared to the Ca^{2+} content reported for vesicles isolated from other types of smooth muscle (Table II). The

Table II. Comparison of Ca^{2+} Content of Sarcoplasmic Reticulum in Different Preparations

Tissue	Ca^{2+} content (μmoles/g protein)	pCa	ATP, Mg (mM)	Temp. (°C)	Reference
Microsomal fraction, guinea pig ileum	25	4.7	3, 5	37	Godfraind (10)
Microsomal fraction, guinea pig taenia coli	14	6.3	5, 5	25	Raeymaekers (18)
Microsomal fraction, rabbit myometrium	20	6.0	5, 5	—	Batra (1)
Saponin skinned smooth muscle, rat caudal artery	≈5–15	6.3	2, 2	20	Stout and Diecke (23)
Intact smooth muscle, rat caudal artery	≈30–90	≈7.0	—	37	Stout (unpublished)

upper limit of the Ca^{2+} content measured for skinned caudal artery agrees favorably with that reported for vesicles isolated from other smooth muscle types in spite of differences in temperature, free Ca^{2+}, MgATP, and the absence of an ATP-regenerating system.

Because of the variation in the amount of SR in various types of smooth muscle, the Ca^{2+} content of intact and skinned caudal artery strips was compared. The intracellular Ca^{2+} content of intact strips desaturated in La^{3+}-containing solution extrapolates to 260 μmoles/kg wet tissue (Stout, unpublished). This value converts to about 30–90 μmoles/g protein. The skinned preparation sequesters only one-sixth the amount of Ca^{2+} estimated for the SR of the intact preparation. Since the uptake for intact strips was measured at 37°C, the increased temperature level may account for a large part of the difference.

^{45}Ca Release

A final measure of the functional integrity of the preparation was to demonstrate that the sequestered Ca^{2+} could be released with Ca^{2+} agonists. Test solutions were applied during the slow component of Ca^{2+} washout when the rate coefficients had converged. Figure 5A shows a small but statistically significant release of Ca^{2+} induced by caffeine. In Fig. 5B a substantial Ca^{2+}-induced Ca^{2+} release is illustrated. The magnitude of the efflux in each of these cases depends on the Ca^{2+} concentration of the loading medium and the concentration of caffeine or Ca^{2+} in the test solution. Epinephrine had no effect on Ca^{2+} efflux (Fig. 5C). Norepinephrine did not cause a contraction in skinned guinea pig mesenteric artery [13]. This

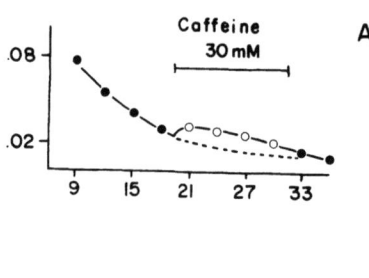
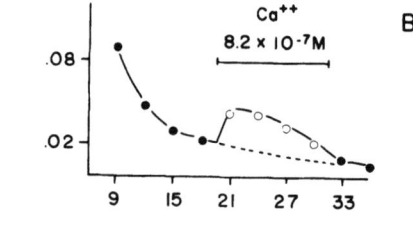
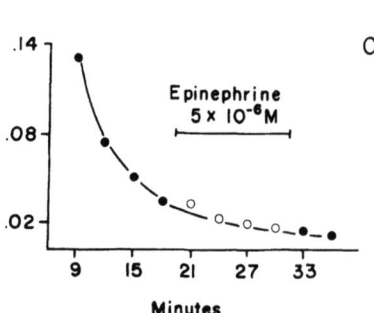

Figure 5. Effect of Ca^{2+}-releasing agents on ^{45}Ca efflux from skinned caudal artery. Tissues were loaded at 5.5×10^{-7} M Ca^{2+} and desaturated in 4.7×10^{-9} M Ca^{2+}. The first 9 min of desaturation is not shown. The horizontal bars and open circles represent the time interval the Ca^{2+} agonist was applied. Each point represents the mean of four tissues. Standard errors are 10% or less. [Reproduced from *J. Pharmacol. Exp. Ther.* **255**:108, 1983.]

contrasts with the reported contractile response to NE in saponin skinned mesenteric artery of rabbit [11]. Epinephrine responses have been observed in caudal artery strips skinned with low concentrations of saponin. The difference, therefore, may be due to incompletely skinned muscle strips.

Summary and Conclusions

A method has been developed which enables Ca^{2+} uptake and release from the SR to be isolated and studied in skinned vascular smooth muscle strips. Under optimal skinning conditions, force responses to increasing Ca^{2+} and ATP indicated that the contractile and regulatory proteins were intact and that molecules as large as ATP entered the cytosolic space. Caffeine-induced contractions demonstrated that the Ca^{2+}-sequestering system of the SR was functional.

Although the plasma membranes of skinned arterial tissues are hyperpermeable, the caudal artery strip still represents a multicompartmental system for tracer measurements. Three components were identified in this system and the compartmental Ca^{2+} distribution and rate of release for each were analyzed. The data indicate that the fast component represents the extracellular space, while the intermediate component corresponds in every aspect to the cytosolic space made accessible to solutes and dyes having molecular weights of at least 600. The slow component represents the Ca^{2+}-sequestering compartment. The Ca^{2+} uptake is dependent on ATP and is stimulated by oxalate. The energy-dependent Ca^{2+} uptake is destroyed by Triton X-100. The sequestered Ca^{2+} can be released by Ca^{2+} and by caffeine. These characteristics are consistent with Ca^{2+} sequestration and releases described for the SR of other muscle preparations. On the basis of these observations, this appears to be a suitable preparation to study SR Ca^{2+} uptake and release as well as drug interactions in smooth muscle.

ACKNOWLEDGMENT. The author wishes to thank Dr. F. P. J. Diecke for helpful suggestions on the manuscript.

References

1. Batra, S. The importance of calcium binding by subcellular components of smooth muscle in excitation contraction coupling. In: *Excitation–Contraction Coupling in Smooth Muscle*, R. Casteels, T. Godfraind, and J. C. Ruegg, eds., Amsterdam, Elsevier/North-Holland, 1977, pp. 225–232.
2. Colbeau, A.; Nachbaur, J.; Vignais, P. M. Enzymic characterization and lipid composition of rat liver subcellular membranes. *Biochim. Biophys. Acta* **249**:462–492, 1971.
3. Comte, J.; Maïsterrena, B.; Gautheron, D. C. Lipid composition and protein profiles of outer and inner membranes from pig heart mitochondria: Comparison with microsomes. *Biochim. Biophys. Acta* **419**:271–284, 1976.
4. Devine, C. E.; Somlyo, A. V.; Somlyo, A. P. Sarcoplasmic reticulum and excitation–contraction coupling in mammalian smooth muscle. *J. Cell Biol.* **52**:690–718, 1972.
5. Diecke, F. P. J.; Stout, M. A. Calcium distribution and transport in myelinated nerve. *J. Comp. Physiol.* **141**:319–326, 1981.
6. Endo, M.; Iino, M. Specific perforation of muscle cell membranes with preserved SR functions by saponin treatment. *J. Muscle Res. Cell Motil.* **1**:89–100, 1980.
7. Endo, M.; Kitazawa, T.; Yagi, S.; Iino, M.; Kakuta, Y. Some properties of chemically skinned smooth muscle fibers. In: *Excitation–Contraction Coupling in Smooth Muscle*, R. Casteels, T. Godfraind, and J. C. Ruegg, eds., Amsterdam, Elsevier/North-Holland, 1977, pp. 199–209.

8. Endo, M.; Yagi, S.; Iino, M. Tension–pCa relation and sarcoplasmic reticulum responses in chemically skinned smooth muscle fibers. *Fed. Proc.* **41:**2245–2250, 1982.
9. Glauert, A. M.; Dingle, J. T.; Lucy, J. A. Action of saponin on biological cell membranes. *Nature (London)* **196:**952–955, 1962.
10. Godfraind, T.; Sturbois, X.; Verbeke, N. Calcium incorporation by smooth muscle microsomes. *Biochim. Biophys. Acta* **455:**254–268, 1976.
11. Haeusler, G.; Richards, J. G.; Thorens, S. Noradrenaline contractions in rabbit mesenteric arteries skinned with saponin. *J. Physiol. (London)* **321:**537–556, 1981.
12. Hess, M. L.; Ford, G. D. Calcium accumulation by subcellular fractions from vascular smooth muscle. *J. Mol. Cell. Cardiol.* **6:**275–282, 1974.
13. Itoh, T.; Suzuki, H.; Kuriyama, H. Effects of sodium depletion on contractions evoked in intact and skinned muscles of the guinea-pig mesenteric artery. *Jpn. J. Physiol.* **31:**831–847, 1981.
14. Jacquez, J. A. *Compartmental Analysis in Biology and Medicine: Kinetics of Distribution of Tracer-Labeled Materials,* New York, American Elsevier, 1972.
15. Jones, A. W. Content and fluxes of electrolytes. In: *Handbook of Physiology,* Section 2, *The Cardiovascular System,* Volume 2, *Vascular Smooth Muscle,* D. F. Bohr, A. P. Somlyo, and H. V. Sparks, Jr., eds., Baltimore, Williams & Wilkins, 1980, pp. 253–299.
16. Levy, M.; Toury, R.; Sauner, M.-T.; Andre, J. Recent findings on the biochemical and enzymatic composition of the two isolated mitochondrial membranes in relation to their structure. In: *Mitochondria Structure and Function,* L. Ernster and Z. Drahota, eds., New York, Academic Press, 1969, pp. 33–42.
17. Martonosi, A. Sarcoplasmic reticulum. V. The structure of sarcoplasmic reticulum membranes. *Biochim. Biophys. Acta* **150:**694–704, 1968.
18. Raeymaekers, L.; Wuytack, F.; Batra, S.; Casteels, R. A comparative study of the calcium accumulation by mitochondria and microsomes isolated from the smooth muscle of the guinea-pig taenia coli. *Pfluegers Arch.* **368:**217–223, 1977.
19. Saida, K.; Nonomura, Y. Characteristics of Ca^{+2}- and Mg^{+2}-induced tensions development in chemically skinned smooth muscle fibers. *J. Gen. Physiol.* **72:**1–14, 1978.
20. Seeman, P. Transient holes in the erythrocyte membrane during hypotonic hemolysis and stable holes in the membrane after lysis by saponin and lysolecithin. *J. Cell Biol.* **32:**55–70, 1967.
21. Seeman, P.; Cheng, D.; Iles, G. H. Structure of membrane holes in osmotic and saponin hemolysis. *J. Cell Biol.* **56:**519–527, 1973.
22. Sloane, B. F. Isolated membranes and organelles from vascular smooth muscle. In: *Handbook of Physiology,* Section 2, *The Cardiovascular System,* Volume 2, *Vascular Smooth Muscle,* D. F. Bohr, A. P. Somlyo, and H. V. Sparks, Jr., Baltimore, Williams & Wilkins, 1980, pp. 121–132.
23. Stout, M. A.; Diecke, F. P. J. ^{45}Ca distribution and transport in saponin skinned vascular smooth muscle. *J. Pharmacol. Exp. Ther.* **255:**102–111, 1983.

45

Cellular and Subcellular Approaches to the Mechanism of Action of Calcium Antagonists

T. Godfraind

The concept of Ca^{2+} antagonism has been developed on the basis of two different and parallel experimental approaches, one being mainly supported by studies on smooth muscle [14,18], the other on cardiac muscle [6]. In previous cardiac muscle studies, not all of the effects of Ca^{2+} antagonists could be attributed to inhibition of Ca^{2+} entry, as some effects were still observed after Ca^{2+} withdrawal [7]. The present chapter will summarize the development of studies in my laboratory which deal with the mechanisms of action of Ca^{2+} antagonists in smooth muscle.

In the 1960s, we analyzed the pharmacological action of some drugs acting as polyvalent antagonists, mainly lidoflazine used as an antianginal drug and cinnarizine, at that time considered an antihistiminic. Lidoflazine and cinnarizine inhibit smooth muscle contraction evoked by several agonists, including norepinephrine, angiotensin, acetylcholine, and vasopressin. Because the active concentration was similar for these various agonists, we proposed that lidoflazine and cinnarizine interfered with a common mechanism activated by various spasmogens [18]. Cinnarizine inhibited smooth muscle contraction in the presence of Ca^{2+}, but was devoid of any effect on contraction evoked in the absence of Ca^{2+}. This suggested that cinnarizine does not act directly on the contractile machinery but rather on the translocation of Ca^{2+} evoked by stimuli. This conclusion was supported by experiments showing that cinnarizine inhibits in a dose-dependent manner smooth muscle contraction evoked by a cumulative increase of Ca^{2+} in a K^+-depolarizing solution [14] (Fig. 1).

In the early 1960s, Schild and colleagues [11] showed that depolarization of smooth muscle increases membrane permeability to extracellular Ca^{2+}. A most convenient way to depolarize smooth muscle is to immerse the preparation in K^+-rich solution where increasing Ca^{2+} concentration results in a proportional increase in smooth muscle tone. After pretreatment with cinnarizine, the response to added Ca^{2+} is reduced in a way that is dependent on the preparations used. In rabbit arteries, the maximum response to calcium is depressed in a dose-dependent manner (Fig. 1A). However, with rat isolated aorta, the

---

T. Godfraind • Laboratory of Pharmacodynamics and Pharmacology, Catholic University of Louvain, B-1200 Brussels, Belgium.

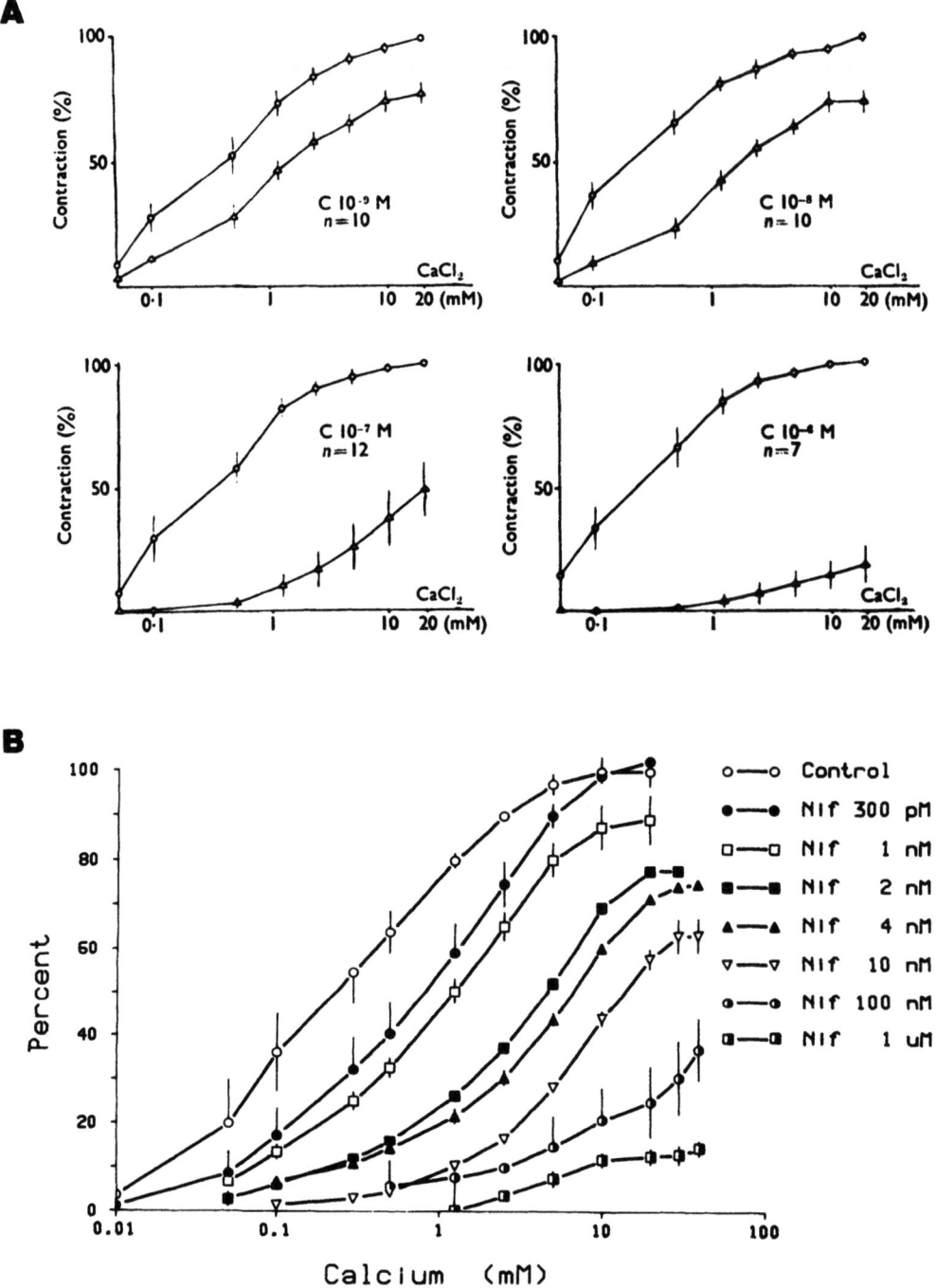

Figure 1. Illustration of the action of Ca^{2+} antagonists on Ca^{2+}-evoked contraction in depolarized vascular smooth muscle showing typical families of dose–effect curves that can be observed in two different tissue preparations. The difference is not due to the drug used but to the tissue used: (A) Cumulative dose–effect curves evoked by Ca^{2+} in K^+-depolarized rabbit mesenteric arteries before (○) and after (△) the addition of cinnarizine at concentrations indicated on the graphs. [From Ref. 14, with permission of the *British Journal of Pharmacology*.] (B) Cumulative dose–effect curves evoked by Ca^{2+} in K^+-depolarized rat aorta before and after the addition of nifedipine at concentrations indicated on the graph. [Modified from Ref. 12.]

inhibitory action of cinnarizine may be partly surmounted and Ca^{2+} dose–effect curves are shifted to the right. The same is true with other Ca^{2+} antagonists such as nifedipine (Fig. 1B).

Because such a family of dose–effect curves was similar to the one observed when studying specific neurotransmitter antagonists (e.g., atropine), the term *Ca^{2+} antagonist* was proposed to characterize cinnarizine and other drugs having the same action on Ca^{2+}-induced contractions [18]. From their experimental work, Godfraind and Kaba [14] concluded that the increase in cell membrane permeability to Ca^{2+} that was evoked by K^+ depolarization was reduced by cinnarizine. They observed that cinnarizine reduces the contraction evoked by noradrenaline, but that a part of this contraction was resistant to the Ca^{2+} antagonists. This resistant portion of the contraction was of the same magnitude as that evoked in the absence of extracellular Ca^{2+} and unaffected by cinnarizine. A similar observation was made with other Ca^{2+} antagonists such as flunarizine, D600, nifedipine [17], and nisoldipine (Godfraind, unpublished). However, diltiazem, another Ca^{2+} antagonist, can completely block contraction evoked by the noradrenaline-induced release of intracellular Ca^{2+} pools [27], an indication of possible important differences between Ca^{2+} antagonists. Therefore, understanding the mode of action of those drugs requires characterization of processes responsible for cellular Ca^{2+} homeostasis.

Subcellular Characterization of Cellular Ca^{2+} Regulator Processes

In a resting muscle, the cytoplasmic free Ca^{2+} concentration is not more than 0.1 μM, so that the ratio of extracellular to intracellular Ca^{2+} exceeds 10^4. The Ca^{2+} electrochemical

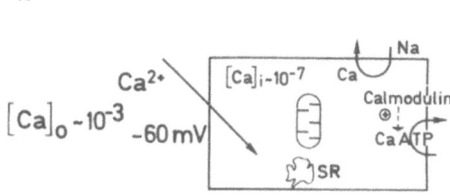

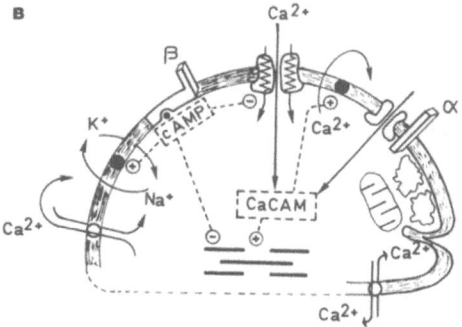

Figure 2. Schematic representation describing cellular processes controlling cellular Ca^{2+} metabolism in smooth muscle at rest (A) and during stimulation (B).

(A) Although the Ca^{2+} electrochemical gradient is oriented inward, the low intracellular Ca^{2+} concentration is maintained by the low permeability of the cell membrane to Ca^{2+} and by the activity of Ca^{2+} buffering systems described in the text. The various mechanisms illustrated here are insensitive to specific Ca^{2+} entry blockers.

(B) Illustration of the various processes activated by smooth muscle stimuli. From left to right: (1) Na^+-Ca^{2+} exchange mechanism. (2) Na^+-K^+ pump. (3) β-Adrenoceptor coupled to adenylate cyclase responsible for the cyclization of ATP into cAMP; the latter has a stimulatory action on the Na^+-K^+ pump, a negative action on the contractile machinery, and has been proposed to impair the opening of the potential-operated channels. (4) Potential-operated Ca^{2+} channels; their opening is dependent on the level of the membrane potential. Once they are open, Ca^{2+} entry occurs as a function of the Ca^{2+} electrochemical gradient. Inside the cell, Ca^{2+} forms a complex with calmodulin (CaCAM); this complex activates the contractile machinery and the Ca^{2+} extrusion pumps. (5) ATP-dependent Ca^{2+} pump stimulated by calmodulin. (6) α-Adrenoceptor associated with receptor-operated channel and with intracellular Ca^{2+} stores, most likely the endoplasmic reticulum. Once the channel is open, Ca^{2+} entry is also passive (see 4). (7) The association formed by mitochondria and endoplasmic reticulum, responsible for intracellular Ca^{2+} buffering. (8) $Ca^{2+}-Ca^{2+}$ exchange mechanism occurring through a leak channel insensitive to Ca^{2+} entry blockers. Other explanations are given in the text.

gradient is oriented inward (Fig. 2), so that Ca^{2+} tends to enter the cell. The normally low intracellular Ca^{2+} concentration is due to a low permeability of the polarized plasma membrane for Ca^{2+} and to cellular Ca^{2+} buffering processes. Cytoplasmic proteins such as parvalbumins have Ca^{2+}-binding properties. Mitochondria are able to take up large amounts of Ca^{2+} and may serve as Ca^{2+} sinks; they can also rapidly release Ca^{2+} in the presence of metabolic inhibitors or Na^+. Although these properties permit mitochondria to regulate cellular Ca^{2+} metabolism, a role for mitochondria in regulation of muscle tension has not been established.

Endoplasmic (or sarcoplasmic) reticulum (ER) appears to be more efficient than mitochondria for normal regulation of Ca^{2+} levels in the sarcoplasm. This is illustrated by experiments performed in visceral smooth muscle showing that ER vesicles have a higher affinity for Ca^{2+} than do mitochondria, whereas the latter organelles are able to take up much more Ca^{2+}. Intracellular binding of Ca^{2+} may, therefore, rapidly buffer changes in free Ca^{2+}. The constancy of the cellular Ca^{2+} content over a protracted period requires that Ca^{2+} efflux exactly balanced Ca^{2+} influx. This can only be achieved by a Ca^{2+} pump located in the plasma membrane. At least two types of mechanism are feasible: a Ca^{2+} pump that actively extrudes Ca^{2+} at the expense of ATP hydrolysis, and a Na^+–Ca^{2+} exchange process that utilizes the Na^+ electrochemical gradient [8].

To characterize plasmalemmal Ca^{2+} transport systems in smooth muscle, it would be desirable to use highly purified plasma membranes, and efforts have been made to obtain such a preparation [20,23]. An alternative approach is to investigate Ca^{2+} transport activities in relatively crude microsomal fractions containing a fair proportion of the plasma membrane present in the homogenate; the subcellular location of those activities will be established by analytical density gradient centrifugation.

The selective effect of digitonin on plasmalemmal elements has proven to be particularly useful in this type of study [26,33,34]. Digitonin forms a highly insoluble, equimolecular complex with cholesterol, and this reaction has long been used for cholesterol determinations. When added at low concentration to a subcellular fraction from smooth muscle, this compound binds preferentially to the plasmalemmal elements that contain cholesterol and markedly increases their equilibrium density in sucrose gradients, whereas other types of membranes such as ER or mitochondrial membranes (which are poor in cholesterol) are not affected [1,33,34]. According to studies on rat liver fractions, the densities of the membrane entities derived from the Golgi complex are also modified by this method, but to a lesser extent than those from plasma membranes. The marked density shift of plasma membranes at least partly reflects the binding of large amounts of digitonin to these cholesterol-rich membranes. Mitochondrial or ER membranes are not shifted, presumably because they contain little, if any, cholesterol. Summarized below are the main results obtained with vascular muscle from rat aorta [26].

When microsomal fractions from rat aorta were incubated in a ^{45}Ca-uptake medium devoid of oxalate, ATP-dependent ^{45}Ca accumulation plateaued at 10–15 nmoles/mg protein after 15–20 min. In fractions previously washed with 1 mM EGTA, Ca^{2+} transport activity was reduced by about 50%. Addition of calmodulin more than restored the original level of ^{45}Ca uptake. Moreover, EGTA treatment, which presumably removes endogenous calmodulin, decreased the apparent affinity of the transport system for Ca^{2+} (half-maximal uptake at 50 μM Ca^{2+}), and upon addition of calmodulin, this system regained a Ca^{2+} affinity similar to that of untreated microsomes (half-maximal uptake at 10 μM Ca^{2+}).

The subcellular localization of the calmodulin-sensitive Ca^{2+} pump was investigated by density gradient centrifugation of EGTA-treated microsomal fractions (Fig. 3). The distribu-

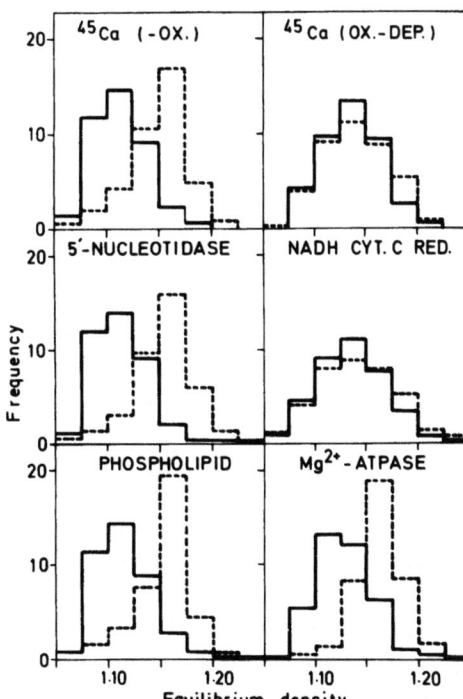

Figure 3. Density distribution patterns of enzymes and ^{45}Ca uptake activity in untreated (solid line) and digitonin-treated (broken line) microsomal samples from smooth muscle. The microsomal preparation had been treated with EGTA before addition of digitonin (0.25 mg/mg protein) [see 33].

tion profile of the ^{45}Ca-uptake activity measured in the presence of calmodulin was virtually superimposable on that of 5'-nucleotidase, a plasmalemmal enzyme marker; and both distribution profiles were shifted to the same extent by digitonin. A similar behavior was observed for both the ^{45}Ca-uptake activity measured in the absence of calmodulin and the major part of the protein. In contrast, two constituents of ER, the enzyme NADH-cytochrome c reductase and sulfatase C, showed distinct density distribution patterns that were hardly modified after addition of digitonin (Fig. 3). Thus, in the microsomal fraction from vascular smooth muscle, the bulk of the ATP-dependent ^{45}Ca uptake seems to be located in plasmalemmal vesicles. Also, marked calmodulin dependency, such as demonstrated in microsomal fractions from rat aorta, seems to be a distinctive property of plasmalemmal Ca^{2+} pumps [4,19]. It is likely that the ATP-dependent Ca^{2+} transport identified in smooth muscle plasma membrane corresponds to the still hypothetical Ca^{2+} pump of the plasma membrane.

Na^+-Ca^{2+} exchange is likely to occur through a membrane carrier process; it is not sensitive to calmodulin and is of variable importance among diverse cell types. A comparison of the relative efficacy of the Na^+-Ca^{2+} exchange mechanism and of the $Ca^{2+}-$ATP pump to transport Ca^{2+} has been made in plasma membrane vesicles prepared from heart, arteries, and intestinal smooth muscle [25] (Fig. 4). It is obvious that a Na^+-Ca^{2+} exchange process is most efficient in heart, whereas a $Ca^{2+}-$ATP pump is of primary importance in smooth muscle.

When Ca^{2+} antagonists are added to test tubes, only calmodulin antagonists (Table I) show a prominent action on the process activated by the Ca^{2+}-calmodulin complex. This indicates that they do not interfere with cellular Ca^{2+} buffering processes but, instead, with processes responsible for Ca^{2+} entry. Therefore, contraction and Ca^{2+} fluxes were compared in intact tissues.

Table I. Main Calmodulin Antagonists

Phenothiazines: trifluoperazine, chlorpromazine
Naphthalene derivatives: W7
Local anesthetics: dibucaine
Dopamine antagonists: pimozide, haloperidol
R 24571

Action of Ca^{2+} Antagonists on Contraction and Ca^{2+} Fluxes

Direct estimation of transmembrane Ca^{2+} fluxes in smooth muscle is obscured by exchange occurring at extracellular sites. Several attempts have been made to measure changes in the level of cytoplasmic Ca^{2+} during excitation of smooth muscle; the most successful have been made with the use of La^{3+} [9,30]. La^{3+} replaces Ca^{2+} at superficial binding sites, and prevents Ca^{2+} from entering the cell. Because La^{3+} blocks transmembrane fluxes of Ca^{2+}, the Ca^{2+} content of a muscle washed in a La^{3+} solution could provide an estimate of cellular Ca^{2+}. Evidence that La^{3+} blocks the binding of Ca^{2+} to superficial sites, as well as Ca^{2+} influx, is based on the observation that the Ca^{2+} diffusion space in the presence of La^{3+} is similar to the insulin diffusion space [30]. However, La^{3+} does not completely block Ca^{2+} efflux from several types of smooth muscle.

Fluxes of Ca^{2+} across the smooth muscle cell membrane are estimated by measuring Ca^{2+} turnover in the La^{3+}-resistant Ca^{2+} fraction. This fraction is defined as that portion not displaced when the tissue is washed in 50 mM $LaCl_3$. In this solution most of the Ca^{2+} displacement occurs during the first 5 min, and Ca^{2+} efflux from intracellular stores is apparently delayed, allowing measurement of Ca^{2+} transmembrane fluxes. Such studies

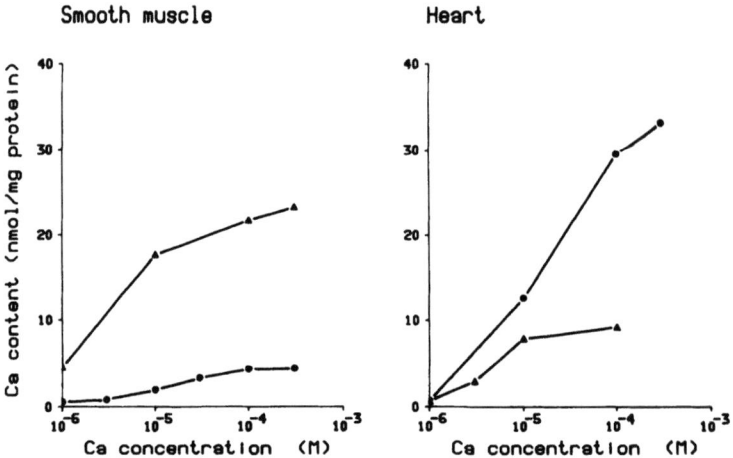

Figure 4. Na^+-induced uptake (●) and ATP-dependent Ca^{2+} uptake (▲) activities of intestinal smooth muscle microsomes (left panel) and cardiac plasma membrane-rich fraction (right panel) as a function of free Ca^{2+} concentration in the medium. Na^+-induced Ca^{2+} uptake was measured 3 min after diluting Na^+-loaded veisicles into Na^+-free 20 mM Tris maleate buffer (pH 7.4) containing $^{45}CaCl_2$ and 150 mM KCl. ATP-dependent Ca^{2+} uptake was measured after 15-min incubation of the membrane in 20 mM Tris maleate buffer (pH 7.4) with 3 mM ATP, 5 mM $MgCl_2$, 100 mM KCl, and $^{45}CaCl_2$. Free Ca^{2+} concentration was controlled by addition of various concentrations of EGTA. Each value is a mean of at least four determinations. [Adapted with modifications from Ref. 25.]

have shown that 10 μM norepinephrine (which evokes a maximum contraction) produces a large increase in the rate of Ca^{2+} influx. There is no net gain in tissue Ca^{2+}, and Ca^{2+} efflux increases in a similar manner. The increase in Ca^{2+} uptake, which is due to α-adrenoceptor activation, is concentration-dependent. In Ca^{2+}-free solution, norepinephrine is still able to evoke a contraction that is most likely supported by Ca^{2+} release from intracellular Ca^{2+} stores [3,9,16].

The observation of a contractile response to increasing extracellular Ca^{2+} concentrations in a muscle depolarized in high-K^+ solution or stimulated by an agonist is a strong indication that, under such conditions, there is an increase in membrane permeability to Ca^{2+}. The depression of this Ca^{2+}-induced effect by a drug may be interpreted as a reduction of membrane permeability to Ca^{2+}. However, because drugs may have more than one action and could also interfere with several cellular processes, more direct evidence is required. This is only possible by measuring Ca^{2+} fluxes or currents and by quantitative demonstrations of a similar depression of Ca^{2+} entry and cellular response. In other words, to attribute the depression of a cellular response to an agonist or a depolarizing stimulus to a blockade of Ca^{2+} entry, dose–inhibition curves for Ca^{2+} entry and response must be superimposed [10].

Direct measurements of ^{45}Ca fluxes have revealed that activation of α-adrenoceptors or K^+ depolarization evokes a fast increase in Ca^{2+} entry into smooth muscle cells, indicating that these stimuli open Ca^{2+} channels in the cell membrane [9,29]. Because of the disparate sensitivity of these two processes to Ca^{2+} antagonists, receptor-operated channels and potential-operated channels may have different sensitivities to blockade by Ca^{2+} antagonists [2,5]. Indeed, the I_{50} (concentration of drug inhibiting 50% of stimulus-evoked calcium entry) and IC_{50} values (concentration of drug inhibiting 50% of stimulus-evoked contraction supported by extracellular calcium) are similar (see Table III)—a confirmation that blockade of contraction may be attributable to a depression of Ca^{2+} entry [10,12]. Studies of ^{45}Ca fluxes have also shown that in Ca^{2+}-free solution, an agonist such as norepinephrine releases Ca^{2+} from intracellular stores, this Ca^{2+} release is insensitive to blockade by diphenylpiperazines and dihydropyridines as well as to verapamil and D600. This is not the case for chlorpromazine, papaverine, and high doses of diltiazem. Entry of Ca^{2+} occurring at a slow rate in resting preparations is not greatly affected by those drugs. Ca^{2+} entry evoked by depolarization appears to be more sensitive to Ca^{2+} channel blockers than does entry associated with receptor–response coupling (Table II). Such experimental evidence supports the hypothesis of potential-operated and receptor-operated Ca^{2+} channels [2,5,10,32]. The interaction between Ca^{2+} entry blockers and potential-operated channels shows use-dependency, i.e., the degree of Ca^{2+} entry blockade for a given concentration of the drug increases with the duration of the depolarization up to a steady state. This use-dependency, which has been observed with radiochemical and electrophysiological techniques [10,24], could be due to an increase in affinity of the open channel for the blocker; it has not been observed when Ca^{2+} channel opening is operated by receptor activation. It must be emphasized that use-dependency has been determined after a preincubation time sufficient to reach an equilibrium between medium and tissue concentration. The rate at which it reaches its steady state depends on the drug studied and follows the sequence: nifedipine, D600, diltiazem, flunarizine [17].

Thus, for a given drug and tissue, Ca^{2+} gating mechanisms show differences in sensitivity. Also, for a given gating mechanism (agonist or depolarization activation) and Ca^{2+} entry blocker, differences in sensitivity to Ca^{2+} entry blockers may be observed among different tissues. This tissue selectivity has to be examined with reference to histological and anatomical classifications.

Comparison of the concentrations of Ca^{2+} antagonists required to reduce the contrac-

Table II. Major Ca^{2+} Antagonists and Their Therapeutic Applications

---

Specific Ca^{2+} entry blockers
  Angina
    Lidoflazine, verapamil, diltiazem, nifedipine
  Hypertension
    Verapamil, nifedipine, diltiazem
  Arrhythmia
    Verapamil
  Peripheral vascular disorders
    Cinnarizine
  Cerebral ischemia and postapoplectic states
    Flunarizine, nimodipine
  Under experimental study
    Nisoldipine, niludipine, nicardipine, tiapamil
Ca^{2+} entry blockers with an additional action
  On catecholamine receptors
    Chlorpromazine
    Rauwolscine and corynanthine
    Pimozide
  On fast Na^+ channel
    Angina: bepridil, perhexiline, prenylamine, fendiline, bencyclane
  On Ca^{2+} calmodulin
    Felodipine
  On phosphodiesterase
    Papaverine
    Amrinone

---

tion of myocardium and smooth muscle by 50% reveals that the myocardium is much less sensitive. Furthermore, the ratio of activity between heart and smooth muscle varies among Ca^{2+} entry blockers; this ratio is equal to 3–6 for verapamil and to 210 for flunarizine [31]. There are also variations in sensitivity to Ca^{2+} antagonists among smooth muscles, as is shown by the resistance to the myogenic contraction of rat portal vein to flunarizine and its high sensitivity to D600 and nifedipine [31]. A comparison between flunarizine and nifedipine may help to illustrate some aspects of specificity among arteries. Figure 5 illus-

Table III. Comparison of the Concentrations of Ca^{2+} Entry Blockers Required to Reduce to 50% Agonist-Evoked ^{45}Ca Influx ($I_{C_{50}}$) and Contraction (IC_{50})

Drug	Tissue	Agonist	I_{50} (M)	IC_{50} (M)
Flunarizine[a]	Rat aorta	K^+	3.7×10^{-8}	1.9×10^{-8}
		Norepinephrine	7×10^{-7}	5×10^{-7}
	Rat mesenteric artery	K^+		2×10^{-9}
		Norepinephrine	5×10^{-8}	2.4×10^{-8}
		$PGF_{2\alpha}$	1.3×10^{-7}	1.4×10^{-7}
Nifedipine[b]	Rat aorta	K^+	1.6×10^{-9}	1.3×10^{-9}
		Norepinephrine	1.9×10^{-8}	1.7×10^{-8}
Diltiazem[c]	Rabbit aorta	K^+	5.0×10^{-7}	3.0×10^{-7}

[a]Data from Refs. 13 and 17.
[b]Data from Ref. 12.
[c]Data from Ref. 30.

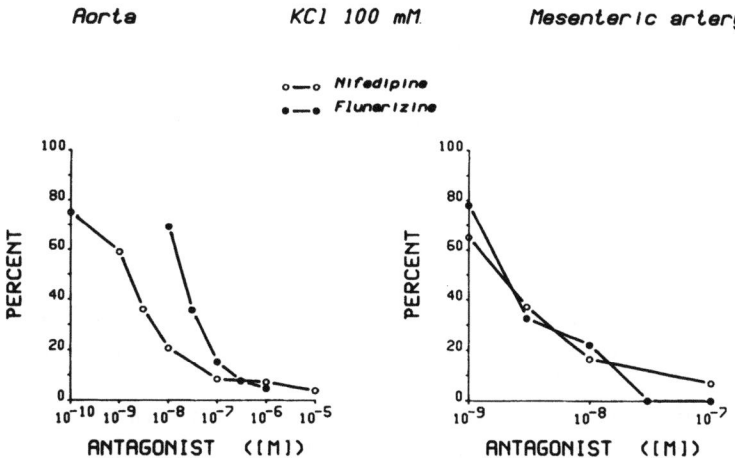

Figure 5. Illustration of tissue selectivity for the action of flunarizine as compared to nifedipine (which has similar potency in both arteries). Ordinate: contractile response to K^+-depolarizing solution in percent of control; abscissa: concentration of the antagonist [see 12 and 13].

trates dose–inhibition curves of contraction of rat aorta and mesenteric artery evoked by K^+-induced depolarization. Nifedipine is much more potent than flunarizine in aorta, but the two drugs are equipotent in mesenteric artery. The identical IC_{50} values for nifedipine in the two tissues indicate that the flunarizine inhibitory potency was greater in mesenteric artery than in aorta. Such changes in sensitivity to Ca^{2+} antagonist blockade among various arteries was initially observed by Godfraind *et al.* [15], who studied depression of the K^+-depolarization-evoked contractions in isolated rabbit arteries pretreated with cinnarizine. This Ca^{2+} antagonist was inactive in thoracic aorta, although it abolished contraction in small mesenteric arteries. This tissue selectivity is not only found with the diphenylpiperazine group but also with the dihydropyridines. This is clearly the case for nimodipine, which at 2×10^{-6} M inhibits the serotonin-induced contraction in rabbit basilar artery, whereas the response of the saphenous artery is only slightly affected [22,28].

The degree of inhibition of smooth muscle contraction evoked by an agonist is governed by two main factors: (1) the occupancy of Ca^{2+} channels by a given Ca^{2+} antagonist and (2) the relative contribution of extracellular and intracellular Ca^{2+} pools in the control contractions. At maximum inhibition when occupancy of Ca^{2+} channels is 100%, receptor activation no longer promotes Ca^{2+} entry. Receptor–response coupling is entirely dependent on release of the intracellular Ca^{2+} pool insensitive to most Ca^{2+} antagonists with the exception of at least diltiazem, papaverine, and chlorpromazine. Therefore, differences in sensitivity to Ca^{2+} antagonists may be due to the relative importance of the contractions sensitive and resistant to Ca^{2+} antagonists, as well as to the affinities of receptor-operated channels for the drugs. The comparative study done in this laboratory between aortae and mesenteric arteries shows that these two possibilities may coexist. The relative contribution of each factor is dependent on the specific drug; only dose–effect curves will characterize the nature of the vascular specificity.

Concluding Summary

From experimental data summarized in this report it is apparent that so-called Ca-antagonists act to reduce smooth muscle contraction by a mechanism allowing us to consider these drugs as calcium entry blockers. However, for some of them, one may consider multiple sites of action [32] and only experimental observation showing superimposition of dose–effect curves for inhibition of stimulus-dependent calcium entry and inhibition of contraction indicate that the action on the latter may be attributed to an interaction with receptor operated or potential operated calcium channels. An additional evidence should come from analysis of agonist-evoked Ca efflux. In physiological solutions, this process is due to a stimulation of the calcium pumping of plasma membrane. This process is governed by free calcium concentration in the cytoplasm and is dependent upon the formation of a Ca-calmodulin complex responsible for the activation of the Ca-ATPase of the pump. This process may be altered by calmodulin antagonists and by drugs interfering with the release of calcium from internal stores. Studies with subcellular preparations may help in differentiating between these two possibilities.

Characterization of the pharmacological profile of calcium entry blockers shows that they present Ca-channel specificity and tissue selectivity. Ca channel specificity means that calcium entry blockers are more active at lower concentration on potential operated channels than on receptor operated channels. Furthermore, the interaction with the former shows use-dependency. Tissue selectivity means that some of these drugs could have a low affinity for the slow Na–Ca channel of heart as indicated by a weak negative inotropic effect. In addition, some are more active in some smooth muscles than in others, for example, a drug such as nifedipine is equipotent in every arterial bed so far studied, but cinnarizine, flunarizine and nimodipine show a predilective action in some arterial territories (such as the cerebral circulation). From a therapeutic point of view, this tissue selectivity is important for the rationale of the clinical use of calcium antagonists.

References

1. Amar-Costesec, A.; Wibo,M.; Thines-Sempoux, D.; Beaufay, H.; Berthet, J. Analytical study of microsomes and isolated subcellular membranes from rat liver. IV. Biochemical, physical, and morphological modifications of microsomal components induced by digitonin, EDTA, and pyrophosphate. *J. Cell Biol.* **62**:717–745, 1974.
2. Bolton, T. B. Mechanisms of action of transmitters and other substances on smooth muscle. *Physiol. Rev.* **59**:606–718, 1979.
3. Broekaert, A.; Godfraind, T. A comparison of the inhibitory effect of cinnarizine and papaverine on the noradrenaline- and calcium-evoked contraction of isolated rabbit aorta and mesenteric arteries. *Eur. J. Pharmacol.* **53**:282–288, 1979.
4. Caroni, P.; Carafoli, E. The Ca^{++}-pumping ATPase of heart sarcolemma. *J. Biol. Chem.* **256**:3263–3270, 1981.
5. Cauvin, C.; Loutzenhiser, R.; van Breemen, C. Mechanisms of calcium antagonist-induced vasodilatation. *Annu. Rev. Pharmacol. Toxicol.* **23**:373–396, 1983.
6. Fleckenstein, A. Specific pharmacology of calcium in myocardium, cardiac pacemakers, and vascular smooth muscle. *Annu. Rev. Pharmacol. Toxicol.* **17**:149–166, 1977.
7. Fleckenstein, A. (ed.) *Calcium Antagonism in Heart and Smooth Muscle,* New York, Wiley, 1983.
8. Godfraind-De Becker, A.; Godfraind, T. Calcium transport system: A comparative study in different cells. *Int. Rev. Cytol.* **67**:141–170, 1980.
9. Godfraind, T. Calcium exchange in vascular smooth muscle, action of noradrenaline and lanthanum. *J. Physiol. (London)* **260**:21–35, 1976.
10. Godfraind, T. Mechanisms of action of calcium entry blockers. *Fed. Proc.* **40**:2866–2871, 1981.
11. Godfraind, T. Pharmacology of calcium entry blockers. In: *Calcium Modulators,* T. Godfraind, A. Albertini, and R. Paoletti, eds., Amsterdam, Elsevier/North-Holland, 1982, pp. 51–65.
12. Godfraind, T. Actions of nifedipine on calcium fluxes and contraction in isolated arteries. *J. Pharmacol. Exp. Ther.* **224**:443–450, 1983.
13. Godfraind, T.; Dieu, D. The inhibition by flunarizine of the norepinephrine-evoked contraction and calcium influx in rat aorta and mesenteric arteries. *J. Pharmacol. Exp. Ther.* **217**:510–515, 1981.

14. Godfraind, T.; Kaba, A. Blockade or reversal of the contraction induced by calcium and adrenaline in depolarized arterial smooth muscle. *Br. J. Pharmacol.* **36**:549–560, 1969.
15. Godfraind, T.; Kaba, A.; Polster, P. Specific antagonism to the direct and the indirect action of angiotensin on the isolated guinea-pig ileum. *Br. J. Pharmacol. Chemother.* **28**:93–104, 1966.
16. Godfraind, T.; Miller, R. C. Prostaglandin $F_{2\alpha}$ ($PGF_{2\alpha}$) mediated contraction and ^{45}Ca influx into rat mesenteric arteries: Inhibition by flunarizine, a calcium entry blocker. *Br. J. Pharmacol.* **73**:252, 1981.
17. Godfraind, T.; Miller, R. C. Specificity of action of Ca^{++} entry blockers: A comparison of their actions in rat arteries and in human coronary arteries. *Circ. Res.* **52**:I81–I91, 1983.
18. Godfraind, T.; Polster, P. Etude comparative de médicaments inhibant la réponse contractile de vaisseaux isolés d'origine humaine ou animale. *Therapie* **23**:1209–1220, 1968.
19. Gopinath, R. M.; Vincenzi, F. F. Phosphodiesterase protein activator mimics red blood cell cytoplasmic activator of (Ca^{++}-Mg^{++})ATPase. *Biochem. Biophys. Res. Commun.* **77**:1203–1209, 1977.
20. Grover, A. K.; Kwan, C. Y.; Crankshaw, J.; Crankshaw, D. J.; Garfield, R. E.; Daniel, E. E. Characteristics of calcium transport and binding by rat myometrium plasma membrane subfractions. *Am. J. Physiol.* **239**:C66–C74, 1980.
21. Janis, R. A.; Triggle, D. J. New developments on Ca^{2+} channel antagonists. *J. Med. Chem.* **26**:775–785, 1983.
22. Kazda, S.; Towart, R. Differences in the effect of the calcium antagonists nimodipine (BAY e 9736) and bencyclan on cerebral and peripheral vascular smooth muscle. *Br. J. Pharmacol.* **72**:582–583, 1980.
23. Kwan, C. Y.; Garfield, R.; Daniel, E. E. An improved procedure for the isolation of plasma membranes from rat mesenteric arteries. *J. Mol. Cell. Cardiol.* **11**:639–659, 1979.
24. Wibo, M.; Tuyet Duong, A. T.; Godfraind, T. Subcellular location of semicarbazide-sensitive amine oxidase in rat aorta. *Eur. J. Biochem.* **112**:87–94, 1980.
25. Morel, N.; Godfraind, T. Na–Ca exchange in heart and smooth muscle microsomes. *Arch. Int. Pharmacodyn. Ther.* **258**:319–321, 1982.
26. Morel, N.; Wibo, M.; Godfraind, T. A calmodulin-stimulated Ca^{2+} pump in rat aorta plasma membranes. *Biochim. Biophys. Acta* **644**:82–88, 1981.
27. Saida, K.; van Breemen, C. Mechanism of Ca^{2+} antagonist-induced vasodilation: Intracellular actions. *Circ. Res.* **52**:137–142, 1983.
28. Towart, R. The selective inhibition of serotonin-induced contractions of rabbit cerebral vascular smooth muscle by calcium antagonistic dihydropyridines: An investigation of the mechanism of action of nimodipine. *Circ. Res.* **48**:650–657, 1981.
29. Triggle, D. J.; Swamy, V. C. Calcium antagonists: Some chemical-pharmacologic aspects. *Circ. Res.* **52**:I17–I28, 1983.
30. van Breemen, C.; Hwang, O. K.; Meisheri, K. D. The mechanism of inhibitory action of diltiazem on vascular smooth muscle contractility. *J. Pharmacol. Exp. Ther.* **218**:459–463, 1981.
31. Van Nueten, J. M. Selectivity of calcium entry blockers. In: *Calcium Modulators,* T. Godfraind, A. Albertini, and R. Paoletti, eds., Amsterdam, Elsevier/North-Holland, 1982, pp. 199–208.
32. Weiss, G. (ed.) Sites of action of calcium antagonists. In: *New Perspectives on Calcium Antagonists,* Bethesda, American Physiological Society, 1981, pp. 83–94.
33. Wibo, M.; Morel, N.; Godfraind, T. Differentiation of Ca^{++} pumps linked to plasma membrane and endoplasmic reticulum in the microsomal fraction from intestinal smooth muscle. *Biochim. Biophys. Acta* **649**:651–660, 1981.

46

Calcium Antagonist Effects on Vascular Muscle Membrane Potentials and Intracellular Ca^{2+}

Kent Hermsmeyer

One of the first hypotheses for the mechanism of action of a Ca^{2+} antagonist that should be tested is the blockade of Ca^{2+} entry [10]. Because block of Ca^{2+} entry would remove a depolarizing influence, hyperpolarization, or at least block of part of the depolarization, would result. A test of the Ca^{2+} entry block hypothesis would be measurement of membrane potential (E_m), with the predictions that E_m would be more electronegative in the presence of a Ca^{2+} antagonist than in the absence, and that the greatest effects would be observed at higher concentrations of stimuli, because there would be more Ca^{2+} influx to inhibit. Experiments with verapamil are confusing because verapamil (and gallopamil, or D600) does not block norepinephrine (NE) depolarization of muscle cells of the rat caudal artery [16]. In contrast, nitrendipine (a newer Ca^{2+} antagonist) causes an immediate and significant hyperpolarization of both resting and stimulated caudal artery [16]. Because there is more hyperpolarization elicited with nitrendipine at rest than during activation, a mechanism other than block of Ca^{2+} influx is likely to be involved.

Considering other possible mechanisms of hyperpolarization, one likely candidate is the active electrogenic transport of Ca^{2+} across the cell membrane. In support of this hypothesis, earlier experiments showed that hyperpolarization by nitrendipine is dependent on extracellular K^+. Thus, with a K^+ concentration in the extracellular fluid of zero, hyperpolarization and relaxation of the rat caudal artery by nitrendipine is eliminated [16]. While it is not clear why there should be an interaction between extracellular K^+ and Ca^{2+} transport, a K^+-sensitive electrogenic Ca^{2+} transport could be involved, and data from rat caudal artery are consistent with such a hypothesis. The mechanism for Ca^{2+} transport out of a cell has only recently been approached at the molecular level [3] and many details (perhaps including K^+ interactions) are missing. In the experiments reviewed here, the relationship between relaxation and E_m in the presence of nitrendipine will be examined and related to Ca^{2+} in rat caudal arterial vascular muscle cells. Interpretation of data in accord with depolarization and Ca^{2+} release will be chosen rather than less conventional interpretations recently presented [2,5].

Kent Hermsmeyer • Department of Pharmacology, The Cardiovascular Center, University of Iowa, Iowa City, Iowa 52242.

Evidence for the Action of Ca^{2+} Antagonists on the Cell Membrane

Nitrendipine and other Ca^{2+} antagonists inhibit the influx of ^{45}Ca into vascular muscle cells, with a proportional decrease in the tension developed [32]. The second phase of contraction of certain arteries, thought to depend heavily on intracellular Ca^{2+} [28], is selectively inhibited by Ca^{2+} antagonists [29]. Furthermore, when the surface membrane is disrupted by the cell skinning process, nitrendipine has little or no effect on contractions induced by K^+ depolarization (F. P. J. Diecke, personal communication). Therefore, the site of the predominant immediate action of nitrendipine, and at least some other Ca^{2+} antagonists, is believed to be located at the cell surface membrane.

These observations support the Ca^{2+} entry hypothesis. However, other possible mechanisms might also fit the observations and not be distinguishable from blockade of Ca^{2+} entry, especially if only ^{45}Ca flux were measured. For example, nitrendipine might act by stimulation of a Ca^{2+} pump [16]. In ^{45}Ca flux experiments, such an action would appear identical to inhibition of Ca^{2+} influx. The possibility that nitrendipine stimulates a Ca^{2+} pump could not be differentiated by radioactive ion flux measurements because of the limited time resolution, but could be detected by specific intracellular indicators of Ca^{2+} activity.

Association of Membrane Potential Effects with Relaxation

Having established that a surface membrane action of nitrendipine is likely, it is logical to consider the roles of hyperpolarization and relaxation. We have emphasized in recent reviews the close correlation between membrane potential and tension in at least the caudal artery [15,17,19]. For stimulation from the normal resting state by K^+ or NE, it is likely that membrane depolarization is the major event initiating contraction in the rat caudal artery. However, it is entirely possible to modify the voltage–tension relationship in caudal artery, as in other kinds of muscle [1]. In particular, drugs acting on excitation–contraction coupling, like Ca^{2+} antagonists, can shift the voltage–tension relationship to the right. While examples of such shifts are given by verapamil and related drugs, nitrendipine and other dihydropyridines appear to have a potent direct action on E_M, causing relaxation without a shift in the voltage–tension relationship. When nitrendipine is added to the resting caudal artery, there is a rapid and significant hyperpolarization of at least 5 mV [16]. When nitrendipine is used with NE, relaxation correlates with E_m only at lower NE concentrations. At higher NE concentrations, there is a pronounced shift of tension downward and to the right [16].

In considering the possibilities for E_m action, it is important to remember that if nitrendipine acts by stimulation of electrogenic Ca^{2+} transport, there would be a dual effect to decrease contraction. Hyperpolarization (caused by stimulating Ca^{2+} transport out of the cell) would by itself reduce the release of intracellular Ca^{2+} by moving down the voltage–tension relationship. In addition, movement of Ca^{2+} out of the cell would directly reduce intracellular Ca^{2+}. Thus, divergence from the voltage–tension curve would be explained by the Ca^{2+} pump mechanism. However, much more work will be needed to determine whether other mechanisms, including intracellular processes, might not also play a role.

Intracellular Actions of Nitrendipine

There is specific evidence that nitrendipine stimulates Ca^{2+}-ATPase and Ca^{2+} uptake in sarcoplasmic reticulum isolated from skeletal and cardiac muscle [6]. Although higher concentrations of nitrendipine are needed to produce intracellular Ca^{2+} pump stimulation, nitrendipine in cell membranes may reach a concentration more than 10 times that in the extracellular fluid [21]. Some action on Ca^{2+} transport, perhaps including both surface membrane and intracellular sequestration sites, may be stimulated by nitrendipine.

Furthermore, other actions of nitrendipine on calmodulin and related phosphodiesterase proteins have been proposed [9]. Myosin light-chain phosphorylation appears to be involved in maintained contractions [25]. The explanation of contraction by light-chain phosphorylation is probably complex [30], and nitrendipine and other Ca^{2+} antagonists may interfere with the phosphorylation process to decrease tension. Further studies of possible intracellular actions of Ca^{2+} antagonists are unquestionably needed.

ROCs, a Misleading Concept

It must be recognized at the outset that a difficulty in understanding the mechanism of action of Ca^{2+} antagonists on vascular muscle is the lack of detailed knowledge about excitation–contraction coupling in vascular muscle. The ideal approach to studying the mechanism of action of drugs which cause relaxation of vascular muscle might be to examine each of the sequential steps in the normal activation pathway, testing whether each drug has an effect on each step, and quantitating the magnitude of each effect found. Furthermore, since vascular muscle appears to be activated differently by various agonists, the interaction of Ca^{2+} antagonists with stimulation produced by diverse agents should be examined. Since there is overwhelming evidence that excitation and subsequent steps in the activation of vascular muscle are markedly disparate in different blood vessels within an animal (and even within a given organ), such studies should be carried out and the conclusions drawn properly limited for each vessel studied [17]. Generalizations with respect to all blood vessels, to all Ca^{2+} antagonists, to all stimulation mechanisms, or even to all experimental conditions, should not be made unless there are experimental data to justify such inductive reasoning. It is particularly illogical to use data from nonvascular muscle experiments (e.g., from vas deferens or intestinal muscle) to draw conclusions about the mechanisms of action of Ca^{2+} antagonists on vascular muscle, since visceral muscle membranes possess many properties that differ from those of vascular muscle membranes [14].

One concept that has been offered to explain Ca^{2+} antagonists which appears to have little predictive value is the "receptor-operated channel" (ROC). With the deficit in understanding the excitation–contraction coupling in vascular muscle, a new concept may indeed have been useful. The terminology, at least the acronym ROC, introduced by Bolton [2], has been used by many authorities on Ca^{2+} antagonism [e.g., 10]. Presumably, the concept would explain early data on aorta in which stimulation by KCl was antagonized to a greater extent by Ca^{2+} antagonists than was stimulation by NE [5]. However, such a generalization simply does not hold when one considers almost any vessel except the aorta. Indeed, small resistance arteries are readily dilated by Ca^{2+} antagonists in the presence of NE [16].

It has been assumed that KCl specifically depolarizes the surface membrane of vascular muscle to open voltage-dependent ion channels. This hypothesis is an implicit assumption that has never been demonstrated in a single blood vessel, and may well vary from one vessel to another. Furthermore, stimulation by NE (which undoubtedly releases intracellular Ca^{2+}) can hardly act independently of membrane depolarization [15,17,19]. In my opinion, stimulation of at least some vascular muscle by K^+, NE, or any other agonist will cause a combination of opening of voltage-dependent ion channels and of intracellular Ca^{2+} release. The essence of the question is to understand the sources and control mechanisms for intracellular Ca^{2+} release in each type of vascular muscle. This is really the main question that should be the focus of research on excitation–contraction coupling in vascular muscle. Since we do not know the answer (and there are probably several answers), it is not possible to make categorical statements about how vasodilator drugs cause relaxation.

These questions about the steps in normal excitation–contraction coupling of vascular muscle, considered together with the questions about the mechanism of action of Ca^{2+} antagonists, form a complicated network of interdependent unknowns in which use of circular reasoning is almost inevitable. In order to understand the mechanism of action of Ca^{2+} antagonists, one should understand all of the important steps in excitation–contraction coupling (in a given vessel under the experimental circumstances). And in order to use Ca^{2+} antagonists to define the excitation–contraction coupling process, one needs to know the mechanism of action of a given Ca^{2+} antagonist. Investigators often draw conclusions based on the assumption that one or more factors are known, without making that assumption sufficiently explicit.

In order to state that NE causes both voltage-dependent ion channels and "receptor-operated channels" to open, it would be necessary to measure each of those events independently. In vascular muscle, voltage-clamp quantitation is not sufficient to accurately establish the amount of voltage-dependent ionic current [14]. Thus, reasoning has had to rely largely upon measurements of radioactive Ca^{2+} fluxes. Since the time resolution of ion flux data is hundreds of times too slow to allow measurement of a single excitation process and the ion flux experiments on vascular muscle are complicated by multiple pools of unknown location, it is not possible to separate what is due to depolarization from what might hypothetically be due to "receptor-operated channels." The only circumstance where any differentiation can be made is when cells are completely depolarized in high K^+ concentrations (usually greater than 100 mM), so there is little possibility for changes in membrane potential. However, this condition is by no means representative of normally polarized vascular muscle, and there is substantial indication that intracellular Ca^{2+} has a very different distribution in the depolarized cell than in the polarized resting state [20].

In fact, a critical test of the ROC hypothesis is presently available only through patch-clamp types of experiments. The recording of single ion channels using gigaseal pipette technology and voltage-clamp techniques has allowed identification of acetylcholine-operated channels [12]. When patch-clamp electrodes are applied to single vascular muscle cells, membrane potential can be held at normal resting values and NE added to ascertain whether there are ion channels directly responding to interaction of NE with its receptors (without the influence of membrane potential). To directly test this question, I have carried out such experiments using single vascular muscle cells cultured from rat azygous veins, which have been extensively characterized. This characterization includes demonstration of NE sensitivity, spontaneous electrical activity and contraction, and membrane potentials and the electrical spikes underlying contraction [13,18,22]. Individual vascular muscle cells were identified and gigaseal pipettes were attached to the cells giving seal resistances in excess of 3 gigaohms. Spontaneous channel opening and closing occurred in these preparations with amplitudes similar to slow channels that have previously been reported [12,27]. In unpublished studies of six cells so far regarded as technically acceptable, no response to NE was elicited in the range from 10 pM to 10 μM. There was never any indication of any difference in number, amplitude, or duration of channel openings with or without NE. This range of concentrations of NE includes that which we have previously demonstrated causes increased contraction frequency and duration. At least in the rat azygous vein, experiments which should have offered the most unequivocal support for "receptor-operated channels" have argued against it.

Even overlooking a direct test of ROCs, the concept has proved to be difficult to understand. The recent review by Cauvin et al. [5] develops the concept of ROCs to attempt to explain the selectivity of Ca^{2+} antagonists, and then concludes that ROCs do not provide a valid generalization. Integral to the concept of ROCs is the proposition that normally polarized vascular muscle contracts to NE without any change in membrane potential [5]. In fact, the concept of contraction to NE without any depolarization is an error of experimental design [31]. The very same preparation (rabbit ear artery) used by Droogmans et al. [7] does in fact markedly depolarize under the influence of NE [31]. Rabbit main pulmonary artery, also claimed not to depolarize [4] with NE, does indeed depolarize [11]. It is therefore inaccurate to suggest [5] that the close correlation between membrane potential and tension in the rat caudal artery [15,17,19] does not exist in the rabbit ear artery and, by implication, in other arteries. In fact, a strong argument may be made for the association between membrane potential and tension in all vascular muscle so far studied [15,17].

Considering the failure to find ROCs by direct measurement (patch-clamp recording) and with correction of the mistake in the relationship between E_m and tension in rabbit ear artery, the concept of ROCs seems inappropriate. Indeed, the concept of ion channels not controlled by E_m could have been formulated in any kind of muscle. One might have proposed that any deviation from a perfect relationship between E_m and tension in skeletal muscle is due to ROCs. In striated muscle (as well as in smooth muscles), that concept adds more confusion than explanation because the more conservative hypothesis incorporates not only surface membrane excitation due to voltage-operated channels, but also intracellular Ca^{2+} release [20]. In fact, the concept of Ca^{2+}-induced Ca^{2+} release still enjoys the attention of some investigators [8,26] and should be explored in vascular muscles.

My hypothesis is that excitation in vascular muscle occurs by an initial depolarization step followed by the release of intracellular Ca^{2+} that is probably due to: (1) depolarization of the surface membrane, (2) release of trigger Ca^{2+} from a location near the periphery of the cell, and (3) interdependent ion movements within the cell as a result of Ca^{2+}, K^+, Na^+, and possibly Cl^- movements through the cell surface membrane. Contraction initiated by any agent is thus likely to be a complicated event, which will need to be studied by detailed methods that allow separation of individual ion movements, loci, and events. Promising techniques for furthering our understanding of excitation–contraction coupling in vascular muscle and, therefore, enhancing our insight into the mechanism of action of Ca^{2+} antagonists, would appear to be intracellular indicators that allow continuous, millisecond resolution studies of living cells. Calcium ion can be monitored, for example, by the photoprotein aequorin [23], or by the metallochromic dye arsenazo III [15]. In each case, direct indications of intracellular Ca^{2+} concentration can be continuously monitored for studies of the effect of Ca^{2+} antagonists. These studies appear most promising, and that is why I have chosen preliminary experiments that monitor intracellular Ca^{2+} to conclude this presentation.

Measurements of Intracellular Ca^{2+} Activity

A recent development is the use of Ca^{2+} indicators to directly examine intracellular Ca^{2+} activity during contraction. We have developed methods for introducing arsenazo III into single vascular muscle cells with liposome techniques [24]. In previous experiments on single cells, a close correlation has been demonstrated between intracellular Ca^{2+} activity and contraction within a portion of a single cell [15]. With liposome techniques, we are consistently loading the cells to concentrations greater than 100 μM. This is one instance where the high surface-to-volume ratio of vascular muscle cells actually represents an advantage for the experiments.

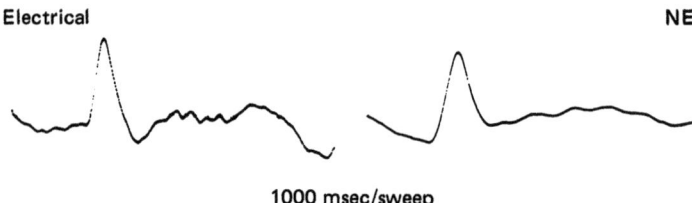

Figure 1. Optical records of single contractions of a single vascular muscle cell were stimulated electrically (left) or by pulse application of 300 nM norepinephrine (right). The record shown is the light intensity at 660 nm after subtraction of the 580-nm signal. The increase in light intensity represents a change of 0.006 of the resting light intensity in each record. Signal-to-noise enhancement was carried out by a Princeton Applied Research light chopper and lockin amplifiers, and the record was filtered to eliminate frequencies above 30 Hz. Notice that there is a strong initial Ca^{2+} release phase followed by rapid reduction in Ca^{2+} level to below the resting value in the left panel, and to slightly above the resting value in the right panel. The duration of the contraction, as monitored by the 580-nm signal, was about 0.3 sec longer in each case.

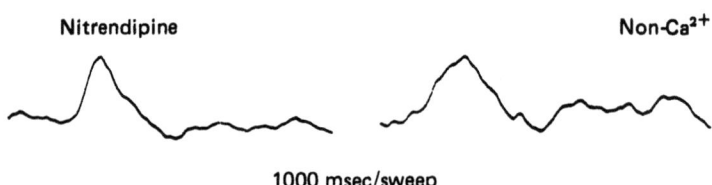

Figure 2. Optical recordings similar to those shown in Fig. 1 after treatment with 100 nM nitrendipine (left) or in low-Ca^{2+} solution (right). The vertical amplitude in both records is 0.001 of the resting level. Contractions were stimulated by pulse application of norepinephrine. Notice that the initial phase of Ca^{2+} release was strongly reduced by nitrendipine, with a lesser effect on the later stage of elevated intracellular Ca^{2+}. The signal was also reduced in the solution which had no added Ca^{2+} (concentration measured to be 20 μM Ca^{2+}), with a lesser reduction in the initial phase of intracellular Ca^{2+} release (right). The contraction signal at 580 nm gave a duration about 0.3 sec longer in each case than the 660- to 580-nm signal (representing intracellular Ca^{2+}).

To provide the severest test for the hypothesis that there is a close correlation between intracellular Ca^{2+} and contraction, we have chosen the rapidly contracting preparation of neonatal rat azygous vein [13,18]. These cells contract and relax in less than 1 sec at 37°C. A very close correlation exists between the electrical spike underlying the contraction and the contractile element shortening [13]. Measurement of Ca^{2+} by changes in arsenazo III absorbance at 660 nm is similar to the contractile time course, without a fall in Ca^{2+} after the initial peak (Fig. 1) [as reported by Morgan and Morgan, (23)], but with a duration about 0.3 sec shorter than contraction.

Nitrendipine strongly inhibited the elevation of intracellular Ca^{2+} in single cells, especially in the initial phase (first 100 msec), with a proportional decrease in contraction (Fig. 2). Nitrendipine reduced the initial phase more than in the absence of added Ca^{2+} (Fig. 2). Furthermore, immediately (within 1 min) after treatment with 0.1 μM nitrendipine, the resting Ca^{2+} concentration fell to levels lower than those observed in Ca^{2+}-deprived solutions. These data are consistent with the hypothesis that nitrendipine stimulates a Ca^{2+} pump, producing both hyperpolarization and a decrease in intracellular Ca^{2+} activity.

Certainly, these preliminary data are only a suggestion of the detailed mechanisms of nitrendipine action that will be explored in subsequent experiments. However, the data suggest that nitrendipine stimulates Ca^{2+} extrusion from the cell, supporting the hypothesis formulated on the basis of E_m experiments [16].

ACKNOWLEDGMENT. This research was supported by Grants HL-14388 and HL-16328 from the National Institutes of Health.

References

1. Bianchi, C. Drugs affecting excitation coupling mechanisms. In: *Fundamentals of Cell Pharmacology*, S. Dikstein, ed., Springfield, Charles C. Thomas, 1973, pp. 454–468.
2. Bolton, T. B. Mechanisms of action of transmitters and other substances on smooth muscle. *Physiol. Rev.* **3**:606–718, 1979.
3. Carafoli, E.; Zurini, M. The Ca^{2+}-pumping ATPase of plasma membranes: Purification, reconstitution and properties. *Biochim. Biophys. Acta* **683**:279–301, 1982.
4. Casteels, R.; Kitamura, K.; Kuriyama, H.; Suzuki, H. Excitation–contraction coupling in the smooth muscle cells of the rabbit main pulmonary artery. *J. Physiol. (London)* **271**:63–79, 1977.
5. Cauvin, C.; Loutzenhiser, R.; VanBreemen, C. Mechanisms of calcium antagonist-induced vasodilation. *Annu. Rev. Pharmacol. Toxicol.* **23**:373–396, 1983.
6. Colvin, R. A.; Pearson, N.; Messineo, F. C.; Katz, A. M. Effects of Ca channel blockers on Ca transport and Ca ATPase in skeletal and cardiac sarcoplasmic reticulum vesicles. *J. Cardiovasc. Pharmacol.* **4**:935–941, 1982.
7. Droogmans, G.; Raeymaekers, L.; Casteels, R. Electro- and pharmacomechanical coupling in the smooth muscle cells of the rabbit ear artery. *J. Gen. Physiol.* **70**:129–148, 1977.
8. Endo, M. Calcium release from the sarcoplasmic reticulum. *Physiol. Rev.* **57**:71–108, 1977.
9. Epstein, P. M.; Fiss, K.; Hachisu, R.; Adrenyak, D. M. Interaction of calcium antagonists with cyclic AMP phosphodiesterases and calmodulin. *Biochem. Biophys. Res. Commun.* **105**:1142–1149, 1982.
10. Fleckenstein, A. History of calcium antagonists. *Circ. Res.* **52**(Suppl. I):3–16, 1983.
11. Haeusler, G. Contraction, membrane potential, and calcium fluxes in rabbit pulmonary arterial muscle. *Fed. Proc.* **42**:263–268, 1983.
12. Hamill, O. P.; Marty, A.; Neher, E.; Sakmann, B.; Sigworth, F. J. Improved patch-clamp techniques for high-resolution current recording from cells and cell-free membrane patches. *Pfluegers Arch.* **391**:85–100, 1981.
13. Hermsmeyer, K. High shortening velocity of isolated single arterial muscle cells. *Experientia* **35**:1599–1602, 1979.
14. Hermsmeyer, K. Electrophysiology of vascular muscle. In: *Microcirculation*, Volume III, G. Kaley and B. M. Altura, eds., Baltimore, University Park Press, 1980, pp. 439–460.

15. Hermsmeyer, K. Electrogenic ion pumps and other determinants of membrane potential in vascular muscle [the 1982 Henry Pickering Bowditch Lecture]. *Physiologist* **25**:454–465, 1982.
16. Hermsmeyer, K. Might nitrendipine enhance Ca^{2+} transport in vascular muscle? In: *Ca^{2+} Entry Blockers, Adenosine, and Neurohumors*, G. F. Merrill and H. R. Weiss, eds., Baltimore, Urban & Schwarzenberg, 1983, pp. 51–61.
17. Hermsmeyer, K. Sodium pump hyperpolarization-relaxation in rat caudal artery. *Fed. Proc.* **42**:246–252, 1983.
18. Hermsmeyer, K.; Mason, R. Norepinephrine sensitivity and desensitization of cultured single vascular muscle cells. *Circ. Res.* **50**:627–632, 1982.
19. Hermsmeyer, K.; Trapani, A.; Abel, P. W. Membrane-potential dependent tension in vascular muscle. In: *Vasodilatation*, P. M. Vanhoutte and I. Leusen, eds., New York, Raven Press, 1981, pp. 273–284.
20. Johansson, B.; Somlyo, A. P. Electrophysiology and excitation–contraction coupling. In: *Handbook of Physiology*, Section 2, *The Cardiovascular System*, D. F. Bohr, A. P. Somlyo, and H. V. Sparks, Jr., eds., Bethesda, American Physiological Society, 1980, pp. 301–324.
21. Keef, D. L.; Kates, R. E. Myocardial disposition and cardiac pharmacodynamics of verapamil in the dog. *J. Pharmacol. Exp. Ther.* **220**:91–96, 1982.
22. Marvin, W.; Robinson, R.; Hermsmeyer, K. Correlation of function and morphology of neonatal rat and embryonic chick cultured cardiac and vascular muscle cells. *Circ. Res.* **45**:528–540, 1979.
23. Morgan, J. P.; Morgan, K. G. Vascular smooth muscle: The first recorded Ca^{2+} transients. *Pfluegers Arch.* **395**:75–77, 1982.
24. Mueller, T. M.; Marcus, M.; Mayer, H.; Williams, J.; Hermsmeyer, K. Liposome concentration in canine ischemic myocardium and depolarized myocardial cells. *Circ. Res.* **49**:405–415, 1981.
25. Murphy, R. A.; Gerthoffer, W. T. Cell calcium and contractile system regulation in arterial smooth muscle. In: *Calcium Antagonists in Cardiovascular Disease*, L. H. Opie and R. Krebs, eds., New York, Raven Press, 1984, pp. 75–84.
26. Oetliker, H. An appraisal of the evidence for a sarcoplasmic reticulum membrane potential and its relation to Ca^{2+} release in skeletal muscle. *J. Muscle Res. Cell Motil.* **3**:247–272, 1982.
27. Reuter, H.; Stevens, C. F.; Tsien, R. W.; Yellen, G. Properties of single calcium channels in cardiac cell culture. *Nature (London)* **297**:501–504, 1982.
28. Steinsland, O. S.; Furchgott, R. F.: Kirpekar, S. M. Biphasic vasoconstriction of the rabbit ear artery. *Circ. Res.* **32**:49–58, 1973.
29. Steinsland, O. S.; Berkowitz, C. J.; Scriabine, A. Effects of nitrendipine (N) and verapamil (V) on adrenergic neurotransmission in rabbit ear artery. *Pharmacologist* **23**:198, 1981.
30. Stull, J. T.; Silver, P. J.; Miller, J. R.; Blumenthal, D. K.; Botterman, B. R.; Klug, G. A. Phosphorylation of myosin light chain in skeletal and smooth muscles. *Fed. Proc.* **42**:21–26, 1983.
31. Trapani, A. J.; Matsuki, N.; Abel, P. W.; Hermsmeyer, K. Norepinephrine produces tension through electromechanical coupling in rabbit ear artery. *Eur. J. Pharmacol.* **72**:87–91, 1981.
32. Weiss, G. B. Sites of action of calcium antagonists in vascular smooth muscle. In: *New Perspectives on Calcium Antagonists*, G. B. Weiss, ed., Bethesda, American Physiological Society, 1981, pp. 83–94.

47

Suppression of Experimental Coronary Spasms by Major Calcium Antagonists

Gisa Fleckenstein-Grün

Maintenance of coronary vascular tone and phasic contractility requires free Ca^{2+}. This requirement applies even more to supernormal contractile activity which manifests itself in the form of coronary spasms. The discovery of the new pharmacological family of specific Ca^{2+} antagonists during the years 1964–1969 in our laboratory* has demonstrated a novel way to interfere with the transmembrane Ca^{2+} supply to the contractile systems of both myocardial fibers and various types of smooth muscle cells. Particularly the Ca^{2+} transport systems in heart and vascular smooth muscle membranes have crucial features in common regarding responsiveness to Ca^{2+}-antagonist drugs. This chapter focuses specifically on arterial smooth muscle of the large extramural coronary trunks and on the major antispastic actions that are exerted by four highly specific Ca^{2+} antagonists [verapamil, gallopamil (D600), nifedipine, and diltiazem]. This subject has been considered in our preceding publications [2–13]. There are, however, a number of more recent findings of general interest that are provided in the present report.

Clinical Significance of Extramural Coronary Vasculature

Pathophysiologically, the large subepicardial stem arteries are the most important functional part of the coronary system. In fact, more than 95% of the stenosing atherosclerotic processes are located in the proximal part of the large extramural coronary arteries. Therefore, a major aim in clinical therapy is to improve blood flow in this particular section of the coronary bed. Under normal physiological conditions, changes in the diameter of the coronary stem arteries do not exert a major influence on either total or local cardiac flow resistance. But this situation radically changes in patients with occlusive coronary heart

---

*A comprehensive survey of this research work has recently been given in a monograph by Fleckenstein [1].

Gisa Fleckenstein-Grün • Physiological Institute, University of Freiburg, 7800 Freiburg, West Germany.

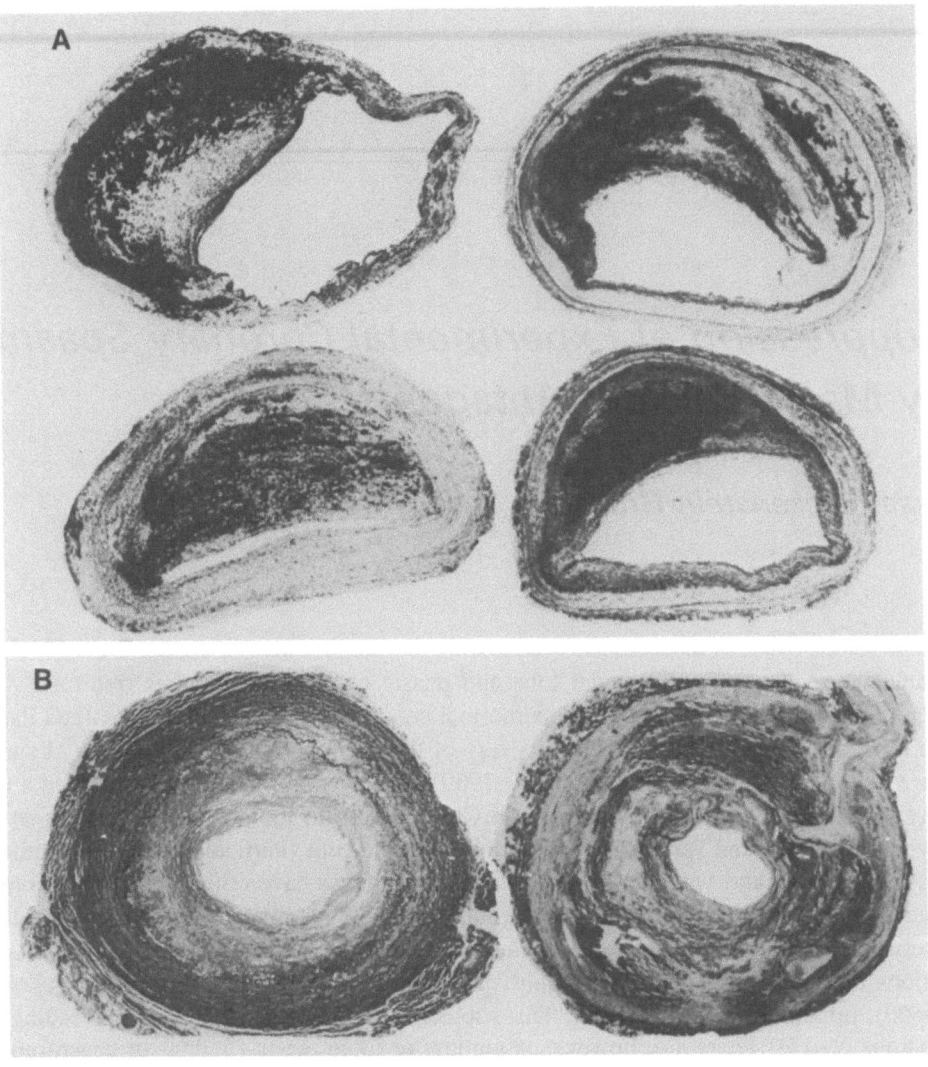

Figure 1. Collection of eccentric (A) and concentric (B) stenoses of human coronary stem arteries. [From Ref. 1.]

disease, since the blood supply to the myocardium will then be determined primarily by the residual diameter of the narrowed extramural stem arteries.

Obviously, pharmacotherapy with coronary vasodilators will only be successful if the atherosclerotic extramural arteries (or collaterals that may act as a physiological bypass) are still able to respond. This prerequisite is often fulfilled when the stenosing process develops eccentrically so that, at least for a time, a portion of the vascular wall remains functionally intact (Fig. 1). In this situation, local spasms in a narrowed arterial segment may dramatically decrease myocardial perfusion, whereas, conversely, significant circulatory improvement will be attained with suitable vasodilators. The modern use of Ca^{2+} antagonists represents a landmark in coronary therapy because, apart from their oxygen-saving and cardioprotective effects on heart muscle, these drugs also produce a long-lasting relaxation of extramural coronary smooth muscle. This relaxation is particularly prominent in cases of angina with a spastic component [for pertinent literature see Fleckenstein (1)].

Possible Pathways of Calcium-Dependent Extramural Coronary Smooth Muscle Activation

The fundamental mechanism of action of Ca^{2+} antagonists in coronary smooth muscle relaxation (excitation–contraction uncoupling due to blockade of transmembrane Ca^{2+} influx) can be most clearly demonstrated on K^+-depolarized strips excised from great extramural stem arteries. Figure 2 shows log dose–response curves that summarize the results of a comparative study with several Ca^{2+} antagonists on approximately 1000 K^+-depolarized pig coronary strips. In comparison with papaverine, nifedipine is approximately 3000 times more potent, gallopamil is intermediate, and verapamil and diltiazem exceed the coronary relaxing potency of papaverine by 50 to 100 times.

The Ca^{2+} antagonists are generally believed to be less effective against vasospasm induced by organic vasoconstrictor agonists than against that elicited by high K^+ [14–17]. However, our results obtained on coronary smooth muscle of pigs and rabbits are not fully consistent with this view. First, our studies indicate that the contractile responses of coronary

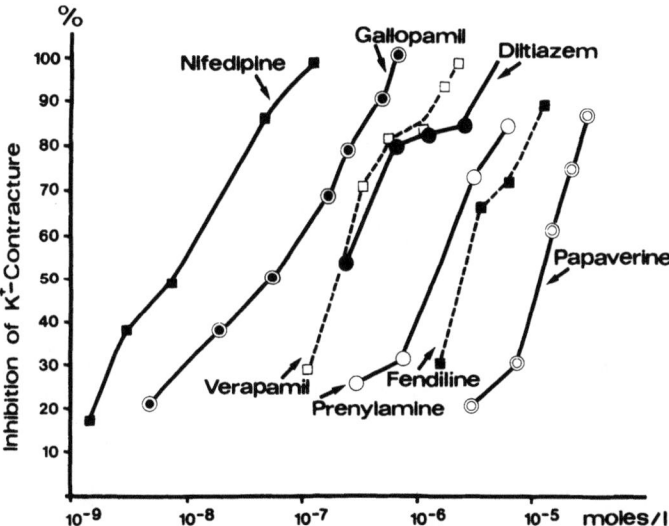

Figure 2. Suppression of K^+-induced contractures of pig coronary strips by Ca^{2+}-antagonistic inhibitors of excitation–contraction coupling. The strips were depolarized with a K^+-rich Tyrode solution (43 mM KCl) for 40 min to produce full contractures. Then different concentrations of the Ca^{2+}-antagonistic compounds were added, so that, depending on the dose applied, the tension development was more or less inhibited. The degree of relaxation obtained is expressed as percent of peak tension just before addition of the Ca^{2+}-antagonistic drugs. Each point represents the average relaxation calculated from at least 15 individual experiments for each concentration, S.E. not exceeding ± 2%. Before administration of the K^+-rich Tyrode solution, the coronary strips (wet weight: 12–15 mg) were kept in a Tyrode solution with a normal K^+ content (concentrations in mM: NaCl 155; KCl 4; $NaHCO_3$ 11.9; $CaCl_2$ 1.0; NaH_2PO_4 0.48; Glucose 5.6) for a period of 60 min under a load of 2.0 g. Throughout the experiment a gas mixture of 97% O_2 and 3% CO_2 was used for oxygenation of the bath at a constant temperature of 35°C and at a pH of 7.4. Isometric tension was continuously recorded with the use of a mechano-electronic transducer [8,37].

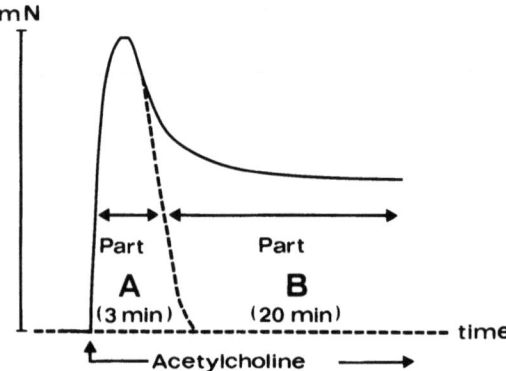

Figure 3. A mixed phasic/tonic contractile response of a pig coronary strip to acetylcholine. With histamine and serotonin, similar inhomogeneous contractures are also produced.

smooth muscle to the vasoconstrictor agonists acetylcholine, histamine, and serotonin are not homogeneous. They consist of an initial fast (phasic) component A which ends within a few minutes and a subsequent tonic component B that may last for as long as several hours (see Fig. 3). The tonic contractile phase B was somewhat more sensitive to Ca^{2+} antagonists than the initial phase A. Thus, in our experiments, the tonic phase B could be suppressed by concentrations of Ca^{2+} antagonists similar to those inhibiting K^+-induced coronary spasm [18].

On the other hand, similar increases in the external Ca^{2+} concentration could counteract both the relaxing effect of Ca^{2+} antagonists on the tonic phase B and on K^+-induced contractures to an equal extent. These observations indicate that the basic mechanisms of

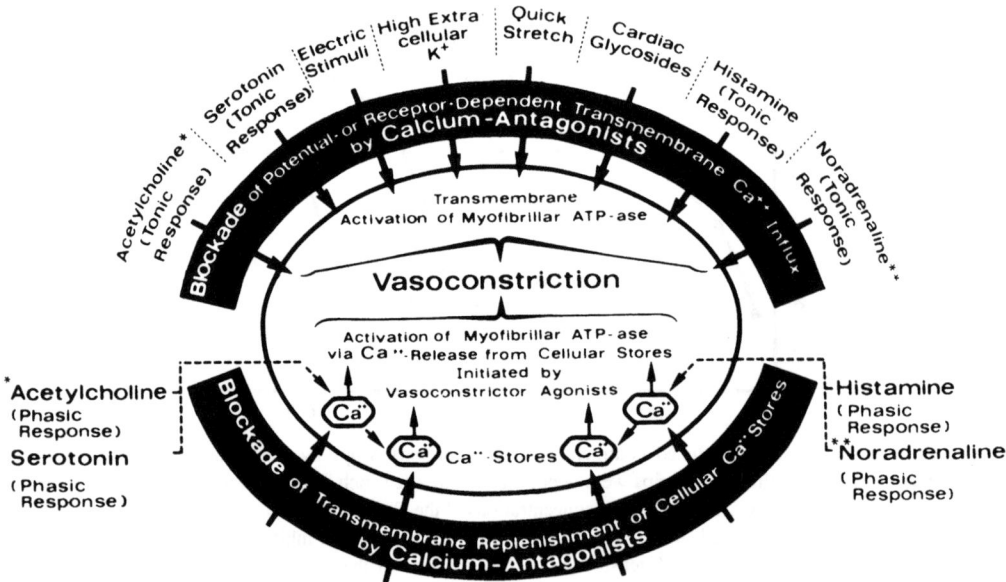

Figure 4. Dual source of activator Ca^{2+} for tonic and phasic vascular smooth muscle contractions, i.e., (1) potential- or receptor-dependent transmembrane Ca^{2+} supply for tonic responses, and (2) release of Ca^{2+} from cellular stores for phasic responses to vasoconstrictor agonists. *Vasculature originating from systemic arteries is not activated by acetylcholine. **Noradrenaline produces contraction of extramural coronary vasculature only after previous β-receptor blockade.

both the tonic responses to vasoconstrictor agonists and the K^+-induced contracture of coronary smooth muscle are closely related or even identical. A similar biphasic response and a higher Ca^{2+} antagonistic drug susceptibility of the tonic contraction phase B has been reported on the vasculature of systemic arteries by Godfraind and Kaba [19] with epinephrine and by Bevan et al. [20] with norepinephrine, histamine, and serotonin.

Figure 4 explains the dualistic concept of phasic and tonic contractile responses. The free Ca^{2+} required for activation of vascular smooth muscle could be supplied by different pathways [21–26]. Tonic responses preferentially occur by Ca^{2+} influx through potential-dependent or receptor-operated plasma membrane channels (Fig. 4, upper). Electrical stimuli, high extracellular K^+, quick stretch, or cardiac glycosides increase coronary vascular tone by enhancing transmembrane Ca^{2+} uptake. The same is true of acetylcholine, serotonin, histamine, or norepinephrine (after β-receptor blockade) when these vasoconstrictor agonists produce long-lasting tonic responses. Very low concentrations of Ca^{2+} antagonists block this transmembrane pathway for Ca^{2+}. On the other hand, rapid phasic responses of coronary smooth muscle may be more closely related to liberation of activator Ca^{2+} from cellular storage sites (Fig. 4, lower). Serotonin, histamine, or acetylcholine mobilizes cellular Ca^{2+} deposits for short-lasting phasic responses. Ca^{2+} antagonists do not seem to interfere directly with this Ca^{2+} release mechanism, but they probably impair the subsequent transmembrane replenishment of cellular Ca^{2+} stores.

Peculiarities of the Coronary Vasculature Treated with Vasoconstrictor Agents, Ca^{2+} Antagonists, and Cardiac Glycosides

Although the scheme of Fig. 4 adequately meets didactic requirements, in the coronary vasculature the dualistic concept which differentiates between phasic activation and tonic activation is not as clear as in certain systemic arteries. In fact, we can conclude from thousands of observations on coronary arteries of pigs and rabbits that the two mechanisms have considerable overlap under practically all circumstances. Moreover, a number of unexpected results require special consideration.

Staircase phenomena during phasic activity of coronary smooth muscle. Figure 5 shows repeated 3-min exposures of pig coronary strips to high K^+ and to the vasoconstrictor agonists histamine, serotonin, and acetylcholine in medium containing 1 mM Ca^{2+}. Because both high K^+ and the agonists were added each time for only 3 min and then washed out, the contractions merely reflect the initial phasic portion A of the mechanical response. The amplitudes of the initial (phasic) contractile responses to high K^+ (43 mM) or to specific concentrations of histamine (1.4×10^{-6} M), serotonin (1.4×10^{-6} M), or acetylcholine (3.6×10^{-7} M) steadily grow with each successive exposure of the coronary vasculature to the activating media (Fig. 5). This unexpected reaction represents a typical *positive staircase phenomenon,* similar to that described by Bowditch [27] on heart muscle more than 100 years ago. In heart muscle, the positive staircase effect can be rapidly abolished or converted into a negative one by low concentrations of Ca^{2+} antagonists. Although the occurrence of an analogous phenomenon in coronary smooth muscle has hitherto not been considered, the positive staircase effects in the experiments of Fig. 5 are also highly susceptible to Ca^{2+} antagonist treatment.

It is generally accepted that a positive staircase phenomenon reflects a progressive saturation of cellular stores due to a repetitive Ca^{2+} uptake during a series of successive membrane excitations. Thus, our observations on K^+-depolarized coronary smooth musculature indicate that repeated Ca^{2+} fluxes across the voltage-dependent membrane chan-

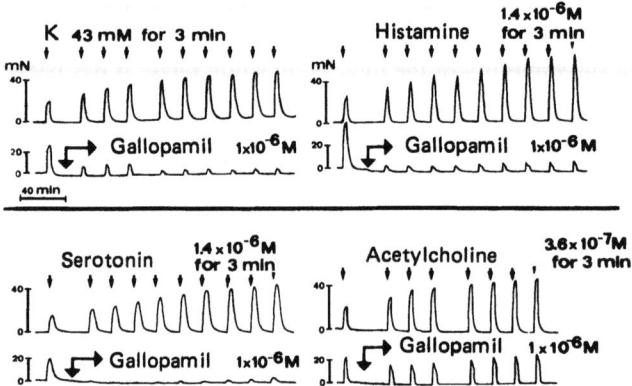

Figure 5. Repeated administration of a depolarizing K$^+$-rich Tyrode solution (43 mM K$^+$) and of Tyrode solutions containing different coronary vasoconstrictor agonists (histamine, serotonin, acetylcholine) to pig coronary strips in the presence of 1 mM Ca^{2+}. The duration of each exposure to the contracture-producing media was 3 min so that only the phasic part (A) of the contractile responses was assessed, followed by a washout period of 17 min. There is in all experiments a progressive increase in contraction amplitude ("positive staircase phenomenon") with each exposure of the coronary strips to the activating media. With gallopamil, the phasic responses and the staircases are greatly depressed.

nels, apart from directly activating the contractile machinery, may also favor a transient storage of Ca^{2+} ions in cellular depots. Then, during a series of successive depolarizations, increasing amounts of additional activator Ca^{2+} could be delivered from such stores as supplements to direct transmembrane sources of Ca^{2+}.

Is there a Ca^{2+}-triggered cellular Ca^{2+} release during phasic activity of coronary smooth muscle? The Ca^{2+} antagonist gallopamil, at 1×10^{-6} M, is highly efficient, not only in suppressing the contractures of K$^+$-depolarized coronary smooth muscle strips, but also in reducing the phasic responses elicited by histamine, serotonin, and acetylcholine. Verapamil, nifedipine, and diltiazem cause the same inhibitory effects as gallopamil (see Figs. 5–7). These observations are at variance with the model in Fig. 4, in which merely a Ca^{2+} release from cellular depots accounts for the rapid type of vascular smooth muscle activation by organic vasoconstrictor agonists. But the acute inhibitory effects of Ca^{2+} antagonists and of Ca^{2+} withdrawal (Figs. 6 and 7) indicate that the phasic responses also depend on Ca^{2+} availability from *extracellular* sources. To explain these findings, we propose that the liberation of Ca^{2+} from cellular stores (according to mechanism A) requires transmembrane influx of a small amount of Ca^{2+} which then, in turn, serves as a trigger for Ca^{2+} release from internal binding sites. This would represent an analogy to the Ca^{2+}-triggered Ca^{2+} release from the sarcoplasmic reticulum in heart muscle [28–33].

Comparison of Ca^{2+} dependency and drug susceptibility of phasic and tonic components in experimental coronary spasms. The scheme in Fig. 4 emphasizes, mainly for

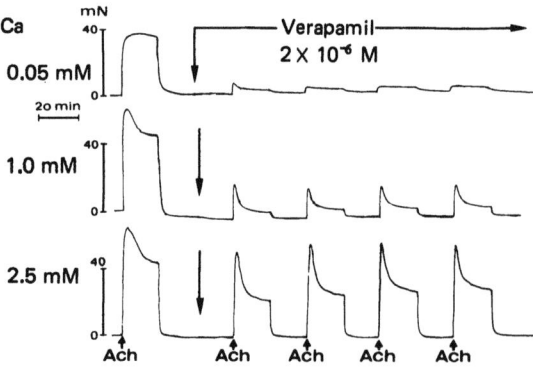

Figure 6. Contractile responses of pig coronary strips to repeated 20-min exposures to a Tyrode solution with 50 μg/liter acetylcholine (Ach). Influence of extracellular Ca^{2+} concentration on the degree of coronary relaxation produced by verapamil: Whereas the verapamil effect is potentiated by a low Ca^{2+} concentration (0.05 mM), verapamil loses most of its relaxing potency in the presence of a high Ca^{2+} concentration (2.5 mM). The tonic part (B) of the acetylcholine-induced contractile responses is slightly more inhibited by verapamil than is the phasic part (A).

Figure 7. Evaluation of the inhibitory influence of various Ca^{2+} antagonists on the phasic part (A), measured after 3 min, and on the tonic part (B), measured after 20 min, during five successive exposures of pig coronary strips to a Tyrode solution containing 50 µg/liter acetylcholine. Whereas part A and part B of the acetylcholine-induced contractile responses exhibit clear staircase phenomena in the control strips, both parts are greatly diminished or abolished in the presence of different Ca^{2+} antagonists added to the experimental media after the first exposure to acetylcholine. However, (1) the degree of contractile inhibition of the tonic part (B) of the acetylcholine-induced contractures is somewhat greater than is that of the phasic part (A), and (2) all Ca^{2+} antagonists inhibit the acetylcholine-induced contractures much more at a low (0.05 mM) than at a high (2.5 mM) extracellular Ca^{2+} concentration [Experimental techniques as in Fig. 6. Isometric tension development is calculated in absolute units (mN/mm^2 cross section of cylindric coronary smooth muscle strip).]

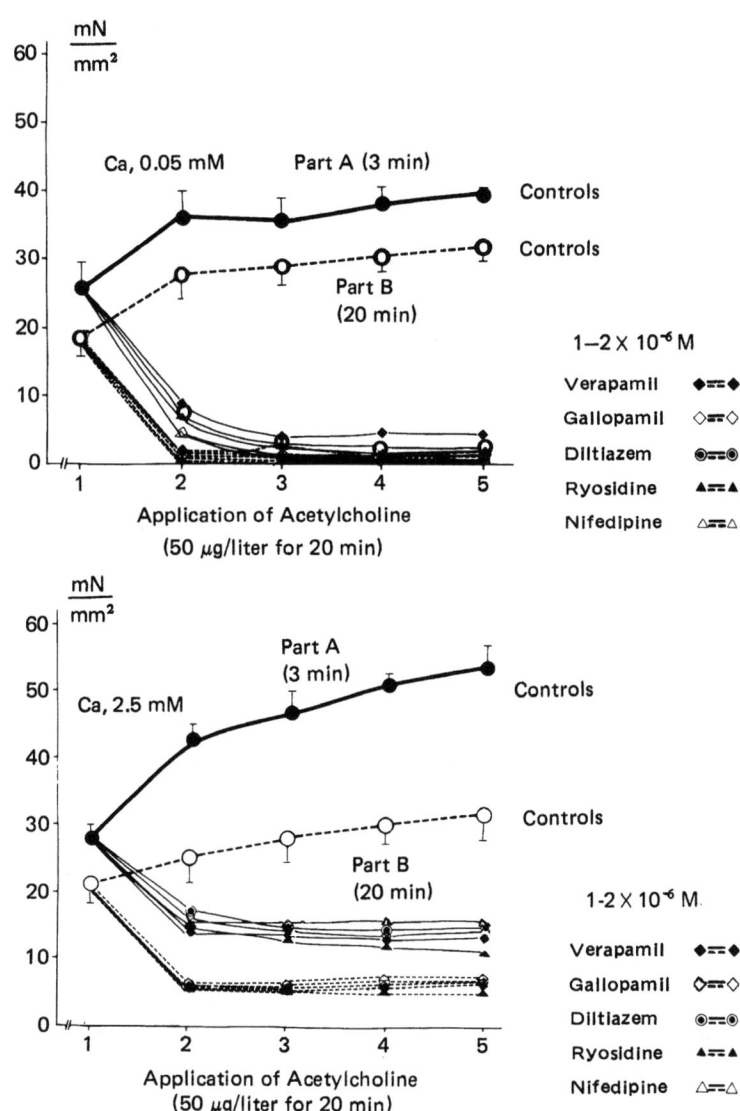

analytical reasons, the existing dissimilarities between the pathways of Ca^{2+} supply for phasic and tonic vascular smooth muscle activation. However, one should not overemphasize this theoretical differentiation between the two mechanisms because, at least in the coronary vasculature, Ca^{2+} antagonists and Ca^{2+} withdrawal always interfere with *both types* of Ca^{2+}-dependent contractile activation. For example, Fig. 6 represents a series of experiments in which a constant dose of acetylcholine (50 µg/liter) has been repeatedly applied for 20 min. By this procedure, a separate assessment of tension development during the initial phasic peak A (measured after 3 min) and the subsequent tonic part B (measured after 20 min, according to Fig. 3) is possible. Three conclusions can be drawn from the data in Fig. 6. (1) In pig coronary smooth muscle, the Ca^{2+} antagonist verapamil affects both the phasic and the tonic responses to acetylcholine simultaneously and to a similar extent. (2) The degree of depression of both phasic and tonic acetylcholine-induced contractile responses by a given dose of verapamil (2×10^{-6} M) is strongly dependent on extracellular

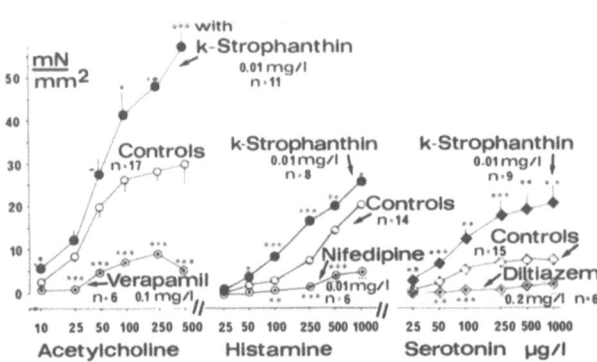

Figure 8. Potentiation by k-strophanthin and suppression by Ca^{2+} antagonists of the phasic parts (A) of the contractile responses of pig coronary strips to acetylcholine, histamine, and serotonin. Increasing doses of these vasoconstrictor agonists were successively added to Tyrode solutions containing k-strophanthin (0.01 mg/liter) or Ca^{2+} antagonists (verapamil: 0.1 mg/liter; nifedipine: 0.01 mg/liter; diltiazem: 0.2 mg/liter). Isometric tension development is calculated as mN/mm^2 cross section of cylindric coronary smooth muscle strip. ***$p < 0.001$, **$p < 0.005$, *$p < 0.05$.

Ca^{2+}. Thus, at a low extracellular Ca^{2+} concentration (0.05 mM), verapamil (as well as other Ca^{2+} antagonists) produces a more potent contractile inhibition. Conversely, the effects of Ca^{2+} antagonists are always decreased at an elevated Ca^{2+} concentration (2.5 mM). (3) There is only a small quantitative difference between the influence of verapamil on the two parts of the acetylcholine-induced contractile response. The tonic part is slightly more depressed than the phasic part.

These conclusions are representative of *all* Ca^{2+} antagonists tested so far (verapamil, gallopamil, diltiazem, ryosidine, nifedipine). In fact, these drugs always produce (in coronary smooth muscle) the same pattern of a "mixed" inhibition of phasic and tonic activity without a strong preference for suppression of tonic responses (Fig. 7). Experiments with histamine or serotonin, instead of acetylcholine, produce similar results (see also Fig. 8).

In this context, it should be noted that an increased transmembrane Ca^{2+} supply induced with cardiac glycosides, alters coronary contractile performance in the opposite direction from the Ca^{2+} antagonists [1,6–8,34–36]. In fact, cardiac glycosides facilitate transmembrane Ca^{2+} uptake and thereby enhance the contractile responses of coronary smooth muscle to mechanical, electrical, and pharmacological stimuli. Accordingly, k-strophanthin markedly increases the strength of serotonin-, histamine-, or acetylcholine-induced coronary contractions [10,18]. However, this potentiation by cardiac glycosides always manifests itself on both phasic and tonic portions of the response, although the tonic portion of the vasoconstrictor responses usually increases to a slightly greater extent.

An evaluation of our pertinent experiments, referring only to opposite changes of the phasic component A by Ca^{2+} antagonists (verapamil, nifedipine, diltiazem) and k-strophanthin, is given in Fig. 8. Here, definitive evidence is provided that the phasic responses of coronary smooth muscle to acetylcholine, histamine, and serotonin are almost as susceptible to alterations of the transmembrane Ca^{2+} conductivity produced by Ca^{2+} antagonists or cardiac glycosides as are the tonic responses. Obviously, phasic and tonic contractile performance is potentiated almost in parallel when transmembrane Ca^{2+} influx into the coronary vasculature is enhanced by cardiac glycosides. Conversely, both types of coronary contractile performance are inhibited if the transmembrane Ca^{2+} supply is decreased by Ca^{2+} antagonists. Our observations do not negate the possibility that more of the activator Ca^{2+} important for phasic contractions of coronary smooth muscle originates from cellular stores than via a transmembrane pathway, whereas the supply of activator Ca^{2+} for tonic responses depends to a greater degree on direct transmembrane influx rather than on cellular release. In any case, the model in Fig. 4 must be modified for the coronary vasculature. The practical implication is that Ca^{2+} antagonists have a particularly wide array of vasodilatory and spasmolytic actions against both *tonic and phasic* constrictions of the extramural coronary vasculature, irrespective of the specific spasmogenic factors involved.

In summary, both the phasic and the tonic activation of coronary smooth muscle from extramural stem arteries requires a transmembrane supply of Ca^{2+}. Any interruption of the transmembrane Ca^{2+} influx by Ca^{2+} antagonists simultaneously inhibits the two types of contractile response, although the tonic responses are always affected to a slightly greater extent than are the phasic components. Cardiac glycosides, on the other hand, enhance transmembrane Ca^{2+} uptake and thereby potentiate both phasic and tonic contractility simultaneously. The inhibitory effects of Ca^{2+} antagonists are enhanced if availability of extracellular Ca^{2+} is reduced; conversely, a high extracellular Ca^{2+} concentration counteracts the effects of Ca^{2+} antagonists. Coronary smooth muscle, analogous to the myocardium, exhibits positive staircase phenomena with respect to phasic and (to a slightly lesser degree) tonic activation, whereas Ca^{2+} antagonists abolish these positive staircase effects or convert them into negative ones.

References

1. Fleckenstein, A. *Calcium Antagonism in Heart and Smooth Muscle—Experimental Facts and Therapeutic Prospects*, New York, Wiley, 1983.
2. Grün, G.; Fleckenstein, A.; Byon, Y. K. Ca-antagonism, a new principle of vasodilation. In: *Proc. 25th Congr. Int. Union Physiol. Sci.*, Munich 1971, Vol. 9, 1971, p. 221.
3. Grün, G.; Fleckenstein, A.; Byon, Y. K. Blockierung der Ca^{++}-Effekte auf Tonus und Autoregulation der glatten Gefässmuskulatur durch Ca^{++}-Antagonisten (Verapamil, D 600, Prenylamin, Bay a 1040 u.a.). In: *Vascular Smooth Muscle* E. Betz, ed., Berlin, Springer-Verlag, 1972, pp. 69–70.
4. Grün, G.; Fleckenstein, A. Ca-antagonism, a new principle of vasodilation. *Naunyn-Schmiedebergs Arch. Exp. Pathol. Pharmakol.* **270**:(Suppl.):R48, 1971.
5. Grün, G.; Fleckenstein, A. Die elektromechanische Entkoppelung der glatten Gefässmuskulatur als Grundprinzip der Coronardilatation durch 4-(2'-Nitrophenyl)-2,6-dimethyl-1, 4-dihydropyridin-3,5-dicarbonsäure-dimethylester (Bay a 1040, Nifedipin). *Arzneim. Forsch.* **22**:334–344, 1972.
6. Fleckenstein, A.; Nakayama, K.; Fleckenstein-Grün, G.; Byon, Y. K. Interactions of vasoactive ions and drugs with Ca-dependent excitation–contraction coupling of vascular smooth muscle. In: *Calcium Transport in Contraction and Secretion*, E. Carafoli, F. Clementi, W. Drabikowski, and A. Margreth, eds., Amsterdam, North-Holland, 1975, pp. 555–566.
7. Fleckenstein, A.; Nakayama, K.; Fleckenstein-Grün, G.; Byon, Y. K. Mechanism and sites of action of calcium antagonistic coronary therapeutics. In: *Coronary Angiography and Angina Pectoris*, P. R. Lichtlen, ed., Stuttgart, Thieme, 1976, pp. 297–316.
8. Fleckenstein, A.; Nakayama, K.; Fleckenstein-Grün, G.; Byon, Y. K. Interactions of H ions, Ca-antagonistic drugs and cardiac glycosides with excitation–contraction coupling of vascular smooth muscle. In: *Ionic Actions on Vascular Smooth Muscle*, E. Betz, ed., Berlin, Springer-Verlag, 1976, pp. 117–123.
9. Fleckenstein-Grün, G.; Fleckenstein, A. Ca-dependent changes in coronary smooth muscle tone and the action of Ca-antagonistic compounds with special reference to Adalat. In: *New Therapy of Ischemic Heart Disease*, W. Lochner, W. Braasch, and G. Kroneberg, eds., Berlin, Springer-Verlag, 1975, pp. 66–75.
10. Fleckenstein-Grün, G.; Fleckenstein, A. Calcium-Antagonismus, ein Grundprinzip der Vasodilatation. In: *Calcium-Antagonismus*, A. Fleckenstein and H. Roskamm, eds., Berlin, Springer-Verlag, 1980, pp. 191–207.
11. Fleckenstein-Grün, G.; Fleckenstein, A. Calcium antagonism, a basic principle in vasodilation. In: *Calcium Antagonism in Cardiovascular Therapy—Experience with Verapamil*, A. Zanchetti and D. M. Krikler, eds., Amsterdam, Excerpta Medica, 1981, pp. 30–48.
12. Fleckenstein-Grün, G.; Fleckenstein, A. Blockierung der Ca^{++}-abhängigen bioelektrischen Automatie und elektromechanischen Koppelung glatter Muskelzellen durch Gallopamil (D 600). In: *Gallopamil—Pharmakologisches und klinisches Wirkungsprofil eines Kalziumantagonisten*, M. Kaltenbach and R. Hopf, eds., Berlin, Springer-Verlag, 1983, pp. 35–51.
13. Fleckenstein, A.; Späh, F.; Fleckenstein-Grün, G.; Byon, Y. K.; Frey, M.; von Witzleben, H. Wirkungsspektrum und Spezifität des Calciumantagonisten Diltiazem. 1. Dilzem Symposium, Copenhagen, Amsterdam, Excerpta Medica, 1982, pp. 3–45.

14. Haeusler, G. Differential effect of verapamil on excitation–contraction coupling in smooth muscle and on excitation–secretion coupling in adrenergic nerve terminals. *J. Pharmacol. Exp. Ther.* **180**:672–682, 1972.
15. Bilek, I.; Peiper, U. The influence of verapamil on the noradrenaline activation of the isolated aorta and portal vein of the rat. *Pfluegers Arch. Gesamte Physiol. Menschen Tiere* **343**(Suppl.):R57, 1973.
16. Massingham, R. A study of compounds which inhibit vascular smooth muscle contraction. *Eur. J. Pharmacol.* **22**:75–82, 1973.
17. Schümann, H. J.; Görlitz, B. D.; Wagner, J. Influence of papaverine, D 600 and nifedipine on the effects of noradrenaline and calcium on the isolated aorta and mesenteric artery of the rabbit. *Naunyn-Schmiedebergs Arch. Exp. Pathol. Pharmakol.* **289**:409–418, 1975.
18. Fleckenstein-Grün, G. Control of coronary spasms by calcium antagonists. In: *Calcium Modulators*, T. Godfraind, A. Albertini, and R. Paoletti, eds., Amsterdam, Elsevier, 1982, pp. 141–154.
19. Godfraind, T.; Kaba, A. Actions phasiques et tonique de l'adrénaline sur un muscle lisse vasculaire et leur inhibition par des agents pharmacologiques. *Arch. Int. Pharmacodyn. Ther.* **178**:488–491, 1969.
20. Bevan, J. A.; Garstka, W.; Su, C.; Su, M. O. The bimodal basis of the contractile response of the rabbit ear artery to norepinephrine and other agonists. *Eur. J. Pharmacol.* **22**:47–53, 1973.
21. Hinke, J. A. M.; Wilson, M. L.; Burnham, S. C. Calcium and the contractility of arterial smooth muscle. *Am. J. Physiol.* **206**:211–217, 1964.
22. Hudgins, P. M.; Weiss, G. B. Differential effects of calcium removal upon vascular smooth muscle contraction induced by norepinephrine, histamine and potassium. *J. Pharmacol. Exp. Ther.* **159**:91–97, 1968.
23. Van Breemen, C.; Lesser, P. The absence of increased membrane permeability during norepinephrine stimulation of arterial smooth muscle. *Microvasc. Res.* **3**:113–114, 1971.
24. Van Breemen, C.; Farinas, B. R.; Gerba, P.; McNaughton, E. D. Excitation–contraction coupling in rabbit aorta studied by the lanthanum method for measuring cellular calcium influx. *Circ. Res.* **30**:44–54, 1972.
25. Karaki, H.; Weiss, G. B. Alterations in high and low affinity binding of ^{45}Ca in rabbit aortic smooth muscle by norepinephrine and potassium after exposure to lanthanum and low temperature. *J. Pharmacol. Exp. Ther.* **211**:86–92, 1979.
26. Karaki, H.; Weiss, G. B. Effects of stimulatory agents on mobilization of high and low affinity site ^{45}Ca in rabbit aortic smooth muscle. *J. Pharmacol. Exp. Ther.* **213**:450–455, 1980.
27. Bowditch, H. P. Über die Eigenthümlichkeiten der Reizbarkeit, welche die Muskelfasern des Herzens zeigen. *Arb. Physiol. Anstalt Leipzig* **6**:139–176, 1871.
28. Endo, M.; Tanaka, M.; Ogawa, Y. Calcium induced release of calcium from the sarcoplasmic reticulum of skinned skeletal muscle fibres. *Nature (London)* **228**:34–36, 1970.
29. Ford, L. E.; Podolsky, R. J. Regenerative calcium release within muscle cells. *Science* **67**:58–59, 1970.
30. Fabiato, A.; Fabiato, F. Contractions induced by a calcium-triggered release of calcium from the sarcoplasmic reticulum of single skinned cardiac cells. *J. Physiol. (London)* **249**:469–495, 1975.
31. Fabiato, A.; Fabiato, F. Calcium release from the sarcoplasmic reticulum. *Circ. Res.* **40**:119–129, 1977.
32. Fabiato, A.; Fabiato, F. Calcium-induced release of calcium from the sarcoplasmic reticulum of skinned cells from adult human, dog, cat, rabbit, rat and frog hearts and from fetal and newborn rat ventricles. *Ann. N.Y. Acad. Sci.* **307**:491–522, 1978.
33. Endo, M. Calcium release from the sarcoplasmic reticulum. *Physiol. Rev.* **57**:71–108, 1977.
34. Fleckenstein, A.; Byon, Y. K. Prevention by Ca-antagonistic compounds (verapamil, D 600) of coronary smooth muscle contractures due to treatment with cardiac glycosides. *Naunyn-Schmiedebergs Arch. Exp. Pathol. Pharmacol.* **282**(Suppl.):R20, 1974.
35. Grün, G.; Fleckenstein, A.; Weder, U. Changes in coronary smooth muscle tone produced by Ca, cardiac glycosides and Ca-antagonistic compounds (verapamil, D 600, prenylamine, etc.). *Pfluegers Arch. Gesamte Physiol. Menschen Tiere* **347**(Suppl.):R1, 1974.
36. Fleckenstein, A.; Fleckenstein-Grün, G. Further studies on the neutralization of glycoside-induced contractures of coronary smooth muscle by Ca-antagonistic compounds (verapamil, D 600, prenylamine, nifedipine, fendiline or nitrites). *Nauny-Schmiedebergs Arch. Exp. Pathol. Pharmakol.* **287**(Suppl.):R38, 1975.
37. Fleckenstein, A. Specific pharmacology of calcium in myocardium, cardiac pacemakers, and vascular smooth muscle. *Annu. Rev. Pharmacol. Toxicol.* **17**:149–166, 1977.

48

CGP 28392, a Dihydropyridine Ca^{2+} Entry Stimulator

Arnold G. Truog, Hellmut Brunner, Leoluca Criscione,
Marc Fallert, Hans Kühnis, Max Meier, and Harald Rogg

Agents termed Ca^{2+} antagonists or Ca^{2+} channel blockers (e.g., nifedipine, verapamil, diltiazem) reduce Ca^{2+} influx through specific channels located in the membranes of vascular smooth muscle and myocardial cells. This action results in vasodilation and decreased myocardial contractility [1–5]. Consequently, this group of agents has clinical value in the therapy of angina pectoris, arrhythmias, and hypertension [6]. Compounds that act to increase Ca^{2+} influx through specific membrane channels might be expected to possess pharmacological activities opposite to those of the Ca^{2+} antagonists (e.g., vasoconstriction and increased myocardial contractility) [7]. The first Ca^{2+} entry stimulator (or Ca^{2+} agonistic compound) structurally related to nifedipine was YC-170 [8]. Like nifedipine, YC-170 is a 1,4-dihydropyridine derivative, but produced vasopressor effects in anesthetized rats and dogs as well as vasoconstriction in the isolated rabbit aorta. Based on indirect evidence, the authors suggested that these effects were due to an enhanced Ca^{2+} influx into vascular smooth muscle cells.

This chapter presents experimental information about a new dihydropyridine derivative, CGP 28392 (Fig. 1), which also belongs to this new class of Ca^{2+} entry stimulators and possesses prominent positive inotropic effects.

Inotropic Effects

The effects of CGP 28392 on myocardial contractility were assessed in isolated guinea pig atria electrically driven at 2.5 Hz in Krebs–Henseleit solution at 32°C (2.5 mM Ca^{2+}). Under these conditions, nifedipine reduced the peak developed tension by 25 and 50% at concentrations of 0.09 and 0.35 µM, respectively. In contrast to this well-known negative

---

Arnold G. Truog, Hellmut Brunner, Leoluca Criscione, Marc Fallert, Hans Kühnis, Max Meier, and Harald Rogg • Biological Research Laboratories, Pharmaceuticals Division, Ciba-Geigy Corporation, CH 4000 Basel, Switzerland.

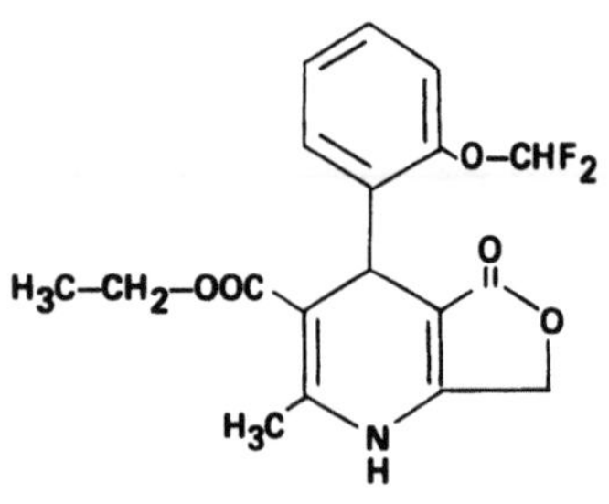

Figure 1. Chemical structure of CGP 28392, 4-[2-(Difluoromethoxy)-phenyl]-1,4,5,7-tetrahydro-2-methyl-5-oxo-furo[3,4-b]pyridine-3-carboxylic acid ethylester.

inotropic effect of a prototype Ca^{2+} antagonist, CGP 28392 increased peak developed tension in the concentration range of 0.03 to 30 μM with a half-maximal effect (EC_{50}) at 0.7 μM (Fig. 2). The inotropic efficacy of CGP 28392 was about 70% of that of a maximal stimulation induced by the β-adrenergic agonist isoproterenol. The efficacy was similar to that of a high concentration of ouabain (0.8 μM); about 2.5 times higher than that of the cardioselective $β_1$-adrenergic agonist prenalterol; and about 30% higher than that of the bipyridine derivative amrinone. Except for ouabain, all compounds were tested in cumulative concentrations. Due to the occurrence of arrhythmias after the establishment of its inotropic effect, ouabain had to be tested in single concentrations. All other compounds did not produce rhythm disturbances in these experiments. The initial values of the peak developed tension were consistently between 2 and 2.5 g.

As estimated from equi-effective concentrations, CGP 28392 proved to be about 300–600 times less active than 1-isoproterenol; 2–5 times less active than prenalterol and ouabain; and about 500 times more active than amrinone. Compared to CGP 28392, the Yamanouchi compound (YC-170) was about 70 times less active and its efficacy was roughly two-thirds that of CGP 28392.

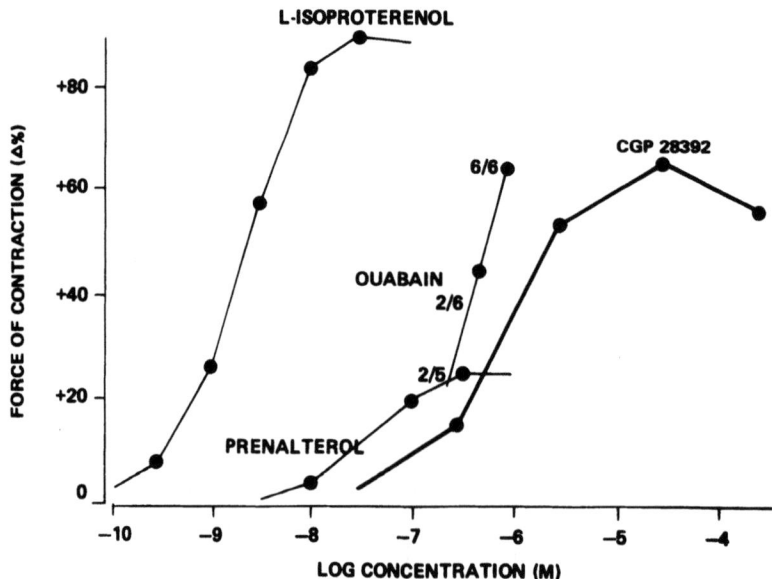

Figure 2. Inotropic dose–response curves in guinea pig left atria. Given are mean values of percentage increases in peak developed tension in 5 to 8 experiments (CGP 28392, ouabain, prenalterol) and 30 experiments (isoproterenol). CGP 28392, prenalterol, and L-isoproterenol were tested in cumulative concentrations; ouabain was tested in single concentrations. (Figures indicate the relative occurrence of arrhythmias.)

In electrically stimulated human right atrial strips obtained from patients undergoing open heart surgery, CGP 28392 induced clearly defined positive inotropic effects. As in guinea pig atria, the maximal peak developed tension (efficacy) of CGP 28392 was again slightly lower than that of isoproterenol. In human atrial tissue, the potency of CGP 28392 was about 10 times lower (EC_{50} 8 μM, $n = 6$) and that of l-isoproterenol was slightly higher (EC_{50} 1 nM, $n = 8$) than in the guinea pig atria.

CGP 28392 increased left ventricular aortic dP/dt_{max}, mean arterial blood pressure, and total peripheral vascular resistance in preliminary hemodynamic studies in anesthetized cats (3 mg/kg/5 min i.v., $n = 2$) and dogs (1 and 10 mg/kg/20 min i.v., $n = 4$).

Mechanism of Action

Displacement of [³H]nitrendipine. In guinea pig heart membranes, [³H]nitrendipine (0.15 nM) can be displaced from specific binding sites by dihydropyridine Ca^{2+} entry blockers such as nifedipine (IC_{50} 2 nM) (Fig. 3). CGP 28392 also displaced [³H]nitrendipine with an IC_{50} value of 0.4 μM. Therefore, CGP 28392 has about 1/200 of the affinity of nifedipine for dihydropyridine-binding sites. The displacement of the ligand by CGP 28392 occurred in the same concentration range as did the positive inotropic effects. Thus the actions of CGP can be attributed to stimulation of calcium influx into myocardial cells on a dihydropyridine sensitive site.

Lack of adrenergic stimulant properties. The inotropic dose–response curve of CGP 28392 in guinea pig left atria remained unchanged in the presence of a high concentration of the β-adrenergic antagonist oxprenolol (1 μM, added to the bathing solution 30 min before CGP 28392). The EC_{50} values of CGP 28392 amounted to 1 and 0.7 μM, respectively, in the presence and absence of oxprenolol; and the maximal increases in peak developed tension were +68 ± 5 and +66 ± 6%, respectively. Since oxprenolol produces a rightward shift of

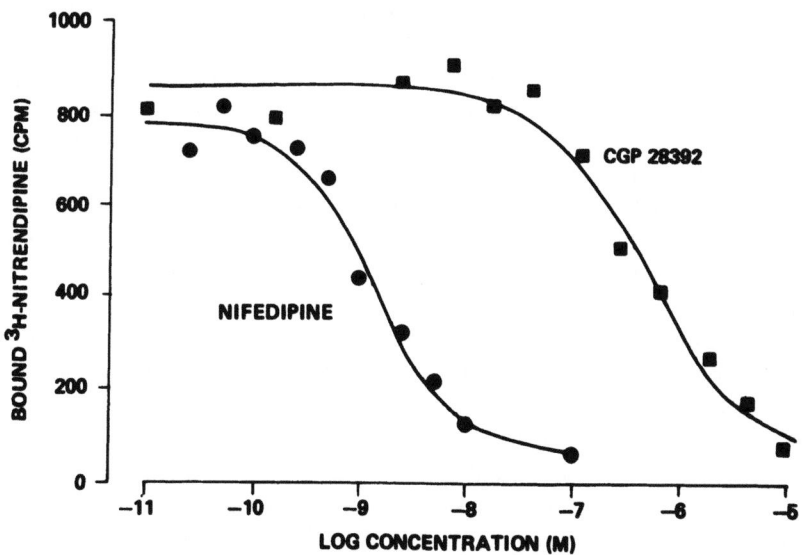

Figure 3. Displacement of [³H]nitrendipine by CGP 28392 and nifedipine in guinea pig heart membranes. The data given are representative of four to five experiments, each performed in duplicate with an [³H]nitrendipine concentration of 0.15 nM. The mean IC_{50} values obtained were 2 nM (nifedipine) and 0.4 μM (CGP 28392).

the inotropic dose–response curve of isoproterenol (by a factor of 280), the existence of β-adrenergic stimulant properties of CGP 28392 can be excluded. Similarly, in anesthetized cats, phentolamine (0.3 mg/kg, i.v.) did not inhibit the pressor effect of CGP 28392 (0.1 mg/kg, i.v.), so this compound is also devoid of α-adrenergic stimulant properties.

Lack of histaminergic stimulant properties. In guinea pig left atria, the inotropic dose–response curve of CGP 28392 remained unchanged in the presence of 1 μM pyribenzamine (which produces a rightward shift of the histamine dose–response curve by a factor of about 80). Therefore, CGP 28392 does not exhibit histaminergic stimulant properties.

Na^+,K^+-ATPase. In the concentration range from 10 nM to 0.1 mM, CGP 28392 did not affect enzyme activity of dog kidney Na^+,K^+-ATPase, whereas 0.1 mM ouabain completely inhibited this enzyme (IC_{50} 2 μM). Therefore, the positive inotropic effect of CGP 28392 is not mediated by inhibition of Na^+,K^+-ATPase.

Chronotropic effects. In the same concentration range in which the inotropic effects occurred (0.03 to 30 μM), CGP 28392 also slightly increased the rate of contraction by about 40 beats/min in spontaneously beating guinea pig right atria (initial rate of contraction 150 beats/min). The EC_{50} value was 1 μM, and the chronotropic efficacy of the compound was substantially lower than that of isoproterenol (about one-third) and approximately 50% of that of prenalterol, ouabain, and amrinone. Elevating extracellular Ca^{2+} also increased the rate of contraction of the spontaneously beating right atria by about 60 beats/min.

At a very high concentration (300 μM), CGP 28392 totally blocked spontaneous contractile activity in two out of six right atria. Also, pacemaker cells are highly dependent on extracellular Ca^{2+}, and guinea pig atria ceased to beat at higher concentrations of nifedipine (7 out of 12 atria at 0.3 μM) [9]. CGP 28392, therefore, seems to possess some additional Ca^{2+} antagonistic properties at very high concentrations (see also Vascular Effects section).

Influence of CGP 28392 on Inotropic Dose–Response Curves for Ca^{2+} in Guinea Pig Left Atria

In these studies, the general experimental conditions were the same as those used for assessment of the inotropic effects of CGP 28392. Since atria had attained stable peak developed tension levels at a normal extracellular Ca^{2+} concentration of 2.5 mM, the Ca^{2+} concentration in the solution was reduced for 20 min in order to reach the nadir of the inotropic dose–response curve. In control experiments, a low Ca^{2+} concentration (0.5 mM) decreased contractility from 2.3 to 0.17 g ($n = 18$). Subsequent stepwise increases in extracellular Ca^{2+} restored contractility, and these increases were continued until no further augmentation in contractility occurred. The maximum response was obtained at 8–10 mM Ca^{2+}, and the peak developed tension was 4.1 g. This value approximates that which can be attained by a maximal stimulation with isoproterenol at normal Ca^{2+} concentrations. Thus, temporary reduction of extracellular Ca^{2+} concentration did not damage the atria.

In the presence of single concentrations of CGP 28392 (0.3 and 3 μM, added 40 min before reduction of extracellular Ca^{2+}) the inotropic dose–response curves to Ca^{2+} were shifted to the left in a parallel manner by factors of 1.5 and 3.1, respectively (Fig. 4). At concentrations of 0.1, 1, and 10 μM, nifedipine shifted the Ca^{2+} dose–response curve to the right by factors of 1.3, 2.1, and 3.2, respectively. At the same concentrations, verapamil produced almost identical rightward shifts (1.5, 2.3, and 3.4). Compared to control experiments, the maximal contractility reached at the peak of the Ca^{2+} dose–response curves remained virtually unaffected by CGP 28392, nifedipine, and verapamil.

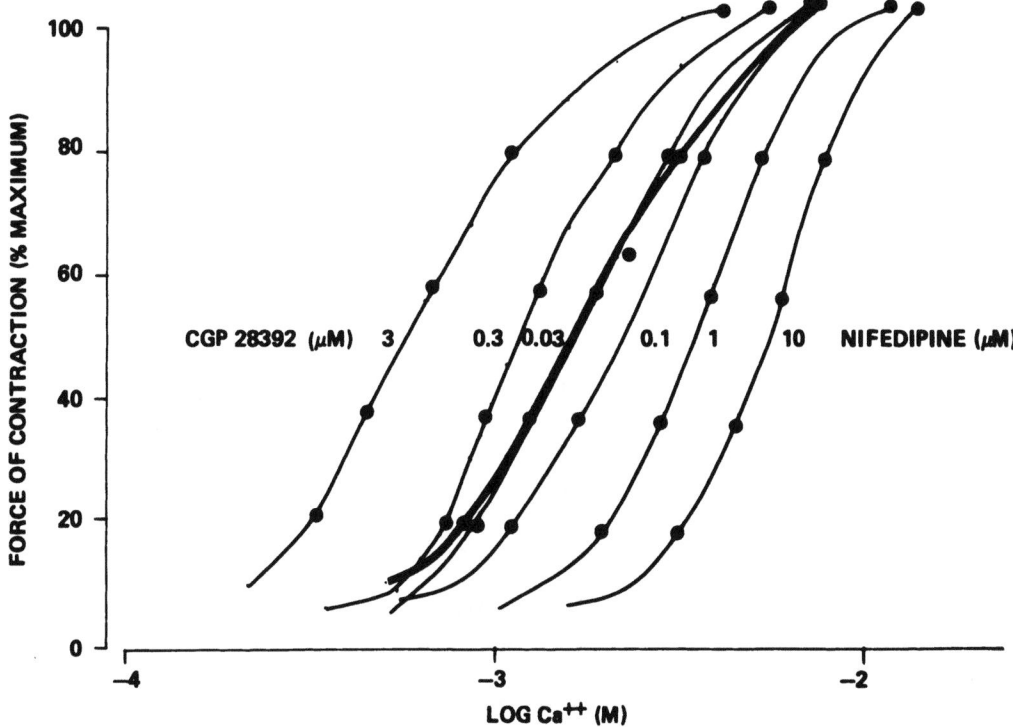

Figure 4. Influence of CGP 28392 and nifedipine on inotropic Ca^{2+} dose–response curves in guinea pig left atria. Given are mean dose–response curves ($n=6$ each, except for controls, $n=18$). The inotropic effects during the 40-min incubation period at normal (2.5 mM) Ca^{2+} concentration amounted to $-7 \pm 4\%$, $+20 \pm 5\%$, and $+56 \pm 6\%$ (0.03, 0.3, and 3 μM CGP 28392), $-24 \pm 3\%$, $-66 \pm 3\%$, and $-90 \pm 1\%$ (0.1, 1, and 10 μM nifedipine).

Competitive Antagonism of the Inotropic Effects of CGP 28392 by Verapamil

In electrically stimulated guinea pig atria, inotropic dose–response curves of CGP 28392 were determined in the absence or presence of various verapamil concentrations ranging from 0.3 to 10 μM. In each experiment only one dose–response curve was determined; verapamil was added to the bathing solution 60 min before addition of the first concentration of CGP 28392.

Verapamil depressed myocardial contractility in a concentration-dependent manner during the incubation period, and the threshold of the CGP 28392 dose–response curves was consequently observed at successively lower levels of contractility. The maximum contractile response attained by the agonist remained unchanged up to 1 μM verapamil, but decreased slightly at higher concentrations of the antagonist. However, conversion of the data from Δabs. into percentage maximum effect resulted in virtually parallel dose–response curves.

From the Schild plot (Fig. 5), it can be concluded that verapamil competitively antagonizes the effects of CGP 28392 with a pA_2 value of 6.8 (0.16 μM). At this concentration, verapamil reduced contractility of the guinea pig atria by one-third. The limited solubility of the agonist and/or its antagonistic properties at high concentrations may be responsible for

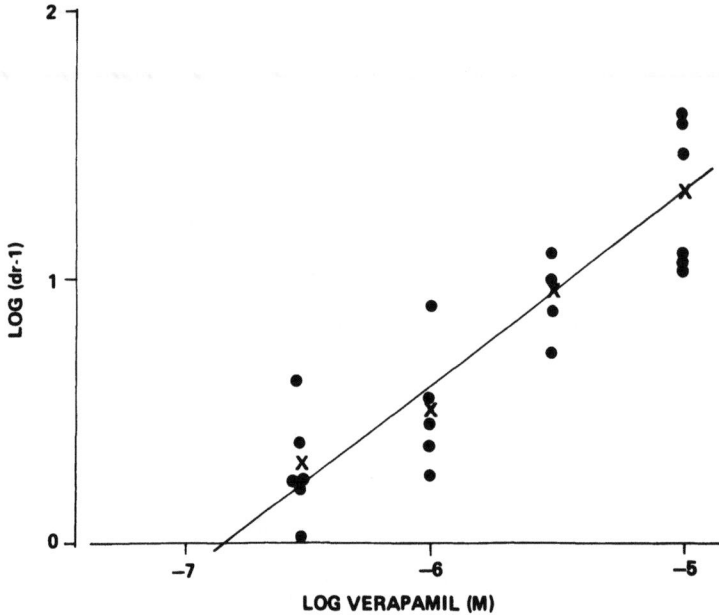

Figure 5. Competitive antagonism of the positive inotropic effects of CGP 28392 by verapamil in guinea pig left atria (Schild plot). ●, log (dr − 1) from single experiments calculated by using the mean EC_{50} value from eight controls. X, mean values of log (dr − 1) for each concentration of verapamil. The pA_2 value was 6.8 (0.16 μM), the slope about 0.8.

the fact that the slope in the Schild plot was slightly less than the theoretical value (0.8 instead of 1) [10,11].

Several attempts were also made to determine the pA_2 value of nifedipine. Since both the agonist and the antagonist have limited solubility in the Krebs–Henseleit solution, highly variable and inconsistent rightward shifts of the agonist dose–response curves were obtained.

Effects on the Action Potential

Electrophysiological studies in guinea pig papillary muscles at normal K^+ concentration (4.7 mM) indicate that CGP 28392 (0.1 to 10 μM) clearly prolongs the upper part of the action potential (20 and 40% repolarization) (Fig. 6A). At high concentrations (10 μM), the lower part of the action potential (90% repolarization) was also prolonged. In the same concentration range, the compound increased the mechanical activity (peak developed tension and dT/dt) of the papillary muscle, whereas maximum diastolic potential and action potential amplitude remained unchanged up to 10 μM. In the same test system, nifedipine (0.1 to 10 μM) again had effects opposite to those of CGP 28392 (Fig. 6B). In the partially depolarized preparation (K_0^+ = 27 mM, Ba_0^{2+} = 0.5 mM), however, prolongation of the action potential could not be observed. Slight prolongation of action potential duration was observed in automatically discharging calf Purkinje cells; the steepness of diastolic depolarization was increased and the diastolic interval shortened.

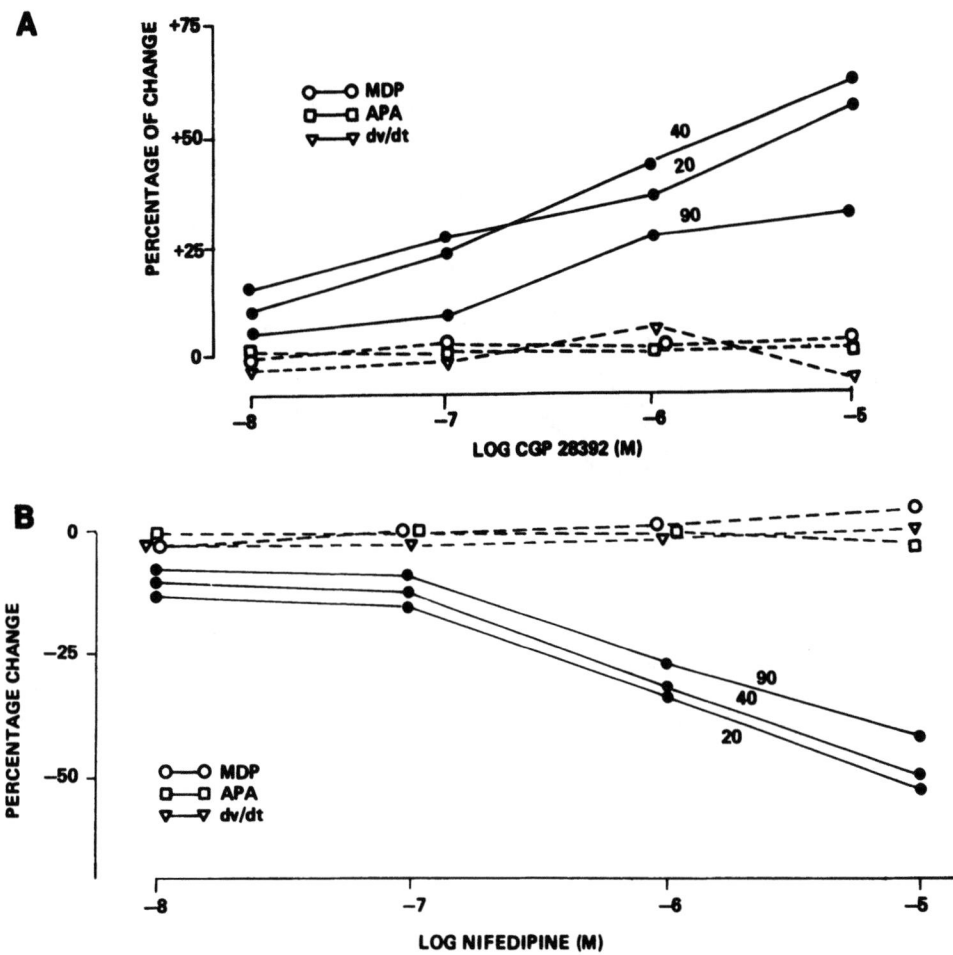

Figure 6. Effects of CGP 28392 and nifedipine on the action potential in guinea pig papillary muscle. Given are mean percent changes from initial values ($n=9-15$). Shown are the effects of CGP 28392 (A) and nifedipine (B) on the action potential duration at 20, 40, and 90% of repolarization. ○, maximal diastolic potential (MDP); □, action potential amplitude (APA); ▽, maximal upstroke velocity (*dv/dt*).

Vascular Effects

In the rat isolated perfused mesenteric artery preparation, vasoconstrictor Ca^{2+} dose–response curves were determined in the absence and presence of CGP 28392 (Fig. 7). CGP 28392 was dissolved in ethanol, and the solvent had no significant effects on the control dose–response curve. CGP 28392 (3 to 300 nM) shifted the vasoconstrictor Ca^{2+} dose–response curves to the left and, in contrast to the inotropic effects observed in guinea pig atria, also increased the maximum effects of Ca^{2+}. CGP 28392 was approximately 120 times more active in rat vascular tissue than in guinea pig myocardial tissue. Assuming that the reduction of the leftward shift observed at 3 and 9 μM indicates Ca^{2+} antagonistic activity, this action is more pronounced in vascular tissue. Therefore, lower concentrations of CGP 28392 appear to have vascular-selective Ca^{2+} entry stimulant activities, and higher concentrations of CGP 28392 have additional vascular-selective Ca^{2+} antagonistic proper-

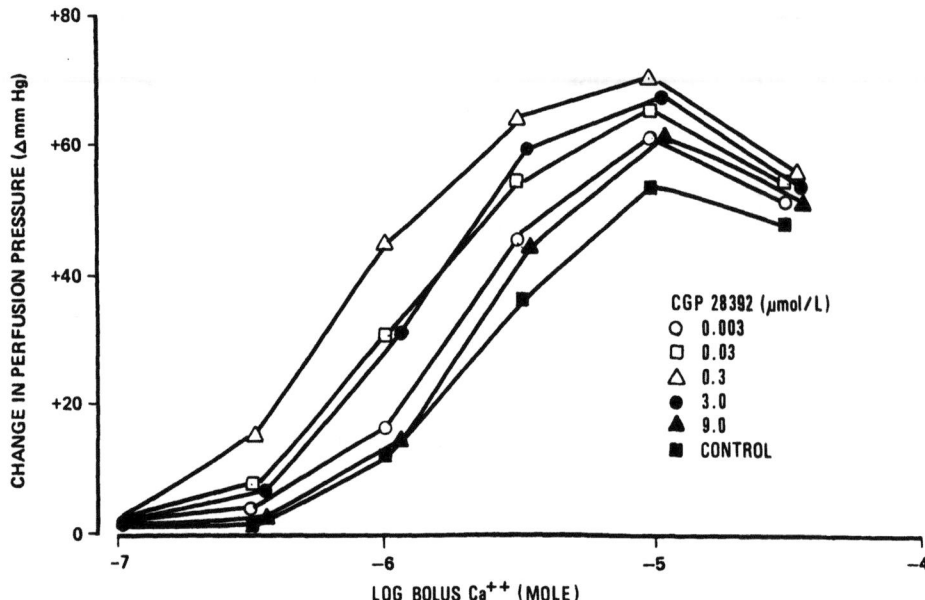

Figure 7. Effects of CGP 28392 on vasoconstrictor Ca^{2+} dose–response curves in the rat isolated perfused mesenteric artery preparation. The perfusate was Tyrode's solution containing 90 mM K^+, 0.1 mM EDTA, and no added Ca^{2+}. Incubation time 45 min. In control experiments, cumulative increases in the extracellular Ca^{2+} concentration resulted in a maximum perfusion pressure increase of about 53 ± 3 mm Hg. Given are mean values for six experiments.

ties (for *in vivo* vasoconstrictor effects of CGP 28392, see preliminary hemodynamic studies in Inotropic Effects section).

Summary

CGP 28392 is a dihydropyridine derivative with pharmacological activities opposite to those of Ca^{2+} antagonists but with affinity for the same [^{3}H]nitrendipine-binding sites (IC_{50} of 0.4 μM as compared to an IC_{50} of 2 nM for nifedipine). These findings indicate that CGP 28392 is a member of a new class of Ca^{2+} entry stimulators. The presence of agonist activity at low concentrations and antagonist activity at higher concentrations in a single molecule, taken together with an affinity for dihydropyridine-binding sites and competitive antagonism by Ca^{2+} entry blockers, characterizes CGP 28392 as a partial agonist. Hence, a similarity exists between the properties of this Ca^{2+} entry stimulator and those of partial β-adrenergic receptor agonists (e.g., prenalterol).

References

1. Fleckenstein, A. (ed.) Discovery and mechanism of action of specific calcium-antagonistic inhibition of excitation–contraction coupling in the mammalian myocardium. In: *Calcium Antagonism in Heart and Smooth Muscle*, New York, Wiley, 1983, pp. 34–41.
2. Fleckenstein, A. (ed.) Selective suppression of the transsarcolemmal inward current by calcium antagonists. In: *Calcium Antagonism in Heart and Smooth Muscle*, New York, Wiley, 1983, pp. 51–56.
3. Fleckenstein, A.; Tritthart, H.; Döring, H.-J.; Byon, K. Y. BAY a 1040—A highly potent Ca^{++} antagonistic

inhibitor of excitation–contraction coupling in the mammalian ventricular myocardium. *Drug Res.* **22**:22–33, 1972.

4. Nakajima, H.; Hoshiyama, M.; Yamashita, K.; Kiyomoto, A. Effect of diltiazem on electrical and mechanical activity of isolated cardiac ventricular muscle of guinea-pig. *Jpn. J. Pharmacol.* **25**:383–392, 1975.

5. Reuter, H. Calcium channel modulation by neurotransmitters, enzymes and drugs. *Nature (London)* **301**:569–574, 1983.

6. Fleckenstein, A. (ed.) The practical significance of calcium antagonists in cardiovascular therapy. In: *Calcium Antagonism in Heart and Smooth Muscle,* New York, Wiley, 1983, pp. 286–320.

7. Evans, D. B.; Weishaar, R. E.; Kaplan, H. R. Strategy for the discovery and development of a positive inotropic agent. *Pharmacol. Ther.* **16**:303–330, 1982.

8. Takenaka, T.; Maeno, H. A new vasoconstrictor 1,4-dihydropyridine derivative, YC-170. *Jpn. J. Pharmacol.* **32**(Suppl.):139P, 1982.

9. Fleckenstein, A. (ed.) General involvement of calcium ions in sinoatrial or atrioventricular automaticity and intranodal impulse conduction: Differentiation between fast- and slow-channel-mediated excitatory events. In: *Calcium Antagonism in Heart and Smooth Muscle,* New York, Wiley, 1983, pp. 165–185.

10. Schild, H. O. pA, a new scale for the measurement of drug antagonism. *Br. J. Pharmacol.* **2**:189–206, 1947.

11. Tallarida, R. J.; Cowan, A.; Adler, M. W. pA_2 and receptor differentiation: A statistical analysis of competitive antagonism. *Life. Sci.* **25**:637–654, 1979.

XI

Calcium Entry Blockers and Disease

49

Role of Ca^{2+} in and Effect of Ca^{2+} Entry Blockers on Excitation–Contraction and Excitation–Secretion Coupling

R. Casteels and G. Droogmans

It is now generally accepted that, in a variety of cells, a change in cytoplasmic Ca^{2+} concentration is a primary signal for cell function modulation (contraction, secretion, metabolic pathway changes). Ca^{2+} enters the cytoplasm by release from a cellular compartment or by influx from the extracellular medium. The latter process depends on the large inwardly directed electrochemical gradient for Ca^{2+}. A notable exception to this dependence on external Ca^{2+} is skeletal muscle [6,12]. The activation cycle of the contractile proteins in skeletal muscle depends to a large extent on Ca^{2+} movements between cytoplasm and the sarcoplasmic reticulum (SR); the amount of Ca^{2+} supplied from the external medium is rather limited. Other cells (e.g., smooth muscle cells, several types of secretory cells) require a continuous influx of Ca^{2+} from the outer medium for maintained activation. In these cells, Ca^{2+} release from the intracellular store in a Ca^{2+}-free medium can only elicit a transient response [2,11,20,31].

The transient nature of this response might be due to a higher affinity of the plasmalemmal Ca^{2+} extrusion pump than the Ca^{2+} reuptake system of the SR. These Ca^{2+} pumps and a low value for Ca^{2+} permeability of the plasma membrane in quiescent cells are of primary importance not only for preserving the modulatory role of low Ca^{2+} concentrations but also to prevent a compromise of cell function by excessively high cytoplasmic Ca^{2+} concentrations. Under normal quiescent conditions, cellular Ca^{2+} exchanges with external ^{45}Ca, which results in labeling of an important cellular Ca^{2+} fraction [11]. This slow Ca^{2+} entry does not interfere with steady-state cytoplasmic $[Ca^{2+}]$ because such slow Ca^{2+} leakage is completely neutralized by the Ca^{2+} extrusion pump.

A maintained increase in cytoplasmic Ca^{2+} can be induced either by an increase in Ca^{2+} influx (by opening Ca^{2+} channels in the cell membrane) or by an inhibition of the Ca^{2+} extrusion pump. The first mechanism is well known and has been validated by

---

R. Casteels and G. Droogmans • Laboratory of Physiology, Catholic University of Louvain, Gasthuisberg Campus, B3000 Louvain, Belgium.

considerable experimental evidence. The second one is currently of theoretical importance and has been proposed in relation to the action of oxytocin [1,28]. It could also explain the increased force development in smooth muscle induced by an agonist in Ca^{2+}-free solution at lower temperatures [10].

Ca^{2+} Channels of the Plasma Membrane

The regulation of Ca^{2+} influx depends on the presence of Ca^{2+} channels in the cell membrane. We may postulate that Ca^{2+} channels are membrane structures [14] which can be modulated by such membrane signals as a change of membrane potential, binding of some agonist to its receptor [19,22,26], or cytoplasmic ionic (e.g., Ca^{2+}) composition changes [5,29]. Much less is known about the nature of these Ca^{2+} channels than about Na^+ channels [15]. The apparent classification of Ca^{2+} channels as potential-dependent pores and receptor-operated channels could not be maintained because of observations on Ca^{2+} currents in *Aplysia* neurons [33] and in cardiac muscle cells [25,30], thus adducing the existence of multiple regulatory channels [25]. The Ca^{2+} current through these channels depends on changes in the membrane potential and the action of an agonist.

The introduction of organic substances (Ca^{2+}-blocking agents) which reduce Ca^{2+} entry into cells [13] has further stimulated interest in Ca^{2+} channels in various tissues. However, the availability of a wide variety of Ca^{2+}-blocking agents (although supporting the concept of Ca^{2+} channels) has not yet contributed to a better understanding of the nature of these membrane pores. There are important differences in the sensitivity of various tissues for the same Ca^{2+} antagonists. Smooth muscle and cardiac muscle are much more sensitive than neurons or secretory cells [27,34]. Insight into the mechanism of action of Ca^{2+} antagonists on Ca^{2+} channels is further confounded by the wide variety of molecular structures that limit Ca^{2+} entry into cells. Known Ca^{2+}-blocking actions are not only exerted by such compounds as verapamil, dihydropyridines, and diltiazem, but also by naturally occurring substances such as reserpine [8], batrachitoxin, and veratridine [27]. This variety of active compounds contrasts with the high specificity of tetrodotoxin for the Na^+ channel [15] and might indicate that a wider variety of molecular structures with which organic compounds may interact is present in the Ca^{2+} channel than in the Na^+ channel.

Ca^{2+} Channels in Smooth Muscle

The amount and quantitative precision of data obtained in smooth muscle regarding the Ca^{2+} current are meager relative to cardiac muscle. In cardiac muscle the study of Ca^{2+} channels by patch clamp [26] and in single dialyzed heart cells [23] has greatly contributed to our knowledge. By contrast, the concept of Ca^{2+} channels in smooth muscle depends only on indirect experimental evidence [3]. At least for certain types of vascular smooth muscle, this evidence is still compatible with classifying the Ca^{2+} channels into voltage-dependent and receptor-operated channels. In these cells, the direct action of the agonist would increase Ca^{2+} influx through receptor-operated channels. The ensuing changes in cytosolic Ca^{2+} elicit a contraction and lead to a secondary modification of membrane permeability to monovalent ions and an accompanying change in the resting potential (either depolarizing or hyperpolarizing).

An increase in the external $[K^+]$ depolarizes the cell membrane and activates voltage-dependent channels. These voltage-dependent channels are much more sensitive to the organic Ca^{2+}-blocking agents than are the receptor-operated channels [7,32]. Such differences in sensitivity also support the hypothesis that two different types of Ca^{2+} channels exist in some vascular smooth muscle cells. However, it is not yet clear whether in some of these vascular smooth muscle cells (which show a secondary membrane depolarization in response to agonist–receptor interactions), an opening of the potential-dependent pores (in addition to the receptor-operated channels) could further increase Ca^{2+} influx. The portion of the Ca^{2+} entry occurring through voltage-dependent channels may be blocked by Ca^{2+} antagonists.

In visceral smooth muscle the above classification of Ca^{2+} channels as potential-dependent pores and receptor-operated channels cannot be readily maintained. Stimulation induced by acetylcholine or histamine (H_1), as well as by high $[K^+]_o$, depolarizes the cells and induces a contraction dependent on an increased Ca^{2+} influx through Ca^{2+} channels [4]. However, both types of excitation are equally sensitive to Ca^{2+} antagonists [24]. Although agonists such as acetylcholine and histamine also release Ca^{2+} from intracellular stores [9], their main action consists of increasing membrane permeability to Ca^{2+}. In visceral smooth muscle, it is not clear whether there exist the usual two types of Ca^{2+} channels, only voltage-dependent Ca^{2+} channels, or multiple-regulated channels.

The clinical use of Ca^{2+}-blocking agents is largely concerned with treatment of vascular and cardiac pathology, whereas their use in intestinal and myometrial pathology has been rather limited. It is, therefore, useful to summarize the effects of Ca^{2+} antagonists on vascular smooth muscle Ca^{2+} channels and their possible actions on perivascular nerve terminal function. These two types of tissue may be important for the regulation of vascular tone.

Activation of Vascular Smooth Muscle Cells by Perivascular Nerve Stimulation

It can be assumed that constriction of arteries *in vivo* depends to a large extent on release of norepinephrine from the perivascular nerve fibers. This agonist would, according to *in vitro* findings, act by activating receptor-operated channels, which are at least 10 times less sensitive to blockade by Ca^{2+} antagonists than are voltage-dependent channels [32]. It is difficult to reconcile these observations with the *in vivo* activity of Ca^{2+} antagonists in relaxing blood vessel tone. If the action of norepinephrine on arteries and arterioles *in situ* were also dependent on receptor-operated channel activation, much higher concentrations of Ca^{2+} antagonists would have to be employed to prevent the action of norepinephrine. For various arterial smooth muscle cells, there is an important difference between activation by exogenous and endogenous norepinephrine [17,18]. Stimulation of the perivascular nerves elicits excitatory junction potentials (ejp) in the smooth muscle cells, and, by summation of these ejp, an active response or an action potential occurs. This membrane response is accompanied by a contraction of the cells, which is also the case in visceral smooth muscle. Thus, these arterial smooth muscle cells respond in a different way to exogenous norepinephrine and to norepinephrine released from nerve fibers.

The first type of norepinephrine stimulus could be classified as pharmacomechanical coupling and the second type as electromechanical coupling. It has been proposed that norepinephrine added to the organ bath affects smooth muscle cells through an interaction with α-receptors, while norepinephrine released by nerve fibers activates a different type of adrenergic receptor, which Hirst and Neild called a γ-receptor [16]. Because the membrane signal during perivascular nerve stimulation consists of an active membrane response, it is not surprising that Ca^{2+} antagonists reduce the amplitude of this response and the concomitant force development [21].

Although the most likely explanation for this inhibitory action is a direct effect of Ca^{2+} antagonists on voltage-dependent Ca^{2+} channels of the smooth muscle cells, Ca^{2+} antagonists may also act on nerve terminals to reduce Ca^{2+} influx and diminish transmitter release. A reduction in transmitter release would decrease the postsynaptic response and contraction, a result similar to experimentally observed changes. In order to ascertain which mechanism is responsible for the Ca^{2+} antagonist-induced depression of the vascular response, the effect of these compounds on transmitter release during nerve stimulation was determined. After exposing tissues to [^{3}H]norepinephrine, efflux of radioactivity was determined in the absence and presence of Ca^{2+} antagonists (nicardipine, diltiazem, and flunarizine) under basal conditions and during electrical stimulation of the nerves. Addition of Ca^{2+} antagonists to the perfusion fluid increased the loss of radioactivity in quiescent tissues due to an augmented release of norepinephrine metabolites ([^{3}H]-DOPEG). Nerve stimulation elevated [^{3}H]norepinephrine release in the presence or absence of Ca^{2+} antagonists. In the presence of antagonists, enhanced norepinephrine release was superimposed on the increased DOPEG efflux.

These findings lead to the conclusion that Ca^{2+} antagonists do not interfere with norepinephrine release in these acute *in vitro* experiments. Instead, they exert their major effect on the smooth muscle membrane by affecting the voltage-dependent Ca^{2+} channels. These channels probably play a fundamental role in the cellular response to stimulation of the perivascular nerves. It is not clear how Ca^{2+} antagonists interfere with the metabolism of norepinephrine in these nerve fibers or whether the augmented DOPEG release during prolonged treatment *in vivo* diminishes evoked transmitter release. The absence of an effect on stimulated transmitter release might be related to a difference in sensitivity of the Ca^{2+} channels of nerve fibers as compared to those of smooth muscle cells.

Additional studies are necessary to establish whether prolonged *in vivo* treatment with Ca^{2+} antagonists affects nerve terminal metabolism to the extent of interfering with the amount of transmitter released during stimulation. The nature and the properties of the various Ca^{2+} channels present in smooth muscle also require more fundamental research not only to obtain more effective Ca^{2+}-blocking agents but also to develop drugs with greater tissue specificity.

References

1. Ackerman, K. E.; Wikstrom, M. K. F. (Ca^{2+} + Mg^{2+})-stimulated ATPase activity of rabbit myometrium plasma membrane is blocked by oxytocin. *FEBS Lett.* **97**:283–287, 1979.
2. Bohr, D. F. Vascular smooth muscle: Dual effect of calcium. *Science* **139**:597–599, 1963.
3. Bolton, T. Mechanism of action of transmitters and other substances on smooth muscle. *Physiol. Rev.* **59**:606–718, 1979.
4. Bolton, T. B. Cholinergic mechanisms in smooth muscle. *Br. Med. Bull.* **35**:275–283, 1979.
5. Brehm, P.; Eckert, R.; Tillotson, D. Calcium-mediated inactivation of calcium current in *Paramecium*. *J. Physiol. (London)* **306**:193–203, 1980.
6. Caputo, C. Excitation and contraction processes in muscle. *Annu. Rev. Biophys. Bioeng.* **7**:63–84, 1978.
7. Casteels, R.; Droogmans, G. Exchange characteristics of the noradrenaline-sensitive calcium store in vascular smooth muscle cells of rabbit ear artery. *J. Physiol. (London)* **317**:263–279, 1981.
8. Casteels, R.; Login, I. Reserpine has a direct action as a calcium antagonist on mammalian smooth muscle cells. *J. Physiol. (London)* **340**:403–414, 1983.
9. Casteels, R.; Raeymaekers, L. The action of acetylcholine and catecholamines on an intracellular calcium store in the smooth muscle cells of the guinea-pig taenia coli. *J. Physiol. (London)* **294**:51–68, 1979.

10. Droogmans, G.; Casteels, R. Temperature-dependence of ^{45}Ca fluxes and contraction in vascular smooth muscle cells of rabbit ear artery. *Pfluegers Arch.* **391**:183–189, 1981.
11. Droogmans, G.; Raeymaekers, L.; Casteels, R. Electro- and pharmacomechanical coupling in the smooth muscle cells of rabbit ear artery. *J. Gen. Physiol.* **70**:129–148, 1977.
12. Endo, M. Calcium release from the sarcoplasmic reticulum. *Physiol. Rev.* **57**:71–105, 1977.
13. Fleckenstein, A. Specific pharmacology of calcium in myocardium, cardiac pacemakers and vascular smooth muscle. *Annu. Rev. Pharmacol. Toxicol.* **151**:491–501, 1977.
14. Hagiwara, S.; Byerly, L. Calcium channel. *Annu. Rev. Neurosci.* **4**:69–125, 1981.
15. Hille, R. Ionic selectivity of Na and K channels of nerve membranes. In: *Membranes: A Series of Advances*, Volume 3, G. Eisenman, ed., New York, Dekker, 1975, pp. 255–323.
16. Hirst, G. D. S.; Neild, T. D. Evidence for two populations of excitatory receptors for noradrenaline on arteriolar smooth muscle. *Nature (London)* **283**:767–768, 1980.
17. Hirst, G. D. S.; Neild, T. D. Localization of specialized noradrenaline receptors at neuromuscular junctions on arterioles of the guinea-pig. *J. Physiol. (London)* **313**:343–350, 1981.
18. Holman, M. E.; Surprenant, A. Some properties of the excitatory junction potentials recorded from saphenous arteries of rabbits. *J. Physiol. (London)* **287**:331–351, 1979.
19. Horn, J. P.; McAfee, D. A. Alpha-adrenergic inhibition of calcium-dependent action potentials in rat sympathetic neurones. *J. Physiol. (London)* **301**:191–204, 1980.
20. Hudgins, P. M.; Weiss, G. B. Differential effects of calcium removal upon vascular smooth muscle contraction induced by norepinephrine, histamine and potassium. *J. Pharmacol. Exp. Ther.* **159**:91–97, 1968.
21. Kajiwara, M.; Casteels, R. Effects of Ca-antagonists on neuromuscular transmission in the rabbit ear artery. *Pfluegers Arch.* **396**:1–7, 1983.
22. Klein, M.; Kandel, E. R. Presynaptic modulation of voltage-dependent Ca^{2+} current: Mechanism for behavioral sensitization in *Aplysia californica*. *Proc. Natl. Acad. Sci. USA* **75**:3512–3516, 1978.
23. Lee, K. S.; Tsien, R. W. Mechanism of calcium channel blockade by verapamil, D600, diltiazem and nitrendipine in single dialyzed heart cells. *Nature (London)* **302**:790–794, 1983.
24. Ohashi, H.; Takewadi, T.; Shibata, N.; Okada, T. Effects of calcium antagonists on contractile response of guinea-pig taenia caecum in a calcium deficient, potassium rich solution. *Jpn. J. Pharmacol.* **25**:214–216, 1975.
25. Reuter, H. Properties of two inward membrane currents in the heart. *Annu. Rev. Physiol.* **41**:413–424, 1979.
26. Reuter, H. Calcium channel modulation by neurotransmitters, enzymes and drugs. *Nature (London)* **301**:569–574, 1983.
27. Romey, G.; Lazdunski, M. Lipid-soluble toxins thought to be specific for Na^+ channels block Ca^{2+} channels in neuronal cells. *Nature (London)* **297**:79–80, 1982.
28. Soloff, M. S.; Sweet, P. Oxytocin inhibition of $(Ca^{2+} + Mg^{2+})$-ATPase activity in rat myometrial plasma membranes. *J. Biol. Chem.* **257**:10687–10693, 1982.
29. Tillotson, D. Inactivation of Ca conductance dependent on entry of Ca-ions in molluscan neurons. *Proc. Natl. Acad. Sci. USA* **76**:1497–1500, 1979.
30. Tsien, R. W. Cyclic AMP and contractile activity in heart. *Adv. Cyclic Nucleotide Res.* **8**:363–420, 1977.
31. Van Breemen, C.; Farinas, B. R.; Gerba, P.; McNaughton, E. D. Excitation–contraction coupling in rabbit aorta studied by the lanthanum method for measuring cellular calcium flux. *Circ. Res.* **30**:44–54, 1972.
32. Van Breemen, C.; Wang, O.; Meisheri, K. D. The mechanism of inhibitory action of diltiazem on vascular smooth muscle contractility. *J. Pharmacol. Exp. Ther.* **218**:459–463, 1981.
33. Wilson, W. A.; Wachtel, H. Prolonged inhibition in burst firing neurons: Synaptic inactivation of the slow regenerative inward current. *Science* **202**:772–775, 1978.
34. Wollheim, C. B.; Kikuchi, M.; Renol, A. E.; Sharp, G. W. G. The roles of intracellular and extracellular Ca^{++} in glucose-stimulated biphasic insulin release by rat islets. *J. Clin. Invest.* **62**:451–458, 1978.

50

Calcium Antagonists in the Treatment of Arrhythmias

A. Fleckenstein

Involvement of Ca^{2+} in Cardiac Pacemaker Function—Susceptibility to Ca^{2+} Antagonists

The crucial role that Ca^{2+} may play in spontaneous impulse discharge from a cardiac pacemaker can be visualized in an isolated frog ventricle. For example, the experiment in Fig. 1 illustrates the following three important facts (1):

1. Withdrawal of Ca^{2+} impairs excitation–contraction coupling so that contractile force declines. However, the action potential persists.
2. Ca^{2+} deficiency, in addition to inhibiting mechanical performance, reduces the pacemaker activity of the preparation. Thus, withdrawal of Ca^{2+} slows the heart rate and, finally, causes cardiac arrest.
3. Epinephrine, by virtue of its function as a Ca^{2+} promoter, restores both contractile force and automaticity of the Ca^{2+}-deprived myocardium.

However, in these hearts, epinephrine and other β-adrenergic substances are only effective as long as traces of Ca^{2+} are available. If Ca^{2+} is completely withdrawn, the β-adrenergic agents are not effective. Thus, β-receptor stimulation enhances both cardiac contractility and pacemaker function by mobilization of Ca^{2+}

In mammalian hearts, contractile activation as well as spontaneous or evoked SA- and AV-nodal excitation is linked with transmembrane Ca^{2+} influx. Moreover, the Ca^{2+} transport systems that operate in nodal cell membranes closely resemble the "slow membrane channels" of ordinary myocardial fiber. Accordingly, in both nodal cells and ordinary myocardial fibers, these channels respond to virtually the same activators and inhibitors of the inward Ca^{2+} current. Thus, β-adrenergic agents, by promoting Ca^{2+} transfer through slow channels, cause not only an increase in contractile force but also an analogous rise in heart rate and AV conduction velocity. Conversely, organic Ca^{2+} antagonists exert negative

---

A. Fleckenstein • Physiological Institute, University of Freiburg, 7800 Freiburg, West Germany.

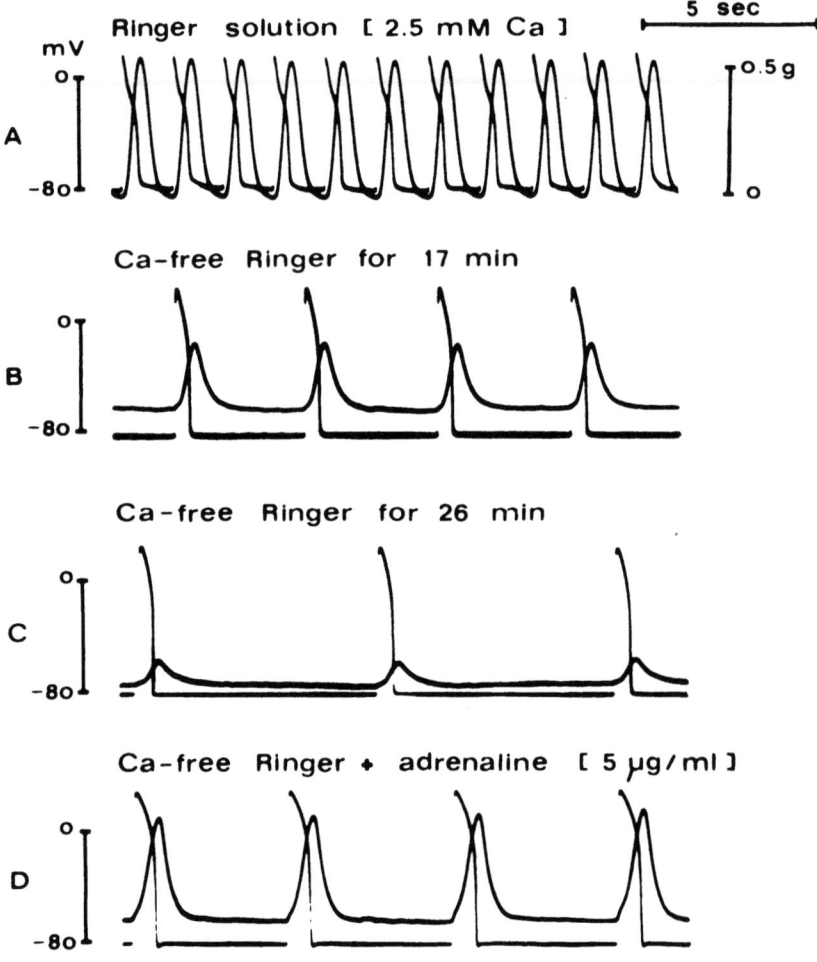

Figure 1. Simultaneous impairment of excitation–contraction coupling and pacemaker activity in a spontaneously beating isolated frog ventricle upon Ca^{2+} withdrawal during an observation period of 26 min (A–C). Thereafter, a considerable recovery of excitation–contraction coupling and heart rate took place within 6 min following addition of epinephrine to the Ca^{2+}-deficient solution (see D). This restitution of contractile force and pacemaker activity by epinephrine or by other β-mimetic agents occurs only as long as at least traces of extracellular (or membrane-bound) Ca^{2+} are available. [From Antoni et al. [1].]

inotropic, chronotropic, and dromotropic influences simultaneously (For details see the surveys of Cranefield [2] and Fleckenstein [3,4]).

Table I includes the most potent and specific Ca^{2+} antagonists. All were identified in our laboratory [4] with the exception of diltiazem, which was identified by Japanese workers [5]. In experiments on guinea pigs and rats, all of these Ca^{2+} antagonists have distinct inhibitory effects on nodal excitation. However, there are important species differences. For example, nifedipine and other 1,4-dihydropyridines exert a relatively weak depression of sinus node automaticity and AV conduction in dogs, monkeys, and humans. On the other hand, such species-dependent differences between the relative strengths of nodal and contractile inhibition are absent with verapamil, gallopamil, or diltiazem. For example, verapamil shows a close quantitative correlation between decrease in SA node frequency and reduction of contractile force of isolated atria in all laboratory animals tested so far.

50. Arrhythmias

Table I. Calcium Antagonists of Outstanding Specificity (Group A)[a]

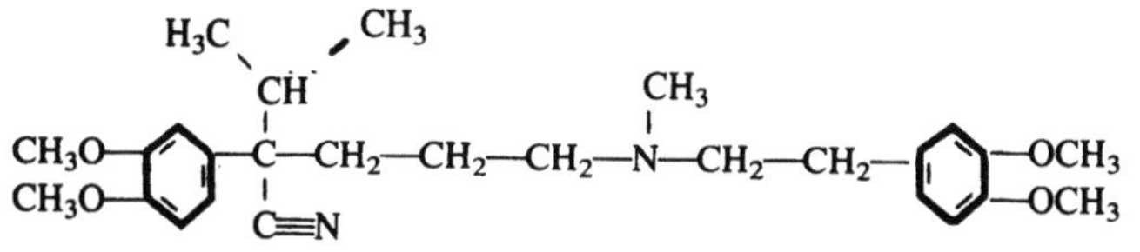

Verapamil (M_r 454.59)

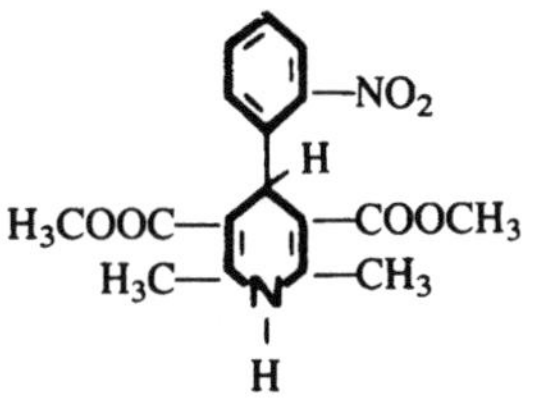

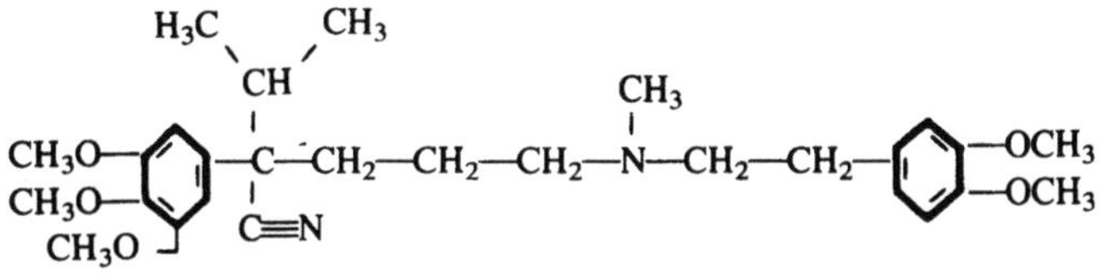

Compound D600 (Gallopamil, M_r 485.59)

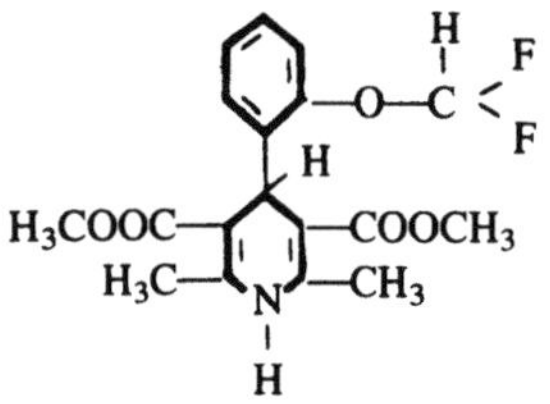

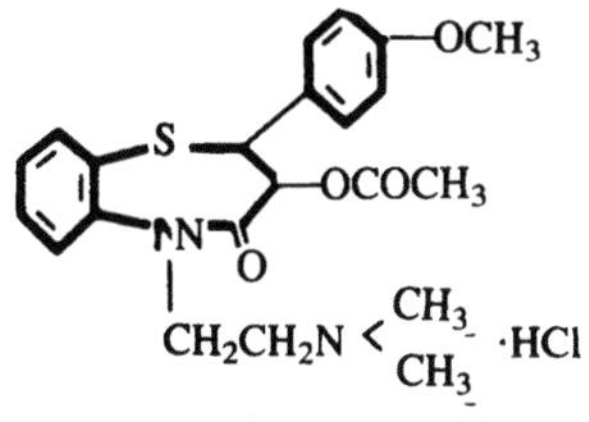

Diltiazem (M_r 414.52)

Nifedipine (M_r 346.34)

Ryosidine (M_r 367.35)

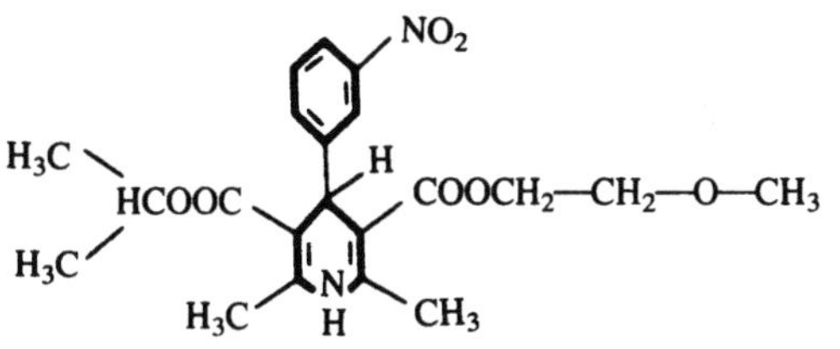

Nimodipine (M_r 418.45)

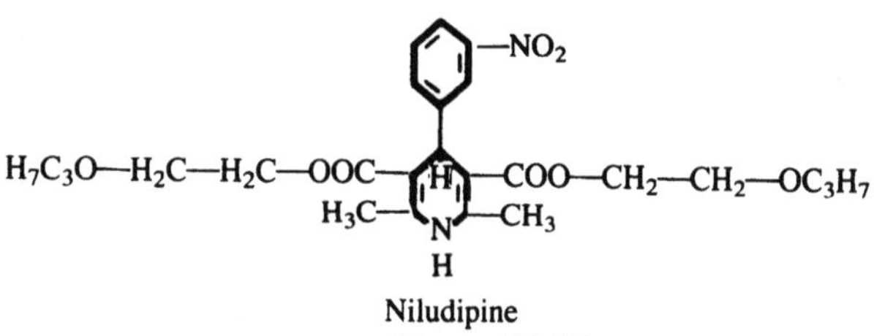

Niludipine (M_r 490.55)

[a] Criterion: Inhibition by 90 to 100% of slow inward Ca^{2+} current *without* concomitant influence on electrogenic Na^+ or Mg^{2+} effects on ventricular myocardium.

Electrophysiological Assessment of Ca^{2+} Antagonist Actions on Nomotopic Automaticity

Electrophysiological studies have clearly established that the mechanism of membrane excitation of nodal cells widely differs from that of ordinary myocardial fibers (Table II). Thus, the two contrasting types of cardiac excitation can be easily distinguished from each other by their different sensitivities to drugs and by a number of bioelectric criteria. There is

Table II. Different Types of Excitation in Ordinary Myocardium and Cardiac Pacemakers

	Ordinary myocardial fibers	SA and AV nodal cells
Resting potential or maximal diastolic potential	−80 to −90 mV	−50 to −60 mV
Upstroke velocity (dV/dT_{max})	170–180 V/sec	2–7 V/sec
Conduction velocity	50–120 cm/sec	2–6 cm/sec
Operating transmembrane carrier systems	Fast channel for Na^+	Slow channel for Ca^{2+} and Na^+
Drug susceptibilities of transmembrane currents	Blockade by TTX No influence of β-adrenergic drugs[a] No influence of Ca^{2+} antagonists[a]	No influence of TTX Promotion of Ca^{2+} flux by β-adrenergic drugs Blockade of Ca^{2+} flux by Ca^{2+} antagonists

[a]The only changes in action potential are a slight prolongation of the plateau phase by β-adrenergic drugs and a slight abbreviation by Ca^{2+}-antagonistic agents.

evidence of a fast, obviously Na^+-dependent type of excitation which occurs in atrial and ventricular myocardium as well as in His bundle and Purkinje fibers. This Na^+-dependent mechanism of excitation does not respond to changes in transmembrane Ca^{2+} conductivity produced by Ca^{2+} antagonists or β-adrenergic catecholamines, but is blocked by the fast-channel inhibitor tetrodotoxin (TTX). The other obviously Ca^{2+}-dependent slow type of excitation underlies impulse generation and conduction in the SA and AV nodes and is TTX-insensitive. This type of excitation is suppressed by Ca^{2+} antagonists, and, conversely, enhanced by β-receptor stimulation (or by dibutyryl cAMP). Other criteria on which the dualistic concept of a fast Na^+-carried and a slow Ca^{2+}-mediated form of excitation is based, consist of large differences in upstroke velocity and propagation rate of the respective action potentials. Upstroke velocity in ordinary myocardium and nodal tissue is 170–180 and 2–7 V/sec, respectively. Propagation rate is 50–120 cm/sec in ordinary myocardium and 2–6 cm/sec in sinus and AV nodes, respectively [for details see 3,4].

Perhaps the most obvious difference between transmembrane potentials of ordinary myocardial fibers and those of pacemaker cells is that in pacemaker cells, automaticity is linked with an instability of resting potential. Ordinary atrial and ventricular fibers maintain a steady level of resting potential during diastole. The pacemaker cells, on the other hand, undergo a slow diastolic depolarization, which is usually large enough to reach threshold potential. Then, the slow diastolic depolarization gives rise to the upstroke of a propagated pacemaker action potential (Fig. 2). If, however, the slow diastolic depolarization fails to reach this critical level and remains confined to the primary pacemaker region, the intrinsic automaticity of the nodal cells appears as continuous local oscillations of membrane potential.

The intrinsic pacemaker automaticity proved to be extremely insensitive to depolarization. With K^+-depolarized pacemaker cells of the frog AV node, there were still persistent bioelectric oscillations at a membrane potential of less than −20 mV (at which the fast Na^+ channel no longer operates). The same local oscillation of membrane potential appeared in largely depolarized SA nodal cells of rabbits (Fig. 3). Also in these studies, the isolated

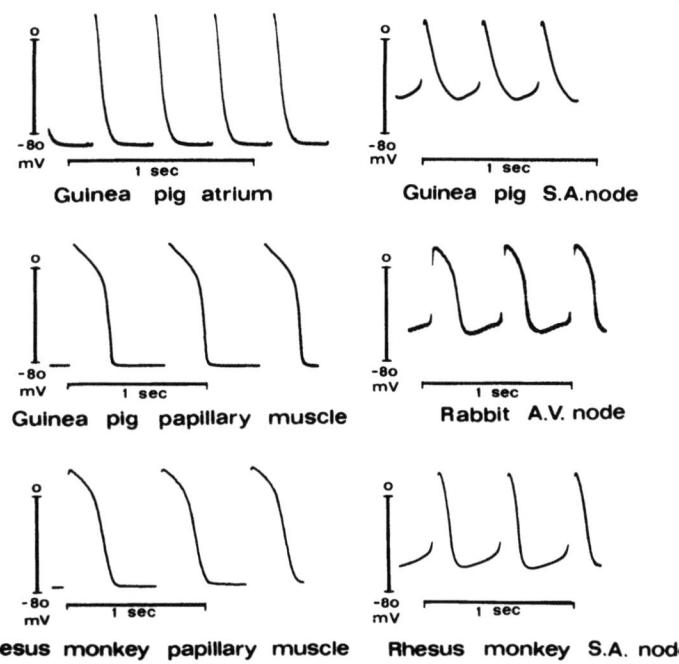

Figure 2. Comparison of transmembrane action potentials of electrically stimulated nonautomatic myocardial fibers from atrial or papillary muscles with those of spontaneously discharging SA- or AV-nodal cells. The characteristic feature of spontaneous active nodal cells is a slow depolarization that initiates a propagated pacemaker action potential as soon as the membrane potential is lowered to a critical level (threshold potential). [From a collection of transmembrane action potentials recorded by Antoni et al. [6]; see Fleckenstein [4].]

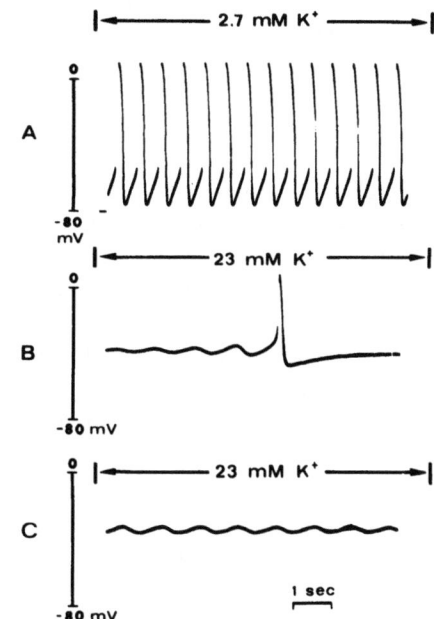

Figure 3. Persistence of rhythmic oscillations of membrane potential in a largely depolarized rabbit SA node after transfer from a normal (2.7 mM K^+) to a K^+-rich Tyrode solution containing 23 mM K^+. In B at a membrane potential of around −40 mV, one of the spontaneous rhythmic oscillations still elicited a small propagated action potential. In C at a membrane potential of about −30 to −35 mV, the pacemaker activity was reduced to local oscillations only. Temperature: 36°C. [Observations of Antoni et al. [6]; see Fleckenstein [4,7].]

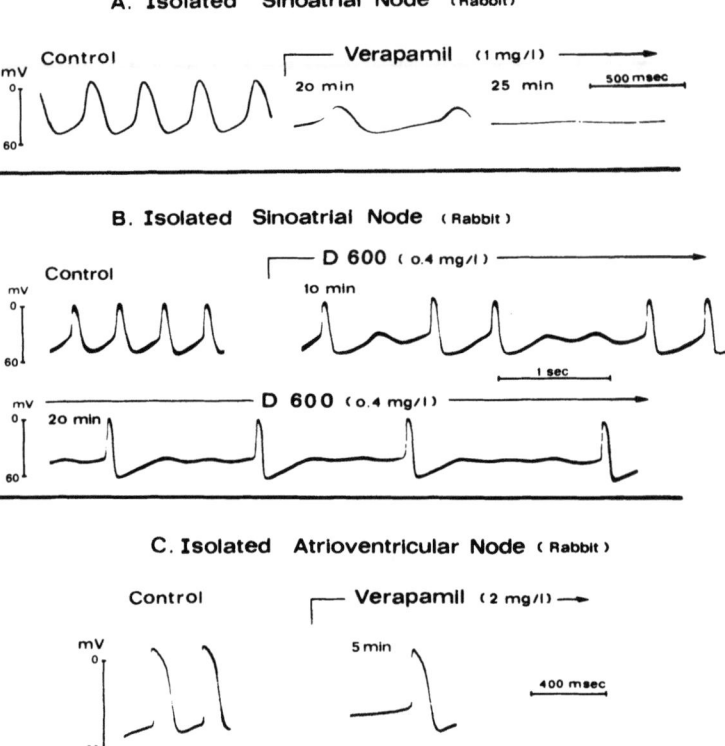

Figure 4. Inhibition by verapamil and D600 of the pacemaker function of isolated SA and AV nodes from rabbits. Verapamil and D600 reduce both the steepness of the slow diastolic depolarization (thereby producing a negative chronotropic effect) and the velocity of upstroke of the propagated nodal action potentials. The height of the overshoot also decreases upon prolonged exposure to the drugs, whereas there is an increasing tendency toward hyperpolarization during diastole. As shown in B a nodal cell treated with the Ca^{2+} antagonist D600 increasingly fails to elicit propagated pacemaker action potentials. This is due to (1) a reduction in the amplitude of the intrinsic oscillatory potential changes and (2) a shift of threshold potential toward zero. Thus, an increasing number of spontaneous local depolarization waves remained unpropagated. Each record in experiments A, B, and C was taken by steady impalement of a single cell. [A from Kohlhardt *et al.* [8]; C from Tritthart *et al.* [9]; see Fleckenstein [10].]

pacemaker tissue was exposed to a K^+-rich (23 mM) Tyrode solution. Consequently, at a membrane potential level of about -35 mV, the production of propagated pacemaker action potentials was no longer possible. Instead, permanent local subthreshold oscillations demonstrated the persistence of intrinsic nodal automaticity. In a series of four successive swings, there was (by chance) an increase in amplitude so that the depolarizing phase of the fourth oscillation reached threshold; by this manner, one single pacemaker propagated action potential was elicited (Fig. 3B).

However, both local oscillatory depolarizations and upstrokes of superimposed pacemaker propagated action potentials are Ca^{2+}-dependent; thus, they are greatly enhanced by β-receptor stimulation, which increases the inward Ca^{2+} current. Conversely, local oscillations as well as nodal propagated action potentials are readily blocked by Ca^{2+} withdrawal or Ca^{2+} antagonists. For example, Fig. 4 shows the typical inhibition by verapamil and D600 of the pacemaker function of isolated SA and AV nodes. Verapamil and D600 reduce the steepness of slow diastolic depolarization, thereby producing a negative chronotropic effect. Moreover, the upstroke velocity and height of the pacemaker propagated action potentials also decrease upon prolonged exposure to the drugs. A nodal cell, treated with the Ca^{2+} antagonist D600, increasingly fails to elicit pacemaker propagated action potentials (Fig. 4B). This is due both to a reduction of amplitude of the intrinsic oscillatory potential changes and to a shift of threshold potential toward zero. Thus, an increasing number of spontaneous depolarization waves remain unpropagated.

Akiyama and Fozzard [11] have calculated that, during excitation of AV nodal cells, the transmembrane permeability that allows the entry of Ca^{2+} is about 70 times greater than that for Na^+. Because there is no appreciable involvement of fast-channel activity, TTX does not exert an inhibitory effect on nodal excitation. Nevertheless, recent observations indicate that during nodal excitation, Na^+ ions can pass through the slow channels together with Ca^{2+} ions in a proportion that may even exceed the ratio of 1 : 1. However, the contribution of Na^+ appears to diminish as the range of membrane potential declines and oscillatory local depolarizations develop.

Suppression by Ca^{2+} Antagonists of Ectopic Cardiac Autorhythmicity

Every ordinary myocardial fiber possesses the latent capability of adopting, under certain conditions, the abnormal role of an ectopic pacemaker. One of the most suitable

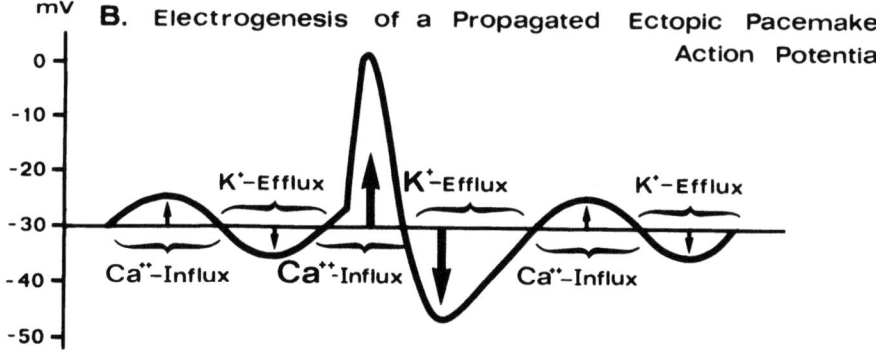

Figure 5. Scheme of the coupled $Ca^{2+}-K^+$ exchange that underlies the electrogenesis of ectopic subthreshold oscillations (A) and propagated ectopic pacemaker action potentials (B) in a low range of membrane potential (below −50 mV). In both cases A and B, the depolarizing slow-channel-mediated inward current is carried by Ca^{2+}, while the subsequent repolarization is brought about by the efflux of an equivalent amount of K^+ (I_K), which drives the membrane potential back to diastolic values. The greater influx of Ca^{2+} in case of a propagated ectopic pacemaker action potential is apparently compensated by an enhancement of I_K, so that an overshooting repolarization is produced. [From Fleckenstein [4,10].]

mechanisms for experimentally inducing ectopic pacemaker activity consists of the application of soluble Ba^{2+} salts. This functional change ensues from an alteration of the bioelectric membrane properties in that the regular stable resting potential declines from its high level until, after partial depolarization, spontaneous subthreshold oscillations appear and, finally, automaticity becomes manifest. However, the essence of this myocardial fiber transformation is that the excitatory process, with the development of ectopic automaticity, progressively converts from the fast-channel-mediated Na^+-dependent type at a high membrane potential into the predominantly Ca^{2+}-dependent slow form in a lower potential range [4].

The electrophysiological features of ectopic automaticity closely resemble the well-known pattern of spontaneous impulse discharge in the SA or AV nodes. The scheme of Fig. 5 illustrates the electrogenesis of ectopic automaticity in a low range of membrane potential (from -30 to -50 mV). In fact, Ca^{2+} influx is the primary step responsible for spontaneous depolarization. This is true of both the depolarizing phase of the local oscillations and the calcium-dependent upstroke of the propagated ectopic pacemaker action potentials. In a second step, after the entry of Ca^{2+}, a delayed K^+ outward current is elicited and drives the membrane potential back to the higher diastolic values. Oscillations of this type, as well as ectopic automaticity, can be elicited by a multitude of factors such as myocardial anoxia or ischemia, cardiac glycosides, K^+ deficiency, hypercalcemia, and particularly Ba^{2+}. The Ca^{2+} antagonists interrupt this fundamental $Ca^{2+}-K^+$ exchange cycle and thereby abolish the uncontrolled firing of Ca^{2+}-dependent atrial or ventricular ectopic foci [4].

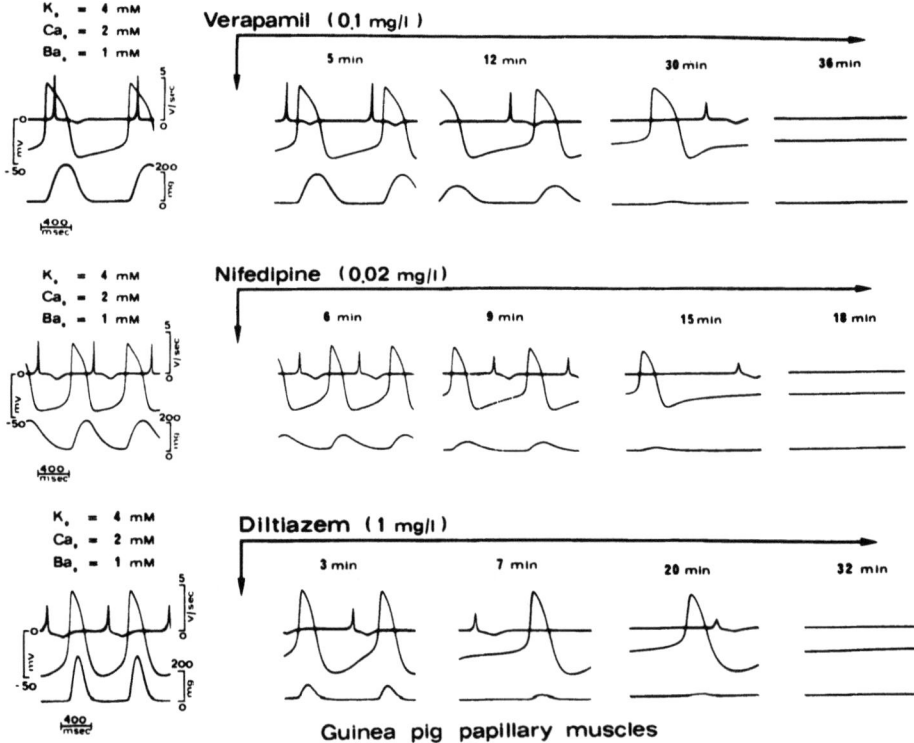

Figure 6. Suppression of Ba^{2+}-induced ectopic pacemaker activity by minute amounts of the specific Ca^{2+} antagonists verapamil (0.1 mg/liter), nifedipine (0.02 mg/liter), and diltiazem (1.0 mg/liter). Experiments on isolated guinea pig papillary muscles immersed in Tyrode solution with 1 mM Ba^{2+}, 2 mM Ca^{2+}, and 4 mM K^+ for 30 min at 30°C until sustained rhythmic automaticity had developed. Then the Ca^{2+} antagonists were added. The intracellular records originated in each experiment from one and the same myocardial fiber. [From Fleckenstein and Späh [unpublished]; see Fleckenstein [4].]

β-Adrenergic agonists, by stimulating the inward Ca^{2+} current, promote the spontaneous impulse discharge not only in SA or AV nodal cells, but also in ectopic atrial or ventricular pacemakers. On the other hand, minute amounts of the Ca^{2+} antagonists verapamil, nifedipine, or diltiazem suppress the propagated ectopic action potentials in Ba^{2+}-treated guinea pig papillary muscles (Fig. 6). The Ba^{2+}-induced ectopic foci operate in a particularly low range of membrane potential (where Ca^{2+} ions are the only charge carriers for depolarization). This explains why such ectopic Ba^{2+}-induced pacemakers are even more effectively suppressed by Ca^{2+} antagonists than is normal nodal automaticity. Conversely, Ba^{2+}-induced ectopic autorhythmicity is completely resistant even to high doses (50–100 mg/liter) of Na^+-antagonistic drugs, such as procaine amide or lidocaine. Moreover, even high doses of TTX (10 mg/liter) are completely devoid of any effect on the Ba^{2+}-induced spontaneous action potentials [4].

Clinical Aspects

Verapamil was the first Ca^{2+} antagonist introduced into human therapy in 1967 for the treatment of *supraventricular tachyarrhythmias* such as atrial fibrillation or flutter [12]. Subsequent clinical studies definitely established this new therapy [13–16]. Using intracardiac recordings of electric activity and programmed electric stimulation of the heart, Krikler [14–16] demonstrated that the pivotal effects of verapamil consist of retardation of AV-impulse conduction and prolongation of the AV-node recovery time. In this manner, the maximum frequency of atrial impulses that effectively reach the His–Purkinje system is lowered. Most striking results in cases with paroxysmal supraventricular reentry tachycardias were obtained by rapid intravenous injection of 5–10 mg of the drug. Paroxysmal supraventricular tachycardia is either due to a reciprocating mechanism within the AV node or related to a reentry loop associated with an anomalous bypass tract as in patients with the Wolff–Parkinson–White syndrome. In the latter case, the anterograde pathway of the reentry loop consists mainly of the normal conduction system (including the AV node), whereas the retrograde impulses use the anomalous AV-bypass tract for return. Verapamil, by inhibiting Ca^{2+}-dependent impulse conduction through the AV node, dramatically interrupts such circus movements regardless of whether they are based on an intranodal circuit or on an accessory AV connection. In humans, verapamil is the most effective inhibitor of AV-nodal excitation of the major Ca^{2+} antagonists. Thus, verapamil has, up to now, maintained a primary role in the treatment of supraventricular reentry tachycardias.

However, the situation is more complex as far as the efficacy of Ca^{2+} antagonists in the treatment of *ventricular tachyarrhythmias* is concerned. There are certainly cases of premature ventricular beats and ventricular tachycardia that respond as promptly to Ca^{2+} antagonists as do the Ba^{2+}-induced ectopic pacemakers. This is particularly true for patients who suffer from ventricular autorhythmicity in connection with ventricular ischemia. Patients with Prinzmetal's "variant angina" belong to this group [4,17].

But there is an alternative category of ventricular automaticity that is more sensitive to the inhibitory actions of agents such as quinidine, procaine amide, or lidocaine which exert predominantly Na^+-antagonistic membrane effects. The theoretical explanation is that, in these cases, ectopic automaticity arises from Na^+-dependent foci in Purkinje fibers. The peculiarity of Purkinje fibers is that they tend to operate as ectopic pacemakers in a rather high range of membrane potential where the impulse production depends more on the fast Na^+ than on the slow Ca^{2+} influx. For instance, frequent spontaneous impulse discharges can be elicited in isolated Purkinje fibers by simple stretch [18] even at a diastolic potential of about -75 mV (at which the fast Na^+ channel is still fully active). Under such conditions no effective damping of automaticity by Ca^{2+} antagonists can be expected.

Final Remarks

During the last decade Ca^{2+} antagonists, serving as research tools, have contributed invaluably to elucidating the vital roles of Ca^{2+} ions in nomotopic and ectopic pacemaker activity. Furthermore, these drugs have considerably enriched the therapeutic armamentarium for the treatment of *supraventricular tachyarrhythmias*. However, with respect to ectopic automaticity arising from the *ventricular walls*, the efficacy of Ca^{2+} antagonists is less predictable because of the differential responsiveness of the two types of foci (Na^+- or Ca^{2+}-driven). Possibly, even intermediate forms of automaticity occur that operate with both Na^+ and Ca^{2+} ions as electric charge carriers. Thus, the therapy of human ventricular arrhythmias will continue to be empirical as long as the underlying ionic mechanisms are not accessible to direct local measurement.

References

1. Antoni, H.; Engstfeld, G.; Fleckenstein, A. Inotropic effects of ATP and epinephrine on hypodynamic frog myocardium following excitation–contraction uncoupling by Ca withdrawal [in German]. *Pfluegers Arch. Gesamte Physiol. Menschen Tiere* **272**:91–106, 1960.
2. Cranefield, P. F. *The Conduction of the Cardiac Impulse—The Slow Response and Cardiac Arrhythmias*, New York, Futura, 1975.
3. Fleckenstein, A. Specific pharmacology of calcium in myocardium, cardiac pacemakers, and vascular smooth muscle. *Annu. Rev. Pharmacol. Toxicol.* **17**:149–166, 1977.
4. Fleckenstein, A. *Calcium Antagonism in Heart and Smooth Muscle—Experimental Facts and Therapeutic Prospects*. New York, Wiley, 1983.
5. Nakajima, H.; Hoshiyama, M.; Yamashita, K.; Kiyomoto, A. Effect of diltiazem on electrical and mechanical activity of isolated cardiac ventricular muscle of guinea pig. *Jpn. J. Pharmacol.* **25**:383–392, 1975.
6. Antoni, H.; Herkel, K.; Fleckenstein, A. Neutralization by epinephrine of cardiac pacemaker blockade in a high K^+ solution: Electrophysiological studies on isolated SA nodes (guinea pig, rhesus monkey), and Purkinje fibres (rhesus monkey) [in German]. *Pfluegers Arch. Gesamte Physiol. Menschen Tiere* **277**:633–649, 1963.
7. Fleckenstein, A. Experimental restitution of cardiac automaticity, and impulse conduction by sympathomimetic amines [in German]. *Verh. Dtsch. Ges. Kreislaufforsch.* **30**:102–113, 1964.
8. Kohlhardt, M.; Figulla, H. R.; Tripathi, O. The slow membrane channel as the predominant mediator of the excitation process of the sinoatrial pacemaker cell. *Basic Res. Cardiol.* **71**:17–26, 1976.
9. Tritthart, H.; Fleckenstein, B.; Fleckenstein, A. Some fundamental actions of antiarrhythmic drugs on the excitability and the contractility of single myocardial fibres. *Naunyn-Schmiedebergs Arch. Exp. Pathol. Pharmakol.* **269**:212–219, 1971.
10. Fleckenstein, A. Pharmacology and electrophysiology of calcium antagonists. In: *Calcium Antagonism in Cardiovascular Therapy: Experience with Verapamil*, A. Zanchetti and D. M. Krikler, eds., Amsterdam, Excerpta Medica, 1981, pp. 10–29.
11. Akiyama, T.; Fozzard, H. A. Ca and Na selectivity of the active membrane of rabbit A.V. nodal cells. *Am. J. Physiol.* **236**:C1–C8, 1979.
12. Bender, F. Therapeutic use of Isoptin (verapamil) in the treatment of supraventricular tachyarrhythmias [in German]. *Med. Klin. (Munich)* **62**:634–636, 1967.
13. Schamroth, L. Immediate effects of intravenous verapamil on atrial fibrillation. *Cardiovasc. Res.* **5**:419–424, 1971.
14. Krikler, D. M. A fresh look at cardiac arrhythmias. *Lancet* pp. 851, 913, 974, 1034, 1974.
15. Krikler, D. M. Verapamil in cardiology. *Eur. J. Cardiol.* **2**:3–10, 1974.
16. Krikler, D. M.; Spurrell, R. A. J. Verapamil in the treatment of paroxysmal supraventricular tachycardia. *Postgrad. Med. J.* **50**:447–453, 1974.

17. Mizuno, K.; Tanaka, K.; Honda, Y.; Kimura, E. Suppression of repeatedly occurring ventricular fibrillation with nifedipine in variant form of angina pectoris. In: *International Nifedipine (Adalat) Panel Discussion*, P. R. Lichtlen, E. Kimura, and N. Taira, eds., Amsterdam, Excerpta Medica, 1979, pp. 61–67.
18. Kaufmann, R.; Theophile, U. Stretch-induced automaticity of Purkinje fibres, papillary muscles, and atrial trabeculae of rhesus monkeys [in German]. *Pfluegers Arch. Gesamte Physiol. Menschen Tiere* **297**:174–189, 1967.

51

Calcium Entry Blockers in Coronary Artery Disease

Jay N. Cohn, Robert J. Bache, and Jeffrey S. Schwartz

Coronary artery disease may be complicated by a variety of symptomatic manifestations of inadequate myocardial perfusion. These include exertional angina, rest or variant angina, unstable angina, acute myocardial infarction, sudden death, and chronic congestive heart failure. Although these manifestations of coronary disease are quite divergent, all may in some way relate to an imbalance between myocardial oxygen demand and myocardial oxygen delivery. The dynamics of this imbalance are most thoroughly understood in the syndrome of exertional angina. This review will therefore focus on the pathophysiology of angina and the mechanism of effect of Ca^{2+} entry blockers.

An understanding of the response of pharmacologic interventions in angina requires an analysis of the physiology of the coronary vascular tree. The total coronary vascular resistance that determines localized coronary perfusion is made up of at least four independent resistances in series (Fig. 1). The resistance in the conductance vessels, identified as R_1, is normally extremely low but may become critically increased in the presence of proximal coronary stenoses. The arteriolar resistance (R_2) is independently determined in the subendocardial and subepicardial vessels, and these are therefore depicted as resistances in parallel (R_{2a} and R_{2b}). R_3 represents the subendocardial compressive force transmitted from diastolic pressure in the left ventricle. This compressive force may be negligible in the presence of a normal left ventricular diastolic pressure but may become an important determinant of subendocardial perfusion when the diastolic pressure is elevated in the presence of heart failure. R_4 refers to the coronary venous resistance including drainage into the coronary sinus that may be impeded in the presence of a high right atrial pressure. An additional collateral resistance (R_C) may be an important determinant of flow in the presence of a proximal stenosis.

The arteriolar resistance is acutely responsive to metabolic factors. but the distribution of resistance across the wall of the left ventricle also is determined by hemodynamic factors. Since endocardial flow requires perfusion from vessels that perforate from the epicardium

---

Jay N. Cohn, Robert J. Bache, and Jeffrey S. Schwartz • Cardiovascular Division, Department of Medicine, University of Minnesota Medical School, Minneapolis, Minnesota 55455.

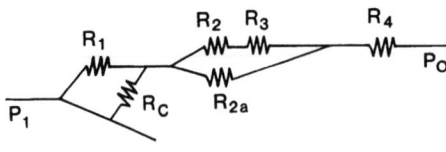

Figure 1. Electrical analog of coronary vascular resistances in series. Total coronary flow (Q) is directly related to pressure drop across the coronary bed and inversely related to the sum of all series resistances. R_1, conductance (stenosis); R_2, arteriolar (endocardial); R_{2a}, arteriolar (endocardial); R_3 compressive; R_t venous; R_c collateral.

$$Q = \frac{P_1 - P_0}{\Sigma R}$$

through the myocardial wall, a fall in perfusion pressure may deprive the more distally perfused subendocardium while flow is maintained in the more superficial epicardial layers [2]. A fall in distal coronary artery pressure resulting in subendocardial ischemia is a prominent manifestation of increased myocardial oxygen demand in the presence of a proximal coronary stenosis [7]. The effect of mild restriction of coronary blood flow on the distribution of flow in response to exercise is shown in Fig. 2. Thus, when a proximal stenosis increases R_1, even a small change in total coronary flow may be associated with considerable maldistribution of flow across the wall of the myocardium [4].

The therapeutic goal in exercise-induced angina is either to increase blood flow to potentially ischemic myocardium during exertion or to decrease myocardial oxygen requirements during exertion. Many of the drugs utilized to treat angina have been classified as coronary dilators because the concept was held that a drug which produced arteriolar dilation might increase blood flow and thus relieve exertional angina. In contrast, however, recent evidence suggests that coronary arteriolar dilators may actually reduce blood flow to ischemic areas rather than increase it. In experiments carried out in our laboratory by Schwartz et al. [16] coronary arteriolar dilation in the presence of a severe proximal stenosis reduced flow beyond the stenosis (Fig. 3). This response can be demonstrated to result from pacing [17], from exercise [18], from reactive hyperemia [16], and from local administration of an

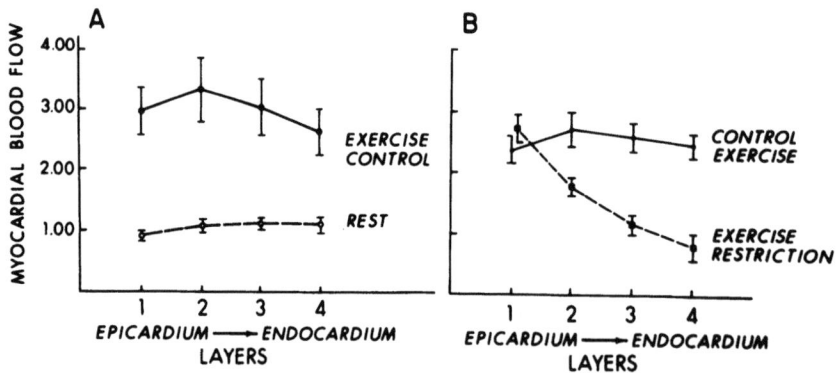

Figure 2. The effect of a subcritical coronary stenosis which became flow limiting during exercise on transmural myocardial perfusion. Blood flow to four transmural myocardial layers from epicardium (layer 1) to endocardium (layer 4) was measured with radioactive microspheres in nine chronically instrumented dogs at rest (heart rate = 82 ± 2 beats/min) and during treadmill exercise (heart rate = 224 ± 12 beats/min) during unrestricted coronary artery inflow (control exercise) and in the presence of a proximal coronary artery stenosis which did not impede resting flow but which limited the increase in coronary blood flow during exercise to 66% of the normal response (exercise restriction). During exercise the coronary stenosis resulted in redistribution of perfusion to maintain normal subepicardial blood flow at the expense of severe subendocardial perfusion.[Copyright by Grune & Stratton, reproduced with permission.]

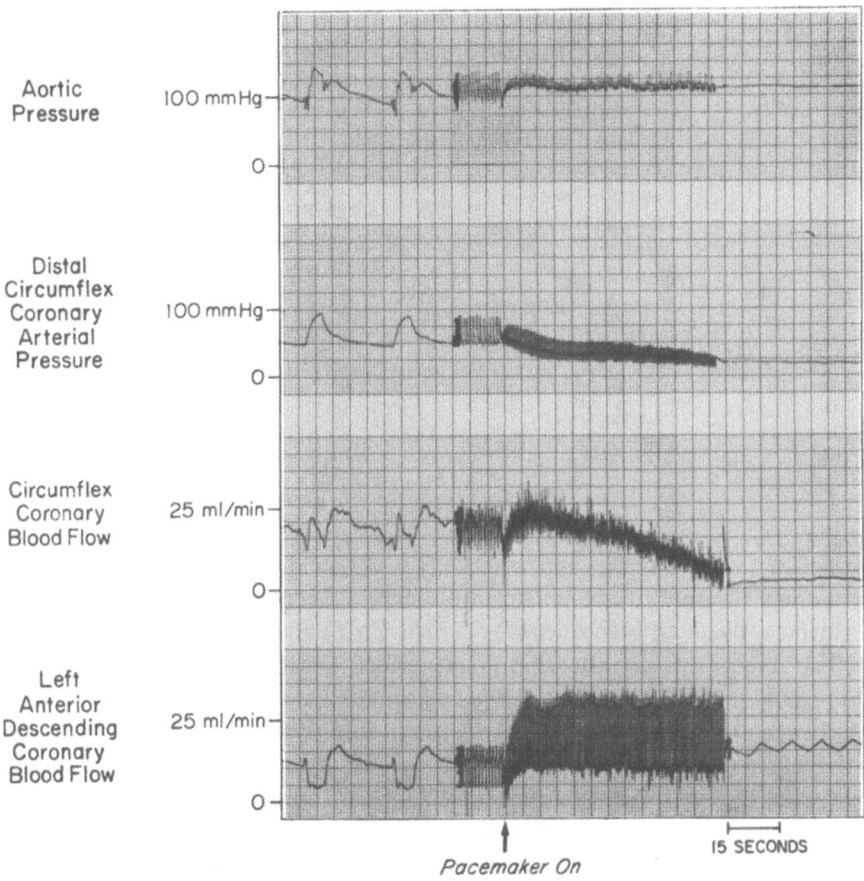

Figure 3. Effects of rapid ventricular pacing in the presence of a severe circumflex stenosis. Phasic pressures and flows are shown followed by mean pressures and flows. Rapid pacing resulted in a fall in distal circumflex pressure and circumflex blood flow. Flow in the normal left anterior descending artery increased and aortic pressure did not change. Calculated resistance across the circumflex stenosis increased.

arteriolar dilator agent [16]. Dependence of this phenomenon on reactivity of the stenosis itself was demonstrated by comparing several different types of stenoses. When the stenosis was created by a wire snare or by plication of one wall of the proximal coronary artery, dilation of the distal bed resulted in a fall in pressure distal to the stenosis and an increase in resistance across the stenosis. When the stenosis was created by utilizing a small segment of polyethylene tubing inserted into the coronary artery, the resultant fixed stenosis was not altered during interventions that lowered distal pressure. Therefore, most of the physical stenoses utilized experimentally and many that occur naturally in men [9] apparently remain pliable enough so that hemodynamic forces, especially coronary artery pressure, may significantly alter the stenosis resistance and paradoxically influence coronary flow. Thus, drugs that have a prominent coronary arteriolar dilator effect may aggravate angina in patients with severe proximal coronary stenosis. The relative dilator and constrictor activity of pharmacologic agents on the coronary arteriolar bed is shown in Table I. The most potent dilators, including dipyridamole, probably are ineffective for angina and may aggravate the syndrome [5].

Table I. Coronary Arteriolar Effects of Various Pharmacologic Agents in Order of Decreasing Dilator and/or Increasing Constrictor Activity

Adenosine, chromallin, dipyridamole
Lidoflazine
Captopril
Prazosin, hydralazine
Nifedipine
Nitroprusside
Verapamil, diltiazem
Nitroglycerin
Atenolol
Propranolol
Phenylephrine
Ergonovine, vasopressin

Since Ca^{2+} entry blockers appear on the list as potent coronary arteriolar dilators, how can these drugs be useful in the treatment of angina? Recent studies by Bache et al. [1–3] shed light on a potential mechanism by which Ca^{2+} entry blockers may relieve angina despite their direct coronary arteriolar dilator effect. During ischemia local dilation in the area of impaired perfusion might be viewed as initiating a positive feedback vicious circle in which ischemia produces dilation which results in further ischemia. Studies with nifedipine and diltiazem have demonstrated that pretreatment of the dog with these drugs significantly attenuates the hyperemia that follows a 10-sec occlusion of a coronary artery [1,3] (Fig. 4). By inhibiting reactive hyperemia, therefore, the Ca^{2+} entry blockers may break the vicious circle which may contribute to subendocardial hypoperfusion during exertion. Therefore

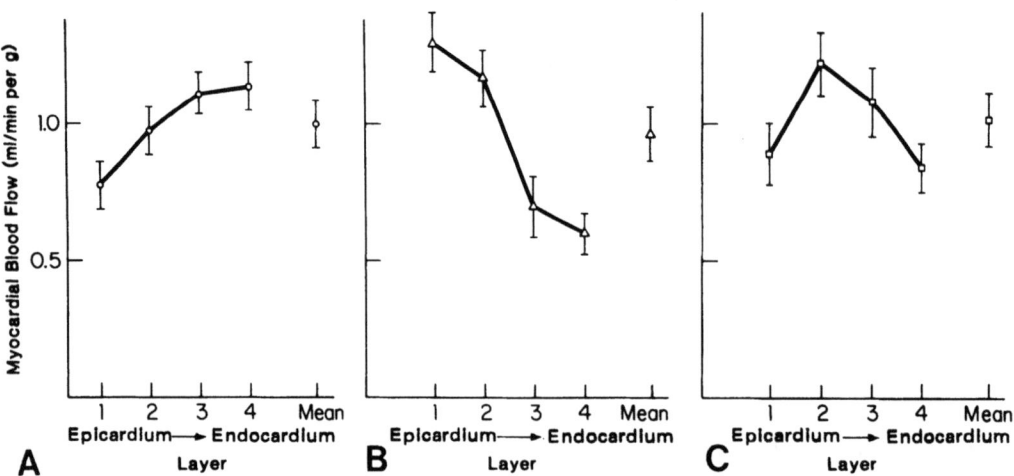

Figure 4. Myocardial blood flow to four transmural layers from epicardium to endocardium during control conditions with unrestricted arterial inflow (A), in the presence of a coronary stenosis which prevented blood flow from increasing after a 10-sec period of myocardial ischemia (B), and during a similar stenosis 30 min after administration of nifedipine (C). Although the total volume of blood flow was unaltered following nifedipine administration, the transmural distribution of perfusion was improved with reduction of the excess blood flow delivered to the subepicardium and increased subendocardial blood flow. [Modified from Bache and Tockman [3].]

these drugs may improve the anginal syndrome not by virtue of their basal coronary arteriolar dilator effect but by virtue of their effect to inhibit more intense coronary arteriolar dilation during exertion.

These studies on the mechanism and pharmacologic response in angina make it necessary to focus attention on dynamic changes in the more proximal coronary arteries as important in altering the balance between oxygen demand and supply. Recent evidence from numerous clinical studies raises the possibility that dynamic changes in large coronary artery caliber may play an important role in the genesis of angina. Whereas coronary artery spasm may be the predominant mechanism of angina in patients with the variant anginal syndrome accompanied by transient chest pain and ST segment elevation occurring at rest [14], even in the more traditional forms of exertional angina it has recently been suggested that spasm or changes in caliber may play a role in the syndrome [10,19,23]. Caliber changes in the coronary artery may be induced by a number of phenomena. A passive change in stenosis resistance because of dynamic pressure changes distal to the stenosis occurs in response to a number of physiologic events that have been described above [15–18]. More active changes in coronary caliber may occur either by virtue of a generalized change in the caliber of the coronary artery or a more localized change in the vicinity of an atherosclerotic plaque. In either case the plaque that may have been producing little or no obstruction to flow might be converted into a severe resistance by virtue of a change in the caliber of the artery in the vicinity of the plaque. The mechanism of these active changes in coronary artery tone is not entirely understood but may at least in some patients involve activation of the sympathetic nervous system [21], cholinergic stimulation [24], thromboxane release [11], decreased synthesis of prostacyclin in the local vessel wall [13], a membrane defect for Ca^{2+} [8], or alkalosis [22].

Drugs that relax large coronary arteries should therefore be effective in the treatment of both variant angina and at least some patients with exertional angina. Both the nitrates and the Ca^{2+} entry blockers have been demonstrated to relax large coronary arteries. This has been demonstrated both in experimental animals [20] and more recently in human coronary arteries using quantitative angiography [6].

Another important factor that may influence coronary perfusion in ischemic areas is the coronary collateral resistance. A rich supply of collaterals beyond the stenosis may maintain distal coronary pressure and thus perfusion during exertion. In experiments in the open-chest dog, the effect of collateral flow on stenosis resistance was studied by using a collateral circuit from the carotid artery to the coronary artery distal to a wire snare occluder. When the snare was tightened to reduce blood flow in the coronary artery a marked fall in distal coronary artery pressure occurred. When the collateral channel was opened to maintain pressure distal to the stenosis, flow through the proximal coronary artery actually rose rather than fell as the distal bed perfusion was increased [15]. It is clear then that part of the mechanism of the decrease in flow in that coronary artery was passive change in the caliber of the artery in the area of the stenosis. Maintenance of the distal coronary pressure was able to prevent the passive collapse of the coronary artery and better maintain forward flow. Therefore, development of a collateral network and drugs which might reduce collateral resistance could play a role not only in providing flow to the distal coronary arteries but in maintaining proximal flow as well. The nitrates and Ca^{2+} entry blockers both have been demonstrated to relax collateral channels [12]. β blockers probably also may improve collateral flow but in this case it may be due largely to a prolongation of diastolic interval.

The demand side of the equation is influenced predominantly by changes in systolic blood pressure, heart rate, and ventricular volume during exertion. Drugs which attenuate the rise in systolic blood pressure and slow the heart rate response to exercise or cause peripheral blood pooling tend to reduce myocardial oxygen demand during exertion and thus have proved to be therapeutically useful in the management of angina. In this regard all of the antihypertensive agents may be effective, but the nitrates, the β blockers, and the Ca^{2+} entry blockers all are particularly effective. The bradycardic effect of diltiazem and verapamil are particularly useful in this regard.

In summary, therefore, the Ca^{2+} entry blockers may be effective in relieving angina by a number of possible mechanisms. They attenuate ischemic arteriolar vasodilation and block the dilation-induced fall in distal pressure that aggravates proximal stenosis. They relax the conduit arteries and thus minimize the effect of proximal stenoses on resistance. They may relax collateral vessels which maintain distal perfusion and at the same time reduce proximal stenosis resistance. They may reduce myocardial oxygen demand by lowering exertional systolic blood pressure and attenuating exertional tachycardia. When combined with nitrates and at times with β-adrenoceptor blockers these drugs appear to be particularly effective in the prevention of exertion-induced myocardial ischemia. Their role in the other syndromes of coronary disease, including acute myocardial infarction, congestive heart failure, and sudden death, requires further study.

References

1. Bache, R. J.; Dymek, D. J. Effect of diltiazem on myocardial blood flow. *Circulation* **65**:I19–I26, 1982.
2. Bache, R. J.; Schwartz, J. S. Effect of perfusion pressure distal to a coronary stenosis on transmural myocardial blood flow. *Circulation* **65**:928–935, 1982.
3. Bache, R. J.; Tockman, B. A. Effect of nitroglycerin and nifedipine on subendocardial perfusion in the presence of a flow-limiting coronary stenosis in the awake dog. *Circ. Res.* **50**:678–687, 1982.
4. Ball, R. M.; Bache, R. J. Distribution of the myocardial blood flow in the exercising dog with restricted coronary artery inflow. *Circ. Res.* **38**:60–66, 1976.
5. Brown, B. C.; Josephson, M. A.; Peterson, R. B.; Pierce, C. D.; Wong, M.; Hecht, H. S.; Bolson, E.; Dodge, H. T. Intravenous dipyridamole combined with isometric handgrip for near maximal aortic increase in coronary flow in patients with coronary artery disease. *Am. J. Cardiol.* **48**:1077–1085, 1981.
6. Chew, C. Y. C.; Brown, B. G.; Singh, B. N.; Wong, M. M., Pierce, C.; Petersen, R. Effects of verapamil on coronary hemodynamic function and vasomobility relative to its mechanism of antianginal action. *Am. J. Cardiol.* **51**:699–705, 1983.
7. Flameng, W.; Wusten, B.; Schaper, W. On the distribution of myocardial flow. Part II. Effects of arterial stenosis and vasodilation. *Basic Res. Cardiol.* **69**:435–446, 1974.
8. Fleckenstein, A. Specific pharmacology of calcium in myocardium, cardiac pacemakers and vascular smooth muscle. *Annu. Rev. Pharmacol. Toxicol.* **17**:149–166, 1977.
9. Freudenberg, H.; Lichtlen, P. R. The normal wall segment in coronary stenoses: A postmortem study. *Z. Kardiol.* **70**:863–869, 1981.
10. Fuller, C. M.; Raizner, A. E.; Chahine, R. A.; Nahormek, P.; Ishimori, T.; Verani. M.; Nitishin, A.; Mokotoff, D.; Luchi, R. J. Exercise-induced coronary arterial spasm: Angiographic demonstration, documentation of ischemia by myocardial scintigraphy and results of pharmacologic intervention. *Am. J. Cardiol.* **46**:500–506, 1980.
11. Levy, R. I.; Wiener, L.; Walinsky, P.; Lefer, A. M.; Silver, M. J.; Smith, J. B. Thromboxane release during pacing-induced angina pectoris: Possible vasoconstrictor influence on the coronary vasculature. *Circulation* **61**:1165–1171, 1980.
12. Nagao, T.; Mureta, S.; Sato, M. Effects of diltiazem on developed coronary collaterals in the dog. *Jpn. J. Pharmacol.* **25**:281–288, 1975.
13. Needleman, P.; Kaley, G. Cardiac and coronary prostaglandin synthesis and function. *N. Engl. J. Med.* **298**:1122–1128, 1978.
14. Oliva, P. B.; Potts, D. E.; Pluss, R. G. Coronary arterial spasm in Prinzmetal angina: Documentation by coronary arteriography. *N. Engl. J. Med.* **288**:745–751, 1973.
15. Schwartz, J. S. Effects of distal coronary pressure on rigid and compliant coronary stenoses. *Am. J. Physiol.* **245**:H1054–H1060, 1983.

16. Schwartz, J. S.; Carlyle, P. F.; Cohn, J. N. Effect of dilation of the distal coronary bed on flow and resistance in severely stenotic coronary arteries in the dog. *Am. J. Cardiol.* **43**:219–224, 1979.
17. Schwartz, J. S.; Carlyle, P. F.; Cohn, J. N. Decline in blood flow in stenotic coronary arteries during increased myocardial energetic demand in response to pacing-induced tachycardia. *Am. Heart J.* **101**:435–440, 1981.
18. Schwartz, J. S.; Tockman, B.; Cohn, J. N.; Bache, R. J. Exercise-induced decrease in flow through stenotic coronary arteries in the dog. *Am. J. Cardiol.* **50**:1409–1413, 1982.
19. Specchia. G.; deServi, S.; Falcone, C.; Bramucci, E.; Angoli, L.; Mussini, A.; Marinoni, G. P.; Montemartini, C.; Bobba, P. Coronary arterial spasm as a cause of exercise-induced ST segment elevation in patients with variant angina. *Circulation* **59**:948–954, 1979.
20. Vatner, S. F.; Heintze, T. H. Effects of a calcium-channel antagonist on large and small coronary arteries in conscious dogs. *Circulation* **66**:579–588, 1982.
21. Vatner, S. F.; Macho, P. Regulation of large coronary vessels by adrenergic mechanisms. *Basic Res. Cardiol.* **76**:508–517, 1981.
22. Yasue, H.; Nagao, M.; Omote, S.; Takizawa, A.; Miwa, K.; Tanaka, S. Coronary arterial spasm and Prinzmetal's variant form of angina induced by hyperventilation and Tris-buffer infusion. *Circulation* **58**:56–62, 1978.
23. Yasue, H.; Omote, S.; Takizawa, A.; Nagao, M.; Miwa, K.; Tanaka, S. Exertional angina pectoris caused by coronary arterial spasm: Effects of various drugs. *Am. J. Cardiol.* **43**:647–652, 1979.
24. Yasue, H.; Touyama, M.; Shimamota, M.; Kato, H.; Tanaka, S.; Akiyama, F. Role of autonomic nervous system in the pathogenesis of Prinzmetal's variant form of angina. *Circulation* **50**:534–539, 1974.

52

The Use of Calcium Entry Blockers in Congestive Heart Failure

Robert Zelis and Edward Toggart

Patients with severe congestive heart failure operate on a depressed ventricular function curve [28]. They experience symptoms and signs due to: (1) an inadequate cardiac output (especially with exercise) and/or (2) an increase in ventricular filling pressure. Low-output symptoms and signs result from redistribution of an inadequate cardiac output and include fatigue, pallor, and a decreased urinary output. Congestive symptoms and signs result from an increase in pulmonary and systemic venous pressure and fluid accumulation in tissues. The following may occur: shortness of breath from pulmonary congestion (pulmonary edema), edema in dependent portions of the body, hepatic enlargement and tenderness due to congestion, as well as ascites and pleural effusions.

Vasoconstriction in Heart Failure

Peripheral vasoconstriction, which is a prominent feature in heart failure [26], can be considered a favorable compensatory mechanism when it helps to maintain systemic perfusion pressure in the face of an inadequate cardiac output. This helps preserve blood flow to the heart and brain, circulations which are relatively spared from neurohumoral constrictor influences. However, systemic vasoconstriction may also be excessive, and can impair ventricular function further by presenting an increased aortic impedance to the failing heart [28]. There has been considerable interest recently in the use of vasodilators to minimize the deleterious effects of the vasoconstrictor response to heart failure [4,6]. These drugs can be classified according to either the category of vessels on which they exert their major action, or the vasconstrictor mechanism their primary action overcomes.

According to the former classification system, drugs can be considered either arteriolar dilators (e.g., hydralazine), venodilators (e.g., nitroglycerin), or mixed vasodilators having a balanced effect on arteries and veins (e.g., nitroprusside, prazosin, and captopril). Al-

---

Robert Zelis and Edward Toggart • Cardiology Division, The Milton S. Hershey Medical Center, The Pennsylvania State University, Hershey, Pennsylvania 17033.

though some vasodilators are considered to have a "direct" effect on blood vessels (e.g., the nitrates, nitroprusside, and hydralazine), other drugs may specifically interrupt a vasoconstrictor mechanism activated by heart failure. The type of vasoconstrictor mechanism and the extent to which it is employed varies with the type and severity of heart failure, and varies with time as heart failure develops [26].

However, in general, three major abnormalities may be seen. First there appears to be hyperactivity of the sympathetic nervous system, which may result in increased neurogenic vasomotor tone and increased circulating catecholamines, especially during exercise. At least three mechanisms may be responsible for increased activity of the sympathetic nervous system, including (1) a depression of the baroreceptor (cardiac and arterial) buffering of cardiovascular vasoconstrictor reflexes [26], (2) activation of somatic afferent receptors due to relative skeletal muscle ischemia during exercise [26], and (3) enhanced ability of the sympathetic nerves to release norepinephrine perhaps through down-regulation of presynaptic α_2 receptors [29].

A second general vasoconstrictor mechanism operative in heart failure is an increase in circulating vasoconstrictor hormones (e.g., angiotensin and vasopressin) [26]. The third postulated mechanism is an increased stiffness of arterial conductance vessels [26]. This may be related to increased vascular sodium, which limits the skeletal muscle vasodilator response to a metabolic stimulus.

Therefore, vasodilators can be classified according to which of these mechanisms they antagonize. The vasoconstrictor effects of norepinephrine are antagonized by the α_1 blocker prazosin, and the vasoconstrictor effects of angiotensin are limited by the angiotensin-converting enzyme inhibitor captopril. Although diuretics are not generally classified as "vasodilators," they may reduce vascular sodium and improve vasodilator capacity.

As a group, calcium entry blockers also dilate blood vessels and might be considered for use in congestive heart failure, but they also depress myocardial contractility [2,13,27]. Before considering their use in heart failure, their complex effects on isolated cardiac and vascular tissues will be reviewed. Since the response of the heart and circulation in an intact animal depends on many factors, the integrated cardiocirculatory response to these drugs will be considered. Only three drugs presently approved for use in humans, diltiazem, verapamil, and nifedipine, will be reviewed in detail. At present, they are approved for use in angina, but not in heart failure.

The Cardiovascular Effects of Ca^{2+} Blockers

Influx of Ca^{2+} is responsible for the action potentials of the sinoatrial and atrioventricular (AV) nodes and for excitation–contraction coupling in cardiac and vascular muscle. Thus, impairment of Ca^{2+} influx might be expected to slow sinus node firing rate, depress AV conduction, reduce myocardial contractility, and relax vascular smooth muscle [2,13]. All these effects are readily apparent when isolated tissues are evaluated (Table I). In an isolated right atrial preparation the order of potency for decreasing sinus node frequency is: diltiazem > nifedipine > verapamil [13]. In terms of negative inotropy, the order of potency is: nifedipine > verapamil > diltiazem [13]. In aortic vascular preparations, the Ca^{2+} blockers appear to be more potent in inhibiting contraction mediated by the activation of potential-dependent Ca^{2+} channels than they are in inhibiting contraction produced by the opening of receptor-operated Ca^{2+} channels [13,27]. However, this differential effect is not seen in resistance vessels of most other circulatory systems [13]. There, Ca^{2+} blockers may be equally potent in blocking contractions initiated by opening of both types of Ca^{2+}

Table I. The Different Effects of Calcium Blockers on Isolated Cardiovascular Tissue and Cardiovascular Function in Intact Animals and Humans

	Diltiazem	Verapamil	Nifedipine
Isolated tissue			
Frequency (SA node)	− − −	− −	−
Myocardial contractility	−	− −	− − −
Vascular contraction	−	− −	− − −
Intact animal			
Heart rate	−	− +	+
AV conduction	− −	− −	+
Contractility indices	0	−	+
Coronary blood flow	+ + +	+ +	+ + +
Limb blood flow	+ +	+ + +	+ + +

channels. In vascular smooth muscle the order of potency in decreasing order in inhibiting vascular contraction is: nifedipine > verapamil > diltiazem [13].

When evaluating the effects of these drugs on persons or animals with an intact circulation and active baroreceptor reflexes, a different type of cardiocirculatory response is observed [2,13] (Table I). The marked systemic vasodilation produced by nifedipine results in a reflex increase in heart rate and a reflex increase in AV conduction. The systemic vasodilation occurs at a clinical dose which is well below that which produces myocardial depression. Therefore, reflex sympathetic activation is the predominant effect on the heart, and a slight increase in contractility results [13]. The systemic vasodilation produced by nifedipine is widespread and not selective for any vascular bed [13].

Diltiazem is a less potent generalized systemic arterial dilator than nifedipine or verapamil but appears to selectively increase coronary blood flow [13]. Systemic administration of the drug generally results in a decrease in heart rate and AV conduction, whereas negative inotropic effects are rarely seen [13].

Verapamil is a potent systemic vasodilator, whose effects on nonvisceral resistance vessels are greater than those on vessels in the coronary circulation [13]. Thus, vasodilation and activation of baroreceptor reflexes partially offset the direct depressant effects of the drug on the sinus and AV nodes, as well as the ventricular myocardium. Generally, verapamil does not alter heart rate and myocardial contractility, but does slow AV conduction [13]. However, at higher doses a negative inotropic effect is observed which is particularly troublesome for patients with markedly depressed baseline myocardial function caused by disease or other cardiodepressant drugs (e.g., β blockers and disopyramide).

The Use of Ca^{2+} Blockers in Heart Failure

It is clear from the above discussion that the peripheral vasodilator capacity of Ca^{2+} blockers might be of potential use in heart failure. Equally clear are the risks that might be posed by the negative chronotropic, dromotropic, and inotropic properties of certain Ca^{2+} blockers. For example, the use of diltiazem in patients with sinus node disease (e.g., sick sinus syndrome) is contraindicated. Similarly, both diltiazem and verapamil should not be administered to patients who have significant problems with AV conduction. Lastly, the negative inotropic effects of verapamil make it contraindicated for patients with heart failure and significantly depressed myocardial contractility. However, acute intravenous administration of verapamil is well tolerated by most patients with mild ventricular dysfunction.

Specific studies evaluating the hemodynamic changes that occur after administration of Ca^{2+} blockers to patients with heart disease will be reviewed next. Chew et al. studied the effects of intravenous verapamil in 22 patients with acute myocardial infarction or symptomatic coronary artery disease [5] and found that the drug reduced systemic vascular resistance 23%; cardiac index rose 18% and systolic arterial pressure fell 15%. Heart rate was essentially unchanged and left ventricular filling pressure increased 13%; of note is that these patients all had normal ventricular function even though they had heart disease. Their mean ejection fraction was 56% and they displayed a normal cardiac index and left ventricular filling pressure. In three patients with severely depressed ventricular function and mean ejection fraction of 26%, a comparable dose of intravenous verapamil (145 μg/kg followed by 5 μg/kg/min) increased left ventricular filling pressure by 30% (from 25.7 mm Hg) and decreased cardiac index 15% (from 2.8 liters/min/m^2). Recently, Vlietstra et al. compared the acute effects of intravenous verapamil (200 μg/kg) with placebo in patients with near-normal ventricular function and mean ejection fraction of 58% [25]. They, too, noted minimal changes in heart rate, a reduction in systemic arterial pressure and vascular resistance, and an increase in cardiac index. Left ventricular end-diastolic pressure was virtually unchanged; however, there was a modest increase in left ventricular end-diastolic volume (from 94 to 102 ml/m^2). Thus, for patients with heart disease, but without severe left ventricular failure, the acute administration of verapamil appears to be well tolerated. However, in patients with significant congestive heart failure, it is a potent negative inotrope and is contraindicated.

Although diltiazem is not as potent a systemic vasodilator as nifedipine or verapamil, it produces a slight but significant increase in left ventricular ejection fraction in patients with moderate ventricular dysfunction [20]. Diltiazem increased urinary sodium excretion, without a significant change in renal plasma flow, glomerular filtration, or urinary volume in nine patients with moderate to severe congestive heart failure [18]. These authors suggested that diltiazem either alters intrarenal blood flow distribution to favor natriuresis or decreases renin release. We have preliminary unpublished data in a rat model of chronic congestive heart failure produced by coronary ligation which suggest that diltiazem attenuates the renal vasoconstrictor response to exercise. This effect is similar to that reported when intravenous nitroglycerin was given to another chronic rat model of heart failure produced by an aortocaval fistula [12].

Since the hemodynamic profile of nifedipine closely resembles that of arteriolar dilators, the Ca^{2+} blocker has been most often evaluated for the treatment of heart failure (Table II). Matsumoto et al. observed an increase in cardiac index of 16% and a decrease in systemic vascular resistance of 20% with minimal change in left ventricular filling pressure in eight patients with congestive heart failure who were given nifedipine [22]. These patients, however, were not severely ill, and as a group had a normal baseline cardiac index and left ventricular filling pressure. In another study nifedipine was administered sublingually to 11 patients with congestive heart failure and they had significantly compromised hemodynamics at rest, with a depressed mean cardiac index (2.1 liters/min/m^2) and an increased pulmonary capillary wedge pressure (25 mm Hg) [19]. Nifedipine elicited a 44% reduction in systemic vascular resistance, a 34% reduction in pulmonary capillary wedge pressure, and a 47% increase in cardiac index.

The above studies examined the effects of nifedipine in patients who were stable. Polese et al. measured hemodynamics in 24 patients with acute pulmonary edema from a variety of causes (hypertension, cardiomyopathy, aortic and mitral regurgitation) before and after

Table II. The Effects of Nifedipine on Hemodynamics in Patients with Congestive Heart Failure (CHF)[a]

Reference	Patients	Dose		HR	BP	CI	LVFP	SVR
Matsumoto et al. [22]	Chronic CHF (n = 8)		C	73	113	3.5	11	15.6
		20 mg SL	N	81	101	4.1	12	12.4
Klugman et al. [19]	Chronic CHF (n = 11)		C	78	95	2.1	25	24.6
		20 mg SL	N	80	75	3.1	17	13.9
Polese et al. [23]	Pulmonary edema		C	88	221	2.5	28	31.6
	HT (n = 7)	10 mg PO	N	96	177	3.4	18	19.3
			C	107	132	1.9	39	28.7
	CM (n = 7)	10 mg PO	N	103	116	2.6	26	17.7
			C	82	173	2.5	26	22.7
	AR (n = 5)	10 mg PO	N	91	149	3.5	17	15.0
			C	90	145	2.8	31	20.8
	MR (n = 5)	10 mg PO	N	92	118	3.6	23	13.6
Brooks et al. [3]	Chronic CHF (n = 4)		C	91	110	4.0[†]	31	25.5
		20–30 mg PO	N	89	90	5.2[†]	24	15.8
	(n = 2)		C	90	91	4.5[†]	35	16.5
		20,30 mg PO	N	76	43	3.0[†]	24	8.5
Bellocci et al. [1]	Chronic CHF (n = 10)		C	86	86*	1.8	25	26.4
		20 mg PO	NA	88	74*	3.0	19	13.6
		80 mg/day	NC	86	76*	3.0	21	13.8
		0 × 48 hr	NW	87	85*	2.0	24	24.7

[a] Abbreviations: HR, heart rate (min^{-1}); BP, systemic systolic arterial pressure (mm Hg) (* = mean pressure); CI, cardiac index (liters/min/m^2) ([†] = cardiac output, liters/min); LVFP, left ventricular filling pressure (mm Hg); SVR, systemic vascular resistance (units); SL, sublingual; PO, oral; HT, hypertension; CM, cardiomyopathy; AR, aortic regurgitation; MR, mitral regurgitation; NA, NC, NW are, respectively, nifedipine acute, chronic, and 48 hr after withdrawal.

nifedipine (Table II) [23]. All patients were in significant heart failure and demonstrated elevated left ventricular filling pressures (26 to 39 mm Hg) and most had depressed cardiac indices (1.9 to 2.8 liters/min/m^2). In this group nifedipine reduced left ventricular filling pressure 26–35%, and increased cardiac index 29–40%, while reducing systemic vascular resistance 34–39%.

In all of the above studies, acutely administered nifedipine led to a favorable response. However, there are two reports in the literature of significant unfavorable responses. One patient with severe aortic valvular stenosis (peak systolic aortic valve gradient of 124 mm Hg), who was given 10 mg of nifedipine orally for increasing angina pectoris, developed acute pulmonary edema [15]. However, patients with this degree of aortic stenosis do not normally receive vasodilator therapy. In another study, two patients given nifedipine exhibited severe systemic hypotension and reduction in cardiac output [3] (Table II), while four other patients had a favorable response.

The chronic effects of nifedipine in patients with significant heart failure have been evaluated in only one study [1] (Table II). Bellocci et al. recorded hemodynamics at four time points: before and immediately after giving nifedipine 20 mg orally; 2 months later during chronic nifedipine therapy (80 mg/day); and 48 hr after withdrawing the drug. They noted that acutely administered nifedipine increased cardiac index 66%, reduced systemic vascular resistance 48%, and reduced left ventricular filling pressure 23%. When the patients were studied during chronic therapy, the improvement in cardiac index, systemic vascular resistance, and left ventricular filling pressure were all sustained. After stopping nifedipine therapy for 48 hr, these values returned to the level noted prior to therapy.

It is clear from these studies that nifedipine is a potent systemic vasodilator with predominant effects on the arterial side of the circulation. In most patients to whom nifedipine is given acutely, significant improvement in cardiac index occurs, which appears to be sustained during chronic drug therapy. But on occasion, patients with severe heart failure may experience an adverse reaction to large doses of nifedipine given acutely. Thus, nifedipine can be administered safely, and improves systemic hemodynamics in patients with congestive heart failure.

One important question remains: do Ca^{2+} blockers have any *unique* advantages over other vasodilators? There is a suggestion that, in certain situations, they do.

Unique Advantages of Ca^{2+} Blockers over Other Vasodilators

In hypertropic cardiomyopathy, verapamil reduces angina pectoris more effectively than β blockers [17,24]. Since Ca^{2+} influx may be an initial component for the development of myocardial hypertrophy, verapamil might also reduce left ventricular mass in these patients. Unfortunately, patients with hypertropic cardiomyopathy show progression, regression, or no effect of the drug on myocardial mass [9]. In an interesting preliminary report by French et al., chronic verapamil administration to normal dogs produced an increase in myocardial weight [14]. Thus, the initial hope that Ca^{2+} blocker therapy might reduce myocardial hypertrophy has apparently not been borne out.

However, certain Ca^{2+} blockers possess some unique properties that might be uniquely beneficial. One mechanism postulated to be responsible for increased sympathetic nerve activity in heart failure is a depressed baroreceptor sensitivity [26], and one salutary effect of digitalis in heart failure is the enhancement of baroreceptor sensitivity. Nifedipine also enhances baroreceptor activity, but verapamil depresses it [8]. The former agent might provide the patient with heart failure a greater degree of baroreceptor modulation of the sympathetically mediated vasoconstriction that normally accompanies exercise. Therefore, one potentially unique advantage for nifedipine in heart failure is enhanced baroreceptor sensitivity.

A second potentially unique mechanism for Ca^{2+} blocker therapy is in modulating norepinephrine release. An influx of Ca^{2+} is necessary for excitation–secretion coupling in sympathetic nerves. Previously, it was thought that nerves are relatively resistant to the effects of Ca^{2+} blockers, which do not significantly alter neurotransmitter release in the brain and release of norepinephrine from cardiac sympathetic nerves. However, recently it has been noted that diltiazem inhibits norepinephrine release from vascular sympathetic nerves at a concentration which is only twice that which inhibits vascular contraction [30]. Whether or not this will prove to be a unique benefit from Ca^{2+} blocker therapy of heart failure remains to be determined.

A third suggested unique effect of Ca^{2+} blockers is in modifying ventricular diastolic function in heart failure. The effects of sublingual nifedipine (20 mg) on 19 patients with various degrees of left ventricular dysfunction were evaluated [21]. An increase in cardiac index and a reduction in systemic vascular resistance occurred for the group as a whole. In patients whose left ventricular filling pressure was elevated, nifedipine produced a significant reduction in this variable. The authors evaluated a number of indices of left ventricular diastolic function and noted that peak negative *dp/dt* was significantly reduced by nifedipine. However, tau, the left ventricular isovolumic relaxation constant, was unaltered by nifedipine. The authors concluded that nifedipine did not exert a primary effect on left ventricular diastolic compliance but, rather, enhanced relaxation by altering the loading conditions of the heart. Dash evaluated the effects of intravenous diltiazem on left ventricular diastolic function and noted that the drug reduced tau as well as peak negative dp/dt [7]. Some of this effect might have been related to altering of loading condition of the heart as well.

Lastly, there appears to be a small subgroup of patients in whom acute left ventricular failure is the primary manifestation of myocardial ischemia. In these individuals, acute shortness of breath is an "anginal equivalent" and minimal or no cardiac pain occurs. In most patients with severe congestive heart failure secondary to coronary artery disease, anginal symptoms commonly abate. The role played by continuing myocardial ischemia in aggravating heart failure in these patients is unclear. Diltiazem may improve exercise hemodynamics in some angina patients by improving myocardial perfusion [16]. Concordant with this conclusion is the report by Dash, who noted in a group of patients given diltiazem intravenously that there were subtle but definite improvements in regional systolic function in areas of myocardium compromised by atherosclerotic coronary arteries [7]. On the other hand, in other studies it was concluded that the major mechanism by which verapamil improved angina was by reducing myocardial oxygen requirements, rather than by increasing coronary blood flow [10,11].

In summary, the negative inotropic effects of verapamil preclude its use as a vasodilator for heart failure therapy. The more potent systemic arteriolar dilator, nifedipine, clearly produces a sustained improvement in cardiac output, reduction in systemic vascular resistance, and reduction in left ventricular filling pressure. In patients with left ventricular dysfunction there are a number of potential mechanisms by which nifedipine and perhaps diltiazem may have additional unique advantages over conventional vasodilators. These include an enhancement in baroreceptor sensitivity, inhibition of norepinephrine release from vascular nerves, and improvement in myocardial blood flow during exercise. Whether or not these advantages will prove significant enough to warrant more widespread use of Ca^{2+} blockers for heart failure therapy remains to be determined. It seems reasonable to consider nifedipine or diltiazem as being potentially useful for patients with ventricular dysfunction secondary to coronary artery disease in whom attacks of acute shortness of breath might be related to transient myocardial ischemia. However, this probably represents a relatively small proportion of patients with congestive heart failure.

ACKNOWLEDGMENTS. Supported in part by a grant from the American Heart Association York–Adams Chapter and a gift from The Grand Chapter of Pennsylvania. Order of the Eastern Star. ET. is the recipient of NIH Research Service Award HL-06239.

References

1. Bellocci, F.; Ansalone, G.; Santarelli. P.; Loperfido, F.; Scabbia, E.; Zecchi, P.; Manzoli, U. Oral nifedipine in the long-term management of severe chronic heart failure. *J. Cardiovasc. Pharmacol.* **4**:847–855, 1982.
2. Braunwald, E. Mechanisms of action of calcium-channel-blocking agents. *N. Engl. J. Med.* **26**:1618–1627, 1982.
3. Brooks, N.; Cattell, M.; Pidgeon, J.; Balcon, R. Unpredictable response to nifedipine in severe cardiac failure. *Br. Med. J.* **281**:1324, 1980.
4. Chatterjee, K.; Parmley, W. W. Vasodilator therapy for chronic heart failure. *Annu. Rev. Pharmacol. Toxicol.* **20**:475–512, 1980.
5. Chew, C. Y. C.; Hecht, H. S.; Collett, J. T.; McAllister, R. G.; Singh, B. N. Influence of severity of ventricular dysfunction on hemodynamic responses to intravenously administered verapamil in ischemic heart disease. *Am. J. Cardiol.* **47**:917–922, 1981.

6. Cohn, J. N.; Franciosa, J. A. Vasodilator therapy of chronic cardiac failure. *N. Engl. J. Med.* **297**:27–31, 1977.
7. Dash, H. Regional wall motion response to diltiazem in coronary artery disease. *Circulation* **68**:iii–18, 1983.
8. Dorsey, J. K.; Ferguson, D. W.; Tracy, J.; Mark, A. L. Effects of calcium blockers on baroreceptor control of vascular resistance in humans. *Clin. Res.* **31**:523A. 1983.
9. Epstein, S. E.; Rosing, R. D. Verapamil: Its potential for causing serious complications in patients with hypertrophic cardiomyopathy. *Circulation* **64**:437–441, 1981.
10. Ferlinz, J.; Stavens, C. S. Effects of intracoronary versus intravenous verapamil on exercise performance of patients with coronary artery disease. *Clin. Res.* **31**:182A, 1983.
11. Ferlinz, J.; Turbow, M. E. Antianginal and myocardial metabolic properties of verapamil in coronary artery disease. *Am. J. Cardiol.* **46**:1019–1026, 1980.
12. Flaim, S. F.; Weitzel, R. L.; Zelis. R. Mechanism of action of nitroglycerin during exercise in a rat model of heart failure: Improvement of blood flow to the renal, splanchnic, and cutaneous beds. *Circ. Res.* **49**:458–468, 1981.
13. Flaim, S. F.; Zelis, R. *Calcium Channel Blockers: Mechanisms of Action and Clinical Applications.* Baltimore, Urban & Schwarzenberg, 1982.
14. French, W. J.; Adomian, G. E.; Averill, W. K.; Garner, D.; Laks, M. M. Chronic infusion of verapamil produces increased heart weight in conscious dogs. *Clin. Res.* **31**:184A, 1983.
15. Gillmer, D. J.; Kark, P. Pulmonary oedema precipitated by nifedipine. *Br. Med. J.* **280**:1420, 1980.
16. Hossack, K. F.; Bruce, R. A.; Ritterman, J. B.; Kusumi, F.; Trimble, S. Divergent effects of diltiazem in patients with exertional angina. *Am. J. Cardiol.* **49**:538–546, 1982.
17. Kaltenbach, M.; Hopf, R.; Kober, G.; Bussman, W. D.; Keller. M.; Petersen, Y. Treatment of hypertrophic obstructive cardiomyopathy with verapamil. *Br. Heart J.* **42**:35–42, 1979.
18. Kinoshita, M.; Kusakawa, R.; Shimono, Y.; Motomura, M.; Tomoroga, G.; Hoshino, T. Effect of diltiazem hydrochloride on sodium diuresis and renal function in chronic congestive heart failure. *Arzneim. Forsch.* **29**:676–681, 1979.
19. Klugmann, S.; Salvi, A.; Camerini, F. Haemodynamic effects of nifedipine in heart failure. *Br. Heart J.* **43**:440–446, 1980.
20. Low, R. I.; Takeda, P.; Lee, G.; Mason, D. T.; Awan, N. A.; DeMaria, A. N. Effects of diltiazem-induced calcium blockade upon exercise capacity in effort angina due to chronic coronary artery disease. *Am. Heart J.* **101**:713–718, 1981.
21. Ludbrook, P. A.; Tiefenbrunn, A. J.; Sobel, B. E. Influence of nifedipine on left ventricular systolic and diastolic function. Relationships to manifestations of ischemia and congestive failure. *Am. J. Med.* **71**:683–692, 1981.
22. Matsumoto, S.; Ito, T.; Sada, T.; Takahashi, M.; Su, K.-M.; Ueda, A.; Okabe, F.; Sato, M.; Sekine, I.; Ito, Y. Hemodynamic effects of nifedipine in congestive heart failure. *Am. J. Cardiol.* **46**:476–479, 1980.
23. Polese, A.; Fiorentini, C.; Olivari, M. T.; Guazzi, M. D. Clinical use of a calcium antagonistic agent (nifedipine) in acute pulmonary edema. *Am. J. Med.* **66**:825–830, 1979.
24. Rosing, D. R.; Kent, K. M.; Borer, J. S.; Seides, S. F.; Maron, B. K.; Epstein, S. E. Verapamil therapy: a new approach to the pharmacologic treatment of hypertrophic cardiomyopathy (parts I and II). *Circulation* **60**:1201–1213, 1979.
25. Vlietstra, R. E.; Farias, M. A. C.; Frye, R. L.; Smith, H. C.; Ritman, E. L. Effect of verapamil on left ventricular function: A randomized, placebo-controlled study. *Am. J. Cardiol.* **51**:1213–1217, 1983.
26. Zelis, R.; Flaim, S. F. Alterations in vasomotor tone in congestive heart failure. *Prog. Cardiovasc. Dis.* **24**:437–459, 1982.
27. Zelis, R.; Flaim, S. F. Calcium blocking drugs for angina pectoris. *Annu. Rev. Med.* **33**:465–478, 1982.
28. Zelis, R.; Flaim, S. F.; Liedtke, A. J.; Nellis, S. H. Cardiocirculatory dynamics in the normal and failing heart. *Annu. Rev. Physiol.* **43**:455–476, 1981.
29. Zelis, R.; Starke, K.; Brunner, H. Abnormal vascular sympathetic nerve function in heart failure. *Clin. Res.* **31**:464A, 1983.
30. Zelis, R.; Wichmann, T.; Starke, K. Inhibition of vascular noradrenaline release by diltiazem. *Circulation* **66**:II139, 1982.

C

NUTRITIONAL AND PATHOPHYSIOLOGIC ASPECTS OF CALCIUM ACTION

XII

Vitamin D and Other Calcemic Agents

53

The Vitamin D–Calcium Axis—1983

Hector F. DeLuca

This chapter is devoted to a brief review of the vitamin D endocrine system coupled with the enumeration of new information on the role of vitamin D and its metabolites in calcium metabolism. Additionally, the most recent outlook for new applications of vitamin D hormonal substances in the treatment of metabolic bone diseases such as postmenopausal osteoporosis is presented.

Photobiogenesis of Vitamin D

It has been known that vitamin D is not a true vitamin [1,2], since the independent studies of Huldshinsky [3] and Chick et al. [4] showed that higher animals exposed to sufficient amounts of ultraviolet light do not require vitamin D. The isolation of 7-dehydrocholesterol and the demonstration that this intermediate in cholesterol biosynthesis is a precursor of vitamin D_3 ultimately led to the assumption that vitamin D_3 is produced in skin by ultraviolet irradiation [5]. More recently, two laboratories isolated vitamin D_3 from the skin of vitamin D-deficient animals irradiated with ultraviolet light [6,7].

The conversion of 7-dehydrocholesterol to vitamin D_3 in skin is strictly a photolysis reaction quite similar to that occurring in organic solvents in the laboratory [8] and involves a two-step conversion. The first is a photochemical eruption of the B-ring followed by a 5,7-sigmatropic shift to yield previtamin D. Previtamin D then slowly equilibrates to form vitamin D_3. So far, the photochemical conversion has not been shown to be regulated, although pigmentation of skin and tanning of skin appears to reduce the conversion of 7-dehydrocholesterol to vitamin D_3 probably by blocking a fraction of ultraviolet light from reaching the 7-dehydrocholesterol [9,10].

Vitamin D is not required in the diet if the subject is exposed to sufficient amounts of ultraviolet light. However, vitamin D can be absorbed with fats in the distal small intestine through the lacteal system, ultimately making its appearance in the liver [2]. Hence, because of climatic conditions, industrial pollution of the atmosphere, and life style, vitamin D_3 can

---

Hector F. DeLuca • Department of Biochemistry, University of Wisconsin, Madison, Wisconsin 53706.

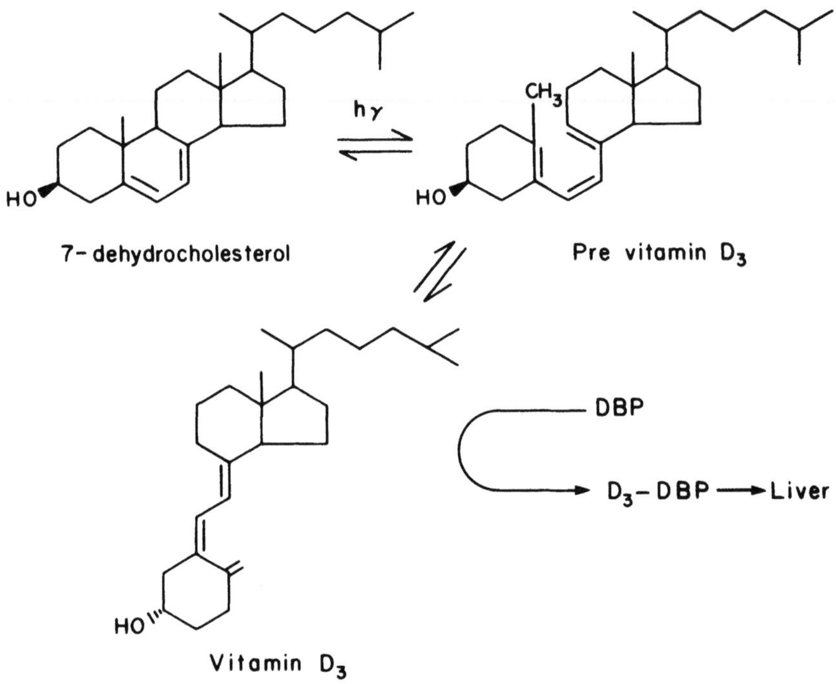

Figure 1. Photobiogenesis of vitamin D_3 and its transport from skin to liver.

be required in the diet, thus relegating it to the category of the vitamins [2]. In a true sense, however, vitamin D is a prohormone normally prepared in the skin by photochemical conversion (Fig. 1).

Physiologic Functions of Vitamin D

Vitamin D was originally discovered as being necessary for the mineralization of bone [11,12]. In the absence of vitamin D, the deposition of calcium and phosphorus on collagen fibrils of the bone matrix does not take place [13]. The major reason for this failure is an insufficient supply of calcium and phosphorus to the calcification sites [14,15]. In vitamin D deficiency the product of calcium × phosphorus in the plasma is very low as compared to normal. Normally, blood is supersaturated with regard to the hydroxyapatite mineral form in bone and with regard to calcium and phosphorus concentration [15], while the plasma is undersaturated in vitamin D deficiency [16]. Thus, the essence of vitamin D action is to elevate plasma calcium and phosphorus levels to support mineralization of newly forming bone (Fig. 2).

Vitamin D stimulates both active calcium transport from the lumen of the intestine across the enterocyte into the plasma compartment [17,18] and the independent active phosphate transport in the enterocyte [19–21]. However, to support the critical functions of calcium in nerves, muscle, and other tissues, there must be a more readily available source of calcium. The mineralized skeleton not only serves a structural role but also as a reservoir for calcium to be called upon when required [22,23].

The calcified skeleton is always separated from the plasma compartment by a membrane of cells composed of osteoblasts, osteocytes, bone-lining cells, and perhaps osteoclasts [24]

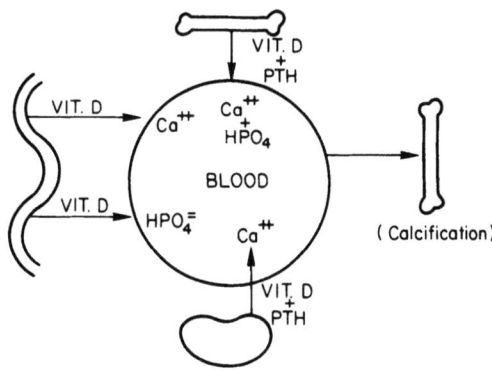

Figure 2. Sites of action of Vitamin D and parathyroid hormone in the elevation of plasma calcium and phosphorus concentrations to levels required for mineralization of bone. Note that vitamin D activates intestinal calcium and phosphorus transport. It, together with parathyroid hormone, activates the mobilization of calcium and phosphorus from the bone fluid compartment, and stimulates renal reabsorption of calcium in the distal tubule.

forming a bone fluid compartment whose calcium level is dominated by the solubility product of hydroxyapatite. The calcium level in that compartment is therefore below that in the plasma or extracellular fluid compartment. If calcium is required, it must be transported from the bone fluid compartment to the plasma compartment. Two hormones are required in this process, the vitamin D and parathyroid hormone [25,26]. The mobilization of calcium from bone is compromised unless both hormones are present [26]. The reabsorption of calcium in the distal renal tubule is also likely to be under control of both parathyroid and the vitamin D hormone [27], although the nature of the interaction of these two substances in the renal conservation of calcium has not been defined.

There has been considerable debate especially among clinical investigators as to whether vitamin D might have a further action at the mineralization site [28,29]. It has been suggested that some form of vitamin D might be necessary for the transfer of mineral from the plasma compartment to the calcification sites [28,29], or that it might be required in preparing the calcification sites [30–32]. We have recently completed a major effort in examining this question. J. Underwood, a Ph.D. student, developed techniques whereby rats raised from completely vitamin D-deficient mothers were cannulated through a skull cap into the jugular veins and infusion pumps were used to infuse concentrated calcium and phosphorus solutions to maintain the plasma calcium and phosphorus levels in the normal range for a period of 9 days. Similar infusions with saline were done with animals receiving vitamin D, and they were compared to animals without infusion. The results are illustrated in Table I. This technique was effective in maintaining normal plasma calcium and phosphorus levels in vitamin D-deficient rats for the 9-day period, whereas in the noninfused vitamin D-deficient controls plasma calcium and phosphorus levels were both low. The maintenance of plasma calcium and phosphorus in the normal range stimulated bone growth to the same degree as in animals receiving vitamin D and infused with saline. Most importantly, bones from vitamin D-deficient animals exposed to normal levels of calcium and phosphorus calcified normally. In fact, the ash content of their bones was higher than that of the + D animals receiving a saline infusion. Preliminary results obtained from bones labeled with tetracycline suggest that the mineralization brought about by calcium and phosphorus infusions are normal histologically and histomorphometrically [33]. These results show quite clearly that vitamin D is not required for the mineralization of cartilage and bone. In fact, there is also no evidence that some form of vitamin D is required for the growth of cartilage as has been suggested by *in vitro* techniques [33].

Although these experiments and those of others [34,35] reveal that vitamin D is not required at the mineralization sites, this does not rule out an action of vitamin D on bone. Indeed, vitamin D plays an important role, together with parathyroid hormone, in the

*Table I. Mineralization and Bone Growth Does Not Require Vitamin D
If Bone Is Supplied with Calcium and Phosphorus[a]*

	Not infused		Infused	
	− D	+ D	− D	+ D
Serum Ca (mg/dl)	5.1*	10.2	10.4	10.4
Serum P (mg/dl)	6.2*	8.5*	9.8	9.7
Femur length (nm)	22.2*	24.5	25.1	25.8
Femur ash (%)	36.7	46.2	52.1[†]	47.5
Femur ash (mg)	52.3*	89.6	117.5[†]	95.1

[a] Vitamin D-deficient pups were obtained from vitamin D-deficient mothers and were then fed the normal calcium, normal phosphorus vitamin D-deficient diet until they weighed 100 g at the end of 5 weeks. At that time, one-half the rats were given 75 IU of vitamin D_3 every third day, while the remaining animals were maintained in the vitamin D-deficient state. The rats were then either maintained as controls (not infused) or were infused continuously for 9 days with solutions of calcium and phosphorus in the case of the − D animals and with saline in the case of the + D animals. At the end of the 10-day period, all animals were killed and the indicated measurements were made.
*Lower than others in group, $p < 0.001$.
†Higher than corresponding + D, $p < 0.001$.

mobilization of calcium from bone [25,26]; hence, osteoblasts, osteocytes, and bone-lining cells are likely sites of action of the vitamin D hormone [36]. Quite apart from calcium homeostasis, vitamin D plays an important role in the modeling and remodeling sequences [37]. Thus, when bone is being turned over, the resorption process is quite likely to be under direct control of the vitamin D hormone [38,39].

Physiologic Metabolism of Vitamin D

For vitamin D to carry out its functions in intestine, bone, and kidney, it must be metabolically activated [1,2] (Fig. 3). Vitamin D must be hydroxylated in the 25 position primarily in the liver to form 25-hydroxyvitamin D (25-OH-D) or calcidiol. There are two systems capable of 25-hydroxylating vitamin D. A microsomal system carries out 25-hydroxylation at physiologic doses of vitamin D [40]. This is a two-component, mixed-function monooxygenase, dependent on flavoprotein and a cytochrome P-450 [41,42]. A mitochondrial system, which also hydroxylates vitamin D and other sterols in the 25 position [43], begins to function when large doses of vitamin D are given. Some 25-hydroxylation also

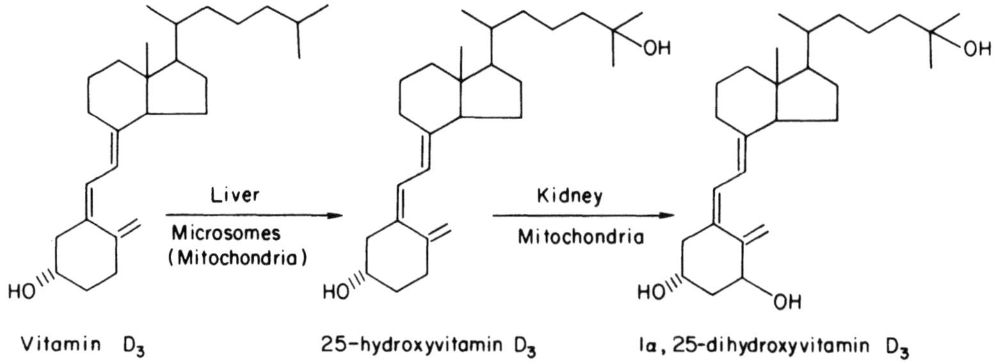

Figure 3. Metabolism of vitamin D_3 required for its functions.

occurs in intestine and kidney [44,45]. The 1α-hydroxylation of 25-OH-D_3 in the nonpregnant mammal takes place in the kidney [46,47], but in the pregnant mammal, some conversion also takes place in the placenta [48,49]. The renal mixed-function monooxygenase system, which is quite similar to the adrenal steroidogenic system [50,51], is dependent on a flavoprotein, an iron sulfur protein, and a cytochrome P-450; it places a hydroxyl group in the 1α position to form 1α,25-dihydroxyvitamin D_3 (calcitriol) [1,25-$(OH)_2D_3$] [52].

Recently, it has been argued that the kidney is not the exclusive site of synthesis of the 1,25-$(OH)_2D_3$ or calcitriol [53–55], which, if true, would severely compromise the endocrine nature of the kidney in the vitamin D system. We, therefore, considered a reexamination of this question [56]. If kidneys are removed from vitamin D-deficient animals, 25-OH-D_3 is unable to stimulate the target organ responses of bone calcium mobilization [57] and intestinal calcium transport at physiologic concentrations [58,59]. On the other hand, calcitriol produces a clear response in both intestine and bone whether the kidneys are present or not. This indicates that calcitriol or a metabolite is the active form and that the kidney is essentially the exclusive site of calcitriol synthesis. Since bone cells did not respond to physiologic doses of calcidiol, one can infer that a significant amount of extrarenal 1α-hydroxylation does not occur.

Nevertheless, using newly synthesized radiolabeled calcidiol of 160 Ci/mmole as the most sensitive method of detecting the presence of calcitriol, we reexamined the question of whether anephric rats can synthesize even small amounts of calcitriol [56]. We nephrectomized vitamin D-deficient rats, or tied their ureters or sham-operated them. All animals were injected with a physiologic dose of [^{3}H]calcidiol (160 C/mmole) and were sacrified after 20 hr. The radioactivity was extracted, chromatographed through Sephadex LH-20, straight-phase HPLC, and reverse-phase HPLC. In no case did we detect any radiolabeled calcitriol in bone, intestine, or other organs. Similar findings were obtained by Schultz *et al.* [60]. We conclude that *in vivo*, significant amounts of calcitriol synthesis do not take place in organs other than kidney and placenta.

A number of metabolites of vitamin D have recently been isolated and identified, including 24-oxo vitamin D compounds, 23-oxo compounds, and various hydroxylated side chain derivatives [61–64]. Some of these compounds which have been isolated and identified from the plasma of animals given enormous doses of vitamin D [65,66] cannot be regarded as physiologic metabolites, but are likely to be either pharmacologic products or artifacts of *in vitro* production. Certainly, they should not be regarded as physiologically significant metabolities of vitamin D until their significant concentration can be demonstrated *in vivo* following physiological doses of vitamin D. Therefore, these metabolites will not be discussed. Instead, we will focus on the metabolites of the vitamin produced *in vivo* under physiologic circumstances (cf. Fig. 4).

The major metabolic product of calcitriol is the 23-carboxylic acid (or calcitroic acid) isolated and identified in our laboratory [67]. This compound has been synthesized and shown to be biologically inactive [68]. It is likely that the calcitroic acid pathway is the major pathway of degradation of calcitriol. So far no product of calitriol has been shown to be biologically active except 1,24R,25-trihydroxyvitamin D_3 [1,24,25-$(OH)_3D_3$] [69], which is a minor metabolic product having approximately 1/10th the biologic activity of calcitriol itself [70]. Similarly, 1,25,26-trihydroxyvitamin D_3 is approximately 1/10th as active as calcitriol and is even more minor than the 1,24,25-$(OH)_3D_3$ [71].

A new pathway has recently been discovered by our research group, which leads ultimately to the compound, calcidiol 26,23-lactone [72]. This lactone has been isolated, synthesized, and the stereochemical configuration of the 23 and 25 positions elucidated (23*S*, 25 *R*) [73]. It is derived from a 23*S*-hydroxylation of calcidiol, followed by functionalization

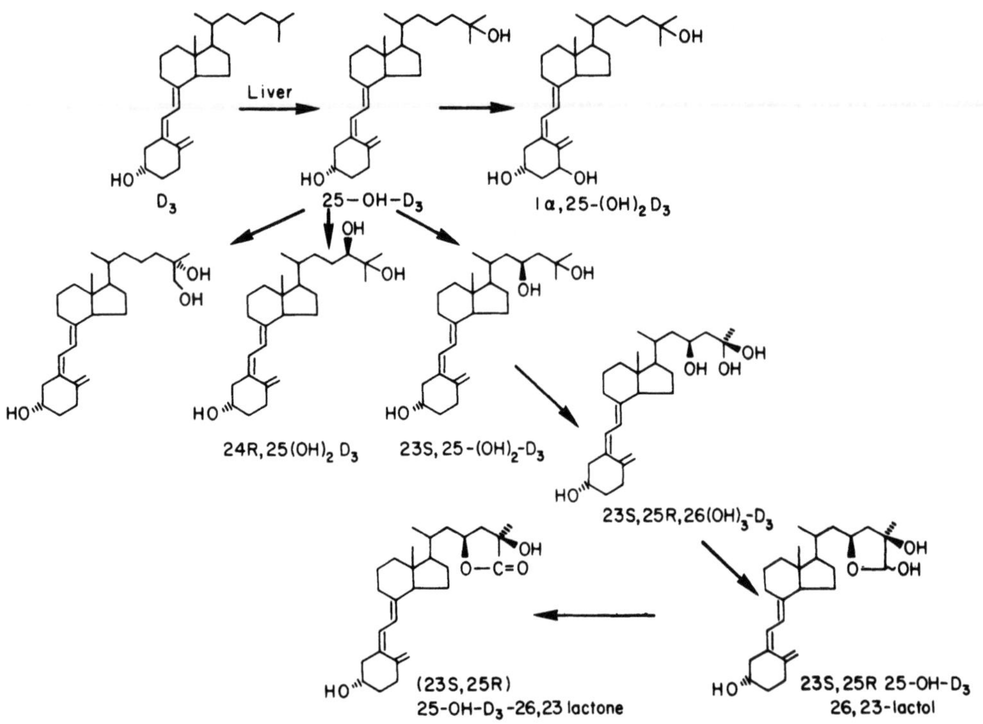

Figure 4. The physiologic pathways of vitamin D_3 metabolism. Note that $1,25\text{-}(OH)_2D_3$, the vitamin hormone, is metabolized in a series of reactions in intestine and liver to the C_{23} carboxylic acid called calcitroic acid (not shown here). All other pathways shown except the 25-hydroxylase have been located in the kidney but not necessarily exclusively.

of the 26 position, likely hydroxylation followed by aldehyde formation, lactol formation. and oxidation to the corresponding lactone (Yamada, personal communication). A peroxy lactone has been suggested as an intermediate [74], although evidence is not convincing that it is a true intermediate. This pathway apparently is quantitatively of significant magnitude with increasing doses of vitamin D, although its products are biologically inactive and the biologic significance of this pathway remains to be determined.

A pathway discovered in 1970 is 26-hydroxylation of calcidiol [75], which involves two 26-hydroxylase activities. One hydroxylase inserts the hydroxyl giving rise to a $25S$ configuration and another gives rise to a $25R$ configuration. The $25S,26\text{-}(OH)_2D_3$ has very little biologic activity, and the significance of this pathway remains unknown [76]. It has recently been shown that natural $25,26\text{-}(OH)_2D_3$ is an epimeric mixture of $25S$ and $25R$ compounds [77].

A major pathway receiving the greatest interest besides 1α-hydroxylation is 24-hydroxylation, which was discovered in 1972 [78]. When calcitriol synthesis is suppressed, this pathway is activated. Conversely, when calcitriol synthesis is stimulated, this pathway is shut off [79]. Additionally $24R,25\text{-}(OH)_2D_3$ has significant biologic activity in mammals but less than its precursor [80], and it possesses weak biologic activity in birds [81,82]. Nevertheless, the availability of this material as a consequence of the efforts of the Hoffman–La Roche Company has led to the suggestion that it is required for suppression of parathyroid hormone secretion [83], mineralization of bone [84,85], embryonic development in the chicken [86], and cartilage growth and development [87].

In conjunction with two groups of Japanese chemists, we devised experiments to test this hypothesis which are based on the synthesis of 24,24-difluoro-25-OH-D_3 (24,24-F_2-calcidiol) [88,89]. In addition, 24,24-F_2-calcitriol has also been synthesized [90,91]. No $24R,25$-$(OH)_2D_3$ is produced from 24,24-F_2-25-OH-D_3 *in vivo;* furthermore, binding characteristics of this compounds are quite similar to 25-OH-D_3 but quite dissimilar from $24R,25$-$(OH)_2D_3$, especially in the chick intestinal cytosol receptor binding test [92]. Electronically, the fluoro group is less like the hydroxyl group than hydrogen and thus cannot be regarded as an analog of the hydroxyl radical. It has been clearly demonstrated that 24,24-F_2-25-OH-D_3 is equally active relative to 25-OH-D_3 in intestinal calcium transport, bone calcium mobilization, bone mineralization, cartilage mineralization, and in the maintenance of plasma calcium and phosphorus levels [93–95]. Thus, these reponses do not require the presence of the 24-hydroxyl position or even the capability of hydroxylating in this position. Nevertheless, the claims that $24R,25$-$(OH)_2D_3$ is required for mineralization of bone have persisted.

We, therefore. devised experiments in which animals born to vitamin D-deficient mothers were maintained exclusively on: (1) 25-OH-D_3 delivered either by mouth or by osmotic minipump, (2) 1,25-$(OH)_2D_3$ delivered by osmotic minipump, or (3) 24,24-F_2-25-OH-D_3 given by mouth [96,97]. These animals were grown to maturity, the males were studied, and the females were then mated with normal males and their offspring maintained again on the same treatment schedule. At the end of this time the animals were examined. In every case, all of the animals given the vitamin D compounds were equivalent in their ability to grow, the maintenance of serum calcium and phosphorus, and bone mineralization. Furthermore, in rats given either 24,24-F_2-calcidiol or calcitriol as their only form of vitamin D for two generations, there was no apparent defect. The animals appeared entirely normal. The bones following tetracycline labeling were sent in coded fashion to Michael Parfitt and his associate, who demonstrated that mineralization was normal in all of the vitamin D treatment groups [98] as compared to animals that were maintained in the vitamin D-deficient state throughout. Thus, no evidence was found for a significant function of 24-hydroxylation of vitamin D; and the claim that 24,25-$(OH)_2D_3$ plays an important functional role cannot be supported.

With Professors Ikekawa and Kobayashi, we have also prepared and studied 26,26,26,27,27,27-F_6-25-OH-D_3 [99,100], which was also used to prepare 26,26,26,27,27,27-F_6-1,25-$(OH)_2D_3$ [101]. Experiments with this compounds are in complete agreement with those using 24,24-F_2-25-OH-D_3 in which no necessity for the 26-carbon modification or

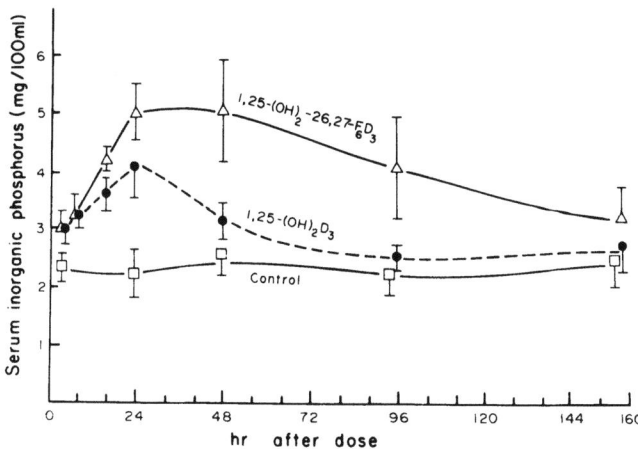

Figure 5. The response of serum inorganic phosphorus of rachitic, vitamin D-deficient rats to a single injection of vehicle (control), 1,25-$(OH)_2D_3$, or 26,27-F_6-1,25-$(OH)_2D_3$. The dose (650 pmoles) was injected intrajugularly in 50 μl of ethanol. The control vitamin D-deficient animals received the 50 μl of ethanol without compound.

lactone formation to the function of vitamin D can be demonstrated. Thus, there is no evidence that any of the side chain-modified metabolites of vitamin D except 1,25-(OH)$_2$D$_3$ plays an important role in the functions of the vitamin. We can, therefore, focus our attention on the true vitamin D hormone, calcitriol.

Of considerable interest is the fact that the biopotency of 26,26,26,27,27,27-F$_6$-1,25-(OH)$_2$D$_3$ is at least 5–10 times that of the native hormone, calcitriol (Fig. 5). Similar results are obtained with the 24,24-F$_2$-1,25-(OH)$_2$D$_3$ [102]. These compounds may be regarded as a second generation of vitamin D therapeutic agents, since they are more active and have a longer duration of action than 1,25-(OH)$_2$D$_3$.

Mechanism of Action of Calcitriol

The time course of response of intestinal calcium transport to calcitriol as illustrated in Fig. 6 is quite complex [103], involving two independent mechanisms. The first mechanism is an initial rapid response of villous cells to calcitriol; the second, more prolonged response is regarded as a programming of the crypt cells to transport calcium ions once they move into the villous region.

The remaining portion of this section will be devoted to considering available information with regard to the first or 6-hr response to calcitriol. Using chemically synthesized [26,27-^{3}H]calcitriol prepared in our laboratory, we were able to carry out freeze–thaw autoradiography both in our own group [104] and with the kind collaboration of Dr. Walter Stumpf and his laboratory in North Carolina [105]. Within ½ hr postinjection, [^{3}H]calcitriol appeared selectively in the villous cell nuclei without localizing in any other section of the cell. Radioactivity also did not localize in the nuclei of goblet, submucosa, or smooth muscle cells. The nuclear localization of [^{3}H]calcitriol is noted not only in bone [106] and kidney [107], but also in other cell types not previously appreciated as target organs [105,108]. This opens the possibility that calcitriol has multiple functions, some of which may be more subtle than mineralization or regulation of calcium levels in the plasma. The nuclear localization of calcitriol does not represent localization of its degradation product, calcitroic acid, since this product would have lost the tritium label. Furthermore, nuclear localization supports previously obtained biochemical data using cell fractionation techniques which demonstrated that the nucleus represents the functional site at which calcitriol exerts its effects [109,110].

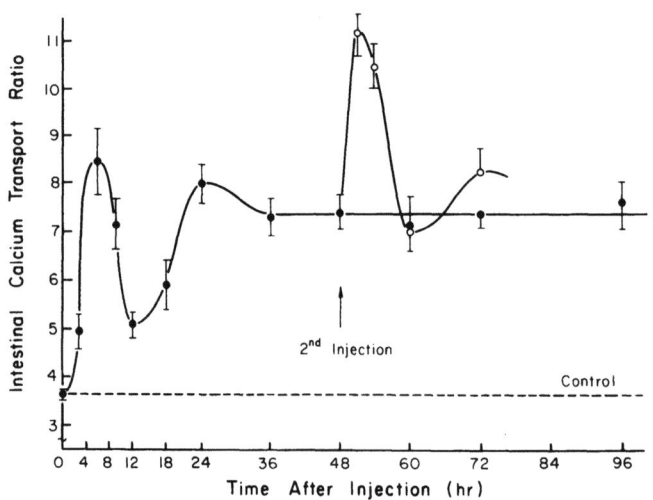

Figure 6. Intestinal calcium transport response to a single injection of 1,25-(OH)$_2$D$_3$ as described in Halloran and DeLuca [103].

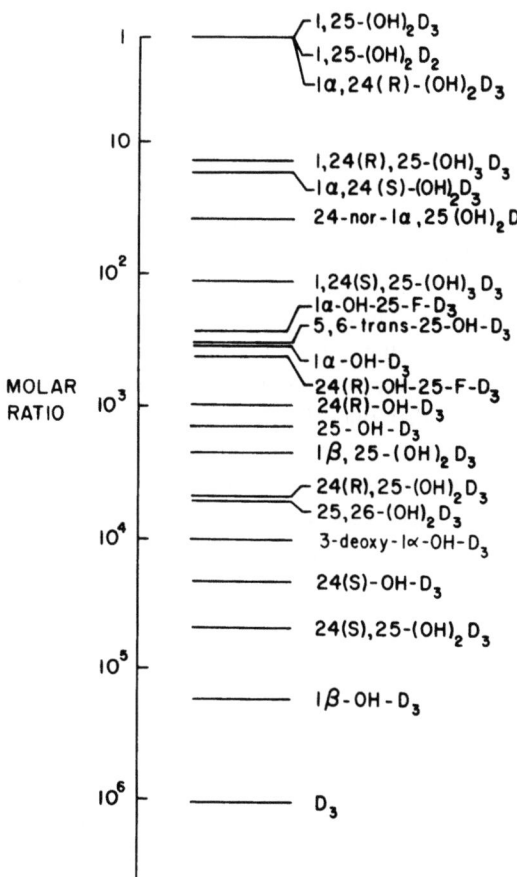

Figure 7. Diagrammatic representation of the relative potency of vitamin D metabolites and analogs to compete for the binding site for $1,25\text{-}(OH)_2D_3$ on the chick intestinal cytosol receptor. The diagram illustrates the molar ratio of unlabeled analog required to displace 50% of labeled $1,25\text{-}(OH)_2D_3$ from the chick intestinal cytosol receptor.

Because of its steroidal nature and the proposed nucleus-mediated mechanism of the steroid hormones, it is natural to suspect that the vitamin D hormone would act in a similar fashion. Indeed, a 3.7 S receptor protein for calcitriol was first demonstrated in chick intestinal cytosol [111,112], rat intestinal cytosol [113], bone [114], skin [115], kidney [116], and other sites within the body [117]. In addition, the calcitriol-specific receptor molecule has been found in a number of neoplastic cells and cell lines [118]. Exactly how this receptor functions in all of these tissues has not yet been determined. The intestinal receptor of the chicken, which has been the best characterized, is quite specific in its ability to bind to calcitriol (Fig. 7). Besides the synthetic $1,24R\text{-}(OH)_2D_3$ or $1,25\text{-}(OH)_2D_2$ and $24,24\text{-}F_2\text{-}1,25\text{-}(OH)_2D_3$ [102,119], no metabolite or analog approaches the potency of calcitriol in displacement of label (Fig. 7). The metabolite $1,24R,25\text{-}(OH)_3D_3$ is approximately one order of magnitude less active; significant displacement activity is apparent from $25\text{-}OH\text{-}D_3$ (calcidiol) when present in a 1000-fold excess [120]. This interaction with the receptor probably accounts for the fact that calcidiol will cause a target organ response in the absence of kidneys when it is present in pharmacologic amounts. This fortunate circumstance permitted doctors to treat metabolic bone disease with large amounts of ordinary vitamin D long before calcitriol was known. The receptor is nevertheless quite specific for calcitriol but there have been questions raised as to whether this is an obligatory mechanism for calcitriol in mediating intestinal calcium transport [121].

What evidence can be offered, therefore, that the interaction of $1,25\text{-}(OH)_2D_3$ with the receptor is required for calcitriol action? In the neonatal development of rat pups, intestinal calcium transport does not appear until 16–18 days postpartum [122], and the intestine is not sensitive to vitamin D or calcitriol until this time. In examining the basis for this deficiency, we have found that the neonatal rat intestine lacks the calcitriol receptor during the first 16–18 days of life [123]. At the time the receptor appears, the intestine also assumes the ability to carry out active calcium transport. Adrenalectomy delays both events, while hydrocortisone stimulates both. Finally, explantation of intestines in organ culture at 14 days postpartum does not by itself result in the appearance of the receptor in a 24-hr period, but if the explant is incubated with hydrocortisone, $1,25\text{-}(OH)_2D_3$-specific binding activity develops [124,125]. Thus, the correlation between intestinal calcium transport, sensitivity to calcitriol, and the appearance of the receptor provides strong evidence that an interaction of calcitriol with the receptor is required for the small intestine to carry out its function.

Another line of evidence has been the discovery of the autosomal recessive disorder called vitamin D-dependency rickets type II [126,127]. This disorder in children is one in which the subjects present with severe rickets despite high circulating levels of calcitriol and exhibit a lack of sensitivity to calcitriol. Eil et al. [128] at the National Institutes of Health have provided evidence that some of these patients lack the receptor for calcitriol.

As a consequence of the fact that the receptor plays an important role in the function of calcitriol, we have recently succeeded in isolating in homogeneous form by two conventional series of protein fractionation methods the receptor from chick intestine [129,130] (Fig. 8). The final isolate is a protein with a molecular weight of approximately 64,000 which has a single polypeptide chain that binds calcitriol. Unfortunately, we have not yet generated antibodies to this receptor, although Pike et al. [131] have reported recently the development of monoclonal antibodies to the chick intestinal receptor. We believe purified receptor and its antibodies will be a useful adjunct in unraveling the molecular mechanism of action of calcitriol in the small intestine and in other target organs.

There has been considerable question as to whether gene expression is involved in calcitriol action in the small intestine [121]. Using the Corradino organ culture technique, we

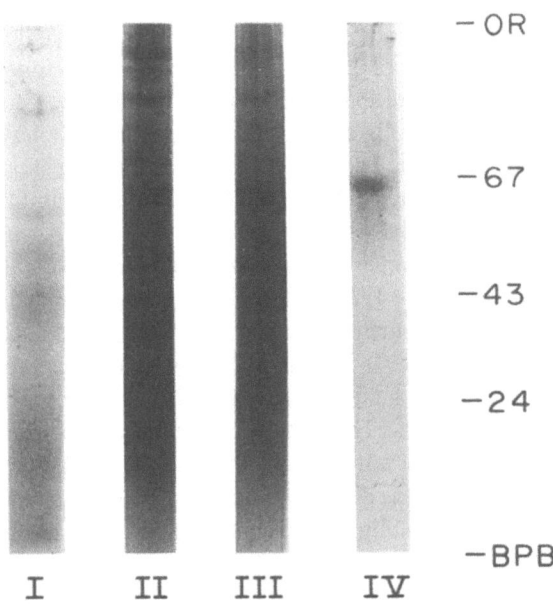

Figure 8. SDS polyacrylamide gel electrophoresis of the chick intestinal nuclear receptor at various stages of purification. I, II, and III are various stages of purification as described in Simpson et al., whereas IV represents the final purified chick nuclear binding protein for $1,25\text{-}(OH)_2D_3$ (Simpson et al., [130]).

have provided evidence favoring the concept that the action of calcitriol on the intestinal calcium transport involves gene expression [132,133]. It has been reported that actinomycin D fails to block the intestinal calcium transport response to calcitriol *in vivo* [134]. Unfortunately, the concentration of 1 μg actinomycin D/100 g body wt utilized in these experiments is not sufficient to totally block mRNA synthesis. Larger doses of actinomycin D result in death of the animals. Therefore, conclusions from these *in vivo* experiments are not possible.

Using the organ culture system of Corradino, one can add calcitriol directly to the cultures and observe the calcium uptake response. At 6 hr following *in vitro* addition of calcitriol, a clear transport response can be seen. This response can be blocked by the addition of cycloheximide, a protein synthesis inhibitor; actinomycin D, an RNA synthesis inhibitor; anisomycin, and other RNA-blocking agents [133]. This is not merely the result of compromised viability, since one can restore the response after cycloheximide treatment by placing the cultures in fresh medium. Thus, the molecular mechanism of action of calcitriol in the small intestine appears to involve gene expression. A diagrammatic sketch of the postulated mechanism of action of calcitriol is illustrated in Fig. 9. Calcitriol must first interact with a receptor either in the cytosol or in the nucleus resulting ultimately in the appearance of the receptor–ligand complex in the nucleus. This brings about expression of at least one and perhaps several genes that code for calcium and phosphorus transport proteins. These proteins could function either at the brush border surface, in the cytosol, and/or at the basal lateral membrane. At the basal lateral membrane, sodium is required for the ultimate expulsion of calcium into the serosal fluid [135].

The outlines of action of calcitriol are now becoming clear although much still remains to be learned. One of the major unsolved questions is the nature of the gene products that bring about the transport processes. At least one of these proteins is the calcium-binding protein originally discovered by Wasserman and his colleagues [136]. In the chick, this is a protein of 28,000 molecular weight, whereas in mammals, it is of the order of 8000–12,000 [137]. There has been considerable debate whether this protein makes its appearance at an appropriate time to participate in transport [138]. We have recently developed a two-dimen-

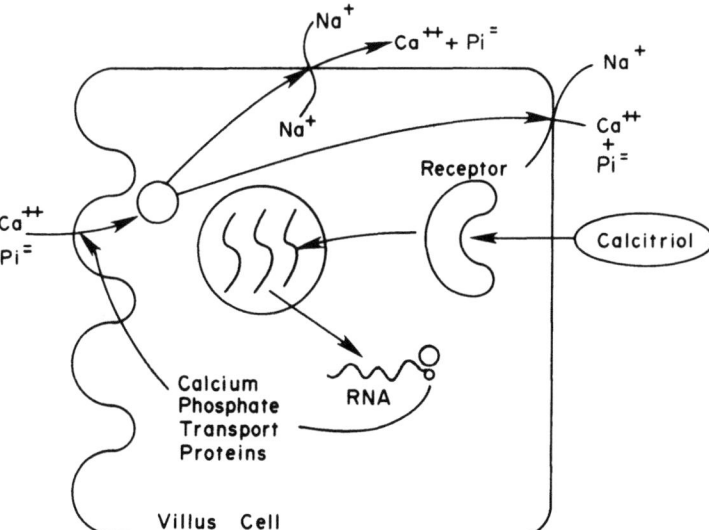

Figure 9. Diagrammatic representation of the mechanism of action of 1,25-$(OH)_2D_3$ (calcitriol) in the intestinal villous cell.

sional gel electrophoresis–electrofocusing technique for detecting proteins in large complex mixtures. Using the double-label technique in which ^{3}H-labeled amino acids are supplied the calcitriol-stimulated intestines, while ^{14}C-labeled amino acids are supplied to to the vitamin D-deficient intestines, one can carry out a separation of these combined preparations and determine the ^{3}H and ^{14}C contents of these spots. In addition, by means of computer analysis of the fluorographs and autoradiographs of these gels, one can calculate and integrate the amount of ^{3}H and ^{14}C in each of the spots [139,140]. Following addition of calcitriol to cultures of chick embryonic intestine the calcium-binding protein appears within 2 hr [140], which is far in advance of the first observable calcium uptake response occurring at 4–6 hr. In addition, the double-label and two-dimensional gel techniques will undoubtedly reveal the presence of at least one other protein that must play a role in the calcium transport process. With the development of appropriate methods, it is likely that within the next 2 or 3 years the gene products required for the calcium and phosphorus transport processes will become known.

Regulation of the 1α-Hydroxylase

Following the discovery of the regulation of 1α-hydroxylase by Boyle, Gray, and DeLuca in 1971 [79,141] and the role played by the parathyroid glands therein [142,143], a description of the vitamin D endocrine system was possible [144]. A simplified diagram of the calcium homeostatic mechanism is shown in Fig. 10. Thus, when plasma calcium falls below normal, the parathyroid gland secretes parathyroid hormone, which binds to the kidney and to bone but not to intestine [145]. In the kidney, the parathyroid hormone stimulates calcitriol formation in the proximal convoluted tubules [146] and facilitates calcium reabsorption in the distal tubule [27]. The calcitriol formed not only stimulates renal reabsorption of calcium together with parathyroid hormone, but also the bone calcium mobilization system in conjunction with the parathyroid hormone. Most importantly, calcitriol itself stimulates intestinal calcium transport. The three sources of calcium raise the

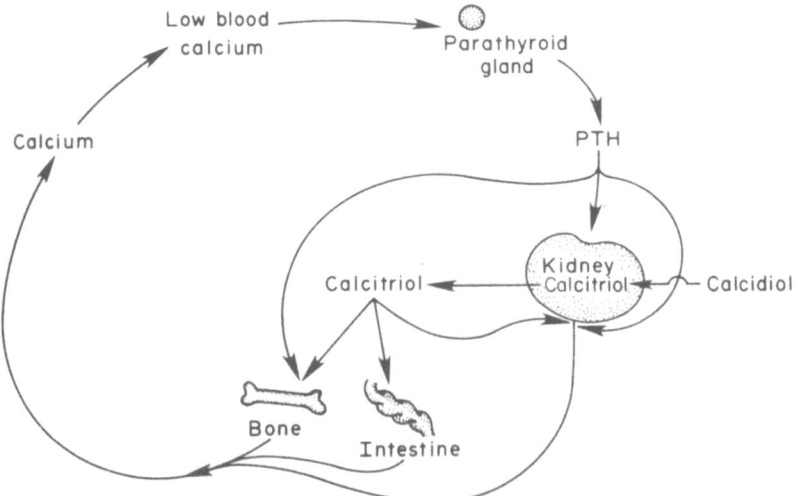

Figure 10. Diagrammatic representation of the calcium homeostatic action of the vitamin D endocrine system involving the kidney, bone, intestine, and parathyroid glands.

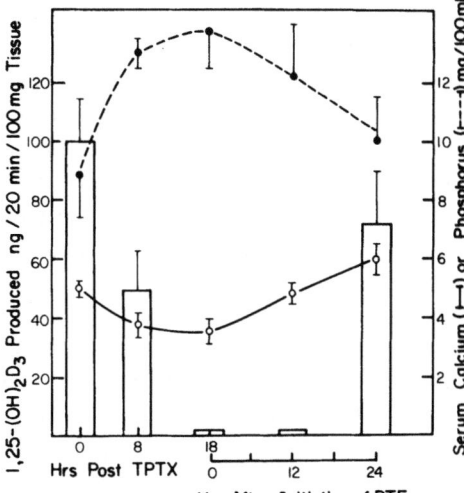

Figure 11. Response of 25-OH-D-1-hydroxylase and serum, calcium, or phosphorus of rats on a low-calcium diet following thyroparathyroidectomy and after initiation of parathyroid hormone injections. The bars represent the 1α-hydroxylase measurement; the dashed and solid lines represent serum calcium and serum phosphorus concentrations [148].

plasma calcium to normal, thus terminating parathyroid hormone secretion and completing the calcium homeostatic loop mechanism.

The time course of response to calcitriol is of relatively long duration [103]. Thus, a minute-to-minute regulation of plasma calcium by the actions of calcitriol on intestine, bone, and kidney would not be expected [147]. However, calcitriol is required for parathyroid hormone to effect bone calcium mobilization and possibly renal tubular reabsorption of calcium. We have examined the question of whether calcitriol as well as parathyroid hormone play a role in the minute-to-minute regulation of plasma calcium levels. Thyroparathyroidectomy results in a rapid fall in plasma calcium and a much slower fall in the renal 1α-hydroxylase (Fig. 11) [148]. Similarly, plasma phosphate levels rise quite rapidly in response to parathyroidectomy. The changes in plasma calcium and phosphorus cannot be the result of alterations in 1α-hydroxylase activity. Furthermore, when parathyroid hormone is administered to animals that have no 1α-hydroxylase activity as a result of parathyroidectomy, some 18–24 hr is required before enzyme activity is restored (Fig. 11) [148]. This contrasts sharply with the rapid response of plasma calcium and plasma phosphorus. These results reveal two important facts. (1) Changes in plasma calcium and phosphorus are not directly regulated by renal 1α-hydroxylase activity. (2) Parathyroid hormone carries out the short-term regulation of plasma calcium and phosphorus *in vivo* by acting on bone and kidney that was previously primed by calcitriol. Continued lack of calcium increases calcitriol levels which would raise the sensitivity of bone and possibly kidney to parathyroid hormone. Most importantly, the increase in calcitriol would stimulate the intestinal absorption of calcium, thereby providing maximum use of environmental calcium [149].

Vitamin D Metabolism and Bone Disease

Although calcitriol is extremely useful in the treatment of renal osteodystrophy [150–152], certain types of vitamin D-resistant rickets [153–156], and hypocalcemic disorders such as hypoparathyroidism [157,158], considerable debate has raged as to whether calcitriol is of any use in the treatment of osteoporosis. One problem has been that clinicians have equated calcitriol to vitamin D. Vitamin D_3 itself will not correct osteoporosis [159], and therefore it is natural to assume that calcitriol, being a derivative of vitamin D, will also not be effective. However, calcitriol is the functional form of vitamin D, and if it is not produced when required, insufficient amounts of calcium absorption will take place. If calcium is not absorbed, the organism will rely on bone calcium for its soft tissue needs, resulting in a chronic loss of bone. A continuous thinning of bones, especially trabecular bone, will occur, ultimately predisposing to fractures. Two questions must therefore be raised: Is a lack of calcitriol involved in the genesis of osteoporosis? Is it possible to reverse the osteoporotic process by stimulating intestinal calcium transport and bone remodeling, regardless of its pathogenesis?

The fact that calcium absorption diminishes with age has been demonstrated in a number of species including man [160–162]. This fall closely correlates with plasma calcitriol levels in both man and rat [159,160]. Furthermore, a single dose of calcitriol stimulates intestinal calcium absorption in 65-year-old subjects [161]. Another important fact is that if subjects above age 65 are placed on a low-calcium diet, neither their calcitriol levels nor their calcium absorption increases, in contrast to younger subjects [163]. Similarly, postmenopausal osteoporotic women show a failure in adapting their intestinal calcium absorption to dietary calcium [162]. Futhermore, analysis of plasma of postmenopausal women with osteoporosis reveals a 30% reduction in calcitriol levels beyond that which occurs as a result of age [162]. Therefore, a contributing factor to osteoporosis may be a diminished supply of calcitriol. The primary insult that results in calcitriol deficiency remains unknown. In the postmenopausal woman, this may result from estrogen lack. This deficiency would cause premature loss of calcium from bone that in turn shuts off the calcium-mobilizing hormones, parathyroid hormone and calcitriol. Whatever the mechanism, intervention with calcitriol treatment might be effective in these disorders. One study reported that a 0.25 µg/day dose of calcitriol was ineffective [164]. However, this dose is far below the normal amount synthesized in an average adult and is not likely to produce a measurable response. However, if the dose is raised to 0.5 µg/day, a response becomes evident [165]. In this latter study (Table II) postmenopausal women with osteoporosis were divided into two groups. One group received calcitriol, 0.5 µg/day; the other received placebo. After 6 months, the placebo group failed to show an improvement in calcium absorption or calcium balance, whereas the calcitriol-treated group showed clear improvement in both parameters. It was therefore necessary to shift the control group to the treatment group. The results of this study, which has now proceeded for 3 years, reveal improved calcium absorption and calcium balance at the end of 2 years, and a significant increase in trabecular bone as revealed by bone biopsies. In untreated osteoporosis a decrease in trabecular bone volume would be expected. Bone resorption was not a factor, since hydroxyproline measurements reveal that calcitriol caused a decrease in bone resorption. Drs.

Table II. Postmenopausal Osteoporosis: Treatment with 1,25-$(OH)_2D_3$ (0.5 µg/Day)[a]

	Baseline	6–8 months	2 years
Ca absorption (%)	7	27[†]	27[†]
Ca balance (mg/day)	−59	2*	−27
Trabecular bone volume (%)	11.3	13.4	16.2*
Urinary hydroxyproline (mg/day)	26.3	21.1*	22.0*

[a]Data from Gallagher et al. [165].
*$p < 0.01$.
[†]$p < 0.001$.

Gallagher, Recker, and Riggs will soon be reporting the results of two independent studies, one at Creighton and one at the Mayo Clinic, that show that the fracture incidence is markedly reduced even after 1 year of calcitriol treatment in these patients and is dramatically reduced by the end of 2–3 years. Thus, the effectiveness of calcitriol in the treatment of osteoporosis is becoming increasingly apparent, and this compound and its analogs will ultimately prove useful as therapeutic agents in this bone disease.

In conclusion, insight into how vitamin D works has yielded not only increased understanding of the pathophysiology of metabolic bone disease but has provided therapeutic tools which now have great promise in the treatment of osteoporosis and other forms of bone disease.

ACKNOWLEDGMENT. This work was supported by Program Project Grant AM-14881 from the National Institutes of Health and the Harry Steenbock Research Fund of the Wisconsin Alumni Research Foundation. This chapter is based on the 3M Life Sciences Award Lecture delivered at the FASEB Meetings in Chicago, April 1983.

References

1. DeLuca, H. F. The transformation of a vitamin into a hormone: The vitamin D story. In: *The Harvey Lectures*, Series 7, New York, Academic Press, 1981, pp. 333–379.
2. DeLuca, H. F.; Schnoes, H. K. Vitamin D: Recent advances. *Annu. Rev. Biochem.* 52:411–439, 1983.
3. Huldshinsky, K. Heilung von rachitis durch künstliche höhensonne. *Dtsch. Med. Wochenschr.* 45:712–713, 1919.
4. Chick, H.; Palzell, E. J.; Hume, E. M. Studies of rickets in Vienna 1919–1922. Medical Research Council, Special Report No. 77, 1923.
5. Windaus, A.; Bock. F. Über das provitamin aus dem sterin der schweineschwarte. *Z. Physiol. Chem.* 245:168–170, 1937.
6. Esvelt, R. P.; Schnoes, H. K.; DeLuca, H. F. Vitamin D_3 from rat skins irradiated *in vitro* with ultraviolet light. *Arch. Biochem. Biophys.* 188:282–286, 1978.
7. Holick. M. F.; Richtand, N. M.; McNeill, S. C.; Holick, S. A.; Frommer, J. E.; Henley, J. W.; Potts, J. T., Jr. Isolation and identification of previtamin D_3 from the skin of rats exposed to ultraviolet irradiation. *Biochemistry* 18:1003–1008. 1979.
8. Velluz, L.; Amiard, G. Chimie organique-equilibre de reaction entre precalciferol et calciferol. *C. R. Acad. Sci.* [D] (Paris) 228:853–855, 1949.
9. Clemens, T. L.; Henderson, S. L.; Adams, J. S.; Holick, M. F. Increased skin pigment reduces the capacity of skin to synthesise vitamin D_3. *Lancet* 9:74–76, 1982.
10. Holick, M. F.; MacLaughlin, J. A.; Doppelt, S. H. Regulation of cutaneous previtamin D_3 photosynthesis in man: Skin pigment is not an essential regulator. *Science* 211:590–593, 1981.
11. Mellanby, E. An experimental investigation on rickets. *Lancet* 1:407–412. 1919.
12. McCollum, E. V.; Simmonds, N., Becker, J. E.; Shipley, P. G. Studies on experimental rickets. XXI. An experimental demonstration of the existence of a vitamin which promotes calcium deposition. *J. Biol. Chem.* 53:293–312, 1922.
13. Kramer, B.; Knof, A. Chemical pathology and pharmacology. In: *The Vitamins*, W. H. Sebrell, Jr., and R. S. Harris, eds., New York, Academic Press, 1954, pp. 248–253.
14. DeLuca, H. F. Mechanism of action and metabolic fate of vitamin D. *Vitam. Horm. (N.Y.)* 25:315–367, 1967.
15. Lamm, M.; Neuman, W. F. On the role of vitamin D in calcification. *Arch. Pathol.* 66:204–209, 1958.
16. Shipley, P. G.; Kramer, B.; Howland, J. Calcification of rachitic bones *in vitro*. *Am. J. Dis. Child.* 30:37–39, 1925.
17. DeLuca, H. F. Vitamin D and calcium transport. In: *Calcium Transport and Cell Function*, A. Scarpa and E. Carafoli, eds., New York, New York Academy of Sciences, 1978, pp. 356–376.
18. Wasserman, R. H. Vitamin D and the intestinal absorption of calcium and phosphate. In: *Membranes and Transport*, A. N. Martonosi, ed., New York, Plenum Press, 1982, pp. 665–673.

19. Harrison, H. E.; Harrison, H. C. Intestinal transport and phosphate: Action of vitamin D, calcium, and potassium. *Am. J. Physiol.* **201**:1007–1012, 1961.
20. Chen, T. C.; Castillo, L.; Korycka-Dahl, M.; DeLuca, H. F. Role of vitamin D metabolites in phosphate transport of rat intestine. *J. Nutr.* **104**:1056–1060, 1974.
21. Walling, M. W. Effects of 1α,25-dihydroxyvitamin D_3 on active intestinal inorganic phosphate absorption. In: *Vitamin D: Biochemical, Chemical, and Clinical Aspects Related to Calcium Metabolism*, A. W. Norman, K. Schaefer, J. W. Coburn, H. F. DeLuca, D. Fraser, H.-G. Grigoleit, and D. von Herrath, eds., Berlin, de Gruyter, 1977, pp. 321–330.
22. Carlsson, A. Tracer experiments on the effect of vitamin D on the skeletal metabolism of calcium and phosphorus. *Acta Physiol. Scand.* **26**:212–220, 1952.
23. Omdahl, J. L.; DeLuca, H. F. Regulation of vitamin D metabolism and function. *Physiol. Rev.* **53**:327–372, 1973.
24. Talmage. R. V. Morphological and physiological considerations in a new concept of calcium transport in bone. *Am. J. Anat.* **129**:467–476, 1970.
25. Rasmussen, H.; DeLuca, H.; Arnaud, C.; Hawker, C.; von Stedingk, M. The relationship between vitamin D and parathyroid hormone. *J. Clin. Invest.* **42**: 1940–1946, 1963.
26. Garabedian, M.; Tanaka, Y.; Holick, M. F.; DeLuca, H. F. Response of intestinal calcium transport and bone calcium mobilization to 1,25-dihydroxyvitamin D_3 in thyroparathyroidectomized rats. *Endocrinology* **94**:1022–1027, 1974.
27. Sutton, R. A. L.; Dirks, J. H. Renal handling of calcium. *Fed. Proc.* **37**:2112–2119, 1978.
28. Eastwood, J. B.; Harris, E., Stamp, T. C. B.; Dewardener, H. E. Vitamin D deficiency in the osteomalacia of chronic renal failure. *Lancet* **2**:1209–1211, 1976.
29. Bordier, P.; Miravet, L.; Tun Chot, S.; Hioco, D. Evidences of a direct effect of vitamin D_3 or 25-hydroxycholecalciferol upon human adult bone mineralization. In: *Cellular Mechanisms for Calcium Transfer and Homeostasis*, G. Nichols, Jr., and R. H. Wasserman, eds., New York, Academic Press, 1971, pp. 459–460.
30. Dickson, I. R.; Kodicek, E. Effect of vitamin D deficiency on bone formation in the chick. *Biochem. J.* **182**:429–435, 1979.
31. Russell, J. E.; Avioli, L. V. 25-Hydroxycholecalciferol-enhanced bone maturation in the parathyroprivic state. *J. Clin. Invest.* **56**:792–798, 1975.
32. Mechanic, G. L.; Toverud, S. U.; Ramp, W. K.; Gonnerman. W. A. The effect of vitamin D on the structural crosslinks and maturation of chick bone collagen. *Biochim. Biophys. Acta* **383**:419–425, 1975.
33. Garabedian, M.; Lieberherr, M.; Nguyen, T. M.; Corvol, M. T.; Dubois, M. B.; Balsan, S. *In vitro* production and activity of 24,25-dihydroxycholecalciferol in cartilage and calvarium. *Clin. Orthop. Relat. Res.* **135**:241–248, 1978.
34. Holtrop, M. F.; Cox. K. A.; Clark, M. B.; Holick, M. F.; Anast, C. S. 1,25-Dihydroxycholecalciferol stimulates osteoclasts in rat bones in the absence of parathyroid hormone. *Endocrinology* **108**:2293–2301, 1981.
35. Howard, G. A.; Baylink, D. J. Matrix formation and osteoid maturation in vitamin D-deficient rats made normocalcemic by dietary means. *Miner. Electrolyte Metab.* **3**:44–50, 1980.
36. Stumpf, W. E.; Sar, M.; DeLuca, H. F. Sites of action of 1,25 $(OH)_2$ vitamin D_3 identified by thaw-mount autoradiography. In: *Hormonal Control of Calcium Metabolism*, D. V. Cohn, R. V. Talmage, and J. L. Matthews, eds., Amsterdam, Excerpta Medica, 1981, pp. 222–229.
37. Frost, H. M. Bone dynamics in osteoporosis and osteomalacia, Henry Ford Hospital Surgical Monograph Series, Springfield, Ill., Thomas, 1966.
38. Trummel. C. L.; Raisz, L. G.; Blunt, J. W.; DeLuca, H. F. 25-Hydroxycholecalciferol: Stimulation of bone resorption in tissue culture. *Science* **163**:1450–1451, 1969.
39. Stern. P. H.; Mavreas, T.; Trummel. C. L.; Schnoes, H. K.; DeLuca, H. F. Bone-resorbing activity of analogues of 25-hydroxycholecalciferol and 1,25-dihydroxycholecalciferol: Effects of side chain modification and stereoisomerization on responses of fetal rat bones *in vitro*. *Mol. Pharmacol.* **12**:879–886, 1977.
40. Bhattacharyya, M.; DeLuca, H. F. Subcellular location of rat liver calciferol-25-hydroxylase. *Arch. Biochem. Biophys.* **160**:58–62, 1974.
41. Madhok, T. C.; DeLuca, H. F. Characteristics of the rat liver microsomal enzyme system converting cholecalciferol into 25-hydroxycholecalciferol: Evidence for the participation of cytochrome P-450. *Biochem. J.* **184**:491–499, 1979.
42. Yoon, P. S.; DeLuca, H. F. Resolution and reconstitution of soluble components of rat liver microsomal vitamin D_3 25-hydroxylase. *Arch. Biochem. Biophys.* **203**:529–541, 1980.
43. Bjorkhem, I.; Holmberg, I.; Oftebro, H.; Pedersen, J. I. Properties of a reconstituted vitamin D_3 25-hydroxylase from rat liver mitochondria. *J. Biol. Chem.* **255**:5244–5249, 1980.

44. Tucker, G., III; Gagnon, R. E.; Haussler, M. R. Vitamin D_3-25-hydroxylase: Tissue occurrence and apparent lack of regulation. *Arch. Biochem. Biophys.* **155**:47–57, 1973.
45. Bhattacharyya, M. H.; DeLuca, H. F. The regulation of calciferol-25-hydroxylase in the chick. *Biochem. Biophys. Res. Commun.* **59**:734–741, 1974.
46. Fraser, D. R.; Kodicek, E. Unique biosynthesis by kidney of a biologically active vitamin D metabolite. *Nature (London)* **228**:764–766, 1970.
47. Gray, R.; Boyle, I.; DeLuca, H. F. Vitamin D metabolism: The role of kidney tissue. *Science* **172**:1232–1234, 1971.
48. Tanaka, Y.; Halloran, B.; Schnoes, H. K.; DeLuca, H. F. In vitro production of 1,25-dihydroxyvitamin D_3 by rat placental tissue. *Proc. Natl. Acad. Sci. USA* **76**:5033–5035, 1979.
49. Gray, T. K.; Lester, G. E.; Lorenc, R. S. Evidence for extra-renal 1α-hydroxylation of 25-hydroxyvitamin D_3 in pregnancy. *Science* **204**:1311–1313, 1979.
50. Ghazarian, J. G.; Jefcoate, C. R.; Knutson. J. C.; Orme-Johnson, W. H.; DeLuca, H. F. Mitochondrial cytochrome P_{450}: A component of chick kidney 25-hydroxycholecalciferol-1α-hydroxylase. *J. Biol. Chem.* **249**:3026–3033, 1974.
51. Yoon, P. S.; DeLuca, H. F. Purification and properties of chick renal mitochondrial ferredoxin. *Biochemistry* **19**:2165–2171, 1980.
52. Holick, M. F.; Schnoes, H. K.; DeLuca, H. F.; Suda, T.; Cousins, R. J. Isolation and identification of 1,25-dihydroxycholecalciferol: A metabolite of vitamin D active in intestine. *Biochemistry* **10**:2799–2804, 1971.
53. Howard, G. A.; Turner, R. T.; Sherrard, D. J.; Baylink, D. J. Human bone cells in culture metabolize 25-hydroxyvitamin D_3 to 1,25-dihydroxyvitamin D_3 and 24,25-dihydroxyvitamin D_3. *J. Biol. Chem.* **256**:7738–7740, 1981.
54. Puzas, J. E.; Turner, R. T.; Howard, G. A.; Baylink, D. J. Cells isolated from embryonic intestine synthesize 1,25-dihydroxyvitamin D_3 and 24,25-dihydroxyvitamin D_3 in culture. *Endocrinology* **112**:378–382, 1983.
55. Lambert, P. W.; Stern, P.; Avioli, R. C.; Brockett, N. C.; Turner, R. P.; Green, A.; Fu, I. Y.; Bell, N. H. Evidence for extrarenal production of 1,25-dihydroxyvitamin D_3 in man. *J. Clin. Invest.* **69**:722–725, 1982.
56. Reeve, L.; Tanaka, Y.; DeLuca, H. F. Studies on the site of 1,25-dihydroxyvitamin D_3 synthesis in vivo. *J. Biol. Chem.* **258**:3615–3617, 1983.
57. Holick. M. F.; Garabedian, M.; DeLuca, H. F. 1,25-Dihydroxycholecalciferol: Metabolite of vitamin D_3 active on bone in anephric rats. *Science* **176**:1146–1147, 1972.
58. Boyle, I. T.; Miravet, L.; Gray, R. W.; Holick, M. F.; DeLuca, H. F. The response of intestinal calcium transport to 25-hydroxy and 1,25-hydroxy vitamin D in nephrectomized rats. *Endocrinology* **90**:605–608, 1972.
59. Wong, R. G.; Norman, A. W.; Reddy, C. R.; Coburn, J. W. Biologic effects of 1,25-dihydroxycholecalciferol (a highly active vitamin D metabolite) in acutely uremic rats. *J. Clin. Invest.* **51**:1287–1291, 1972.
60. Shultz, T. D.; Fox, J.; Heath. H., III; Kumar, R. Do tissues other than the kidney produce 1,25-dihydroxyvitamin D_3 in vivo? A reexamination. *Proc. Natl. Acad. Sci. USA* **80**:1746–1760, 1983.
61. Takasaki, Y.; Suda, T.; Yamada, S.; Takayama, H.; Nishii, Y. Isolation, identification, and biological activity of 25-hydroxy-24-oxovitamin D_3: A new metabolite of vitamin D_3 generated by in vitro incubations with kidney homogenates. *Biochemistry* **20**:1681–1686, 1981.
62. Yamada, S.; Ohmori, M.; Takayama, H.; Takasaki, Y.; Suda, T. Isolation and identification of 1α- and 23-hydroxylated metabolites of 25-hydroxy-24-oxovitamin D_3 from *in vitro* incubates of chick kidney homogenates. *J. Biol. Chem.* **258**:457–463, 1983.
63. Ohnuma, N.; Kruse, J.; Popjak, G.; Norman, A. W. Isolation and chemical characterization of two new vitamin D metabolites produced by the intestine: 1,25-Dihydroxy-23-oxo-vitamin D_3 and 1,25,26-trihydroxy-23-oxo-vitamin D_3. *J. Biol. Chem.* **257**:5097–5102, 1982.
64. Ohnuma, N.; Norman, A. W. Identification of a new C-23 oxidation pathway of metabolism for 1,25-dihydroxyvitamin D_3 present in intestine and kidney. *J. Biol. Chem.* **257**:8261–8271, 1982.
65. Wichmann, J. K.; Schnoes, H. K.; DeLuca, H. F. 23,24,25-Trihydroxyvitamin D_3, 24,25,26-trihydroxyvitamin D_3. 24-keto-25-hydroxyvitamin D_3, and 23-dehydro-25-hydroxyvitamin D_3: New in vivo metabolites of vitamin D_3. *Biochemistry* **20**:7385–7391. 1981.
66. Wichmann, J. K.; Schnoes, H. K.; DeLuca, H. F. Isolation and identification of 24(R)-hydroxyvitamin D_3 from chicks given large doses of vitamin D_3. *Biochemistry* **20**:2350–2353, 1981.
67. Esvelt, R. P.; Schnoes, H. K.; DeLuca, H. F. Isolation and characterization of 1α-hydroxy-tetranor-vitamin D-23-carboxylic acid: A major metabolite of 1,25-dihydroxyvitamin D_3. *Biochemistry* **18**:3977–3983, 1979.
68. Esvelt, R. P.; DeLuca, H. F. Calcitroic acid: Biological activity and tissue distribution studies. *Arch. Biochem. Biophys.* **206**:403–413, 1981.
69. Holick, M. F.; Kleiner-Bossaller, A.; Schnoes, H. K.; Kasten, P. M.; Boyle, I. T.; DeLuca, H. F. 1,24,25-

Trihydroxyvitamin D_3: A metabolite of vitamin D_3 effective on intestine. *J. Biol. Chem.* **248**:6691–6696, 1973.
70. Castillo, L.; Tanaka, Y.; DeLuca, H. F.; Ikekawa, N. On the physiological role of 1,24,25-trihydroxyvitamin D_3. *Miner. Electrolyte Metab.* **1**:198–207, 1978.
71. Tanaka, Y.; Schnoes, H. K.; Smith, C. M.; DeLuca, H. F. 1,25,26-Trihydroxyvitamin D_3: Isolation, identification, and biological activity. *Arch. Biochem. Biophys.* **210**:104–109, 1981.
72. Wichmann, J. K.; DeLuca, H. F.; Schnoes, H. K.; Horst, R. L.; Shepard, R. M.; Jorgensen, N. A. 25-Hydroxyvitamin D_3 26,23-lactone: A new *in vivo* metabolite of vitamin D. *Biochemistry* **18**:4775–4780, 1979.
73. Eguchi, T.; Takatsuto, S.; Ishiguro, M.; Ikekawa, N.; Tanaka, Y.; DeLuca, H. F. Synthesis and determination of configuration of natural 25-hydroxyvitamin D_3 26,23-lactone. *Proc. Natl. Acad. Sci. USA* **78**:6579–6583, 1981.
74. Ishizuka, S.; Ishimoto, S.; Norman, A. W. Isolation and identification of 25-hydroxyvitamin D_3-26,23-peroxylactone: A novel *in vivo* metabolite of vitamin D_3. *J. Biol. Chem.* **257**:14708–14713, 1982.
75. Suda, T.; DeLuca, H. F.; Schnoes, H. K.; Tanaka, Y.; Holick, M. F. 25,26-Dihydroxycholecalciferol, a metabolite of vitamin D_3 with intestinal calcium transport activity. *Biochemistry* **9**:4776–4780, 1970.
76. Lam, H.-Y.; Schnoes, H. K.; DeLuca, H. F. Synthesis and biological activity of 25ξ,26-dihydroxycholecalciferol. *Steroids* **25**:247–256, 1975.
77. Ikekawa, N.; Koizumi, N.; Ohshima, E.; Ishizuka, S.; Takeshita, T.; Tanaka, Y.; DeLuca, H. F. Natural 25,26-dihydroxyvitamin D_3 is an epimeric mixture. *Proc. Natl. Acad. Sci. USA* **80**:5286–5288, 1983.
78. Hollick, M. F.; Schnoes, H. K.; DeLuca, H. F.; Gray, R. W.; Boyle, I. T.; Suda, T. Isolation and identification of 24,25-dihydroxycholecalciferol: A metabolite of vitamin D_3 made in the kidney. *Biochemistry* **11**:4251–4255, 1972.
79. Boyle, I. T.; Gray, R. W.; DeLuca, H. F. Regulation by calcium of *in vivo* synthesis of 1,25-dihydroxycholecalciferol and 21,25-dihydroxycholecalciferol. *Proc. Natl. Acad. Sci. USA* **68**:2131–2134, 1971.
80. Tanaka, Y.; DeLuca, H. F.; Ikekawa, N.; Morisaki, M.; Koizumi, N. Determination of stereochemical configuration of the 24-hydroxyl group of 24,25-dihydroxyvitamin D_3 and its biological importance. *Arch. Biochem. Biophys.* **170**:620–626, 1975.
81. Holick, M. F.; Baxter, L. A.; Schraufrogel, P. K.; Tavela, T. E.; DeLuca, H. F. Metabolism and biological activity of 24,25-dihydroxyvitamin D_3 in the chick. *J. Biol. Chem.* **251**:397–402, 1976.
82. Boris, A.; Hurley, J. F.; Trmal, T. Relative activities of some metabolites and analogs of cholecalciferol in stimulation of tibia ash weight in chicks otherwise deprived of vitamin D. *J. Nutr.* **107**:194–198, 1977.
83. Henry, H. L.; Taylor, A. N.; Norman, A. W. Response of chick parathyroid glands to the vitamin D metabolites 1,25-dihydroxyvitamin D_3 and 24,25-dihydroxyvitamin D_3. *J. Nutr.* **107**:1918–1926, 1977.
84. Ornoy, A.; Goodwin, D.; Noff, D.; Edelstein, S. 24,25-Dihydroxyvitamin D is a metabolite of vitamin D essential for bone formation. *Nature (London)* **276**:517–519, 1978.
85. Rasmussen, H.; Bordier, P. Vitamin D and bone. *Metab. Bone Dis. Relat. Res.* **1**:7–13, 1978.
86. Henry, H. L.; Norman, A. W. Vitamin D: Two dihydroxylated metabolites are required for normal chicken egg hatchability. *Science* **201**:835–837, 1978.
87. Garabedian, M.; Corvol, M. T.; Nguyen, T. M.; Balsan, S. Metabolism and activity of 25-hydroxycholecalciferol in chondrocyte culture. *Ann. Biol. Anim. Biochim. Biophys.* **18**:175–180, 1978.
88. Kobayashi, Y.; Taguchi, T.; Terada, T.; Oshida, J.; Morisaki, M.; Ikekawa, N. Synthesis of 24,24-difluoro- and 24ξ-fluoro-25-hydroxyvitamin D_3. *Tetrahedron Lett.* **22**:2023–2026, 1979.
89. Yamada, S.; Ohmori, M.; Takayama, H. Synthesis of 24,24-difluoro-25-hydroxyvitamin D_3. *Tetrahedron Lett.* **21**:1859–1862, 1979.
90. Tanaka, Y.; DeLuca, H. F.; Schnoes, H. K.; Ikekawa, N.; Kobayashi, Y. 24,24-Difluoro-1,25-dihydroxyvitamin D_3: *In vitro* production, isolation, and biological activity. *Arch. Biochem. Biophys.* **199**:473–478, 1980.
91. Yamada, S.; Ohmori, M.; Takayama, H. Synthesis of 24,24-difluoro-1α,25-dihydroxyvitamin D_3. *Chem. Pharm. Bull.* **27**:3196–3198, 1979.
92. Tanaka, Y.; Wichmann, J. K.; DeLuca, H. F.; Kobayashi, Y.; Ikekawa, N. Metabolism and binding properties of 24,24-difluoro-25-hydroxyvitamin D_3. *Arch. Biochem. Biophys.* **225**:649–655, 1983.
93. Tanaka, Y.; DeLuca, H. F.; Kobayashi, Y.; Taguchi, T.; Ikekawa, N.; Morisaki, M. Biological activity of 24,24-difluoro-25-hydroxyvitamin D_3: Effect of blocking of 24-hydroxylation on the functions of vitamin D. *J. Biol. Chem.* **254**:7163–7167, 1979.
94. Okamoto, S.; Tanaka, Y.; DeLuca, H. F.; Yamada, S.; Takayama, H. 24,24-Difluoro-25-hydroxyvitamin D_3-enhanced bone mineralization in rats: Comparison with 25-hydroxyvitamin D_3 and vitamin D_3. *Arch. Biochem. Biophys.* **206**:8–14, 1981.

95. Halloran, B. P.; DeLuca, H. F.; Barthell, E.; Yamada, S.; Ohmori, M.; Takayama, H. An examination of the importance of 24-hydroxylation to the function of vitamin D during early development. *Endocrinology* **108**:2067–2071, 1981.
96. Jarnagin, K.; Brommage, R.; DeLuca, H. F.; Yamada, S.; Takayama, H. 1- but not 24-hydroxylation of vitamin D is required for growth and reproduction in rats. *Am. J. Physiol.* **244**:E290–E297, 1983.
97. Brommage, R.; Jarnagin, K.; DeLuca, H. F.; Yamada, S.; Takayama, H. 1- but not 24-hydroxylation of vitamin D is required for skeletal mineralization in rats. *Am. J. Physiol.* **244**:E298–E304, 1983.
98. Parfitt, A. M.; Mathews, C. H. E.; Brommage, R.; Jarnagin, K.; DeLuca, H. F. Calcitriol but no other metabolite of vitamin D is essential for normal bone growth and development in the rat. *J. Clin. Invest.* **73**:576–586, 1984.
99. Kobayashi, Y.; Taguchi, T.; Kanuma, N.; Ikekawa, N.; Oshida, J.-I. Synthesis of 26,26,26,27,27,27-hexafluoro-25-hydroxyvitamin D_3. *J. Chem. Soc. Chem. Commun.* **10**:459–460, 1980.
100. Tanaka, Y.; Pahuja, D. N.; Wichmann, J. K.; DeLuca, H. F.; Kobayashi, Y.; Taguchi, T.; Ikekawa, N. 25-Hydroxy-26,26,26,27,27,27-hexafluorovitamin D_3: Biological activity in the rat. *Arch. Biochem. Biophys.* **218**:134–141, 1982.
101. Kobayashi, Y.; Taguchi, T.; Mitsuhashi, S.; Eguchi, T.; Ohshima, E.; Ikekawa, N. Synthesis of 1,25-dihydroxy-26,26,26,27,27,27-hexafluorovitamin D_3. *Chem. Pharm. Bull.* **30**:4297–4301, 1982.
102. Okamoto, S.; Tanaka, Y.; DeLuca, H. F.; Kobayashi, Y.; Ikekawa, N. Biological activity of 24,24-difluoro-1,25-dihydroxyvitamin D_3. *Am. J. Physiol.* **7**:E159–E163, 1983.
103. Halloran, B. P.; DeLuca, H. F. Intestinal calcium transport: Evidence for two distinct mechanisms of action of 1,25-dihydroxyvitamin D_3. *Arch. Biochem. Biophys.* **208**:477–486, 1981.
104. Zile, M.; Bunge, E. C.; Barsness, L.; Yamada, S.; Schnoes, H. K.; DeLuca, H. F. Localization of 1,25-dihydroxyvitamin D_3 in intestinal nuclei *in vivo*. *Arch. Biochem. Biophys.* **186**:15–24, 1978.
105. Stumpf, W. E.; Sar, M.; Reid, F. A.; Tanaka, Y.; DeLuca, H. F. Target cells for 1,25-dihydroxyvitamin D_3 in intestinal tract, stomach, kidney, skin, pituitary and parathyroid. *Science* **206**:1188–1190, 1979.
106. Narbaitz, R.; Stumpf, W. E.; Sar, M.; Huang, S.; DeLuca, H. F. Autoradiographic localization of target cells for 1α-dihydroxyvitamin D_3 in bones from fetal rats. *Calcif. Tissue Int.* **35**:177–182, 1983.
107. Stumpf, W. E.; Sar, M.; Narbaitz, R.; Reid, F. A.; DeLuca, H. F.; Tanaka, Y. Cellular and subcellular localization of 1,25-$(OH)_2$-vitamin D_3 in rat kidney: Comparison with localization of parathyroid hormone and estradiol. *Proc. Natl. Acad. Sci. USA* **77**:1149–1153, 1980.
108. Stumpf, W. E.; Sar, M.; Clark, S. A.; DeLuca, H. F. Brain target sites for 1,25-dihydroxyvitamin D_3. *Science* **215**:1403–1405, 1982.
109. Chen, T. C.; Weber, J. C.; DeLuca, H. F. On the subcellular location of vitamin D metabolites in intestine. *J. Biol. Chem.* **245**:3776–3780, 1970.
110. Tsai, H. C.; Wong, R. G.; Norman, A. W. Studies on calciferol metabolism. IV. Subcellular localization of 1,25-dihydroxy-vitamin D_3 in intestinal mucosa and correlation with increased calcium transport. *J. Biol. Chem.* **247**:5511–5519, 1972.
111. Kream, B. E.; Reynolds, R. D.; Knutson, J. C.; Eisman. J. A.; DeLuca, H. F. Intestinal cytosol binders of 1,25-dihydroxyvitamin D_3 and 25-hydroxyvitamin D_3. *Arch. Biochem. Biophys.* **176**:779–787, 1976.
112. Brumbaugh, P. F.; Haussler, M. R. 1α,25-Dihydroxyvitamin D_3 receptor: Competitive binding of vitamin D analogs. *Life Sci.* **13**:1737–1746, 1973.
113. Kream, B. E.; DeLuca, H. F. A specific binding protein for 1,25-dihydroxyvitamin D_3 in rat intestinal cytosol. *Biochem. Biophys. Res. Commun.* **76**:735–738, 1977.
114. Kream, B. E.; Jose, M.; Yamada, S.; DeLuca, H. F. A specific high-affinity binding macromolecule for 1,25-dihydroxyvitamin D_3 in fetal bone. *Science* **197**:1086–1088, 1977.
115. Simpson, R. U.; DeLuca, H. F. Characterization of a receptor-like protein for 1,25-dihydroxyvitamin D_3 in rat skin. *Proc. Natl. Acad. Sci. USA* **77**:5822–5826, 1980.
116. Simpson, R. U.; Franceschi, R. T.; DeLuca, H. F. Characterization of a specific, high affinity binding macromolecule for 1α,25-dihydroxyvitamin D_3 in cultured chick kidney cells. *J. Biol. Chem.* **255**:10160–10166, 1980.
117. Eisman, J. A.; MacIntyre, I.; Martin, T. J.; Frampton, R. J.; King, R. J. B. Normal and malignant breast tissue is a target organ for 1,25-$(OH)_2$ vitamin D. *Clin. Endocrinol.* **13**,267–272, 1980.
118. Eisman, J. A.; Martin, T. J.; MacIntyre, I. 1,25-Dihydroxyvitamin D_3 receptors in cancer. *Lancet* **1**:1188–1191, 1980.
119. Franceschi, R. T.; Simpson, R. U.; DeLuca, H. F. Binding proteins for vitamin D metabolites: Serum carriers and intracellular receptors. *Arch. Biochem. Biophys.* **210**:1–13, 1981.
120. Eisman, J. A.; DeLuca, H. F. Intestinal 1,25-dihydroxyvitamin D_3 binding protein: Specificity of binding. *Steroids* **30**:245–257, 1977.

121. Rasmussen, H.; Fontaine, O.; Matsumoto, T. Liponomic regulation of calcium transport by 1,25-$(OH)_2D_3$. *Ann. N.Y. Acad. Sci.* **372**:518–523, 1981.
122. Halloran, B. P.; DeLuca, H. F. Calcium transport in the small intestine during early development: The role of vitamin D. *Am. J. Physiol.* **239**:G473–G479, 1980.
123. Halloran, B. P.; DeLuca, H. F. Appearance of the intestinal cytosolic receptor for 1,25-dihydroxyvitamin D_3 during neonatal development in the rat. *J. Biol. Chem.* **256**:7338–7342, 1981.
124. Massaro, E. R.; Simpson, R. U.; DeLuca, H. F. Glucocorticoids and appearance of 1,25-dihydroxyvitamin D_3 receptor in rat intestine. *Am. J. Physiol.* **244**:E230–E235, 1983.
125. Massaro, E. R.; Simpson, R. U.; DeLuca, H. F. Stimulation of specific 1,25-dihydroxyvitamin D_3 binding protein in cultured postnatal rat intestine by hydrocortisone. *J. Biol. Chem.* **257**:13736–13737, 1982.
126. Bell, N. H.; Hamstra, A. J.; DeLuca, H. F. Vitamin D-dependent rickets type II: Resistance of target organs to 1,25-dihydroxyvitamin D. *N. Engl. J. Med.* **298**:996–999, 1978.
127. Rosen, J. F.; Fleischman, A. R.; Finberg, L.; DeLuca, H. F. A new type of rickets: Unresponsiveness of bone and intestine to high levels of endogenously synthesized 1,25-dihydroxyvitamin D_3. *Am. Pediatr. Soc.* 1978 (abstract).
128. Eil, C.; Liberman, U. A.; Rosen, J. F.; Marx, S. J. A cellular defect in hereditary vitamin D-dependent rickets type II: Defective nuclear uptake of 1,25-dihydroxyvitamin D in cultured skin fibroblasts. *N. Engl. J. Med.* **304**:1588–1591, 1981.
129. Simpson, R. U.; DeLuca, H. F. Purification of chicken intestinal receptor for $1\alpha,25$-dihydroxyvitamin D_3 to apparent homogeneity. *Proc. Natl. Acad. Sci. USA* **79**:16–20, 1982.
130. Simpson, R. U.; Hamstra, A.; Kendrick, N. C.; DeLuca, H. F. Purification of the receptor for $1\alpha,25$-dihydroxyvitamin D_3 from chicken intestine. *Biochemistry* **22**:2586–2594, 1983.
131. Pike, J. W.; Marion, S. L.; Donaldson, C. A.; Haussler, M. R. Serum and monoclonal antibodies against the chick intestinal receptor for 1,25-dihydroxyvitamin D_3. *J. Biol. Chem.* **258**:1289–1296, 1983.
132. Franceschi, R. T.; DeLuca, H. F. The effect of inhibitors of protein and RNA synthesis on $1\alpha,25$-dihydroxyvitamin D_3 dependent calcium uptake in cultured embryonic chick duodenum. *J. Biol. Chem.* **256**:3848–3852, 1981.
133. Franceschi, R. T.; DeLuca, H. F. Characterization of 1,25-dihydroxyvitamin D_3-dependent calcium uptake in cultured embryonic chick duodenum. *J. Biol. Chem.* **256**:3840–3847, 1981.
134. Bikle, D. D.; Zolock, D. T.; Morrissey, R. L.; Herman, R. H. Independence of 1,25-dihydroxyvitamin D_3-mediated calcium transport from *de novo* RNA protein synthesis. *J. Biol. Chem.* **253**:484–488, 1978.
135. Martin, D. L.; DeLuca, H. F. Influence of sodium on calcium transport by the rat small intestine. *Am. J. Physiol.* **216**:1351–1359, 1969.
136. Wasserman, R. H.; Taylor, A. N. Vitamin D_3-induced calcium-binding protein in chick intestinal mucosa. *Science* **152**:791–793, 1966.
137. Wasserman, R. H.; Feher, J. J. Vitamin D-dependent calcium-binding proteins. In: *Calcium Binding Proteins and Calcium Function*, R. H. Wasserman, R. A. Corradino, E. Carafoli, R. H. Kretsinger, D. H. MacLennan, and S. L. Siegel, eds., Amsterdam, Elsevier/North-Holland, 1977, pp. 292–302.
138. Spencer, R.; Charman, M.; Wilson, P.; Lawson, E. Vitamin D-stimulated intestinal calcium absorption may not involve calcium binding protein directly. *Nature (London)* **263**:161–163, 1976.
139. Kendrick, N. C.; Barr, C. R.; Moriarity, D.; DeLuca, H. F. Effect of vitamin D-deficiency on in vitro labeling of chick intestinal proteins: Analysis by two-dimensional electrophoresis. *Biochemistry* **20**:5288–5294, 1981.
140. Bishop, C. W.; Kendrick, N. C.; DeLuca, H. F. Induction of calcium binding protein before 1,25-dihydroxyvitamin D_3 stimulation of duodenal calcium uptake. *J. Biol. Chem.* **258**:1305–1310, 1982.
141. Omdahl, J. L.; Gray, R. W.; Boyle, I. T.; Knutson, J.; DeLuca, H. F. Regulation of metabolism of 25-hydroxycholecalciferol by kidney tissue *in vitro* by dietary calcium. *Nature (London)* **237**:63–64, 1972.
142. Garabedian, M.; Holic, M. F.; DeLuca, H. F.; Boyle, I. T. Control of 25-hydroxycholecalciferol metabolism by the parathyroid glands. *Proc. Natl. Acad. Sci. USA* **69**:1673–1676, 1972.
143. Fraser, D. R.; Kodicek, E. Regulation of 25-hydroxycholecalciferol-1-hydroxylase activity in kidney by parathyroid hormone. *Nature (London)* **241**:163–166, 1973.
144. DeLuca, H. F. Vitamin D: The vitamin and the hormone. *Fed. Proc.* **33**:2211–2219, 1974.
145. Zull, J. E.; Repke, D. W. The tissue localization of tritiated parathyroid hormone in thyroparathyroidectomized rats. *J. Biol. Chem.* **247**:2195–2199, 1972.
146. Brunette, M. G.; Chan, H.; Ferriere, C.; Roberts, K. D. Site of 1,25-dihydroxyvitamin D_3 synthesis in the kidney. *Nature (London)* **276**:287–289, 1978.
147. DeLuca, H. F. The cardinal role of 1,25-dihydroxyvitamin D_3 in mineral homeostasis. In: Proceedings of the Symposium on Clinical Disorders of Bone and Mineral Metabolism, Detroit, 1983.
148. Tanaka, Y.; DeLuca, H. F. Stimulation of 1,25-dihydroxyvitamin D_3 production by 1,25-dihydroxyvitamin D_3 in the hypocalcaemic rat. *Biochem. J.* **214**:893–897, 1983.

149. Boyle, I. T.; Gray, R. W.; Omdahl, J. L.; DeLuca, H. F. Calcium control of the *in vivo* biosynthesis of 1,25-dihydroxyvitamin D_3: Nicolaysen's endogenous factor. In: *Endocrinology 1971*, S. Taylor, ed., London, Heinemann, 1972, pp. 468–476.
150. Silverberg, D. S.; Bettcher, K. B.; Dossetor, J. B.; Overton, T. R.; Holick, M. F.; DeLuca, H. F. Effect of 1,25-dihydroxycholecalciferol in renal osteodystrophy. *Can. Med. Assoc. J.* **112**:190–195, 1975.
151. Brickman, A. S.; Coburn, J. W.; Norman, A. W. Action of 1,25-dihydroxycholecalciferol, a potent, kidney-produced metabolite of vitamin D_3, in uremic man. *N. Engl. J. Med.* **287**:891–895, 1972.
152. Brickman, A. S.; Sherrard, D. J.; Jowsey, J.; Singer, F. R.; Baylink D. J.; Maloney, N.; Massry, S. G.; Norman, A. W.; Coburn, J. W. 1,25-Dihydroxycholecalciferol: Effect on skeletal lesions and plasma parathyroid hormone levels in uremic osteodystrophy. *Arch. Intern. Med.* **134**:883–888, 1974.
153. Fraser, D.; Kooh, S. W.; Kind, H. P.; Holick, M. F.; Tanaka, Y.; DeLuca, H. F. Pathogenesis of hereditary vitamin D-dependent rickets: An inborn error of vitamin D metabolism involving defective conversion of 25-hydroxyvitamin D to 1α,25-dihydroxyvitamin D. *N. Engl. J. Med.* **289**:817–822, 1973.
154. Glorieux, F. H.; Marie, P. J.; Pettifor, J. M.; Delvin, E. E. Bone response to phosphate salts, ergocalciferol, and calcitriol in hypophosphatemic vitamin D-resistant rickets. *N. Engl. J. Med.* **303**:1023–1031, 1980.
155. Chesney, R. W.; Mazess, R. B.; Rose, P.; Hamstra, A. J.; DeLuca, H. F.; Breed, A. L. Long-term influence of calcitriol (1,25-$(OH)_2$-vitamin D) and supplemental phosphate in X-linked hypophosphatemic rickets. *Pediatrics* **71**:559–567, 1983.
156. Rasmussen, H.; Pecket, M.; Anast, C.; Mazur, A.; Gertner, J.; Broadus, A. E. Long-term treatment of familial hypophosphatemic rickets with oral phosphate and 1α-hydroxyvitamin D_3. *J. Pediatr.* **99**:16–25, 1981.
157. Neer, R. M.; Holick, M. F.; DeLuca, H. F.; Potts, J. T., Jr. Effects of 1α-hydroxyvitamin D_3 and 1,25-dihydroxyvitamin D_3 on calcium and phosphorus metabolism in hypoparathyroidism. *Metabolism* **24**:1403–1413, 1975.
158. Kooh, S. W.; Fraser, D.; DeLuca, H. F.; Holick, M. F.; Belsey, R. E.; Clark, M. B.; Murray, T. M. Treatment of hypoparathyroidism and pseudohypoparathyroidism with metabolites of vitamin D: Evidence for impaired conversion of 25-hydroxyvitamin D to 1α,25-dihydroxyvitamin D. *N. Engl. J. Med.* **293**:840–844, 1975.
159. Riggs, B. L.; Hodgson, S. F.; Hoffman, D. L.; Kelly, P. J.; Johnson, K. A.; Taves, D. Treatment of primary osteoporosis with fluoride and calcium: Clinical tolerance and fracture occurrence. *J. Am. Med. Assoc.* **243**:446–449, 1980.
160. Horst, R. L.; DeLuca, H. F.; Jorgensen, N. A. The effect of age on calcium absorption and accumulation of 1,25-dihydroxyvitamin D_3 in intestinal mucosa of rats. *Metab. Bone Dis. Relat. Res.* **1**:29–33, 1978.
161. Gallagher, J. C.; Riggs, B. L.; Eisman, J.; Hamstra, A.; Arnaud, S. B.; DeLuca, H. F. Intestinal calcium absorption and serum vitamin D metabolites in normal subjects and osteoporotic patients: Effect of age and dietary calcium. *J. Clin. Invest.* **64**:729–736, 1979.
162. Bullamore, J. R.; Gallagher, J. C.; Wilkinson, R.; Nordin, B. E. C. Effect of age on calcium absorption. *Lancet* **2**:535–537, 1970.
163. Gallagher, J. C.; Riggs, B. L.; Eisman, J.; Arnaud, S.; DeLuca, H. F. Impaired intestinal calcium absorption in postmenopausal osteoporosis: Possible role of vitamin D metabolites and PTH. *Clin. Res.* **24**:360A, 1976.
164. Christiansen, C.; Cristensen, M. S.; Rodbro, P.; Hagen, C.; Transbol, I. Effect of 1,25-dihydroxyvitamin D_3 in itself or combined with hormone treatment in preventing postmenopausal osteoporosis. *Eur. J. Clin. Invest.* **11**:305–311, 1981.
165. Gallagher, J. C.; Jerpbak, C. M.; Jee, W. S. S.; Johnson, K. A.; DeLuca, H. F.; Riggs, B. L. 1,25-Dihydroxyvitamin D_3: Short- and long-term effects on bone and calcium metabolism in patients with postmenopausal osteoporosis. *Proc. Natl. Acad. Sci. USA* **79**:3325–3329, 1983.

54

Vitamin D and the Intestinal Membrane Calcium-Binding Protein

David Schachter and Szloma Kowarski

The importance of vitamin D in calcium homeostasis is well recognized, but the molecular basis of its action remains to be characterized. In this chapter we describe the identification, biochemical purification, and preliminary quantification in various tissues of a membrane-bound Ca^{2+}-binding protein which is dependent on vitamin D. Inasmuch as the protein was identified initially in membranes of small intestinal mucosal homogenates, we named it IMCAL, an acronym for intestinal membrane calcium-binding protein [11]. The results to be described, however, indicate that the protein is more widely distributed in the body tissues and it is also appropriate to consider IMCAL an acronym for integral membrane calcium-binding protein. Perspectives on the possible role of this protein in Ca^{2+} transport and homeostasis derive from a body of information concerning the mechanism of Ca^{2+} absorption across the small intestinal enterocyte.

Working Model of Ca^{2+} Transport across the Rat Intestinal Enterocyte

Vitamin D is required for the intestinal absorption of Ca^{2+}, a facultative process which varies with the needs of the organism for the mineral. In recent years a general understanding of how the sterol vitamin functions in relation to Ca^{2+} transport has come from the work of many laboratories. Experiments with intestinal preparations *in vitro* have demonstrated that the transcellular transfer of Ca^{2+} is an active cation transport which occurs against electrochemical potential gradients [4,7,9,17,19–25]. The pump mechanism is relatively specific for Ca^{2+} as compared to other divalent cations, dependent on vitamin D, and localized mainly in the proximal (duodenal) region of the intestine in the rat, chicken, and a number of other species. Moreover, the active transport studied *in vitro* varies predictably with the needs of the organism for calcium and is enhanced during growth, pregnancy, and low Ca^{2+} intake.

David Schachter and Szloma Kowarski • Department of Physiology, Columbia University College of Physicians and Surgeons, New York, New York 10032.

At the cellular level, the active Ca^{2+} transport involves at least two steps: entry into the cell by translocation across the microvillous membrane and exit from the cell across the basolateral membrane [20,21,23,24]. These two steps were first identified by transport experiments with intestinal segments *in vitro*, and many of the features of each step have subsequently been characterized by studies of microvillous and basolateral membrane vesicles [6,13,14,15,26]. The entry flux across the microvillous membrane exhibits saturation kinetics, is competitively inhibited by certain hexoses, and is vitamin D-dependent [21]. Attempts to show dependence of this flux on ATP have not been successful to date, and since the entry of Ca^{2+} would normally occur *in vivo* along an electrochemical potential gradient, it is reasonable to hypothesize that the translocation is a facilitated diffusion. The exit step via the basolateral membrane was initially found to be rate-limited and dependent on oxidative phosphorylation and Na^+ in the extracellular fluid. Moreover, the exit flux occurred against concentration gradients of Ca^{2+} and seemed clearly, therefore, to be the uphill, pumping step in the sequence [20,21,23,24]. Basolateral membrane vesicle studies have demonstrated directly a Ca^{2+}-ATPase activity, active flux of Ca^{2+} dependent on ATP [6,14,15,26], and evidence for a possible Ca^{2+}–Na^+ exchange (antiport) mechanism. Recently, the basolateral exit flux has also been shown to require vitamin D [26].

Evidence that inhibitors of protein synthesis effectively block the action of vitamin D on the small intestine of vitamin-deficient rats [16,22,30] provided the first clue that an important mechanism of the sterol's action is to induce the synthesis or maintain the level of specific proteins required for the translational. After hydroxylation of the sterol vitamin in the liver and kidney [1,8,12] (see also Chapter 53), the active metabolites are taken up and bound to a cytosol receptor [2,3] in the intestinal enterocyte. To what extent the subsequent effects involve the control of transcriptional or translational modifications of proteins or enzymes is not yet clear.

In summary, a working model which incorporates the evidence above is as follows. Vitamin D acts via its hydroxylated active metabolites to induce or maintain specific cellular proteins required for the active transport of Ca^{2+} across the small intestinal enterocyte. The overall translocation occurs via a vitamin D-dependent entry mechanism at the luminal membrane, presumably a facilitated diffusion, followed by active pumping out of the cell across the basolateral membrane via a Ca^{2+}-ATPase and possibly a Ca^{2+}–Na^+ exchange mechanism. It is noteworthy that translocation of the cation within the cell may also involve membrane-mediated transfers by intracellular organelles, as suggested by Freedman *et al.* [5] for the Golgi apparatus. Moreover, it is possible that additional mechanisms of vitamin D action which do not require new protein synthesis may occur in the chicken intestine, as suggested by Rasmussen *et al.* [18].

Inasmuch as the characterization of the molecular mechanisms of action of vitamin D requires the identification of the mucosal proteins regulated by the sterol, there has been considerable interest in this problem. Wasserman and his colleagues [27–29] described a soluble Ca^{2+}-binding protein (CaBP) of relatively low molecular weight which has subsequently been studied extensively. Properties and tissue distribution of this class of cytosolic proteins are considered by Parkes (see Chapter 55). Since it seemed unlikely that a soluble, cytosolic protein could mediate the translocation of Ca^{2+} across cell membranes, we initiated studies to identify more appropriate candidates, i.e., membrane particulate fractions with activities dependent on vitamin D.

Calcium-Binding Complex (CaBC) of Rat Intestinal Mucosa

A particulate fraction of rat duodenal homogenates was shown to contain at least three activities dependent on vitamin D: Ca^{2+} binding of high affinity; *p*-nitrophenylphosphatase ("alkaline phosphatase") and Ca^{2+}-ATPase dependent on millimolar Ca^{2+} concentrations (i.e., distinguishable from the basolateral Ca^{2+}-ATPase, which requires Ca^{2+} in the micromolar range). The preparation of this particulate fraction, CaBC, has been described [10]. Of particular significance is the evidence that the Ca^{2+}-binding activity of the particulate varies concordantly with the intestinal Ca^{2+} transport mechanism. The activity is greater in the duodenal than the jejunal segment; in rats maintained on low-calcium diets; and in young, growing rats. Cycloheximide, an inhibitor of protein synthesis, decreases both the CaBC activity and Ca^{2+} transport in rat intestine. Moreover, the compound blocks the restoration of both Ca^{2+} transport and the CaBC activity when vitamin D is administered to deficient rats. The close correlation of the CaBC activity and Ca^{2+} transport supports a working hypothesis that the membrane particulate contains one or more essential components of the translocation mechanism. This hypothesis led to the solubilization and purification of the calcium-binding membrane protein, IMCAL, described below.

It is noteworthy that the three vitamin D-dependent activities of the CaBC (calcium binding, p-nitrophenylphosphatase, and Ca^{2+}-ATPase) vary concordantly with vitamin D status, segment of the intestine, and level of dietary Ca^{2+} [11]. They diverge, however, with respect to age. We did not observe a decrease in the Ca^{2+}-ATPase activity of the CaBC prepared from old rats [10], even though both intestinal Ca^{2+} transport and the Ca^{2+}-binding activity of the CaBC decrease in the aged animals. These results portended the subsequent demonstration that the Ca^{2+}-binding and enzyme activities of the CaBC are dissociable (see below). Of additional interest, the observations suggest that the Ca^{2+}-binding activity is more directly implicated in the Ca^{2+} transport process and is more likely to play a role in the rate-limiting step of the transcellular flux.

Identification and Purification of IMCAL

The procedure for isolating IMCAL was developed by following and comparing the Ca^{2+}-binding activity of $-D$ and $+D$ extracts; details of the procedure can be found in Ref. 11. In a typical procedure butanol extracts prepared from the CaBC are resolved by gel filtration on Sephadex G-150 and by spheroidal hydroxylapatite column chromatography. In the column chromatography step the Ca^{2+}-binding activity is purified and dissociated from p-nitrophenylphosphatase and Ca^{2+}-ATPase. As described previously [11] the product contains 1–2% of the total protein of the crude butanol extract and is purified approximately 32- to 110-fold in relation to the Ca^{2+}-binding activity of the crude mucosal homogenate.

The purified native IMCAL in 0.1% Triton X-100 has an apparent M_r of 200,000 as judged by gel filtration. The native protein migrates as a single band on polyacrylamide gel electrophoresis and after dissociation in 1% SDS, a monomer of M_r 20,450 ± 450 (mean ± S.E.) is observed. The IMCAL monomer is clearly separated from the rat duodenal CaBP ($M_r \sim 11,500$) on SDS-polyacrylamide gels. We have also described differences in amino acid composition between the two proteins [11].

Immunochemical Quantification and Tissue Distribution

Recent studies in our laboratory focus on the immunochemical quantification and tissue distribution of IMCAL. A sensitive and specific assay for the IMCAL SDS-monomer has been developed, using rabbit antibodies prepared against a highly purified, SDS-polyacrylamide gel fraction of the monomer. This assay has a detection limit of approximately 10 ng of IMCAL and is linear to at least 300 ng; as little as 0.01% of a typical sample of 100 μg

Table I. Assay of Soluble Calcium-Binding Protein (CaBP) Preparations for the IMCAL SDS-Monomer[a]

Preparation	Ca^{2+}-binding activity (cpm/μg protein)	IMCAL SDS-monomer (% of total protein)
− vitamin D	11	<0.01%
+ vitamin D	128	<0.01%

[a]The supernatant fractions obtained on centrifugation of duodenal homogenates were assayed for Ca^{2+}-binding activity and for the IMCAL monomer. To ensure complete specificity, a combined polyacrylamide gel separation and enzyme immunoassay were employed, with inclusion of [^{125}I]-IMCAL as an internal reference for recovery. The tissue sample and a known quantity of [^{125}I]-IMCAL were mixed and solubilized in 2% SDS by heating at 100°C for 3 min. The SDS proteins were then resolved by polyacrylamide gel electrophoresis and the proteins transferred to a nitrocellulose sheet electrophoretically. The IMCAL band was cut out and assayed for (1) ^{125}I and (2) total monomer by enzyme immunoassay, using the rabbit primary antibody, a goat anti-rabbit, peroxidase-labeled secondary antibody, and o-phenylenediamine as substrate for the enzyme reaction. From the final specific radioactivity (^{125}I cpm/mg monomer), the initial specific radioactivity and IMCAL content were calculated. Groups of seven vitamin D-depleted and -repleted (20,000 IU vitamin D_3 given i.p.) rats were used.

of protein applied to the gel can be quantified. Further, the assay does not detect the soluble intestinal Ca^{2+}-binding protein (CaBP), as indicated by the results in Table I.

The results of immunochemical assays of intestinal mucosa have confirmed that the content of IMCAL monomer is highly correlated with the activity of the Ca^{2+} transport mechanism. For example, the IMCAL monomer content (percent of total particulate protein) in normal duodenal, jejunal, and ileal mucosal preparations was 0.501, 0.017, and <0.01%, respectively. The IMCAL monomer content of duodenal mucosa of vitamin D-deficient and -repleted (20,000 IU of vitamin D_2 given i.p. 18 hr before excision) rats was $0.135 \pm 0.019\%$ (mean ± S.E.) and $0.206 \pm 0.013\%$, respectively $p < 0.01$). Repletion with vitamin D increased the mucosal IMCAL content significantly in 1 hr. Moreover, the time course of repletion with vitamin D was very similar for Ca^{2+} transport and IMCAL content (Table II). Recently, we have also demonstrated that the duodenal mucosal content of IMCAL is increased by a low-calcium as compared to a high-calcium diet.

Using groups of vitamin D-deficient and -repleted rats, the IMCAL content of a number of organs has been assayed as a function of vitamin D. Increments dependent on vitamin D

Table II. Time Course of Effects of Vitamin D_3 on Intestinal Calcium Transport and on Duodenal Mucosal IMCAL Content[a]

Hours after vitamin D_3	IMCAL content (% of whole particulate protein)	IMCAL content relative to zero time[b]	Ca^{2+} transport relative to zero time[c]
0	0.130	100%	100%
1	0.218	168%	152%
4	0.202	155%	157%
24	0.375	288%	210%

[a]At zero time 20,000 IU of vitamin D_3 was injected i.p. IMCAL content is in percent of the total protein in the particulate fraction of the duodenal mucosal homogenate.
[b]Zero time value is taken as 100%.
[c]Data taken from Schachter et al. [21].

repletion have been demonstrated in cecal mucosa, kidney, and tibial metaphysis, as expected for a membrane protein involved in Ca^{2+} transport. Unexpectedly, we have also observed vitamin D-dependent increments in brain (cerebral cortex and cerebellum), heart muscle, lung, spleen, and testis. Liver and skeletal muscle contain small quantities of immunologically reactive material which does not vary with vitamin D treatment.

Lastly, we have assayed various organellar fractions of the duodenal mucosa of normal rats. Values of the IMCAL content of the microsomal, mitochondrial–lysosomal, and nuclear fractions, respectively, were 0.70, 0.22, and 0.13%. Isolated duodenal microvillous and basolateral plasma membranes both contain the monomer, with relative contents of 0.88 and 0.31%, respectively.

Summary and Conclusions

The foregoing observations support the working hypothesis that IMCAL is involved in the vitamin D-dependent transport of Ca^{2+} across the intestinal mucosa. The higher content of the protein in the luminal (microvillous) membrane would suggest that it mediates or modulates the vitamin D-dependent entry mechanism. On the other hand, vitamin D also influences the active transport of the cation across the basolateral membrane [26] and IMCAL is present in this membrane. It seems reasonable to suggest that the protein is either part of the translocation apparatus itself, e.g., forms a Ca^{2+} channel, or that it regulates other membrane components of the translocation apparatus. The widespread tissue distribution, and the vitamin D-dependency of the IMCAL content in such organs as the brain, lung, heart muscle, and testis are difficult to account for at present. One possibility is that the membrane protein regulates Ca^{2+} flux in many cell types. On the other hand, the immunologically cross-reactive proteins in these organs may not be identical to the native IMCAL originally isolated from duodenal mucosa, and further investigation is needed to clarify the interrelationships.

ACKNOWLEDGMENT. We acknowledge with gratitude the support of Grant AM-01483 from the National Institutes of Health (NIADDK).

References

1. Blunt, J. W.; DeLuca, H. F.; Schnoes, H. K. 25-Hydroxycholecalciferol: A biologically active metabolite of vitamin D. *Biochemistry* **7**:3317–3322, 1968.
2. Brumbaugh, P. F.; Haussler, M. R. 1α,2-Dihydroxycholecalciferol receptors in intestine. II. Temperature dependent transfer of the hormone to chromatin via a specific cytosol receptor. *J. Biol. Chem.* **249**:1258–1262, 1974.
3. Brumbaugh, P. F.; Haussler, M. R. Specific binding of 1α,25-dihydroxycholecalciferol to nuclear components of chick intestine. *J. Biol. Chem.* **250**:1588–1594, 1975.
4. Dowdle, E. B.; Schachter, D.; Schenker, H. Requirement for vitamin D for the active transport of calcium by the intestine. *Am. J. Physiol.* **198**:269–274, 1960.
5. Freedman, R. A.; Weiser, M. M.; Isselbacher, K. J. Calcium translocation by Golgi and lateralbasal membrane vesicles from rat intestine: Decrease in vitamin D-deficient rats. *Proc. Natl. Acad. Sci. USA* **74**:3612–3616, 1977.
6. Ghijsen, W. E. J. M.; De Jong, M. D.; van Os, C. H. ATP-dependent calcium transport and its correlation with Ca^{2+}-ATPase activity in basolateral plasma membranes of rat duodenum. *Biochim. Biophys. Acta* **689**:327–336, 1982.
7. Harrison, H. E.; Harrison, H. C. Transfer of Ca^{45} across intestinal wall *in vitro* in relation to action of vitamin D and cortisone. *Am. J. Physiol.* **199**:265–271, 1960.

8. Holick, M. F.; Schnoes, H. K.; DeLuca, H. F. Identification of 1,25-dihydroxycholecalciferol, a form of vitamin D, metabolically active in the intestine. *Proc. Natl. Acad. Sci. USA* **68**:803–804, 1971.
9. Kimberg, D. V.; Schachter, D.; Schenker, H. Active transport of calcium by intestine: Effects of dietary calcium. *Am. J. Physiol.* **200**:1256–1262, 1961.
10. Kowarski, S.; Schachter, D. Vitamin D-dependent, particulate calcium-binding activity and intestinal calcium transport. *Am. J. Physiol.* **229**:1198–1204, 1975.
11. Kowarski, S.; Schachter, D. Intestinal membrane calcium-binding protein: Vitamin D-dependent membrane component of the intestinal calcium transport mechanism. *J. Biol. Chem.* **255**:10834–10840, 1980.
12. Lawson, D. E.; Fraser, D. R.; Kodicek, E.; Morris, H. R.; Williams, D. H. Identification of 1,25-dihydroxycholecalciferol, a new kidney hormone controlling calcium metabolism. *Nature (London)* **230**:228–230, 1971.
13. Moriuchi, S.; DeLuca, H. F. The effect of vitamin D_3 metabolites on membrane proteins of chick duodenal brush borders. *Arch. Biochem. Biophys.* **174**:367–372, 1976.
14. Murer, H.; Hildmann, B. Transcellular transport of calcium and inorganic phosphate in the small intestinal epithelium. *Am. J. Physiol.* **240**:G409–G416, 1981.
15. Nellans, H. N.; Popovitch, J. E. Calmodulin-regulated, ATP-driven calcium transport by basolateral membranes of rat small intestine. *J. Biol. Chem.* **256**:9932–9936, 1981.
16. Norman, A. W. Actinomycin D and the response to vitamin D. *Science* **149**:184–186, 1965.
17. Rasmussen, H. The influence of parathyroid extract upon the transport of calcium in isolated sacs of rat small intestine. *Endocrinology* **65**:517–519, 1959.
18. Rasmussen, H.; Fontaine, O.; Matsumoto, T. Liponomic regulation of calcium transport by $1,25(OH)_2D_3$. *Ann. N.Y. Acad. Sci.* **372**:518–529, 1981.
19. Schachter, D. Toward a molecular description of active transport. In: *Biological Membranes*, R. M. Dowben, ed., Boston. Little, Brown, 1969, pp. 157–176.
20. Schachter, D.; Dowdle, E. B.; Schenker, H. Active transport of calcium by the small intestine of the rat. *Am. J. Physiol.* **198**:263–268, 1960.
21. Schachter, D.; Kimberg, D. V.; Schenker, H. Active transport of calcium by intestine: Action and bioassay of vitamin D. *Am. J. Physiol.* **200**:1263–1271, 1961.
22. Schachter, D.; Kowarski, S. Radioactive vitamin D: Preparation, metabolism and mechanism of action. *Bull. N.Y. Acad. Med.* **41**:241, 1965.
23. Schachter, D.; Kowarski, S.; Finkelstein, J. D.; Ma, R.-I. W. Tissue concentration differences during active transport of calcium by intestine. *Am. J. Physiol.* **211**:1131–1136, 1966.
24. Schachter, D.; Kowarski, S., Reid, P. Active transport of calcium by intestine: Studies with a calcium activity electrode. In: *A Symposium on Calcium and Cellular Function*, A. W. Cuthbert, ed., London, Macmillan & Co., 1969, pp. 108–123.
24. Schachter, D.; Rosen, S. M. Active transport of Ca^{45} by the small intestine and its dependence on vitamin D. *Am. J. Physiol.* **196**:357–362, 1959.
26. van Os, C. H.; Ghijsen, W. E. J. M. Calcium transport mechanisms in rat duodenal basolateral plasma membranes: Effects of $1,25(OH)_2D_3$. In: *Vitamin D: Chemical, Biochemical and Clinical Endocrinology of Calcium Metabolism*, A. W. Norman, K. Schaefer, D. V. Herrath, H.-G. Grigoleit, eds., Berlin, de Gruyter, 1982, pp. 295–297.
27. Wasserman, R. H.; Corradino, R. A.; Taylor, A. N. Vitamin D-dependent calcium binding protein: Purification and some properties. *J. Biol. Chem.* **243**:3978–3986, 1968.
28. Wasserman, R. H.; Taylor, A. N. Vitamin D-induced calcium-binding protein in chick intestinal mucosa. *Science* **152**:791–793, 1966.
29. Wasserman, R. H.; Taylor, A. N. Vitamin D-dependent calcium-binding protein: Response to some physiological and nutritional variables. *J. Biol. Chem.* **243**:3987–3993, 1968.
30. Zull, J. E.; Czarnowski-Misztal, E.; DeLuca, H. F. Actinomycin D inhibition of vitamin D action. *Science* **149**:182–184, 1965.

55

Calcium-Binding Protein in the Central Nervous System and Other Tissues

C. Owen Parkes

The first report of vitamin D-dependent calcium-binding protein (D-CaBP) was by Wasserman and Taylor in 1966 [22]. The small intestine of the chick contains this 28,000-dalton protein in remarkably high concentrations (15 µg/mg soluble protein in the duodenum). Subsequent reports from Wasserman's laboratory [23] described D-CaBP in high concentrations in a number of other chick organs. A majority of the sites are clearly involved in calcium translocation, e.g., shell gland, kidney, and intestine. However, large amounts of immunoreactive D-CaBP are also present in chick cerebellum.

The relationship between D-CaBP synthesis and vitamin D metabolites has been studied extensively in the chick both *in vivo* and *in vitro*. The presence of some form of vitamin D is mandatory for D-CaBP synthesis by embryonic chick duodenum. This tissue poses an interesting question about the control of gene expression. *In ovo*, chick duodenum produces no CaBP, but a surge of CaBP synthesis occurs on hatching [10]. However, the embryo itself contains adequate levels of 1,25-dihydroxyvitamin D [$1,25(OH)_2D_3$], the active metabolite of vitamin D; and the kidneys and cerebellum of the embryo contain D-CaBP. The embryonic duodenum, removed from the chick and maintained *in vitro*, will respond to exogenous $1,25(OH)_2D_3$ by synthesizing CaBP. We have been unable to duplicate the effect of vitamin D steroids with any of a wide range of other hormones, both steroid and peptide, although the effect of vitamin D metabolites is enhanced by PTH and thyroxine [13].

While chick duodenal and renal D-CaBP levels are vitamin D-dependent, those of the cerebellum are not. Immunohistochemical studies by Jande *et al.* [6] reveal that CaBP is found exclusively in Purkinje cells, disseminated throughout the cytoplasm. The cerebellar cortex is by far the major site of CaBP (7 µg/mg soluble protein) but smaller quantities were detected by specific radioimmunoassay from the medulla to the forebrain [2]. Kainic acid treatment reduced the CaBP content of both cerebellum and hippocampus. As this glutamate analog preferentially destroys large cortical neurons, we conclude that both cerebellar and hippocampal CaBP are present in large neurons.

C. Owen Parkes • Department of Physiology, University of British Columbia, Vancouver, British Columbia V6T 1W5, Canada.

Mammalian CaBP

All chick tissues appear to contain the same vitamin D-dependent C-CaBP. Mammals, on the other hand, exhibit a more complex pattern. At least three different vitamin D-dependent CaBPs have been described.

1. 28,000-dalton L-CaBP. Antibodies raised to chick 28,000-dalton D-CaBP will cross-react with material from mammalian cerebellum and kidney. We used this feature in the purification of human cerebellar L-CaBP and the development of a specific RIA [3]. The RIA, which has a sensitivity of 1 ng/assay tube, may be used to measure L-CaBP in a variety of mammalian species, including rat, baboon, mouse. and cat. It is not, however, suitable for the quantification of chick CaBP.
2. 7500-dalton S-CaBP. While very little immunoreactive L-CaBP can be detected in the intestine of mammals, there is most certainly vitamin D-dependent calcium-binding activity present [17]. Marche et al. [9] purified a small CaBP (S-CaBP) from rat duodenum and developed a specific RIA for this CaBP. In contrast to the L-CaBP RIA, this assay is absolutely specific for rat duodenal CaBP and is unable to detect other mammalian S-CaBPs. Murray et al. [11] reported a similarly restricted specificity for their pig intestinal S-CaBP RIA.
3. 11,000-dalton skin CaBP. Yet a third vitamin D-dependent CaBP has been reported to occur in the basal proliferative layer of the rat epidermis [14]. This protein is immunologically quite distinct from either of the other mammalian CaBPs.

Tissue Distribution of Mammalian CaBPs

Each CaBP has a characteristic distribution pattern which does not markedly overlap with that of other CaBPs (Table I). Skin CaBP is largely confined to the proliferative layer of stratified Malpighian-type epithelia, but it has also been detected in other proliferating epithelia, such as the lens epithelium and the ependymal cells of the brain lateral ventricles.

Table I. Tissue Distribution of Three Vitamin D-Dependent CaBPs in the Rat

Tissue	S-CaBP (μg/mg soluble protein)[a]	L-CaBP (μg/mg soluble protein)[a]	Skin CaBP[b]
Esophagus	14 ± 4	12 ± 5	++
Duodenum	18,250 ± 1620	95 ± 6	--
Jejunum	2,930 ± 430	35 ± 5	nd[c]
Ileum	430 ± 10	39 ± 7	nd
Cecum	3,080 ± 679	76 ± 2	nd
Pancreas	10 ± 7	67 ± 14	nd
Parathyroid	nd	nd	nd
Kidney cortex	35 ± 3	8,680 ± 600	nd
Bone	209 ± 7	210 ± 9	nd
Cerebellum	4 ± 1	17,950 ± 109	---
Skin	1 ± 0.3	1 ± 0.3	++
Brain ventricle ependyma	—	—	++

[a]Data from Ref. 22.
[b]Data from Ref. 14.
[c]nd, not detectable.

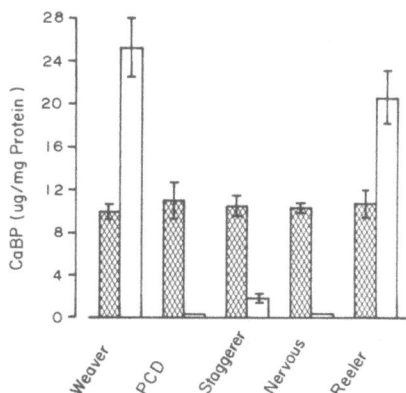

Figure 1. Concentrations (mean ± S.E.) of L-CaBP in the cerebella of mutant mice. Hatched bars, heterozygous control; white bars, homozygous mutant. PCD, Purkinje-cell-degenerate mice. L-CaBP was measured by RIA.

While its function in these epithelia is unknown, it appears to be characteristic of undifferentiated cells.

On the other hand, duodenal S-CaBP contained within the enterocytes increases in concentration as these cells mature and travel up the villus. The small intestine is by far the major site of S-CaBP levels, but even in this tissue the distribution is not constant. The concentration of calmodulin is the same in all three segments of the rat small intestine, but the duodenal content of S-CaBP is 6-fold greater (18 μg/mg soluble protein) than that of the adjacent jejunum (3 μg/mg), and 40-fold greater than that of the ileum [19]. However, the more distal cecum has an S-CaBP level of 3 μg/mg. The major concentration of L-CaBP in the nervous system resides in the cerebellum [20], where measured levels range from 8 μg/mg in mouse and man to 15 μg/mg in rat. Independent studies agree that the protein is restricted to the Purkinje cells [6,7].

To correlate the presence of CaBP with cerebellar function, J. Mariani (Institut Pasteur) and I have measured the cerebellar L-CaBP content of mutant mice (Fig. 1). It is clear that mutants lacking Purkinje cells (Purkinje-cell-degenerate and nervous) possess negligible amounts of cerebellar CaBP. Homozygous recessive staggerer mice, which are devoid of Purkinje and granular cells, have reduced CaBP. Those mutants that showed increased levels of CaBP, weaver and reeler, suffer from a loss of other cerebellar components. For example, weavers are missing their granular cells. As these cells form a significant portion of the cerebellar mass in the normal mouse, their absence leads to a *de facto* increase in L-CaBP content per unit mass. It, therefore, appears that L-CaBP is an intrinsic component of Purkinje cells. As long as these cells are present, protein can be detected. Furthermore, the L-CaBP concentration does not appear dependent on the activity of other cerebellar neuronal components.

There is also a specific and discontinuous distribution of L-CaBP in the hippocampus. The cell bodies of the large pyramidal cells of the CA1 region stain strongly for L-CaBP, but the cells of the adjoining region remain unstained [1]. The dentate gyrus granule cells also stain strongly, as does the mossy fiber tract between CA1 and the dentate region. Here, as in the cerebellum, L-CaBP seems to be distributed throughout the cytoplasm of certain neurons.

Our immunohistochemical studies of L-CaBP in the mammalian kidney [15] confirm the intracellular nature of CaBP. As in the brain, renal L-CaBP was found in the cytoplasm, with occasional nuclei stained. Although all stained cells are in the region of the distal convoluted tubule–proximal collecting duct, not all cells in that region stain. Clearly distinguishable unstained cells are interspersed among the L-CaBP-positive cells. Monkey kidney is unique in that cells distributed throughout the length of the collecting tubule as far as the papillary ducts stain for L-CaBP.

Relationship between S-CaBP and L-CaBP

As both L-CaBP and S-CaBP bind calcium with similar affinities, and are intracellular proteins synthesized in response to vitamin D, they may have a common ancestry and share major structural features. We examined this possibility by extracting the mRNAs coding for L-CaBP and S-CaBP and translating them in a rabbit reticulocyte lysate system [21]. Poly(A)-rich mRNA from duodenum produced CaBP immunoprecipitable with antibody to S-CaBP only. No newly synthesized L-CaBP was detected. Conversely, mRNA from kidney and cerebellum coded for CaBP precipitable with anti-L-CaBP antibodies. There was no indication that the syntheses of the two CaBPs were interdependent, and we concluded that there are at least two CaBP-producing genes which are vitamin D-sensitive. We also found evidence to support the exclusively intracellular distribution of both CaBPs. When translation was carried out in the presence of microsomal membrane, there was no difference in the migration of the newly synthesized CaBP on SDS-PAGE from that found for CaBPs synthesized in the absence of microsomal membrane. They, therefore, appear to lack the leader sequence characteristic of extracellular proteins.

It is of interest to compare the translational efficiencies of mRNAs from each of the three tissues. The incorporation of [^{35}S]methionine into S-CaBP reaches almost 10% of the total TCA-precipitable material for the duodenal mRNA, while comparable levels for kidney and cerebellar mRNAs are 1 and 0.4%, respectively. These data are consonant with the histochemical results and with the relative rates of cell turnover in the three tissues.

Function of CaBPs

Numerous attempts have been made, beginning with Wasserman and Taylor [22], to correlate the appearance and distribution of CaBP with calcium transport. However, published reports are about equally divided for and against the concept. Significant levels of CaBP in rat cecum and recent studies on chick duodenum [4] indicate such a correlation. But the absence of measurable CaBP at the onset of calcium absorption following 1,25(OH)D treatment of rachitic chicks and rats, respectively [16,18], engenders a powerful argument against the direct involvement of CaBP in calcium translocation.

Two possibilities thus remain. Vitamin D-dependent CaBPs may act as intracellular calcium buffers. High calcium loads stress duodenal enterocytes, renal distal tubule cells, and the cells of the chick shell gland as the egg shell is formed. CaBPs are found under these conditions. The cerebellar Purkinje cells produce voltage-dependent calcium spikes [8]. The resulting calcium surges may be stabilized by the high (10^{-4} M) CaBP level, thereby preventing disruption of the internal architecture of the cell.

Although there are extensive reports of the modulatory interaction of calmodulin with specific target proteins, little evidence exists to suggest that CaBP has a similar function. CaBP binds to rat enterocyte brush border [5] and bovine alkaline phosphatase [12] and will stimulate rat alkaline phosphatase activity [6]. This, of course, is a most attractive role for a group of proteins whose synthesis is so tightly regulated and whose distribution is so strikingly specific. Nevertheless, experimental support for this idea is quite meager.

There is little doubt that the vitamin D-dependent CaBPs are important intracellular proteins. Although their exact functions remain obscure, their widespread but specific occurrence raises new questions about the role of vitamin D. The steroid homones derived from this vitamin may not be restricted in their effects on the traditional target organs of gut, kidney, and bone. Instead, they may be intimately concerned, through the action of CaBPs, with the regulation of calcium activity and, hence, cell function in a diversity of body tissues.

ACKNOWLEDGMENT. Supported by Grant MA-6566 from the Medical Research Council, Canada.

References

1. Baimbridge, K. G.; Miller, J. J. Immunohistochemical localization of calcium-binding protein in the cerebellum, hippocampal formation and olfactory bulb of the rat. *Brain Res.* **245**:223–229, 1982.
2. Baimbridge, K. G.; Parkes, C. O. Vitamin D-dependent CaBP in the chick brain. *Cell Calcium* **2**,65–76, 1981.
3. Baimbridge, K. G.; Miller, J. J.; Parkes, C. O. CaBP distribution in the rat brain. *Brain Res.* **239**:519–525, 1982.
4. Bishop, C. W.; Kendrick, N. C.; DeLuca, H. F. Induction of CaBP before 1,25(OH)D stimulation of duodenal calcium uptake. *J. Biol. Chem.* **258**:1305–1310, 1983.
5. Freund, T. S. Vitamin D-dependent intestinal CaBP as an enzyme modulator. In: *Vitamin D: Chemical, Biochemical and Clinical Endocrinology of Calcium Metabolism*, A. W. Norman, K. Schaefer, D. von Herrath, and H.-G. Grigoleit, eds., Berlin, de Gruyter, 1982, pp. 248–251.
6. Jande, S. S.; Tolnai, S.; Lawson, D. E. M. Immunohistochemical localization of vitamin D-dependent CaBP in duodenum, kidney, uterus and cerebellum of chickens. *Histochemistry* **71**:99–116, 1981.
7. Legrand, C.; Thomasset, M.; Parkes, C. O.; Clavel, M. C.; Rabie, A. CaBP in the developing rat cerebellum: An immunohistochemical study. *Cell Tissue Res.* **233**:389–402, 1983.
8. Llinas, R.; Hess, R. Tetrodotoxin-resistant dendritic spikes in avian Purkinje cells. *Proc. Natl. Acad. Sci. USA* **73**:2520–2523, 1976.
9. Marche, P.; Pradelles, P.; Gros, C.; Thomasset. M. Radioimmunoassay for a vitamin D-dependent calcium-binding protein in rat duodenal mucosa. *Biochem. Biophys. Res. Commun.* **76**:1020–1026, 1977.
10. Moriuchi, S.; DeLuca, H. F. Metabolism of vitamin D in the chick embryo. *Arch. Biochem. Biophys.* **164**:165–171, 1974.
11. Murray, T. M.; Arnold, B. M.; Tam, W. H.; Hitchman, A. J. W.; Harrison, J. E. A radioimmunoassay for a porcine intestinal calcium-binding protein. *Metabolism* **23**:829–837, 1974.
12. Norman, A. W.; Leathers, V. Preparation of a photoaffinity probe for the vitamin D-dependent intestinal CaBP: Evidence for a calcium-dependent specific interaction with intestinal alkaline phosphatase. *Biochem. Biophys. Res. Commun.* **108**:220–226, 1982.
13. Parkes, C. O.; Reynolds, J. J. The effect of PTH on duodenal CaBP synthesis. *Horm. Metab. Res.* **10**:75–77, 1978.
14. Pavlovich, J. H.; Didierjean, L.; Rizk, M.; Balsan, S.; Saurat, J. H. Skin calcium-binding protein: Distribution in other tissues. *Am. J. Physiol.* **244**:C50–C57, 1983.
15. Schreiner, D. S.; Jande, S. S.; Parkes, C. O.; Lawson, D. E. M.; Thomasset, M. Immunohistochemical demonstration of two vitamin D-dependent calcium-binding proteins in mammalian kidney. *Acta Anat.* **117**:1–14, 1983.
16. Spencer, R.; Charman, M.; Wilson, P.; Lawson, D. E. M. Vitamin D-stimulated intestinal calcium absorption may not involve calcium-binding protein directly. *Nature (London)* **263**:161–163, 1976.
17. Thomasset, M.; Cuisinier-Gleizes, P.; Mathieu, H. Duodenal CaBP and phosphorus depletion in growing rats. *Biomedicine* **25**:345–349, 1976.
18. Thomasset, M.; Cuisinier-Gleizes, P.; Mathieu, H. 1,25-Dihydroxycholecalciferol: Dynamics of the stimulation of duodenal calcium-binding protein, Ca transport and bone calcium mobilization in vitamin D and calcium-deficient rats. *FEBS Lett.* **107**:91–94, 1979.
19. Thomasset, M.; Molla, A.; Parkes, C. O.; Demaille, J. G. Intestinal calmodulin and calcium-binding protein differ in their distribution and in the effect of vitamin D steroids on their concentrations. *FEBS Lett.* **127**:13–16, 1981.
20. Thomasset, M.; Parkes, C. O.; Cuisinier-Gleizes, P. Rat calcium-binding proteins: Distribution, development and vitamin D dependence. *Am. J. Physiol.* **243**:E483–E488, 1982.

21. Thomasset, M.; Desplan, C.; Parkes, C. O. Rat vitamin D-dependent CaBP's: Specificity of mRNA's coding for the 7500Mr from duodenum and 28000Mr protein from the kidney and cerebellum. *Eur. J. Biochem.* **129:**519–524, 1983.
22. Wasserman, R. H.; Taylor, A. N. Vitamin D induced calcium-binding protein in chick intestinal mucosa. *Science* **152:**793–797, 1966.
23. Wasserman, R. H.; Fulmer, C. S.; Taylor, A. N., The vitamin D-dependent calcium-binding proteins. In: *Vitamin D*, D. E. M. Lawson, ed., New York, Academic Press, 1978, pp. 133–136.

56

The Vitamin K-Dependent Bone Protein and the Action of 1,25-Dihydroxyvitamin D_3 on Bone

Paul A. Price

This chapter considers the evidence that the vitamin K-dependent bone protein [bone Gla protein (BGP)] mediates an aspect of the action of 1,25-dihydroxyvitamin D_3 on bone. BGP is a 49-residue protein of known structure [1] which contains three residues of the vitamin K-dependent Ca^{2+}-binding amino acid, γ-carboxyglutamic acid (Gla). A general review on BGP has recently been published [2], so only the most salient of its properties will be described here. BGP is one of the most abundant bone proteins [3,4] and has been detected in all vertebrates examined to date. It is also present in serum [5]. Its structure is highly conserved [6], with over 50% of the amino acid residues unchanged in BGP from swordfish and calf bone, species which diverged about 400 million years ago. The most prominent *in vitro* properties of BGP are its high affinity for hydroxyapatite and its ability to retard the crystallization of hydroxyapatite from supersaturated solutions of calcium phosphate [4]. Both of these properties are lost upon thermal decarboxylation of γ-carboxyglutamic acid to glutamic acid [7], a specific chemical modification which yields an abnormal BGP apparently identical to that produced in animals treated with a vitamin K antagonist such as warfarin [8]. BGP is synthesized in bone [9] and by cultured osteosarcoma cells with an osteoblastic phenotype [10].

Effects of 1,25(OH)$_2$D$_3$ on BGP Synthesis by Rat Osteosarcoma Cells

Initial evidence that BGP synthesis is regulated by 1,25(OH)$_2$D$_3$ came from studies in cultured rat osteosarcoma cells [11], showing that 1,25(OH)$_2$D$_3$ produced a sixfold elevation in intracellular levels of BGP, followed by a similar increase in media BGP levels (Fig. 1). The 1,25(OH)$_2$D$_3$ concentration which half-stimulated BGP (0.04 ng/ml) is approximately the level normally present in rat serum.

The time course of the response to 1,25(OH)$_2$D$_3$ is consistent with the hypothesis that 1,25(OH)$_2$D$_3$ activates the gene for BGP synthesis in the same manner postulated for the

---

Paul A. Price • Department of Biology, University of California at San Diego, La Jolla, California 92093.

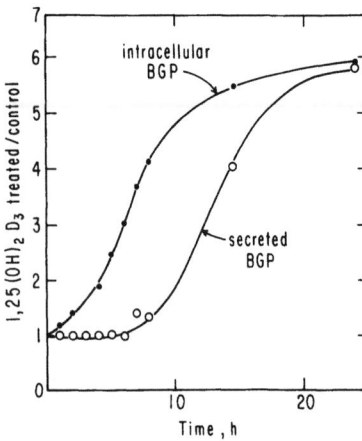

Figure 1. Stimulation of BGP synthesis and secretion by 1,25 (OH)$_2$D$_3$. The media of confluent 60-mm culture plates were exchanged for the same media or media containing 1 ng/ml of 1,25(OH)$_2$D$_3$ at time zero. Two experimental and two control plates were terminated at each time point and analyzed for intracellular and media levels of BGP. Each point is the average experimental value divided by the average control [From Price and Baukol [11] with permission.]

1,25(OH)$_2$D$_3$-dependent regulation of the cytosolic Ca^{2+}-binding protein of intestine. Regulation of BGP synthesis at the message level is also supported by the observation that actinomycin D totally blocks the increase in secreted levels of BGP in ROS 17/2 cells treated with 1 ng/ml of 1,25(OH)$_2$D$_3$ (unpublished). Secreted levels of BGP are, as expected, dependent on ongoing synthesis and therefore were reduced in ROS 17/2 cells treated with the protein synthesis inhibitor cycloheximide.

Effect of 1,25(OH)$_2$D$_3$ on Serum BGP Levels in the Rat

The evidence that 1,25(OH)$_2$D$_3$ also stimulates BGP synthesis in the rat came from studies on serum levels of BGP following either administration of 1,25(OH)$_2$D$_3$ [12] or a period of dietary Ca^{2+} deficiency [13,14]. Serum BGP arises from new cellular synthesis rather than from the release of bone matrix BGP during bone resorption [8]. It is cleared rapidly by kidney filtration, with a half-time of 5 min. Since the administration of warfarin produces a complete change in the nature of serum BGP (from normally γ-carboxylated to non-γ-carboxylated) within 3 hr, the maximal interval between a change in BGP synthesis at the ribosomal level and the reflection of this change in altered concentrations of serum BGP is less than 3 hr [8].

It is clear that a single dose of 1,25(OH)$_2$D$_3$ enhances the rate of BGP synthesis, as reflected by serum BGP levels, and that the time course of this response closely parallels that of the BGP synthesis response in cultured osteosarcoma cells (Fig. 2). The dosage of 1,25(OH)$_2$D$_3$ required to elicit the maximal BGP synthesis response was 350 ng 1,25(OH)$_2$D$_3$/180-g rat, a dosage considered to be pharmacologic rather than physiologic. However, comparable dosages of 1,25(OH)$_2$D$_3$ were also required to elicit a maximal increase in synthesis of the duodenal Ca^{2+}-binding protein in calcium-replete rats of this age [15]. The physiologic relevance of the regulation of BGP synthesis by 1,25(OH)$_2$D$_3$ is reinforced by the observation that dietary Ca^{2+} deficiency, a circumstance which normally elevates serum 1,25(OH)$_2$D$_3$ levels over 10-fold [16], produced a 3- to 4-fold elevation in serum BGP levels [13]. The elevation in serum BGP levels was abolished in rats on a diet deficient in vitamin D as well as Ca^{2+} and was restored upon subsequent daily oral administration of vitamin D$_3$ [13,14].

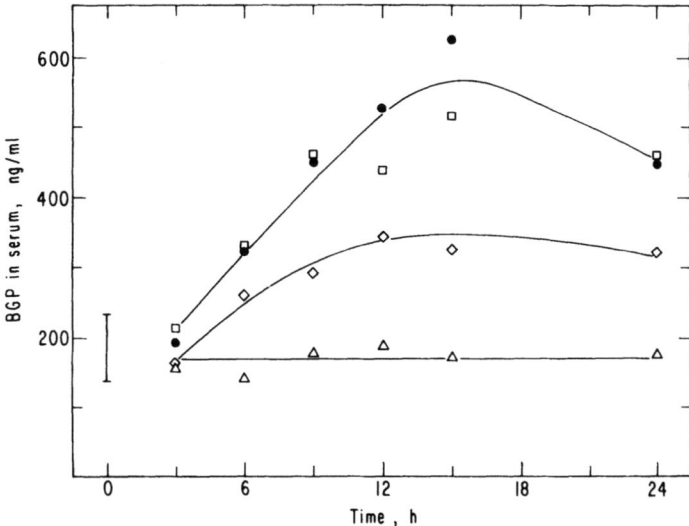

Figure 2. Dose dependence of serum BGP levels on the amount of $1.25(OH)_2D_3$ injected. Forty-seven-day-old male rats were injected intrajugularly with 80 ($\Diamond$), 350 ($\bullet$) and 1500 ($\Box$) ng of $1,25(OH)_2D_3/180$-g body wt. Control rats ($\triangle$) were injected with ethanol vehicle alone. [From Price and Baukol [12] with permission.]

Effect of Warfarin on the Response of Bone to $1,25(OH)_2D_3$

To assess possible functions of BGP as a mediator of $1,25(OH)_2D_3$ action on bone, daily dosages of $1,25(OH)_2D_3$ have been administered to rats concurrently treated with the vitamin K antagonist warfarin [17]. Weanling animals were fed a calcium- and phosphate-replete diet and received sufficient $1,25(OH)_2D_3$ to achieve nearly maximal stimulation of BGP synthesis. The warfarin dosages employed in these studies are sufficient to reduce levels of BGP in newly synthesized bone to 2% of normal [18]. In the vitamin K-replete animals, $1,25(OH)_2D_3$ treatment dramatically reduced bone density in the tibial diaphysis while not significantly affecting bone density in the proximal tibial metaphysis (Fig. 3). In contrast, the administration of $1,25(OH)_2D_3$ to animals treated concurrently with warfarin reduced bone density dramatically in the metaphysis as well as the diaphysis.

To provide a quantitative picture of the effects of warfarin on the response to $1,25(OH)_2D_3$, right tibias were sectioned into proximal epiphysis and two serial 0.5-cm sections from the proximal growth plate. The first 0.5-cm section, the metaphysis, includes the area of greater radiological density seen in the $1,25(OH)_2D_3$-treated vitamin K-replete rat (Fig. 3). The second 0.5-cm section is the diaphysis. As anticipated from radiologic data, warfarin treatment reduced total calcium content of the proximal metaphysis in the $1,25(OH)_2D_3$-treated rat by 40% ($p < 0.01$). Warfarin treatment did not produce a significant effect on radiologic density or on the calcium content of any tibial section in the absence of $1,25(OH)_2D_3$ treatment (Figs. 3 and 4).

The efficacy of the warfarin protocol in antagonizing the action of vitamin K was assessed by measuring BGP levels in demineralization extracts from each bone segment. This analysis revealed that warfarin treatment reduced BGP levels to the greatest extent in the metaphysis and epiphysis [17]. This result is consistent with the observation [18] that overall BGP levels in bone reflect the rate at which high BGP content matrix, formed prior to warfarin treatment, is replaced with the low (2% of normal) BGP content matrix formed during warfarin treatment.

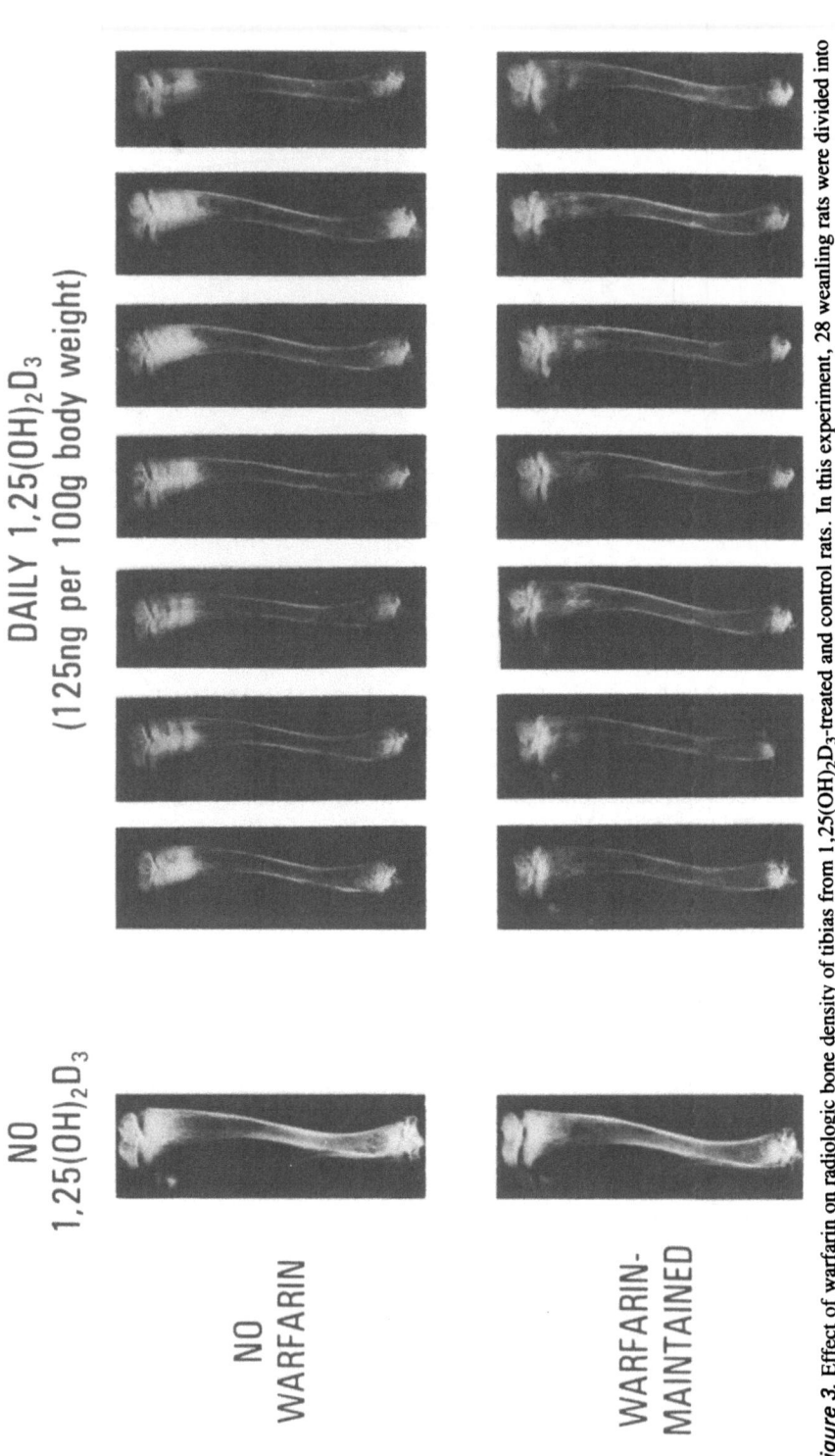

Figure 3. Effect of warfarin on radiologic bone density of tibias from $1,25(OH)_2D_3$-treated and control rats. In this experiment, 28 weanling rats were divided into four treatment groups of seven each and given subcutaneous injections of $1,25(OH)_2D_3$, warfarin, or vehicle for 11 days as described [17]. All right tibias from the seven animals in the $1,25(OH)_2D_3$ and the $1,25(OH)_2D_3$ plus warfarin treatment groups are compared with representative right tibias from the control and the warfarin-only groups. Magnification × 1.4. [From Price and Sloper [17] with permission.]

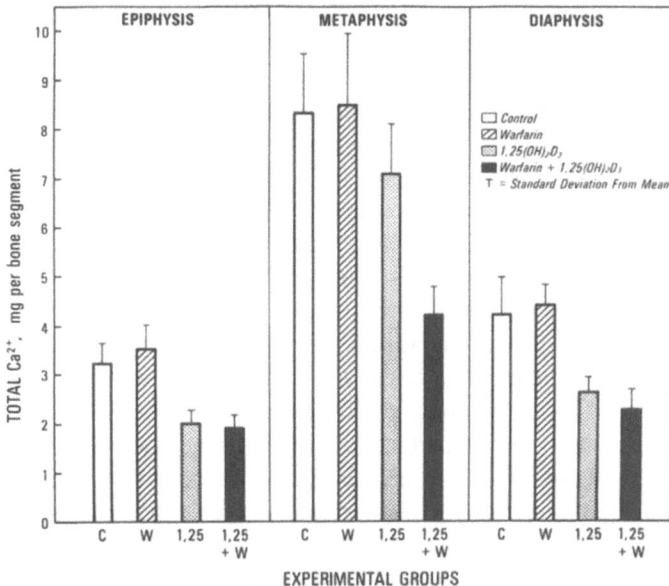

Figure 4. Effect of warfarin on the mean calcium content of bone segments from the tibias of 1,25(OH)$_2$D$_3$-treated and control rats. Right tibias were removed at autopsy and sectioned into proximal epiphysis and two serial 0.5-cm transverse segments from the proximal growth plate. Segments were demineralized and analyzed for calcium as described [17]. White bars, untreated controls ($n = 15$); striped bars, warfarin-treated ($n = 13$); dotted bars, 1,25(OH)$_2$D$_3$-treated ($n = 15$); solid bars, 1,25(OH)$_2$D$_3$- and warfarin-treated ($n = 13$). [From Price and Sloper [17] with permission.]

The efficacy of the warfarin protocol was also evaluated by measuring the fraction of normally γ-carboxylated BGP in serum by the hydroxyapatite binding assay. Serum was obtained from animals treated with 1,25(OH)$_2$D$_3$ plus warfarin for 3 days and analyzed by the hydroxyapatite binding assay as described [8]. Warfarin treatment reduced the levels of serum BGP which binds to hydroxyapatite, and therefore presumably accumulates in bone, to 2% of normal.

To better assess the nature of the response of metaphyseal bone to 1,25(OH)$_2$D$_3$, the insoluble metaphyseal residue which remained after demineralization for the calcium analysis shown in Fig. 4 was analyzed for 4-hydroxyproline and aspartic acid with an amino acid analyzer. These analyses showed 1,25(OH)$_2$D$_3$ increased total 4-hydroxyproline content of the demineralized metaphyseal bone residue from 2.6($\pm$ 0.6) to 4.3($\pm$ 0.6) μmoles ($p < 0.001$) and the total aspartic acid content from 4.0($\pm$ 0.7) to 4.7($\pm$ 0.06) μmoles ($p < 0.01$). These results indicate that the collagen content of the isoluble metaphyseal residue was increased by over 60% in the 1,25(OH)$_2$D$_3$-treated rat. Since the total calcium content of the metaphyseal bone segments was not elevated by 1,25(OH)$_2$D$_3$ treatment (Fig. 4), the metaphyseal matrix in the 1,25(OH)$_2$D$_3$-treated rat appears to be poorly mineralized.

General Discussion

It seems probable the effect of warfarin on the response of bone to 1,25(OH)$_2$D$_3$ is due to its effect on Gla synthesis in BGP. Not only is the activity of warfarin as a vitamin K antagonist well established, the warfarin protocol itself was developed to reduce bone levels of BGP to

2% of normal [18]. In addition, the dosage of 1,25(OH)$_2$D$_3$ used in these experiments was chosen so as to elicit a maximal serum BGP response [12], which serum measurements at the end of the experiment showed were in fact achieved [17]. If BGP is not the protein responsible for the warfarin effect on metaphyseal bone levels in the 1,25(OH)$_2$D$_3$-treated rat, it is necessary to postulate the existence of another vitamin K-dependent bone protein whose synthesis is also stimulated by the 1,25(OH)$_2$D$_3$ dosages employed in these experiments.

Two general mechanisms could account for BGP's action to enhance the amount of metaphyseal bone of the 1,25(OH)$_2$D$_3$-treated rat. BGP could stimulate metaphyseal bone synthesis perhaps by recruiting or activating bone osteoblasts. If the degree of increased metaphyseal bone formation exceeded the rate of 1,25(OH)$_2$D$_3$-induced bone resorption, then at the end of the 1,25(OH)$_2$D$_3$ treatment period the metaphyseal mineral levels would be higher than those in the diaphysis, where the 1,25(OH)$_2$D$_3$-induced increased resorption would not be opposed by a concomitant increase in new synthesis. Partial support for this model is provided by the observation that collagenous bone matrix levels in the metaphysis of the 1,25(OH)$_2$D$_3$-treated rat were increased by over 60% (see above). The alternative mechanism is that BGP selectively blocks osteoclastic resorption of bone in the metaphysis without affecting 1,25(OH)$_2$D$_3$-induced bone resorption in the diaphysis.

Both mechanisms imply that BGP has an effect on the activity of bone cells, which could either be direct, with BGP itself interacting with the bone cell, or indirect, with the change in bone cell activity a secondary consequence of a BGP-induced alteration in the bone matrix. There is presently no clear evidence to distinguish between these possibilities. BGP is a chemotactic signal for monocytes [19], the presumed precursor for osteoclasts, as well as for an osteosarcoma cell line with a highly osteoblastic phenotype [20]. These observations suggest that the protein directly interacts with bone cells in enhancing the amount of bone in the 1,25(OH)$_2$D$_3$-treated rat. An indirect action of BGP is also possible. The major effect of 1,25(OH)$_2$D$_3$ on the metaphysis is to increase the total levels of bone matrix without a concomitant increase in the amount of bone mineral. The role of BGP, therefore, might be to retard mineralization of this bone matrix. If the degree of mineralization is important for osteoclastic bone resorption, then warfarin treatment could, by removing the BGP block to mineralization, permit metaphyseal bone to be mineralized to an extent sufficient to enable osteoclasts to resorb it.

This explanation has the advantage of linking the present warfarin action with the previously reported effects of warfarin on the proximal tibial growth plate of 8-month-old rats [21]. These effects include the excessive mineralization of the proximal tibial growth cartilage and complete cessation of longitudinal growth. The mechanism postulated to explain this growth plate closure disorder was that warfarin, by antagonizing the vitamin K-dependent formation of Gla in BGP, released a BGP block to the spontaneous propagation of hydroxyapatite in the growth cartilage [21]. This hypothesis was supported by the earlier observation that BGP is a potent inhibitor of hydroxyapatite formation [4]. This inhibition is abolished when the vitamin K-dependent modification of BGP is reversed by the specific thermal decarboxylation of Gla to glutamic acid [7].

If the elevated levels of BGP in the 1,25(OH)$_2$D$_3$-treated animals reduce the degree of bone mineralization, the Ca^{2+} normally deposited in bone would then be available to restore serum Ca^{2+} levels. BGP would then indeed be implicated as a mediator of a calcemic hormone in bone.

ACKNOWLEDGMENT. This research was supported in part by USPHS Grants AM-25921 and AM-27029.

References

1. Price, P. A.; Poser, J. W.; Raman, N. Primary structure of the gamma-carboxyglutamic acid-containing protein from bovine bone. *Proc. Natl. Acad. Sci. USA* **73**:3374–3375, 1976.
2. Price, P. A. Osteocalcin. In: *Bone and Mineral Research Annual*, Volume 1, W. A. Peck, ed., Amsterdam, Excerpta Medica, 1983, pp. 157–190.
3. Hauschka, P. V.; Lian, J. B.; Gallop, P. M. Direct identification of the calcium-binding amino acid, γ-carboxyglutamate, in mineralized tissue. *Proc. Natl. Acad. Sci. USA* **72**:3925–3929, 1975.
4. Price, P. A.; Otsuka, A. S.; Poser, J. W.; Kristaponis, J.; Raman, N. Characterization of a γ-carboxyglutamic acid-containing protein from bone. *Proc. Natl. Acad. Sci. USA* **73**:1447–1451, 1976.
5. Price, P. A.; Nishimoto, S. K. Radioimmunoassay for the vitamin K-dependent protein of bone and its discovery in plasma. *Proc. Natl. Acad. Sci. USA* **77**:2234–2238, 1980.
6. Poser, J. W.; Esch, F. S.; Ling, H. C.; Price, P. A. Isolation and sequence of the vitamin K-dependent protein from human bone: Undercarboxylation of the first glutamic acid residue. *J. Biol. Chem.* **255**:8685–8691, 1980.
7. Poser, J. W.; Price, P. A. A method for decarboxylation of gamma-carboxyglutamic acid in proteins: Properties of the decarboxylated gamma-carboxyglutamic acid protein from calf bone. *J. Biol. Chem.* **254**:431–436, 1979.
8. Price, P. A.; Williamson, M. K.; Lothringer, J. W. Origin of the vitamin K-dependent bone protein found in plasma and its clearance by kidney and bone. *J. Biol. Chem.* **256**:12760–12766, 1981.
9. Nishimoto, S. K.; Price, P. A. Proof that the γ-carboxyglutamic acid-containing bone protein is synthesized in calf bone. *J. Biol. Chem.* **254**:437–441, 1979.
10. Nishimoto, S. K.; Price, P. A. Secretion of the vitamin K-dependent protein of bone by rat osteosarcoma cells: Evidence for an intracellular precursor. *J. Biol. Chem.* **255**:6579–6583, 1980.
11. Price, P. A.; Baukol, S. A. 1,25-Dihydroxyvitamin D_3 increases synthesis of the vitamin K-dependent bone protein by osteosarcoma cells. *J. Biol. Chem.* **255**:11660–11663, 1980.
12. Price, P. A.; Baukol, S. A. 1,25-Dihydroxyvitamin D_3 increases serum levels of the vitamin K-dependent bone protein. *Biochem. Biophys. Res. Commun.* **99**:928–935, 1981.
13. Price, P. A.; Williamson, M. K.; Baukol, S. A. The vitamin K-dependent bone protein and the biological response of bone to 1,25 dihydroxyvitamin D_3. In: *The Chemistry and Biology of Mineralized Connective Tissues*, A. Veis, ed., Amsterdam, Elsevier/North-Holland, 1981, pp. 327–335.
14. Price, P. A.; Williamson, M. K.; Baukol, S. A. The vitamin K-dependent bone protein and the vitamin D-dependent biochemical response of bone to dietary calcium deficiency. In: *Vitamin D: Chemical, Biochemical and Clinical Endocrinology of Calcium Metabolism*, A. W. Norman, K. Schaefer, D. von Herrath, and H.-G. Grigoleit, eds., Berlin, de Gruyter, 1982, pp. 351–361.
15. Buckley, M; Bronner, F. Calcium-binding protein biosynthesis in the rat: Regulation by calcium and 1,25-dihydroxyvitamin D_3. *Arch. Biochem. Biophys.* **202**:235–241, 1980.
16. Rader, J. I.; Baylink, D. J.; Hughes, M. R.; Safilian, E. F.; Haussler, M. R. Calcium and phosphorus deficiency in rats: Effects of PTH and 1,25-dihydroxyvitamin D_3. *Am. J. Physiol.* **236**:E118–E122, 1979.
17. Price, P. A.; Sloper, S. A. Concurrent warfarin treatment further reduces bone mineral levels in 1,25-dihydroxyvitamin D_3-treated rats. *J. Biol. Chem.* **258**:6004–6007, 1983.
18. Price, P. A.; Williamson, M. K. Effects of warfarin on bone: Studies on the vitamin K-dependent protein of rat bone. *J. Biol. Chem.* **256**:12754–12759, 1981.
19. Malone, J. D.; Teitelbaum, S. L.; Griffin, G. L.; Senior, R. M.; Kahn, A. J. Recruitment of osteoclast precursors by purified bone matrix constituents, *J. Cell Biol.* **92**:227–235, 1982.
20. Mundy, G. R.; Poser, J. W. Chemotactic activity of the γ-carboxyglutamic acid containing protein in bone, *Calcif. Tissue Int.* **35**:164–169, 1983.
21. Price. P. A.; Williamson, M. K.; Haba, T.; Dell, R. B.; Jee, W. S. S. Excessive mineralization with growth plate closure in rats on chronic warfarin treatment. *Proc. Natl. Acad. Sci. USA* **79**:7734–7738, 1982.

57

Prostaglandins as Mediators of Bone Cell Metabolism

Rosemary Dziak

Prostaglandins are unsaturated, hydroxylated, 20-carbon fatty acids consisting of a five-membered ring and two aliphatic side chains. They are formed from essential polyunsaturated fatty acids and appear to be ubiquitous throughout the body [28]. Most of the prostaglandin biosynthetic activity occurs within or adjacent to the cellular membrane. The prostaglandins synthesized are not stored in the cells but appear to be released immediately after synthesis.

Prostaglandin Effects in Bone

Bone can metabolize both endogenous and exogenous arachidonic acid via the cyclooxygenase pathway to the prostaglandins E_2 (PGE_2) and F_2 as well as to prostacyclin (PGI_2) [15,30]. These prostaglandins have been widely implicated in the regulation of bone cell metabolism. They stimulate osteoclastic bone resorption directly by a process that is inhibitable by calcitonin [2,11] and at relatively high concentrations decrease bone collagen formation [14]. Prostaglandins have been proposed as mediators for hypercalcemia and increased bone resorption in certain human and animal malignancies [26,27], bone loss in chronic periodontal disease [7], and rheumatoid arthritis [19].

Prostaglandin Synthesis in Bone

Increased endogenous synthesis of prostaglandins occurs in fetal rat long bone cultured in the presence of complement and antibody to cell surface antigens [22], in neonatal mouse calvaria treated with phorbol esters, mellitin, as well as epidermal growth factor [25], and may be responsible for the subsequent resorption of bone that occurs with these treatments

---

Rosemary Dziak • Department of Oral Biology, State University of New York at Buffalo, Buffalo, New York 14226.

Table I. Prostaglandin Synthesis in Bone Cell Populations[a]

	PGE_2 released into media (pg/10^6 cells)	
	Control	Indomethacin (10^{-6} M)
Osteoclastic	469.5 ± 2.4	222.8 ± 2.7**
Osteoblastic	925.0 ± 98.3*	473.4 ± 43.7**

[a]Calvaria from 1- to 2-day newborn rats are subjected to four 15-min collagenase digestions at 37°C [23]. The cells released during the first two digestions were characterized osteoclastic on the basis of high acid phosphatase levels and ability to resorb bone *in vitro* [23]. The cells released during the last two digestions were characterized osteoblastic on the basis of their alkaline phosphatase levels [23], their ability to synthesize collagen and form bone *in vitro* [24]. Following isolation and washing, cells were suspended in Eagle's MEM with bovine serum albumin (fraction V) (4 mg/ml) and maintained as suspension cultures for 48 hr at 37°C. Cells were separated from media by centrifugation. Prostaglandins were extracted from aliquots of the cell-free media by a mixture of ethyl acetate, isopropanol, and hydrochloric acid (3:3:1). PGE_2 levels were determined using a commercially available radioimmunoassay employing an antibody specific for PGE_2 (Seragen Inc., Boston, Mass.). Values are the mean ± S.E. of three samples. *, significantly greater than osteoclastic cell levels; **, significantly different from control levels of respective cell type: $p < 0.005$, Student's *t* test, difference between means.

[25]. Mechanical forces also increase prostaglandin levels in bone and thus these agents may also have a role in remodeling [9].

Information on the nature of the cells synthesizing prostaglandins and those responding to the increased levels has been difficult to obtain from intact bone because of the cellular heterogeneity of the tissue. In recent years, however, studies with isolated, characterized cell populations have helped to elucidate the possible mechanism of prostaglandin action on bone. Osteosarcoma cells with osteoblastic characteristics synthesize relatively large amounts of PGE_2 which can function as a bone resorptive factor [21]. We have been studying prostaglandin synthesis and action in enriched populations of osteoblastic and osteoclastic cells obtained by an adaptation of the sequential collagenase digestion technique of Luben *et al.* [12]. Our studies reveal that osteoblastic cells synthesize significantly greater amounts of PGE_2 than the osteoclastic-enriched populations (Table I).

PGE_2 Binding in Bone Cells

Although calvarial cells with osteoclastic characteristics have a relatively low prostaglandin synthetic ability, they possess specific binding sites for exogenous PGE_2 [5]. Binding studies performed by incubating cells with [^{3}H]-PGE_2 reveal that osteoclastic calvarial cells have a significantly greater binding capacity for exogenous PGE_2 than the osteoblastic cells (Fig. 1). This binding in calvarial cells satisfies the basic requirements of saturability, reversibility. and specificity generally recognized for specific receptors. The increased PGE_2 binding capacity in the osteoclastic-enriched cell preparation is consistent with the relative effects of PGE_2 on resorption and collagen synthesis in bone organ culture studies. PGE_2 at a concentration of 10^{-9} M induces osteoclastic resorption in fetal rat bone but at least a 1000-fold greater concentration is required before inhibition of osteoblastic collagen synthesis occurs [2].

PGE_2 Effects on Bone Cell cAMP and Calcium Uptake

The cellular events that occur after binding of a prostaglandin to its specific receptor are not completely understood but appear to involve calcium and cAMP. PGE_2 increases cAMP

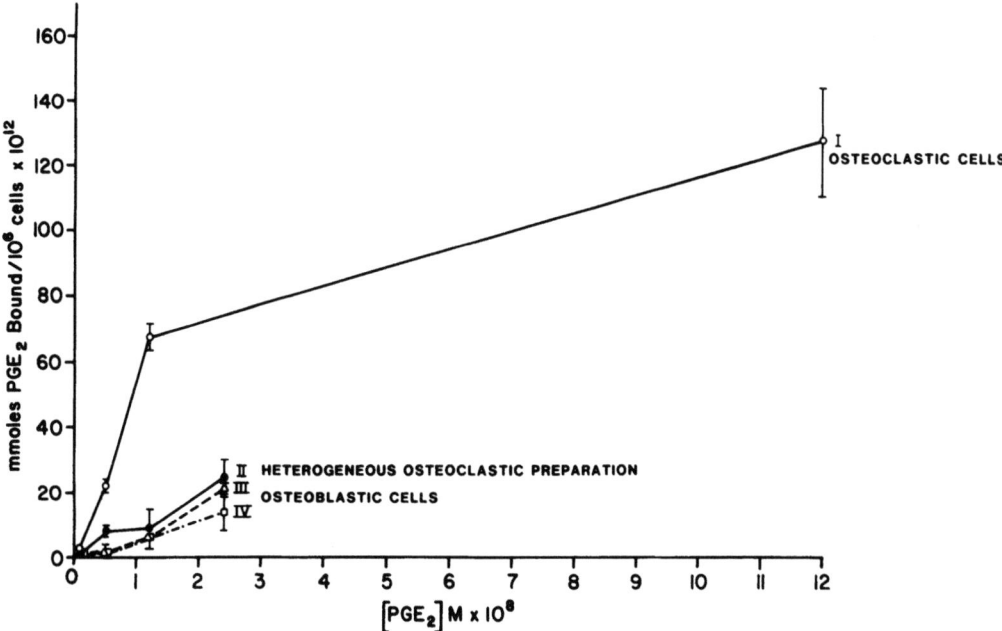

Figure 1. Binding of PGE_2 in bone cell populations. Osteoclastic and osteoblastic bone cell preparations were incubated with [^3H]-PGE_2 at various concentrations for 5 min at 37°C. Parallel incubations were done with 10^{-6} M nonradioactive PGE_2 or control solvent (0.5% ethanol) to assess specific binding. Values are the mean ± S.E. of three samples. Osteoclastic cells bound significantly more PGE_2 than osteoblastic cells at all the concentrations tested.

in both osteoclastic and osteoblastic cells, with osteoclastic cells manifesting a greater responsivity [5]. Osteoclastic-enriched populations respond significantly with an increase in cAMP to concentrations of PGE_2 as low as 10^{-9} M, while osteoblastic cells require concentrations approximately 1000-fold greater for a significant response. Likewise, the effect in osteoclastic cells is significantly greater than in osteoblastic cells at the relatively high concentration of 10^{-6} M PGE_2. PGI_2, another major prostanoid synthesized in bone, appears to act in a similar manner. Osteoclastic cells respond with an increase in cAMP to lower concentrations of PGI_2 than do osteoblastic cells [17]. As with PGE_2, the doses of PGI_2 (2.3×10^{-8}–3×10^4 M) which elicit an increase in cAMP in osteoclastic cells correlate quite well with those that are effective in stimulating bone resorption *in vitro* [15].

Calcium also appears to be involved in the initial response of bone cells to prostaglandins [6]. Under normal basal conditions, PGE_2 (10^{-7}–10^{-5} M) increases calcium uptake in osteoclastic-enriched preparations (Table II). The effect occurs within 5 min after the simultaneous addition of ^{45}Ca and PGE_2 and appears to involve calcium influx. The increase in influx is transient, and no change in the steady-state level of calcium is induced. PGI_2, PGE_1 and $PGF_{2\alpha}$ also stimulate osteoclastic calcium uptake in a similar manner, but with slightly smaller effects than PGE_2. No effects of the prostanoids have been observed in osteoblastic cells under identical incubation conditions. Inhibition of prostaglandin synthesis by incubation with a cyclooxygenase antagonist, indomethacin or flufenamic acid (cf. Table I), has no effect on calcium uptake in osteoclastic cells and causes a significant decrease in osteoblastic cells. Moreover, after treatment with these inhibitors, a highly significant calcium response to PGE_2 can be observed in osteoblastic cells (Fig. 2).

These studies suggest that endogenous prostaglandin synthesis may regulate calcium

Table II. Effect of PGE_2 on Calcium Uptake Dose-Response[a]

	Ca^{2+} uptake (nmoles $Ca^{2+}/10^6$ cells)	
	Osteoclastic cells	Osteoblastic cells
$PGE_2 10^{-5}$ M	3.59 ± 0.05* (28.2%)	1.48 ± 0.26
10^{-6} M	3.45 ± 0.03* (23.2%)	1.26 ± 0.11
10^{-7} M	3.33 ± 0.06* (18.9%)	1.18 ± 0.12
10^{-8} M	2.92 ± 0.15	1.09 ± 0.13
Control	2.79 ± 0.09	1.08 ± 0.21

[a]Bone cells were preincubated for 15 min at 37°C before the simultaneous addition of the indicated PGE_2 and ^{45}Ca or ethanol (final concentration 1%) and ^{45}Ca (control). Incubation continued for 5 min at 37°C. Cells were collected on Millipore filters and counted by liquid scintillation spectrometry. Values are the mean ± S.E. of three samples.
*$p < 0.005$, Student's t test, difference between means. Numbers in parentheses indicate the percent increase compared to control means. Significant PGE_2-induced increases in calcium uptake occurred only in osteoclastic cells.

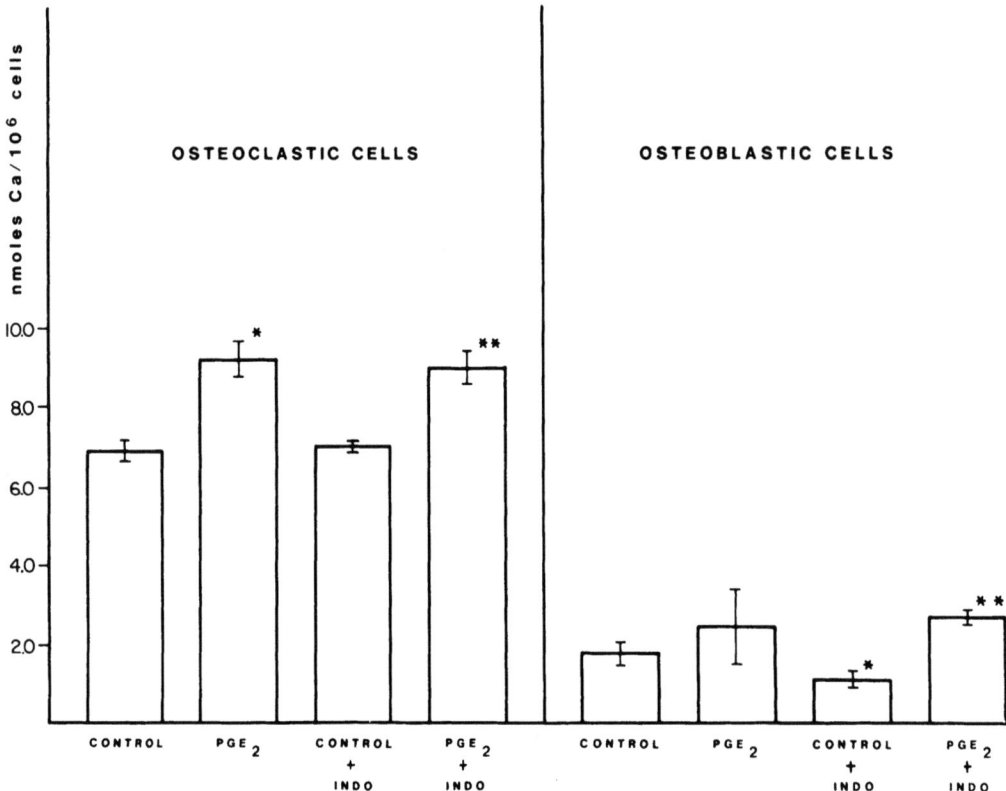

Figure 2. Effect of indomethacin on PGE_2-stimulated calcium uptake. Bone cells were preincubated for 1 hr at 37°C in the presence or absence of indomethacin (INDO) at 10^{-6} M. Controls received an appropriate amount of ethanol. Cells were then incubated for 5 min at 37°C in the presence of PGE_2 (10^{-6} M) and ^{45}Ca or ethanol and ^{45}Ca (controls). Values are the mean ± S.E. of three samples. *, significantly different from controls; **, significantly different from controls + INDO = $p < 0.005$, Student's t test, difference between means. In osteoclastic cells, INDO had no effect on calcium uptake and PGE_2 induced its characteristic increase even in the presence of the cyclooxygenase antagonist. In osteoblastic cells, incubation with INDO decreased control calcium uptake. After treatment with INDO, exogenous PGE_2 produced an increase in osteoblastic cell calcium uptake but no effect of PGE_2 was seen in cells of this type not treated with the INDO. Preincubation with 10^{-6} M flufenamic acid produced results similar to those with INDO (data not shown).

movements across osteoblastic cell membranes. A negative feedback control may exist between the effect on calcium uptake of exogenous prostaglandin and that of endogenous prostaglandin synthesis. When endogenous prostaglandin synthesis is low, the cells respond to exogenous prostaglandin with an increase in calcium uptake; but when endogenous synthesis is high, the response of the calcium regulatory system to exogenous levels of prostaglandins is low. However, studies using cyclooxygenase antagonists suggest that endogenous prostaglandin synthesis does not influence cAMP levels in either osteoclastic or osteoblastic cells [6]. These data support our previous observations [3] that calcium and cAMP metabolism in bone cells are not interdependent processes.

The calcium iontophoretic effects of the prostaglandins observed in both osteoclastic and osteoblastic cells are consistent with the observations that these agents modify cellular calcium movements in several tissues including the heart, skin [10], stomach [16], and uterus [29]. Thus, prostaglandin-induced increase in calcium uptake may be involved in mediating the resorptive effects of these compounds [4]. This concept is supported by a study with cultured mouse calvaria showing that the calcium antagonist D600 can inhibit the resorptive effects of PGE_2 [31]. Furthermore, another bone resorber, parathyroid hormone, induces increases in osteoclastic calcium uptake similar to those caused by the prostaglandins [23], while inhibitors of resorption, such as calcitonin [23] and BL-5583 [18], depress this parameter. The mechanism by which changes in calcium transport control bone cell metabolism has not yet been elucidated. However, the intracellular calcium regulatory protein calmodulin [13] appears to be involved in the mediation of the effects of PGE_2 on osteoclastic cell calcium. Preincubation with the calmodulin antagonist trifluoperazine prevents the characteristic effect of PGE_2 on calcium uptake without affecting basal calcium levels [1].

Conclusions

Although the osteoclast appears to be directly responsible for bone resorption [8], osteoblasts may also participate in the regulation of this process [20]. Osteoblasts may act as message-transducing cells which coordinate hormone signals and release resorbing factors. Our studies with osteoblastic calvarial cells, and those of Rodan and his colleagues with osteoblastic osteosarcoma cells [21] suggest that endogenous synthesis of PGE_2 may be an important regulator of resorption. The factors controlling the production of PGE_2 by osteoblastic cells *in situ* still remain elusive. However, the data reviewed herein suggest that prostaglandins act as a coupling factor between osteoblastic and osteoclastic activity and serve as potent regulators of bone cell metabolism.

ACKNOWLEDGMENTS. Supported by NIH Grant AM-25271 and an award from the Arthritis Foundation (W.N.Y. Chapter). The author is the recipient of a Research Career Development Award, K04-AM-00726-01, from the NIH.

References

1. Brown, M. J.; Weinfeld, N.; Dziak, R. Effect of calmodulin antagonists on bone cell calcium and cyclic AMP. In: Proceedings of the 5th annual meeting, American Society for Bone and Mineral Research, San Antonio, 1983.
2. Dietrich, J. W. Stimulation of bone resorption by various prostaglandins in organ culture. *Prostaglandins* **10**:231–240, 1975.

3. Dziak, R.; Stern, P. H. Calcium transport in isolated bone cells. III. Effects of parathyroid hormone and cyclic 3',5' AMP. *Endocrinology* **97:**1281–1287, 1975.
4. Dziak, R.; Stern, P. H. Response of fetal rat bone cells and bone organ cultures to the ionophore, A23187. *Calcif. Tissue Res.* **22:**137–147, 1976.
5. Dziak, R. M.; Hurd, D.; Miyasaki, K. T.; Brown, M.; Weinfeld, N.; Hausmann, E. Prostaglandin E_2 binding and cyclic AMP production in isolated bone cells. *Calcif. Tissue Int.* **35:**243–249, 1983.
6. Farr, D.; Pochal, W.; Brown, M.; Shapiro, E.; Weinfeld, N.; Dziak, R. Effects of prostaglandins on bone cell calcium. *Arch. Oral Biol.* in press.
7. Goodson, J. M.; Dewhirst, F. E.; Bruneth, A. Prostaglandin E_2 levels and human periodontal disease. *Prostaglandins* **6:**81–85, 1974.
8. Hancox, N. H. The Osteoclast. In: *The Biochemistry and Physiology of Bone*, 2nd ed., Volume 1, G. H. Bourne, ed., New York, Academic Press, 1972, pp. 45–46.
9. Harell, A.; Dekel, S.; Binderman, I. Biochemical effect of mechanical stress on cultured bone cells. *Calcif. Tissue Res. Suppl.* **22:**202, 1977.
10. Klaus, W.; Piccini, F. Uber die wirkung von prostaglandin E_1 auf der Ca-Haushalt isoberter Meerschweinherzen. *Experientia* **23:**556–557, 1976.
11. Klein, D. C.; Raisz, L. G. Prostaglandin: Stimulation of bone resorption in tissue culture. *Endocrinology* **86:**1436–1440, 1970.
12. Luben, R. A.; Wong, G. L.; Cohn, D. V. Biochemical characterization with parathormone and calcitonin of isolated bone cells: Provisional identification of osteoclasts and osteoblasts. *Endocrinology* **99:**526–534, 1976.
13. Means, A. R.; Dedman, J. R. Calmodulin—an intracellular Ca receptor. *Nature (London)* **285:**73–77, 1980.
14. Raisz, L. G.; Koolemans-Beynen, A. R. Inhibition of bone collagen synthesis by prostaglandin E_2 in organ culture. *Prostaglandins* **8:**377–385, 1974.
15. Raisz, L. G.; Vanderhoek, J. V.; Simmons, H. A.; Kream, B. E.; Nicolaou, K. C. Prostaglandin synthesis by fetal rat bone in vitro: Evidence for a role of prostacyclin. *Prostaglandins* **17:**905–914, 1979.
16. Ramwell, P. W.; Shaw, J. E. Biological significance of the prostaglandins. *Recent Prog. Horm. Res.* **26:**139–187, 1970.
17. Robin, J. C.; Brown, M. J.; Weinfeld, N.; Dziak, R. Prostacyclin: Effects on cyclic AMP in bone cells. *Res. Commun. Chem. Pathol. Pharmacol.* **35:**43–49, 1982.
18. Robin, J. C.; Brown, M. J.; Weinfeld, N.; Dziak, R. M. Benzo (B) thiophene-2-carboxylic acid: Calcium uptake and cyclic AMP production in isolated bone cells. *Calcif. Tissue Int.* **36:**194–199, 1984.
19. Robinson, D. R.; Tashjian, A. H., Jr.; Levine, L. Prostaglandin-stimulated bone resorption by rheumatoid synovia: A possible mechanism for bone destruction in rheumatoid arthritis. *J. Clin. Invest.* **56:**1181–1188, 1975.
20. Rodan, G. A.; Martin, T. J. The role of osteoblasts in the hormonal control of bone resorption: A hypothesis. *Calcif. Tissue Int.* **33:**349–351, 1981.
21. Rodan, S. B.; Rodan, G. A.; Simmons, H. A.; Walenga, R. A.; Feinstein, M. B.; Raisz, L. G. Bone resorptive factor produced by osteosarcoma cells with osteoblastic features is PGE_2. *Biochem. Biophys. Res. Commun.* **102:**1358–1365, 1981.
22. Sandberg, A. L.; Raisz, L. G.; Goodson, J. M.; Simmons, H. A.; Mergenhagen, S. E. Inhibition of bone resorption by the classical and alternative C pathways and its mediation by prostaglandins. *J. Immunol.* **119:**1378–1381, 1977.
23. Shlossman, M.; Brown, M.; Shapiro, E.; Dziak, R. Calcitonin effects on isolated bone cells. *Calcif. Tissue Int.* **34:**190–196.
24. Simmons, D.; Kent, G. N.; Jilka, R. L.; Scott, D. M.; Fallon, M.; Cohn, D. V. Formation of bone by isolated, cultured osteoblasts in Millipore diffusion chambers. *Calcif. Tissue Int.* **34:**291–294, 1982.
25. Tashjian, A. H., Jr. Prostaglandin as local mediators of bone resorption. In: *Proceedings, Mechanisms of Localized Bone Loss*, J. R. Horton, T. M. Tarpley, and W. F. Davis, eds., Supplement to Calcified Tissue Abstracts, Arlington, IRL, 1978, pp. 173–179.
26. Tashjian, A. H.; Voelkel, E. F.; Levine, L.; Goldhaber, P. Evidence that the bone resorption stimulating factor produced by mouse fibrosarcoma cells is prostaglandin E_2: A new model for the hypercalcemia of cancer. *J. Exp. Med.* **136:**1329–1343, 1972.
27. Tashjian, A. H., Jr.; Voelkel, E. F.; Goldhaber, P.; Levine, L. Prostaglandins, calcium metabolism and cancer. *Fed. Proc.* **33:**81–86, 1974.
28. Van Dorp, V. A.; Beerthius, R. K.; Nugteren, D. H.; Vonkeman, H. The biosynthesis of prostaglandins. *Biochim. Biophys. Acta* **90:**204–206, 1974.
29. Villani, F.; Chiarro, A.; Cristalli, S.; Piccini, F. Effect of PGE_2 on the turnover of calcium in rat uterus. *Experientia* **303:**532–534, 1974.

30. Voelkel, E. F.; Tashjian, A. H.; Levine, L. Cyclooxygenase products of arachidonic acid metabolism by mouse bone in organ culture. *Biochim. Biophys. Acta* **620**:418–428, 1980.
31. Yu, J.; Wells, H.; Ryan, W. J.; Lloyd, W. S. Effects of prostaglandins and other drugs on the cyclic AMP content of cultured bone cells. *Prostaglandins* **12**:501–504, 1976.

58

Interactions of Calcemic Hormones and Divalent Cation Ionophores on Fetal Rat Bone in Vitro

Paula H. Stern

The ionophore A23187, which has been widely used to assess the role of divalent cations in physiologic processes [1,2], has been a difficult agent to work with due to the variability in response [2–5] and the narrow range of stimulatory concentrations [6,7]. We previously found that A23187 mimics some of the effects of parathyroid hormone (PTH) on bone, increasing the release of previously incorporated ^{45}Ca and promoting the appearance of multinucleated osteoclasts [7–9]. Higher concentrations of the ionophore inhibited PTH-induced resorption. We now have observed a second and somewhat different pattern of response, a marked potentiation by the ionophore of the effects of low concentrations of calcemic hormones. Both response patterns are shown here and a possible explanation for the variable effects of the agent is proposed.

Additive Interactions

Experiments were carried out on fetal rat limb bones, prelabeled with ^{45}Ca and maintained in culture for 48–72 hr. Details of procedure are described elsewhere [10]. A23187 stimulated bone resorption in some bone cultures, and parallel effects on ^{45}Ca release, bone ^{40}Ca, and bone collagen hydroxyproline were obtained (Fig. 1). Although resorption was stimulated by A23187, PTH, and 1,25(OH)$_2$D$_3$ in these studies (Table I), there was no evidence for synergism when A23187 and PTH or 1,25(OH)$_2$D$_3$ were added in combination, suggesting that the stimulators were acting independently in a purely additive manner.

Potentiation

In other experiments a different pattern of response was observed. In experiment A, Table II, for example, neither A23187 (10^{-7} to 3×10^{-7} M) nor 10^{-9} M PTH alone

---

Paula H. Stern • Department of Pharmacology, Northwestern University Medical and Dental Schools, Chicago, Illinois 60611.

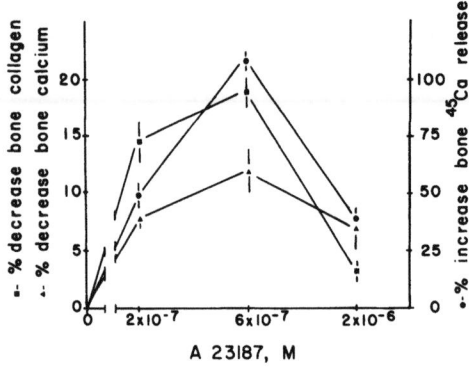

Figure 1. Responses to A23187 of fetal rat limb bones prelabeled with ^{45}Ca, and maintained in culture for 72 hr. Values are means of the responses of six bones per point. Parameters measured were: ●, % increase in bone ^{45}Ca release; ■, % decrease in bone collagen; ▲, % decrease in bone calcium. ^{45}Ca release was measured by liquid scintillation spectrometry and total calcium was determined by a Fiske calcium titrator. Changes in bone collagen were assessed colorimetrically after hydrolysis [11].

stimulated resorption, whereas combinations of PTH with these concentrations of ionophore gave significant stimulation. A23187 (2×10^{-7} and 3×10^{-7} M) also enhanced the stimulatory effects of $1,25(OH)_2D_3$. Potentiation of subthreshold concentrations of PTH and PGE_2 by 2×10^{-7} M A23187 is also shown in Table II, experiment B. When higher hormone concentrations were used, potentiation was generally not observed (Fig. 2). When A23187 itself produced resorption as in Table I, potentiation was usually not seen. Two other divalent cation ionophores, ionomycin and 4Br-A23187, had similar potentiating effects on PTH- and $1,25(OH)_2D_3$-induced ^{45}Ca release at concentrations in the 6×10^{-8} to 10^{-6} M range (data not shown).

Further experiments were carried out to determine whether the potentiation was reflected in intermediate events in bone metabolism. β-Glucuronidase was used to assess effects on lysosomal enzyme release. Parallel potentiation was observed on ^{45}Ca release and

Table I. Additive Interactions between A23187 and PTH or $1,25(OH)_2D_3$[a]

Expt.	Treatment	% bone ^{45}Ca released	Δ from control Observed	Expected
A	None	21 ± 2		
	A23187, 10^{-7} M	36 ± 4*	15	
	A23187, 2×10^{-7} M	46 ± 7**	25	
	PTH, 3×10^{-9} M	48 ± 6**	27	
	$1,25(OH)_2D_3$, 10^{-11} M	44 ± 6**	23	
	PTH + A23187, 10^{-7} M	61 ± 9***	40	42
	$1,25(OH)_2D_3$ + A23187, 10^{-7} M	52 ± 9***	31	38
	$1,25(OH)_2D_3$ + A23187, 2×10^{-7} M	68 ± 10***	47	48
B	None	26 ± 1		
	A23187, 2×10^{-7} M	31 ± 3	5	
	A23187, 6×10^{-7} M	33 ± 3*	7	
	$1,25(OH)_2D_3$, 2×10^{-11} M	39 ± 4*	13	
	$1,25(OH)_2D_3$ + A23187, 2×10^{-7} M	44 ± 3**	18	18
	$1,25(OH)_2D_3$ + A23187, 6×10^{-7} M	40 ± 3*	14	20

[a]Values are means ± S.E.; medium was BGJ + albumin. Experiment A, $n = 4$; experiment B, $n = 6$. *$p < 0.05$, **$p < 0.01$, ***$p < 0.001$, compared to control.

Table II. Potentiation of Effects of Calcemic Hormones by A23187[a]

Expt.	Treatment	% bone ^{45}Ca released	Δ from control Observed	Δ from control Expected
A	Control	19 ± 2		
	A23187, 10^{-7} M	18 ± 2	−1	
	A23187, 2×10^{-7} M	16 ± 1	−3	
	A23187, 3×10^{-7} M	16 ± 1	−3	
	PTH, 10^{-9} M	29 ± 3	10	
	1,25(OH)$_2$D$_3$, 10^{-10} M	51 ± 7***	32	
	PTH + A23187, 10^{-7} M	32 ± 6*	13	9
	PTH + A23187, 2×10^{-7} M	54 ± 5***,†††	35	7
	PTH + A23187, 3×10^{-7} M	51 ± 6***,†††	32	7
	1,25(OH)$_2$D$_3$ + A23187, 10^{-7} M	42 ± 6***	23	31
	1,25(OH)$_2$D$_3$ + A23187, 2×10^{-7} M	83 ± 4***,†††	64	29
	1,25(OH)$_2$D$_3$ + A23187, 3×10^{-7} M	61 ± 5***,†	42	29
B	Control	23 ± 1		
	A23187, 2×10^{-7} M	24 ± 1	1	
	PTH, 10^{-9} M	31 ± 3	8	
	PGE$_2$, 10^{-9} M	23 ± 1	0	
	1,25(OH)$_2$D$_3$, 10^{-11} M	42 ± 2**	19	
	PTH + A23187	55 ± 4***,††	32	9
	PGE$_2$ + A23187	37 ± 1*,†	14	1
	1,25(OH)$_2$D$_3$ + A23187	53 ± 4***	30	20

[a]Experiment A, DMEM + 15% horse serum, $n = 6$; experiment B, BGJ + 1 mg/ml albumin, $n = 4$. Significantly different from control: *$p < 0.05$, ***$p < 0.001$; significantly different from stimulator [PTH, 1,25(OH)$_2$D$_3$, or PGE$_2$] alone: †$p < 0.05$, ††$p < 0.01$, †††$p < 0.001$.

β-glucuronidase activity in the culture medium when PTH and A23187 were employed in combination (Table III). In contrast, potentiation of ^{45}Ca release observed when A23187 was added together with PTH or parathyroid extract (PTE) was not reflected in changes in medium cAMP (Table IV).

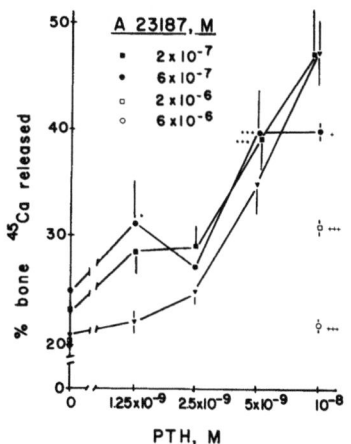

Figure 2. Effect of A23187 on resorption produced by parathyroid hormone (PTH). Results are composite findings from two experiments. Concentrations of 1.25×10^{-9} to 5×10^{-9} M PTH were studied in the first experiment. The studies with 10^{-8} H PTH derived from a second experiment. Values are means ± S.E. of the responses from six bones per point (first experiment) or three bones per point (second experiment) after 72 hr of incubation. ▼, no ionophore; ■, 2×10^{-7} M A23187; ●, 6×10^{-7} M A23187; □, 2×10^{-6} M A23187; ○, 6×10^{-6} M A23187. Significant increase vs. PTH, *$p < 0.05$ ***$p < 0.001$ significant decrease vs. PTH, †$p < 0.05$ †††$p < 0.001$.

Table III. Effects of Parathyroid Hormone and A23187 on ^{45}Ca Release and β-Glucuronidase in Limb Bones[a]

Treatment	% bone ^{45}Ca released	β-Glucuronidase (U/bone)
Control	25 ± 2	0.052 ± 0.001
A23187, 2 × 10^{-7} M	26 ± 4	0.055 ± 0.002
A23187, 3 × 10^{-7} M	25 ± 2	0.051 ± 0.002
PTH, 10^{-9} M	35 ± 2	0.058 ± 0.002
PTE, 0.09 U/ml	42 ± 5*	0.063 ± 0.004**
PTH + A23187, 2 × 10^{-7} M	43 ± 4*	0.060 ± 0.003*
PTH + A23187, 3 × 10^{-7} M	37 ± 11	0.060 ± 0.004
PTE + A23187, 2 × 10^{-7} M	60 ± 4***,[†]	0.070 ± 0.005***,[†]
PTE + A23187, 3 × 10^{-7} M	65 ± 3***,[††]	0.075 ± 0.004***,[††]

[a]Medium: BGJ + 1 mg/ml albumin; n = 5. Significantly different from control: *$p < 0.05$, **$p < 0.01$, ***$p < 0.001$; significantly different from PTH or PTE alone: [†]$p < 0.05$, [††]$p < 0.01$.

Discussion

Diverse evidence suggests that the mechanism of action of bone-resorbing agents involves a calcium-dependent step. Calcium antagonists, including verapamil, D-600, cobalt, and manganese, prevent the bone-resorbing effects of PTH, PGE_2, or $1,25(OH)_2D_3$ [12–15]. Increased calcium mimics or potentiates actions of PTH on enzymatic activity of bone cells [16] or bone (Stern, unpublished). An alternative approach has been the use of divalent cation-selective ionophores. Although these agents have provided some evidence in support of a calcium-dependent mechanism in bone resorption, the findings have been somewhat equivocal. Our initial studies [7–9] showed that A23187 had a biphasic effect on resorption of fetal rat limb bones, stimulating resorption over a narrow concentration range and inhibiting at higher concentrations. Similar findings have been reported by others [17–19]. A23187 also promoted osteoclast proliferation in fetal rat limb bones [9]. However, only inhibitory effects were observed when neonatal mouse calvaria were treated with A23187 [20]. Combinations of A23187 and submaximal concentrations of calcemic hormones failed to produce potentiation in the calvarial system [20].

Table IV. Lack of Potentiation of Medium cAMP Levels by Combined Treatment with PTH and A23187[a]

Treatment	% bone ^{45}Ca released	Medium cAMP (pmole/bone)
Control	15 ± 1	0.18 ± 0.02
A23187, 2 × 10^{-7} M	16 ± 1	0.19 ± 0.01
PTH, 10^{-9} M	22 ± 2	0.26 ± 0.04*
PTH + A23187	28 ± 6*	0.27 ± 0.02*
Control	16 ± 1	0.11 ± 0.01
A23187, 3 × 10^{-7} M	16 ± 2	0.17 ± 0.01
PTE, 0.025 U/ml	32 ± 6*	0.27 ± 0.03***
PTE, 0.025 U/ml + A23187	51 ± 4***,[††]	0.29 ± 0.02***
PTE, 0.1 U/ml	76 ± 8***	0.39 ± 0.02***

[a]Medium was DMEM + 15% horse serum; n = 4–6. Significantly different from control: *$p < 0.05$, ***$p < 0.001$; significantly different from PTE alone: [††]$p < 0.01$.

In the present studies, we have shown that low concentrations of PTH, PTE, PGE$_2$, or 1,25(OH)$_2$D$_3$ in combination with divalent cation ionophores can significantly potentiate resorption in fetal rat limb bones. Effects were seen with A23187 and with the more calcium-selective ionophores, ionomycin, and 4Br-A23187 [21,22]. In some studies, subthreshold concentrations of the calcemic hormones produced significant responses when added together with ionophore. In other experiments, calcemic hormones were effective and addition of ionophore resulted in potentiation. The observations would be consistent with a rate-limiting calcium-dependent step in the action of the hormones on bone. The potentiation of ^{45}Ca release was reflected in parallel changes in β-glucuronidase in the culture medium. The PTH-induced increase in medium cAMP, however, was not enhanced. The lack of potentiation of PTH-induced increases in medium cAMP suggests that the calcium-dependent step occurs subsequent to or is independent of changes in cAMP. Studies on the effects of A23187 on bone cAMP have given disparate results. The ionophore failed to increase cAMP in neonatal mouse calvaria [20]; however, in other studies A23187 mimicked or potentiated the effect of PTH on cAMP in cells isolated from neonatal mouse calvaria [13,23].

The potentiating effect of the ionophores would seem to be qualitatively different from the additive responses which we and other investigators have reported previously and which are shown in Fig. 1 and Table I. One interpretation would be that in the studies in which the ionophore per se stimulates resorption, potentiation of the effects of an endogenous bone-resorbing agent is being observed. Subtle differences in the status of the animals or the isolated tissues may affect the concentrations of this resorbing agent. We cannot explain why potentiation has not been observed in experiments in which the ionophore itself is stimulatory. Perhaps ionophore action involves acceleration of an initiating step in the resorption process, and once activated, this step cannot be stimulated further.

It is perhaps relevant that the literature relating to studies with A23187 also reveals inconsistencies between laboratories with respect to effects of A23187 on vasopressin secretion from the neurohypophysis [24–26] and on contractility of guinea pig atrium [4,5]. Cochrane et al. emphasized the inconsistency of the effects of A23187 in the adrenal medulla by presenting individually the diverse responses obtained in 13 separate experiments [3].

Studies at the biochemical and membrane level serve to emphasize the complex actions of the divalent cation ionophores. Short-term exposure to 10^{-6}–10^{-5} M ionomycin or A23187 stimulated synthesis of specific proteins in chicken pectoralis muscle [27,28]. Concentrations of A23187 less than one order of magnitude greater resulted in muscle protein degradation [29]. Concentrations in the micromolar range increased [^{3}H]thymidine incorporation and stimulated DNA synthesis and mitosis in lymphocytes [6]. However, similar concentrations inhibited thymidine incorporation in cultured bone [18], and two- to threefold increases in A23187 inhibited thymidine incorporation in lymphocytes [6]. These findings probably reflect the small degree of alteration needed to perturb the homeostatic control exercised by intracellular calcium. In addition, compartmentalization of calcium may be affected, as reflected in flux studies [30,31]. Bone, which has a substantial reservoir of extracellular calcium, may be unusually sensitive to small changes in ionophore concentration.

We have as yet no explanation for the divergent results in calvaria and limb bones. Although calcemic hormones [PTH, PGE$_2$, and 1,25(OH)$_2$D$_3$] have essentially the same effects on the two bone types [32], the two systems may respond differently to some agents. Potassium, for example, stimulates resorption in calvaria, but not in limb bones [33]. This may be in part secondary to prostaglandin release from the calvaria. In recent studies we have found that manganese (3×10^{-6}–3×10^{-4} M) stimulates or potentiates resorption in limb bones [15], whereas higher concentrations (10^{-3} M) are inhibitory. In contrast, a given concentration of manganese stimulates resorption in calvaria when present alone but inhibits the bone-resorbing effects of calcemic hormones when added in combination [15]. Clearly, there are differences between the two systems that require further study.

The mechanism by which A23187 stimulates or potentiates resorption in limb bones is not yet known. Conceivably, it could enhance calcium fluxes from intracellular or extracellular compartments. A recent study demonstrated that A23187 potentiates the effects of the neural calcium-binding peptide S-100 and enables it to release amino acids and catecholamines from neural tissue [34]. The authors postulate that the S-100 peptide may function physiologically by interacting with an endogenous ionophore. A similar mechanism of action may be considered for calcemic hormones.

ACKNOWLEDGMENTS. These studies were supported by USPHS Grant AM-11262. The excellent technical assistance of Thalia Mavreas and Shirley Snerling is gratefully acknowledged.

References

1. Pressman, B. C. Biological applications of ionophores. *Annu. Rev. Biochem.* **45**:501–530, 1976.
2. Stern, P. H. Ionophores: Chemistry, physiology and potential applications to bone biology. *Clin. Orthop.* **122**:273–298, 1977.
3. Cochrane, D. E.; Douglas, W. W.; Mouri, T.; Nakazato, Y. Calcium and stimulus–secretion coupling in the adrenal medulla: Contrasting stimulating effects of the ionophores X-537A and A23187 on catecholamine output. *J. Physiol. (London)* **252**:363–378, 1975.
4. Holland, D. R.; Steinberg, M. I.; Armstrong, W. M. D. A23187: A calcium ionophore that directly increases cardiac contractility. *Proc. Soc. Exp. Biol. Med.* **148**:1141–1145, 1975.
5. Schwartz, A.; Lewis, R. M.; Flanley, H. G.; Munson. R. G.; Dial, F. D.; Ray, M. V. Hemodynamic and biochemical effects of a new positive inotropic agent: Antibiotic ionophore RO2-2985. *Circ. Res.* **34**:102–111, 1974.
6. Manio, V. C.; Green, M. M.; Crumpton, M. J. The role of calcium ions in initiating transformation of lymphocytes. *Nature (London)* **251**:324–327, 1974.
7. Dziak, R.; Stern, P. H. Parathyromimetic effects of the ionophore, A23187 on bone cells and organ cultures. *Biochem. Biophys. Res. Commun.* **65**:1343–1349, 1975.
8. Dziak, R.; Stern, P. J. Responses of fetal rat bone cells and bone organ cultures to the ionophore A23187. *Calcif. Tissue Res.* **22**:137–147, 1976.
9. Stern. P. J.; Orr, M. F.; Brull, E. Ionophore A23187 promotes osteoclast formation in bone organ culture. *Calcif. Tissue Int.* **34**:31–36, 1982.
10. Stern, P. H.; Phillips, T. E.; Mavreas, T. Bioassay of 1,25-dihydroxy-vitamin D in human plasma purified by partition, alkaline extraction and high-pressure chromatography. *Anal. Biochem.* **102**:22–30. 1980.
11. Cheng, P.-T. H. An improved method for the determination of hydroxyproline in rat skin. *J. Invest. Dermatol.* **53**:112–115, 1969.
12. Herrmann-Erlee, M. P. M.; Gaillard, P. J.; Hekkelman, J. W.; Nijewide, P. J. The effect of verapamil on the action of parathyroid hormone on embryonic bone *in vitro. Eur. J. Pharmacol.* **46**:51–58, 1977.
13. Yu, J. H.; Wells, H.; Ryan, W. J., Jr.; Lloyd. W. Effects of prostaglandins and other drugs on the cyclic AMP content of cultured bone cells. *Prostaglandins* **12**:501–513, 1976.
14. Lerner, U.; Gutafson, T. T. Inhibition of 12-hydroxy-vitamin D_3 stimulated bone resorption in tissue culture by the calcium antagonist verapamil. *Eur. J. Clin. Invest.* **12**:185–190, 1982.
15. Stern, P. Cationic agonists and antagonists of bone resorption. Proc. 8th Int. Conf. Calcium Regulating Hormones, Kobe, Japan, 1983.
16. Wong, G. L.; Kent, G. N.; Ku, K. Y.; Cohn, D. U. The interaction of parathormone and calcium on the hormone-regulated synthesis of hyaluronic acid and citrate decarboxylation in isolated bone cells. *Endocrinology* **103**:2274–2282, 1978.
17. Hahn, T. J.; Debartolo, T. F.; Halstead, L. R. Ouabain effects on hormonally-stimulated bone resorption and cyclic AMP content in cultured fetal rat bones. *Endocrin. Res. Commun.* **7**:189–200, 1980.

18. Lorenzo, J.; Raisz, L. G. Divalent cation ionophores stimulate resorption and inhibit DNA synthesis in cultured fetal rat bone. *Science* **212**:1157–1159, 1981.
19. Eilon, G.; Raisz, L. G. Comparison of the effects of stimulators and inhibitors of resorption on the release of lysosomal enzymes and radioactive calcium from fetal bone in organ culture. *Endocrinology* **103**:1969–1975, 1978.
20. Ivey, J. L.; Wright, D. R.; Tashjian, A. H., Jr. Bone resorption in organ culture: Inhibition by the divalent cation ionophores A23187 and X-537A. *J. Clin. Invest.* **58**:1327–1338, 1976.
21. Toeplitz, B. K.; Cohen, A. I.; Funka, P. T.; Parker, W. L.; Gougoutas, J. Z. Structure of ionomycin—A novel diacidic polyether antibiotic having high affinity for calcium ions. *J. Am. Chem. Soc.* **101**:3344–3353, 1981.
22. DeBono, M.; Molloy, R. M.; Dorman, D. E.; Paschal, J. W.; Babcock, D. F.; Deber, C. M.; Pfeiffer, D. R. Synthesis and characterization of halogenated derivatives of the ionophore A23187: Enhanced calcium ion transport specificity by the 4-bromo-derivative. *Biochemistry* **20**:6865–6872, 1981.
23. Peck, W. A.; Kohler, G.; Barr, S. Calcium-mediated enhancement of the cyclic AMP response in cultured bone cells. *Calcif. Tissue Int.* **33**:409–416, 1981.
24. Nakazato, Y.; Douglas, W. W. Vasopressin release from the isolated neurohypophysis induced by a calcium ionophore X-537A. *Nature (London)* **249**:479–481, 1974.
25. Nordmann, J. J.; Currell, G. A. The mechanism of calcium ionophore-induced secretion from the rat neurohypophysis. *Nature (London)* **253**:646–647, 1975.
26. Russell, J. T.; Hansen, E. L.; Thorn, N. A. Calcium and stimulus–secretion coupling in the neurohypophysis. III. Ca^{2+} ionophore (A23187) induced release of vasopressin from isolated rat neurohypophyses. *Acta Endocrinol. (Copenhagen)* **77**:443–450, 1974.
27. Roufa, D.; Wu. F. S.; Martonosi, A. N. The effect of Ca^{2+} ionophores upon the synthesis of proteins in cultured skeletal muscle. *Biochim. Biophys. Acta* **674**:225–237, 1981.
28. Wu, F. S.; Park, Y.-C.; Roufa, D.; Martonosi, A. N. Selective stimulation of the synthesis of an 80,000-dalton protein by calcium ionophores. *J. Biol. Chem.* **256**:5309–5312, 1981.
29. Sugden, P. The effects of calcium ions, ionophore A23187 and inhibition of energy metabolism on protein degradation in the rat diaphragm and epitrochlearis muscle *in vitro*. *Biochem. J.* **190**:593–603, 1980.
30. Warner, W.; Carchman, R. A. Effect of ionophore A23187 on calcium fluxes from cultured adrenal cells. *Biochim. Biophys. Acta* **645**:346–350, 1981.
31. Babcock, D. F.; First, N. L.; Lardy, H. A. Action of ionophore A23187 at the cellular level: Separation of effects at the plasma and mitochondrial membranes. *J. Biol. Chem.* **251**:3881–3886, 1976.
32. Stern. P. H.; Krieger. N. S. Comparison of fetal rat limb bones and neonatal mouse calvaria: Effects of parathyroid hormone and 1,25-dihydroxyvitamin D_3. *Calcif. Tissue Int.* **35**:172–176, 1983.
33. Krieger, N. S.; Stern, P. H. Potassium effects on bone: Comparison of two model systems. *Am. J. Physiol.* **245**:E303–E307, 1983.
34. Gallo, V.; Levi, G.; Raiteri, M.; Coletti, A. A nervous system specific protein potentiates the biological effects of the calcium ionophore A23187. *Life Sci.* **27**:761–769, 1980.

XIII

Alterations in Calcium Metabolism and Homeostasis

59

Altered Cell Calcium Metabolism and Human Diseases

Howard Rasmussen and Genaro M. A. Palmieri

Both genetic and environmental factors are usually considered to play interacting roles in the pathogenesis of human disease. Scant attention, however, has been paid to the possible role that evolutionary events may play in health and disease. Life on this planet began in an anaerobic atmosphere and in a sea rich in K^+, Mg^{2+}, and phosphate, but low in Na^+ and Ca^{2+}. Within this milieu, ATP appeared, along with the machinery for metabolizing glucose to lactate, alcohol, or acetate. The development of the process of oxidative phosphorylation provided the cell with large amounts of ATP for its work function. In the case of Na^+ and Ca^{2+}, these new extracellular ions became the means of providing the cell with two major new attributes: the property of excitability of its surface membrane (the Na^+ gradient), and the ability to couple events occurring at this surface membrane to the metabolic events within the cell (the calcium messenger).

Yet evolutionary and biological solutions are never perfect. A mammalian cell is susceptible to oxidative damage if its protective systems fail. It is also susceptible to ionic damage if its ability to maintain a proper distribution of calcium within its various subcellular compartments is impaired. It is this latter consideration which will be the focus of our present discussion.

The magnitude of the challenge faced by each mammalian cell to exclude calcium can be illustrated by depicting cellular calcium metabolism in its simplest aspect. The overriding fact of cellular life is that a 10,000-fold gradient of $[Ca^{2+}]$ exists across the plasma membrane of every cell [6] (Fig. 1). Even more impressive than the magnitude of this gradient is the fact that less than 1% of the basal metabolism of the cell is required to maintain it. This economical solution is achieved primarily by having a plasma membrane which is only very slightly permeable to calcium. If calcium entry into the cell were not balanced by an equal rate of efflux, the cell would become calcium overloaded in less than 10 hr [20]. This does

---

Howard Rasmussen • Departments of Cell Biology and Internal Medicine, Yale University, New Haven, Connecticut 06510.   *Genaro M. A. Palmieri* • Department of Internal Medicine, University of Tennessee Center for the Health Sciences, Memphis, Tennessee 38163.

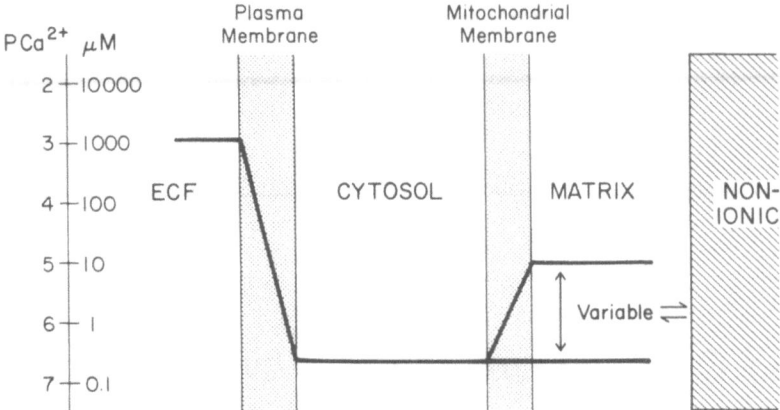

Figure 1. A schematic representation of cellular calcium metabolism. The points of emphasis are: (1) the cytosolic free calcium is 0.1 μM or less but the extracellular free calcium is 100 μM: a 10,000-fold calcium concentration gradient exists across the plasma membrane; (2) there is a cytosolic buffer pool which may be largely in the endoplasmic reticulum; and (3) there is an exchangeable calcium pool in the mitochondria which is largely nonionic and has a considerable capacity to store calcium.

not normally occur because of energy-dependent calcium efflux balances influx so that over a prolonged interval the cell neither gains nor loses calcium.

There are two mechanisms by which the cell extrudes calcium across its plasma membrane. The first is a primary calcium pump or calcium:nH$^+$-ATPase that exchanges Ca^{2+} for H$^+$ and catalyzes the uphill transport of calcium out of the cell [22]. The second is a process of secondary active transport of calcium catalyzed by a Ca^{2+}–3Na$^+$ exchange (intracellular calcium for extracellular Na$^+$) and the activity of the Na$^+$-K$^+$-ATPase [3]. In addition, to minimize risk of cellular calcium intoxication, the cell has evolved intracellular storage sites (mitochondria and endoplasmic reticulum) in which calcium can be sequestered during periods of cell activation. There is another important relationship between the plasma membrane and cellular calcium metabolism. As will be discussed below, when cells employ calcium as a second messenger in hormone or neurotransmitter action, a common event is a four- to fivefold increase in the calcium permeability of the plasma membrane [6]. This type of change is not simply a nonspecific change in permeability but represents the opening of potential- and receptor-dependent channels in the membrane. In addition to whatever messenger function this calcium influx serves, it also poses the potential threat of calcium overload and cell dysfunction [12,20].

Dyshomeostasis

Cellular calcium dyshomeostasis can occur as a consequence of several different types of change, and can lead to either acute or chronic, dramatic or subtle alterations in cell function. Such alterations can occur as a result of an excess of a physiological stimulus, an inappropriate response to a physiological stimulus, an acute toxic injury, a dietary excess, or an inherited disease (Table I).

Excess of a normal physiological stimulus. Since one common means by which hormones act is by altering cellular calcium metabolism, states of hormone excess may lead to cellular calcium dyshomeostasis. For example, cellular calcium intoxication may result from

*Table I. Representative Examples
of Altered Cell Calcium
and Human Disease*

Excess of normal physiological stimulus
   Catecholamines and the heart
   Parathyroid hormone and the kidney
   Thyroid hormone and skeletal muscle
Inappropriate response to a normal stimulus
Acute toxic injury
   Tissue anoxia
   Acute pancreatitis
   Chemical injury to hepatocytes
   Acute respiratory alkalosis
Dietary excess
   Hypertension—Na^+
   Atherosclerosis—lipids
Inherited diseases
   Duchenne's muscular dystrophy
   Cystic fibrosis

---

excessive stimulation of the heart with large doses of catecholamine [12]. Under these conditions, rates of influx and efflux of calcium are increased, but influx exceeds efflux, so there is a net accumulation of calcium [12]. The limited capacity of the calcium pool in the sarcoplasmic reticulum (SR) is soon exceeded and the bulk of the extra calcium is stored in the mitochondrial pool. If the mitochondrial pool becomes overloaded, oxidative phosphorylation is uncoupled, calcium-dependent proteases and phospholipases are activated, and autolysis of intracellular components takes place causing cell death. Anoxia and pharmacological injury can also precipitate a similar sequence of cell calcium overload (see below). This means that in a clinical situation, cardiac cell damage can result from the interplay of several of these factors [9].

Parathyroid hormone (PTH) stimulates calcium uptake and retention in cultured kidney cells [6], and the experimental induction of secondary hyperparathyroidism elevates the total renal cell content of calcium [7]. Since the acute infusion of large doses of PTH into thyroparathyroidectomized rats leads to the rapid appearance of massive nephrocalcinosis and renal failure [19], it is likely that high concentrations of PTH, together with excess phosphate, increase cell calcium uptake, and contribute to progressive cell destruction and renal failure.

A recent study of the action of thyroid hormone on muscle focuses on changes in cellular calcium metabolism as an important factor in thyroid thermogenesis [24]. Dantrolene, a drug that blocks calcium release from SR, was reported to have no significant effect on the basal rate of O_2 consumption in muscle from hypothyroid animals, but caused a 20% inhibition of O_2 consumption in muscles from euthyroid animals, and a 35% inhibition in muscles from hyperthyroid animals [24]. The increase in metabolic rate, seen either when thyroid hormone corrected the hypothyroid state to a euthyroid state, or in going from hypothyroid to hyperthyroid state, was depressed by dantrolene. The findings of this study suggested that over one-half of the thermogenic response of thyroid hormone in an intact muscle may be accounted for by increases in calcium-activated ATPase of one type or another, e.g., an increased cycling of calcium into and out of SR and/or mitochondria, or an increase in a calcium-activated futile metabolic cycle similar to that seen in bumble bee flight muscle. The point of interest in this discussion is that an abnormality of cellular calcium metabolism other than abnormal accumulation may play a crucial pathogenic role in thyrotoxicosis.

Inappropriate response to a normal stimulus. Certain disease states develop because of unrecognized and inappropriate actions of normal concentrations of a given hormone leading to another type of cell calcium intoxication. In order to understand this mechanism, it is necessary to review our current view of hormone action and our current knowledge of hormone receptor distribution (Fig. 2). Many hormones regulate cell function by stimulating, directly or indirectly, the uptake of calcium into the cell by acting principally, if not exclusively, via specific receptors on the cell surface. However, many cells, other than the classic target cells, possess receptors for a particular hormone, and many of these receptors have a lower affinity for the hormone that those receptors located specifically on the target cells. There are at least two situations in which these low-affinity, non-target-tissue receptors may become functional. The first is simply a pathological increase in hormone concentration. For example, the rise in PTH levels seen in patients with chronic renal failure could be sufficient to activate receptors in a variety of tissues, e.g., heart and skeletal muscle, bone, and kidney. The effects of the increased concentrations of PTH on these "nontarget" tissues might be unrecognized.

A second potential situation is even more subtle. A change in a nontarget cell could lead to the appearance of a population of high-affinity surface receptors for a particular hormone. Such a situation might lead to chronic or progressive changes in the function of these cells. Such a possibility is not simply theoretical. In acromegaly, TRH receptors paradoxically appear on the surface of the GH-secreting cells, and when this receptor interacts with TRH, there is an anomalous increase in GH secretion. Other examples of abnormalities in the tissue distribution of specific types of hormone receptors will probably be found that lead to cellular dysfunction.

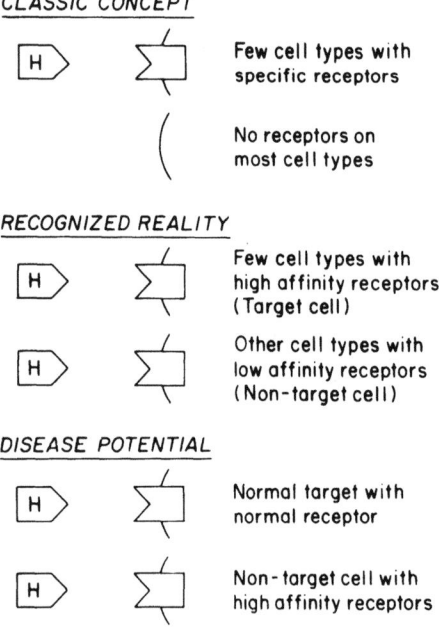

Figure 2. Abnormal hormone receptors and human disease. Depicted in the top panel is the classic view of hormone receptors in which there are only high-affinity receptors on target tissues and none on nontarget tissues. The middle panel is a representation of the new reality that receptors of lower affinity exist on the cell surfaces of cells in "nontarget" tissues. The lower panel represents a hypothetical situation in which the receptors on a classic "nontarget" tissue have a high rather than lower affinity for the particular hormone.

Acute Toxic Injury

In considering the role of calcium overload in acute toxic injury, it is important to recall that there is an extremely large [Ca^{2+}] gradient across the cell membrane and that cells have only a limited capacity to store excess calcium in a biologically inactive form. In addition, energy (in the form of ATP) is absolutely essential for maintaining cellular calcium homeostasis. Hence, an acute toxic injury to the plasma membrane can lead to calcium overload. In fact, cellular calcium overload may be a final common pathway to cell death in response to ischemia, viral infection, radiation, or chemical injury [11,21,25]. It is not possible to consider all situations in which calcium has been postulated to play such a role. Cardiac cell anoxia, chemical injury to hepatocytes, acute pancreatitis, and acute respiratory alkalosis will be discussed as examples of acute toxic injury.

Tissue anoxia. Myocardial injury secondary to tissue anoxia is widely studied because of its immediate relevancy to the process of myocardial infarction. The effect of hypoxia on cardiac cell function can be considered to result in three successive changes, depending on the duration of hypoxia and the metabolic state of the heart [16]. During phase 1 (the first 5 min), there is a rapid fall in peak developed (systolic) tension, without any significant change in resting (diastolic) tension. During phase 2 (the next 5–10 min) peak developed tension continues to fall slowly and resting tension begins to rise. During phase 3 (the next 15–30 min), active tension development no longer takes place but resting tension increases rapidly. If the muscle is reoxygenated during phase 1, complete recovery occurs; during phase 2, partial to nearly complete recovery occurs, with the extent of recovery being inversely proportional to the duration of the hypoxia; and during phase 3, recovery does not usually occur.

In phase 1, the major alteration in cellular calcium metabolism is a lack of activator calcium for initiating large contractile responses in cell in which no appreciable changes in total cell calcium occur. This change in cellular calcium metabolism can be viewed as a protective device conserving a vanishing supply of metabolic energy needed for cell survival.

During the second phase of hypoxic stress, it is postulated that energy stores become so depleted that even though calcium entry into the cell is reduced, the calcium normally stored in nonionic form in SR and mitochondria can no longer be retained by these organelles. Hence, the [Ca^{2+}]$_c$ begins to rise and contracture of the actin–myosin complex follows.

During the third phase of hypoxic stress, membrane potential falls, calcium begins to enter the cell via the reversal of the normal pathway of Na^+–Ca^{2+} exchanges, and maximal resting tension develops. Reexposure to normal O_2 concentrations at this point leads to a marked increase in the rate of calcium entry into the cell. Even though the capacity to synthesize ATP is restored, the cell is overloaded with calcium. So, an uncoupling of oxidative phosphorylation results, calcium-dependent proteases and phospholipases are activated, and cell death follows. Treatment during periods of hypoxia and recovery from hypoxia with agents that block calcium entry into the heart are effective in minimizing the consequences of calcium overload [8]. These drugs will reduce the degree or extent of irreversible damage caused by prolonged tissue hypoxia.

Acute pancreatitis. One of the most common systemic manifestations of acute pancreatitis is hypocalcemia. The exact pathogenesis of this fall in serum calcium concentration has been a matter of considerable debate [1]. Hypotheses have ranged from a lack of parathyroid hormone secretion and/or an excess of calcitonin secretion, to an accumulation of calcium in areas of fat necrosis and/or a diffuse extracellular sequestration of calcium. In dogs with experimentally induced acute pancreatitis a fall in the serum calcium concentration

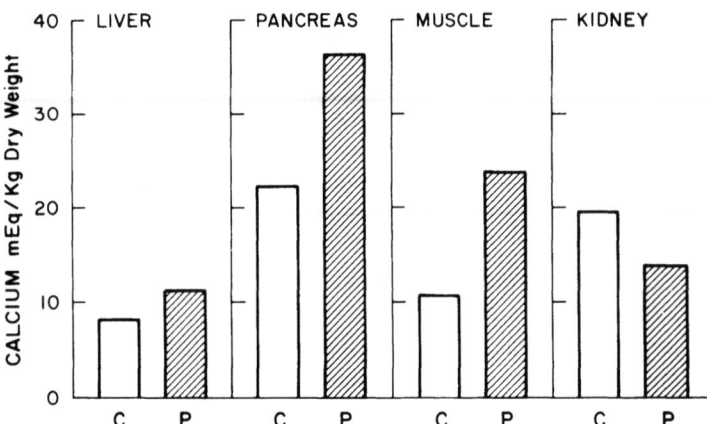

Figure 3. The change in the content of calcium in a variety of tissues following the induction of acute pancreatitis in the dog. C, control animals with pancreatitis; P, animals with pancreatitis. [Replotted from Ref. 2.]

occurs within 6 hr [2]. This fall in serum calcium is associated with an increase in the calcium content of pancreas, liver, and muscle, but a reduction in the kidney (Fig. 3). Biopsies of pancreatic tissue reveal calcium accumulation, whether or not necrosis is evident, suggesting that calcium accumulation precedes rather than occurs as a consequence of necrosis. Either some substance released from the pancreas, or a degradation product (e.g., peptide, phospholipid) resulting from the action of liberated pancreatic enzymes brings about an alteration in muscle cell membrane permeability that, in turn, leads to an accumulation of calcium by the muscle.

Chemical injury to hepatocytes. Livers of experimental animals exposed to toxic amounts of carbon tetrachloride accumulate large quantities of calcium that is sequestered in mitochondria. Recent studies with hepatocytes grown in tissue culture have shown that toxic effects of such agents as CCl_4, galactosamine, and phalloridin are diminished if calcium is excluded from the medium during the time of exposure of the cells to otherwise lethal concentrations of toxin [11]. These results have been interpreted to mean that in chemically induced hepatic cell necrosis, the uptake of abnormal quantities of calcium plays a critical role in the irreversible process of cell injury.

Although chemical injury often leads to plasma membrane injury and cellular accumulation, there are some hepatic toxins, such as menadione, which appear to interfere specifically with the retention of calcium by mitochondria and possibly the endoplasmic reticulum [23]. The loss of calcium from organelles is thought to be associated with a rise in $[Ca^{2+}]_c$ and to play a critical role in the pathogenesis of menadione-induced cytotoxicity. Thus, a common denominator in hepatic cytotoxicity appears to be an increase in the $[Ca^{2+}]_c$. This can occur because of an alteration in the structure and function of the plasma membrane or intracellular membranes such that intracellular sequestration of calcium is impaired.

Acute respiratory alkalosis. Patients who are chronic hyperventilators often have dramatic falls in pCO_2 and a consequent rise in blood pH. Associated with this change in acid–base balance are a variety of neurological and cardiovascular abnormalities, including anginalike symptoms. Coronary vasoconstriction can be induced *in vitro* by an elevation of medium pH [26]. Elevation of extracellular pH may lead to an increase in the rate of calcium influx, which, in turn, augments vascular tone. In harmony with this supposition is the finding that diltiazem, a calcium-channel blocker, prevents the alkalosis-induced increase in coronary artery tone.

In a broader context, it is now recognized that coronary artery spasm caused by a sudden increase in the $[Ca^{2+}]$ in the cell is a common precipitating cause of anginal attacks and even myocardial infarctions. Use of calcium-channel blockers has provided a new therapeutic modality in these states, where rather transient changes in cellular calcium metabolism appear to be responsible for the pathological events.

Dietary Excess

Hypertension and atherosclerosis are two of the most common diseases which afflict man, and alterations in cellular calcium metabolism are considered of great importance in their pathogenesis. Although change in the $[Ca^{2+}]_c$ is the common mechanism by which contraction is initiated in vascular smooth muscle [5], a high uptake of dietary Na^+ is thought to be a contributing factor in the pathogenesis of essential hypertension in man. As discussed previously [18] the control of $[Ca^{2+}]_c$ and of vascular tone are both extremely complex processes in which there is no direct relationship between $[Ca^{2+}]_c$ and degree of tone. So, at the present our knowledge is insufficient to define the equally complex relationship of $[Ca^{2+}]_c$ to Na^+ intake, and link this relationship to changes in vascular tone.

It has been known for many years that an accumulation of calcium salts occurs as a component of atherogenic plaques [27]. Recent indirect evidence suggests that changes in cell calcium play an early and critical role in the pathogenesis of the initial changes in arterial intima and media. Rabbits fed an atherogenic diet fail to develop atheromata in their aortae if they are simultaneously fed lanthanum chloride or other agents which block calcium uptake into cells [14]. Likewise, rats fed large doses of vitamin D develop calcific deposits in their smaller arteries and arterioles, which are suppressed by verapamil, a calcium-channel blocker [13]. The exact step at which calcium participates in the sequence of events that occur in developing atheromata is not known. However, platelets and vasoactive substances stimulate calcium influx into vascular endothelial cells, and bring about a calcium-dependent stimulation of growth [10]. Hence, it is possible that an early event in the vascular wall injury is an influx of calcium into the cell across an altered plasma membrane barrier.

Inherited Diseases

Duchenne's muscular dystrophy (DMD) and cystic fibrosis are two of several inherited diseases in which alterations in cellular calcium metabolism have been postulated to play a pathogenic role. DMD represents a disorder in which cellular calcium metabolism may be of considerable pathogenic importance [25], since muscles of patients and animals with similar dystrophies contain abnormally large concentrations of calcium, primarily localized in mitochondria. Moreover, a variety of agents or maneuvers that reduce the rate of calcium uptake will ameliorate the severity of the muscle lesions *pari passu* with a decline in calcium accumulation within the muscle cells. In addition, a rise in muscle $[Ca^{2+}]$ leads to the activation of neutral proteases which cause the disruption and degradation of the proteins constituting the contractile system. An increase in $[Ca^{2+}]$ may also lead to partial uncoupling of mitochondrial oxidative phosphorylation and a decline in ATP and creatine-P stores [12].

A particularly interesting observation is the fact that ablation of the parathyroid gland in dystrophic hamsters leads to a marked reduction in the accumulation of excessive calcium in both heart and skeletal muscle, which is associated with a decline in the serum creatine kinase, a biochemical marker of disease activity [17]. However, in the hamster, parathyroidectomy does not lead to permanent hypocalcemia. Thus, parathyroidectomized, dystrophic animals have normal serum calcium concentrations in spite of which their muscles accumulate less calcium. PTH may act directly upon the muscle cells of the dystrophic hamster to increase calcium uptake. If so, dystrophy in these animals could represent a situation in which a hormone produces an abnormal response in a nontarget tissue (see above). Alternatively, the effect of parathyroidectomy on the muscles of dystrophic hamsters may be explained by a change in vitamin D metabolism; this metabolic change may be responsible for the observed alteration in calcium accumulation by muscle.

These various observations, though interesting, do not address a fundamental question, i.e., the nature of the primary defect in dystrophic muscle. The primary defect may be an abnormality of plasma membrane structure which leads to aberrant cellular calcium metabolism causing muscle cell destruction. Alternatively, the primary defect may lie in some intracellular membrane system that causes an initial tissue injury or damage to the plasma membrane, and the changes in cellular calcium metabolism are a response of the muscle cell to this injury.

Cystic fibrosis is an inherited disease in which changes in cellular calcium metabolism may play a pathogenic role. One of the major factors contributing to a change in the viscosity of secretions in this disorder is the interaction of calcium with glycoproteins of these secretions. Elevated amounts of cellular calcium have been detected in fibroblasts, lymphocytes, granulocytes, and parotid acinar cells from patients with cystic fibrosis [4,15]. Of particular interest is the discovery of a factor in the serum of cystic fibrosis patients which promotes calcium uptake into and potassium efflux from isolated cells. This factor released into the blood may cause subtle changes in cellular calcium metabolism, which may ultimately lead to clinical symptoms.

A primary abnormality in plasma membrane calcium exchange could lead to persistently high levels of intracellular calcium, which, in turn, could eventuate in cellular dysfunction and/or cell death. Alternatively, a primary abnormality in some other cellular system leading to tissue injury could, in turn, effect a secondary disturbance in cellular calcium metabolism and the net accumulation of cellular calcium. At present, there are no clues to distinguish between these alternatives in the cause of cystic fibrosis. However, one of the difficulties of accepting the postulate that either muscular dystrophy or cystic fibrosis represents defects of primary membrane calcium transport is the restriction of lesions in both diseases to a limited number of specific tissues.

Conclusions

Along with an understanding of the central role of calcium in stimulus–response coupling, and the recognition of the complex means by which cellular calcium homeostasis is maintained has emerged the recognition that cellular calcium dyshomeostasis plays a pathogenic role in both acute and chronic disease states. In this brief review, an attempt has been made to point out certain cellular calcium disorders, and their relationship to disease. A better understanding of cellular calcium metabolism will provide new insights into the pathogenesis of many diseases, and foster new approaches to their study and management.

ACKNOWLEDGMENTS. Work in the laboratory of H. R. has been supported by a grant from the National Institutes of Health (AM-09650) and a grant from the Muscular Dystrophy

Association of America. Work in the laboratory of G. P. has been supported by a grant from the National Institutes of Health (CA-20939) and a grant from the Muscular Dystrophy Association.

References

1. Balait, L. A.; Ferrante, W. A. Pathophysiology of acute and chronic pancreatitis. *Arch. Intern. Med.* **142**:113–117, 1982.
2. Palmieri, G. M. A.; Bertorini, T. E.; Notting, D. S.; Bhattacharya, S. K.; Luther, R. W.; Pate, J. W. Muscle accumulation in muscular dystrophy and acute pancreatitis: A common pathogenic mechanism. *Acta Endocrinol. Suppl.* **256**:103–112, 1983.
3. Blaustein, M. P. The interrelationship between sodium and calcium fluxes across cell membranes. *Rev. Physiol. Biochem. Pharmacol.* **70**:33–82, 1974.
4. Bogart, B. I.; Conod, E. J.; Conover, J. H. The biologic activities of cystic fibrosis serum. I. The effects of cystic fibrosis sera and calcium ionophore A23187 on rabbit tracheal explants. *Pediatr. Res.* **11**:131–134, 1977.
5. Bolton, T. B. Mechanism of action of neurotransmitters and other substances on smooth muscle. *Physiol. Rev.* **59**:609–718, 1979.
6. Borle, A. B. Control, modulation and regulation of cell calcium. *Rev. Physiol. Biochem. Pharmacol.* **90**:13–153, 1981.
7. Borle, A. B.; Clark, I. Effects of phosphate-induced hyperparathyroidism and parathyroidectomy on rat kidney calcium *in vivo*. *Am. J. Physiol.* **241**:E136–E141, 1981.
8. Braunwald, E. Mechanisms of action of calcium-channel blocking agents. *N. Engl. J. Med.* **307**:1618–1627, 1982.
9. Chapman, R. W. Excitation–contraction coupling in cardiac muscle. *Prog. Biophys. Mol. Biol.* **35**:1–52, 1979.
10. D'Amore, P.; Shepro, D. Stimulation of growth and calcium influx in cultured, bovine, aortic endothelial cells by platelets and vasoactive substances. *J. Cell. Physiol.* **92**:117–183, 1977.
11. Farber, J. L. The role of calcium in cell death. *Life Sci.* **29**:1289–1295, 1981.
12. Fleckenstein, A. Drug-induced changes in cardiac energy. *Adv. Cardiol.* **12**:183–197, 1974.
13. Frey, M.; Keidel, J.; Fleckenstein, A. Verhutung experimenteller Gelassverkalkungen (Monckeberts type der arterosclerosis) durch calcium-antagonisten bei Ratten. In: *Calcium Antagonismus*, A. Fleckenstein and H. Roskamon, eds., Berlin, Springer-Verlag, 1980, pp. 258–269.
14. Kramsch, D. M.; Aspen, A. J.; Apstein, C. S. Suppression of experimental atherosclerosis by the calcium-antagonist lanthanum. *J. Clin. Invest.* **65**:967–981, 1980.
15. Mangos, J. A.; Donnelly, W. H. Isolated parotid acinar cells from patients with cystic fibrosis: Morphology and composition. *J. Dent. Res.* **60**:19–25, 1981.
16. Naylor, W. G.; Polle-Wilson, P. A.; Williams, A. Hypoxia and calcium. *J. Mol. Cell. Cardiol.* **11**:683–706, 1979.
17. Palmieri, G. M. A.; Nutting, D. F.; Bhattacharya, S. K.; Bertorini, T. E.; Williams, J. C. Parathyroid ablation in dystrophic hamsters. *J. Clin. Invest.* **68**:646–654, 1981.
18. Rasmussen, H. Cellular calcium metabolism. *Ann. Intern. Med.* **98**:809–816, 1983.
19. Rasmussen, H.; Tenenhouse, A. Thyrocalcitonin, osteoporosis and osteolysis. *Am. J. Med.* **43**:711–726, 1967.
20. Rasmussen, H.; Waisman, D. M. Modulation of cell function in the calcium messenger system. *Rev. Physiol. Biochem. Pharmacol.* **95**:111–148, 1982.
21. Schanne, F. A. X.; Kane, A. B.; Young, E. E.; Farber, J. L. Calcium dependence of toxic cell death: A final common pathway. *Science* **206**:700–702, 1979.
22. Smallwood, J.; Waisman, D. M.; Lefreniere, D.; Rasmussen, H. The erythrocyte calcium pump catalyzes a Ca^{2+}:nH^+ exchange. *J. Biol. Chem.* **258**:11092–11097, 1983.
23. Thor, H.; Smith, M. T.; Hartzell, P.; Bellomo, G.; Jewell, S. A.; Orrenius, S. The metabolism of menadione (2-methyl-1,4-napthoquione) by isolated hepatocytes. *J. Biol. Chem.* **257**:12419–12425, 1982.
24. Van Hardeveld, C.; Kassenaar, A. A. H. A possible role for calcium in thyroid hormone-dependent oxygen consumption in skeletal muscle of the rat. *FEBS Lett.* **121**:349–351, 1980.
25. Wrogemann, K.; Pena, S. D. J. Mitochondrial calcium overload: A general mechanism for cell necrosis in muscle disease. *Lancet* **1**:672–673, 1976.

26. Yasue, H.; Omote, S.; Takizawa, A.; Nagoa, M.; Nosaka, H.; Nakajima, H. Alkalosis-induced coronary vasoconstriction: Effect of calcium, diltiazem, nitroglycerine, and propranolol. *Am. Heart J.* **102**:206–210, 1981.
27. Yu, S. Y.; Blumenthal, H. T. The calcification of elastic fibers. I. Biochemical studies. *J. Gerontol.* **18**:119–126, 1963.

60

Dietary Calcium in the Pathogenesis and Therapy of Human and Experimental Hypertension

David A. McCarron

The thesis that disorders of calcium metabolism and homeostasis are a factor in the pathogenesis of increased arterial pressure in a subset of patients with essential hypertension will be the focus of this review. An understanding of the evidence drawn from basic and clinical investigations supporting this postulate is essential to appropriate application of new therapeutic options in the management of human hypertension. Pharmacologic interventions that modify an individual's calcium balance, and hence cellular physiology, represent treatment options that will be potentially specific for selected patients with essential hypertension.

Examples of how calcium and calcium-dependent events contribute to the regulation of cardiac output and peripheral resistance will be addressed first. Laboratory and clinical observations which suggest that experimental and human hypertension are associated with disturbances of calcium metabolism will then be summarized. Proposed mechanisms whereby abnormal Ca^{2+} metabolism might adversely affect normal cardiovascular physiology are set forth. Finally, specific therapeutic options, including dietary and pharmacologic, that reverse these calcium-related abnormalities of blood pressure control will be assessed, as well as the patient populations most likely to benefit from the application of these advances in cardiovascular pharmacology.

Calcium and Cardiovascular Physiology

Intraarterial pressure is a function of two cardiovascular parameters: cardiac output and peripheral resistance. Calcium is essential to the normal regulation and function of both of these components [1]. Calcium directly modifies the critical determinants of cardiac output: left ventricular ejection, heart rate, and intravascular volume. Likewise, peripheral resistance

---

David A. McCarron • Division of Nephrology and Hypertension, Oregon Health Sciences University, Portland, Oregon 97201.

is altered by calcium's effects on intrinsic tone, release of vasoactive substances, and membrane reception/transmission. Via its action on intracellular function and stabilization of the cell membrane, calcium sets the intrinsic tone of resistance arterioles. Furthermore, this cation directly controls vascular reactivity of these same vessels. Both synthesis and release of the humoral substances that mediate both vasodilator and vasoconstrictor cardiovascular reflexes are regulated by calcium [1,2]. The ability of these substances to affect the appropriate vascular response is also calcium-dependent. Calcium influences the binding of humoral compounds to the vascular smooth muscle membrane and then the transmission of their signal internally via the movement of calcium down the "slow channels" and/or the release of membrane-bound calcium [3,4].

The action of calcium alters not only cardiac function and the peripheral vasculature, but also other organs essential for normal cardiovascular activity: central nervous system, kidney, and adrenal gland. While reported abnormalities linked with human hypertension are diverse and encompass many of the organs and cellular mechanisms outlined above, abnormally elevated peripheral resistance remains the hallmark of this disorder. Both resting vascular tone and vascular reactivity are accentuated in human and experimental hypertension. Defining whether primary or secondary alterations in the cellular metabolism of calcium are central to the manifestation of increased vascular resistance is an area of current research.

Disturbances of Calcium Metabolism Associated with Hypertension

In both human and experimental hypertension, disordered calcium metabolism has been noted in a variety of organ functions, biochemical parameters, and cellular fractions [5–7]. Organ involvement has included the peripheral vasculature, kidney, intestines, adipose tissue, erythrocytes, and bone. In the laboratory, animal vascular tissue exhibits increased membrane permeability, altered binding kinetics in the cell membranes, and calcium accumulation in subcellular fractions [5–7]. Both Na^+-dependent and genetic models of hypertension are reported to possess these abnormalities [7]. In addition, similar disturbances of cellular calcium have been identified in adipocytes and erythrocytes of the spontaneously hypertensive rat (SHR). The renal defects include enhanced urinary calcium excretion in the adult SHR, a failure of the young animal to appropriately diminish his urinary calcium excretion when placed on a calcium-deficient intake, and a blunted stimulation of urinary cAMP generation under conditions of metabolic stress [8]. Intestinal handling of calcium has also been noted to be abnormal in the SHR. The reports to date have been conflicting as both increased and decreased absorption of ingested calcium has been suggested [9,10].

Disturbances of calcium metabolism linked to human hypertension include altered renal and bone metabolism [11–13]. The latter encompasses increased urinary calcium excretion, enhanced excretion of an acutely administered calcium load, increased urinary cAMP excretion, and elevated phosphate clearance. Bone disturbances have been implied by the increased prevalence of hypertension reported in osteoporotic women [13].

The biochemical disorders associated with calcium metabolism that have been reported in both human and experimental hypertension are summarized in Table I. While serum total calcium is reported to be normal, free or ionized calcium is decreased [14]. Parathyroid hormone values are elevated and serum phosphorus concentrations are low. Vitamin D precursors and intestinal calcium transport have also been noted to be abnormal in the SHR [9,10].

*Table I. Theoretical Mechanisms of
Cellular Calcium Defects*

Altered binding kinetics
   Membrane
   Cytosol
Abnormal kinetics of membrane fluxes
   Cell membrane
   Vesicle membranes
Depletion
   Inadequate storage
   Inadequate exposure

The defects in calcium metabolism in vascular smooth muscle of SHR include: (1) increased permeability of the extracellular membrane to calcium [15]; (2) increased membrane calcium content [16]; (3) decreased binding of ^{45}Ca to the cell membrane [17]; (4) increased mitochondrial calcium content with delayed mitochondrial uptake of cytosolic calcium [18]; and (5) altered calcium transport across the cell membrane [19]. Recent observations have indicated that at least some, if not all, of the cellular alterations in calcium metabolism are primary in origin and not simply the result of the increased arterial pressure that evolves in experimental models [20].

Dietary Calcium: Epidemiologic and Intervention Data in Experimental Hypertension

The functional importance of disordered calcium homeostasis in the pathogenesis of high blood pressure has been suggested by the epidemiologic observations in humans that decreased dietary [21,22] and environmental exposure [23,24] to calcium is associated with an increased risk of developing hypertension. Reduced dietary calcium consumption is the nutritional pattern that most closely follows the demographics of hypertension in the United States [25]. The data depicted in Fig. 1 are based upon the Health and Nutrition Examination Survey I (HANES I) from the National Center for Health Statistics. We have used that data base to identify adult Americans between the ages of 35 and 74 who were free of known hypertensive cardiovascular disease. Their reported daily dietary calcium consumption was analyzed and a mean value calculated for those subjects whose blood pressures were normal, borderline, or elevated and who were then further subdivided by sex and by two age groups of 35-54 years and 55-74 years.

The data indicate that hypertensive Americans consume less calcium and this difference is greatest at the extremes of body mass index, i.e., the leanest 30% and the heaviest 10% of hypertensives. Age, race, sex, and weight characteristics do not account for these differences. These observations are consistent with previous reports that calcium consumption is reduced in subjects with essential hypertension and that environmental exposure to calcium in several areas of the world is a correlate of blood pressure and hypertensive cardiovascular disease [21-25]. Reduced calcium intake from either food or water is associated with an increased probability of an abnormally high blood pressure and thereby ultimately increasing the risk of cardiovascular disease.

Acute and chronic studies in the SHR and its genetic [5,26], normotensive control, the

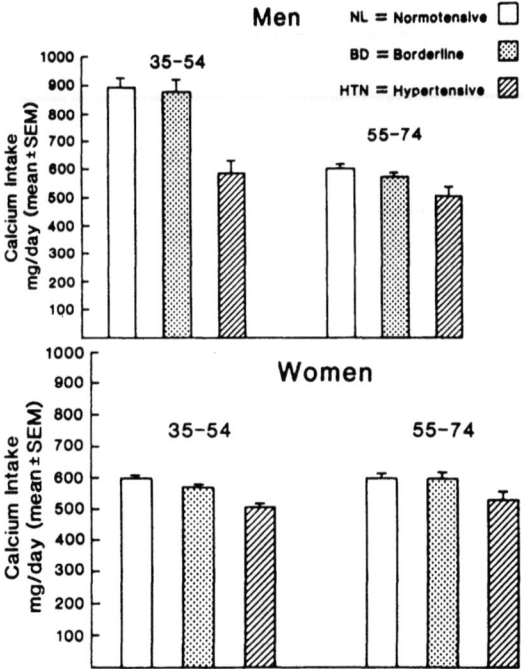

Figure 1. Dietary calcium intake in males and females, aged 35–74, from the Health and Nutrition Examination Survey I of the National Center for Health Statistics, based on level of blood pressure.

Wistar–Kyoto rat [27], have provided additional evidence that calcium homeostasis is a factor in blood pressure control. Dietary supplementation of the SHR with calcium results in a marked attenuation of what had previously been called fixed hypertension (Fig. 2). In contrast, removal or reduction of dietary calcium accelerates hypertension. In the adult SHR (24 weeks of age or older), the influence of dietary exposure to calcium may be evident within 2 weeks. Reducing dietary calcium results in a further rise in blood pressure, while supplementation with calcium significantly lowers pressure within 2 weeks (Fig. 3). Observations in the Wistar–Kyoto rat are consistent with those in the SHR. The normotensive rat's blood pressure will track inversely depending on the calcium content of the diet. Increased dietary exposure lowers pressure in the adult animal, while calcium restriction produces borderline hypertension in this putative normotensive animal. Again, dietary calcium manip-

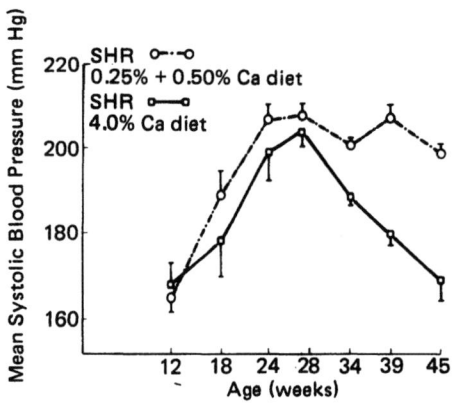

Figure 2. Mean systolic blood pressure in the SHR fed a normal diet (1% Ca^{2+}) and a diet supplemented with 4% Ca^{2+}, beginning at 12 weeks of age. After 26 weeks of age, the Ca^{2+}-supplemented SHR experienced a significant ($p < 0.0001$) attenuation of blood pressure.

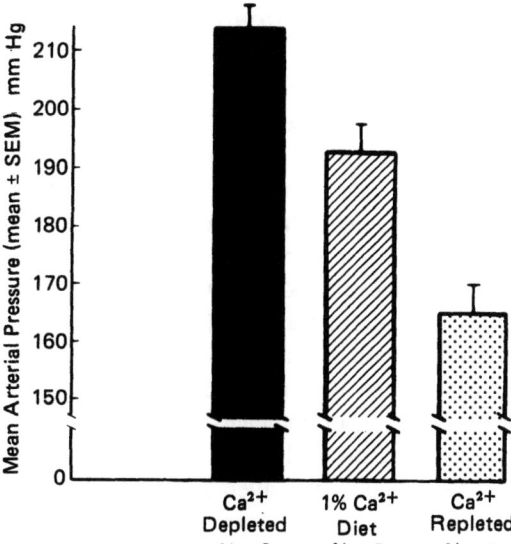

Figure 3. Mean intraarterial blood pressure (MAP) of the SHR ($n = 18$) raised on a standard (1% Ca^{2+}) diet until 24 weeks of age. Six were switched to a low-Ca^{2+} (0.02%) diet, while six were placed on a 4% Ca^{2+} diet. The remaining six were maintained on the 1% diet. MAP increased ($p < 0.001$) on the 0.02% diet, while it decreased ($p < 0.001$) on the 4% within 2 weeks.

ulations no longer than 2 weeks in duration modify the arterial pressure of adult Wistar–Kyoto rats.

Theoretical Relation of Calcium Metabolism to Hypertension

The metabolic pathways by which disturbed calcium homeostasis produces a rise in arterial pressure remain conjectural. Factors that reduce exposure of vascular smooth muscle to calcium (Table I) or impair storage and mobilization of calcium increase peripheral resistance and vascular reactivity. When any one or more of these factors exist, the muscle membrane is less stable, calcium permeability is increased ("slow channels" are open), and an increase in transmembrane calcium flux ensues. As a consequence, smooth muscle tone, reactivity, and contractility are all enhanced. As proposed earlier in this review, calcium serves a diverse role in the regulation of smooth muscle function, not only by initiating contraction, but also by stabilizing the membrane and reducing the kinetics of the interaction of the contractile proteins. The latter mechanisms are essential to the maintenance of the relaxation phase of the contractile process [28,29].

A genetic-based alteration in the binding kinetics of membrane calcium whereby calcium is not readily mobilized is one such putative mechanism that could result in vascular smooth muscle dysfunction. Such an abnormality is apparent in the extracellular space of both humans and experimental animals with high blood pressure, as serum total calcium levels are normal in hypertensives, but ionic calcium levels are depressed. Thus, increased calcium binding exists in the extracellular space of a subset of hypertensive humans and animals. A defect in calcium transport across cell membranes is a second pathogenetic mechanism that might alter the kinetics of calcium compartmentalization and thereby adversely modify smooth muscle function. Reduced dietary exposure to calcium may effectively produce similar metabolic consequences in the smooth muscle cell, i.e., enhanced calcium fluxes by depleting calcium from membrane storage sites. As a consequence, the vascular smooth muscle cell membrane is less stable, and vascular tone and reactivity are increased with a resultant elevation of peripheral resistance.

Therapeutic Implications: Dietary Calcium and Pharmacologic Agents

As noted above, dietary modification of calcium intake in experimental animals influences the development and maintenance of blood pressure levels. In both the genetically hypertensive (SHR) and normotensive Wistar–Kyoto rat, blood pressure is lowered by addition of calcium to the diet. For the SHR, the reversal of fixed hypertension simply by calcium supplementation represents the only reported nonpharmacologic maneuver that lowers blood pressure.

Two recent reports now suggest that comparable effects of calcium supplementation on blood pressure are attainable in both normotensive and hypertensive humans [30,31]. Belizan *et al.* demonstrated a 6–10% reduction in the blood pressure of normotensive males and females given 1000 mg of calcium as the carbonate salt for 6 to 20 weeks. A more recent, preliminary report indicates that 40–45% of hypertensive subjects will normalize their blood pressure. In this study, which was double-blind, placebo-controlled, 1000 mg of calcium for 8 weeks produced an average of 17 mm Hg reduction in blood pressure, principally systolic blood pressure. The current evidence from basic laboratory investigation indicates that calcium present in adequate concentrations acts as a channel blocker, controlling its own flux rate across the cell membrane [32,33]. Consequently, oral calcium and calcium channel blockers may be interchangeable in many patients being treated for high blood pressure.

The use of dietary calcium supplementation or calcium channel blockers, of which nifedipine is the clinical prototype in the United States, has been proposed as specific therapy for many patients with either essential or secondary forms of hypertension [34]. This premise is based on: (1) the association of calcium fluxes through the "slow channels," with initiation of vascular smooth muscle contraction; and (2) the recent evidence that one or more defects in cellular handling of calcium contribute in either a primary or a secondary fashion to the pathogenesis and maintenance of increased vascular tone and thereby elevated arterial pressure. By inhibiting the initial calcium influx, membrane calcium and compounds such as nifedipine may represent relatively specific therapy. Their action may either reverse primary abnormalities of calcium fluxes across the vascular smooth muscle cell or prevent the manifestation of increased smooth muscle contractility due to other hypothesized defects in blood pressure regulation.

Clinical experience with calcium channel blockers and/or dietary calcium in the management of hypertension is still limited [34,35]. Experiments in animal models have demonstrated the short- and long-term ability of supplemental calcium and calcium channel blockers to lower blood pressure. The acute intravenous administration of nifedipine to either calcium-replete or -deficient SHRs produced an immediate reduction and virtual normalization of arterial pressure [36]. The injection of nifedipine produced the same degree of hypotension, independent of the animal's calcium status. A chronic hypotensive effect has been demonstrated in both sodium-dependent and genetic forms of experimental hypertension without adverse effects on cardiac output, heart rate, or renal function.

Future clinical investigation will define subsets of hypertensive patients and their clinical characteristics that are associated with a favorable response to dietary calcium or calcium channel blockers. Based on the experience in animal models and the observations summarized above regarding calcium homeostasis in human hypertension, several categories of hypertensives are more likely to respond to these agents including elderly, black, older female, and obese patients; systolic-, pregnancy-, and osteoporosis-related hypertensives; and diuretic responders. Patient populations with primary increases in systemic vascular resistance and/or those hypertensives who by virtue of their demographic characteristics risk a long-standing exposure to low calcium intake are included. These individuals are likely to experience the most favorable responses to nifedipine because of abnormal vascular responsiveness or predisposition to the calcium fluxes being enhanced. An example of the former group are the elderly, because increased resistance is a hallmark of their systolic hypertension. From the latter group, older women are an excellent example of long-standing, low dietary calcium intake producing calcium depletion. In vascular tissue, this may produce an activation of the calcium channels, and increased cation fluxes and smooth muscle contractility.

Hypertensive patients with selected, associated medical disorders may represent additional subjects who would benefit from antihypertensive therapy with calcium channel blockade. In such cases, either the associated condition will also respond, e.g., coronary artery spasm, or adverse effects which might occur with other commonly used antihypertensives will be avoided, e.g., β-adrenergic antagonists exacerbating reactive airway disease.

Based on currently available data from clinical trials in essential hypertension, the chronic administration of oral calcium or calcium channel blockers does not produce significant adverse metabolic effects or substantive changes in biochemical parameters. This is to be contrasted with many of the currently used antihypertensive agents, such as thiazide diuretics and β blockers, which frequently induce alterations in serum chemistry.

In summary, nutritional and/or pharmacologic interventions that specifically modify calcium metabolism appear to be specific means of reversing abnormalities of calcium metabolism associated with essential hypertension. Such therapeutic measures, by either inhibiting calcium flux or directly stabilizing cell membranes of vascular smooth muscle, should promote vasorelaxation, thereby lowering peripheral resistance and arterial pressure. These emerging alternatives to the classic pharmacologic and/or dietary management of human hypertension may ultimately prove applicable to a substantial portion of the adult American population currently being treated for, or at risk of developing, arterial hypertension.

References

1. Kuriyama, H.; Yushi, I.; Suzuki, H.; Kitamura, K.; Itoh, T. Factors modifying contraction–relaxation cycle in vascular smooth muscles. *Am. J. Physiol.* **243**:H641–H662, 1982.
2. Rasmussen, H. Calcium and cAMP in stimulus–response coupling. *Ann. N.Y. Acad. Sci.* **356**:346–353, 1980.
3. Rasmussen, H. Cellular calcium metabolism. *Ann. Intern. Med.* **98**:809–816, 1983.
4. Gagnon, G.; Regoli, D.; Rioux, F. Studies on the mechanism of action of various vasodilators. *Br. J. Pharmacol.* **70**:219–227, 1980.
5. McCarron, D. A.; Yung, N. N.; Ugoretz, B. A.; Krutzik, S. Disturbances of calcium metabolism in the spontaneously hypertensive rat. *Hypertension* **3**:1162–1167, 1981.
6. McCarron, D. A. Calcium and magnesium nutrition in human hypertension. *Ann. Intern. Med.* **98**:800–805, 1983.
7. Webb, R. C.; Bohr, D. F. Mechanism of membrane stabilization by calcium in vascular smooth muscle. *Am. J. Physiol.* **235**:C227–C232, 1978.
8. Grady, J. R.; Dorow, J.; McCarron, D. A. Urinary calcium excretion and cAMP response of the spontaneously hypertensive rat to Ca^{2+} deprivation. *Clin. Res.* **31**:330A, 1983.
9. Toraason, M. A.; Wright, G. L. Transport of calcium by duodenum of spontaneously hypertensive rat. *Am. J. Physiol.* **241**:G344, 1981.
10. Scheld, H.; Miller, D.; Pape, J.; Horst, R. Vitamin D metabolism and intestinal calcium transport are abnormal in the spontaneously hypertensive rat. *Clin. Res.* **31**:488A, 1983.
11. McCarron, D. A.; Pingree, P.; Rubin, R. J.; Gaucher, S. M.; Molitch, M.; Kurtzik, S. Enhanced parathyroid function in essential hypertension: A homeostatic response to a urinary calcium leak. *Hypertension* **2**:162, 1980.
12. Strazzullo, P.; Nunziata, V.; Cirillo, M.; Giannattasio, R.; Ferrara, L. A.; Mattioli, P. L.; Mancini, M. Abnormalities of calcium metabolism in essential hypertension. *Clin. Sci.* **65**:137–141, 1983.

13. McCarron, D. A.; Chestnut, C. H., III; Cole, C.; Baylink, D. J. Blood pressure response to the pharmacologic management of osteoporosis. *Clin Res.* **29**:274A, 1981.
14. McCarron, D. A. Low serum concentrations of ionized calcium in patients with hypertension. *N. Engl. J. Med.* **307**:226–228, 1982.
15. Noon, J. P.; Rice, P. J.; Baldessarini, R. J. Calcium leakage as a cause of the high resting tension in vascular smooth muscle from the spontaneously hypertensive rat. *Proc. Natl. Acad. Sci. USA* **75**:1605, 1978.
16. Devynck, M. A.; Pernollet, M. G.; Nunez, A. M.; Meyer, P. Analysis of calcium handling in erythrocyte membranes of genetically hypertensive rats. *Hypertension* **3**:397, 1981.
17. Postnov, Y. V.; Orlov, S. N.; Pokudin, N. I. Decrease in calcium binding by the red blood cell membranes in spontaneously hypertensive rats and in essential hypertension. *Pfluegers Arch.* **379**:191, 1979.
18. Webb, R. C.; Bhalla, R. C. Altered calcium sequestration by subcellular fractions of vascular smooth muscle from spontaneously hypertensive rats. *J. Mol. Cell. Cardiol.* **8**:651, 1976.
19. Bhalla, R. C.; Webb, R. C.; Singh, D.; Ashley, T.; Broch, T. Calcium fluxes, calcium binding and adenosine cyclic 3; 5-monophosphate-dependent protein kinase activity in the aorta of spontaneously hypertensive and Kyoto Wistar normotensive rats. *Mol. Pharmacol.* **14**:468, 1978.
20. Mulvany, H. J.; Korsgaard, N.; Nyborg, N. Evidence that the increased calcium sensitivity of resistance vessels in spontaneously hypertensive rats is an intrinsic defect of the vascular smooth muscle. *Clin. Exp. Hypertens.* **3**:749, 1981.
21. McCarron, D. A.; Morris, C. D.; Cole, C. Dietary calcium in human hypertension. *Science* **217**:267–269, 1982.
22. Ackley, S.; Barrett-Conner, E.; Suarez, L. Dairy products, calcium and blood pressure. *Am. J. Clin. Nutr.* **38**:457–461, 1983.
23. Stitt, F. W.; Crawford, M. D.; Clayton, D. G.; Morres, J. N. Clinical and biochemical indicators of cardiovascular disease among men living in hard and soft water areas. *Lancet* **1**:122–126, 1973.
24. Belizan, J. M.; Villar, J. The relationship between calcium intake and edema-, proteinuria-, and hypertension-gestosis: An hypothesis. *Am. J. Clin. Nutr.* **33**:2202–2210, 1980.
25. McCarron, D. A.; Stanton, R. J.; Henry, H. J.; Morris, C. D. Assessment of nutritional correlates of blood pressure. *Ann. Intern. Med.* **98**:715–719, 1983.
26. McCarron, D. A.; Yung, N. N.; Ugoretz, B. A.; Krutzik, S. Disturbances of calcium metabolism in the spontaneously hypertensive rat. *Hypertension* **3**:1162–1167, 1981.
27. McCarron, D. A. Blood pressure and calcium balance in the Wistar–Kyoto rat. *Life Sci.* **30**:683–689, 1982.
28. Bohr, D. F. Vascular smooth muscle: Dual effect of calcium. *Science* **139**:597–599, 1963.
29. Piascik, M. T.; Babich, M.; Rush, M. E. Calmodulin stimulation and calcium regulation of smooth muscle adenylate cyclase activity. *J. Biol. Chem.* **258**:10913–10918, 1983.
30. Belizan, J. M.; Villar, J.; Pineda, O.; Gonzalez, A. E.; Sainz, E.; Garrera, G.; Sibrian R. Reduction of blood pressure with calcium supplementation in young adults. *J. Am. Med. Assoc.* **249**:1161–1165, 1983.
31. Morris, C. D.; Henry, H. J.; McCarron, D. A. Randomized, placebo-controlled trial of oral Ca^{2+} in human hypertension. Abstract presented at the 16th Annual Meeting of the American Society of Nephrology, December 1983.
32. Eckert, R.; Ewald, D. Residual calcium ions depress activation of calcium-dependent current. *Science* **216**:730–733, 1982.
33. Hurwitz, L.; McGuffee, L. J.; Smith, P. M.; Little, S. A. Specific inhibition of calcium channels by calcium ions in smooth muscle. *J. Pharmacol. Exp. Ther.* **220**:382–388, 1982.
34. Guazzi, M. D.; Polese, A.; Fiorentini, C.; Bartorelli, A.; Moruzzi, P. Treatment of hypertension with calcium antagonism [Review]. *Hypertension* **5**:97–102, 1983.
35. Pederson, O. L. Calcium blockade in arterial hypertension [Review]. *Hypertension* **5**:1174–1179, 1983.
36. Grady, J. R.; McCarron, D. A. Divergent effects of Ca^{2+} balance on the vasodilating response to PTH and nifedipine in the SHR. *Clin. Res.* **32**:36a, 1984.

61

Calcium Intake and Bone Loss in Population Context

Stanley M. Garn and Victor M. Hawthorne

A quarter of a century ago, adult bone loss was often described as "senile, postmenopausal osteoporosis," and was thought to be characteristic of Caucasian women over 50 who were predominantly of northwest European ancestry [16]. As we now know, adult bone loss is by no means an exclusively female complaint. Both men and women lose bone over the years, though the rate of loss (mg/day) is both absolutely and relatively greater in the female. The onset of adult bone loss also begins far earlier than menopause, though the incidence of fractures does not rise until bone mass becomes somewhat diminished and there are slips, slides, and falls to stress them beyond the breaking point [4].

Adult bone loss is by no means unique to American and European whites. National and international comparative studies involving tens of thousands of participants have revealed that adult bone loss is an inevitable part of the human condition [3,8]. Japanese and Chinese lose bone at a rate that may be exceptionally high [7]. Eskimos lose bone, whether on a traditional Eskimo or an "Alaskan" diet, and so do Central American Indians. Mexican-Americans in the United States also lose bone and so do the people in Nicaragua and Panama. American blacks lose bone (again in both sexes), though the rate of loss in American black women may be below the sex-specific average [3]. In fact, taking all survey data into account, there is no known population in which adult bone loss does not take place, regardless of activity, diet, or way of life (Table I). Whether ingesting grain, meat, milk, tubers, or the Mongongo nut, bone will be lost with increasing age [3,8,9].

Though adult bone loss has been described as a liability of modern man, radiogrammetric studies (akin to those conducted on the living) and sawn-bone sections of major tubular bones indicate that past populations also lost bone. Estimating age from tooth wear and other age-indicating conditions, it is clear that hunting-and-gathering peoples of the past, digging and grubbing populations, and those who later adopted wheat, barley, or other cereal agricultures all lost bone. Adult bone loss is therefore hardly a recent phenomenon, unique to sedentary television-watchers or limited to groups that consume 60 to 100 g of quality protein

---

Stanley M. Garn and Victor M. Hawthorne • Center for Human Growth and Development and the Epidemiology Department of the School of Public Health, University of Michigan, Ann Arbor, Michigan 48109.

Table I. Adult Bone Loss as Reported in 20 North American Studies

Group	Location
1. Ovolacto vegetarians	Berrien Springs, Michigan
2. American-born and China-born Chinese	Dayton, Ohio
3. American-born and Japan-born Japanese	Cincinnati, Ohio
4. Skeletalized American blacks	St. Louis, Missouri
5. Skeletalized American whites	St. Louis, Missouri
6. Ohio-born whites	Dayton, Xenia, and Fairborn, Ohio
7. Ten-State whites	Ten states
8. Ten-State Nutrition Survey Puerto Ricans	Primarily New York City
9. Ten-State Nutrition Survey blacks	Ten states
10. Ten-State Nutrition Survey Mexican-Americans	Primarily Texas and California
11. Tecumseh Community Survey whites	Tecumseh, Michigan
12. Omaha nuns	Omaha, Nebraska
13. Canadian whites	Toronto, Canada
14. Wisconsin whites	Madison, Wisconsin
15. American Indians	Mukleshoot Reservation, Washington
16. Obese American blacks and whites	Ten states
17. National Aging Center white males	Baltimore, Maryland
18. Wainwright Eskimos	Alaska
19. Sadlermiut Eskimos	Alaska
20. California insured subjects	San Francisco, California

per day [1,2]. People and their pets lose bone, and there is no reason to believe that this age-associated liability began after the Pleistocene.

The early orthopedists, concerned with their bone-losing female patients, made two clinical observations. These patients were deemed inactive and rarely attained the then-recommended dietary allowance for calcium. (These patients also rarely attained the recommended dietary allowance for calories.) Since calcium is a major constituent of bone, it seemed reasonable to ascribe bone loss to an inadequate calcium intake and to recommend an increased calcium intake as the obvious solution. A few workers suggested supplemental phosphate rather than calcium because milk drinkers tend to have lower phosphorus levels relative to calcium—in contrast to consumers of cereal and beans. A few investigators have advocated a higher protein intake to preserve the bone matrix [6]. There are also some who perceive activity as a solution to the problem, since activity decreases with age (and bone is certainly lost in the immobilization state).

Still, dietary supplementation has attracted much attention in the nutritional community and elsewhere, because it is easier to persuade older people to take a supplement than to exercise. Calcium supplementation has been most favored, following the "Nordin hypothesis" (though Nordin is not an adamant followers of his earlier claims).

However, comparative studies do not generate strong evidence favoring the calcium hypothesis. A review of all available radiogrammetric data and the data from direct-photon absorptiometry, fails to reveal any human population that does not lose bone and it is questionable whether there are any large population differences in absolute loss-rates or percent bone loss on an annualized basis. A study in the mid-1960s of the six Central American nations, which involved nearly 20,000 radiographs, demonstrated that none of the national groups failed to lose bone with age [3,8]. In the United States, using the massive data of the Ten-State Nutrition Survey of 1968–1970, blacks, whites, Puerto Ricans, and Mexican-Americans all lost bone in much the same fashion as did prehistoric populations

[3,4]. Nordin's multinational WHO studies of the early 1960s, the Central American Nutritional Survey of the mid-1960s, the Ten-State Nutrition Survey of the late 1960s, and such longitudinally-followed American groups as in the Fels Institute Studies and the total-community Tecumseh Community Health Survey all agree that people lose bone (Table II).

Yet, populations differ markedly in calcium intakes. The highest calcium levels are found among Guatemalan Indians subsisting on yellow maize that is high in calcium. While Guatemalan orthopedists have reported that Indian women do not develop osteoporosis, radiographs reveal that they do. Men and women in Panama lose bone much as they do elsewhere and at low levels of calcium intake (often below 400 mg/day). These national and population comparisons, combining radiogrammetry and dietary-intake methodology, provide no evidence that high calcium intakes are bone-sparing and that low calcium intakes are bone-losing. From the large national surveys and comparisons, the evidence justifies the Scottish assize verdict "not proven" insofar as calcium intake and bone loss are concerned (see Table I).

There are, moreover, additional studies involving smaller or selected samples. They are inconclusive not only as to relationship between calcium intakes and bone loss but also as to the benefits or risks of other life-styles with respect to bone loss in later years. Among ovolacto-vegetarians from Berrien Springs in Michigan, the skeletal mass of the older women seems to be above the Wisconsin absorptiometric standards [13], but no such difference appears among the ovolacto-vegetarian men [12]. In either event, their diets were reasonably high in calcium—primarily from milk and milk products—but they lost bone. In the Dakota studies, in which a high fluoride intake apparently improved the fracture picture among females, the corresponding fluorotic Dakota males had increased rather than decreased fracture rates. Data on Eskimos, some suggesting increased loss rates, are equivocal with respect to calcium intakes (and whether they consume fish bones along with the fish). One matched-population comparison from Yugoslavia is in the expected direction but other studies are not [14,15]. The Dayton area Chinese, characterized by relatively low skeletal masses, also had surprisingly high calcium intakes [7]. (They showed marked food aculturation with respect to milk and milk products but little food aculturation when it came to rice!) So the data comparing calcium intake levels on the one hand and radiogrammetric measurements or absorptiometric measurements on the other, provide little support for the salutary effect of calcium ingestion.

Evidence is lacking that a high calcium intake is bone-protective, or that a low calcium

Table II. A Partial Listing of Countries Reporting Bone Loss

Canada	India
Costa Rica[b]	Jamaica[a]
Denmark	Japan[a]
Dominica	Nicaragua[b]
El Salvador[b]	Panama[b]
Finland[a]	South Africa
Gambia[a]	Sweden
Greece	Uganda[a]
Guatemala[b]	United Kingdom[a]
Honduras[b]	United States

[a]Nordin's WHO report.
[b]Central American Nutrition Survey.

intake accelerates bone loss. Notably, American blacks enjoy the combination of a greater skeletal mass and a lower calcium intake. American black women have a less-than-average rate of bone loss—only half that of women in general and quite comparable to that in black men. The lower calcium intake of black females may reflect lactose intolerance and lactase insufficiency. However, comparative and epidemiological studies fail to provide conclusive evidence favoring either side of the calcium intake controversy (cf. Table III) [11].

Another approach is to look *within* populations and contrast the bone quality of those individuals whose calcium intakes are highest with those whose intakes are lowest for their group. Admittedly, this is no easy task for both methodological and statistical reasons [5]. Calcium intake is easily underestimated by simplified methodologies that are overly dependent on foods of higher calcium content, or overestimated by the food-frequency method. Again, the integrity of the skeletal mass may be imperfectly represented by the second metacarpal or the distal aspect of the radius and the ulna. Simply to get a minimum of 150 low- and 150 high-calcium consumers of a sex and an age demands a starting sample of 1000 and no dropouts whatsoever if a five-decade period is to be covered. Very few studies attain these minimal criteria, particularly those based on patient material and derived from highly selected groups of "hospital normals" [5]. Given the limitations of methodology and sample size, it is doubtful that we have more than one or two major surveys that meet these necessary expectations.

A survey of the truly longitudinal studies and those of a cross-sectional survey nature, reveal no monotonic relationship between calcium intake and bone mass when the data are examined age by age, i.e., by decade intervals. Those whose calcium consumption is 400 mg/day or less (a low calcium intake) are neither more nor less prone to osteopenia. Similarly, those whose calcium intakes exceed 1 g/day, either by heavy use of milk products or by antacid supplementation, are not hyperostotic. While it may be that the calcium "requirement" for bone is either below 400 mg/day or above 1000/day, no benefits or detriments can be envisioned over a 2:1 range of calcium intakes. In the massive data from the Ten-State Nutrition Survey, senior citizens whose calcium intakes were high for their age had no better or worse bones than those at low levels of calcium intake for sex, age, and race, even though the highest intakes were associated with the largest individuals and with elevated caloric intakes as well [10,11]. If calcium were truly bone-protective under the conditions mentioned and if low calcium intakes were associated with increased rates of loss, large samples should certainly reveal it (see Table IV).

Table III. Calcium Intake Percentiles for Older Black and White Americans[a]

		Ca intake percentiles (mg/day)							
		Black				White			
Sex	Midpoint age	N	p15	p50	p85	N	p15	p50	p85
Males	60	27	172	447	813	115	317	765	1308
	70	52	148	531	1033	260	321	664	1365
	80	19	175	429	620	129	302	625	1130
Females	60	37	158	320	1085	159	218	471	991
	70	103	131	326	790	361	249	513	984
	80	23	242	739	1127	150	216	541	1063

[a]From the Ten-State Nutrition Survey of the USA and Ref. 11.

Table IV. Calcium, Calorie, and Protein Intake as Related to Bone Mass and Bone Quality[a]

	Correlation with	
Intake variable	Cortical area	Percent cortical area
Calcium (mg/day)	0.07	0.08
Protein (g/day)	0.06	0.01
Caloric intake (kcal)	0.05	0.00
Calorie-corrected calcium intake	0.05	0.09[b]

[a]$N = 879$ older adults. All correlations computed from sex-, age-, and race-specific normalized Z-scores [11].
[b]Partial correlation.

If the calcium intake–bone integrity relationship exists, it may be of low magnitude and obscured by methodological limitations and sample sizes. Available measurements of bone mass and bone density may be inadequate for the task. We may have simply failed to accumulate sufficient two- and three-decade follow-up data that would really demonstrate that higher calcium intakes are associated with diminished bone loss. We hope to explore this possibility in the Tecumseh Community Health Survey after a lapse of 25 years and in complete family-line context as well.

The major population comparisons we have cited so far, the inter- and intrapopulation comparisons, as well as the special human situations, are in full agreement in documenting the inevitability of bone loss with age, regardless of the calcium intake [8]. Bone minerals and bone matrix are rather easily lost in adults (as we can demonstrate in returned Indochina-area prisoners-of-war). Once lost, bone minerals and matrix are very difficult to restore, except perhaps in growing children and adolescents. Regardless of the calcium intakes, bone tissue is not easily rebuilt following kidney transplants, in chronic renal disease (CRD) or end-state renal disease patients. Unlike teeth, where mineral and matrix tightly sequester calcium, bone represents a readily available reservoir of this cation. It is doubtful whether encouraging more calcium deposition will benefit the bone-bank, while the door of the bank is left open (cf. Table V).

Daily calcium intakes in excess of 1 g are high for older people (who may ingest less than 1500 kcal/day) and are implausible by world standards, as reflected in the RDA for

Table V. Contradictory Evidence from Calcium Balance Studies and from Population Studies

Description of findings	Source of information
Calcium balance can be achieved in adults at calcium intakes of 200, 400, 600, or 1000 mg/day	Balance studies using various markers and corrections for calcium loss in sweat[a]
Adult bone loss occurs in all populations regardless of the level of calcium intake and in individuals at > 1000 mg/day	Population studies using radiogrammetric and absorptiometric measurements[b]

[a]Includes data of Hegsted, Moscosco, and Collazos, Malm, Outhouse, and others.
[b]See Tables I and II.

calcium of the World Health Organization. Such levels could be met in the USA by calcium supplements such as calcium carbonate, lactate, or gluconate. It is possible to modify the dietary habits of Americans by increasing the calcium intake of all Americans beginning at the onset of bone loss which occurs at the age of 39.

However, the evidence in favor of an increased calcium intake to minimize bone loss is not as compelling as the evidence favoring dietary restrictions to minimize atherosclerosis. Multinational population comparisons are not available to indicate the value of a high-calcium intake. Moreover, intrapopulation comparisons as with railroad workers and Minneapolis executives or the Framinghams, Muscatines, and Bogalusas are not yet completed. The potential values of exercise, vitamin D, fluoride, phosphorus, or calcium have not yet been unequivocally established for the over-forty group. As with protein supplementation (which does not prevent age-associated muscle loss), the population data do not support the notion that promoting calcium will prevent, forestall, or ameliorate bone loss.

ACKNOWLEDGMENT. Supported in part by NIH Grant HD-13823.

References

1. Dewey, J. R.; Armelagos, G. J.; Bartley, M. H. Femoral cortical involution in three Nubian archeological populations. *Hum. Biol.* **41**:13–28, 1968.
2. Ericksen, M. F. Cortical bone loss with age in three native American populations. *Am. J. Phys. Anthropol.* **45**:443–452, 1976.
3. Garn, S. M. *Earlier Gain and the Later Loss of Cortical Bone: In Nutritional Perspective*, Springfield, Ill., Thomas, 1970.
4. Garn, S. M. Bone loss and aging. In: *Nutrition of the Aged*, W. W. Hawkins, ed., Calgary, Nutrition Society of Canada, 1978, pp. 73–89.
5. Garn, S. M. Sampling, sample size, correlations and causality in studies of adult bone loss. In: *Osteoporosis: Recent Advances in Pathogenesis and Treatment*, H. F. DeLuca, H. M. Frost, W. S. S. Jee, C. C. Johnston, Jr., and A. M. Parfitt, eds., Baltimore, University Park Press, 1981, pp. 17–18.
6. Garn, S. M.; Kangas, J. Protein intake, bone mass, and bone loss. In: *Osteoporosis: Recent Advances in Pathogenesis and Treatment*, H. F. DeLuca, H. M. Frost, W. S. S. Jee, C. C. Johnston, Jr., and A. M. Parfitt, eds., Baltimore, University Park Press, 1981, pp. 257–263.
7. Garn, S. M.; Pao, E. M.; Rohmann, C. G. Calcium intake and compact bone loss in adult subjects. *Fed. Proc.* **24**:2415, 1965.
8. Garn, S. M.; Rohmann, C. G.; Wagner, B. Bone loss as a general phenomenon in man. *Fed. Proc.* **76**:1729–1736, 1967.
9. Garn, S. M.; Rohmann, C. G.; Wagner, B.; Davila, G. H.; Ascoli, W. Population similarities in the onset and rate of adult endosteal bone loss. *Clin. Orthop.* **65**:51–60, 1969.
10. Garn, S. M.; Solomon, M. A. Do the obese have better bones? *Ecol. Food Nutr.* **10**:195–197, 1981.
11. Garn, S. M.; Solomon, M. A.; Friedl, J. Calcium intake and bone quality in the elderly. *Ecol. Food Nutr.* **10**:131–133, 1981.
12. Marsh, A. G.; Sanchez, T. V.; Chaffee, F. L.; Mayor, G. H.; Mickelsen. O. Bone mineral mass in adult lacto-ovo-vegetarian and omnivorous males. *Am. J. Clin. Nutr.* **37**:453–456, 1983.
13. Marsh, A. G.; Sanchez, T. V.; Mickelsen, O.; Keiser, J.; Mayor, G. Cortical bone density of adult lacto-ovo-vegetarian and omnivorous women. *J. Am. Diet. Assoc.* **76**:148–151, 1976.
14. Matkovic, V.; Ciganovic, M.; Tominac, C.; Kostial, K. Osteoporosis and epidemiology of fractures in Croatia. *Henry Ford Hosp. Med. J.* **28**:116–125, 1980.
15. Matkovic, V.; Kostial, K.; Simonovic, I.; Buzina, R.; Broderec, A.; Nordin, B. E. C. Bone status and fracture rates in two regions of Yugoslavia. *Am. J. Clin. Nutr.* **32**:540–549, 1979.
16. McClean, F. C.; Urist, M. R. *Bone*, 3rd ed., Chicago, University of Chicago Press, 1973.

62

Similarities and Differences in the Response of Animals and Man to Factors Affecting Calcium Needs

H. H. Draper

Evidence accumulated over the past 35 years indicating that there are various hormonal, nutritional, genetic, and occupational factors which influence aging osteopenia and the development of osteoporotic bone disease in humans has prompted a search for animal models which might be useful in the study of specific predisposing factors. Aging bone loss in animals has been systematically studied only in recent years and there is still a paucity of information regarding the histological changes involved. There is no known animal condition which faithfully mimics human osteoporosis, but there are animal models which are useful in the study of some factors, such as nutrition and exercise, which influence certain aspects of the disease [22].

The use of animal models is complicated by persistent uncertainties about the nature of the human disease itself, including the relative importance of trabecular versus compact bone loss [21] and by evidence for as many as four distinct subgroups, differing in bone composition, among a clinically homogeneous cohort of osteoporotic subjects [5]. Nevertheless, research has revealed a number of similarities, as well as differences, in aging osteopenia in man and animals, and in the skeletal response to nutritional factors.

Aging Osteopenia in Animals

Age-related osteopenia has been documented in nonhuman primates, carnivores, and rodents, but whether it results in osteoporotic bone fractures is unclear. The macaque monkey undergoes cortical thinning of the second metacarpal and changes in tibial mineralization similar to those seen in man [4]. Adult dog bone resembles human bone histologically except for a much faster turnover rate and a faster rate of bone loss [13]. In both

---

H. H. Draper • Department of Nutrition, College of Biological Science, University of Guelph, Guelph, Ontario N1G 2W1, Canada.

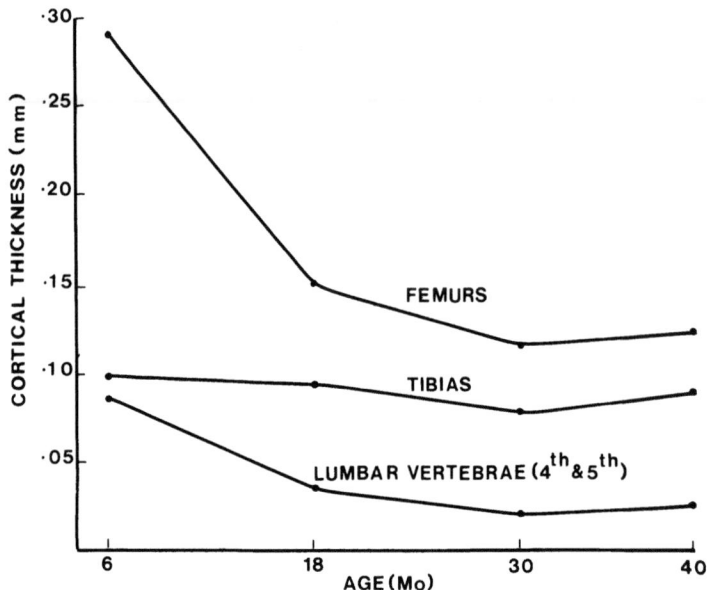

Figure 1. Cortical thickness as a function of adult age in mice fed a stock diet [16].

species the female is more susceptible to bone loss, and a similar fraction of bone consists of cortical bone. The increased risk of fracture implied by a faster bone loss in the dog may be counterbalanced by a greater bone mass at maturity (a key factor in fracture risk in man). Aging osteopenia in rats and mice results in cortical thinning and bone fragility, but the changes have not been well characterized histologically and the incidence of fractures, if any, is unknown. Cortical thinning in female mice of the $B6D1F_1$ Bar Harbor strain fed a stock diet apparently adequate in osteogenic nutrients is illustrated in Fig. 1 [16]. Extensive thinning of the lumbar vertebral and femoral cortices (the bones most susceptible to fracture in the human) occurs by 18 months of age with evidence of further thinning at 30 months. The incidence of cortical porosities, however, is low. A progressive increase with age in femoral width is indicative of periosteal bone apposition; hence, cortical thinning evidently occurs as a result of an overriding increase in endosteal bone resorption. As in the human and dog, female rodents are more susceptible to bone loss than males. The main distinction between rodent and human skeletons is that a lack of lamellar bone confers on rodents a limited capacity for bone remodeling. This characteristic of rodent bone renders it sensitive to depletion caused by factors which accelerate bone resorption.

The state of knowledge regarding the response of adult animals to those nutrients which are most frequently associated with human osteoporosis (calcium, phosphorus, protein, vitamin D, and fluoride) is summarized in the following sections.

Calcium

Although inadequate calcium intake is the nutritional factor most frequently implicated in human osteoporosis, there are major inconsistencies in the evidence for this association which make it difficult to compare the responses of animals and man. Garn *et al.* [7] failed to find a significant relationship between aging bone loss and the calcium content of national food supplies. By contrast, Matkovic *et al.* [20] found that bone mass and fracture rate in two

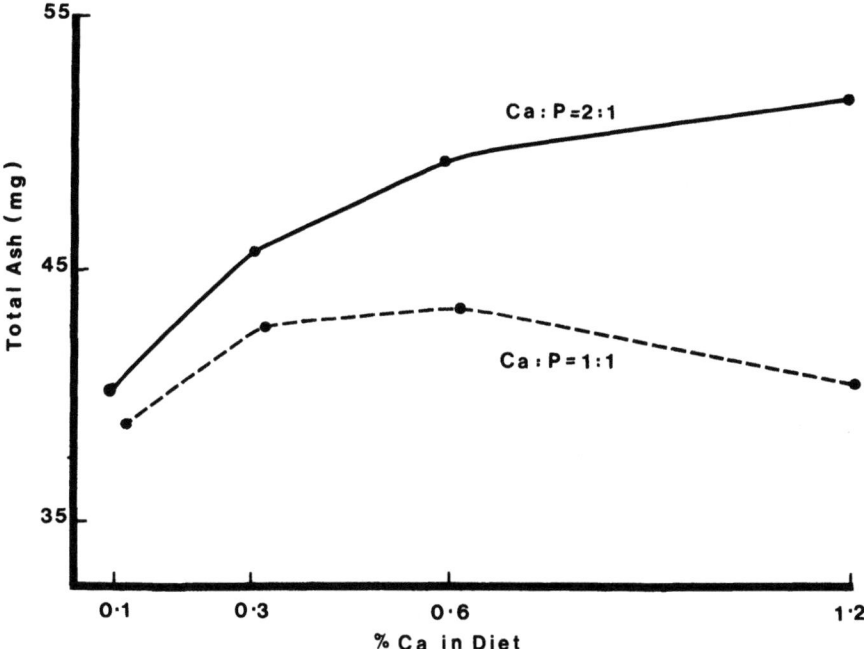

Figure 2. Femur ash content in mice fed diets containing various concentrations and calcium and phosphorus from 10 to 26 months of age [23].

population cohorts of Yugoslavian adults were significantly correlated with calcium intake, and Heaney *et al.* [8] found that a high calcium intake was an effective deterrent of postmenopausal bone loss in a sample of American women. Calcium deficiency, particularly in the presence of excess dietary phosphate, causes parathyroid stimulation and a condition resembling human osteoporosis in adult cats and dogs [14,18]. This condition responds to calcium administration. The fact that human osteoporosis does not respond similarly has been used as an argument against a role of calcium in the human disease.

Adaptation to a reduction in calcium intake, now known to be mediated by a parathyroid hormone-dependent increase in calcitriol synthesis and in calcium absorption, has been demonstrated in adult rats [11] and humans [19]. Calcium balance in men has been observed on intakes as low as 200 mg/day [9]. However, adaptation to a low calcium intake is not uniformly successful [19].

There is evidence of a reduced efficiency of calcium absorption in aged humans, whereas no decrement could be found in aged rats [12]. Bone mass in aged mice is not significantly influenced by the calcium intake during adult life unless the calcium content of the diet is very low (0.1%) [23] (Fig. 2). Skeletal homeostasis in this species is more sensitive to dietary phosphate than to calcium. Rats self-regulate their phosphorus intake within narrow limits when given the opportunity and thereby maintain serum calcium homeostasis [24].

Phosphorus

The recognition that bone metabolism in animals is impaired by either a deficiency of calcium or a relative excess of phosphorus is reflected in recommended ratios of calcium to phosphorus in rations for domestic species. Excess serum phosphate is associated with a depression of serum calcium and increased calcium secretion into the gut of adult humans and animals [6]. Parathyroid activity is stimulated by the decrease in serum calcium ("secondary hyperparathyroidism"). These events have been seen in both adult animals [6,15] and adult human subjects consuming high-phosphate diets [1,31].

Bone loss associated with the consumption of excess phosphate has been observed in several species of laboratory animals and has been cited as a cause of some bone diseases of domestic livestock. The adverse effect of high-phosphate diets on the skeleton is not limited to growing animals, as sometimes alleged [8], and has been observed, for example, in adult mice fed excess phosphate between the ages of 18 and 30 months [17].

The concept of an optimal Ca:P ratio that is generally applicable to diets for either animals or humans is compromised by the fact that any such ratio depends on the absolute intake of both elements. The ratio of absorbed calcium to absorbed phosphorus changes progressively when the intake of both elements is increased proportionately. Calcium absorption in adult man and animals is a curvilinear function of intake, whereas phosphorus absorption is a linear function of intake (about 70%) over a very broad range [30]. Consequently, the ratio of absorbed phosphorus to absorbed calcium (the important consideration as far as any effect of the dietary Ca:P ratio on serum calcium homeostasis is concerned) shifts in favor of phosphorus when the intake of both elements is increased at a fixed ratio. This shift is accentuated when phosphorus intake increases without a concomitant increase in calcium (Fig. 3). The dietary Ca:P ratio has no significant effect on bone homeostasis in adult mice when the concentration of both nutrients is relatively low [23] (Fig. 2). The greater efficiency of phosphorus absorption at high intakes, however, necessitates increases in the ratio of dietary calcium to phosphorus to prevent a depression of serum calcium and osteopenia [2,23].

Despite evidence of parathyroid stimulation in human adults consuming high-phosphate diets, calcium balance has been reported to be maintained, indicating that increased bone resorption is not associated with bone loss [25,31]. Even at a relatively low intake of calcium (500 mg) and at a phosphorus intake as high as 2525 mg, bone resorption appears to remain

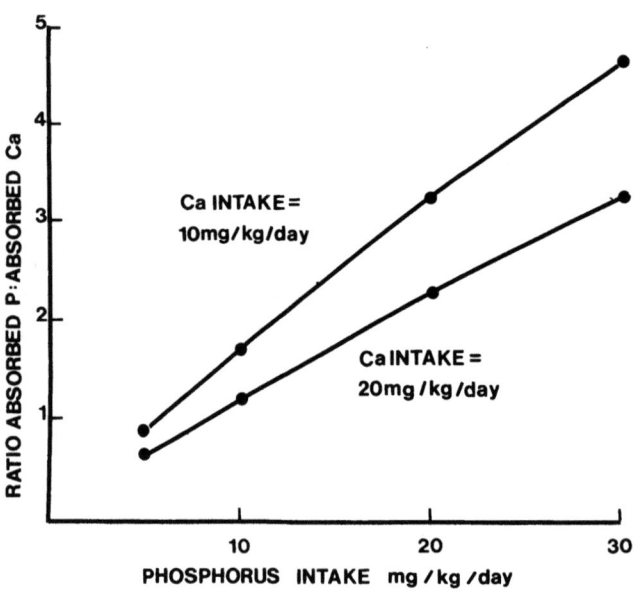

Figure 3. Ratio of absorbed phosphorus to absorbed calcium in man as a function of phosphorus intake at two intakes of calcium. Calculated from the data of Wilkinson [30].

closely coupled to bone formation [10]. These studies indicate that human adults are more adaptable to high-phosphate intakes than adult rodents. A lower remodeling capacity may predispose to bone loss in rodents, but this does not explain the osteopenia produced by excess phosphate in the dog, which remodels bone faster than the human.

Protein

Consumption of experimental diets containing an excess of isolated proteins (but within the range of protein intakes in North America) produces calciuria in human adults and increases the amount of calcium required to maintain calcium balance [10]. Calciuria is due mainly to a decrease in the fractional reabsorption of calcium from the renal tubules and in part to an increase in glomerular filtration rate. Mature rats fed high-protein diets exhibit a similar fractional increase in urinary calcium. However, radiocalcium studies and long-term feeding experiments have shown that there is no net loss of calcium from the skeleton induced by high-protein intakes in this species, the increment in urinary calcium loss being offset by a small decrease in endogenous fecal calcium excretion [29].

The calciuretic effect of a high-protein diet is attributable mainly (though not entirely) to increased acid production, as sulfuric acid, which arises from the oxidation of excess sulfur-containing amino acids [28]. There appears to be a significant difference in the capacity of the rat and man to buffer an exogenous or endogenous acid load. Long-term studies on rats fed excess protein, organic acids, mineral acids, or ammonium chloride [27,28] indicate that this species is capable of buffering acid loads well beyond normal physiological requirements without any adverse effect on bone. Only when acid loads are sufficient to depress serum pH and cause a reduction in food intake is a reduction in bone mass observed [27]. The loss of calcium by human adults consuming large quantities of purified proteins [10] indicates that the resulting excess of endogenous acid is buffered in part at the expense of neutralizing equivalents obtained from bone. Carbonates and phosphates released from bone act as extracellular buffers.

Protein–Phosphorus Interaction

The close association between protein and phosphorus in the human diet and the opposite effects of excesses of these nutrients on urinary calcium raise a question regarding the skeletal response to high-protein diets as normally consumed. Consumption of a high-meat diet, in which phosphorus intake is allowed to rise with protein intake, causes no significant change in urinary calcium, fecal calcium, or calcium balance [26]. The calciuretic effect and negative calcium balance caused by a high intake of purified protein (150 g/day) are also ameliorated by a high intake of phosphorus (2525 mg), even at a modest intake of calcium (500 mg) [10]. These findings provide further evidence that bone homeostasis in human adults is less vulnerable to the increase in parathyroid activity caused by excess phosphate than it is in some adult animals. The resulting increase in renal tubular reabsorption of calcium reduces the impact of the acid load imposed by excess protein on the human skeleton. Whether the phosphorus naturally associated with high-protein diets fully offsets the calciuretic effect of the protein is uncertain. These findings on human adults raise a novel consideration in the assessment of phosphorus needs, i.e., that the intake of phosphorus required to maintain calcium homeostasis rises with the intake of protein.

The protein–phosphorus interaction is not amenable to study in the rat because feeding excess protein has no effect on calcium balance. This difference between rat and human is attributable to the greater capacity of the rat to buffer endogenous acid and to the much smaller fraction (about 5%) of absorbed calcium that is excreted in the urine. Hence, a similar fractional increase in urinary calcium produced by high-protein feeding has a smaller impact on total calcium excretion. In rodents, the adverse effect of excess phosphorus is unaffected by the level of dietary protein. Thus, adult rodents and adult humans differ with respect to the skeletal response to both excess protein and excess phosphorus.

Vitamin D

It has been suggested that the decline in calcium absorption which accompanies aging in some human populations may be related to hypovitaminosis D associated with a decreased exposure to sunlight [8]. Although there is evidence that vitamin D status declines with adult age in some geographic areas, it is uncertain whether this decline is associated with predisposition to osteoporosis. There is no information on any possible relationship between vitamin D and aging osteopenia in animals.

Fluoride

Fluoride administration leads to increased bone mass in osteoporotic subjects and there is evidence that there may be less bone loss in areas with a naturally high fluoride content in the drinking water [3]. The fluoride level in fluoridated water seems to be without effect. High-fluoride drinking water (10 ppm) has been found to have no effect on cortical thinning in aging mice. There is controversy over the dietary essentiality of fluoride in animals and there is no evidence that it has any effect on aging bone loss.

References

1. Bell, R. R.; Draper, H. H.; Tzeng, D. Y. M.; Shin, H. K.; Schmidt, G. R. Physiological responses of human adults to foods containing phosphate additives. *J. Nutr.* **107**:42–50, 1977.
2. Bell, R. R.; Tzeng, D. Y. M.; Draper, H. H. Long-term effects of calcium, phosphorus and forced exercise on the bones of mature mice. *J. Nutr.* **110**:1161–1168, 1980.
3. Bernstein, D.; Sadowski, N.; Hegsted, D. M.; Guri, D.; Stare. F. J. Prevalence of osteoporosis in high and low fluoride areas in North Dakota. *J. Am. Med. Assoc.* **198**:499–504, 1966.
4. Bowden, D. M.; Teets, C.; Witkin, J.; Young, D. M. Long bone calcification and morphology. In: *Aging in Nonhuman Primates*, D. M. Bowden, ed., Princeton, N.J., Van Nostrand–Reinhold, 1979, pp. 335–347.
5. Burnell, J. M.; Baylink, D. J.; Chestnut, C. H.; Mathews, M. W.; Teubner, E. J. Bone matrix and mineral abnormalities in postmenopausal osteoporosis. *Metabolism* **31**:1113–1119, 1982.
6. Draper, H. H.; Sie, T. L.; Bergan, J. G. Osteoporosis in aging rats induced by high phosphorus diets. *J. Nutr.* **102**:1133–1142, 1972.
7. Garn, S. M.; Rohmann, C. G.; Wagner, B. Bone loss as a general phenomenon in man. *Fed. Proc.* **26**:1729–1736, 1967.
8. Heaney, R. P.; Gallagher, J. C.; Johnston, C. C.; Neer, R.; Parfitt, A. M.; Whedon, G. D. Calcium nutrition and bone health in the elderly. *Am. J. Clin. Nutr.* **36**:986–1013, 1982.
9. Hegsted, D. M.; Moscosco, I.; Collazos, C. Study of minimum calcium requirements of adult men. *J. Nutr.* **46**:181–201, 1981.
10. Hegsted, M.; Schuette. S. A.; Zemel. M. B.; Linkswiler, H. M. Urinary calcium and calcium balance in young men as affected by level of protein and phosphorus intake. *J. Nutr.* **111**:553–563, 1981.
11. Henry, K. M.; Kon, S. K. The relationship between calcium retention and body stores of calcium in the rat: Effect of age and vitamin D. *Br. J. Nutr.* **7**:147–159, 1953.

12. Hironaka, R.; Draper, H. H.; Kastelic, J. Physiological aspects of aging. III. The influence of aging on calcium metabolism in rats. *J. Nutr.* **71**:356–360, 1960.
13. Jee, W. S. S.; Kimmel, D. B.; Hashimoto, E. G.; Dell, R. B.; Woodbury, L. A. Quantitative studies of beagle lumbar vertebral bodies. In: *Bone Morphology*, Z. F. G. Jaworski, ed., Ottawa, University of Ottawa Press, 1976, pp. 110–117.
14. Jowsey, J.; Gershon-Cohen, J. Effect of dietary calcium levels on production and reversal of experimental osteoporosis in cats. *Proc. Soc. Exp. Biol. Med.* **116**:437–441, 1964.
15. Jowsey, J.; Reiss, E.; Canterbury, J. M. Long-term effects of high phosphate intake on parathyroid hormone levels and bone metabolism. *Acta Orthop. Scand.* **45**:801–808, 1974.
16. Krishnarao, G. V. G.; Draper, H. H. Age-related changes in the bones of the adult mice. *J. Gerontol.* **24**:149–151, 1969.
17. Krishnarao, G. V. G.; Draper, H. H. Influence of dietary phosphate on bone resorption in senescent mice. *J. Nutr.* **102**:1143–1146, 1972.
18. Krook, L.; Lutwak, L.; Henrikson, P. A.; Kallfetz, F.; Hirsch, C.; Romanus, B.; Belanger, L. F.; Marier, J. R.; Sheffy, B. E. Reversibility of nutritional osteoporosis: Physicochemical data on bones from an experimental study in dogs. *J. Nutr.* **101**:233–246, 1971.
19. Malm, O. J. Adaptation to alterations in calcium intake. In: *The Transfer of Calcium and Strontium across Biological Membranes*, R. H. Wasserman, ed., New York, Academic Press, 1963, pp. 143–173.
20. Matkovic, V.; Kostial, K.; Simonovic, I.; Buzina, R.; Broderec, A.; Nordin, B. E. C. Bone status and fracture rates in two regions of Yugoslavia. *Am. J. Clin. Nutr.* **32**:540–549, 1979.
21. Mazess, R. B. On aging bone loss. *Clin. Orthop. Relat. Res.* **165**:239–252, 1982.
22. National Research Council. Committee on Animal Models for Research on Aging, Washington, D.C., National Academy Press, 1981.
23. Shah, B. G.; Krishnarao. G. V. G.; Draper, H. H. The relationship of calcium and phosphorus nutrition during adult life and osteoporosis in aged mice. *J. Nutr.* **92**:30–42, 1967.
24. Siu, G. M.; Hadley, M.; Draper, H. H. Self-regulation of phosphate intake by growing rats. *J. Nutr.* **111**:1681–1685, 1981.
25. Spencer, H.; Kramer, L.; Osis, D.; Norris, C. Effect of phosphorus on the absorption of calcium and on calcium balance in man. *J. Nutr.* **108**:447–457, 1978.
26. Spencer, H.; Kramer, L.; Osis, D.; Norris. C. Effect of a high protein (meat) diet on calcium balance in man. *Am. J. Clin. Nutr.* **31**:2167–2180, 1978.
27. Upton, P. K.; L'Estrange, J. L. Effects of chronic hydrochloric and lactic acid administrations on food intake, blood acid–base balance and bone composition in the rat. *Q. J. Exp. Physiol.* **62**:223–235, 1977.
28. Whiting, S. J.; Draper, H. H. Role of sulfate in the calciuria of high protein diets. *J. Nutr.* **110**:212–222, 1980.
29. Whiting, S. J.; Draper, H. H. Effect of chronic high protein feeding on bone composition in the adult rat. *J. Nutr.* **111**:178–183, 1981.
30. Wilkinson, R. Absorption of calcium, phosphorus and magnesium. In: *Calcium, Phosphate and Magnesium Metabolism*, B. E. C. Nordin, ed., Edinburgh, Churchill Livingstone, 1976, pp. 36–112.
31. Zemel, M. B.; Linkswiler, H. M. Calcium metabolism in the young adult male as affected by level and form of phosphorus intake and level of calcium intake. *J. Nutr.* **111**:315–324, 1981.

63

Factors Influencing Calcium Balance in Man

Herta Spencer and Lois B. Kramer

Calcium balance studies are time-consuming and tedious to perform. They also require strictly controlled study conditions, namely a constant dietary intake, completeness of collections of excreta, analyses of calcium intake as well as urinary and fecal excretions. These balances have to be determined for relatively long periods of time in order to obtain interpretable data because of the time required for adaptation to a given calcium intake. Another reason for employing long-term balance studies is the fact that passage of minerals, including calcium, through the intestine may be considerably delayed [2,38]. Despite these difficulties and the unavoidable errors inherent in determining calcium balances, the state of calcium metabolism can be quite reliably assessed by determining the intake and excretion of calcium in urine and stool. The loss of calcium in sweat is usually not monitored, and calcium balances must therefore be considered as maximal values.

Most of the ingested calcium passes unabsorbed through the intestine, irrespective of the calcium intake. The urinary calcium varies from subject to subject even during a constant intake and depends on many factors, such as the intestinal absorption, previous intake, deposition of absorbed calcium in bone, rate of bone resorption, state of mobilization, renal and parathyroid function, other hormonal functions, interactions of calcium with other minerals, and probably other still unknown factors. Aside from the effect of age [1], state of activity [9,10,48], and hormonal status [2], dietary factors and drugs can affect calcium balance [42].

The present communication is concerned with calcium balance studies carried out in various age groups and during the intake of a variety of minerals, drugs, and high protein intake. These studies were performed in adult males under strictly controlled conditions in the Metabolic Research Unit. The diet was kept constant throughout the relatively long-term studies and complete collections of urine and stool were obtained. The diet and the excretions in urine and stool were analyzed for calcium and phosphorus throughout the studies. Calcium was analyzed by atomic absorption spectroscopy [50,51] and phosphorus by a modification of the method of Fiske and SubbaRow [12].

Table I shows mean values for calcium balance determined in 10 adult males and 10

Herta Spencer and Lois B. Kramer • Metabolic Section, Veterans Administration Hospital, Hines, Illinois 60141.

Table I. Calcium Excretions and Balances during a Low Calcium Intake

	Age (years)	Study days	Calcium (mg/day)				Phosphorus (mg/day)			
			Intake	Urine	Stool	Balance	Intake	Urine	Stool	Balance
Female patients (10)										
Average	63	30	162	92	144	−75	659	460	183	+35
S.E.M.	± 2.5	± 4.4	± 6.9	± 16.3	± 10.4	± 13.6	± 26.6	± 29.5	± 10.4	± 28.4
Male patients (10)										
Average	63	37	214*	104	196*	−86	866*	547	239†	+80
S.E.M.	± 2.8	± 5.0	± 10.0	± 15.4	± 17.2	± 12.3	± 27.3	± 32.9	± 20.5	± 16.4

*$p < 0.001$; †$p < 0.05$.

adult females during low calcium intake (approximately 200 mg/day). The studies were carried out for 18 to 72 days in male subjects and for 18 to 56 days in females. The calcium balances of both males and females of similar age were negative, −86 mg/day for males and −75 mg/day for females.

We have observed in extensive long-term studies that there is no adaptation to a low calcium intake with time. For example, the calcium balance of a patient who received a low calcium intake of 230 mg/day for 108 days was as negative at the end of the study as it was at 18 or 60 days of low calcium intake. Calcium balance at 18 days was −88 mg/day and at the end of 60 and 108 days, −43 and −92 mg/day, respectively. We have also observed that the intestinal absorption of calcium, determined in Ca^{47} absorption studies, does not increase with time in order to compensate for the prolonged low calcium intake. The lack of adaptation to a low calcium intake has been reported [3] but is in contrast to reports of other investigators [11,15].

Figure 1 shows data on calcium balance during different calcium intakes obtained in 88 studies. During a low calcium intake of approximately 230 mg/day, the calcium balance was

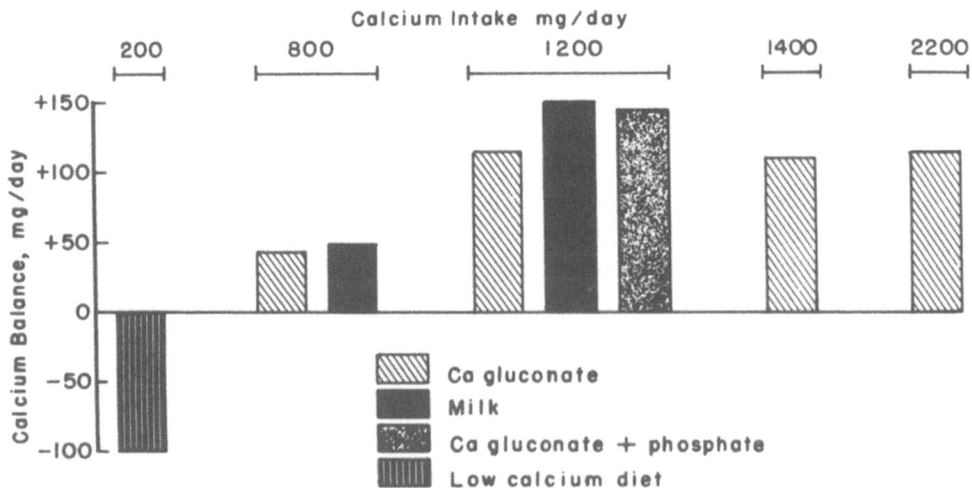

Figure 1. Calcium balances during different calcium intakes. Bars below the line denote negative calcium balance, those above the line represent positive balance.

negative, as expected. During a normal calcium intake of 800 mg/day, the calcium balance was slightly positive, +43 mg/day. This balance was slightly more positive when 800 mg calcium was given as whole milk rather than as calcium gluconate. Increasing the calcium intake to 1200 mg/day, resulted in a further increase in the positivity of calcium balance, to an average of +110 mg/day. Here again, the calcium balance was more positive during the intake of milk than during the intake of calcium gluconate. This improvement of the calcium balance during the intake of milk was not due to increased absorption of calcium from the intestine, but due to a lowering of urinary calcium. Since phosphorus is known to decrease urinary calcium [14,44] and the phosphorus content of milk is high, an additional study was carried out in which the same amount of calcium as calcium gluconate and the same amount of phosphorus as glycerophosphate was given as that contained in milk at the 1200 mg calcium intake. The calcium balance was similar to the balance during the intake of milk, strongly indicating that the lowering of the urinary calcium during milk intake is due to the phosphorus content of milk.

The increase in calcium intake to higher levels, i.e., to 1400 and 2200 mg calcium/day, failed to improve calcium balance further, indicating that a plateau is reached at the 1200 mg calcium intake level. Further analysis of the data obtained in this study has shown that a relatively high percentage of patients receiving a calcium intake of 800 mg/day were in negative calcium balance. These subjects had either latent, asymptomatic osteoporosis or roentgenographic evidence of this bone condition, indicating that this group requires a higher calcium intake than 800 mg/day. In fact, most of these osteoporotic patients were in positive calcium balance at 1200 mg calcium intake. The observation for the higher calcium requirement than 800 mg/day for patients with osteoporosis is in agreement with studies carried out by Heaney and Recker [17] who have shown that postmenopausal women require a calcium intake of 1500 mg/day. A number of investigators who have reported on the calcium requirement [3,15,17,21,37] state that various amounts of calcium are needed. An earlier report indicates that the calcium requirement of adults is much lower [20] than that recommended by the Food and Nutrition Board of the National Academy of Sciences [32]. Recently, other factors such as smoking and caffeine have been reported to affect the calcium requirement [16].

Table II shows the effect of various minerals on calcium balance. Increasing phosphorus

Table II. Effect of Minerals on the Calcium Balance

Patient	Study	Days	Calcium (mg/day)			
			Intake	Urine	Stool	Balance
1	Control[a]	40	1217	206	908	+103
	High phosphorus	36	1204	131	921	+152
2	Control[b]	30	812	214	592	+6
	High magnesium	42	818	200	605	+13
3	Control[c]	24	1972	114	1668	+190
	High fluoride	96	1944	66	1688	+190
4	Control[d]	30	230	93	299	−162
	High zinc	94	233	45	350	−163

[a]Phosphorus intake in control study = 800 mg/day and in high phosphorus study = 2000 mg/day due to the addition of sodium glycerophosphate.
[b]Magnesium intake in control study = 264 mg/day and in high magnesium study = 826 mg/day due to the addition of MgO.
[c]Fluoride intake in control study ≃ 4.5 mg/day and in high fluoride study = 50 mg/day due to the addition of sodium fluoride.
[d]Zinc intake in control study = 15 mg/day and in high zinc study = 155 mg/day due to the addition of zinc as sulfate.

intake by a factor of approximating 2.5, i.e., from 800 to 2000 mg/day during a calcium intake of 1200 mg/day, decreased the urinary calcium; the fecal calcium remained essentially unchanged and the calcium balance became slightly more positive. Increasing the magnesium intake from 264 mg/day to 826 mg/day by adding magnesium oxide to the constant diet did not produce changes in urinary or fecal calcium nor in the calcium balance during a calcium intake of 800 mg/day. Changing the fluoride intake from 4.5 mg/day, contained in the diet and in the drinking water consumed, to 50 mg/day, by giving sodium fluoride supplements, during a high calcium intake of approximately 2000 mg/day resulted in a decrease in urinary calcium, while fecal calcium and the calcium balance remained unchanged. Increasing the zinc intake from 15 mg/day, contained in the diet, to 155 mg/day by adding zinc sulfate supplements, elicited a slight to moderate increase in fecal calcium, which was associated with a decrease of the urinary calcium. Thus, overall calcium balance was unchanged during the high zinc intake.

With regard to the effect of phosphorus on calcium metabolism, a decrease in urinary calcium has previously been reported [14]. The lack of change in calcium balance in our studies is in agreement with an earlier report [26]. In view of our observations that intestinal absorption of calcium does not change during high phosphorus intake, irrespective of calcium intake [44], the dietary Ca/P ratio does not appear to play an important role. This viewpoint has also been expressed by the World Health Organization [52] and by other investigators [18]. In our studies the addition of magnesium to the diet did not affect calcium balance (Table II), which is in agreement with another report [36]. However, in some of our studies, the high magnesium intake was associated with an increase in urinary calcium, probably due to exchange of magnesium with calcium in bone. Large amounts of fluoride are used in the treatment of osteoporosis, because of its reported ability to improve calcium balance and to increase intestinal absorption of calcium [33]. Our strictly controlled studies on the effect of fluoride on calcium metabolism could not confirm either of these findings [45]. A consistent finding was the decrease in urinary calcium during high fluoride intake, which is a desirable effect in osteoporosis. This decrease in urinary calcium is most likely the result of diminished bone resorption during intake of fluoride.

A great deal of information is available on the effect of calcium on availability of zinc, particularly in animals [28,29], but only a few studies have been carried out in man [30,41]. On the other hand, to our knowledge, no studies have been reported on the effect of zinc on calcium metabolism in man. In studies carried out in this Research Unit, large amounts of zinc significantly decreased intestinal absorption of calcium if zinc supplements were given during a low calcium intake [46]. The increase in fecal calcium as well as the decrease in urinary calcium during the high zinc intake (Table II) reflect the decrease in calcium absorption.

Table III shows the effects of various drugs on calcium balance. When corticosteroids were given (Aristocort 40 mg/day) daily for 60 days, urinary and fecal calcium increased, and the calcium balance changed from a positive to a negative value. The administration of thyroid extract increased urinary and fecal calcium markedly, and calcium balance became distinctly negative. During long term use of isoniazid, an antituberculous drug, urinary calcium was very high in relation to the calcium intake; the fecal calcium was relatively low and the calcium balance was highly negative, −284 mg/day. The administration of the antibiotic tetracycline (250 mg four times daily) led to an increase in urinary and fecal calcium, and calcium balance changed from a positive to a negative value. The use of the antacid Maalox containing aluminum hydroxide and magnesium hydroxide, used in small doses of 30 ml three time daily, produced an increase in urinary and fecal calcium, and the calcium balance became more negative. When a large dose of the same antacid was given

Table III. Effect of Drugs on the Calcium Balance

Patient	Study	Study days	Calcium (mg/day)			
			Intake	Urine	Stool	Balance
1	Control	36	1955	90	1733	+132
	Corticosteroids (Aristocort, 40 mg/day)	60	1950	209	1877	−136
2	Control	30	158	63	72	+23
	Thyroid extract gr III	36	160	179	190	−209
3	Isoniazid 100 mg three times daily long term (3 years)	84	240	414	110	−284
4	Control	36	1475	157	1263	+55
	Tetracycline 250 mg four times daily	24	1488	211	1422	−145
5	Control	30	231	89	229	−87
	Maalox 30 ml three times daily	36	225	123	264	−162
6	Control	30	254	86	240	−72
	Maalox 30 ml (15 doses = 450 ml/day)	28	279	421	380	−522

therapeutically (30 ml every hour for 15 doses/day), urinary and fecal calcium increased and calcium balance became very negative (−522 mg/day).

With regard to the effect of drugs on calcium metabolism, corticosteroids [25], as well as thyroid extract or increased function of the thyroid gland [5,8], increase urinary and fecal calcium excretion similar to data reported herein (Table III). The increase in urinary calcium induced by long-term use of isoniazid and tetracycline could lead to considerable calcium loss [42]. Little is known about the mechanism of the hypercalciuria induced by isoniazid. Isoniazid has been shown to inhibit lysyl oxidase and to interfere with deamination of lysyl residues in collagen [4]. With regard to the effect of tetracycline, X-ray diffraction studies of teeth have shown that hydroxyapatite crystal is hypomineralized in areas of tetracycline localization [27]. Large doses of aluminum-containing antacids induce calcium loss [24,31]. This loss results from phosphorus depletion induced by the complexation of dietary phosphorus in the intestine with the aluminum in the antacids. Phosphorus depletion increases bone resorption [24], leading to the removal of phosphorus and calcium from bone and to increased excretion of urinary and fecal calcium. Even relatively small doses of these antacids can have a similar effect [40].

The effect of a high-protein diet given as meat on calcium balance is shown in Table IV. These studies were carried out during different calcium intakes and the duration of the high protein intake ranged from 36 to 132 days. Irrespective of calcium intake, the high protein intake given as meat did not affect urinary or fecal calcium excretion, or calcium balance.

Purified proteins are reported to increase urinary calcium in animals [6] and man [19,23,47]. The mechanism of this increase appears to be based on decreased tubular reabsorption of calcium caused by the sulfur amino acid content of these high-protein diets [35,49]. The differences observed in other studies using purified proteins [6,19,23,35,49] and in our studies in man using red meat as the protein source [39,43] appear to be due to the high phosphorus content of meat. Phosphorus decreases urinary calcium [14,44] and any increase in urinary calcium which might have been induced by the high meat intake could

Table IV. Effect of a High Protein (Meat) Intake on Calcium Balance

Patient	Study	Protein (g/kg)	Phosphorus intake (mg/day)	Study days	Calcium (mg/day)			
					Intake	Urine	Stool	Balance
1	Control	1.0	874	30	217	93	218	−94
	High protein[a]	2.0	1167	36	205	110*	177	−82
2	Control	1.0	751	48	848	187	707	−46
	High protein[a]	2.0	1274	132	830	156	719	−45
3	Control	1.0	859	24	1113	69	863	+181
	High protein[a]	2.0	1189	42	1109	50	849	+210

[a]High protein intake given as red meat.
*Not significant.

have been counteracted by the high phosphorus content of meat. A recent study in animals using beef as the source of high protein also reported that urinary calcium did not increase [7]. The addition of phosphorus to high-protein diets containing purified proteins also led to a decrease of the elevated urinary calcium in man [34].

In conclusion, although metabolic balance studies are limited in scope because of problems inherent in this type of study [13,22], the results obtained under controlled conditions reflect the state of calcium metabolism of the individual. Among various dietary factors, calcium intake plays a major role in maintaining calcium balance. Phosphate, zinc, and fluoride ingested in relatively large amounts cause only minor changes of calcium balance depending on the decrease in urinary calcium. Magnesium supplements have a similar effect but may lead to an increase of the urinary calcium, probably on the basis of exchange of magnesium and calcium in bone. A high-protein diet given as red meat does not induce hypercalciuria or change the intestinal absorption of calcium and thus does not affect calcium balance. Several commonly used drugs increase urinary calcium and cause a negative calcium balance. It is conceivable that the calcium loss induced by the long-term use of such drugs may contribute to skeletal demineralization.

References

1. Albanese, A. A. Principles and applications for nutrition and diet in aging. In *Handbook of Geriatric Nutrition*, J. M. Hsu and R. L. Davis, eds., Park Ridge, N.J., Noyes Publications, 1981, pp. 219–249.
2. Albright, F.; Reifenstein, E. D., Jr. *The Parathyroid Glands and Metabolic Bone disease*, Baltimore, Williams & Wilkins, 1948.
3. Allen, L. H. Calcium bioavailability and absorption: A review. *Am. J. Clin. Nutr.* 35:783–808, 1982.
4. Arem, J. H.; Misiorowski, R. Lathyritic activity of isoniazid. *J. Med. (N.Y.)* 7:239–248, 1976.
5. Aub, J. C.; Bauer, W.; Heath, C.; Ropes, M. Studies of calcium and phosphorus metabolism. III. The effects of the thyroid hormone and thyroid disease. *J. Clin. Invest.* 7:97–137, 1929.
6. Bell, R. R.; Engelman, D. T.; Sie, T.; Draper, H. H. Effect of a high protein intake on calcium metabolism in the rat. *J. Nutr.* 105:475–483, 1975.
7. Calvo, M. S.; Bell, R. R.; Forbes, R. M. Effect of protein-induced calciuria on calcium metabolism and bone status in adult rats. *J. Nutr.* 112:1401–1413, 1982.
8. Cook, P. B.; Nassim, J. R.; Collins, J. The effects of thyrotoxicosis upon the metabolism of calcium, phosphorus, and nitrogen. *Q. J. Med.* 28:505–529, 1959.
9. Deitrick, J. E.; Whedon, G. D.; Shorr, E. Effects of immobilization upon various metabolic and physiologic functions of normal men. *Am. J. Med.* 4:3–36, 1948.
10. Donaldson, C. L.; Hulley, S. B.; Vogel, J. M.; Hattner, R. S.; Bayers, J. H.; McMillian, D. E. Effect of prolonged bed rest on bone mineral. *Metabolism* 19:1071–84, 1970.

11. Draper, H. H.; Scythes, C. A. Calcium, phosphorus, and osteoporosis. *Fed. Proc.* **40**:2434–2438, 1981.
12. Fiske, C. H.; SubbaRow, Y. T. The colorimetric determination of phosphorus. *J. Biol. Chem.* **66**:375–400, 1925.
13. Forbes, G. B. Another source of error in the metabolic balance method. *Nutr. Rev.* **31**:297, 1973.
14. Goldsmith, R. S.; Ingbar, S. H. Inorganic phosphate treatment of hypercalcemia of diverse etiologies. *N. Engl. J. Med.* **274**:1–7, 1966.
15. Heaney, R. P.; Gallagher, J. C.; Johnston, C. C.; Neer, R.; Parfitt, A. M.; Whedon, G. D. Calcium nutrition and bone health in the elderly. *Am. J. Clin. Nutr.* **36**:986–1013, 1982.
16. Heaney, R. P.; Recker, R. R. Effects of nitrogen, phosphorus, and caffeine on calcium balance in women. *J. Lab. Clin. Med.* **99**:46–55, 1982.
17. Heaney, R. P.; Recker, R. R.; Saville, P. D. Calcium balance and calcium requirements in middle-aged women. *Am. J. Clin. Nutr.* **30**:1603–1611, 1977.
18. Hegsted, D. M. Mineral intake and bone loss. *Fed. Proc.* **26**:1747–1754, 1967.
19. Hegsted, M.; Linkswiler, H. M. Long-term effects of level of protein intake on calcium metabolism in young adult women. *J. Nutr.* **111**:244–251, 1981.
20. Hegsted, D. M.; Moscosco, I.; Collazos. C. H. C. Study of minimum calcium requirements by adult men. *J. Nutr.* **46**:181–201, 1952.
21. Irwin, M. I.; Kienholz, E. W. A conspectus of research on calcium requirements of man. *J. Nutr.* **103**:1019–1095, 1973.
22. Isaksson, B.; Sjorgren, B. A critical evaluation of the calcium balance technique. I. Variation in fecal output. *Metabolism* **16**:295–302, 1967.
23. Linkswiler, H. M.; Zemel, M. B.; Hegsted, D. M.; Schuette, S. Protein-induced hypercalciuria. *Fed. Proc.* **40**:2429–2433, 1981.
24. Lotz, M.; Zisman, E.; Bartter, F. C. Evidence for a phosphorus-depletion syndrome in man. *N. Engl. J. Med.* **278**:409–415, 1968.
25. Lukert, B. P.; Adams, J. S. Calcium and phosphorus homeostasis in man: Effect of corticosteroids. *Arch. Intern. Med.* **136**:1249–1253, 1976.
26. Malm, O. J. On phosphates and phosphoric acid as dietary factors in the calcium balance of man. *Scand. J. Clin. Lab. Invest.* **5**:75–84, 1953.
27. Nonomura, E.; Okamoto, M.; Sobue, S.; Moriwaki, Y. X-ray microbeam diffraction analyses on a tooth discolored by tetracycline. *J. Dent. Res.* **56**:447–450, 1977.
28. Oberleas, D.; Muhrer, M. E.; O'Dell, B. L. Effects of phytic acid on zinc availability and parakeratosis in swine. *J. Anim. Sci.* **21**:57–61, 1962.
29. O'Dell, B. L.; Savage, J. E. Effect of phytic acid on zinc availability. *Proc. Soc. Exp. Biol. Med.* **103**:304–306, 1960.
30. Pecoud, A.; Donzel, P.; Schelling, J. L. Effect of foodstuffs on the absorption of zinc sulfate. *Clin Pharmacol. Ther.* **17**:469–474, 1975.
31. Pronove, P.; Bell, N. H.; Bartter, F. C. Production of hypercalciuria by phosphorus deprivation on a low calcium intake: A new clinical test for hyperparathyroidism. *Metabolism* **10**:364–371, 1961.
32. Recommended dietary allowances, 8th rev. ed., Washington, D.C., National Academy of Sciences, 1980.
33. Rich, C.; Ensinck, J.; Ivanovich, P. The effects of sodium fluoride on calcium metabolism of subjects with metabolic bone diseases. *J. Clin. Invest.* **43**:545–556, 1964.
34. Schuette, S. A.; Linkswiler, H. M. Effects on Ca and P metabolism in humans by adding meat, meat plus milk, or purified proteins plus Ca and P to a low protein diet. *J. Nutr.* **112**:338–349, 1982.
35. Schuette, S. A.; Zemel, M. B.; Linkswiler, H. M. Studies on the mechanism of protein-induced hypercalciuria in older men and women. *J. Nutr.* **110**:305–315, 1980.
36. Schwartz, R.; Woodcock, N. A.; Blakely, J. D.; MacKellar, I. Metabolic responses of adolescent boys to two levels of dietary magnesium and protein. II. Effect of magnesium and protein level on calcium balance. *Am. J. Clin. Nutr.* **26**:519–523, 1973.
37. Spencer, H. Osteoporosis: Goals of therapy. *Hosp. Pract.* **17**:131–151, 1982.
38. Spencer H.; Friedland, J. A.; Ferguson, V. Human balance studies in mineral metabolism. In: *Biological Mineralization*, I. Zipkin, ed., New York, Wiley, 1973, pp. 689–727.
39. Spencer, H.; Kramer, L.; DeBartolo, M.; Norris, C.; Osis, D. Further studies of the effect of a high protein diet as meat on calcium metabolism. *Am. J. Clin. Nutr.* **37**:924–929, 1983.
40. Spencer, H.; Kramer, L.; Norris, C.; Osis, D. Effect of small doses of aluminum-containing antacids on calcium and phosphorus metabolism. *Am. J. Clin. Nutr.* **36**:32–40, 1982.
41. Spencer, H.; Kramer, L.; Osis, D. Zinc balances in humans. In: *Clinical, Biochemical and Nutritional Aspects of Trace Elements*, A. S. Prasad, ed., New York, Liss, 1982, pp. 103–115.

42. Spencer, H.; Kramer, L.; Osis, D. Factors contributing to calcium loss in aging. *Am. J. Clin. Nutr.* **36**:776–787, 1982.
43. Spencer, H.; Kramer, L.; Osis, D.; Norris, C. Effect of a high protein (meat) intake on calcium metabolism in man. *Am. J. Clin. Nutr.* **31**:2167–2180, 1978.
44. Spencer, H.; Kramer, L.; Osis, D.; Norris, C. Effect of phosphorus on the absorption of calcium and on the calcium balance in man. *J. Nutr.* **108**:447–457, 1978.
45. Spencer, H.; Lewin, I.; Osis, D.; Samachson, J. Studies of fluoride and calcium metabolism in patients with osteoporosis. *Am. J. Med.* **49**:814–822, 1970.
46. Spencer, H.; Osis, D.; Kramer, L. Metabolic effects of pharmacologic doses of zinc in man. *Am. J. Clin. Nutr.* **30**:611, 1977.
47. Walker, P. M.; Linkswiler, H. M. Calcium retention in the adult human male as affected by protein intake. *J. Nutr.* **102**:1297–1302, 1972.
48. Whedon, G. D.; Shorr, E. Metabolic studies in paralytic acute anterior poliomyelitis. II. Alterations in calcium and phosphorus metabolism. *J. Clin. Invest.* **36**:966–981, 1957.
49. Whiting, S. J.; Draper, H. H. Effect of chronic high protein feeding on bone composition in the adult rat. *J. Nutr.* **111**:178–183, 1981.
50. Willis, J. B. Determination of metals in blood serum by atomic absorption spectroscopy. I. Calcium. II. Magnesium. *Spectrochim. Acta* **16**:259–272, 1960.
51. Willis, J. B. Determination of calcium and magnesium in urine by atomic absorption spectroscopy. *Anal. Chem.* **33**:556–559, 1961.
52. World Health Organization Technical Report Series No. 230. Calcium requirements, Geneva, WHO, 1962, pp. 16–18.

64

Role of Dietary Calcium and Vitamin D in Alveolar Bone Health
Literature Review Update

Gary S. Rogoff, Roger B. Galburt, and Abraham E. Nizel

Two alveolar bone disorders of major concern to the oral health researcher and practitioner are periodontal disease in dentulous persons and resorption of the residual ridges in the edentulous. Since the health of the skeletal system in general has been correlated with adequacy of dietary intake and efficient utilization of mineralizing nutrients such as calcium, phosphorus, and vitamin D, it follows that a similar relationship probably exists between the utilization of these nutrients and alveolar bone health. The postulates are that (1) deficiencies or ratio imbalances in dietary calcium, phosphorus, and vitamin D may condition adversely the resistance of the alveolar bone to resorption from pathological processes associated with local and/or systemic risk factors, and that (2) the density of alveolar bone can be correlated with the resorption of alveolar bone.

Under normal conditions, alveolar bone formation and resorption are in equilibrium. If osteoid synthesis is insufficient to maintain alveolar homeostasis, osteoporosis or osteopenia will result. This osteoporotic state in the alveolus may be the "negative bone factor" that Glickman refers to as a possible systemic conditioning factor that can promote periodontal disease in dentulous persons [12], and that Atwood refers to as a metabolic factor responsible for the rapid reduction of residual alveolar ridges in the edentulous [6].

The hypothesis of interest here is that dietary supplementation with calcium, phosphorus, and vitamin D in excess of the current recommended dietary allowances should be ingested by dental patients—particularly postmenopausal females who are susceptible to systemic osteoporosis. This supplementation will mitigate the progression of periodontal disease in dentulous patients and that of residual ridge resorption in edentulous patients.

The two-fold purpose of this chapter is: (1) to present an overview of the major research to date which deals with the effects on alveolar bone density of calcium, phosphorus, and

---

Gary S. Rogoff • Department of Complete Denture Prosthetics, Tufts University, School of Dental Medicine, Boston, Massachusetts 02111.   *Roger B. Galburt* • Department of Restorative Dentistry, Tufts University, School of Dental Medicine, Boston, Massachusetts 02111.   *Abraham E. Nizel* • Department of Oral Health Services, Tufts University, School of Dental Medicine, Boston, Massachusetts 02111.

vitamin D dietary deficiencies, imbalances, or supplements, and (2) to make recommendations for future studies to acquire more basic knowledge and to rationalize therapeutic measures for prevention of alveolar bone loss.

Calcium, Phosphorus, and Vitamin D Nutrition and Periodontal Disease

Fifty-six years ago, Jones and Simonton reported "retrograde changes" in the alveolar bone of dogs fed diets low in calcium [15]. They suggested that dietary calcium deficiencies affect alveolar bone more rapidly and to a greater extent than other skeletal structures. Shortly thereafter, Mellanby described from a histologic and radiographic perspective the loss of alveolar bone in dogs fed a vitamin D-deficient ricketic diet [22].

The relationship between osteoporosis found in the general skeletal system and alveolar bone was reported by Becks and Weber [9] in 1931, who fed dogs diets with varying amounts of calcium and vitamin D. Bony changes were more extensive in the paradentium (i.e., periodontal alveolar bone) than in the general skeletal system. This could be interpreted to mean that alveolar osteoporosis serves as an early sign of systemic osteoporosis.

Interest in the interrelationship between dietary calcium deficiency and periodontal disease waned until the report by Henrikson in 1968 [13] that the high incidence of periodontal disease in natives of India might be attributed, in part, to a relatively low dietary calcium intake. Further, he produced secondary hyperparathyroidism (NSH) in beagle dogs by feeding them a low-calcium, high-phosphorus diet and noted that this experimental group suffered more than the control group from gingival recession, marked reduction in alveolar crest height, and loss of interradicular bone—cardinal signs of periodontal disease. Included in the same report were the observations that alveolar bone was most susceptible to osteoporosis, followed in order by vertebrae and long bones.

After reviewing Henrikson's epidemiological data, Alfano reaffirmed that natives of India have calcium-deficient diets and are susceptible to periodontal disease; but this correlation was not observed when similar studies were conducted in other countries [3]. Moreover, Svanberg et al. could not confirm Henrikson's periodontal disease findings when the plaque and calculus (local irritants) on beagle dogs' teeth were thoroughly removed prior to placing them on the dietary calcium deficiency regimen and when oral hygiene was maintained throughout the experimental period [26]. After 18 months on the calcium-deficient diet, their experimental group of six dogs showed no greater periodontal bone loss or gingival inflammation than control animals. Svanberg et al. concluded that dogs made hyperparathyroid on calcium-deficient diets were not more susceptible to gingivitis and that osteoporosis induced by the hyperparathyroidism did not produce marginal periodontal disease. However, Alfano, in reviewing the data of Svanberg et al., has pointed out that the dogs fed the calcium-deficient diet did, in fact, experience significant increases in some periodontal disease indices [2]. Also, the lack of statistically significant differences between control and experimental diet groups may have simply been due to an inadequate sample size.

In a two-part study, Krook et al. compared periodontal histology in humans and adult dogs, both ingesting calcium-deficient diets [17]. The periodontal tissues of both dogs and humans were qualitatively similar. In the second part, they reported an improvement in periodontal health when calcium intake was supplemented. Two key variables in this study were not standardized, i.e., local dental treatment (periodic scaling and plaque control procedures) and interpretation of data used for diagnosing periodontal bone changes.

Similar beneficial effects of dietary calcium supplements on periodontal health have been reported [10,20,25]. In two of these studies [10,20], the criterion for improvement was increased density of mandibular alveolar bone. Although alveolar bone density may be a factor in periodontal health, decreased height of the alveolar crest (a morphological change) and apical migration of the epithelial attachment are the two major sequelae of periodontal disease.

Calcium Nutrition and Residual Alveolar Ridge Resorption

The alveolar process in the edentulous person no longer serves its primary function of tooth support. Chronic progressive alveolar resorption inevitably occurs, producing a significant reduction in maxillary and mandibular residual ridge heights [5]. Residual ridge resorption limits the edentulous patient's potential for successful restoration with a dental prosthesis [7, 14].

Wical and Swoope [30] conducted a survey in which they correlated residual alveolar ridge height in edentulous persons with their respective intake of dietary calcium. Complete denture patients were categorized into two groups: those with minimal alveolar ridge resorption, and those with extreme alveolar ridge resorption. Those patients in the minimal resorption group ingested almost twice as much calcium daily as the extreme resorption group, implying a correlation between low dietary calcium intake and severe residual ridge resorption.

Wical and Brussee [29] reported a double-blind study on the effect of calcium and vitamin D supplements on the rate of residual ridge resorption in immediate denture patients (patients who have their complete denture protheses inserted at the same time as surgery for extraction of the remaining natural teeth). After a 12-month supplementation period (750 mg calcium and 375 U vitamin D), those patients whose diets included the supplementation had 36% less alveolar bone loss than those given a placebo.

Residual Ridge Resorption and Osteoporosis Interrelationships

Atwood, while affirming that evidence is lacking for a direct cause-and-effect relationship between the rate of residual ridge reduction in the edentulous person and osteoporosis [7], cites evidence for osteoporosis in the mandible [8,13,21,27,28] to support the proposition that osteoporosis is a contributing factor to the rate of residual ridge reduction [6]. However, Mercier and Inoue [23] found no relationship between the density of the radius and atrophy of the alveolar ridge, although mandibular density was not measured directly. Also, Kribbs et al. [16] found no correlation between skeletal osteopenia and vertical alveolar ridge resorption in osteoporotic females. Thus, mandibular density changes may only precede the morphological changes evidenced by ridge resorption. Also, the lack of correlation within existing cross-sectional data does not necessarily preclude correlation of longitudinal events for a particular patient.

Relative Susceptibility of Alveolar Bone to Systemic Nutrition

Alveolar bone appears to be more labile than the axial or appendicular skeleton when exposed to systemic nutritional inadequacies, particularly calcium and vitamin D [1,9,11, 15,18,19,24]. In patients with a history of prolonged dietary calcium deficiency, the loss of bone from the mandible is far greater than that from finger phalanx 5-2 [1]. Furthermore, supplementation with calcium, vitamin D, and vitamin C produced an increased density in alveolar bone which was even greater than that in finger phalanx 5-2.

Directions of Future Research

The primary question yet unresolved is whether an adequate dietary intake of mineralizing nutrients such as calcium and vitamin D is effective in decreasing susceptibility to periodontal disease in dentulous persons and in decreasing the rate of residual alveolar ridge resorption in the edentulous.

Critical analysis of the studies briefly reviewed here indicates that valid conclusions concerning the effects of dietary calcium and vitamin D supplementation on alveolar bone health will only be possible when (1) more precise diagnostic criteria for periodontal disease are used (e.g., crestal bone height and apical migration of the epithelial attachments; (2) more comprehensive experimental designs are used (i.e., longitudinal in addition to cross-sectional designs, double-blind and placebo-controlled designs); (3) experimental populations are substantially larger (large enough to produce statistically significant data); and (4) reliable quantitative analyses of alveolar bone density are utilized *in vivo*.

The significance of studying the effect of nutritional factors on alveolar bone health lies in putting into proper perspective the relative roles of metabolic, anatomic, mechanical, and microbial factors in the etiology of the two major alveolar bone diseases, periodontal disease and resorption of the residual ridges. If dietary calcium is able to influence the resistance of tooth-supporting bone to periodontal disease and/or the rate of residual ridge resorption, significant advances in therapeutic and preventive dentistry may be realized.

The prophylactic, rather than therapeutic, approach to osteoporosis in general and alveolar osteoporosis in particular seems most promising. One significant goal is to improve the alveolar bone prognosis for the elderly, the group most prone to osteoporosis, so that they will be able to enjoy their natural teeth for a lifetime. If removal of teeth, however, does become necessary, maintenance of a healthy residual ridge will improve the potential for satisfactory retention and function of dental prostheses.

ACKNOWLEDGMENTS. The authors wish to acknowledge the joint support of the National Dairy Council and the United States Department of Agriculture Human Nutrition Research Center on Aging.

References

1. Albanese, A. A.; Edelson, A. H.; Wein, E. H.; Carroll, L. *Calcium Nutrition and Dental Health*, A Scientific Exhibit, the Burke Rehabilitation Center, White Plains, N.Y., 1982, pp. 1–11.
2. Alfano, M. C. Controversies, perspectives, and clinical implications of nutrition in periodontal disease. *Dent. Clin. North Am.* **20**:519–548, 1976.
3. Alfano, M. C. Nutrition in periodontal disease. In: *Periodontics*, 5th ed., D. A. Grant, ed., St. Louis, Mosby, 1979, pp. 171–197.
4. Alfano, M. C. Nutrition in dental practice—A plea for sanity. *N.Y. State Dent. J.* **47**:450–452, 1981.
5. Atwood, D. A. Reduction of the residual ridges: A major oral disease entity. *J. Prosthet. Dent.* **26**:266–278, 1971.
6. Atwood, D. A. The problem of reduction of residual ridges. In: *Essentials of Complete Denture Prosthodontics*, S. Winkler, ed., Philadelphia, Saunders, 1979, pp. 38–59.
7. Atwood, D. A. Bone loss of edentulous alveolar ridges. Eighth English Symposium, *J. Periodontol.* Spec. Ed. 11–21, 1979.

8. Baylink, D. J.; Wergedal, J. E.; Yamamoto, K.; Manzke, E. Systemic factors in alveolar bone loss. *J. Prosthet. Dent.* **31**:486–505, 1974.
9. Becks, H.; Weber, M. The influence of diet on the bone system with special reference to the alveolar process and the labyrinthine capsule. *J. Am. Dent. Assoc.* **18**:197–264, 1931.
10. Coulston, A.; Lutwak, L. Bone, calcium and fluoride II. *Fed. Proc.* Abstract No. 2845, 1972.
11. Ericcson, Y.; Ekberg, O. Dietetically provoked general and alveolar osteopenia in rats and its prevention or cure by calcium and fluoride. *J. Periodontal Res.* **10**:256–269, 1975.
12. Glickman, I. (ed.) The "bone factor" concept of periodontal disease. In: *Clinical Periodontology*, 4th ed., Philadelphia, Saunders, 1972, pp. 432–439.
13. Henrikson, P.-Å. Periodontal disease and calcium deficiency—An experimental study on the dog. *Acta Odontol. Scand.* **26**(Suppl. 50):1–132, 1968.
14. Henrikson, P.-Å.; Wallenius, K.; Astrand, K. The mandible and osteoporosis. *J. Oral Rehab.* **1**:67–84, 1974.
15. Jones, M. R.; Simonton, F. V. Mineral metabolism in relation to alveolar atrophy in dogs. *J. Am. Dent. Assoc.* **15**:881–911, 1928.
16. Kribbs, P. J.; Smith, D. E.; Chesnut, C. H. Oral findings in osteoporosis. II. Relationship between mandibular and generalized skeletal osteopenia. *J. Prosthet. Dent.* **50**:719–724, 1983.
17. Krook, L.; Lutwak, L.; Whalen, J. P.; Henrikson, P.-Å.; Lesser, G.; Uris, R. Human periodontal disease: Morphology and response to calcium therapy. *Cornell Vet.* **61**:32–53, 1970.
18. Krook, L.; Lutwak, L.; Henrikson, P.Å.; Kallfelz, F.; Hirsch, C.; Romanus, B.; Belanger, L. F.; Marier, J. R.; Shefry, B. E. Reversibility of nutritional osteoporosis: Physicochemical data on bones from an experimental study in dogs. *J. Nutr.* **101**:233–246, 1971.
19. Krook, L.; Whalen, J. P.; Lesser, G. V.; Lutwak, L. Human periodontal disease and osteoporosis. *Cornell Vet.* **62**:371–391, 1972.
20. Lutwak, L.; Krook, L.; Henrikson, P.-Å.; Uris, R.; Whalen, A. C.; Lesser, G. Calcium deficiency and human periodontal disease. *Isr. J. Med. Sci.* **7**:504–505, 1971.
21. Manson, J. P.; Lucas, R. B. A microradiographic study of age changes in the human mandible. *Arch. Oral Biol.* **7**:761–769, 1962.
22. Mellanby, M. Experiments on dogs, rabbits and rats and investigations on man which indicate the power of certain food factors to prevent and control dental diseases. *J. Am. Dent. Assoc.* **17**:1456–1480, 1930.
23. Mercier, P.; Inoue, S. Bone density and serum minerals in cases of residual alveolar ridge atrophy. *J. Prosthet. Dent.* **46**:250–255, 1981.
24. Person, P. Metabolic studies of human alveolar bone disease. *Oral Surg. Oral Med. Oral Pathol.* **12**:610–625, 1959.
25. Spiller, W. A clinical evaluation of calcium therapy for periodontal disease. *Dent. Dig.* September 1971, pp. 522–526.
26. Svanberg, G.; Lindhe, J.; Hugoson, A.; Grondahl, H. G. Effect of nutritional hyperparathyroidism on experimental periodontitis in the dog. *Scand. J. Dent. Res.* **81**:155–162, 1973.
27. Von Wowern, N. Histoquantitation of ground sections of human mandibles. *Scand. J. Dent. Res.* **81**:567–571, 1973.
28. Von Wowern, N.; Stolze, K. Sex and age-difference in bone structure in human mandible. IADR Abstract No. 229, 1977.
29. Wical, K.; Brussee, P. Effects of a calcium and vitamin D supplement on alveolar ridge resorption in immediate denture patients. *J. Prosthet. Dent.* **41**:4–11, 1979.
30. Wical, K.; Swoope, C. C. Studies of residual ridge resorption. Part II. The relationship of dietary calcium and phosphorus to residual ridge resorption. *J. Prosthet. Dent.* **32**:13–22, 1974.

XIV

Normal Biological Calcification

65

Normal Biological Mineralization
Role of Cells, Membranes, Matrix Vesicles, and Phosphatase

H. Clarke Anderson

Biological mineralization is a phenomenon of great importance through the vertebrate and invertebrate kingdoms. Calcification allows man to stand erect, chew food, and maintain calcium and phosphate homeostasis (since bones serve as a storehouse of calcium and phosphate ions). The reason for the selective deposition of rock-hard, insoluble crystalline salts of calcium phosphate only within certain supportive tissues has been a fascination to generations of scientists interested in calcification. Although the reason for selective calcification is still not totally understood, it is an article of faith among many investigators that the skeletal cells (of bone, cartilage, and tooth) somehow initiate and control calcification, keeping it confined to tissues in which hardness is useful to the host. Only under pathological conditions does the normal process of calcium phosphate crystal deposition break free of homeostatic control, producing disastrous consequences [6].

My presentation will focus mainly on the mechanism by which skeletal cells initiate calcification and will introduce areas of investigation to be addressed by subsequent authors. The growth plate or epiphyseal plate will be used to illustrate the calcification process, but it is likely that the same mechanism of calcification also occurs in newly formed bone [14] and dentin [13,23,42,43].

Growth Plate Calcification

The growth plate, sometimes called epiphyseal plate or physis, is located at ends of long bones and is the site at which growth in length of the bone occurs (Fig. 1). It is conventional to subdivide the growth plate into structural–functional layers with reserve zone at the top (a relatively inactive zone); overlying proliferative zone (where most cell divisions occur and chondrocytes lie flattened in vertical columns); overlying hypertrophic zone (with enlarging

---

H. Clarke Anderson • Department of Pathology and Oncology, University of Kansas Medical Center, Kansas City, Kansas 66103.

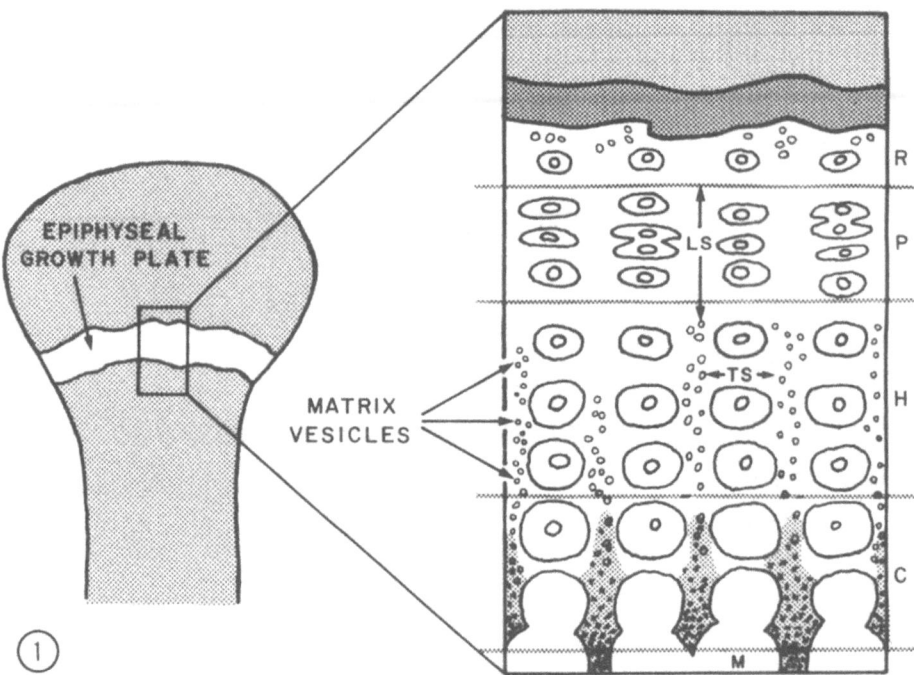

Figure 1. Diagram of the epiphyseal growth plate of a long bone, the site at which growth in length occurs. The growth plate is subdivided into the following anatomical regions: The *reserve zone* (R), at the top of the growth plate, contains apparently inactive chondrocytes. The *proliferative zone* (P) is a zone of active cell division where cell columns first appear, thus allowing the matrix to be anatomically subdivided into transverse matrix septa (TS), separating cells within a column, and longitudinal septa (LS), separating adjacent cell columns. The *hypertrophic zone* (H) contains enlarging chondrocytes and many matrix vesicles, found in clusters in the longitudinal septa. The first mineral crystals arise within matrix vesicles of the hypertrophic zone (see Fig. 2). The *calcifying zone* (C) contains degenerating chondrocytes. This is the level at which proliferating mineral spreads from matrix vesicles radially outward to infiltrate the interstices of the longitudinal septal matrix. At the base of the growth plate lies the bony *metaphysis* (M) with small vessels which remove the uncalcified transverse matrix septa and degenerate cells, and leaving calcified longitudinal septa upon which osteoblasts from the marrow will deposit new bone (the primary spongiosa).

chondrocytes, actively synthesizing and secreting matrix); overlying calcifying zone (with degenerate cells and calcified matrix located almost exclusively in the longitudinal septa). The kinetics of the growth plate are such that daughter cells, derived from stem cells of the proliferative zone, stack up in columns, undergo hypertrophy, and ultimately die and degenerate. Probably most chondrocytes are no longer viable by the time they are approached by small blood vessels of the advancing metaphysis, which erode away the uncalcified transverse matrix septa and degenerate cells. Calcified longitudinal septa are left as a lattice upon which osteoblasts from the marrow space will deposit new medullary bone (the primary spongiosa).

Examination of the calcification process in growth plate by light microscopy reveals two important general features: (1) the process is extracellular; therefore, if cells are to control calcification, they must do so at a distance from the main cell body; and (2) calcification is restricted to longitudinal septa, suggesting that this area of matrix is somehow selectively predisposed to calcify. How calcification of selective sites is brought about seems to be related to the pattern of distribution of matrix vesicles, described below.

Matrix Vesicle Calcification

About 15 years ago, electron microscopy revealed an important difference between longitudinal septa and transverse septa of the growth plate of precalcified hypertrophic cartilage, namely, the presence of clusters of tiny, extracellular matrix vesicles located only in the longitudinal septa [3]. Matrix vesicles are membrane-invested particles, measuring about 200 nm in diameter, which were shown to contain the first electron microscopically recognizable deposits of apatite bone mineral (Fig. 2) [3,16]. These deposits are constituted of a highly insoluble crystalline compound comprised of $(Ca)_6(PO_4)_{10}(OH)_2$. The first crystals of apatite usually appear in apposition to the inner leaflet of the matrix vesicle membrane, a favored site for the formation of such mineral.

There are literally hundreds of reports confirming matrix vesicles as the initial site of mineralization in cartilage, bone, and dentine [4]. However, most of these studies utilized conventional techniques to prepare thin sections which expose calcified tissues to aqueous solutions. The latter might cause loss or translocation of small mineral crystals. Thus, in a recent report, Landis and Glimcher [32], using cryoultramicrotomy to produce anhydrous sections of growth plate, were unable to identify mineral crystals within or even in associa-

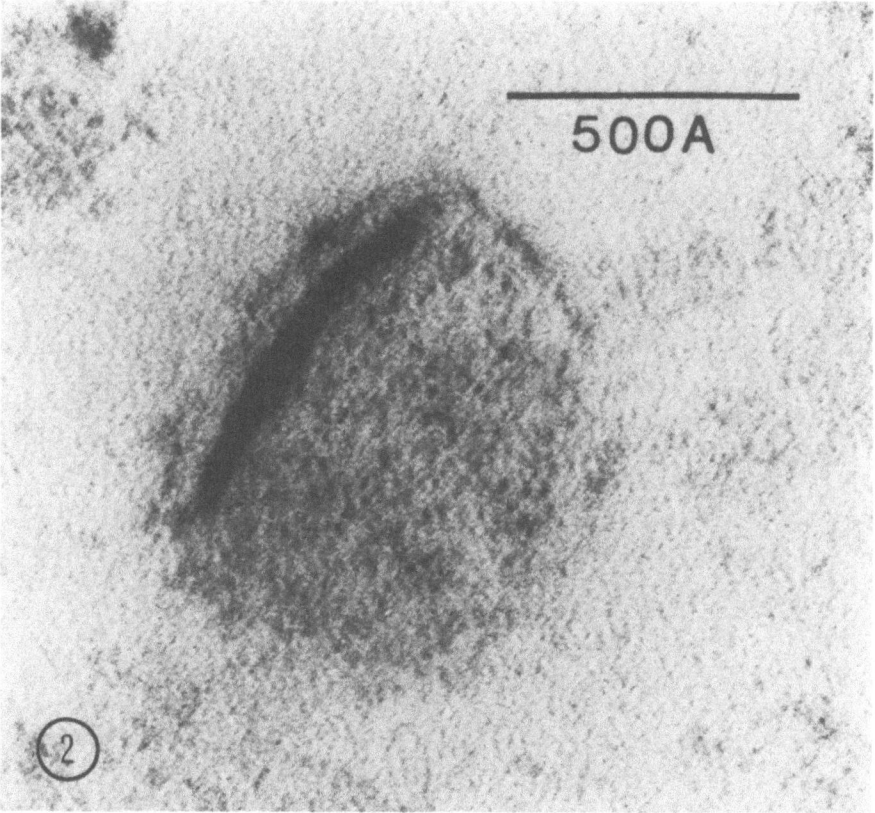

Figure 2. Electron micrography of a calcifying matrix vesicle in the hypertrophic zone of rat growth plate. The first electron microscopically observable crystalline apatite mineral is seen as a needlelike, dense precipitate within the matrix vesicle, often in apposition to the inner leaflet of the vesicle membrane. Stained with lead and uranium, × 512,000. [Reprinted from *Proceedings of the Workshop on Cell-Mediated Calcification—Matrix Vesicles,* Elsevier, Amsterdam, 1983.]

tion with matrix vesicles, and questioned the proposed role of the calcification of matrix vesicles in tissue mineralization.

In an effort to resolve the issue of whether matrix vesicles contain solid-phase calcium phosphate, David Morris has recently carried out an electron microscopic study, in our laboratory, in which rat growth plate cartilage was prepared and sectioned anhydrously by three separate methods: cryoultramicrotomy, similar to the method of Landis and Glimcher [32]; freeze substitution [29]; and ethylene glycol fixation [31]. Solid-phase calcium phosphate was identified within and in association with matrix vesicles by all three techniques, and its presence was confirmed in all three types of preparation by nondispersive X-ray microanalysis [36]. Therefore, we must take issue with the findings of Landis and Glimcher. In so doing it should be pointed out that our results are only the most recent in a series of prior reports [2,12,25,26,37], all of which used anhydrous techniques to demonstrate the presence of solid-phase calcium phosphate mineral associated with matrix vesicles of growth plate and newly formed bone. The critical point that matrix vesicles are the initial site of biological mineralization would now seem to be convincingly established.

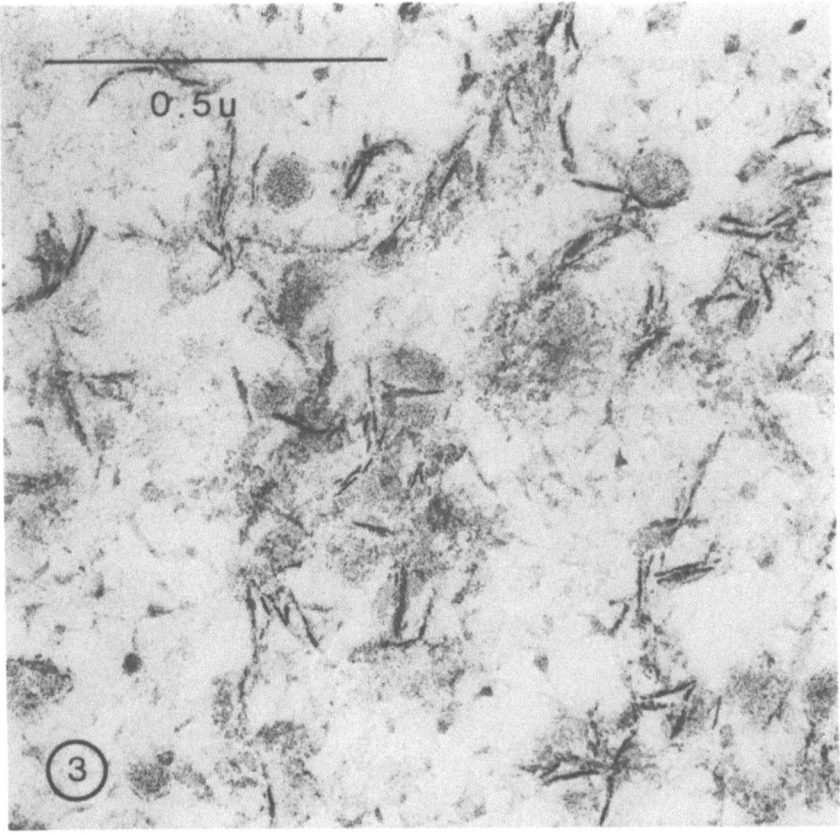

Figure 3. Electron micrograph of calcifying zone. Proliferation of apatite needles occurs at the surfaces of matrix vesicles to form radial clusters of mineral. The latter enlarge and coalesce with adjacent clusters to form contiguous mineral. Matrix vesicles are consumed in the process. Stained with lead and uranium, × 105,000. [Reprinted from *Pathology Annual*, Part 2, Vol. 15, 1980.]

Progressive Calcification

Shortly after the appearance of the first mineral within matrix vesicles, similar apatite crystals begin to accumulate at the surfaces of matrix vesicles (Fig. 3) to form calcospherites of sufficient size to be recognized by light microscopy. The calcospherites then grow by the proliferation and accumulation of tiny apatite crystals, extending into and eventually infiltrating the surrounding matrix. The matrix contains collagen, proteoglycan, noncollagenous proteins, water, electrolytes, and other substituents in variable amounts depending on the tissue, i.e., whether matrix of cartilage, bone, or dentin. An excellent description of the pattern of mineral extension from matrix vesicles, through the extracellular space to involve fibrils, has recently been provided for bone [34], and the same pattern of mineral extension from matrix vesicles, through the extracellular space to involve fibrils, has recently been provided for bone [34], and the same pattern of mineral extension has been repeatedly observed in growth plate cartilage [3,12,16,29,37].

Matrix vesicles not only initiate calcification during fetal bone development [14]. but also postnatally whenever rapid bone growth occurs as, for example, in fracture callus [40], in normal rat alveolar bone and after tooth removal [11,41], in woven and lamellar endochondral bone of newborn mice [14], in postnatal avian bone stimulated to grow by steroids [25], in induced heterotopic bone stimulated by human cells and urinary bladder [5,17], in woven bone of exostoses developing in osteolathyrism [34], in phenotypic cultures of osteoblasts [22,44], and in the osteoid of experimental and pathologically occurring osteomalacic bone [10,20]. Therefore, it is incorrect to assert that matrix vesicles are present in only the very early stages of embryonic chick bone development (Glimcher, this volume). However, electron microscopic evidence strongly suggests that matrix vesicles are involved only in the formation of initial mineral crystals. Following this, the crystals proliferate autocatalytically; thus, their rate of proliferation would seem to be controlled by matrix factors other than vesicles, including collagen and noncollagenous proteins (see Chapter 66) and proteoglycans (see Chapter 67).

Two-Phase Scheme of Calcification

The calcification process probably should be envisioned as having two major phases: (1) initiation of the first crystals (phase 1), and (2) regulation of crystal proliferation (phase 2). A great emphasis has been placed in calcification research upon "initiating" factors, but there seems to be little justification for such a priority of emphasis since 99% of mineral deposition is under the control of regulatory factors rather than true initiators. Thus, an understanding of regulatory factors may actually be the key to successful therapeutic manipulation of the system.

In initiation (phase 1), the matrix vesicle membrane seems to be critically involved (Fig. 2). Calcium, which is attracted by acidic phospholipids concentrated in the matrix vesicles [38,47]. probably accumulates prior to phosphate [1]. Phosphate is concentrated in the matrix vesicle by phosphatases (alkaline phosphatase, ATPase, AMPase, pyrophosphatase, etc.), which are localized to the matrix vesicle membrane [30,35], and are activated as the vesicle approaches the calcification front [35]. Vesicle membrane phosphatase(s) may function as phosphotransferase [18,30], rather than indiscriminately hydrolyzing and liberating orthophosphate. High calcium and high phosphate within the protected microenvironment of the matrix vesicle would be sufficient to promote deposition of solid-phase calcium phosphate mineral, probably at first as an amorphous or poorly crystalline compound [45], but later as highly insoluble hydroxyapatite. The inner leaflet of the vesicle membrane may provide a favorable environment for formation and growth of initial mineral deposits, possibly because of the selective localization of phosphatidylserine (PS) in the inner portion of the vesicle membrane [33]. Yaari has recently developed an intriguing experimental model for membrane calcification in which PS, dissolved in an organic solvent, transports calcium from one aqueous compartment to another, across the nonaqueous, PS-containing compartment. Transport of calcium is promoted by a pH gradient across the organic solvent and the presence of phosphate in the solvent phase [48] (see Chapter 68).

Phase 2, the phase of crystal proliferation, is brought about by exposure of preformed apatite crystals to the extracellular fluid. Cartilage fluid contains insufficient calcium and phosphate to initiate mineral deposition in the absence of apatite [28], but when presented with preformed apatite from vesicles, crystal proliferation would be anticipated, given normal levels of Ca^{2+} and PO_4^{3-} in cartilage fluid. The rate of phase 2 calcification is controlled by factors residing outside of vesicles in the extracellular space. Natural inhibitors of crystal proliferation include: (1) inadequate Ca^{2+} and/or PO_4^{3-} as seen in rickets or osteomalacia [7–10]; (2) certain organic phosphate compounds such as pyrophosphate (PP_i) [24], and ATP [15,45], which prevent hydroxyapatite crystal growth; (3) certain noncollagenous phosphoproteins [19] and γ-carboxyglutamic acid-containing proteins of bone which impede hydroxyapatite deposition from metastable calcium phosphate solutions *in vitro* [39]. Factors seen as promoting phase 2 mineralization would include: (1) elevated Ca^{2+} and/or PO_4^{3-} in the extracellular fluid; (2) collagen, which not only spatially orients hydroxyapatite in bone [21], but also initiates apatite deposition from metastable calcium phosphate solutions [27]; and (3) the recently described protein of bone called "osteonectin," which may serve as a link protein between collagen and hydroxyapatite mineral deposits [46]. As indicated above, the role of these regulating factors is to control the process of mineralization once initiated. Control over one or more regulating factors may permit management of the calcification cascade.

In summary, cells control skeletal mineralization, not only by regulating the composition of the matrix in which calcification occurs but also by depositing tiny, extracellular matrix vesicles at the exact site where calcification will begin. Matrix vesicles contain lipids and enzymes within their membranes, that promote initial calcification. An integrative understanding of the structure and function of the matrix vesicle membrane, which I believe is the "business part" of the matrix vesicle, will lead to a clearer understanding of mineral initiation.

References

1. Ali, S. Y. Analysis of matrix vesicles and their role in the calcification of epiphyseal cartilage. *Fed. Proc.* **35:**135–142, 1976.
2. Ali, S. Y.; Wisby, A.; Craig-Gray, J. Electronprobe analysis of cryosections of epiphyseal cartilage. *Metab. Bone Dis. Relat. Res.* **1:**97–103, 1978.
3. Anderson, H. C. Vesicles associated with calcification in the matrix of epiphyseal cartilage. *J. Cell Biol.* **41:**59–72, 1969.
4. Anderson, H. C. Introduction to the first international conference on matrix vesicle calcification. *Fed. Proc.* **35:**105–108, 1976.
5. Anderson, H. C. Osteogenetic epithelial mesenchymal cell interactions. *Clin. Orthop. Relat. Res.* **119:**211–224, 1976.
6. Anderson, H. C. Calcific diseases. *Arch. Pathol. Lab. Med.* **107:**341–348, 1983.

7. Anderson, H. C.; Hsu, H. H. T. A new method to measure ^{45}Ca accumulation by matrix vesicles in slices of rachitic growth plate cartilage. *Metab. Bone Dis. Relat. Res.* **1**:193–198, 1978.
8. Anderson, H. C.; Sajdera, S. W. Calcification of rachitic cartilage to study matrix vesicle function. *Fed. Proc.* **35**:148–153, 1976.
9. Anderson, H. C.; Cecil, R.; Sajdera, S. W. Calcification of rachitic rat cartilage *in vitro* by extracellular matrix vesicles. *Am. J. Pathol.* **79**:237–255, 1975.
10. Anderson, H. C.; Johnson, T. F.; Avramides, A. Matrix vesicles in osteomalacic bone. *Metab. Bone Dis. Relat. Res.* **25**:79–86, 1980.
11. Bab, I. A.; Muhlrad, A.; Sela, J. Ultrastructural and biochemical study of extracellular matrix vesicles in normal alveolar bone of rat. *Cell Tissue Res.* **202**:1–7, 1979.
12. Barckhaus, R. H.; Krefting, E. R.; Althoft, J.; Quint, P.; Höhling, H. J. Electron-microscopic microprobe analyses on the initial stages of mineral formation in the epiphyseal growth plate. *Cell Tissue Res.* **217**:661–666, 1981.
13. Bernard, G. W. Ultrastructural observations of initial calcification in dentine and enamel. *J. Ultrastruct. Res.* **41**:1–17, 1972.
14. Bernard, G. W.; Pease, D. C. An electron microscopic study of initial intramembranous osteogenesis. *Am. J. Anat.* **125**:271–290, 1969.
15. Betts, F.; Blumenthal, N. C.; Posner, A. S.; Becker, G. L.; Lehninger, A. L. The atomic structure of intracellular amorphous calcium phosphate deposits. *Proc. Natl. Acad. Sci. USA* **72**:2088–2090, 1975.
16. Bonucci, E. Fine structure and histochemistry of calcifying of globules in epiphyseal cartilage. *Z. Zellforsch. Mikrosk. Anat.* **103**:192–217, 1970.
17. Callis, P. D. Bone development following transplants of urinary bladder wall: A quantitative histological and ultrastructural study. *J. Anat.* **135**:53–63, 1982.
18. Cyboron, G. W.; Vijims, M. S.; Wuthier, R. E. Activity of epiphyseal cartilage membrane alkaline phosphatase and effects of its inhibitors at physiological pH. *J. Biol. Chem.* **257**:4141–4146, 1982.
19. Desteno, C. V.; Feagin, F. F. Effect of matrix bound phosphate and fluoride on mineralization of dentin. *Calcif. Tissue Int.* **17**:151–159, 1975.
20. Dopping-Hepenstal, P. J. C.; Ali, S. Y.; Stamp, T. C. B. Matrix vesicles in the osteoid of human bone. In: *Proc. 3rd Int. Conf. Matrix Vesicles*, A. Ascenzi, E. Bonucci, and B. deBernard, eds., Milano, Wichtig Editore, 1981, pp. 229–234.
21. Dudley, H. R.; Spiro, D. The fine structure of bone cells. *J. Biophys. Biochem. Cytol.* **11**:627–649, 1961.
22. Ecarot-Charrier, B.; Glorieux, F. H.; van der Rest, M.; Pereira, G. Osteoblasts isolated from mouse calvaria initiate matrix mineralization in culture. *J. Cell Biol.* **96**:639–643, 1983.
23. Eisenman. D. R.; Glick, P. L. Ultrastructure of initial crystal formation in dentine. *J. Ultrastruct. Res.* **41**:18–28, 1972.
24. Fleisch, H.; Bisaz. S. Mechanism of calcification: Inhibitory role of pyrophosphate. *Nature (London)* **195**:911, 1962.
25. Gay, C. B. The ultrastructure of the extracellular phase of bone as observed in frozen thin sections. *Calcif. Tissue Res.* **23**:215–223, 1977.
26. Gay, C. V.; Schraer, H.; Hargest, T. E., Jr. Ultrastructure of matrix vesicles and mineral in unfixed embryonic bone. *Metab. Bone Dis. Relat. Res.* **1**:105–108, 1978.
27. Glimcher, M. J.; Hodge, A. J.; Schmidt, F. O. Macromolecular aggregation states in relation to mineralization: The collagen–hydroxyapatite system as studied *in vitro*. *Proc. Natl. Acad. Sci. USA* **43**:860–866, 1957.
28. Howell, D. S.; Pita, J. C.; Marquez, J. F.; Madruga, J. E. Partition of calcium, phosphate and protein in the fluid phase aspirated at calcifying sites in epiphyseal cartilage. *J. Clin. Invest.* **47**:1121–1132, 1968.
29. Huntziker, E. B.; Hermann, R. K.; Schenk, R. K.; Marti, T.; Muller, M.; Moor, H. Structural integration of matrix vesicles in calcifying cartilage after cryofixation and free substitution. In: *Proceedings of the Third International Conference on Matrix Vesicles*, A. Ascenci, E. Bonucci, and B. de Bernard, eds., Milan, Wichtig Editore srl, 1981, pp. 25–32.
30. Kanabe, S.; Hsu, H. H. T.; Cecil, R. N. A.; Anderson, H. C. Electron microscopic localization of adenosine triphosphate (ATP) hydrolyzing activity in isolated matrix vesicles and reconstituted vesicles from calf cartilage. *J. Histochem. Cytochem.* **31**:462–470, 1983.
31. Landis, W. J.; Paine, M. D.; Glimcher, M. J. Electron microscopic observations on bone tissue prepared anhydrously in organic solvents. *J. Ultrastruct. Res.* **59**:1–30, 1977.
32. Landis, W. J.; Glimcher, M. J. Electron optical and analytical observations of rat growth plate cartilage prepared by ultracryomicrotomy: The failure to detect a mineral phase in matrix vesicles and the identification of heterodispersed particles as the initial solid phase of calcium phosphate deposited in the extracellular matrix. *J. Ultrastruct. Res.* **78**:227–268, 1982.

33. Majeska, R.; Holwerda, D. L.; Wuthier, R. E. Localization of phosphatidyl serine in isolated chick epiphyseal cartilage matrix vesicles with trinitrobenzenesulfonate. *Calcif. Tissue Int.* **27**:4–56, 1979.
34. Martino, L. J.; Yaeger, V. L.; Taylor, J. J. An ultrastructural study of the role of calcification nodules in the mineralization of woven bone. *Calcif. Tissue Int.* **27**:57–64, 1979.
35. Matsuzawa, T.; Anderson, H. C. Phosphatases of epiphyseal cartilage studied by electron microscopic cytochemical methods. *J. Histochem. Cytochem.* **19**:801–808, 1971.
36. Morris, D. C.; Anderson, H. C. Electron microscopy of matrix vesicle calcification in anhydrous preparations of rat epiphyseal cartilage. *Proc. Am. Soc. Bone Miner. Res.* A-38, 1983.
37. Ozawa. H.; Yamamoto, T. An application of energy-dispersive X-ray microanalysis for the study of biological calcification. *J. Histochem. Cytochem.* **31**:210–213, 1983.
38. Peress, N. S.; Anderson, H. C.; Sajdera, S. W. The lipids of matrix vesicles from bovine fetal epiphyseal cartilage. *Calcif. Tissue Res.* **14**:275–281, 1974.
39. Price, P. A.; Otsuka, A. S.; Poser, J. P.; Kristaponis, J.; Romand, N. Characterization of a gamma-carboxyglutamic acid-containing protein from bone. *Proc. Natl. Acad. Sci. USA* **73**:1447–1451, 1976.
40. Schenk, R. K.; Miller, J.; Zinkernagel, R.; Willenegger, H. Ultrastructure of normal and abnormal bone repair. *Calcif. Tissue Res.* **41**(Suppl.):110–111, 1970.
41. Sela, J.; Bab, I. A. Correlative transmission and scanning electron microscopy of the initial mineralization of healing alveolar bone in rats. *Acta Anat.* **105**:401–408, 1979.
42. Sisca, R. F.; Provenza, D. V. Initial dentin formation in human deciduous teeth: An electron microscopic study. *Calcif. Tissue Res.* **9**:1–16, 1972.
43. Slavkin, H. C.; Bringas, L., Jr.; Croissant, R.; Bavetta, L. A. Epithelial–mesenchymal interactions during odontogenesis. II. Intercellular matrix vesicles. *Mech. Ageing Dev.* **1**:1–23, 1972.
44. Sudo, H.; Kodama, H.-A.; Amagai, Y.; Yamamoto, S.; Kasai, S. *In vitro* differentiation and calcification in a new clonal osteogenic cell line derived from newborn mouse calvaria. *J. Cell Biol.* **96**:191–198, 1983.
45. Termine, J. D.; Conn, K. M. Inhibition of apatite formation by phosphorylated metabolites and macromolecules. *Calcif. Tissue Res.* **22**:149–157. 1976.
46. Termine, J. D.; Kleinman, H. K.; Whitson, W. S.; Conn, K. M.; McGarvey, M. L.; Martin, G. R. Osteonectin, a bone specific protein linking mineral to collagen. *Cell* **26**:99–105, 1981.
47. Wuthier, R. E. Lipid composition of isolated cartilage cells, membranes and matrix vesicles. *Biochim. Biophys. Acta* **409**:128–143, 1975.
48. Yaari, A.; Shapiro I. M. Effect of phosphate on phosphatidyl serine-mediated calcium transport. *Calcif. Tissue Int.* **34**:43–48, 1982.

66

The Role of Collagen and Phosphoproteins in the Calcification of Bone and Other Collagenous Tissues

Melvin J. Glimcher

Collagens and Tissue Calcification

The mineralization of any *tissue* includes the potential calcification of a number of independent, spatially distinct, intra- and extracellular tissue components (Fig. 1) [11] each of which represents an independent series of physical-chemical events beginning with the nucleation of a solid phase of calcium–phosphate from solution. In this presentation, I will attempt to review some of the evidence which supports the role of collagen and the phosphoproteins in the calcification of bone, dentine, and other collagenous mineralized tissues.

Electron micrographs and electron diffraction studies of cross-sections of fish and embryonic chick bones [7,19,21–23] show collagen fibrils impregnated with the calcium–phosphorus apatitic mineral phase. The solid calcium–phosphate mineral phase is not randomly distributed with the collagen fibrils. but is deposited preferentially in specific regions of the fibrils, namely the hole zone region, which has sufficient volume to accommodate the crystals without completely disrupting and destroying the fibrils [7–9]. Equally important is the observation that even at the earliest stages of *collagen* calcification (the emphasis on collagen is to distinguish the calcification of collagen fibrils from the general term *tissue calcification*, which includes calcification of all of the intra- and extracellular components— (see Fig. 1) the individual crystallites are highly oriented with their crystallographic c-axis (long physical axis) roughly parallel to the long axes of the collagen fibrils in which they are located [7,9,15,16]. The localization of the earliest deposited mineral phase to the hole zone regions of the collagen fibrils by electron microscopy has been confirmed by the elegant low-angle neutron diffraction study of intact tissue [3,43] and by the reconstruction of the location of the mineral phase in collagen fibrils by optical transforms of low-angle X-ray diffraction data [5]. In addition, longitudinal and cross-sectional views of the collagen fibrils

---

Melvin J. Glimcher • Laboratory for the Study of Skeletal Disorders and Rehabilitation, Department of Orthopedic Surgery, Harvard Medical School, Children's Hospital Medical Center, Boston, Massachusetts 02115.

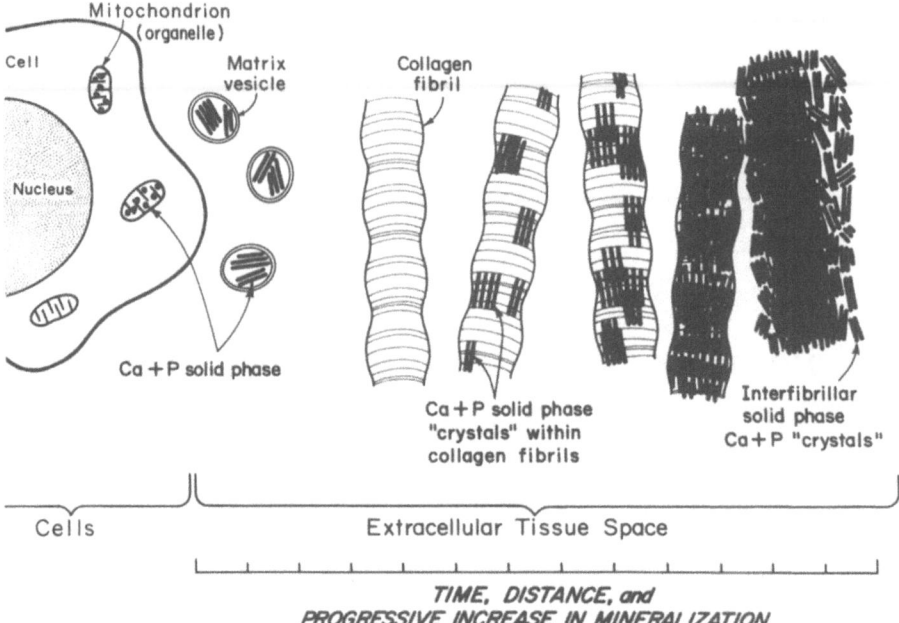

Figure 1. Schematic of tissue calcification. [From Ref. 12.]

of the extracellular matrix of chick [19,21–23] and fish bone [6] in the newly deposited and just-beginning-to-be-mineralized osteoid portion reveal many regions in which there exist widely spaced collagen fibrils in various stages of mineralization *without* an intervening mineral phase between the fibrils. These observations indicate that mineralization is *initiated* within each individual collagen fibril and that the process is independent of any *direct* physical contact of a solid phase of calcium–phosphate located between the fibrils. The physical-chemical implications of these morphological findings are considerable and must be explained by any theory of mineralization.

The *de novo* formation of a solid mineral phase of calcium–phosphate within collagen fibrils free of mineral-phase particles must clearly begin with the nucleation of crystals from a solution phase. The question thus arises whether the formation of the initial crystals within the fibrils results simply from an increase in the metastability of the circulating extracellular fluids above the level of spontaneous precipitation (homogeneous nucleation), in which case the holes and other spaces of the collagen fibrils are viewed simply as *passively* providing the volume within which the crystals are deposited. Alternatively, the collagen fibrils may play an *active role* in the formation of the inorganic crystals (heterogeneous nucleation).

The passive theory is discounted from both basic physical-chemical considerations and the morphological observations already described. First, in all solutions of ions, it is almost impossible to reach the level of metastability at which homogeneous nucleation occurs (point of unstable equilibrium). Instead, as the level of metastability is increased, heterogeneous nucleation almost always occurs either on the surface of the containers or on particles within the solution phase. As has been pointed out previously [7,9,16], it is therefore highly improbable that the metastability of extracellular fluid could be increased to the point of spontaneous or homogeneous nucleation, inasmuch as biological tissues offer so many discrete components, the surface of any one of which could easily serve as a nucleation catalyst. Second, increasing the metastability of the extracellular fluids to a point where spontaneous precipitation of mineral (homogeneous nucleation) occurs would result in a massive and random deposition of the mineral phase throughout the extracellular tissue spaces. This is not observed and thus could not account for the unique localization of the mineral phase to the hole zone region of the collagen fibrils. Moreover, as has already been pointed out, mineralized collagen fibrils can be observed during early deposition of the solid calcium–phosphate phase at a time when little or no mineral phase is present in the extracellular spaces between certain fibrils.

Similarly, the proposal by Anderson [1] and Wuthier [44] that crystals within matrix vesicles perforate the vesicles, propagate large numbers of additional crystals (multiplication) [9,11] in the intervening extracellular spaces by secondary nucleation, and eventually reach the hole zone region of the collagen fibers is equally improbable from a strictly physical-chemical point of view and again is in conflict with the actual morphology of the process as observed by electron microscopy. No observer has ever reported that in most instances the spaces between the collagen fibrils are first filled with a solid mineral phase of calcium–phosphate before mineralization of the collagen fibrils begins. The striking feature of early mineralization is that deposition of the initial and subsequent crystals within the collagen fibrils occurs selectively within the hole zone region at a time when there is generally *little or no mineral between the collagen fibrils*. Moreover, not only would progressive secondary nucleation and propagation of mineral crystals in the extracellular tissue spaces lead to a random dispersal of the crystallites between the collagen fibrils, which is also not observed directly by electron microscopy or X-ray diffraction, but it would also result in randomly organized crystals on the surface of the fibrils. This is also not the case as observed by electron microscopy, electron diffraction [7,19,21–23], or low-angle neutron diffraction [3,43].

Another very important point is the fact that matrix vesicles are observed in only the very early stages of embryonic chick bone development. In later stages of embryonic development, matrix vesicles either are not observed at all or are only few in number. These data first reported by Anderson [1] and reviewed in Chapter 65 are consistent with our own observations. Thus, the collagen fibrils of all the new bone laid down after this early embryonic bone has been resorbed [11] are calcified *in the absence of matrix vesicles* (1, see Chapter 65; Landis and Glimcher, unpublished data). Matrix vesicles are therefore *not obligatory* for the calcification of bone tissue. Indeed, if they do play a role in the calcification of bone, their action must be limited to a brief period during the early stages of embryonic development since bone which is synthesized after this very early stage calcifies in the absence of matrix vesicles.

Proposals that postulate that nucleation of the mineral phase in early embryonic bone occurs in matrix vesicles and that subsequent crystal growth takes place in collagen fibrils of the bone subsequently deposited [18] do not take into account that the early bone containing matrix vesicles is eventually completely resorbed during growth and development [11]. Second, if the crystals are first deposited in matrix vesicles, they must grow or multiply in the vesicles, i.e., they cannot grow where they are not deposited [9,11]. If they somehow were able to traverse the membrane of the matrix vesicles and multiply in the intervening extracellular space, eventually reaching and calcifying the collagen fibrils, the process would still be self-limited to that tissue and those collagen fibrils present in it at a time when the matrix vesicles are also present. When the bone tissue containing matrix vesicles is resorbed, all subsequently synthesized bone tissue, both embryonic and postnatal, is calcified in the absence or relative absence of matrix vesicles.

As far as calcification of epiphyseal cartilage is concerned, we are unable to find inorganic crystals of calcium–phosphate within the matrix vesicles in tissue processed anhydrously [20]. In fact, the vesicles do not contain either a solid phase of calcium–phosphate or an increased concentration of calcium and phosphorus, as determined by high-resolution microprobe analysis [20]. The initial solid-phase mineral particles of calcium and phosphorus were found by electron probe microanalysis to be spatially distinct from the matrix vesicles. Similarly, a recent stereographic study of epiphyseal cartilage [32] revealed that the distribution of matrix vesicles differs completely from the distribution of the mineral phase. Other groups [31,33,37] have also failed to observe any unique relationship between the calcium–phosphate mineral particles in calcifying epiphyseal cartilage and the matrix vesicles. The solid-phase particles of calcium–phosphate observed in the matrix vesicles and mitochondria cannot themselves play a *direct* physical role in the calcification of relatively remote compartments or components [9,11]. If matrix vesicles or mitochondria produce an increase in metastability of the extracellular fluid permeating the collagen fibrils, then their role in the calcification of collagen fibrils could be one of *facilitating* nucleation of calcium and phosphorus crystals within the collagen fibrils. In other words, these organelles may have an important regulatory role [11].

There appears to be one workable hypothesis to explain the findings that the calcium–phosphate crystals are initially formed *de novo* within collagen fibrils from a solution phase of calcium–phosphate and that these crystals do not form spontaneously in the extracellular fluid in the intervening space between the fibrils. The collagen fibrils in their native configuration act as a heterogeneous nucleation catalyst and are themselves responsible for *de novo* formation of the initial calcium–phosphate solid-phase particles in hole zones of the fibrils. The thesis that native collagen fibrils are heterogeneous nucleation catalysts was tested by exposing partially purified, reconstituted collagen fibrils to operationally metastable solutions of calcium–phosphate [2,7,8,14,36]. Native-type collagen fibrils were found to be potent nucleation catalysts for the heterogeneous formation of apatite crystallites *in vitro* [7,14], whereas other proteins, including muscle fibrils, failed to nucleate apatite crystals from metastable solutions under identical conditions.

Thus, the specificity of collagen as a nucleation catalyst was established. Even more striking, however, was the finding that the heterogeneous nucleation catalytic ability of collagen resided in the specific macromolecular aggregation state of native collagen fibrils and not in the collagen molecule per se, i.e., the collagen molecules had to be aggregated in the specific staggered fashion of native fibrils in order for the fibrils to heterogeneously nucleate apatite crystals from metastable solutions [7,14]. Electron micrographs of the *in vitro* nucleation of apatite crystals by reconstituted native-type collagen fibrils and decalcified bone collagen showed that the initial crystals were not randomly dispersed but were highly organized within the collagen fibrils within the hole zone regions [7]. Thus, aggregation of the collagen molecules in native-type fibrils must provide the necessary space for deposition of the crystals and also the stereochemical charge distribution and configuration required for nucleation to occur. These *in vitro* observations have been extended to *in vivo* experiments in which reconstituted native collagen, placed back in animals intraperitoneally and subcutaneously (30; Glimcher *et al.*, unpublished), bring about the deposition of apatite crystals within the hole zone regions of only the native-type collagen fibrils (Figs. 2 and 3) [11].

Thus, not only are native-type collagen fibrils potent heterogeneous catalysts *in vitro* and *in vivo*, but the high selectivity of sites within the collagen fibrils is identical in native tissues and in collagen fibrils calcified *in vitro* and *in vivo*. This specificity establishes that collagen fibrils are potent nucleation catalysts because they contain specific regions within them (nucleation sites) where fibril calcification is initiated (Fig. 4) [11]. It is reasonable to

66. Calcification of Collagenous Tissues 611

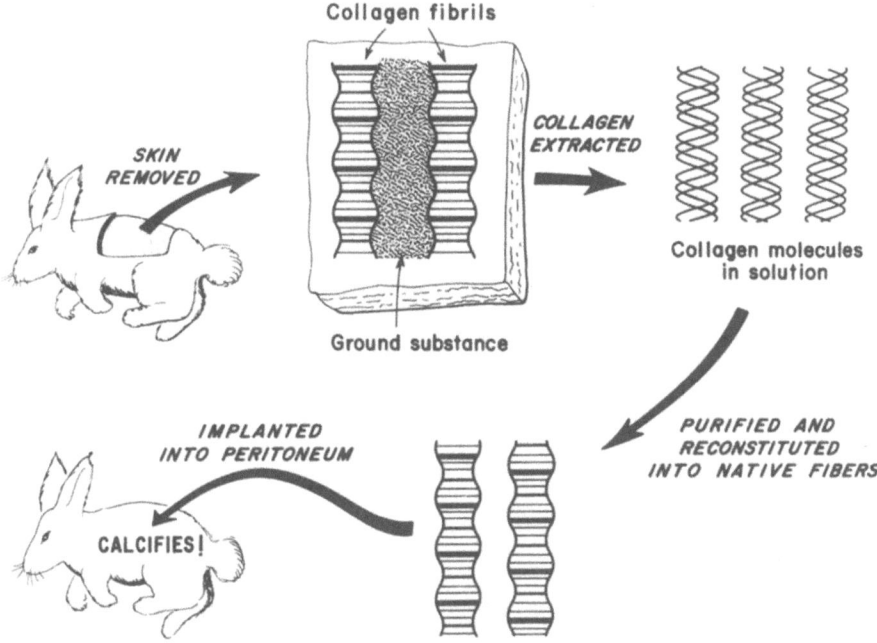

Figure 2. Schema of *in vitro–in vivo* experiments to show ability of native-type reconstituted collagen fibrils to calcify *in vivo*. [From Ref. 11.]

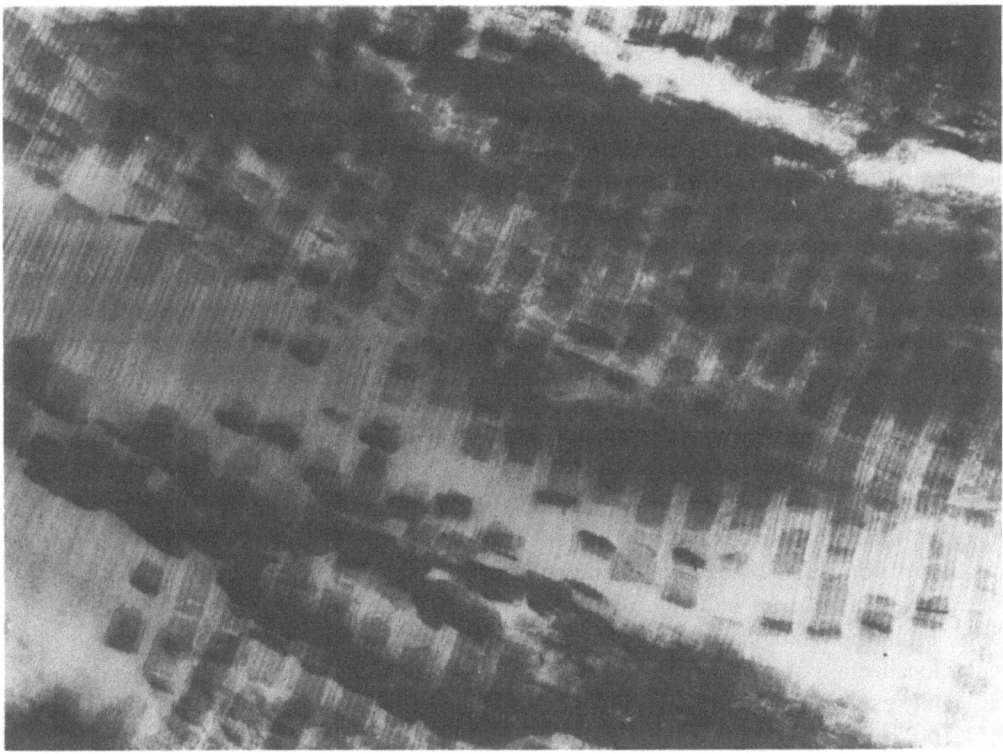

Figure 3. Electron micrograph of *in vivo* calcified reconstituted skin collagen implanted into peritoneum. (Courtesy of Dr. Marie Nylen, NIDR.)

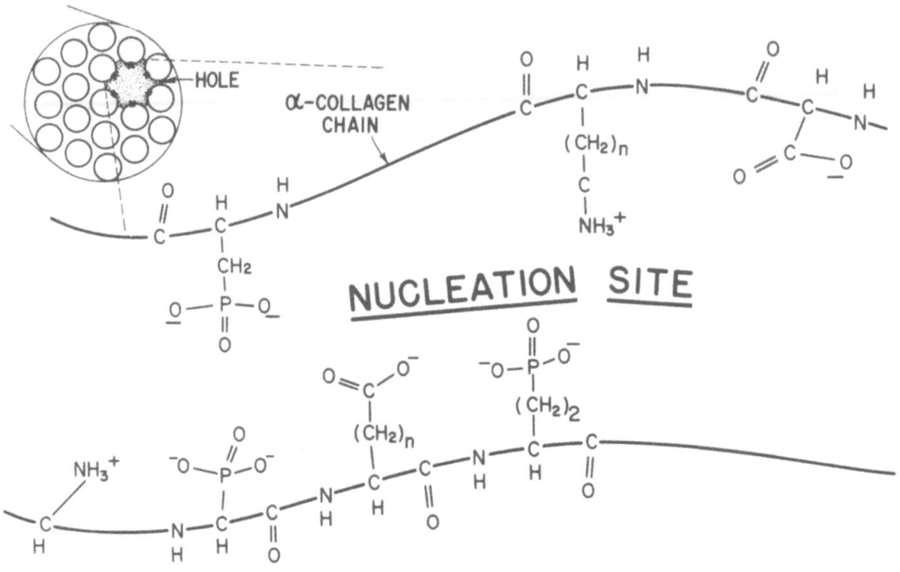

Figure 4. Diagrammatic representation of a nucleation site in collagen fibril. [From Ref. 11.]

conclude that the native aggregation state of collagen fibrils is necessary for their calcification [7,9,14,16].

Thus, calcification of mitochondria, matrix vesicles, collagen, and other spatially distinct intra- and extracellular tissue components are independent physical-chemical events. In each of the components, crystals or particles of calcium–phosphate are formed *de novo* from a solution phase of calcium and inorganic orthophosphate ions permeating the particular component. The formation of a solid phase of calcium–phosphate from a solution phase of calcium and phosphate can therefore be described as a physical change in state [7]. This change in state may occur by heterogeneous nucleation, the nucleation substrate consisting of specific organic macromolecules which, through particular stereochemical and electrochemical array of their side chains, provide sites where the heterogeneous nucleation begins. The particular organic constituents which initiate mineralization in any of these components may vary, but the basic physical-chemical mechanism would be the same.

Potential Role of Bone Matrix Phosphoproteins

Organically bound phosphate groups have the unique chemical and physical chemical properties which make it likely that they play a critical role in the first stages of calcification in collagen fibrils and other tissue components [8,9,11,16]. The organically bound phosphate groups may be an integral part of the nucleation sites at the hole zone of collagen fibrils. From the biological standpoint, this would permit precise cellular control and molecular localization of the process. Chemical compositional, morphological, physical-chemical, and experimental pathophysiological data are consistent with this hypothesis [9,11,16,27, 28,41,42].

Phosphoproteins, almost all of which are soluble in EDTA, were first isolated and purified from chicken and bovine bone in the early 1970s [34,35]. The partially purified proteins from eight species were shown to contain both O-phosphoserine [Ser(P)] and O-phosphothreonine [Thr(P)], which together account for essentially 100% of the total organic phosphorus content [4]. Further work on chicken bone identified two homogeneous phosphoproteins of M_r 12,000 and 30,000, respectively [25,26]. Both proteins contained Ser(P) and Thr(P), establishing that the two phosphorylated amino acids were present in individual bone matrix proteins. ^{31}P NMR studies confirmed that the organic phosphorus was entirely in the form of monoesters [26]. The composition of the proteins was characterized by their richness in dicarboxylic aspartic and glutamic acids which constitute approximately 40% of the total amino acid content of the proteins, thus providing abundant sites for calcium binding in addition to phosphate groups.

There was considerable uncertainty whether the bone matrix proteins were synthesized in bone or were, like albumin and α_2HS-glycoproteins, for example, synthesized in the liver and transported to and bound to bone matrix [38–40]. This issue was settled by experiments utilizing bone cultured *in vitro* which established that phosphoproteins are synthesized by bone tissue (13; Glimcher *et al.*, in preparation). In order to explore whether the phosphoproteins of bone are synthesized by bone cells per se, we isolated fractions rich in osteoblasts and osteoclasts from young mouse calvaria. Using the incorporation of [^{3}H]serine into [^{3}H]-Ser(P) as an indication of phosphoprotein synthesis [26,34,35], it was found that the osteoblast-rich fraction was the most active of the isolated fractions in synthesizing phosphoproteins containing Ser(P). The behavior of these phosphoproteins on molecular sieving and ion-exchange chromatography was similar to that of phosphoproteins isolated from young mouse calvaria—indeed, from the same bone from which the bone cells were derived [17]. The synthesis of phosphoproteins containing Ser(P) and Thr(P) was confirmed by autoradiographic studies utilizing ^{33}P injected into embryonic chicks. ^{33}P was concentrated first in the preosteoblasts and osteoblasts and then transported to the mineralized organic matrix. Localization of the phosphoproteins at extracellular sites of calcification was also established, strongly supporting the role of phosphoproteins in calcification [24].

To explore certain important physical-chemical properties of phosphoproteins related to their postulated role in calcification, we examined by ^{31}P NMR spectroscopy the homogeneous 100,000-M_r phosphoprotein of embryonic bovine dentine. The single resonance at 3.7 ppm at pH 10 and its chemical shift during acid titration established the phosphomonoester nature of the organic phosphorus moiety. During titration of the phosphoproteins with $CaCl_2$ in the presence of inorganic orthophosphate ions, line broadening for the orthophosphate resonance was both phosphoprotein- and calcium-dependent, indicating ternary complex formation. These data indicate that the phosphoprotein of fetal calf dentine binds both calcium and inorganic orthophosphate ions and therefore has the requisite physical-chemical properties to facilitate the heterogeneous nucleation of a calcium–phosphate solid phase from solution during tissue mineralization [27].

In order to further test the hypothesis that bone phosphoprotein plays a significant role in mineralization, vitamin D-deficient rickets was produced in young chicks. The resulting decrease in mineralization of whole bone and of fractions separated by density centrifugation was accompanied by a decrease in Ser(P) and Thr(P) content. The total amount of Ser(P) and Thr(P), as well as the concentrations of these phosphoamino acids, in EDTA extracts and in fractions obtained by molecular sieving was reduced. These data provide the first *in vivo* evidence that phosphoproteins may be critically involved in the calcification of bone [28].

In summary, phosphoproteins have been found in all normal and pathologically mineralized tissues and deposits in vertebrates [10,11] but have *not* been identified in unmineralized soft tissues. They thus appear to be true calcified-tissue-specific proteins. For example, phosphoproteins are not present in significant amounts in normal skin or tendon but appear in increasing amounts during progressive pathologic calcification of these tissues [12,18].

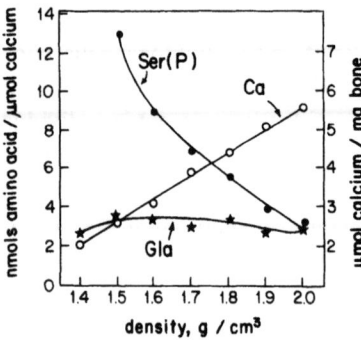

Figure 5. Change in Gla (★) and Ser(P) (●) in the density fractions of bone plotted as a function of the calcium content of the tissue. Total calcium content (○) of each fraction is also shown. [From Ref. 29.]

During bone calcification, the concentration of Ser(P) relative to calcium is highest in the earliest stages of mineralization and progressively falls with increased mineralization (Fig. 5) [29]. This is consistent with the proposal that Ser(P) and Thr(P) (when the latter is present) are necessary for the nucleation of the first mineral particles, but that further mineralization occurs by secondary nucleation from the heterogeneously nucleated mineral particles. Thus, during the initial stages, the ratio of Ser(P) to calcium will be high, but since Ser(P) is not involved in the crystal multiplication process that occurs by secondary nucleation, its concentration relative to calcium will fall as the number of calcium–phosphate particles progressively increases.

The strong interaction which occurs between collagen and phosphoproteins may reflect specific binding of the phosphoproteins to particular sites at hole zone regions of the collagen fibrils. This combined system of natively aggregated collagen fibrils and strongly interacting phosphoproteins is absent in normally soft, unmineralized, connective tissues. But this system which is present in all calcifying collagenous tissues that have been studied, including pathologic tissues, probably serves as the nucleation site for collagen fibril calcification *in vivo*. In the presence of normal extracellular fluid conditions, aggregated collagen fibrils plus phosphoproteins bound to specific regions of fibrils in bone may constitute a system both necessary and sufficient for calcification of collagen fibrils and thus for normal tissue calcification.

ACKNOWLEDGMENTS. This work was supported in part by grants from the National Institutes of Health (AM-15671) and The New England Peabody Home for Crippled Children, Inc.

References

1. Anderson, H. C. Evolution of cartilage. In: *The Comparative Molecular Biology of Extracellular Matrices*, H. C. Slavkin, ed., New York, Academic Press, 1972, pp. 200–205.
2. Bachra, B. N.; Sobel, A. E. Calcification. XXV. Mineralization of reconstituted collagen. *Arch. Biochem. Biophys.* **85**:9–18, 1959.
3. Berthet-Colominas, C.; Miller, A.; White, S. W. Structural study of the calcifying collagen in turkey leg tendons. *J. Mol. Biol.* **134**:431–445, 1979.
4. Cohen-Solal, L.; Lian, J. B.; Kossiva, D.; Glimcher, M. J. The identification of O-phosphothreonine in the soluble non-collagenous phosphoproteins of bone matrix. *FEBS Lett.* **89**:107–110, 1978.
5. Engstrom, A. Apatite–collagen organization in calcified tendon. *Exp. Cell Res.* **43**:241–245, 1966.
6. Geraudie, J.; Landis, W. J. The fine structure of the developing pelvic fin dermal skeleton in the trout *Salmo gairdneri*. *Am. J. Anat.* **163**:141–156, 1982.
7. Glimcher, M. J. Molecular biology of mineralized tissues with particular reference to bone. *Rev. Mod. Phys.* **31**:359–393, 1959.

8. Glimcher, M. J. Specificity of the molecular structure of organic matrices in mineralization. In: *Calcification in Biological Systems*, R. F. Sognnaes, ed., Washington, D.C., Am. Assoc. Adv. Sci., 1960, pp. 421–487.
9. Glimcher, M. J. Composition, structure, and organization of bone and other mineralized tissues and the mechanism of calcification. In: *Handbook of Physiology: Endocrinology*, Volume VII, R. O. Greep and E. B. Astwood, eds., Washington, D.C., American Physiological Society, 1976, pp. 25–116.
10. Glimcher, M. J. Phosphopeptides of enamel matrix. *J. Dent. Res.* **58B**:790–806, 1979.
11. Glimcher, M. J. On the form and function of bone: From molecules to organs. Wolff's law revisited, 1981. In: *The Chemistry and Biology of Mineralized Connective Tissues*, A. Veis, ed., Amsterdam, Elsevier/North-Holland, 1981, pp. 618–673.
12. Glimcher, M. J.; Brickley-Parsons, D.; Kossiva, D. Phosphopeptides and γ-carboxyglutamic acid-containing peptides in calcified turkey tendons: Their absence in uncalcified tendon. *Calcif. Tissue Int.* **27**:281–284, 1979.
13. Glimcher, M. J.; Brickley-Parsons, D.; Kossiva, D. Proof that the phosphoproteins containing O-phosphoserine and O-phosphothreonine are synthesized in bone. *Trans. Orthop. Res. Soc.* **7**:36, 1982.
14. Glimcher, M. J.; Hodge, A. J.; Schmitt, F. O. Macromolecular aggregation states in relation to mineralization: The collagen–hydroxyapatite system as studied *in vitro*. *Proc. Natl. Acad. Sci. USA* **43**:860–867, 1957.
15. Glimcher, M. J.; Katz, E. P.; Travis, D. F. The organization of collagen in bone: The role of noncovalent forces in the physical properties and solubility characteristics of bone collagen. In: *Symp. Int. Biochim. Physiol. Tissu Conjonctif, 1966*, P. Comte, ed., Lyon, Societe Ormeco et Imprimerie du Sud-Est, 1966, pp. 491–503.
16. Glimcher, M. J.; Krane, S. M. The organization and structure of bone, and the mechanism of calcification. In: *Treatise on Collagen*, Volume IIB, G. N. Remachandran and B. S. Gould, eds., New York, Academic Press, 1968, pp. 68–251.
17. Gotoh, Y.; Sakamoto, M.; Sakamoto, S.; Glimcher, M. J. Biosynthesis of O-phosphoserine-containing phosphoproteins by isolated bone cells of mouse calvaria. *FEBS Lett.* **154**:116–120, 1983.
18. Irving, J. T. Theories of mineralization of bone. *Clin. Orthop.* **97**:225–236, 1973.
19. Landis, W. J.; Glimcher, M. J. Electron diffraction and electron probe microanalysis of the mineral phase of bone tissue prepared by anhydrous techniques. *J. Ultrastruct. Res.* **63**:188–223, 1978.
20. Landis, W. J.; Glimcher, M. J. Electron optical and analytical observations of rat growth plate cartilage prepared by ultracryomicrotomy: The failure to detect a mineral phase in matrix vesicles and the identification of heterodispersed particles as the initial solid phase of calcium phosphate deposited in the extracellular matrix. *J. Ultrastruct. Res.* **78**:227–268, 1982.
21. Landis, W. J.; Hauschka, B. T.; Rogerson, C. A.; Glimcher, M. J. Electron microscopic observations of bone tissue prepared by ultracryomicrotomy. *J. Ultrastruct. Res.* **59**:185–206, 1977.
22. Landis, W. J.; Paine, M. C.; Glimcher, M. J. Electron microscopic observations of bone tissue prepared anhydrously in organic solvents. *J. Ultrastruct. Res.* **59**:1–30, 1977.
23. Landis, W. J.; Paine, M. C.; Glimcher, M. J. Use of acrolein vapors for the anhydrous preparation of bone tissue for electron microscopy. *J. Ultrastruct. Res.* **70**:171–180, 1980.
24. Landis, W. J.: Sanzone, C. F.; Brickley-Parsons, D.; Glimcher, M. J. Radioautographic visualization and biochemical identification of O-phosphoserine- and O-phosphothreonine-containing phosphoproteins in mineralizing embryonic chick bone. *J. Cell Biol.* **98**:986–990, 1984.
25. Lee, S. L.; Glimcher, M. J. Bone matrix phosphoprotein from adult avian metatarsals. *J. Cell Biol.* **83**:464a, 1979.
26. Lee, S. L.; Glimcher, M. J. The purification, composition and ^{31}P spectroscopic properties of a non-collagenous phosphoprotein isolated from chicken bone matrix. *Calcif. Tissue Int.* **33**:385–394, 1981.
27. Lee, S. L.; Glonek, T.; Glimcher, M. J. ^{31}P nuclear magnetic resonance spectroscopic evidence for ternary complex formation of fetal phosphoprotein with calcium and inorganic orthophosphate ions. *Calcif. Tissue Int.* **35**:815–818, 1983.
28. Lian, J. B.; Cohen-Solal, L.; Kossiva, D.; Glimcher, M. J. Changes in phosphoproteins of chicken bone matrix in vitamin D-deficient rickets. *FEBS Lett.* **149**:123–125, 1982.
29. Lian, J. B.; Roufosse, A. H.; Reit, B.; Glimcher, M. J. Concentrations of osteocalcin and phosphoprotein as a function of mineral content and age in cortical bone. *Calcif. Tissue Int.* **34**:S82–S87, 1982.
30. Mergenhagen, S. E.; Martin, G. R.; Rizzo, A. A.; Wright, D. N.; Scott, D. B. Calcification *in vivo* of implanted collagen. *Biochim. Biophys. Acta* **43**:563–565, 1960.
31. Poole, A. R.; Pidoux, I.; Reiner, A.; Choi, H.; Rosenberg, L. C. The association of a matrix protein, proteoglycan and link protein with mineralization in endochondral bone formation: An immunohistochemical study. *Trans. Orthop. Res. Soc.* **8**:146, 1983.
32. Reinholt. F. P.; Engfeldt. B.; Hjerpe, A.; Jansson, K. Stereological studies on the epiphyseal growth plate with special reference to the distribution of matrix vesicles. *J. Ultrastruct. Res.* **80**:270–279, 1982.
33. Shepard, N. L.; Mitchell, N. S. Proteoglycan aggregates in the longitudinal septa of rat growth plate. *Trans. Orthop. Res. Soc.* **7**:76, 1982.

34. Spector, A. R.; Glimcher, M. J. The extraction and characterization of soluble anionic proteins from bone. *Biochim. Biophys. Acta* **263**:593–603, 1972.
35. Spector, A. R.; Glimcher, M. J. The identification of O-phosphoserine in the soluble anionic phosphoproteins of bone. *Biochim. Biophys. Acta* **303**:360–362, 1973.
36. Strates, B.; Neuman, W. F.; Levinskas, G. L. The solubility of bone mineral. II. Precipitation of near-neutral solutions of calcium and phosphate. *J. Phys. Chem.* **61**:279–282, 1957.
37. Thyberg, J. Electron microscopic studies on the initial phases of calcification in guinea pig cartilage. *J. Ultrastruct. Res.* **46**:206–218, 1974.
38. Triffitt, J. T.; Gebauer, U.; Ashton, B. A.; Owen, M. E. Origin of plasma α_2HS-glycoprotein and its accumulation in bone. *Nature (London)* **262**:226–227, 1976.
39. Triffitt, J. T.; Owen, M. E. Studies on bone matrix glycoproteins: Incorporation of [1-^{14}C]glucosamine and plasma [^{14}C]glycoprotein into rabbit cortical bone. *Biochem. J.* **136**:125–134, 1973.
40. Tiffitt, J. T.; Owen, M. E.; Ashton, B. A.; Wilson, J. M. Plasma disappearance of rabbit α_2HS-glycoproteins and its uptake by bone tissue. *Calcif. Tissue Int.* **26**:155–161, 1978.
41. Veis, A. The role of acidic proteins in biological mineralization. In: *Ions in Macromolecular and Biological Systems*, D. H. Everett and V. Vincent, eds., Bristol, Scientechnia, 1978, pp. 259–272.
42. Veis, A.; Stettler-Stevenson, W.; Takagi, Y.; Sabsay, B.; Fullerton, R. The nature and localization of the phosphorylated proteins of mineralized dentin. In: *The Chemistry and Biology of Mineralized Connective Tissue*, A. Veis, ed., Amsterdam, Elsevier/North-Holland, 1981, pp. 377–393.
43. White, S. W.; Hulmes, D. J. S.; Miller, A.; Timmins, P. A. Collagen–mineral axial relationship in calcified turkey leg tendon by X-ray and neutron diffraction. *Nature (London)* **266**:421–425, 1977.
44. Wuthier, R. E. A review of the primary mechanisms of endochondral calcification with special emphasis on the role of cells, mitochondria and matrix vesicles. *Clin. Orthop.* **169**:219–242, 1982.

67

Biological Processes Involved in Endochondral Ossification

Lawrence C. Rosenberg, Haing U. Choi, and A. Robin Poole

Cartilage growth plate consists of the zones of resting cartilage cells, proliferating chondrocytes, hypertrophic chondrocytes; as well as the metaphysis, where new bone is laid down on longitudinal bars of calcified cartilage. Several crucial biological processes occur in these regions which result in the longitudinal growth of bone. The progress and completion of each of these biological processes depends not only on the functions of the cells, but on the structure and properties of the extracellular matrix, and the biochemical events which take place in the extracellular matrix surrounding the cells in a particular region. Thus, in the hypertrophic zone, mineralization occurs in the extracellular matrix of the longitudinal septa between the lowermost hypertrophic chondrocytes. Anderson has already described how matrix vesicles initiate the mineralization process which occurs in the hypertrophic zone of cartilage growth plate (see Chapter 65).

Two additional biological processes occur in the lowermost hypertrophic zone which may facilitate the spread of mineralization from matrix vesicles throughout the extracellular matrix. The first process is the structural modification of proteoglycan aggregates. In their native state, proteoglycan aggregates from epiphyseal cartilage or the zone of proliferating chondrocytes possess the capacity to inhibit hydroxyapatite formation. The question is raised as to how these proteoglycan aggregates are structurally modified in the zone of hypertrophic chondrocytes so that their capacity to inhibit mineralization is abolished; as a result, mineralization can easily spread from matrix vesicles throughout the extracellular matrix. The second process is the deposition and selective concentration of a 35,000-M_r protein with a strong affinity for hydroxyapatite in the extracellular matrix of the longitudinal septa of the lowermost hypertrophic zone. It takes place in the identical region where mineralization of the cartilage growth plate first occurs. The question is raised whether selective concentration of the 35K protein is directly related to the mineralization which occurs in this region.

Lawrence C. Rosenberg and Haing U. Choi • Orthopedic Research Laboratories, Montefiore Medical Center, Bronx, New York 10467. *A. Robin Poole* • Joint Diseases Research Laboratory, Shriners Hospital for Crippled Children, Montreal, Quebec H3G 1A6, Canada.

Proteoglycans in Endochondral Ossification

Cartilage proteoglycan aggregates are huge macromolecules formed by the noncovalent association of proteoglycan monomers, link protein and hyaluronic acid. The structure and molecular architecture of these aggregates have been described in detail [2,6,13,17,18]; they are reversibly deformable, elastic molecules, which tend to expand into a large domain of solution, imbibe water into their domain, and resist compression into a smaller volume of solution. In the proliferating zone, chondrocytes divide at a rapid rate and surround themselves with an extracellular matrix composed of type II collagen, cartilage proteoglycans, and a variety of noncollagenous proteins and glycoproteins. The rate of longitudinal bone growth is determined in part by the rate of cell proliferation and the formation of extracellular matrix with which the newly formed cells surround themselves. Proteoglycans are major structural components of the extracellular matrix in the growth plate, although they may also serve other special functions.

Unhampered appositional growth in the zone of proliferating chondrocytes requires that these rapidly dividing cells be surrounded by an extracellular matrix which can accommodate and adjust to changes in shape and volume associated with cell division. The extracellular matrix in this region must be elastic, malleable, compliant, and noncalcifiable. Proteoglycan aggregates possess elastic properties [12] that contribute to the elasticity and compliance of the extracellular matrix in the zone of proliferating chondrocytes, which is essential in this region.

In epiphyseal cartilage and the zone of proliferating chondrocytes, proteoglycan aggregates are also believed to act to prevent mineralization in the zone of proliferating chondrocytes and the upper hypertrophic zone *in vivo* [4,8-10,14,15]. Since native, unmodified proteoglycan aggregates inhibit hydroxyapatite formation [1], the question is raised as to how these proteoglycan aggregates may be altered in the region of the lowermost hypertrophic chondrocytes and in the zone of provisional calcification, so that their capacity to inhibit mineralization is abolished. In what way are proteoglycan aggregates structurally modified during endochondral ossification so that a noncalcifiable matrix is transformed into a calcifiable matrix?

One widely held hypothesis is that proteoglycans are degraded by proteases and removed prior to the onset of mineralization in the lowermost hypertrophic zone [5,7,11], prior to or during mineralization. This results in the transformation of a noncalcifiable matrix into a calcifiable matrix. Our recent studies (16) have led us to question this hypothesis and suggest that proteoglycans are structurally modified by a different, more specific mechanism. If proteoglycans are extensively degraded by proteases and removed prior to mineralization, the concentration of proteoglycan should decrease in the longitudinal septa of the lowermost hypertrophic zone where mineralization occurs, and within the spicules of calcified cartilage in the metaphysis. However, using cationic dyes which stain proteoglycan (provided that the specimens are gently decalcified in the cold with EDTA in the presence of protease inhibitors) no such decrease is demonstrable [16]. Moreover, when immunofluorescence microscopy was carried out using monospecific antibodies to proteoglycan monomer core protein, or to link protein, proteoglycan monomer core protein and link protein were demonstrable without detectable loss throughout the extracellular matrix of the longitudinal septa of the hypertrophic zone and in the calcified cartilage of the metaphysis [16].

These results indicate that there is no net loss of proteoglycans during mineralization, but that the proteoglycans become entombed in calcified cartilage and remain there until the entire trabeculum is eroded from its surface and removed by osteoclasts. Electron microscopic studies of the lengths of the core proteins of proteoglycan monomers isolated from epiphyseal cartilage, uncalcified growth plate, and from the calcified cartilage of the metaphysis further support the concept that proteoglycan monomer core protein is not degraded by proteases prior to or during mineralization.

These studies of J. Buckwalter and L. Rosenberg (unpublished results) demonstrated that the lengths of core protein in proteoglycan monomers from epiphyseal cartilage, uncalcified growth plate, or calcified cartilage are essentially the same and do not reveal degradation of proteoglycan monomer core protein. These investigations also reveal a large increase in the spacing of monomers on hyaluronic acid, and fewer monomers per aggregate, in proteoglycan aggregates from calcified cartilage, suggesting that the proteoglycan monomers are in some way being dissociated from the central filament of hyaluronic acid. The biochemical mechanism responsible for this change in the structure of the proteoglycan aggregates has not been elucidated.

Isolation, Characterization, and Immunohistochemical Localization of a 35K Subunit Protein in Mineralizing Growth Plate Cartilage

Developing epiphyseal and growth plate cartilage contain a variety of noncollagenous extracellular matrix proteins and glycoproteins, few of which have been isolated and characterized. We recently studied the noncollagenous proteins extracted from bovine fetal epi-

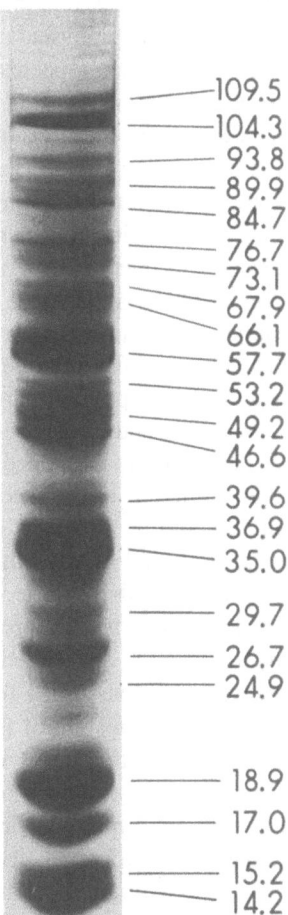

Figure 1. Protein species extracted from the epiphyseal cartilage of a 189-day-old early third trimester bovine fetus. Noncollagenous proteins and proteoglycans were extracted from cartilage in 4 M guanidine hydrochloride containing protease inhibitors. The noncollagenous proteins were separated from proteoglycans by equilibrium density gradient centrifugation under associative conditions. The fraction containing the extracted proteins was examined by SDS-PAGE under reducing conditions.

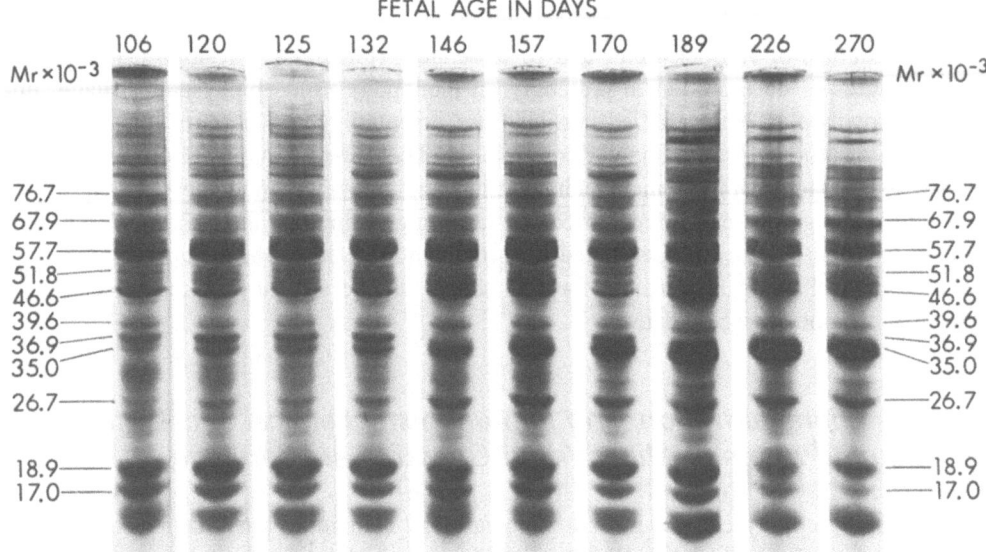

Figure 2. Protein species from 10 bovine fetal epiphyseal cartilages, ranging in age from 106 to 270 days, demonstrated by SDS-PAGE on 7.5% gels under reducing conditions.

physeal cartilages ranging in age from 106 to 270 days [3]. The protein species extracted from the fetal epiphyseal cartilages of different ages were identified by SDS-PAGE. Figure 1 shows the protein species from a 189-day-old, early third trimester, fetal epiphyseal cartilage. Gel scans showed that the 35K protein was the protein species present in highest concentration. Figure 2 shows the protein species from epiphyseal cartilages of bovine fetuses ranging in age from 106 to 270 days. The 35K protein is present in relatively low concentration in the epiphyseal cartilages from the younger fetuses. The relative amounts of the 35K protein then increase progressively in the second and third trimesters. This progressive increase in the concentration of the 35K protein as a function of fetal age is directly related to the appearance and increase in the size of the secondary center of ossification, which is surrounded by a cartilage growth plate.

During early fetal development, before the secondary ossification center has appeared within the epiphyseal cartilage, the 35K subunit protein is present in low concentrations. Toward the end of the second trimester, at about the same time as the secondary ossification center appears, the concentration of the 35K subunit protein greatly increases. During the third trimester, as the secondary ossification center increases in size, grows into and replaces the epiphyseal cartilage, the 35K subunit protein is present in the highest concentration relative to that of any other protein species.

Motivated by these observations, we decided to isolate, characterize, and study the properties of the 35K protein. The isolated 35K protein [3] exhibited a strong affinity for hydroxyapatite and eluted at 0.6 M phosphate. On SDS-PAGE under reducing conditions, the protein exists in the form of its 35K subunit (Fig. 3). Under nondenaturing conditions, the molecular weight of the protein was 69K based on analytical gel chromatography (Fig. 4), indicating that under nondenaturing conditions, the protein exists as a dimer of its 35K subunit. Table I shows the amino acid and carbohydrate composition of the protein.

An antiserum was prepared to the purified cartilage matrix protein, which was used to establish the homogeneity and the immunological identity of the protein by immunodiffusion, crossed immunoelectrophoresis, Laurell rocket electrophoresis, and ELISA [3]. Based

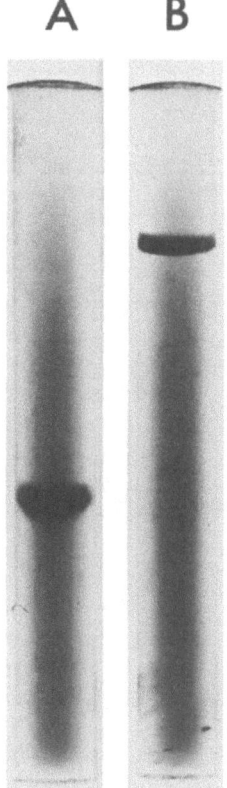

Figure 3. Homogeneity of the isolated cartilage matrix protein following gel chromatography on Sephacryl S-200, demonstrated on SDS-PAGE under reducing conditions (left) and nonreducing conditions (right).

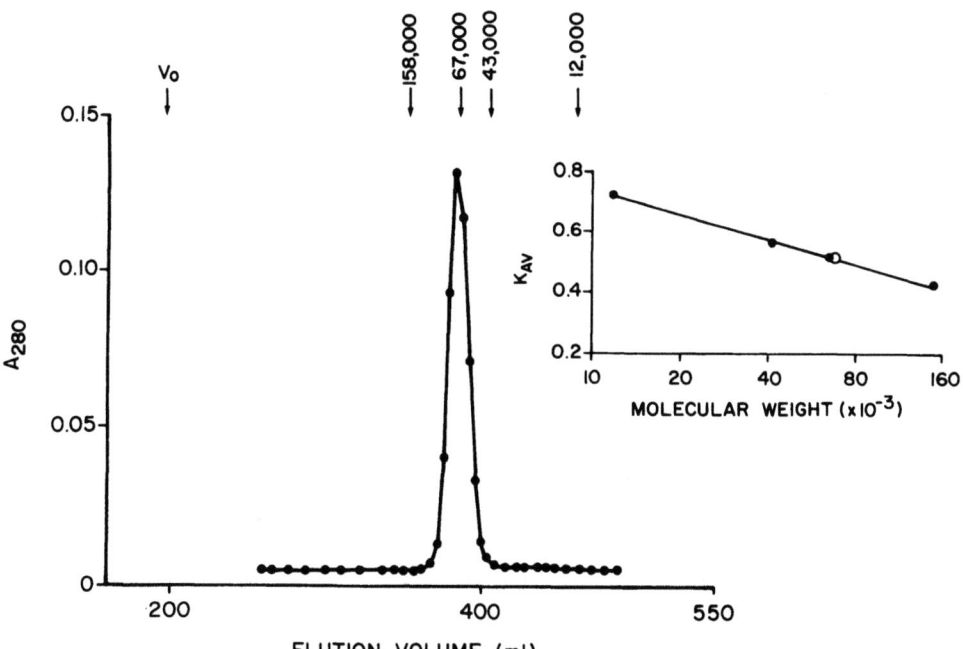

Figure 4. Determination of the molecular weight of the cartilage matrix protein under nondenaturing conditions in 0.15 M NaCl, 0.01 M Mes, pH 7, by gel filtration on Sephacryl S-200. The insert shows that plot of K_{av} versus molecular weight plotted on a logarithmic scale. The open circle represents the cartilage matrix protein.

Table I. Amino Acid and Carbohydrate Composition of the 35K Cartilage Matrix Protein

	Residues/1000 residues	Residues/mole
Asp	127	42
Thr	51	17
Ser	136	45
Glu	100	33
Pro	42	14
Gly	118	39
Ala	72	24
Cys	26	9
Val	39	13
Met	9	3
Ile	43	14
Leu	66	22
Tyr	23	8
Phe	27	9
His	25	8
Lys	67	22
Arg	32	11
	% dry weight	Residues/mole
Man	3.04	5.6
GlcN	0.78	1.4
GalN	tr	
Glc	tr	
Gal	ND	
Fuc	ND	
Xyl	ND	
NeuAc	ND	

on the immunological analyses, the purified matrix protein did not contain other protein species present in the mixture of proteins extracted from epiphyseal cartilage. There was no immunological relation between the matrix protein and link protein or proteoglycan monomer.

A monospecific antiserum prepared to the matrix protein was used to demonstrate the immunofluorescent localization of the matrix protein in epiphyseal cartilage and cartilage growth plate. These studies showed that the protein is present throughout the extracellular matrix of fetal epiphyseal cartilage at relatively low concentrations compared to mineralizing cartilage growth plate. The concentration of the matrix protein in the cartilage growth plate rapidly and strikingly increased in the same region where mineralization occurs. Figure 5A shows a von Kossa stained undecalcified section of the cartilage growth plate from the proximal tibia of the bovine fetus. Mineralization was initiated in the perilacunar and territorial matrix of the longitudinal septa of the lowermost hypertrophic chondrocytes. Figure 5B shows the localization of the matrix protein by immunofluorescence microscopy using a monospecific antibody to the matrix protein. The matrix protein is selectively concentrated in the perilacunar and territorial matrix of the longitudinal septa of the lowermost hypertrophic chondrocytes, in the same region where mineralization of the extracellular matrix first occurs. The protein also binds strongly to hydroxyapatite [3], eluting at 0.6 M potassium

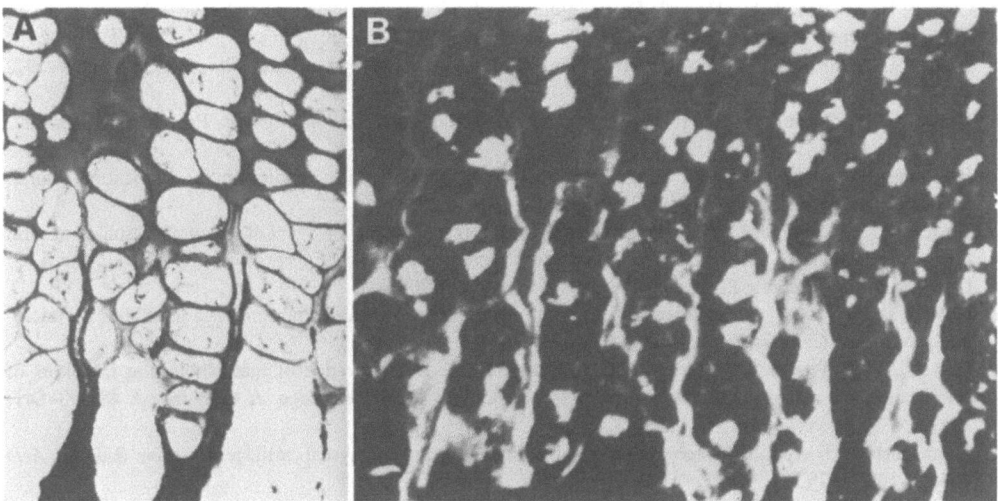

Figure 5. An undecalcified section of bovine fetal cartilage growth plate stained with the von Kossa stain (A) and by immunofluorescence microscopy using monospecific antibody to the cartilage matrix protein (B).

phosphate. Osteocalcin, a known calcium-binding protein, is eluted from hydroxyapatite at 0.1 to 0.15 M potassium phosphate. These observations suggest that the 35K subunit matrix protein may be a calcium-binding protein that has a role in hydroxyapatite deposition in the extracellular matrix of the hypertrophic zone during endochondral ossification.

Since the 35K matrix protein is present at lower concentrations throughout the extracellular matrix of fetal epiphyseal cartilage, an additional biochemical event probably occurs in the lowermost hypertrophic zone, before the 35K matrix protein can exercise its role in mineralization. As noted above, proteoglycan aggregates, such as those present in fetal epiphyseal cartilage, possess the capcity to inhibit hydroxyapatite formation. The matrix protein may not exercise its role in mineralization until these proteoglycans are in some way structurally modified in the lowermost hypertrophic zone, so that they lose their capacity to inhibit hydroxyapatite formation. This would explain why mineralization occurs only in the lowermost hypertrophic zone, despite the fact that the matrix protein is present throughout the extracellular matrix of fetal epiphyseal cartilage. The mechanism by which proteoglycan aggregates are structurally modified in the lowermost hypertrophic zone remains to be elucidated.

ACKNOWLEDGMENT. This work was supported by United States Public Health Service Grants AM HD-21498 and CA AM-23945.

References

1. Blumenthal, N. C.; Posner, A. S.; Silverman, L. D.; Rosenberg, L. C. Effect of proteoglycans on *in vitro* hydroxyapatite formation. *Calcif. Tissue Int.* 27:75–82, 1979.
2. Buckwalter, J. A.; Rosenberg, L. C. Electron microscopic studies of cartilage proteoglycans: Direct evidence for the variable length of the chondroitin sulfate-rich region of proteoglycan subunit core protein. *J. Biol. Chem.* 257:9830–9839, 1982.
3. Choi, H. U.; Tang, L.-H.; Johnson, T. L.; Pal, S.; Rosenberg, L. C.; Reiner, A.; Poole, A. R. Isolation and characterization of a 35,000 molecular weight subunit fetal cartilage matrix protein. *J. Biol. Chem.* 258:655–661, 1983.

4. Cuervo, L. A.; Pita, J. C.; Howell, D. S. Inhibition of calcium phosphate mineral growth by proteoglycan fractions in synthetic lymph. *Calcif. Tissue Res.* **13**:1–10, 1973.
5. Granda, J. L.; Posner, A. S. Distribution of four hydrolases in the epiphyseal plate. *Clin. Orthop.* **74**:269–272, 1971.
6. Hascall, V. C.; Kimura, J. H. Proteoglycans: Isolation and characterization. *Methods Enzymol.* **82**:769–800, 1982.
7. Hirschman, A.; Dziewiatkowski, D. D. Proteinpolysaccharide loss during endochondral ossification: Immunochemical evidence. *Science* **154**:393–395, 1966.
8. Howell, D. S.; Madruga, J.; Pita, J. C. Evidence for an organic nucleation agent of calcium phosphate mineral forms in endochondral plates. In: *Extracellular Matrix Influences on Gene Expression*, H. C. Slavkin and R. C. Greulich, eds., Academic Press, New York, 1975, pp. 701–706.
9. Howell, D. S.; Pita, J. C. Calcification of growth plate cartilage with special reference to studies on micropuncture fluids. *Clin. Orthop.* **118**:208–229, 1976.
10. Howell, D. S.; Pita, J. C.; Marquez; J. F.; Gatter, R. A. Demonstration of macromolecular inhibitors of calcification and nucleational factors in fluid from calcifying sites in cartilage. *J. Clin. Invest.* **48**:630–641, 1969.
11. Lohmander, S.; Hjerpe, A. Proteoglycans of mineralizing rib and epiphyseal cartilage. *Biochim. Biophys. Acta* **404**:93–109, 1975.
12. Mow, V. C.; Mak, A. F.; Lai, W. M.; Rosenberg, L. C.; Tang, L.-H. Visceolastic properties of proteoglycan subunits and aggregates in varying solution concentration. *J. Biomech.* **17**:325–338, 1984.
13. Pal, S.; Tang, L.; Choi, H.; Habermann, E.; Rosenberg, L.; Roughley, P.; Poole, A. R. Structural changes during development in bovine fetal epiphyseal cartilage. *Coll. Relat. Res.* **1**:151–176, 1981.
14. Pita, J. C.; Cuervo, L. A.; Madruga, J. E.; Muller, F. J.; Howell, D. S. Evidence for the role of proteinpolysaccharides in regulation of mineral phase separation in calcifying cartilage. *J. Clin. Invest.* **49**:2188–2197, 1970.
15. Pita, J. C.; Muller, F.; Howell, D. S. Disaggregation of proteoglycan aggregate during endochondral ossification: Physiological role of cartilage lysozyme. In: *Dynamics of Connective Tissue Macromolecules*, P. M. C. Burleigh and A. R. Poole, eds., Amsterdam, North-Holland, 1975, pp. 247–258.
16. Poole, A. R.; Pidoux, I.; Rosenberg, L. C. Role of proteoglycans in endochondral ossification: Immunofluorescent localization of link protein and proteoglycan monomer in bovine fetal epiphyseal growth plate. *J. Cell Biol.* **92**:249–260, 1982.
17. Rosenberg, L. C. Structure of cartilage proteoglycans. In: *Dynamics of Connective Tissue Macromolecules*, P. M. C. Burleigh and A. R. Poole, eds., Amsterdam, North-Holland, 1975, pp. 105–128.
18. Rosenberg, L.; Hellmann, W.; Kleinschmidt, A. K. Electron microscopic studies of proteoglycan aggregates from bovine articular cartilage. *J. Biol. Chem.* **250**:1877–1883, 1975.
19. Tang, L.-H.; Rosenberg, L. C.; Reiner, A.; Poole, A. R. Proteoglycans from bovine nasal cartilage: Properties of a soluble form of link protein. *J. Biol. Chem.* **254**:10523–10531, 1979.

68

Role of Lipids in Mineralization
An Experimental Model for Membrane Transport of Calcium and Inorganic Phosphate

Abraham M. Yaari and Charles Eric Brown

During mineralization of the epiphyseal growth plate, mineral crystals are first seen within extracellular membranous structures termed *matrix vesicles* [1,4]. Matrix vesicles accumulate calcium from the cartilage lymph and concentrate calcium and inorganic phosphate (P_i) to levels at which hydroxyapatite crystal formation is promoted [2]. Mineral formation within the vesicles is associated with an increase in the concentration of extracellular P_i [5] and the presence of acidic phospholipids [9,12]. It has been reported that a stable nondissociable calcium–phospholipid–PO_4 complex isolated from calcifying tissues can nucleate hydroxyapatite crystals from metastable solutions [3,7]. This complex is considered the site of mineral nucleation during calcification [6,13].

Aside from acting as a site for mineral binding and crystallization, phospholipids may have a second function. It has been demonstrated that phosphatides can act as ionophores to promote calcium transport across a lipid barrier [8,11]. How calcium and P_i are transported through the phospholipid complex and the process by which P_i modulates either the transport or the deposition step are not understood.

The intent of this investigation is to demonstrate that phosphatidylserine (PS), a major acidic phospholipid of matrix vesicles, mediates calcium transport nonenzymatically and that this process is enhanced by P_i. Further, the use of ^{31}P and ^1H NMR spectroscopy revealed that complexes of PS, calcium, and P_i are formed which permit transport of calcium across membrane bilayers; additionally, P_i promotes formation of complex aggregates, thus enhancing calcium transport.

PS-Mediated Calcium Transport

The rate of calcium transport from an aqueous donor compartment to an aqueous receiver compartment, separated by a nonaqueous phospholipid phase, was measured in a

---

Abraham M. Yaari • Department of Developmental Dentistry, University of Southern California School of Dentistry, Los Angeles, California 90089-0641. *Charles Eric Brown* • Department of Biochemistry, The Medical College of Wisconsin, Milwaukee, Wisconsin 53226.

*Table I. Calcium Translocation
in the Three-Phase System[a]*

Time (hr)	Calcium translocation	
	Organic phase (μmole/μmole PS)	Receiver buffer[b] (μmole/μmole PS)
0	0.000	0.000
1	0.090 ± 0.004	0.116 ± 0.005
2	0.100 ± 0.015	0.258 ± 0.012
3	0.125 ± 0.017	0.418 ± 0.015
4	0.120 ± 0.020	0.501 ± 0.018
5	0.128 ± 0.010	0.639 ± 0.016

[a]The buffer used in the donor compartment was 25 mM tetramethylammonium Pipes at pH 6.4 and the receiver buffer was 25 mM tetramethylammonium citrate at pH 4.0. The organic phase contained 200 μM PS dissolved in presaturated chloroform/methanol/water in a volume ratio of 2.0/2.0/1.8. Donor buffer contained 1.0 mM calcium labeled with ^{45}Ca.
[b]Mean of eight experiments ± S.E.M.

modified Pressman cell [10]. The details of this three-compartment lipid–aqueous phase model system are provided in a recent publication [14]. The time course for the transport of calcium across the bulk organic phase in the reaction chamber revealed that calcium accumulates in the aqueous receiver compartment at a linear rate (0.13 mole calcium/hr per mole PS) and that accumulation of calcium in the receiver compartment does not limit the translocation rate of calcium (Table I). The calcium level in the organic phase reached a steady state after about 1 hr, and in the absence of PS less than 0.006% of donor calcium was translocated into the receiver buffer. The rate of PS-mediated calcium transport from the donor to the receiver through the lipid phase was modulated by the donor P_i concentration; a threefold increase in the rate of calcium transport across the lipid barrier was observed between 1.0 and 2.0 mM P_i (Table II).

PS-Mediated Calcium Partitioning

Since P_i accelerates calcium translocation either by increasing the rate by which calcium enters the PS phase or by facilitating calcium release from the ionophore, the aqueous-lipid

*Table II. Effect of P_i on Calcium
Translocation Rate in the
Three-Phase System[a]*

Donor P_i (mM)	Calcium translocation rate[b] (μmole/μmole PS per hr)
0.0	0.116 ± 0.015
0.5	0.113 ± 0.017
1.0	0.138 ± 0.012
2.0	0.306 ± 0.016
5.0	0.345 ± 0.005

[a]Experimental conditions were the same as in Table I.
[b]Mean of three experiments ± S.E.M.

Table III. The Effect of P_i on Calcium Extraction
and Release at Equilibrium[a]

Donor P_i (mM)	Calcium extracted[b] (μmole/μmole PS)	Calcium reextracted[b] (μmole/μmole PS)
0.0	0.275	0.299
1.0	0.604	0.433
2.0	0.808	0.789
5.0	1.375	1.028

[a]Two compartment methods were employed to measure calcium uptake and release [14]. To measure calcium uptake, calcium was extracted from the donor buffer into the organic phase in the presence of 0–5.0 mM P_i. When calcium release was determined, calcium was reextracted from the organic phase into a receiver buffer. PS concentration was 200 μM and donor buffer contained 1.0 mM calcium.
[b]Mean of three experiments, S.E.M. was less than 0.1 μmole.

partition assay was employed to measure calcium binding separately from calcium release at equilibrium [14]. Calcium uptake and release from PS were studied by a partitioning technique in which calcium was first extracted into an organic phase containing PS and then reextracted from the organic phase [14].

Calcium uptake by the lipid phase increased with increasing P_i concentrations up to 5 mM (Table III). A decrease in the amount of calcium binding over 5 mM P_i was a consequence of the precipitation of calcium as an insoluble salt. The increase in calcium binding was accompanied by an enhancement in calcium release (Table III). These data indicate that the elevation in the calcium transport rate in the presence of P_i is the result of an increase in calcium binding to PS and that almost all of the bound calcium could be released (Table III).

At low concentrations of calcium and in the presence of P_i, partitioning of calcium into the lipid phase was almost completely reversible. However, as the concentration of calcium was increased above 2 mM, and in the presence of 5.0 mM P_i, less calcium partitioned out of the organic phase into the aqueous buffer (Table IV).

Based on these experiments, it may be concluded that PS-mediated calcium uptake into the lipid phase is increased in the presence of P_i. Furthermore, at low calcium concentrations, and in the presence of P_i, calcium uptake is increased and the process is reversible. However, at high calcium levels (above 2 mM), while calcium uptake is increased in the presence of P_i, the release step is inhibited. In this latter situation, calcium accumulates in the lipid phase.

Table IV. Effect of Calcium Concentration on Partitioning
of Calcium between Aqueous and Organic Phases

Donor calcium[a] (mM)	Calcium extracted[b] (μmole/μmole PS)	Calcium reextracted[b] (μmole/μmole PS)
0.2	0.045	0.045
0.5	0.500	0.230
1.0	1.740	1.760
2.0	3.480	3.350
5.0	10.000	3.660

[a]Donor buffer contained 5.0 mM P_i.
[b]Mean of three experiments, S.E.M. was less than 0.10 μmole.

PS–Ca–P_i Complex Structure

Partitioning of calcium into the phospholipid phase involves binding of calcium to the phosphate head group of PS to form a PS–Ca complex. The kinetics of calcium transport in the three-phase system, and partitioning studies indicate that P_i modulates this process. To investigate the structure of the PS–Ca complex and the interaction of P_i with PS and calcium, ^{31}P and ^{1}H NMR spectroscopy was employed [15]. The effect of calcium and P_i on the NMR spectra of PS, in solvents of different polarity, indicates that lipid complexes and larger lipid aggregates are formed. In a polar solvent system that contains chloroform/methanol/D_2O, the addition of calcium to PS shifted the ^{31}P resonance of the PS upfield about 1.5 ppm. When P_i was added to this complex, only a single ^{31}P resonance from both PS and P_i was observed. This single ^{31}P resonance occurred at the same chemical shift as PS–Ca in the absence of P_i. These results suggest that P_i and the phosphate head group of PS bind to the calcium to form PS–Ca and PS–Ca–P_i complexes. Since the linewidths of the ^{31}P resonances are relatively narrow in this polar solvent, one can conclude that the PS–Ca and PS–Ca–P_i complexes exhibit little constraint on the thermal motion of the phosphate groups in the relatively polar chloroform/methanol/D_2O solvent system.

When a nonpolar solvent, deuterochloroform, was used, calcium had a marked effect on the mobility of P_i and the head group of PS. In this solvent, PS yielded a narrow ^{31}P resonance similar to that observed in the polar solvent, but the ^{31}P resonances from PS–Ca and PS–Ca–P_i were broadened. This suggests that these complexes aggregated in a nonpolar environment. However, the ^{1}H resonances of the fatty acid side chains of these complexes in deuterochloroform exhibited similar linewidths as PS. It seems likely that in nonpolar environments the PS–Ca and PS–Ca–P_i complexes form aggregates in which the PS phosphate head groups and calcium and P_i are centrally sequestered while the fatty acid side chains interact with the solvent.

Possible Mechanism of PS-Mediated Calcium Transport

The results of these experiments demonstrate that calcium can be translocated across a lipid phase by PS and that P_i stimulates calcium transport. The experimental data indicate that calcium translocation by PS is a two-step process in which calcium partitions from an aqueous phase into the phospholipid and then repartitions from the phospholipid to the receiver aqueous phase. P_i increases the capacity of the phospholipid to accept calcium and thereby elevates the rate at which calcium can be transported. When the calcium concentration is raised (above 2 mM), the second step is inhibited and the maximum rate at which calcium can be transported is limited.

Partitioning of calcium into the phospholipid of membrane bilayer might involve binding of calcium to the phosphate head group of PS, followed by aggregation of several of these complexes to form a "micelle" in the center of which calcium ions are sequestered. This "micelle" would be lipophilic on its surface and thus able to traverse the core of a lipid bilayer. Combination of this "micelle" with the opposite head group region of the membrane bilayer would complete the transport of calcium across the membrane. P_i would be expected to facilitate transport by neutralizing the charge on the calcium, thereby facilitating closure of the micelle and also possibly permitting transport of additional calcium.

It is interesting to relate the findings of this model system to the mineralization of vesicles *in vivo*. The early accumulation of calcium within matrix vesicles could be related to the presence of a proton gradient across the membrane. Such a gradient might exist since the intravesicular pH would be expected to reflect the intracellular pH and be lower than that of the extracellular cartilage lymph [7]. Following the initial uptake of calcium, subsequent calcium accumulation may be related to the appearance of P_i in the cartilage matrix [5]. Elevated levels of calcium in the vesicle interior, in the presence of P_i, could result in retention of calcium in the lipid bilayer and subsequent nucleation of calcium phosphates in the vesicle membrane.

ACKNOWLEDGMENTS. Supported by NIDR Grant DE-02623, NIH Grant RR-05337-21, and a Cottrell research grant from the Research Corporation.

References

1. Anderson, H. C. Vesicles associated with calcification in the matrix of epiphyseal cartilage. *J. Cell Biol.* **41:**59–72, 1969.
2. Anderson, H. C.; Sajdera, S. W. Calcification of rachitic cartilage to study matrix vesicle function. *Fed. Proc.* **35:**148–153, 1976.
3. Boskey, A. L.; Posner, A. S. The role of synthetic and bone extracted Ca–phospholipid–PO_4 complexes in hydroxyapatite formation. *Calcif. Tissue Res.* **23:**251–258, 1977.
4. Bounucci, E. Fine structure and histochemistry of calcifying globules in epiphyseal cartilage. *Z. Zellforsch. Mikrosk. Anat.* **130:**192–217, 1970.
5. Boyde, A.; Shapiro, I. M. Energy dispersive X-ray elemental analysis of isolated growth plate chondrocyte fragments. *Histochemistry* **69:**85–94, 1980.
6. Cotmore, J. M.; Nichols, G.; Wuthier, R. E. Phospholipid–calcium–phosphate complex: Enhanced calcium migration in the presence of phosphate. *Science* **172:**1339–1341, 1971.
7. Cuervo, L. A.; Pita, J. C.; Howell, D. S. Ultramicroanalysis of pH, P_{CO_2} and carbonic anhydrase activity at calcifying sites in cartilage. *Calcif. Tissue Res.* **7:**220–231, 1971.
8. Green, D. E.; Fry, M.; Blondin, G. A. Phospholipids as the molecular instrument of ion and solute transport in biological membranes. *Proc. Natl. Acad. Sci. USA* **77:**257–261, 1980.
9. Peress, N. S.; Anderson, H. C.; Sajdera, S. W. The lipids of matrix vesicles from bovine fetal epiphyseal cartilage. *Calcif. Tissue Res.* **14:**275–281, 1974.
10. Pressman, B. C. Properties of ionophores with broad range cation selectivity. *Fed. Proc.* **32:**1698–1703, 1973.
11. Tyson, C. A.; Vande Lande, H.; Green, D. E. Phospholipids as ionophores. *J. Biol. Chem.* **251:**1326–1332, 1976.
12. Wuthier, R. E. Zonal analysis of phospholipids in the epiphyseal cartilage and bone of normal and rachitic chickens and pigs. *Calcif. Tissue Res.* **8:**36–53, 1971.
13. Wuthier, R. E.; Gore, S. T. Partition of inorganic ions and phospholipids in isolated cell, membrane and matrix vesicle fractions: Evidence for Ca–Pi–acidic phospholipid complexes. *Calcif. Tissue Res.* **24:**163–171, 1977.
14. Yaari, A. M.; Shapiro, I. M. Effect of phosphate on phosphatidylserine-mediated calcium transport. *Calcif. Tissue Int.* **34:**43–48, 1982.
15. Yaari, A. M.; Shapiro, I. M.; Brown, C. E. Evidence that phosphatidylserine and inorganic phosphate may mediate calcium transport during calcification. *Biochem. Biophys. Res. Commun.* **105:**778–784, 1982.

XV

Pathological Calcification

69

Factors Contributing to Intracavitary Calcification

Jane B. Lian

This section on soft tissue implant and intracavitary calcification includes six investigators of different disciplines who will focus on an understanding of the circumstances and pathology involved in these calcifications. In order to control or intervene in the pathologic calcification process, both general and specific factors contributing to calcium deposition must be defined. These include physiologic and biochemical-initiating factors, as well as those factors involved in growth and maturation of ectopic calcified deposits. The emphasis of this section concerns the problem encountered with the advent of successful implantation of cardiac assist devices, including xenograft heart valves (R. Levy, H. Harasaki, and M. Dewanjee), left ventricular assist pumps (J. Lian), and total artificial hearts (D. Coleman). Calcification of these bioprostheses leads to dysfunction, the final complication of uncontrollable calcium deposition. Recent studies have evaluated the calcification problem as related to the mechanics of the device, clinical properties of the biomaterials used in such devices, and the blood–material interactions.

The first paper will consider pathologic calcification in relation to postulated mechanisms of bone mineralization and the role of calcium-binding proteins in contributing to calcification in blood pumps which retain an organized thrombus. Adele Boskey's presentation will provide some insight, from mineral and organic analysis of various calcified deposits, to support the hypothesis that common factors may exist in nucleation of and growth and maturation of mineral deposits, despite differences in the etiology of the calcific disease or origin of plaque formation. The following papers present specific aspects of cardiovascular implants which relate to calcification. Dennis Coleman considers the charge, porosity, and lipid adsorption properties of the polymeric biomaterial used in blood pumps. Robert Levy describes the experimental model of subcutaneous implantation to define chemical alterations in bioprosthetic values and certain physiologic factors contributing to their calcification. Hiroaki Harasaki's studies evaluate three different bioprosthetic tissue valves

---

Jane B. Lian • Department of Biological Chemistry (Orthopedic Surgery), Laboratory for the Study of Skeletal Disorders and Rehabilitation, Harvard Medical School, Children's Hospital Medical Center, Boston, Massachusetts 02115.

in the circulation and emphasize mechanical stress and hemodynamic factors contributing to calcification. An approach to the calcification problem would be deficient without a consideration of the means for detecting and evaluating the onset of the problem in patients. Noninvasive diagnosis techniques are being applied to this problem by Mrinal Dewanjee. In the final chapter, Dieter Kramsch considers potential modes of inhibition and/or prevention of the calcification process.

Pathologic Calcification: General Considerations

Pathologic calcification is seen in a broad spectrum of diseases. The clinical disorders which involve ectopic calcification have recently been reviewed [12,51]. Metastatic calcifications, initiated in tissues when the [Ca] × [PO$_4$] product of the tissue milieu exceeds 60 mg/dl [52], are distinguished from dystrophic calcifications which have diverse associations: insoluble deposits, scarring, injury, necrosis, persistent foreign bodies, structural disorders of collagen and elastin, tumors and calcifications related to microorganisms (dental plaques) [42], intrauterine devices [17].

Calcifications produced by biomaterial implants in soft tissues (e.g., sutures, poly-HEMA sponges, mammary prosthesis) are not a common occurrence [56]. They are generally dystrophic in nature, although the porosity of the biomaterial can be a determinant for heterotopic bone formation [51,56]. On the other hand, devices in the circulation have a high incidence of calcification. These include valve prostheses in children [14], and cardiac assist pumps [20,29,50] or total artificial heart implants in cattle [11].

The ectopic deposition of hydroxyapatite can occur as either ossification, i.e., the induction of heterotopic bone formation (e.g., myositis ossificans), or direct calcinosis, be it due to a systemic etiology (e.g., dermatomyositis, scleroderma) or of local origin (aortic and mitral valve calcification). In dystrophic calcification, degenerating cells or tissue become foci for calcific deposits. Plasma membranes, rich in calcium and phosphate ion-binding sites, are persistent remnants of degenerating cells. Intracellular organelles, typically mitochondria, become calcified, e.g., in ischemic heart injury [19,23]. Needlelike crystals of hydroxyapatite have also been found in association with endoplasmic reticulum in tumoral calcinosis [8].

The question arises as to whether at the molecular level there are factors common to the deposition of calcium and phosphate in pathological calcification and normal mineralization. In recent years, the mineralization mechanism of normal bone formation has come under scrutiny [1,7,15,25], and several laboratories are examining pathologic calcifications for factors which have an apparent role in the deposition of calcium and phosphate in bone [2,7,30]. New facts are emerging regarding ultrastructural and biochemical analyses of pathologic calcifications [2,32] which could portray ectopic calcification as a phenomenon similar to calcified cartilage formation and bone mineralization. For example, nucleation events have been considered in terms of calcium–phosphate–lipid complexes [17] and matrix vesicle formation [57]. Matrix vesicles, membrane-bound structures formed from hypertrophic cells identified in cartilage, dentine, and enamel [1,57], sequester calcium and enzymes for precipitation of calcium phosphate. Such structures have been identified as mineralizing foci in arteriosclerosis tissue [24], atherosclerotic plaque [33], subcutaneous calcinosis [53], and tumors, particularly skeletal tumors, e.g., chondrosarcomas [41]. However, matrix vesicles have not been identified in numerous other pathologic calcifications including calcified deposits in scleroderma, dermatomyositis, intracavitary implants, and renal calcifications to name only a few.

The role of collagen and noncollagenous extracellular bone matrix proteins in mineralization has been elaborated in great detail [15]. With respect to pathologic calcifications, similar matrix proteins can also contribute and promote epitaxial growth of a calcified deposit. Following necrosis of connective tissue, collagen and elastin fibers persist and remain as recognizable and exposed structures for secondary nucleation. The participation of elastin and collagen fibers becomes apparent in vascular calcific diseases [14], pseudoxanthoma elasticum, scleroderma, and dermatomyositis.

Role of Calcium-Binding Proteins in Mineral Deposition

Circulating and tissue proteins with affinities for hydroxyapatite surfaces need to be considered in ectopic calcium deposition. These include phosphorproteins [54], osteonectin [47], osteocalcin [22], and serum proteins, albumin [13] and α_2-glycoprotein [48] which accumulate in bone and could similarly concentrate at sites of ectopic calcification. In bone and dentine, acidic noncollagenous proteins enriched in O-phosphoserine and O-phosphothreonine bind calcium and are associated with mineralization [54]. The most abundant bone-synthesized noncollagenous protein, osteocalcin [22] or bone Gla protein [39], is a vitamin K-dependent protein containing 2-3 residues/5700-dalton molecule of the calcium-binding amino acid, γ-carboxyglutamic acid (Gla). This protein has a unique structural affinity for hydroxyapatite [21].

Comparison of O-phosphoserine, Gla, and osteocalcin concentrations in normal bone as a function of bone age and mineral content has revealed distinctive distributions [30]. Phosphoprotein/calcium ratio is highest in the least mineralized or newly synthesized fraction of bone, while Gla/calcium ratio remains constant at all stages of mineralization. These findings suggest that phosphate groups are more involved in the initial events of mineralization. The highest Gla and osteocalcin concentrations occur in bone diaphysis or most mineralized fraction of bone, suggesting a function which may be related to the mineral phase after its formation. Osteocalcin is found in the circulation and becomes elevated with increased bone turnover. In children, serum osteocalcin concentrations are 10-15 times higher than adults [18] and it is of interest to note that children and adolescents incur more severe calcification problems than adults with the same disease; as in juvenile dermatomyositis [35] and with xenograft valve prostheses [43]. It thus became apparent to us that proteins with affinities for hydroxyapatite surfaces, particularly circulating proteins, should be considered as factors contributing to ectopic calcium deposition.

In pathologic calcifications, O-phosphoserine is found in significant concentrations in kidney stones (8 residues/10^5 amino acids) and plaque of cardiovascular calcifications (1.2-4.0 residues /10^5 amino acids), including atherosclerotic tissue and assist devices in cattle; but it is not detectable in uncalcified host tissue. These levels are higher than those found in adult human bone and heterotopic bone formation (e.g., myositis ossificans). Trace levels of O-phosphoserine (0.15-0.4/10^5 amino acids) were measured in human subcutaneous calcific deposits (scleroderma and dermatomyositis), but more significant quantities were found in rat experimental calcinosis (1.5-1.8/10^5 amino acids) [16]. The differentiating feature in observing high levels of O-phosphoserine appears to be in calcific deposits associated more with thrombus, degenerating cells, and plasma membranes [9,32]. Fibrinogen and numerous intracellular proteins and fibrinogen contain O-phosphoserine [40], and at sites of injury in disintegrating cells, these will denature, unfold, and could thereby provide calcium-binding sites. Thus, the concentration of O-phosphoserine may be attributed to intracellular phosphoproteins.

Gla content of hydroxyapatite ectopic deposits is significant, and correlations of Gla and osteocalcin with different mineral types from human specimens are elaborated in Chapter 70. We have shown in two experimental models of subcutaneous calcifications, calcergy and calciphylaxis, that the appearance of Gla and osteocalcin accumulation coincides with first formation of hydroxyapatite crystals, as observed by electron microscopy [32]. In several pathologic calcifications, including calcergy and calciphylaxis [32], atherosclerostic tissue [26], and calcified valves [27], a correlation of calcium deposition with the amino acid Gla has been demonstrated. We have also observed increased urine excretion of Gla in juvenile dermatomyositis [31], scleroderma patients with massive calcinosis [28], and patients with paraosteoarthropy who develop heterotropic bone from severe skeletal loss [44]. Such findings suggest a relation between calcification and turnover of Gla-containing proteins.

Other Gla-containing proteins include the vitamin K-dependent clotting proteins. The function of Gla residues in prothrombin and liver-synthesized serum proteins is to specifically bind calcium for interaction with phospholipid membrane vesicles [45]. The properties of the Gla-containing peptides are such that with the presence of phospholipid membranous fragments and the extravasation of blood in damaged soft tissues, circulating osteocalcin and Gla-containing clotting factor fragments could promote the calcification process.

Evaluation of the Calcification Problem in Left Ventricular Assist Devices

The use of cardiac assist devices requires a biomaterial interface compatible with blood. Two approaches have evolved to meet this requirement. One relies on nonthrombogenic smooth surfaces [37]; the other utilizes a biomaterial with a rough or textured surface [5,49] to promote deposition of a biologically compatible layer, pseudoneointima. The use of smooth surfaces has the advantages of minimal thrombosis, fibrin deposition, and minimal anticoagulation requirements. However, this approach has the disadvantage of a limited flex life of the biomaterials due to microscopic defects produced during fabrication; this can lead to degradation of the material and calcium deposition within the surface defects [10]. The rough or textured surface approach has the advantage of a biologically derived and highly flexible blood-contacting layer, the pseudoneointima. The major disadvantages are, however, the necessity to control layer thickness with anticoagulation and calcification leading to the loss of compliance.

Our calcification studies have focused on textured surfaces which retain an organized thrombus, using the Model XI left ventricular assist device (LVAD) developed by the ThermoElectron Corporation (Waltham, Mass.) [4]. The device has been successfully implanted in calves for periods up to 1 year and is currently used clinically [5]. It is an axisymmetrical cylindrical blood pump with a polyurethane chamber fabricated from Biomer (Ethicon, Inc., Sommerville, N.J.), which is interfaced on the blood-contacting surface with either internal surfaces of the same polyurethane Biomer and termed "integrally textured" surfaces, or flocked with Dacron fibers [46]. When biomaterials contact the circulating blood a sequence of events is elicited: protein adsorption, inflammatory response, platelet–fibrin interactions, thrombus organization, and pseudoneointima (PNI) formation. The occurrence of each event is dependent on the nature of the biomaterial. Both Biomer and Dacron textured surfaces promote the formation of a fibrinized matrix which contains platelets, leukocytes, monocytic and macrophagic cells (Fig. 1a). Phagocytic and multinucleated giant cells are found on the polymer surface wrapped around the protruding fibrils. We observe that such cells are more apparent on Dacron flocked surfaces (Fig. 1b) than integrally

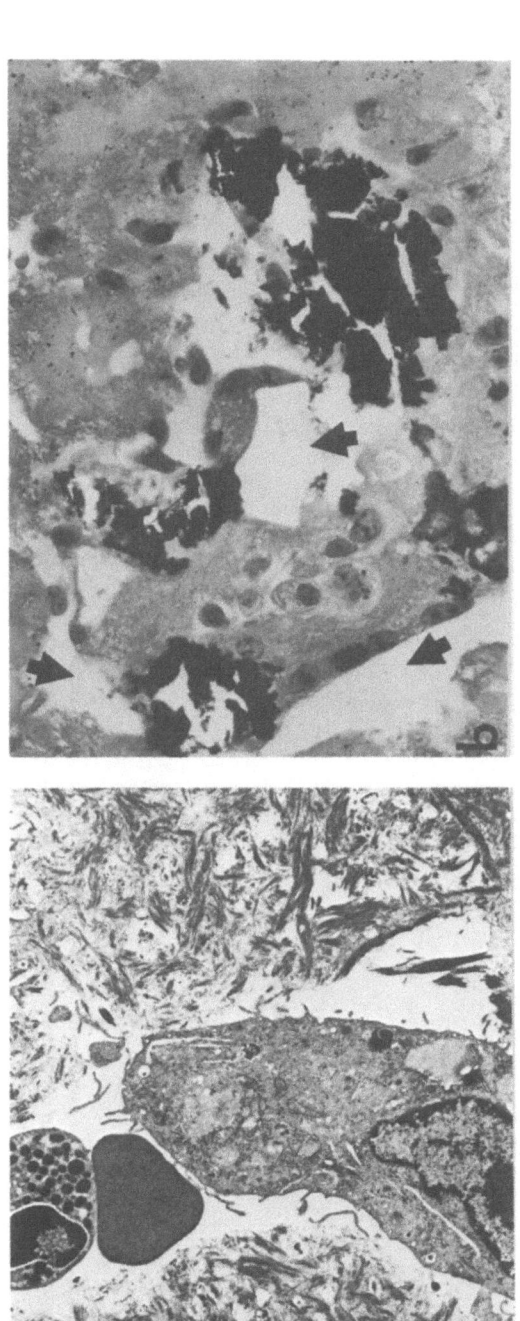

Figure 1. Cells of the pseudoneointima. (a) Loosely organized fibrin matrix, characteristic of the blood-contacting surface where a macrophage, red blood cell, and leukocyte are seen. × 3600. (b) Multinucleated cells surrounding protruding Dacron fibrils (arrow) and in close apposition to patches of mineral. Note the dense, compact organization of surrounding pseudoneointima in this area. von Kossa/H & E. × 1200.

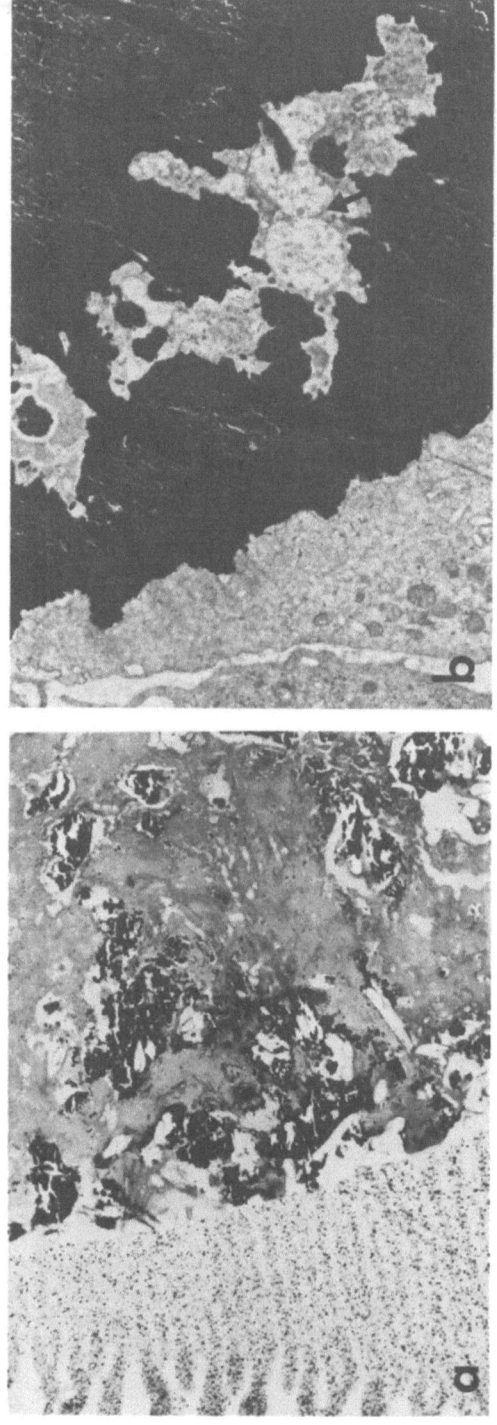

Figure 2. Calcification of left ventricular assist device. (a) Overview of polymer–pseudoneointima (PNI) interface of a Dacron flocked Biomer bladder implanted 184 days showing a heavy density of calcium deposition at polymer–PNI interface with disrupted areas of PNI containing mineral throughout the approaching blood-contacting surface. von Kossa/H & E. × 190. (b) Cell remnants trapped within mineral (arrow) with viable phagocytic cell attached to mineralized area. × 7200.

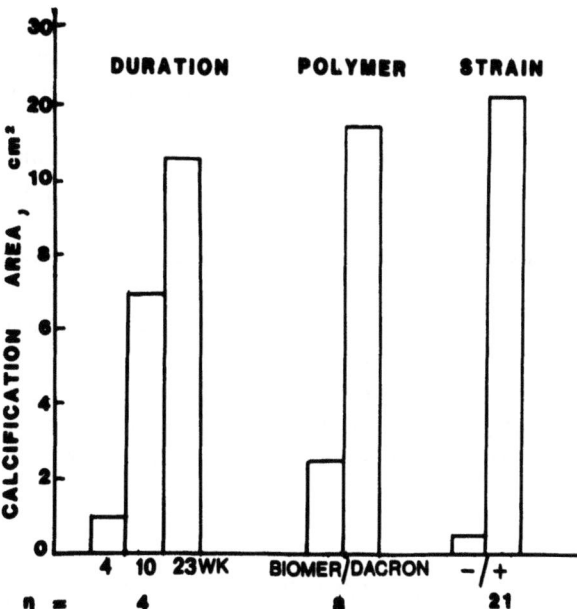

Figure 3. Some contributing factors in left ventricular assist device calcification. Measurements of calcium quantitation were performed on radiographs of the entire explanted bladder (Kodak X-O-mat film). Duration of bypass was assessed in 12 explants of Dacron flocked textured surfaces (four calves at each time point). Polymer Biomer and Dacron were compared in 16 calves who were matched for time of bypass. Strain: On all explants the area of calcification on the three flexing axes and junctions of the pumps (+) were compared to remaining area (−).

textured Biomer (J. B. Lian, S. Dethlefsen, W. Bernhard, and F. Schoen, unpublished observations).

Calcification on Dacron flocked surfaces invariably initiates at the polymer–PNI interface, and not from the blood-contacting surface (Fig. 2a), and occurs with greater incidence (70%) than integrally textured Biomer surfaces. Mineral appears to be associated with phagocytic vesicles, membranous fragments, and other cell debris (Fig. 2b). Calcification is not as apparent on integrally textured Biomer as at the polymer–PNI interface, but occurs within breaks or tears within the PNI, close to the polymer surface. Calcium deposits have never been seen to be initiated from the blood-contacting surface.

Our studies ([29] and Fig. 3) and observations in other types of assist devices [49,55] indicate that calcification predominates on flexing and mechanically stressed areas of the pump. It has been hypothesized that local conditions lead to necrosis of cells and precipitation of calcium salts [34]. Loss of adherence of the fibrin matrix to the bladder surface in such areas is evident [29] and for this reason more stable PNI with strong adherent properties are being developed [6]. The calcification problem can be significantly controlled by seeding the surface with fibroblasts before implant, to favor the development of a dense stable collagenous matrix, rather than a laminated fibrin matrix [46].

Analysis of Calcific Deposits in LVAD

Mineral content of the plaques vary from 40 to 65% ash, with calcium ranging from 14 to 22% of total dry weight and phosphorus 11 to 18%. X-ray diffraction patterns show a poorly crystalline hydroxyapatite with molar ratios of calcium to phosphate indicative of a hydroxyapatite that is seen in embryonic bone (J. B. Lian, W. J. Landis, and L. Bonar, unpublished observations). Protein components were extracted from noncalcified PNI and calcific deposits in 0.5 M EDTA, pH 8.0. Comparison by disc gel electrophoresis and gel filtration liquid chromatography revealed a remarkable similarity of proteins, although quantitative differences occurred. Less than 3% of the total dry starting weight was recovered from the 0.5 M EDTA pH 8.0 extracts of noncalcified PNI, whereas almost 22% of the starting dry weight could be solubilized by this demineralizing solution from calcific materials. Our data from profiles of 22 different calves (J. Lian and W. Bernhard, unpublished observations) suggest that the protein components of the plaque are also present in the PNI, either adsorbed from the circulation or as remnants of degenerating cells that were not phagocytized (Fig. 2b). During the calcification process the proteins become accumulated components of the calcific plaque.

Specific proteins of the calcific deposits have been identified and quantitated by immunochemical techniques. Immunoelectrophoresis of extracted proteins showed that albumin accounts for 8–12% of the total protein concentration. Measurements of the Gla-containing proteins by radioimmunoassay reveal significant levels of F_1 prothrombin fragment in the plaque (from 3 to 11 µg/mg) and osteocalcin (from 30 to 400 ng/mg) with the latter correlating ($r = 0.82$) to the calcium content of the extract. This, in fact, was a surprising result since our cows are maintained on sodium warfarin throughout the implant period.

Effect of Anticoagulation on Gla-Containing Proteins and Calcification

Anticoagulation is necessary to limit the amount of thrombus in bovine tissue with assist pumps which promote PNI formation. Sodium warfarin exerts its effect by inhibiting synthesis of Gla residues with a consequent decrease in calcium binding [45]. However, two factors may contribute to the accumulation of osteocalcin in calcific deposits during therapeutic anticoagulation. One, the anticoagulation regimen increases circulating osteocalcin; and, second, a fraction of this osteocalcin is partially carboxylated. Partially carboxylated bovine osteocalcin maintains its calcium-binding properties to hydroxyapatite (J. B. Lian and C. M. Gundberg, unpublished data), and this fraction accumulates in calcific deposits.

Measurements of plasma osteocalcin are shown in two anticoagulated calves monitored during the implant period (Fig. 4). Up to 10-fold increases above the pretreated values occur within 4 weeks. This phenomenon of increased serum osteocalcin has previously been observed in young rats which were subjected to pharmacological dosage of sodium warfarin [38]. These calves, on the other hand, were maintained at therapeutic prothombin times of twice control values. We were thus concerned with the effects of sodium warfarin on circulating osteocalcin concentrations in patients who are candidates for cardiac assist devices. In collaboration with Drs. R. Blanchard and B. Furie (Tufts New England Medical Center, Boston) 24 patients who were maintained on sodium warfarin for periods of weeks to 6 years were examined for circulating osteocalcin concentrations. All patients were adults ranging in age from 22 to 58 years and all had serum concentrations in the normal range from 4 to 10 ng/ml. Only 25% of these patients were in fact in the upper limit of the normal range: 10 ± 2.3 ng/ml.

It has previously been reported [36] that sodium warfarin regimen during implant of a similar device in calves limited the deposition of calcified plaque. This observation may be attributed to inhibition of synthesis of Gla residues directly or to decreased thrombus formation which limits the environment for potential calcium deposition. It appears that sodium warfarin may retard calcification but not totally prevent it, based on the observations in our calves, who required anticoagulation throughout the implant period. Although the observations of Pierce [36] and others [3] are promising, further quantitative study is necessitated before any conclusions are reached regarding the therapeutic effect of sodium warfarin anticoagulation on ectopic calcium deposition.

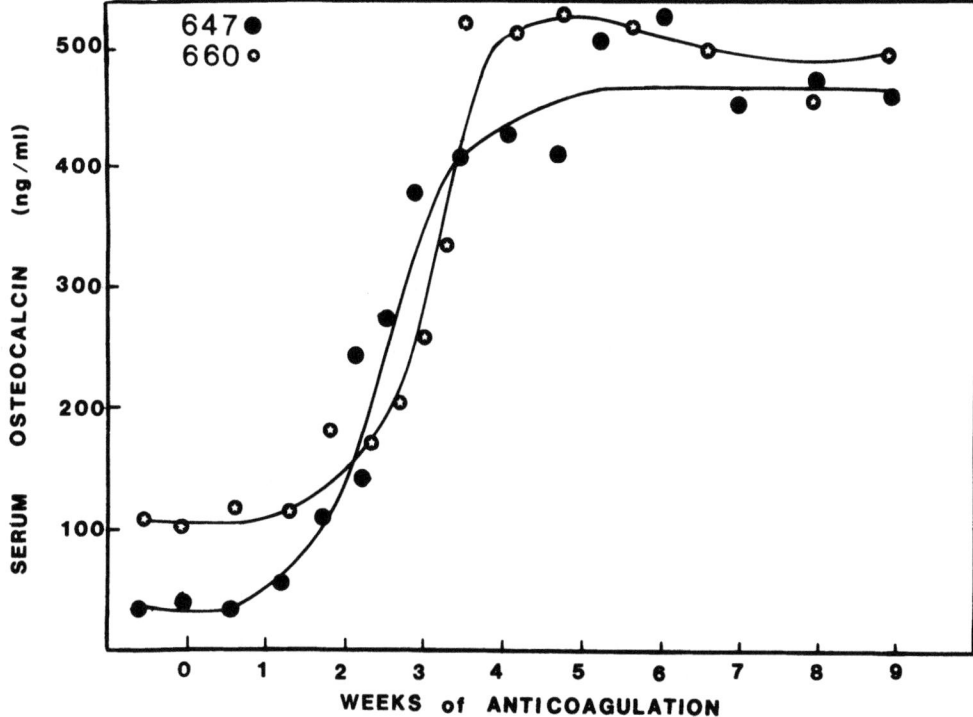

Figure 4. Effect of sodium warfarin on circulating osteocalcin. Two cows [647,660] were monitored every 4 days before and after implant of an LVAD. Cows are maintained twice above control prothrombin throughout the bypass period. Elevations above pretreatment are seen after 2 weeks and increase to a steady-state value after 4 weeks.

ACKNOWLEDGMENTS. This work was supported by National Institutes of Health Grants HL-24029 and AM-26333.

References

1. Anderson, H. C. Calcification processes. *Pathol. Annu.* **15**:45–75, 1980.
2. Anderson, H. C. Calcific disease: A concept. *Arch. Pathol. Lab. Med.*, **107**:341–348, 1983.
3. Berger, R. G.; Hadler, M. M. Treatment of calcinosis universalis secondary to dermatoyositis or scleroderma with low dose warfarin. *Arthritis Rheum.* **26**:511, 1983.
4. Bernhard, W. F.; Poirier, V.; Carr, J. C. A new method for temporary left ventricular bypass. *J. Thorac. Cardiovasc. Surg.* **70**:880–890, 1975.
5. Bernhard, W. F.; LaFarge, G.; Liss, R. H.; Szycher, M.; Berger, R.; Poirier, V. An appraisal of blood trauma and the blood prosthetic interface during left ventricular bypass in calf and human. *Ann. Thorac. Surg.* **26**:427–437, 1978.
6. Bernhard, W. F.; Colo, N. A.; Szycher, M.; Wesolowski, B. S.; Haudenschild, M. D.; Franzblan, C. F.; Parkman, R.; Liss, R. Development of a nonthrombogenic collagenous blood–prosthetic interface. *Ann. Surg.* **192**:369–381, 1980.
7. Boskey, A. L. Current concepts of the biochemistry and physiology of calcification. *Clin. Orthop.* **157**:165–196, 1981.
8. Boskey, A. L.; Vigorita, V. J., Spencer, O.; Stuchin, S. A.; Lane, J. M. Chemical, microscopic and ultrastructural characterization of the mineral deposits in tumoral calcinosis. *Clin. Orthop.* **178**:258–269, 1983.
9. Caufield, J. B.; Schrag, P. E. Electron microscopic study of renal calcification. *Am. J. Pathol.* **44**:365–381, 1964.
10. Coleman, D.; Lawson, J.; Kolff, W. Scanning electron microscopic evaluation of the surfaces of artificial hearts. *Artif. Organs* **2**:275–280, 1978.

11. Coleman, D. L.; Lim, D.; Kessler, T.; Andrade, J. Calcification of nontextured implantable blood pumps. *Trans. Am. Soc. Artif. Intern. Organs* **27**:708-713, 1981.
12. Dalinka, M. K.; Melchior, E. L. Soft tissue calcifications in systemic disease. *Bull. N.Y. Acad. Med.* **56**:539-563, 1980.
13. Dickson, I.; Bagga, K. Changes with age in non-collagenous proteins of human bone. *Connect. Tissue Res.* in press.
14. Ferrans, V. J.; Boyce, S. W.; Billingham, M. E.; Jones, M.; Ishihara, T.; Robert W. C. Calcific deposits in bioprostheses: Structure and pathogenesis. *Am. J. Cardiol.* **46**:721-734, 1980.
15. Glimcher, M. J. Recent studies of the mineral phase in bone and its linkage to the organic matrix by protein bound phosphate bonds. *Philos. Trans. R. Soc. (London B.)* **304**:479-508, 1984.
16. Glimcher, M. J.; Reit, B.; Kossiva, D. Serine phosphate, threonine phosphate and γ-carboxyglutamic acid in normal and experimentally induced pathologically calcified rat skin. *Calcif. Tissue Int.* **33**:185-190, 1983.
17. Gonzales, E. R. Medical news: Calcium deposits in IUD may play role in infections. *J. Am. Med. Assoc.* **245**:1625-1626, 1980.
18. Gundberg, C. M.; Lian, J. B.; Gallop, P. M. Urinary measurements of γ-carboxyglutamate and circulating osteocalcin in normal children and adults. *Clin. Chim. Acta* **128**:1-8, 1983.
19. Hagler, H. K.; Lopez, L. E.; Murphy, M. E.; Greico, C. A.; Briya, L. M. Quantitative x-ray microanalysis of mitochondrial calcification in damaged myocardium. *Lab. Invest.* **45**:241-247, 1981.
20. Harasaki, H.; Gerrity, R.; Kiraly, R.; Jacobs, G.; Nose, H. Calcification in blood pumps. *Trans. Am. Soc. Artif. Intern. Organs* **25**:305-310, 1979.
21. Hauschka, P. V.; Carr, S. A. Calcium-dependent α-helical structure in osteocalcin. *Biochemistry* **21**:638-642, 1982.
22. Hauschka, P. V.; Lian, J. B.; Gallop, P. M. Direct identification of the calcium binding amino acid γ-carboxyglutamate, in mineralized tissue. *Proc. Natl. Acad. Sci. USA* **72**:3925-3929, 1975.
23. Jennings, R. B.; Ganote, C. E.; Reimer, K. A. Ischemic tissue injury. *Am. J. Pathol.* **811**:179-198, 1975.
24. Kim, K. M. Calcification of matrix vesicles in human aortic valve and aortic media. *Fed. Proc.* **35**:156-162, 1976.
25. Landis, W. J.; Glimcher, M. J. Electron optical and analytical observations of rat growth plate cartilage prepared by ultracryomicrotomy: The failure to detect a mineral phase in matrix vesicles and the identification of heterodispered particles as the initial solid phase of calcium phosphate deposited in the extracellular matrix. *J. Ultrastruct. Res.* **78**:227-268, 1982.
26. Levy, R. J.; Lian, J. B.; Gallop, P. M. Atherocalcin, a γ-carboxyglutamic acid-containing protein from atherosclerotic plaque. *Biochem. Biophys. Res. Commun.* **91**:41-49, 1979.
27. Levy, R. J.; Zenker, J. A.; Lian, J. B. Vitamin K dependent calcium binding proteins in aortic valve calcification. *J. Clin. Invest.* **65**:563-566, 1980.
28. Lian, J. B.; Glimcher, M. J.; Gallop, P. M. Identification of γ-carboxyglutamic acid in urinary proteins. In: *Calcium Binding Proteins and Calcium Function*, R. H. Wasserman, ed., Amsterdam, Elsevier/North-Holland, 1977, pp. 379-381.
29. Lian, J. B.; Levy, R. J.; Bernhard, W.; Szycher, M. LVAD mineralization and γ-carboxyglutamic acid containing proteins in normal and pathologically mineralized tissues. *Trans. Am. Soc. Artif. Intern. Organs* **27**:683-689, 1981.
30. Lian, J. B.; Roufosse, A. H.; Reit, B.; Glimcher, M. J. Concentrations of osteocalcin and phosphoprotein as a function of mineral content and age in cortical bone. *Calcif. Tissue Int.* **34**:S82-S87, 1982.
31. Lian, J. B.; Pachman, L. M.; Gundberg, C. M.; Partridge, R.; Maryjowski, M. C. Gamma-carboxyglutamate excretion and calcinosis in juvenile dermatomyositis. *Arthritis Rheum.* **25**:1094-1100, 1982.
32. Lian, J. B.; Boivin, G.; Patterson-Allen, P.; Grynpas, M.; Walzer, C. Calcergy and calciphylaxis: Timed appearance of γ-carboxyglutamic acid and osteocalcin in mineral deposits. *Calcif. Tissue Int.* **35**:555-561, 1983.
33. McGregor, D. H.; Tanimura, A.; Anderson, H. C. Morphogenesis of calcification in atherosclerosis. *Fed. Proc.* **41**:918, 1982.
34. Nose, Y.; Harasaki, H.; Murray, J. Mineralization of artificial surfaces that contact blood. *Trans. Am. Soc. Artif. Intern. Organs* **27**:714-719, 1981.
35. Pachman, L. M.; Cooke, N. Juvenile dermatomyositis: A clinical and immunologic study. *J. Pediatr.* **96**:226-234, 1980.
36. Pierce, W. S.; Donarchy, J. H.; Rosenberg, R.; Baier, R. Calcification inside artificial hearts; inhibition by warfarin sodium. *Science* **208**:601-603, 1980.
37. Pierce, W. S.; Myers, J. L.; Donarchy, J. H.; Rosenberg, G.; Landis, D. L.; Prophet, G. A.; Snyder, A. J. Approaches to the artificial heart. *Surgery* **90**:137-148, 1981.
38. Price. P. A.; Williamson, M. K. Effects of warfarin on bone. *J. Biol. Chem.* **257**:12754-12759, 1981.

39. Price, P. A.; Otsuka, A. S.; Poser, J. W.; Kristaponis, J.; Raman, N. Characterization of γ-carboxyglutamic acid-containing protein from bone. *Proc. Natl. Acad. Sci. USA* **73**:1447–1451, 1976.
40. Rosen, O. N.; Krebs, E. G. (eds.) *Protein Phosphorylation*. Volume 8, Book A, Cold Spring Harbor, N.Y., Cold Spring Harbor Laboratory, 1981.
41. Schajowicz, M. D.; Cabrini, M. D.; Simes, R. J.; Klein-Szanto, A. J. P. Ultrastructure of chondrosarcoma. *Clin. Orthop.* **100**:378–386, 1974.
42. Sidaway, D. A. A microbiological study of dental calculus. 4. An electron microscopic study of *in vitro* calcified organisms. *J. Periodontal Res.* **15**:240–254, 1980.
43. Silver, M. M.; Pollock, J.; Silver, M. D.; Williams, W. G.; Truster, G. A. Calcification of porcine xenograft valves in children. *Am. J. Cardiol.* **45**:685–690, 1980.
44. Steinberg, J. T.; Rossier, A. B.; Lian, J. B.; Gundberg, C. M.; Gallop, P. M. γ-carboxyglutamate excretion as an index of ectopic bone formation in post-traumatic myelopathy. *Trans. Orthop. Res. Soc.* **6**:269, 1981.
45. Stenflo, J.; Suttie, J. W. Vitamin K-dependent formation of γ-carboxyglutamate acids. *Annu. Rev. Biochem.* **46**:157–172, 1977.
46. Szycher, M.; Poirier, V.; Bernard, W. F.; Franzblau, C.; Haudenschild, C. C.; Toselli, P. Integrally textured polymeric surfaces for permanently implantable cardiac assist devices. *Trans. Am. Soc. Artif. Intern. Organs* **25**:493–499, 1980.
47. Termine, J. D.; Belcourt, A. B.; Conn, K. M.; Kleinman, H. K. Mineral and collagen binding proteins of fetal calf bone. *J. Biol. Chem.* **256**:10403–10408, 1981.
48. Triffit, J. T.; Gebauer, U.; Ashton, B. A.; Owen, M. E. Origin of plasma α2HS-glycoprotein and its accumulation in bone. *Nature (London)* **262**:226–227, 1976.
49. Turner, S. A.; Milton, L. T.; Poirier, V. L.; Norman, J. C. Sequential studies of pseudoneointimae within long-term THI E-type ALVAD's: Thickness, calcification, and compositional analyses. *Artif. Organs* **5**:18–27, 1981.
50. Turner, S. A.; Bossart, M. I.; Milam, J. D.; Fugua, J. M.; Ago, S. R. Calcification in chronically-implanted blood pumps: Experimental results and review of the literature. *Tex. Heart Inst. J.* **9**:195–205, 1982.
51. Urist, M. R. (ed.) Heterotopic bone formation. In: *Fundamental and Clinical Bone Physiology*, Philadelphia, Lippincott, 1980, pp. 369–393.
52. Valentzas, C. Visceral calcification and the Ca × P product. *Adv. Exp. Med. Biol.* **103**:195–201, 1978.
53. Waltzer, C.; Boivin, G.; Schonborner, A. A.; Baud, C. A. Ultrastructural and cytochemical aspects of an experimental cutaneous calcinosis in rat. *Cell Tissue Res.* **212**:185–202, 1980.
54. Weinstock, M.; LeBlond, C. P. Radioautographic visualization of the deposition of a phosphoprotein at the mineralization front in the dentin of rat incisor. *J. Cell Biol.* **56**:838–845, 1973.
55. Whalen, R. L.; Snow, J. L.; Harasaki, H.; Nose, Y. Mechanical strain and calcification in blood pumps. *Trans. Am. Soc. Artif. Intern. Organs* **26**:487–492, 1980.
56. Woodward, S. C. Mineralization of connective tissue surrounding implanted devices. *Trans. Am. Soc. Artif. Intern. Organs* **27**:697–701, 1981.
57. Wuthier, R. E. A review of the primary mechanism of endochondral calcification with special emphasis on the role of cells, mitochondria and matrix vesicles. *Clin. Orthop.* **169**:219–242, 1982.

70

Mineral, Lipids, and Proteins Associated with Soft Tissue Deposits

Adele L. Boskey

A wide variety of crystalline deposits occur as a result of pathophysiologic disturbances in soft tissues of humans [3–5,43,44,38,56]. These include mineral deposits, e.g., calcium phosphates and carbonates, the calcium salts of organic molecules, and crystals of non-mineral materials. Many examples of each of these categories exist, and a list of the most commonly encountered soft tissue crystalline deposits is presented in Table I. The biochemical disturbances associated with crystal deposition in soft tissues differ, but in general, abnormal crystal deposition can be attributed to one or more of the following mechanisms: increased local or systemic supersaturations of the tissue, exposure of agents that promote crystal deposition (nucleators), and loss or destruction of agents that in healthy tissue prevent mineral deposition (inhibitors).

Lipids and Proteins Associated with Mineral Deposits

Molecules commonly found associated with soft tissue crystalline deposits include nucleators and inhibitors, as well as macromolecules which play no direct role in the process of crystal deposition. These molecules were derived from surrounding tissues and circulating fluids. They include components which were adsorbed to or trapped within the deposit during its formation, as well as those whose presence in the soft tissue facilitated crystal formation. Insight into the function of these associated materials has come from studies of their effect on *in vitro* crystal deposition and growth. Most studies have focused on the calcified deposits (the calcium phosphates, calcium pyrophosphates, etc.), with an emphasis on the molecules associated with normal and heterotopic bone mineral (hydroxyapatite).

Table II lists the macromolecules associated with mineralized deposits and their putative functions. Of these, the phosphorproteins [25,38,45.60], osteocalcin [5,50,57], and proteoglycans, or their component glycosaminoglycans [32,35,49], all have properties which

---

Adele L. Boskey • Department of Biochemistry, Cornell University Medical College, and Department of Ultrastructural Biochemistry, The Hospital for Special Surgery, New York, New York 10021.

Table I. Crystalline Deposits in Human Soft Tissues

Deposit	Site	Reference
Hydroxyapatite	Cartilage	26, 59
	Synovium	26, 58
	Synovial fluid	26, 59
	Subcutaneous tissues	47
	Bursa	59, 48
	Muscle	47
	Ligament and tendon	59
	Aorta	24
	Bioprostheses	29
Brushite	Fibrocartilage	4, 34
	Joint fluid	41
Calcium pyrophosphate	Cartilage	41, 57
	Synovium	41, 57
	Dura[a]	57
	Tendon insertions[a]	57
Calcium carbonate	Bronchial liths	53
	Pancreatic stones	45
Calcium oxalate	Granulomas	41
Sodium oxalate	Synovium	41
	Cartilage	41
	Skin	41
Monosodium urate	Synovium	44
	Skin	44
	Cartilage	44
	Other connective tissues	44
Cholesterol	Joints	54

[a]Less commonly.

allow them to inhibit hydroxyapatite *in vitro* growth. Such *in vitro* systems, however, may not reflect the *in vivo* participation of these molecules in mineral formation [31]. On the other hand, calcium–acidic phospholipid–phosphate complexes, which are found in association with soft tissue hydroxyapatite deposits, can, as nucleators, cause hydroxyapatite deposition both *in vitro* and *in vivo* [22,42]. This chapter is devoted to a consideration of these complexes and their possible role in calcification.

Acidic Phospholipids and Mineralization

The observation in normal (bone, calcified cartilage, dentine) and ectopic hydroxyapatite depositions, that acidic phospholipids accumulated at the mineralization front [36], led to the discovery of calcium–acidic phospholipid–phosphate complexes [13]. These complexes, which were later shown to be membrane components [22,63], contain, in decreasing proportion: calcium (50 mole%), acidic phospholipids (38–46 mole%), and inorganic phosphate (3–12 mole%). *In vitro*, these complexes cause hydroxyapatite deposition from metastable calcium phosphate solutions [14,15]. These studies indicate that all acidic phospholipids capable of causing hydroxyapatite deposition do so via the formation of such complexes.

Further studies revealed that levels of the complexes are highest in actively mineralizing tissues [10,11,16,20]. More detailed studies revealed that calcium–acidic phospholipid–

Table II. Lipids and Proteins Associated with Mineral Deposits

Molecule	Found in	Function
Ca–PL–PO$_4$	HA[a] deposits	Promotes HA deposition
Proteolipids	Atherosclerotic plaques and other HA deposits	Promotes HA deposition
Collagen	All connective tissue deposits (HA, monosodium urate, calcium pyrophosphate dihydrate)	Oriented support for mineral crystals
Proteoglycans	Articular cartilage: kidney stones, atherosclerotic plaques	Inhibits mineralization, promotes flocculation
Osteocalcin	HA deposits	Controls mineral growth?
Phosphoproteins	HA deposits, pancreatic stones	Control mineral nucleation and/or growth
IgG	Monosodium urate deposits, silicate deposits, calcium pyrophosphate dihydrate deposits	?
Enzymes[b]	Monosodium urate deposits, silicate deposits, calcium pyrophosphate dihydrate deposits	?

[a] HA, hydroxyapatite.
[b] Enzymes found associated with these deposits include acid phosphatase, β-glucuronidase, and 5'-nucleotidase.

phosphate complex concentrations increase prior to the appearance of mineral [19], suggesting that the formation of these complexes prepares the tissue for mineral deposition. Thus, complex levels are elevated in bone matrix-induced endochondral ossification [18], in nonmineralizing rachitic rat growth plates [19], and in noncalcified aortic plaques [27]. Implanted in Millipore diffusion chambers within abdominal rabbit muscle these calcium–acidic phospholipid–phosphate complexes cause *in vivo* hydroxyapatite accumulation [55]. These complexed acidic phospholipids, as well as their component acidic phospholipids, have also been shown to facilitate calcium transport *in vitro* (see Chapter 68) [64].

In both physiologic and nonphysiologic calcifications, the first hydroxyapatite crystals appear in association with extracellular, membrane-bound bodies, known as matrix vesicles [2,3,7,17] (see Chapter 65). These extracellular matrix vesicles, derived from cells, are thought to facilitate mineral deposition by concentrating calcium and phosphate, and by providing an appropriate milieu for hydroxyapatite nucleation and growth. The membranes of extracellular matrix vesicles of cartilage contain calcium–acidic phospholipid–phosphate complexes [63] probably associated with membrane [30] proteolipids [22,28]. In addition, each of the tissues shown to contain elevated levels of these complexes (growth cartilage, osteomalacic cortical bone, aortic plaques, osteoarthritic cartilage, etc.) also possesses abundant extracellular matrix vesicles [1,2,6,37]. We therefore believe that calcium–acidic phospholipid–phosphate complexes are associated with the extracellular matrix vesicles, and that their formation precedes hydroxyapatite deposition, preparing the matrix for calcification. Our studies [18,19] indicate that the acidic phospholipid complexes will be associated with the first deposits of hydroxyapatite, and that all hydroxyapatite deposits derived from cells involve these complexes.

Table III. Complexed Acidic Phospholipid Content Is Elevated in Soft Tissues in Which Hydroxyapatite Deposits

Disease in which HA deposits	N^a
Calcific tendonitis	2
Arthritis	5
Calcinosis	4
(associated with scleroderma, polymyositis, milk alkali syndrome, vitamin D toxicity)	
Synovial calcification	1
Heterotopic bone formation	2
(traumatic myositis ossificans)	
Tumoral calcinosis	5
Atherosclerosis	13
LVAD[b] calcification	1

[a] Number of patients with disease in which Ca–PL–PO$_4$ levels were elevated relative to noncalcified controls.
[b] Left ventricular assist device implanted in a calf.

Calcium–Acidic Phospholipid Complexes in Soft Tissue Deposits

Calcium–acidic phospholipid–phosphate complexes in concentrations greater than those found in healthy, nonmineralizing tissues have been isolated from the matrices of all hydroxyapatite deposits examined to date. These include stones, such as sialoliths [12,17], and nephroliths [9], normally calcified tissues, and the soft tissue deposits listed in Table III. Thus, in all cases of hydroxyapatite deposition, whether normal or diseased, and regardless of the etiology of the disease [21,23,34,46,47], calcium–acidic phospholipid–phosphate complexes appear within the soft tissue matrix. Moreover, the amount of the complex present in all of the tissues examined is significantly greater than that found in noncalcified tissues [21,27,29]. For example, the amount of calcium–acidic phospholipid–phosphate

Table IV. Complexed Acidic Phospholipid, Osteocalcin, and Gla Content of Soft Tissue Deposits[a]

	Ca–PL–PO$_4$ (μg/mg demin. dry wt)	Osteocalcin (ng/mg demin. dry wt)	Gla (residues/ 100,000)
HA ($n = 66$)	10 ± 2	13 ± 8	6 ± 4
CPPD ($n = 26$)	1.7 ± 0.1	0.4 ± 0.2	0.2 ± 0.1
MSU ($n = 5$)	0.39 ± 0.03	ND[b]	ND
Oxalates ($n = 5$)	0.3 ± 0.2	ND	ND

[a] Abbreviations: HA, hydroxyapatite; CPPD, calcium pyrophosphate dihydrate; MSU, monosodium urate; oxalates, sodium and calcium oxalates.
[b] ND, not determined.

complex in dermis of normal individuals and patients with tumoral calcinosis [21] was found to be 1 and 8 µg/mg dry weight, respectively.

Table IV summarizes the calcium–acidic phospholipid–phosphate complex concentrations of 66 soft tissue deposits containing hydroxyapatite, and compares these to calcium pyrophosphate dihydrate, monosodium urate, and sodium oxalate deposits. The concentration of complexed phospholipids in the matrices associated with the pyrophosphate deposits was identical to that found in comparable, normal, nonmineralized tissues. The inclusion of these lipids within the deposit suggests that membranes were trapped in the deposit during the growth and development of these crystals. This is a reasonable assumption, since there is ample data connecting these deposits with membranolysis [43,61]. In contrast, the monosodium urate and oxalate deposits contain no complexed acidic phospholipids in their matrices. These deposits, which generally form due to increased local supersaturations [41], and which are due to their extreme insolubility, form quite easily *in vivo* and *in vitro,* most likely precipitate directly within the matrix, rather than in contact with cell or other organelle membranes.

Interaction of Proteins and Lipids in Soft Tissue Calcification

Complexed acidic phospholipids and the extracellular matrix vesicles in which they may be situated are not the only etiologic agents in mineral deposition. We believe that the calcification process in general, and the formation by hydroxyapatite in particular, results from the interaction of numerous factors that promote crystal nucleation and growth, or have an inhibitory, regulatory function. The nature of the factors controlling bone mineralization has recently been reviewed [8,31,62]. These factors exist in a complex matrix, and therefore have the capacity to interact. For this reason, we have begun to investigate the interaction of the acidic phospholipids with several of the proteins involved in the control of the mineralization process.

Osteocalcin, or bone Gla protein [52], by analogy with the γ-carboxyglutamic acid (Gla)-containing blood-clotting proteins, should have the capacity to interact with phospholipids (see Chapter 69). Osteocalcin and Gla-containing proteins are found concentrated in densely mineralized bone [40], and in apatite-containing pathologic deposits [39]. Table IV includes the mean osteocalcin content found in the mineral deposits examined in our complexed phospholipid study, along with the number of Gla residues per 100,000 amino acid residues of matrix protein. Only the apatite-containing deposits contain osteocalcin. However, there is no correlation between concentrations of osteocalcin and complexed acidic phospholipids (Table V). In fact, the only significant correlations found between parameters measured in the mineralized deposits are between mineral composition and complexed acidic phospholipid content, mineral composition and Gla content, site of deposit and nature of the primary disease, complexed acidic phospholipid content and Gla content (but not osteocalcin content), and the content of osteocalcin and the nature of the mineral. It is of interest to note that in developing matrix-induced endochondral bone [51], and in embryonic bone [40], the amino acid Gla correlates with calcium content, but osteocalcin content lags behind and is present in proportionately lower concentrations than in mature bone.

Osteocalcin inhibits hydroxyapatite growth [50], presumably by binding to the surface of the mineral, thus preventing access of incoming ions to growth sites. Preliminary studies in our laboratory confirm these results, while indicating that osteocalcin does not affect the rate of lipid-induced calcification until a significant amount of hydroxyapatite crystals are formed. Additionally, osteocalcin does not appear to either interact with calcium–acidic

Table V. Correlations between Parameters Measured in Soft Tissue Deposits

Variable 1	Variable 2	r
Mineral phase	Ca–PL–PO_4 content	0.6
Mineral phase	Gla content	0.8
Site of deposit	Primary disease	0.8
Ca–PL–PO_4 content	Gla content	0.8
Ca–PL–PO_4 content	Mineral phase	0.6
Ca–PL–PO_4 content	Ca/P ratio of ash	0.8
Gla content	Total lipid content	0.5
Gla content	Mineral phase	0.8
Osteocalcin content	Mineral phase	0.5
Osteocalcin content	Primary disease	0.6

phospholipid complexes, or bind tightly to acidic phospholipids (F. Wians, personal communication). This finding suggests that osteocalcin peptide and complexed acidic phospholipids do not interact to regulate initial apatite deposition. However, since Gla and phospholipid-complex contents correlate, another larger Gla-containing bone protein [33] in pathologic calcifications may interact with phospholipids, analogous to the vitamin K-dependent clotting proteins.

In conclusion, the results presented in this chapter suggest that while calcium–acidic phospholipid–phosphate complexes are formed prior to mineral deposition, preparing the matrix for calcification, osteocalcin probably adsorbs to mineral after its deposition. Osteocalcin and the complexed acidic phospholipid contents are increased in mineralized soft tissue deposits, although they do not appear to be associated with other mineral deposits. However, the nature of their association with hydroxyapatite deposits, and their functions within the hydroxyapatite-containing tissues appear to be quite different.

ACKNOWLEDGMENTS. Supported by NIH Grant DE-04141. We are grateful to Dr. Peter G. Bullough, Dr. Vincent Vogorita, Dr. Foster Betts, and Dr. Jane Lian for their assistance in the performance of these studies. This is publication No. 169 from the Laboratory of Ultrastructural Biochemistry.

References

1. Ali, S. Y. Matrix vesicles and apatite nodules in arthritic cartilage. In: *Perspectives in Inflammation*, D. A. Willoughby, J. P. Giroud, and G. P. Vello, eds., Lancaster, M.T.P. Press, 1977, pp. 211–223.
2. Anderson, H. C. Calcification processes. *Pathol. Annu.* 15:45–75, 1980.
3. Anderson, H. C. Calcific diseases: A Concept. *Arch. Pathol. Lab. Med.* 107:341–348, 1983.
4. Bigi, A.; Foresti, E.; Incerti, A.; Roverti, N.; Borea, P. A.; Zavagli, G. Chemical and structural study of the mineral phase associated with a human subcutaneous ectopic calcification. *Inorg. Chim. Acta* 46:271–274, 1980.
5. Boivin, G.; Lian, J. B.; Grynpas M. D.; Glimcher, M. J. The appearance of γ-carboxyglutamic acid in subcutaneous calcinosis. *Calcif. Tissue Int.* 33:314, 1981.
6. Bonucci, E.; Dearden, L. C. Matrix vesicles in aging cartilage. *Fed. Proc.* 35:163–168, 1976.
7. Boskey, A. L. Models of matrix vesicle calcification. *Inorg. Perspect. Biol. Med.* 2:51–92, 1978.
8. Boskey, A. L. Current concepts of the biochemistry and physiology of calcification. *Clin. Orthop.* 157:165–196, 1981.

9. Boskey, A. L. The role of Ca-phospholipid-PO$_4$ complexes, membrane lipids and matrix vesicles in calcification. In: *Matrix Vesicles*, A. Ascenzi, E. Bonucci, and B. de Bernard, eds., Milan, Wichtig Editore, 1981, pp. 161–167.
10. Boskey, A. L.; Bullough, P. G.; Dmitrovsky, E. D. The biochemistry of the mineralization front. *Metab. Bone Dis. Relat. Res.* 15:61–67, 1980.
11. Boskey, A. L.; Bullough, P. G.; Posner, A. S. Calcium acidic phospholipid phosphate complexes in diseased and normal human bone. *Bone Dis. Relat. Res.* 4:151–156, 1982.
12. Boskey, A. L.; Mandel, I.; Boyan-Salyers, B. D. Lipids and salivary stone calcification. *Arch. Oral Biol.* 26:779–785, 1981.
13. Boskey, A. L.; Posner, A. S. Extraction of a calcium-phospholipid-phosphate complex from bone. *Calcif. Tissue Res.* 19:273–283, 1976.
14. Boskey, A. L.; Posner, A. S. The role of synthetic and bone extracted Ca-phospholipid-phosphate complexes in hydroxyapatite formation. *Calcif. Tissue Res.* 23:251–258, 1977.
15. Boskey, A. L.; Posner, A. S. Optimal conditions for Ca-acidic phospholipid-PO$_4$ complex formation. *Calcif. Tissue Int.* 34:S1–S7, 1982.
16. Boskey, A. L.; Posner, A. S.; Lane, J. M.; Goldberg, M. R.; Cordella, D. M. Distribution of lipids associated with mineralization in the bovine epiphyseal growth plate. *Arch. Biochem. Biophys.* 199:305–311, 1980.
17. Boskey, A. L.; Posner, A. S.; Mandel, I. Phospholipids associated with human parotid sialoliths. *Arch. Oral Biol.* 28:655–657, 1981.
18. Boskey, A. L.; Reddi, A. H. Changes in lipids during matrix-induced endochondral bone formation. *Calcif. Tissue Int.* 35:549–554, 1983.
19. Boskey, A. L.; Timchak, D. M. Phospholipid changes in the bones of the vitamin D deficient P deficient immature rat. *Metab. Bone Dis. Relat. Res.* 5:81–85, 1984.
20. Boskey, A. L.; Timchak, D. M.; Lane, J. M.; Posner, A. S. Phospholipid changes during fracture healing. *Proc. Soc. Exp. Biol. Med.* 165:368–373, 1980.
21. Boskey, A. L.; Vigorita, V.; Stuchin, S.; Sencer, O.; Lane, J. M. Chemical characterization of the mineral deposits in tumoral calcinosis. *Clin. Orthop.* 178:258–269, 1983.
22. Boyan-Salyers, B. D.; Boskey, A. L. Relationship between proteolipid and Ca-phospholipid-phosphate complexes in bacterionema matruchotti calcification. *Calcif. Tissue Int.* 30:167–174, 1980.
23. Burnett, C. H.; Commons, R. R.; Albright, F.; Howard, J. E. Hypercalcemia without hypercalciuria or hyperphosphatemia, calcinosis and renal insufficiency: A syndrome following prolonged intake of milk and alkali. *N. Engl. J. Med.* 240,787–794, 1949.
24. Carlstom, D.; Engfeldt, B.; Engstrom, A.; Ringertz, N. Studies on the chemical composition of normal and abnormal blood vessel walls. I. Chemical nature of vascular calcified deposits. *Lab. Invest.* 2:325–335, 1953.
25. De Caro, A.; Lohse, J.; Sarles, H. Characterization of a protein isolated from pancreatic calculi of men suffering from chronic calcifying pancreatitis. *Biochem. Biophys. Res. Commun.* 87:1176–1182, 1979.
26. Dieppe, P. A.; Crocker, P.; Huskisson, E. C.; Willoughby, D. A. Apatite deposition disease: A new arthropathy. *Lancet* 1:266–268, 1976.
27. Dmitrovsky, E. D.; Boskey, A. L.; Minick, C. R. Complexed acidic phospholipids associated with aortic calcification. *Fed. Proc.* 41:918, 1982.
28. Ennever, J.; Vogel, J. J.; Riggan, L. J. Calcification by proteolipid from atherosclerotic aorta. *Atherosclerosis* 35:209–213, 1980.
29. Ferrans, V. J.; Boyce, S. W.; Billingham, M. E.; Jones, M.; Ishihara, T.; Roberts, W. C. Calcific deposits in porcine bioprostheses: Structure and pathogenesis. *Am. J. Cardiol.* 46:721–734, 1980.
30. Folch, J.; Lees, M. Proteolipids, a new type of tissue lipoproteins. *J. Biol. Chem.* 91:807–817, 1951.
31. Glimcher, M. J. On the form and function of bone: From molecules to organs. Wolf's law revisited, 1981. In: *The Chemistry and Biology of Mineralized Connective Tissues*, A. Veis, ed., Amsterdam: Elsevier/North-Holland, 1981, pp. 617–673.
32. Hascall, V. C.; Kimura, J. H. Proteoglycans–Isolation and characterization. *Methods Enzymol.* 82:769–800, 1982.
33. Hauschka, P. V.; Frenkel, J.; DeMuth, R.; Gundberg, C. M. Presence of osteocalcin and related higher molecular weight 4-carboxyglutamic acid-containing proteins in developing bone. *J. Biol. Chem.* 258:176–182, 1983.
34. Howell, D. S. Biochemical studies of osteoarthritis. In: *Arthritis and Allied Conditions*, 9th ed., D. J. McCarty, ed., Philadelphia, Lea & Febiger, 1979, pp. 1154–1160.
35. Howell, D. S.; Pita, J. C. Calcification of growth plate cartilage with special reference to studies on micropuncture fluids. *Clin. Orthop.* 118:208–229, 1976.
36. Irving, J. T.; Wuthier, R. E. Histochemistry and biochemistry of calcification with special reference to the role of lipids. *Clin. Orthop.* 56:237–260, 1968.

37. Kim, K. M. Calcification of matrix vesicles in human aortic valve and aortic media. *Fed. Proc.* **35**:156-162, 1976.
38. Lee, S. L.; Glimcher, M. J. The purification, composition and 31-P spectroscopic properties of a noncollagenous phosphoprotein isolated from chicken bone matrix. *Calcif. Tissue. Int.* **33**:385-394, 1981.
39. Levy, R. J.; Lian, J. B.; Gallop, P. Atherocalcin, a γ-carboxyglutamic acid containing protein from atherosclerotic plaque. *Biochem. Biophys. Res. Commun.* **91**:41-49, 1979.
40. Lian, J. B.; Roufosse, A. H.; Reit, B.; Glimcher, M. J. Concentrations of osteocalcin and phosphorprotein as a function of mineral content and age in cortical bone. *Calcif. Tissue Int.* **34**:S82-S87, 1982.
41. McCarty, D. J. (ed.) Pathogenesis and treatment of crystal-induced inflammation. In: *Arthritis and Allied Conditions*, 9th ed., Philadelphia, Lea & Febiger, 1979, pp. 1245-1261.
42. Mackel, A. M.; DeLustro, F.; Hasper, F. E.; LeRoy, E. C. Antibodies to collagen in scleroderma. *Arthritis Rheum.* **25**:522-531, 1982.
43. Mandel, N. S. The structural basis of crystal-induced membranolysis. *Arthritis Rheum.* **19**:439-445, 1976.
44. Mandel, N. S. Mandel, G. S. Monsodium urate monohydrate, the gout culprit. *J. Am. Chem. Soc.* **98**:2319-2323, 1976.
45. Multigner, L.; DeCaro, A.; Lombardo, D.; Sarles, H. Pancreatic stone protein: A phosphoprotein which inhibits calcium carbonate precipitation from human pancreatic juice. *Biochem. Biophys. Res. Commun.* **110**:69-74, 1983.
46. Mundy, G. R.; Raisz, L. G. Disorders of bone resorption. In: *Disorders of Mineral Metabolism*, Volume III, F. Bronner and N. W. Coburn, eds., New York, Academic Press, 1981, pp. 1-66.
47. Ogilvie-Harris, D. J.; Hons, C. B.; Fornasier, V. L. Pseudomalignant myositis ossificans: Heterotopic newbone formation without a history of trauma. *J. Bone Jt. Surg. Am. Vol.* **62**:1274-1283, 1980.
48. Perugia, I.; Sadun, R. New trends in muscle calcification. *Int. Orthop.* **1**:165-170, 1977.
49. Posner, A. S.; Boskey, A. L.; Chen, C.-C. Regulation of apatite proliferation by phospholipid complexes and proteoglycans. *Excerpta Med. Int. Congr. Ser.* **589**:44-49, 1982.
50. Poser, J. W.; Price, P. A. A method for decarboxylation of γ-carboxyglutamic acid in protein. *J. Biol. Chem.* **254**:431-436, 1979.
51. Price, P. A.; Lothringer, J. W.; Baukol, S. A.; Reddi, A. H. Developmental appearance of the vitamin K-dependent protein of bone during calcification: Analysis of mineralizing tissues in human, calf, and rat. *J. Biol. Chem.* **257**:3781-3784, 1982.
52. Price, P. A.; Otsuka, A. S.; Poser, J. P.; Kristanponis, J.; Raman, N. Characterization of a γ-carboxyglutamic acid-containing protein from bone. *Proc. Natl. Acad. Sci. USA* **73**:1447-1451, 1976.
53. Pritzker, K. H.; Desai, S. D.; Patterson, M. C.; Cheng, P.-T. Calcite sputum lith: Characterization by analytic scanning electron microscopy and X-ray diffraction. *Am. J. Clin. Pathol.* **75**:253-257, 1981.
54. Pritzker, K. H.; Fam, A. G.; Omar, S. A.; Gertzbein, S. D. Experimental cholesterol crystal arthropathy. *J. Rheumatol.* **8**:281-290, 1981.
55. Raggio, C. L.; Boskey, A. L.; Boyan, B. D.; Urist, M. R. *In vitro* induction of hydroxyapatite formation by lipid macromolecules. *Trans. Annu. ORS* **6**:20, 1983.
56. Reginato, A. J.; Schumacher, H. R. Synovial calcification in a patient with collagen vascular disease. *J. Rheumatol.* **4**:261-267, 1977.
57. Rynes, R. I. Calcium pyrophosphate dihydrate crystal deposition. *J. Rheumatol.* **7**:5-8, 1980.
58. Sarkar, K.; Uhthoff, H. K. Ultrastructural localization of calcium in calcifying tendinitis. *Arch. Pathol. Lab. Med.* **102**:266-269, 1968.
59. Schumacher, H. R.; Somlyo, A. P.; Tse, R. L.; Maurer, K. Arthritis associated with apatite crystals. *Ann. Intern. Med.* **87**:411-416, 1977.
60. Termine, J. D.; Eanes, E. D.; Conn, K. M. Phosphorprotein modulation of apatite crystallization. *Calcif. Tissue Int.* **31**:247-251, 1980.
61. Weissman, G.; Rita, G. A. Molecular basis of gouty inflammation: Interaction of monosodium urate crystals with lysosomes and liposomes. *Nature (London)* **240**:167-172, 1972.
62. Wuthier, R. E. A review of the primary mechanism of endochondral calcification with special emphasis on the role of cells: Mitochondria and matrix vesicles. *Clin. Orthop.* **169**:219-242, 1982.
63. Wuthier, R. E.; Gore, S. T. Partition of inorganic ions and phospholilipids in isolated cell, membrane and matrix vesicle fractions: Evidence for Ca-Pi-acidic phospholipid complexes. *Calcif. Tissue Res.* **24**:163-171, 1977.
64. Yaari, A. M.; Shapiro, I. M.; Brown, C. E. Evidence that phosphatidylserine and inorganic phosphate may mediate calcium transport during calcification. *Biochem. Biophys. Res. Commun.* **105**:778-784, 1982.

71

Polymer Properties Associated with Calcification of Cardiovascular Devices

Dennis L. Coleman, Hwei-Chuen Hsu, David E. Dong, and Donald B. Olsen

Nucleation and growth of calcium phosphate crystals on blood pumps designed to assist or replace the natural heart is a limiting factor in the long-term survival of experimental animals. This phenomenon, first reported by Olsen *et al.* [24] in 1975, is an acknowledged problem in all cardiovascular implant centers with routine animal survival times greater than 100 days [8]. The exact cause of the deposition of calcium phosphate, crystallization, and growth is not known, but several factors have been identified as important participants in the process. These factors will be discussed in some detail below.

The search for hemocompatible materials is several decades old. However, this term takes on new meaning with the discovery that long-term implants will fail or become severely impaired due to mineral deposition. The two basic approaches to blood-compatible surfaces, i.e., textured surfaces to promote pseudoneointima formation and nontextured surfaces to prevent the attachment of thrombus, both exhibit calcification when implanted in calves [8,9,16,36]. Methods to prevent mineralization are dependent on understanding the mechanism of induction and propagation of calcification. Unfortunately, this information is not well defined, and data supporting various hypotheses are difficult to obtain in a well-controlled environment. *In vivo* experiments are expensive and highly variable, and it is not well established that *in vitro* calcification experiments reproduce the same pattern or morphology seen *in vivo*.

Various investigational approaches have attempted to identify the cause of mineralization. Biochemical hypotheses have emphasized the importance of calcium-binding proteins [20,22] or the role of lipids [4,40]. Pathologists have stressed the mechanism of dystrophic calcification and the role of matrix vesicles [2,44]. Anderson [3] has reviewed the current hypotheses of abnormal mineralization in reference to implants. Properties of implant materials are also considered important variables to the induction of mineralization of car-

---

Dennis L. Coleman, Hwei-Chuen Hsu, and David E. Dong • Department of Pharmaceutics, University of Utah, Salt Lake City, Utah 84112.   *Donald B. Olsen* • Department of Surgery, Division of Artificial Organs, University of Utah, Salt Lake City, Utah 84112.

diovascular devices [8,9]. Absorption of biological components such as lipids may alter the physical and mechanical properties of elastomers, making them more prone to stress aging and/or mineral deposition [5,25].

General Properties of Polymers

A wide variety of polymers are currently used in the construction of cardiovascular implants. Implantable blood pumps are generally constructed of polyurethanes that have a polyether soft segment, a diisocyanate hard segment, and a diol or diamine chain extender. The most common polyurethane for this purpose is Biomer (Ethicon, Inc.) which is thought to have the chemical structure shown in Fig. 1. The hard segments affect the modulus, hardness, and tear strength of polyurethanes, while the flexible or soft segments control the elastic properties and also contribute to the ultimate tear strength, modulus, and hardness of the material [1].

Other polymers exposed to biological environments develop mineral deposits on the surface after extended use. These materials include poly(dimethyl siloxane) [38], ethylene-vinyl acetate [30], poly(hydroxyethyl methacrylate) [18], glutaraldehyde cross-linked gelatin [23], polysulfone-coated ceramic [31], expanded polytetrafluoroethylene [28], poly(ethylene terephthalate) [15], and poly(vinyl alcohol) [19]. In relation to mineralization, the common denominator among all of these materials is not entirely clear because of the wide variety of bulk and surface characteristics.

Hydrophobicity/Hydrophilicity

Polymers vary greatly in hydrophobic or hydrophilic character. This variation contributes in part to the resulting interfacial energetics thought to be important in hemocompatibility [10]. The importance of interfacial energy to the nucleation of calcium phosphate on a foreign surface has been considered by Garside [13]. He suggests that the contact angle between the crystal and the solid substrate can be used to predict the affinity between the crystalline solid and the foreign surface.

Mineralization has been associated with hydrophilic and hydrophobic polymers. Poly(hydroxyethyl methacrylate), a hydrogel with about 40% water, is reported to calcify in different biological environments including the eye [35], subcutaneous implants in rats, guinea pigs, and hamsters [18], clinically in breast augmentation [6] and plastic reconstruction of the nose [41]. Porous hydrogels calcify much more rapidly than homogeneous gels [6,32], suggesting the possibility of two distinct mechanisms. The porosity issue will be addressed in more detail below. Homogeneous gels are permeable to ion diffusion allowing ionic species to penetrate the bulk material [27]. Precipitation or crystallization in the bulk

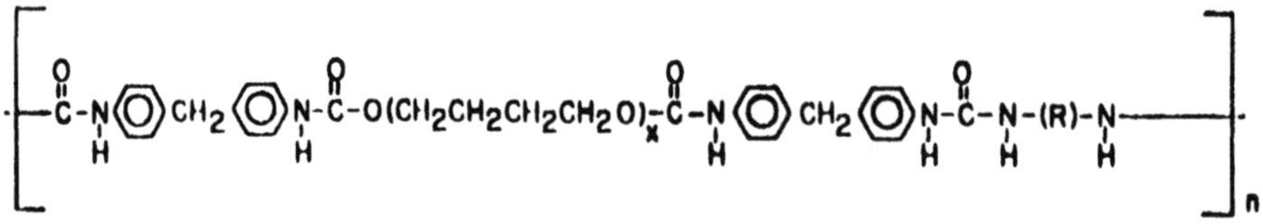

Figure 1. The chemical structure of Biomer, a polyurethane commonly used in the construction of cardiac support devices, has a polyether soft segment, a diisocyanate hard segment, and a diol or diamine chain extender.

gel can occur given optimum calcium and phosphate concentrations [26]. This mechanism is not consistent with mineralization associated with low-water-content materials.

In 1971 Urry [37] proposed the "neutral site binding/charge neutralization" mechanism to explain elastin mineralization. This hypothesis suggests that hydrophobic neutral peptide moieties in elastin attract and bind calcium ions with subsequent attraction of phosphate to neutralize the resulting positive charge. The increase in phosphate is followed by the migration of calcium into the matrix resulting in a self-perpetuating mineralization process [21]. An analogous situation exists with many of the elastomers used in cardiovascular devices. Hydrophobic and hydrophilic microdomains have been identified using transmission electron microscopy [34]. It is tempting to speculate that hydrophobic domains have an affinity for calcium ions, thus inducing mineralization by the "neutral site binding/charge neutralization" mechanism.

Surface Charge

Most biomedical polymers are thought to be neutral in terms of surface charge. Methods to determine surface charge are not routine but evidence is available to suggest that hydrophobic materials have a negative interfacial charge [39]. This is very likely due to specific ion adsorption at the solid–liquid interface [29]. The importance of charge in the calcification of copolymers of hydroxyethyl methacrylate, with either trimethylaminoethyl methacrylate-chloride to provide a positive charge or methacrylic acid to vary the amount of negative charge [14], can be demonstrated *in vitro*. Samples of these polymers placed in calcium-enhanced barbital buffer (pH 7.4, containing 1.5 mM Ca^{2+} and 2.0 mM PO_4^{2-}) and maintained at 37°C for 7 days have shown significant differences in calcium uptake (Fig. 2). This effect can also be demonstrated with sera from different species. The mechanism for

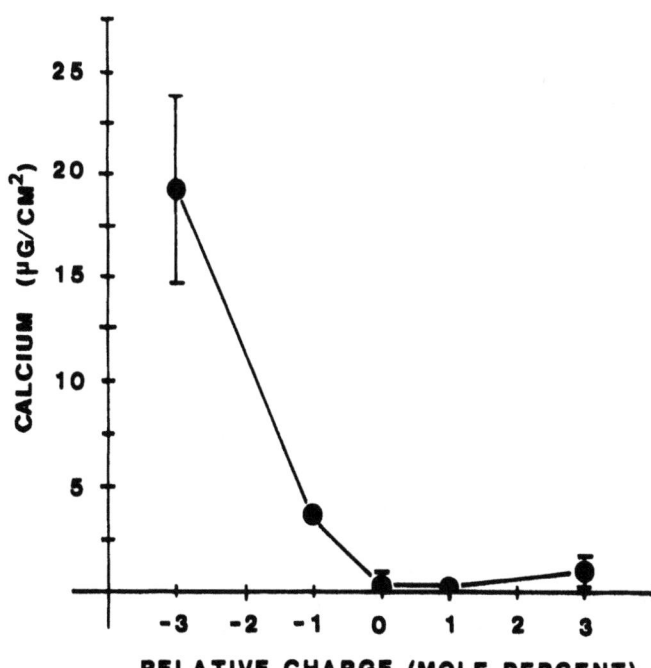

Figure 2. Calcium uptake by hydrogels, as determined by atomic absorption, is significantly greater in gels with a negative charge than gels that are neutral or positively charged.

calcium uptake by hydrogel materials may involve ion diffusion into the bulk gel. Mineralization of biomedical elastomers cannot occur by a diffusion mechanism because of the nonporous nature of the membrane. However, it is difficult to evaluate the importance of interfacial charge using elastomeric polymer systems.

Porosity

A strong case can be made for the importance of polymer porosity in mineralization. Wesolowski et al. [42] evolved a calcification index based on the porosity of prosthetic vascular grafts and noted that calcification increased as biologic porosity decreased below a critical value. This phenomenon has also been demonstrated for porous hydrogels implanted in soft tissue of experimental animals [6,18,32], as well as in clinical applications [6]. Microbubbles or surface fractures may act like porous sites to induce mineralization of smooth elastomers. Degenerative changes of biological components trapped in pores or voids of the implant material will result in what have been termed "metabolic shadows," which have been identified as loci for lipid and calcium accumulation [7]. Microbubble defects and surface fractures have been noted on smooth polyurethane bladders, and microscopic evidence suggests that mineralization is associated with these defects [8,9,17]. Removal of microscopic calcium phosphate deposits (Fig. 3A) using a decalcifying solution reveals microbubble defects (Fig. 3B) that provide a site for mineral deposition on smooth polyurethane artificial hearts. However, neither microbubble defects nor calcium deposits were detected on a heart implanted in a human for 112 days.

Degradative changes occur even in polymers thought to be stable in a biological environment [25,43]. Mineral deposits on mechanically active polymers accelerate this process by propagating surface fractures in the material [9]. Taguchi et al. [33] have reported surface fractures and microbubbles on the surface of cardiac assist devices implanted for more than a year. They suggest that microbubbles trapped in the bulk polymer migrate to the surface with time. However, data to support this hypothesis have not been forthcoming.

The role of mechanical stress in polymer mineralization has been a concern especially for artificial trileaflet valves and glutaraldehyde-treated porcine valves [11]. While evidence supports the concept that mineral deposits occur in areas of high stress, it is not clear if the stress causes microfractures which act as sites for mineralization, or if lipid absorption or thrombus formation is the precursor to mineral deposition.

Lipid Absorption

Normal biologic calcification consists of an inorganic mineral phase and a calcium-bound organic component which contains both proteins [20,22] and lipids [4,40]. Experience with select Silastic rubber formulations demonstrates absorption of various lipid fractions into the bulk polymer [5,38]. The hypothesis that absorbed lipids initiate mineralization is a natural consequence of these observations.

Adsorption of low-density lipoproteins (LDL) onto various polymers demonstrates that hydrophobic materials generally possess enhanced LDL adsorption [12]. Experiments to determine if LDL adsorption leads to lipid absorption are under way. However, direct evidence linking mineral deposition to LDL adsorption or absorption has not been obtained in preliminary experiments. Absorbed lipids may also act in an indirect way to alter the surface and initiate sites for mineral deposition. Evidence supporting this hypothesis is also

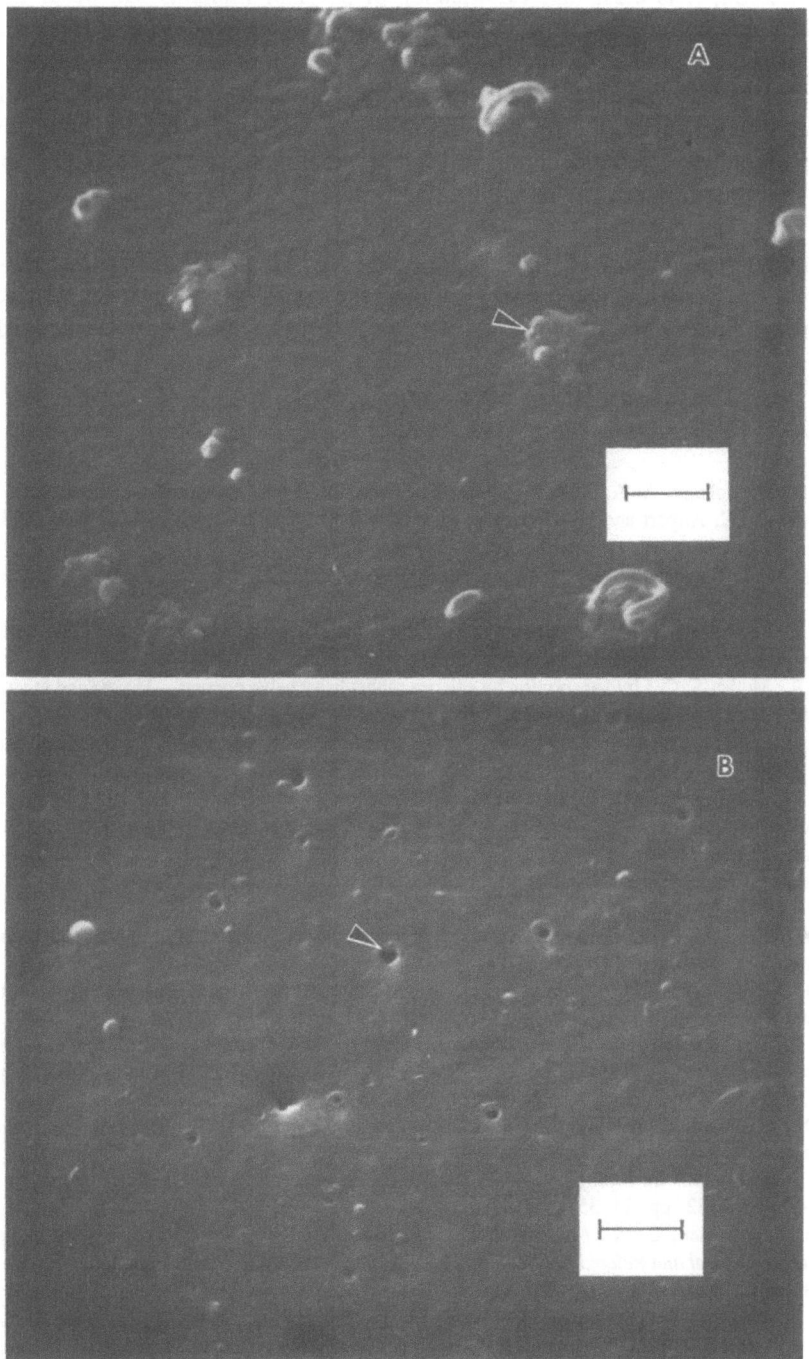

Figure 3. A scanning electron micrograph of the pumping bladder from an artificial heart implanted in a sheep for 278 days reveals calcium phosphate deposits (arrowhead, A). After treatment with a decalcifying solution, microbubble defects are clearly visible (arrowhead, B). Scale bars = 20 μm (A); 10 μm (B).

not presently available, but this problem should provide a fruitful area of research in the future.

In conclusion, it is clear that calcification of polymer implants can occur in a wide variety of biologic environments. In the case of hydrogel materials, ion diffusion into the bulk material is an important factor. However, calcification of elastomers does not involve diffusion into the bulk material but appears to be dependent on microbubble defects or microfractures on the surface of the materials acting as voids which provide an environment for calcium phosphate crystallization. The latter situation is somewhat analogous to mineralization of necrotic tissues or dystrophic calcification.

ACKNOWLEDGMENTS. This work was supported by NIH Grants HL-27747 and HL-24561. The authors thank Drs. D. E. Gregonis, J. D. Andrade, and W. I. Higuchi for their scientific discussions and advice. The technical assistance of Pam Dew is greatly appreciated.

References

1. Allport, D. C.; Mohajer, A. A. Property-structure relationships in polyurethane block copolymers. In: *Block Copolymers*, D. C. Allport and W. H. James, eds., New York, Applied Science Publishers, 1973. p. 443.
2. Anderson, H. C. Matrix vesicle calcification. *Fed. Proc.* 35:105–108, 1976.
3. Anderson, H. C. Normal and abnormal mineralization in mammals. *Trans. Am. Soc. Artif. Intern. Organs* 27:702–708, 1981.
4. Boskey, A. L.; Posner, A. S. The role of synthetic and bone extracted Ca-phospholipid-PO_4 complexes in hydroxyapatite formation. *Calcif. Tissue Res.* 23:251–258, 1977.
5. Carmen, R.; Mutha, S. C. Lipid absorption by silicone rubber heart valve poppets—In vivo and in vitro results. *J. Biomed. Mater. Res.* 6:327–346, 1972.
6. Cerveny, L.; Sprincl, L. The calcification of poly(glycol methacrylate) gel in experimental and clinical practice. *Polym. Med.* 11:71–77, 1981.
7. Chvapil, M. Development of degenerative changes in relation to the porosity of implanted vessel prosthesis: Role of diffusion, O_2, metabolic shadows. In: *Connective Tissue and Aging*, Volume 1, H. G. Vogel, ed., Amsterdam, Excerpta Medica, 1973.
8. Coleman, D. L. Mineralization of blood pump bladders. *Trans. Am. Soc. Artif. Intern. Organs* 27:708–713, 1981.
9. Coleman, D. L., Lim, D.; Kessler, T.; Andrade, J. D. Calcification of nontextured implantable blood pumps. *Trans. Am. Soc. Artif. Intern. Organs* 27:97–103, 1981.
10. Coleman, D. L.; Gregonis, D. E.; Andrade, J. D. Blood-materials interactions: The minimum interfacial free energy and the optimum polar/apolar ratio hypotheses. *J. Biomed. Mater. Res.* 16:381–398, 1982.
11. Deck, J. D.; Thubrikar, M.; Nolan, S. P.; Aouad, J. Role of mechanical stress in calcification of bioprostheses. In: *Proceedings of the Second International Symposium on Cardiac Bioprostheses*, L. H. Cohn and V. Gallucci, eds., New York, York Medical Books, 1982, pp. 293–305.
12. Dong, D. E.; Andrade, J. D.; Colemen, D. L. Low density lipoprotein (LDL) adsorption to cardiovascular implant materials. *ACS Polym. Prepr.* 23:40–42, 1983.
13. Garside, J. Nucleation. In: *Biological Mineralization and Demineralization*, G. H. Nancollas, ed., Berlin, Springer-Verlag, 1982, pp. 23–35.
14. Gregonis, D. E.; Chen, C. M.; Andrade, J. D. The chemistry of some selected metacrylate hydrogels. In: *Hydrogels for Medical and Related Applications*, J. D. Andrade, ed., ACS Symposium Series 31, Washington, D.C., American Chemical Society, 1976, pp. 88–104.
15. Guidoin, R.; Gosselin, C.; Domurado, D.; Marois, M.; Levaillant, P. A.; Awad, J.; Rouleau, C., Levasseur, L. Dacron as arterial prosthetic material: Nature, properties, brands, fate and perspectives. *Biomater. Med. Devices Artif. Organs* 5:177–203, 1977.
16. Harasaki, H.; Gerrity, R.; Kiraly, R.; Jacobs, G.; Nose, Y. Calcification in bloodpumps. *Trans. Am. Soc. Artif. Intern. Organs* 25:305–309, 1979.
17. Hennig, E.; Keilbach, H.; Hoder, D.; Bucherl, E. S. Calcification of artificial heart values and artificial hearts. *Proc. Eur. Soc. Artif. Intern. Organs* 8:76–80, 1981.
18. Imai, Y.; Masuhara, E. Long-term in vivo studies of poly(2-hydroxyethyl methacrylate). *J. Biomed. Mater. Res.* 16:609–617, 1982.

19. Kojima, K.; Imai, Y.; Masuhara, E. Reaction between poly(vinyl alcohol) graft copolymers and tissue. *Artif. Organs Jpn.* 3:443–448, 1974.
20. Lian, J. B.; Levy, R. J.; Bernhard, W.; Szycher, M. LVAD mineralization and γ-carboxyglutamic acid containing proteins in normal and pathologically mineralized tissues. *Trans. Am. Soc. Artif. Intern. Organs* 27:683–689, 1981.
21. Long, M. M.; Urry, D. W. On the molecular mechanism of elastic fiber calcification. *Trans. Am. Soc. Artif. Intern. Organs* 27:690–696, 1981.
22. Nelsestuen, G. L. Interactions of vitamin K-dependent proteins with calcium ions and phospholipid membranes. *Fed. Proc.* 37:2621–2629, 1978.
23. Nose, Y.; Harasaki, H.; Murray, J. Mineralization of artificial surfaces that contact blood. *Trans. Am. Soc. Artif. Intern. Organs* 27:714–719, 1981.
24. Olsen, D. B.; Unger, F.; Oster, H.; Lawson, J.; Kessler, T.; Kolff, W. J. Thrombus generation within the artificial heart. *J. Thorac. Cardiovasc. Surg.* 70:248–255, 1975.
25. Parins, D. J.; McCoy, K. D.; Horvath, N.; Olson, R. W. *In vivo* degradation of a polyurethane: Pre-clinical studies. *ASTM Tech. Bull.*, in press.
26. Pokric, B.; Pucar, Z. Precipitation of calcium phosphates under conditions of double diffusion in collagen and gels of gelatin and agar. *Calcif. Tissue Int.* 27:171–176, 1979.
27. Ratner, B. D.; Miller, I. F. Transport through crosslinked poly(2-hydroxyethyl methacrylate) hydrogel membranes. *J. Biomed. Mater. Res.* 7:353–368, 1973.
28. Selman, S. H.; Rhodes, R. S.; Anderson, J. M.; DePalma, R. G.; Clowes, A. W. Atheromatous changes in expanded polytetrafluoroethylene grafts. *Surgery* 87:630–637, 1980.
29. Shaw, D. J. *Electrophoresis*, New York, Academic Press, 1969, p. 4.
30. Sheppard, B. L.; Bonnar, J. Scanning and transmission electron microscopy of material adherent to intrauterine contraceptive devices. *Br. J. Obstet. Gynecol.* 87:155–159, 1980.
31. Spector, H.; Wigger, W. B.; Buse, M. G. Radionuclide bone imaging of femoral prostheses with porous coatings. *Clin. Orthop.* 160:242–249, 1981.
32. Sprincl, L.; Kopecek, J.; Lim, D. Effect of the structure of poly(glycol monomethacrylate) gel on the calcification of implants. *Calcif. Tissue Res.* 13:63–72, 1973.
33. Taguchi, K.; Hasegawa, T.; Fukunaga, S.; Tagami, S.; Hironaka, T.; Iwase, K.; Otsubo, A.; Kajihara, H. An analysis of assisted heart implantation in calves for more than one year. *Trans. Am. Soc. Artif. Intern. Organs* 28:584–588, 1982.
34. Thomas, D. A. Morphology characterization of multiphase polymers by electron microscopy. *J. Polym. Sci. Polym. Symp.* 60:189–200, 1977.
35. Tripathi, R. C.; Tripathi, B. J. The role of the lids in soft lens spoilage. *Contact Intraocul. Lens Med. J.* 7:234–240, 1981.
36. Turner, S. A.; Bossart, M. I.; Milam, J. D.; Fuqua, J. M.; Igo, S. R.; McGee, M. G.; Frazier, O. H. Calcification in chronically-implanted blood pumps: Experimental results and review of the literature. *Tex. Heart Inst. J.* 9:195–205, 1982.
37. Urry, D. W. Neutral sites for calcium ion binding to elastin and collagen: A charge neutralization theory for calcification and its relationship to atherosclerosis. *Proc. Natl. Acad. Sci. USA* 68:810–814, 1971.
38. van Noort, R.; Black, M. M.; Harris, B. Developments in the biomedical evaluation of silicone rubber. *J. Mater. Sci.* 14:197–204, 1979.
39. Van Wagenen, R. A.; Coleman, D. L.; King, R. N.; Triolo, P.; Brostrom, L.; Smith, L. M.; Gregonis, D. E.; Andrade, J. D. Streaming potential investigations: Polymer thin films. *J. Colloid Interface Sci.* 84:155–162, 1981.
40. Vogel, J. J.; Boyan-Salyers, B.; Campbell, M. M. Protein–phospholipid interactions in biologic calcification. *Metab. Bone Dis. Relat. Res.* 1:149–153, 1978.
41. Voldrich, Z.; Tomanek, Z.; Vacik, J.; Kopeck, J. Long-term experience with the poly(glycol monomethacrylate) gel in plastic operations of the nose. *J. Biomed. Mater. Res.* 9:675–685, 1975.
42. Wesolowski, S. A.; Fries, C. C.; Karlson, K. E.; De Bakey, M.; Sawyer, P. N. Porosity: Primary determinant of ultimate fate of synthetic vascular grafts. *Surgery* 50:91–96, 1961.
43. Williams, D. F. Review: Biodegradation of surgical polymers. *J. Mater. Sci.* 17:1233–1246, 1982.
44. Woodward, S. C. Mineralization of connective tissue surrounding implanted devices. *Trans. Am. Soc. Artif. Intern. Organs* 27:697–702, 1981.

72

Calcification of Cardiac Valve Bioprostheses
Host and Implant Factors

Robert J. Levy, Frederick J. Schoen, Susan L. Howard, Judith T. Levy, Lauren Oshry, and Marguerite Hawley

Glutaraldehyde-preserved stent-mounted porcine aortic valve bioprostheses are widely used in the surgical management of valvular heart disease [3,13,16,17]. Since 1971, several hundred thousand have been implanted in patients undergoing cardiac valve replacement. In most cases, bioprostheses offer the distinct advantages of freedom from chronic anticoagulation and favorable hemodynamic performance. However, calcification of bioprostheses is frequent after long-term function [16,17]. Mineral deposits often lead to clinically significant valvar dysfunction, which usually necessitates reoperation, but is occasionally fatal.

Clinicopathologic studies of bioprosthetic heart valves have identified several important features of the mineralization process. Younger patients are at a greater risk than older recipients for bioprosthesis calcification [16]. The principal morphologic abnormality in valve failures is intrinsic mineralization of the cusps, primarily of the valvar spongiosa with localization to collagen fibrils and residual porcine connective tissue cells [5]. Calcification also occurs, but less frequently, as a secondary process in superficial thrombi and the vegetations of bacterial endocarditis. The principal associated biochemical alteration [12] in primary valve failures is the deposition of calcium-binding proteins containing the vitamin K-dependent amino acid, γ-carboxyglutamic acid (Gla). Gla-containing proteins are not present either in unimplanted bioprosthetic valve leaflets or in noncalcified explants. Osteocalcin, a vitamin K-dependent bone protein, is one of the Gla-containing proteins occurring in the bioprosthestic calcifications [8]. Osteocalcin is the most abundant noncollagenous protein normally present in bone, where its function may involve the regulation of mineral density [7,14].

Robert J. Levy • Department of Cardiology and Laboratory of Human Biochemistry, Children's Hospital Medical Center and Department of Pediatrics, Harvard Medical School, Boston, Massachusetts 02115. *Frederick J. Schoen* • Department of Pathology, Brigham and Women's Hospital, Boston, Massachusetts 02115. *Susan L. Howard and Marguerite Hawley* • Department of Cardiology and Laboratory of Human Biochemistry, Children's Hospital Medical Center, Boston, Massachusetts 02115. *Judith T. Levy and Lauren Oshry* • Department of Chemistry, Wellesley College, Wellesley, Massachusetts 02025.

Experimental Models

Calcification of porcine aortic valve bioprostheses in sheep [1] and calves [11] is dramatically accelerated, occurring in several months, compared generally to at least several years in human valve replacements [3,5,13,16,17]. Orthotopic or bypass-situated glutaraldehyde-preserved porcine xenograft valve replacements in these animals develop progressive calcification, with a morphology comparable to that noted in clinically retrieved specimens. Analyses of retrieved valves from calves reveal the progressive accumulation of Gla-containing proteins [11], including osteocalcin. However, models utilizing implants in the circulatory system of large animals are expensive, and yield limited quantities of calcified tissue for analysis.

The morphologic and biochemical abnormalities (including Gla-protein deposition) associated with clinical bioprosthesis calcification are also stimulated when bioprosthetic valve leaflets are implanted subcutaneously in rabbits [6] and rats [10]. Extensive calcification occurs in only 3 weeks in the rat, with a total calcium and phosphorus accumulation of the same order of magnitude as that noted in long-term clinical retrievals.

Morphology and Comparisons

Light microscopic examination of rat-derived explants reveals that early diffuse mineral deposition is present in implants analyzed as early as 3 days postimplantation [10]. The

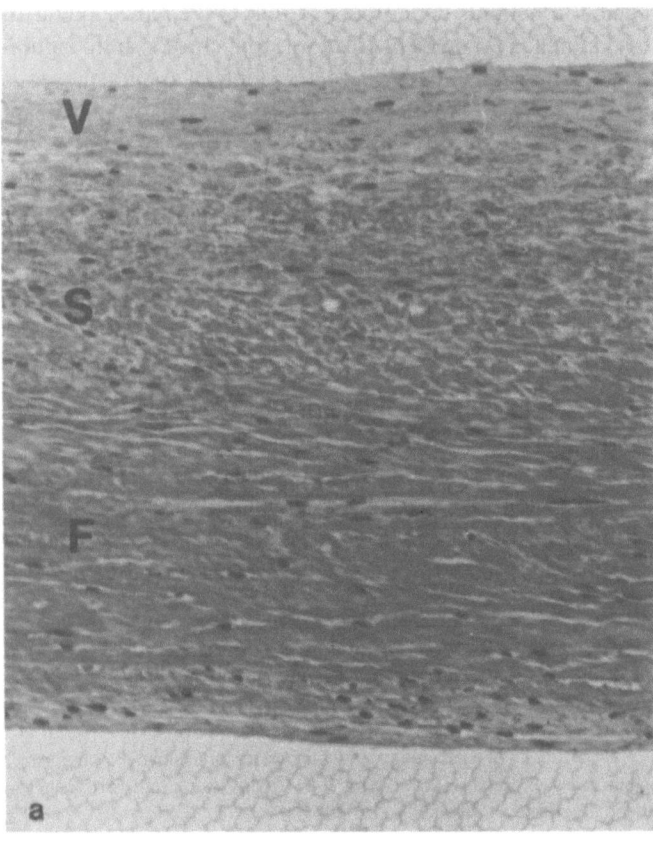

Figure 1. Typical histologic appearance of unimplanted and retrieved experimental porcine aortic valve bioprostheses. (a) Unimplanted glutaraldehyde-treated leaflet; (b) leaflet implanted subcutaneously in rat for 21 days; (c) leaflet from mitral valve implant in calf for 92 days. Calcification in (b) and (c) is seen to be virtually identical in morphology and is greater in spongiosa (S) than fibrosa (F) or ventricularis (V), a dense fibrous layer similar in structure to the fibrosa. Enhanced mineralization with coalescence into nodules (arrows) occurs at spongiosa–ventricularis and spongiosa–fibrosa junction in (b) and (c), respectively. The mineral deposits in both (b) and (c) are localized to cells (enclosed in boxes) and diffusely related to the background collagen, confirmed by transmission electron microscopy [13]. Glycol-methacrylate-embedded, H & E stain; × 375.

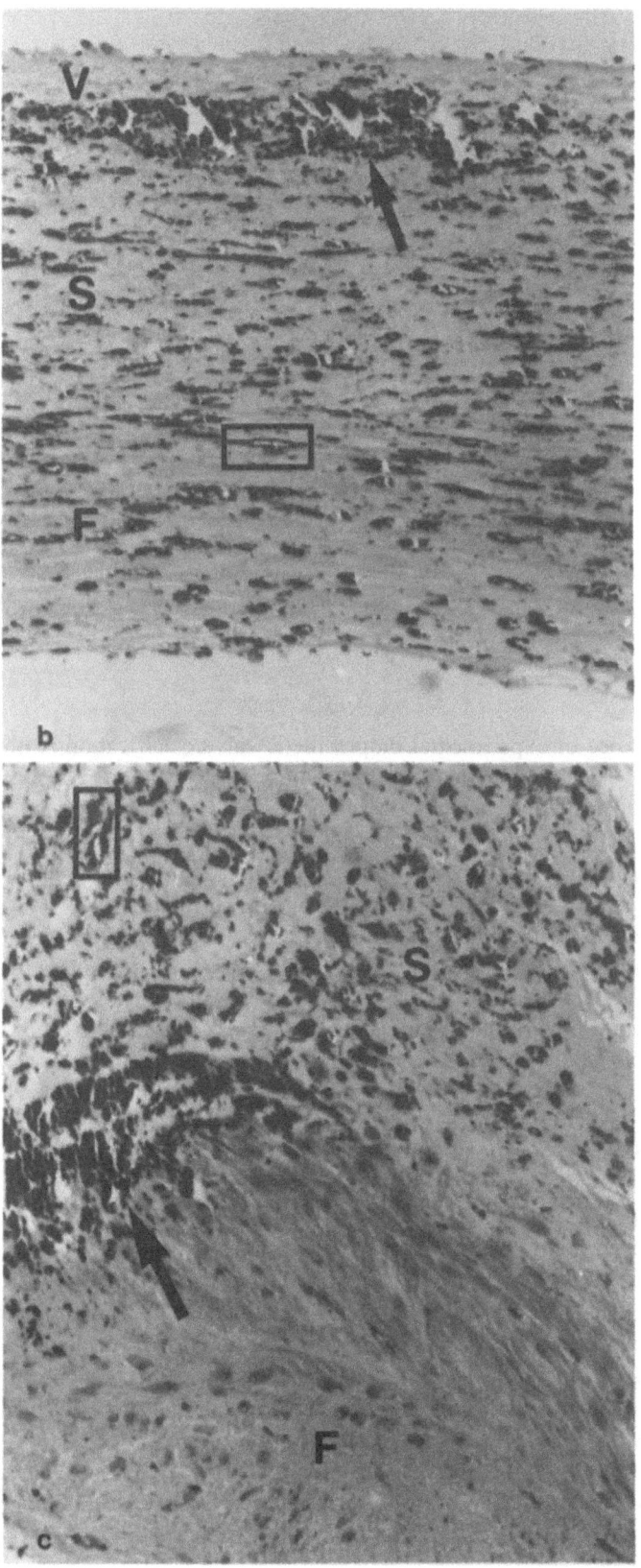

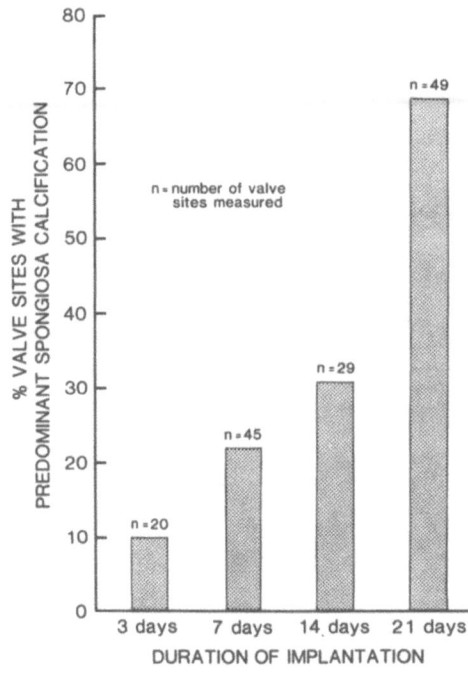

Figure 2. Prevalence of calcific deposits in valvar spongiosa in experimental bioprosthetic valves. Early deposition of mineral in the fibrosa in rat subcutaneous implants is followed by subsequent strong predominance in the spongiosa.

histologic appearance of experimental bioprosthetic valve calcification is illustrated in Fig. 1. Both diffuse and focal nodular mineralization are observed after 3 weeks (Fig. 1). Nodularity, occasionally noted at the spongiosa–fibrosa junction, appears to be due to a coalescence of diffuse deposits. The same mineralization pattern occurs in rabbit subcutaneous explants [6]. Furthermore, an accentuation of calcific deposits in tissue bends and folds occurs in rat explants.

Calcification was seen primarily in the valvar fibrosa at the earliest time periods examined; however, as time progressed, a clear predominance of deposition in the spongiosa evolved (Fig. 2). While only 10% of randomly selected valve sites in rat explants examined at 3 days had greater mineral deposition in the spongiosa, at 21 days 69% of sites had this localization. In an additional 22% of these sites measured at 21 days, deposition was equivalent in the cuspal layers. Preliminary analysis of these same specimens using scanning electron microscopy with energy-dispersive X-ray analysis suggests that this distribution results from predominent early mineral deposition in the fibrosa, with subsequent deposition (after 3–7 days) virtually limited to the spongiosa. This aspect of the morphology is under further investigation.

By comparison, porcine aortic bioprostheses implanted in calves also demonstrate diffuse spongiosa calcification (Fig. 1), with focal confluent nodularity at the spongiosa–fibrosa junction. This pattern is identical to that seen in subcutaneous implants as described above. Previous work emphasized the finding by light microscopy of nodular calcification of the valvar spongiosa in clinical retrievals [5]. Although subcutaneous implants initially develop a diffuse pattern, studies of the calf retrievals demonstrate that in this model the diffuse pattern is an early developmental stage in the formation of the clinically observed nodular calcific deposits. Valves retrieved from circulatory [5,13] and subcutaneous implantations [6,10], when analyzed by transmission electron microscopy, demonstrate calcification of collagen fibrils and mineralization of devitalized porcine connective tissue cells, regardless of the pattern seen by light microscopy.

Implant Factors

Subcutaneous implants in rats of leaflets not pretreated with glutaraldehyde do not accumulate significant amounts of calcium, as compared to unimplanted leaflets (Table I). Examination of these retrievals by light microscopy reveals necrotic valve tissue undergoing organization by host cells [10]. In contrast, glutaraldehyde-preserved implants develop severe calcification. Thus, it is clear that chemical alteration by glutaraldehyde is a prerequisite for calcification. Although the reasons for this are not understood, it is hypothesized that a critical extent of structural protein cross-links is induced by the glutaraldehyde pretreatment and that these cross-links attract an anion influx. This results in hydroxyapatite deposition associated with both structural proteins and proteolipid membrane components. Glutaraldehyde is known to induce cationic pyridinium linkages [2] between proteins which could contribute to a net phosphate influx according to our hypothesis. The foci of phosphate attraction could then become hydroxyapatite nucleation sites.

Bioprosthetic valves are typically prepared by incubating porcine aortic leaflets in phosphate-buffered glutaraldehyde. Since calcium phosphate salts compose the mineral phase of valve calcifications, preincubation of the bioprosthetic leaflets in phosphate solutions could theoretically enhance mineralization. The importance of buffering reagents in valve pretreatment was examined by comparing two sets of bioprosthetic leaflet subcutaneous rat implants, one fixed in phosphate-buffered glutaraldehyde, and the other pretreated in glutaraldehyde buffered at pH 7.40 in a monophosphate-containing buffer, 0.05 M Hepes. The results of these studies (Table I) indicate that the two sets of implants calcified to a similar degree, and suggest that the nature of the buffering agent is not a critical determinant in the calcification process.

Species Differences

The kinetics of calcification of porcine aortic bioprosthetic leaflets varies markedly depending on the recipient animal used in the implant experiment. To date, the most rapidly progressive response has been noted in the young rat (Table I). Implants in 3-week-old male

Table I. Porcine Aortic Valve Leaflet Subcutaneous Implants[a]

Aortic valve implants	N	Valve[b]	Ca^{2+} (μg/mg)	P (μg/mg)
Unimplanted	7	P	2.8 ± 0.7	2.5 ± 0.5
Untreated in 3-week-old rat	14	—	5.6 ± 1.0	9.4 ± 1.0
In 8-month-old rat	16	P	11.1 ± 2.7	7.2 ± 1.4
In 8-month-old castrated rat	18	P	3.4 ± 0.9	5.5 ± 0.4
In 3-week-old rat	54	P	122.9 ± 6.0*	64.1 ± 0.1*
In 3-week-old rat	9	H	121.9 ± 9.7*	75.0 ± 5.9*

[a]Values are mean ± S.E.
[b]P = prepared in 0.2% glutaraldehyde, in 0.05 M NaH_2PO_4 (pH 7.40); H = prepared in 0.2% glutaraldehyde, in 0.05 M Hepes (pH 7.40).
*$p < 0.001$ compared to unimplanted.

CD rats demonstrate a level of calcium accumulation comparable to that noted after 6 months [6] in subcutaneous rabbit explants, several months to a year in calf circulatory explants [11], and several years in retrieved human bioprostheses [16]. Failure to detect calcification in animal implant experiments may be the result of either species variation or age-dependent effects (see below).

Age Effects

Explants of glutaraldehyde-treated bioprosthetic valve leaflets from 8-month-old rats were calcified significantly less than explants from 3-week-old animals (Table I). These results clearly indicate that accelerated calcification of bioprosthetic leaflet implants occurs in younger animals, in a direct parallel to clinical observations. The basis for this is not obvious, but age-related differences in calcium and phosphorus homeostasis may in part explain the phenomenon since the serum phosphorus level is significantly higher in younger animals [10,15]. Since castration does not restore accelerated calcification in the mature rats (Table I), gonadal effects are thereby ruled out as the cause of the age-related differences in mineralization kinetics. Age-related differences in vitamin D metabolism [4,15] may also be important in this regard.

Host Response

The host inflammatory response to the bioprosthetic tissue probably does not play a role in either the deposition or the resorption of mineral. Morphological studies of both circulatory [5,13] and subcutaneous implants have noted a classic foreign body reaction to bioprostheses, with mononuclear phagocyte attachment to the surface without cellular invasion of the cusps. Two studies using subcutaneous implants support the hypothesis that neither nonspecific nor immunologic cell-mediated processes play a role in bioprosthesis calcification [9,10]. Subcutaneous valve implants enclosed in Millipore diffusion chambers with a 0.45μm pore size (thereby excluding cells, but not extracellular fluid) have the same degree of calcification as that found in direct tissue implants [10]. Furthermore, bioprosthetic leaflets implanted subcutaneously into congenitally athymic (nude) mice that have essentially no T-lymphocyte function, develop calcification which does not significantly differ from that of implants in control (immunologically competent) mice [9].

Osteocalcin

Osteocalcin, the principal vitamin K-dependent, calcium-binding protein of bone [7,14], is thought to function as a regulatory inhibitor of mineralization [14]; it may serve the same function in bioprosthesis mineralization. In valves retrieved from the rat, osteocalcin is present in all calcified bioprosthetic leaflets, but is not detectable in unimplanted leaflets (Table II). Osteocalcin radioimmunoassays reveal that virtually all of the Gla-containing proteins present in the bioprosthetic valve calcifications may be accounted for as osteocalcin. Inhibition of osteocalcin biosynthesis induced by warfarin administration (80 mg/kg per 24 hr) produces immeasurable prolongation of the one-stage prothrombin times; however, this does not influence the extent of calcification in 72-hr subcutaneous implants. As expected, tissue osteocalcin content is significantly reduced in the treated animals (Table II).

Table II. The Effect of Warfarin (80mg/kg per day) on Bioprosthetic Valve Leaflet (GPV) Subcutaneous Implant Calcification and Osteocalcin Accumulation

GPV	N	Ca^{2+} (μg/mg)	P (μg/mg)	Osteocalcin (ng/mg)
Unimplanted	7	2.8 ± 0.7	2.5 ± 0.5	0.0
72-hr control	12	39.6 ± 2.9*	23.4 ± 2.2*	69.0 ± 13.7*
72-hr warfarin	8	40.8 ± 4.8*	22.3 ± 2.6*	15.3 ± 1.2*

^aValues are mean ± S.E.
*$p < 0.001$ compared to unimplanted.

These results, where short-term suppression of osteocalcin biosynthesis is induced, do not preclude the possibility that osteocalcin deposition may influence bioprosthetic leaflet calcification by a more chronic mechanism. Furthermore, since the osteocalcin content of subcutaneously calcified bioprosthetic valve implants correlates closely with the extent of calcium accumulation [10], noninvasive measurements of valvar osteocalcin content may potentially provide a means of monitoring progression of mineralization.

Clinical Implications

No known valve preparation procedures exist which will completely abolish bioprosthetic valve calcification. Pretreatment of the leaflets with detergents, which probably remove phospholipids, or chemical reduction of glutaraldehyde-induced cross-links, may be promising approaches. Furthermore, there is at present no established therapeutic regime to prevent progression of pathologic calcification or promote its regression in this setting. Nevertheless, the diphosphonate compound, ethanehydroxydiphosphonate (EHDP), may be beneficial in preventing both the onset and the progression of bioprosthetic valve calcification. Bioprosthetic leaflets retrieved from EDHP-treated rats have reduced calcification compared to those from controls, although EHDP-treated animals demonstrate significant growth retardation. Nevertheless, EDHP administration may be feasible, either locally via extended controlled release from the valve annulus or systemically at low dosage for a limited treatment course.

ACKNOWLEDGMENT. Supported in part by NHLBI Grant 24463, and a Grant-in-Aid from the American Heart Association (81760).

References

1. Barnhardt, G. R.; Jones, M.; Ishihara, T.; Rose, D.; Chavez, A. M.; Ferrans, V. J. Degeneration and calcification of bioprosthetic cardiac valves. *Am. J. Pathol.* 106:136–139, 1982.
2. Bowes, J. H.; Carter, C. W. The interaction of aldehydes with collagen. *Biochim. Biophys. Acta* 168:341–352, 1968.
3. Carpentier, A.; Lemaigre, G.; Robert, L.; Carpentier, S.; DuBost, C. Biological factors affecting long term results of valvular heterografts. *J. Thorac. Cardiovasc. Surg.* 58:467–483, 1969.
4. DeLuca, H. F.; Schnoes, H. K. Metabolism and mechanism of action of vitamin D. *Annu. Rev. Biochem.* 45:631–666, 1977.
5. Ferrans, V. J.; Boyce, S. W.; Billingham, M. E.; Jones, M.; Ishihara, T.; Roberts, W. C. Calcific deposits in porcine bioprostheses: Structure and pathogenesis. *Am. J. Cardiol.* 46:721–734, 1980.
6. Fishbein, M.; Levy, R. J.; Nashef, A.; Ferrans, V. J.; Dearden, L. C.; Goodman, A. P.; Carpentier, A.

Calcification of cardiac valve bioprostheses: Histologic, ultrastructural, and biochemical studies in a subcutaneous implantation model system. *J. Thorac. Cardiovasc. Surg.* **83**:602–609, 1982.

7. Hauschka, P. V.; Lian, J. B.; Gallop, P. M. Direct identification of the calcium binding amino acid, γ-carboxyglutamate, in mineralized tissue. *Proc. Natl. Acad. Sci. USA* **72**:3925–3929, 1975.

8. Levy, R. J.; Gundberg, C.; Scheinman, R. The identification of the vitamin K-dependent bone protein, osteocalcin as one of the γ-carboxyglutamic acid containing proteins present in calcified atherosclerotic plaque and mineralized heart valves. *Atherosclerosis* **46**:49–56, 1983.

9. Levy, R. J.; Schoen. F. J.; Howard, S. L. Calcification of porcine bioprosthetic aortic valve leaflets is not mediated by immunological processes. *Am. J. Cardiol.* **52**:829–831, 1983.

10. Levy, R. J.; Schoen, F. J.; Levy, J. T.; Nelson, A. C.; Howard, S. L.; Oshry, L. J. Biological determinants of dystrophic calcification and osteocalcin deposition in glutaraldehyde-preserved porcine aortic valve leaflets implanted in glutaraldehyde-preserved porcine aortic valve leaflets implanted subcutaneously in rats. *Am. J. Pathol.* **113**:143–155, 1983.

11. Levy, R. J.; Zenker, J. A.; Bernhard, W. F. Porcine bioprosthetic valve calcification in bovine left ventricle to aorta shunts: Studies of the deposition of vitamin K-dependent proteins. *Ann. Thorac. Surg.* **36**:187–192, 1983.

12. Levy, R. J.; Zenker, J. A.; Lian, J. B. Vitamin K-dependent calcium binding proteins in aortic valve calcification. *J. Clin. Invest.* **65**:563–566, 1980.

13. Oyer, P. E.; Miller, D. C.; Stinson, E. B.; Reitz, B. A.; Moreno-Carbral, R. J.; Shumway, N. E. Clinical durability of the Hancock porcine bioprosthetic valve. *J. Thorac. Cardiovasc. Surg.* **82**:127–137, 1980.

14. Price, P. A.; Williamson, M. K.; Haba T.; Dell, R. B.; and Jee, W. S. S. Excessive mineralization with growth plate closure in rats on chronic warfarin treatment. *Proc. Natl. Acad. Sci. USA* **79**:7734–7738, 1982.

15. Round, J. M. Plasma calcium, magnesium, phosphorus, and alkaline phosphtase in normal British school children. *Br. Med. J.* **3**:137–140, 1972.

16. Sanders, S. P.; Levy, R. J.; Freed, M. D.; Norwood, W. I.; Castaneda, A. R. Use of Hancock porcine xenografts in children and adolescents. *Am. J. Cardiol.* **46**:429–438, 1980.

17. Schoen, F. J.; Collins, J. J.; Cohn, L. H. Long-term failure rate and morphologic correlations in porcine bioprosthetic heart valves. *Am. J. Cardiol.* **51**:957–964, 1983.

73

Pathogenesis of Valve Calcification
Comparison of Three Tissue Valves

Hiroaki Harasaki, James T. McMahon, and Yukihiko Nose

Since glutaraldehyde was introduced in preservation of tissue valve prostheses in 1969 [5], the durability of these valves has greatly improved. However, the problem of degeneration and calcification continues to complicate the use of various glutaraldehyde-fixed tissue valves [2,3,7,9,10,15,16,18,22,25], especially following implantation in children and adolescents [4,6,8,11,17,24,26–28]. In the younger patient group, it is indicated that less than 50% of valves remain functional after 5 years of implantation [6]. Elevated heart rate and calcium turnover appear to be the causative factors in early calcification, although no clearcut evidence has been offered. Renal failure may also render the patient more vulnerable to calcific complication [9,15,17,19], although its etiology has not been established.

Dunn [6] reported on 20 cases of valve calcification among 227 late survivors. Calcification was most common in the systemic valve positions (aortic and mitral), while valves implanted in either tricuspid or pulmonary position showed no calcification. A hemodynamic effect on calcification was suggested in a report of five cases [32], where the tissue valves in the mitral position showed more extensive calcification than those in the aortic position in the same patients for the same duration. However, the link between calcification phenomenon and mechanical stress has not been clarified. The present overview is mainly concerned with our use of various tissue valves as components of artificial hearts in calves with special emphasis on: (1) comparison of three tissue valves of different origin and preservation, (2) hemodynamic effect of valve calcification, comparing multiple valves used under different hemodynamic conditions in the same animal, and (3) the mechanism of calcification taking place in tissue valves.

Tissue Valves

Three different valves were utilized: 0.45% glutaraldehyde-treated bovine aortic valves (BAV, $n=31$), 98% glycerol-preserved human dura mater valves (HDV, $n=189$) and com-

---

Hiroaki Harasaki and Yukihiko Nose • Department of Artificial Organs, Cleveland Clinic Foundation, Cleveland, Ohio 44106.   James T. McMahon • Department of Pathology, Cleveland Clinic Foundation, Cleveland, Ohio 44106.

mercial, 0.5% glutaraldehyde-treated bovine pericardial valves (BPV, $n=4$). These valves were used with various combinations in the inflow and outflow ports of artificial hearts, which were implanted in 3- to 4-month-old male calves. The total artificial heart (TAH) is composed of a pair of right and left blood pumps, which incorporate inflow and outflow valves in the locations corresponding to the natural atrioventricular and semilunar valves. Thus, four valves can be tested simultaneously in a single animal. The only variables include the degree of oxygen saturation and mechanical stresses due to pressure and velocity of circulating blood. In the left ventricular assist devices (LVAD), and in both intrathoracic left ventricular assist device (ILVAD) and parathoracic left ventricular assist device (PVAD), two valves can be tested in inflow and outflow positions, respectively. The LVAD pump is driven pneumatically in synchrony with natural heart in a counterpulsation manner, bypassing most of the stroke volume of the natural left ventricle from its apex to descending aorta.

Table I shows the incidence of calcification seen in the BAVs used in the outflow position of TAHs and LVADs. The presence of calcification was detected histochemically with Von Kossa staining. In selected cases, needlelike crystals were observed by transmission electron microscopy (TEM) and shown to be calcium phosphate using energy-dispersive X-ray microanalysis. Typical apatitic diffraction was also confirmed by X-ray diffraction. All of the BAVs were used less than 5 months. The incidence of calcification was quite high in TAHs (55.6%) and in LVADs (92.3%). The difference in incidences between these two groups appears merely to reflect the difference in implantation duration, regardless of the type of artificial heart, since all BAVs utilized more than 27 days exhibited macroscopic calcified foci. In the TAH group, calcification occurred with the same incidence in the inflow and outflow positions, although the extent of calcification was always more severe on the left side. By contrast, HDVs showed a lesser incidence of calcification (25.9%) than the BAVs (Table II), even though the valves were used for an equivalent duration and in the same manner. Again, no significant difference was observed between right and left positions or between inflow and outflow positions. In this study only four commercially available BPVs were tested: two in the inflow and two in the outflow position of LVAD. Each of these four valves was macroscopically calcified within 3 weeks. X-ray photographs of valves indicated that initial calcification occurred in the flexing area and in the free edge of the cusp and that calcification was more prominent in the inflow valve.

It appears from these results that both glutaraldehyde-treated BAVs and BPVs show higher propensity to calcification than HDVs treated only with 98% glycerol. Glutaraldehyde definitely diminishes tissue degeneration and calcification, which were frequent and early complications in fresh tissue valves or in valves treated with formaldehyde [5]; however, it has not obviated valve degeneration and calcification. Perhaps optimization of preservation

Table I. Calcification in Bovine Aortic Valve

Expt	Location	No. of valves	Duration (days)	Calcification Right pump	Calcification Left pump	Earliest occurrence	Incidence
TAH[a]	Outflow	18	6–145 (Av. 37)	5	5	40 days	10/18 (55.6%)
ILVAD and PVAD	Outflow	13	14–105 (Av. 53)		12	19 days	12/13 (92.3%)
Total		31		5	17		22/31 (71.0%)

[a]TAH, total artificial heart; ILVAD, intrathoracic left ventricular assist device; PVAD, parathoracic left ventricular assist device.

Table II. Calcification in Human Dura Mater Valve

Expt	Location	No. of valves	Duration (days)	Calcification Right pump	Calcification Left pump	Earliest occurrence	Incidence
TAH	Inflow	40	6–145 (Av. 41)	6	6	45 days	12/40 (30%)
	Outflow	22	7–109 (Av. 36)	2	3	45 days	5/22 (22.7%)
LVAD	Inflow	70	7–316 (Av. 55)		18	19 days	18/70 (25.7%)
	Outflow	57	7–316 (Av. 56)		14	51 days	14/57 (24.6)
Total		189		8	41		49/189 (25.9%)

conditions, including the selection of appropriate buffers and glutaraldehyde concentration, will ameliorate these complications. Ultrastructurally, human dura mater has a more uniform and denser collagen fiber structure than pericardium. The dura is also devoid of a loose connective tissue layer equivalent to the tunica spongiosa, which appears to be the initiation site of calcification in aortic valve xenografts [14]. Thus, the structural integrity of the material may be partly responsible for the difference in vulnerability among the three valves tested.

Mechanical Stress

Although only a few articles have been published regarding the initial site of calcification loci in tissue valves [31,33], this subject seems relevant in analyzing the mechanism of calcification. Most of the reported valve calcifications have occurred in left valves (mitral and aortic), partially reflecting the fact that tricuspid and pulmonary valve replacements are far outnumbered by those of the left side. In a study of 27 valves used in children, 5 out of 20 left-side valves were calcified, while none of 7 valves in the right side showed calcification [11]. Dunn [6] also reported no calcification on the right side in his study of 227 valve replacement patients. A recent study [32] in 5 double-replacement patients revealed more severe calcification in mitral than aortic valves. Higher and uneven mechanical stress distribution in the mitral position was the explanation for this phenomenon.

In our present study, multiple valves were tested in a single animal and for the same duration under different hemodynamic conditions. The left valves were always more severely affected than the right counterparts in the TAHs. Moreover, the valves in the inflow position were calcified more extensively than those in the outflow position, both in TAHs and in LVADs, indicating that mechanical stress affects the extent of calcification. Our previous observation that calcification starts in the flexing area of the cusps [12,14] supports the findings of the present study. *In vivo* stress analysis of the porcine xenograft revealed that the flexing area is subjected to large compressive stresses during valve opening [29].

It is, however, necessary to emphasize that the present study shows no difference in incidence of calcification among different valve locations. In addition, the pericardial layer lining the left ventricular cannula and pump housing backed with rigid materials, were calcified frequently despite the fact that they were not subjected to mechanical stress [13].

These observations suggest that mechanical stress is one of the promoting factors, but may not be a prerequisite for initiation of calcification.

Mechanism of Valve Calcification

The key questions regarding valve calcification appear to be: How are the initial crystallizations generated in an environment where spontaneous precipitation is not expected? Which components of valvular tissue are the primary targets of calcium precipitation? What is the role played by mechanical stresses in the calcification process? Why does calcification occur so rapidly in a younger population?

Our previous study in 1980 indicated that initial calcium deposits in BAVs occur in the myocardial cell at the interface to the tunica spongiosa, and in fibroblasts and collagen bundles of the tunica spongiosa in the flexing area [14]. In addition, host blood elements (platelets, leukocytes, and erythrocytes) accumulated in valve tissue as a result of fiber

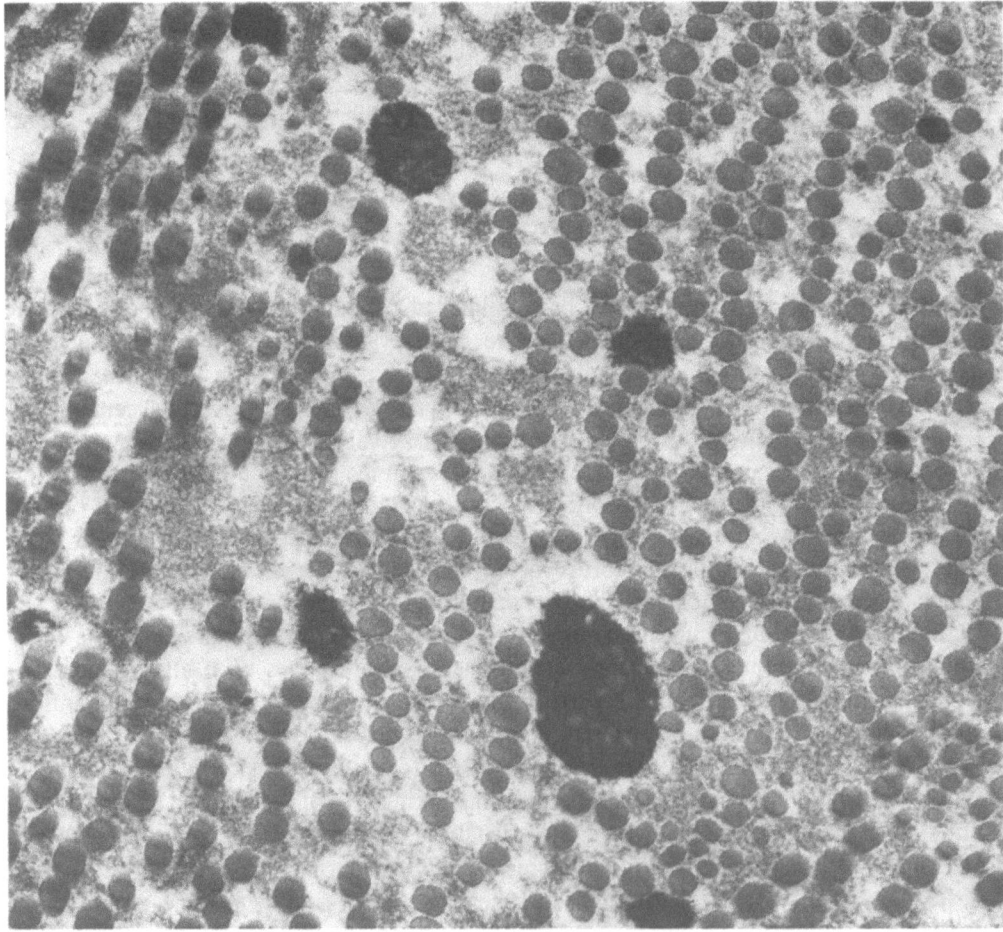

Figure 1. Transmission electron micrograph showing an initial calcification focus directly involving collagen. Calcium phosphate crystals were identified by X-ray microanalysis. × 11,900.

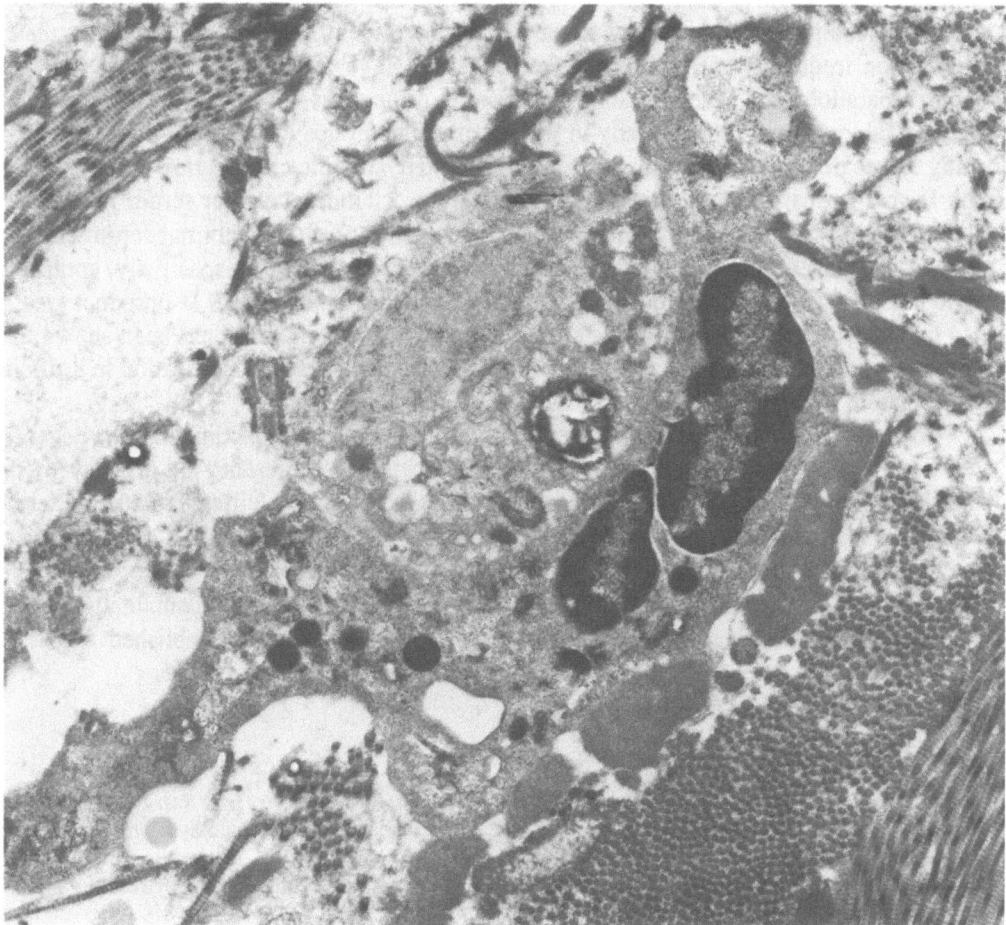

Figure 2. Calcification in subcellular components in a mononuclear cell. Note that needlelike crystals are observed on the membrane-bound structure. × 4200.

separation and void formation in valvular tissue. This insudation was preferentially seen in the flexing area of the cusp, regardless of the type of tissue valve. In HDVs and PVBs, insudation was also observed along the free edge of the cusp. These blood elements entrapped in the valve appeared to undergo a degenerative process, resulting in calcification.

The plasma membrane and subcellular components appear to be the initial site of calcification. Figures 1 and 2 show the early stages of calcification in the dura mater valve. Calcium phosphate crystals are seen directly involving the collagen bundles and degraded leukocytes, suggesting that valvular calcification is dystrophic. However, proteoglycans and glycoproteins in the ground substance may mask collagen and inhibit its calcification by blocking ε-amino groups of lysine (or hydroxylysine), which might otherwise react with calcium or phosphorus [1]. Repeated bending of the cusp may cause the gradual disappearance of this ground substance, which is revealed as void formation, particularly in the flexing area. The role of immunological mechanisms in tissue valve calcification remains in doubt since no leukocyte reaction has been observed in the calcification front.

Our proposed mechanism involving degraded cells and intracellular components, either constituting native valve tissue itself or blood components entrapped in the valve tissue as a result of insudation, is supported by ample evidence (e.g., Fig. 2). This mechanism explains calcification frequently seen along the free edge of HDVs and BPVs, where the collagen bundle separation and space formation may readily occur. It is not known why calcification in tissue valves occurs preferentially in the younger age groups or in renal failure, although young calves do, indeed, have a strong propensity to dystrophic calcification [13,23]. Common factors in these three groups are normal serum calcium levels but serum phosphorus levels 2–3 times higher than normal [23]. Moreover, in our studies, normal serum calcium and phosphorus levels from 28 normal calves are 9.7 and 6.6 mg/dl, respectively, contributing to a [Ca] × [P] product of 64. In renal failure, patients with a [Ca] × [P] product greater than 60 manifest a tendency to metastatic calcification. Furthermore, it has been shown that serum phosphorus levels are significantly higher in groups showing calcification in artificial hearts than in animals not having calcification [23].

Thus, it is our belief that high serum phosphorus levels in conjunction with the degenerated tissues may be key factors in valvular calcification. At present, the efficacy of diphosphonate and C_{12} alkyl sulfate (T6) in retarding tissue valve calcification is being tested [20]. The experimental results to date on T6 are somewhat controversial [20,30], and further study along these lines is warranted. Of recent interest has been the discovery of proteins with γ-carboxyglutamic acid in calcified tissue valves [21]. Whether the accumulation of this family of proteins is the cause or result of calcification has not been established.

References

1. Anderson, J. C. Glycoprotein of the connective tissue matrix. *Int. Rev. Connect. Tissue Res.* 7:251–267, 1976.
2. Angell, W. W.; Angell, J. D.; Kosek, J. C. Twelve-year experience with glutaraldehyde-preserved porcine xenografts. *J. Thorac. Cardiovasc. Surg.* 83:493–502, 1982.
3. Borkon, A. M.; McIntosh, C. L.; Von Reuden, T. J.; Morrow, A. G. Mitral valve replacement with the Hancock bioprosthesis: Five-to-ten year follow up. *Ann. Thorac. Surg.* 32:127–137, 1981.
4. Brown, J. W.; Dunn, J. M.; Spooner, E.; Kirsh, M. M. Late spontaneous disruption of a porcine xenograft mitral valve: Clinical, hemodynamic, echocardiographic, and pathological findings. *J. Thorac. Cardiovasc. Surg.* 75:606–611, 1978.
5. Carpentier, A.; Lemaigre, G.; Robert, L.; Carpentier, S.; Dubost, C. Biological factors affecting long-term results of valvular heterografts. *J. Thorac. Cardiovasc. Surg.* 58:467–483, 1969.
6. Dunn, J. M. Porcine valve durability in children. *Ann. Thorac. Surg.* 32:357–368, 1981.
7. Ferrans, V. J.; Boyce, S. W.; Billingham, M. E.; Jones, M.; Ishihara, T.; Roberts, W. C. Calcific deposits in porcine bioprostheses: Structure and pathogenesis. *Am. J. Cardiol.* 46:721–734, 1980.
8. Fiddler, G. I.; Gerlis, L. M.; Walker, D. R.; Scott, O.; Williams, G. J. Calcification of glutaraldehyde-preserved porcine and bovine xenograft valves in young children. *Ann. Thorac. Surg.* 35:257–261, 1983.
9. Fishbein, M. C.; Gissen, S. A.; Collins, J. J. Jr.; Barsamian, E. M.; Cohn, L. H. Pathologic findings after cardiac valve replacement with glutaraldehyde-fixed porcine valves. *Am. J. Cardiol.* 40:331–337, 1977.
10. Forfar, J. C.; Cotter, L.; Morritt, G. N. Severe and early stenosis of porcine heterograft mitral valve. *Br. Heart J.* 40:1184–1187, 1978.
11. Geha, A. S.; Laks, H.; Stansel, R. C., Jr.; Cornhill, J. F.; Kilman, J. W.; Buckley, M. J.; Roberts, W. C. Late failure of porcine valve heterografts in children. *J. Thorac. Cardiovasc. Surg.* 78:351–364, 1979.
12. Harasaki, H.; Snow, J. L.; Kiraly, R. J.; Nose, Y. The dura mater valve: In vitro characteristics and pathological changes after implantation in calves. *Artif. Organs* 3:176–183, 1979.
13. Harasaki, H.; Gerrity, R.; Kiraly, R.; Jacobs, G.; Nose Y. Calcification in blood pumps. *Trans. Am. Soc. Artif. Intern. Organs* 25:305–309, 1979.
14. Harasaki, H.; Kiraly, R. J.; Jacobs, G. B.; Snow, J. L.; Nose, Y. Bovine aortic and human dura mater valves. *J. Thorac. Cardiovasc. Surg.* 79:125–137, 1980.
15. Hetzer, R.; Hill, J. D.; Kerth, W. J.; Wilson, A. J.; Adappa, M. G.; Gerbode, F. Thrombosis and degeneration of Hancock valves: Clinical and pathological findings. *Ann. Thorac. Surg.* 26:317–322, 1978.
16. Ishihara, T.; Ferrans, V. J.; Jones, M.; Cabin, H. S.; Roberts, W. C. Calcific deposits developing in a bovine pericardial bioprosthetic valve 3 days after implantation. *Circulation* 63:718–723, 1981.

17. Kutsche, L. M.; Oyer, P.; Shumway, N.; Baum, D. An important complication of Hancock mitral valve replacement in children. *Circulation* **60**(Suppl. 1):I98–I103, 1979.
18. Lakier, J. B.; Khaja, F.; Magilligan, D. J., Jr.; Goldstein, S. Porcine xenograft valves: Long-term (60–89 months) follow up. *Circulation* **62**:313–318, 1980.
19. Lamberti, J. J.; Wainer, B. H.; Fisher, K. A.; Karunaratne, H. B.; Al-Sadir, J. Calcific stenosis of the porcine heterograft. *Ann. Thorac. Surg.* **28**:28–32, 1979.
20. Lentz, D. J.; Pollock, E. M.; Olsen, D. B.; Andrews, E. J. Prevention of intrinsic calcification in porcine and bovine xenograft materials. *Trans. Am. Soc. Artif. Intern. Organs* **28**:494–497, 1982.
21. Levy, R. J.; Zenker, J. A.; Lian, J. B. Vitamin K-dependent calcium binding proteins in aortic valve calcification. *J. Clin. Invest.* **65**:563–566, 1980.
22. McIntosh, C. L.; Michaelis, L. L.; Morrow, A. G.; Itscoitz, S. B.; Redwood, D. R.; Epstein, S. E. Artrioventricular valve replacement with the Hancock porcine xenograft: A five year clinical experience. *Surgery* **78**:768–775, 1975.
23. Nose, Y.; Harasaki, H.; Murray, J. Mineralization of artificial surfaces that contact blood. *Trans. Am. Soc. Artif. Intern. Organs* **27**:714–719, 1981.
24. Oyer, P. E.; Stinson, E. B.; Reitz, B. A.; Miller, D. C.; Rossiter, S. J.; Shumway, N. E. Long-term evaluation of the porcine xenograft bioprosthesis. *J. Thorac. Cardiovasc. Surg.* **78**:343–350, 1979.
25. Platt, M. R.; Mills, L. J.; Estrera, A. S.; Hillis, L. D.; Buja, L. M.; Willerson, J. T. Marked thrombosis and calcification of porcine heterograft valves. *Circulation* **62**:862–869, 1980.
26. Rose, A. G.; Forman, R.; Bowen, R. M. Calcification of glutaraldehyde-fixed porcine xenograft. *Thorax* **33**:111–114, 1978.
27. Silver, M. M.; Pollock, J.; Silver, M. D.; Williams, W. G.; Trusler, G. A. Calcification in porcine xenograft valves in children. *Am. J. Cardiol.* **45**:685–689, 1980.
28. Thandroyen, F. T.; Whitton, I. N.; Pirie, D.; Rogers, M. A.; Mitha, A. S. Severe calcification of glutaraldehyde-preserved porcine xenografts in children. *Am. J. Cardiol.* **45**:690–696, 1980.
29. Thubrikar, M. J.; Skinner, J. R.; Eppink, R. T.; Nolan, S. P. Stress analysis of porcine bioprosthetic heart valves in vivo. *J. Biomed. Mater. Res.* **16**:811–826, 1982.
30. Thubrikar, M. J.; Nolan, S. P.; Deck, J. D.; Aouad, J.; Levitt, L. C. Intrinsic calcification of T6-processed and control porcine and bovine bioprostheses in calves. *Trans. Am. Soc. Artif. Intern. Organs Abstr.* **12**:25, 1983.
31. Wallace, R. B.; Londe, S. P.; Titus, J. L. Aortic valve replacement with preserved aortic valve homografts. *J. Thorac. Cardiovasc. Surg.* **67**:44–52, 1974.
32. Warnes, C. A.; Scott, M. L.; Silver, G. M.; Smith, C. W.; Ferrans, V. J.; Roberts, W. C. Comparison of late degenerative changes in porcine bioprostheses in the mitral and aortic valve position in the same patient. *Am. J. Cardiol.* **51**:965–968, 1983.
33. Yarbrough, J. W.; Roberts, W. C.; Reis, R. L. Structural alterations in tissue cardiac valves implanted in patients and in calves. *J. Thorac. Cardiovasc. Surg.* **65**:364–375, 1973.

74

Noninvasive Imaging of Dystrophic Calcification

Mrinal K. Dewanjee

Several modalities are used for noninvasive evaluation of bone abnormality, e.g., radiography, dual photon absorptiometry, neutron activation analysis, and scintigraphy with ^{99m}Tc-labeled disphosphonates. Although radiography is still the most widely used imaging technique, radiographs of bone are difficult to interpret, and radiographic evidence of bone destruction does not appear until more than 50% of the skeleton is destroyed [3,19]. In addition, a dense area of active bone formation may appear as identical to an old healed process. However, the distribution pattern of intravenously administered ^{99m}Tc-labeled disphosphates depends on bone vascularity and the extent of bone remodeling. Thus, increased sensitivity of early and subtle bone changes makes bone scintigraphy a potentially invaluable imaging tool in the evaluation of bone abnormality. The kinetics of tracer uptake [1,4], synthesis of ^{99m}Tc-polyphosphates and applications in imaging the skeleton in health and disease [6,14], and the mechanism of tracer uptake in normal and heterotopic bone have been discussed [15,21]. This chapter will address the use of γ-emitting radionuclide-labeled tracers for imaging bone abnormalities and consider the use of its application in examining dystrophic calcification. Table I summarizes the physical decay properties of radionuclides, amount administered, and radiation dose to whole body and bone.

Evaluation of Tracers for Skeletal Uptake Measurements

The metabolic activity of bone was first recognized in 1935 by Chiecwitz and Hevesy [2] who reported that 22 days after administration of $Na_3{}^{32}PO_4$ to rats, 25% was sequestered in bone. The radionuclides of calcium, i.e., ^{45}Ca, ^{47}Ca, and ^{49}Ca, have been used for radiocalcium kinetic studies, and autoradiography of bone has contributed significantly to our understanding of skeletal metabolism. Compartmental analysis of these simple ions shows the similarity of Ca^{2+} uptake, although the rate of exchange between extracellular fluid and bone, the size of the pool, and amount of extraction are different. A similar attempt was made for Tc-complexes. Due to the presence of several species of ^{99m}Tc-complexes in

---

Mrinal K. Dewanjee • Section of Nuclear Medicine, Mayo Clinic and Foundation, Rochester, Minnesota 55905.

Table I. Physical Decay Properties of Bone-Scanning Tracers, Amount Administered, Bone Uptake, and Radiation Exposure

Tracer	$t_{1/2}$	% dose bone	Diagnostic dose (mCi)	Absorbed dose (mrad) Bone	Absorbed dose (mrad) Whole body	Comments
Phosphorus (phosphate) ^{32}P	14.3 days	25	0.2	6000		No gamma. Therapeutic use only; can image bremsstrahlung
Calcium						
^{45}Ca	165 days		0.005	250–650	75	No gamma
^{47}Ca	4.7 days		0.01	300	30–70	Multiple high-energy gammas
Strontium						
^{85}Sr	65 days	20–50	0.05–0.2	1500–10,000	250–4000	Relatively high rad dose
^{87m}Sr	2.8 hr	20–50	1–5	100–3000	7–100	Short half-life and high fecal content; generator system
Barium						
^{135m}Ba	1.2 days		3	3500	900	Difficult to obtain
Dysprosium						
^{157}Dy	8.1 hr	50	10	1270		Difficult to obtain, costly
Gallium						
^{67}Ga	78 hr	20	2.5	800	600	Multiple gammas; protein bound, high stool content, marrow and soft tissue uptake
Fluorine (fluoride)						
^{18}F	1.85 hr	50–75	2–10	200–3000	50–500	Possible ^{3}H contamination, high-energy gamma; potential for coincidence detection
Technetium						
^{99m}Tc-PP	6 hr	40–50	10–20	350–700	100–220	Not biodegradable
^{99m}Tc-MDP	6 hr	40–50	10–20	350–700	90–220	Most rapidly cleared ^{99m}Tc-phosphate

bone, the rate and extent of uptake and renal excretion of each are different, and hence only average rate constants could be obtained.

Most of the mathematical models of radiocalcium kinetic data evaluated the two important parameters: the transfer rate of radiocalcium from plasma into bone and the mass of rapidly exchangeable pool of calcium; this latter amounts to 5–10 g in normal individuals. About 60% of rapidly exchangeable calcium is in soft tissue and the remainder is in bone. Autoradiographic studies have shown that the skeletal component of this pool is located on the bone surface. After an intravenous injection of ^{45}Ca in man, there is a slow buildup of radioactivity on bone surface to a maximal value at 2–3 days, at which time bone surfaces are in equilibrium with ^{45}Ca in plasma [11,12]. Afterwards, the specific activities in plasma and bone surfaces decrease in a parallel fashion, conforming to a single compartment of exchange between bone and plasma. The second parameter, often called the bone accretion rate, has a magnitude of 350–450 mg/day in normal individuals. ^{45}Ca autoradiography indicates that in adult man the lowest ^{45}Ca uptake occurs in the oldest bone, while subsequent generations of remodeled bone have progressively higher rates of ^{45}Ca uptake, with the highest uptake in the most recently formed bone.

The changes in bone remodeling and resorption are affected by a variety of factors, e.g., stage of skeletal growth, endocrine diseases of parathyroid and thyroid, hormones generated by these glands and that of growth hormones and estrogen, renal disorders, and intestinal absorption of vitamin D. A few of these factors and local stress also affect the rate of ectopic calcification of cardiovascular prostheses [20,21]. The bone remodeling and the variation with site, age, and hormone may be visualized by scintigraphy with ^{99m}Tc-disphosphonate. The most aesthetic scintigraphs are obtainable in the pediatric and young adult, with a wide gradation of uptake in different parts of the skeleton.

Synthesis of ^{99m}Tc-Polyphosphate, Pyrophosphate, and Diphosphonate

The discovery of ^{99m}Tc-complexes in synthetic inorganic chemistry is a milestone in diagnostic nuclear medicine; the ^{99m}Tc-phosphonates are used in millions of patients for noninvasive imaging of a variety of bone diseases, especially in the survey of metastases in malignant cancers in bone. In 1971 Subramanian and McAfee [18] first reported the synthesis of ^{99m}Tc-polyphosphate and applied it successfully for skeletal scintigraphy in animals and humans. The stable chemical form of Tc is pertechnetate ion (^{99m}TcO$_4^-$). Mixtures of ^{99m}Tc-diphosphonate complexes are prepared using ^{99m}Tc-pertechnetate and Sn (II) = diphosphonate. The structure of ^{99m}Tc-diphosphonate as suggested by X-ray crystallography indicates the presence of (Tc-0)$^{3+}$ cation chelated to two units of diphosphonate (Fig. 1).

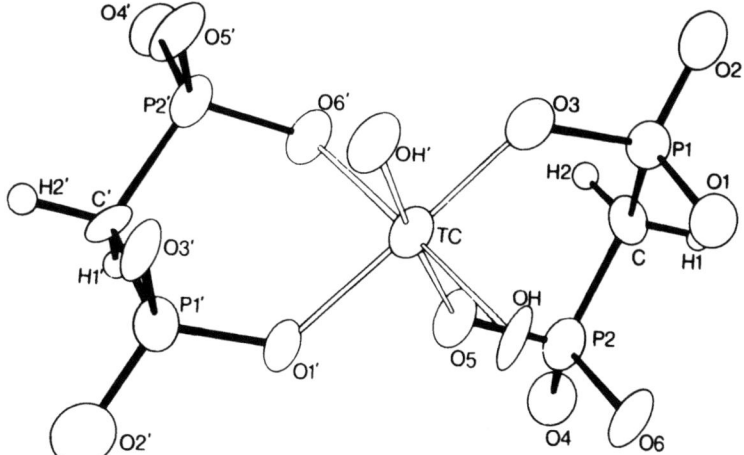

Figure 1. Structure of Tc-(methylene diphosphonate)$_2$ with a central oxotechnetium cation. [Courtesy of E. Deutsch.]

Mechanism of Localization of ^{99m}Tc-Pyrophosphate and ^{99m}Tc-Diphosphonate

To reach the mineral matrix, radioactive cations, anions, and metal complexes must permeate from plasma via (1) the wall of blood vessels made of discontinuous endothelial cell junction, (2) through the extracellular fluid space of the osteon, (3) into the hydration shell of hydroxyapatite, (4) where these tracers undergo ion exchange or chemisorption on the surface, and (5) enter the interior of the osteon. The first three steps are very rapid, with half-times measured in minutes. Step 4 is probably measured in hours, and the actual incorporation of ions into the deep bone crystal takes a much longer time. Once the radioactive ions or molecules enter the water-bound shell for ion exchange or chemisorption, they can be considered a constituent of bone. There are numerous small blood vessels in the metaphyseal ends of tubular bones where there is greater Ca^{2+} turnover and osteogenesis than in the diaphysis. The disparity in vasculature helps create a metabolic gradient among the different parts of the same skeleton and is easily reflected by gradations of uptake of [^{3}H]diphosphonate (Figs. 2, 3), and ^{99m}Tc-pyrophosphate and ^{45}Ca (Fig. 4) [9,10,17].

Hughes et al. [13] used a canine tibia model to study the extraction of tracers by arterial injection. The apparent fractional retentions by bone of ^{99m}Tc-polyphosphates, ^{99m}Tc-diphosphonate, ^{85}Sr-chloride, and ^{18}F-fluoride are 0.42, 0.27, 0.65, and 0.64, respectively. The larger molecular size of ^{99m}Tc-complexes is presumably responsible for the reduction in extraction of tracer.

Among the radiopharmaceuticals, ^{99m}Tc-labeled methylene diphosphonate and ^{99m}Tc-

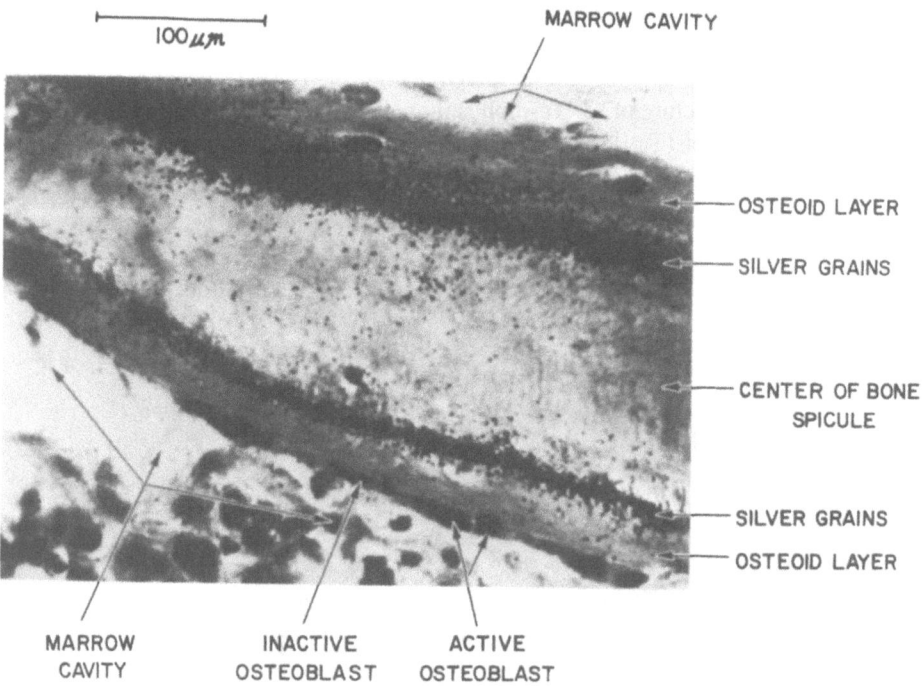

Figure 2. Microautoradiograph of a spicule of rabbit bone 24 hr postadministration of a single dose of [^{3}H]hydroxyethylene diphosphonate. The silver grains indicate the distribution of the tracer along the calcified surface of the spicule and below the lamellar osteoid layer. [Courtesy of M. D. Francis.]

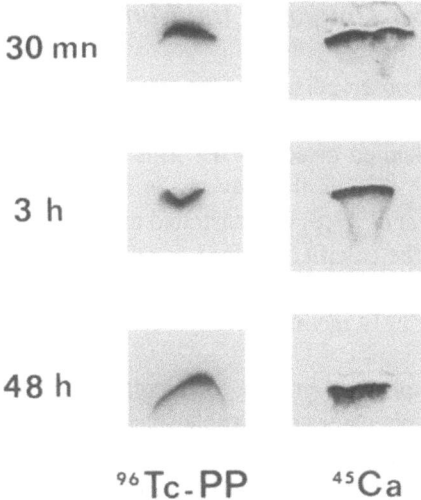

Figure 3. Autoradiograph of longitudinal section of rabbit epiphysis (femur) 30 min, 3 hr, and 48 hr postadministration of ^{96}Tc-pyrophosphate and ^{45}Ca. [Courtesy of A. Guillemart.]

labeled hydroxymethylene diphosphonate appear to be the best agents for skeletal scintigraphy [18,19]. About 10–20 mCi of ^{99m}Tc-diphosphonate is administered intravenously without requiring any patient preparation. The optimum time for imaging is 90–120 min postadministration, when 35–40% of the radioactivity localizes in the skeleton and the rest of the radioactivity is excreted in the urine. At this time, the bone-to-blood and bone-to-muscle activity ratios reach maximum values. Radioactivity distribution of the whole body is recorded on film using a gamma camera, and additional spot views of suspected bone or extraosseous lesions are obtained for clarification. Since the radiation dose to the bone and whole body is minimal, studies may be repeatedly performed, even in children.

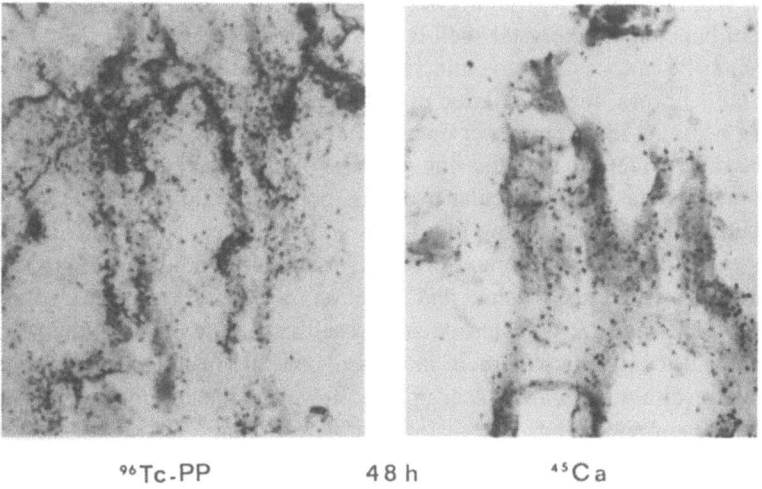

Figure 4. Microautoradiograph of rabbit diaphysis (500 μm beneath calcified cartilage) 48 hr postadministration of ^{96}Tc-pyrophosphate and ^{45}Ca. [Courtesy of A. Guillemart.]

Specific Site of Tracer Uptake in the Mineral Phase of Bone

Guillemart et al. [10] observed significant accumulation of ^{99m}Tc-diphosphonate in the cytoplasm of osteoclast and in the network of vascular buds below epiphyseal plate, which are also the sites of accelerated osteogenesis. This cytoplasmic radioactivity in osteoclast may represent minerals or crystals that have been resorbed from the apposed mineralized bone. In addition, Francis et al. [7,8] demonstrated the presence of ^{3}H-labeled hydroxyethylene diphosphonate along the calcified surface of rabbit bone spicule. These studies clearly demonstrate that diphosphonate (Fig. 2) and ^{99m}Tc-labeled diphosphonate localize mainly in the mineral surface, with ^{99m}Tc having a piggyback ride on diphosphonate to the osteogenic site.

In addition, the phosphatase enzymes and collagen in bone are probable sites of localization of ^{99m}Tc-diphosphonates and pyrophosphate [16]. Dewanjee [5] showed increased protein binding of ^{99m}Tc-pyrophosphate with respect to that of ^{99m}Tc-diphosphonate. Francis et al. [8] found that a maximum of 20% of total ^{99m}Tc-diphosphonate in bone may bind to collagen; the rest is adsorbed to calcium phosphate. ^{99m}Tc-pyrophosphate localizes in parenchymal tissues during such disease states as amyloidosis and myocardial infarction. The nature and extent of binding of Tc-pyrophosphate and diphosphonate with protein and calcium phosphate may depend on their abundance in these tissues and may account for the uptake of these bone-scanning agents in several soft tissue pathologies of skeletal muscle, lung, and stomach. On a competitive basis hydroxyapatite binds 40 times more ^{99m}Tc-diphosphonate than organic matrix of bone. In spite of this drawback of protein binding, the enormous surface area of apatite crystals (200–300 m^2/g or 3×10^6 m^2/person) permits extensive chemisorption and makes these agents extremely sensitive for detection of bone accretion and resorption.

Noninvasive Imaging of Bone Pathology with ^{99m}Tc-Diphosphonate

The five predominant varieties of bone pathology are: (1) osteoporosis, (2) osteomalacia, (3) osteitis fibrosa, (4) osteosclerosis, and (5) avascular necrosis of bone, and uptake patterns in scintigraphy are different. In osteoporosis, fewer bone crystals are spread over a wide area of bone, and the crystal shell is poorly hydrated, resulting in less tracer uptake. In osteomalacia, the matrix does not mineralize, and in severe osteomalacia, tracer uptake is low. In osteitis fibrosa, the stimulation of bone cells caused by hyperparathyroidism leads to remodeling resulting in higher tracer uptake. In osteosclerosis, overmineralization in certain parts of the skeleton results in increasing numbers of apatite crystals with greater surface area for higher tracer uptake. In avascular necrosis, bone cells die. Due to fewer blood vessels, minimal perfusion occurs resulting in an area poor in tracer (cold spot). The destruction of blood vessels after radiation therapy produces a similar condition. In dystrophic calcification, tissue necrosis due to ischemia and infection results in local calcification of organelles and vesicles. These dispersed calcified sites result in diffuse or focal uptake, as in tumoral calcinosis. In osteomyelitis, there is increased bone perfusion due to increased vascular permeability resulting in greater tracer uptake; however, greater uptake does not necessarily imply more bone remodeling.

In summary, the development of ^{99m}Tc-labeled pyrophosphate and diphosphonate has simplified the noninvasive diagnoses of sites of dystrophic calcification, and has even begun to aid us in our understanding of ectopic calcification after cardiovascular prosthesis [21]. ^{99m}Tc is an ideal radioisotope for imaging with a gamma camera in diagnostic nuclear medicine, since the change in bone turnover as measured by imaging is more sensitive than the change in bone density as observed by radiography.

References

1. Bernett, B. L.; Strickland, L. C. Structure of disodium dihydrogen 1-hydroxyethylidene diphosphonate tetrahydrate; a bone growth regulator. *Acta Crystallogr. Sect. B* **35**:1212–1214, 1979.
2. Chiecwitz, O.; Hevesy, G. Radioactive indicators in the study of phosphorus metabolism in the rats. *Nature (London)* **136**:754, 1935.
3. Davis, M. A.; Jones, A. G. Comparison of ^{99m}Tc-labeled phosphate and phosphonate agents for skeletal imaging. *Semin. Nucl. Med.* **6**:19–31, 1976.
4. Deutsch, E. Inorganic radiopharmaceuticals. Proc. 2nd Int. Symp. Radiopharmaceuticals, Seattle, 1979, pp. 129–146.
5. Dewanjee, M. K. Binding of diagnostic radiopharmaceuticals to human serum albumin by sequential and equilibrium dialysis. *J. Nucl. Med.* **23**:753–754, 1982.
6. Dewanjee, M. K.; Prince, E. W. Cellular necrosis model in tissue culture: Uptake of ^{99m}Tc-tetracyclin and the pertechnetate ion. *J. Nucl. Med.* **15**:577–581, 1974.
7. Francis, M. D.; Russell, R. G. G.; Fleisch, H. Diphosphonates inhibit formation of calcium phosphate crystals *in vitro* and pathological calcification *in vivo*. *Science* **165**:1264–1266, 1969.
8. Francis, M. D.; Davis, T. L.; Benedict, J. J.; Tofe, J. J. Disphosphonates: *In vitro* adsorption and desorption studies on hydroxyapatite and diffusion in bone. Etidronate. Proc. 1st Int. Symp. Diphosphonate in Therapy, Instituto Gentili, Italy, 1980, pp. 33–50.
9. Genant, H. K.; Bautovich, G. J.; Singh, M.; Lathrop, L. A.; Harper, P. V. Bone-seeking radionuclides: An *in vivo* study of factors affecting skeletal uptake. *Radiology* **113**:373–382, 1974.
10. Guillemart, A.; LePape, A.; Galy, G.; Besnard, J. C. Bone kinetics of calcium-45 and pyrophosphate labeled with technetium-96: An autoradiographic evaluation. *J. Nucl. Med.* **21**:466–470, 1980.
11. Harrison, G. E.; Carr, T. E. F.; Sutton, A. Distribution of radioactive calcium, strontium, barium and radium following intravenous injection into a healthy man. *Int. J. Radiat. Biol.* **13**:235–247, 1957.
12. Heaney, R. P. Evaluation and interpretation of calcium kinetic data in man. *Clin. Orthop.* **31**:153–183, 1963.
13. Hughes, S. P. F.; Davies, D. R.; Bassingthwaighte, J. B.; Knox, F. G.; Kelly, P. J. Bone extraction and blood clearance of disphosphonate in the dog. *Am. J. Physiol.* **232**:H341–H347, 1977.
14. Makler, T. P.; Charkes, N. D. Studies of skeletal tracer kinetics. IV. Optimum time delay for Tc-99m (Sn) methylene disphosphonate bone imaging. *J. Nucl. Med.* **21**:641–645, 1980.
15. Marshall, J. H. Measurements and models of skeletal metabolism. In: *Mineral Metabolism: An Advanced Treatise*, Volume III, C. L. Comar and F. Bronner, eds., New York, Academic Press, 1969, pp. 1–122.
16. Rosenthall, L.; Kaye, M. Observations on the mechanisms of ^{99m}Tc-labeled phosphate complex uptake in metabolic bone disease. *Semin. Nucl. Med.* **6**:59–67, 1976.
17. Stevenson, J. S.; Bright, R. W.; Dusson, G. L.; Nelson, F. R. Technetium-99m phosphate bone imaging: A method for assessing bone graft healing. *Radiology* **110**:391–394, 1974.
18. Subramanian, G.; McAfee, J. G. A new complex of ^{99m}Tc for skeletal imaging. *Radiology* **99**:192, 1971.
19. Subramanian, G.; McAfee, J. G.; Blair, R. F.; Rosenstreich, M.; Coco, M.; Duxbury, C. E. ^{99m}Tc-labeled stannous imido-disphosphate, a new radiodiagnostic agent for bone scanning: Comparison with other ^{99m}Tc complexes. *J. Nucl. Med.* **16**:1137–1143, 1975.
20. Urist, M. R. (ed.) Heterotopic bone formation. In: *Fundamental and Clinical Bone Physiology*, Philadelphia, Lippincott, 1982, pp. 369–393.
21. Wahner, H. W.; Dewanjee, M. K. Drug-induced modulation of Tc-99m pyrophosphate tissue distribution: What is involved. *J. Nucl. Med.* **22**:556–559, 1981.

75

Prevention of Arterial Calcium Deposition with Diphosphonates and Calcium Entry Blockers

Dieter M. Kramsch

The Role of Calcium in the Formation of Atherosclerotic Lesions

The arterial disease most commonly associated with an increase in arterial calcium is atherosclerosis. In the past, the deposition of calcium in atherosclerotic lesions has been considered to be an end-stage of advanced atheromata formation only [1]. While such end-stage calcification of plaques certainly is one of the prominent features of late atherosclerosis, it has long been recognized that discrete calcium mineral deposition may occur, especially on intimo-medial elastica, in otherwise normal-appearing arteries [2] and, therefore, may be considered an early event in atherogenesis. In fact, in recent years there has been mounting evidence that localized increases in arterial ionic calcium may play a key role in early atherogenesis, stimulating cellular functions of arterial smooth muscle cells and macrophages such as increased cell migration, mitosis, endocytosis of lipoproteins, as well as excessive synthesis and/or secretion of connective tissue macromolecules [3,4]. Focal increases in arterial calcium content also have been implicated in the early pathobiochemistry of the intercellular matrix molecules. Calcium ions are required for the complexing of low-density (LDL) and very-low-density lipoproteins (VLDL) to sulfated glycosaminoglycans. Calcium-dependent mechanisms may be responsible for the excessive degradation of arterial connective tissue such as the activation of elastolysis by macrophage elastase [5] or the release of collagenolytic, elastolytic, and mucolytic enzymes from platelets [6].

In addition, calcium appears to play an important role in the binding of other acidic proteins to the elastin of atherosclerotic lesions resulting in a marked increase in the elastin content of polar amino acids [7]. This, in turn, appears to lead to a firm binding of large amounts of cholesteryl ester to the altered elastin protein [8], presumably from lipoproteins. Modification in the amino acid (and lipid) composition of lesion elastin, which is associated with an increased content in elastin calcium, is rather characteristic and of a similar nature in atherosclerotic arteries of man (fatty streak and advanced fibrous plaque) as well as rabbit

---

Dieter M. Kramsch • Cardiovascular Institute, Boston University School of Medicine, Boston, Massachusetts 02118.

and monkey [9]. These elastin changes occur in rabbits and monkeys within a few weeks after starting an atherogenic diet, and are invariably found before other abnormalities can be demonstrated in the arteries by light microscopic or biochemical analysis (Kramsch, unpublished observations).

The binding of acidic proteins to lesion elastin presumably occurs through calcium bridges or simply by incrustation of the whole elastic fiber with calcium minerals [7]. Proposed nucleation sites for calcium mineralization of elastin include: the polar microfibrillar glycoproteins [10] or the proteoglycans [11,12] normally associated with the elastin, or neutral peptide groups of the elastin itself [13]. Binding of calcium ions to normal elastin *in vitro* causes configurational changes of this protein, exposing hydrophobic sites and giving rise to an increased adsorption of hydrophobic molecules such as cholesterol [14].

Lastly, it should be noted that the principal tissue for gross calcification of arteries in atherosclerosis and aging appears to be the arterial elastica [15–17]. The main constituents of minerals associated with elastins appear to be calcium and phosphorus [16,18], which are present in the late stages of calcium mineralization in the form of hydroxyapatite [16,19]. In still later stages, plaque collagen can also calcify, presumably with associated glycoproteins serving as nucleation sites. A γ-carboxyglutamic acid-containing, noncollagenous protein which binds calcium also has been detected in calcified plaques [20].

Prevention of Calcium-Induced Arterial Changes by Anticalcium Drugs

Our laboratory has been experimenting for the last decade with agents that have the capacity to prevent or even reverse calcium mineralization and atherosclerotic disease in rabbits and *Macaca fascicularis* monkeys on fibrogenic atherogenic diets (Tables I and II). These substances, which do not affect elevated serum cholesterol levels, initially included ethane-1-hydroxy-1,1-diphosphonic acid (EHDP), known to prevent the deposition of calcium into soft tissues including into arteries [21,24]. Other and newer diphosphonates of similar properties [24] were later employed and included the disodium salts of amino-1-hydroxy-propane-1,1-diphosphonic acid (APDP), azacycloheptane-2,2-diphosphonic acid

Table I. Components of Intima Media from Rabbits on Control Diets and Atherogenic Diets with and without Treatment[a]

Experimental group	Calcium	Nonlipid phosphorus	Collagen	Elastin	Cholesterol
Control diet	0.03 ± 0.007[b]	0.04 ± 0.02	2.8 ± 0.5	10.3 ± 2.8	0.3 ± 0.2
Atherogenic diet, untreated	0.08 ± 0.012*	0.15 ± 0.03*	6.5 ± 1.3*	27.1 ± 1.7*	3.5 ± 1.2*
Atherogenic diet + EHDP	0.06 ± 0.013*	0.11 ± 0.06**	4.3 ± 1.1**	21.5 ± 1.6*	1.3 ± 0.5*
Atherogenic diet + APDP	0.06 ± 0.015**	0.13 ± 0.08	3.7 ± 0.4**	15.3 ± 1.6*	0.7 ± 0.3**
Atherogenic diet + AHDP	0.04 ± 0.017	0.11 = 0.05	2.9 ± 0.8	12.4 ± 1.4	0.5 ± 0.1**
Atherogenic diet + IPDP	0.02 ± 0.009	0.06 ± 0.05	2.6 ± 0.9	7.3 ± 2.3	0.3 ± 0.3
Atherogenic diet + LaCl$_3$	0.03 ± 0.008	0.05 ± 0.02	2.9 ± 0.8	13.6 ± 4.7	0.6 ± 0.1**

[a]The fibrogenic atherogenic diet contained 8% peanut oil and 2% cholesterol by weight (27). The control diet was Purina rabbit chow. Nine groups of 10 male New Zealand white rabbits were fed the fibrogenic atherogenic or control diets for 8 weeks. One group of rabbits was fed the control diet and one group the atherogenic diet alone, while seven groups were given the atherogenic diet plus drug. The daily oral dose of each drug was 40 mg/kg body wt.
[b]Values are absolute amounts in mg/whole aorta per kg body wt; mean ± S.D.
*Highly significant changes from control values ($p < 0.01$).
**Significant changes from control values ($p < 0.05$).

Table II. Constituents of Elastin from Control Monkeys and Monkeys on the Atherogenic Diet with and without Drugs[a]

Artery	Calcium				Nonlipid phosphorus			
	Control diet	Athero diet	Athero + La^{3+}	Athero + EHDP	Control diet	Athero diet	Athero + La^{3+}	Athero + EHDP
Thoracic aorta	3.6 ± 0.3[b]	97.1 ± 21.3*	5.9 ± 0.4*	7.0 ± 0.5*	3.1 ± 1.5	50.7 ± 16.3*	3.5 ± 0.9	3.0 ± 1.1
Abdominal aorta	2.8 ± 1.1	77.4 ± 19.0*	3.2 ± 1.3	4.2 ± 1.6	3.7 ± 1.3	14.0 ± 3.8*	2.2 ± 1.5	3.2 ± 0.8
Subclavian	3.9 ± 0.8	22.9 ± 6.5*	2.0 ± 1.2	5.3 ± 0.6**	2.8 ± 0.7	8.3 ± 2.2*	3.5 ± 1.4	4.5 ± 1.1
Carotid	3.7 ± 1.2	22.5 ± 7.7*	3.8 ± 0.4	5.4 ± 1.2	1.2 ± 0.4	10.2 ± 4.1*	1.2 ± 0.2	1.4 ± 0.3
Iliac	4.3 ± 1.5	16.9 ± 4.9*	3.8 ± 0.5	7.7 ± 2.6	5.3 ± 1.5	11.9 ± 2.6*	3.2 ± 1.7	5.4 ± 1.4
Femoral	2.6 ± 0.7	6.4 ± 1.8*	3.1 ± 0.6	3.9 ± 1.3	—	—	—	—

[a] The fibrogenic atherogenic diet contained 10% butter and 0.1% cholesterol (28). The control diet was Purina monkey chow. Altogether, six groups of eight adult male *Macaca fascicularis* (cynomolgus) monkeys were fed the diets for 24 months. One group of monkeys was fed the control diet and one group the atherogenic diet alone, while two groups were given the atherogenic diet plus EHDP or LaCl$_3$, APDP. The daily oral dose of each drug was 120 mg/kg body wt for the first 6 months followed by a maintenance dose of 40 mg thereafter. The large initial doses in monkeys were given to saturate the tissues.
[b] Values are μg/g elastin; mean ± S.D.
*Highly significant changes from control values ($p < 0.01$).
**Significant changes from control values ($p < 0.05$).

(AHDP), and 2-imino-pyrrolidone-5,5-diphosphonate (IPDP). Successful treatment with these calcium-regulating agents suggests that atherosclerosis and associated calcification may be beneficially influenced by regulating the availability of ionic calcium. We, therefore, also treated animals on an atherogenic diet with lanthanum, a calcium entry blocker that inhibits calcium influx into cells at receptor-operated channels [25,26].

As compared to animals on the control diets, rabbits (Table I) and monkeys [4] on the fibrogenic atherogenic diet manifested drastic changes in the components of aortic intima media: a marked increase in the arterial content of calcium was associated with drastic rises in the arterial content of nonlipid phosphorus, collagen, elastin, and cholesterol. As noted previously [8], the arterial elastin fraction of untreated atherogenic diet animals of both species also was markedly altered, revealing marked increases in calcium and phosphorus, as well as of polar amino acids and cholesterol (Tables I and II).

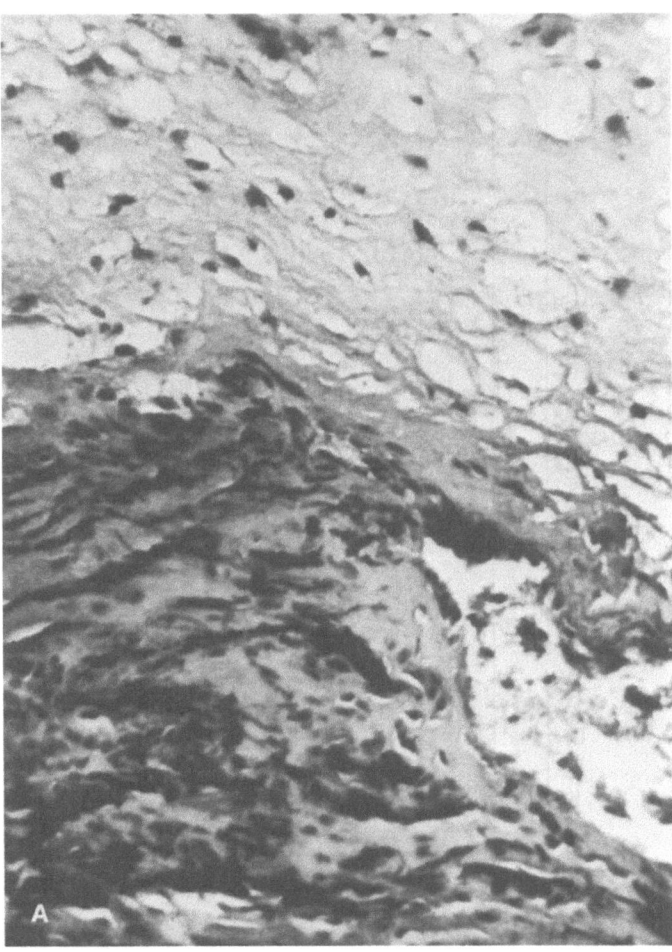

Figure 1. Sections through characteristic lesions of thoracic aorta of rabbits on a fibrogenic atherogenic diet for 8 weeks without drugs (A, B = sequential sections of the same lesion) and treated simultaneously with IPDP (C). (A) Note proliferation of cells and accumulation of collagen (gray) in intima and deranged elastica (black); Verhoeff's-Van Gieson, × 85. (B) Note deposition of calcium (black) on deranged elastica; Yasue's-light green, × 85. (C) Note absence of collagen accumulation and calcium deposition, normal elastica, raising of intima by a thin layer of (lipid-rich) foam cells; Yasue's-light green, × 70.

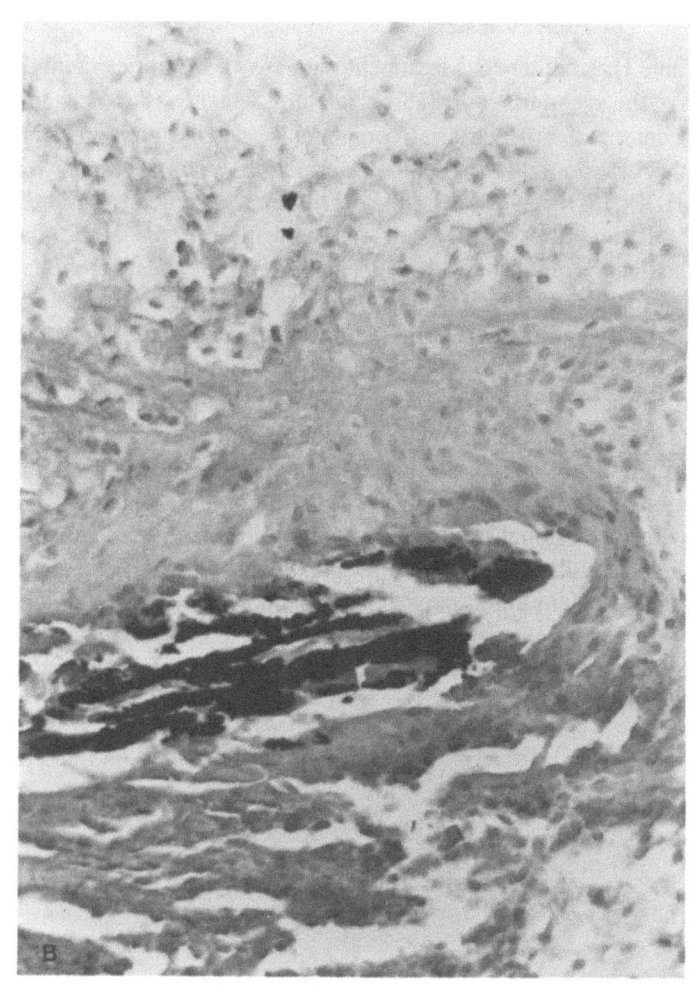

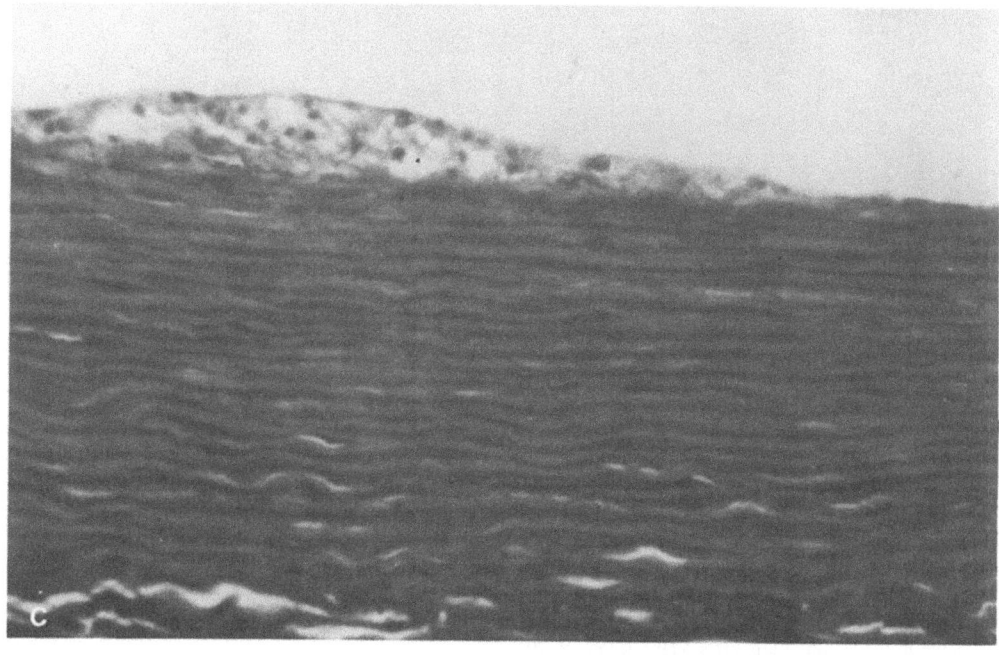

In rabbits (Table I), simultaneous treatment with any of the anticalcium drugs produced marked inhibition of the accumulation of aortic calcium, which was associated with a marked reduction in the content of other arterial components, despite continued presence of the atherogenic stimulus of unmitigated hypercholesterolemia. Optimal results were obtained by treatment with AHDP, IPDP, and $LaCl_3$ (40 mg), which completely suppressed the aortic accumulation of calcium, phosphorus, collagen, and elastin; the aortic accumulation of cholesterol was drastically reduced, with the remaining lipids presumably contained in the thin layer of intimal foam cells (Fig. 1C) that was observed as the only atherogenic alteration in these animals. The calcium, phosphorus, amino acid, and cholesterol composition was also normal in rabbits treated with these three drugs. In monkeys [4] treatment with lanthanum and diphosphonates resulted in similar reductions in the content of aortic calcium, as well as the other arterial components, with the changes in calcium, phosphorus, collagen, and elastin being completely suppressed in abdominal aorta and other major systemic arteries. The exception again was the arterial cholesterol content, which was greatly reduced but not normal, with the residual lipids presumably being mainly contained in the intimal cells of the few small cellular lesions present. Treatment with EHDP and $LaCl_3$ also suppressed the atherogenic calcium and phosphorus changes of elastin in abdominal aorta, as well as in other systemic major arteries of monkeys (Table II). Arterial elastin of APDP- and AHDP-treated monkeys was not similarly analyzed, but presumably would have shown similar results as with EHDP.

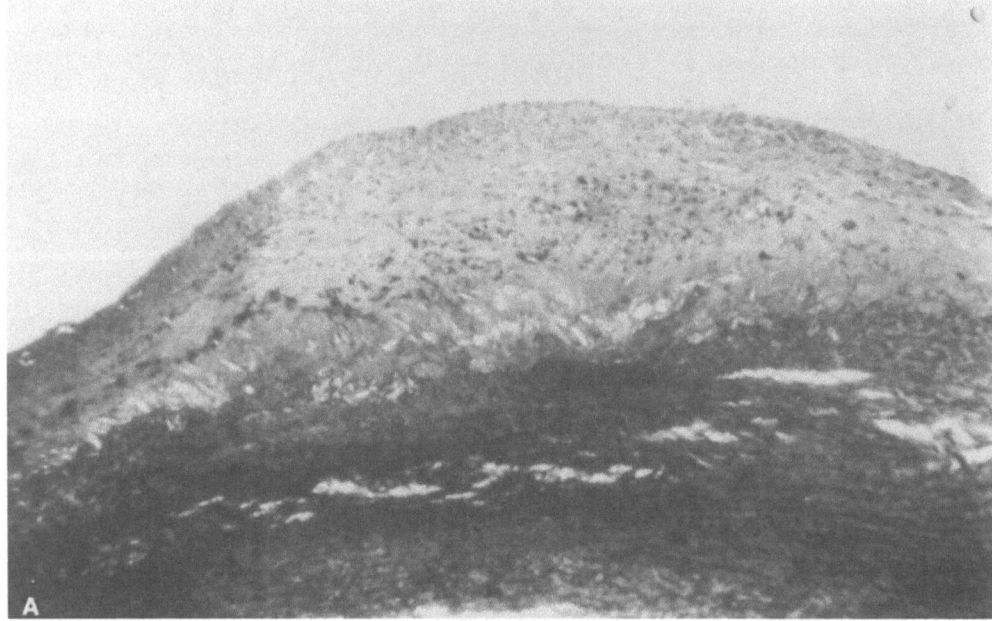

Figure 2. Sections through characteristic lesions of thoracic aorta of monkeys on a fibrogenic atherogenic diet for 24 months witout drugs (A, B = sequential sections through the same lesion) and treated simultaneously with AHDP (C). (A) Note collagenous plaque capsule (gray) surrounding a necrotic core with cholesterol clefts; the intimo-medial elastica (black) is fragmented and deranged; Verhoeff's-Van Gieson, × 70. (B) Note calcium deposition (black), including on intimo-medial elastica; Yasue's-light green, × 70. (C) The residual lesion consists of a few layers of (lipid-filled) foam cells in the intima. No connective tissue changes (or calcifications) were detectable in the lesion; Verhoeff's-Van Gieson, × 70.

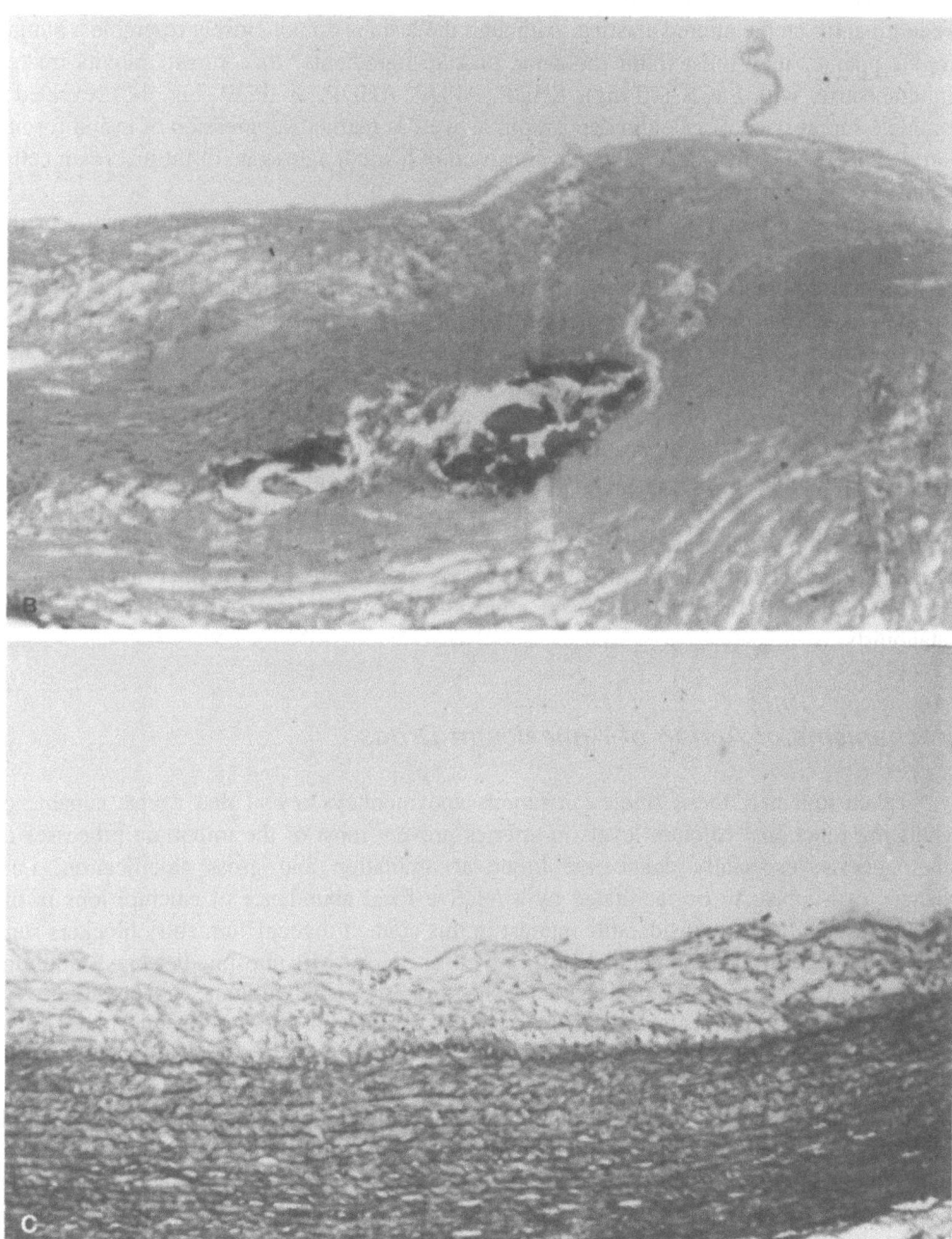

Morphological Findings

The biochemical results were in agreement with morphological findings. Typical lesions of thoracic aorta are shown in Fig. 1 for rabbits and in Fig. 2 for monkeys. In untreated rabbits, the fibrogenic atherogenic diet elicited numerous characteristic fibrous-atheromatous plaques (Fig. 1A), consisting of a markedly raised intima which contained marked accumulations of collagen, many proliferated cells (some of which were lipid-filled foam cells), fragmented and deranged intimo-medial elastica, as well as depositions of calcium minerals predominantly on the altered elastica. Although the lesions do not closely resemble a human fibrous plaque, they did exhibit the same plaque ingredients. In contrast, rabbits treated simultaneously with $LaCl_3$ (40 mg), EHDP, APDP, AHDP, or IPDP (Fig. 1C) revealed a complete suppression of calcium deposition, as well as marked suppression of lesion formation, leaving a few small lesions which consisted of a small aggregate of intimal foam cells, mounted on an otherwise normal arterial wall.

In untreated monkeys, the fibrogenic atherogenic diet employed elicited many of the same human-type fibrous plaques (Fig. 2A), as demonstrated in an earlier study [29]. These markedly raised plaques were characterized by a massive accumulation of intimal collagen, especially a fibrous cap, necroses and cholesterol clefts deep in the raised intima, derangement of intimo-medial elastica, but very few intimal foam cells. Calcium mineralization also was present which included deranged elastica. By contrast, no visible calcium deposits were detectable in arteries of monkeys on the same atherogenic diet that were treated with either lanthanum or any of the diphosphonates. In addition, the accumulation of collagen, the elastica alterations, necroses, and cholesterol crystal deposition were also suppressed; the few small residual lesions consisted of a few layers of lipid-filled foam cells (Fig. 2C), giving the lesion the appearance of a mild human fatty streak, especially in monkeys treated with $LaCl_3$ and AHDP. Similar microscopic observations were made in other major systemic arteries, including coronary arteries of control monkeys and those treated with calcium antagonists.

Mechanisms of Action of Anticalcium Drugs

Taken together, these studies in rabbits and monkeys reveal that agents capable of regulating functional calcium levels in arteries prevent most of the important processes of atherogenesis, especially connective tissue accumulation and gross calcification, i.e., changes that appear to be facilitated by a relative focal abundance of calcium ions in the arterial tissues. It is of considerable interest in this context that calcium entry blockers such as nifedipine [30] and verapamil [31], as well as the cAMP phosphodiesterase inhibitor trimazosin [32], possess marked antiatherogenic properties, including the capacity to prevent abnormal deposition of calcium and of connective tissue in arteries, despite unmitigated hypercholesterolemia.

At a more general level, these findings indicate that promotion of calcium homeostasis at the arterial tissue level may be an important element in the beneficial effects of all "calcium antagonists." It appears to be particularly relevant in this context that the formation of immunogenic arterial lesions in rabbits, as well as calcification of these lesions, can be suppressed by daily injections of thyrocalcitonin, the hormone that plays a pivotal role in calcium homeostasis [33]. Likewise, diet-induced rabbit atherosclerosis and arterial calcification is markedly inhibited by oral administration of thiophenecarboxylic acid [34], which is thought to exert a thyrocalcitonin-like effect on maintenance of calcium homeostasis [35].

The precise mechanisms by which these anticalcium agents inhibit calcification of soft tissue is known only for the diphosphonates [21–24]. These agents bind specifically to minerals normally existing in tissues, thereby preventing the binding of additional calcium ions to these minerals. However, this hardly explains the marked inhibitory effects on most other cellular and extracellular processes of atherogenesis. On the other hand, the diphosphonates, like other calcium antagonists, inhibit tissue calcium influx and transport, including transport across cellular membranes [36]. These latter properties appear instrumental in suppressing excessive cellular synthesis and/or secretion of calcifiable arterial components (collagen, altered elastin, proteoglycans) in response to atherogenic stimuli. Suppression of the overproduction of dangerous connective tissue components of plaques, and their calcification, appears to be the most prominent effect common to all anticalcium agents studied so far. It also is of interest to note that treatment with methyl-thiophenecarboxylic acid [37,38], as well as AHDP and lanthanum, promotes regression of connective tissue, calcium, and cholesterol components of preestablished lesions in rabbits and monkeys after cessation of a fibrogenic atherogenic diet. However, the possible mechanisms for promoting the reversal of atherosclerosis by these anticalcium agents are at present unclear.

Therapeutic Outlook

From our studies and those of others, it appears that at least some of the various calcium antagonists may provide therapeutic tools for the prevention (and perhaps even reversal) of the most important and life-threatening aspects of human atherosclerosis—fibrosis and calcification. The most suitable candidates as therapeutic agents may be the calcium entry blockers (perhaps including lanthanum) and trimazosin, because of their low incidence of serious side effects. The diphosphonates, AHDP and APDP, also appear to be good candidates, since these agents appear to have a wider safety margin than does EHDP [39,40], causing no bone demineralization (osteomalacia) even at higher doses. In fact, APDP (APD) has been used clinically for the treatment of osteoporosis in Paget's disease [40]. No clinical data are currently available concerning the newest diphosphonate, IPDP. It is noteworthy to add in conclusion that all of these calcium antagonists appear to exert their beneficial effects without altering the unfavorable serum lipid and lipoprotein concentrations that predominate in Western man.

ACKNOWLEDGMENT. Supported by U.S. Public Health Service Research Grants HL-15512 and HL-13262.

References

1. National Heart, Lung and Blood Institute Task Force on Arteriosclerosis. DHEW Publ. (NIH) 72-137. Vol 1971.
2. Meyer, W. W. The mode of calcification in atherosclerotic lesions. *Adv. Exp. Med. Biol.* **82**:786-792, 1977.
3. Kramsch, D. M.; Aspen, A. J.; Apstein, C. S. Suppression of experimental atherosclerosis by the Ca-antagonist lanthanum: Possible role of calcium in atherogenesis. *J. Clin. Invest.* **65**:967-981, 1980.
4. Kramsch, D. M.; Aspen, A. J.; Rozler, L. J. Atherosclerosis: Prevention by agents not affecting abnormal levels of blood lipids. *Science* **213**:1511-1512, 1981.
5. Werb, Z.; Gordon, S. Elastase secreted by stimulated macrophages. *J. Exp. Med.* **142**:361-377, 1975.
6. Packham, M. A.; Cazenave, J. P.; Kinlough-Rathbone, R. L.; Mustard, J. F. Drug effects on platelet adherence to collagen and damaged vessel walls. *Adv. Exp. Med. Biol.* **109**:253-276, 1978.
7. Keeley, F. W.; Partridge, S. M. Amino acid composition and calcification of human aortic elastin. *Atherosclerosis* **19**:287-296, 1974.
8. Kramsch, D. M.; Hollander, W. The interaction of serum and arterial lipoproteins with elastin of the arterial intima and its role in the lipid accumulation in atherosclerotic plaques. *J. Clin. Invest.* **52**:236-247, 1973.
9. Kramsch, D. M. Biochemical changes of the arterial wall in atherosclerosis with special reference to connective tissue: Promising experimental avenues for their prevention. In: *Connective Tissues in Arterial and Pulmonary Disease*, T. F. McDonald and A. B. Chandler, eds., Berlin, Springer-Verlag, 1981, pp. 95-151.

10. Ross, R.; Bornstein, P. The elastic fiber. I. Separation and partial characterization of its macromolecular components. *J. Cell Biol.* **40**:366–381, 1969.
11. Gotte, L.; Meneghali, V.; Castellani, A. Electron microscope observations and chemical analyses of human elastin. In: *Structure and Function of Connective and Skeletal Tissue,* S. Fitton-Jackson, R. D. Harkness, and G. R. Tristram, eds., London, Butterworths, 1965, pp. 93–101.
12. Moczar, M.; Robert, L. Extraction and fractionation of the media of thoracic aorta. *Atherosclerosis* **11**:7–25, 1970.
13. Urry, D. W.; Cunningham, W. D.; Ohnishi, T. A neutral polypeptide–calcium ion complex. *Biochim. Biophys. Acta* **292**:853–857, 1973.
14. Hornebeck, W.; Partridger, S. M. Confirmational changes in fibrous elastin due to calcium ions. *Eur. J. Clin. Invest.* **51**:73–78, 1975.
15. Weissman, G.; Weissman, S. X-ray diffraction studies of human aortic elastin residues. *J. Clin. Invest.* **39**:1657–1666, 1960.
16. Yu, S. Y.; Blumenthal, H. T. The calcification of elastic fibers. I. Biochemical studies. *J. Geront.* **18**:119–126, 1963.
17. Haust, M. D.; Geer, J. C. Mechanism of calcification in spontaneous artheriosclerotic lesions of the rabbit. *Am. J. Pathol.* **60**:329–338, 1970.
18. Yu, S. Y. Cross-linking of elastin in human atherosclerotic aorta. *Lab. Invest.* **25**:121–125, 1971.
19. Bladen, H. A.; Martin, G. R. Preferential mineralization of elastin in a matrix containing collagen. In: *Fifth International Congress on Electron Microscopy,* New York, Academic Press, 1962, pp. QQ-5.
20. Lian, J. B.; Skinner, M.; Glimcher, M. J.; Gallop, P. The presence of gamma-glutamic acid in proteins associated with ectopic calcification. *Biochem. Biophys. Res. Commun.* **73**:349–355, 1976.
21. Fleisch, H.; Russel, R. G. G.; Bisaz, S.; Muehlbauer, R. C.; Williams, D. A. The inhibitory effect of diphosphonates on the formation of calcium phosphate crystals in vitro and on aortic and kidney calcification in vivo. *Eur. J. Clin. Invest.* **1**:12–18, 1970.
22. Rosenblum, I. Y.; Flora, L.; Eisenstein, R. The effect of sodium ethane-1-hydroxy-1,1-diphosphonate (EHDP) on a rabbit model of arterio-atherosclerosis. *Atherosclerosis* **22**:411–421, 1975.
23. Wagner, W. D.; Clarkson, T. B. Slowly miscible cholesterol pools in progressing and regressing atherosclerotic aortas. *Proc. Soc. Exp. Biol. Med.* **143**:804–809, 1973.
24. Potocar, M.; Schmidt-Dunker, M. The effect of new diphosphonic acids on aortic and kidney calcifications in vivo. *Atherosclerosis* **30**:313–320, 1978.
25. Weiss, G. B. Cellular pharmacology of lanthanum. *Annu. Rev. Pharmacol.* **14**:343–354, 1974.
26. Weiss, G. B.; Goodman, F. R. Distribution of lanthanide (^{174}Pm) in vascular smooth muscle. *J. Pharmacol. Exp. Ther.* **198**:366–374, 1974.
27. Kritchevsky, D.; Tepper, S. A.; Vesselinovitch, D.; Wissler, R. W. Cholesterol vehicle in experimental atherosclerosis. Part II (peanut oils). *Atherosclerosis* **14**:53–64, 1971.
28. Kramsch, D. M.; Hollander, W.; Renand, S. Induction of fibrous plaques versus foam cell lesions in *Macaca fascicularis* by varying the composition of dietary fats. *Circulation* **48**:(Suppl. IV):41, 1973.
29. Kramsch, D. M.; Aspen, A. J.; Abramowitz, B. M.; Kreimendahl, T.; Hood, W. B. Jr. Reduction of coronary atherosclerosis by moderate conditioning exercise in monkeys on an atherogenic diet. *N. Engl. J. Med.* **305**:1483–1489, 1981.
30. Henry, P. D.; Bentley, K. I. Suppression of atherogenesis in cholesterol-fed rabbit treated with nifedipine. *J. Clin. Invest.* **68**:1366–1396, 1981.
31. Rouleau, J. L.; Parmley, W. W.; Stevens, J.; Wilkman-Coffelt, J.; Sievers, R.; Mahley, R.; Havel, R. J. Verapamil suppresses atherosclerosis in cholesterol-fed rabbits. *Am. J. Cardiol.* **49**:889, 1982.
32. Kramsch, D. M.; Aspen, A. J.; Swindell, A. C. Trimazosin suppresses fibrosis of atherosclerotic plaques. *Fed. Proc.* **42**:808, 1983.
33. Robert, A. M.; Moczar, M.; Brechamier, D.; Godeau, G.; Miskulin, M.; Robert, L. Biosynthesis and degradation of matrix molecules of the arterial wall: Regulation by drug action. In: *International Symposium: State of Prevention and Therapy in Human Atherosclerosis and in Animal Models,* W. H. Hauss, R. W. Wissler, and R. Lehmann, eds., Abh. Rhein.-Westf. Akad. Wiss., Volume 3, Opladen, Germany, Westdeutscher Verlag, 1978, pp. 301–312.
34. Chan, C. T.; Wells, H.; Kramsch, D. H. Suppression of fibrous-fatty plaque formation in rabbits by agents not affecting elevated serum cholesterol levels: The effect of thiophene compounds. *Circ. Res.* **43**:115–125, 1978.
35. Lloyd, W.; Fang, W. S.; Wells, H.; Tashjian, A. H. 2-Thiophene carboxylic acid: A hypocalcemic, antilipolytic agent with hypocalcemic and hypophosphatemic effects in rats. *Endocrinology* **85**:763–768, 1969.
36. Guilland, D. F., Salis, J. D., Fleisch, H. The effect of two diphosphonates on the handling of calcium by rat kidney mitochondria in vitro. *Calcif. Tissue Res.* **15**:303–314, 1974.

37. Kramsch, D. M.; Chan, C. T.; Wells, H. Effects of thiophene compounds on progression and regression of atherosclerosis in rabbits and cynomolgus monkeys. *Abstr. Counc. Arterioscler. Am. Heart Assoc.*, 1976, p. 24.
38. Kramsch, D. M. Role of connective tissue in atherosclerosis. *Adv. Exp. Med. Biol.* **109**:155–194, 1978.
39. Fleisch, H.; Felix, R. Diphosphonates. *Calcif. Tissue Res. Int.* **27**:91–94, 1979.
40. Frijlink, W. B.; TeVelde, J.; Bijvoet, O. L. M.; Heynen, G. Treatment of Paget's disease with (3-amino-1-hydroxypropylidene)-1,1-biphosphonate (APD). *Lancet* **1**:799–803, 1979.

XVI

Crystal Deposition

76

Arthritis and Calcium-Containing Crystals
An Overview

Daniel J. McCarty

In 1961, I found that aspiration of the acutely inflamed knee joints of two patients with acute arthritis yielded crystals having the following properties by compensated polarized light microscopy: weak positive birefringence, inclined extinction, and resistance to uricase [1]. By contrast, the crystals contained within chalk from a gouty tophus possessed: a needlelike shape, strong negative birefringence, axial extinction, and disappeared when incubated with highly purified uricase. This prompted the conclusion that joints could harbor two different kinds of crystals, one of which was not sodium urate. Both types of crystals were found predominantly within polymorphonuclear leukocytes during an acute attack [2]. As the acute episodes associated with the nonurate crystals were goutlike, they were called "pseudogout" [3].

Characterization of CPPD Crystal Deposition

Much has been learned over the years about this condition which we now call "calcium pyrophosphate dihydrate (CPPD) crystal deposition disease." In England it is called "pyrophosphate arthropathy." On the European continent, "chondrocalcinosis" still seems to be preferred, and this term is widely used in North America as a relatively nonspecific reference to the radiologic appearance of calcified articular cartilages.

The pathologic calcifications from various tissues have been characterized. Most are basic calcium phosphates, such as hydroxy- or carbonate-apatite or Whitlockite [$Ca_3(PO_4)_2$] [4]. CPPD crystals were found in tendons, ligaments, synovium, and articular capsules, in addition to hyaline and fibrocartilages. Three crystal species have been detected in menisci removed from knee joints [5]. CPPD deposits were found in 3.2% and brushite [dicalcium phosphate dihydrate (DCPD), $CaHPO_4 \cdot 2H_2O$] in 2–3% of the joints analyzed. Both DCPD and CPPD crystals were deposited in multiple sites in multiple cartilages, i.e., they repre-

---

Daniel J. McCarty • Division of Rheumatology, Department of Medicine, Medical College of Wisconsin, Milwaukee, Wisconsin 53226.

Table I. Classification of CPPD Crystal Deposition Disease[a]

Hereditary
Caucasian
(Hungarian, Dutch, Swedish, French, French-Canadian, German, Mexican, Swiss-German, English)
Oriental
(Japanese)
Sporadic (idiopathic)
Metabolic disease-associated (Table II)

[a]For additional information see Ref. 8.

sented a "primary" type of calcification. Lastly, the ubiquitous apatite was found, usually as a solitary deposit or as vascular calcification, in the peripheral third of menisci.

Our tentative classification of cases of CPPD crystal deposition as shown in Table I separates three main groups: (1) hereditary, (2) metabolic disease-associated, and (3) sporadic. Hereditary disease had been described exclusively in Caucasians (Hungarians, Dutch, Swedish, Spanish, French, German, French-Canadian, Mexican, and Americans with Swiss-German and English ethnic backgrounds), but a Japanese family with CPPD crystal deposition has been reported recently [6]. Most familial studies have shown an autosomal dominant pattern with complete penetrance after midlife.

A Pandora's box of metabolic diseases and multiple clinical presentations has been reported in association with CPPD crystal deposits [7,8]. It is difficult to ascertain whether a putative association is "real" in the sense of cause and effect, or whether it represents chance coincidence. CPPD deposition is sufficiently common that associations with other common conditions might be predicted. However, except for some abnormalities in calcium metabolism in several Swedish subjects [9], no metabolic disease associations have been described in familial cases. Cases of CPPD crystal deposition without recognizable metabolic abnormality or affected relatives are classified as "sporadic."

An incomplete list of putative disease associations is shown in Table II. Sustained

Table II. Conditions Associated with CPPD Crystal Deposition

Primary hyperparathyroidism
Familial hypocalciuric hypercalcemia
Hemochromatosis
Hemosiderosis
Hypophosphatasia
Hypomagnesemia
Bartter's syndrome
Hypothyroidism
Gout (MSU monohydrate crystal deposition)
Neuropathic joints
Amyloidosis
Localized trauma
Surgery for osteochondritis dissecans
Hypermobility syndrome
Aging

hypercalcemia appears important, and there are many reports describing an association with tissue iron overload, especially hemachromatosis. As with hypercalcemia, older patients are most often afflicted and the crystal deposits persist despite treatment of the primary condition. Hypothyroidism is a common association, and hypomagnesemia and hypophosphatasia are rare but associated conditions. CPPD crystal deposition in patients receiving corticosteroids over a prolonged period for arthritis symptoms has also been recently described, but it is likely that this is not a true association [10].

Pathogenesis of Acute Attacks

Inflammation experimentally induced by either sodium urate or calcium pyrophosphate crystals is dose-dependent and mediated by phagocytosis by polymorphonuclear leukocytes (summarized in Ref. 11). Crystal phagocytosis by leukocytes is accompanied by release of prostaglandins, leukotrienes [12], reactive oxygen species, lysosomal enzymes, and neutral proteases, including neutrophil collagenase. A cell-derived chemotactic factor, an 8500-dalton glycopeptide, is released from leukocytes during and after phagocytosis of urate crystals [13]. There are approximately 5×10^5 receptors on each leukocyte for this cell-induced chemotactic factor [14], which is released by CPPD and diamond crystals also [15]. Finally, various humoral factors, such as fibronectin, fibrin, and complement components, adsorb to urate crystals and may be responsible for many of the biologic effects attributable to crystals [16].

How do crystals gain access to joint space? We have postulated that crystals are shed into the joint space as a result of increased solubility, causing the crystals to loosen within their matrix. Once inflammation begins, shedding might be further accentuated by additional factors such as: (1) a lowered joint fluid level of inorganic pyrophosphate (PP_i) due to increased synovial blood flow causing increased PP_i efflux, and (2) enzymatic digestion of the matrix encasing the crystals, resulting in "strip-mining." In this context, CPPD crystals have been released readily from articular cartilage by incubation *in vitro* with partially purified synovial cell collagenase [17]. Such a mechanism may explain the presence of CPPD (or urate) crystals in association with other forms of arthritis, including gout or sepsis.

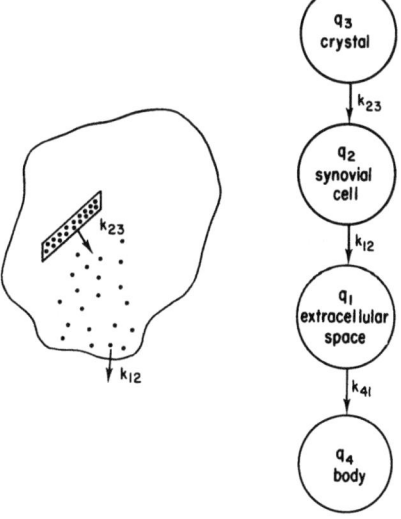

Figure 1. Kinetic model used to analyze labeled CPPD crystal clearance from lapine joints. Nuclide flux was calculated from the known specific activity of the injected crystals and the rate constants for nuclide moving from compartment to compartment as determined by the clearance rate (serial external counting) and the percentage of nuclide found in crystal and cell compartments in a synovial biopsy sample. The crystals were doubly labeled with ^{45}Ca and the ^{85}Sr/^{45}Ca was the same in the crystal and cell compartments, suggesting identical ^{85}Sr and ^{45}Ca efflux from the cells. [Reproduced with permission of *Annals of the Rheumatic Diseases*.]

It appears that virtually all CPPD crystal dissolution occurs within fixed synovial cells after endocytosis as evidenced by: (1) the rapid disappearance of labeled crystals from the joint space with concentration in the synovium [18,19]; (2) the microscopic localization of crystals within synovial cells [20]; (3) the relative insolubility of crystals in body fluids; (4) the lack of effect on CPPD crystal clearance rate of manipulation of variables controlling solubility in extracellular fluid; and (5) a definite effect on crystal clearance of manipulating the intracellular environment. Therefore, we postulated a "traffic" of CPPD crystals from cartilaginous sites of origin through the joint fluid into the synovium. The paradigm developed is shown in Fig. 1.

"Milwaukee Shoulder" and the Pathogenesis of Destructive Arthropathy

Insight into the mechanism of the destructive arthropathies accompanying CPPD crystal deposition was gained by discovery of a peculiar condition associated with calcium phosphate crystals ("Milwaukee shoulder" syndrome). "Hydroxyapatite-like" crystals have been identified in synovial fluid of these patients by transmission electron microscopy (TEM) [21,22]. Moreover, a semiquantitative technique utilizing ^{14}C-labeled ethane-1-hydroxy-1,1-diphosphonate (EHDP) was employed in our laboratory to detect crystals in human synovial fluid samples [23]. Fluids from joints affected with "Milwaukee shoulder" syndrome show: (1) microspheroidal aggregates of calcium phosphates; (2) particulate collagens types I, II, and III suggesting origin both from synovial membrane (types I and III) and from articular cartilage (type II); (3) collagenase activity, often evident even before "activation" with trypsin; (4) neutral protease activity; (5) leukocyte concentrations usually less than 500/mm^3, nearly all mononucleated cells [24,25]. Histologic examination of the synovium of our index case showed typical chondromatosis [26]. TEM showed extracellular masses of crystals enmeshed in a collagen network as well as within synovial cells. Some of these particles appeared to be shedding from the tissue through areas denuded of synovial lining cells.

The nucleation and growth of CPPD crystals in gels is discussed in Chapter 78. As cartilage is essentially a gel at neutral pH, the work considered in this chapter provides a model for this phenomenon *in vivo*. The initial stimulus for formation of basic calcium phosphate mineral phase in the periarticular tissues is unknown. Chapter 79 focuses on the histopathology of calcific deposits in tendons which provides important insights into this question. Reports of increased secretion of collagenase and neutral protease from cultured synovial cells after phagocytosis of latex beads [27] led us to speculate that these findings might be equivalent to those observed *in vivo*. Chapter 79 focuses on this hypothesis. We envision crystal uptake by synovial cells—analogous to the "traffic" of CPPD crystals—to involve augmented protease release, and further release of crystals and particulate collagens into the joint space by the "strip-mining" mechanism. Calcium phosphate crystal-containing microaggregates were released from minced synovial tissue when incubated with partially purified mammalian synovial cell collagenase [17]. All but one of the crystal populations in these shoulder joint fluids contained partially carbonate-substituted hydroxyapatite, octacalcium phosphate (OCP), and collagens [28]. The one exception, crystals from our index case, contained tricalcium phosphate instead of OCP. These crystal populations are now referred to generically as basic calcium phosphates (BCP).

Increased intracellular levels of PP$_i$ in cultured skin fibroblasts and in Epstein–Barr virus-transformed peripheral blood lymphocytes taken from French patients with familial CPPD crystal deposition were reported recently [29]. Histologic, electron microscopic, and biochemical studies on cartilage biopsy samples from Swedish patients with familial CPPD crystal deposition imply an underlying abnormality of the matrix [30], with CPPD crystal deposition an epiphenomenon. Information relative to the generation of PP_i by cartilage is covered in Chapter 77. Although these studies need confirmation and extension, CPPD crystals, like monosium urate crystals in gout, may represent a final common pathway of a number of different primary metabolic aberrations.

ACKNOWLEDGMENT. Supported by USPHS Grants AM-26062 and AM-31395.

References

1. McCarty, D. J.; Hollander, J. L. Identification of urate crystals in gouty synovial fluid. *Ann. Intern. Med.* **54**:452–460, 1961.
2. McCarty, D. J. Phagocytosis of urate crystals in gouty synovial fluid. *Am. J. Med. Sci.* **243**:288–295, 1962.
3. McCarty, D. J.; Kohn, N. N.; Faires, J. S. The significance of calcium phosphate crystals in the synovial fluid of arthritis patients: The "pseudogout syndrome": I. Clinical aspects. *Ann. Intern. Med.* **56**:711–737, 1962.
4. Gatter, R. A.; McCarty, D. J. Pathological tissue calcifications in man. *Arch. Pathol.* **84**:346–353, 1967.
5. McCarty, D. J.; Hogan, J. M.; Gatter, R. A.; Grossman, M. Studies on pathological calcifications in human cartilages. I. Prevalence and types of crystal deposits in the menisci of two hundred fifteen cadavers. *J. Bone Jt. Surg.* **48**,309–325, 1966.
6. Sakaguchi, M.; Kitagawa, T.; Ishikawa, K.; Numata, T. Familial pseudogout with destructive arthropathy in Japan. *Arthritis Rheum.* **25**:S34 (abstract), 1982.
7. McCarty, D. J. Crystals, joints and consternation. The Heberden Oration–1982, London, England. *Ann. Rheum. Dis.* **42**:243–253, 1983.
8. McCarty, D. J. (ed.) Calcium pyrophosphate dihydrate crystal deposition disease; pseudogout; articular chondrocalcinosis. In: *Arthritis and Allied Conditions*, 9th ed., Philadelphia, Lea & Febiger, 1979, pp. 1276–1299.
9. Bjelle, A. Pyrophosphate arthropathy in two Swedish families. *Arthritis Rheum.* **25**:66–74, 1982.
10. Alexander, G. J. M.; Dieppe, P. A.; Doherty, M.; Scott, D. G. I. Pyrophosphate arthropathy: A study of metabolic associations and laboratory parameters. *Ann. Rheum. Dis.* **41**:377–381, 1982.
11. McCarty, D. J. Crystal-induced inflammation and its treatment. In: *Arthritis and Allied Conditions*, 9th ed., D. J. McCarty, ed., Lea & Febiger, Philadelphia, 1979, pp. 1244–1261.
12. Rae, S. A.; Davidson, E. M.; Smith, M. J. H. Leukotriene B_4, an inflammatory mediator in gout. *Lancet* **2**:1122–1124, 1982.
13. Phelps, P. Polymorphonuclear leukocyte motility in vitro. IV. Colchicine inhibition of chemotactic activity formation after phagocytosis of urate crystals. *Arthritis Rheum.* **13**:1–9, 1970.
14. Spilberg, I.; Mehta, J. Demonstration of a specific neutrophil receptor for a cell-derived chemotactic factor. *J. Clin. Invest.* **63**:85–88, 1979.
15. Spilberg, I.; Mehta, J.; Simchowitz, L. Induction of a chemotactic factor from human neutrophils by diverse crystals. *J. Lab. Clin. Med.* **100**:399–404, 1982.
16. Terkeltaub, R.; Kozin, F.; Ginsberg, M. Plasma protein binding by monosodium urate crystals (MSUC): Analysis by 2-dimensional gel electrophoresis. *Arthritis Rheum.* **25**:S77 (abstract), 1982.
17. Halverson, P. B.; Cheung, H. S.; McCarty, D. J. Enzymatic release of microspheroids containing hydroxypatite crystals from synovium and of calcium pyrophosphate crystals from cartilage. *Ann. Rheum. Dis.* **41**:527–531, 1982.
18. McCarty, D. J.; Palmer, D. W.; Halverson, P. B. Clearance of calcium pyrophosphate dihydrate (CPPD) crystals in vivo. I. Studies using ^{169}Yb labeled triclinic crystals. *Arthritis Rheum.* **22**:718–727, 1979.
19. McCarty, D. J.; Palmer, D. W.; James, C. Clearance of calcium pyrophosphate dihydrate (CPPD) crystals in vivo. II. Studies using triclinic crystals doubly labeled with ^{45}Ca and ^{85}Sr. *Arthritis Rheum.* **22**:1122–1131, 1979.
20. McCarty, D. J., Palmer, D. W.; Garancis, J. C. Clearance of calcium pyrophosphate dihydrate crystals in vivo. III. Effects of synovial hemosiderosis. *Arthritis Rheum.* **24**:706–710, 1981.
21. Dieppe, P. A.; Huskisson, E. C.; Crocker, P.; Willoughby, D. A. Apatite deposition disease. *Lancet* **1**:266–269, 1976.
22. Schumacher, H. R.; Somlyo, A. P.; Tse, R. L.; Maurer, K. Arthritis associated with apatite crystals. *Ann. Intern. Med.* **87**:411–416, 1977.

23. Halverson, P. B.; McCarty, D. J. Identification of hydroxyapatite crystals in synovial fluid. *Arthritis Rheum.* **22**:389–395, 1979.
24. Halverson, P. B.; Cheung, H. S.; McCarty, D. J.; Garancis, J.; Mandel, N. "Milwaukee shoulder": Association of microspheroids containing hydroxyapatite crystals, active collagenase and neutral protease with rotator cuff defects. II. Synovial fluid studies. *Arthritis Rheum.* **24**:474–483, 1981.
25. McCarty, D. J.; Halverson, P. B.; Carrera, G. F.; Brewer, B. J.: Kozin, F. "Milwaukee shoulder": Association of microspheroids containing hydroxyapatite crystals, active collagenase and neutral protease with rotator cuff defects. I. Clinical aspects. *Arthritis Rheum.* **24**:464–473, 1981.
26. Garancis, J. C.; Cheung, H. S.; Halverson, P. B.; McCarty, D. J. "Milwaukee shoulder": Association of microspheroids containing hydroxyapatite crystals, active collagenase and neutral protease with rotator cuff defects. III. Morphologic and biochemical studies of an excised synovium showing chondromatosis. *Arthritis Rheum.* **24**:484–491, 1981.
27. Werb, Z.; Reynolds, J. J. Stimulation by endocytosis of the secretion of collagenase and neutral proteinase from rabbit synovial fibroblasts. *J. Exp. Med.* **140**:1482–1497, 1974.
28. McCarty, D. J.; Lehr, J. R.; Halverson, P. B. Crystal populations in human synovial fluid: Identification of apatite, octacalcium phosphate and tricalcium phosphate. *Arthritis Rheum.* **26**:1220–1224, 1983.
29. Lust, G.; Faure, G.; Netter, P.; Gaucher, A.; Seegmiller, J. E. Evidence of generalzed metabolic defect in patients with hereditary chondrocalcinosis: Increased inorganic pyrophosphate in cultured fibroblasts and lymphoblasts. *Arthritis Rheum.* **24**:1517–1521, 1981.
30. Bjelle, A. Cartilage matrix in hereditary pyrophosphate arthropathy. *J. Rheumatol.* **8**:959–964, 1981.

77

Nucleoside Triphosphate (NTP) Pyrophosphohydrolase in Chondrocalcinotic and Osteoarthritic Cartilages

David S. Howell, Jean Pierre Pelletier, Johanne Martel-Pelletier, Sara Morales, and Ofelia Muniz

Previous studies have led to the view that pyrophosphate ion metabolism is disturbed in diseased articular cartilages, particularly chondrocalcinotic and osteoarthritic cartilages [1–4]. In organ culture, a consistent output of pyrophosphate by normal and diseased articular cartilages has been found [1–3], and all ulcerated osteoarthritic cartilages studied so far have elaborated pyrophosphate into the surrounding medium [2,5]. These findings were not the result of cell death [1–3]. Variable fluctuations in the synthesis by chondrocytes of glycosaminoglycans [3–6], discussed elsewhere, also seem an unlikely cause of the sustained synovial fluid pyrophosphate elevations and of the large accumulations of calcium pyrophosphate salts seen clinically.

The ability of growth cartilage extracts and cartilage slices [7–11,14] to generate pyrophosphate from cold or labeled ATP indicates that a nucleoside triphosphate (NTP) pyrophosphohydrolase is present in growth plate and articular cartilages. In this chapter, data are presented on the properties of this enzyme isolated from human articular cartilage, as well as on its tissue cellular distribution. Fresh cartilage was excised from the tibial plateaus of seven patients who had primary osteoarthritis (OA) by X-ray criteria [12], six patients who manifested symptoms of chondrocalcinosis calcium pyrophosphate deposition disease (CPDD) [13], and three normal subjects. Tissue preparations and histological studies were carried out as previously described [14–18].

Distribution of enzyme activities in whole extracts (Table I) and in subcellular fractions from mechanically disrupted OA and calcium pyrophosphate dihydrate crystal deposition disease (CPDD) articular cartilage are shown (Fig. 1). In comparison to whole cartilage

---

David S. Howell • Department of Medicine, University of Miami, School of Medicine, and U.S. Veterans Administration, Miami, Florida 33101.   *Jean Pierre Pelletier and Johanne Martel-Pelletier.* • Rheumatic Disease Unit, Notre Dame Hospital, University of Montreal, Montreal, Quebec H2L 4K8, Canada.   *Sara Morales and Ofelia Muniz* • Department of Medicine, University of Miami, School of Medicine, Miami, Florida 33101.

Table I. Total Activity of NTP Pyrophosphohydrolase and Other Hydrolases Based on [8-14C]-ATP Assay, pH 9.8, as Well as [λ-32]-ATP Assays Performed on These Samples[a]

Diagnostic group	No. of patients	Alkaline phosphatase	5'-Nucleotidase (U/g)[b]	NTP pyrophosphohydrolase[c,d]	DNA (μg/g)
1. Normal	3	1.2 ± 3	151 ± 48	19.1 ± 8.7	19.8 ± 10
2. OA	7	50.3 ± 14	143 ± 51	33.1 ± 9	17 ± 6
3. CPDD	6	25.9 ± 17	461 ± 39	41.0 ± 10	22 ± 11

[a]Assays for alkaline phosphatase, 5'-nucleotidase, nucleoside pyrophosphohydrolase, and DNA were performed as previously described (14). Histological examination of the OA and CPDD cartilage revealed cell cloning, deep fissures, and by histological criteria grade 7 to 11 (15).
[b]g wet weight, starting sample of cartilage; U = 1 nmole product formed per minute. Significance levels for differences among diagnostic groups for 5'-nucleotidase and alkaline phosphatase: CPDD vs. OA $p < 0.001$.
[c]It was necessary to use levamisole, 1 mM, to inhibit other hydrolytic enzyme activities.
[d]For NTP pyrophosphohydrolase: CPDD vs. OA, $p > 0.05$; CPDD vs. normal, $p < 0.01$; OA vs. normal, $p < 0.01$.

extracts, roughly 100% recoveries of alkaline phosphatase and NTP pyrophosphohydrolase were obtained in the subsequent subcellular fractions by a recycling method. Approximately 75% of the total activity of NTP pyrophosphohydrolase was recovered in pellet 1, a fraction enriched in plasma membranes (Fig. 1). The next highest proportion of NTP pyrophosphohydrolase was in the microsome–vesicle fraction. Lesser quantities of the enzyme were detected in mitochondrial and ER fractions. These results are in approximate agreement with a recent careful study of Hsu characterizing apparently the same enzyme in epiphyseal fetal bovine cartilage [9].

Analysis of Table II reveals that the enzyme probably could function in either an intra- or an extracellular milieu. Although enzyme activity was greatest in the alkaline range, substantial activity at neutral pH in whole slices was evident from other studies [9,10]. The apparent K_m of our cartilage enzyme was higher than that found by Hsu [9].

To assess further whether a plasma membrane ectoenzyme was involved, an experiment was performed in which NTP pyrophosphohydrolase activity was compared in articular

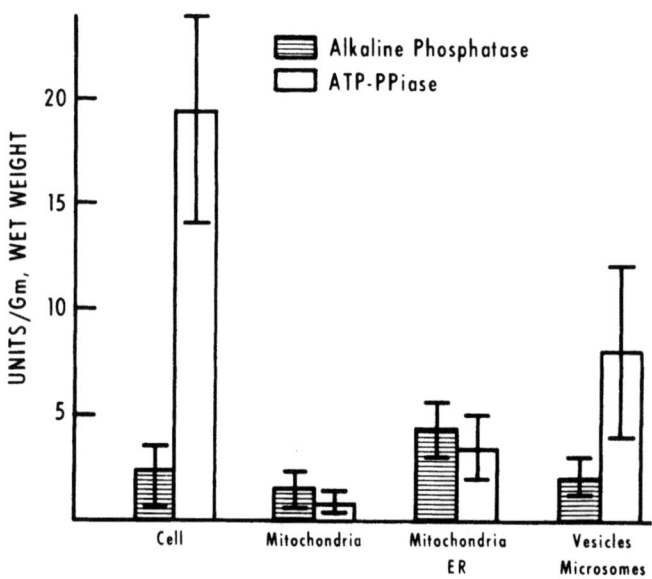

Figure 1. Subcellular distribution of enzymes (mean ± S.D.) in patients with both OA and CPDD. Cartilage–Triton X-100 mixtures were homogenized and the homogenate centrifuged at 23,000g for 20 min [16]. The supernatant was used for biochemical assays. Subcellular fractionation was performed by the grinding method of Wuthier et al. [16], for this partition. A similar partition was obtained following a collagenase method for subcellular fractionation [11].

Table II. Some Properties of NTP Pyrophosphohydrolase in Human Cartilage

Substrates: ATP, CTP, UTP, GTP
ph range 7–11; optimum 9.5–9.8
K_m 2.3 ± 0.52 × 10^{-5} M

Inhibited by[a] dithiothreitol, cysteine
Not inhibited by[b] levamisole, tetramisole, bromotetramisole
Metal ion effects: strong inhibition by Zn^{2+}, EDTA, EGTA
80–137% of control activity
 with 0.1–1.0 mM Mg^{2+} or Ca^{2+}

[a]1 mM.
[b]All activity lost in tissue treated 2 hr with Pronase.

cartilages of three OA patients, submitted to Triton X-100 extraction. Table III reveals that enzyme activity was approximately the same in slices as in cartilage extracts. In contrast, about half the activity for alkaline phosphatase was found in the latter samples. These current findings strongly support the view that NTP pyrophosphohydrolase is an ectoenzyme in cartilage [10]. The loss of all enzyme activity by brief treatment with pronase also favors an ectoenzymatic function (Table II) [10].

Alkaline phosphatase activity is found in high concentrations in chondrocytes along the tidemark in rabbits [19], and in human OA articular cartilage [20]. This high level in normal adult cartilage has been ascribed to the presence of bone remodeling and capillary invasion associated with remodeling. These remodeling activities may be increased at sites of experimental cartilage cavitation and in OA joints [20]. This raises the question as to whether NTP pyrophosphohydrolase is concentrated in such remodeling sites. Thus, the activities of alkaline phosphatase in the deep zone of OA marginal lesions were compared to the superficial two-thirds of the hyaline cartilage and the activity of NTP pyrophosphohydrolase was correlated with that of alkaline phosphatase. Higher activity in alkaline phosphatase was found in the deep third of "cartilage" adjacent to the tidemark or Nelaton's line in comparison to the superficial cartilage (Table IV), where only a trace of activity was detected. In contrast, there was no difference in NTP pyrophosphohydrolase levels between these anatomical sites, thus supporting the view that this enzyme may be distributed ubiquitously.

A membrane-bound ATP pyrophosphohydrolase (EC 3.6.1.8) associated with liver cell plasma membranes has also been observed [19–23]. The generation of pyrophosphate in such membranes is thought to be associated with energy transfers needed for optimal calcium transport [20]. The recent evidence indicating that it is an ectoenzyme points to the need for exploring alternative possible functions, e.g., this enzyme might either supply energy for plasma membrane surface biochemical reactions or conserve nucleosides liberated by cell

Table III. Results of Enzyme Assays on Slices vs. Triton X-100 Extracts of OA Cartilages[a,b]

No. of patients	NTP pyrophosphohydrolase (U/g)		Alkaline phosphatase (U/g)	
	Whole cartilage extract	Slice	Whole cartilage extract	Slice
3	35 ± 4	36 ± 6.1	6.9 ± 1.5	3.7 ± 1.4

[a]Values are mean ± S.D. Cell counts: 3.4 ± 0.7 × 10^3.
[b]Of each cartilage sample, half was used for extraction and the rest for slice assays.

Table IV. Distribution of Enzyme Activities between Superficial and Deep Layers of Articular Cartilage in Patients with CPDD[a]

No. of patients	Cartilage size	Alkaline phosphatase[b]	5'-Nucleotidase	NTP pyrophos-phohydrolase	Cell counts/mg wet wt
6	Superficial layer	17.2 ± 4.6	162 ± 31	29.6 ± 10.2	$3.3 \pm 0.7 \times 10^3$
6	Deep layer	64.1 ± 30	79 ± 44	24.1 ± 8.2	$3.6 \pm 1.5 \times 10^3$

[a] Values are mean ± S.D.
[b] Alkaline phosphatase activities in superficial vs. deep layer: significantly different, $p < 0.05$.

injury. A role for the exceptionally high levels of 5' nucleotidase is equally enigmatic. Finally, the question as to whether the new enzyme contributes ions to production of calcium pyrophosphate mineral deposits remains for future study.

ACKNOWLEDGMENTS. Supported by the Research Service, U.S. Veterans Administration, Grant AM-08622 from the National Institutes of Health, the Medical Research Council of Canada and Canadian Arthritis Society Fellowship Awards, and the Kroc Foundation.

Refernces

1. Howell, D. S.; Muniz, O.; Pita, J. C. Extrusion of pyrophosphate into extracellular media by cartilage incubates. In: *Normal and Osteoarthrotic Articular Cartilage*, S. Y. Ali, M. W. Elves, D. H. Leaback, eds., London, Kingswood Press, 1974, pp. 177-178.
2. Howell, D. S.; Muniz, O. E.; Pita, J. C.; Enis, J. E. Extrusion of pyrophosphate into extracellular media by osteoarthritic cartilage incubates. *J. Clin. Invest.* **56**:1473-1480, 1975.
3. Ryan, L. M.; Cheung, H. S.; McCarty, D. J. Release of pyrophosphate by normal mammalian articular hyaline and fibrocartilage in organ culture. *Arthritis Rheum.* **24**:1522-1527, 1981.
4. Lust, G.; Nuki, G.; Seegmiller, J. E. Inorganic pyrophosphate and proteoglycan metabolism in cultured human articular chondrocytes and fibroblasts. *Arthritis Rheum.* **19**:479, 1976.
5. Howell, D. S. Crystal deposition disease. In: *Topical Reviews in Rheumatic Disorders*, V. Wright, ed., London, Wright, 1982.
6. McGuire, M. K. B.; Bayliss, M.; Baghat, N.; Colman, C. H.; Russell, R. G. G. Pyrophosphate metabolism in human articular chondrocytes. *J. Rheum.* **8**:1016, 1981.
7. Cartier, P.; Picard, J. La mineralization du cartilage ossifiable III le mecanise de la reaction ATPasique du cartilage. *Bull. Soc. Chem. Biol.* **37**:1159-1176, 1955.
8. Howell, D. S.; Muniz, O. E.; Morales, S. 5'Nucleotidase and pyrophosphate (PPi) generating activities in articular cartilage extracts in calcium pyrophosphate deposition disease (CPDD) and in primary osteoarthritis (OA). In: *Epidemiology of Osteoarthritis*, J. G. Peyron, ed., Paris, Ciba Geigy. 1980, pp. 99-108.
9. Hsu, H. H. T. Purification and partial characterization of ATP pyrophosphohydrolase from fetal bovine epiphyseal cartilage. *J. Biol. Chem.* **258**:3463-3468, 1983.
10. Ryan, L. M.; Wortmann, R. L.; Karas, B.; McCarty, D. J. Nucleoside triphosphate (NTP) pyrophosphohydrolase activity in canine articular cartilage. *Arthritis Rheum.* **25**:S30 (abstract), 1982.
11. Howell, D. S.; Pelletier, J.; Martel-Pelletier, J. P.; Morales, S.; Muniz, O. E. Nucleoside triphosphate pyrophosphohydrolase in chondrocalcinotic and osteoarthritic cartilages. *Arthritis Rheum.* **27**:193-199, 1984.
12. Kellgren, J. H.; Lawrence, J. S. Radiological assessment of osteoarthrosis. *Ann. Rheum. Dis.* **16**:494-502, 1957.
13. Resnick, D.; Niwayama, G.; Goergen, T. G.; Utsinger, P. D.; Shapiro, R. F.; Haselwood, D. H.; Wiesner, K. B. Clinical radiographic and pathologic abnormalities in calcium pyrophosphate dihydrate deposition disease (CPDD): Pseudogout. *Radiology* **122**:1-15, 1977.
14. Tenenbaum, J.; Muniz, O. E.; Howell, D. S. Comparison of pyrophosphohydrolase activities from articular cartilage in calcium pyrophosphate-deposition disease and primary osteoarthritis. *Arthritis Rheum.* **24**:492-500, 1981.

15. Mankin, H. J.; Dorfman, H.; Lipiello, L.; Zarins, A. Biochemical and metabolic abnormalities in articular cartilage from osteoarthritic human hips. II. Correlation of morphology with biochemical and metabolic data. *J. Bone Jt. Surg. Am. Vol.* **53**:523–537, 1971.
16. Wuthier, R. E.; Linder, E.; Warner, G. P.; Gore, S. T.; Borg, T. K. Nonenzymatic isolation of matrix vesicles: Characterization and initial studies on ^{45}Ca and ^{32}P-orthophosphate metabolism. *Metab. Bone Dis. Relat. Res.* **1**:125–136, 1978.
17. Stockwell, R. The chondrocytes. In: *Adult Articular Cartilage*, M. A. E. Freeman, ed., London, Pitman, 1979, pp. 51–99.
18. Ali, S. Y.; Bayliss, M. T. Enzymic changes in human osteoarthrotic cartilage. In: *Normal and Osteoarthrotic Articular Cartilage*, S. Y. Ali, M. W. Elves, and D. H. Leaback, eds., London, Kingswood Press, 1974, pp. 189–205.
19. Lieberman, I.; Lansing, A. I.; Lynch, W. E. Nucleoside triphosphate pyrophosphohydrolase of the plasma membrane of the liver cell. *J. Biol. Chem.* **242**:736–739, 1967.
20. Torp-Pedersen, C.; Flodgaard, H.; Saermark, T. Studies on Ca^{2+} dependent nucleoside triphosphate pyrophosphohydrolase in rat liver plasma membranes. *Biochim. Biophys. Acta* **571**:94–104, 1979.
21. Franklin, J. E.; Tramz, E. G. Metabolism of coenzyme A and related nucleotides by liver plasma membranes. *Biochim. Biophys. Acta* **230**:105–116, 1971.
22. House, P. D. R.; Poulis, P.; Weidemann, M. J. Isolation of a plasma membrane subfraction from rat liver containing an insulin sensitive cyclic-AMP phosphodiesterase. *Eur. J. Biochem.* **24**,429–437, 1972.
23. Flodgaard, H.; Torp-Pedersen, C. A calcium ion-dependent adenosine triphosphate pyrophosphohydrolase in plasma membrane from rat liver. *Biochem. J.* **171**:817–820, 1978.

78

Nucleation and Growth of CPPD Crystals and Related Species in Vitro

Neil S. Mandel and Gretchen S. Mandel

Chondrocalcinosis is characterized clinically by the deposition of crystalline calcium pyrophosphate dihydrate (CPPD) in joint tissues [1,2]. Although crystals appear both in synovial fluid and in intraarticular cartilage, the extremely low pyrophosphate (PP_i) level in synovial fluids [3–6] strongly implicates the cartilage as the primary crystal growth site. Although the mechanism of calcium pyrophosphate crystal formation in articular cartilage has not been elucidated, it has been proposed that acute inflammatory episodes are the result of a process whereby CPPD crystals are shed from the cartilage into the joint space [7].

In vitro preparations of CPPD crystals necessitate a pH of 3 [8], and attempts are being made to develop a physiological model of CPPD deposition using both solution [9–11] and gel-mediated [12,13] crystallization media. Comparison of these studies with the current data indicates that collagen matrix has a role in regulating calcium and pyrophosphate ionic mobilities which would predispose to the formation of CPPD crystals.

Generation of a Gelatin Matrix Model

The experimental design is based on an ionic gradient controlled crystallization process, where the collagen matrix is derived from biological-grade gelatin, which is 80% collagen and contains all physiological ions at approximately 10–25% of their normal concentrations. The gels are prepared by pouring sequential layers of hot gel, allowing each layer to cool before another addition. The calcium- and pyrophosphate-containing gels are separated by a layer of nondoped gelatin, and formation of crystals occurs when the calcium and pyrophosphate ions migrate toward each other and interact. This system allows for the analysis of numerous parameters in the crystal formation process and the quantitation of calcium and pyrophosphate as a function of time and type of crystal formed.

Neil S. Mandel and Gretchen S. Mandel • Department of Medicine and Research Service, Veterans Administration Medical Center, Medical College of Wisconsin, Milwaukee, Wisconsin 53193.

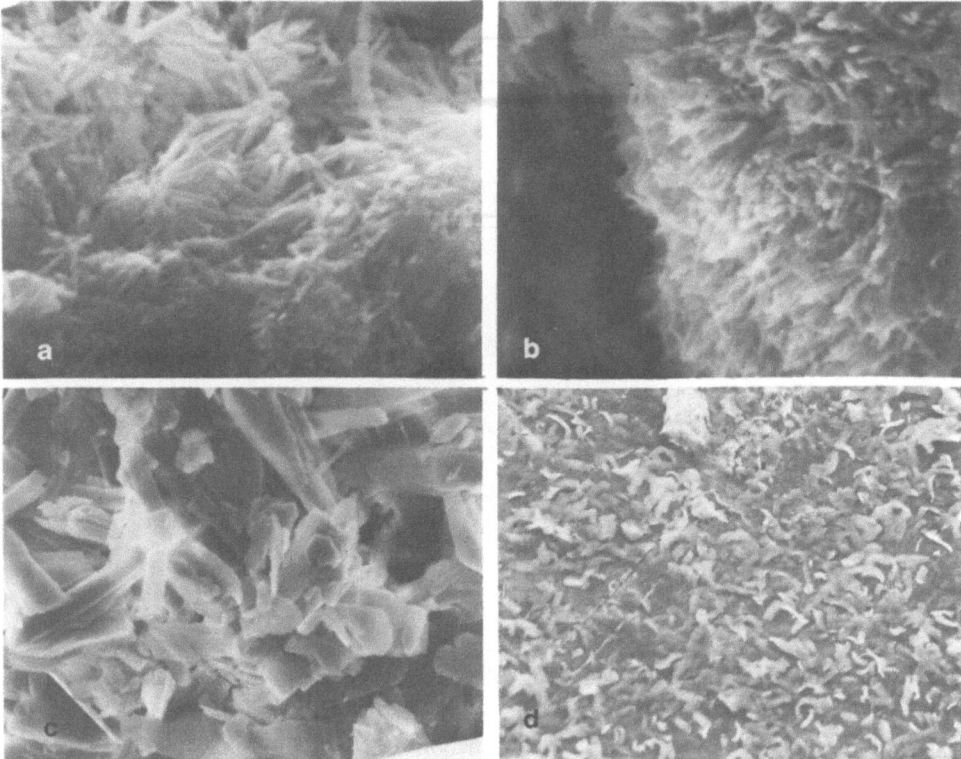

Figure 1. The different crystal morphologies are seen in the scanning electron microscope pictures of (a) triclinic calcium pyrophosphate dihydrate (t-CPPD), (b) monoclinic calcium pyrophosphate dihydrate (m-CPPD), (c) orthorhombic calcium pyrophosphate tetrahydrate (o-CPPT), and (d) amorphous calcium pyrophosphate. The Ca : P ratios are all 1 : 1, as assessed by energy-dispersive X-ray microanalysis, with t, m, and o having a standard deviation of approximately 5% and amorphous a standard deviation of approximately 30% (measured at five different locations). Crystal growth tubes were prepared at pH 7 with a 5% gelatin concentration at room temperature. The tubes were characterized at periodic intervals by gently heating the gelatin, sliding out the intact column, dissecting the gel at 1-cm sections, and separating the crystals from the geletin. The crystals were then analyzed by high-resolution X-ray powder diffraction.

In total, the collagen matrix crystal growth studies have yielded six crystalline species and an amorphous calcium pyrophosphate. Crystals of triclinic CPPD (t), monoclinic CPPD (m), orthorhombic calcium pyrophosphate tetrahydrate (o), and amorphous calcium pyrophosphate (a) are shown in Fig. 1. Only t-CPPD and m-CPPD crystals have been observed *in vivo*. In addition, three mixed calcium/sodium pyrophosphate crystals were isolated, which had varying amounts of sodium in their crystalline lattices. All six crystals are structurally diverse, based on their X-ray powder diffraction patterns. A visible precipitate with insufficient material for an X-ray diffraction pattern was termed unknown (u). Sufficient crystalline material isolated from the gel, but exhibiting no diffraction pattern was termed amorphous. The amorphous material had a sickled appearance and a calcium to phosphorus ratio of 1 : 1 with a standard error of 30%, indicative of noncrystalline material.

Isolation of Crystalline Intermediates

After 4 weeks of incubation with an initial calcium and pyrophosphate concentration of 50 mM, an intact crystal growth tube was examined (Fig. 2). The growth band labeled I is in the center of the tube, and was the site of the first interaction of calcium and PP_i, and therefore, the first crystal growth site. During the first 5 days of incubation, this band contained a significant amount of amorphous calcium pyrophosphate. This band did not change morphologically during the first 4 weeks, but after 4 weeks, the crystals were orthorhombic calcium pyrophosphate tetrahydrate. At 6 weeks (II), the milky band had dissolved, and larger crystals, which were a combination of o and m, had formed. At 14 weeks (III), the milky band had dissolved further, and the crystals were now primarily triclinic CPPD, with some o and m.

Thus, it appears that the amorphous milky band rapidly changes to crystalline o, and that o is dissolved before either m or t is formed. The formation and dissolution of o may be an important step in the formation of the localized ionic gradients which are necessary for the formation of both m and t. It is noteworthy that m-CPPD formed much earlier than the clinically predominant t-CPPD crystals. These findings might parallel calcium phosphate supersaturation solution studies [15] where amorphous calcium phosphate precedes the formation of crystalline calcium phosphate.

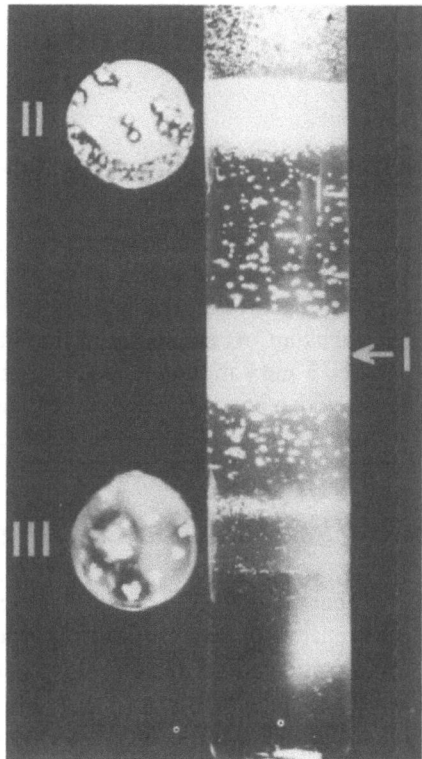

Figure 2. A crystal growth tube is shown intact after 4 weeks of incubation with the amorphous and o-CPPD band being labeled I. This band was excised from identical tubes and studied in cross-section at 6 weeks (II), and at 14 weeks (III). The dissolution of amorphous material and o-CPPD, the diffuse milky band, is followed by the formation of both m-CPPD and t-CPPD, the larger clumps of crystals visible at both 6 and 14 weeks.

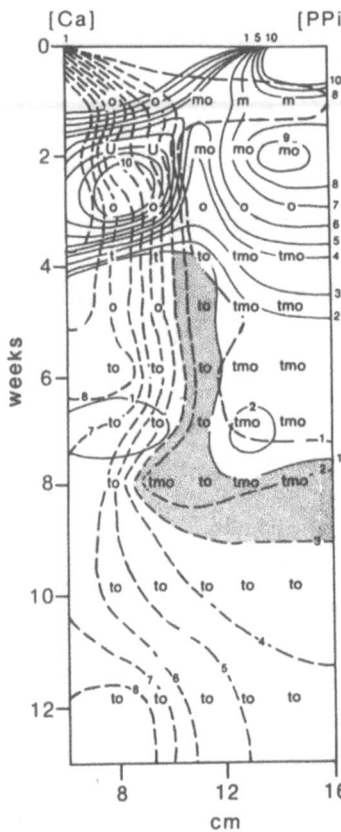

Figure 3. The kinetic data on the migration of the calcium and pyrophosphate ionic concentration gradients have been contoured as a function of time and tube position. The [Ca^{2+}] has been contoured in 1 mM intervals in dashed lines originating from the upper left of the plot, the top of the tube. The [PP$_i$] has also been contoured in 1 mM intervals but in solid lines originating from the upper right of the plot, the bottom of the tube. Crystal types have been superimposed on the plot. The stippled area represents the near-physiological region with [Ca^{2+}] between 1 and 3 mM and [PP$_i$] < 1 mM. The small numerals indicate the actual ionic millimolar concentrations. Noncrystalline [Ca^{2+}] was determined by atomic absorption spectrophotometry and [PP$_i$] by spectrophotometric assay after PP$_i$ hydrolysis [14].

The Kinetics of Crystallization

The multiparameter kinetic data can be visualized in a contour map of the calcium and PP$_i$ concentration gradients as a function of time, tube position, and type of crystal formed (Fig. 3). The crystal types have been superimposed on the kinetic map using t, m, o, a, and u nomenclature. At the center of the tube, crystals o formed at 7 days when the calcium was between 3 and 7 mM and the PP$_i$ was between 0.5 and 2 mM. Crystal o rapidly dissolved at 2 weeks with a burst of available PP$_i$ and reprecipitated at 3 to 4 weeks as t and o. At 8 weeks when the ionic gradients were closer to physiological conditions, the crystal types included t, m, and o.

Gel versus Solution Crystallization

The dependence of ionic concentrations is more easily visualized in Fig. 4, where occurrence of the various crystal types has been plotted as a function of calcium and PP$_i$ concentrations. Triclinic CPPD crystals tended to form at low to medium PP$_i$ concentrations and throughout the range of calcium concentrations. There was more t at relatively high calcium concentrations. In comparison, monoclinic CPPD formation was strongly favored at lower calcium concentrations and at somewhat higher PP$_i$ levels. There did not appear to be an absolute calcium or PP$_i$ concentration range that predisposed to o, m, or t. Crystalline o was always a precursor of both m and t and formed under most conditions, except at very high calcium or PP_i levels. The no-crystal panel (Fig. 4, lower right) shows that calcium pyrophosphate crystals were not favored at either high calcium or high PP_i levels. This finding is in agreement with solubility constant studies [16], which emphasize the competitive role of soluble calcium pyrophosphate complexes in crystal formation.

Solution crystallization studies [9–11] have been conducted with initial ionic concentrations between 1 and 2 mM calcium and between 0.01 and 100 mM PP_i. Crystal o has been found at both low and high PP_i levels. t-CPPD formed at high PP_i levels (31.6 mM) [10]; however, it did not form at low PP_i levels (< 1 mM) in the presence of physiological sodium levels. m-CPPD was not observed below 3.2 mM PP_i [10].

The effect of collagen matrix on crystal growth becomes most striking with ionic concentrations approaching physiological levels (Fig. 5). In all solution studies conducted at or near physiological conditions ($[Ca^{2+}]$ 1–2 mM, $[PP_i]$ 2–20 μM), the clinically observed t and m crystals were not grown. By contrast, in the collagen-based system, t, m, and o

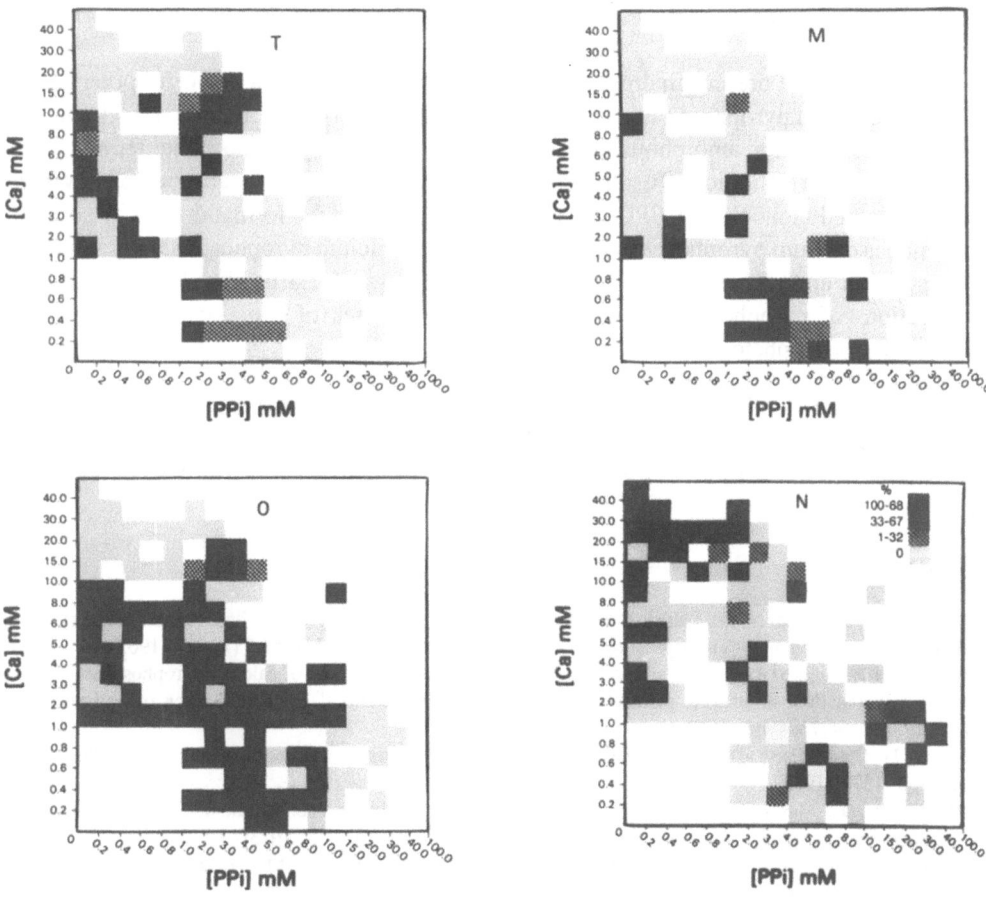

Figure 4. A composite plot of the frequency of occurrence in the gels of t-CPPD, m-CPPD, o-CPPD, and no crystals (N) as a function of the respective $[Ca^{2+}]$ and $[PP_i]$ in all sections of gel highlights the dependence of crystal type on ionic concentrations. The % occurrence scale is defined in the N panel.

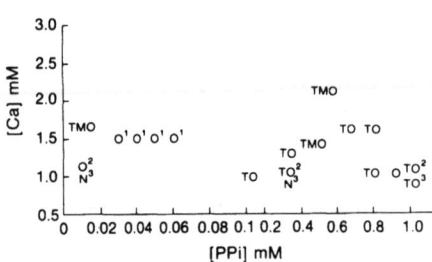

Figure 5. A comparison of the results of various solution and gel-based crystallization model studies shows the effect of collagen matrix on crystal growth. The entries superscripted are all solution studies. The clinically observed crystals t and m grew only at physiological [PP$_i$] in the collagen matrix system. Solution studies have been denoted with superscripted crystal types. Entries with superscript 1 are from Ref. 9. Those with 2 are from Ref. 10 and were conducted in the presence of low [Na$^+$]. Those with 3 are also from Ref. 10, but were derived using physiological [Na$^+$]. N indicates that no crystals formed in the presence of physiological [Na$^+$].

crystals were grown. Further, monoclinic-CPPD could be grown in the collagen-based system at PP$_i$ concentrations 300 times smaller than those required in the solution studies.

It is, therefore, clear that collagen matrix modifies the type of crystal formed at specific calcium and PP$_i$ concentrations, suggesting that the crystals nucleate on collagen molecules. Recent studies have verified the presence of hydroxyproline in the crystals grown in the first 6 days of incubation. The intracrystalline hydroxyproline concentration appears to recede as the crystallization process continues with the dissolution of o and the subsequent formation of m and then t.

In summary, the present findings reveal that collagen matrix is a powerful potentiating force in the calcium pyrophosphate crystal growth process, serving both as a crystal nucleating site, particularly for amorphous and orthorhombic calcium pyrophosphate tetrahydrate, and as a moderator of ionic diffusion. The deposition of the two crystal types observed *in vivo*, triclinic and monoclinic CPPD, was always preceded by the formation and dissolution of amorphous calcium pyrophosphate and orthorhombic calcium pyrophosphate tetrahydrate. The dissolution and recrystallization process, which allows for a localized release of high concentrations of pyrophosphate, is consistent with the lack of a major biochemical imbalance in pyrophosphate metabolism in chondrocalcinosis.

ACKNOWLEDGMENTS. This work was supported in part by grants from the Veterans Administration, the Kroc Foundation for Medical Research, the Arthritis Foundation, the National Institutes of Health (AM-30579), and the Wisconsin Chapter of the Arthritis Foundation.

References

1. McCarty, D. J.; Kohn, N. N.; Faires, J. S. The significance of calcium pyrophosphate crystals in the synovial fluid of arthritic patients—The "pseudogout syndrome." *Ann. Intern. Med.* **56:**711–737, 1962.
2. Kohn, N. N.; Hughes, R. E.; McCarty, D. J.; Faires, J. S. The significance of calcium pyrophosphate crystals in the synovial fluid of arthritic patients: The "pseudogout syndrome." II. Identification of crystals. *Ann. Intern. Med.* **56:**738–745, 1962.
3. Silcox, D. C.; McCarty, D. J. Measurement of inorganic pyrophosphate in biological fluids. *J. Clin. Inves.* **52:**1863–1870, 1973.
4. Altman, R. D.; Muniz, O. E.; Pita, J. C.; Howell, D. S. Articular chondrocalcinosis: Microanalysis of pyrophosphate in synovial fluid and plasma. *Arthritis Rheum.* **16:**171–178, 1973.
5. Russell, R. G. G.; Bisaz, S.; Fleisch, H.; Currey, H. L. F.; Rubinstein, H. M.; Dietz, A. A.; Boussina, I.; Micheli, A.; Fallet, G. Inorganic pyrophosphate in plasma, urine and synovial fluid of patients with pyrophosphate arthropathy (chondrocalcinosis or pseudogout). *Lancet* **1:**899–902, 1970.
6. Ryan, L. M.; Kozin, F.; McCarty, D. J. Quantification of human plasma inorganic pyrophosphate. *Arthritis Rheum.* **22:**886–891, 1979.
7. Bennett, R. M.; Lehr, J. R.; McCarty, D. J. Crystal shedding and acute pseudogout. *Arthritis Rheum.* **19:**93–97, 1976.

8. Brown, E. H.; Lehr, J. R.; Smith, A. W. Preparation and characterization of some calcium pyrophosphates. *Agric. Food Chem.* **11**:214–222, 1963.
9. Hearn, R. P.; Russell, R. G. G. Formation of calcium pyrophosphate crystals in vitro: Implications for calcium pyrophosphate crystal deposition disease (pseudogout). *Ann. Rheum. Dis.* **39**:222–227, 1980.
10. Cheng, P. T.; Pritzker, K. P. H.; Adams, M. E.; Nyburg, S. C.; Omar, S. A. Calcium pyrophosphate crystal formation in aqueous solutions. *J. Rheum.* **7**:609–616, 1980.
11. Cheng, P. T.; Pritzker, K. P. H. The effect of calcium and magnesium ions on calcium pyrophosphate crystal formation in aqueous solutions. *J. Rheum.* **8**:772–782, 1981.
12. Pritzker, K. P. H.; Cheng, P. T.; Adams, M. E.; Nyburg, S. C. Calcium pyrophosphate dihydrate crystal formation in model hydrogels. *J. Rheum.* **5**:469–473, 1978.
13. Pritzker, K. P. H.; Cheng, P. T.; Omar, S. A.; Nyburg, S. C. Calcium pyrophosphate crystal formation in model hydrogels. II. Hyaline articular cartilage as a gel. *J. Rheum.* **8**:451–455, 1981.
14. Fiske, C. H.; SubbaRow, Y. The colorimetric determination of phosphorus. *J. Biol. Chem.* **66**:375–403, 1925.
15. Boskey, A. L.; Posner, A. S. Formation of hydroxyapatite at low super saturation. *J. Phys. Chem.* **80**,40–45, 1976.
16. Brown, W. E.; Gregory, T. M. Calcium pyrophosphate crystal chemistry. *Arthritis Rheum.* **19**:446–463, 1976.

79

Biological Effects of Calcium-Containing Crystals on Synoviocytes

Herman S. Cheung and Daniel J. McCarty

Recently, we described a clinical condition affecting the shoulder joints termed "Milwaukee shoulder syndrome." Bilateral rotator cuff defects, glenohumeral osteoarthritis, and joint stiffness or instability were accompanied in each case by the finding of hydroxyapatite (HA) crystal clumps (microspheroids), activated collagenase, neutral protease, and particulate collagen types I, II, and III in a nearly acellular synovial fluid [1–3]. Some crystals were detected within synoviocytes and others were observed shedding into the joint space through areas denuded of synovial cells [3]. These observations focused our interest on the role of synovial cells in the pathogenesis of the devolutionary joint changes induced by calcium-containing crystals.

Stimulation of Collagenase, Neutral Protease, and Prostaglandin Secretion

In 1975, Werb and Reynolds reported that endocytosis of latex particles by cultured rabbit synovial cells stimulates secretion of collagenase and neutral protease into the surrounding medium [4]. On the basis of these findings, we proposed that HA crystals shed from synovial depots into the joint fluid might be phagocytized and stimulate synovial cell secretion of enzymes. Enzymatic tissue digestion might then release more crystals, completing a pathogenetic cycle.

To test the hypothesis, various concentrations of calcium pyrophosphate dihydrate (CPPD) and HA crystals were added to canine synovial fibroblasts [5], and the release of collagenase [6] and neutral protease [5] was determined after 7 days. Both types of crystals, which were phagocytized by synovial cells [5], stimulated collagenase and neutral protease release in a dose-related fashion (Table I). HA crystals (100 µg/ml) induced a sixfold increase in collagenase release and a sevenfold increase in neutral protease release. CPPD

Herman S. Cheung and Daniel J. McCarty • Division of Rheumatology, Department of Medicine, Medical College of Wisconsin, Milwaukee, Wisconsin 53226.

Table I. Cumulative Release of Collagenase and Neutral Protease Activities from Cultured Canine Synovial Cells after Exposure to CPPD or HA Crystals for 7 Days[a]

Addition	Concentration (μg/ml)	Collagenase (U/10^6 cells)[b]	Neutral protease (U/10^6 cells)[b]
Control	—	4 ± 1	11 ± 5
CPPD	10	10 ± 1	13 ± 2
CPPD	100	14 ± 2	25 ± 6
CPPD	500	10 ± 0.5	40 ± 6
HA	10	18 ± 3	31 ± 4
HA	50	22 ± 2	65 ± 8
HA	100		76 ± 9

[a] Various concentrations of CPPD and HA crystals were suspended in Dulbecco's modified Eagle's medium (DMEM), 10% horse serum, and added to a confluent monolayer of canine synovial fibroblasts [5]. After 24 hr, the medium was replaced by DMEM–0.2% lactoalbumin hydrolysate [7]. Medium was then replaced daily for 7 days.
[b] Values are mean ± S.E.M.; $n = 3$.

crystals (500 μg/ml) stimulated collagenase and neutral protease release by approximately fourfold.

Stimulation of collagenase synthesis and release by crystals appeared to be quite specific. Collagenase secretion from lapine chondrocytes in response to increasing doses of HA crystals was stimulated more than 10-fold, while synthesis of other secreted proteins increased by only 2-fold [8]. The recent report of the induction of mRNA translating for rabbit synovial collagenase in a cell-free RNA-dependent system after stimulation of rabbit synovial fibroblasts with monosodium urate crystals also suggests that such stimulated cells synthesize collagenase out of proportion to other macromolecular species [9].

Using identical experimental conditions, the effect of calcium-containing crystals on prostaglandin E_2 (PGE_2) production was also examined in canine synovial cells (Table II). CPPD (500 μg/ml) and HA (100 μg/ml) crystals increased PGE_2 levels by 18-fold and over 100-fold, respectively. Moreover, increased PGE_2 production correlated with increasing crystal dose, and the cyclooxygenase inhibitor indomethacin (10^{-6} M) abolished the stim-

Table II. Prostaglandin E_2 Production by Canine Synovial Cells Exposed to CPPD and HA Crystals for 24 hr

Addition	Concentration (μg/ml)	Prostaglandin E_2 (pg/mg cell protein)[a,b]
Control	—	1,251 ± 85
CPPD	10	1,058 ± 69
CPPD	100	10,623 ± 771
CPPD	500	20,893 ± 989
CPPD + indomethacin (10^{-6} M)	500	0
HA	10	20,455 ± 2011
HA	50	84,556 ± 9125
HA	100	127,831 ± 8333
HA + indomethacin (10^{-6} M)	100	117 ± 29

[a] Values are mean ± S.E.; $n = 3$.
[b] PGE_2 was measured by radioimmunoassay [10]. From 0.5 to 3.0 ml of conditioned medium was extracted for each determination to provide the desired sensitivity.

ulatory effect. However, Dayer et al. [11] have reported that indomethacin blocks PGE_2 production, but stimulates collagenase synthesis, indicating that PGE_2 does not mediate collagenase production. The more potent action of HA crystals as an inducer of proteolytic enzyme and PGE_2 release relative to CPPD crystals may be due to the smaller size of the HA crystals, in which the number of particles per unit mass and the surface area are orders of magnitude smaller.

Mitogenic Properties of Calcium-Containing Crystals

In the experiments relating crystals to enzyme secretion by synovial cells, we also noted that those cultures exposed to calcium-containing crystals consistently contained more cells than did control cultures, suggesting that these particles were mitogenic. Recently, we found that synthetic HA crystals (50–100 μg/ml) in 1% serum stimulated [^{3}H]thymidine incorporation into quiescent canine synovial fibroblasts and human foreskin fibroblast cultures to the same extent as did 10% serum. Both the onset and the peak incorporation of thymidine after crystal addition were delayed by 2 to 3 hr, as compared to the effect of 10% serum.

Stimulation of [^{3}H]thymidine uptake by canine synovial cells exposed to various concentrations of CPPD, HA, calcified material isolated from a patient with calcinosis, and synthetic calcium urate crystals is shown in Table III. HA crystals (10 μg/ml) stimulated incorporation by approximately twofold; further stimulation occurred as the dose of crystals was increased up to 100 μg/ml. CPPD crystals (10–20 μm) also elicited a dose–response relationship. The mixture of naturally occurring basic calcium phosphate crystals [octacalcium phosphate and carbonate-substituted apatite (12)] isolated from a patient with calcinosis was as mitogenic as pure synthetic HA (Table III). Calcium urate likewise markedly stimulated nuclide uptake (Table III). However, not all particulates were effective mitogenic agents. Latex beads and diamond crystals had no effect, and monosodium urate modestly

Table III. Incorporation of [^{3}H]Thymidine by Canine Synovial Cells Exposed to Various Concentrations of CPPD, HA, Calcified Material from Patients with Calcinosis (C), and Calcium Urate Crystals (CaU)

Addition	Concentration (μg/ml)	[^{3}H]Thymidine incorporation (% of control)[a]
Control	—	100 ± 11
CPPD	50	210 ± 18
CPPD	100	260 ± 15
CPPD	200	320 ± 14
HA	10	188 ± 20
HA	50	390 ± 22
HA	100	417 ± 24
C	50	260 ± 17
C	100	344 ± 15
CaU	50	318 ± 29
CaU	100	415 ± 21
CaU	200	520 ± 16

[a] $n = 4$ for all cases.

stimulated [^{3}H]thymidine incorporation only at certain critical concentrations (data not shown). The range of concentrations of synthetic and natural calcified crystal is of the same order of magnitude as the crystal concentrations found in joint fluid of patients with the Milwaukee shoulder syndrome, i.e., 10 to 25 μg [2].

Stimulated [^{3}H]thymidine uptake into mouse 3T3 cells in culture by precipitates of calcium and inorganic pyrophosphate (PP_i) has been described [13,14]. Orthophosphate could substitute for PP_i at higher concentrations. Stimulation of thymidine uptake was partially blocked by increasing magnesium concentrations; strontium or barium could substitute for calcium, although they manifested weaker stimulation. On the basis of these findings, divalent cation was designated as the critical factor in promoting thymidine uptake [13,14].

It seems likely that the mitogenic effects of the calcium-containing crystals on cultured synovial cells described here are due to their ability to confer "competence" to stimulate DNA synthesis [15]. If HA or CPPD crystals are indeed "competence"-promoting factors, they may be responsible for synovial cell proliferation in the crystal deposition disease, since the plasma factors needed for "progression" are present in synovial fluid [16,17], which is largely a plasma dialysate [18].

In Vitro Dissolution of HA Crystals

To confirm the suggestion that *in vivo* dissolution of CPPD crystals takes place within synovial cells [19], the mechanism by which HA crystals are degraded by synovial cells *in vitro* was examined [20]. At 21 hr of incubation, there were greater amounts of ^{45}Ca solubilization in cultures with synovial cells than with the cell-free culture media or conditioned media (Table IV). Since there was no difference in the amount of free ^{45}Ca in the conditioned media and the cell-free control, the solubilization of crystals was not due to secretion of unknown factors but required physical contact with cells. As time-lapse photography, scanning and transmission electron microscopy all showed crystal endocytosis, we speculate that intracellular dissolution of crystals raises intracellular free calcium and renders the cells competent to synthesize DNA [21].

In summary, calcium-containing crystals such as HA, CPPD, and calcium urate have several biological effects on synovial cells including enzyme production and secretion, PGE_2 synthesis, and cell division. The production of collagenase and PGE_2 by synovial cells may relate to devolutionary changes in arthritic joints. The mechanism by which calcium crystals

Table IV. Dissolution of ^{45}Ca-Labeled HA Crystals by Cultured Canine Synovial Fibroblasts

	Time (hr)	Culture conditions	^{45}Ca HA solubilized[a,b]
A	0	7.5% serum, no cells	0.25 ± 0.01%
B	21	7.5% serum, no cells	0.48 ± 0.02%
C	21	7.5% serum, DSC	1.17 ± 0.07%
D	21	Conditioned media	0.51 ± 0.06%

[a]Values are mean ± S.E.M.; $n = 6$.
[b]^{45}Ca-labeled HA crystals were added to confluent monolayers of cultured canine synovial cells, cell-free control media, and conditioned media from the same synovial cell cultures. At time 0 and 21 hr after crystal addition, media were removed, centrifuged at 15,000g for 1 min, and a sample of media was monitored for free ^{45}Ca.

induce cell division or enzyme production and secretion is unknown but may involve changes in the intracellular calcium concentration. Synovial cells are also capable of degrading calcium-containing crystals although the mechanism of HA and CPPD crystal solubilization is not known. Demonstration of both types of crystals inside synovial cell phagolysosomes suggests that they might be solubilized by the acidic conditions found in the phagolysosomes [3,5]. These data thus support the proposition that synovial cells play a pivotal role in the pathogenesis of the joint changes induced by calcium-containing crystals.

ACKNOWLEDGMENTS. Supported by USPHS Grant AM-26062, a grant from the Kroc Foundation, and NIH Research Career Development Award AM-01065 (H.S.C.).

References

1. McCarty, D. J.; Halverson, P. B.; Carrera, G. F.; Brewer, B. J.; Kozin, F. "Milwaukee shoulder": Association of microopheroids containing hydroxyapatite crystals, active collagenase and neutral protease with rotator cuff defects. I. Clinical aspects. *Arthritis Rheum.* **24**:464–473, 1981.
2. Halverson, P. B.; Cheung, H. S.; McCarty, D. J.; Garancis, J. C.; Mandel, N. "Milwaukee shoulder": Association of microspheroids containing hydroxyapatite crystals, active collagenase and neutral protease with rotator cuff defects. II. Synovial fluid studies. *Arthritis Rheum.* **24**:474–483, 1981.
3. Garancis, J. C.; Cheung, H. S.; Halverson, P. B.; McCarty, D. J. "Milwaukee shoulder": Association of microspheroids containing hydroxyapatite, active collagenase and neutral protease with rotator cuff defects. III. Morphologic and biochemical studies of an excised synovium showing chondromatosis. *Arthritis Rheum.* **24**:484–491, 1981.
4. Werb, Z.; Reynolds, J. J. Stimulation of endocytosis of the secretion of collagenase and neutral proteinase from rabbit synovial fibroblasts. *J. Exp. Med.* **140**:1482–1497, 1974.
5. Cheung, H. S.; Halverson, P. B.; McCarty, D. J. Release of collagenase, neutral protease and prostaglandins from cultured mammalian synovial cells by hydroxyapatite and calcium pyrophosphate dihydrate crystals. *Arthritis Rheum.* **24**:1338–1344, 1981.
6. Evanson, M; Jeffrey, J. J.; Krane, S. M. Studies on collagenase from rheumatoid synovium in tissue culture. *J. Clin. Invest.* **47**:2639–2651, 1968.
7. Werb, Z.; Mainardi, C. L.; Vater, C. A.; Harris, E. D., Jr. Endogenous activation of collagenase by rheumatoid synovial cells: Evidence for a role of plasminogen activator. *N. Engl. J. Med.* **296**:1017–1023, 1977.
8. Cheung, H. S.; Halverson, P. B.; McCarty, D. J. Phagocytosis of hydroxyapatite or calcium pyrophosphate dihydrate crystals by rabbit articular chondrocytes release of collagenase, neutral protease and prostaglandin E_2 and F_2. *Proc. Soc. Exp. Biol. Med.* **173**:181–189, 1983.
9. Brinckerhoff, C. E.; Gross, R. H.; Nagase, H.; Sheldon, L.; Jackson, R. C.; Harris, E. D. Increased level of translatable collagenase in RNA in rabbit synovial fibroblasts treated with phorbol myristate acetate or crystals of monosodium urate monohydrate. *Biochemistry* **21**:2674–2679, 1982.
10. Dunn, M. J.; Laird, J. F.; Dray, F. Basal and stimulated rates of renal secretion and excretion of prostaglandins F_2, $F_{2\alpha}$ and 13,14-dihydro 15 beta F_2 in the dog. *Kidney Int.* **13**:136–143, 1978.
11. Dayer, J. M.; Krane, S. M.; Russell, R. G. G.; Robinson, D. R. Production of collagenase and prostaglandins by isolated adherent rheumatoid synovial cells. *Proc. Natl. Acad. Sci. USA* **73**:945–949, 1976.
12. McCarty, D. J.; Lehr, J. R.; Halverson, P. B. Crystal populations in human synovial fluid: Identification of apatite, octacalcium phosphate and beta tricalcium phosphate. *Arthritis Rheum.* **26**: 1220–1224, 1983.
13. Rubin, H.; Sanui, H. Complexes of inorganic pyrophosphate orthophosphate and calcium as stimulants of 3T3 cells multiplication. *Proc. Natl. Acad. Sci. USA* **74**:5026–5030, 1977.
14. Barnes, D. W.; Colowick, S. P. Stimulation of sugar uptake and thymidine incorporation in mouse 3T3 cells by calcium phosphate and other cellular particles. *Proc. Natl. Acad. Sci. USA* **74**:5593–5597. 1977.
15. Scher, D. C.; Shepard, R. C.; Antoniades, H. H.; Stiles, C. D. Platelet derived growth factor and the regulation of the mammalian fibroblast cell cycle. *Biochim. Biophys. Acta* **560**:217–241, 1979.
16. Coates, C. L.; Burwell, K. G.; Buttery, P. J.; Walker, C. T.; Woodward, P. M. Somatomedin activity in synovial fluid. *Ann. Rheum. Dis.* **36**:50–55, 1977.
17. Coates, C. L.; Burwell, T. G.; Lloyd-Jones, K.; Swannell, A. J.; Walker, C. T.; Shelby, C. Somatomedin activity in synovial fluid from patients with joint diseases. *Ann. Rheum. Dis.* **37**:303–314, 1978.

18. Ropes, M. W.; Bauer, W. *Synovial Fluid Changes in Joint Diseases*, Cambridge, Mass., Harvard University Press, 1953, p. 92.
19. McCarty, D. J.; Palmer, D. W.; James, C. Clearance of calcium pyrophosphate dihydrate (CPPD) crystals in vivo. II. Studies using triclinic crystals doubly labeled with ^{45}Ca and ^{85}Sr. *Arthritis Rheum.* **22:**1122–1131, 1979.
20. Evans, R. W.; Cheung, H. S.; McCarty, D. J. Cultured canine synovial cells solubilized ^{45}Ca labeled hydroxyapatite crystals. *Arthritis Rheum.* in press.
21. Nykyforiak, C. J.; Young, R. B.; Phillips, T. A. Changes in intracellular Ca^{2+} distribution during the transition from the proliferating to the stationary state. *Biochem. Biophys. Res. Commun.* **93:**585–587, 1980.

80

Rotator Cuff Tendinopathies with Calcifications

Kiriti Sarkar and Hans K. Uhthoff

Rotator cuff tendinopathies are the commonest cause of painful shoulder syndromes in persons of middle age. In primary conditions, the disease process begins in the tendon, while the pathologic changes in secondary tendinopathies are either an extension of diseases from contiguous structures or a manifestation of a systemic disease. Not all of the tendinopathies are necessarily associated with calcification. In Table I, we have classified rotator cuff tendinopathies indicating the mechanisms which induce them, their sites of predilection, and associations with calcification. Calcification is the event that characterizes the condition in calcifying tendinitis. Among the degenerative forms of primary tendinopathies, one of the cardinal features of enthesopathy is deposition of calcium at the area of tendon attachment to the bone that may eventually become bony excrescences or enthesophytes [1]. On the other hand, the reported incidence of calcification in rotator cuff tear varies from less than 10% to 23% [2].

A significant degree of calcification is one of the principal components of Milwaukee shoulder syndrome (see Chapter 79). The phagocytic action of synovial cells, brought about by the generation of calcium phosphate crystals, induces the release of collagenase and neutral protease which attack periarticular tissues including tendon, causing rotator cuff loss [3]. The other instance of secondary rotator cuff tendinopathy, due to spread of systemic diseases such as gout, may sometimes manifest calcification. In the present account, discussion will be limited to elucidating differences in the morphological characteristics of the calcification process in reactive and degenerative forms of primary rotator cuff tendinopathies.

Self-Healing Tendinopathy and Reactive Calcification

The site of calcification in calcifying tendinitis is more than 1.0 cm away from the bony insertion of the supraspinatus tendon. The calcification site corresponds to an area of pre-

---

Kiriti Sarkar • Department of Pathology, School of Medicine, Faculty of Health Sciences, University of Ottawa, Ottawa, Ontario K1H 8M5, Canada.   *Hans K. Uhthoff* • Department of Surgery, School of Medicine, Faculty of Health Sciences, University of Ottawa, and Division of Orthopaedics, Ottawa General Hospital, Ottawa, Ontario K1H 8M5, Canada.

Table I. Rotator Cuff Tendinopathies and Calcification

Type	Mechanism	Condition	Site	Calcification
Primary	Traumatic	Acute avulsion	Insertion	−
	Reactive	Calcifying tendinitis	1.5–2 cm from insertion	+
	Degenerative	Enthesopathy	Insertion	+
		Rotator cuff tear	At or near insertion	±
Secondary	Contiguous structures	Impingement syndrome	Midtendon (bursal aspect)	−
		Milwaukee shoulder	Insertion (articular aspect)	+
	Systemic diseases	Gout, etc.	Not area specific	±

carious blood supply to the tendon and is known as the "critical zone" [4]. Through morphological examinations of biopsy tissues obtained during surgery, we have been able to reconstruct the events that occur during the evolution of calcifying tendinitis, which shows a remarkable tendency to heal spontaneously. The events seem to occur in three distinctive stages. In the precalcific stage, the tendon undergoes fibrocartilaginous transformation at multiple foci within the "critical zone." The transformed areas contain typical chondrocytes

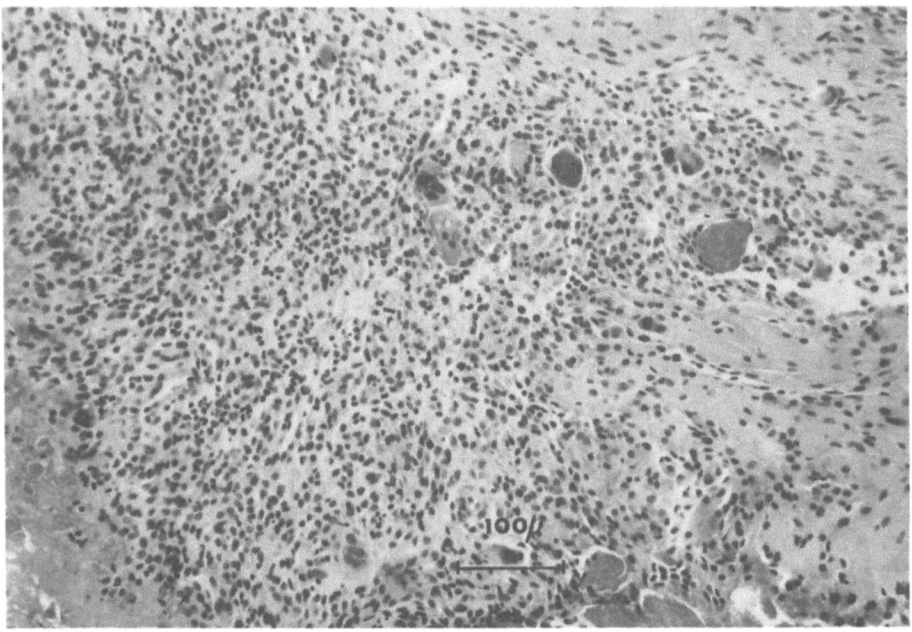

Figure 1. The area of calcium resorption shows intense cellular infiltration, consisting mainly of macrophages and several multinucleated giant cells. Only a few polymorphonuclear leukocytes are present. H & E.

and the matrix shows metachromasia. The ensuing calcific stage can be divided into formative and resorptive phases [5].

In the formative phase, rectangular crystalline structures representing apatites are deposited initially within membrane-bound matrix vesicles [6]. The chondrocytes that appear responsible for the generation of the matrix vesicles show the ultrastructural features of ruffled plasma membrane, intracytoplasmic glycogen, and well-developed rough endoplasmic reticulum, but no pericellular matrix. Matrix vesicles filled with crystals lose their membranes and coalesce to form large nodular deposits. There is hardly any cellular infiltrate in the surrounding tissue and vascular channels are sparse. The chondrocytes in the later period of the formative phase have a broad pericellular matrix and show all of the features described above except containing glycogen. Vacuolations of various sizes are common in the cytoplasm and pericellular lacunae are present in many of these cells.

The clinical onset of acute pain signals the initiation and maintenance of the resorptive phase of the calcific stage [7]. Both macrophages and multinucleated giant cells actively participate in phagocytizing calcium, and the proliferation of these cells continues as more and more of the calcium is resorbed (Fig. 1). Polymorphonuclear leukocytes constitute no more than 2–3% of the total number of cells, and the leukocyte population is even less remarkable in the subacromial bursa when calcium is detected within that structure, secondary to calcifying tendinitis. Ultrastructurally, the phagocytic cells are seen to engulf large portions of the calcific deposits. Once within the phagocytic vacuoles of the cytoplasm, the crystalline structure of the deposits seems to disintegrate. The postcalcific stage is characterized by the reconstitution of the tendon when the removal of calcium is complete.

Degenerative Tendinopathies and Dystrophic Calcification

In degenerative tendinopathies, the structural alterations have no propensity to heal without sequelae, as the lesions may spontaneously do in calcifying tendinitis. For example, the majority of patients with rotator cuff tear have satisfactory functional recovery with conservative management, while the defect in the rotator cuff persists and osteoarthritic changes may develop in the joint. When calcification occurs in degenerative rotator cuff tendinopathies, evidence of resorption is usually negligible.

Because of a lack of surgically obtainable tissues, we studied the area of supraspinatus tendon insertion in postmortem cases of enthesopathy; the age of these cases ranged from the second to the ninth decade. At the area of tendon insertion, the tidemark appears as a blue line in routine H & E stain, demarcating the calcifying from the noncalcifying part of the fibrocartilage consisting of Sharpey's fibers. A broadening or reduplication of the tidemark begins to appear as early as the fourth decade, indicative of excess calcification in this area. Localized calcific excrescences on the tidemark, very commonly found in later decades of life, may be accompanied by small calcareous deposits in close proximity (Fig. 2).

Although the reported incidence of calcification in rotator cuff tear varies widely, the presence of calcific deposits has great prognostic significance because it adversely affects satisfactory functional recovery. Radiologically visible calcification as well as microscopic calcification preclude a favorable outcome [8]. It is uncertain whether all tears begin as incomplete ruptures which eventually develop into full-thickness tears. When postmorten specimens of the supraspinatus tendon are examined, small ruptures at the articular aspect of the tendon insertion are not unusual in older individuals. Torn fragments of tendinous tissue with villous configuration often show signs of repair. However, we have never found any evidence of calcification in these small, incomplete tears.

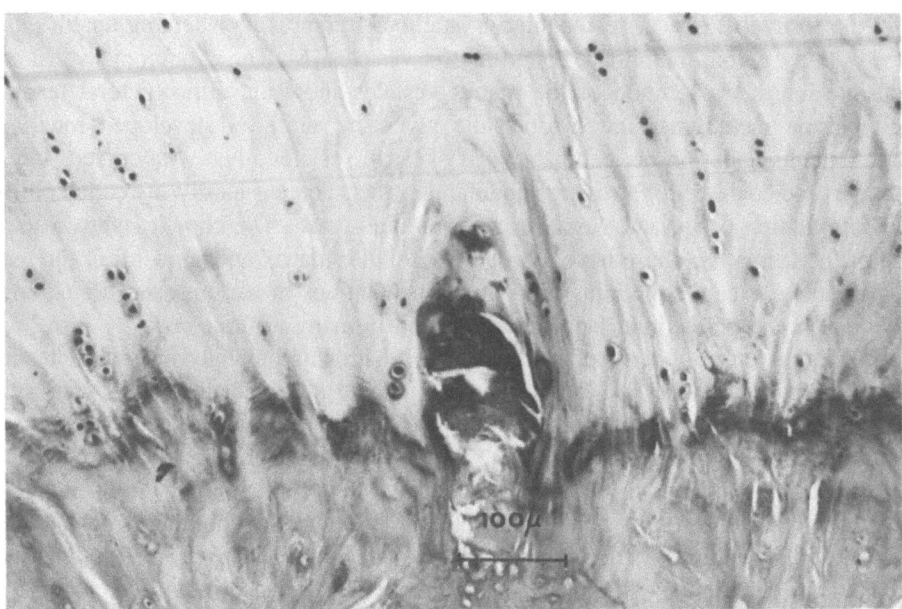

Figure 2. The tidemark at the bony insertion of supraspinatus tendon from a 50-year-old male shows calcific excrescence. H & E.

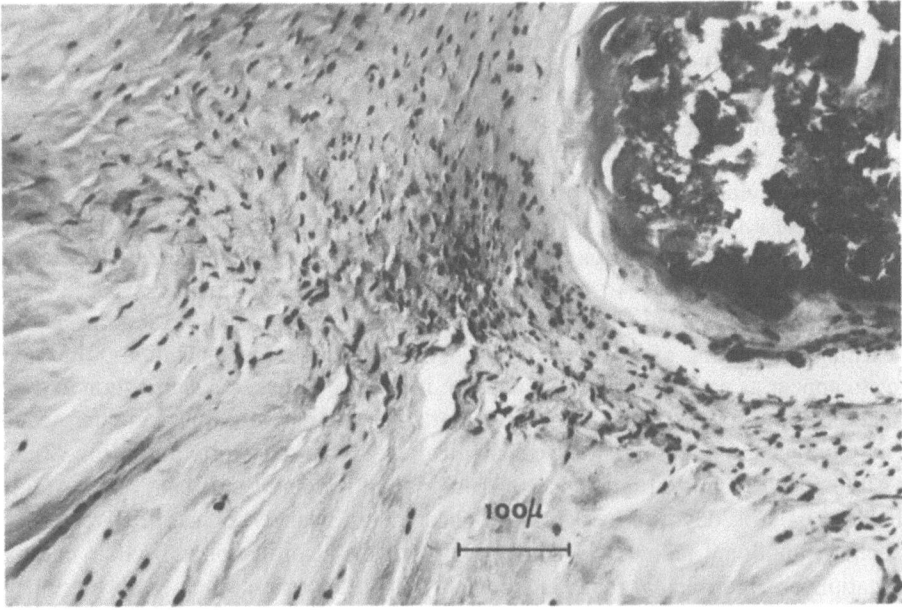

Figure 3. A large calcific deposit in a rotator cuff tear is encroaching on Sharpey's fibers. The adjacent area of the deposit shows numerous mesenchymal cells. H & E.

Complete rotator cuff tear tends to occur across the junctional area where tendon fiber proper merges with Sharpey's fibers. When calcification occurs in rotator cuff tear, the site is somewhat distant from the tidemark, although with continued expansion the lobulated deposits may extend toward the tidemark. The effect of calcification on the tendon varies according to the size of the deposit. Histologically identifiable microscopic deposits hardly evoke a tissue reaction. On the other hand, large lobulated deposits not only cause fraying of the adjacent tissue, but may elicit a marked cellular reaction, involving what appear to be undifferentiated mesenchymal cells (Fig. 3). Seldom do we see any attempt to resorb the calcium in rotator cuff tear, regardless of the intensity of the cellular reaction surrounding the calcification.

In conclusion, it should be emphasized that it is of great importance to determine the localization of calcification in rotator cuff tendinopathies, because the site not only suggests pathogenesis but indicates prognosis as well.

References

1. Niepel, G. A.; Sit'aj, S. Enthesopathy. *Clin. Rheum. Dis.* **5**:857–872, 1979.
2. Wolfgang, G. L. Surgical repair of tears of the rotator cuff of the shoulder: Factors influencing the result. *J. Bone Jt. Surg. Am. Vol.* **56**:14–26, 1974.
3. McCarty, D. J.; Halverson, P. B.; Carrera, G. F.; Brewer, B. J.; Kozin, F. "Milwaukee shoulder": Association of microspheroids containing hydroxyapatite crystals, active collagenase, and neutral protease with rotator cuff defects. I. Clinical aspects. *Arthritis Rheum.* **24**:464–473, 1981.
4. Moseley, H. F.; Goldie, I. The arterial pattern of the rotator cuff of the shoulder. *J. Bone Jt. Surg. Br. Vol.* **45**:780–789, 1963.
5. Uhthoff, H. K.; Sarkar, K.; Maynard, J. A. Calcifying tendinitis: A new concept of its pathogenesis. *Clin. Orthop.* **118**:164–168, 1976.
6. Sarkar, K.; Uhthoff, H. K. Ultrastructural localization of calcium in calcifying tendinitis. *Arch. Pathol. Lab. Med.* **102**:266–269, 1978.
7. Uhthoff, H. K.; Sarkar, K. Calcifying tendinitis: Its pathogenetic mechanism and a rationale for its treatment. *Int. Orthop.* **2**:187–193, 1978.
8. Earnshaw, P.; Desjardins, D.; Sarkar, K.; Uhthoff, H. K. Rotator cuff tears: The role of surgery. *Can. J. Surg.* **25**:60–63, 1982.

Index

Acetylcholine, 61, 67, 73, 74, 411, 434, 436, 437, 438, 455
  release, 110–117
Actin, 137, 275, 279, 280
Actinomycin, 276
Action potentials, calcium-dependent, 227–233, 464
Actomyosin, 143, 166, 167, 363
Adenosine, 110, 111, 165
  diphosphate, 46, 159, 167, 168, 169, 381
Adenylate cyclase, 114–115, 413
  regulation by calmodulin, 10, 283–288
Adrenergic
  agonists, 442, 448
  antagonists, 257, 259, 567
Aequorin, 356, 362, 370, 371, 374
Alcohols, 201 (*see also* Ethanol)
Alkaline phosphatase, 515, 603, 706, 707
Amiloride, 138, 174, 175
Aminopyridine, 228
Amrinone, 442, 444
Anesthesia, theory of, 7, 197, 198
Angina, 366, 418, 433, 441, 467, 471, 472, 473, 475, 480, 485
Angiotensin, 38, 411, 480, 557
Anionic sites, 8, 27–28
Anticalcium therapy, 686–693
Antigen, cellular activation by, 130, 173, 175
Antimycin, 151
Antipyrylazo, 339–348
Apamin, 202–204
Apatite, 607, 610, 582, 699, 700
  hydroxy, 529, 530, 601, 604, 617, 623, 634, 636, 639, 645–650, 665, 682, 686, 702, 719–723
Arachidonic acid
  actions of, 37, 46, 140–141
  metabolism of, 10, 41–42, 45, 50–51, 129, 140–141, 147, 155, 168, 533

Arrhythmias, 366, 441, 459–468
Arsenazo, 29–30, 326, 339, 427, 429
Artery
  aorta, 411, 418, 441
  ear, 393, 427
  caudal, 400, 423, 424
  coronary, 431–439, 471–476, 692
  mesenteric, 395, 412, 418
Arthritis, 699–703
Arthropathy, 699–703
Aspirin, 47, 150, 157, 158, 161
Atherogenic diet, 557, 688, 691, 693
Atherosclerosis, 475, 553, 557, 634, 685–693
Atomic absorption spectrometry, 84, 583
ATPase Na-K$^+$, 173–177, 265
Azide, 405

Barbiturates, effects of, 215, 221, 227–233, 325
Barium, 5, 105, 112, 324, 325, 326, 446, 466, 467, 678
Basophils, 129–136
Batrachitoxin, 454
Benzodiazepine(s), 193, 196, 217, 260
  antagonist, 197
Bioprosthesis, 633, 661–667
Blood pumps, implantable, 654–658
Bone, 7–8
  alveolar, disorders of, 591–594
  density, 570, 572, 573, 591–594
  Gla protein, 635, 636, 640, 648 (*see also* Carboxyglutamic acid)
  loss (resorption), 494, 533, 541, 544, 569–574, 575–580, 679
Brushite (dicalcium phosphate dihydrate), 646, 699

C-kinase (*see* Protein kinase, phospholipid-dependent)
Cadmium, 183, 229, 257, 325, 335

731

Caffeine, 115, 332, 356, 400, 408, 409
Calcidiol (see Vitamin D)
Calcification, 493, 599–604, 607–614, 640–641, 648, 649–650, 653, 656, 658, 661–667, 669–674, 682, 693, 725–729
  arterial, 692
  dystrophic, 653, 674, 677–683, 727–729
  intracavitary, 633–641
Calcineurin, 307–313
Calcinosis, 636, 648, 649
Calcitonin, 533, 537, 555, 692
Calcitriol (see Vitamin D_3)
Calcium
  as coupling factor, 13, 23
  as mercurial messenger, 15–18
  as minatory messenger, 18–20
  as synarchic messenger, 14
  absorption, 25, 492, 513–517, 522
  antagonists, 366, 411–419, 417, 419, 423–429, 431–439, 448, 459–468, 544
  ATPase, 7, 8, 27, 63, 94, 139, 167, 198, 216, 217, 260, 265–272, 276, 277, 313, 363, 377–386, 424, 514, 515, 552, 553, 603
  balance, 561–567, 583–588
  binding, 8, 27, 139, 143, 203, 215, 216, 237, 256, 405
    complex of rat intestinal mucosa (CaBC), 514, 515
  binding protein, 10, 24, 25, 30, 198, 501, 514, 519–523, 635–636, 653
  buffer, 24–27
  –Ca^{2+} exchange, 93, 379, 382, 405, 413, 453
  and cardiac ultrastructure, 362, 365
  and cell death, 8, 553, 558
  channels, 24, 28, 90, 112, 115, 321–328, 377, 453, 459
  compartmentalization, 400–408
  current, 322
  dietary, 552, 561–567, 569, 591–594
  deficiency, 504, 515, 526, 593
  deprivation, 83, 96, 115, 159, 162, 195, 365, 428, 436, 437, 453, 454, 459, 460
  efflux, 93, 139, 216, 248, 323, 327, 377–385, 408, 416, 453, 553
  electrochemical gradient of, 24, 30
  entry blockers (see individual compounds and Channels (ionic), blockers)
  entry stimulators, 441–448
  extracellular, 18, 83, 112, 391, 392, 396, 411, 413, 417, 419, 434, 438, 444
  extrusion, 27
  fecal, 586, 587
  and fertilization, 8
  flux, 173, 193, 215, 377–386, 416
  gating mechanism, 377–386
  homeostasis, 413, 494, 502, 555, 558, 578, 666
  hydrogen exchange, 266
  independent events, 40, 157
  induced Ca^{2+} release, 369–374

Calcium (cont.)
  intake, 569–574, 571, 576, 583–588
  and intercellular communication, 8
  intoxication, 552–558
  intracellular, 7, 37, 84, 129, 163, 413
  medicated potassium conductance, 8, 181–191, 193–198, 465
  membrane, 7, 58, 85
  and membrane permeability, 6, 207–210
  mobilization, 48, 57, 58, 61, 75, 76, 84, 142, 147, 504
  and myocardial metabolism, 361–362
  on membrane permeability, 5–6, 125, 181–211
  phosphate crystals, 602, 607, 609, 610, 629, 653, 656, 702
  phosphate ratio, 492, 578, 585–586, 587, 591, 674, 712
  phospholipid-phosphate complexes, 646, 647, 648, 649–650
  protein interactions, 10, 16, 579, 588
  pump, 166, 340, 346, 363, 378–386, 413, 415, 424, 429, 453
  pyrophosphate dihydrate, 645, 699–703, 705–706, 708, 711–716, 719–723
  release, 8, 15, 53, 85, 93, 143, 156, 334–336, 351–357, 392, 396, 400, 408–409, 417, 423, 425, 428, 436, 453, 455, 542, 626, 627
  transport, 24, 25, 26, 198, 221–225, 399–409, 414, 423, 424–431, 492, 495, 498, 514, 517, 522, 562, 604, 625–629, 647, 693, 707
  trigger, 427
  uptake, 84, 90, 93, 101–107, 112, 139, 156, 250, 362, 377–386, 400, 677, 679
  urinary, 409, 502, 562, 579, 580, 583, 586, 588, 627
Caldesmon, binding to F-actin, 275–276
Calmodulin, 10, 16–17, 21, 24, 30, 88, 106, 139, 363, 413, 313, 537
  binding to
    calspectin, 277–280
    F-actin and tubulin, 275–281
    tau factor, 280
  and brain, 283–288
  dependent kinase(s), 10, 166, 276, 291–302
  effect on GTPase activity, 286–288
  inhibitors of, 255–262
  interaction with
    ATPase, 269
    guanyl nucleotides, 283–288
  modulation of adenylate cyclase, 283–288
  -phenothiazine complex, 255–262
  -sensitive phosphodiesterase, 258–262
  stimulation of phosphatase, 307–313
Calsequestrin, 355
Calspectin, 277–280
Captopril, 479, 480
Carboxyglutamic acid (Gla), 525–530, 604, 649, 661, 662, 666, 674, 686
Cardiac assist devices, 633, 653–658

Index

Cardiac glycosides, 7, 435, 438, 439, 466
Cartilage, 599, 603, 617–623, 629, 634, 647, 701, 705–708
Catecholamine
    effects of, 366, 392, 395, 400, 408, 411, 413, 417–418, 423, 424, 462, 553
    secretion, 101–107, 546
Cesium, 229, 233, 324, 326
CGP 28392, 441–448
Channels (ionic)
    blockers, 229, 238, 239, 325, 326, 393, 396, 413, 417, 441, 453–456, 471–476, 479–485, 567
    calcium, 24, 104–105, 110, 112, 115, 223, 321–328, 456, 567
    fast, 223
    inactivation of, 323
    porin, 121
    potassium, 181–191, 242, 321
    receptor-operated, 61, 104–105, 143, 396, 413, 417, 419, 425–427, 454, 455, 480, 552, 565, 566
    slow, 222, 455, 466, 480, 552, 565, 566
    sodium, 84–105, 321, 454, 466, 468
    spare, 110, 114
    stretch-operated, 396
    voltage-dependent, 95, 105, 181–191, 321, 396, 426, 435, 454, 455, 456, 552
Chelating agents, 84
    EDTA, 124, 266, 612, 618, 639
    EGTA, 54–55, 56, 62, 64, 69, 174, 245, 256, 284, 285, 321, 322, 333, 371, 400, 403, 414
Chemotactic factor(s), 137, 701
Chloride, 30, 83, 196, 197, 221, 354, 427
    induced Ca^{2+} release, 331–334
2-chloroadenosine, 111–117
Chlorotetracycline, 84
Chlorpromazine, 223, 259, 417, 419
Cholesterol, 399, 414, 646, 686, 688, 690, 692, 693
Choline, 333
Chondrocalcinosis, 699, 705–708, 716
Chondrocytes, 599–600, 707, 726–727
Chromaffin cells, 101–108
Cinnarizine, 411, 413
Cobalt, 257, 325, 544
Collagen, 48, 160–161, 162, 534, 541, 604, 607–614, 634, 635, 661, 672, 682, 688, 690, 692, 702
    matrix, 530, 711, 715, 716
Collagenase, 701, 702, 719–721, 722
Concanavalin, 38, 131
Congestive heart failure, 476, 479–485
Corticosteroids, 581, 700
CPPD crystals, 699–703
Crystal deposition, 607–614, 645, 699–703, 719–723, 725
    crystallization, 625, 653, 654, 711–716
Cyclic AMP, 138, 562
    as a second messenger, 13–14, 53, 313, 534, 545
    –calcium interactions, 250, 543, 545

Cyclic AMP (cont.)
    dependent protein kinase, 10, 86, 88, 165, 291, 308, 313
    dependent phosphorylation, 21, 86, 117
    inhibitory effects of, 114–117
    prostaglandin interactions, 48, 534–535
Cyclic GMP, 37, 40, 166, 250
    as a second messenger, 140
    -dependent protein kinase, 165, 291
    effects of, 87
Cyclooxygenase, 10, 41, 47, 156, 161, 535, 537, 720
Cystic fibrosis, 557, 558
Cytochrome P-450, 494
Cytoskeleton (see also Microfilaments and Microtubules), 137, 275–281

D-600 (methoxyverapamil), 84, 105, 238, 239, 327, 413, 417, 418, 423, 464, 465, 537, 544
Dantrolene, 553
Dark current, 245–250
7-dehydrocholesterol, 491
Dentin, 599–604, 634, 646
Deoxyglucose, 151
Dermatomyositis, 634
Diacylglycerol
    action of, 155–163, 165
    kinase, 46, 147, 151
    lipase, 47
    production of, 37, 40, 49, 54, 61, 68, 74, 140, 147, 155–163, 168
Diazepam, 242
Differentiation, cellular, 173–177
    effect of Ca^{2+} on, 175
Dihydropyridines, 417, 419, 441, 448, 454, 460, 474
Diltiazem, 396, 413, 417, 419, 431, 438, 441, 454, 460, 461, 467, 480–485, 556
Dimethylsulfoxide, 174, 175
Diphosphonate(s), 667, 674, 679, 680, 682, 685–693
Dipyridamole, 473
DNA synthesis, 8, 545
Dopamine
    antagonists, 416
    effects on adenylate cyclase, 283–288

EDHP (ethanehydroxyphosphonate), 667, 686, 690, 692, 693
Elastin, 634, 656
Electron probes, 351–357, 685, 686, 688, 690, 702
Endoplasmic reticulum (see also Microsomes), 9, 24, 706
    Ca^{2+} binding to, 24, 25–26, 85, 156, 188, 198, 556
Endoperoxides, 48
Enkephalins, 231, 232
Epinephrine, actions of, 57, 62, 161, 408, 435, 459
Erythrocytes, 201–204, 266, 277
Erythropoesis, 173
Ethanol, 193–198, 201–204, 215, 224, 237, 238–241
Ether, 197

N-ethylmaleimide, 107
Excitation-contraction coupling, 326, 361, 424, 425, 426, 453–456, 480
Excitatory junction potentials, 455
Exocytosis, 8, 109, 121, 129, 135, 156, 237

Fatty acids, 628
FCCP, 176
Filamin, induced gelation of actin filaments, 276
Flunarizine, 413, 417, 419
Fluphenazine, 262
Fluoride, 576, 580, 586, 588, 678
f-Met-Leu-Phe, 137–143

GABA, action of, 193–198, 228, 230
Gallopamil, 423, 431, 433, 436, 438, 460, 461
Ganglion cells
  bullfrog, 181–191
  dorsal root, 227
  stellate, 228
Gating mechanism(s), 374, 377, 380, 417
Gelation, 711–712
Glands
  exocrine, 73–78, 83–91
  endocrine, 93–98, 101–107
Glucose
  and insulin release, 93–99
  3-O-methyl, 98
  utilization of, 379
Glutamic acid, 230
Glutaraldehyde, 661, 665, 669–671
Glycerol, treatment of muscle, 321
Glycoproteins, 673, 686
Glycosaminoglycans, 645
Gout, 701, 703, 725
GppNHP, stimulation of adenylate cyclase, 283, 285, 288
Granules (secretory), 95, 137
Guanylate cyclase, 37
Guanyl nucleotides, 283–288

Haloperidol, 416
Halothane, 197
Hanes I survey, 563
Hemodynamics, 479–485
Hippocampal, slices, 193, 198, 228, 237
Histamine, 38, 129, 434, 435, 438, 455
Hydralazine, 474, 479
Hydroxyproline, 529, 541
Hypercalcemia, 466, 533
Hyperparathyroidism, 592, 682, 700
Hyperpolarization, 187, 190, 193–198, 423, 424
Hypertension, 441, 553, 557, 561–567
Hypoparathyroidism, 503
Hypoxia, 555

IMCal, 513–517
Immunoglobulin, and cell differentiation, 173–177
Imidazole, 115, 335–336, 400

Implants, 38, 57, 62, 64, 653, 661–667, 669
Indomethacin, 47, 150, 167, 535, 536, 720
Infarction, 482, 682
Inositol, phosphates, 40–42, 56, 61, 68, 69–70, 74, 76, 162
Insulin, release of, 93–99, 175
Ionophores, 141, 144, 152, 175, 541–546, 625, 626
  A23187, 40, 54, 68, 69, 87, 150, 175
  ionomycin, 156, 542, 545
  valinomycin, 30
Ischemia, myocardial, 471–476
Isoproterenol, 53, 442, 443, 444

Joint inflammation, 699–703

Lanthanum, 6, 110, 362, 392, 408, 416, 690, 692, 693
Leukocytes (*See* Neutrophils)
Leukotrienes
  action of, 10, 37, 140–141
  synthesis and release of, 10, 37, 129, 140, 701
Lidocaine, 467
Lidoflazine, 411, 474
Lipopolysaccharide, 173
Lipoproteins, 693
  low density (LDL), 656, 685
  very low density (VLDL), 685
Liposomes, 112, 269, 427
Lipoxygenase, 10, 41, 140
Lithium
  biochemical effects of, 42
  physiological effects of, 209, 324
  therapeutic action of, 42
Local anesthetics, 6, 240, 241, 257, 258, 325–326, 416
Lymphocyte(s), 173–177, 545, 666
Lysophospholipids, 45
Lysosomal enzyme(s), 366, 542–543, 701

Magnesium, 6, 125, 151, 269, 326, 332, 339–340, 346, 348, 353, 354, 357, 551, 586, 588, 707
  as an inhibitor, 5
  competition with Ca^{2+}, 27, 28, 30, 105, 115, 257, 324, 347, 355–356
  –ATP as a secretagogue, 102
Manganese, 324, 325, 392, 544, 546
Mast cells, 129
Matrix vesicles, 599–604, 609, 612, 617, 628, 634, 647, 653, 727
Mellitin, 258, 533
Membrane
  fusion, 121–126
  permeability, 6, 207–210
  stabilization, 6–7
Messenger, role of Ca^{2+}, 13–21, 23–30
Microfilament(s), 8
Microsomes, 706
Microtubule(s), 8
  depolymerization, 8, 169

Index

Midazolam, 196, 198
Milwaukee Shoulder syndrome, 702, 719, 722
Mineralization, 530, 599–604, 607–614, 617–623, 635–636, 653, 654, 656, 661, 664, 686
Mitochondria, 706
   binding of Ca^{2+}, 9, 15, 18, 19, 25, 41, 85, 95, 112, 176, 198, 216, 237, 405, 406, 413–414, 553, 612
Monensin, 176
Morphine, 215
   -like-receptors, 231
Muscle
   cardiac, 5, 321, 326, 361–367, 369–374, 377–386, 411, 418, 424, 431, 436, 439, 441–448, 454, 459–468, 471–476, 558
   skeletal, 321–328, 331–336, 339–348, 369, 381, 382, 424, 480, 558
   smooth, 391, 396, 399–409, 411–419, 423–429, 431–439
Muscular dystrophy, 557, 558
Myosin, light chain kinase, 10, 24, 143, 166, 169, 276, 292, 302, 454, 565

Naloxone, actions of, 232
Nerve terminal, 69, 109
Neurotransmitter release, 215, 217, 221, 228, 229, 230, 238
Neutrophil(s), 137–143, 673, 699, 727
Nickel, 324, 325, 327
Nicotine, 101–107
Nifedipine, 325, 327, 412, 417, 431, 433, 438, 441, 444, 446, 474, 480–485, 567, 692
Niludipine, 461
Nimodipine, 396, 419, 460, 461, 467
Nisoldipine, 413
Nitrates, 475
Nitrendipine, 327, 423–429, 443, 448
Nitroglycerin, 479, 482
Nitroprusside, 166, 474, 479
Norepinephrine, 55, 392, 408, 411, 413, 417, 423, 424, 425, 426, 427, 428, 435, 455, 456, 480, 484
Nucleosides, 111–117, 165
Nucleoside triphosphate pyrophosphohydrolase, 705–708

Opiates
   action of, 215, 227–233, 287
   receptor, 217
Osmotic gradients, 121–126
Ossification, endochondrial, 617–623
Osteoarthritis, 705–708
Osteoblasts, 492, 530, 534, 535, 536, 537
Osteocalcin, 623, 635, 636, 640–641, 645, 647, 648, 649–650, 661, 666–667
Osteoclasts, 492, 530, 534, 535, 536, 537, 541, 682
Osteodystrophy, 503
Osteonectin, 604

Osteopenia, 575, 579, 580, 591
Osteoporosis, 504, 562, 569, 575–580, 586, 591, 594, 682, 693
Ouabain, 173, 176, 332, 333–334, 442, 444
Oxalate, 166, 405, 409, 646, 648, 649
Oxygen consumption in myocardium, 471–476

Pacemaker potentials, 394
Paget's disease, 693
Pancreas
   endocrine, 93–99
   exocrine, 78, 83–91
Papaverine 417, 419, 433
Parathyroid hormone, 38, 493, 502, 504, 537, 541–546, 553, 554, 557, 558, 562, 577, 578, 679
Parvalbumin, 24, 30, 340, 346, 348, 353, 355, 357, 414
Pentobarbital, 193, 197, 223, 325
Peptides, action of, 257, 546
Periodontal disease, 533, 591, 592–593, 594
Peripheral resistance, 561, 567
Permeabilized cells
   by digitonin, 102, 105
   by high voltage discharge, 156
Phenothiazines, 106, 225, 257, 258, 309, 416
Phorbol ester(s), phorbol myristate acetate, 40, 140, 157, 162, 533
Phosphatase(s), 90, 165, 166, 599–604
   stimulation by calmodulin, 307–313
Phosphate, 551, 562, 577, 588, 603, 604, 612, 613, 625–629, 655, 677, 686, 713
   Ca^{2+} binding to, 30, 628, 634
   incorporation into phospholipids, 49, 53–55, 61, 67, 71, 74, 75
Phosphatidic acid, 40, 45–51, 61, 67, 74, 86, 140, 147–152
Phosphatidyl
   choline, 140
   ethanolamine, 45, 140
   inositol, 38, 45, 53, 61, 67, 74, 76, 86, 147
   kinase, 40, 68, 75
   serine, 88, 604, 625–629
Phosphodiesterase, 16, 68, 250, 307
   activation by Ca^{2+}, 16–17, 257
   inhibition of, 115, 259, 260, 692
Phospholamban, 296, 363
Phospholipases, 62, 365
   A_2, 10, 45–51, 140
   C, 10, 45–51, 54, 56, 58, 63, 69, 75–76, 86, 140, 148
Phospholipid(s), 395
   acidic, 10, 603, 625, 646, 647, 649, 650
   bilayer membranes, 121
   methylation of, 142
Phosphoproteins, 292–302, 604, 612, 613, 635, 645, 647
Phosphorus, 562, 574, 576, 577–579, 585, 587, 591, 592, 662, 666, 673, 688
   –protein interaction, 579

Phosphorylase
  activation of, 54, 61
  kinase, 90, 313
  phosphoserine, 613, 614, 635
  phosphothreonine, 613, 614
Photoexcitation, 245–250
Protein phosphorylation, 10, 86–88, 106, 165–170, 174, 238
  calcium-dependent, 10, 86, 116, 279, 291–302
  cyclic AMP-dependent, 117, 166, 216, 363
  40K protein, 49, 157, 159, 168
Pimozide, 416
Plaque formation, 475, 633, 634, 639, 640, 647, 686, 693
Plasma membrane, 409, 474, 706
  Ca^{2+} pump, 18, 19, 265, 414
  Ca^{2+} release from, 143
  receptor mediated PI depletion from, 55, 58, 62
Platelet activating factor, 49, 141, 159, 162
Platelets, 45–51, 141, 147–153, 155–163, 165–170, 557, 685
Polymers, 653, 658
Polyphosphoinositide(s)
  hydrolysis of, 37–40, 45, 53–58, 61–62, 68–71, 86, 140, 147, 156, 162
  tris-, 69, 71, 75, 76, 142
Polyurethane(s), as implantable blood pumps, 654–658
Pore, model of a Ca^{2+} channel, 324
Potassium, 30
  contractures, 326, 327, 356, 400, 411, 417, 418, 434
  currents in cells, 138, 181–191, 193, 324, 353
  deprivation, 466
  efflux induced by calcium, 24, 53, 181–191, 193–198, 201–204, 216, 217, 238, 326, 465, 466, 558
  excess, depolarization by, 94, 104, 138, 221, 248, 333, 354, 395, 396, 419, 426, 433, 435, 436, 463
  release by, 104
Prazosin, 56, 474, 479
Procaine, 325, 467
Propionate, 332, 333
Propranolol, 474
Prostacyclin, 47, 475, 533, 535
Prostaglandins
  actions of, 10, 37, 165
  PGE_2, 533–537, 542, 543, 544, 545, 722
  $PGF_2\alpha$, 418, 533
  release of, 533–537, 545, 701
  synthesis of, 719–721
Proteases(s), 365, 618, 702, 719–721
Protein
  contractile, 167, 168, 321, 400, 565
  intestinal membrane calcium binding (see IMCal), 513–517
Protein kinase(s), 165
  and calcium, 10, 14, 88–90
  and cyclic nucleotides, 10, 363

Protein kinase(s) (cont.)
  phospholipid-dependent (protein kinase C), 37, 40, 49, 140, 155, 157, 159, 163, 165, 169
Proteoglycan(s), 617, 618–619, 645, 647, 673
Protons, 699
  movements of, 355
Pyrophosphate, 679, 701, 705, 711

Quin-2, as a fluorescent indicator, 84, 96, 140, 155, 159
Quinidine, effect on Na^+ transport, 207, 467

Receptors, 8
  alpha-adrenergic, 37, 53, 54, 62, 73, 417, 455
  beta-adrenergic, 53, 459, 462, 467, 475, 481, 567
  Ca^{2+}-mobilizing, 37–44, 53, 73
  desensitization of, 134–135
  histaminergic, 37
  IgE, 129
  Muscarinic, 37, 68, 73, 75
  Peptidergic, 37, 73, 78, 84, 137
Red blood cells (see Erythrocytes)
Renal tubule, 207–211, 493
Reserpine, 454
Residual ridge resorption, 591, 593
Retina, calcium extrusion by, 245–250
Rickets, 500, 503, 604, 613
Ringer's experiment
  in amphibian heart, 5, 361, 369
  in mammalian heart, 369
RNA, synthesis, 501, 720
Rubidium
  efflux, 73–74, 76, 201
  uptake, 174, 324
Ryanodine, 381, 461
Ryosidine, 438, 461

Saponin, 399
Sarcolemma, 362, 363, 365, 366, 369, 373
Sarcoplasmic reticulum, 266, 326, 339, 340, 399, 400, 405–407, 409
  Ca^{2+} binding to, 24, 363, 366, 413, 424, 553
  Ca^{2+} release from, 326, 328, 331–336, 351–357, 369–374, 377–386, 436, 453
Scintography, 677, 679, 681, 682
SDS electrophoresis, 620
Secretion, 8, 83–119, 125, 129–135, 137–143, 147–152, 155, 163, 166, 453
Sedative-hypnotic drugs, 193–198, 201–204, 215–217, 221–225, 237–242
Serotonin
  action of, 419, 434, 435, 438
  release of, 47, 129, 167, 168
Skinned fibers, 331–336, 354, 370, 399–409
Sodium, 30, 480
  ATPase, 173, 174, 265, 444, 552
  –calcium exchange, 26, 27, 94, 174, 198, 201–203, 207, 209, 216, 237–242, 246, 362, 414, 415, 501, 514, 555
  currents, 138, 427, 465

Sodium *(cont.)*
  cytosolic activity of, 210
  flux, 173–177, 209, 324
  pump, 173–177, 208, 246, 413
  transport, 207–211
  uptake, 105
  urate, 646, 647, 649, 701, 721
Spectrin, 277
Spontaneously hypertensive rats (SHR), 562–564, 566
Stimulus-secretion coupling, 83–91, 484
Strontium, 5, 105, 110, 114, 257, 324, 326, 678, 680
Strophanthin, 438
Substance P, 38, 73, 76
Surface charge, 655–656
Synapsin
  phosphorylation of, 292
  membranes, 215, 237, 241, 279, 291
Synaptic vesicles, 109, 112
Synaptosomes, 69–70, 203, 221, 228, 291
Synoviocytes, 719–723

Technetium, 467
  labeled polyphosphate, 677–683
Tendinopathy, calcific, 725–729
Terminal cisternae, 353–354
Tetanus, 353–357
Tetracaine, 325, 493, 587
Tetracyclines (*see* Chlorotetracycline)
Tetraethylammonium, 183, 187, 188, 228, 326
Tetrodotoxin, 193, 195, 196, 197, 454, 462, 467
Theophylline, 111
Thrombin, 40, 45, 49, 147, 159, 168
Thromboxane, 41, 47, 155–156, 161, 162, 475
Thyroid, 679
Tolerance, to depression of Ca^{2+} uptake by drugs, 222
Transformed cells, 173–178
Transmitter release, 6–7, 24, 109–117, 190, 456, 484
  feedback inhibition of, 111
Transverse tubules, 321, 331, 351
Trifluoperazine, 49, 88, 107
Trimazosin, 692, 693
Tropomyosin, 363, 364
Troponin, 10, 13, 24, 30, 353, 357, 364
Trypsin, 309–310, 702
  splitting of membranous Ca^{2+} ATPase, 266–271
Tubulin, 272, 302

Vaccenic acid, 241
Valinomycin, 354–355
Valve
  aortic, 661, 662, 671
  biprosthetic leaflets, 656

Valve *(cont.)*
  calcification, 636, 669–674
  mitral, 671
  pericardial, 670
  prosthesis, 635, 661–667, 669–674
  pulmonary, 671
  tricuspid, 671
Vascular
  reactivity, 562
  resistance, 561, 562
  tone, 391–396, 455, 562, 566
Vasoconstrictor mechanism(s), 479–485
Vasodilation, 488, 567
Vasopressin, 38, 57, 62, 411, 474, 545
Ventricular assist device(s), calcific deposits in, 636–641, 648, 670
Verapamil, 238, 239, 242, 417, 423, 431, 436, 437, 441, 444, 445–446, 454, 460, 461, 464, 467, 474, 480, 485, 544, 649
Veratridine, 101, 454
Vesicles
  calcium pumping, 166, 377–386, 399, 405
  phospholipid, 121–126, 636
Vitamin D, 491–503, 562, 574, 576, 580, 591–594, 666, 679
  deficiency, 493, 516, 613
  -dependent Ca^{2+} binding proteins, 501, 513–517, 519–523
  25-hydroxy (calcidiol), 494
  and intestinal absorption, 492
  metabolites, 494–498, 558
  and phosphate metabolism, 493, 497, 503
  synthesis, 491–492
Vitamin D_3, 491–503, 525–530, 541–546
  induced Ca^{2+}-binding protein, 24, 30, 502
  24,24-difluoro-25-OH, 497
  1,24R,25-trihydroxy, 497
  1,25-dihydroxy (calcitriol), 495–505
  26,26,26,27,27,27,F_6-25-OH(D_3), 497
Vitamin K, 650
  -dependent bone protein, 525–530, 661, 666–667
Voltage clamp, 121, 182, 193, 326, 328, 339

Warfarin, 527–530, 640, 641, 666–667
Whitlockite, 669

X-ray
  crystallography, 711–712
  diffraction, 587, 609, 639
  microanalysis, 25

Zinc, 257, 586, 588

MIX
Papier aus verantwortungsvollen Quellen
Paper from responsible sources
FSC® C105338

If you have any concerns about our products,
you can contact us on
ProductSafety@springernature.com

In case Publisher is established outside the EU,
the EU authorized representative is:
**Springer Nature Customer Service Center GmbH
Europaplatz 3, 69115 Heidelberg, Germany**

Printed by Libri Plureos GmbH
in Hamburg, Germany